ELECTRICAL WIRING RESIDENTIAL

Based on the 2014 National Electrical Code®

18TH EDITION

ELECTRICAL WIRING
RESIDENTIAL

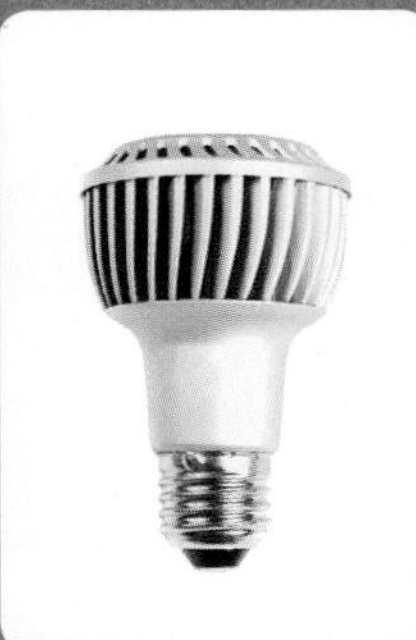

Based on the 2014 National Electrical Code®

18TH EDITION

RAY C. MULLIN
PHIL SIMMONS

CENGAGE Learning®

Australia • Brazil • Mexico • Singapore • United Kingdom • United States

Electrical Wiring Residential, 18E
Ray C. Mullin, Phil Simmons

Vice President, GM Skills & Product Planning: Dawn Gerrain

Product Team Manager: James DeVoe

Senior Director, Development: Marah Bellegarde

Senior Product Development Manager: Larry Main

Senior Content Developer: John Fisher

Product Assistant: Andrew Ouimet

Vice President, Marketing Services: Jennifer Baker

Market Manager: Linda Kuper

Senior Production Director: Wendy A. Troeger

Production Manager: Mark Bernard

Senior Content Project Manager: Kara A. DiCaterino

Senior Art Director: David Arsenault

Technology Project Manager: Joe Pliss

Media Editor: Debbie Bordeaux

Cover and Interior Design Image(s):
© iStockphoto/piovesempre
© iStockphoto/tarczas

For product information and technology assistance, contact us at
Cengage Learning Customer & Sales Support, 1-800-354-9706
For permission to use material from this text or product,
submit all requests online at **www.cengage.com/permissions**
Further permissions questions can be e-mailed to
permissionrequest@cengage.com

Library of Congress Control Number: 2013947429

ISBN-13: 978-1-285-17095-4

ISBN-10: 1-285-17095-4

Cengage Learning
200 First Stamford Place, 4th Floor
Stamford, CT 06902
USA

Cengage Learning is a leading provider of customized learning solutions with office locations around the globe, including Singapore, the United Kingdom, Australia, Mexico, Brazil, and Japan. Locate your local office at: **www.cengage.com/global**

Cengage Learning products are represented in Canada by Nelson Education, Ltd.

To learn more about Cengage Learning, visit **www.cengage.com**

Purchase any of our products at your local college store or at our preferred online store **www.cengagebrain.com**

Printed in the United States of America
3 4 5 6 7 18 17 16 15

Contents

CHAPTER 7

CHAPTER 8

CHAPTER 9

Plans for Single-Family Dwelling

THE IMPORTANCE OF PROPER TRAINING

Now that I have retired after 48 years in the electrical industry, it has become even more evident that a good solid education about the world of electricity is of utmost importance.

Accurate materials and training are the two sides of the electrical safety coin. This coin is spent every day by various persons intrinsically involved in the electrical construction industry. Unfortunately, some spend it less wisely than others. Usually, the unwise spenders are those who rush to career, having neglected to acquire accurate materials and to focus on training themselves to a high level of proficiency.

Ray Mullin, coauthor of this book, *Electrical Wiring—Residential*, has often stated, "The cost of education is small when compared to the price paid for ignorance." All too often, we, the citizens, pay the price for others' ignorance—ignorance of the codes, ignorance of proper wiring methods, ignorance of proper installation procedures, ignorance of design requirements, ignorance of product evaluations. This price becomes dear when our friends and family lose health or life or when our homes are destroyed.

It is exciting to see that Phil Simmons has joined with Ray as coauthor of *Electrical Wiring—Residential*. Phil has served the electrical industry with distinction for many years. His ability to express complex electrical issues clearly and to illustrate them accurately is unparalleled among his peers.

Fortunately, accurate materials are so easy to obtain. Ray Mullin and Phil Simmons are both technical writers who have paid their dues in the electrical industry. Each has put in many years as an apprentice, a journeyman, and then as a master electrician before beginning to write about his trade. Phil was additionally a professional in the electrical inspection arena and managed the International Association of Electrical Inspectors (IAEI) for several years. Both have served or are serving on *NEC* Code Making Panels. *Electrical Wiring—Residential* contains accurate, up-to-date information about all aspects of residential wiring.

When installers and inspectors don't keep abreast of installation procedures and code requirements, things like cables across scuttle access to attics; improper spacing of receptacle outlets; improper short-circuit and ground-fault protection; and improper grounding of electrical systems, phone system, and CATV systems can lead to hazardous situations causing electrical shocks and fire. Not just anybody can install or inspect safe electrical systems. Trained professionals can, but even they must be constantly improving their knowledge and skills.

Because Ray Mullin and Phil Simmons care about the electrical safety coin, they have striven to provide the most accurate information possible. It is up to each of us, however, to focus on the training. Some training can be acquired simply by reading the best books in our trade; some training can come through the online programs available; and other training, through participation in classes and seminars. In each instance, though, motivation and desire come from within—to know everything involved in our trade, to be totally proficient, to focus continually on improvement. As we seek both accurate information and training, we learn to spend the coin of safety to benefit others as well as ourselves. I commend you for acquiring *Electrical Wiring—Residential*; now I challenge you to make it part of yourself. I challenge you to spend the electrical safety coin wisely.

James W. Carpenter
Former CEO and Executive Director,
International Association of Electrical Inspectors
Past Chair of the *NEC* Technical Correlating Committee

Preface

INTENDED USE AND LEVEL

STOP . . . Don't read any further . . . yet. Take a moment to familiarize yourself with how to use this text to get the most benefit from it. Think of it as a three-legged stool. One leg is this text, the second leg is the 2014 edition of the *National Electrical Code*®, and the third leg is the set of Plans that are in the packet in the inside back cover. If any one of the legs is missing, the stool will collapse. Stated another way, you will not get as much out of this course. When you have completed all of the chapters in *Electrical Wiring—Residential*, you will have virtually wired a typical house according to the requirements of the 2014 *National Electrical Code*, an accomplishment you can be proud of!

The *NEC*® defines a "qualified person" as One who has skills and knowledge related to the construction and operation of the electrical equipment and installations and has received safety training to recognize and avoid the hazards involved.*

Electrical Wiring—Residential is intended for use in residential wiring courses at high schools, two-year and four-year colleges, and apprenticeship training programs. This comprehensive book guides readers, room by room, through the wiring of a typical residence and builds a foundation of knowledge by starting with the basic requirements of the *National Electrical Code* (*NEC*), then continuing on to the more advanced wiring methods. Each *Code* rule is presented through text, illustrations, examples, and wiring diagrams. In addition, an accompanying set of Plans at the back of the book guides the reader through the wiring process by applying concepts learned in each chapter to an actual residential building in order to understand and meet the requirements set forth by the *NEC*.

An Important Note about Safety

In the educational field, it is pretty much a given that "Society will pay for education . . . one way or another." Proper training of a skilled trade is much better than hit-or-miss learning. Having to do the job over, having a house burn down, or having someone get electrocuted because of improper wiring is costly!

It really doesn't take any longer to do it right the first time than to have to do it over. You probably have heard the phrase "Measure twice . . . cut once. Measure once . . . cut twice." How true!

Electrical wiring is a skilled trade. Wiring should not be done by anyone not familiar with the hazards involved. It is a highly technical skill that requires much training. This text

**National Electrical Code*® and *NEC*® are registered trademarks of the National Fire Protection Association, Inc., Quincy, MA 02169.

*Reprinted with permission from NFPA 70-2014.

provides all of the electrical codes and standards information needed to approach house wiring in a safe manner. In fact, *Electrical Wiring—Residential* has been adopted as the core text by the major electrical apprenticeship programs across the country. Their residential curriculum program directors and committee members made this text their top choice for their residential wiring training.

Electrical Wiring—Residential will provide you with the know-how so you can wire houses that "Meet *Code*."

Electrical Wiring—Residential has become an integral part of approved (accredited) training programs by an increasing number of states that require residential electricians to have a residential license if they are going to wire homes and small apartments.

The *NEC* has one thing in mind—safety! There is too much at stake to do less than what the *NEC* requires. Anything less is unacceptable! The *NEC* in *90.1(A)* makes it pretty clear. It states that *The purpose of this Code is* The practical safeguarding of persons and property from hazards arising from the use of electricity.*

Do not work on live circuits! Always de-energize the system before working on it! There is no compromise when it comes to safety! Many injuries and deaths have occurred when individuals worked on live equipment. The question is always: "Would the injury or death have occurred had the power been shut off?" The answer is "No!"

All mandatory safety-related work practices are found in the Federal Regulation Occupational Safety and Health Administration (OSHA), Title 29, Subpart S—Electrical, Sections 1910.331 through 1910.360.

SUBJECT AND APPROACH

The 18th edition of *Electrical Wiring—Residential* is based on the 2014 *National Electrical Code* (*NEC*). The *NEC* is used as the basic standard for the layout and construction of residential electrical systems. In this text, thorough explanations are provided of *Code* requirements as they relate to residential wiring. To gain the greatest benefit from this edition, the student must use the *NEC* on a continuing basis.

It is extremely difficult to learn the *NEC* by merely reading it. This text brings together the rules of the *NEC* and the wiring of an actual house. You will study the rules from the *NEC* and apply those rules to a true-to-life house wiring installation.

Take a moment to look at the Table of Contents. It is immediately apparent that you will not learn such things as how to drill a hole, tape a splice, fish a cable through a wall, use tools, or repair broken plaster around a box. These things you already know or are learning on the job. The emphasis of this text is to teach you how to wire a house that "Meets *Code*." Doing it right the first time is far better than having to do it over because the electrical inspector turned down your job.

The first seven chapters in this book concentrate on basic electrical code requirements that apply to house wiring. This includes safety when working with electricity; construction symbols, plans, and specifications; wiring methods; conductor sizing; circuit layout; wiring diagrams; numerous ways to connect switches and receptacles; how to wire recessed luminaires; ground-fault circuit interrupters (GFCIs); arc-fault circuit interrupters (AFCIs); and surge suppressors.

The remaining chapters are devoted to the wiring of an actual house—room by room, circuit by circuit. All of these circuits are taken into account when calculating the size of the main service. Because proper grounding is a key safety issue, the subject is covered in detail.

You will also learn about security systems, fire and smoke alarms, low-voltage remote-control wiring, swimming pools, and standby generators, and you will be introduced to structured wiring for home automation.

You will find this text unique in that you will use the text, an actual set of Plans and specifications, and the *NEC*—all at the same time. The text is perfect for learning house wiring and makes an excellent reference source for looking up specific topics relating to house wiring. The blueprints serve as the basis for the wiring schematics, cable layouts, and discussions provided in the text. Each chapter dealing with a specific type of wiring is referenced to

*Reprinted with permission from NFPA 70-2014.

the appropriate plan sheet. All wiring systems are described in detail—lighting, appliance, heating, service entrance, and so on.

The house selected for this edition is scaled for current construction practices and costs. Note, however, that the wiring, luminaires, appliances, number of outlets, number of circuits, and track lighting are not all commonly found in a home of this size. The wiring may incorporate more features than are absolutely necessary. This was done to present as many features and *Code* issues as possible, to give the student more experience in wiring a residence. Also included are many recommendations that are above and beyond the basic *NEC* requirements.

Note: The *NEC* (NFPA 70) becomes mandatory only after it has been adopted by a city, county, state, or other governing body. Until officially adopted, the *NEC* is merely advisory in nature. State and local electrical codes may contain modifications of the *NEC* to meet local requirements. In some cases, local codes will adopt certain more stringent regulations than those found in the *NEC*. For example, the *NEC* recognizes nonmetallic-sheathed cable as an acceptable wiring method for house wiring. Yet, the city of Chicago and surrounding counties do not permit nonmetallic-sheathed cable for house wiring. In these areas, all house wiring is done with electrical metallic tubing (thinwall).

There are also instances where a governing body has legislated action that waives specific *NEC* requirements, feeling that the *NEC* was too restrictive on that particular issue. Such instances are very rare. The instructor is encouraged to furnish students with any local variations from the *NEC* that would affect this residential installation in a specific locality.

THE ELECTRICAL TRADE—TRAINING PROGRAMS

As you study *Electrical Wiring—Residential*, study with a purpose—to become the best residential wireman possible.

There will always be a need for skilled electricians! Qualified electricians almost always have work. It takes many hours of classroom and on-the-job training to become a skilled electrician. The best way to learn the electrical trade is through a training program approved by the U.S. Department of Labor (http://www.dol.gov). Many times an apprenticeship program is called "Earn while You Learn." These programs offer the related classroom training and the advantage of working on the job with skilled journeymen electricians. Completion of a registered apprenticeship program generally leads to higher pay, job security, higher quality of life, recognition across the country, and the opportunity for college credit and future degrees.

As a rule, these training programs require 144 to 180 hours of classroom-related technical training and 2000 hours of on-the-job training per year. Some programs have day classes and some have night classes. An electrical apprenticeship training program might run four to five years. The end result—becoming a full-fledged licensed journeyman electrician capable of doing residential, commercial, and industrial electrical work. A residential electrician training program might run two to three years, with the training limited to the wiring of single- and multifamily dwellings. The end result—receiving a license limited to residential wiring.

To get into an apprenticeship program, the individual usually must have a high school education, with at least 1 year of high school algebra; be at least 18 years old; must be physically in shape to perform the work electricians are called upon to do (e.g., climbing, lifting, work in inclement weather); and, most importantly, be drug free. There generally is a qualifying aptitude test to make sure the applicant has the ability to take on the responsibility of a rigid apprenticeship training program. In some areas, passing the high school equivalency General Education Development (GED) test is acceptable in place of a high school diploma.

What does it take to make a good apprentice and journeyman electrician? In no particular order: commitment to master the electrical field, willingness to study and understand the training material, strong math skills, ability to think clearly and logically to analyze and solve problems, ability to work indoors and outdoors, comfortable working with your head and hands, good mechanical skills, ability to communicate and work with others, good verbal skills, ability to follow directions, strong work and personal ethics, and being a self-starter.

Following completion of an apprenticeship program, continuing education courses are available to keep the journeyman up to date on codes and other related topics and skills.

Journeymen electricians who have an interest in teaching apprentices will usually have to take instructor training courses. In certain programs, satisfactory completion of the required courses can lead to an associate degree. Others will go on to become crew leaders, supervisors, and contractors.

There are some areas where a "pre-apprenticeship" program is offered. To learn more about the careers possible in the electrical field, chat with your instructor; your local high school's guidance counselor; your vocational, technical, and adult education schools; and electricians and electrical contractors. Go online and search for electrical apprenticeship programs.

Your future is in your hands.

Some very important two-letter words that you should remember are

IF IT IS TO BE, IT IS UP TO ME!

Job Titles

Most building codes and standards contain definitions for the various levels of competency of workers in the electrical industry. Here are some examples of typical definitions:

Apprentice shall mean a person who is required to be registered, who is in compliance with the provisions of this article, and who is working at the trade in the employment of a registered electrical contractor and is under the direct supervision of a licensed master electrician, journeyman electrician, or residential wireman.

Residential Wireman shall mean a person having the necessary qualifications, training, experience, and technical knowledge to wire for and install electrical apparatus and equipment for wiring one-, two-, three-, and four-family dwellings. A residential wireman is sometimes referred to as a *Class B Electrician.*

Journeyman Electrician shall mean a person having the necessary qualifications, training, experience, and technical knowledge to wire for, install, and repair electrical apparatus and equipment for light, heat, power, and other purposes, in accordance with standard rules and regulations governing such work.

Master Electrician means a person having the necessary qualifications, training, experience, and technical knowledge to properly plan, lay out, and supervise the installation and repair of wiring apparatus and equipment for electric light, heat, power, and other purposes, in accordance with standard codes and regulations governing such work, such as the *NEC.*

Electrical Contractor means any person, firm, partnership, corporation, association, or combination thereof who undertakes or offers to undertake for another the planning, laying out, supervising and installing, or the making of additions, alterations, and repairs in the installation of wiring apparatus and equipment for electrical light, heat, and power.

Most electrical inspectors across the country are members of the International Association of Electrical Inspectors (IAEI). This organization publishes one of the finest technical bimonthly magazines devoted entirely to the *NEC* and related topics, and it is open to individuals who are not electrical inspectors. Electrical instructors, vo-tech students, apprentices, electricians, consulting engineers, contractors, and distributors are encouraged to join the IAEI so they can stay up to date on all *NEC* issues, changes, and interpretations. An application form that explains the benefits of membership in the IAEI can be found in the Appendix of this text.

NEW TO THIS EDITION

Continuing in the tradition of previous editions, this edition thoroughly explains how *Code* changes affect house wiring installations. New and revised full-color illustrations supplement the explanations to ensure that electricians understand the new *Code* requirements. New photos reflect the latest wiring materials and components available on the market. Revised review questions test student understanding of the new content. New tables that summarize *Code* requirements offer a quick reference tool for students. Other reference aids are the tables reprinted directly from the 2014 edition of the *NEC.* The extensive revisions for the eighteenth edition make

Electrical Wiring—Residential the most up-to-date and well-organized guide to house wiring. Coverage of the *NEC* has been expanded to well over 1000 *Code* references.

This text focuses on the technical skills required to perform electrical installations. It covers such topics as calculating conductor sizes, calculating voltage drop, determining appliance circuit requirements, sizing service, connecting electric appliances, grounding service and equipment, installing recessed luminaires (fixtures), and much more. These are critical skills that can make the difference between an installation that "Meets *Code*" and one that does not. The electrician must understand the reasons for following *Code* regulations to achieve an installation that is essentially free from hazard to life and property.

Note: Symbols have been added to indicate changes in the 2014 *National Electrical Code* from the 2011 *National Electrical Code*.▶◀

This text might be called "Work in Progress." The authors stay in touch with the latest residential wiring trends and the *National Electrical Code*. Because the *NEC* is revised every three years, this text follows the same cycle. *Electrical Wiring—Residential* has been carefully reviewed to editorially simplify, streamline, and improve its readability. Many diagrams have been simplified. Some units were reorganized so the *Code* requirements for the various applications are more uniform.

Much rewriting was done. The 2014 *NEC* contains many editorial changes as well as renumbering and relocation of numerous *Code* references. All of these have been addressed in this edition of *Electrical Wiring—Residential*. Some text has been condensed and reformatted for ease in reading. Many diagrams have been simplified for clarity.

The Objectives have been fine-tuned for easier readability.

- Thousands of proposals are submitted each cycle 3-year to update the *NEC*. Additional comments are submitted to revise the action taken by the Code Panels on the proposals. The end result is the publishing of a new edition of the *National Electrical Code*.
- In *Electrical Wiring—Residential*, all *Code* requirements have been updated to the 2014 edition of the *NEC*. These have been revised throughout the text, wiring diagrams, and illustrations.
- Illustrations have been enhanced for improving clarity and ease in understanding.
- Emphasis was given to making the wiring of the residence conform to energy saving standards. In other words, the residence in *Electrical Wiring—Residential* is "green."
- A most significant new rule in the 2011 *NEC* that a grounded conductor be taken to every switch location has been revised for the 2014 edition. This revision has been addressed in *Electrical Wiring—Residential*, with all wiring diagrams revised accordingly. In some locations 3-wire and 4-wire switch loops, and possibly larger boxes, will be required.
- The requirement for arc-fault circuit protection has been expanded to include kitchens and laundry areas. In addition, devices that are located in the rooms where AFCI protection is required now require that same AFCI protection. See *NEC 210.12(A)*.
- The outlets or devices in several rooms or areas now require both GFCI and AFCI protection. See *NEC 210.8(A)* and *210.12(A)*.
- Wiring for receptacle outlets in an attached garage or detached garage with electric power is not permitted to supply outlets outside the garage. See *NEC 210.52(G)(1)*.
- At least one receptacle outlet is required for each car space in the garage. The location relative to the car space is not specified. See *NEC 210.52(G)(1)*.
- All of the wiring diagrams have been updated to show the latest system of electrical symbols. This is based on the National Electrical Contractors Association's *National Electrical Installation Standard*.
- Major revisions of many diagrams and figures have made to improve the clarity and ease of understanding the *Code* requirements.
- Many new full-color illustrations have been added.

FEATURES OF THIS TEXT

This text may be used as a classroom text, as a learning resource for an individual student, or as a reference text for technicians on the job.

Objectives Objectives are listed at the beginning of each unit. The objective statements have been stated clearly and simply to give students direction.

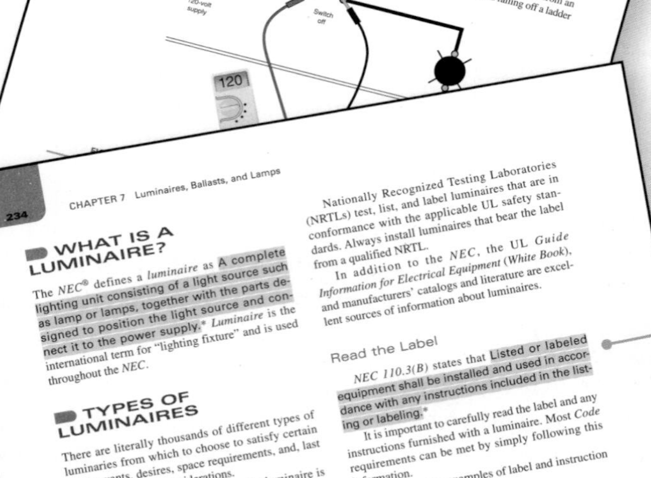

CHAPTER 7 Luminaires, Ballasts, and Lamps

234

WHAT IS A LUMINAIRE?

The *NEC*® defines a *luminaire* as A complete lighting unit consisting of a light source such as lamp or lamps, together with the parts designed to position the light source and connect it to the power supply.* *Luminaire* is the international term for "lighting fixture" and is used throughout the *NEC*.

TYPES OF LUMINAIRES

There are literally thousands of different types of luminaires from which to choose to satisfy certain needs, wants, desires, space requirements, and, last but not least, price considerations.

Note in Table 7-1 that whether the luminaire is incandescent or fluorescent, the basic categories are surface mounted, recessed mounted, and suspended ceiling mounted.

The *Code* Requirements

Article 410 sets forth the requirements for installing luminaires. The electrician must "Meet *Code*" with regard to mounting, supporting, grounding, live-parts exposure, insulation clearances, supply conductor types, maximum lamp wattages, and so forth.

Probably the two biggest contributing factors to fires caused by luminaires are installing lamp wattages that exceed that for which the luminaire has been designed, and burying recessed luminaires under thermal insulation if the luminaire has not been designed for such an installation.

Nationally Recognized Testing Laboratories (NRTLs) test, list, and label luminaires that are in conformance with the applicable UL safety standards. Always install luminaires that bear the label from a qualified NRTL.

In addition to the *NEC*, the UL *Guide Information for Electrical Equipment* (*White Book*), and manufacturers' catalogs and literature are excellent sources of information about luminaires.

Read the Label

NEC 110.3(B) states that Listed or labeled equipment shall be installed and used in accordance with any instructions included in the listing or labeling.*

It is important to carefully read the label and any instructions furnished with a luminaire. Most *Code* requirements can be met by simply following this information.

Here are a few examples of label and instruction information:

- Maximum lamp wattage
- Type of lamp
- For supply connections, use wire rated for at least ________°C
- Type-IC
- Type Non-IC
- Suitable for damp, dry, or wet locations
- Thermally protected

Installing and Connecting Luminaires

The circuit conductors in a wall or ceiling box where luminaires are to be installed are usually

- white—the "identified" grounded conductor,

Safety Alert Safety is emphasized throughout the book and is fully covered in the first chapter. Special considerations in working with electricity provide readers with an overview of what dangers are to be expected on the job.

Direct Quotes from the *NEC* Direct quotes from the *National Electrical Code* are set in roman type with shaded background, enabling the reader to clearly distinguish direct *NEC* content.

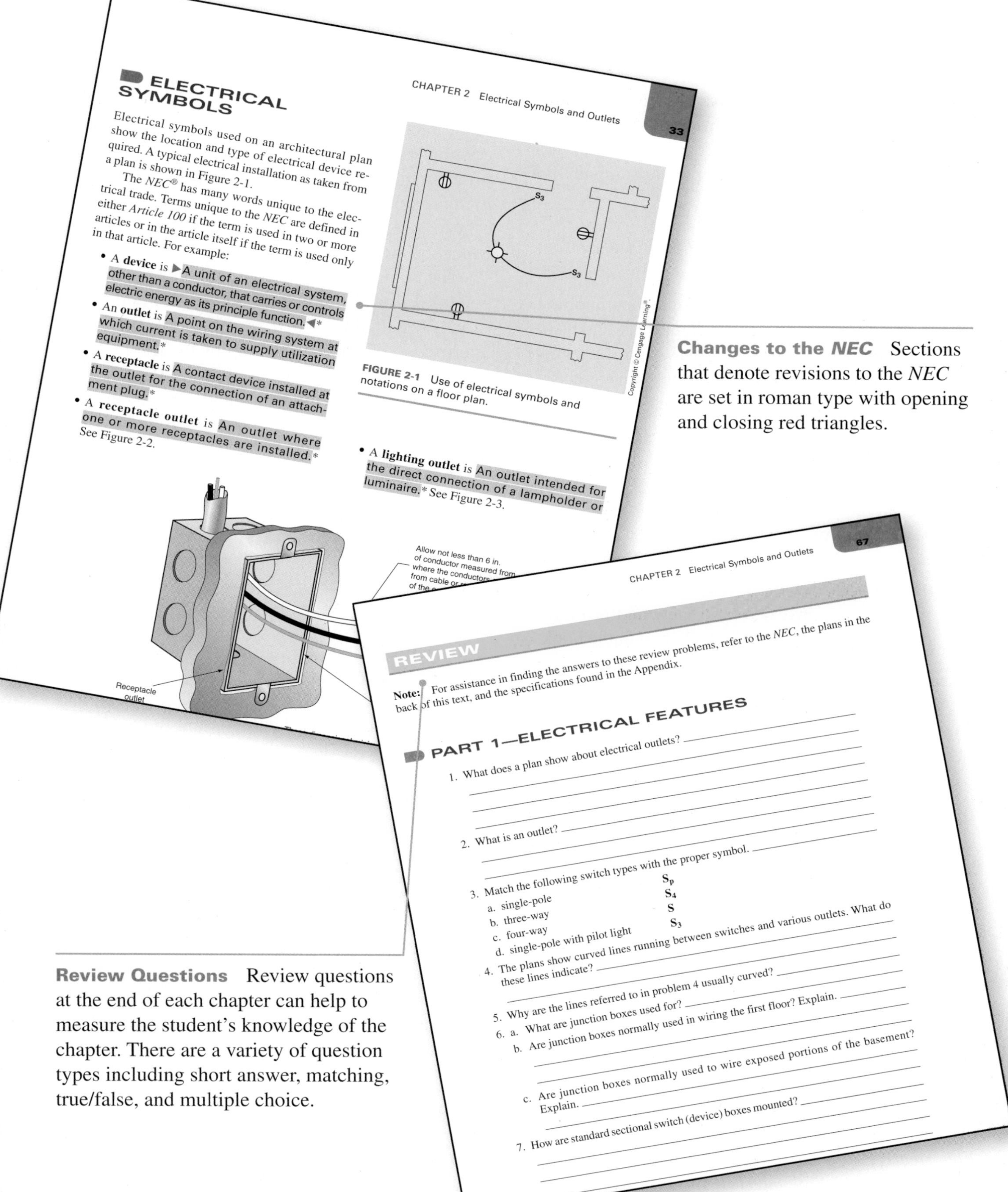

Changes to the *NEC* Sections that denote revisions to the *NEC* are set in roman type with opening and closing red triangles.

Review Questions Review questions at the end of each chapter can help to measure the student's knowledge of the chapter. There are a variety of question types including short answer, matching, true/false, and multiple choice.

SUPPLEMENT PACKAGE

An ***Instructor's Guide*** contains answers to all Review questions included in the book and a blank service-entrance calculation form, which is also available in electronic format on the accompanying *Instructor Resources CD*.
(Order #: 978-1-285-17099-2)

An *Instructor Resources CD* provides instructors with valuable classroom materials on CD-ROM:

- ***PowerPoint Presentations*** outline the important concepts covered in each chapter. Extensively illustrated with photos, tables, and diagrams from the book, the presentations enhance classroom instruction. PowerPoint presentations also allow instructors to tailor the course to meet the needs of their individual class.
- An ***Image Gallery*** contains nearly all the images in the book and can be used to enhance the PowerPoint presentation or to create transparency masters and handouts.
- ***Electronic Instructor's Guide*** in Microsoft Word enables instructors to view and print answers to review questions contained in the book.
- ***Electronic Blueprints*** provide an online version of the drawings that are included at the back of the book, allowing instructors to project and reference in classroom presentations.
- ***Video Clips*** from the accompanying video series visually highlight important concepts presented in the book

(Order #: 978-1-285-17100-5)

A ***Lab Manual*** provides over twenty exercises to aid students in learning both basic and complex wiring circuits. Each lab consists of a hands-on wiring exercise as well as *NEC* drill problems. Students are also required to draw an electrical layout of wiring booths to familiarize themselves with electrical symbols. Allowing for instructor verification and student self-assessment, this manual is essential to applying important wiring concepts.
(Order #: 978-1-285-17112-8)

Electrical Wiring Residential DVD Series: *Electrical Wiring—Residential Video Series* correlates directly with the *Electrical Wiring—Residential* book. Shot on various construction sites and enhanced with quality animations, each video is devoted to a specific, specialized topic. In addition to high-quality animations, questions have also been incorporated into each tape at strategic points to promote discussion aimed at enhancing viewers' understanding and preparing them for successful attainment of all learning objectives. Each video is approximately 20 minutes in length, making it easy to incorporate into a comprehensive electrician education and training program.

The complete set includes: Video #1: Introduction to Electrical Installation, Video #2: Planning for Circuit Installation, Video #3: Ground-Fault Circuit Interrupters, Video #4: Lighting by the Room, Video #5: Special Circuits, Video #6: Special-Purpose Outlets, Video #7: Miscellaneous and Custom Installations, Video #8: Service Entrances.
(DVD 1, Videos 1–4 Order #: 978-1-4354-9530-2)
DVD 2, Videos 5–8 Order #: 978-1-4354-9525-8)

COURSEMATE

CourseMate complements the text and course content with study and practice materials. Cengage Learning's CourseMate brings course concepts to life with interactive learning, study, and exam preparation tools that support the printed textbook. Watch student comprehension soar as your class works with the printed textbook and the textbook-specific website.

CourseMate includes an integrated MindTap reader; interactive teaching and learning tools including quizzes, flashcards, videos, and Engagement Tracker, a first-of-its-kind tool that monitors student engagement in the course; and more. CourseMate goes beyond the book to deliver what you need!

To access additional course materials, including CourseMate, please visit www.cengagebrain.com. At the CengageBrain home page, search for the ISBN (from the back cover of the book), using the search box at the top of the page. This will take you to the product page where these resources can be found.

INSTRUCTOR SITE

An Instructor Companion website containing supplementary material is available. This site contains an Instructor Guide, Cognero test banks, an image gallery of text figures, and chapter presentations done in PowerPoint. Contact Cengage Learning or your local sales representative to obtain an instructor account.

Accessing an Instructor Companion Website from SSO Front Door

1. Go to http://login.cengage.com and log in using the instructor e-mail address and password.
2. Enter author, title, or ISBN in the **Add a title to your bookshelf** search.
3. Click **Add to my bookshelf** to add instructor resources.
4. At the Product page, click the **Instructor Companion site** link.

Also Available from Cengage Learning

Videos. The *House Wiring DVD Series* is an integrated part of the *Residential Construction Academy House Wiring* package. The series contains a set of eight 20-minute videos that provide step-by-step instruction for wiring a house. All the essential information is covered in this series, beginning with the important process of reviewing the plans and following through to the final phase of testing and troubleshooting. Need-to-know *NEC* articles are highlighted, and Electrician's Tips and Safety Tips offer practical advice from the experts.

The complete set includes the following: Video #1: Safety and Safe Practices, Video #2: Hardware, Video #3: Tools, Video #4: Initial Review of Plans, Video #5: Rough-In, Video #6: Service Entrance, Video #7: Trim-Out, Video #8: Testing & Troubleshooting.

For more information, visit http://www.residentialconstructionacademy.com. Visit us at www.cengage.com/community/electrical, now LIVE for the 2014 Code cycle!

This newly designed website provides information on other learning materials offered by Cengage Learning, as well as industry links, career profiles, job opportunities, and more!

ABOUT THE AUTHORS

This text was coauthored by Ray C. Mullin and Phil Simmons.

Mr. Mullin is a former electrical instructor for the Wisconsin Schools of Vocational, Technical, and Adult Education. He is a former member of the International Brotherhood of Electrical Workers. He is a member of the International Association of Electrical Inspectors, the Institute of Electrical and Electronic Engineers, and the National Fire Protection Association, Electrical Section, and has served on Code-Making Panel 4 of the *National Electrical Code*.

Mr. Mullin completed his apprenticeship training and worked as a journeyman and supervisor for residential, commercial, and industrial installations. He has taught both day and night electrical apprentice and journeyman courses, has conducted engineering seminars, and has conducted many technical *Code* workshops and seminars at International Association of Electrical Inspectors Chapter and Section meetings, and has served on their code panels.

He has written many technical articles that have appeared in electrical trade publications. He has served as a consultant to electrical equipment manufacturers regarding conformance of their products to industry standards, and on legal issues relative to personal injury lawsuits resulting from the misuse of electricity and electrical equipment. He has served as an expert witness.

Mr. Mullin presents his knowledge and experience in this text in a clear-cut manner that is easy to understand. This presentation will help students to fully understand the essentials required to pass the residential licensing examinations and to perform residential wiring that "Meets *Code*."

Mr. Mullin is the author of *House Wiring with the NEC*—a text that focuses entirely on the *National Electrical Code* requirements for house wiring. He is coauthor of *Electrical Wiring—Commercial*, *Illustrated Electrical Calculations*, and *The Smart House*. He contributed technical material for Cengage Learning's *Electrical Grounding and Bonding* and to the International Association of Electrical Inspectors' texts *Soares' Book On Grounding* and *Ferm's Fast Finder*.

He served on the Executive Board of the Western Section of the International Association of Electrical Inspectors and on their Code Clearing Committee, and, in the past, served as Secretary/Treasurer of the Indiana Chapter of the IAEI.

Mr. Mullin is past Chairman of the Electrical Commission in his hometown.

Mr. Mullin is past Director, Technical Liaison for a major electrical manufacturer. In this position, he was deeply involved in electrical codes and standards as well as contributing and developing technical training material for use by this company's field engineering personnel.

Mr. Mullin attended the University of Wisconsin, Colorado State University, and the Milwaukee School of Engineering.

Phil Simmons is self-employed as Simmons Electrical Services. Services provided include consulting on the *National Electrical Code* and other codes; writing, editing, illustrating, and producing technical publications; and inspection of complex electrical installations. He develops training programs related to electrical codes and safety and has been a presenter on these subjects at numerous seminars and conferences for universities, the NFPA, IAEI, Department of Defense, and private clients. Phil also provides plan review of electrical construction documents. He has consulted on several lawsuits concerning electrical shocks, burn injuries, and electrocutions.

Mr. Simmons is coauthor and illustrator of *Electrical Wiring—Residential* (17th and 18th editions), coauthor and illustrator of *Electrical Wiring—Commercial* (14th and 15th editions), and author and illustrator of *Electrical Grounding and Bonding*, all published by Cengage Learning. While at IAEI, Phil was author and illustrator of several books, including the *Soares' Book on Grounding of Electrical Systems* (five editions), *Analysis of the NEC* (three editions), and *Electrical Systems in One- and Two-Family Dwellings* (three editions). Phil wrote and illustrated the National Electrical Installation Standard (NEIS) on *Types AC and MC Cables* for the National Electrical Contractors Association.

Phil presently serves NFPA on Code-Making Panel 5 of the *National Electrical Code* Committee (grounding and bonding). He previously served on the *NEC* CMP-1 (*Articles 90*, *100*, and *110*), as Chair of CMP-19 (articles on agricultural buildings and mobile and manufactured buildings), and member of CMP-17 (health care facilities). He served six years on the NFPA Standards Council, as NFPA Electrical Section President, and on the *NEC* Technical Correlating Committee.

Phil began his electrical career in a light-industrial plant. He is also a master electrician in the state of Washington and was owner and manager of Simmons Electric Inc., an electrical contracting company. He is also a licensed journeyman electrician in Montana. Phil passed the certification examinations for Electrical Inspector General, Electrical Plan Review, and Electrical Inspector One- and Two-Family.

He previously served as Chief Electrical Inspector for the State of Washington from 1984 to 1990 as well as an Electrical Inspector Supervisor, Electrical Plans Examiner, and field Electrical Inspector. While employed with the State, Phil performed plan review and inspection of health care facilities, including hospitals, nursing homes, and boarding homes.

Phil served the International Association of Electrical Inspectors as Executive Director from 1990 to 1995 and as Education, Codes, and Standards Coordinator from 1995 through June 1999. He was International President in 1987 and has served on local and regional committees.

He served Underwriters Laboratories as a Corporate Member and on the Electrical Council from 1985 to 2000. He served on the UL Board of Directors from 1991 to 1995 and is a retired member of the International Brotherhood of Electrical Workers.

IMPORTANT NOTE

Every effort has been made to be technically correct, but there is always the possibility of typographical errors. If changes in the *NEC* do occur after the printing of this text, these changes will be incorporated in the next printing.

The National Fire Protection Association has a standard procedure to introduce changes between *Code* cycles after the actual *NEC* is printed. These are called "Tentative Interim Amendments," or TIAs. NFPA also publishes errata or changes of an editorial nature to each edition of the *NEC*. TIAs and errata documents can be downloaded from the NFPA website, http://www.nfpa.org, to make your copy of the *Code* current.

Acknowledgments

Ray Mullin wishes to again thank his wife, Helen, for her understanding and support while he devoted unlimited time attending meetings and working many hours on revising this edition of *Electrical Wiring—Residential*. Major revisions such as this take somewhere between 1000 and 1500 hours! Patience is a virtue!

Phil Simmons once again wants to express his appreciation to his wife, Della, for her generosity in allowing him to devote so much time and effort to updating this book as well as *Electrical Wiring Commercial* and *Electrical Grounding and Bonding* to the new *NEC* during the year. Time after time she picked up the ball and ran with it on projects Phil would customarily attend to.

As always, the team at Cengage Learning has done an outstanding job in bringing this edition to press. Their drive, dedication, and attention to minute details ensure that this text, without question, is the country's leading text on house wiring. They sure know how to keep the pressure on!

Special thanks to our good friend Jimmy Carpenter, former Executive Director of the International Association of Electrical Inspectors, for his inspiring Foreword to this text regarding the "Importance of Proper Training."

We are so appreciative of our friends and colleagues in the electrical industry who have provided assistance and information. Several are respected as *Code* experts. These include Madeline Borthick, David Dini, Joe Ellwanger, David Hittinger, Mike Johnston, Robert Kosky, Richard Loyd, Bill Neitzel, Cliff Rediger, Gordon Stewart, Clarence Tibbs, Charlie Trout, and David Williams. Over the years, technical reviewers have provided invaluable suggestions to make *Electrical Wiring Residential* the best it can be! Special thanks again to Robert Boiko for his technical input on water heaters and their safety related controls.

We wish we could name all our friends in the electrical industry, but there are so many, it would take many pages to include all of their names. Thanks to all of you for your input. We apologize if we missed anyone.

The authors gratefully acknowledge the contribution of the chapter on Residential Utility Interactive Photovoltaic Systems by Pete Jackson, electrical inspector for the City of Bakersfield, CA.

The coauthors and publisher would like to thank the following reviewers for their contributions:

William Dunakin
Independent Electrical Contractors
West Hartford, CT

Chad Kurdi
Dunwoody College of Technology
Minneapolis, MN

Orville Lake
Augusta Technical College
Augusta, GA

Marvin Moak
Hinds Community College
Raymond, MS

Richard Moore
Wake Tech Community College
Raleigh, NC

Bill F. Neitzel
Madison, WI

Michael Ross
San Jacinto College
Pasadena, TX

Alan W. Stanfield
Southern Crescent Technical College
Griffin, GA

CHAPTER 1

General Information for Electrical Installations

OBJECTIVES

After studying this chapter, you should be able to

- understand the basic safety rules for working on electrical systems.
- access the Internet to obtain a virtual unlimited source of safety and technical related information.
- become familiar with important electrical codes, safety codes, and building codes such as NFPA 70, 70A, 70B, 70E, 73, OSHA, NIOSH, ADA, NRTL, and the ICC.
- learn about licensing, permits, plans, specifications, symbols, and notations.
- understand the role of the electrical inspector and the International Association of Electrical Inspectors.
- understand the metric system of measurement.
- understand the role of nationally recognized testing laboratories (NRTLs) and the necessity for listed equipment.

SAFETY IN THE WORKPLACE

Electricity is great when it is doing what it is intended to do, and that is to stay in its intended path and do the work intended. But electricity out of its intended path can be dangerous, often resulting in fire, serious injury, or death.

Before getting into residential wiring and the *National Electrical Code* (*NEC*®), we need to discuss on-the-job safety. Safety is not a joke! Electricians working on new construction, remodel work, maintenance, and repair work find that electricity is part of the work environment. Electricity is all around us, just waiting for the opportunity to get out of control. Repeat these words: **Safety First . . . Safety Last . . . Safety Always!**

The voltage level in a home is 120 volts between one "hot" conductor and the "neutral" conductor, or grounded surface. Between the two "hot" conductors (line-to-line), the voltage is 240 volts.

An electrical shock is received when electrical current passes through the body. From basic electrical theory, you learned that line voltage appears across an open in a series circuit. Getting caught "in series" with a 120-volt circuit will give you a 120-volt shock. For example, open-circuit voltage between the two terminals of a single-pole switch on a lighting circuit is 120 volts when the switch is in the OFF position and the lamp(s) are in place. See Figure 1-1. Likewise, getting caught "in series" with a 240-volt circuit will give you a 240-volt shock.

Working on equipment with the power turned on can result in death or serious injury, either as a direct result of electricity (electrocution or burns) or from an indirect secondary reaction, such as falling off a ladder

SAFETY ALERT

Working on switches, receptacles, luminaires, or appliances with the power turned on is dangerous. Turn off the power! In addition, check with a voltmeter to be sure the power is off. Safety procedures require that you test the voltage tester on a known live source, then test for absence of voltage, and finally test the voltage tester again on a live source to make certain it is operational.

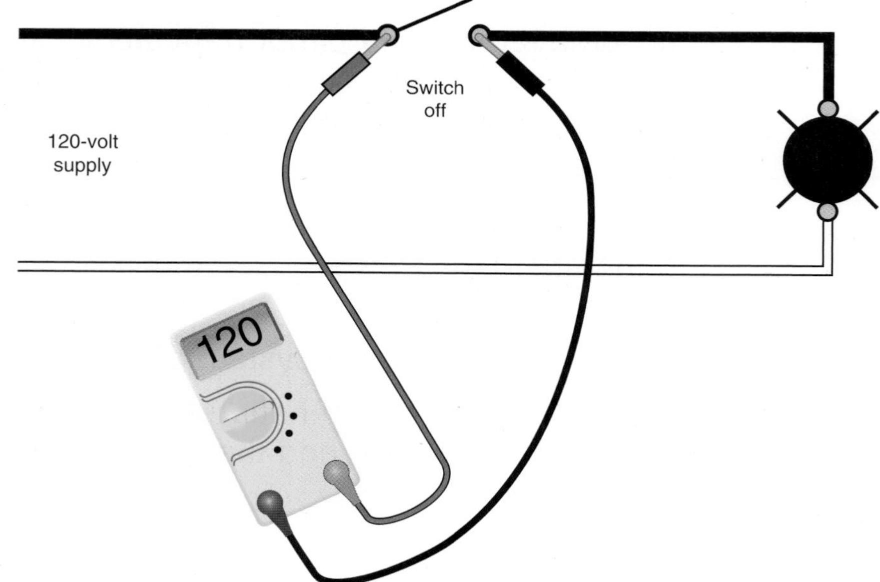

FIGURE 1-1 The voltage across the two terminals of the single-pole switch is 120 volts.

or jerking away from the "hot" conductor into moving parts of equipment such as the turning blades of a fan. For example: A workman was seriously injured while working a live circuit that supplied a piece of equipment. He accidentally came into contact with a "hot" terminal, and reflex action caused him to pull his hand back into a turning pulley. The pulley cut deeply into his wrist, resulting in a tremendous loss of blood.

Dropping a metal tool onto live parts, allowing metal shavings from a drilling operation to fall onto live parts of electrical equipment, cutting into a live conductor and a neutral conductor at the same time, or touching the live wire and the neutral conductor or a grounded surface at the same time can cause injury directly or indirectly.

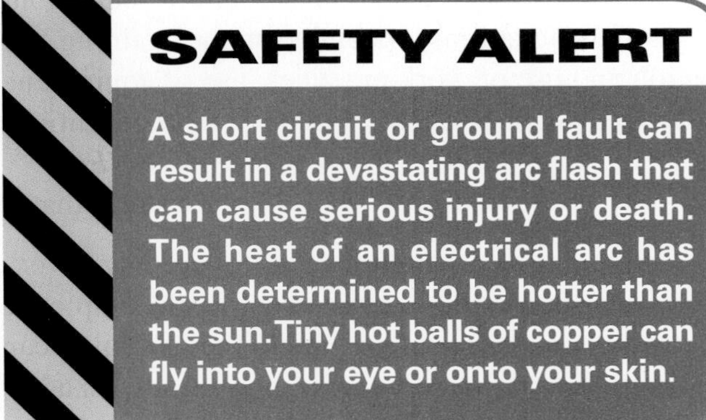

SAFETY ALERT

A short circuit or ground fault can result in a devastating arc flash that can cause serious injury or death. The heat of an electrical arc has been determined to be hotter than the sun. Tiny hot balls of copper can fly into your eye or onto your skin.

FIGURE 1-2 A typical disconnect switch with a lock and a tag attached to it. In the OSHA, ANSI, and NFPA standards, this is referred to as the lockout/tagout procedure.

Figure 1-2 shows a disconnect switch that has been locked and tagged. Figure 1-3 shows a circuit-breaker panelboard with a lock-off accessory installed. After the circuit breaker handle is moved to the off position, a padlock can be easily added to prevent the breaker from being turned on.

Lockout/tagout (sometimes called LOTO) is the physical restraint of all hazardous energy sources that supply power to a piece of equipment. It simply means putting a padlock on the switch or circuit breaker and attaching a warning tag.

Dirt, debris, and moisture can also set the stage for equipment failure and personal injury. Neatness and cleanliness in the workplace are a must.

FIGURE 1-3 A lock-off accessory that can be added to circuit breakers allows for adding a padlock and tag to individual circuit breakers in a panelboard.

What about Low-Voltage Systems?

Although circuits of less than 50 volts generally are considered harmless, don't get too smug when working on so-called low voltage. Low-voltage circuits are not necessarily low hazard. A slight tingle might cause a reflex. A capacitor that is discharging can give you quite a jolt, causing you to jump or pull back.

In commercial work, such as telephone systems with large battery banks, there is extreme danger even though the voltage is "low." Think of a 12-volt car battery. If you drop a wrench across the battery terminals, you will immediately see a tremendous and dangerous arc flash.

It is the *current* that is the harmful component of an electrical circuit. *Voltage* pushes the current through the circuit. If you're not careful, you might become part of the circuit.

Higher voltages can push greater currents through the body. Higher voltages like 240, 480, and 600 volts can cause severe skin burns and possibly out-of-sight injuries such as internal bleeding and/or destruction of tissues, nerves, and muscles.

It's the Law!

Not only is it a good idea to use proper safety measures as you work on and around electrical equipment, it is **required** by law. Electricians and electrical contractors need to be aware of these regulations. Practicing safety is a habit—like putting on your seat belt as soon as you get into your car.

The *NEC* is full of requirements that are safety related. For example, *430.102(B)* requires that a disconnecting means be located in sight from the motor location and the driven machinery location. This section also has in-sight and lock-off requirements, discussed in detail in Chapter 19.

The *NEC* in *Article 100* defines a qualified person as: One who has skills and knowledge related to the construction and operation of the electrical equipment and installations and has received safety training to recognize and avoid the hazards involved.* Merely telling someone or being told to be careful does not meet the definition of proper training and does not make the person qualified. An individual qualified in one skill might very well be unqualified in other skills.

According to NFPA 70E, *Electrical Safety in the Workplace*, circuits and conductors are not considered to be in an electrically safe condition until all sources of energy are removed, the disconnecting means is under lockout/tagout, and the absence of voltage is verified by an approved voltage tester.

The US Department of Labor **Occupational Safety and Health Administration (OSHA)** regulations (Standards–29 CFR) is the law! This entire standard relates to safety in the workplace for general industry. Specifically, Part 1910, Subpart S, involves electrical safety requirements. The letters *CFR* stand for *Code of Federal Regulations.*

Key topics in the standard are electric utilization systems, wiring design and protection, wiring methods, components and equipment for general use, specific purpose equipment and installations, hazardous (classified) locations, special systems, training, selection and use of work practices, use of equipment, safeguards for personnel protection, and definitions (a mirror image of definitions found in the *NEC*).

A direct quote from 1910.333(a)(1) states that

> "Live parts to which an employee may be exposed shall be de-energized before the employee works on or near them, unless the employer can demonstrate that de-energizing introduces additional or increased hazards or is infeasible due to equipment design or operational limitations. Live parts that operate at less than 50 volts to ground need not be de-energized if there will be no increased exposure to electrical burns or to explosion due to electric arcs."**

OSHA 1910.333(c)(2) states that:

> "Only qualified persons may work on electric circuit parts or equipment that have not been de-energized under the procedures of paragraph (b) of this section. Such persons shall be capable of working safely on energized circuits and shall be familiar with the proper use of special precautionary techniques, personal protective equipment, insulating and shielding materials, and insulated tools."**

OSHA 1910.399 defines a *qualified person* as "One familiar with the construction and operation of the equipment and the hazards involved," almost the same definition as that in the *NEC.*

*Reprinted with permission from NFPA 70-2014.

**Source: US Department of Labor Occupational Safety and Health Administration (OSHA).

For the most part, turning the power off and then locking and tagging the disconnecting means is the safest practice. As the OSHA regulations state:

> "A lock and a tag shall be placed on each disconnecting means used to de-energize circuits and equipment."

Part 1926 in the OSHA regulation (Standards–29 CFR) deals with *Safety and Health Regulations for Construction*. Here we find the rules for anyone working in the construction industry, not just in the electrical field. A few of the topics are medical services and first aid, safety training and education, recording and reporting injuries, housekeeping, personal protective equipment, means of egress, head protection, hearing protection, eye and face protection, ladders, scaffolds, rigging, hand and power tools, electrical requirements (a repeat of Part 1910, Subpart S), fall protection, and required signs and tags.

PERSONAL PROTECTIVE EQUIPMENT

Safety courses refer to personal protective equipment (PPE). These include such items as rubber gloves, insulating shoes and boots (footwear suitable for electrical work is marked with the letters "EH"), face shields, safety glasses, hard hats, ear protectors, Nomex™, and similar products. OSHA 1910.132(f)(1) requires that

> "The employer shall provide training to each employee who is required by this section to use PPE."

Working on electrical equipment while wearing rings and other jewelry is not acceptable. OSHA states that

> "Conductive articles of jewelry and clothing (such as watch bands, bracelets, rings, key chains, necklaces, metalized aprons, cloth with conductive thread, or metal headgear) may not be worn if they might contact exposed energized parts. However, such articles may be worn if they are rendered nonconductive by covering, wrapping, or other insulating means."**

**Source: US Department of Labor Occupational Safety and Health Administration (OSHA).

ARC FLASH

Don't get too complacent when working on electrical equipment.

A major short circuit or ground fault at the main service panel or at the meter cabinet or base can deliver a lot of energy. On large electrical installations, an arc flash (also referred to as an arc blast) can generate temperatures of 35,000°F (19,427°C). This is hotter than the surface on the sun. This amount of heat will instantly melt copper, aluminum, and steel. The blast will blow hot particles of metal and hot gases all over, resulting in personal injury, fatality, and/or fire. An arc flash also creates a tremendous air pressure wave that can cause serious hearing damage and/or memory loss due to the concussion. The blast might blow the victim away from the arc source.

Don't be fooled by the size of the service. Typical residential services are 100, 150, and 200 amperes. Larger services are found on large homes. Electricians seem to feel out of harm's way when working on residential electrical systems and seem to be more cautious when working on commercial and industrial electrical systems. A fault at a small main service panel can be just as dangerous as a fault on a large service. The available fault current at the main service disconnect for all practical purposes is determined by the kVA rating and impedance of the transformer. Other major limiting factors for fault current are the size, type, and length of the service-entrance conductors. Available fault current can easily reach 22,000 amperes, as is evident by panels that have a 22,000/10,000-ampere series rating.

Short-circuit calculations are discussed in Chapter 28 of this text.

Don't be fooled into thinking that if you cause a fault on the load side of the main disconnect that that main breaker will trip off and protect you from an arc flash. An arc flash will release the energy that the system is capable of delivering for as long as it takes the main circuit breaker to open. How much current (energy) the main breaker will let through is dependent on the available fault current and the breaker's opening time.

Although not required for house wiring, *NEC 110.16* requires that ▶Electrical equipment, such as switchboards, switchgear, panelboards, industrial control panels, meter socket enclosures, and motor control centers, that are in other than dwelling units, and are likely to require examination, adjustment, servicing, or maintenance while energized shall be field or factory marked to warn qualified persons of potential electric arc flash hazards. The marking shall meet the requirements in 110.21(B) and shall be located so as to be clearly visible to qualified persons before examination, adjustment, servicing, or maintenance of the equipment.◄* More information on this subject is found in NFPA 70E and in the ANSI Standard Z535.4, *Product Safety Signs and Labels.*

SAFETY ALERT

When turning a standard disconnect switch ON, don't stand in front of the switch. Instead, stand to one side. For example, if the handle of the switch is on the right, then stand to the right of the switch, using your left hand to operate the handle of the switch, and turn your head away from the switch. That way, if an arc flash occurred when you turned the disconnect switch ON, you would not be standing in front of the switch. You would not have the switch's door fly into your face, and the molten metal particles resulting from the arc flash would fly past you.

Classifying Electrical Injuries

OSHA recognizes these as the four main types of electrical injuries:

- Electrical shock (touching live line-to-line or line-to ground conductors) (ground-fault circuit interrupters are discussed in Chapter 6)
- Electrocution (death due to severe electrical shock)
- Burns (from an arc flash)
- Falls (an electrical shock that might cause a person to lose balance, pull back, jump, or fall off a ladder)

What to Do If You Are Involved with a Possible Electrocution

The following is taken in part from the OSHA, NIOSH, NSC regulations, and the American Heart Association recommendations. These are steps that should be taken in the event of a possible electrocution (cardiac arrest). Refer to the actual cardiopulmonary resuscitation (CPR) instructions for complete and detailed requirements, and to take CPR training.

- "First, you must recognize that an emergency exists. Timing is everything. The time between the accident and arrival of paramedics is crucial. Call 911 immediately. Don't delay.
- Don't touch the person if he or she is still in contact with the live circuit.
- Shut off the power.
- Stay with the person while someone else contacts the paramedics, who have training in the basics of life support. In most localities, telephoning 911 will get the paramedics.
- Have the caller verify that the call was made and that help is on the way.
- Don't move the person.
- Check for bleeding; stop the bleeding if it occurs.
- If the person is unconscious, check for breathing.
- The *ABCs* of CPR are: *a*irway must be clear; *b*reathing is a must, either by the victim or the rescuer; and *c*irculation (check pulse).
- Perform CPR if the victim is not breathing—within 4 minutes is critical. If the brain is deprived of oxygen for more than 4 minutes, brain damage will occur. If it is deprived of oxygen for more than 10 minutes, the survival rate is 1 in 100. CPR keeps oxygenated blood flowing to the brain and heart.
- Defibrillation may be necessary to reestablish a normal heartbeat. Ventricular fibrillation is common with electric shock, which causes the

*Reprinted with permission from NFPA 70-2014.

heartbeat to be uneven and unable to properly pump blood.

- By now, the trained paramedics should have arrived to apply advanced care.
- When it comes to an electrical shock, *timing is everything!"***

Who Is Responsible for Safety?

You are!

The electrical inspector inspects electrical installations for compliance to the *NEC*. He or she is not really involved with on-the-job safety.

For on-the-job safety, OSHA puts the burden of responsibility on the employer. OSHA can impose large fines for noncompliance with its safety rules. But because it's your own safety that we are discussing, you share the responsibility to apply safe work practices, use the proper tools and PPE equipment the contractor furnishes, and install listed electrical equipment. Be alert to what's going on around you! Do a good job of housekeeping!

Tools

Using the proper tools for a job is vital to on-the-job safety.

OSHA Standard 1926.302 specifically covers the requirements for hand and power tools. The American National Standards Institute (ANSI) also has standards relating to tools.

If you want to learn more about tools, visit the website of the Hand Tools Institute at www.hti.org. The institute has a number of excellent safety education materials available. Of particular interest is its 90-plus-page publication *Guide to Hand Tools: Selection, Safety Tips, Proper Use and Care* that includes topics for selecting, proper use, maintaining, and avoiding hazards, as well as special emphasis on eye protection using all types of hand tools.

Electrical Power Tools

You will be using portable electric power tools on the job. Electricity on construction sites is usually in the form of temporary power, covered by *Article 590* of the *NEC*.

Courtesy Hubbell Lighting Outdoor & Industrial.

FIGURE 1-4 Two types of portable plug-in cord sets that have built-in GFCI protection.

NEC 590.6(A)(1) requires that All 125–volt, single-phase, 15-, 20-, and 30-ampere receptacle outlets that are not a part of the permanent wiring of the building or structure and that are in use by personnel shall have ground-fault circuit interrupter protection for personnel.*

Because this requirement is often ignored or defeated on job sites, you should carry and use as part of your tool collection a portable GFCI of the types shown in Figure 1-4—an inexpensive investment that will protect you against possible electrocution. Remember, *"The future is not in the hands of fate, but in ours."****

*Reprinted with permission from NFPA 70-2014.

**Source: US Department of Labor Occupational Safety and Health Administration (OSHA).

***Jean Jules Jusserand, http://www.inspirationalstories.com/quotes/t/on-future/page/48/

Digital Multimeters

Some statistics show that more injuries occur from using electrical meters than from electric shock.

For safety, electricians should use quality digital multimeters that are *category rated*. The International Electrotechnical Commission (IEC) Standard 1010 for *Low Voltage Test, Measurement, and Control Equipment* rates the ability of a meter to withstand voltage transients (surges or spikes). This standard is very similar to UL Standard 3111. When lightning strikes a high line; when utilities are performing switching operations; or when a capacitor is discharging, a circuit can "see" voltage transients that greatly exceed the withstand rating of the digital multimeter. The meter could explode, causing an arc flash (a fireball) that in all probability would result in personal injury. A properly selected category-rated digital multimeter is able to withstand the spike without creating an arc blast. The leads of the meter are also able to handle high transient voltages.

Digital multimeters also are category rated based on the location of the equipment to be tested, because the closer the equipment is to the power source, the greater the danger from transient voltages.

Cat IV multimeters are used where the available fault current is high, such as a service entrance, a service main panel, service drops, and the house meter.

Cat III multimeters are used for permanently installed loads such as in switchgear, distribution panels, motors, bus bars, feeders, short branch circuits, and appliance outlets where branch-circuit conductors are large and the distance is short.

Cat II multimeters are used on residential branch circuits for testing loads that are plugged intoreceptacles.

Cat I multimeters are used where the current levels are very low, such as electronic equipment. Note that the lower the category rating, the lower is the meter's ability to withstand voltage transients. If you will be using the multimeter in all of the above situations, select the higher category rating.

Category-rated digital multimeters also contain fuses that protect against faults that happen when the meter is accidentally used to check voltage while it is inadvertently set in the current reading position.

To learn more about meters, visit the website of Fluke Corporation, http://www.fluke.com, for a wealth of technical information about the use of meters and other electrical and electronic measuring instruments.

Ladders

To learn more about ladders, visit the website of Werner Ladder Company, http://www.wernerco.com. You can download their pamphlet entitled *Ladder Safety Tips*. You will learn about the right and wrong ways to use a ladder: Never work on a stepladder in which the spreaders are not fully locked into position; the 4:1 ratio, which means that the base of an extension ladder should be set back (S) one-fourth the length (L) of where the upper part of the ladder is supported (S = ¼ L); the duty ratings, such as do not stand higher than the second step from the top for stepladders, and do not stand higher than the fourth rung from the top for extension ladders; plus many more safety tips.

Ladders are labeled with their duty rating. Medium-duty commercial (Type II—225#), heavy-duty industrial (Type I—250#), and extra-heavy-duty (Type IA—300#) ladders bear an OSHA compliance label. Light-duty household (Type III—200#) ladders do not bear an OSHA logo.

Hazardous Chemicals

Increasingly, hazardous chemicals are found on the job. What do you do if you get a spilled chemical on your skin or in your eyes, or if you breathe the fumes?

Every manufacturer of these products is required to publish and make available a comprehensive data sheet called the Material Safety Data Sheet (MSDS). There are supposedly over 1.5 million of these data sheets, containing product identification, ingredients, physical data, fire and explosion hazard data, health-hazard data, reactive data, spill or leak procedures, protection information, and special precautions.

The least you can do is to be aware that this information is available. Apprenticeship programs include some training about MSDSs.

You can learn more about MSDSs by checking any search engine for the letters MSDS.

TRAINING

If you want to learn more, visit manufacturers' websites. For example, Bussmann's website is http://www.bussmann.com. It is easy to use and has an application for "smartphones" that can be downloaded for making arc-flash and fault-current calculations. This website also has a technical publication, *Selecting Protective Devices*, or bulletin SPD, a 268-plus-page publication about overcurrent protection selection, application, the *NEC*, and safety.

The OSHA Training Institute offers outreach training programs of interest to electricians, contractors, and instructors. The basic safety courses for general construction safety and health are the OSHA 10-hour and OSHA 30-hour courses. Instructors interested in becoming an outreach trainer for the 10- and 30-hour courses must complete the OSHA 500 course entitled "Trainer Course in Occupational Safety and Health Standards for the Construction Industry." To become an outreach trainer, you must pass a test. Before the end of 4 years, outreach trainers must take the OSHA 502 update course for the construction industry or the OSHA 502 update course "Update for Construction Industry Outreach Trainers." Completion cards are issued on completion of these courses.

Other courses, publications, "free loan" videos, schedules of upcoming safety training seminars, and other important information relating to safety on the job are available from OSHA for electricians, contractors, and trainers.

Visit the OSHA website at http://www.OSHA.gov for everything there is to know about OSHA safety requirements in the workplace. The OSHA website is a virtual gold mine of information relating to safety on the job.

Another valuable source of safety information is the National Institute for Occupational Safety and Health (NIOSH), a division of the Department of Health and Human Services Centers for Disease Control and Prevention (CDC). Check out its website at http://www.cdc.gov/niosh. NIOSH offers an excellent downloadable 80-plus-page manual on *Electrical Safety*.

The National Safety Council has a vast amount of information relative to all aspects of safety. Check out its website at http://www.nsc.org.

The Consumer Product Safety Commission offers many safety publications for downloading. Visit its website at http://www.cpsc.gov, click on Library, click on CPSC Publications, click on By General Category, and then click on Electrical Safety. Here you will see a list of CPSC publications about GFCIs, AFCIs, metal ladder hazards, home wiring hazards, repairing aluminum wiring, and others.

The National Fire Protection Association offers many publications, videos, and a training course relating to safety. Browse its website at http://www.nfpa.org.

NFPA 70E, *Standard for Electrical Safety in the Workplace and NFPA 70B, Recommended Practices for Electrical Equipment Maintenance*, present much of the same text regarding electrical safety as does the OSHA regulation.

Accredited apprenticeship training programs incorporate safety training as an integral part of their curriculum.

SAFETY CANNOT BE COMPROMISED!

It is impossible to put a dollar value on a life.

Don't take chances! Use the right tools! Turn off the power. Follow a lockout/tagout procedure. Mark the tag with a description of exactly what that particular disconnect controls.

How many times have we heard "The person would not have been injured (or electrocuted) had he turned the power OFF"? How many more times can we say it? **Turn OFF the power before working on the circuit!**

Visit the websites of the various organizations mentioned earlier. The website list can also be found in the back of this text. These organizations have a wealth of information about on-the-job safety educational material and safety training courses.

Check out the website of the Electrical Safety Foundation International (ESFI) at http://www.electrical-safety.org. This organization has a tremendous amount of down-to-earth, simple-to-understand electrical safety material. Some of their educational material is free; other items are priced. Certain items are downloadable. The bottom line is to reduce deaths and injuries from preventable electrical accidents.

LICENSING AND PERMITS

Several communities, counties, and/or states require electricians and electrical contractors to be licensed. This usually means they have taken and passed a test. Often, these regulatory agencies have minimum qualifications that must be met before the applicant is permitted to take an examination. To maintain a valid license, many licensing agencies require electricians and electrical contractors to attend and satisfactorily complete approved continuing education courses consisting of a specified number of classroom hours over a given period of time. Quite often, a community will have a "Residential Only" license for electricians and contractors that limits their activity to house wiring.

Permits are a means for a community to permanently record electrical work to be done and who is doing the work, and to schedule inspections during and after the rough-in stage and in the final stages of construction. Usually, permits must be issued prior to starting an electrical project. In most cases, homeowners are allowed to do electrical work in their own home where they live, but not in other properties they might own.

Figure 1-5 is a simple application for an electrical permit form. Some permit application forms are much more detailed.

APPLICATION FOR ELECTRICAL PERMIT

VILLAGE OF ANYWHERE, USA 1-234-567-8900, EXT. 1234

Date ______ Permit No. ______
Owner ______ Job Address ______
Telephone No. ______ Job Start Date ______

CONTRACTOR INFORMATION AND SIGNATURE

Electrical Contractor ______ Tel. No. ______
Address ______ City ______ State ______ Zip ______
Registration No. ______ City of Registration ______
Supervising Electrician (Please Print Name) ______
Supervising Electrician's Signature ______
Insurance Bond ______ Village Business License ______

SERVICE INSPECTIONS OR REVISIONS

Existing Service Size: Amps ______ Volts ______ No. of Circuits ______ No. Added ______
New Service Size: Amps ______ Volts ______ No. of Circuits ______
Type: Overhead ______ Underground ______
Service Installation Fees: 100–200 amps $50 ______ over 200 amps $75 ______

MOTORS AND AIR-CONDITIONING EQUIPMENT

No. of Motors:up to 1 HP @ $50 ______ Over 1–10 HP @ $50 ______
11–25 HP @ $25 ______ Over 25 HP @ $25 ______
Air Conditioner/Heat Pump:
No. of Tons ______ ($20 for first ton, $5 for each additional ton) ______
Furnace (electric): kW ______ Amps ______ ($25) ______
Dryer (electric): kW ______ Amps ______ ($10) ______
Range, oven, cooktop (electric): Total kW ______ Amps ______ ($10 each) ______
Water Heater (electric): kW ______ Amps ______ ($10) ______

TYPE OF MISCELLANEOUS ELECTRIC WORK

______
______

Minimum Inspection Fee: $40.00 ______
Escrow Deposit, if applicable ______
TOTAL DUE ______

FIGURE 1-5 A typical Application for Electrical Permit.

If you are not familiar with licensing and permit requirements in your area, it makes sense to check this out with your local electrical inspector or building department before starting an electrical project. Not to do so could prove to be very costly. Many questions can be answered: you will find out what tests, if any, must be taken; which permits are needed; which electrical code is enforced in your community; minimum size electrical service; and so on. Generally, the electrical permit is taken out by an electrical contractor who is licensed and registered as an electrical contractor in the jurisdictional area.

For new construction or for a main electrical service change, you will also need to contact the electric utility.

Temporary Wiring

There is an ever-present electrical shock hazard on construction sites. The *NEC* addresses this in *Article 590*. This is covered in Chapter 6 of this text.

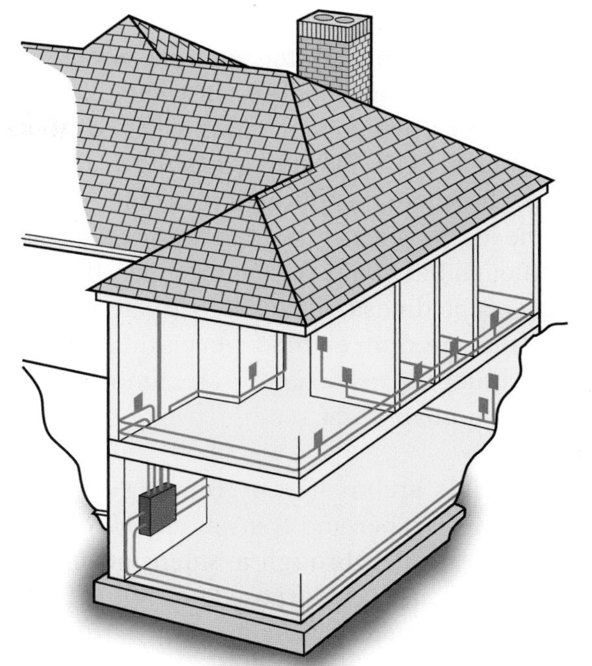

FIGURE 1-6 Three-dimensional view of house wiring.

Construction Terms

Electrical Wiring—Residential covers all aspects of typical residential wiring, with focus on the *NEC*. Electricians work with others on construction sites. Knowing construction terms and symbols is therefore a key element to getting along with the other workers. A rather complete dictionary of construction terms can be found on http://www.constructionplace.com/glossary.asp. Architectural symbols are found in this text in Chapter 2 and in the Appendix.

PLANS

An architect or electrical engineer prepares a set of drawings that shows the necessary instructions and details needed by the skilled workers who are to build the structure. These are referred to as plans, prints, blueprints, drawings, construction drawings, or working drawings. The sizes, quantities, and locations of the materials required and the construction features of the structural members are shown at a glance. These details of construction must be studied and interpreted by each skilled construction craftsperson—masons, carpenters, electricians, and others—before the actual work is started.

The electrician must be able to (1) convert the 2-dimensional plans into an actual electrical installation and (2) visualize the many different views of the plans and coordinate them into a 3-dimensional picture, as shown in Figure 1-6.

The ability to visualize an accurate 3-dimensional picture requires a thorough knowledge of blueprint reading. Because all of the skilled trades use a common set of plans, the electrician must be able to interpret the lines and symbols that refer to the electrical installation and also those used by the other construction trades. The electrician must know the structural makeup of the building and the construction materials to be used.

Plans might be black line prints, which are simply photocopies; diazo prints, referred to as white line prints; or computer-aided drawings (CAD), which are most commonly used today. The need for the draftsman has for the most part become history. We now need CAD technicians and engineers.

Today's high-tech computers and wide-frame printers allow for storage and backup of clear and concise drawings, standardization of lines in symbols, ease of making revisions, the ability to zoom in and out when viewed on a monitor, simplicity of printing (black or color), and in many cases the ability to view the structure in 2D or 3D, turn a drawing to any angle, and view the more complicated plans in layers. The larger the job, the greater the need for more detail.

Although blueprints for the most part have gone the way of the dinosaur, they are still around. Electricians that are from an older generation are familiar with blueprints. Those of you coming into the trade will probably not see blueprints on the jobs you work on.

Blueprints are created by running yellowish, light-sensitive paper and transparent Mylar® with black images on it through a blueprint machine, where the two sheets are exposed to a bright light for a short period of time. The light-sensitive paper turns white in all places except where the black images are. Where the black images are, the light-sensitive paper turns blue. That is why it is called a blueprint.

A tear-out/fold-out set of full-size plans for the residence referred to throughout this text is included in the back of the book.

A few of the more common sizes for blueprints are

- Size C (17 inches by 22 inches),
- Size B (22 inches by 34 inches), and
- Size E (34 inches by 44 inches).

SPECIFICATIONS

The specifications (specs, for short) for the electrical work indicated on the plans for the residence discussed throughout this text are found in the back of the book, before the Appendix.

Working drawings are usually complex because of the amount of information that must be included. To prevent confusing detail, it is standard practice to include with each set of plans a set of detailed written specifications prepared by the architect.

These specifications provide general information to be used by all trades involved in the construction. In addition, specialized information is given for the individual trades. The specifications include information on the sizes, the type, and the desired quality of the standard parts to be used in the structure.

Typical specifications include a section on "General Clauses and Conditions," which is applicable to all trades involved in the construction. This section is followed by detailed requirements for the various trades—excavating, masonry, carpentry, plumbing, heating, electrical work, painting, and others.

In the electrical specifications, the listing of standard electrical parts and supplies frequently includes the manufacturers' names and the catalog numbers of the specified items. Such information ensures that these items are of the correct size, type, and electrical rating, and that the quality meets a certain standard. To allow for the possibility that the contractor will not always be able to obtain the specified item, the phrase "or equivalent" is usually added after the manufacturer's name and catalog number.

The specifications are also useful to the electrical contractor in that all of the items needed for a specific job are grouped together, and the type or size of each item is indicated. This information allows the contractor to prepare an accurate cost estimate without having to find all of the data on the plans.

If there is a difference between the plans and specifications, the specifications will take preference. The electrical contractor should discuss the matter with the homeowner, architect, and engineer. The cost of the installation might vary considerably because of the difference(s), so obtain any changes to the plans and/or specifications in writing.

SYMBOLS AND NOTATIONS

The architect uses **symbols** and **notations** to simplify the drawing and presentation of information concerning electrical devices, appliances, and equipment. For example, an electric range outlet symbol is shown in Figure 1-7.

Most symbols have a standard interpretation throughout the country as adopted by ANSI. Symbols are described in detail in Chapter 2.

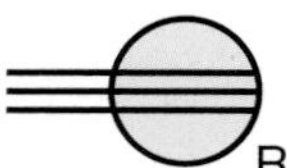

FIGURE 1-7 An electric range outlet symbol.

A notation is generally found on the plans (blueprints) next to a specific symbol calling attention to a variation, type, size, quantity, or other necessary information. In reality, a symbol might be considered to be a notation because symbols represent words, phrases, numbers, quantities, and so on.

Another method of using notations to avoid cluttering up a blueprint is to provide a system of symbols that refer to a specific table. For example, the written sentences on plans could be included in a table referred to by a notation. Figure 2-11 on page 39 is an example of how this could be done. The special symbols that refer to the table would have been shown on the actual plan.

NATIONAL ELECTRICAL CODE (NEC)

The *NEC* is the electrical *Code* standard recognized by everyone in the electrical industry.

The first sentence in the *NEC* is found in *90.1*. This sentence lays the foundation for all electrical installations. It states, ▶The purpose of this *Code* is the practical safeguarding of persons and property from hazards arising from the use of electricity. This *Code* is not intended as a design specification or an instruction manual for untrained persons.◀*

It goes on to state, This *Code* contains provisions that are considered necessary for safety. Compliance therewith and proper maintenance will result in an installation that is essentially free from hazard but not necessarily efficient, convenient, or adequate for good service or future expansion of electrical use.*

As you study this text, you will learn how to "Meet *Code*" for wiring a typical house, not "Beat *Code*."

As *NEC* requirements are discussed throughout this text, the sheer number of *Code* references can become mind-boggling. To simplify using this text as a reference, in addition to the conventional subject index at the back of the text, there is a Cross Index, making it easy for you to pinpoint specific *Code* sections and articles found in this text.

*Reprinted with permission from NFPA 70-2014.

The *NEC* is published by the National Fire Protection Association and is referred to as NFPA 70. The *NEC* was first published in 1897. It is revised every 3 years so as to be as up to date as possible. The *NEC* does not become law until adopted by official action of the legislative body of a city, municipality, county, or state. Because of the ever-present danger of fire or shock hazard through some failure of the electrical system, the electrician and the electrical contractor must use listed materials and must perform all work in accordance with recognized standards.

Other Electrical Codes

In addition to the *NEC*, you must also consider local and state electrical regulations and codes. Many agencies create local amendments or additional regulations when they adopt the *NEC*. These local regulations supercede the requirements of the *NEC* and become the minimum level of installation of electrical equipment. Don't even think about planning the electrical installation project without learning the local regulations!

In addition, most every electrical serving utility has requirements for the electrical supply to the service equipment. These requirements must be complied with as a condition for receiving electrical energy from the power company. Many electric utilities have a booklet that can be downloaded from their Internet site. These booklets usually include drawings that illustrate their requirements.

Code Arrangement

It is important to understand the arrangement of the *NEC*, which is laid out in a very efficient and precise manner, as stated in *90.3* and illustrated in Figure 1-8.

Wiring of dwellings is covered mostly in *Chapters 1* through *4* of the *NEC*. The requirements for grounding and bonding of spas, hot tubs, and swimming pools in *Article 680* is an example of how the rules in *Chapter 6* modify the rules in *Article 250*, which is in *Chapter 2*.

Table 1-1 provides additional information on the content of the chapters and *Informative Annexes* of the *NEC*.

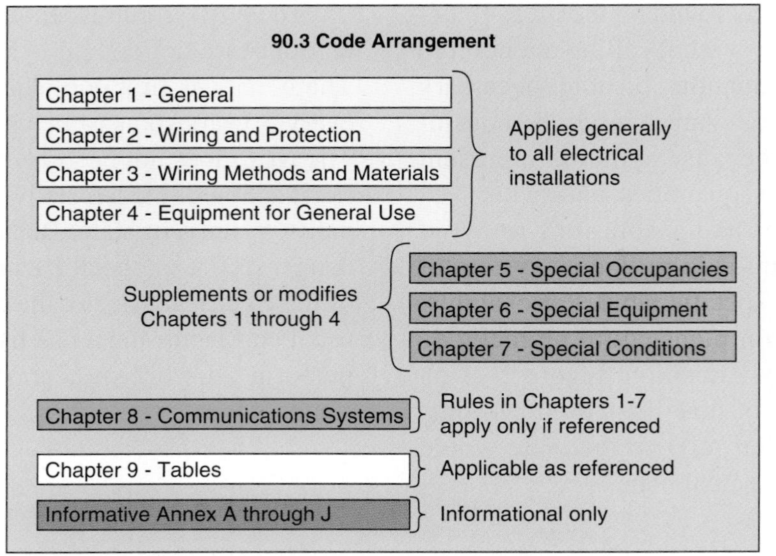

FIGURE 1-8 NEC Figure 90.3 illustrates the arrangement of the *NEC*.

TABLE 1-1

Arrangement of the *NEC*.

Article 90	An introduction to the *NEC*. Explains what is and what is not covered in the *NEC*, the arrangement of the *NEC*, who has the authority to enforce the *Code*, what mandatory rules are, what permissive rules are, and the basics of metric vs. inch-pound measurements found throughout the *NEC*.
Chapter 1	General *Article 100: Definitions* *Article 110: Requirements for Electrical Installations* Applies to all electrical installations.
Chapter 2	*Wiring and Protection* *Articles 200–285* Applies to all electrical installations.
Chapter 3	*Wiring Methods* *Articles 300–399* Applies to all electrical installations.
Chapter 4	*Equipment for General Use* *Articles 400–490* Applies to all electrical installations.
Chapter 5	*Special Occupancies* *Articles 500–590*
Chapter 6	*Special Equipment* *Articles 600–695*
Chapter 7	*Special Conditions* *Articles 700–770*
Chapter 8	*Communications Systems* *Articles 800–840* The articles in *Chapter 8* are not subject to the requirements of *Articles 1* through *7* unless specifically stated.
Chapter 9	Tables showing dimensional data for raceways and conductors, resistance and reactance values of conductors, Class 2 and Class 3 circuit power limitations.

(*continues*)

TABLE 1-1 *(Continued)*

Informative Annex A	Provides a comprehensive list of product safety standards from Underwriters Laboratories (UL).
Informative Annex B	Provides data for determining conductor ampacity where there is engineering supervision.
Informative Annex C	Has tables showing the maximum number of conductors permitted in various types of raceways.
Informative Annex D	Examples of load calculations.
Informative Annex E	Types of building construction.
Informative Annex F	*Availability and Reliability of Critical Operations Power Systems*
Informative Annex G	*Supervisory Control and Data Acquisitions (SCADA)*
Informative Annex H	*Administration and Enforcement. A* Comprehensive suggested typical electrical ordinance that could be adopted on a local level.
Informative Annex I	*Recommended Tightening Torque Tables from UL Standard 486A-B*
Informative Annex J	*ADA Standards for Accessible Design*
Index	The alphabetical index for the *NEC.*

Note: *Annexes* and *Informational Notes* are informational only. They are not mandatory.

NEC 90.3 covers the arrangement of the *NEC* and reads, This *Code* is divided into the introduction and nine chapters, as shown in Figure 90.3. Chapters 1, 2, 3, and 4 apply generally; Chapters 5, 6, and 7 apply to special occupancies, special equipment, or other special conditions. These latter chapters supplement or modify the general rules. Chapters 1 through 4 apply except as amended by Chapters 5, 6, and 7 for the particular conditions.

Chapter 8 covers communications systems and is not subject to the requirements of Chapters 1 through 7 except where the requirements are specifically referenced in Chapter 8.

Chapter 9 consists of tables that are applicable as referenced.

Informative annexes are not part of the requirements of this *Code* but are included for informational purposes only.*

LANGUAGE CONVENTIONS

The *National Electrical Code* is intended for mandatory adoption by authorities having jurisdiction. As such, it is very important that the language used in the *Code* be suitable for mandatory enforcement. *NEC 90.5* provides an explanation of mandatory rules, permissive rules, and explanatory material. Other requirements for writing the *National Electrical Code* are contained in the *NEC Style Manual*. These rules help ensure uniformity throughout the *NEC*.

- Mandatory rules identify what is required or prohibited, and use the term *shall* or *shall not*.
- Permissive rules are actions that are allowed, but not required. Permissive rules use the term *shall be permitted* or *shall not be required*.
- Explanatory material is identified as an *Informational Note*. *Informational Notes* may make reference to other important rules or provide helpful information related to the *Code* itself. These *Informational Notes* are not intended to be enforceable. If more than one *Informational Note* is applicable to a *Code* rule, they are numbered sequentially.

Exceptions

The *NEC Manual of Style* gives instructions on how Exceptions are to be used in the *NEC*. Although there has been an effort in recent years to reduce the number of exceptions used in the *NEC*, in some cases they remain the best method of rule construction. When exceptions are used, the general requirement is stated first, followed by one or more modifications of the general requirement. Often the exception contains a condition that must be met for the exception to apply.

*Reprinted with permission from NFPA 70-2014.

- Exceptions are required to immediately follow the main rule to which they apply. If exceptions are made to items within a numbered list, the exception must clearly indicate the items within the list to which it applies. Exceptions containing the mandatory terms *shall* or *shall not* are to be listed first in the sequence. Permissive exceptions containing *shall be permitted* are to follow any mandatory exceptions and should be listed in their order of importance as determined by the Code-Making Panel.
- If used, exceptions are to convey alternatives or differences to a basic Code rule. The terms *shall* and *shall not* are used to specify a mandatory requirement that is either different from the rule or diametrically opposite to the rule. The term *shall be permitted* designates a variance from the main rule that is permitted but not required.
- See *250.110* for an example of three exceptions to the general rule. The exceptions present a different set of conditions for providing relief from the general rule.

Citing *Code* References

Every time an electrician makes a decision concerning the electrical wiring, the decision should be checked by reference to the *Code*. This may be done from memory, without actually using the *Code* book. Since requirements in the Code change from time to time, the *Code* should be referenced directly—just to make sure. When the *Code* is referenced, it is a good idea to record the location of the information in the *Code* book—this is referred to as "citing the *Code* reference." Electrical inspectors should always give a reference, preferably in writing, for any correction they ask be made. If they cannot cite the site of the rule, they should not cite the installation!

There is a very exact way that the location of a *Code* item is to be cited. The various levels of *Code* referencing are shown in Table 1-2. Starting at the top of the table, each step becomes a more specific reference. If a person references *Chapter 1*, this reference includes all the information and requirements that are set forth in several pages. When citing a specific *Section* or an *Exception*, only a few words may be included in the citation. The electrician and inspector should be as specific as possible when citing the *Code*. For the most part, the word *section* does not precede the section numbers in the *Code*.

Common Numbering

Finding similar requirements in different *Articles* in the current edition of the *NEC* is much easier

TABLE 1-2

Citing the *NEC*.

DIVISION	DESIGNATION	EXAMPLE
Chapter	1–9	*Chapter 1*
Article	90 through 840	*Article 250*
Part	Roman numeral	*Article 250, Part II*
Section	Article number, a dot (period), plus one, two, or three digits	*250.20*
Paragraph	Section designation, plus uppercase letter in ( ), followed by digit in ( ), followed by a lowercase letter in ( ) as is required	*250.119(A)(1)*
List	Usually follows an opening paragraph or section	*285.23(B), (1), (2), (3), and (4)*
Exception to	Follows a rule that applies generally and applies under the conditions included in the Exception	*Exception No. 1* to *250.24(B)* or *250.61(B) Exception 3a*
Informational Note	Explanatory material, such as references to other standards, references to related sections, or information related to a Code rule. Such notes are informational only and are not enforceable.	*Informational Note*
Informative Annex	A, B, C, D, E, F, G, H, I, and J (are not part of the *NEC* and are not enforceable)	*Informative Annex A*

than in past editions. All articles in *Chapter 3* of the *NEC* cover wiring methods. In each article, the same section number has been assigned for a particular requirement. Here are a few examples:

- *Scope* is found in *XXX.1 such as 320.1, 330.1 and 344.1*
- *Definitions* (if present in the *Article*) are found in *XXX.2* such as *240.2, 517.2* and *680.2.*
- *Permitted Uses* are found in *XXX.10.*
- *Uses Not Permitted* are found in *XXX.12.*
- *Securing and Supporting* is found in *XXX.30.*

This is referred to as "the parallel numbering system."

How to Spot Changes in the 2014 *NEC*

Each new edition of the *NEC* includes the following usability features as aids to the user. Changes other than editorial are highlighted with gray shading within sections and with vertical ruling for large blocks of changed or new text and for new tables and changed or new figures. Where one or more complete paragraphs have been deleted, the deletion is indicated by a bullet (•) between the paragraphs that remain. The previous edition of the *NEC* must be consulted to determine what text was deleted. Large blocks of new text such as a new article are identified by a vertical rule in the margin.

Individual pages have dictionary-style headers that indicate beginning and ending sections.

Who Writes the *Code*?

For each *Code* cycle, the NFPA solicits proposals from anyone interested in electrical safety. Anyone may submit a proposal to change the *NEC* using the Proposal Form found in the back of the *NEC*. A form may also be downloaded from www.nfpa.org. Proposals received are then given to a specific Code-Making Panel (CMP) to accept, reject, accept in part, accept in principle, or accept in principle in part. These actions are published in the Report on Proposals (ROP) and may be downloaded from www.nfpa.org. Individuals may send in their comments on these actions using the Comment Form found in the ROP. The CMPs meet again to review and take action on the comments received. These actions are published in the Report on Comments (ROC) and also may be downloaded from www.nfpa.org. Final action (voting) on proposals and comments is taken at the NFPA annual meeting.

However, before the *NEC* is published, if there is disagreement on any specific *Code* requirement adopted through the above process, the NFPA will consider an *Appeal* that is reviewed and acted upon by the NFPA Standards Council about six weeks after the annual meeting. After an *Appeal* is acted upon by the Standards Council, should there still be controversy, another step not often used in the *Code* adoption process is a petition that is reviewed and acted upon by the NFPA Board of Directors.

Individuals who serve on CMPs are electrical inspectors, electrical contractors, electricians, electrical engineers, individuals from utilities, manufacturers, testing laboratories, the Consumer Product Safety Commission, insurance companies, and similar organizations. CMP members are appointed by the NFPA. The CMPs have 10 to 20 principal members, plus a similar number of alternate members. All of the CMPs have a good balance of representation.

Which Edition of the *NEC* to Use

This text is based on the 2014 edition of the *NEC*. Some municipalities, cities, counties, and states have not yet adopted this edition and may continue using older editions. Check with your local electrical inspector to find out which edition of the *Code* is in force.

Copies of the latest edition of the *NEC* NFPA 70 may be ordered from the following:

National Fire Protection Association
1 Batterymarch Park
Quincy, MA 02269-9101
Phone: 617-770-3000
http://www.nfpa.org

International Association of Electrical Inspectors
901 Waterfall Way, Suite 602
Richardson, TX 75080-7702
Phone: 800-786-4234
http://www.iaei.org

Electrical Inspection Code for Existing Dwellings

This code is published by the NFPA and is referred to as *NFPA 73*. It is a brief nine-page code that provides requirements for evaluating installed electrical systems within and associated with existing dwellings to identify safety, fire, and shock hazards—such as improper installations, overheating, physical deterioration, and abuse. This code lists most of the electrical things to inspect in an existing dwelling that could result in a fire or shock hazard if not corrected. It only points out things to look for that are visible. It does not get into examining concealed wiring that would require removal of permanent parts of the structure. It also does not get into calculations, location requirements, and complex topics, as does the *NEC* NFPA 70.

This code can be an extremely useful guide for electricians doing remodel work and for electrical inspectors wanting to bring an existing dwelling to a reasonably safe condition. Many localities require that when a home changes ownership, the wiring must be brought up to some minimum standard, but not necessarily as extensively as would be the case for new construction.

Homes for the Physically Challenged

The *NEC* now covers the wiring of homes for the physically or mentally challenged and disabilities associated with the elderly in Informative Annex J. Because this information is in an Informative Annex, it is not a requirement.

The preface to *Informative Annex J* reads, ▶The provisions cited in Informative Annex J are intended to assist the users of the *Code* in properly considering the various electrical design constraints of other building systems and are part of the 2010 ADA Standards for Accessible Design. They are the same provisions as those found in ANSI/ICC A117.1-2009, *Accessible and Usable Buildings and Facilities.* ◀*

Each installation for the physically challenged must be based on the specific need(s) of the individual(s) who will occupy the home. Some physically challenged people are bedridden, require the mobility of a wheelchair, have trouble reaching or bending, and so forth. There are no hard and fast rules that *must* be followed—only many suggestions to consider.

Some of these follow:

- Install more ceiling luminaires instead of switching receptacles. Cords and lamps are obstacles.
- Install luminaires having more than one bulb.
- Go "overboard" in the amount of lighting for all rooms, entrances, stairways, stairwells, closets, pantries, bathrooms, and so on.
- Use higher wattage bulbs—not to exceed the wattage permitted in the specific fixture.
- Consider installing luminaires in certain areas (such as bathrooms and hallways) to be controlled by motion detectors.
- Consider installing exhaust fans in certain areas (such as laundries, showers, and bathrooms) that turn on automatically when the humidity reaches a predetermined value.
- Consider the height of switches and thermostats, usually 42 in. (1.0 m) or lower, instead of the standard 46–52 in. (1.15–1.3 m).
- Consider rocker-type switches instead of toggle type.
- Install pilot light switches.
- Consider "jumbo" switches.
- Be sure stairways and stairwells are well lit.
- Consider stair tread lighting.
- Position lighting switches so as not to be over stairways or ramps.
- Locate switches and receptacles to be readily and easily accessible—not behind doors or other hard-to-reach places.
- Consider installing wall receptacle outlets at a height of 24–27 in. (600–675 mm) above the finished floor rather than the normal 12 in. (300 mm) height.
- Install lighted doorbell buttons.
- Chimes: Consider adding a strategically located "dedicated" lamp(s) that turn on when doorbell buttons are pushed. The wiring diagram for this is found in Figure 25-25.

*Reprinted with permission from NFPA 70-2014.

- Telephones: Consider adding visible light(s) strategically located that flash at the same time the telephone is ringing.
- Consider installing receptacle outlets and switches on the face of kitchen cabinets. Wall outlets and switches can be impossible for the physically challenged person to reach.
- Consider fire, smoke, and security systems, directly connected to a central office for fast response to emergencies, that do not depend on the disabled to initiate the call.
- Consider installing the fuse box or breaker panel on the first floor instead of in the basement.
- Consider the advanced home systems concept of remotely controlling lighting, receptacles, appliances, television, telephones, and so on. The control features can make life much easier for a disabled person.

When involved with multifamily dwellings, check with your local building authority. They have copies of the Americans with Disabilities Act (ADA) from the U.S. Department of Justice and the Fair Housing Act from the U.S. Department of Housing and Urban Development (HUD); these basically require that all units must have accessible light switches, electrical outlets, thermostats, and other environmental controls.

The ANSI publication *ANSI A117.1–2009, Standard for Accessible and Useable Buildings and Facilities*, contains many suggestions and considerations for buildings and facilities for the physically challenged. This and other standards can be ordered from

American National Standards Institute
25 West 43rd Street
New York, NY 10036
Tel: 212-642-4900
Fax: 212-398-0023

The same standard is available from the International Code Council (ICC) under the publication number *ICC/ANSI A117.1–2009*.

Another virtually unlimited source of standards is

Global Engineering Documents
15 Inverness Way East
Englewood, CO 80112
Phone: 800-854-7179
303-397-7956
Fax: 303-397-2740
http://www.global.ihs.com
e-mail: globalcustomerservice@ihs.com

BUILDING CODES

The majority of the building departments across the country have for the most part adopted the *NEC* rather than attempting to develop their own electrical codes. As you study this text, you will note numerous references to the *NEC*, the electrical inspector, or the authority having jurisdiction.

An authority's level of knowledge varies. The electrical inspector may be full-time or part-time and may also have responsibility for other trades, such as plumbing or heating. The heads of the building departments in many communities are typically called the Building Commissioners or Directors of Development. Regardless of title, they are responsible for ensuring that the building codes in their communities are followed.

International Code Council (ICC)

Rather than writing their own codes, most communities adopt the codes of the International Code Council (ICC). Over the years, the ICC has developed comprehensive and coordinated model construction codes. These include building, mechanical, plumbing, fire, energy conservation, existing building, fuel gas, sewage disposal, property maintenance, zoning, residential, and electrical codes. The ICC provides technical, educational, and informational products. Its address is

ICC Headquarters
500 New Jersey Avenue, NW, 6th Floor
Washington, DC 20001
Phone: 888-422-7233
http://www.iccsafe.org

The *ICC International Residential Code*, Chapters 33–42, contains electrical provisions that were written and produced under the guidance of the NFPA. The material in these chapters is copyrighted by the NFPA. These provisions are similar to the *NEC* other than the layout and numbering system.

This text, *Electrical Wiring—Residential*, is based on the *NEC* and does not include local amendments to the *NEC*.

Because local electrical codes may differ from the *NEC*, you should check with the local inspection authority to determine which edition of the *NEC* is enforced, and what, if any, local requirements or amendments take precedence over the *NEC*.

American National Standards Institute

The **American National Standards Institute (ANSI)** is an organization that coordinates the efforts and results of the various standards-developing organizations, such as those mentioned in previous paragraphs. Through this process, ANSI approves standards that then become recognized as American National Standards. There is a lot of similarity among the technical information found in ANSI standards, the UL standards, the International Electronic and Electrical Engineers standards, and the *NEC*.

International Association of Electrical Inspectors

The International Association of Electrical Inspectors (IAEI) is a nonprofit organization. The IAEI membership consists of electrical inspectors, building officials, electricians, engineers, contractors, and manufacturers throughout the United States and Canada. The major goal of the IAEI is to improve the understanding of the *NEC*. Representatives of this organization serve as members of the various CMPs of the *NEC* and share equally with other members in the task of reviewing and revising the *NEC*.

The IAEI publishes a bimonthly magazine—*The IAEI News*. It is devoted entirely to electrical code topics. Anyone in the electrical industry is welcome to join the IAEI. An application form is found after the Appendix of this text. Its address is

International Association of Electrical Inspectors
901 Waterfall Way, Suite 602
Richardson, TX 75080-7702
Phone: 800-786-IAEI
http://www.iaei.org

Code Definitions

The electrical industry uses many words (terms) that are unique to the electrical trade. These terms need clear definitions to enable the electrician to understand completely the meaning intended by the *Code*.

Article 100 of the *NEC* is a "dictionary" of these terms. *Article 90* also provides further clarification of terms used in the *NEC*. Here are a few examples:

Ampacity: The maximum current, in amperes, that a conductor can carry continuously under the conditions of use without exceeding its temperature rating.*

Approved: Acceptable to the authority having jurisdiction.*

Authority Having Jurisdiction (AHJ): An organization, office, or individual responsible for enforcing the requirements of a code or standard, or for approving equipment, materials, an installation, or a procedure. See *NEC 90.4*.*

Dwelling Unit: A single unit, providing complete and independent living facilities for one or more persons, including permanent provisions for living, sleeping, cooking, and sanitation.*

Listed: Equipment, materials, or services included in a list published by an organization that is acceptable to the authority having jurisdiction and concerned with evaluation of products or services, that maintains periodic inspection of production of listed equipment or materials or periodic evaluation of services, and whose listing states that either the equipment, material, or services meets appropriate designated standards or has been tested and found suitable for a specified purpose.*

Shall: Indicates a mandatory rule, *90.5*. As you study the *NEC*, think of the word "shall" as meaning "must." Some examples found in the *NEC* where the word "shall" is used in combination with other words are: *shall be, shall have, shall not, shall be permitted, shall not be permitted*, and *shall not be required*.

Refer to Key Terms in the Appendix of this text.

*Reprinted with permission from NFPA 70-2014.

Read and Follow *110.3(B)* Carefully!

One of the most far-reaching *NEC* rules is *110.3(B)*. This section states that Listed or labeled equipment shall be installed and used in accordance with any instructions included in the listing or labeling.* This means that an entire electrical system and all of the system's electrical equipment must be *installed* and *used* in accordance with the *NEC* and the applicable standards to which the electrical equipment has been tested.

The phrase In accordance with any instructions included in the listing or labeling* is commonly interpreted to include the Guide Card information included in the UL *White Book*. The Guide Card information contains hundreds, if not thousands, of installation instructions that must be carefully followed to achieve a safe and *NEC*-compliant installation. Here are a few examples of installation requirements contained in the UL *White Book,* followed by the UL four alpha character key:

- "Enclosure Type 1 is suitable for only indoor locations (AALZ)"**
- "Enclosure Type 3 is suitable for outdoor use, and will be undamaged by the formation of ice on the enclosure (AALZ)"**
- "Except as noted in the general Guide Information for some product categories, most terminals, unless marked otherwise, are for use only with copper wire. If aluminum or copper-clad aluminum wire can be used, marking to indicate this fact is provided. Such marking is required to be independent of any marking on terminal connectors, such as on a wiring diagram or other visible location. The marking may be in an abbreviated form, such as "AL-CU." (AALZ)"**
- "Except as noted in the following paragraphs or in the general Guide Information for some product categories, the termination provisions are based on the use of 60°C ampacities for wire size Nos. 14-1 AWG, and 75°C ampacities for wire size Nos. 1/0 AWG and larger, as specified in Table 310.15(B)(16) of the NEC. (AALZ)"**

Manufacturers of equipment may include specific instructions such as these:

- Supply conductors must have an insulation rated not less than 90°C.
- Do not locate this baseboard heater below a receptacle outlet.
- This appliance must be supplied from a receptacle connected to an equipment grounding conductor.

The UL *White Book* is sometimes referred to as "the other Code book." Some have emphasized the importance of the *White Book* by stating, "Don't leave home without it!"

METRICS (SI) AND THE *NEC*

The United States is the last major developed country in the world not using the metric system of weights and measures as the primary system. For most of our lifetime, we have used the English system of weights and measures, also referred to as inch-pound and U.S. Customary. But this is changing!

Manufacturers are now showing both inch-pound and metric units in their catalogs. By law, plans and specifications for new governmental construction and renovation projects have used the metric system since January 1, 1994.

You may not feel comfortable with metric measurements, but metric measurements are here to stay. You might just as well become familiar with the metric system.

All measurements in the *NEC* are shown in both inch-pound and metric values.

For more information about metrics, refer to the "Metric System of Measurement" section found in the Appendix of this text.

Trade Sizes

A unique situation exists. Strange as it may seem, what electricians have been referring to for years has not been correct!

*Reprinted with permission from NFPA 70-2014.

TABLE 1-3

Comparison of trade size vs. actual inside diameters.

TRADE SIZE	INSIDE DIAMETER (I.D.)
½ Electrical Metallic Tubing	0.622 in.
½ Electrical Nonmetallic Tubing	0.560 in.
½ Flexible Metal Conduit	0.635 in.
½ Rigid Metal Conduit	0.632 in.
½ Intermediate Metal Conduit	0.660 in.

Raceway sizes have always been an approximation. For example, there has never been a ½ in. raceway. Measurements taken from the *NEC* for a few types of raceways are shown in Table 1-3.

You can readily see that the cross-sectional areas, critical when determining conductor fill, are different. It makes sense to refer to conduit, raceway, and tubing sizes as *trade sizes*. The *NEC* in *90.9(C)(1)* states that Where the actual measured size of a product is not the same as the nominal size, trade size designators shall be used rather than dimensions*. This edition of *Electrical Wiring—Residential* uses the term *trade size* when referring to conduits, raceways, and tubing. For example, a ½ in. Electrical Metallic Tubing (EMT) is referred to as trade size ½ EMT. EMT is also referred to in the trade as "thinwall."

The *NEC* also uses the term *metric designator*. A trade size ½ EMT is shown as *metric designator 16 (½)*. A trade size 1 EMT is shown as *metric designator 27 (1)*. The numbers 16 and 27 are the metric designator values; the (½) and (1) are the trade sizes. The metric designator is the raceway's inside diameter—in rounded-off millimeters (mm). Table 1-4 shows some of the more common sizes of conduits, raceways, and tubing. A complete listing is found in *NEC Table 300.1(C)*.

For ease in understanding, this text uses only the term *trade size* when referring to conduit and raceway sizes.

Conduit knockouts in boxes do not measure up to what we call them. Table 1-5 shows some examples.

Outlet boxes and device boxes use their nominal measurement as their *trade size*. For example, a 4 in. × 4 in. × 1½ in. box does not have an internal cubic-inch volume of 4 in. × 4 in. × 1½ in. = 24 in^3. *Table 314.16(A)* shows this size box as having a 21-cubic in. volume. This table shows *trade sizes* in two columns—millimeters and inches.

TABLE 1-4

Trade sizes of raceways and their metric designator identification.

METRIC DESIGNATOR AND TRADE SIZE

Metric Designator	Trade Size
12	⅜
16	½
21	¾
27	1
35	1¼
41	1½
53	2
63	2½
78	3

TABLE 1-5

Comparison of knockout trade size vs. actual measurement.

TRADE SIZE KNOCKOUT	ACTUAL MEASUREMENT
½	⅞ in.
¾	1³⁄₃₂ in.
1	1⅜ in.

In this text, a square outlet box is referred to as trade size 4 × 4 × 1½. Similarly, a single-gang device box would be referred to as a trade size 3 × 2 × 3 box.

Trade sizes for construction material will not change. A 2 × 4 is really a *name*, not an actual dimension. A 2 × 4 will keep its name forever. This is its *trade size*.

In this text, most measurements directly related to the *NEC* are given in both inch-pound and metric units. In many instances, only the inch-pound units are shown. This is particularly true for the examples of raceway and box fill calculations, load calculations for square foot areas, and on the plans (drawings).

*Reprinted with permission from NFPA 70-2014.

Because the *NEC* rounded off most metric conversion values, a calculation using metrics results in a different answer when compared to the same calculation done using inch-pounds. For example, load calculations for a residence are based on 3 volt-amperes per square foot, or 33 volt-amperes per square meter.

For a 40 ft × 50 ft dwelling: 3 VA × 40 ft × 50 ft = 6000 volt-amperes.

In metrics, using the rounded-off values in the *NEC*: 33 VA × 12 m × 15 m = 5940 volt-amperes.

The difference is small; nevertheless, there is a difference.

To show calculations in both units throughout this text would be very difficult to understand and would take up too much space. Calculations in either metrics or inch-pounds are in compliance with the *NEC, 90.9(D)*. In *90.9(C)(3)* we find that metric units are not required if the industry practice is to use inch-pound units.

It is interesting to note that the examples in *Chapter 9* of the *NEC* use inch-pound units, not metrics.

Guide to Metric Usage

The metric system is a base-10, or decimal system in that values can be easily multiplied or divided by 10 or powers of 10. The metric system as we know it today is known as the International System of Units (SI), derived from the French term "le Système International d'Unités."

In the metric system, the units increase or decrease in multiples of 10, 100, 1000, and so on. For instance, one megawatt (1,000,000 watts) is 1000 times greater than one kilowatt (1000 watts).

By assigning a name to a measurement, such as a *watt*, the name becomes the unit. Adding a prefix to the unit, such as *kilo*, forms the new name *kilowatt*, meaning 1000 watts. Refer to Table 1-6 for prefixes used in the metric system.

The prefixes used most commonly are *centi*, *kilo*, and *milli*. Consider that the basic unit is a meter (one). Therefore, a centimeter is 0.01 meter, a kilometer is 1000 meters, and a millimeter is 0.001 meter.

Some common measurements of length and equivalents are shown in Table 1-7.

Electricians will find it useful to refer to the conversion factors and their abbreviations shown in Table 1-8.

Refer to the Appendix of this text for a comprehensive metric conversion table. The table includes information on how to round off numbers for practical use on the job.

TABLE 1-6

Metric prefixes, symbols, multipliers, powers, and values.

PREFIX	SYMBOL	MULTIPLIER	SCIENTIFIC NOTATION (POWERS OF TEN)	VALUE
tera	T	1 000 000 000 000	10^{12}	one trillion (1 000 000 000 000/1)
giga	G	1 000 000 000	10^{9}	one billion (1 000 000 000/1)
mega	M	1 000 000	10^{6}	one million (1 000 000/1)
kilo	k	1 000	10^{3}	one thousand (1 000/1)
hecto	h	100	10^{2}	one hundred (100/1)
deka	da	10	10^{1}	ten (10/1)
unit		1	—	one (1)
deci	d	0.1	10^{-1}	one-tenth (1/10)
centi	c	0.01	10^{-2}	one-hundredth (1/100)
milli	m	0.001	10^{-3}	one-thousandth (1/1 000)
micro	m	0.000 001	10^{-6}	one-millionth (1/1 000 000)
nano	n	0.000 000 001	10^{-9}	one-billionth (1/1 000 000 000)
pico	p	0.000 000 000 001	10^{-12}	one-trillionth (1/1 000 000 000 000)

TABLE 1-7

Some common measurements of length and their equivalents.

one inch	=	2.54	centimeters
	=	25.4	millimeters
	=	0.025 4	meter
one foot	=	12	inches
	=	0.304 8	meter
	=	30.48	centimeters
	=	304.8	millimeters
one yard	=	3	feet
	=	36	inches
	=	0.914 4	meter
	=	914.4	millimeters
one meter	=	100	centimeters
	=	1000	millimeters
	=	1.093	yards
	=	3.281	feet
	=	39.370	inches

TABLE 1-8

Useful conversions and their abbreviations.

inches (in.) × 0.0254	= meters (m)
inches (in.) × 0.254	= decimeters (dm)
inches (in.) × 2.54	= centimeters (cm)
centimeters (cm) × 0.393 7	= inches (in.)
inches (in.) × 25.4	= millimeters (mm)
millimeters (mm) × 0.039 37	= inches (in.)
feet (ft) × 0.304 8	= meters (m)
meters (m) × 3.280 8	= feet (ft)
square inches (in.^{2}) × 6.452	= square centimeters (cm^2)
square centimeters (cm^2) × 0.155	= square inches (in.^{2})
square feet (ft^2) × 0.093	= square meters (m^2)
square meters (m^2) × 10.764	= square feet (ft^2)
square yards (yd^2) × 0.836 1	= square meters (m^2)
square meters (m^2) × 1.196	= square yards (yd^2)
kilometers (km) × 1000	= meters (m)
kilometers (km) × 0.621	= miles (mi)
miles (mi) × 1.609	= kilometers (km)

LISTED EQUIPMENT AND NATIONALLY RECOGNIZED TESTING LABORATORIES (NRTL)

Electrical Equipment

For safety, it is very important that electrical equipment be listed by a Nationally Recognized Testing Laboratory (NRTL). Concepts about listed electrical equipment are found in *NEC 90.7*.

NEC 90.7: Examination of Equipment for Safety. For specific items of equipment and materials referred to in this *Code*, examinations for safety made under standard conditions provide a basis for approval where the record is made generally available through promulgation by organizations properly equipped and qualified for experimental testing, inspections of the run of goods at factories, and service/value determination through field inspections. This avoids the necessity for repetition of examinations by different examiners, frequently with inadequate facilities for such work, and the confusion that would result from conflicting reports on the suitability of devices and materials examined for a given purpose.

It is the intent of this *Code* that factory-installed internal wiring or the construction of equipment need not be inspected at the time of installation of the equipment, except to detect alterations or damage, if the equipment has been listed by a qualified electrical testing laboratory that is recognized as having the facilities described in the preceding paragraph and that requires suitability for installation in accordance with this *Code*.*

OSHA rules state that:

"All electrical products installed in the work place shall be listed, labeled, or otherwise determined to be safe by a Nationally Recognized Testing Laboratory (NRTL)."

Testing Electrical Equipment Safety

How does one know if a product is safe to use? Manufacturers, consumers, regulatory authorities, and others recognize the importance of independent, third-party testing of products in an effort

*Reprinted with permission from NFPA 70-2014.

to reduce safety risks. Unless you have all of the necessary test equipment and knowledge of how to properly test a product for safety, the surest way is to accept the findings of a third-party testing agency.

Nationally Recognized Testing Laboratories (NRTL) have the knowledge, wherewithal, and test equipment to test and evaluate products for safety. The term Nationally Recognized Testing Laboratories (NRTL) is not used in the *National Electrical Code*. The term is used by the Occupational Safety and Health Administration and is used to classify and recognize qualified electrical product testing laboratories. Additional information about the OSHA NRTL program can be found on the Internet at http://www.osha.gov/dts/otpca/nrtl/.

A NRTL performs tests on a product based on a specific nationally recognized safety standard. After the product has been tested and found to comply with the safety standard, the product is considered to be free from reasonably foreseeable risk of fire, electric shock, and related hazards. The product is then "listed" to have a listing mark (label) attached by the manufacturer of the electrical product.

As you work with electrical products, make sure the product has a listing mark on it from a NRTL. If the product is too small to have a listing mark on the product itself, then look for the mark on the carton the product came in.

The following laboratories do a considerable amount of testing and listing of electrical equipment:

Underwriters Laboratories Inc. (UL®)

Underwriters Laboratories Inc. (UL), founded in 1894, is a highly qualified, nationally recognized testing laboratory with several testing laboratories in the United States and service locations in several other countries. UL develops product safety standards and performs tests to these standards. Most reputable manufacturers of electrical equipment submit representative samples of their products to UL, where the equipment is subjected to numerous tests. These tests determine whether the product can perform safely under normal and abnormal conditions to meet published standards. After UL tests and evaluates product samples, and determines that the product samples comply with the specific

FIGURE 1-9 The Underwriters Laboratories listing mark.

standard, the manufacturer is then permitted to *label* its product with the UL Mark (Figure 1-9). The products are then *listed* in UL's Online Certification Directory.

The UL Mark. The UL mark is required to be on the product! The UL mark almost always consists of four elements—UL in a circle, the word "LISTED" in capital letters, the product identity, and a unique alphanumeric control or issue number. If the product is too small, or has a shape or is made of a material that will not accept the UL Mark on the product itself, the marking is permitted on the smallest unit carton or container that the product comes in. Marking on the carton or box is nice but does not ensure that the product is UL listed! For additional information on the listing marks of Underwriters Laboratories, see the information at the front of the UL *White Book*.

The listing mark shown in Figure 1-9 indicates the product is in compliance with the applicable product safety standards in the United States and in Canada. A listing mark with only UL in a circle indicates the product has been evaluated only to US standards.

When UL tests and lists products that comply only to the requirements of a particular Canadian standard, the UL mark shown in Figure 1-9 appears with a "C" outside of and to the left of the circle. This means the product has been tested and evaluated for compliance *only* with Canadian requirements. Product standards are being harmonized in North America as a result of the North American Free Trade Act (NAFTA). Discussions are also going on with Mexico. When all of this is finalized, electrical equipment standards may be the same in the United States, Canada, and Mexico.

Additional efforts are being made to harmonize North American standards with those of Europe.

SAFETY ALERT

Counterfeit electrical products may present a significant safety hazard as there is no assurance the construction complies with a safety standard.

FIGURE 1-10 The *recognized components* mark.

Counterfeit Products. Be on the lookout for counterfeit electrical products. These products have not been tested and listed by a recognized testing laboratory. They can present a real hazard to life and property. Counterfeit products continue to come from China as well as many other countries. Counterfeit electrical products might also be referred to as "black market products."

Look for unusual logos or wording. For example, the UL mark might be illustrated in an oval instead of a circle, or the UL mark might not be encircled with anything, or the wording might say *approved* instead of LISTED. UL doesn't approve anything! It tests representative samples of the product. If samples of the product meet UL standards, the product is then LISTED.

Federal legislation was passed by Congress and is now law that makes it a criminal offense to traffic in counterfeit products and counterfeit trademarks. The law makes it mandatory that the counterfeit products and any tools that make the products or markings be seized and destroyed.

To learn more about counterfeits, check out http://www.ul.com, then search on (type in) the word "counterfeit". Also, check out http://www.nema.org, then type in the word "counterfeit".

Do not confuse a UL marking in a circle with the markings found on *recognized components*. Recognized components that have passed certain tests are marked with the letters "RU" printed in mirror image (Figure 1-10). By themselves, recognized components are not to be field-installed. They are intended for use in end-use products or systems that would ultimately be tested and listed, with the final assembly becoming a UL-listed product. Some examples of recognized components are relays, ballasts, insulating materials, special switches, and so on.

UL previously produced several directories including: *Electrical Construction Equipment Directory (Green Book)* and the *Electrical Appliance and Utilization Equipment Directory (Orange Book)*. The information previously provided in these directories is now in UL's Online Certifications Directory. UL continues to publish the *Guide Information for Electrical Equipment Directory (White Book)*. More on the UL *White Book* follows.

It is extremely useful for an electrician, an electrical contractor, and/or an electrical inspector to refer to these directories when looking for specific requirements, permitted uses, limitations, and others for a certain product. Here's the link to the UL Online Certifications Directory: http://database.ul.com/cgi-bin/XYV/template/LISEXT/1FRAME/index.htm.

Many times the answer to a product-related question that cannot be found in the *NEC* can be found in the *White Book* or the UL Online Certifications Directory.

The best companion to the *NEC* is probably the UL *White Book* (Figure 1-11). As mentioned earlier in this text, the UL *White Book* contains invaluable safety information necessary to make a NEC-compliant installation. It can be downloaded from http://www.ul.com.

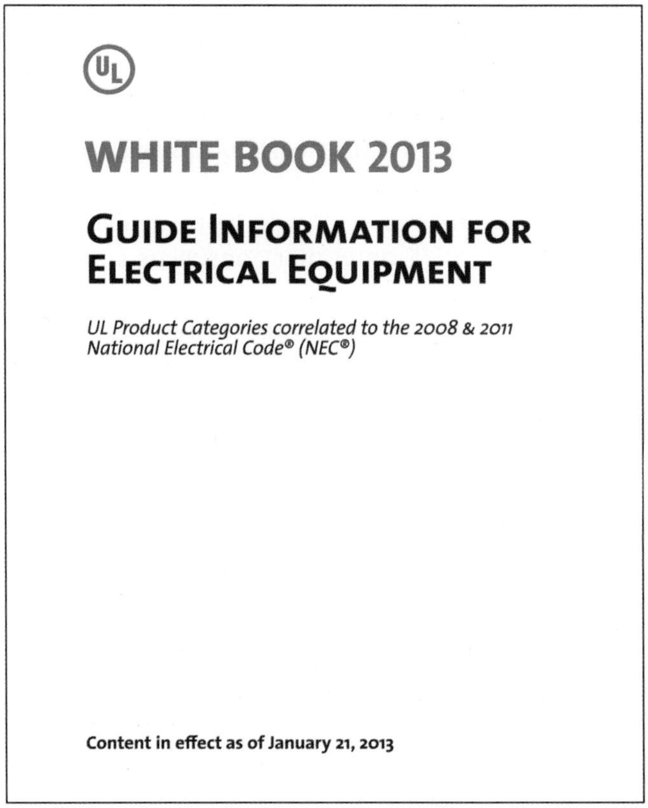

WHITE BOOK 2013

GUIDE INFORMATION FOR ELECTRICAL EQUIPMENT

UL Product Categories correlated to the 2008 & 2011 National Electrical Code® (NEC®)

Content in effect as of January 21, 2013

FIGURE 1-11 The Underwriters Laboratories *White Book*.

FIGURE 1-12 The listing mark of Canadian Standards Association.

Courtesy CSA Group

Underwriters Laboratories Inc., and the CSA Group have worked out an agreement whereby either agency can test, evaluate, and list equipment for the other agency. For example, UL might test and list an air conditioner unit to the requirements of UL Standard 1995 (Heating and Cooling Equipment) because the Canadian Standard C22.2 No. 236-M90 is a mirror image of UL Standard 1995. One by one, the UL and CSA Group Standards are becoming harmonized.

CSA Group

In ways, the CSA Group is the Canadian counterpart of Underwriters Laboratories, Inc., in the United States. CSA Group develops the *Canadian Electrical Code (CEC)* and of the Canadian Standards for the testing, evaluation, and listing of electrical equipment in Canada. The *Canadian Electrical Code* is quite different from the *NEC*. A Canadian version of *Electrical Wiring—Residential* is available in Canada.

Figure 1-12 is a representation of the CSA Group listing mark. Like the UL listing mark, the appearance of "C" and "US" at the approximately 4:00 and 8:00 o'clock positions indicates to what nation's standards the equipment has been evaluated.

The *Canadian Electrical Code* and CSA Group standards can be obtained by contacting

CSA Group
178 Rexdale Boulevard
Toronto, Ontario, CANADA
M9W 1R3
Phone: 416-747-4044; 800-463-6727
Fax: 416-747-2510
http://www.csa-group.org

Intertek

Intertek is a nationally recognized testing laboratory. Its Products division provides testing, evaluation, labeling, listing, and follow-up service for the safety testing of electrical products. This is done in conformance to nationally recognized safety standards.

The Intertek ETL listing mark, like the UL and CSA Group listing marks, indicates which nation's product safety standards the equipment has been found to be in compliance with. See Figure 1-13.

Courtesy Intertek

FIGURE 1-13 The listing mark of Intertek.

Information can be obtained by contacting

Intertek
3933 US Route 11
Cortland, NY 13045
Phone: 800-345-3851
http://www.intertek.com

National Electrical Manufacturers Association

The National Electrical Manufacturers Association (NEMA) represents nearly 600 manufacturers of electrical products. NEMA develops electrical equipment standards, which, in many instances, are very similar to UL and other consensus standards. NEMA has representatives on the CMPs for the *NEC*. Additional information can be obtained by contacting

National Electrical Manufacturers Association
1300 North 17th Street, Suite 1847
Rosslyn, VA 22209
Phone: 703-841–3200
http://www.nema.org

REVIEW

Note: Refer to the *NEC* or the plans in the back of this text where necessary.

1. What is the purpose of specifications? ______________________________

____

____

2. Refer to the specifications in the back of this text following Chapter 33.
   a. What electrical codes must be conformed to?

   ____

   ____

   ____

   b. What is the wiring method to be used in the workshop?

   ____

   c. What size conductors are to be used for
   the lighting branch circuits? ____________ AWG
   the small-appliance branch circuits? ____________ AWG
   d. What section of the specifications tells us the size of the service entrance to be installed for this residence?
3. What is done to prevent a plan from becoming confusing because of too much detail?

____

____

____

4. Name three requirements contained in the specifications regarding material.

   a. ______________________ c. ______________________

   b. ______________________

5. The specifications state that all work shall be done ______________________

   ____

6. What phrase is used when a substitution is permitted for a specific item? ______________________

   ____

7. What is the purpose of an electrical symbol? ______________________

   ____

8. What is a notation? ______________________

   ____

9. Where are notations found? ______________________

10. List at least 12 electrical notations found on the plans for this residence. Refer to the plans at the back of the text. ______________________

    ____

    ____

    ____

    ____

    ____

11. What three parties must be satisfied with the completed electrical installation?

    a. ______________________ c. ______________________

    b. ______________________

12. What code sets standards for electrical installation work? ______________________

    ____

13. What authority enforces the requirements set by the *NEC*? ______________________

    ____

14. Does the *NEC* provide minimum or maximum standards? ______________________

15. What do the letters "UL" signify?

    ____

16. What section of the *NEC* states that all listed or labeled equipment shall be installed or used in accordance with any instructions included in the listing or labeling?

    ____

17. When the word "shall" appears in a *NEC* reference, it means that it (must)(may) be done. (Underline the correct word.)

18. What is the purpose of the *NEC*?

    ____

    ____

    ____

19. Does compliance with the *NEC* always result in an electrical installation that is adequate, safe, and efficient? Why? ______

20. Name two nationally recognized testing laboratories. ______

21. a. Do Underwriters Laboratories and the other recognized testing laboratories "approve" products? ______
    b. What do these testing laboratories do? ______

22. a. Has the *NEC* been officially adopted by the community in which you live? ______
    b. By the state in which you live? ______
    c. If your answer is YES to (a) or (b), are there amendments to the *NEC*? ______
    d. If your answer is YES to (c), list some of the more important amendments.

23. Does the *NEC* make suggestions about how to wire a house that will be occupied by handicapped persons? ______

24. A junction box on a piece of European equipment is marked 200 cm^3. Convert this to cubic inches. ______

25. Convert 4500 watts to Btu/hour. ______

26. You will learn in Chapter 3 that residential lighting loads are based on 3 volt-amperes per ft^2 (33 volt-amperes per m^2). Determine the minimum lighting load required for an area of 186 m^2. Do calculations for both feet squared and meters squared so you can see the difference in answers. To convert meters squared to feet squared, refer to Table 1-8. ______

SAFETY-RELATED PROBLEMS

27. What federal organization dictates requirements for work-related safety issues? Circle the correct answer.
    a. *NEC*
    b. OSHA
    c. IAEI
28. What can you do to reduce or eliminate the possibility of receiving an electric shock when working on electrical circuits? Circle the correct answer.
    a. Turn off the power on the circuit you are working on.
    b. Turn off the power on the circuit you are working on, tag it, and lock the disconnect in the OFF position.
    c. Turn off the power on the circuit you are working on, then tell everyone else working with you not to turn the power back on.
29. What is the *NEC* definition of a *qualified person*?
    ____
30. Are you a *qualified person*? Explain your answer.
    ____
    ____
31. Are low-voltage systems totally safe? Explain.
    ____
    ____
32. What NFPA standard specifically covers safety in the workplace? ______________
33. What do the letters "PPE" stand for?
    ____
34. Explain briefly what an arc flash is.
    ____
    ____
35. Where might you obtain information about on-the-job safety and safety training? ____
    ____
    ____
    ____
36. a. Is it a safe practice to use a stepladder in which the spreaders have not been fully locked into position? ______________
    b. If the ladder collapses, who is at fault? ______________

CHAPTER 2

Electrical Symbols and Outlets

OBJECTIVES

After studying this chapter, you should be able to

- identify and explain the electrical outlet symbols used in the plans of the single-family dwelling.
- discuss the types of outlets, boxes, luminaires, and switches used in the residence.
- explain the methods of mounting the various electrical devices used in the residence.
- understand the preferred way to position receptacles in wall boxes.
- understand issues involved in remodel work.
- understand how to determine the maximum number of conductors permitted in a given size box.
- understand the concept of fire-resistance rating of walls and ceilings.

ELECTRICAL SYMBOLS

Electrical symbols used on an architectural plan show the location and type of electrical device required. A typical electrical installation as taken from a plan is shown in Figure 2-1.

The *NEC*® has many words unique to the electrical trade. Terms unique to the *NEC* are defined in either *Article 100* if the term is used in two or more articles or in the article itself if the term is used only in that article. For example:

- A **device** is ▶A unit of an electrical system, other than a conductor, that carries or controls electric energy as its principle function.◀*
- An **outlet** is A point on the wiring system at which current is taken to supply utilization equipment.*
- A **receptacle** is A contact device installed at the outlet for the connection of an attachment plug.*
- A **receptacle outlet** is An outlet where one or more receptacles are installed.* See Figure 2-2.

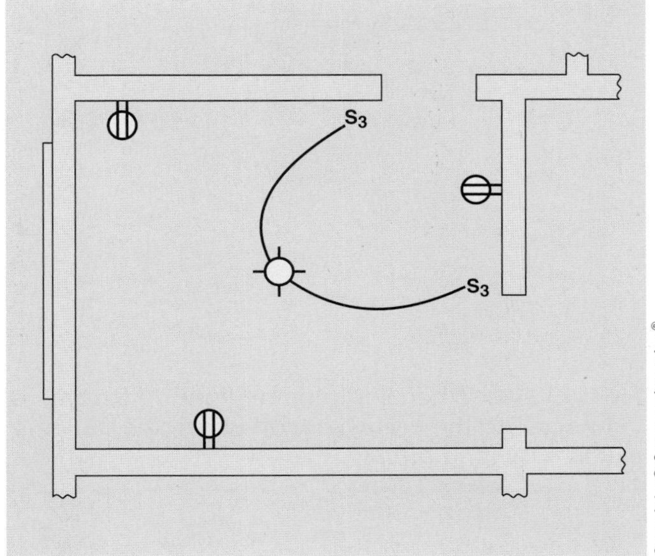

FIGURE 2-1 Use of electrical symbols and notations on a floor plan.

- A **lighting outlet** is An outlet intended for the direct connection of a lampholder or luminaire.* See Figure 2-3.

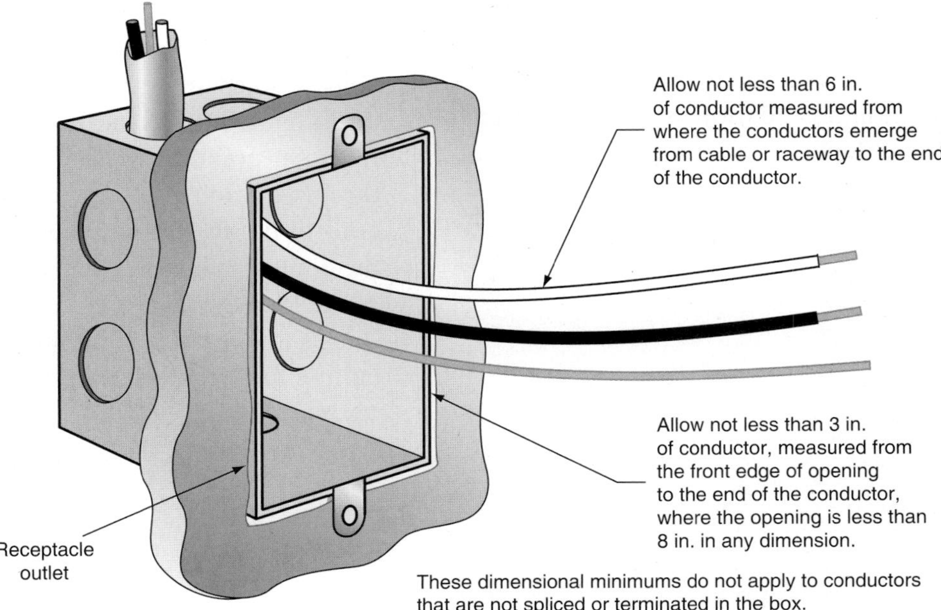

FIGURE 2-2 Where a receptacle is connected to the branch-circuit wires, the outlet is called a *receptacle outlet. NEC 300.14* states the minimum length of conductors at an outlet, junction, or switch point for splices or the connection of luminaires or devices. Leaving the conductors too long can cause crowding of the wires in the box, leading to possible short circuits and/or ground faults.

*Reprinted with permission from NFPA 70-2014.

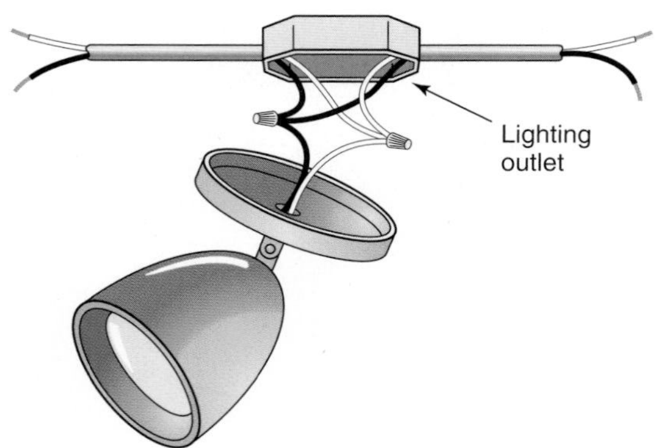

FIGURE 2-3 When a luminaire (lighting fixture) is connected to the branch-circuit wires, the outlet is called a *lighting outlet*.

- A **split-wired receptacle** is electrician's jargon, not an official *NEC* definition. Electricians are very creative in their use of terms. Other terms for the use of these receptacles include *split receptacle*, *split-wired*, *split-switched receptacle*, *switched receptacle*, and *half-switched receptacle*.

To convert a conventional duplex receptacle into a split-wired receptacle, simply remove the tab between the two ungrounded conductor terminals (brass colored). See Figure 2-4. The receptacle can then be used where one receptacle is "hot" at all times, and the other receptacle is switch controlled. Another common application is to connect each receptacle of the duplex to a separate branch circuit.

FIGURE 2-4 Method to create a split-wired receptacle.

By definition, toggle switches, receptacles circuit breakers, fuses, and occupancy sensors are *devices* because they carry or control current as their principle function.

The term *opening* is widely used by electricians and electrical contractors when estimating the cost of an installation. The term *opening* covers all lighting outlets, receptacle outlets, junction boxes, switches, etc. The electrician and/or electrical contractor estimates a job at "X dollars per lighting outlet," "X dollars per switch," "X dollars per receptacle outlet," and so on. These estimates include the time and material needed to complete the job. Each type of electrical *opening* is represented on the electrical plans as a symbol. The electrical openings in Figure 2-1 are shown by the symbols in Figure 2-5.

ANSI recently published a totally revised standard entitled *Symbols for Electrical Construction Drawings*. This was the first revision in over 25 years. Figures 2-6, 2-7, 2-8, 2-9, 2-10, and 2-11 show the electrical symbols most commonly found on architectural and electrical plans. Because some items may have more than one symbol, it is important to check the plans and specifications of any job you are working on for a symbol schedule to make sure you have interpreted the symbols correctly.

The curved lines in Figure 2-1 run from the outlet to the switch or switches that control the outlet. Curved lines are used to differentiate the electrical circuitry from the building construction drawing lines. Several drawings show a curved line between three-way and four-way switches to indicate the control points. Some receptacles have one-half of the duplex switched. A curved line runs from the receptacle to the switch or control point. Other curved lines show the connection of luminaires or the circuiting for receptacles.

Receptacle outlet:

Three-way switch:

Ceiling lighting outlet:

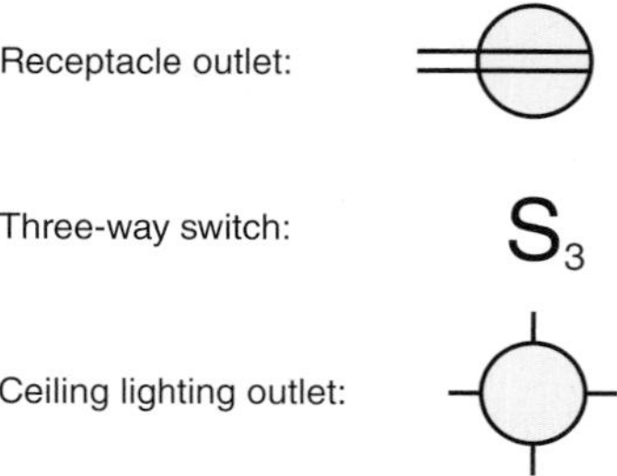

FIGURE 2-5 Some common electrical symbols.

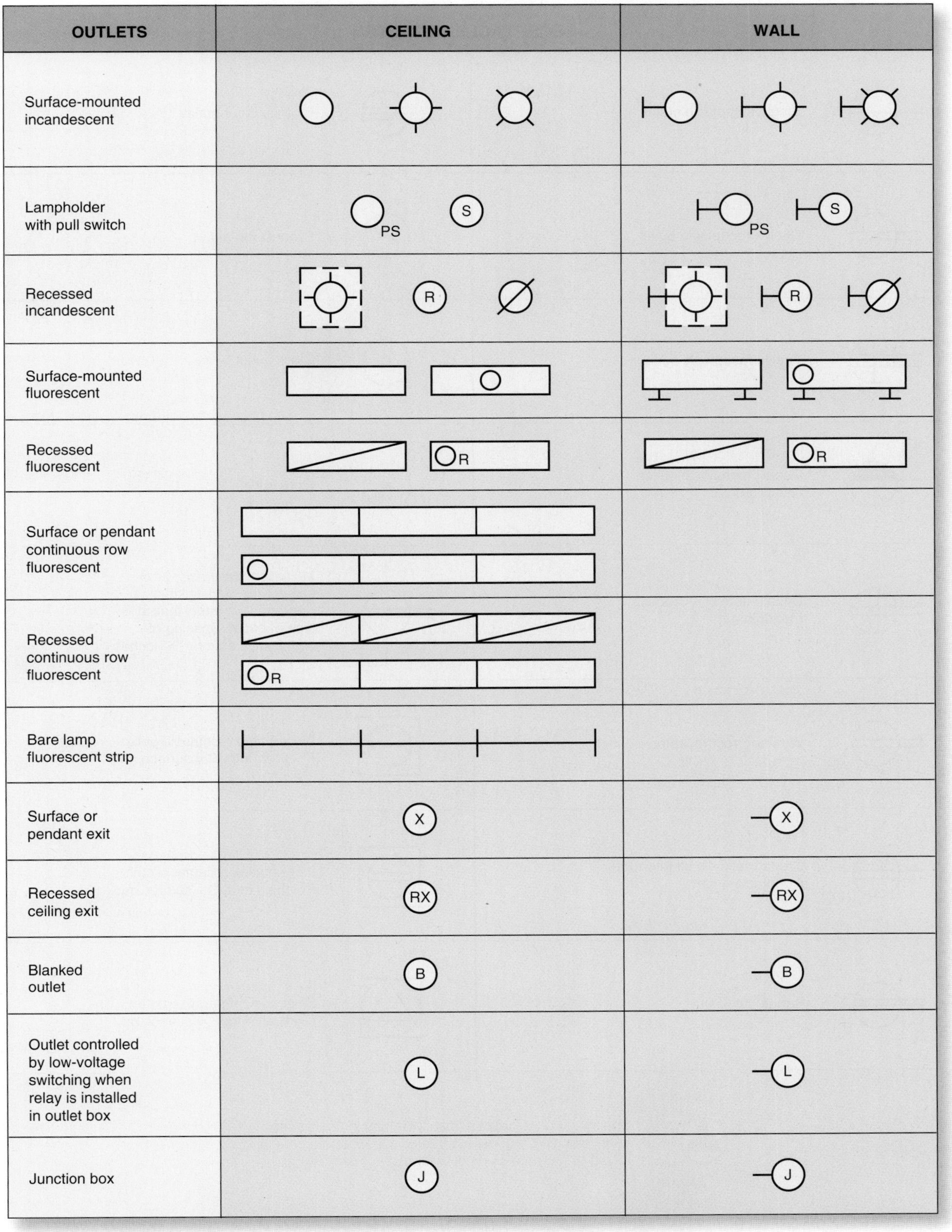

OUTLETS	CEILING	WALL
Surface-mounted incandescent		
Lampholder with pull switch	PS S	PS S
Recessed incandescent	R	R
Surface-mounted fluorescent		
Recessed fluorescent	R	R
Surface or pendant continuous row fluorescent		
Recessed continuous row fluorescent	R	
Bare lamp fluorescent strip		
Surface or pendant exit	X	X
Recessed ceiling exit	RX	RX
Blanked outlet	B	B
Outlet controlled by low-voltage switching when relay is installed in outlet box	L	L
Junction box	J	J

Courtesy of National Electrical Contractors Association.

FIGURE 2-6 Lighting outlet symbols.

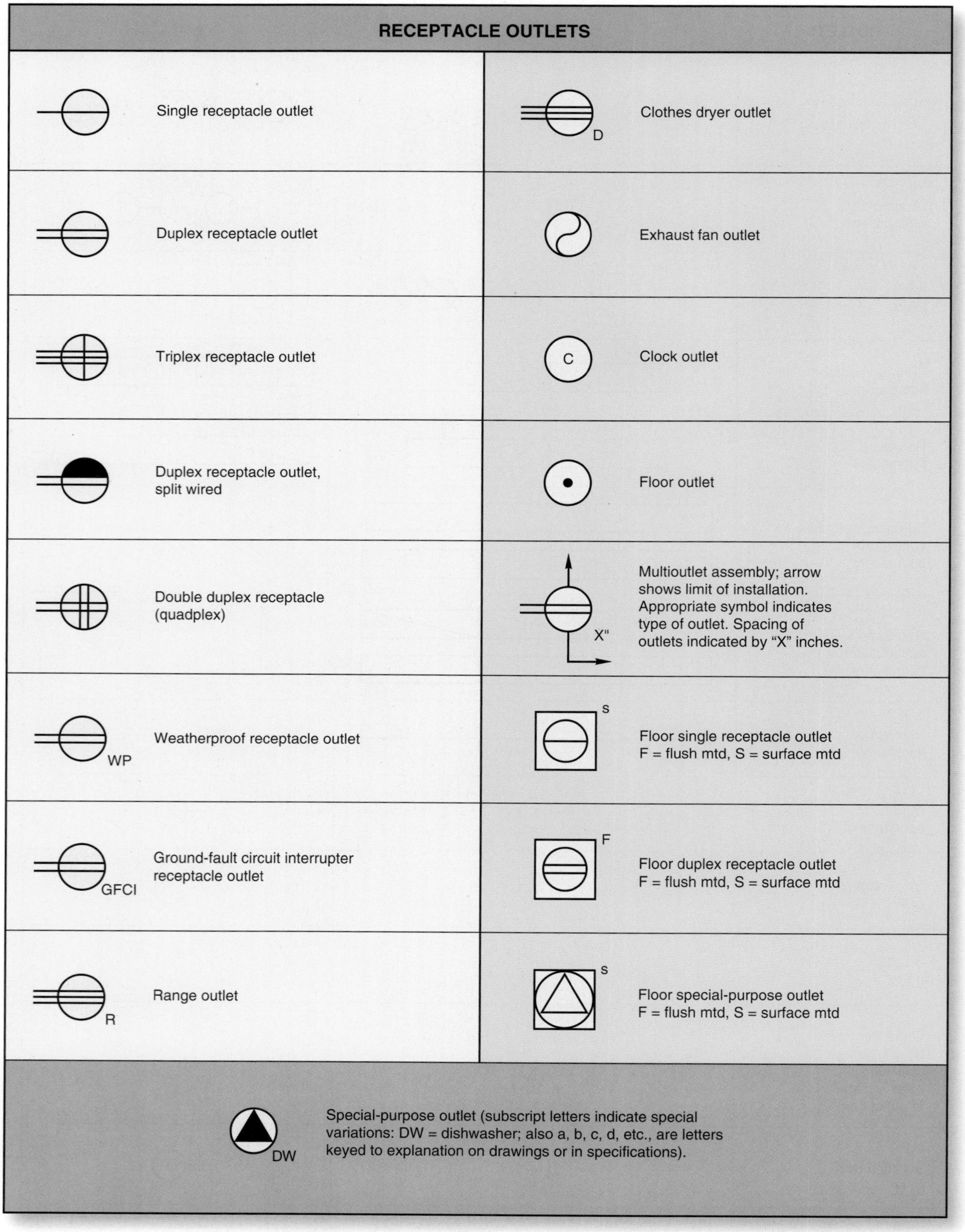

RECEPTACLE OUTLETS			
	Single receptacle outlet	D	Clothes dryer outlet
	Duplex receptacle outlet		Exhaust fan outlet
	Triplex receptacle outlet	C	Clock outlet
	Duplex receptacle outlet, split wired		Floor outlet
	Double duplex receptacle (quadplex)	X"	Multioutlet assembly; arrow shows limit of installation. Appropriate symbol indicates type of outlet. Spacing of outlets indicated by "X" inches.
WP	Weatherproof receptacle outlet	S	Floor single receptacle outlet F = flush mtd, S = surface mtd
GFCI	Ground-fault circuit interrupter receptacle outlet	F	Floor duplex receptacle outlet F = flush mtd, S = surface mtd
R	Range outlet	S	Floor special-purpose outlet F = flush mtd, S = surface mtd
DW	Special-purpose outlet (subscript letters indicate special variations: DW = dishwasher; also a, b, c, d, etc., are letters keyed to explanation on drawings or in specifications).		

Courtesy of National Electrical Contractors Association.

FIGURE 2-7 Receptacle outlet symbols.

SWITCH SYMBOLS	
S	Single-pole switch
S_2	Double-pole switch
S_3	Three-way switch
S_4	Four-way switch
S_D	Door switch
S_{DS}	Dimmer switch
S_G	Glow switch toggle—glows in off position
S_K	Key-operated switch
S_{KP}	Key switch with pilot light
S_{LV}	Low-voltage switch
S_{LM}	Low-voltage master switch
S_{MC}	Momentary-contact switch
M (diamond)	Occupancy sensor—wall mounted with "Off-auto" override switch
M (diamond), P	Occupancy sensor—ceiling mounted; "P" indicates multiple switches wire-in parallel
S_P	Switch with pilot light on when switch is on
S_T	Timer switch
S_R	Variable-speed switch
S_{WP}	Weatherproof switch

Courtesy of National Electrical Contractors Association.

FIGURE 2-8 Switch symbols.

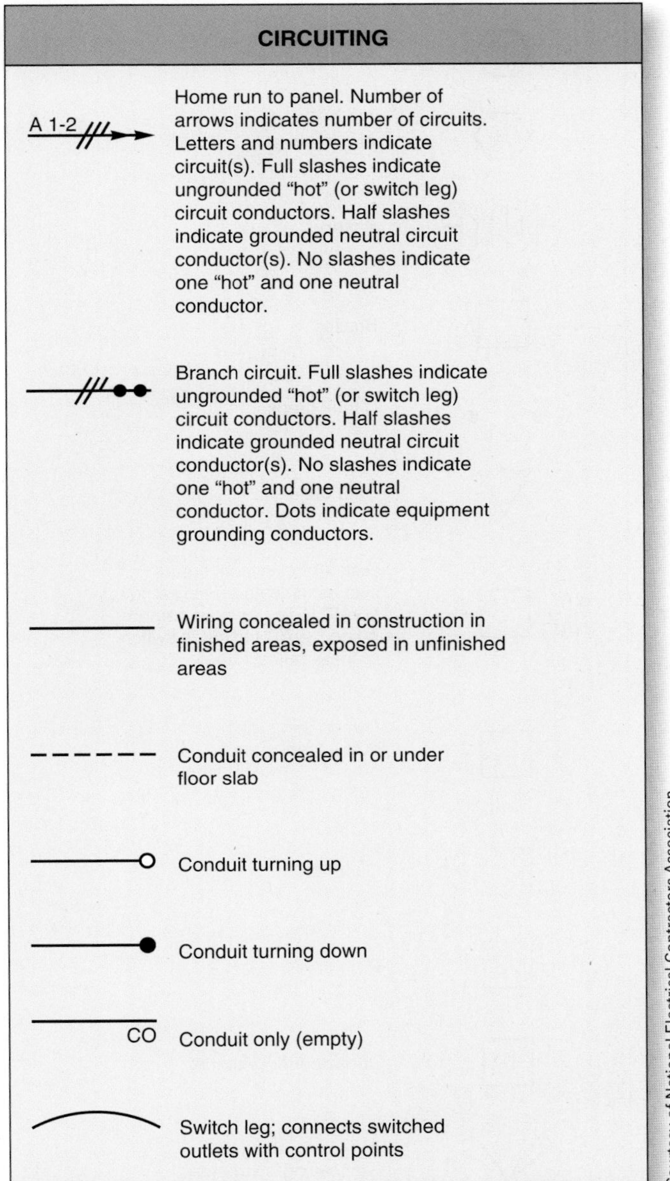

FIGURE 2-9 Circuiting symbols.

A study of the plans for the single-family dwelling shows that many different electrical symbols are used to represent the electrical devices and equipment used in the building.

Be very careful when interpreting symbols because of the similarity between them, as shown in Figure 2-12. For example, the left symbol can be used for a split-wired or a half-switched duplex receptacle. Several of the rooms in this dwelling have duplex receptacles with half of the receptacle switched and the other half remaining hot, or energized, all the time. Table or floor lamps can be plugged into the receptacle that is switched, to allow control of room lighting from near the door. The receptacle remains "hot" all the time for any appliance or fixture that is to remain energized at all times. In some areas these receptacles are referred to as switched receptacles or split-wired receptacles. The concept of split-wired receptacles includes using a branch circuit with two "hot" or energized, conductors and a shared neutral.

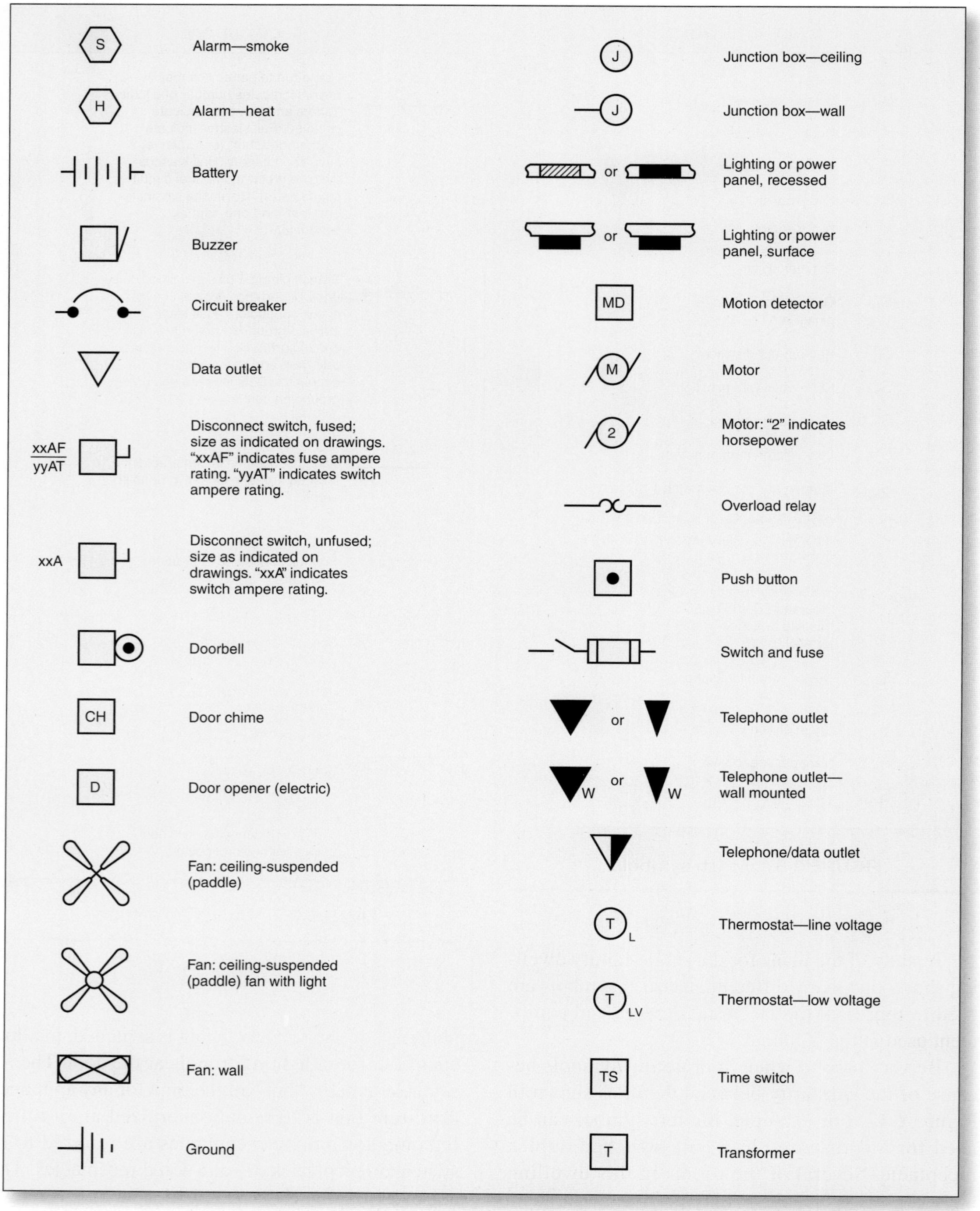

Symbol	Description	Symbol	Description
S	Alarm—smoke	J	Junction box—ceiling
H	Alarm—heat	J	Junction box—wall
	Battery	or	Lighting or power panel, recessed
	Buzzer	or	Lighting or power panel, surface
	Circuit breaker	MD	Motion detector
	Data outlet	M	Motor
xxAF / yyAT	Disconnect switch, fused; size as indicated on drawings. "xxAF" indicates fuse ampere rating. "yyAT" indicates switch ampere rating.	2	Motor: "2" indicates horsepower
xxA	Disconnect switch, unfused; size as indicated on drawings. "xxA" indicates switch ampere rating.		Overload relay
	Doorbell		Push button
CH	Door chime		Switch and fuse
D	Door opener (electric)	or	Telephone outlet
	Fan: ceiling-suspended (paddle)	W or W	Telephone outlet—wall mounted
	Fan: ceiling-suspended (paddle) fan with light		Telephone/data outlet
	Fan: wall	T L	Thermostat—line voltage
	Ground	T LV	Thermostat—low voltage
		TS	Time switch
		T	Transformer

Courtesy of National Electrical Contractors Association.

FIGURE 2-10 Miscellaneous symbols.

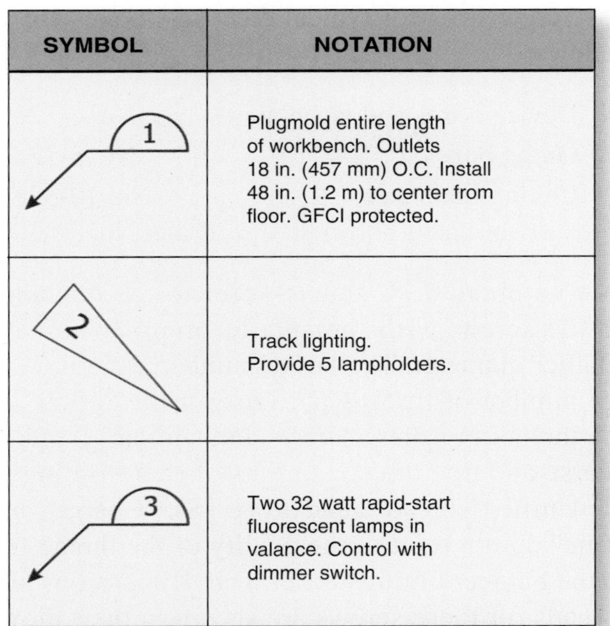

SYMBOL	NOTATION
1	Plugmold entire length of workbench. Outlets 18 in. (457 mm) O.C. Install 48 in. (1.2 m) to center from floor. GFCI protected.
2	Track lighting. Provide 5 lampholders.
3	Two 32 watt rapid-start fluorescent lamps in valance. Control with dimmer switch.

FIGURE 2-11 Example of how a symbol refers to notations. Used if there is insufficient space on the sheet where the notation applies. Sometimes referred to as "sheet notes." The triangular symbol may be rotated to any angle.

For both applications mentioned here, the tab is removed on the side of the duplex receptacle so that each receptacle can be controlled separately or have both ungrounded conductors connected to it.

This switching arrangement satisfies *NEC 210.70(A)(1), Exception No. 1*, which permits switch-controlled receptacles to serve as the lighting outlet in habitable rooms.

The notation for the receptacle, shown at the right in Figure 2-12, indicates the receptacle is to be in an enclosure suitable for a weatherproof location.

When preparing electrical plans for commercial and industrial buildings, most architects and electrical engineers use symbols approved by ANSI wherever possible. For residential projects, usually only the basic symbols are used, showing the location of receptacles, switches, appliances, and luminaires. Specific circuitry, as shown in Figure 2-9, is generally left up to the electrician.

FIGURE 2-12 Variations in significance of outlet symbols.

Later in this text, you will find suggested *cable layouts* for all of the branch circuits in this residence. These *cable layouts* show the required number of conductors between outlets and switches.

LUMINAIRES AND OUTLETS

The term *luminaire* is found in the *NEC*. In the United States, we previously used the term *lighting fixture* for years. But because the *NEC* is an international code, we must become familiar with some new terms such as *luminaire*. By definition, a *luminaire* is A complete lighting unit consisting of a light source such as a lamp or lamps, together with the parts designed to position the light source and connect it to the power supply. It may also include parts to protect the light source, ballast, or distribute the light. A lampholder itself is not a luminaire.*

Simply stated for our purposes, a luminaire and a lighting fixture are one and the same.

The location of lighting outlets is determined by the amount and type of illumination required to provide the desired lighting effects. It is not the intent of this text to describe how proper and adequate lighting is determined. Rather, the text covers the proper methods of installing the circuits for such lighting.

Luminaire manufacturers publish catalogs that provide a tremendous amount of information regarding recommendations for residential lighting. These publications are available at electrical distributors, lighting showrooms, and home centers.

Architects often include a specific amount of money in the specifications, referred to as a *fixture allowance*. Electrical contractors include this amount in their bid, and the choice of luminaires is then left to the homeowners. If the owner selects

*Reprinted with permission from NFPA 70-2014.

luminaires whose total cost exceeds the fixture allowance, the owner is expected to pay the difference between the actual cost and the fixture allowance. The electrician needs to know the details of the luminaire to be installed so that a box designed for the size and weight of the luminaire can be positioned correctly. Figure 2-21 shows several types of boxes that may be appropriate, depending on the size and weight of the luminaire.

How luminaires are handled varies by job and geographical area. Sometimes the recessed luminaire housings are included in the base bid, and the trims are included in the fixture allowance. Always read the specifications and study the plans carefully for important symbols and notations that might reference luminaires.

Surface-mounted luminaires are secured to an outlet box, a raised plaster (adapter) ring, or a device box, using the mounting hardware (straps and No. 8-32 or No. 6-32 screws) furnished with the luminaire. Some outlet box/bar hanger assemblies have a fixture stud on which to fasten the luminaire. See Figure 2-13.

Symbols Used to Represent Lighting Outlets and Switches

Typical lighting outlet symbols are shown in Figure 2-13. Typical switch symbols are shown in Figure 2-14.

OUTLET, DEVICE, AND JUNCTION BOXES

NEC Article 314 contains much detail about outlet boxes, device boxes, pull boxes, junction boxes, and conduit bodies (LB, LR, LL, C, etc.), all of which are used to provide the necessary space for making electrical connections, to contain wiring, and to support luminaires, and so on. *NEC Article 314* is quite lengthy. For the moment, we will focus on some of the basic *Code* requirements for outlet boxes, device boxes, and junction boxes. Additional *Code* requirements are covered on an "as needed" basis in this chapter and in later chapters.

Screw sizes for typical electrical boxes are as follows:

Device boxes and faceplates	No. 6-32
Outlet boxes	No. 8-32
Ceiling fan boxes (two sets of holes)	No. 10-32 and No. 8-32

The numbers 6, 8, and 10 refer to the diameter of the screw, with the smaller number being a smaller diameter. The larger number, 32, refers to the number of threads per inch, or the "pitch" of the thread. For these screws and similar screws, a coarse and fine thread is established according to the Unified Thread Standard (UTS). "Coarse" and "fine" do not refer to the quality of the thread but to the number of threads per inch. Dimensions and threads for these screws are shown in the following table:

UTS Number	Diameter	Coarse Pitch	Fine Pitch
6	0.1380	32	40
8	0.1640	32	36
10	0.1900	24	32

Electricians commonly carry a multitap tool to clean out threaded holes that may have become damaged or filled with paint or other contaminates.

Some Basic Code Requirements

- *NEC 300.11* and *314.23:* Boxes must be securely mounted and fastened in place.
- *NEC 300.15:* Where conduit, electrical metallic tubing, nonmetallic sheathed cable, Type AC cable, or other cables are installed, a box or conduit body must be installed at each conductor splice connection point, outlet, switch point, junction point, or pull point. This requirement is illustrated in Figure 2-15. Exceptions to this are found in *300.15(A)* through *(M)*. Typical exceptions are surface-mounted strip luminaires, as shown in Figure 10-2. Most of these have a back plate that has a trade size ½ knockout intended for the direct entry of a cable without the need for a box behind the luminaire.

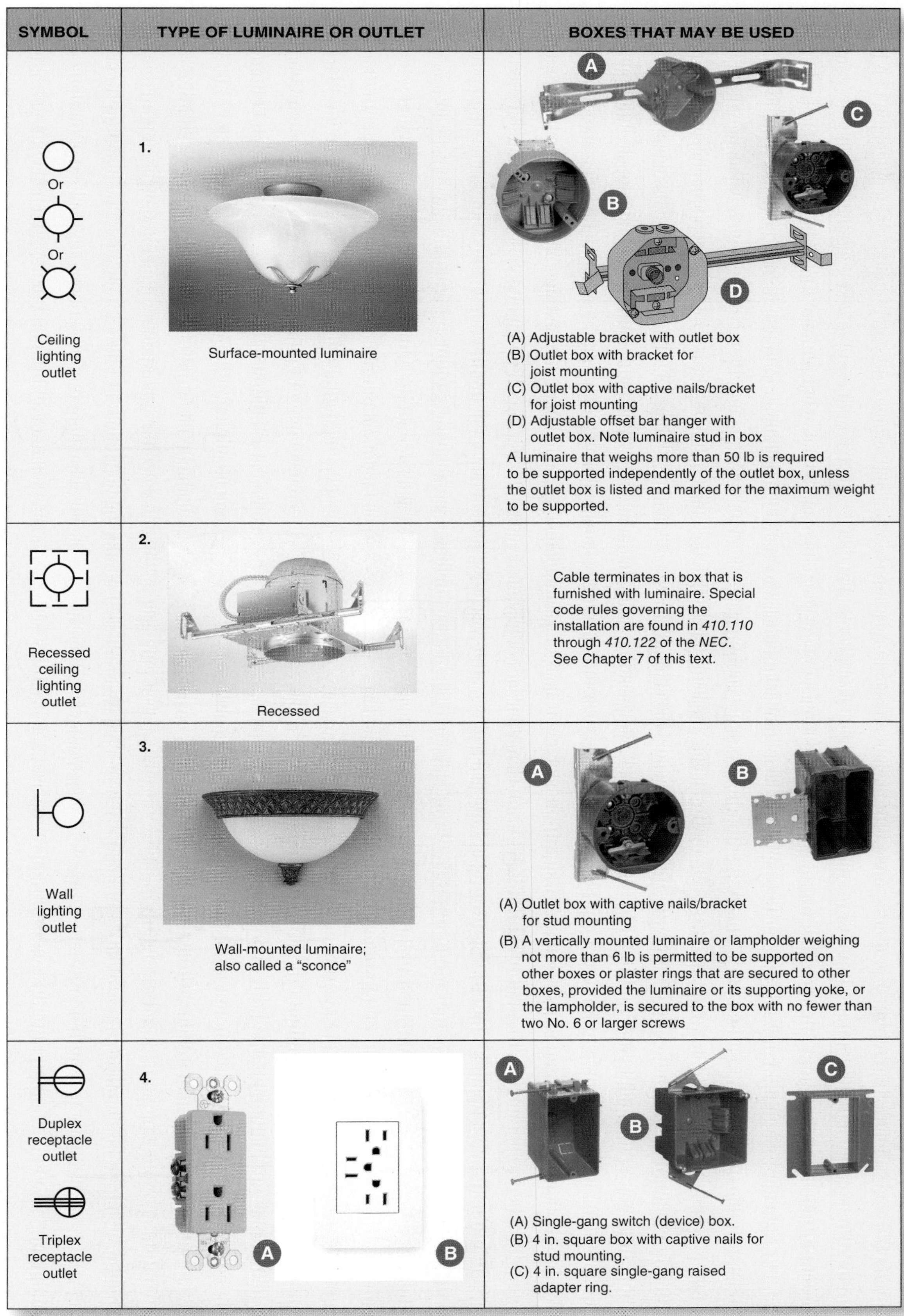

SYMBOL	TYPE OF LUMINAIRE OR OUTLET	BOXES THAT MAY BE USED
Or Or Ceiling lighting outlet	1. Surface-mounted luminaire	(A) Adjustable bracket with outlet box (B) Outlet box with bracket for joist mounting (C) Outlet box with captive nails/bracket for joist mounting (D) Adjustable offset bar hanger with outlet box. Note luminaire stud in box A luminaire that weighs more than 50 lb is required to be supported independently of the outlet box, unless the outlet box is listed and marked for the maximum weight to be supported.
Recessed ceiling lighting outlet	2. Recessed	Cable terminates in box that is furnished with luminaire. Special code rules governing the installation are found in *410.110* through *410.122* of the *NEC*. See Chapter 7 of this text.
Wall lighting outlet	3. Wall-mounted luminaire; also called a "sconce"	(A) Outlet box with captive nails/bracket for stud mounting (B) A vertically mounted luminaire or lampholder weighing not more than 6 lb is permitted to be supported on other boxes or plaster rings that are secured to other boxes, provided the luminaire or its supporting yoke, or the lampholder, is secured to the box with no fewer than two No. 6 or larger screws
Duplex receptacle outlet Triplex receptacle outlet	4.	(A) Single-gang switch (device) box. (B) 4 in. square box with captive nails for stud mounting. (C) 4 in. square single-gang raised adapter ring.

FIGURE 2-13 Typical electrical symbols, items they represent, and boxes that could be used.

Photos courtesy of Progress Lighting [1, 3]; Eaton's Cooper Lighting Business [2]; Leviton Manufacturing Company [4A & 4B].

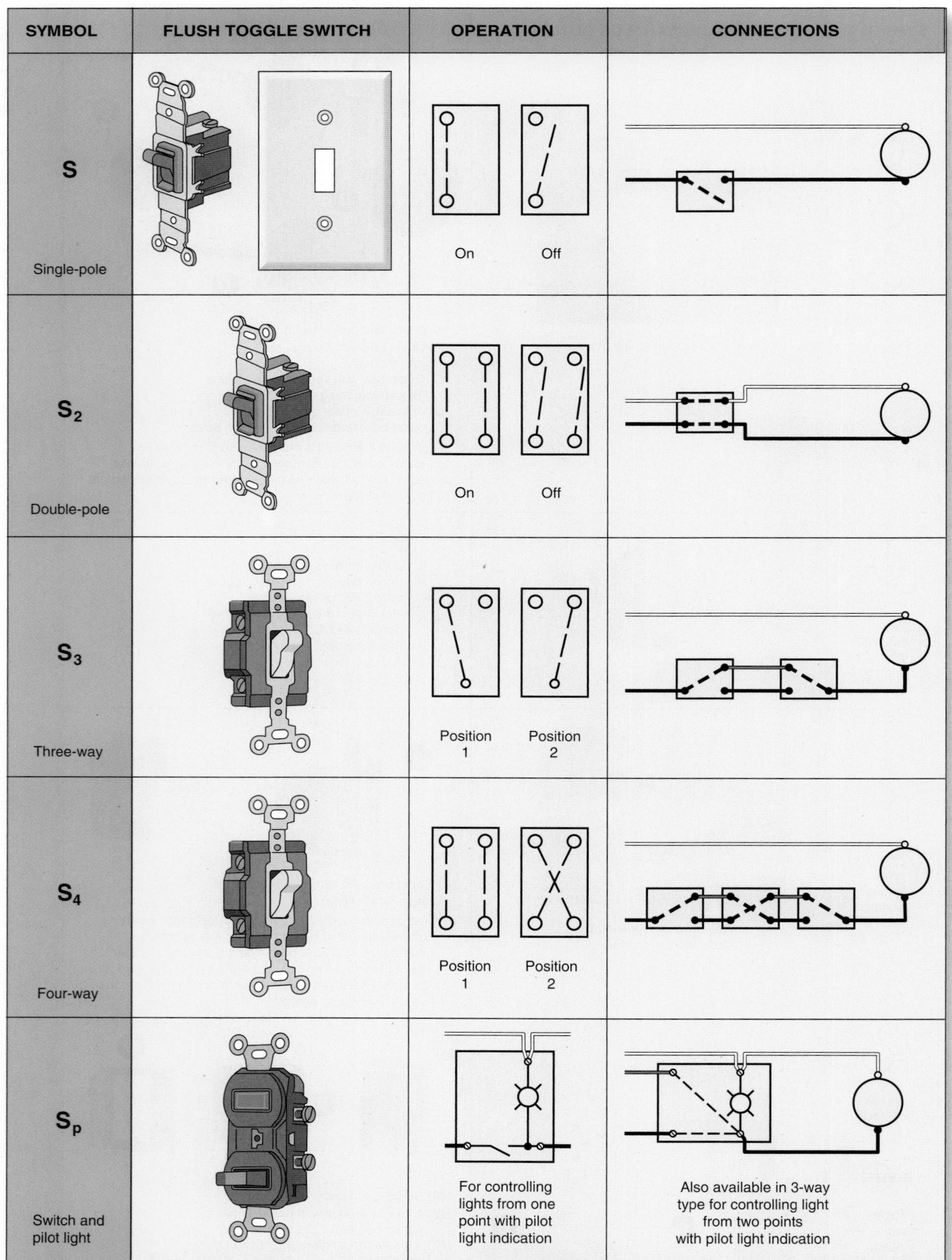

FIGURE 2-14 Standard switches and symbols.

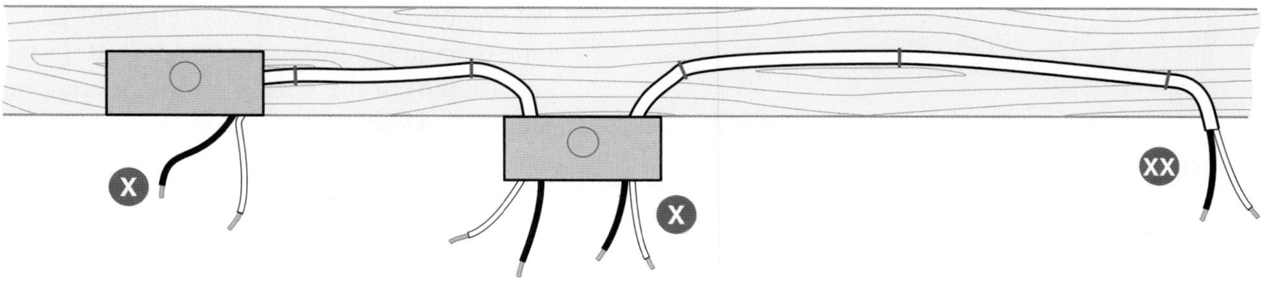

FIGURE 2-15 A box (or fitting) must be installed wherever there are splices, outlets, switches, or other junction points as indicated by the "X." Unless listed for use without a box, such as a typical surface-mounted strip luminaire that has a back plate with a trade size ½ knockout, point XX is a *Code* violation, *300.15.*

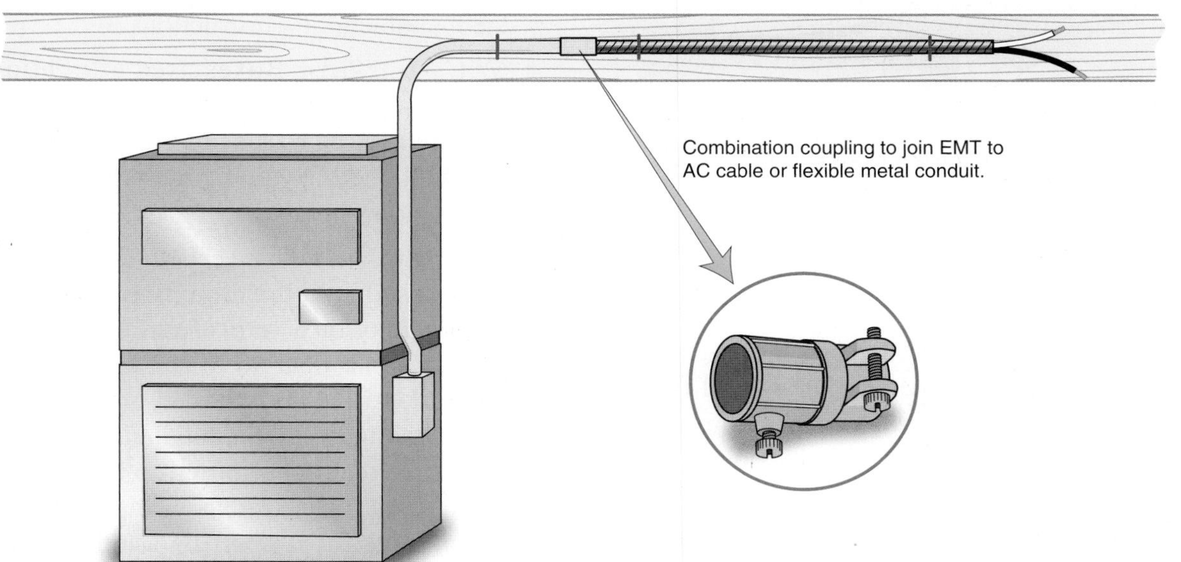

FIGURE 2-16 *NEC 300.15(F)* permits a transition to be made from one wiring method to another wiring method. In this case, the armor of the Type AC cable is removed, allowing sufficient length of the conductors to be run through the EMT. A proper fitting must be used at the transition point, and the fitting must be accessible after installation. The conductors in Type AC cable often have THHN insulation.

- *NEC 300.15(C):* Where cables enter or exit from conduit or tubing that is used to provide cable support or protection against physical damage, a fitting shall be provided on the end(s) of the conduit or tubing to protect the cable from abrasion. Figure 2-16 shows a transition from a cable to a raceway.
- *NEC 314.16:* Boxes must be large enough for all of the enclosed conductors and wiring devices. Installing boxes too small for the number of conductors, wiring devices, luminaire studs, and splices is a real problem! This topic is covered in detail later.
- *NEC 314.17(C):* The exception allows multiple cables to be run through a single knockout opening in a nonmetallic box. This permission is not given for metallic boxes. See Figure 2-17.
- *NEC 314.20:* In walls or ceilings where the surface material is noncombustible, boxes, plaster rings, domed covers, extension rings, or listed extenders shall be mounted so that they will be set back not more than ¼ in. (6 mm). In walls or ceilings where the surface material is combustible, boxes, plaster rings, domed covers, extension rings, or listed extenders shall be set flush with or extend from the surface. Be cautious

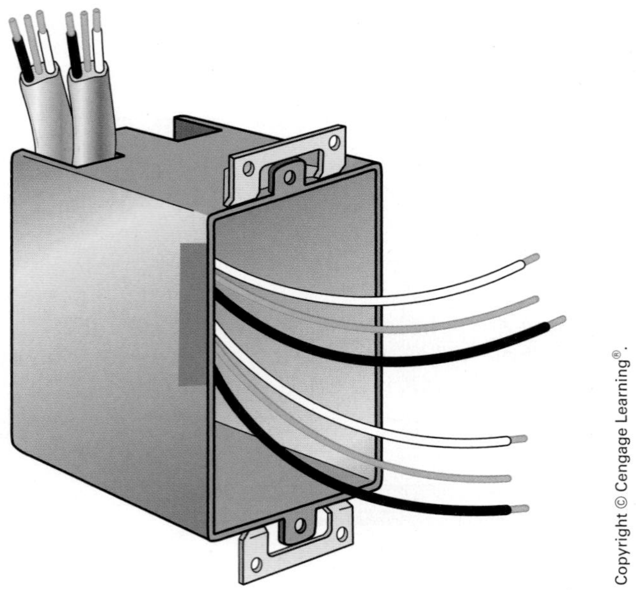

FIGURE 2-17 Multiple cables may run through a single knockout opening in a nonmetallic box, *314.17(C), Exception.*

about boxes or plaster rings that extend from the surface. The owner or electrical contractor is likely to object if the cover plates do not seat against the wall surface.

- *NEC 314.22:* A surface extension from a box of a concealed wiring system is made by mounting and mechanically securing an extension ring over the concealed box. An exception to this is that a surface extension is permitted to be made from the cover of a concealed box if the cover is designed so it is unlikely to fall off, or be removed if its securing means becomes loose. ▶The wiring method must be flexible for an approved length, and permit removal of the cover and provide access to the box interior. It must be arranged so that any required grounding continuity is independent of the connection between the box and cover.◀ This is shown in Figure 2-18.
- *NEC 314.24:* ▶Outlet and device boxes are required to be of an approved depth so that conductors will not be damaged when installing the wiring device or equipment into the box.◀ Merely calculating and providing the proper volume of a box is not always enough. The volume calculation might prove adequate, yet the size (depth) of the wiring device or equipment might be such that conductors behind it may be damaged.
- *NEC 314.25:* In completed installations, all boxes are required to have a cover, faceplate, or luminaire canopy. Do not leave any electrical boxes uncovered. ▶Screws used for the purpose of attaching covers, or other equipment, to the box are required to be either machine screws matching the thread gauge or size that is integral to the box or to be in accordance with the manufacturer's instructions.◀
- *NEC 314.27(A):* Boxes used at lighting outlets must be designed or installed so that a luminaire may be attached to it. This is typically accomplished by the manufacturer providing tapped holes in boxes.
- *NEC 314.27(A)(1)* ▶*Vertical Surface Outlets:* Boxes used at luminaire or lampholder outlets that are in or on a vertical surface, if suitable to support other than 50 lb (23 kg), are required to be identified and marked on the interior of the box to indicate the maximum weight of the luminaire that is permitted to be supported.◀ See Figure 2-13(3) for an example of boxes that are suitable for supporting a luminaire that is mounted on a vertical surface.

  An exception to the rule provides: A vertically mounted luminaire or lampholder weighing not more than 6 lb (3 kg) is permitted to be supported on other boxes or plaster rings that are secured to other boxes, provided the luminaire or its supporting yoke, or the lampholder, is secured to the box with no fewer than two No. 6 or larger screws. See Figure 2-19 for examples of boxes or plaster rings that can be used to support luminaires that are mounted on a vertical surface so long as the luminaire does not weigh more than 6 lb (3 kg). Plaster rings are typically installed on a 4 in. (100 mm) square box to provide a means to secure receptacles, switches, or luminaires and can be thought of as "adapter rings."
- *NEC 314.27(A)(2) Ceiling Outlets:* At every outlet used exclusively for lighting, the box

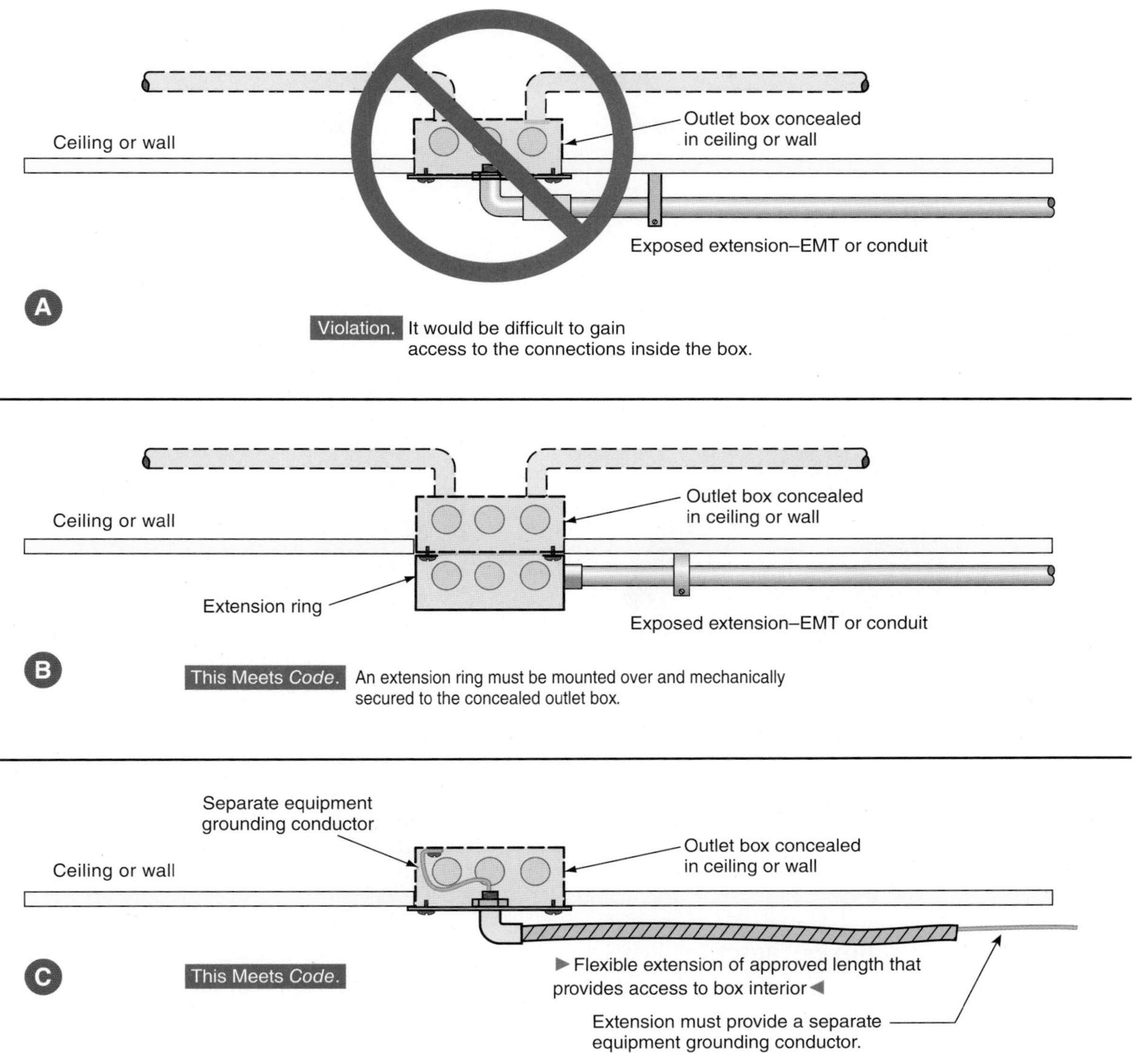

FIGURE 2-18 Making a surface extension from a flush-mounted box.

is required to be designed or installed so that a luminaire or lampholder may be attached. Boxes are required to be suitable to support a luminaire weighing a minimum of 50 lb (23 kg). A luminaire that weighs more than 50 lb (23 kg) is required to be supported independently of the outlet box unless the outlet box is listed and marked on the interior of the box to indicate the maximum weight the box is permitted to support. See Figure 2-13(1) for an example of boxes that are suitable for supporting luminaires that are ceiling mounted.

- *NEC 314.27(C):* Standard outlet boxes must not be used as the sole support for ceiling-suspended (paddle) fans unless they are specifically listed for this purpose. See Chapter 9 for a detailed discussion of this issue.
- *NEC 314.29:* ▶Conduit bodies, junction, pull, and outlet boxes must be installed so the wiring contained in them is accessible without removing any part of the building or structure, or in underground circuits, without excavating sidewalks, paving, earth, or other substance that is to be used to establish the finished grade.◀

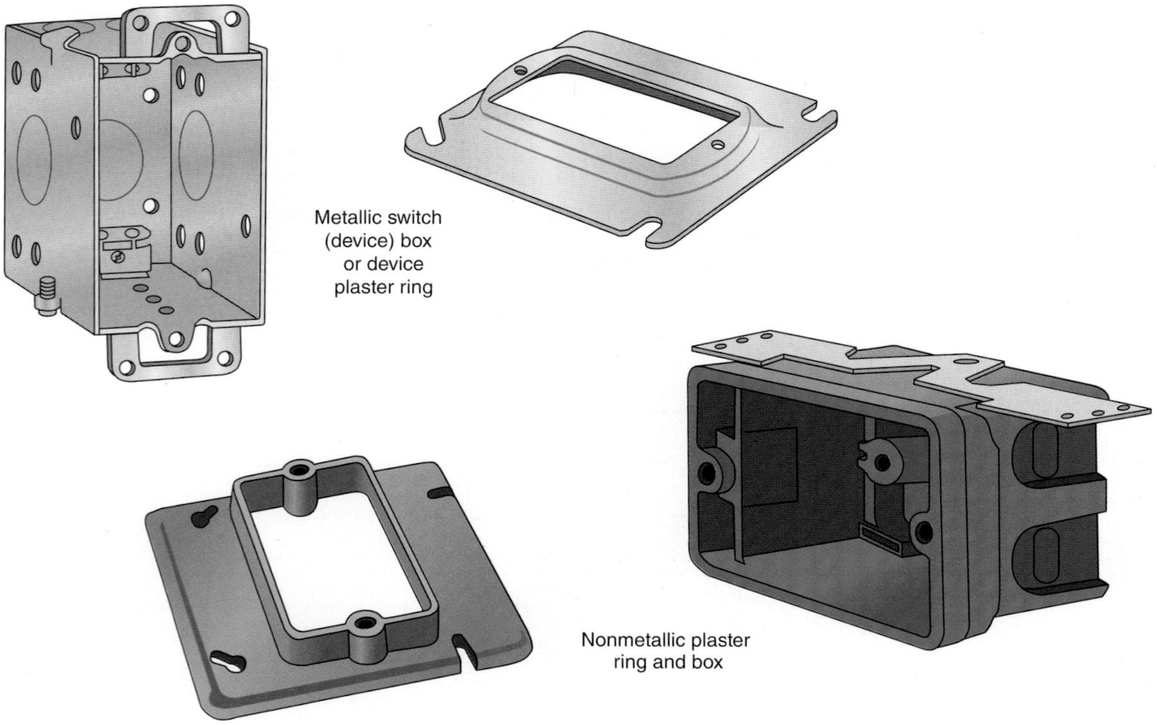

Metallic switch (device) box or device plaster ring

Nonmetallic plaster ring and box

FIGURE 2-19 Luminaires that do not weigh more than 6 lb. (3 kg) and are mounted on a vertical surface such as a wall are permitted to be supported from boxes or plaster rings by not less than two No. 6-32 screws. See *314.27(A)(1), Exception*. Heavier luminaires are required to be supported from a box that is marked for the design weight to be supported, if the box is suitable for supporting less than 50 lb (23 kg).

Never install outlet boxes, junction boxes, pull boxes, or conduit bodies where they will be inaccessible behind or above permanent finished walls or ceilings. Should anything ever go wrong, it would be a nightmare, if not impossible, to troubleshoot and locate hidden splices. Wiring located above "lay-in" ceilings is considered accessible because it is easy to remove a panel(s) to gain access to the space above the ceiling. The *NEC* definition of Accessible (as applied to wiring methods) is: Capable of being removed or exposed without damaging the building structure or finish or not permanently closed in by the structure or finish of the building.*

The same accessibility requirement is true for underground wiring. Boxes must be accessible without excavating sidewalks, paving, earth, or other material used to establish the finished grade.

The exception to *314.29* permits listed boxes to be covered with gravel, light aggregate, or noncohesive granulated soil (not clay) if their location is clearly marked.

- *NEC 314.28:* Pull and junction boxes must provide adequate space and dimensions for the installation of conductors, and they shall comply with the specific requirements of this section.
- *NEC 410.22:* This is another repeat of the requirement that boxes must be covered. The "cover" could be a blank cover, a luminaire canopy, a lampholder, a switch or receptacle, or similar device with faceplate.
- *NEC 410.36(A):* Here we find permission to use outlet boxes and fittings to support luminaires provided the requirements found in *314.27(A)* and *(B)* are met.

*Reprinted with permission from NFPA 70-2014.

What to Do Where Recessed Luminaires Will Be Installed. This topic is covered in detail in Chapter 7 of this text.

NONMETALLIC OUTLET AND DEVICE BOXES

Today, most house wiring is done using nonmetallic boxes, as permitted by *NEC 314.3*, along with nonmetallic sheathed cable Type NM. This is usually because of economics. Although not often done, *NEC 314.3 (Exceptions)* allows metal raceways and metal-jacketed cables (BX) to be used with nonmetallic boxes provided all metal raceways or cables entering the box are bonded together to maintain the integrity of the grounding path to other equipment in the installation.

The house wiring system is formed by a number of specific circuits. Each branch circuit begins at the service or branch-circuit panelboard and consists of runs of cable from outlet to outlet or from box to box. The residence plans show many branch circuits for general lighting, appliances, electric heating, and other requirements. The specific *Code* rules for each of these circuits are covered in later chapters.

GANGED SWITCH (DEVICE) BOXES

Multiple-Gang Boxes

Where more than one wiring device is to be installed at one location, multiple-gang boxes are used. Nonmetallic boxes for wiring devices are available in 2-gang, 3-gang, and 4-gang. Nonmetallic multiple-gang boxes are shown in Figure 2-20. These boxes are marked by the manufacturer with their cubic inch (mm) volume or with the number of conductors permitted in the box. See the discussion on selecting the correct size box later in this Chapter.

Some designs of metal device boxes can be ganged together by removing and discarding one

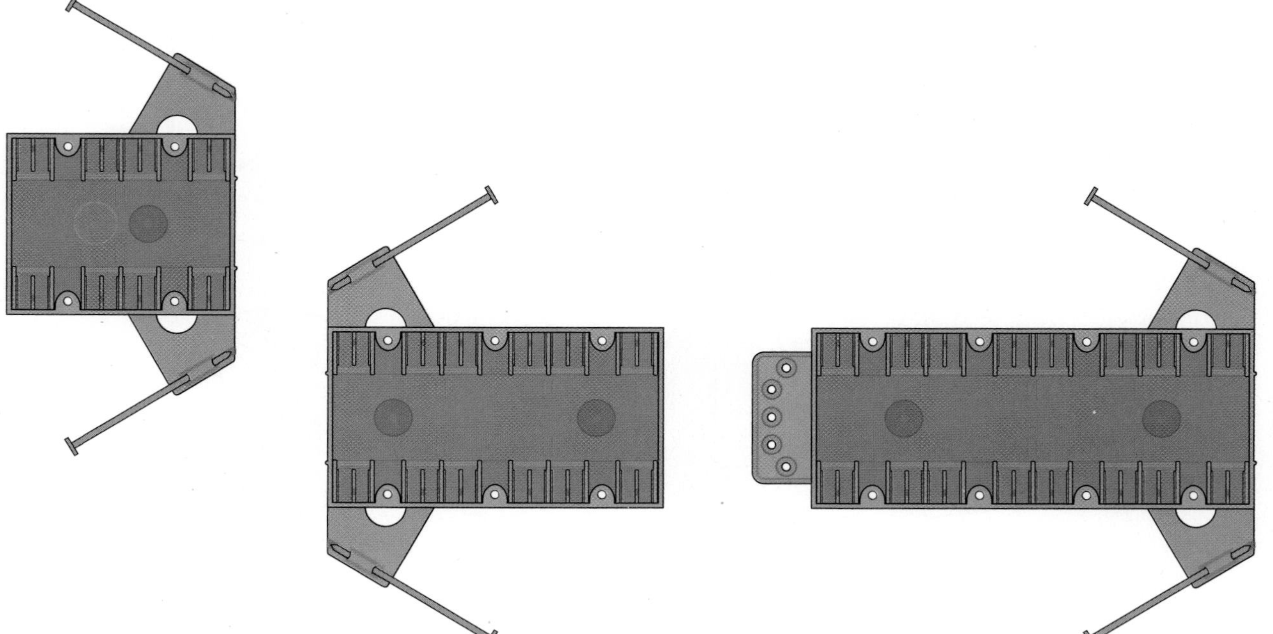

FIGURE 2-20 Examples of multiple-gang device boxes.

side from both the first and third boxes, and both sides from the second (center) box.

BOX MOUNTING

NEC 314.20 states that where noncombustible surface material (tile, gypsum, plaster, and like material) is used on walls or ceilings, then boxes, plaster rings, domed covers, extension rings, or listed extenders must be mounted so that they will be set back not more than ¼ in. (6 mm) from the face of the surface material. Where combustible surface material (wood paneling) is used on walls or ceilings, boxes, plaster rings, domed covers, extension rings, or listed extenders must be set flush with the surface material, Figure 2-21.

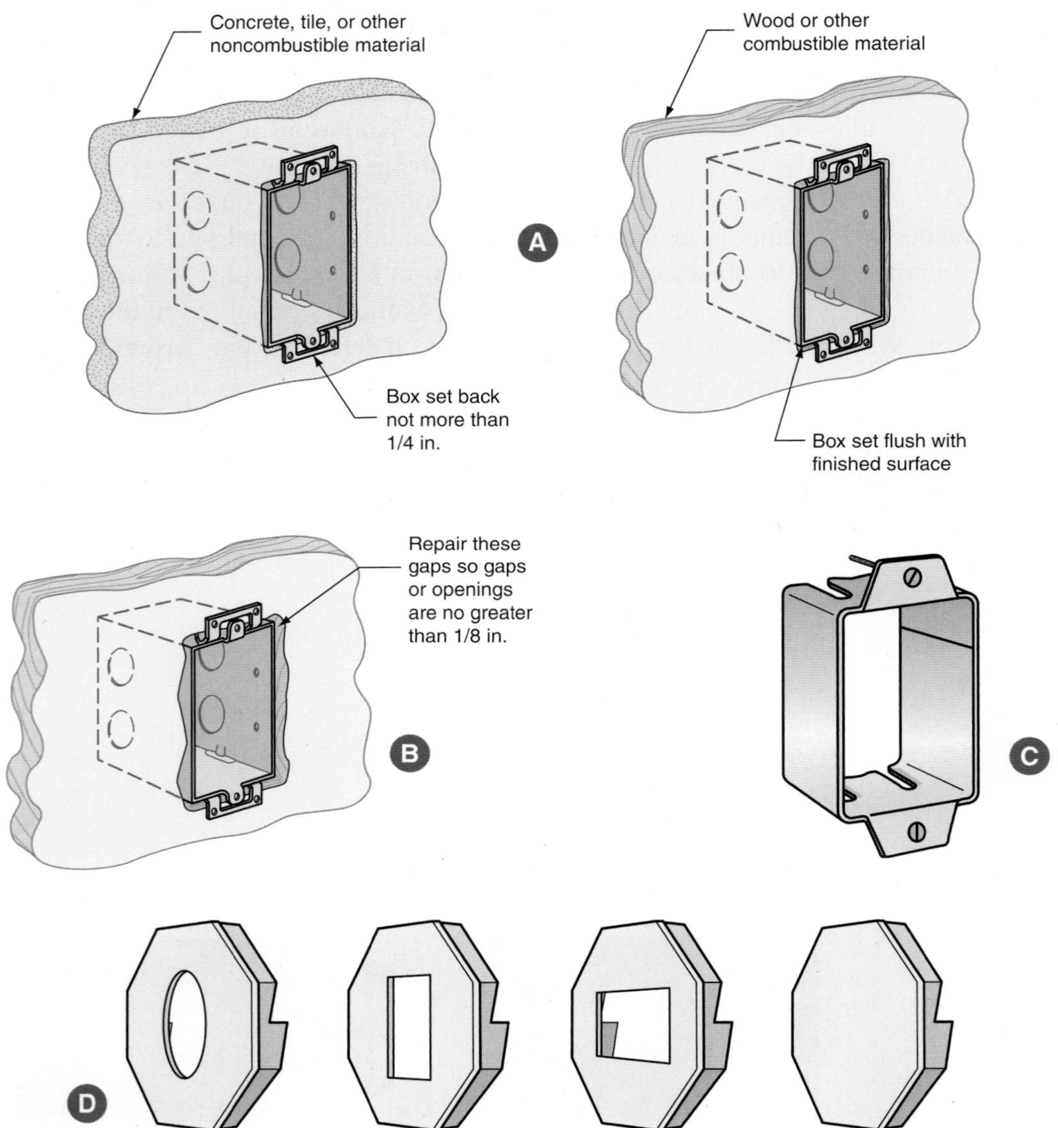

FIGURE 2-21 Box position requirements in walls and ceilings constructed of various materials. If the box is not flush with the finished surface (as a result of improper installation) or if paneling, drywall, tile, or a mirror is added, then install a box extender, sometimes referred to as a "goof ring." One type is shown in (C). Box extenders that are ¼ in., ⅜ in., ½ in., or ¾ in. deep are also available for 1- and 2-gang boxes. Refer to *314.20* and *314.21*. For setback requirements for panelboard cabinets, see *312.3* and *312.4*.

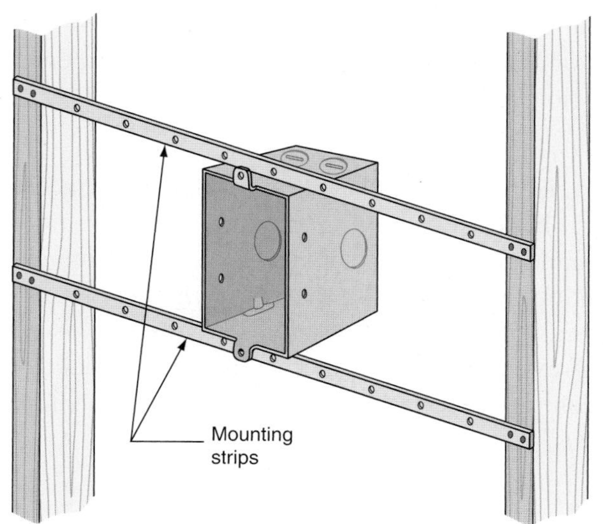

FIGURE 2-22 Switch (device) boxes installed between studs, using metal switch box supports, often referred to as "Kruse strips." If wood strips are used, they must have a cross-sectional dimension of not less than 1 in. × 2 in. (25 mm × 50 mm), *314.23(B)(2)*.

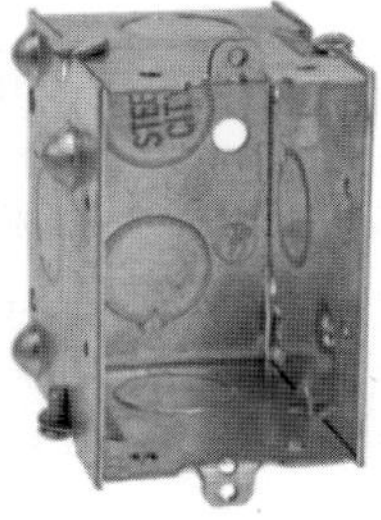

Listed box that complies with the nail-positioning rules.

Listed box that has external nails for securing the box to wooden studs.

FIGURE 2-23 Using nails or screws to install a sectional switch box, *314.23(B)(1)*. UL-listed boxes will have nail mounting holes positioned to meet the requirements of the *NEC*. A better choice is to use boxes that have external brackets or external nail mounting fixtures.

These requirements hold back, retard, and slow down the spread of fire caused by arcing at a loose connection, a short circuit, or a ground fault that might occur within the box. This significantly reduces the supply of oxygen needed to keep a fire burning. Closing the gaps around boxes further reduces the spread of fire.

Ganged sectional switch (device) boxes can be installed using a pair of metal mounting strips. These strips are also used to install a switch box between wall studs, Figure 2-22. When an outlet box is to be mounted at a specific location between joists, as for ceiling-mounted luminaires, an adjustable bar hanger is used, as in Figure 2-12.

NEC 314.23(B)(1) requires that mounting nails or screws through a box shall not be more than ¼ in. (6 mm) from the back or ends of the box, as shown in Figure 2-23. This ensures that the nails or screws will not interfere with the wiring devices in the box. Furthermore, you are not allowed to use screws through a box unless the threads are protected by an approved means to avoid abrasion of conductor insulation.

Why bother with these *Code* requirements that take up valuable wiring space and could lead to conductor damage? It is much better to use boxes that have external mounting means.

Most residences are constructed with wood framing. However, metal studs and joists are being used more and more. Figure 2-24 shows a few types of boxes for use with metal framing members.

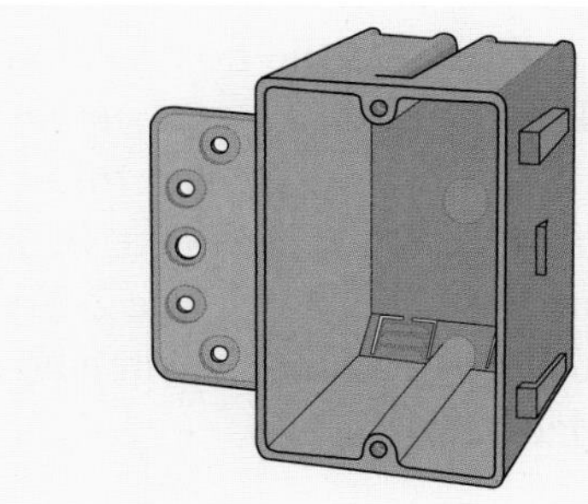

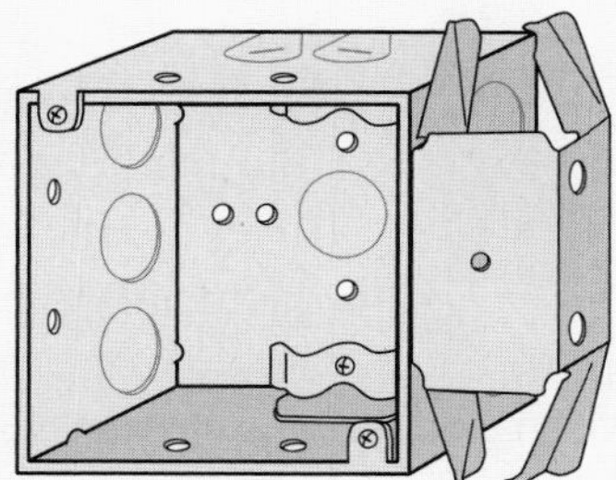

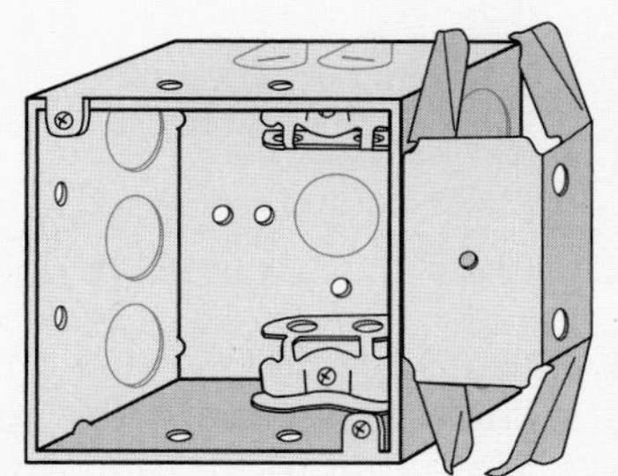

FIGURE 2-24 Types of boxes that can be used with steel framing construction. Note the side clamps that "snap" directly onto the steel framing members.

The first one is a single-gang nonmetallic device box that can be riveted or screwed to the metal stud. The second is a 4 in. square steel box with nonmetallic sheathed cable (Romex) clamps. The third is a 4 in. square steel box with armored cable (BX) clamps. The second and third boxes "snap" onto the metal stud, and do not require drilling the metal studs and fastening the box with nuts and bolts or screws. See Chapter 4 for *Code* requirements about running cables through metal framing members.

Another type of device used for fastening electrical boxes to steel framing members is shown in Figure 2-25. The bracket is screwed to the steel framing members, and then the box or boxes are attached to the bracket.

Damage Control!

One of the most common problems electricians face is what happens to conductors in boxes during the construction stage of a job. It's a major problem. In an attempt to minimize conductor damage, unsuccessful proposals similar to the following have been made to the *NEC:*

> Conductors that are inside electrical boxes and are subject to physical damage from router bits, sheet rock saws, and knives, and nonconductive coatings, such as drywall mud, paint, lacquer and enamel, shall be protected during the construction process by means of a rigid cover, plate, or insert of a thickness and strength as to prohibit penetration by the above-mentioned items.

In the meantime, do everything possible to prevent conductors from being damaged in electrical boxes. Pushing the conductors deep into the box and using deeper boxes can reduce damage to conductors. Installing temporary metal covers on boxes will prevent damage to conductors as well as prevent unwanted material such as sheetrock finishing compounds and paint from entering the boxes.

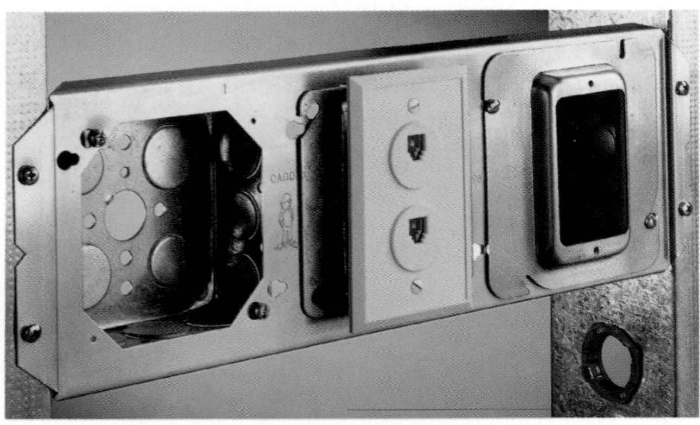

FIGURE 2-25 A bracket for use on steel framing members. The bracket spans the space between two studs. Note that more than one box can be mounted on this bracket.

Photograph is reprinted with permission and provided courtesy of ERICO International Corporation.

Remodel (Old Work)

When installing switches, receptacles, and lighting outlets in remodel work where the walls and ceilings (paneling, drywall, sheetrock and plaster, or lath and plaster) are already in place, first make sure you will be able to run cables to where the switches, receptacles, and lighting outlets are to be installed. After making sure you can get the cables through the concealed spaces in the walls and ceilings, you can then proceed to cut openings the size of the specific type of box to be installed at each outlet location. Cables are then "fished" through the stud and joist spaces behind or above the finished walls and ceilings to where you have cut the openings. Then boxes having plaster ears and snap-in brackets, Figure 2-26, can be inserted from the front, through the holes already cut at the locations where the wall or ceiling boxes are to be installed. After the boxes are snapped into place through the hole from the front, they become securely locked into place.

Another popular method of fastening wall and ceiling boxes in existing walls and ceilings is to use boxes that have plaster ears. Insert the box through the hole from the front. Then position a metal support into the hole on each side of the box. Be careful not to let the metal support fall into the hole. Next, bend the metal support over and into the box (Figure 2-27). Two metal supports are used. These metal switch box supports are known as Madison Hold-Its.[†]

[†]Hold-It™ is a brand name trademark owned by Madison Equipment Co., Cleveland, OH.

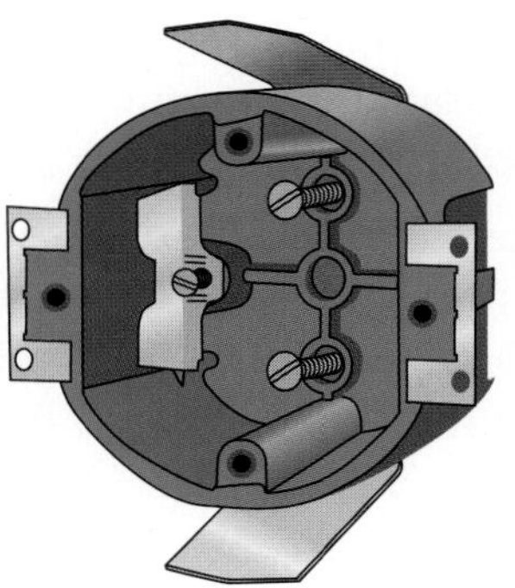

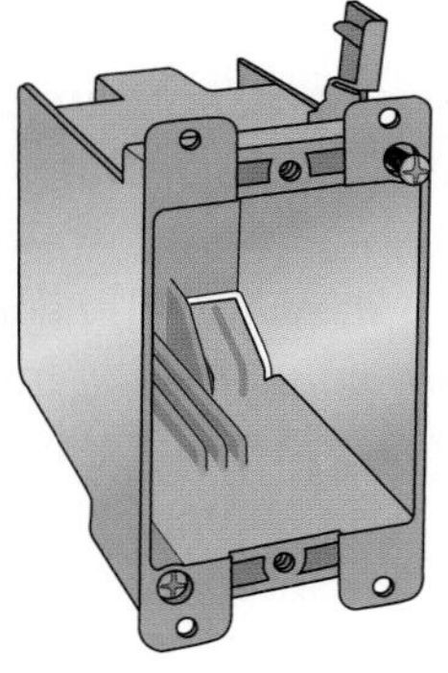

FIGURE 2-26 Nonmetallic boxes are commonly used for remodel work. When boxes are inserted into the hole cut in the drywall, they securely lock into place. This type of box is not for use in new work.

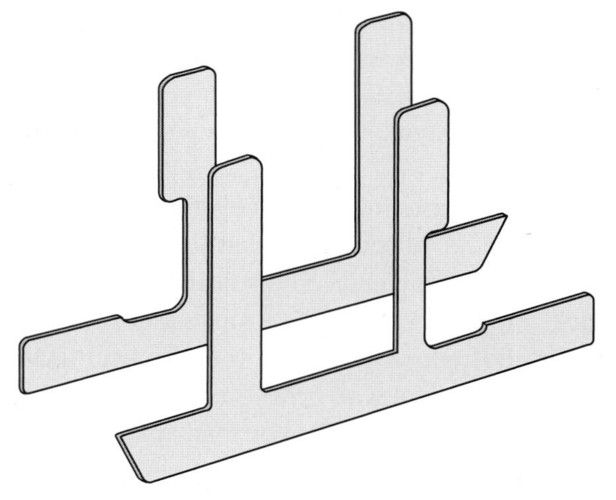

FIGURE 2-27 Metal device box supports used to secure metal boxes in finished surfaces.

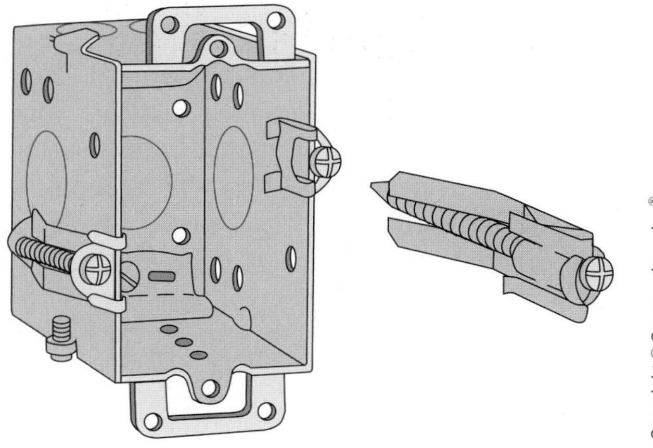

FIGURE 2-28 These boxes are commonly used in remodel work. The box is installed from the front, and then the screws are tightened. The box is adequately supported.

Also available for old work are boxes that have a screw-type support on each side of the box, as in Figure 2-28. This type of box has plaster ears. It is inserted through the hole cut in the wall or ceiling, and then the screws are tightened. This pulls up a metal support tightly behind the wall or ceiling, firmly holding the box in place. The action is very similar to that of Molly screw anchors.

Because these types of remodel boxes are supported by the drywall or plaster, as opposed to new work where the boxes are supported by the framing members, be careful not to hang heavy luminaires from them. Be sure that there is proper support if a ceiling fan is to be installed. This is covered in Chapter 9.

Spread of Fire

To protect lives and property, fire must be contained and not be allowed to spread. In addition to the *NEC*, you must also become familiar with building codes.

Building codes such as the International Code Council (ICC) *International Residential Code* generally do not require fire-resistance-rated walls and ceilings in one- and two-family buildings. Some communities have amendments to the basic building codes, such as requiring ⅝ in. (16 mm) gypsum wallboard [instead of the normal ½ in. (13 mm)] on the ceilings and on the walls between the garage and living areas, and requiring *draft stopping* where ducts, cables, and piping run through bottom (sole) and top plates. Draft stopping is generally done by the insulation contractor. Townhouses, condominiums, and other multi-occupancy buildings do have fire-resistance rating requirements. To avoid costly mistakes, check with your local building officials to determine exactly what the requirements are!

In walls, partitions, and ceilings that are fire-resistance-rated, special consideration must be given when installing wall and ceiling electrical boxes "back to back" or in the same stud or joist space of common walls or ceilings, as might be found in multifamily buildings. Electrical wall boxes installed back to back or installed in the same stud or joist space defeat the fire-resistance rating of the wall or ceiling.

When discussing fire-rated assemblies, the term *membrane penetration* refers to penetrating one side of the fire-rated assembly. The term *through penetration* refers to penetrating entirely through the fire-rated assembly.

What Is Fire-Resistance Rating?

A building assembly's (walls, ceilings, partitions, gypsum, drywall, plaster, etc.) fire-resistance rating refers to the period of time the assembly will serve as a barrier to the spread of fire. It is an indication of how long it can hold back a fully developed blaze before it spreads to adjacent areas of the building and how long the assembly can function structurally after being exposed to a fire.

While undergoing specific tests, passage of flames and/or structural collapse are the determining factors for establishing fire-resistance ratings.

To be truly fire-resistance-rated, the construction and materials must conform to very rigid requirements, as set forth in the UL Standards.

Ratings of fire-resistance materials are expressed in hours. Terms such as a 1-hour fire-resistance rating, 2-hour fire-resistance rating, and so on, are used.

UL Standard 263 Fire Tests of Building Construction and Materials and NFPA 251 cover fire-resistance ratings. In UL Standard 263, and in all building codes, you will find requirements such as these:

> "The surface area of individual metallic outlet or switch boxes shall not exceed 16 square inches" (such as a 4 in. square box is 4 in. × 4 in. = 16 in.^{2}).
>
> "The aggregate surface area of the boxes shall not exceed 100 in.^{2} per 100 ft^2 of wall surface.
>
> Boxes located on opposite sides of walls or partitions shall be separated by a minimum horizontal distance of 24 in. (600 mm).
>
> The metallic outlet or switch boxes shall be securely fastened to the studs and the opening in the wallboard facing shall be cut so that the clearance between the box and the wallboard does not exceed ⅛ in. (3 mm).
>
> The boxes shall be installed in compliance with the *NEC*."

The UL *Fire Resistance Directory* covers fire-resistant materials and assemblies and contains a rather detailed listing of *Outlet Boxes and Fittings Classified for Fire Resistance (QBWY)*, showing all manufacturers' product part numbers for *nonmetallic boxes* to be installed in walls, partitions, and ceilings that meet the fire-resistance standard. This category also covers special-purpose boxes for installation in floors.

When using nonmetallic boxes in fire-resistance rated walls, the restrictions are more stringent than for metallic boxes. Nonmetallic boxes shall be marked with

- UL in a circle (or other listing mark from an NRTL) located in the base of the box
- the hour rating (1 hour or 2 hours)
- the letters F, W, and/or C where F = floor, W = wall, and C = ceiling

Nonmetallic boxes are UL Classified for use in specific fire-resistant rated assemblies. The Classification data shows spacing restrictions and installation details.

Molded fiberglass outlet and device boxes have become available and are listed for use in fire-resistance-rated partitions without the use of putty pads, mineral wool batts, or fiberglass batts, provided a minimum horizontal distance of 3 in. (75 mm) is maintained between boxes and the boxes are not back to back. See Figure 2-31(B).

If you are dealing with fire-resistance-rated construction, be sure to carefully read the box manufacturer's instructions and "Listing" details.

For both metallic and nonmetallic electric outlet boxes, the *minimum* distance between boxes on opposite sides of a fire-rated wall or partition is 24 in. (600 mm). There is an exception, which is discussed later.

For both metallic and nonmetallic electric outlet boxes, the maximum gap between the box and the wall material is ⅛ in. (3 mm). This holds true for fire-resistance-rated walls and non-fire-resistance-rated walls.

To maintain the hourly fire rating of a fire-rated wall that contains electrical outlet and switch boxes, the UL *Fire Resistance Directory* lists an *intumescent* (expands when heated) fire-resistant material that comes in "pads."

These "putty pads" are moldable and can be used to wrap metal electrical boxes or inserted on the inside back wall of metal electrical boxes. When exposed to fire, the material expands and chars, sealing off the opening to prevent the spread of flames and limit the heat transfer from the fire-resistance-rated side of the wall to the non-fire-resistance-rated side of the wall. It is not to be used on nonmetallic boxes.

The ⅛ in. (3 mm) thickness provides a 1-hour fire-resistance rating. A 2-hour rating is obtained with ¼ in. (6 mm) thickness.

When this material is properly installed, the 24 in. (600 mm) separation between boxes is not required. However, the electrical outlet boxes *must not* be installed back to back. Using this material meets the requirements of *NEC 300.21* of the *NEC* and other building codes.

The ICC *International Residential Code* states that "fireblocking" be provided to cut off all concealed draft openings (both vertical and horizontal) and to form an effective fire barrier between stories, and between a top story and the roof space. This includes openings around vents, pipes, ducts, chimneys, and fireplaces at ceiling and floor levels. Oxygen supports fire. Firestopping cuts off or significantly reduces the flow of oxygen. Most communities that have adopted the ICC code require firestopping around electrical conduits. In residential work, this is generally done by the insulation installer, before the drywall installer closes up the walls.

Be careful when roughing in wall boxes where wood paneling is to be installed. Some communities require that ½ in. (13 mm) gypsum board be installed on the studs behind finished wood paneling to establish a fire-resistance rating. Remember that where the finish wall or ceiling surface is combustible material such as wood paneling, the front edge of the box, plaster ring, domed cover, extension ring, or listed extender shall be set flush with the wall surface. Refer to *NEC 314.20* and Figure 2-21.

The previous discussion makes it clear that the electrician must become familiar with certain building codes in addition to the *NEC*. Knowing the *NEC* is not enough!

It is extremely important to check with the electrical inspector and/or building official for complete clarification on the matter of fire-resistance requirements before you find yourself cited with serious violations and lawsuits regarding noncompliance with fire codes. Changes can be costly and will delay completion of the project. Improper wiring is hazardous. The *NEC* addresses this issue in *300.21*.

Figures 2-29, 2-30, 2-31, and 2-32 illustrate the basics for installing electrical outlet and switch boxes in fire-resistance-rated partitions. This is an extremely important issue. If in doubt, check with your local electrical inspector and/or building official.

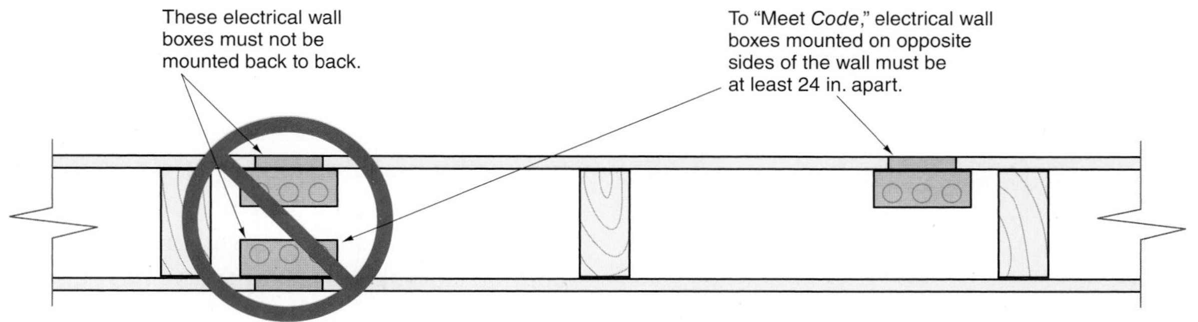

FIGURE 2-29 Building codes prohibit electrical wall boxes to be mounted back to back in fire-resistance-rated walls unless specific installation procedures, as illustrated in Figures 2-30, 2-31, and 2-32, are followed. One of these requirements is that electrical wall boxes must be kept at least 24 in. (600 mm) apart. This is to maintain the integrity of the fire-resistance rating of the wall.

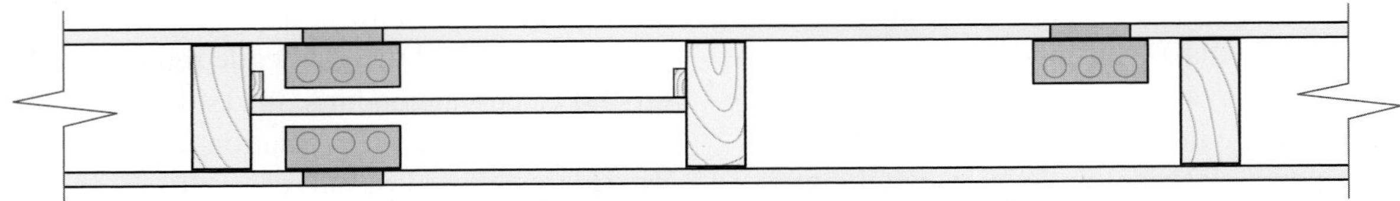

FIGURE 2-30 Building codes permit boxes to be mounted back to back or in the same wall cavity when the fire-resistance rating is maintained. In this detail, gypsum board has been installed in the wall cavity between the boxes, creating a new fire-resistance barrier. Verify that this is acceptable to the building authority in your area.

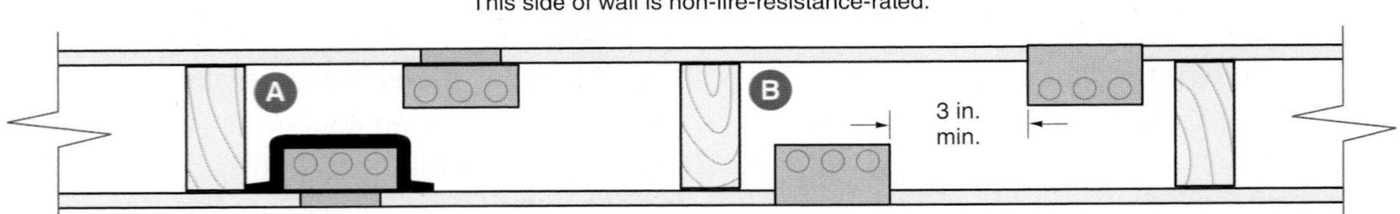

FIGURE 2-31 Building codes permit electrical outlet and device boxes to be mounted in the same stud space under certain conditions. (A) Here we see "Wall Opening Protective Material" packed around one outlet box. (B) Here we see molded fiberglass boxes in the same stud space with a minimum horizontal clearance of 3 in. (75 mm). These boxes are listed for this application. In both (A) and (B), the boxes are not permitted to be back to back.

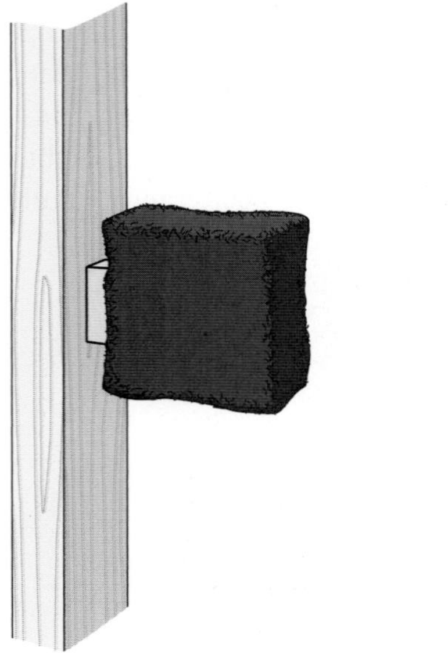

FIGURE 2-32 This illustration shows an electrical outlet box that has been covered with "Listed" fire-resistance "putty pad" to retain the wall's fire-resistance rating per building codes.

BOXES FOR CONDUIT WIRING

Some local electrical ordinances require a metal raceway rather than cable wiring. Conduit wiring is discussed in Chapter 18. Examples of conduit fill (how many wires are permitted in a given size conduit) are presented.

When a metal raceway is installed in a residence, it is quite common to use 4 in. square boxes trimmed with suitable plaster rings or covers for the desired wiring device or luminaire. See Figure 2-33(B). There are sufficient knockouts in the top, bottom, sides, and back of the box to permit a number of EMTs to run to the box. Plenty of room is available for the conductors and wiring devices. Note how easily these 4 in. square outlet boxes can be mounted back to back by installing a small fitting between the boxes. This is illustrated in Figure 2-33(B) and 2-33(C). These boxes are

A

B

C

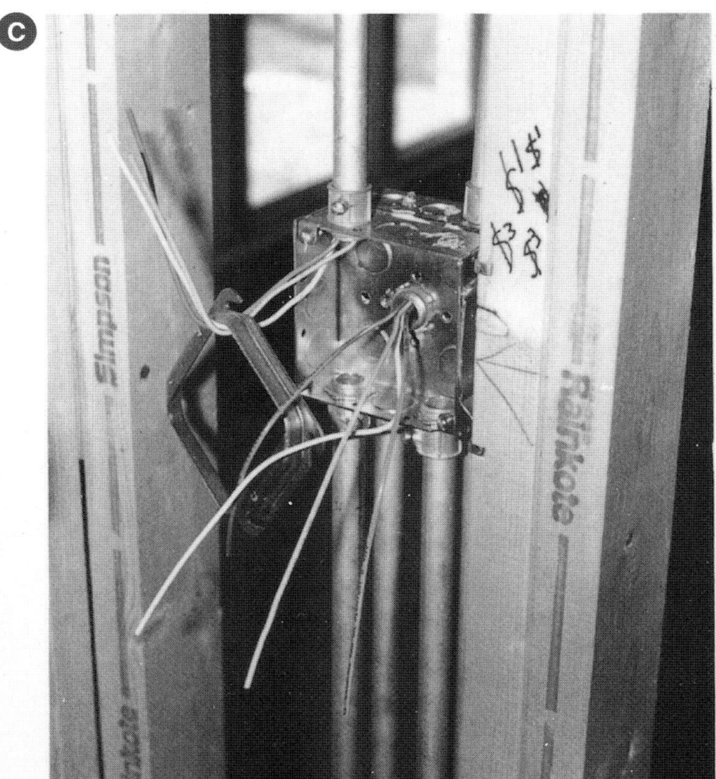

FIGURE 2-33 (A) shows electrical metallic tubing run to a 4 in. square box. (B) and (C) show 4 in. square boxes attached together for "back-to-back" installations. When mounting boxes back to back, be sure to consider possible transfer of sound between the rooms. IMPORTANT: See text "Spread of Fire" for details of fire-resistance ratings and the restrictions related to installing electrical boxes back to back.

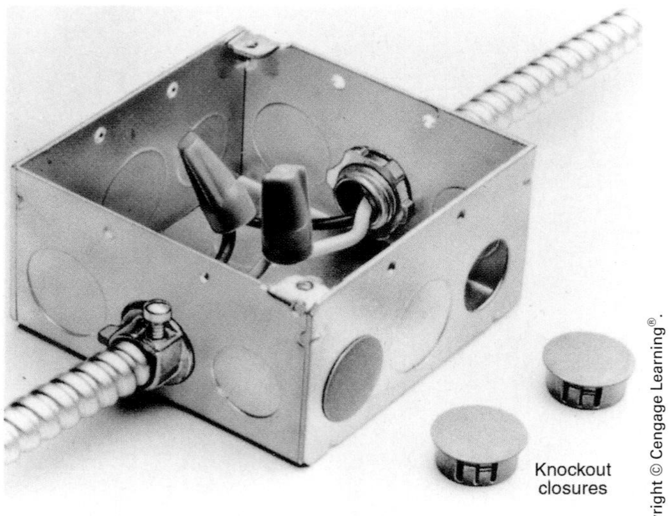

FIGURE 2-34 Unused openings in boxes must be closed according to *110.12(A)*. This is done to contain electrical short-circuit problems inside the box and to keep rodents out. Note that the conductors in the box in the photo *do not* meet the 3 in. (75 mm) and 6 in. (150 mm) conductor length requirements of *300.14*. This is an example of a *Code violation*. See Figure 2-2 for proper installation regarding conductor length at boxes.

available in 1½ in. (38 mm) and 2⅛ in. (54 mm) depths.

Four-inch square outlet boxes can be trimmed with 1-gang or 2-gang plaster rings where wiring devices will be installed. Where luminaires will be installed, a plaster ring having a round opening should be installed.

Any and all unused openings in electrical equipment must be closed! This includes unused knockouts in boxes, spaces for circuit breakers in a panel, meter socket enclosures, and all other similar electrical equipment. The closing material shall provide substantially the same protection as the wall of the equipment. See *110.12(A)* and Figure 2-34.

Openings that are intended for mounting purposes, for the operation of the equipment, or as part of the design of listed equipment do not have to be closed.

Close the Gap around Boxes!

NEC 314.21 requires that Noncombustible surfaces that are broken or incomplete around boxes employing a flush-type cover or faceplate shall be repaired so there will be no gaps or open spaces greater than ⅛ in. (3 mm) at the edge of the box.* This is to minimize the spread of fire. The UL *Fire Resistance Directory* mirrors the *NEC* rule by stating that:

> "The outlet or switch boxes shall be securely fastened to the studs and the opening in the wallboard facing shall be cut so that the clearance between the box and the wallboard does not exceed ⅛ in." See Figure 2-21(B).

This is easier said than done. The electrician installs the wiring (rough-in), which is followed by an inspection by the electrical inspector. Next comes the drywall/plaster/panel installer, who many times cuts out the box openings much larger than the outlet or switch box. In most cases, the walls and ceilings are painted or wallpapered before the electrician returns to install the receptacles, switches, and luminaires. For gaps or openings greater than ⅛ in. (3 mm) around the electrical boxes, should the electrician repair the gap with patching plaster, possibly damaging or marring the finished wall? Whose responsibility is it? Is it the electrician's? Is it the drywall/plaster/panel installer's?

The electrician should check with the general contractor to clarify who is to be responsible for seeing to it that gaps or openings around electrical outlet and switch boxes do not exceed ⅛ in. (3 mm).

YOKE

We use the term *yoke*. The *NEC* uses the term *yoke*. Yet, the *NEC* does not define a *yoke*. What is a yoke? It is the metal or polycarbonate mounting strap on which a wiring device or devices are attached. The yoke in turn is attached to a device box or device cover with No. 6-32 screws. One yoke may have one, two, or three wiring devices attached to it.

Visit an electrical distributor or home center. You will find many combination wiring devices on a single yoke or strap. Selecting one of these combination devices may eliminate the need to use combinations of interchangeable wiring devices.

*Reprinted with permission from NFPA 70-2014.

SPECIAL-PURPOSE OUTLETS

Special-purpose outlets are usually indicated on the plans. These outlets are described by a notation and are usually detailed in the specifications. The plans indicate special-purpose outlets by a triangle inside a circle with subscript letters. In some cases, a subscript number is added to the letter.

When a special-purpose outlet is indicated on the plans or in the specifications, the electrician must check for special requirements. Such a requirement may be a separate circuit, a special 240 volt circuit, a special grounding or polarized receptacle, or other preparation.

A set of plans and specifications for this residence is found at the end of this text. The specifications include a *Schedule of Special-Purpose Outlets*.

NUMBER OF CONDUCTORS IN BOX

NEC 314.16 dictates that outlet boxes, switch boxes, and device boxes should be large enough to provide ample room for the wires in that box, without having to jam or crowd the wires into the box. Jamming the wires into the box can not only damage the insulation on the wires, but can also result in heat buildup in the box that can further damage the insulation on the wires. The *Code* specifies the maximum number of conductors allowed in standard metal outlet, device, and junction boxes; see *Tables 314.16(A)* and *(B)*. Nonmetallic boxes are

NEC TABLE 314.16(A)

Metal Boxes.

Box Trade Size			Minimum Volume		Maximum Number of Conductors* (arranged by AWG size)						
mm	in.		cm^3	in.^{3}	18	16	14	12	10	8	6
100 × 32	(4 × 1¼)	round/octagonal	205	12.5	8	7	6	5	5	5	2
100 × 38	(4 × 1½)	round/octagonal	254	15.5	10	8	7	6	6	5	3
100 × 54	(4 × 2⅛)	round/octagonal	353	21.5	14	12	10	9	8	7	4
100 × 32	(4× 1¼)	square	295	18.0	12	10	9	8	7	6	3
100 × 38	(4 × 1½)	square	344	21.0	14	12	10	9	8	7	4
100 × 54	(4 × 2⅛)	square	497	30.3	20	17	15	13	12	10	6
120 × 32	(4 11/16 × 1¼)	square	418	25.5	17	14	12	11	10	8	5
120 × 38	(4 11/16 × 1½)	square	484	29.5	19	16	14	13	11	9	5
120 × 54	(4 11/16 × 2⅛)	square	689	42.0	28	24	21	18	16	14	8
75 × 50 × 38	(3 × 2 × 1½)	device	123	7.5	5	4	3	3	3	2	1
75 × 50 × 50	(3 × 2 × 2)	device	164	10.0	6	5	5	4	4	3	2
75 × 50 × 57	(3 × 2 × 2¼)	device	172	10.5	7	6	5	4	4	3	2
75 × 50 × 65	(3 × 2 × 2½)	device	205	12.5	8	7	6	5	5	4	2
75 × 50 × 70	(3 × 2 × 2¾)	device	230	14.0	9	8	7	6	5	4	2
75 × 50 × 90	(3 × 2 × 3½)	device	295	18.0	12	10	9	8	7	6	3
100 × 54 × 38	(4 × 2⅛ × 1½)	device	169	10.3	6	5	5	4	4	3	2
100 × 54 × 48	(4 × 2⅛ × 1⅞)	device	213	13.0	8	7	6	5	5	4	2
100 × 54 × 54	(4 × 2⅛ × 2⅛)	device	238	14.5	9	8	7	6	5	4	2
95 × 50 × 65	(3¾ × 2 × 2½)	masonry box/gang	230	14.0	9	8	7	6	5	4	2
95 × 50 × 90	(3¾ × 2 × 3½)	masonry box/gang	344	21.0	14	12	10	9	8	7	4
min. 44.5 depth	FS — single cover/gang (1¾)		221	13.5	9	7	6	6	5	4	2
min. 60.3 depth	FD — single cover/gang (2⅜)		295	18.0	12	10	9	8	7	6	3
min. 44.5 depth	FS — multiple cover/gang (1¾)		295	18.0	12	10	9	8	7	6	3
min. 60.3 depth	FD — multiple cover/gang (2⅜)		395	24.0	16	13	12	10	9	8	4

*Where no volume allowances are required by 314.16(B)(2) through (B)(5).

NEC TABLE 314.16(B)

Volume Allowance Required per Conductor.

Size of Conductor (AWG)	Free Space Within Box for Each Conductor	
	cm³	in.³
18	24.6	1.50
16	28.7	1.75
14	32.8	2.00
12	36.9	2.25
10	41.0	2.50
8	49.2	3.00
6	81.9	5.00

marked by the manufacturer with their cubic-inch capacity.

When conductors are the same size, the proper metal box size can be selected by referring to *Table 314.16(A)*. When conductors are of different sizes and nonmetallic boxes are installed, refer to *Table 314.16(B)* and use the cubic-inch volume for the particular size of wire being used.

Table 314.16(A) and *Table 314.16(B)* do not take into consideration the space taken by luminaire studs, cable clamps, hickeys, switches, pilot lights, or receptacles that may be in the box. These require additional space. Table 2-1 shows the additional allowances that must be considered for different situations.

TABLE 2-1

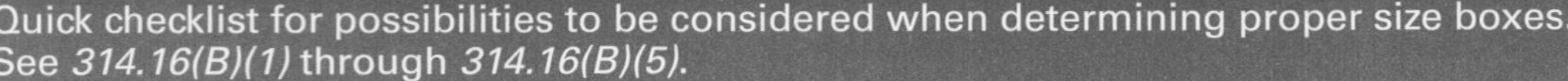

Quick checklist for possibilities to be considered when determining proper size boxes. See *314.16(B)(1)* through *314.16(B)(5)*.

Situation	Allowance
• If box contains no fittings, devices, fixture studs, cable clamps, hickeys, switches, receptacles, or equipment grounding conductors:	• Refer directly to *Table 314.16(A)* or *Table 314.16(B)*.
• **Clamps.** If box contains one or more internal cable clamps:	• Add a single-volume based on the largest conductor in the box.
• **Support Fittings.** If box contains one or more luminaire studs or hickeys:	• Add a single-volume for each type based on the largest conductor in the box.
• **Device or Equipment.** If box contains one or more wiring devices on a yoke:	• Add a double-volume for each yoke based on the largest conductor connected to a device on that yoke. Some large wiring devices, such as a 30-ampere dryer receptacle, require a 2-gang box.
• **Equipment Grounding Conductors.** If a box contains one or more equipment grounding conductors:	• Add a single-volume based on the largest equipment grounding conductor in the box.
• **Isolated Equipment Grounding Conductor.** If a box contains one or more additional "isolated" (insulated) equipment grounding conductors as permitted by *250.146(D)* for "noise" reduction:	• Add a single-volume based on the largest equipment grounding conductor in the additional set.
• For conductors less than 12 in. (300 mm) long that are looped or coiled in the box without being spliced:	• Add a single-volume for each conductor that is looped or coiled through the box.
• For conductors 12 in. (300 mm) or longer that are looped or coiled in the box without being spliced:	• Add a double-volume for each conductor that is looped or coiled through the box.
• For conductors that originated outside of the box and terminate inside the box:	• Add a single-volume for each conductor that originates outside the box and terminates inside the box.
• If no part of the conductor leaves the box, as with a "jumper" wire used to connect three wiring devices on one yoke, or pigtails:	• Don't count this (these). No additional volume required.
• For small equipment grounding conductors or not more than four conductors smaller than 14 AWG that originate from a luminaire canopy or similar canopy (like a fan) and terminate in the box:	• These are not required to be counted. No additional volume required.
• For small fittings, such as locknuts, bushings, and wire connectors:	• These are not required to be counted. No additional volume required.

SELECTING THE CORRECT SIZE BOX

Selecting the correct size box depends on the mix of conductors and wiring devices that will be contained in the particular box. Here are some examples.

Box Fill

Figure 2-35 shows the dimensions of typical metal boxes and the number of conductors permitted in a given size box. This figure replicates *Table 314.16(A)* of the *NEC*.

Nonmetallic boxes are marked with their cubic-inch volume. *Table 314.16(B)* shows the cubic-inch volume allowance required for different size conductors.

When Using a Nonmetallic Box Where All Conductors Are the Same Size. A nonmetallic box is marked as having a volume of 22.8 in.^{3} The box contains no luminaire stud or cable clamps. How many 14 AWG conductors are permitted in this box?

From *Table 314.16(B)*, the volume requirement for a 14 AWG conductor is 2.00 in.^{3} Therefore, the maximum number of 14 AWG conductors permitted in this box is

$$\frac{22.8}{2} = 11.4 \text{ (round down to 11 conductors)}$$

When Using a Metal Device Box Where All Conductors Are the Same Size. A box contains one wiring device and two internal cable clamps. The wiring method is two 12/2 NM with ground cables.

four 12 AWG conductors	4
one wiring device	2
two cable clamps (only count one)	1
two 12 AWG equipment grounding conductors (only count one)	1
Total	8

Referring to *Table 314.16(A)*, a 3 × 2 × 3½ device box is suitable for this example.

When Using a Metal or Nonmetallic Box Where the Conductors Are Different Sizes. What is the minimum cubic-inch volume required for a box that will contain one internal cable clamp, one switch, and one receptacle mounted on one yoke? One 14/2 Type AC and one 12/2 Type AC cable are used. There are no separate equipment grounding conductors involved as the bonding wire in the armored cables are laid back over the armor and do not take up any space within the box.

two 14 AWG conductors = 2 in.^{3} per conductor	= 4.00 in.^{3}
two 12 AWG conductors = 2.25 in.^{3} per conductor	= 4.50 in.^{3}
two cable clamps (only count one) = 2.25 in.^{3}	= 2.25 in.^{3}
one switch and one receptacle, each on a separate yoke = 2.25 in.^{3} × 2	= 4.50 in.^{3}
Total	15.25 in.^{3}

Select a metal box or combination of box and plaster ring from *Table 314.16(A)*. This might consist of a 4 × 1¼ square or deeper box with a 2-gang plaster ring or two 3 × 2 × 2 device boxes that are ganged together. Any other metal box or combination that provides not less than 15.25 in.^{3} area can be used.

For nonmetallic boxes, select a 2-gang box that is marked with a volume not less than 15.25 in.^{3}.

The *Code* in *314.16(A)(2)* requires that all boxes *other* than those listed in *Table 314.16(A)* be durably and legibly marked by the manufacturer with their cubic-inch capacity. Nonmetallic boxes have their cubic-inch volume marked on the box. When sectional boxes are ganged together, the calculated volume is the total cubic-inch volume of the assembled boxes.

When Installing a Box That Will Have a Raised Cover Attached to It. When marked with their cubic-inch volume, the *additional* space provided by plaster rings, domed covers, raised covers, and extension rings is permitted to be used when determining the overall volume. This is illustrated in Figure 2-36.

How many 12 AWG conductors are permitted in a 4 × 1½ in. square box with a

QUIK-CHEK BOX SELECTION GUIDE FOR BOXES GENERALLY USED FOR RESIDENTIAL WIRING

DEVICE BOXES		3x2x1 1/2 (7.5 in.^{3})	3x2x2 (10 in.^{3})	3x2x2 1/4 (10.5 in.^{3})	3x2x2 1/2 (12.5 in.^{3})	3x2x2 3/4 (14 in.^{3})	3x2x3 (16 in.^{3})	3x2x3 1/2 (18 in.^{3})
	Wire size							
	14 AWG	3	5	5	6	7	8	9
	12 AWG	3	4	4	5	6	7	8
		The number of conductors is "per gang"						

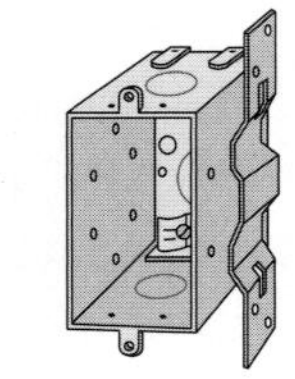

SQUARE BOXES		4x4x1 1/2 (21 in.^{3})	4x4x2 1/8 (30.3 in.^{3})
	Wire size		
	14 AWG	10	15
	12 AWG	9	13

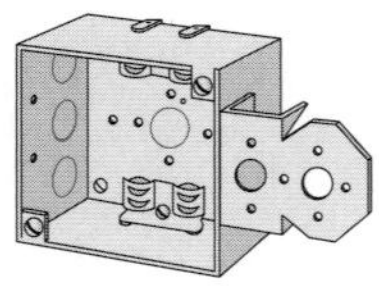

OCTAGON BOXES		4x1 1/2 (15.5 in.^{3})	4x2 1/8 (21.5 in.^{3})
	Wire size		
	14 AWG	7	10
	12 AWG	6	9

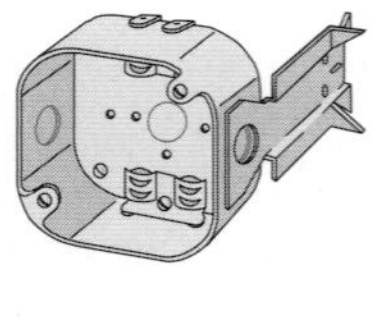

HANDY BOXES		4x2 1/8x1 1/2 (10.3 in.^{3})	4x2 1/8x1 7/8 (13 in.^{3})	4x2x2 1/8 (14.5 in.^{3})
	Wire size			
	14 AWG	5	6	7
	12 AWG	4	5	6

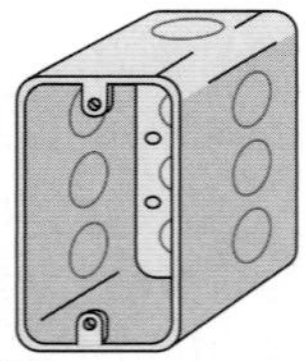

RAISED COVERS

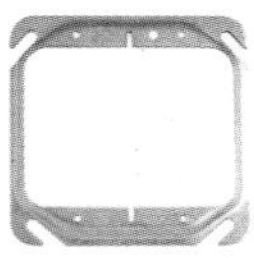

Where raised covers are marked with their volume in cubic inches, that volume may be added to the box volume to determine maximum number of conductors in the combined box and raised cover. Nonmetallic raised covers are available.

MASONRY BOXES	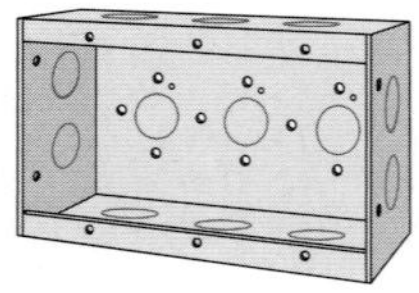	3 3/4x2x2 1/2 (14 in.^{3})	3 3/4x2x3 1/2 (21 in.^{3})
3-gang box	Wire size		
	14 AWG	7	10
	12 AWG	6	9

Masonry boxes have lots of room. Available up to 6 gang. The conductor fill is for each gang. Refer to *Table 314.16(A)* and *(B)*.

Note: Be sure to make deductions from the above maximum number of conductors permitted for wiring devices, cable clamps, fixture studs, and equipment grounding conductors. The cubic-inch (in.^{3}) volume is taken directly from *Table 314.16(A)* of the *NEC*. This table applies to only metal boxes. Nonmetallic boxes are marked with their cubic-inch capacity and are sized using *Table 314.16(B)*.

FIGURE 2-35 Quik-chek selection guide for metal boxes.

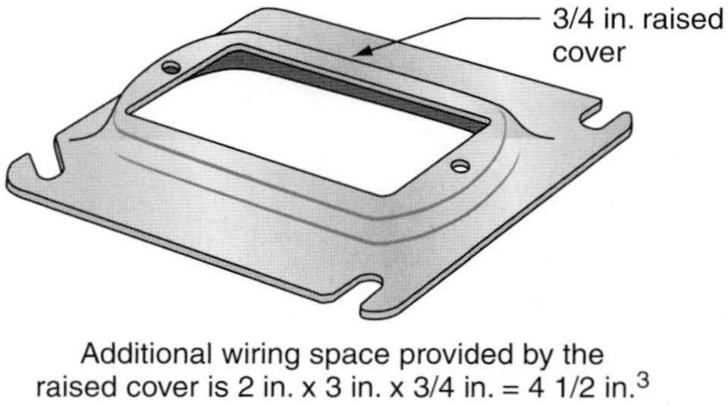

FIGURE 2-36 Raised cover. Raised covers are sometimes called plaster rings.

plaster ring marked with a volume of $4\frac{1}{2}$ in.^{3}? Refer to Figure 2-37 and *Table 314.16(A)* and *Table 314.16(B)*. In *Table 314.16(A)* we find the 4 in. square box that is $1\frac{1}{2}$ in. deep has a volume of 21.5 in.^{3}. Add the volume of $4\frac{1}{2}$ in.^{3} for the plaster ring for a total volume of 25.5 in.^{3}.

$$\frac{25.5 \text{ in.}^3 \text{ of total space}}{2.25 \text{ in.}^3 \text{ per 12 AWG conductor}} = \text{Eleven 12 AWG conductors}$$

This calculation actually resulted in 11.33 conductors. The conductor fill numbers in *Table 314.16(A)* were created by dropping any excess above the whole number after the calculations were made. Following the same procedure, this conductor fill calculation dropped off the excess.

Electricians and electrical inspectors have become very aware of the fact that GFCI receptacles, dimmers, and timers take up a lot more space than regular wiring devices. Therefore, it is a good practice to install boxes that will provide plenty of room for the wires, instead of pushing, jamming, and crowding the wires into the box.

Figure 2-38 shows that transformer leads 18 AWG or larger must be counted when selecting a proper size box.

The minimum required size of equipment grounding conductors is shown in *Table 250.122*. The equipment grounding conductors are the same size as the circuit conductors in cables having 14 AWG, 12 AWG, and 10 AWG circuit conductors. Thus, for normal house wiring, metal box sizes can be calculated using *Table 314.16(A)*. Refer to Figure 2-39.

Figure 2-35, a quick-check box selection guide, shows some of the most popular types of boxes used in residential wiring. This guide also shows the box's cubic-inch volume.

Figure 2-40 illustrates typical wiring using cable or wire and conduit. Note in this figure how the conductor count is determined.

Using electrical metallic tubing (EMT) makes it possible to "loop" conductors through the box, which counts as one conductor only, provided the loops are shorter than 12 in. (300 mm) or as provided in *300.14*. See *314.16(B)(1)*. Looping conductors through a box is not possible when using cable as the wiring method.

The basic rules for box fill are summarized in Table 2-1. Boxes that are intended to support ceiling fans are discussed in detail in Chapter 9.

4 in. x 1 1/2 in.
square box

4 in. square, 3/4 in. deep
raised plaster ring
(raised section measures
2 in. x 3 in. x 3/4 in.)

25.5 in.^{3}
total space

21 in.^{3}

4 1/2 in.^{3}
marked on cover

$$\frac{25.5 \text{ in.}^3 \text{ of total space}}{2.25 \text{ in.}^3 \text{ per 12 AWG conductor}} = \text{Eleven 12 AWG conductors}$$

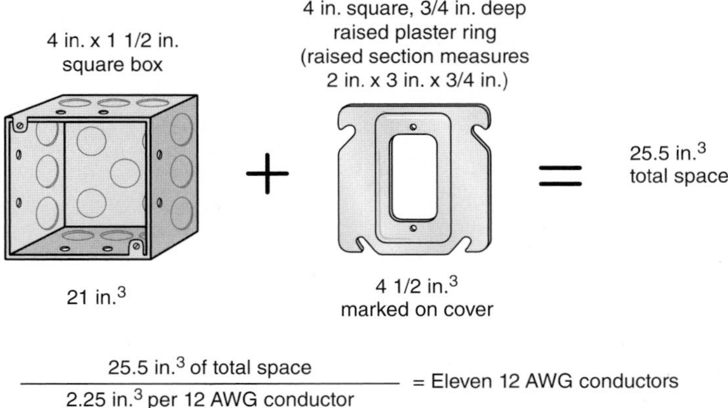

FIGURE 2-37 When marked with their cubic-inch volume, the volume of the cover may be added to the volume of the box to determine the total cubic-inch volume.

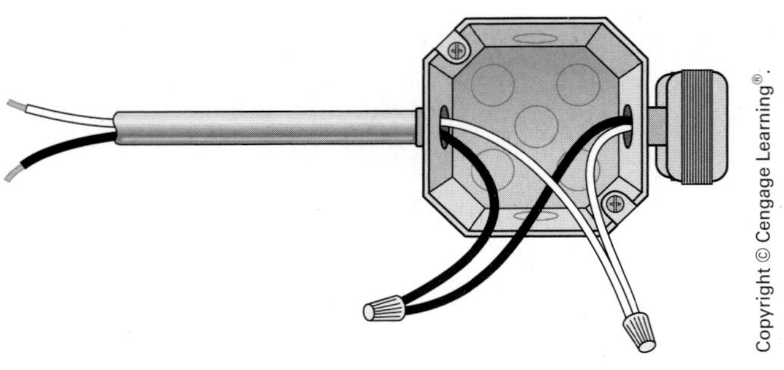

FIGURE 2-38 Transformer leads 18 AWG or larger must be counted when selecting the proper size box, *314.16(B)(1)*. In the example shown, the box contains four conductors.

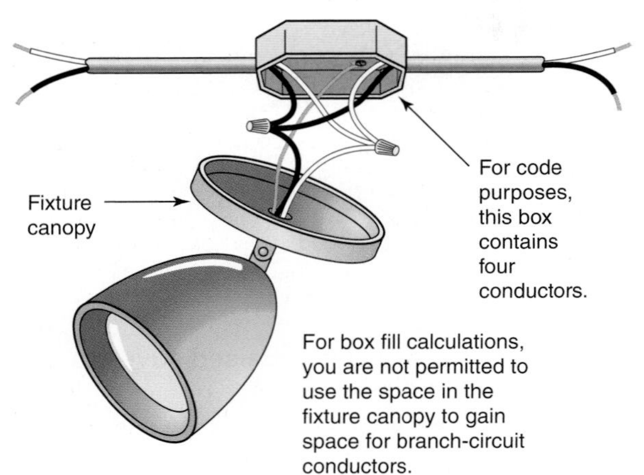

FIGURE 2-39 Four or fewer luminaire conductors, smaller than 14 AWG, and/or small equipment grounding conductors that originate in the luminaire and terminate in the box, need not be counted when calculating box fill, *314.16(B)(1), Exception*.

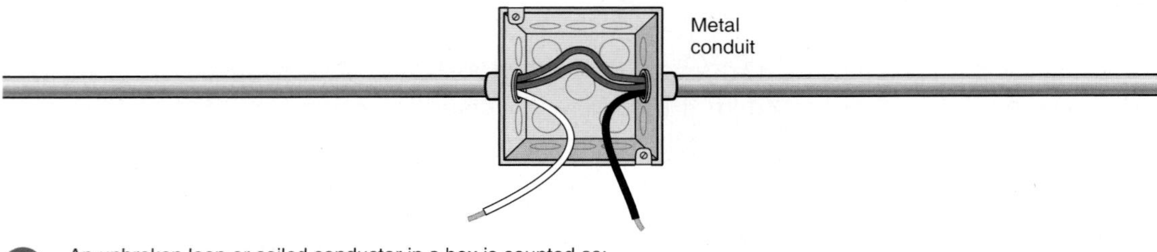

A An unbroken loop or coiled conductor in a box is counted as:
- One conductor (a single-volume count) if less than 12 in. in length.
- Two conductors (a double-volume count) if 12 in. or longer.

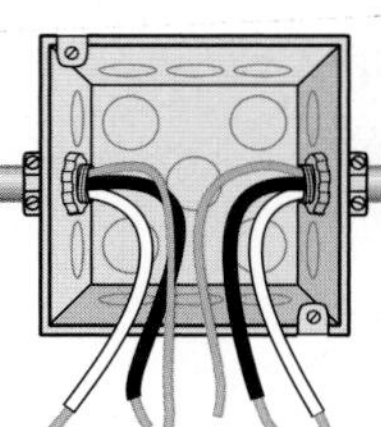

Cable (each with two circuit conductors plus one equipment grounding conductor.)

B When using cable, the conductor count in this illustration is five. The two equipment grounding conductors are counted as one.

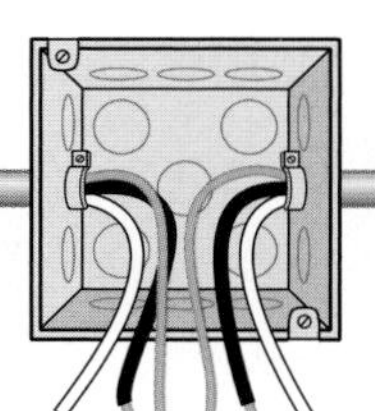

Cable and internal clamps (each cable has two circuit conductors plus one equipment grounding conductor.)

C When using cable and a box that has internal cable clamps, the conductor count in this illustration is six. The two equipment grounding conductors are counted as one. The two cable clamps are counted as one.

FIGURE 2-40 Example of conductor count for both metal conduit and cable installations.

Depth of Box—Watch Out!

Determining the cubic volume of a box is not the end of the story. Outlet and device boxes shall be deep enough so that conductors will not be damaged when installing the wiring device or equipment into the box. Merely calculating and providing the proper volume of a box is not always enough. The volume calculation might prove adequate, yet the size (depth) of the wiring device or equipment might be such that conductors behind it may possibly be damaged. See *NEC 314.24*.

Width of Box—Watch Out!

Large wiring devices, such as a 30-ampere, 3-pole, 4-wire dryer or a 50-ampere-range receptacle will not fit into a single-gang box that is 2 in. (50.8 mm) wide. These receptacles measure 2.10 in. (53.3 mm) in width. Likewise, a 50-ampere, 3-pole, 4-wire receptacle measures 2.75 in. (69.9 mm) in width. Consider using a 4 in. (101.6 mm) square box with a 2-gang plaster device ring for flush mounting or a 2-gang raised cover for surface mounting, or use a 2-gang device box. The center-to-center mounting holes of both the 30-ampere and the 50-ampere receptacles are 1.81 in. (46.0 mm) apart, exactly matching the center-to-center holes of a 2-gang plaster ring, raised cover, or 2-gang device box, so mounting the devices is not a problem. See *NEC 314.16(B)(4)*.

Means for Connecting Equipment Grounding Conductors in a Box

NEC 314.4 requires that all metal boxes be grounded and bonded. The specific requirements for specific applications are found in *Article 250, Parts I, IV, V, VI, VII*, and *X*.

NEC 314.40(D) requires that A means shall be provided in each metal box for the connection of an equipment grounding conductor.* Generally, these are tapped No. 10-32 holes in the box that may be marked GR, GRN, GRND, or similarly identified. The screws used for terminating the equipment grounding conductor must be hexagonal and must be green. Electrical distributors sell green, hexagonal No. 10-32 screws for this purpose.

*Reprinted with permission from NFPA 70-2014.

HEIGHT OF RECEPTACLE OUTLETS

There are no hard-and-fast rules for locating most outlets. A number of conditions determine the proper height for a switch box. For example, the height of the kitchen counter backsplash determines where the switches and receptacle outlets are located between the kitchen countertop and the cabinets.

The residence featured in this text is electrically heated, which is discussed in Chapter 23. The type of electric heat could be

- electric furnace (as in this text).
- electric cable heating buried in ceiling plaster or "sandwiched" between two layers of drywall material.
- electric baseboard heaters.
- wall-mounted, fan-forced heaters
- heat pump.

Let us consider the electric baseboard heaters. In most cases, the height of these electric baseboard units from the top of the unit to the finished floor seldom exceeds 6 in. (150 mm). The important issue here is that the manufacturer's receptacle accessories may have to be used to conform to the receptacle spacing requirements, as covered in *210.52*.

Electrical receptacle outlets are not permitted to be located above an electric baseboard heating unit. Refer to the section "Location of Electric Baseboard Heaters in Relation to Receptacle Outlets," in Chapter 23.

It is common practice among electricians to consult the plans and specifications to determine the proper heights and clearances for the installation of electrical devices. The electrician then has these dimensions verified by the architect, electrical engineer, designer, or homeowner. This practice avoids unnecessary and costly changes in the locations of outlets and switches as the building progresses.

Rooms or areas where specific height for receptacle outlets and switches are often specified include:

- kitchen countertops
- bathroom countertops
- laundry equipment or countertops
- workshop or garage receptacles
- construction for the handicapped, see *Informative Annex J, ADA Standards for Accessible Design* in the *NEC* for additional information.

Typical heights for switches and outlets are shown in Table 2-2. These dimensions usually are satisfactory. However, the electrician must check the blueprints, specifications, and details for measurements that may affect the location of a particular outlet or switch. The cabinet spacing, available space between the countertop and the cabinet, and the tile height may influence the location of the outlet or switch. For example, if the top of the wall tile is exactly 4 ft (1.2 m) from the finished floor line, a wall switch should not be mounted 4 ft (1.2 m) to center. This is considered

TABLE 2-2

Typical heights for switches, receptacles, and wall luminaire outlets. Local height preferences prevail.

	SWITCHES
Regular	46 in. (1.15 m). Some homeowners want wall switches mounted 30–36 in. (750–900 mm) so as to be easy for children to reach and easy for adults to reach when carrying something. The choice is yours!
Between counter and kitchen cabinets.	44–46 in. (1–1.15 m) depending on height of counter and backsplash
	RECEPTACLE OUTLETS
Regular	12 in. (300 mm)
Between counter and kitchen cabinets. Depends on backsplash.	44–46 in. (1–1.15 m)
In garages	46 in. (1.15 m) Minimum 18 in. (450 mm)
In unfinished basements	46 in. (1.15 m)
In finished basements	12 in. (300 mm)
Outdoors (above grade or deck)	18 in. (450 mm)
	WALL LUMINAIRE OUTLETS
Outdoor entrance bracket luminaires	72 in. (1.8 m). If luminaire is "upward," mount wall box lower. If luminaire is "downward," mount wall box higher. A good appearance for typical outdoor wall lanterns is when the center of wall box is approximately 10 in. (0.254 m) below the top of the door. The final choice is yours.
Inside wall brackets	5 ft (1.5 m)
Side of medicine cabinet or mirror above medicine cabinet or mirror	You need to know the measurement of the medicine cabinet. Check the rough-in opening of medicine cabinet measurement of mirror. Mount electrical wall box approx. 6 in. (150 mm) to center above rough-in opening or mirror. Medicine cabinets that come complete with luminaires have a wiring compartment with a conduit knockout(s) in which the supply cable or conduit is secured using the appropriate fitting. Many strip luminaires have a back plate with a conduit knockout in which the supply cable or conduit is secured using the appropriate fitting. Where to bring in the cable or conduit takes careful planning.

Note: All dimensions are from finished floor to center of the electrical box. If possible, try to mount wall boxes for luminaires based on the type of luminaire to be installed. Verify all dimensions before roughing in. If wiring for physically handicapped, the above heights may need to be lowered in the case of switches and raised in the case of receptacles.

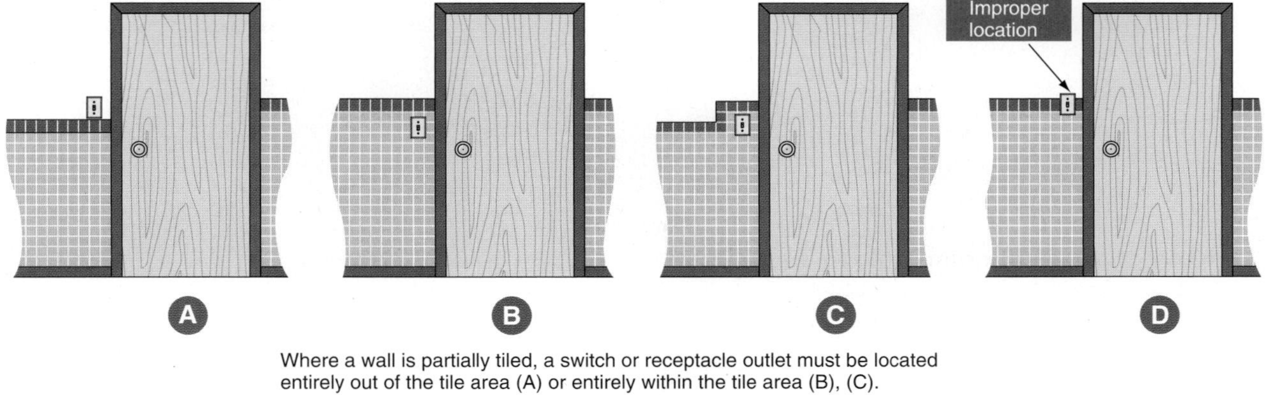

Where a wall is partially tiled, a switch or receptacle outlet must be located entirely out of the tile area (A) or entirely within the tile area (B), (C).

The faceplate in (D) does not "hug" the wall properly. This installation is considered unacceptable by most electricians and is a violation of the last sentence of *NEC 404.9(A).*

FIGURE 2-41 Locating an outlet on a tiled wall.

poor workmanship. The switch should be located entirely within the tile area or entirely out of the tile area, as in Figure 2-41. This situation requires the full cooperation of all craftspersons involved in the construction job.

POSITIONING OF RECEPTACLES

Although no *NEC* rules exist on positioning the grounding slot of receptacles, the electrician should be informed of or consider the following information:

1. **Metal faceplates.** There is always the possibility of a metal wall plate coming loose and falling downward onto the blades of an attachment plug cap that is loosely plugged into the receptacle. This can create sparks that could result in a fire, burns, and potential shock. Positioning the *equipment grounding* conductor slot or the *grounded* conductor slot on top will minimize these hazards.
2. **Length of grounding pin.** The equipment grounding blade on a 3-wire attachment plug cap is longer than the other blades. Thus, when inserted into a receptacle, in agreement with *250.124(A)*, the equipment grounding blade is first to make and last to break.

   There is also the argument that with the equipment grounding slot to the bottom, if a plug cap comes loose and starts to fall out, the longer equipment grounding blade would be the last connection to be disconnected.
3. **Style of flexible cord attachment cap.** Some appliances are supplied with a flexible cord that has an 90-degree angle cord cap (plug) with the ground pin at the bottom. Refrigerators and freezers commonly have cord caps with 90-degree plugs. Some plug strips also have cord caps that are designed this way. The equipment grounding slot on the receptacle should be positioned at the top if the receptacle is mounted vertically and these appliances or plug strips will be supplied.

   See Figures 2-42, 2-43, 2-44, and 2-45.

Faceplates

Faceplates for switches and receptacles are required to be installed so as to completely cover the wall opening and seat against the wall surface, *404.9(A)* and *406.5*.

FIGURE 2-42 Grounding hole to the top.

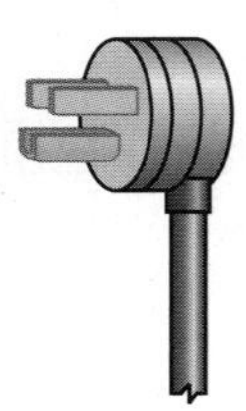

FIGURE 2-43 Grounding hole to the bottom.

FIGURE 2-44 Grounded neutral blades on top. A loose metal faceplate could fall onto these grounded neutral blades. No sparks would fly.

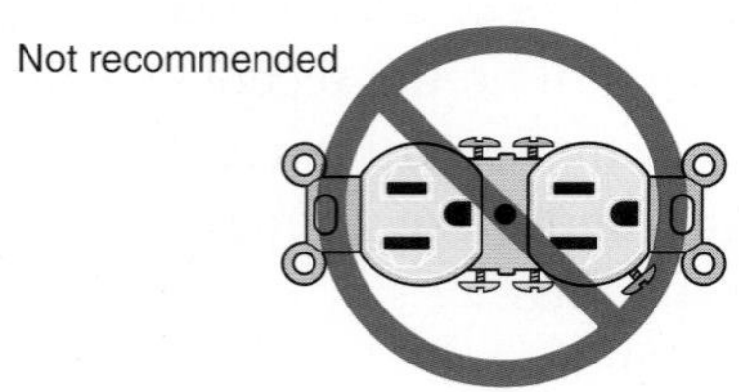

FIGURE 2-45 "Hot" terminal on top. A loose metal faceplate could fall onto these live blades. If this were a split-wired receptacle fed by a 3-wire, 120/240-volt circuit, the short would be across the 240-volt line. Sparks would fly.

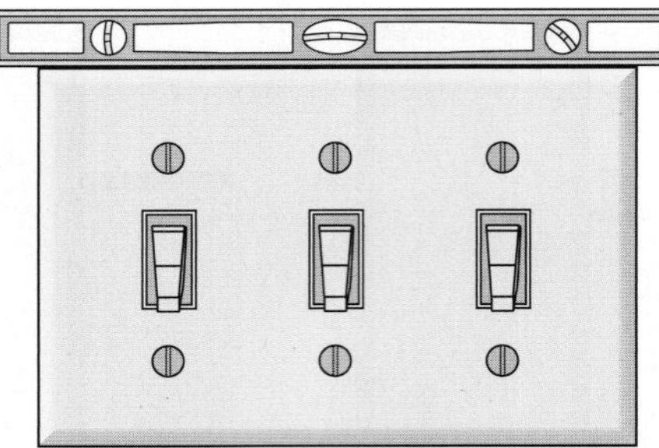

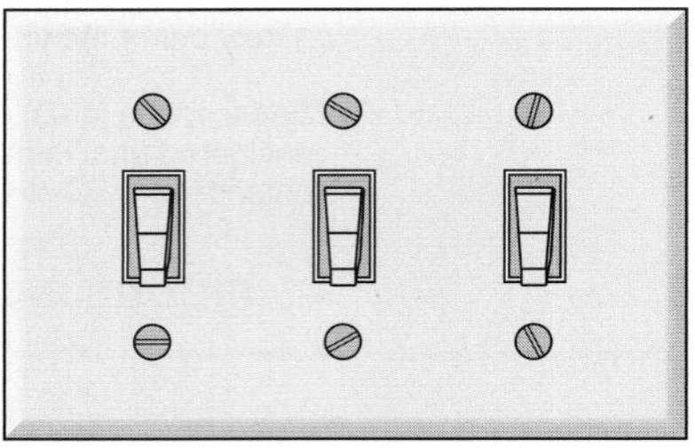

FIGURE 2-46 Aligning the slots of the faceplate. Mounting screws in the same direction makes the installation look neater. Make sure the faceplate is level.

Faceplates should be level! No matter how good the wiring inside the wall is, no matter how good the connections and splices are, no matter how proper the equipment grounding is, the only thing the homeowner will see is the wiring devices and the faceplates. That establishes the impression the homeowner has of the overall wiring installation. Because of the elongated slots in the yoke of a wiring device, it is rather easy to level a wiring device and faceplate on a single-gang box even if the box is not level. This is not true for multigang boxes. It is extremely important to make sure multi-gang boxes are level. If a multigang box is not level, it will be virtually impossible to make the wiring devices and faceplate level.

Another nice detail that can improve the appearance is to align the slots in all of the faceplate mounting screws in the same direction, as in Figure 2-46. This is not a *Code* rule—just a nice little finishing touch.

REVIEW

Note: For assistance in finding the answers to these review problems, refer to the *NEC*, the plans in the back of this text, and the specifications found in the Appendix.

PART 1—ELECTRICAL FEATURES

1. What does a plan show about electrical outlets? ______________________________

____

____

____

2. What is an outlet? ______________________________

____

____

3. Match the following switch types with the proper symbol. ______________________________

a. single-pole	$\mathbf{S_p}$
b. three-way	$\mathbf{S_4}$
c. four-way	**S**
d. single-pole with pilot light	$\mathbf{S_3}$

4. The plans show curved lines running between switches and various outlets. What do these lines indicate? ______________________________

____

5. Why are the lines referred to in problem 4 usually curved? ______________________________

6. a. What are junction boxes used for? ______________________________

b. Are junction boxes normally used in wiring the first floor? Explain. ______________________________

____

____

c. Are junction boxes normally used to wire exposed portions of the basement? Explain. ______________________________

____

7. How are standard sectional switch (device) boxes mounted? ______________________________

____

____

____

8. a. What is an offset bar hanger? ______________________________

____

____

b. What types of boxes may be used with offset bar hangers? ______________

____

____

9. What methods may be used to mount luminaires to an outlet box fastened to an offset bar hanger? ______________________________

____

____

10. What advantage does a 4 in. octagon box have over a 3 ¼ in. octagon box?

____

____

11. What is the size of the opening of a switch (device) box for a single device?

____

12. The space between a door casing and a window casing is 3½ in. (88.9 mm). Two switches are to be installed at this location. What type of switches will be used?

____

____

13. Three switches are mounted in a 3-gang switch (device) box. The wall plate for this assembly is called a ______________________________ plate.

14. The mounting holes in a device (switch) box are tapped for No. 6-32 screws. The mounting holes in an outlet box are tapped for No. 8-32 screws. The mounting holes in metal boxes for attaching equipment grounding conductors are tapped for ______________ screws.

15. a. How high above the finished floor in the living room are switches located?

____

b. How high above the garage floor are switches located?

____

16. a. How high above the finished floor in the living room are receptacles located?

____

b. How high above the garage floor are receptacles located?

____

17. Outdoor receptacle outlets in this dwelling are located ______________ in. above grade.

18. In the spaces provided, draw the correct symbol for each of the descriptions listed in (a) through (r).

a.	________	Lighting panel	j.	________	Special-purpose outlet
b.	________	Clock outlet	k.	________	Fan outlet
c.	________	Duplex outlet	l.	________	Range outlet
d.	________	Outside telephone	m.	________	Power panel
e.	________	Single-pole switch	n.	________	3-way switch
f.	________	Motor	o.	________	Push button
g.	________	Duplex outlet, split-wired	p.	________	Thermostat
h.	________	Lampholder with pull switch	q.	________	Electric door opener
i.	________	Weatherproof outlet	r.	________	Multioutlet assembly

19. The front edge of a box installed in a combustible wall must be ________________ with the finished surface.

20. List the maximum number of 12 AWG conductors permitted in the following metal boxes:

    a. 4 in. × 1½ in. octagon box. ________________

    b. 4 11/16 in. × 1½ in. square box. ________________

    c. 3 in. × 2 in. × 3½ in. device box. ________________

21. When a switch (device) box is nailed to a stud, and the nail runs through the box, the nails must not interfere with the wiring space. To accomplish this, keep the nail:

    a. halfway between the front and rear of the box.

    b. a maximum of ¼ in. (6 mm) from the front edge of the box.

    c. a maximum of ¼ in. (6 mm) from the rear of the box.

22. Hanging a ceiling luminaire directly from a plastic outlet box is permitted only if

    ____

    ____

    ____

    ____

23. It is necessary to count luminaire wires when counting the permitted number of conductors in a box according to *314.16*. True or false? ________________

24. *Table 314.16(A)* allows a maximum 10 ten wires in a certain box. However, the box will have two cable clamps and one fixture stud in it. What is the maximum number of wires allowed in this box? ________________

    ____

25. When laying out a job, the electrician will usually make a layout of the circuit, taking into consideration the best way to run the cables and/or conduits and how to make up the electrical connections. Doing this ahead of time, the electrician determines exactly how many conductors will be fed into each box. With experience, the electrician will probably select two or three sizes and types of boxes that will provide adequate space to "Meet *Code*." *Table 314.16(A)* of the *Code* shows the maximum number of conductors permitted in a given size metal box. In addition to counting the number of conductors that will be in the box, what is the additional volume that must be provided for the following items? Enter *single* or *double* volume allowance in the blank provided.
    a. one or more internal cable clamps: ______________________-volume allowance.
    b. for a fixture stud: ______________________-volume allowance.
    c. for one or more wiring devices on one yoke: ______________________-volume allowance.
    d. for one or more equipment grounding conductors: ______________________-volume allowance.
26. Is it permissible to install a receptacle outlet above an electric baseboard heater?
    ____
    ____
27. What is the maximum weight of a luminaire permitted to be hung directly from an outlet box in a ceiling? ______________________ lb.
28. Two 12-2 AWG and two 14-2 AWG nonmetallic-sheathed cables enter a box. Each cable has an equipment grounding conductor. The 12 AWG conductors are connected to a receptacle. Two of the 14 AWG conductors are connected to a toggle switch. The other two 14 AWG conductors are spliced together because they serve as a switch loop. The box contains two cable clamps. Calculate the minimum cubic-inch volume required for the box.
29. Using the same number and size of conductors as in problem 28, but using electrical metallic tubing, calculate the minimum cubic-inch volume required for the box. There will be no separate equipment grounding conductors, nor will there be any clamps in the box.
30. To allow for adequate conductor length at electrical outlet and device boxes to make up connections, *300.14* requires that not less than [3 in. (75 mm), 6 in. (150 mm), 9 in. (225 mm)] of conductor length be provided. This length is measured from where the

conductor emerges from the cable or raceway to the end of the conductor. For box openings having any dimension less than 8 in. (200 mm), the minimum length of conductor measured from the box opening in the wall to the end of the conductor is [3 in. (75 mm), 6 in. (150 mm), 9 in. (225 mm)]. (Circle the correct answers.)

31. When wiring a residence, what must be considered when installing wall boxes on both sides of a common partition that separates the garage and a habitable room? ________

____

____

32. Does the *NEC* allow metal raceways to be used with nonmetallic boxes?

Yes ____________ No ____________, *NEC* ____________

PART 2—STRUCTURAL FEATURES

Note: Show measurements using inches and feet.

1. To what scale is the basement plan drawn? ____________
2. What is the size of the footing for the steel Lally columns in the basement? Refer to Plan 1 of 10. (Lally is the trademark used for a concrete-filled steel cylinder used as a supporting member for wood or steel girders and beams. The Lally column was named after John Lally, born in Ireland in 1859. The term *lolly* is also used.) ________

   ____
3. To what kind of material will the front porch lighting bracket luminaire be attached?

   ____
4. Give the size, spacing, and direction of the ceiling joists in the workshop. ________

   ____
5. What is the size of the lot on which this residence is located? ____________

   ____
6. The front of the house is facing which compass direction? ____________
7. How far is the front garage wall from the curb? ____________
8. How far is the side garage wall from the property lot line? ____________
9. How many steel Lally columns are in the basement and what size are they?

   ____
10. What is the purpose of the I-beams that rest on top of the steel Lally columns? ______

    ____
11. To make sure that switch boxes and outlet boxes are set properly in the garage walls and ceilings, we need to know the thickness of the gypsum wallboard used in these locations. In the garage, what is the thickness and type of the gypsum wallboard?
    a. On the "warm walls" of the garage? ____________
    b. On the "cold walls" of the garage? ____________
    c. On the ceiling of the garage? ____________
12. Where is access to the attic provided? ____________

13. Give the thickness of the outer basement walls. ______
14. What material is indicated for the foundation walls? ______
15. Where are the smoke detectors located in the basement? ______

    What is the ceiling height in the basement workshop from the bottom of the joists to the floor? ______

    ______
16. Give the size and type of the front door.

    ______
17. What is the stud size for the partitions between the bathrooms in the bedroom area where substantial plumbing is to be installed? ______

    ______
18. Who is to furnish the range hood? ______

    ______
19. Who is to install the range hood? ______

    ______

CHAPTER **3**

Determining the Required Number of Branch Circuits, Lighting Outlets, and Receptacle Outlets

OBJECTIVES

After studying this chapter, you should be able to

- understand the *NEC* requirements for calculating branch-circuit sizing and loading.
- understand the term *volt-amperes per square foot*.
- calculate the occupied floor area of a residence.
- determine the minimum number of lighting and laundry branch circuits.
- know where receptacle outlets and lighting outlets are required.

INTRODUCTION

It is standard practice in the design and planning of dwelling units to permit the electrician to plan the circuits. Thus, the residence plans do not include layouts for the various branch circuits. The electrician may follow the general guidelines established by the architect. However, any wiring systems designed and installed by the electrician must conform to the *NEC*®, as well as local and state code requirements.

This chapter focuses on lighting branch circuits and small-appliance branch circuits. The circuits supplying the electric range, oven, clothes dryer, and other specific circuitry not considered to be a lighting branch circuit or a small-appliance branch circuit are covered later in this text. Refer to the Index for the specific circuit being examined.

BASICS OF WIRE SIZING AND LOADING

The *NEC* establishes some very important fundamentals that weave their way through the decision-making process for an electrical installation. They are presented here in brief form and are covered in detail as required throughout this text.

The *NEC* defines a **branch circuit** as The circuit conductors between the final overcurrent device protecting the circuit and the outlet(s).* See Figure 3-1. In the residence discussed in this text, the wiring to wall outlets, the dryer, the range, and so on, are all examples of a branch circuit.

The *NEC* defines a **feeder** as All circuit conductors between the service equipment, the source of a separately derived system, or other power supply source and the final branch-circuit overcurrent device.* In the residence discussed in this text, the wiring between Main Panel A and Subpanel B is a feeder.

The ampacity (current-carrying capacity) of a conductor must not be less than the rating of the **overcurrent device** protecting that conductor, *NEC 210.19* and *NEC 210.20*. A common exception to this is a motor branch circuit, where it is quite common to have overcurrent devices (fuses or breakers) sized larger than the ampacity of the conductor. Motors and motor circuits are covered specifically in *NEC Article 430*. The **ampere** rating of the

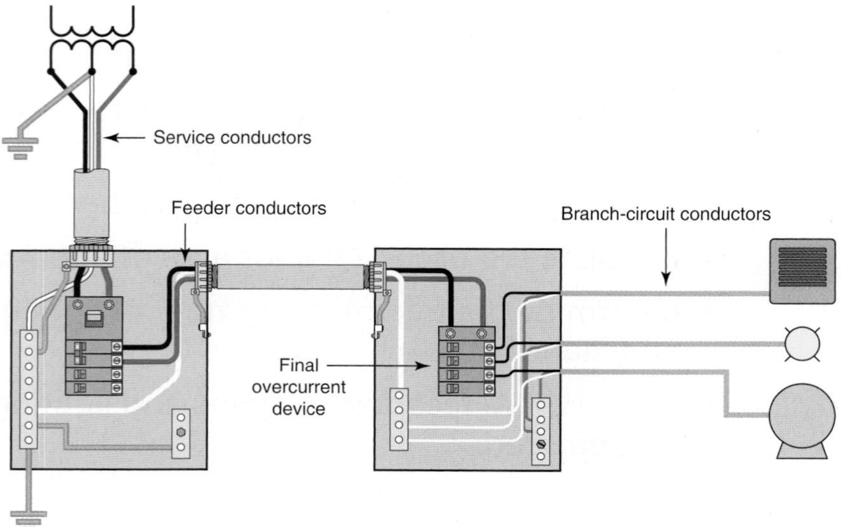

FIGURE 3-1 The branch circuit is that part of the wiring that runs from the final overcurrent device to the outlet. The rating of the overcurrent device, not the conductor size, determines the rating of the branch circuit. Feeder conductors are circuit conductors between service equipment and the final overcurrent device.

*Reprinted with permission from NFPA 70-2014.

branch-circuit overcurrent protective device (fuse or circuit breaker) determines the rating of the branch circuit. For example, if a 20-ampere conductor is protected by a 15-ampere fuse, the circuit is considered to be a 15-ampere branch circuit, *NEC 210.3*.

Standard branch circuits that serve more than one receptacle outlet or more than one lighting outlet are rated 15, 20, 30, 40, and 50 amperes. A branch circuit that supplies an individual load can be of any ampere rating, *NEC 210.3*.

If the ampacity of the conductor does not match up with a standard rating of a fuse or breaker, the next higher standard size overcurrent device may be used, provided the overcurrent device does not exceed 800 amperes, *NEC 240.4(B)*. This deviation is permitted if the circuit supplies fixed loads such as for lighting or not more than one receptacle for cord-and-plug-connected portable loads. Keep in mind a duplex receptacle is two receptacles on one strap or yoke according to the definition of "receptacle" in *NEC Article 100*. So, the round-up rule does not work for branch circuits that supply multiple receptacles, *NEC 240.4(B)(1)*.

For example, a 6 AWG copper conductor has an allowable ampacity of 65 amperes in the 75°C column of *NEC Table 310.15(B)(16)*. Because there is no standard overcurrent device rated 65 amperes, we are permitted to round up to the next standard rating of 70 amperes. However, the load on the conductor cannot exceed 65 amperes.

See Chapter 28 in this text for additional discussion on this topic.

The allowable ampacity of conductors commonly used in residential occupancies is found in *NEC Table 310.15(B)(16)*. This includes Type NM cable. It is required to be manufactured with 90°C insulated conductors. Typically, the insulation is Type THHN. As a result, the cable is limited to use in dry locations. See *NEC 334.10(A)(1)*. *NEC 334.80* allows the 90°C ampacity to be used for derating purposes so long as the final ampacity is selected from the 60°C column of *NEC Table 310.15(B)(16)*.

The ampacities in *Table 310.15(B)(16)* are subject to **correction factors** that must be applied if high ambient temperatures are encountered—for example, in attics; see *NEC Table 310.15(B)(2)(a)*.

Conductor ampacities are also subject to a **derating factor** if more than three current-carrying conductors are installed in a single raceway or cable or if these cables are installed without maintaining separation; see *NEC Table 310.15(B)(3)(a)*. See Chapter 18 of this text for complete coverage of correction and derating factors.

Most general-use receptacle outlets in a residence are included in the general lighting load calculations, *NEC 220.14(J)*.

Receptacle outlets connected to the 20-ampere small-appliance branch circuits in the kitchen, dining room, laundry, and workshop are not considered part of the general lighting load. Additional load values must be added into the calculations for these receptacle outlets. This is discussed later in this text.

The minimum lighting load for dwellings is 3 volt-amperes per square foot. See *NEC 220.12* and *Table 220.12*.

Continuous Loads

The *NEC* defines **continuous load** in *Article 100* as A load where the maximum current is expected to continue for three hours or more.* Continuous loads are not permitted to exceed 80% of the rating of the branch circuit. General lighting outlets and receptacle outlets in residences are not considered to be continuous loads.

Certain loads in homes are considered to be continuous and must be treated accordingly. Examples are electric water heaters (*NEC 422.13*), central electric heating [*NEC 424.3(B)*], snow-melting cables (*NEC 426.4*), and air-conditioning equipment (*NEC 440.32*). For these loads, the branch-circuit rating, the conductors, and the overcurrent device are required to be not less than 125% of the rating of the equipment. Mathematically, sizing the conductors and overcurrent device at 125% of the load is the same as loading the conductors and overcurrent device to 80%.

For example, an electric furnace with a nameplate rating of 40 amperes would require the supply conductors and overcurrent protection to be not less than

$$40 \times 1.25 = 50 \text{ amperes}$$

*Reprinted with permission from NFPA 70-2014.

VOLTAGE

All calculations throughout this text use voltage values of 120 volts and 240 volts. This complies with the requirements of *NEC 220.5*, which lists the various voltage values to use for different electrical systems. This is repeated in the second paragraph of *Annex D* of the *NEC*.

The word *nominal* means "in name only, not a fact." For example, in our calculations, we use 120 volts even though the actual voltage might be 110, 115, or 117 volts. This provides us with a uniformity in making load calculations.

Most public service commissions require that the voltage for residential services be held to ±5% of the nominal voltage as measured at the service point.

CALCULATING LOADS

When wiring a house, it is all but impossible to know which appliances, lighting, heating, and other loads will be turned on at the same time. Different families lead different lifestyles. There is tremendous diversity. There is a big difference between "connected load" and "calculated load." Who knows what will be plugged into a wall receptacle, now or in the future? It's a guess at best. Over the years, the *NEC* has developed procedures for calculating loads in typical one- and two-family homes.

The rules for doing the calculations are found in *NEC Article 220*. For lighting and general-use receptacles, the calculations are based on volt-amperes per square foot. For the small-appliance circuits in kitchens and dining rooms as well as for the 2-wire laundry branch circuit, the basis is 1500 volt-amperes per circuit. For large appliances such as dryers, electric ranges, ovens, cooktops, water heaters, air conditioners, heat pumps, and so on, which are not all used continuously or at the same time, there are demand factors to be used in the calculations. Following the requirements in the *NEC*, the various calculations roll together in steps that result in the proper sizing of branch circuits, feeders, and service equipment.

As you work your way through this text, examples are provided for just about every kind and type of load found in a typical residence.

Inch-pounds versus Metrics When Calculating Loads

As discussed in Chapter 1, converting inch-pound measurements to metric measurements and vice versa results in odd fractional results. Adding further to this problem are the values rounded off when the *NEC* Code-Making Panels did the metric conversions. When square feet are converted to square meters and the unit loads are calculated for each, the end results are different—close, but nevertheless different. To show both calculations would be confusing as well as space consuming. Many of the measurements in this text are shown in both inch-pound units and metric units. Load calculations throughout this text use inch-pound values only, which is in agreement with the *Examples* given in *Annex D* of the 2014 *NEC*.

CALCULATING FLOOR AREA

The general lighting load for a dwelling is based on the square footage of the dwelling. Here is how this is done.

First Floor Area

To estimate the total load for a dwelling, the occupied floor area of the dwelling must be calculated. Note in the residence plans that the first floor area has an irregular shape. In this case, the simplest method of calculating the occupied floor area is to determine the total floor area using the outside dimensions of the dwelling. Then, the areas of the following spaces are subtracted from the total area: open porches, garages, or other unfinished or unused spaces if they are not adaptable for future use, *NEC 220.12*.

Many open porches, terraces, patios, and similar areas are commonly used as recreation and entertainment areas. Adequate lighting and receptacle outlets should be provided for these areas.

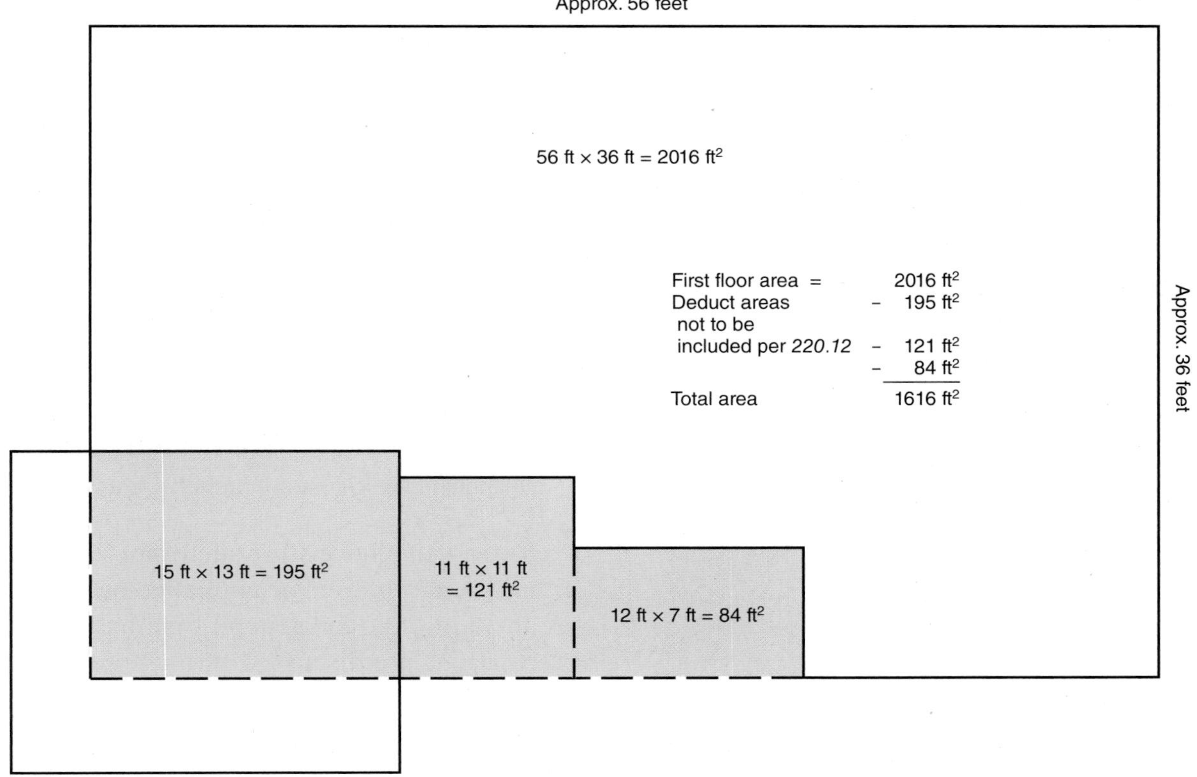

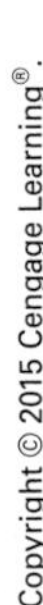

FIGURE 3-2 Simplified method of determining the first floor square footage area. Some dimensions are rounded off to make the calculations simpler and more practical. The areas of the three blocks are subtracted from an overall large rectangle.

For practicality, the author has chosen to round up dimensions for the determination of total square footage and to round down dimensions for those areas (garage, porch, and portions of the inset at the front of the house) not to be included in the calculation of the general lighting load. This produces a slightly larger result as opposed to one on the conservative side. Don't be miserly with your measurements; rather, be generous. Figure 3-2 shows the procedure for calculating the total square footage of this residence.

Basement Area

Although the *NEC* in *220.12* tells us that unused or unfinished spaces not adaptable for future use do not have to be included in calculating the square footage of a dwelling, it makes sense to include some of these spaces.

Nearly all basements in homes today certainly could be considered as being adaptable for future use. A crawl space and most attics would not normally be considered as being adaptable for future use. This is a judgment call based on a close examination of the Plans and Specifications. In this residence, more than half of the total basement area is finished off as a recreation room, which certainly is considered a living area. The workshop area also is intended to be used but is not considered as habitable space.

To simplify the calculation for this residence, we will consider the entire basement as usable space and figure the basement square footage area as being the same as the area of the first floor.

The combined occupied area of the dwelling is found by adding the first floor and basement areas together:

First floor	1616 ft^2 (149 m^2)
Basement	1616 ft^2 (149 m^2)
Total	3232 ft^2 (298 m^2)

DETERMINING THE MINIMUM NUMBER OF LIGHTING BRANCH CIRCUITS

Table 220.12 of the *NEC* tells us that the minimum load requirement for dwelling units is 3 volt-amperes (VA) per square ft (0.093 m^2) of occupied area.

To determine the minimum number of 15-ampere lighting branch circuits required for a residence, here are the simple steps:

$$\text{Step 1:} = \frac{3 \text{ volt-amperes} \times \text{square ft}}{120 \text{ volts}} = \text{amperes}$$

$$\text{Step 2:} = \frac{\text{amperes}}{15} = \begin{array}{l}\text{minimum number of}\\ \text{15-amperes lighting}\\ \text{branch circuits}\end{array}$$

This equates to a minimum of

- one 15-ampere lighting branch circuit for every 600 ft^2 (55.8 m^2).
- one 20-ampere lighting branch circuit for every 800 ft^2 (74.4 m^2).

Note: 20-ampere branch circuits are not commonly used for lighting in residential installations. In this residence, let's calculate the load for the total occupied area of 3232 ft^2 (298 m^2):

$$3232 \times 3 = 9696 \text{ volt-amperes}$$

$$\text{Amperes} = \frac{\text{volt-amperes}}{\text{volts}} = \frac{9696}{120} = 80.8 \text{ amperes}$$

The minimum number of 15-ampere lighting branch circuits is

$$\frac{80.0 \text{ amperes}}{15 \text{ amperes}} = \begin{array}{l}\text{5.4 branch circuit}\\ \text{(round up to 6)}\end{array}$$

Because we cannot have a fraction of a branch circuit, we rounded up to a minimum of six 15-ampere lighting branch circuits.

We get the same answer using 600 ft^2 (55.8 m^2) for each 15-ampere lighting branch circuit as follows:

$$\frac{3232 \text{ ft}^2}{600} = \begin{array}{l}\text{5.4 branch circuit}\\ \text{(round up to 6)}\end{array}$$

No matter which method we use, a minimum of six 15-ampere lighting branch circuits is required.

In the *Informational Annex* of the *NEC*, we find many examples of load calculations. Here, we also find that Except where the calculations result in a major fraction of an ampere (0.5 or larger), such fractions are permitted to be dropped.*

AUTHOR'S TIP: This permission is for load calculations such as in Step 1. This permission to drop fractions is *not* to be used in Step 2 because it is impossible to have a fraction of a branch circuit. That is why we rounded up from 5.4 to arrive at a minimum of six 15-ampere branch circuits. ●

Load calculations for small-appliance branch circuits and other major appliance branch circuits are in addition to the general lighting loads. These calculations are discussed throughout this text as they appear.

The *NEC* in *210.11(B)* states that the calculated load . . . shall be evenly proportioned among multioutlet branch circuits within the panelboard(s).*

Table 3-1 is the schedule of the 15-ampere lighting and 20-ampere branch circuits in this residence.

NEC 240.4(D) is referred to as the "Small Conductor Rule" and tells us that the maximum overcurrent protection is

15 amperes for 14 AWG copper conductors.
20 amperes for 12 AWG copper conductors.
30 amperes for 10 AWG copper conductors.

In conformance to *NEC 210.3*, a branch circuit is rated according to the rating or setting of the overcurrent device. For example, a lighting branch circuit is rated 15 amperes even if 12 AWG copper conductors are installed.

How Many Outlets per Circuit

This is covered in detail in Chapter 8, where you will find *Estimating Loads for Outlets, How Many Outlets per Circuit, Circuit Loading Rules of Thumb,* and the *80% Rule.*

TABLE 3-1

Summary of 15-ampere lighting and 20-ampere receptacle circuits installed in this dwelling.

15-AMPERE CIRCUITS		20-AMPERE CIRCUITS	
A14	Bathrooms, hall lighting	A18	Workbench receptacles
A15	Front entry/porch	A20	Workshop receptacles, window wall
A16	Front bedroom lighting and outdoor receptacle	A22	Master bath receptacle
A17	Workshop lighting	A23	Hall bath receptacle
A19	Master bedroom lighting and outdoor receptacle		
A21	Study/bedroom lighting		
B7	Kitchen lighting	B5	Dishwasher
B9	Wet bar lighting and receptacles	B13	Kitchen north and east wall receptacles
B10	Laundry, rear entry, powder room, attic lighting	B15	Kitchen receptacles, west countertop
B11	Recreation room receptacles	B16	Kitchen receptacles, south countertop, and island
B12	Recreation room lighting	B18	Clothes washer receptacle
B14	Garage lighting and overhead door operator	B19	Food waste disposal
B17	Living room and outdoor receptacle	B20	Laundry room receptacles and outdoor receptacle
		B21	Powder room receptacle
		B22	Refrigerator
		B23	Garage receptacles

Notes: Panelboard "A" is the main panel located in the workshop.
Panelboard "B" is the subpanel located in the corner of the recreation room.

TRACK LIGHTING LOADS

Track lighting loads are considered to be part of the general lighting load for residential installations, based on 3 volt-amperes per square foot as previously discussed. There is no need to add more wattage (volt-amperes) to the load calculations for track lighting in homes.

As required by *410.151(B)*, the connected load on a lighting track is not permitted to exceed the rating of the track. Also, be sure that a lighting track is supplied by a branch circuit having a rating not more than that of the track. There is more on track lighting in Chapters 12 and 13 of this text.

SUMMARY OF WHERE RECEPTACLE AND LIGHTING OUTLETS MUST BE INSTALLED IN RESIDENCES

The following is a recap of the *NEC* requirements for receptacle and lighting outlets. We will look at these *Code* rules, room by room, later in this text.

Remember, the *NEC* provides *minimum* requirements. There is nothing wrong with installing more receptacles than the *NEC minimum*. You will never have enough receptacles! Think about it. Receptacles are needed for computers, cable modems, small UPS devices, printers, stereos, VCRs,

VDRs, CD players, CD burners, shredders, television sets, radios, fax machines, answering machines, typewriters, adding machines, pencil sharpeners, lamps, telephones, and others.

Consider spacing receptacles closer than the minimum. Consider quadplex receptacles or two duplex receptacles at each receptacle outlet location, and consider surge suppression receptacles or surge protection multiple plug-in strips. See Chapter 6 for a discussion of surge protection.

AUTHOR'S TIP: Some electricians install a receptacle below the wall switch in bedrooms, living rooms, family rooms, and similar rooms. As is usually the case, the required receptacles will be behind furniture, making it difficult to reach for plugging in a vacuum cleaner or other temporarily used appliance. A receptacle below the wall switch will be very easy to get to. A real plus! ●

See Figure 3-3 for an explanation of *receptacle* and *receptacle outlet*.

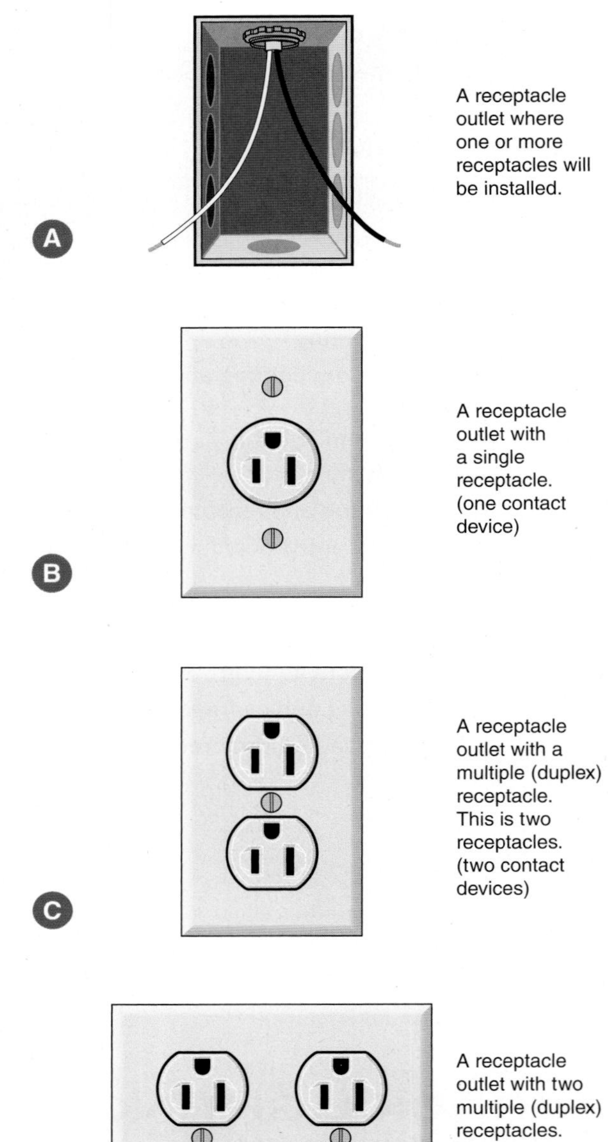

FIGURE 3-3 The branch-circuit wiring and the box where one or more receptacles will be installed is defined by the *Code* as a *receptacle outlet.* The receptacle itself, whether a single or multiple device (one, two, or three receptacles) on one strap (yoke) is defined as a *receptacle.* For house wiring, receptacle loads are considered to be part of the general lighting load calculations, *220.14(J).*

Receptacle Outlets (125 volts, Single-Phase, 15 and 20 Amperes)

Ground-Fault Circuit Interrupter (GFCI). GFCI protection is required in specific locations in a home. GFCIs are discussed in Chapter 6.

Arc-Fault Circuit Interrupter (AFCI). AFCI protection is required in specific locations in a home. AFCIs are discussed in Chapter 6.

The following is a partial list of *Code* requirements where receptacles are required in new homes:

- Receptacles are located throughout a home for the convenience of plugging things in. The following required locations for receptacles are in addition to receptacles that are wall switch controlled, part of a luminaire or appliance, located inside of a cabinet, or located more than 5½ ft (1.7 m) above the floor. A duplex receptacle that has one receptacle switched and one receptacle "hot" continuously meets the receptacle requirements of *210.52* and the switching requirements of *210.70*.

- Wall receptacles must be placed so that no point measured horizontally along the floor line in any wall space is more than 6 ft (1.8 m) from a receptacle outlet. Fixed room dividers and railings are considered to be walls for the purpose of this requirement. See *210.52(A)(1)*.
- Any wall space 24 in. (600 mm) or more in width in kitchens, family rooms, dining rooms, living rooms, parlors, libraries, dens, sunrooms, bedrooms, recreation rooms, or similar room or area of dwelling units must have a receptacle outlet, *210.52(A)(2)(1)*.
- Nonsliding fixed glass panels on exterior walls are considered to be wall space and are to be figured in when applying the rule that no point along the floor line is more than 6 ft (1.8 m) from a receptacle. Sliding panels on exterior walls are not considered to be a wall, *210.52(A)(2)*.
- Receptacle outlets located in the floor more than 18 in. (450 mm) from the wall are not to be counted as meeting the required number of wall receptacle outlets, *210.52(A)(3)*. An example of this would be a floor receptacle installed under a dining room table for plugging in warming trays, coffee pots, or similar appliances.
- Hallways 10 ft (3.0 m) or longer in homes must have at least one receptacle outlet, *210.52(H)*.
- Foyers that have an area that is greater than 60 ft^2 (5.6 m^2) are required to have a receptacle(s) located in each wall space as defined in *210.52(I)*.
- Circuits supplying wall receptacle outlets for lighting loads are generally 15-ampere circuits but are permitted by the *Code* to be 20 amperes.
- Outdoors: One receptacle is required in front and another in back, *210.52(E)*. This is discussed in detail later in this chapter.
- Receptacle(s) for servicing HVAC equipment, *210.63*. This is discussed in detail later in this chapter.
- Outdoor receptacles are required to be GFCI protected, *210.8(A)(3)*. Exempt are receptacles that are not readily accessible, are supplied by a dedicated branch circuit, and serve snow-melting or deicing equipment; see *426.28*. Ground-fault protection of equipment is required for fixed outdoor electric deicing and snow-melting equipment. This equipment protection will typically trip or open at 30 mA or greater leakage current rather than at from 4 to 6 mA for personnel protection.
- At least one receptacle outlet is required in each attached garage and in each detached garage with electric power. ▶The branch circuit supplying this receptacle(s) is not permitted to supply outlets outside of the garage. At least one receptacle outlet must be installed for each car space.◀
- A receptacle outlet is also required for each accessory building that has electric power.
- At least one receptacle outlet is required in each separate unfinished portion of a basement.

Basements

- At least one receptacle outlet must be installed in addition to the laundry outlet, *210.52(G)(3)*.
- Receptacles must be supplied by a 15- or 20-ampere circuit.
- In finished basements, the required number, spacing, and location of receptacles, lighting outlets, and switches are required to be in conformance with *210.52(A)* and *210.70*.
- In finished basements, GFCI protection for receptacles is *not* required unless the receptacles are near a sink.
- Any 125-volt, 15- or 20-ampere receptacle within 6 ft (1.8 m) from the outside edge of a sink shall be GFCI protected. There are no exceptions to this requirement.
- Unfinished basements, or unfinished portions of basements that are not generally "lived in," such as work areas and storage areas, are not subject to the spacing requirements for receptacles as specified in *210.52(A)*.

- In each separate unfinished basement area that is not intended to be "lived in," such as storage areas or work areas, at least one 125-volt, single-phase, 15- or 20-ampere receptacle must be installed, *210.52(G)(3)*. This receptacle(s) must be GFCI protected, *210.8(A)(5)*. This includes receptacles that are an integral part of porcelain or plastic pull chain or keyless lampholders.
- See the "Laundry Area" section in this chapter for receptacle outlet requirements in laundry areas.

Small-Appliance Receptacle Branch Circuits

- At least two 20-ampere small-appliance circuits must supply the receptacles that serve the countertop surfaces in the kitchen, *210.11(C)(1)* and *210.52(B)(3)*. Two small-appliance branch circuits is the minimum. It is advisable to run additional 20-ampere branch circuits to these areas because of the heavy concentration of appliances.
- See Chapter 12 of this text for information on installing lighting and small-appliance branch circuits for kitchens.
- Allow 1500 volt-amperes for each 20-ampere small-appliance branch circuit when calculating loads for determining the size of the service-entrance equipment, *220.52(A)*.
- Refer to *210.8(A)(6)*, *210.11(C)(1)*, *210.52(B)(1)*, *(2)*, and *(3)*, and *220.52(A)*.
- See Chapter 6 of this text for details on GFCI protection.
- See *210.8(A)(6)* and *(7)*, *210.11(C)*, *210.52(C)*, *220.52(A)*, and *406.4(E)* for *NEC* rules on providing GFCI protection, branch circuit requirements, and receptacle outlet location requirements.

Laundry Area

Automatic clothes washers draw a large amount of current during certain operating cycles. Here are the requirements for branch circuits for the laundry area:

- This receptacle(s) must be supplied by an additional 20-ampere branch circuit *NEC 210.11(C)(2)*.
- This branch circuit is not permitted to supply any other outlets outside the laundry area.
- This branch circuit and receptacle(s) are in addition to other required receptacles and branch circuits.
- GFCI protection is required for receptacles in laundry areas, within 6 ft (1.8 m) from the outside edge of a sink, and in unfinished basements. The GFCI device must be readily accessible for monthly testing.
- If you install a single receptacle on a 20-ampere branch circuit, it must be rated 20 amperes.
- If you install a duplex receptacle on a 20-ampere branch circuit, it may be rated 15 amperes or 20 amperes.
- Include 1500 volt-amperes for this laundry branch circuit when calculating the ampacity of the feeder or service-entrance conductors.

Bathroom Branch Circuits

Here is a recap of the *NEC* requirements for branch circuits that supply receptacles in bathrooms:

- The receptacle(s) must be supplied by at least one separate 20-ampere branch circuit and generally must not supply other loads such as lighting outlets *NEC 210.11(C)(3)*.
- The 20-ampere branch circuit and receptacle is in addition to other required outlets and branch circuits.
- The separate 20-ampere branch circuit is permitted to supply more than one bathroom, although it is not recommended because of the use of high-wattage appliances. It is better to run a separate 20-ampere branch circuit to each bathroom.
- If the separate 20-ampere branch circuit supplies only one bathroom, other outlets such as for lighting or exhaust fans in that bathroom are permitted to be connected to it. The other permitted loads are limited to not more than 80% of the rating of the branch circuit if they are cord-and-plug-connected, and 50% if they are fixed or fastened-in-place equipment.
- Instead of the "only one bathroom" exception, it is recommended that you run a separate

20-ampere branch circuit to each bathroom. That's what the plans for the residence in this text call for.

- Don't be left in the dark. Though permitted by the *NEC*, it is best practice not to connect bathroom lighting to this separate 20-ampere receptacle branch circuit.
- No additional load need be added for this separate 20-ampere branch circuit when doing the load calculations. The separate branch-circuit requirement merely redirects the receptacle load off the lighting branch circuit.

Outdoors

- At least one receptacle is required to be installed in the front and another in the back of a one-family dwelling, *210.52(E)(1)*. These receptacles must be readily accessible from grade level and not more than 6½ ft (2.0 m) above grade. See Figure 3-4. The same is true for each unit of a multifamily dwelling where the entrance/egress is at grade level, *210.52(E)(2)*.

Grade level could be grass, sidewalk, or driveway. It is possible that a concrete or brick patio might be built at grade level.

In addition to the "front and back" receptacle requirement, let's take a look at receptacle requirements for balconies, decks, and porches.

- **Balconies, decks, and porches:** We don't want people running extension cords through doors or window openings and pinching the cord. In *210.52(E)(3)*, we find these requirements for balconies, decks, and porches that are attached to the dwelling unit and are accessible from inside the dwelling:
  - At least one receptacle outlet is required that is accessible from the balcony, deck, or porch.
  - The receptacle outlet must not be more than 6½ ft (2.0 m) above the walking surface of the balcony, deck, or porch.

Roughing in receptacle(s) on the front of the house is usually a no-brainer. Plans clearly show the construction of the porch. In the residence in this text, there is one switched receptacle on the front porch and another non-switched receptacle on the outside of the front bedroom wall.

A rear deck/patio/porch is another story. They are generally built after the house is completed. It is easier said than done to figure out where to rough in the required receptacle.

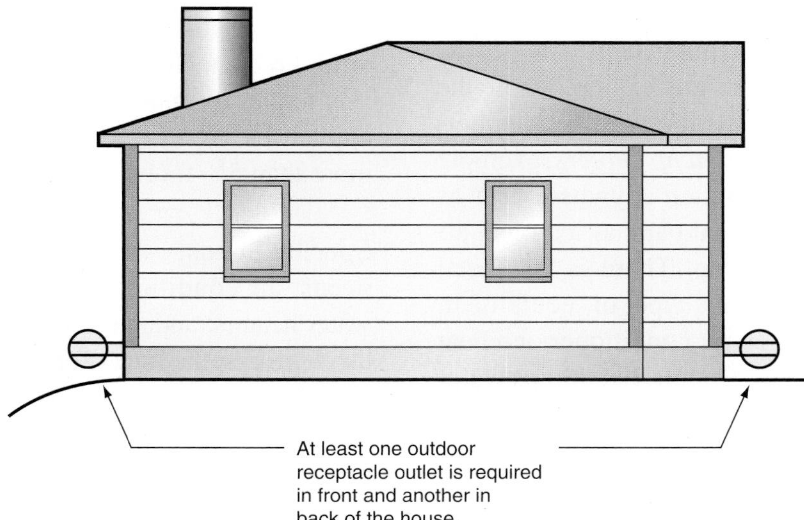

FIGURE 3-4 *NEC 210.52(E)* covers outdoor receptacles for residences and requires that for one-family dwellings and for each unit of a two-family dwelling that is at grade level, at least one receptacle outlet must be installed in the front and another in the back. These receptacle outlets must be readily accessible from grade level and must be mounted not more than 6½ ft (2.0 m) above grade.

The electrician must consider: Where will the deck/patio be built? Is the deck/patio poured concrete or decorative brick/stone at grade level? What is the height if built above grade? Does it have rails or walls? Is it screened in? Sometimes the surest thing to do is

- rough in the one receptacle on the rear of the house close to the means of entrance/egress such as a sliding door not over 6½ ft (2.0 m) above grade.
- rough in a second receptacle somewhere else on the rear of the house outside the footprint of the intended deck/patio/porch.

For the more elaborate deck/patio/porch, consider installing a few additional receptacles around the perimeter of the deck/patio/porch. These would be wired in after the deck/patio/porch is completed.

- Additional receptacles may be installed higher, such as receptacles used for plugging in ice and snow-melting heating cables and/or decorative outdoor lighting.
- For servicing HVAC equipment, *210.63* requires that a 125-volt, single-phase, 15- or 20-ampere-rated receptacle be installed as follows:
  - Be at an accessible location.
  - Be on the same level as the HVAC equipment.
  - Be within 25 ft (7.5 m) of the HVAC equipment.
  - *Not* be connected to the load side of the equipment disconnecting means.

  Evaporative coolers are exempt from the above requirements. Why are evaporative coolers, often referred to as *swamp coolers*, exempt? Because they do not contain anywhere near as many complicated internal parts as a typical air conditioner or heat pump. There is less need for maintenance on this type of equipment. Less expensive than an air conditioner or a heat pump, they contain one motor that pumps water to the top of the unit. The water "falls" downward over porous filter pads where the water is cooled by evaporation. A second motor (fan) pulls in outside air and blows the air over the cooled water into the room. Evaporative coolers work great in hot, dry climates. The higher the humidity, the less efficient is an evaporative cooler.
- The outdoor receptacles required by *210.52(E)(1)* (one in the front, one in the back) might or might not meet the requirements of *210.63*. If the HVAC equipment is located on a roof, in an attic, in a crawl space, or in a similar location, a receptacle must be installed in that location so as to be on the same level as the equipment.
- All outdoor receptacles must be GFCI protected, *210.8(A)(3)*. The GFCI device must be readily accessible for monthly testing. Exempt from this requirement are GFCI outdoor receptacle(s) that are not readily accessible and are supplied by a dedicated branch circuit installed for snow- and ice-melting equipment (e.g., heating cables).
- Ground-fault protection of equipment (GFPE) must be provided for "fixed" outdoor deicing and snow-melting equipment. GFPE can be a circuit breaker or a receptacle, or may be self-contained in the equipment. Don't confuse GFPE with GFCI protection for personnel. GFPE devices for equipment protection trip at approximately 30 milliamperes; see *426.28*.

  When used for deicing and snow melting, mineral-insulated, metal-sheathed cable embedded in noncombustible material such as concrete does not require ground-fault protection.

Receptacles in Other Locations. All of the required receptacles discussed previously are *in addition* to any receptacles that are part of luminaires or appliances, are located inside cabinets, or are located more than 5½ ft (1.7 m) above the floor, *210.52*.

Specific Loads. Appliances such as electric ranges, ovens, air conditioners, electric heat, heat pumps, water heaters, and similar loads are discussed later in this text. See the Index for their location.

Receptacle Definitions. The *NEC* in *Article 100* defines the following:

- **Receptacle:** A receptacle is a contact device installed at the outlet for the connection of an attachment plug.*

*Reprinted with permission from NFPA 70-2014.

- **Single receptacle:** A single receptacle is a single contact device with no other contact device on the same yoke.*
- **Multiple receptacle:** A multiple receptacle is two or more contact devices on the same yoke.*

See Figure 3-3 for an illustration of these definitions.

Receptacle and Branch-Circuit Ratings

Table 3-2 shows the requirements of *210.21(B)* and *Table 210.21(B)(3)* where the branch circuit supplies two or more receptacles or outlets in homes. The table also points out that where the branch circuit supplies a single receptacle, the receptacle must have a rating not less than the ampere rating of the branch circuit.

Arc-Fault Circuit Interrupters (AFCIs). See Chapter 6 of this text for complete information about AFCI requirements in bedrooms.

Replacing Existing Receptacles. See Chapter 6 of this text for detailed discussion and illustrations about replacing existing receptacles.

Lighting Outlets. *(210.70)* The *NEC* contains minimum requirements for providing lighting for dwellings.

TABLE 3-2

Ampere rating for receptacles connected to 15- and 20-ampere branch circuits. See *210.21(B)* and *Table 210.21(B)(3).*

BRANCH-CIRCUIT RATING	RECEPTACLE RATING
15-ampere lighting branch circuit: Circuit has one or more receptacles or outlets.	15-ampere maximum
20-ampere small-appliance branch circuit: Circuit has two or more receptacles.	15 or 20 amperes
Individual 15-ampere branch circuit: Circuit has only a single receptacle connected.	15 amperes
Individual 20-ampere branch circuit: Circuit has only a single receptacle connected.	20 amperes

Switched Lighting Outlets. Per *210.70*, lighting outlets in dwellings must be installed as shown in Figure 3-5. Here is a summary of these requirements.

Habitable Rooms. In every habitable room and in every bathroom, at least one wall switch–controlled lighting outlet must be installed. Switched receptacles meet the intent of this requirement except in kitchens and bathrooms. Quite often, ceiling luminaires are omitted in bedrooms, living rooms, and similar rooms. Instead, the room lighting is dependent on cord-and-plug-connected table, floor, wall, and/or swag lamps.

The recognized definition of habitable room is: A room in a residential occupancy used for living, sleeping, cooking, and eating, but excluding bath, storage and service area, and corridors.

As illustrated in Figure 3-6, swag lamps are not hardwired. They are connected by weaving the cord through the chain.

Wiring split-wired receptacles is the key to providing convenient control of cord-and-plug-connected lighting. These switched receptacles generally should be controlled by ordinary snap switches. Be aware that *NEC 404.14(E)* requires, General-use dimmer switches shall be used only to control permanently installed incandescent luminaires unless listed for the control of other loads and installed accordingly.*

In addition, *NEC 406.15* requires, ▶A receptacle supplying lighting loads shall not be connected to a dimmer unless the plug/receptacle combination is a nonstandard configuration type that is specifically listed and identified for each such unique combination.◀*

The required minimum spacing of receptacles is found in *210.52*, already discussed in this chapter. When laying out and roughing in the wiring for split-wired receptacles, remember that:

- the "always hot" portion of a split-wired receptacle qualifies for the required receptacle.
- the "switched" portion of a split-wired receptacle is not considered to be one of the required receptacles, but rather is considered to be "in addition."

*Reprinted with permission from NFPA 70-2014.

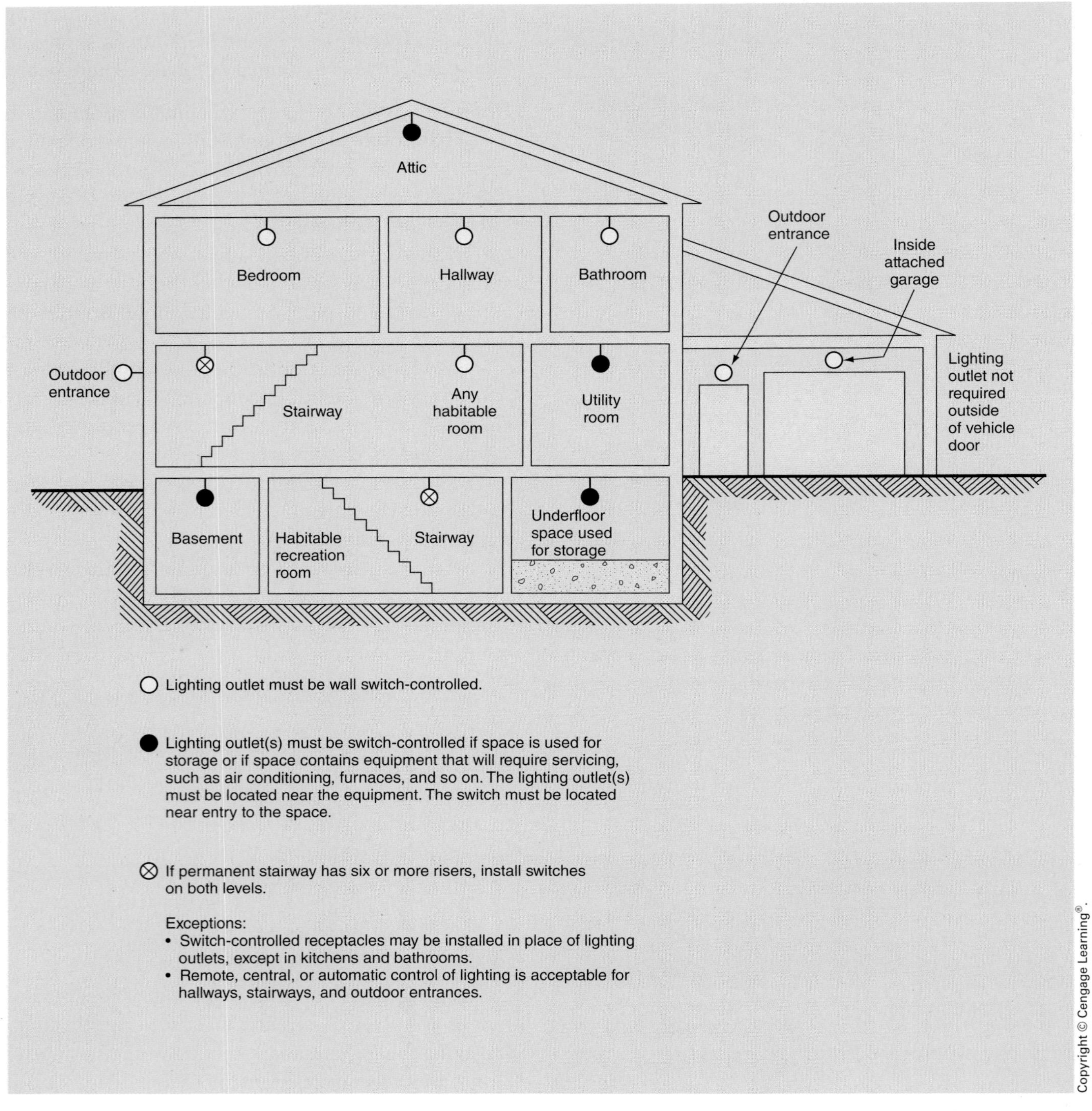

FIGURE 3-5 Lighting outlets required in a typical dwelling unit.

If plug-in swag lighting, as illustrated in Figure 3-6, is used in a dining room, a switched receptacle connected to a general lighting branch circuit must be installed. In other words, you still must install the receptacles supplied by the required 20-ampere small-appliance branch circuits, *210.52(B)(1), Exception No. 1,* Figure 3-7.

Where Do I Learn More about Split-Wired Receptacles? In this residence, split-wired receptacles are used in the Living Room and the bedrooms. The wiring of split-wired receptacles is covered in Chapters 2, 3, 8, 12, 13, 14 of this text, and is shown on Blueprint sheet 9.

Occupancy motion sensors may be used if (1) they are in addition to the required wall switch,

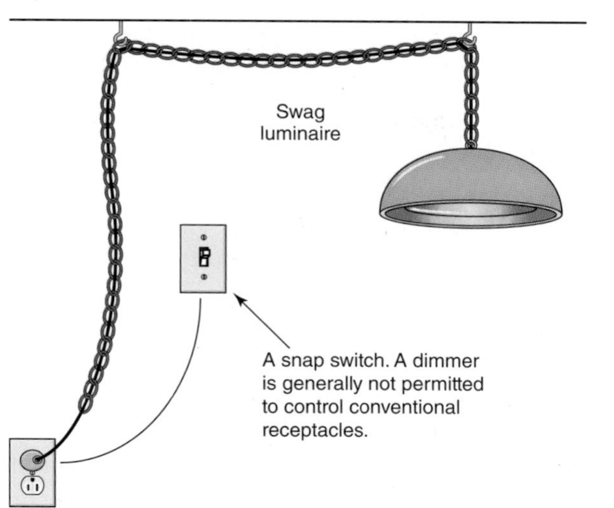

FIGURE 3-6 A receptacle controlled by a wall switch is acceptable as the required lighting outlet in rooms other than a kitchen or bathroom, *210.70(A)(1), Exception.*

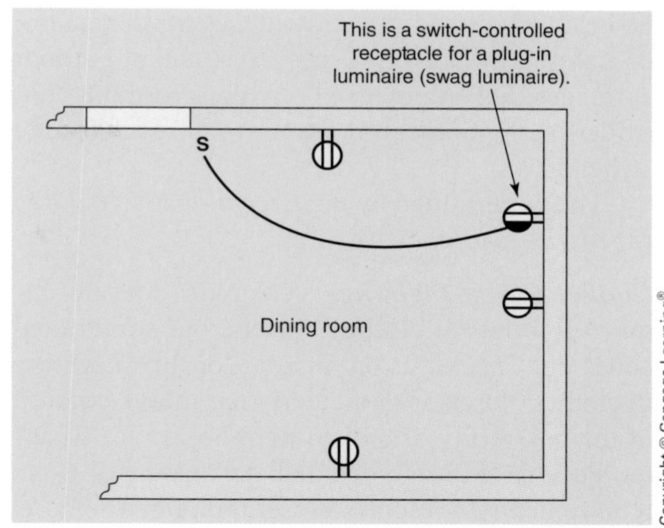

FIGURE 3-7 When a switch-controlled receptacle outlet is installed in rooms where the receptacle outlets normally would be 20-ampere small-appliance circuits (breakfast room, dining room, and so on), it must be in addition to the required small-appliance receptacle outlets, *210.52(B)(1), Exception No. 1.*

or (2) they have a manual override and are located where the wall switch would normally be located.

Additional Locations. At least one wall switch–controlled lighting outlet must be installed in hallways, stairways, attached garages, and detached garages with electric power, *210.70(A)(2)(a)*.

The *NEC* does *not* require that a detached garage have electric power, but if it does have electric power, then it must have at least one GFCI receptacle, and it must have wall switch–controlled lighting.

Some electrical inspectors consider a tool shed or other similar accessory building to be a storage or utility room that comes under the requirements of *210.70(A)(3)*, thus requiring a wall switch–controlled lighting outlet.

At least one wall switch–controlled lighting outlet must be installed to provide illumination on the exterior side of outdoor entrances or exits that have grade-level access to dwelling units, attached garages, and detached garages if they have electric power. A sliding glass door is considered to be an outdoor entrance. A vehicle door is not considered to be an outdoor entrance; see *210.70(A)(2)(b)*.

Although not necessarily good design practice,

- one switch could control the required outdoor lighting at more than one entrance.
- one switch could control both the required outdoor and indoor lighting of a garage.
- one outdoor luminaire, properly located, could meet the required outdoor lighting for both the house entrance and the garage entrance.

Stairways. Where there are six stair risers or more on an interior stairway, the *NEC* requires that wall switch–controlled stairway lighting must be provided on each floor level and any landing level that has an entryway, *210.70(A)(2)(c)*.

This switching requirement includes permanent stairways to basements and attics but does not include fold-down storable ladders commonly installed for access to residential attics.

The *NEC* does not tell us how to illuminate the stairway.

The *International Residential Code* is more specific about stairway lighting. It requires indoor stairways to have lighting that will illuminate the landings and treads, and that the lighting shall be located in the immediate vicinity of each landing at the top and bottom of the stairs. This *Code* also requires that the switching must be accessible at the top and bottom of the stairway, without using any step to reach a switch.

If You Don't Install Switches, What Else Is Acceptable? Remote control, central control, motion sensors, photocells, and other automatic devices may be installed instead of wall switches for the control of lighting in hallways, stairways, and at outdoor entrances. For example, an outdoor post light controlled by a photocell does not require a separate switch.

This is permitted by the *Exception* to *210.70(A)(2)(a)*, *(b)*, and *(c)*.

Clothes Closet Lighting. The *NEC* does not require lighting in clothes closets, but some local codes do. Check this out in your locality. Lighting in clothes closets is particularly hazardous because of the possibility of broken light bulbs. See Chapter 8 of this text for a detailed explanation of *Code* requirements for clothes closet lighting. The *NEC* defines a clothes closet as A nonhabitable room or space intended primarily for storage of garments and apparel.* A utility closet, broom closet, china closet, or "game" closet would not come under the stringent clothes closet requirements found in *410.16*, unless it is obvious that the room or area is primarily for storage of garments and apparel.

Basements, Attics, Storage, and Other Equipment Spaces. At least one switch-controlled lighting outlet must be installed in spaces that are used for storage or contain equipment that requires servicing, such as attics, underfloor spaces, utility rooms, and basements. Examples of equipment that may need servicing are furnaces, air conditioners, heat pumps, sump pumps, and so on. The control is permitted to be a wall switch or a lighting outlet that has an integral switch such as a pull chain switch. The lighting outlet must be at or near the equipment that may require servicing. The control must be at the usual point of entrance to the space. See *NEC 210.70(A)(3)*.

The Number of Branch Circuits Required and How to Determine the Number of Lighting and Receptacle Outlets to Be Connected to One Branch Circuit

This chapter explains the *NEC* rules on where receptacle and lighting outlets are to be installed. The question of *how many* receptacle outlets and lighting outlets should be connected to each branch circuit is covered in detail in Chapter 8 of this text, where we begin the actual layout of the wiring for the residence discussed throughout this text.

Common Areas in Two-Family or Multifamily Dwellings

NEC 210.25 contains specific requirements for circuits that serve common areas of two-family or multifamily dwellings. Be careful when wiring a two-family or multifamily dwelling. For areas such as a common entrance, common basement, or common laundry area, the branch circuits for lighting, central fire/smoke/security alarms, communication systems, or any other electrical requirements that serve the common area *shall not* be served by circuits from any individual dwelling unit's power service. Common areas will require separate circuits, associated panels, and metering equipment. The reasoning for this requirement is that power could be lost in one occupant's electrical system, resulting in no lights or alarms for the other tenant(s) using the common shared area.

For example, a two-family dwelling would probably have two separate stairways to the basement. If only one stairway is provided to serve the basement of a two-family dwelling, then lighting would have to be provided and served from a branch circuit originating from each tenant's electrical system so that each tenant would have control of the lighting. For a common entrance, two luminaires would be installed and connected to each tenant's electrical system.

Likewise, fire/smoke/security systems in a two-family dwelling shall be installed and connected to branch circuits originating from each tenant's electrical panel. This becomes an architectural design issue. This is not generally a problem in two-family dwellings but can become a problem in dwellings having more than two tenants. Rather than having to install a third watt-hour meter and associated electrical equipment on a two-family dwelling, the design of the structure can be such that there are no common areas, Figure 3-8. A fourplex will probably require a fifth watt-hour meter, four watt-hour meters for the tenants and one watt-hour meter to serve common area lighting and the fire/smoke/security systems.

*Reprinted with permission from NFPA 70-2014.

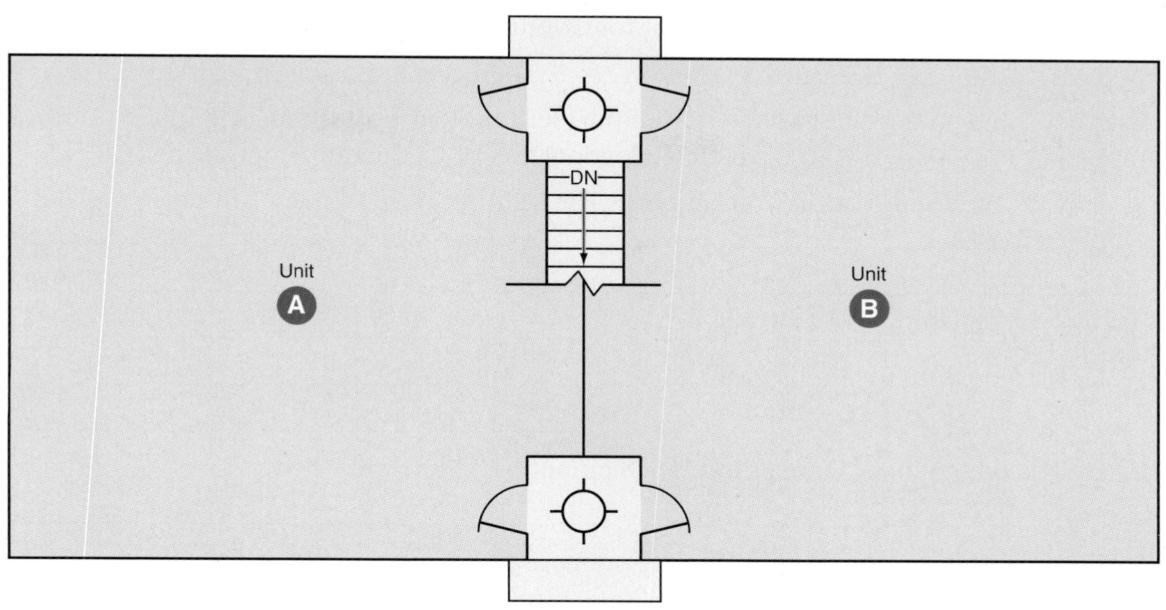

FIGURE 3-8 This outline of a two-family dwelling shows a common rear entry, a common stairway to the basement, and a common front entry. Note that each entry has one luminaire. These common area luminaires are not permitted to be connected to either of the tenants' electrical systems in the individual dwelling units, *210.25.* The choices are (1) the architect/designer must redo the layout of the building to eliminate the common entrances; (2) two luminaires must be provided in each entry, with each luminaire connected to the respective individual tenant's electrical circuit; or (3) a third meter and panel must be installed to supply the electrical loads in the common areas.

REVIEW

Note: For the following problems, you may refer to the *NEC* to obtain the answers. To help you find these answers, references to *Code* sections are included after some of the problems. These references are not given in later reviews, where it is assumed that you are becoming familiar with the *Code*.

1. What is the meaning of calculated load? ______________________________

____

2. How are branch circuits rated? See *NEC 210.3.* ______________________________

____

3. How is the rating of the branch-circuit protective device affected when the conductors used are of a larger size than called for by the *Code*? See *NEC 210.3.* ______________

____

4. What dimensions are used when measuring the area of a building? See *NEC 220.12.*

____

____

5. What spaces are not included in the floor area when calculating the load in volt-amperes per square foot? See *NEC 220.12.* ______________________________

____

6. What is the unit load per square foot for dwelling units? See *Table 220.12.* ________

____

7. According to *NEC 210.50(C)*, a laundry equipment outlet must be placed within ______________ ft (m) of the intended location of the laundry equipment.

8. How is the total load in volt-amperes for lighting purposes determined? See *NEC 220.12.* ____

____

____

9. How is the total lighting load in amperes determined? ________________

____

10. How is the required number of branch circuits determined? ______________

____

11. What is the minimum number of 15-ampere lighting branch circuits required if the dwelling has an occupied area of 4000 ft^2 (368 m^2)? Show all calculations.

12. How many lighting branch circuits are provided in this dwelling? ______________

    a. What is the minimum load allowance for small-appliance circuits for dwellings? See *NEC 220.52(A).* ________________________________

    b. An individual 15-ampere branch circuit is run to the receptacle outlet behind the refrigerator instead of connecting it to one of the two 20-ampere small-appliance branch circuits that are required in kitchens. For this separate circuit, an additional 1500 volt-amperes (shall) (does not have to) be added to the load calculations for dwellings. (Circle the correct answer.)

13. What is the smallest size wire that can be used in a branch circuit rated at 20 amperes?

____

14. How is the load determined for outlets supplying specific appliances? See *NEC 220.14.* ____

____

15. What type of circuits must be provided for receptacle outlets in the kitchen, pantry, dining room, and breakfast room? See *NEC 210.11(C)(1).* ______________

____

16. How is the minimum number of receptacle outlets determined for most occupied rooms? See *NEC 210.52(A).* ________________________________

____

____

____

____

17. In a single-family dwelling, how is overcurrent protection for branch circuits required to be provided? See *NEC 210.20.* ______________________________

____

____

____

____

18. The *NEC* in *Article 100* defines a *continuous load* as, A load where the maximum current is expected to continue for three hours or more.* According to *210.19(A)* and *210.20(A)*, the branch-circuit conductors and overcurrent protection for a continuous load shall be sized at not less than (100%) (125%) (150%) of the continuous load. (Circle the correct answer.)

19. The minimum number of outdoor receptacles for a residence is ______________, *NEC* ______________. State the location. ______________

20. The *Code* indicates the rooms in a dwelling that are required to have switched lighting outlets or switched receptacles. Write "yes" (switch required) or "no" (switch not required) for the following areas [see *NEC 210.70(A)*]:

    a. attic ______________

    b. stairway ______________

    c. crawl space (where used for storage) ______________

    d. hallway ______________

    e. bathroom ______________

    f. clothes closet ______________

21. Is a receptacle required in a bedroom on a 3 ft (900 mm) wall space behind the door? The door is normally left open. ______________

22. The *Code* requires that at least one 20-ampere circuit feeding a receptacle outlet must be provided for the laundry. May this circuit supply other outlets? ______________

23. In a basement, at least one receptacle outlet must be installed in addition to the receptacle outlet installed for the laundry equipment. This additional receptacle outlet and any other receptacle outlets in unfinished areas must be ______________ protected.

24. *NEC 210.8(A)(3)* requires all receptacles installed outdoors to have GFCI protection. Explain the exception and describe the conditions of the exception. ______________

____

____

____

____

25. Define a branch circuit. ____

____

26. Although the *Code* contains many exceptions to the basic overcurrent protection requirements for conductors, in general, the rating of the branch-circuit overcurrent device must (not be less than) (not be more than) the ampacity of a conductor. (Circle the correct answer.)

27. The rating of a branch circuit is based on (Circle the correct answer.)

    a. the rating of the overcurrent device.

    b. the length of the circuit.

    c. the branch-circuit wire size.

28. a. A 25-ampere branch-circuit conductor is derated to 70%. It is important to provide proper overcurrent protection for these conductors. The derated conductor ampacity is ____

    ____

    b. If the connected load is a "fixed" nonmotor, noncontinuous load, the branch-circuit overcurrent device may be sized at (20) (25) amperes. (Circle the correct answer.)

    c. If the above circuits supply receptacle outlets, the branch-circuit overcurrent device must be sized at not over (15) (20) (25) amperes. (Circle the correct answer.)

29. Small-appliance receptacle outlets are (included) (not included) in the 3-volt-ampere-per-square-foot calculations. (Circle the correct answer.)

30. If a homeowner wishes to have a switched receptacle outlet for a swag lamp in a dining room, may this switched receptacle outlet be considered to be one of the receptacle outlets required for the 20-ampere small-appliance circuits? ____________________

31. How many receptacle outlets are required on a 13 ft (4.0 m) wall space between two doors? Refer to *NEC 210.52(H)*. Draw the outlets on the diagram.

32. Is a receptacle required in a hallway in a home when the hallway is 8 ft (2.5 m) long?

____

33. A split-level home has one stairway that has six risers between two levels of the home. Which of the following choices Meets *Code*? (Circle the correct answer.)

    a. Two three-way switches must be installed (one switch at the top of the stairs, one switch at the bottom of the stairs) to control the lighting for the stairway.

    b. One single-pole switch is permitted on either level to control the lighting for the stairway.

    c. No switches are necessary for the stairway lighting because the lighting on the upper level and the lower level provides enough illumination on the stairway.

34. A sliding glass door is installed in a family room that leads to an outdoor deck. The sliding door has one 4 ft (1.22 m) sliding section and one 4 ft (1.22 m) permanently mounted glass section. For inside the recreation room, what does the *Code* say about wall receptacle outlets relative to the receptacle's position near the sliding glass door?

____

____

____

____

____

____

____

____

____

35. When determining the location and number of receptacles required, fixed room dividers and railings (shall be) (need not be) considered to be wall space. (Circle the correct answer.)

36. No point along the wall line shall be more than 6 ft (1.8 m) from a receptacle outlet. This requirement (does apply) (does not apply) to unfinished residential basements. (Circle the correct answer.)

37. In the past, it was common practice to connect the lighting and receptacle(s) in a bathroom to the same circuit. Because of overloads caused by high-wattage grooming appliances, the *Code* now requires that these receptacle(s) be supplied by a separate 20-ampere branch circuit. Where in the *Code* is this requirement found? *NEC* ______________

38. If a residence has two bathrooms, the *Code* states:

    a. The receptacles in each bathroom must be connected to a separate 20-ampere branch circuit. For two bathrooms, this would mean two circuits. (True) (False) (Circle the correct answer.)

    b. The receptacles in both bathrooms are permitted to be served by the same 20-ampere branch circuit. (True) (False) (Circle the correct answer.)

39. In your own words, explain the GFCI exemptions for receptacle outlets installed in unfinished basements. Give some examples.

____

____

____

____

40. *NEC* ______________________ of the *Code* prohibits connecting lighting and/or fire/smoke/security systems to an individual tenant's electrical source where common (shared) areas are present, in which case the lighting provided for both tenants is connected on one tenant's electrical source.

41. When installing weatherproof outdoor receptacles high up under the eaves of a house intended to be used for plugging in deicing and snow-melting cable for the rain gutters, must these receptacles be GFCI protected for personnel protection? (Yes) (No) (Circle the correct answer.)

42. An individual 20-ampere branch circuit supplies a receptacle for a refrigerator. The receptacle shall be or is permitted to be

    a. a duplex receptacle.

    b. a single receptacle.

    Explain why you chose your answer. ______________________________

    ____

    ____

    ____

43. A balcony on the second floor of a new residence measures 4 ft (1.22 m) × 10 ft (3.05 m). The balcony is accessed by a sliding door. Is a receptacle required for this balcony? Circle the correct answer, and give the *NEC* section where the answer is found.

    Yes No *NEC* ______________________________

CHAPTER **4**

Conductor Sizes and Types, Wiring Methods, Wire Connections, Voltage Drop, and Neutral Conductor Sizing for Services

OBJECTIVES

After studying this chapter, you should be able to

- determine the current-carrying capacity (ampacity) of conductors.
- understand overcurrent protection for conductors and maximum loading of branch circuits.
- understand aluminum conductors and the possible fire hazards if they are not properly installed.
- know the *NEC* installation requirements for all types of cables and raceways.
- understand the special ampacity ratings of service-entrance conductors.
- make voltage-drop calculations.
- learn an alternate cost and time-saving method permitted to bring nonmetallic-sheathed cables into the top of a surface-mounted panel.

CONDUCTORS

Throughout this text, all references to conductors are for copper conductors, unless otherwise stated.

Wire Size

The copper wire used in electrical installations is graded for size according to the American Wire Gauge (AWG) Standard. The wire diameter in the AWG standard is expressed as a whole number. The higher the AWG number, the smaller the wire. AWG sizes vary from fine, hairlike wire used in coils and small transformers to very large diameter wire required in industrial wiring to handle heavy loads.

The wire may be a single strand (solid conductor), or it may consist of many strands. Each strand of wire acts as a separate conducting unit. The wire size used for a circuit depends on the maximum current to be carried. The *NEC*® in *Table 210.24* shows that the minimum conductor size for branch-circuit wiring is 14 AWG. Smaller size conductors are permitted for bell wiring, thermostat wiring, communications wiring, intercom wiring, luminaire wires, and similar low-energy circuits.

Don't Get Confused!

In the past, conductor size was shown as, for example, "No. 12 AWG." This same conductor now appears in the *NEC* as "12 AWG." The "12" is a size, not a quantity.

Table 4-1 shows typical applications for different size conductors.

TABLE 4-1

Conductor applications chart.

CONDUCTOR SIZE	OVERCURRENT PROTECTION	TYPICAL APPLICATIONS (Check wattage and/or ampere rating of load to select the correct size conductors based on *Table 310.15(B)(16)*.)
20 AWG	Class 2 circuit transformers provide overcurrent protection; see *Article 725*.	Telephone wiring is usually 20 or 22 AWG.
18 AWG	Class 2 circuit transformers provide overcurrent protection; see *Article 725*. For motor control circuits, 7 amperes, see *Table 430.72(B)*.	Low-voltage wiring for thermostats, chimes, security, remote control, home automation systems, etc. For these types of installations, 18 or 20 AWG conductors can be used depending on the connected load and length of circuit.
16 AWG	Class 2 circuit transformers provide overcurrent protection; see *Article 725*. For motor control circuits, 10 amperes, see *Table 430.72(B)*.	Same applications as above. Good for long runs to minimize voltage drop.
14 AWG	15 amperes	Typical lighting branch circuits.
12 AWG	20 amperes	Small-appliance branch circuits for the receptacles in kitchens and dining rooms. Also laundry receptacles, bathroom and workshop receptacles. Often used as the "home run" for lighting branch circuits. Some water heaters.
10 AWG	30 amperes	Most clothes dryers, built-in ovens, cooktops, some central air conditioners, some water heaters, some heat pumps.
8 AWG	40 amperes	Ranges, ovens, heat pumps, some large clothes dryers, large central air conditioners, heat pumps.
6 AWG	50 amperes	Electric ranges, electric furnaces, heat pumps.
4 AWG	70 amperes	Electric furnaces, feeders to subpanels.
3 AWG and larger	100 amperes	Main service-entrance conductors, feeders to subpanels, electric furnaces.

Solid and Stranded Conductors

For residential wiring, sizes 14, 12, and 10 AWG conductors are generally solid when the wiring method is nonmetallic-sheathed cable or armored cable.

In a raceway, the preference is to use stranded 10 AWG conductors because of their flexibility and ease of handling and pulling.

NEC 310.106(C) requires that unless permitted elsewhere in the *Code*, conductors 8 AWG and larger must be stranded where installed in a raceway.

Ampacity

Ampacity means **The maximum current, in amperes, that a conductor can carry continuously under the conditions of use without exceeding its temperature rating.*** This value depends on the conductor's cross-sectional area, whether the conductor is copper or aluminum, and the type of insulation around the conductor. Ampacity values, also referred to as current-carrying capacity, are found in *Article 310*. The most commonly used conductor ampacities are found in *Table 310.15(B)(16)*.

The allowable ampacity values in the tables are valid where there are no more than three current-carrying conductors in a raceway or cable and where the temperature does not exceed 86°F (30°C). These are considered the *conditions of use* for the conductors.

If there are more than three current-carrying conductors in a raceway or cable, the allowable ampacity values are adjusted according to the factors in *Table 310.15(B)(3)(a)*.

If the ambient temperature exceeds 86°F (30°C), the allowable ampacity values are corrected according to the factors found in *Table 310.15(B)(2)(a)*.

If both conditions are present, then both adjustment and correction factors must be applied. More on conductor ampacities, derating, adjusting, and correction factors is presented in Chapter 18 of this text.

Conductors must have an ampacity not less than the maximum load that they are supplying, as shown in Figure 4-1. All conductors of a specific branch circuit must have an ampacity of the branch circuit's rating, as shown in Figure 4-2. There are exceptions to this rule, such as taps for electric ranges (see Chapter 20 of this text.)

*Reprinted with permission from NFPA 70-2014.

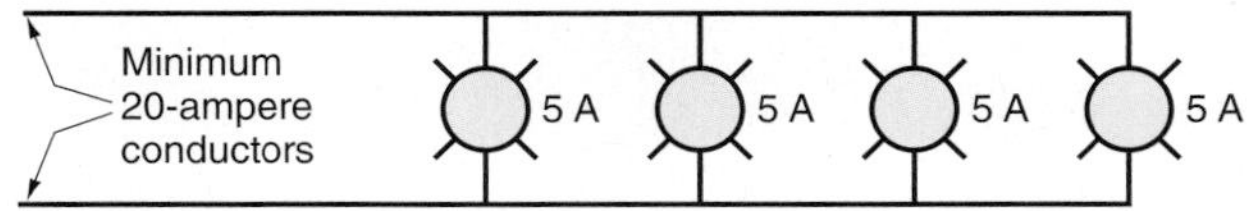

FIGURE 4-1 Branch-circuit conductors must have an ampacity not less than the maximum load to be served, *210.19(A)(1)*. See *210.23* for permissible loads.

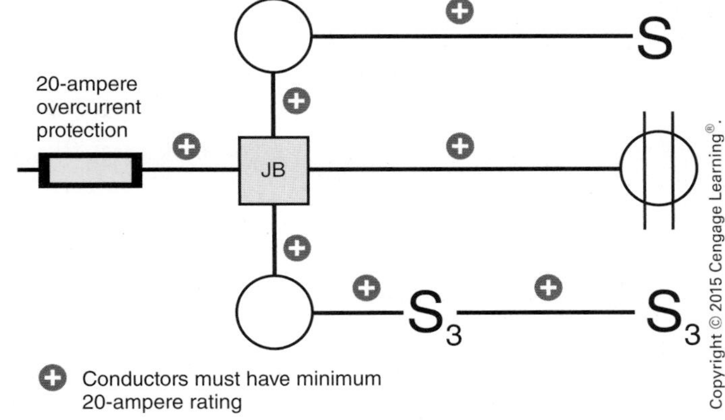

FIGURE 4-2 All conductors in this branch circuit are required to have an ampacity not lower than the rating of the branch circuit. In this example, 20-A conductors must be used including for switch-legs and travelers. See *210.19(A)(2)* and *Table 210.24*.

Ampacity of Flexible Cords

Table 4-2 shows the allowable ampacities for some sizes of flexible cords. Refer to *Table 400.5(A)(1) and (A)(2)* in the *NEC* for other specific types and sizes of flexible cords.

Overcurrent protection requirements for extension cords are found in *NEC 240.5*.

Conductor Sizing

The diameter of conductors is usually given in a unit called a mil. A *mil* is defined as one-thousandth of an inch (0.001 inch). Mils squared are known as circular mils.

Table 8 in *Chapter 9* of the *NEC* clearly shows that conductors are expressed in AWG numbers

TABLE 4-2

Allowable ampacities flexible cords.

CONDUCTOR SIZE AWG COPPER	CORDS IN WHICH THREE CONDUCTORS CARRY CURRENT Example: black, white, red, plus equipment grounding conductor	CORDS IN WHICH TWO CONDUCTORS CARRY CURRENT Example: black, white, plus equipment grounding conductor
18	7	10
16	10	13
14	15	18
12	20	25
10	25	30
8	35	40
6	45	55
4	60	70

Allowable ampacities for some of the more common flexible cords used for residential applications. These values were taken from *Table 400.5(A)(1), NEC*. Refer to the *NEC* table for other specific types and sizes.

from 18 (1620 circular mils) through 4/0 (211,600 circular mils). Wire sizes larger than 4/0 are expressed in circular mils.

Large conductors, such as 500,000 circular mils, are generally expressed as 500 kcmil. Because the letter "k" designates 1000, the term *kcmil* means "thousand circular mils." This is much easier to express in both written and verbal terms.

Older texts used the term *MCM*, which also means "thousand circular mils." The first letter "M" refers to the Roman numeral that represents 1000. Thus, 500 MCM means the same as 500 kcmil. Roman numerals are no longer used in the electrical industry for expressing conductor sizes.

Overcurrent Protection for Conductors

Conductors must be protected against overcurrent by fuses or circuit breakers rated not more than the ampacity of the conductors, *NEC 240.4*.

NEC 240.4(B) permits the use of the next higher standard overcurrent device rating, as shown in *240.6(A)*. This permission is granted *only* when the overcurrent device is rated 800 amperes or less. For example, from *Table 310.15(B)(16)*, a 6 AWG conductor with Type THWN insulation rated for 75°C has an allowable ampacity of 65 amperes. In *240.6(A)*, we find that the next higher standard rating for an overcurrent device is 70 amperes which is permitted to be used to provide overcurrent protection for the conductor with an allowable ampacity of 65 amperes. However, the conductor is not permitted to carry more than 65 amperes. Nonstandard ampere ratings of fuses and breakers are also permitted.

TABLE 4-3

Maximum overcurrent protection for 14, 12, and 10 AWG copper conductors.

CONDUCTOR SIZE	MAXIMUM OVERCURRENT PROTECTION
14 AWG copper	15 amperes
12 AWG copper	20 amperes
10 AWG copper	30 amperes

NEC 240.4(D) spells out the maximum overcurrent protection for small branch-circuit conductors. Table 4-3 illustrates the requirements of *240.4(D)*.

There are exceptions for motor branch circuits and HVAC equipment. Motor circuits are discussed in Chapter 19 and Chapter 22. HVAC circuits are discussed in Chapter 23.

Branch-Circuit Rating

The rating of the overcurrent device (OCD) determines the rating of a branch circuit, as required in *NEC 210.3*. This, except for some motor circuit conductors, determines the minimum ampacity of

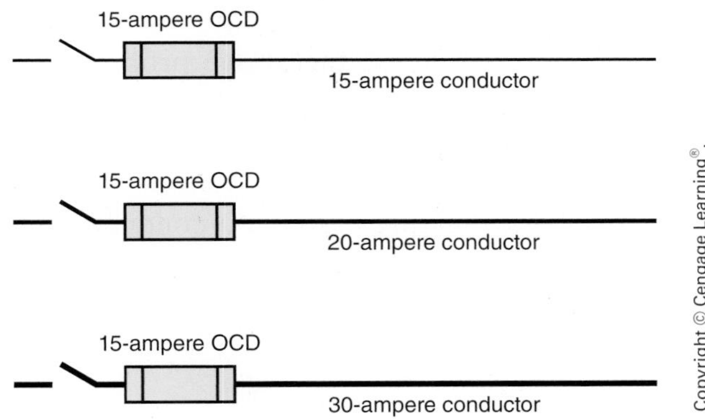

FIGURE 4-3 All three of these branch circuits are rated as 15-ampere circuits, even though larger conductors were used for some other reason, such as solving a voltage-drop problem. The rating of the overcurrent device (OCD) determines the rating of a branch circuit, *210.3*.

branch-circuit conductors. This concept is shown in Figures 4-2 and 4-3.

For example, switch leg and travelers for three- and four-way switch loops are required to have the ampacity of the branch-circuit rating even though the conductors may carry the load of a single luminaire.

PERMISSIBLE LOADS ON BRANCH CIRCUITS *(210.23)*

The *NEC* is very specific about the loads permitted on branch circuits. Here is a recap of these requirements.

- The load must not exceed the branch-circuit rating.
- The branch circuit must be rated 15, 20, 30, 40, or 50 amperes when serving two or more outlets, *NEC 210.3*.
- An individual branch circuit may supply any size load.
- For 15- and 20-ampere branch circuits
  a. they are permitted to supply lighting, other equipment, or both types of loads.
  b. the rating of any one cord-and-plug-connected utilization equipment is not permitted to exceed 80% of the branch-circuit rating.
  c. utilization equipment fastened in place, other than luminaires, is not permitted to exceed 50% of the branch-circuit rating if the branch circuit also supplies lighting, other cord-and-plug-connected equipment, or both types of loads.
- The 20-ampere small-appliance circuits in homes are not permitted to supply other loads, *NEC 210.11(C)(1)*.
- 30-ampere branch circuits may supply equipment such as dryers, cooktops, water heaters, and so forth. Cord-and-plug-connected equipment is not permitted to exceed 80% of the branch-circuit rating.
- 40- and 50-ampere branch circuits may supply cooking equipment that is fastened in place, such as an electric range, as well as HVAC equipment.
- Over 50-ampere-rated branch circuits are for electric furnaces, large heat pumps, air-conditioning equipment, large double ovens, and similar large loads.

ALUMINUM CONDUCTORS

The conductivity of aluminum is not as great as that of copper for a given size. For example, checking *Table 310.15(B)(16)*, an 8 AWG Type THHN copper conductor has an allowable ampacity of 55 amperes. An 8 AWG Type THHN aluminum or copper-clad aluminum conductor has an ampacity of 45 amperes.

In *240.4(D)*, the maximum overcurrent protection for a 12 AWG copper conductor is 20 amperes but only 15 amperes for a 12 AWG aluminum or copper-clad aluminum conductor.

Aluminum conductors have a higher resistance compared to a copper conductor of the same size. When considering voltage drop, a conductor's resistance is a key ingredient. Voltage-drop calculations are discussed later on in this chapter.

Common Connection Problems

Some common problems associated with aluminum conductors when not properly connected may be summarized as follows:

- A corrosive action is set up when dissimilar wires come in contact with one another if moisture is present.

- The surface of aluminum oxidizes as soon as it is exposed to air. If this oxidized surface is not penetrated, a poor connection results. When installing aluminum conductors, particularly in large sizes, an inhibitor (antioxidant) is brushed onto the aluminum conductor, and then the conductor is scraped with a stiff brush where the connection is to be made. The process of scraping the conductor breaks through the oxidation, and the inhibitor keeps the air from coming into contact with the conductor. Thus, further oxidation is prevented. Aluminum connectors of the compression type usually have an inhibitor paste already factory installed inside of the connector.
- Aluminum wire expands and contracts to a greater degree than does copper wire for an equal load. This is referred to as *creep* or *cold flow*. This factor is another possible cause of a poor connection. Crimp connectors for aluminum conductors are usually longer than those for comparable copper conductors, thus resulting in greater contact surface of the conductor in the connector.
- Older technology aluminum such as Alloy 1350 experienced the above problems. The newer technology aluminum such as AA-8000 has much better conductivity, creep resistance, and strength. Still, extreme care must be taken when terminating aluminum conductors. The manufacturer of the aluminum conductor or connector will provide detailed installation instructions.

Proper Installation Procedures

Proper, trouble-free connections for aluminum conductors require terminals, lugs, and/or connectors that are suitable for the type of conductor being installed.

Terminals on receptacles and switches must be suitable for the conductors being attached. Table 4-4 shows how the electrician can identify these terminals. Listed connectors provide proper connection when properly installed. See *NEC 110.14*, *404.14(C)*, and *406.3(C)*.

TABLE 4-4

Terminal identification markings and the acceptable types of conductors permitted to be connected to a specific type of terminal.

TYPE OF DEVICE	MARKING ON TERMINAL OR CONNECTOR	CONDUCTOR PERMITTED
15- or 20-ampere receptacles and switches	CO/ALR	Aluminum, copper, copper-clad aluminum
15- and 20-ampere receptacles and switches	None	Copper, copper-clad aluminum
30-ampere and greater receptacles and switches	AL/CU	Aluminum, copper, copper-clad aluminum
30-ampere and greater receptacles and switches	None	Copper only
Screwless pressure terminal connectors of the push-in type	None	Copper or copper-clad aluminum
Wire connectors	AL	Aluminum
Wire connectors	AL/CU or CU/AL	Aluminum, copper, copper-clad aluminum
Wire connectors	CC	Copper-clad aluminum only
Wire connectors	CC/CU or CU/CC	Copper or copper-clad aluminum
Wire connectors	CU or CC/CU	Copper only
Any of the above devices	COPPER OR CU ONLY	Copper only

The Importance of Making Proper Connections of Aluminum Conductors

Improper connections and terminations of aluminum conductors may cause connections to overheat, which may result in damage to the insulation and be the origination of a fire. Most often, the problem is not the aluminum conductors. The problem is with the connections and terminations.

Many of these problems occurred in the meter base and at the main panel where aluminum service entrance conductors were installed years ago. In the course of time, the connections failed. Failures also happened where aluminum conductors were terminated on receptacles and switches that were not rated for termination of the aluminum conductor.

The Consumer Product Safety Commission (CPSC) (Washington, DC, 20207) publishes an excellent free booklet (#516) entitled *Repairing Aluminum Wiring* that explains what you can do with connections and terminations for the small aluminum wires that were installed years ago for the branch-circuit wiring of receptacles and switches.

The CPSC states that older homes wired with aluminum conductors are 55 times more likely to have one or more connections that will reach "Fire Hazard Conditions" than are homes wired with copper.

Although the *NEC*, notably in *110.5*, *110.14*, *240.4(D)*, *310.106(B)*, *Table 310.15(B)(16)*, and other sections, recognizes aluminum conductors in sizes 12 AWG and larger, there are some local electrical codes that do not allow aluminum conductors to be used in smaller branch-circuit sizes. Check this out before using aluminum conductors.

Some of the signs of potential problems might be

- wall switches and receptacles that feel warm to the touch.
- the smell of burning plastic or rubber at switches, receptacle, main panel, or meter.
- flickering lights.
- lights getting bright or dim.
- circuits that don't work; some circuits are on, others are off.
- TV shifting to another channel for no apparent reason.
- appliances with electronic controls shutting off or changing settings for no apparent reason.
- security system sounding off for no apparent reason.

Old Houses Wired with Aluminum Conductors

If you come across a house wired with aluminum conductors, it was probably wired back in the mid-sixties or early seventies. You will very likely find problems where the conductors are terminated on switches and receptacles. Take special care when changing out switches and receptacles or making splices. Use wiring devices that bear the AL/CU marking. You can also use special "pigtails" designed to make the connection between aluminum and copper conductors and terminals. One such connector is COPALUM—a compression (crimp) connector. Apply an antioxidant to the splice or connection as recommended by the manufacturer of the connector.

The AlumiConn connector is a multi-barrel lug-style wire connector designed as a permanent repair for aluminum wiring in homes and commercial buildings. It is specifically designed for terminating aluminum conductors or for pigtailing a copper conductor from the branch circuit for connection to wiring devices. As an aluminum wire is inserted into the AlumiConn lug, it is coated with a thin layer of silicone grease to provide resistance from oxidation. When the set screw is tightened down it compresses the surface of the wire, which breaks up any surface oxides and provides a secure mechanical connection. In addition, the AlumiConn connector itself is constructed using tin-plated aluminum, which relieves any issues related to dissimilar metals. AlumiConn is UL Listed to meet US and Canadian specifications, declared dependable and economical by independent tests, and now approved by the CPSC for repairing aluminum wiring.

Courtesy of King Innovations.

FIGURE 4-4 A wire connector specifically designed for pigtailing.

Remember, it is not acceptable to connect copper and aluminum conductors in a wire connector where the different metals will be in direct contact with each other unless the connector is specifically listed and marked as being suitable for that type connection.

Wire Connections

Wire connectors are known in the trade by such names as *screw terminal, pressure terminal connector, wire connector, Wing-Nut®, Wire-Nut®, Scotchlok®, Twister®, split-bolt connector, pressure cable connector, solderless lug, soldering lug, solder lug*, and others.

All cartons and manufacturers' literature show the size, number, and types of conductors permitted.

Solder-type lugs and connectors are rarely, if ever, used today. In fact, connections that depend on solder are *not* permitted for connecting service-entrance conductors to service equipment, *230.81*. Nor is solder permitted for grounding and bonding connections, *250.8*. The labor costs and time spent make the use of the solder-type connections prohibitive.

Solderless connectors designed to establish *Code*-compliant electrical connections are shown in Figure 4-5.

As with the terminals on wiring devices (switches and receptacles), wire connectors must be marked "AL" when they are to be used with aluminum conductors. This marking is found on the connector itself, or it appears on or in the shipping carton.

Connectors marked "AL/CU" are suitable for use with aluminum, copper, or copper-clad aluminum conductors. This marking is found on the connector itself, or it appears on or in the shipping carton.

Connectors not marked "AL" or "AL/CU" are for use with copper conductors only.

Unless specially stated on or in the shipping carton, or on the connector itself, conductors made of copper, aluminum, or copper-clad aluminum *may not* be used in combination in the same connector.

When combinations are permitted, the connector will be identified for the purpose and for the conditions when and where they may be used. The conditions usually are limited to dry locations only. There are some "twist-on" wire connectors that have recently been listed for use with combinations of copper and aluminum conductors.

The preceding data are found in *110.14* and in the Underwriters Laboratories (UL) Standards. When terminating large conductors such as service-entrance conductors, be sure to read any instructions that might be included with the equipment where these conductors will be terminated.

Terminations for all electrical connections to devices and equipment are required to be tightened to the torque as required by the manufacturer of such electrical device or equipment, *NEC 110.3(B)*. The tightening torque value is usually marked on the equipment, the carton the connector is shipped in or on the connector itself if it is large enough.

In the absence of a marked torque value, use *NEC Informative Annex I, Recommended Tightening Torque Tables from UL Standard 486-B*.

Connections can fail because they are not tightened correctly.

Connectors used to connect wires together on combinations of 18 AWG through 6 AWG conductors. They are twist-on, solderless, and tapeless. *Wire-Nut,® Wing-Nut,® and Twister® are registered trademarks of Ideal Industries, Inc. Scotchlok® is a registered trademark of 3M.	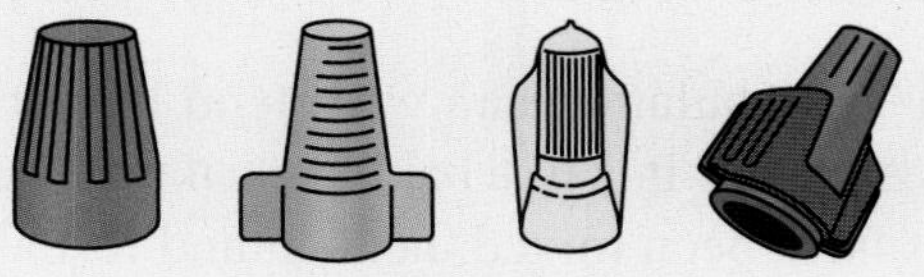 Wire connectors variously known as Wire-Nut, Wing-Nut, Twister, and Scotchlok
Connectors used to connect wires together in combinations of 16, 14, and 12 AWG conductors. They are crimped on with a special tool, then covered with a snap-on insulating cap.	 Crimp-type wire connector and insulating cap
Solderless pressure connectors are available in sizes 4 AWG through 1000 kcmil conductors. They are used for one solid or one stranded conductor only, unless otherwise noted on the connector or on its shipping carton. The screw may be of the standard screwdriver slot type, or it may be for use with an allen wrench or socket wrench.	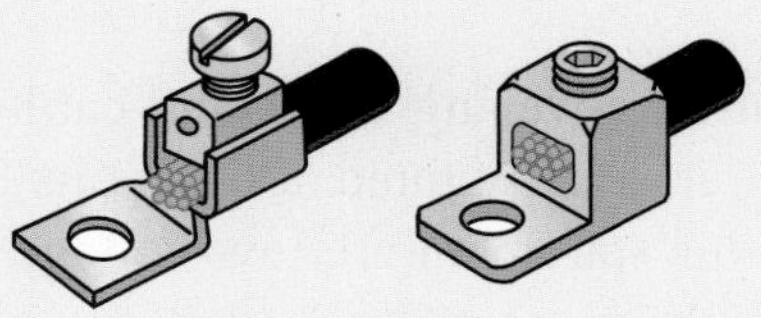 Solderless connectors
Compression connectors are used for 8 AWG through 1000 kcmil conductors. The wire is inserted into the end of the connector, then crimped on with a special compression tool.	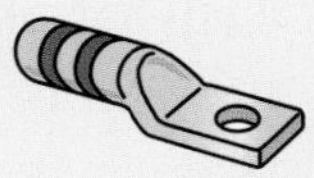 Compression connector
Split-bolt connectors are used for connecting two conductors together, or for tapping one conductor to another. They are available in sizes 10 AWG through 1000 kcmil. They are used for two solid and/or two stranded conductors only, unless otherwise noted on the connector or on Its shipping carton.	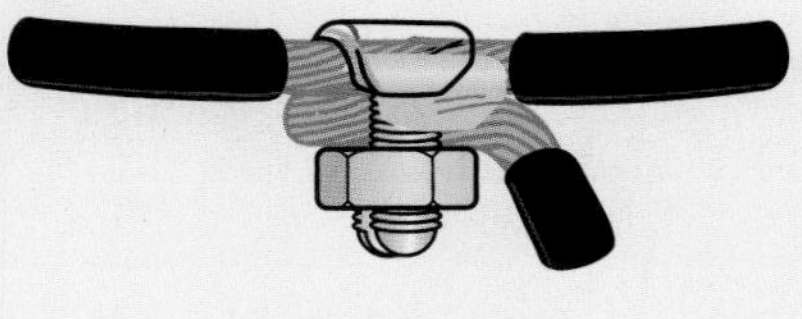 Split-bolt connector

FIGURE 4-5 Types of wire connectors.

CONDUCTOR INSULATION

The *NEC* requires generally that all conductors be insulated, *310.106(D)*. There are a few exceptions, such as the permission to use a bare neutral conductor for services, and bare equipment grounding conductors.

NEC Table 310.104(A) shows many types of conductors, their applications, and insulations. The conductors most commonly used fall into the thermoplastic and thermoset categories. ▶Conductor insulations are typically rated 600 volts but are permitted to be rated up to 1000 volts if listed and marked.◀

What Is Thermoplastic Insulation?

Thermoplastic insulation is the most common. It is like chocolate. It will soften and melt if heated above its rated temperature. It can be heated, melted, and reshaped. Thermoplastic insulation will stiffen at temperatures colder than 14°F (−10°C). Typical examples of thermoplastic insulation are Types THHN, THHW, THW, THWN, and TW.

What Is Thermoset Insulation?

Thermoset insulation can withstand higher and lower temperatures. It is like baking a cake. Once the ingredients have been mixed, heated, and formed, it can never be reheated and reshaped. If heated above its rated temperature, it will char and crack. Typical examples of thermoset insulation are Types RHH, RHW, XHH, and XHHW.

Table 310.104(A) lists the various conductor insulations and their applications.

The allowable ampacities of conductors are given in *Table 310.15(B)(16)* for various types of insulation.

The insulation covering wires and cables used in house wiring is usually rated at 600 volts or less. Exceptions to this statement are low-voltage wiring and luminaire wiring.

Older Types of Conductor Insulation

Although Type THHN/THWN is the most common building wire used today, older types of conductor insulation might still be found, such as Types A, RH, RU, RUH, RUW, T, TW, and THW. Some of these are still shown in the *NEC* but may no longer be manufactured. Some are still manufactured but are difficult, if not impossible, to find at an electrical supply house. Others are available only as a special-order product.

WET, DAMP, DRY, AND SUNLIGHT LOCATIONS

Conductors are listed for specific locations. Be sure that the conductor you use is suitable for the location. Note that the construction of cables often used for wiring residences limits their use to dry locations. See *NEC 334.10* for uses permitted for Type NM cable and *334.12* for uses not permitted. Table 4-5 shows some of the more typical conductors used in house wiring. Here are some definitions taken directly from the *NEC:*

- **Damp Location.** Locations protected from weather and not subject to saturation with water or other liquids, but subject

TABLE 4-5

Typical conductors used for residential wiring.

TRADE NAME	TYPE LETTER	MAXIMUM OPERATING TEMP.	APPLICATION PROVISIONS	INSULATION	AWG	OUTER COVERING
Heat-resistant thermoplastic	THHN	90°C	Dry and damp locations	Flame-retardant, heat-resistant thermoplastic	14–1000 kcmil	Nylon jacket or equivalent
Moisture- and heat-resistant thermoplastic	THHW	75°C 90°C	Wet location Dry location	Flame-retardant, moisture- and heat-resistant thermoplastic	14–1000 kcmil	None
Moisture- and heat-resistant thermoplastic	THWN Note: If marked THWN–2, okay for 90°C in dry or wet locations	75°C	Dry and wet locations	Flame-retardant, moisture- and heat-resistant thermoplastic	14–1000 kcmil	Nylon jacket or equivalent
Moisture- and heat-resistant thermoplastic	THW Note: If marked THW–2, okay for 90°C in dry or wet locations	75°C	Dry and wet locations	Flame-retardant, moisture- and heat-resistant thermoplastic	14–2000 kcmil	None
Underground feeder and branch-circuit single conductor or multiconductor cable. See *Article 339*	UF	60°C 75°C	See *Article 339* See *Article 339*	Moisture-resistant Moisture-resistant	14 through 4/0	Integral with insulation

to moderate degrees of moisture. Examples of such locations include partially protected locations under canopies, marquees, roofed open porches, and like locations; and interior locations subject to moderate degrees of moisture, such as some basements, some barns, and some cold-storage warehouses.* (See Figure 11-4.)

- **Dry Location.** A location not normally subject to dampness or wetness. A location classified as dry may be temporarily subject to dampness or wetness, as in the case of a building under construction.*
- **Wet Location.** Installations underground or in concrete slabs or masonry in direct contact with the earth; in locations subject to saturation with water or other liquids, such as vehicle washing areas; and in unprotected locations exposed to weather.*

  The interior of raceways installed in wet locations is considered a wet location. Insulated conductors and cables in these locations shall be listed for wet locations. See *NEC 300.5(B)*, *300.9*, and *310.10(C)*. Also see Figure 11-4.
- **Locations Exposed to Direct Sunlight.** Insulated conductors and cables used where exposed to direct rays of the sun shall be listed, or listed and marked as being "sunlight resistant." A common application for this requirement is exposed conductors at service masts where the conductors are extended to connect to the service drop.

See *NEC 310.10(D)*, *310.15(A)(3)*, *310.104*, and *Table 310.104(A)* for additional information.

Conductor Insulation Temperature Ratings

Conductors are also rated as to the temperature their insulation system can withstand. Even though the *NEC* provides Fahrenheit temperatures for conductors, they are not used by the electrical industry. Only the Celsius temperatures are used in referring to the temperature rating of insulation.

*Reprinted with permission from NFPA 70-2014.

The temperature rating of conductor insulations commonly used in residential wiring is as follows:

	Celsius	Fahrenheit
Type TW	60°	140°
Type THWN	75°	167°
Type THHN	90°	194°

Without question, conductors with Type THHN/THWN insulation are the most popular and most commonly used, particularly in the smaller sizes, because of their small diameter (easy to handle; more conductors of a given size permitted in a given size raceway) and their suitability for installation where high temperatures are encountered, such as in attics; buried in insulation; and supplying recessed fixtures.

The temperature rating of nonmetallic-sheathed cable is 90°C. However, *NEC 334.80* states that the allowable ampacity is that of the 60°C column in *Table 310.15(B)(16)*. The 90°C ampacity is permitted to be used for derating purposes.

Multiple-Type Designations

Some conductors are listed for more than one application. For example, a conductor marked THHN/THWN is rated 600 volts, 90°C when used in dry locations, *and* 600 volts, 75°C when used in wet locations. As always, read the surface marking on the conductor insulation, the tag, the reel, or the carton. The *NEC* covers *Conductors for General Wiring* in *Article 310*.

The Weakest Link of the Chain

A conductor has two ends!

Selecting a conductor ampacity based solely on the allowable ampacity values found in *Table 310.15(B)(16)* is a very common mistake that can prove costly.

In any given circuit, we have an assortment of electrical components that have different maximum temperature ratings. Our job is to find out which component is the "weakest link" in the system and do our circuit design based on the lowest rated component in the circuit. Maximum temperature ratings are found in the *NEC* and in the UL Standards.

- **Circuit breakers:** UL 489. Generally marked "75°C Only" or "60°C/75°C."

- **Conductors:** *Table 310.104(A)*, *Table 310.15(B)(16)*, and UL 83.
- **Disconnect switches:** UL 98. Generally marked "75°C Only" or "60°C/75°C."
- **Panelboards:** UL 67. Panelboards are generally marked 75°C. The temperature rating of a panelboard has been established as an assembly with circuit breakers in place. Do not use the circuit breakers' temperature marking by itself. It is the temperature marking on the assembly that counts.
- **Receptacles and attachment-plug caps:** UL 498. Most 15- and 20-ampere branch circuits have receptacles that are marked with a wire size and not a temperature rating. As a result, the default temperature rating in *NEC 110.14(C)* applies. This results in a maximum temperature rating of 60°C. Some 30-ampere receptacles are rated 60°C and some are rated for 75°C. Larger 50-ampere receptacles are rated 75°C. You will want to verify the marking on the device.
- **Snap switches:** The safety standard is UL 20. Most 15- and 20-ampere branch circuits have snap switches with a maximum terminal temperature rating of 60°C.
- **Wire splicing devices and connectors:** Wire-Nuts, Wing-Nuts, Twisters, and Scotchloks for copper conductors (UL 486) and for aluminum conductors (UL 486B). These types of connectors are commonly rated 105°C.

Table 4-6 indicates the correct applications for conductors and terminations.

Another key *NEC* requirement is found in *110.14(C)*, "*Termination Limitations*," where we find the following:

For circuits rated 100 amperes or less, or marked for conductor sizes 14 through 1 AWG, unless otherwise identified, the terminals on wiring devices, switches, breakers, motor controllers, and other electrical equipment are based on the ampacity of 60°C conductors. It is acceptable to install conductors having a higher temperature rating, such as 90°C THHN, but you must use the 60°C ampacity values.

For circuits rated over 100 amperes, or marked for conductors larger than 1 AWG, the conductors installed must have a minimum 75°C rating. The ampacity of the conductors is based on the 75°C values. It is acceptable to install conductors having a higher temperature rating, such as 90°C THHN, but you must use the 75°C ampacity values.

TABLE 4-6

Types of conductor temperature ratings permitted for various terminal temperature ratings in conformance to the *NEC* and the UL Standards.

TERMINATION RATING	CONDUCTOR INSULATION RATING		
	60°C	75°C	90°C
60°C	Okay	Okay (at 60°C ampacity)	Okay (at 60°C ampacity)
75°C	No	Okay	Okay (at 75°C ampacity)
60/75°C	Okay	Okay (at 60°C or 75°C ampacity)	Okay (at 60°C or 75°C ampacity)
90°C	No	No	Okay (only if equipment has a 90°C rating)

When using high-temperature insulated conductors where adjustment factors are to be applied, such as *derating* for more than three current-carrying conductors in a raceway or cable, or *correcting* where high temperatures are encountered, the final results of these adjustments (in amperes) must comply with the requirements of *110.14(C)*. Similar requirements are found in the UL Standards.

Examples of applying *derating factors* and *correcting factors* are found in Chapter 18.

Temperature ratings of conductors and cables are discussed later in this chapter.

In *240.4(D)*, there is a built-in code compliance limitation for ampacities of small branch-circuit conductors. Here we find that the maximum overcurrent protection is 15 amperes for 14 AWG, 20 amperes for 12 AWG, and 30 amperes for 10 AWG.

From *Table 310.15(B)(16)* we find that

- 14 AWG copper conductors' allowable ampacity is

  15 amperes in the 60°C column
  20 amperes in the 75°C column
  25 amperes in the 90°C column

- 12 AWG copper conductors' allowable ampacity is

  20 amperes in the 60°C column
  25 amperes in the 75°C column
  30 amperes in the 90°C column

- 10 AWG copper conductors' allowable ampacity is

  30 amperes in the 60°C column
  35 amperes in the 75°C column
  40 amperes in the 90°C column

There are exceptions to this maximum overcurrent protection rule for motor, air-conditioner, and heat pump branch circuits. These exceptions are covered elsewhere in this text.

In summary, we cannot arbitrarily use the ampacity values for conductors as found in *Table 310.15(B)(16)*. Conductor ampacity is not a "stand-alone" issue. We must also consider the temperature limitations of the equipment, such as panelboards, receptacles, snap switches, receptacles, connectors, and so on. The lowest maximum temperature-rated device in an electrical system is the weakest link, and it is the weakest link on which we must base our ultimate conductor ampacity and conductor insulation decision.

Knob-and-Tube Wiring

Older homes were wired with open, individual, insulated conductors supported by porcelain *knobs* and *tubes*. In its day, knob-and-tube wiring served its purpose well. This text does not cover knob-and-tube wiring because this method is no longer used in new construction. On occasion, it is necessary to make some modifications to knob-and-tube wiring when doing remodel work. The *NEC* covers this wiring method in *Article 394*.

Watch out when working on old knob-and-tube wiring. You will possibly find the old rubber insulation on the conductors dried out and brittle from the many years of heat generated by luminaires and from the extreme heat in attics, particularly when the conductors are completely buried in thermal insulation. The conductor insulation might fall apart when touched or moved. This is a potential fire hazard!

If you do find this situation, one fix is to slide insulation (heat shrink) tubing (available from electrical distributors) over the conductors to reinsulate them.

Figure 4-6 is a photo of typical knob-and-tube wiring from the 1930s.

FIGURE 4-6 Photograph of a typical knob-and-tube wiring from the 1930s.

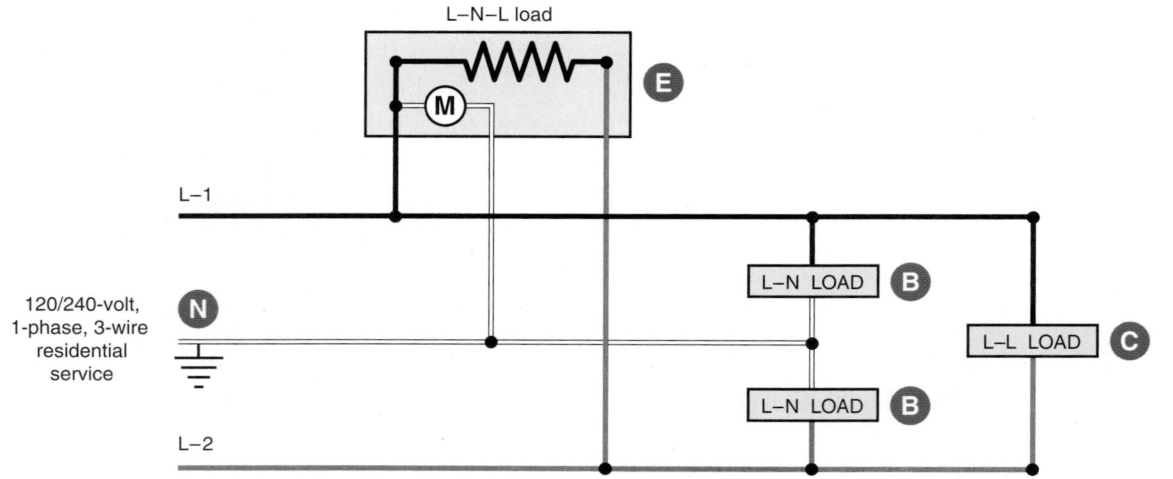

FIGURE 4-7 Diagram showing line-to-line and line-to-neutral loads.

Neutral Conductor Size

Figure 4-7 shows how *line-to-line* loads and *line-to-neutral* loads are connected. The grounded neutral (white) conductor (N) for residential services and feeders is permitted to be smaller than the "hot" ungrounded phase (black, red) conductors L1 and L2 only when the neutral conductor is properly and adequately sized to carry the maximum unbalance loads calculated according to *NEC 215.2*, *220.61*, and *230.42*.

NEC 215.2 relates to feeders and refers us to *Article 220* for calculation requirements.

NEC 230.42 relates to services and refers us to *Article 220* for calculation requirements.

Focusing on the neutral conductor, *220.61* states that The feeder or service neutral load shall be the maximum unbalance of the load determined by this Article.*

NEC 220.61 further states that The maximum unbalanced load shall be the maximum net calculated load between the neutral and any one ungrounded conductor.*

In a typical residence, we find a number of loads that carry little or no neutral current, such as an electric water heater, electric clothes dryer, electric oven and range, electric furnace, and air conditioner. Checking Figure 4-7, you can readily see that load (C) is connected to a 240-volt, 2-wire circuit and has no neutral current. Note that load (E), which is similar to internal connections in an electric clothes dryer, is connected to a 120/240-volt, 3-wire circuit, and in this hookup the120-volt motor will result in current flowing in the neutral conductor. Loads (B) are 120-volt loads that are connected line to neutral.

Thus we find logic in the *NEC*, which permits reducing the neutral conductor size on services and feeders where the calculations prove that the neutral conductor will be carrying less current than the "hot" phase conductors.

See Chapter 20 for sizing neutral conductors for electric range branch circuits and Chapter 29 for sizing neutral service-entrance conductors.

VOLTAGE DROP

Low voltage can cause lights to dim, some television pictures to "shrink," motors to run hot, electric heaters to not produce their rated heat output, and appliances to not operate properly.

Low voltage in a home can be caused by

- wire that is too small for the load being served.
- a circuit that is too long.
- poor connections at the terminals.
- conductors operating at high temperatures having higher resistance than when operating at lower temperatures.

A simple formula for calculating voltage drop on single-phase systems considers only the dc resistance of the conductors and the temperature of the conductor. See *Table 8*, *Chapter 9*, in the *NEC* for the dc resistance values. The more accurate formulas consider ac resistance, reactance, temperature, and spacing, in

*Reprinted with permission from NFPA 70-2014.

metal conduit and in nonmetallic conduit, *NEC Table 9, Chapter 9*. Voltage drop is covered in great detail in *Electrical Wiring—Commercial* 15th Edition (Cengage Learning). The simple voltage-drop formula is more accurate with smaller conductors and gets increasingly less accurate as conductor size increases. It is sufficiently accurate for voltage-drop calculations necessary for residential wiring.

To Find Voltage Drop in a Single-Phase Circuit

$$E_d = \frac{K \times I \times L \times 2}{CMA}$$

To Find Conductor Size for a Single-Phase Circuit

$$CMA = \frac{K \times I \times L \times 2}{E_d}$$

In the above formulae:

E_d = result of voltage drop calculation in volts.

K = approximate resistance in ohms per circular-mil foot at 75°C.
- For uncoated copper wire, use a K factor of 12.9.
- For aluminum wire, use 21.2.

I = current in amperes flowing through the conductors.

L = length in feet from beginning of circuit to the load.

CMA = cross-sectional area of the conductors in circular mils.**

We use the factor of 2 for single-phase circuits because there is voltage drop in both conductors to and from the connected load.

To Find Voltage Drop and Conductor Size for a 3-Phase Circuit

Although residential wiring does not commonly use 3-phase systems, commercial and industrial systems do. To make voltage-drop and conductor sizing calculations for a 3-phase system, substitute the factor of 1.732 (the square root of 3) in place of the factor 2 in the voltage-drop formulae.

Code References to Voltage Drop

For branch circuits, refer to *NEC 210.19(A), Informational Note No. 4*. The recommended maximum voltage drop is 3%, as shown in Figure 4-8.

For feeders, refer to *NEC 215.2(A)(1), Informational Note No. 2*. The recommended maximum voltage drop is 3%. See Figure 4-9.

When both branch circuits and feeders are involved, the total voltage drop should not exceed 5%. This is shown in Figure 4-9.

According to *NEC 90.5(C), Informational Notes* are explanatory only and are not to be enforced as *Code* requirement.

There is nothing in the *NEC* that dictates what the incoming voltage to a home must be. Incoming voltage is really determined by the electric utility and is affected by the type, size, and length of the service conductors, transformer(s), and primary lines. Electric utilities come under the jurisdiction of the *National Electrical Safety Code (NESC)*. A local public service commission, usually at the state level, mandates maximum over- and under-voltages. This is generally within the 5% range of nominal voltage.

Examples of Voltage-Drop Calculations

EXAMPLE

What is the approximate voltage drop on a 120-volt, single-phase circuit consisting of 14 AWG copper conductors where the load is 11 amperes and the distance of the circuit from the panel to actual load is 85 ft (25.91 m)?

Solution:

$$E_d = \frac{K \times I \times L \times 2}{CMA}$$

$$E_d = \frac{12.9 \times 11 \times 85 \times 2}{4110}$$

$$E_d = 5.87 \text{ volts drop}$$

Note: Refer to Table 4-7 for the CMA value.

This exceeds the voltage drop information in the Code, which is

3% of 120 volts = 3.6 volts

****Note:** From Table 4-7 in this text or from *Table 8, Chapter 9, NEC*.

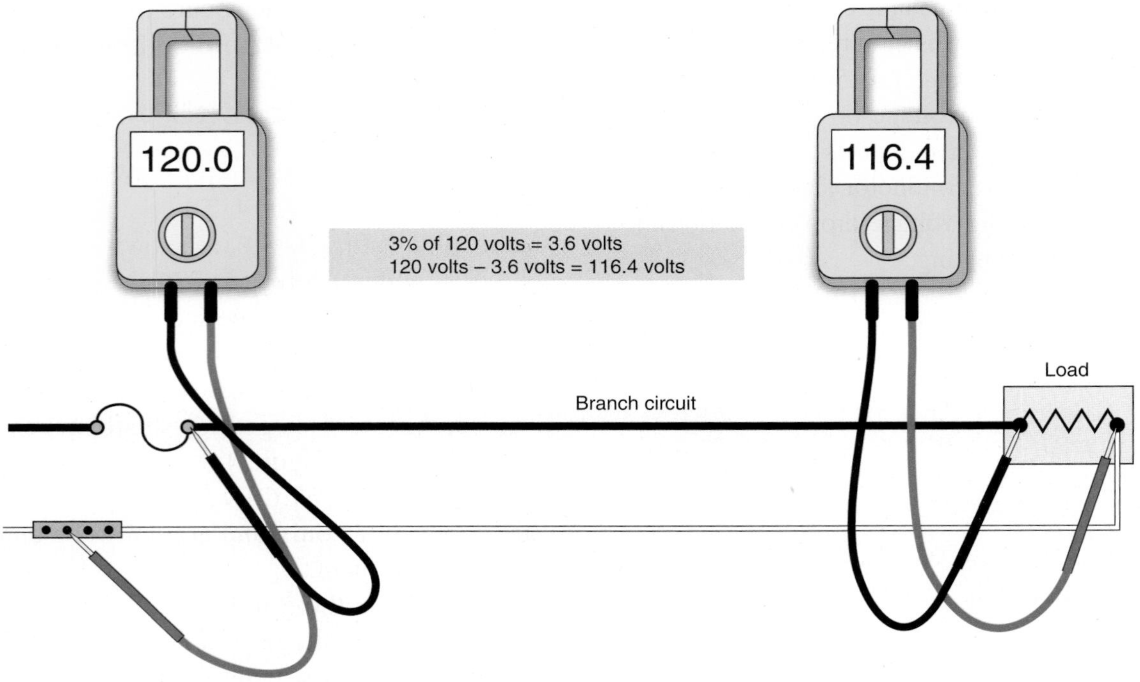

FIGURE 4-8 Maximum recommended voltage drop on a branch circuit is 3%, *210.19(A), Informational Note No. 4.*

TABLE 4-7

Circular mil area for many of the more common conductors.

CONDUCTOR SIZE, AWG	CROSS-SECTIONAL AREA IN CIRCULAR MILS
18	1620
16	2580
14	4110
12	6530
10	10,380
8	16,510
6	26,240
4	41,740
3	52,620
2	66,360
1	83,690
1/0	105,600
2/0	133,100
3/0	167,800
4/0	211,600

Let's try it again using 12 AWG conductors:

$$E_d = \frac{12.9 \times 11 \times 85 \times 2}{6530}$$

$$E_d = 3.69 \text{ volts drop}$$

Since the voltage drop is 3.69 volts, you will want to examine the voltage drop of the feeder (if one exists in your installation) to determine if the wire size is satisfactory or if you will want to install a 10 AWG conductor.

EXAMPLE

Find the wire size needed to keep the voltage drop to no more than 3% on a single-phase, 240-volt, air-conditioner circuit. The nameplate reads: "Minimum Circuit Ampacity 40 Amperes." The circuit originates at the main panel located approximately 65 ft (19.81 m) from the air-conditioner unit. No neutral conductor is required.

Solution: As shown on page 109, we use the following formula to determine the wire size needed to carry a known load for a specific distance without exceeding a desired voltage drop.

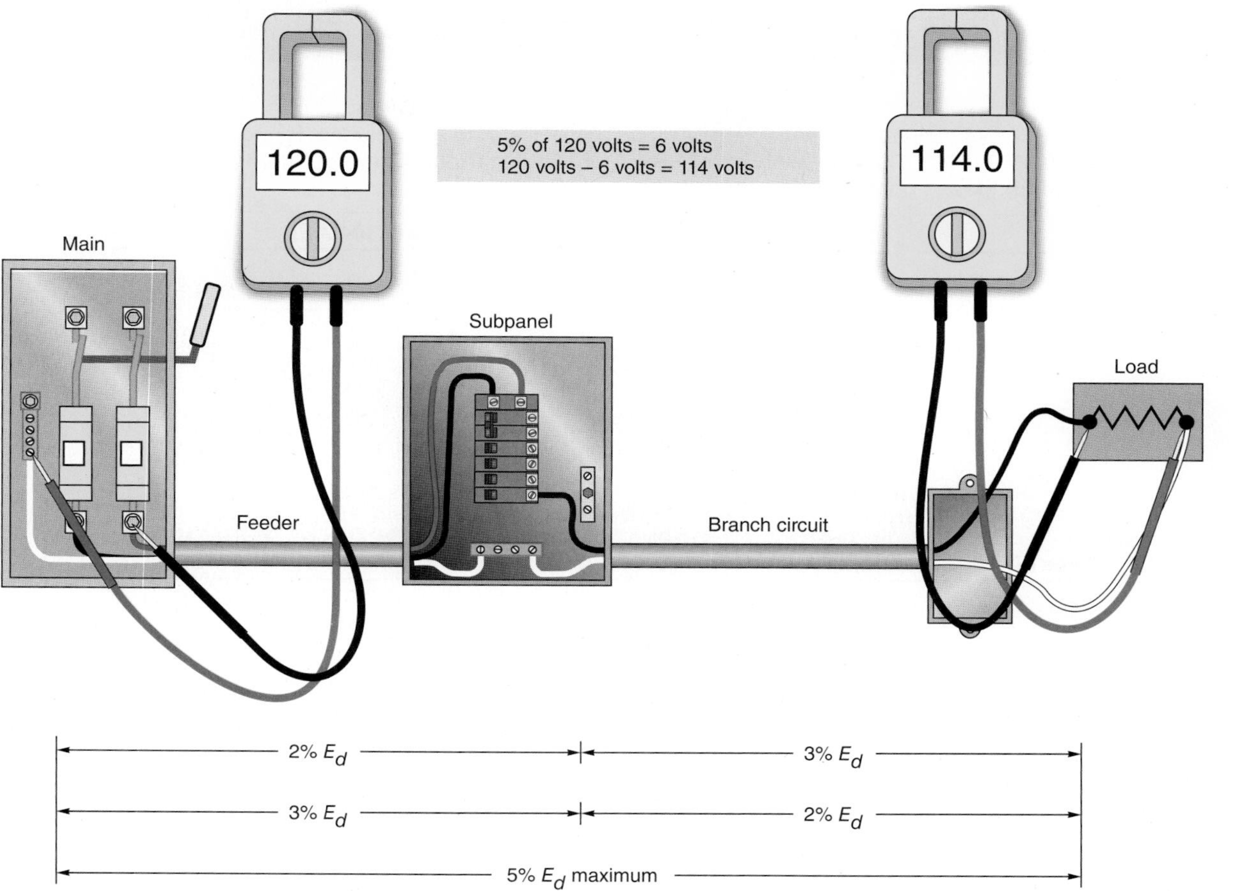

FIGURE 4-9 *NEC 210.19(A), Informational Note No. 4,* and *215.2(A)(1), Informational Note No. 2,* of the *Code* state that the total voltage drop from the beginning of a feeder to the farthest outlet on a branch circuit that does not exceed 5% will provide reasonable efficiency of operation. In this figure, if the voltage drop in the feeder is 3%, then do not exceed 2% voltage drop in the branch circuit. If the voltage drop in the feeder is 2%, then do not exceed 3% voltage drop in the branch circuit.

$$CMA = \frac{K \times I \times L \times 2}{E_d}$$

Substituting the known loads, length of wire and maximum voltage drop of 3% (240 × 3% = 7.2) we get:

$$CMA = \frac{12.9 \times 40 \times 65 \times 2}{7.2}$$

$$CMA = \frac{67{,}080}{7.2}$$

$$CMA = 9{,}317$$

By reference to Table 4-7, we find a 10 AWG to be the smallest copper wire than will result in a 3 percent or lower voltage drop for this circuit. However, to comply with the "Minimum Circuit Ampacity", install not smaller than an 8 AWG copper conductor. [See *NEC Table 310.15(B)(16).*] Obviously, the larger of the two conductors must be installed.

A Word of Caution When Using High-Temperature Conductors

Note in *NEC Table 310.15(B)(16)* that conductor insulations fall into three classes of temperature ratings: 60°C, 75°C, and 90°C. For a given conductor size, we find that the allowable ampacity of a 90°C insulated conductor is greater than that of a 60°C insulated conductor. Therefore, be careful when selecting conductors based on their ability to withstand high temperatures. For example:

- 8 AWG THHN (90°C) copper has an allowable ampacity of 55 amperes.
- 8 AWG THW (75°C) copper has an allowable ampacity of 50 amperes.

NEC TABLE 310.15(B)(16)

Allowable Ampacities of Insulated Conductors.

Table 310.15(B)(16) (formerly Table 310.16) Allowable Ampacities of Insulated Conductors Rated Up to and Including 2000 Volts, 60°C Through 90°C (140°F Through 194°F), Not More Than Three Current-Carrying Conductors in Raceway, Cable, or Earth (Directly Buried), Based on Ambient Temperature of 30°C (86°F)*

	Temperature Rating of Conductor [See Table 310.104(A).]						
	60°C (140°F)	75°C (167°F)	90°C (194°F)	60°C (140°F)	75°C (167°F)	90°C (194°F)	
Size AWG or kcmil	Types TW, UF	Types RHW, THHW, THW, THWN, XHHW, USE, ZW	Types TBS, SA, SIS, FEP, FEPB, MI, RHH, RHW-2, THHN, THHW, THW-2, THWN-2, USE-2, XHH, XHHW, XHHW-2, ZW-2	Types TW, UF	Types RHW, THHW, THW, THWN, XHHW, USE	Types TBS, SA, SIS, THHN, THHW, THW-2, THWN-2, RHH, RHW-2, USE-2, XHH, XHHW, XHHW-2, ZW-2	
	COPPER			ALUMINUM OR COPPER-CLAD ALUMINUM			Size AWG or kcmil
18**	—	—	14	—	—	—	—
16**	—	—	18	—	—	—	—
14**	15	20	25	—	—	—	—
12**	20	25	30	15	20	25	12**
10**	30	35	40	25	30	35	10**
8	40	50	55	35	40	45	8
6	55	65	75	40	50	55	6
4	70	85	95	55	65	75	4
3	85	100	115	65	75	85	3
2	95	115	130	75	90	100	2
1	110	130	145	85	100	115	1
1/0	125	150	170	100	120	135	1/0
2/0	145	175	195	115	135	150	2/0
3/0	165	200	225	130	155	175	3/0
4/0	195	230	260	150	180	205	4/0
250	215	255	290	170	205	230	250
300	240	285	320	195	230	260	300
350	260	310	350	210	250	280	350
400	280	335	380	225	270	305	400
500	320	380	430	260	310	350	500
600	350	420	475	285	340	385	600
700	385	460	520	315	375	425	700
750	400	475	535	320	385	435	750
800	410	490	555	330	395	445	800
900	435	520	585	355	425	480	900
1000	455	545	615	375	445	500	1000
1250	495	590	665	405	485	545	1250
1500	525	625	705	435	520	585	1500
1750	545	650	735	455	545	615	1750
2000	555	665	750	470	560	630	2000

*Refer to 310.15(B)(2) for the ampacity correction factors where the ambient temperature is other than 30°C (86°F).
**Refer to 240.4(D) for conductor overcurrent protection limitations.

- 8 AWG TW (60°C) copper has an allowable ampacity of 40 amperes.

We learned earlier in this chapter that according to *110.14(C)*, we use the 60°C column of *NEC Table 310.15(B)(16)* for circuits rated 100 amperes or less, and the 75°C column for circuits rated over 100 amperes. However, when the equipment is marked as being suitable for 75°C, we can take advantage of the higher ampacity of 75°C conductors. This oftentimes results in a smaller size conductor and a correspondingly smaller raceway.

But installing smaller conductors might result in excessive voltage drop.

Therefore, after selecting the proper size conductor for a given load, it is always a good idea to run a voltage-drop calculation to make sure the voltage drop is not excessive.

What Is the Advantage of High-Temperature Conductors?

They can withstand the high temperatures found in attics and high-temperature climates.

Another big advantage of the higher ampacities of high-temperature insulation is that the *NEC* permits the higher ampacity values to be used as the starting point when applying derating and correcting factors. Example calculations of this are found in Chapter 18.

Effect of Voltage Variation

Chapter 19 has more information on the effect voltage differences have on wattage output of appliances and on electric motors.

APPROXIMATE CONDUCTOR SIZE RELATIONSHIP

There is a definite relationship between the circular mil area and the resistance of conductors. The following rules explain this relationship.

Rule One. For wire sizes up through 4/0, every third size approximately doubles or halves in circular mil area.

Thus, a 1 AWG conductor is about 2 times larger than a 4 AWG conductor (83,690 versus 41,470). A 1/0 wire is about one-half the size of 4/0 wire (105,600 versus 211,600).

Rule Two. For wire sizes up through 4/0, every consecutive wire size is approximately 1.26 times larger or smaller than the preceding wire size.

Thus, a 3 AWG conductor is approximately 1.26 times larger than a 4 AWG conductor (41,740 × 1.26 = 52,592). A 2 AWG conductor is approximately 1.26 times smaller than a 1 AWG conductor (83,690 ÷ 1.26 = 66,420).

TABLE 4-8

"Every third size" and the 1.26 relationship between the circular-mil area (CMA) and the resistance in ohms per 1000 ft of copper conductors sizes 10 AWG through 6 AWG.

WIRE SIZE (AWG)	CMA (IN CIRCULAR MILS)	OHMS PER 1000 FT
10	10,380	1.2
9		
8		
7	20,760	0.6
6	26,158	0.476

Try to fix in your mind that a 10 AWG conductor has a cross-sectional area of 10,380 circular mils and that it has a resistance of 1.2 ohms per 1000 ft (300 m). The resistance of aluminum wire is approximately 2 ohms per 1000 ft (300 m). By remembering these numbers, you will be able to perform voltage-drop calculations without having the wire tables readily available.

EXAMPLE

What is the approximate cross-sectional area, in circular mils, and resistance of a 6 AWG copper conductor?

Solution: Note in Table 4-8 that when the CMA of a wire is doubled, then its resistance is cut in half. Inversely, when a given wire size is reduced to one-half, its resistance doubles.

NONMETALLIC-SHEATHED CABLE (*ARTICLE 334*)

The *NEC* and UL Standard 719 describe the construction details of nonmetallic-sheathed cable. The following is a summary of these details.

Description

Most electricians still refer to nonmetallic-sheathed cable as *Romex*, a name chosen years ago by the Rome Wire and Cable Company. Romex is now a registered trademark of Southwire Company.

Nonmetallic-sheathed cable is A factory assembly of two or more insulated conductors enclosed within an overall nonmetallic jacket.* This cable is most commonly available with two or three current-carrying conductors. The conductors range in size from 14 AWG through 2 AWG for copper conductors, and from 12 AWG through 2 AWG for aluminum or copper-clad aluminum conductors. Two-wire cables contain one black conductor, one white conductor, and one bare equipment grounding conductor. Three-wire cables contain one black, one white, one red, and one bare equipment grounding conductor. Equipment grounding conductors are permitted to have green insulation, but bare equipment grounding conductors are the most common.

NEC 334.108 requires that nonmetallic-sheathed cable have an insulated or a bare equipment grounding conductor. The equipment grounding conductor in Type NM cable is sized to comply with *NEC Table 250.122* based upon the overcurrent device that is typically used ahead of the branch circuit. This sizing is shown in Table 4-9.

Figure 4-10 clearly shows a bare equipment grounding conductor. Sometimes the bare equipment grounding conductor is wrapped with paper

TABLE 4-9

Uses permitted for nonmetallic-sheathed cable.

	TYPE NM-B	TYPE NMC-B	TYPE NMS-B
May be used on circuits of 600 volts or less	Yes	Yes	Yes
May be run exposed or concealed in dry locations	Yes	Yes	Yes
May be installed exposed or concealed in damp and moist locations	No	Yes	No
Has flame-retardant and moisture-resistant outer covering	Yes	Yes	Yes
Has fungus-resistant and corrosion-resistant outer covering	No	Yes	No
May be used to wire one- and two-family dwellings or multifamily dwellings; see *334.10*	Yes	Yes	Yes
May be embedded in masonry, concrete, plaster, adobe, and fill	No	No	No
May be installed or fished in the hollow voids of masonry blocks or tile walls where not exposed to excessive moisture or dampness	Yes	Yes	Yes
May be installed or fished in the hollow voids of masonry blocks or tile walls where exposed to excessive moisture	No	Yes	No
In outside walls of masonry block or tile	No	Yes	No
In inside wall of masonry block or tile	Yes	Yes	Yes
May be used as service-entrance cable	No	No	No
Must be protected against physical damage	Yes	Yes	Yes
May be run in shallow chase of masonry, concrete, or adobe if protected by a steel plate(s) at least 1/16 in. (1.6 mm) thick, then covered with plaster, adobe, or similar finish	No	Yes	No

*Reprinted with permission from NFPA 70-2014.

FIGURE 4-10 A nonmetallic-sheathed Type NM-B cable showing (A) black "ungrounded" (hot) conductor, (B) bare equipment "grounding" conductor, and (C) white "grounded" conductor.

Courtesy of Southwire Company.

or fiberglass, which acts as a filler. The equipment grounding conductor is *not* permitted to be used as a current-carrying conductor.

Size of Equipment Grounding Conductor Relative to Circuit Conductors in Type NM Cable

SIZE OF BRANCH CIRCUIT CONDUCTORS	SIZE OF EQUIPMENT GROUNDING CONDUCTOR
14 AWG	14 AWG
12 AWG	12 AWG
10 AWG	10 AWG
8 AWG	10 AWG
6 AWG	10 AWG
4 AWG	8 AWG

Types of Nonmetallic-Sheathed Cable

UL 719 lists two types of nonmetallic-sheathed cable. The *NEC* shows three types of nonmetallic-sheathed cable.

- **Type NM-B** is the most common type of nonmetallic-sheathed cable in use today. Type NM-B cable has a flame-retardant, moisture-resistant, nonmetallic outer jacket. The conductors are rated 90°C. The ampacity is based on the 60°C column in *NEC Table 310.15(B)(16)*. The conductors in Type NM-B meet all of the requirements of THHN but do not have the identifying marking along the entire length of the individual conductors. In the past, Type NM cable contained conductors having Type TW insulation.
- **Type NMC-B** cable has a flame-retardant, moisture-resistant, fungus-resistant, and corrosion-resistant, nonmetallic outer jacket. The conductors are rated 90°C. The ampacity is based on the 60°C column in *NEC Table 310.15(B)(16)*. Type NMC-B cable is not commercially available. Underground feeder Type UF-B cable can be used as a substitute for NMC.
- **Type NMS-B** cable is a hybrid cable that contains the conventional insulated power conductors as well as telephone, coaxial, home entertainment, and signaling conductors all in one cable. This type of cable is used for home automation systems using the latest in digital technology.

Prior to the 2008 *NEC*, *Article 780* recognized this wiring method for closed loop and programmed power distribution. In the 2008 *NEC*, *Article 780* was deleted. The Type NMS-B cable had a moisture-resistant, flame-retardant, nonmetallic outer jacket. Type NMS-B cable is not commercially available. This type of home automation never did take off. It fell by the wayside. Today, we are seeing a tremendous amount of wireless home automation systems.

Although somewhat confusing, the UL standards use the suffix B, whereas the *NEC* refers to nonmetallic-sheathed cables as Type NM, NMC, and NMS. The conductors in these cables are rated 90°C. Types NM, NMC, and NMS cable, identified by the markings NM-B, NMC-B, and NMS-B, meet this 90°C requirement. See *334.112*.

As of the writing of this edition of *Electrical Wiring—Residential*, there are no UL listings for Type NMS and NMS-B cables.

Why the Suffix "B"?

The suffix "B" means that the conductors have 90°C insulation.

Older nonmetallic-sheathed cable contained 60°C-rated conductors. There were many problems with the insulation becoming brittle and breaking off due to the extremely high temperatures associated with recessed and surface-mounted luminaires: fires resulted. In hot attics, after applying correction factors, the adjusted allowable ampacity of the conductors insulation could very well be zero. To solve this dilemma, since December 17, 1984, UL has required that the conductor insulation be rated 90°C and that the cable be identified with the suffix "B." This suffix makes it easy to differentiate the newer 90°C from the old 60°C cable.

The ampacity of the conductors in nonmetallic-sheathed cable is based on that of 60°C conductors. The 90°C ampacity is permitted to be used for derating purposes, such as might be necessary in attics. The final derated ampacity must not exceed the ampacity for 60°C conductors. More specifics can be found in *NEC 334.80*.

Table 310.15(B)(16) shows the allowable ampacities for conductors.

Overcurrent Protection for Small Conductors

Overcurrent protection for small copper conductors, in conformance to *240.4(D)*, is as follows:

- 14 AWG 15 amperes
- 12 AWG 20 amperes
- 10 AWG 30 amperes

A typical nonmetallic-sheathed cable is illustrated in Figure 4-10.

Color Coding of Type NM-B Jacket

To make installations easier for the electrician and identification of the conductor size easier for the electrical inspector, nonmetallic sheathed cable with a color-coded outer jacket is available. This color-coding is not a requirement of the *NEC* but has become a manufacturer's standard construction.

Conductor Size	Color of Jacket
14 AWG	White
12 AWG	Yellow
10 AWG	Orange
8 and 6 AWG	Black

New Types of NM-B Cable

At least one manufacturer offers a 4-conductor Type NM-B that can be used for two branch circuits. This cable is used where sharing of the neutral is *not wanted*, as in a multiwire branch circuit, or for branch circuits supplying GFCIs and AFCIs, where sharing the neutral is *not permitted*.

This cable contains the following:

- One bare equipment grounding conductor
- Pair 1: one black and one white
- Pair 2: one red and one white (with a red stripe)

Because all four conductors in this cable are current-carrying conductors, the ampacity of the conductors must be adjusted (derated) according to *NEC 310.15(B)(3)(a)*. This is generally not a problem because the *Code* allows the adjustment to be made from the 90°C allowable ampacity of 30 amperes in *NEC Table 310.15(B)(16)*.

$$30 \times 0.8 = 24 \text{ A}$$

This obviously exceeds the 20-ampere rating of the branch-circuit overcurrent device.

Where Nonmetallic-Sheathed Cable May Be Used

Table 4-9 shows the uses permitted for Types NM-B, NMC-B, and NMS-B. This information is extracted from *NEC 334.10*.

Equipment Grounding Conductors (*250.122*)

Equipment grounding conductor (EGC) sizing is based on the rating or setting of the overcurrent device protecting a given branch circuit or feeder. See *NEC Table 250.122*.

The EGC in a nonmetallic-sheathed cable is sized according to *NEC Table 250.122* and is usually bare. Table 4-10 is a short version of *Table 250.122*.

An EGC does not have to be larger than the ungrounded conductors of a given circuit. Why? Because in the event of a ground fault, the amount of ground-fault current returning on the EGC can never be more than the ground-fault current flowing on

TABLE 4-10

Minimum size for equipment grounding conductors is based on the rating or setting of the overcurrent device protecting that particular branch circuit. This table is based on *Table 250.122* of the *NEC*.

MINIMUM SIZE (AWG) EQUIPMENT GROUNDING CONDUCTOR

Rating or Setting of Overcurrent Device Not Exceeding	Copper	Aluminum or Copper-Clad Aluminum
15	14	12
20	12	10
60	10	8
100	8	6
200	6	4

TABLE 4-11

Size of Equipment Grounding Conductor Relative to Circuit Conductors in Type NM Cable

Size of Branch Circuit Conductors	Size of Equipment Grounding Conductors in Cables with 2, 3, and 4 Conductors
14 AWG*	14 AWG**
12 AWG*	12 AWG**
10**, 8***, 6*** AWG	10 AWG**
4*** AWG	8 AWG****
3***, 2*** AWG	8 AWG****

*Solid
**Usually solid, but permitted to be stranded.
***Must be stranded.
**** Permitted to be solid but usually are stranded.

the ungrounded conductor of that circuit that caused the problem. It's a simple series circuit. What goes out—comes back! See *NEC 250.122(A)*.

The Underwriters Laboratories standard for construction of Type NM cable contains requirements for the sizing of equipment grounding conductors that are contained within the cable. This requirement ensures compliance with the *NEC Table 250.122*. As shown in Table 4-11 in this text, 14 and 12 AWG conductors are solid and 10 AWG is typically solid but is permitted to be stranded. Branch-circuit conductors larger than 10 AWG are required to be stranded. Eight AWG equipment grounding conductors are typically stranded but are permitted to be solid.

Installation

Nonmetallic-sheathed cable is probably the least expensive of the various wiring methods. It is relatively lightweight and easy to install. It is widely used for dwelling unit installations on circuits of 600 volts or less. Figure 4-11 shows an example of a stripper used for stripping nonmetallic-sheathed cable. A razor knife also works fine with care. The installation of nonmetallic-sheathed cable must conform to the requirements of *NEC Article 334*. Refer to Figures 4-12 through 4-17.

- The cable must be strapped or stapled not more than 12 in. (300 mm) from a box or fitting.
- Do not staple flat nonmetallic-sheathed cables on edge, Figure 4-16.
- Unless run through horizontal holes or notches in studs and joists, the intervals between straps or staples must not exceed 4½ ft (1.4 m).
- Exposed runs of cables can be subjected to lots of abuse. A child may pull or stand on the cables, and adults may use the cables to hang things on. *NEC 334.15* requires that cables be protected against physical damage where necessary. For example, where exposed in a garage with open studs, run the cables on the sides of the studs. Where cables are run

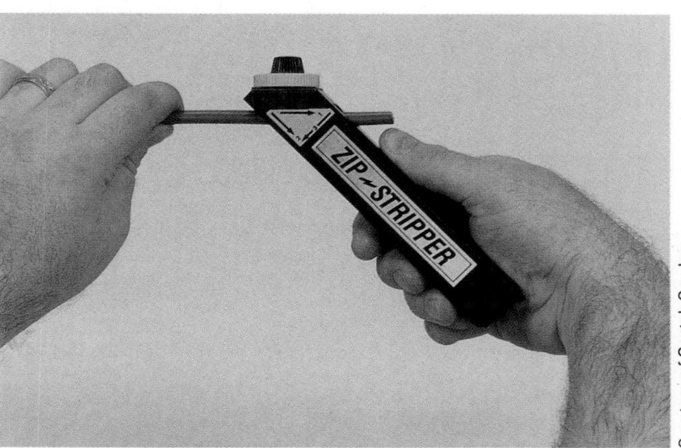

FIGURE 4-11 This stripper is used to remove the outer jacket from nonmetallic-sheathed cable.

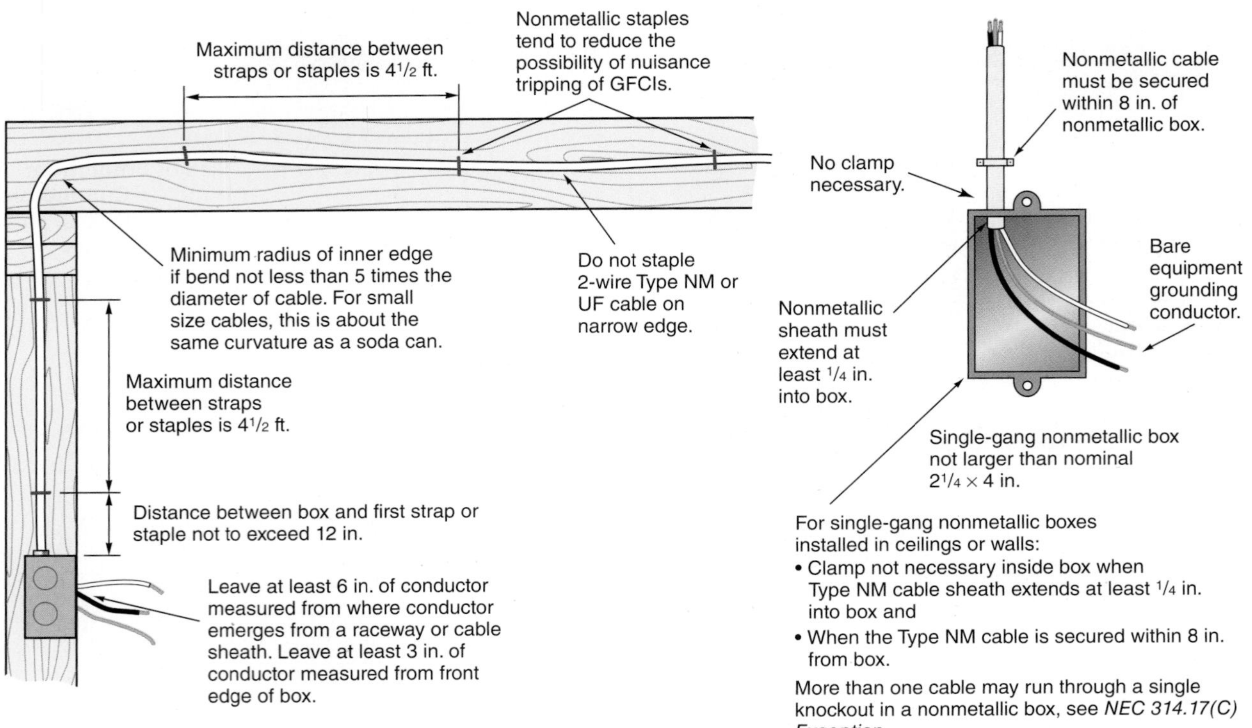

FIGURE 4-12 Installation of nonmetallic-sheathed cable.

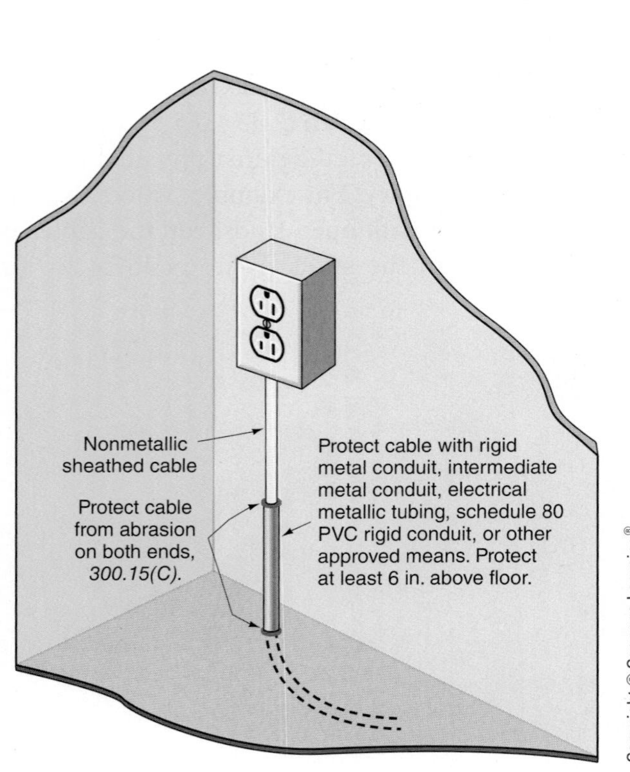

FIGURE 4-13 Installation of exposed nonmetallic-sheathed cable where passing through a floor; see *334.15(B)*.

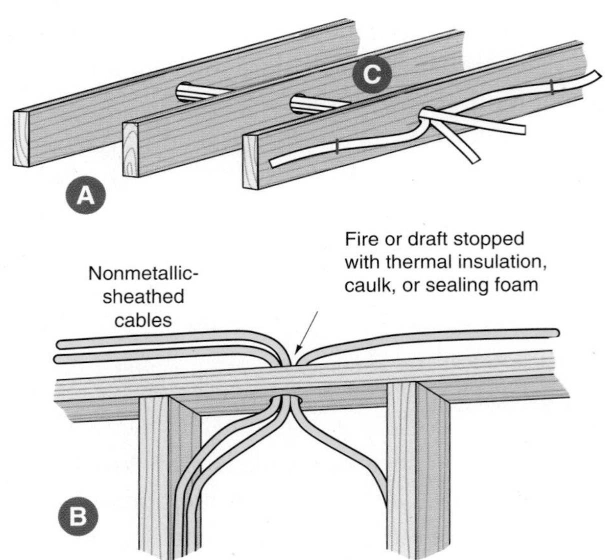

FIGURE 4-14 The conductor ampacity in NM cables shall be adjusted according to *Table 310.15(B)(3)(a):* (A) when NM cables are "bundled" or "stacked" for distances more than 24 in. (600 mm) without maintaining spacing; or (B) when more than two NM cables pass through wood framing members if sealed with thermal insulation, caulk or sealing foam, and there is no spacing between the cables; or (C) when more than two NM cables are installed in contact with thermal insulation. See *310.15(B)(3)(a)* and *334.80.*

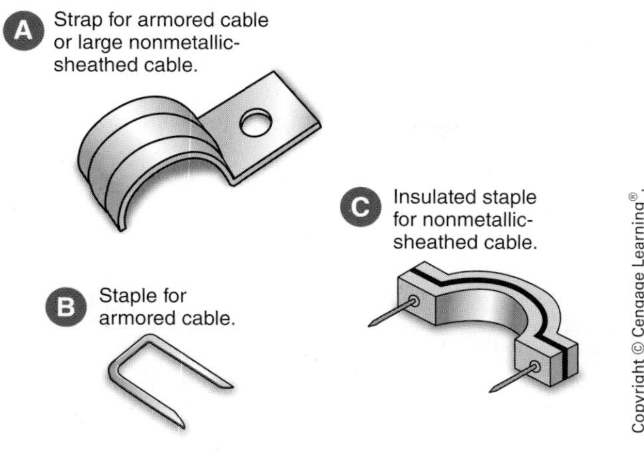

FIGURE 4-15 (A) One-hole strap used for conduit, EMT, large nonmetallic-sheathed cable and armored cable. (B) Metal staple for armored cable and nonmetallic-sheathed cable. (C) Insulated staple for nonmetallic-sheathed cable. Certain types of insulated and noninsulated staples can be applied with a staple gun.

horizontally from or through stud to stud, protection is required. Some inspectors require protective boards over exposed horizontal cable runs, or sheet rock, or other wall finish up to the ceiling, or at least 7 ft (2.13 m) up. For protection of cables in attics, see Chapter 15 of this text.

- The 4½ ft (1.4 m) securing requirement is not needed where nonmetallic-sheathed cable is run *horizontally* through holes in wood or metal framing members (studs, joists, rafters, etc.). The cable is considered to be adequately supported by the framing members, *NEC 334.30(A)*.
- The inner edge of the bend shall have a minimum radius not less than five times the cable diameter, *NEC 334.24*. This prevents damage to the cable jacket and the conductor insulation.
- The cable must not be used in circuits of more than 600 volts.

SAFETY ALERT

Be very careful when installing staples. Driving them too hard can squeeze and damage the insulation on the conductors, causing short circuits and/or ground faults.

- See Figure 4-12 for special conditions when using single-gang nonmetallic device boxes.
- Nonmetallic-sheathed cable must be protected where passing through a floor by rigid metal conduit, intermediate metal conduit, electrical metallic tubing, Schedule 80 rigid PVC conduit not shorter than 6 in. (150 m), or other approved means, *NEC 334.15(B)*. A fitting (bushing or connector) must be used at both ends of the conduit to protect the cable from abrasion, Figure 4-13.
- When nonmetallic-sheathed cables are "bundled" or "stacked" for distances of more than 24 in. (600 mm) without maintaining spacing, their ampacities must be reduced (adjusted, derated) according to *Table 310.15(B)(3)(a)*. When cables are bundled together, the heat

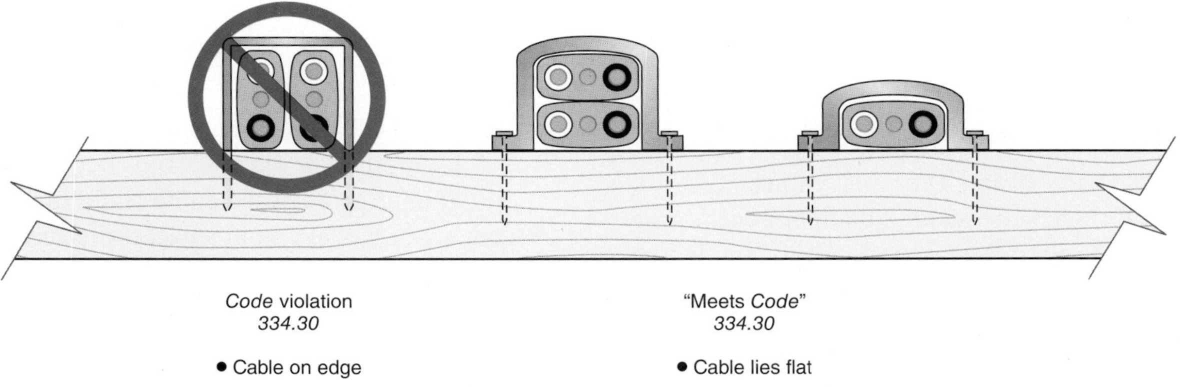

FIGURE 4-16 It is a *Code* violation to staple flat nonmetallic-sheathed cables on edge, *334.30*.

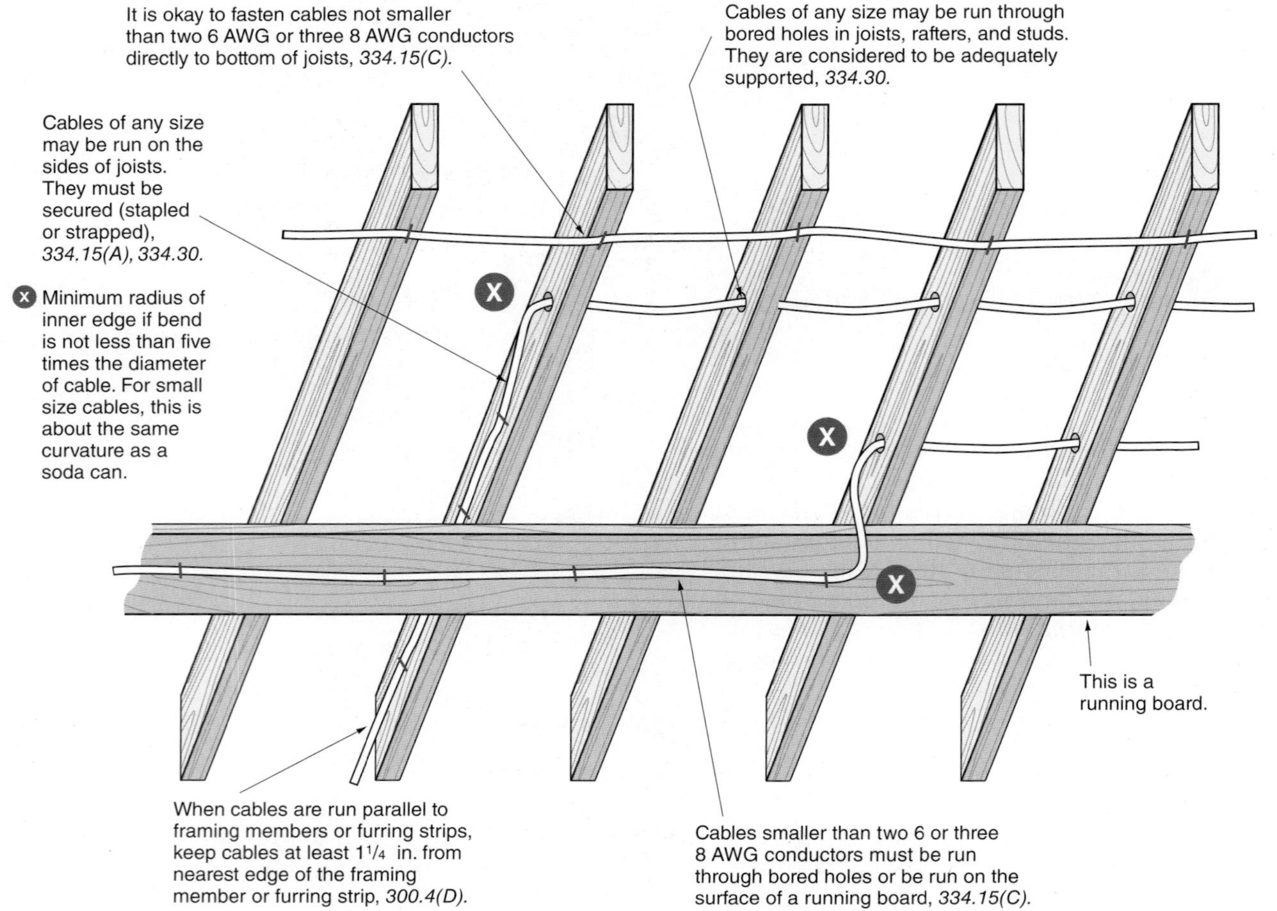

FIGURE 4-17 The requirements for running nonmetallic-sheathed cable in unfinished basements and crawl spaces.

generated by the conductors cannot easily dissipate. See Figure 4-14(A).

- When more than two nonmetallic-sheathed cables are installed through the same opening in wood framing members that are to be fire- or draft-stopped with thermal insulation, caulk, or sealing foam, and there is no spacing between the cables, their conductor ampacities shall be adjusted according to *NEC Table 310.15(B)(3)(a)*. See *NEC 334.80* and Figure 4-14(B).
- When Type AC and Type MC cables are "bundled" or "stacked" without maintaining spacing, derating is not necessary if the following is true:
  a. The cables do not have an overall outer jacket.
  b. There are no more than three current-carrying conductors in each cable.
  c. The conductors are 12 AWG copper.
  d. There are no more than 20 current-carrying conductors in the bundle.

  When there are more than 20 current-carrying conductors in the bundle, for a length greater than 24 in. (600 mm), a 60% adjustment factor must be applied. See *NEC 310.15(B)(3)(a)(4)* and *(B)(3)(a)(5)*.
- Type NM cable may be run in unsupported lengths of not more than 4½ ft (1.4 m) between the last point of support of the cable and a luminaire or other equipment within an accessible ceiling. For this last 4½ ft (1.4 m), the 12 in. (300 mm) and 4½ ft (1.4 m) securing requirements mentioned above are not necessary if the nonmetallic-sheathed cable is used as a luminaire (fixture) whip to connect luminaires of other equipment;

see *NEC 334.30*. Luminaire (fixture) whips are discussed in Chapter 7 and Chapter 17 of this text.

Figure 4-17 shows the installation of nonmetallic-sheathed cable in unfinished basements and crawl spaces, *334.15(C)*.

See Chapter 15 and Figures 15-13 and Figure 15-14 for additional text and diagrams covering the installation of cables in attics.

Figure 4-12 shows the *Code* requirements for securing nonmetallic-sheathed cable when using nonmetallic boxes.

The *Code* in *314.40(D)* requires that all metal boxes have provisions for the attachment of an equipment grounding conductor. Figure 4-18 shows a gang-type switch (device) box that is tapped for a screw by which the grounding conductor may be connected underneath. Figure 4-19 shows an outlet box, also with provisions for attaching grounding conductors. Figure 4-20 illustrates the use of a small grounding clip.

FIGURE 4-18 Metal gangable device box with mounting bracket and tapped hole for ground screw.

Photo courtesy of Thomas and Betts Corp., www.TNB.com.

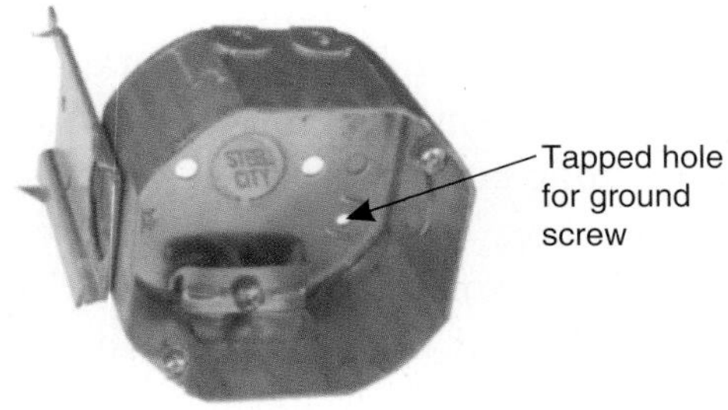

FIGURE 4-19 Metal outlet device box with mounting bracket and tapped hole for ground screw.

Photo courtesy of Thomas and Betts Corp., www.TNB.com.

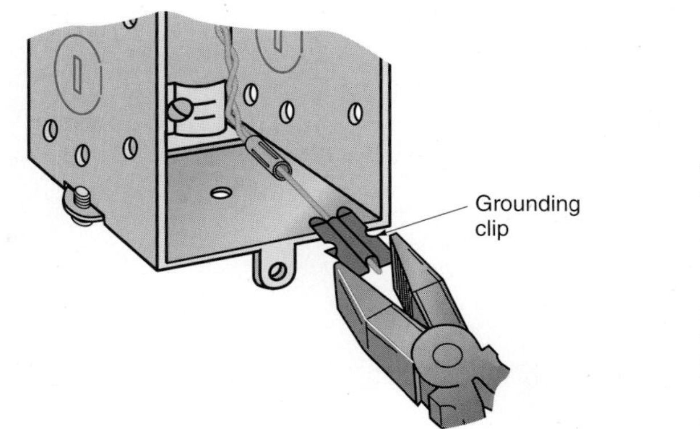

FIGURE 4-20 Method of attaching ground clip to a metal switch (device) box; see *250.148*.

NEC 410.44 requires that there be a provision whereby the equipment grounding conductor can be attached to the exposed metal parts of luminaires.

Are Cables Permitted in Raceways?

The answer is yes, but only if certain conditions are met.

NEC 300.15(C) permits using a listed conduit or tubing without a box where the conduit or tubing is used to protect a nonmetallic-sheathed cable against physical damage. Figure 4-22 shows an example of this in a basement where the wiring method in the ceiling is NM cable, and a switch or receptacle is in a box mounted on the wall. A length of EMT is installed from the box to a point above the ceiling joists, where the EMT is secured to the side of the joists. The NM cable is then run inside the EMT, and an insulating fitting is installed on the upper end of the tubing to protect the cable against abrasion. The cable does not require securing inside the raceway, *334.30*. The finished job is neat.

The conductors inside a nonmetallic-sheathed cable meet all of the UL requirements for Type THHN conductors but are not surface marked in any manner.

NEC 334.15(B) requires protection of nonmetallic-sheathed cable from physical damage where necessary. This section lists different types of raceways permitted to be used for protection of the cable.

FIGURE 4-21 Individual connectors that are listed for the cable are used to connect the cables to the cabinet.

The raceway fill for a cable in a conduit or tubing is based on the allowable percentage fill values specified in *Table 1, Chapter 9*.

Securing Cables to Cabinets and Boxes

The *NEC* generally requires that cables be secured to boxes and cabinets. Keep in mind that the word "cabinet" applies to enclosures for panelboards. As defined in the *NEC*, panelboards are the interior bussing that provides for the installation of circuit breakers. The cabinet is the enclosure that houses the panelboard.

NEC Section 312.5(C) requires Where cable is used, each cable shall be secured to the cabinet, cutout box, or meter socket enclosure.* Most electrical inspectors enforce this rule as though it read, "Where cables are used, each cable shall be individually secured to the cabinet with a connector that is listed for the specific wiring method used, and secures only one cable, unless it is identified for securing more than one." The second paragraph of *NEC* 300.15 reads, Fittings and connectors shall be used only with the specific wiring methods for which they are designed, and listed.*

The requirement for each fitting being suitable to connect only one cable unless it is marked otherwise comes from UL safety standard requirements.

*Reprinted with permission from NFPA 70-2014.

The photographs in Figure 4-21 show a cabinet with individual connectors that secure Type SE and Type NM cables to the enclosure. (The Type SE cable contains the feeder conductors that supply power to the panelboard.) Many different types of connectors are available for securing cable to the box or enclosure. This includes both metal and nonmetallic connectors, as well as snap-in types and those secured with a locknut. One type of metal connector that is secured with a locknut is shown in Figure 4-37.

Cables are generally required to be secured to device and outlet boxes. See *NEC 314.17(C)*. This is illustrated in Figure 4-12. As shown, Type NM cable is generally required to be secured to the framing or other suitable support mechanism within 12 inches of the box or cabinet. It is also required that the cable be secured to the box or cabinet.

An exception to these requirements is also illustrated in Figure 4-12. The exception excludes the requirement for connecting the cable to a single-gang nonmetallic box if the cable is secured within 8 inches of the box. These nonmetallic boxes are designed with a breakout which permits the Type NM cable to be inserted into the box. Note that this exception does not apply to metal boxes or 2-gang, 3-gang or 4-gang nonmetallic boxes.

As shown in several figures in this text, many metal outlet or device boxes are provided with internal cable clamps for securing cable to the box.

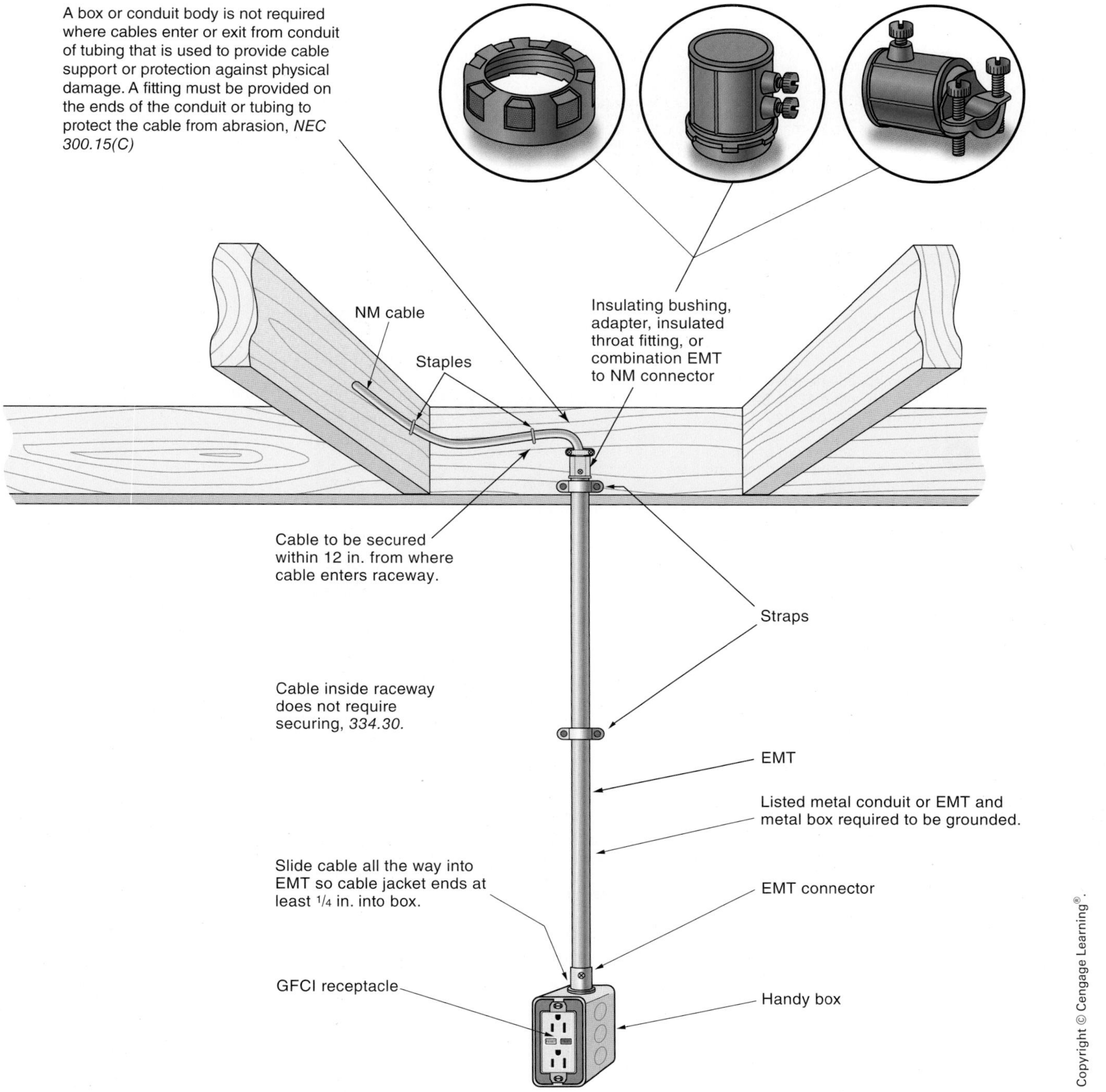

FIGURE 4-22 An example of EMT protecting a nonmetallic-sheathed cable from physical damage, where the cable is run down the wall of an unfinished basement. This also makes a neat installation. The EMT and the metal box must be grounded. Physical protection and grounding are required by *334.15(C). NEC 300.18(A), Exception,* also recognizes this type of installation.

Alternate Method of Installing NM above Panel

Nonmetallic-sheathed cables are usually secured to a panel, box, or cabinet, with a cable clamp or connector. For surface-mounted enclosures only, *312.5(C), Exception*, provides an alternate method for running nonmetallic-sheathed cables, as illustrated in Figure 4-23.

Be aware of a significant rule in *Note 9* to *NEC Table 1* of *Chapter 9*. This note requires that if cables with elliptical cross sections are installed in a

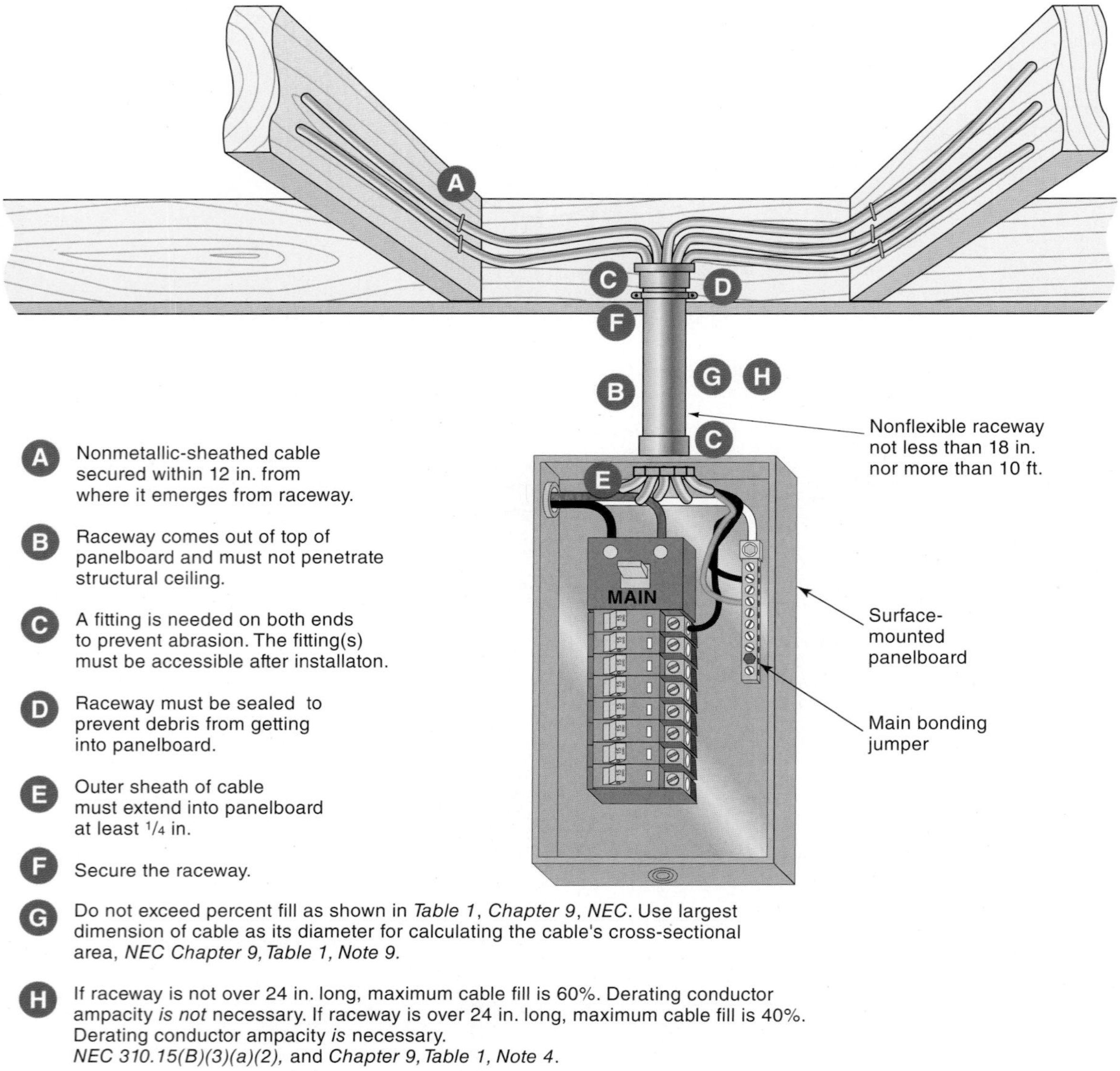

MAIN

Nonflexible raceway not less than 18 in. nor more than 10 ft.

Surface-mounted panelboard

Main bonding jumper

A Nonmetallic-sheathed cable secured within 12 in. from where it emerges from raceway.

B Raceway comes out of top of panelboard and must not penetrate structural ceiling.

C A fitting is needed on both ends to prevent abrasion. The fitting(s) must be accessible after installaton.

D Raceway must be sealed to prevent debris from getting into panelboard.

E Outer sheath of cable must extend into panelboard at least 1/4 in.

F Secure the raceway.

G Do not exceed percent fill as shown in *Table 1*, *Chapter 9*, *NEC*. Use largest dimension of cable as its diameter for calculating the cable's cross-sectional area, *NEC Chapter 9, Table 1, Note 9.*

H If raceway is not over 24 in. long, maximum cable fill is 60%. Derating conductor ampacity *is not* necessary. If raceway is over 24 in. long, maximum cable fill is 40%. Derating conductor ampacity *is* necessary. *NEC 310.15(B)(3)(a)(2),* and *Chapter 9, Table 1, Note 4.*

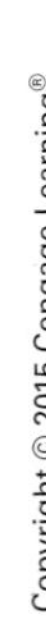

FIGURE 4-23 An alternate method for running nonmetallic-sheathed cables into a surface-mounted panel. This is permitted for surface-mounted panels only, *312.5(C), Exception.*

raceway, the cross-sectional area of the cable is determined by assuming that the greater dimension is the diameter of a circle. This rule is needed because some installers do not straighten the cables before installing them through the raceway, but pull them out of the center of a carton, which results in the cables twisting inside the raceway.

Let's do a calculation using a 12-2 W/G Type NM cable that is 1/4 in. 1/2 in. in size. We will assume the 1/2 in. dimension is the diameter of a circle and perform the calculation using the following formula:

$$A = \pi r^2 \text{ (Area = pi} \times \text{radius squared)}$$
$$A = 3.14159 \times (0.25 \times 0.25)$$
$$A = 3.14159 \times 0.0625$$
$$A = 0.1963 \text{ sq. in.}$$

Let's look at the following table to determine how many of these cables we are permitted to install in various sizes of EMT and whether derating is required.

100% AREA OF EMT	AREA OF 12/2 WG TYPE NM CABLE	NUMBER PERMITTED WITH 60% FILL*	NUMBER PERMITTED WITH 40% FILL**	DERATING REQUIRED?
1 in. = 0.864	0.1963	2	1	No/No
1¼ = 1.496	0.1963	4	3	No/Yes
1½ = 2.036	0.1963	6	4	No/Yes
2 = 3.356	0.1963	10	6	No/Yes

*Maximum 24 in. (600 mm) long

**Longer than 24 in. (600 mm)

Be sure to comply with all the conditions stated in *NEC 312.5(C), Exception*. Also be aware that Type NM cable is limited to installation in dry locations, and thus the installation in EMT as described above is not permitted in damp or wet locations.

Identifying Nonmetallic-Sheathed Cables

One nice thing about nonmetallic-sheathed cable, particularly that with a white or yellow outer jacket, is that at panelboards and device and outlet boxes, you can use a permanent marking pen to mark the outer jacket with a few words indicating what the cable is for. Examples: *FEED TO ATTIC; FEED TO L.R.; FEED TO DINING ROOM RECEPTACLES; FEED TO PANEL; BC#6; TO OTHER 3-WAY SWITCH; SWITCH LEG TO CEILING LIGHT; TO OUTSIDE LIGHT*; and so on.

By marking the cables when they are installed, confusion when making the connections is reduced considerably.

Another great idea is to make a sleeve by cutting off a short 2 in. length of NM cable, pulling out the conductors, then marking an identifying statement on the sleeve. Slide the sleeve over the specific conductors inside the panel or box.

Multifamily Dwellings

Nonmetallic-sheathed cable is permitted to be installed in one- and two-family dwellings and in multifamily dwellings where the building construction is Type III, Type IV, or Type V; see *334 (10)(2)* except as prohibited in *334.12*. Definitions of construction types are found in building codes, the Glossary of this text, and in *Annex E* of the *NEC*.

For years, the *NEC* restricted nonmetallic-sheathed cable to a maximum of three floors above finished grade. This limitation was removed in the *2002 NEC*.

ARMORED CABLE (TYPE AC) AND METAL-CLAD CABLE (TYPE MC)

Type AC and Type MC cables look very much the same (Figure 4-24), yet there are significant differences. Table 4-12 compares some of these differences.

To remove the outer metal armor, a rotary armor-cutting tool like that shown in Figure 4-26 may be used.

If a hacksaw is used, make an angle cut on one of the armor's convolutions as long as the conductors need to be, perhaps 10–12 in. Be careful not to cut too deep or you might cut into the conductor insulation. Bend the armor at the cut and snap it off. Then slide the 10–12 in. of armor off the end of the cable. This method is not recommended by cable manufacturers because of the risk of damage to the insulated conductors. See Figure 4-27.

Most electricians, when using armored cable, will make three cuts with their hacksaw. The middle cut is where the cable and conductors will

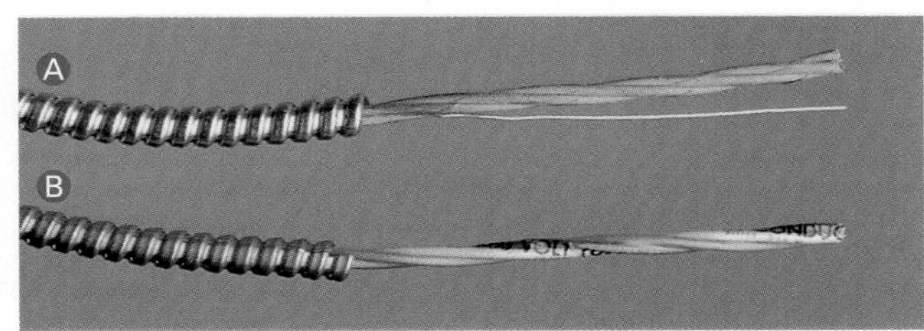

FIGURE 4-24 Flexible Type AC armored (above), Flexible Type MC cable (below).

TABLE 4-12

Comparisons of Type AC and Type MC cables.

TYPE AC (*ARTICLE 320*)	TYPE MC (*ARTICLE 330*)
Article 320: This article contains all of the installation requirements for the installation of Type AC armored cable.	***Article 330:*** This article contains all of the installation requirements for the installation of Type MC metal-clad cable.
***320.1.* Definition:** Type AC cable is a fabricated assembly of insulated conductors in a flexible interlocked metallic armor.	***330.1.* Definition:** Type MC cable is a factory assembly of one or more insulated circuit conductors with or without optical fiber members and enclosed in an armor of interlocking metal tape or a smooth or corrugated metallic sheath.
Number of Conductors: Two to four current-carrying conductors plus bonding wire. It may also have a separate equipment grounding conductor.	**Number of Conductors:** Any number of current-carrying conductors. At least one manufacturer has a "Home Run Cable" that has conductors for more than one branch circuit, plus the equipment grounding conductors.
Conductor Size and Type: 14 AWG through 1 AWG for copper conductors. 12 AWG through 1 AWG for aluminum conductors will be marked "AL" on a tag on the carton or reel. 12 AWG through 1 AWG for copper-clad aluminum conductors will be marked "AL (CU-CLAD)" or "Cu-Clad Al" on a tag on the carton or reel.	**Conductor Size and Type:** 18 AWG through 2000 kcmil for copper conductors. 12 AWG through 2000 kcmil for aluminum or copper-clad aluminum conductors. 12 AWG through 2000 kcmil for copper-clad aluminum conductors will be marked "AL (CU-CLAD)" or "Cu-Clad Al" on a tag on the carton or reel.
Color Coding: Two-conductor: one black, one white. Three-conductor: one black, one white, one red. Four-conductor: one black, one white, one red, one blue. These are in addition to any equipment grounding and bonding conductors.	**Color Coding:** Two-conductor: one black, one white. Three-conductor: one black, one white, one red. Four-conductor: one black, one white, one red, one blue. These are in addition to any equipment grounding and bonding conductors.
Bonding & Grounding: Type AC cable shall provide an adequate path for fault current. Type AC cable has a bonding wire or strip. Bonding wire is in continuous direct contact with the metal armor. The bonding wire and armor act together to serve as the acceptable equipment ground. The bonding wire does not have to be terminated, just folded back over the armor. Figure 4-25 shows how the bonding strip is folded back over the armor. Never use the bonding strip as a grounded neutral conductor or as a separate equipment grounding conductor. It is an internal bonding strip—nothing more!	**Bonding & Grounding:** The "traditional" cable construction provides a separate green insulated equipment grounding conductor. It may have two equipment grounding conductors for isolated ground requirements, such as for computer wiring. (By "traditional," we mean a spiral interlocked armor with mylar over the conductors.) The jacket of the "traditional" Type MC cable is not to be used as an equipment grounding conductor. The sheath of smooth or corrugated tube Type MC is listed as acceptable as the required equipment grounding conductor. Some MC cables have a full-size bare aluminum conductor in continuous, direct contact with the metal armor. This bare conductor and the metal armor act together to serve as the acceptable equipment ground. Cut off the bare conductor at the armor.
Insulation: Type ACTHH has 90°C thermoplastic insulation, by far the most common. Type ACTH has 75°C thermoplastic insulation. This Type AC has thermoset insulation. Type ACHH has 90°C thermoset insulation. Conductors are individually wrapped with a flame-retardant fibrous cover: light-brown Kraft paper. Author's comment: This is an easy way to distinguish Type AC from Type MC.	**Insulation:** Type MC has 90°C thermoplastic insulation. Some Type MC has thermoset insulation. Type MC cable is available with a PVC outer jacket suitable for direct burial and concrete encasement. No individual wrap. Has a polyester (Mylar®) tape over all conductors. Some constructions have the extra layer of Mylar® tape over the individual conductors. Has no bare bond wire but may have a full-size aluminum grounding/bonding conductor. *Author's comment:* These are easy ways to distinguish Type AC from Type MC.

(*continues*)

TABLE 4-12 (*continued*)

TYPE AC (*ARTICLE 320*)	TYPE MC (*ARTICLE 330*)
Armor: Galvanized steel or aluminum. Armored cable with aluminum armor is marked "Aluminum Armor."	**Armor:** Galvanized steel, aluminum, or copper. Three types: *Interlocked (most common).* Requires a separate equipment grounding conductor. The armor by itself is not an acceptable equipment grounding conductor. Steel or aluminum. *Smooth tube.* Does not require a separate equipment grounding conductor. Aluminum only. *Corrugated tube.* Does not require a separate equipment grounding conductor. Copper or aluminum.
Covering over armor: None	**Covering over armor:** Available with PVC outer covering. Suitable for wet locations, for direct burial and concrete encasement if so covered.
Locations: Okay for dry locations. Not okay for damp or wet locations. Not okay for direct burial.	**Locations:** Okay for wet locations if so listed. Okay for direct burial if so listed. Okay to be installed in a raceway.
Insulating bushings: Required to protect conductor insulation from damage. Referred to as "antishorts," these keep any sharp metal edges of the armor from cutting the conductor insulation. Usually supplied in a bag with cable. Figure 4-25(A) shows an antishort being inserted. Figure 4-25(B) shows the recommended way of bending the bonding strip over the antishort to hold the antishort in place.	**Insulating bushings:** Recommended but not required to protect conductor insulation from damage. Usually in a bag supplied with cable.
Minimum radius of bends: 5× diameter of cable. For small size cables, this is about the same curvature as a soda can.	**Minimum radius of bends:** Smooth Sheath: • 10× cable diameter for sizes not over ¾ in. (19 mm) in diameter. • 12× cable diameter for sizes over ¾ in. (19 mm) and not over 1½ in. (38 mm) in diameter. • 15× cable diameter for sizes over 1½ in. (38 mm) in diameter. Corrugated or interlocked: • 7× diameter of cable.
Not available with "super neutral conductor."	Available with "super neutral conductor" where required for computer branch-circuit wiring. For example, three 12 AWG and one 10 AWG neutral conductor.
Not available with fiber-optic cables.	Available with power conductors and fiber-optic cables in one cable.
Support: Not more than 4½ ft (1.4 m) apart. Not more than 12 in. (300 mm) from box or fitting. Support not needed where cable is fished through walls and ceilings. Cables run horizontally through holes in studs, joists, rafters, etc., are considered supported by the framing members. Support not needed when used as a fixture whip within an accessible ceiling for lengths not over 6 ft (1.8 m) from last point of support.	**Support:** Not more than 6 ft (1.8 m) apart. Not more than 12 in. (300 mm) from box or fitting for cables having four or fewer 10 AWG or smaller conductors. Support not needed where cable is fished through walls and ceilings. Cables run horizontally through holes in studs, joists, rafters, etc., are considered supported by the framing members. Support not needed when used as a fixture whip within an accessible ceiling for lengths not over 6 ft (1.8 m) from last point of support.

(*continues*)

TABLE 4-12 *(continued)*

TYPE AC (*ARTICLE 320*)	TYPE MC (*ARTICLE 330*)
Voltage: 600 volts or less.	**Voltage:** 600 volts or less. Some MC has a voltage rating not to exceed 2000 volts.
Ampacity: Use the 60°C column of *Table 310.15(B)(16)*. Derating permitted from 90°C column.	**Ampacity:** Similar to wire pulled in raceways; determine in accordance with *310.15*.
Thermal insulation: If buried in thermal insulation, the conductors must be rated 90°C.	**Thermal insulation:** No mention in *NEC*. Type MC has 90°C thermoplastic insulation.
Connectors: Set screw–type connectors not permitted with aluminum armor. Listed connectors are suitable for grounding purposes.	**Connectors:** Set screw–type connectors not permitted with aluminum armor. Listed connectors are suitable for grounding purposes.
UL Standard 4. Because UL standards have evolved in numerical order, it is apparent that the armored cable standard is one of the first standards to be developed.	**UL Standard 1569.**

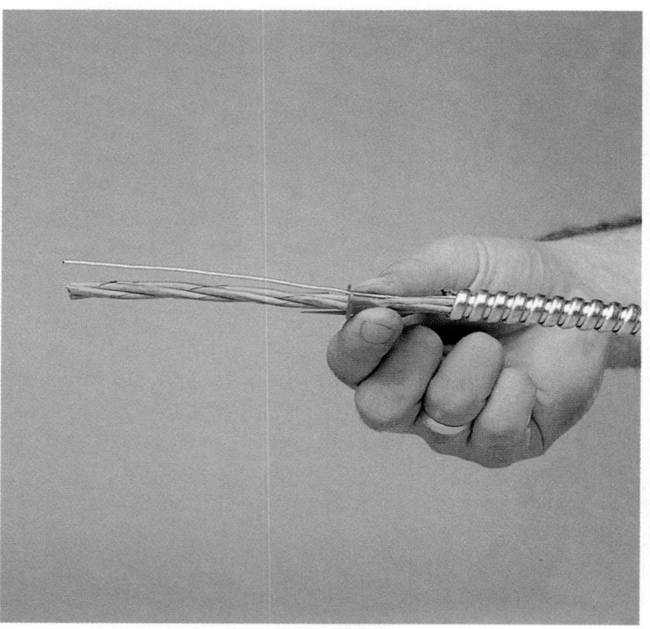

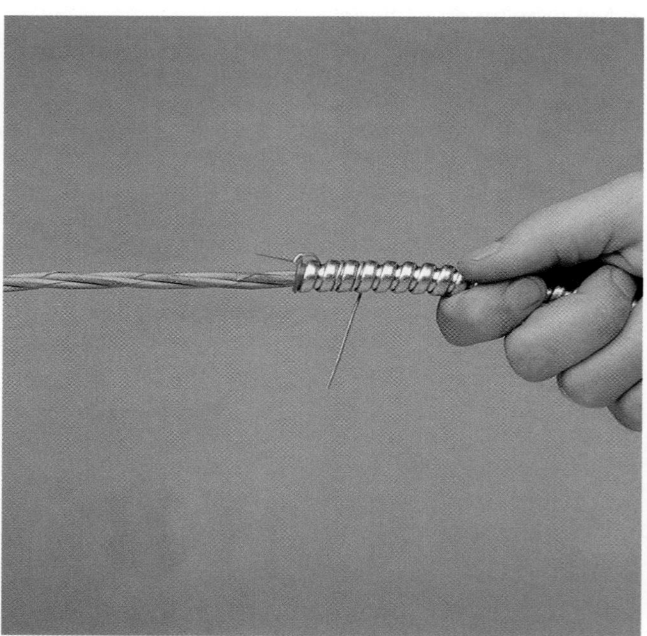

FIGURE 4-25 Antishort bushing prevents cutting of the conductor insulation by the sharp metal armor.

be cut off completely. The first and third cuts are where the armor will be snapped off. This saves time because it actually prepares two ends of the cable.

Most electricians still refer to armored cable as BX, supposedly derived from an abbreviation of the Bronx in New York City, where armored cable was once manufactured. BX was a trademark owned by the General Electric Company, but over the years it has become a generic term.

Figure 4-24 shows armored cable with the armor removed to show the conductors that are wrapped with paper and the bare bonding wire.

Table 4-13 shows the uses and installation requirements for Type AC and Type MC cables for residential wiring.

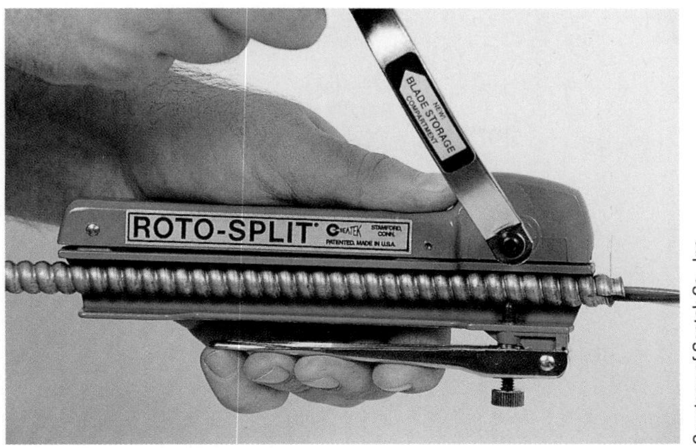

Courtesy of Seatek Co., Inc.

FIGURE 4-26 This is an example of a tool that precisely cuts the outer armor of armored cable, making it easy to remove the armor with a few turns of the handle.

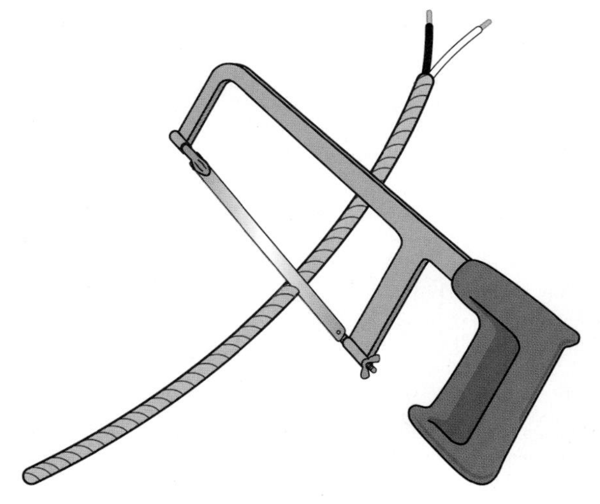

FIGURE 4-27 Using a hacksaw to cut through a raised convolution of the cable armor. This method is not recommended by manufacturers due to the risk of damaging the conductors by cutting too deeply.

TABLE 4-13

Uses and installation in dwellings (Type AC and Type MC).

	TYPE AC	TYPE MC
Branch circuits & feeders	Yes	Yes
Services	No	Yes
Exposed & concealed work in		
• dry locations.	Yes	Yes
• wet locations.	No	Yes (See *300.10(A)(11)* for specifics)
Where subject to physical damage	No	No
May be embedded in plaster finish on masonry walls or run through the hollow spaces of such walls if these locations are not considered damp or wet	Yes	Yes
Run or fished in air voids of masonry block or tile walls where not exposed to excessive moisture or dampness. Masonry in direct contact with earth is considered to be a wet location.	Yes	Yes
In any raceway	No	Yes
Direct burial, embedded in concrete, or in cinder fill	No	Yes, if identified for such use
In accessible attics or roof spaces, must be protected by guard strips at least as high as the cable where run across top of floor joists, or within 7 ft (2.1 m) of the floor or floor joists where the cable is run across the face of studs or rafters. If there is no permanent stairway or ladder, protection is needed only within 6 ft (1.8 m) from edge of scuttle hole or entrance to attic. No guard strips or running boards are necessary for cables run parallel to the sides of joists, rafters, or studs. See Chapter 15 of this text for additional discussion and diagrams regarding physical protection of cables in attics.	Yes	Yes

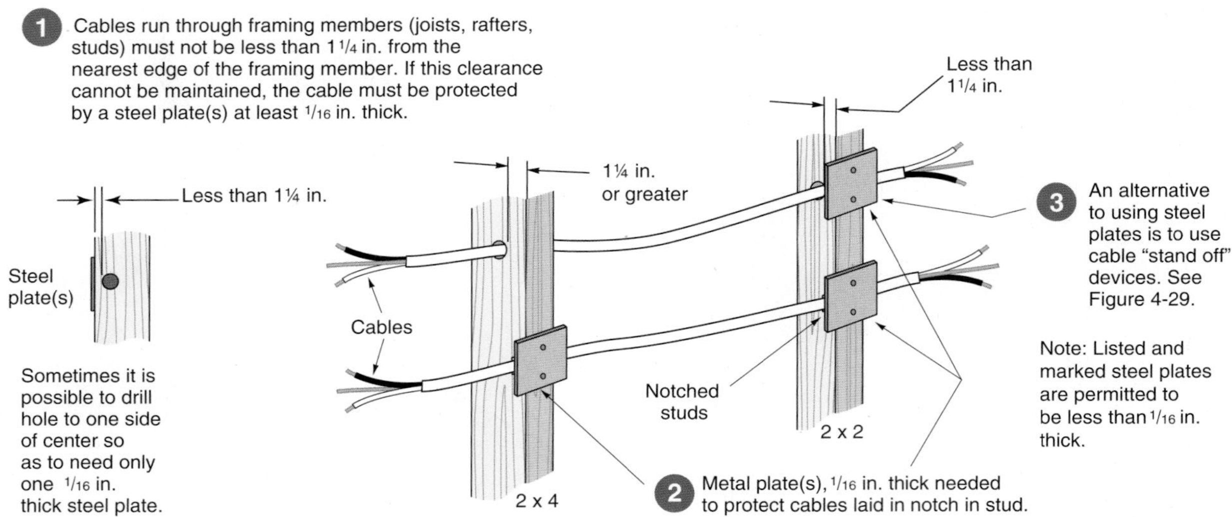

FIGURE 4-28 Methods of protecting nonmetallic and armored cables so nails and screws that "miss" the studs will not damage the cables, *300.4(A)(1)* and *(2)*. Because of their strength, intermediate and rigid metal conduit, PVC rigid conduit, and electrical metallic tubing are exempt from this rule per the *Exceptions* found in *300.4*. See Figure 4-30 for alternative methods of protecting cables.

INSTALLING CABLES THROUGH WOOD AND METAL FRAMING MEMBERS (*300.4*)

Wood Construction

When wiring a house, nonmetallic-sheathed cables are run through holes drilled in studs, joists, and other framing members. Typically, holes are drilled in the approximate centers of wood framing members. To avoid possible damage to the cables, make sure that the edge of the hole is at least 1¼ in. (32 mm) from the nearest edge of the framing member, *300.4(A)(1)*. Refer to Figure 4-28.

You have probably noticed that a 2 × 4 measures 1½ in. × 3½ in., and that a 2 × 3 measures 1½ in. × 2½ in. This can present a real problem!

Steel Plates

If the 1¼ in. (32 mm) from the edge of hole to the edge of framing member cannot be maintained, or if the cable is laid in a notch, a steel plate(s) at least 1⁄16 in. (1.6 mm) thick must be used to protect the cable from errant nails or drywall screws, *300.4(A)(1)*. In all cases where the *NEC* requires 1⁄16 in. (1.6 mm) steel plates, listed and marked steel plates are permitted to be less than 1⁄16 in. (1.6 mm) because they have passed tough mechanical tests. This permission is found in *300.4(A)(1), Exception 2; 300.4(A)(2), Exception 2; 300.4(B)(2), Exception; 300.4(D), Exception 3*; and *300.4(E), Exception 2*. See Figure 4-28.

Instead of using steel plates, Figure 4-29 illustrates devices that can be used to *stand off* the cables from a framing member.

Cables Installed Parallel to Framing Members or Furring Strips

To protect nonmetallic-sheathed cables, armored cables, or nonmetallic raceways that are run parallel to framing members or furring strips from being damaged by drywall screws and nails, you have two choices: (1) keep them more than 1¼ in. (32 mm) from the nearest edge of the framing member or furring strip, or (2) use steel plates. Refer to *300.4(D)*, *320.17*, and *334.17*. See Figures 4-17, 4-28, 4-29, and 4-30.

Metal Construction

Metal studs are being used more and more. They don't burn, and termites don't eat them. They are stronger than wood and can be set 20 in. (500 mm) on

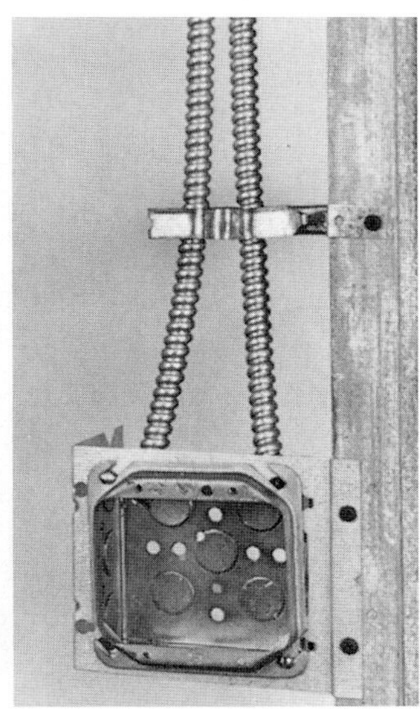

FIGURE 4-29 Devices that can be used to meet the *Code* requirement to maintain a 1¼ in. (32 mm) clearance from framing members.

center instead of the usual 16 in. (400 mm) spacing. Metal studs are 10 times more conductive than wood, so treatment of insulation must be addressed.

Where nonmetallic-sheathed cables pass through holes or slots in metal framing members, the cable must be protected by a listed bushing or listed grommet securely fastened in place prior to installing the cable. The bushing or grommet shall:

- remain in place during the wall finishing process,
- cover the complete opening, and
- be listed for the purpose of cable protection.

This information is found in *300.4(B)(1)* and *334.17*. See Figure 4-31. Bushings that snap into a hole sized for the specific size bushing (½ in., ¾ in., etc.), or a two-piece snap-in bushing that fits just about any size precut hole in the metal framing members are acceptable to use, as shown in Figure 4-31. These types of bushings (insulators) are not required in metal framing members when the wiring method is Type AC, Type MC, EMT, FMC, or electrical nonmetallic tubing.

When running nonmetallic-sheathed cable or electrical nonmetallic tubing through metal framing members where there is a likelihood that nails or screws could be driven into the cables or tubing, a steel plate, steel sleeve, or steel clip at least 1⁄16 in. (1.6 mm) thick must be installed to protect the cables and/or tubing. This is spelled out in *300.4(B)(2)*. See Figures 4-30 and 4-31.

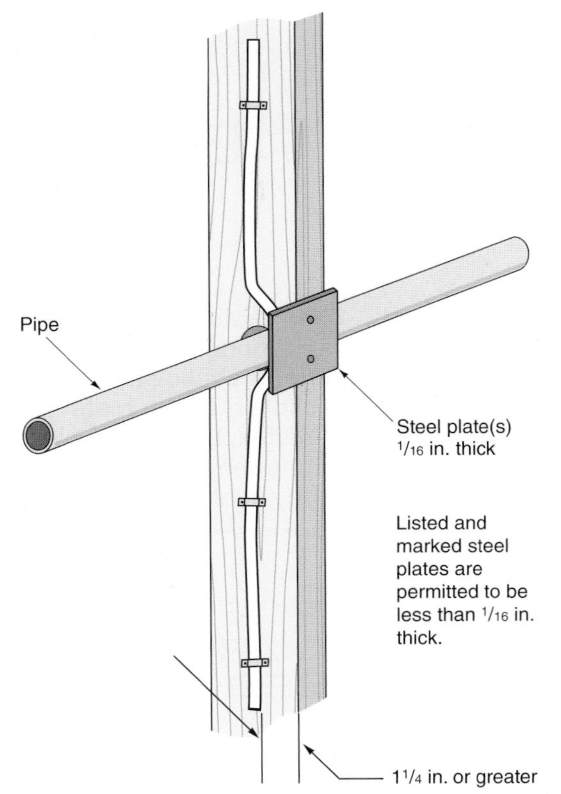

FIGURE 4-30 Where it is necessary to cross pipes or other similar obstructions, making it impossible to maintain at least 1¼ in. (32 mm) clearance from the framing member to the cable, install a steel plate(s) at least 1⁄16 in. (1.6 mm) thick to protect the cable from possible damage from nails or screws. See *300.4(A)(1)* and *(A)(2)*.

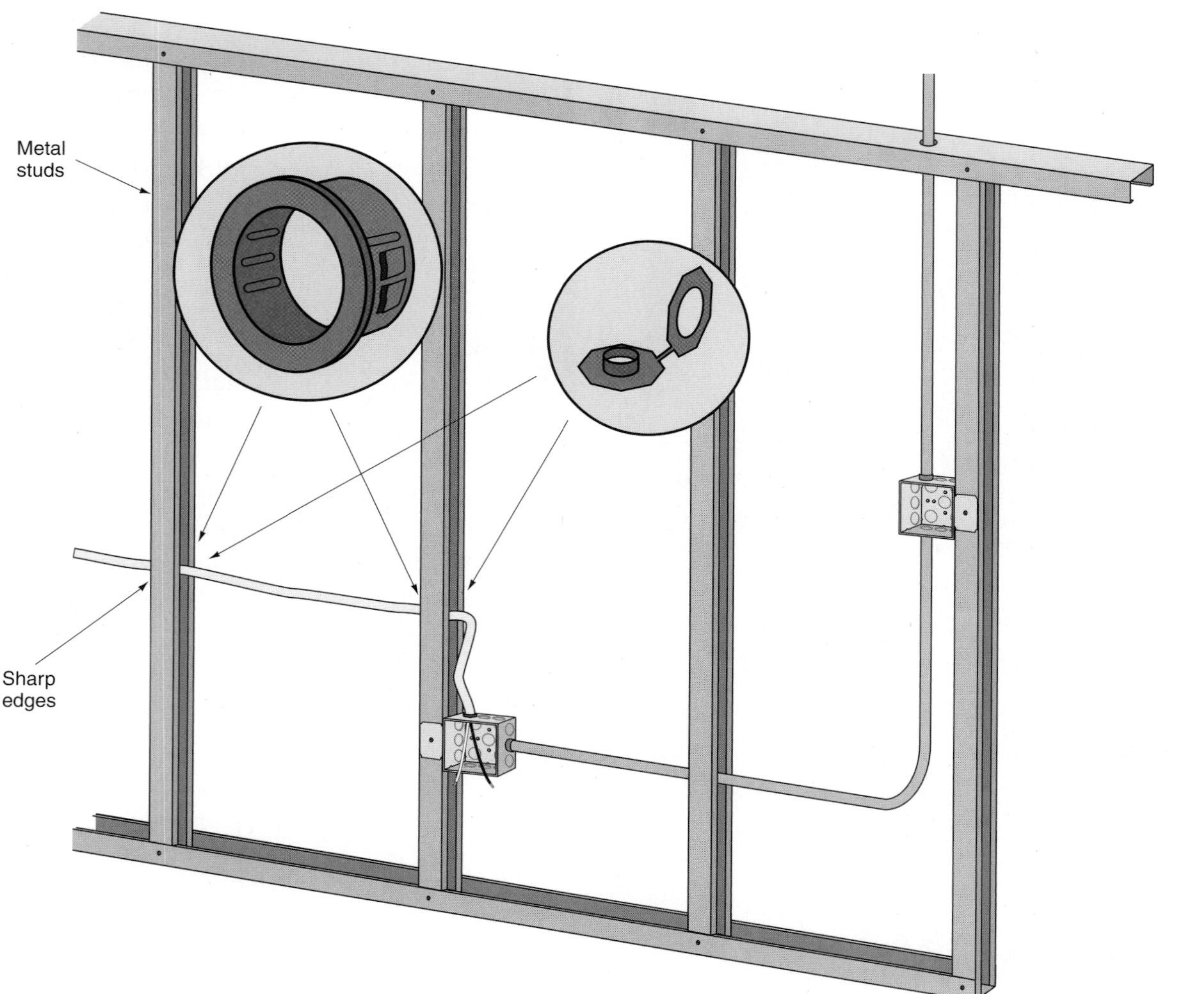

FIGURE 4-31 *NEC 300.4(B)* and *334.17* require protection of nonmetallic-sheathed cable where it runs through holes in metal framing members. Shown are two types of listed bushings (grommets) that can be installed to protect nonmetallic-sheathed cable from abrasion. Electricians generally avoid using nonmetallic-sheathed cable through metal studs and joists. They use electrical metallic tubing, flexible metal conduit, Type AC cable, Type MC cable, or electrical nonmetallic tubing. Where there is a likelihood that nails or screws might penetrate the nonmetallic-sheathed cable, $\frac{1}{16}$ in. (1.6 mm) thick steel plates, sleeves, or clips must be installed to protect the cable or electrical nonmetallic tubing.

CABLES IN SHALLOW GROOVES AND CHASES

NEC 300.4(F) addresses the wiring problems associated with cutting grooves into building material where there is no hollow space for the wiring. Nonmetallic-sheathed cable and some raceways must be protected when laid in a groove. Protection must be provided with a $\frac{1}{16}$ in. (1.6 mm) steel plate(s), sleeve, or equivalent, or the groove must be deep enough to allow not less than 1¼ in. (32 mm) free space for the entire length of the groove. This situation can be encountered where styrofoam insulation building blocks are grooved to receive the electrical cables, then covered with wallboard, wood paneling, or other finished wall material. Another typical example is where solid wood planking is installed on top of wood beams. The top side of the planking is covered with roofing material, and the bottom side is exposed and serves as the finished ceiling. The only way to install wiring for ceiling luminaires and fans is to groove the planking in some manner, as shown in Figure 4-32.

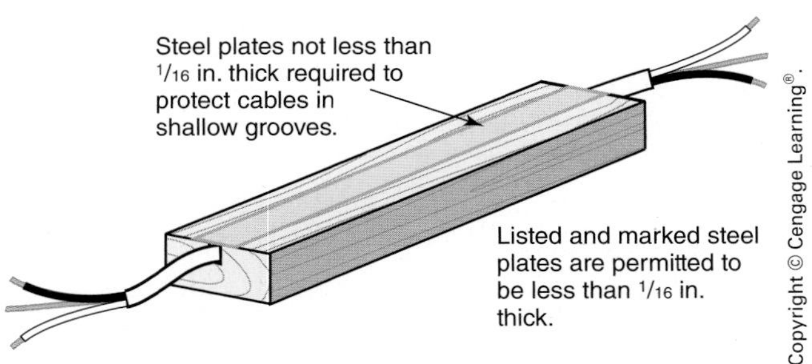

FIGURE 4-32 Example of how a solid wood framing member or Styrofoam insulating block might be grooved to receive the nonmetallic-sheathed cable. After the cable is laid in the groove, a steel plate(s) not less than 1⁄16 in. (1.6 mm) thick must be installed over the groove to protect the cable from damage, such as from driven nails or screws. See *300.4(F)* and *334.15(B)*.

This additional protection is not required when the raceway is intermediate metal conduit *(Article 342)*, rigid metal conduit *(Article 344)*, rigid PVC conduit *(Article 352)*, or electrical metallic tubing *(Article 358)* (see Figure 4-31).

NEC 334.30(B)(1), and *300.4(D), Exception 2*, allow cables to be "fished" between outlet boxes or other access points without additional protection.

In the recreation room of the residence plans, precautions must be taken so there is adequate protection for the wiring. If the carpenter uses 1 × 2 or 2 × 2 furring strips, as in Figure 4-33, nonmetallic-sheathed cable would require the additional 1⁄16 in. (1.6 mm) steel plate(s) protection or the use of *stand-off* devices similar to those illustrated in Figure 4-29.

Some building contractors attach 1 in. insulation to the walls, then construct a 2 × 4 wall in front of the basement structural foundation wall, leaving 1 in. spacing between the insulation and the back side of the 2 × 4 studs. This makes it easy to run cables or conduits behind the 2 × 4 studs and eliminates the need for the additional mechanical protection, as shown in Figure 4-34.

Watch out for any wall partitions where 2 × 4 studs are installed "flat," as in Figure 4-35. In this case, the choice is to provide the required mechanical protection as required by *300.4*, or to install the wiring in EMT.

See Chapter 15 concerning the installation of cables in attics.

Building Codes

When complying with the *NEC*, be sure to also comply with the building code. Building codes are concerned with the weakening of framing members when drilled or notched. Meeting the requirements of the *NEC* can be in conflict with building codes and vice versa. Here is a recap of some of the key building code requirements found in the ICC *International Residential Code*.

Studs.

- Cuts or notches in exterior or bearing partitions shall not exceed 25% of the stud's width.
- Cuts or notches in nonbearing partitions shall not exceed 40% of the stud's width.

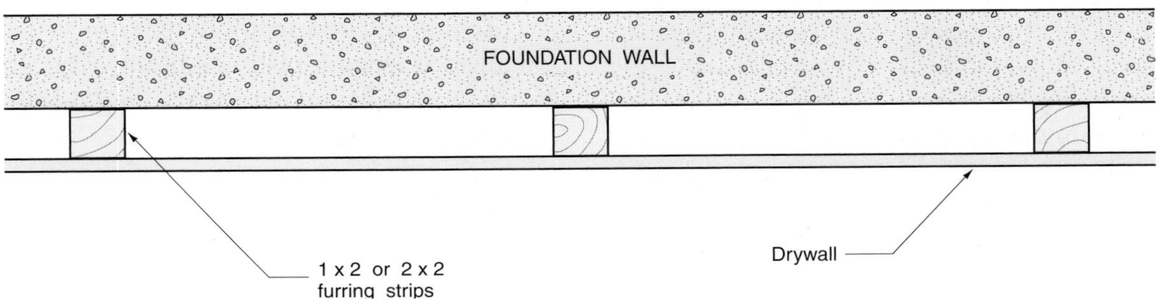

FIGURE 4-33 When 1 × 2 or 2 × 2 furring strips are installed on the surface of a basement wall, additional protection, as shown in Figures 4-28, 4-29, 4-31, and 4-32, must be provided. It is impossible to maintain the minimum distance of 1¼ in. (32 mm) from the nearest edge of the framing member, in which case protection is required the entire length of the cable run, *300.4(D)*.

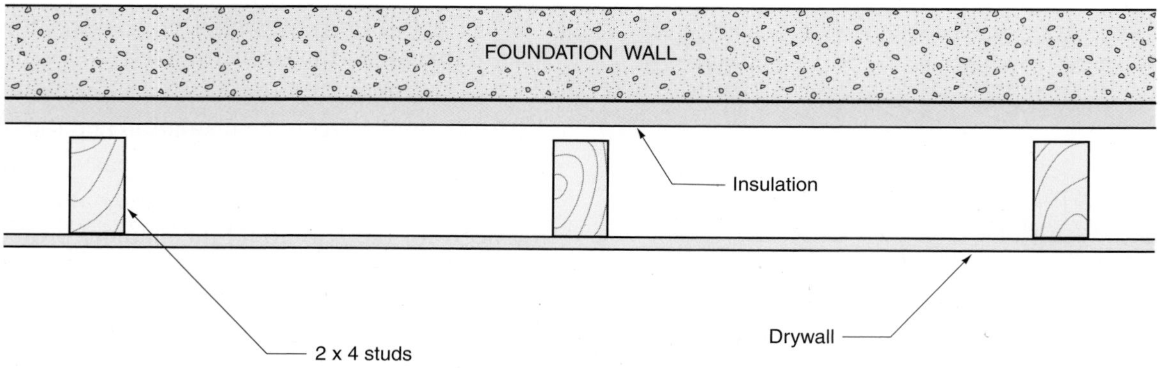

FIGURE 4-34 This sketch shows how to insulate a foundation wall and how to construct a wall that provides a space to run cables and/or raceways that will not require the additional protection as shown in Figures 4-28, 4-29, 4-31, and 4-32.

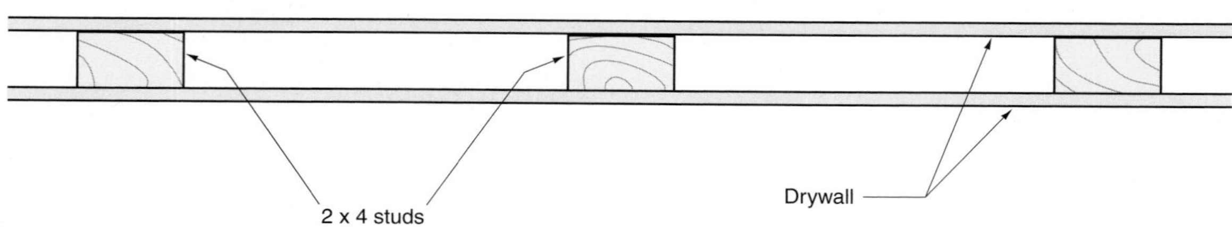

FIGURE 4-35 When 2 × 4 studs are installed "flat," such as might be found in nonbearing partitions, cables running parallel to or through the studs must be protected against the possibility of having nails or screws driven through the cables. This means protecting the cable with 1/16 in. (1.6 mm) steel plates or equivalent for the entire length of the cable, or installing the wiring using intermediate metal conduit, rigid metal conduit, rigid PVC conduit, or EMT. See Figures 4-28, 4-29, 4-30, and 4-31.

- Bored or drilled holes in any stud:
  - shall not be greater than 40% of the stud's width,
  - shall not be closer than 5/8 in. (15.9 mm) to the edge of the stud, and
  - shall not be located in the same section as a cut or notch.

Joists.

- Notches on the top or bottom of joists shall not exceed 1/6 the depth of the joist, shall not be longer than 1/3 the width of the joists, and shall not be in the middle third of the joist.
- Holes bored in joists shall not be within 2 in. (50 mm) of the top or bottom of the joist or any other hole, and the diameter of any such hole shall not exceed 1/3 the depth of the joist. If the joist is also notched, the hole shall not be closer than 2 in. (50 mm) to the notch.

Engineered Wood Products

Today, we find many homes constructed with engineered wood I-joists and beams. The strength of these factory-manufactured framing members conforms to various building codes. It is critical that certain precautions be taken relative to drilling and cutting these products so as not to weaken them. In general:

- Do not cut, notch, or drill holes in trusses, laminated veneer lumber, glue-laminated members, or I-joists unless specifically permitted by the manufacturer.

- Do not cut, drill, or notch flanges. Flanges are the top and bottom pieces.
- Do not cut holes too close to supports or to other holes.
- For multiple holes, the amount of web to be left between holes must be at least twice the diameter of the largest adjacent hole. The web is the piece between the flanges.
- Holes not over 1½ in. (38 mm) usually can be drilled anywhere in the web.
- Do not hammer on the web except to remove knockout prescored holes.
- Check and follow the manufacturer's instructions!

INSTALLATION OF CABLES THROUGH DUCTS

NEC 300.22 is extremely strict as to what types of wiring methods are permitted for installation of cables through ducts or plenum chambers. These stringent rules are for fire safety.

The *Code* requirements are somewhat relaxed by *300.22(C), Exception*. In this exception, permission is given to run nonmetallic-sheathed and armored cables in joist and stud spaces (i.e., cold air returns), but only if they run perpendicular to the long dimensions of such spaces, as illustrated in Figure 4-36.

CONNECTORS FOR INSTALLING NONMETALLIC-SHEATHED AND ARMORED CABLE

The connectors shown in Figure 4-37 are used to fasten nonmetallic-sheathed cable and armored cable to the boxes and panels in which they terminate. These connectors clamp the cable securely to each outlet box. Many boxes have built-in clamps and do not require separate connectors.

The question continues to be asked: May more than one cable be inserted in one connector? Unless the UL listing of a specific cable connector indicates that the connector has been tested for use with more than one cable, the rule is *one cable, one connector*. Connectors suitable for more than one cable will be marked on the carton.

INTERMEDIATE METAL CONDUIT (*ARTICLE 342*), RIGID METAL CONDUIT (*ARTICLE 344*), RIGID PVC CONDUIT (*ARTICLE 352*), AND ELECTRICAL METALLIC TUBING (*ARTICLE 358*)

Some communities do not permit cable wiring. They require that the wiring method be a raceway—metallic or nonmetallic. You will have to check this out before starting a project.

The following comparison chart provides the highlights of the *NEC* requirements for different types of raceways that might be used in a residence. For more detailed information, refer to the manufacturer's installation instructions, the specific *NEC* article, and the UL *White Book*.

Grounding. All listed fittings for listed metallic raceways and listed metal jacketed cables are tested to carry a specified amount of fault current for a specified length of time, making the fittings acceptable for grounding when properly installed.

In this residence, the wiring method for all of the exposed wiring in the workshop is to be electrical metallic tubing (EMT). Wiring with EMT is covered in detail in Chapter 18.

Table 4-14 provides a summary of the requirements for installing intermediate metal conduit (IMC), rigid metal conduit (RMC), rigid polyvinyl chloride conduit (PVC), and electrical metallic tubing (EMT).

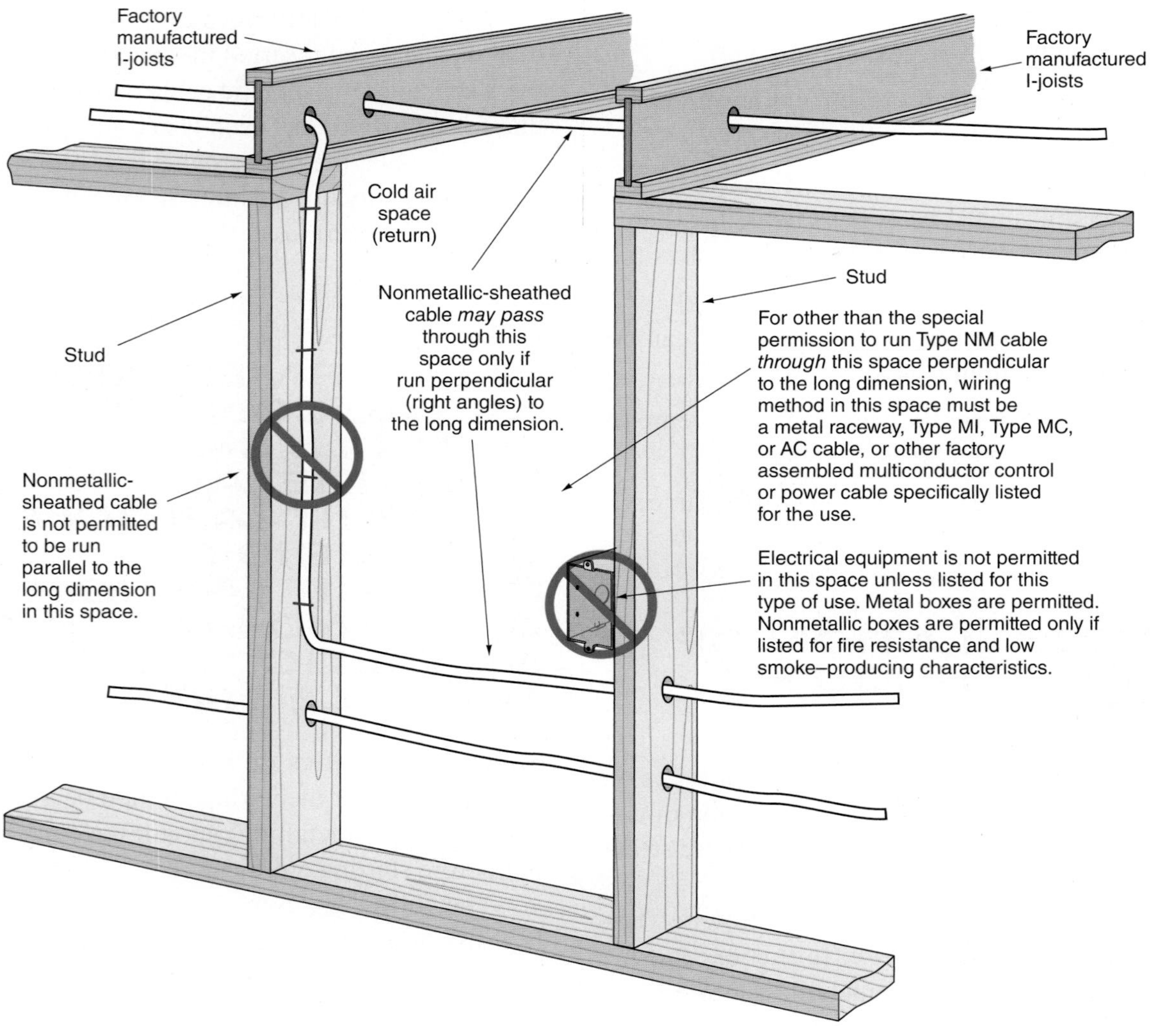

Factory manufactured I-joists

Factory manufactured I-joists

Cold air space (return)

Nonmetallic-sheathed cable *may pass* through this space only if run perpendicular (right angles) to the long dimension.

Stud

Stud

For other than the special permission to run Type NM cable *through* this space perpendicular to the long dimension, wiring method in this space must be a metal raceway, Type MI, Type MC, or AC cable, or other factory assembled multiconductor control or power cable specifically listed for the use.

Nonmetallic-sheathed cable is not permitted to be run parallel to the long dimension in this space.

Electrical equipment is not permitted in this space unless listed for this type of use. Metal boxes are permitted. Nonmetallic boxes are permitted only if listed for fire resistance and low smoke–producing characteristics.

FIGURE 4-36 Nonmetallic-sheathed cable may pass through cold air return, joist, or stud spaces in dwellings, but only if it is run at right angles to the long dimension of the space, *300.22(C), Exception.*

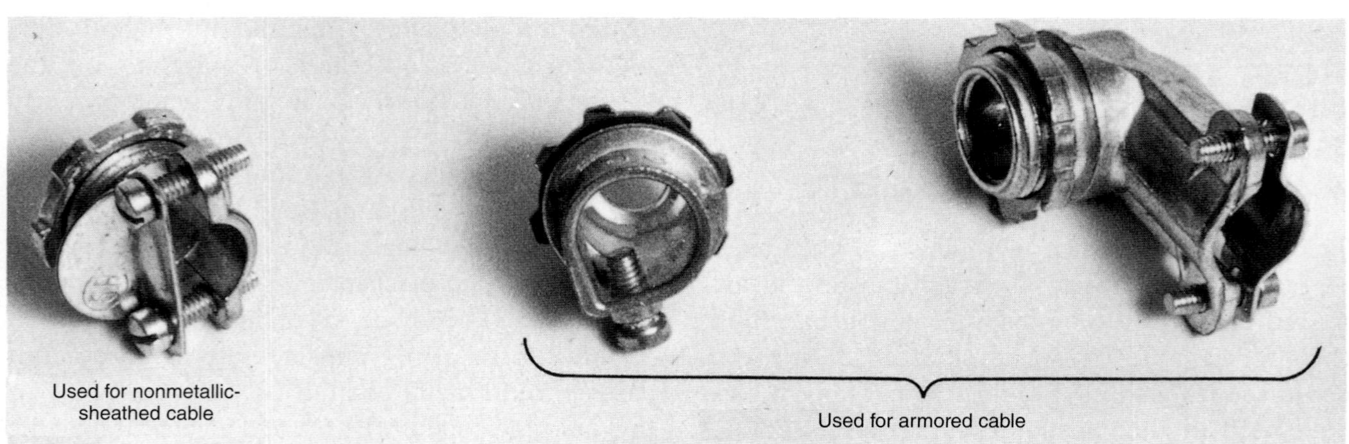

FIGURE 4-37 Cable connectors.

TABLE 4-14

Conduit and tubing.

	INTERMEDIATE METAL CONDUIT (IMC) *ARTICLE 342*	RIGID METAL CONDUIT (RMC) *ARTICLE 344*	RIGID POLYVINYL CHLORIDE CONDUIT: TYPE PVC *ARTICLE 352*	ELECTRICAL METALLIC TUBING (EMT) *ARTICLE 358*
Bends	Not more than the equivalent of four quarter bends between pull points. 360° total. Manufactured bends available. Field bends made with benders.	Not more than the equivalent of four quarter bends between pull points. 360° total. Manufactured bends available. Field bends made with benders.	Not more than the equivalent of four quarter bends between pull points. 360° total. Manufactured bends available. Field bends made with benders that electrically heat the PVC to just the right temperature.	Not more than the equivalent of four quarter bends between pull points. 360° total. Manufactured bends available. Field bends made with benders.
Bend radius	See *Table 2, Chapter 9, NEC.*	See *Table 2, Chapter 9, NEC.*	See *Table 2, Chapter 9, NEC.*	See *Table 2, Chapter 9, NEC.*
Bushings	Required where a conduit enters a box, fitting, or other enclosure unless the design of the box, fitting, or enclosure provides needed protection, *342.46.* Where 4 AWG or larger ungrounded conductors are installed, smoothly rounded insulating fittings or the equivalent must be used, *300.4(G)* and *312.6(C).* See Figure 27-30 and Figure 27-31. Not required where the RMC terminates in a hub or boss. See Figure 27-16.	Required where a conduit enters a box, fitting, or other enclosure unless the design of the box, fitting, or enclosure provides needed protection, *344.46.* Where 4 AWG or larger ungrounded conductors are installed, smoothly rounded insulating fittings or the equivalent must be used, *300.4(G)* and *312.6(C).* See Figure 27-30 and Figure 27-31. Not required where the RMC terminates in a hub or boss. See Figure 27-16.	Required where a conduit enters a box, fitting, or other enclosure unless the design of the box, fitting, or enclosure provides needed protection, *352.46* Where 4 AWG or larger ungrounded conductors are installed, smoothly rounded insulating fittings or the equivalent must be used, *300.4(G)* and *312.6(C).* See Figure 27-30 and Figure 27-31. Not required where the PVC terminates in a hub or boss. See Figure 27-16.	Where 4 AWG or larger ungrounded conductors are installed, smoothly rounded insulating fittings or the equivalent must be used, *300.4(G)* and *312.6(C).* See Figure 27-30 and Figure 27-31. Not required where the PVC terminates in a hub or boss. See Figure 27-16.
Cable permitted inside	Yes, if permitted by the specific cable *Code* article.	Yes, if permitted by the specific cable *Code* article.	Yes, if permitted by the specific cable *Code* article.	Yes, if permitted by the specific cable *Code* article.
Cinder fill	Yes, where subject to permanent moisture if protected on all sides by noncinder concrete not less than 2 in. (50 mm), or if at least 18 in. (450 mm) under the fill, or if it has corrosion protection.	Yes, where subject to permanent moisture if protected on all sides by noncinder concrete not less than 2 in. (50 mm), or if at least 18 in. (450 mm) under the fill, or if it has corrosion protection.	Yes	Yes, where subject to permanent moisture if protected on all sides by noncinder concrete not less than 2 in. (50 mm), or if at least 18 in. (450 mm) under the fill, or if it has corrosion protection.

(*continues*)

TABLE 4-14 (*continued*)

	INTERMEDIATE METAL CONDUIT (IMC) *ARTICLE 342*	RIGID METAL CONDUIT (RMC) *ARTICLE 344*	RIGID POLYVINYL CHLORIDE CONDUIT: TYPE PVC *ARTICLE 352*	ELECTRICAL METALLIC TUBING (EMT) *ARTICLE 358*
Concealed in walls, ceilings, and floors	Yes	Yes	Yes	Yes
Concrete above grade	Yes	Steel: Yes. Aluminum: Yes	Yes	Steel: Yes. Aluminum: Yes, if it has supplemental corrosion protection.
Concrete on grade	Yes	Steel: Yes for galvanized. Steel: No for nongalvanized. Aluminum: Needs supplemental corrosion protection.	Yes	Steel: Yes Aluminum: Yes, if it has supplemental corrosion protection.
Concrete below grade	Yes	Steel: Yes for galvanized. Steel: No for nongalvanized. Aluminum: Needs supplemental corrosion protection.		Steel: Yes, if it has supplemental corrosion protection. Aluminum: Yes, if it has supplemental corrosion protection.
Conductor fill	As determined by *Table 1, Chapter 9, NEC*. Inside interior diameter greater than RMC of same trade size. More room for conductors. See Chapter 18 for examples of conductor fill calculations.	As determined by *Table 1, Chapter 9, NEC*. See Chapter 18 for examples of conductor fill calculations.	As determined by *Table 1, Chapter 9, NEC*. See Chapter 18 for examples of conductor fill calculations.	As determined by *Table 1, Chapter 9, NEC*. See Chapter 18 for examples of conductor fill calculations.
Conductors, when to install?	After complete raceway system is installed, *300.18(A)*.	After complete raceway system is installed, *300.18(A)*.	After complete raceway system is installed, *300.18(A)*.	After complete raceway system is installed, *300.18(A)*.
Corrosive conditions	Yes, if protected by corrosion protection and judged suitable for the condition. Listing will indicate this.	Yes, if protected by corrosion protection and judged suitable for the condition. Listing will indicate this. Aluminum resists most corrosive atmospheres.	Yes. Like those found around swimming pool areas.	Yes, if protected by corrosion protection and judged suitable for the condition. Listing will indicate this.
Dry and damp locations	Yes	Yes	Yes	Yes
Expansion fittings for thermal expansion	Not required.	Not required.	Yes. Refer to *Table 352.44(A)* and *(B)*. Without expansion fittings, RNC in direct sunlight will pull out of hubs of boxes, or will pull the boxes off the wall.	Not required.
Exposed	Yes	Yes	Yes	Yes

(*continues*)

TABLE 4-14 (*continued*)

	INTERMEDIATE METAL CONDUIT (IMC) *ARTICLE 342*	RIGID METAL CONDUIT (RMC) *ARTICLE 344*	RIGID POLYVINYL CHLORIDE CONDUIT: TYPE PVC *ARTICLE 352*	ELECTRICAL METALLIC TUBING (EMT) *ARTICLE 358*
Fittings	Threaded and threadless.	Threaded and threadless compression. For aluminum conduit, aluminum threaded fittings recommended, but cadmium plated or galvanized fittings are satisfactory in most installations.	Attached with solvent-type cement.	Set screw and threadless compression.
Grounding	Is considered to be an equipment grounding conductor.	Is considered to be an equipment grounding conductor.	Where required, a separate equipment grounding conductor must be installed.	Is considered to be an equipment grounding conductor.
Joints	Use listed couplings, connectors, and fittings.	Use listed couplings, connectors, and fittings.	Use proper cement.	Use listed couplings, connectors, and fittings.
Length	10 ft	10 ft	10 ft	10 ft
Listed	Yes	Yes	Yes	Yes
Material	Steel–galvanized. Is as strong as same trade size RMC.	Steel–galvanized. Also aluminum, red brass, and stainless steel.	Polyvinyl chloride (PVC). Schedule 40 (heavy wall thickness). Schedule 80 (extra-heavy wall thickness). O.D. same for both. I.D. of Schedule 40 is greater than that of Schedule 80.	Steel—galvanized. Also aluminum.
Rough edges	Must be properly reamed to remove rough edges.	Must be properly reamed to remove rough edges.	Must be properly reamed to remove rough edges.	Must be properly reamed to remove rough edges.
Support of conduit bodies	Yes	Yes	Yes. Conduit bodies not to contain device or support luminaires or other equipment.	Yes
Securing	Maximum 3 ft (900 mm) from box, panel, cabinet, conduit body, etc.	Maximum 3 ft (900 mm) from box, panel, cabinet, conduit body, etc.	Maximum 3 ft (900 mm) from box, panel, cabinet, conduit body, etc.	Maximum 3 ft (900 mm) from box, panel, cabinet, conduit body, etc.
Supporting	At least every 10 ft (3 m). See *Table 344.30(B) (2)*. See Figure 4-36.	At least every 10 ft (3 m). See *Table 344.30(B)(2)*. See Figure 4-36.	See *Table 352.30* for distance between supports. See Table 4-15 and Figure 4-38.	At least every 10 ft (3 m). See Figure 4-38.
Sizes	Trade sizes ½ through 4.	Trade sizes ½ through 6.	Trade sizes ½ through 6.	Steel: Trade sizes ½ through 4. Aluminum: Trade sizes 2 through 4.
Threaded	Yes. Same threads as RMC. Field cut threads shall be coated with an approved electrically conductive corrosion-resistant compound.	Yes. Field cut threads shall be coated with an approved electrically conductive corrosion-resistant compound.	No	No

(*continues*)

TABLE 4-14 *(continued)*

	INTERMEDIATE METAL CONDUIT (IMC) *ARTICLE 342*	RIGID METAL CONDUIT (RMC) *ARTICLE 344*	RIGID POLYVINYL CHLORIDE CONDUIT: TYPE PVC *ARTICLE 352*	ELECTRICAL METALLIC TUBING (EMT) *ARTICLE 358*
Underground in soil.	Yes	Steel: Yes for galvanized. Steel: No for nongalvanized. Aluminum: Needs supplemental corrosion protection	Yes	Steel: Yes, if it has supplemental corrosion protection. Aluminum: Yes, if it has supplemental corrosion protection.
Weight	Medium. For example, trade size ½: 100 ft (30.5 m) approx. 62 lb (28.1 kg).	Heavy. For example, trade size ½: 100 ft (30.5 m) approx. 82 lb (37.2 kg). Aluminum is approx. ⅓ as heavy as steel. For example, trade size ½: 100 ft (30.5 m) approx. 28 lb (12.7 kg).	Very light. For example, trade size ½ Schedule 40: 100 ft (30.5 m) approx. 17 lb (7.7 kg); Schedule 80 approx. 21 lb (9.53 kg).	Light. For example, trade size ½: 100 ft (30.5 m) approx. 30 lb (13.5 kg).
Wet locations	Yes, if raceway, fittings, bolts, screws, straps, etc., are of corrosion-resistant material.	Yes, if raceway, fittings, bolts, screws, straps, etc., are of corrosion-resistant material.	Yes, if raceway, fittings, bolts, screws, straps, etc., are of corrosion-resistant material.	Yes, if raceway, fittings, bolts, screws, straps, etc., are of corrosion-resistant material.

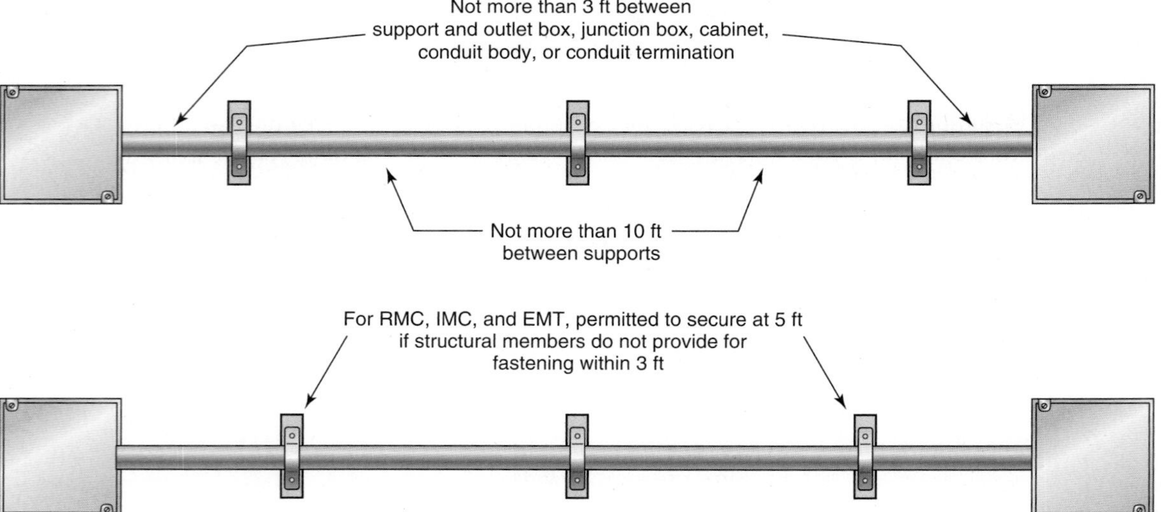

FIGURE 4-38 *Code* requirements for securing and supporting raceways are found in *342.30* (intermediate metal conduit), *344.30* (rigid metal conduit), *352.30* (rigid PVC conduit), and *358.30* (electrical metallic tubing).

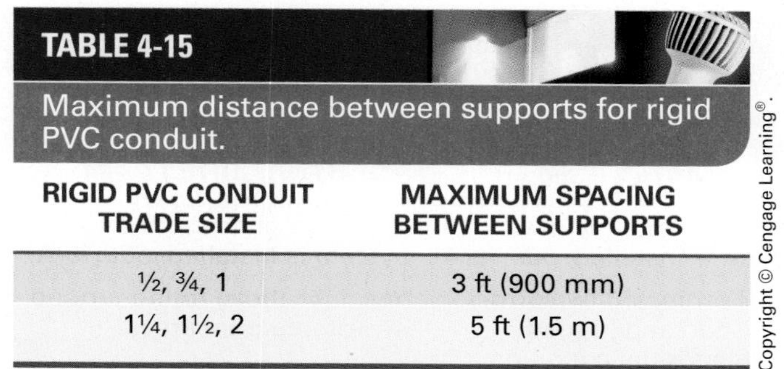

TABLE 4-15

Maximum distance between supports for rigid PVC conduit.

RIGID PVC CONDUIT TRADE SIZE	MAXIMUM SPACING BETWEEN SUPPORTS
½, ¾, 1	3 ft (900 mm)
1¼, 1½, 2	5 ft (1.5 m)

FLEXIBLE CONNECTIONS

The installation of some equipment requires flexible connections, both to simplify the installation and to stop the transfer of vibrations. In residential wiring, flexible connections are used to connect equipment such as attic fans, food waste disposers, dishwashers, air conditioners, heat pumps, and recessed luminaires.

Table 4-16 shows the *NEC* requirements for three types of commonly used flexible conduit.

ELECTRICAL NONMETALLIC TUBING (ENT) (*ARTICLE 362*)

Electrical nonmetallic tubing (ENT) is covered in *Article 362* of the *NEC*. It is a pliable, lightweight, corrugated raceway made of PVC. It requires fittings made specifically for the ENT. Figure 4-46 shows color-coded ENT and some fittings. Color-coded ENT can be used to differentiate power, communication, and fire protection signaling system.

Although not often used in typical house wiring, some electricians find ENT very practical for providing a raceway system for installing structured wiring conductors and cables. Structured wiring is discussed in Chapter 31.

TABLE 4-16

Flexible connections.

	FLEXIBLE METAL CONDUIT (FMC) *ARTICLE 348*	LIQUIDTIGHT FLEXIBLE METAL CONDUIT (LFMC) *ARTICLE 350*	LIQUIDTIGHT FLEXIBLE NONMETALLIC CONDUIT (LFNC) *ARTICLE 356*
	See Figure 4-39(A).	See Figure 4-39(B).	See Figure 4-39(C).
Bends	Not more than the equivalent of four quarter bends between pull points. 360° total.	Not more than the equivalent of four quarter bends between pull points. 360° total.	Not more than the equivalent of four quarter bends between pull points. 360° total.
Bend radius	See *Table 2, Chapter 9, NEC.*	See *Table 2, Chapter 9, NEC.*	See *Table 2, Chapter 9, NEC.*
Bushings	Where 4 AWG or larger ungrounded conductors are installed, smoothly rounded insulating fittings or the equivalent must be used, *300.4(G)* and *312.6(C)*.	Where 4 AWG or larger ungrounded conductors are installed, smoothly rounded insulating fittings or the equivalent must be used, *300.4(G)* and *312.6(C)*.	Where 4 AWG or larger ungrounded conductors are installed, smoothly rounded insulating fittings or the equivalent must be used, 300.4(G) and 312.6(C).
Cable permitted inside	Yes, if permitted by the specific cable *Code* article.	Yes, if permitted by the specific cable *Code* article.	Yes, if permitted by the specific cable *Code* article.
Concealed in walls, ceilings, and floors	Yes.	Yes.	Yes.
Concrete above grade	No.	Yes, if marked "Direct Burial."	Yes, if marked "Direct Burial."
Concrete on grade	No.	Yes, if marked "Direct Burial."	Yes, if marked "Direct Burial."
Concrete	No.	Yes, if marked "Direct Burial."	Yes, if marked "Direct Burial."
Conductor fill	As determined by *Tables 1* and *4, Chapter 9, NEC.* For trade size ⅜, use *Table 348.22.*	As determined by *Table 1, Chapter 9, NEC.* For trade size ⅜, use *Table 348.22.*	As determined by *Table 1, Chapter 9, NEC.* For trade size ⅜, use *Table 348.22.*

(*continues*)

TABLE 4-16 *(continued)*

	FLEXIBLE METAL CONDUIT (FMC) *ARTICLE 348*	LIQUIDTIGHT FLEXIBLE METAL CONDUIT (LFMC) *ARTICLE 350*	LIQUIDTIGHT FLEXIBLE NONMETALLIC CONDUIT (LFNC) *ARTICLE 356*
Conductors, when to install?	After complete raceway system is installed, *300.18(A)*.	After complete raceway system is installed, *300.18(A)*.	After complete raceway system is installed, *300.18(A)*.
Corrosive conditions	Yes, if protected by corrosion protection and judged suitable for the condition. Listing will indicate this.	Yes, if protected by corrosion protection and judged suitable for the condition. Listing will indicate this.	Yes. Like those found around swimming pool areas.
Dry and damp locations	Yes.	Yes.	Yes.
Exposed	Yes.	Yes.	Yes.
Fittings	Must be listed. Sizes larger than trade size ¾ will be marked GRND if acceptable for grounding. Do not conceal angle-type fittings. See Figure 4-40.	Must be listed. Sizes larger than trade size ¾ will be marked GRND if acceptable for grounding. Do not conceal angle-type fittings. See Figure 4-40.	Must be listed. Do not conceal angle-type fittings. See Figure 4-40.
Grounding	When terminated with listed fittings, it is considered to be an EGC • if it is not over 6 ft (1.8 m) long and the overcurrent device does not exceed 20 amperes. See Figure 4-41 and Figure 4-42. Longer than 6 ft (1.8 m), not acceptable as a grounding means. Install a separate equipment grounding conductor sized per *Table 250.122*. If used to connect equipment where flexibility is necessary to minimize the transmission of vibration from equipment or to provide flexibility for equipment that requires movement after installation, an equipment grounding conductor shall be installed. Sized per *250.122*. For grounding, install per *250.134(B)*. For bonding, install per *250.102*. See *250.118(5)*.	When terminated with listed fittings, it is considered to be an EGC • in trade sizes ⅜ and ½ not over 6 ft (1.8 m) long and the overcurrent device does not exceed 20 amperes. • In trade sizes ¾, 1, and 1¼ not over 6 ft (1.8 m) long and the overcurrent device does not exceed 60 amperes. Trade size 1½ and larger, not acceptable for grounding. See Figures 4-41, 4-42, 4-43, and 4-44. Longer than 6 ft (1.8 m), LFMC is not acceptable as a grounding means. Install a separate equipment grounding conductor sized per *Table 250.122*. If used to connect equipment where flexibility is necessary to minimize the transmission of vibration from equipment or to provide flexibility for equipment that requires movement after installation, an equipment grounding conductor shall be installed. Size per *250.122*. For grounding, install per *250.134(B)*. For bonding, install per *250.102*. See *250.118(6)*.	For grounding, install an equipment grounding conductor sized per *250.122*. Install per *250.134(B)*. For bonding, install per *250.102*. See Figures 4-43 and 4-45.
Length	No limit.	No limit.	Type LFNC-B permitted longer than 6 ft (1.8 m) if secured per *356.30*.

(continues)

TABLE 4-16 *(continued)*

	FLEXIBLE METAL CONDUIT (FMC) *ARTICLE 348*	LIQUIDTIGHT FLEXIBLE METAL CONDUIT (LFMC) *ARTICLE 350*	LIQUIDTIGHT FLEXIBLE NONMETALLIC CONDUIT (LFNC) *ARTICLE 356*
Listed	Must be listed. See UL 1.	Must be listed. See UL 360.	Must be listed. See UL 1660.
Material	Steel and aluminum. Must not be used where subject to physical damage.	Steel with nonmetallic outer jacket. Must not be used where subject to physical damage.	Nonmetallic. Three types. See 356.2 and UL *White Book*. Must not be used where subject to physical damage.
Rough edges	Must be properly trimmed to remove rough edges.	Must be properly trimmed to remove rough edges.	Must be properly trimmed to remove rough edges.
Securing and supporting	Maximum 12 in. (300 mm) from box, cabinet, or fitting. Maximum 4½ ft (1.4 m) between supports. Waived if fished through walls and ceilings—where flexibility is needed and the flex is not more than 3 ft (900 mm) from where it is terminated, or as a whip not over 6 ft (1.8 m). See Figures 4-45, 7-2, 7-6, and 7-10. Also waived when supported horizontally through framing members not more than 4½ ft (1.4 m) apart and secured within 12 in. (300 mm) of box, cabinet, or fitting.	Maximum 12 in. (300 mm) from box, cabinet, or fitting. Maximum 4½ ft (1.4 m) between supports. Waived if fished through walls and ceilings—where flexibility is needed and the flex is not more than 3 ft (900 mm) from where it is terminated, or as a whip not over 6 ft (1.8 m). See Figures 4-45, 7-2, 7-6, and 7-10. Also waived when supported horizontally through framing members not more than 4½ ft (1.4 m) apart and secured within 12 in. (300 mm) of box, cabinet, or fitting.	Maximum 12 in. (300 mm) from box, cabinet, or fitting. Maximum 4½ ft (1.4 m) between supports. Waived if fished through walls and ceilings—where flexibility is needed and the flex is not more than 3 ft (900 mm) from where it is terminated, or as a whip not over 6 ft (1.8 m). See Figures 4-45, 7-2, 7-6, and 7-10. Also waived when supported horizontally through framing members not more than 3 ft (900 mm) apart and secured within 12 in. (300 mm) of box, cabinet, or fitting.
Sizes	Steel: Trade sizes ½ through 4. Aluminum: Trade sizes ½ through 3. Trade size ⅜ okay for enclosing leads to a motor or as a luminaire (fixture) whip not longer than 6 ft (1.8 m).	Trade sizes ½ through 4. Trade size ⅜ okay for enclosing leads to a motor or as a luminaire (fixture) whip not longer than 6 ft (1.8 m).	Trade sizes ½ through 4. Trade size ⅜ okay for enclosing leads to a motor or as a luminaire (fixture) whip not longer than 6 ft (1.8 m).
In sunlight	Yes	Yes	Yes, if listed and marked "Outdoor."
Temperature	Limited to the temperature rating of the conductors.	Limited to 60°C unless marked otherwise. Also limited to temperature rating of the conductors.	Limited to 60°C unless marked otherwise. Also limited to temperature rating of the conductors. Can become brittle in extreme cold.
Underground in soil	No.	Yes, if marked "Direct Burial."	Yes, if marked "Direct Burial."
Uses	Where flexibility is required. Do not use where subject to physical damage.	Where flexibility is required. Do not use where subject to physical damage.	Where flexibility is required. Do not use where subject to physical damage.
Wet locations	No.	Yes.	Yes.

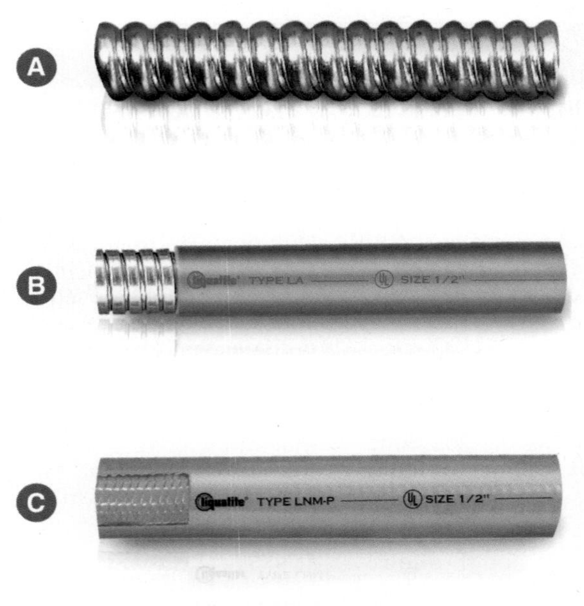

FIGURE 4-39 (A) Flexible metal conduit. (B) Liquidtight flexible metal conduit. (C) Liquidtight flexible nonmetallic conduit.

SERVICE-ENTRANCE CABLE (*ARTICLE 338*)

Service-entrance cable is covered in *Article 338* of the *NEC*.

Service-entrance cable is defined as A single conductor or multiconductor assembly provided with or without an overall covering, primarily used for services.*

As indicated in the definition of service-entrance cable in *NEC* 338.2, service entrance cables are available in two broad types:

- ***Type SE.*** Service-entrance cable having a flame-retardant, moisture-resistant covering.*
- ***Type USE.*** Service-entrance cable, identified for underground use, having a moisture-resistant covering, but not required to have a flame-retardant covering.*

Generally speaking, type SE cables are suitable for installation in or on buildings but are not suitable for installation underground. As shown in Figures 4-47 and 4-48, Type SE cable is typically available in two broad construction types: Types SEU and SER.

Type SEU cable usually has two conductors that are insulated with a bare conductor that is spiraled around the insulated conductors. When used for the service entrance, the bare conductor is usually spiraled around the insulated conductors to form a bare neutral conductor. This construction is also used for feeders and branch circuits where two insulated conductors and a bare equipment ground are used.

Type SER is round in construction and typically has three insulated conductors and one bare equipment grounding conductor. These cables are often used for larger branch circuits as well

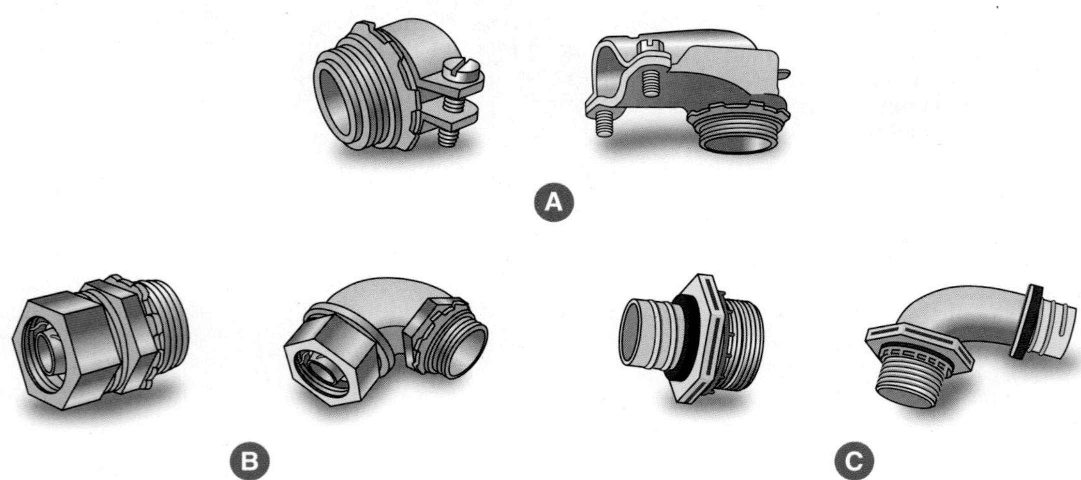

FIGURE 4-40 Various fittings for (A) flexible metal conduit, (B) liquidtight flexible metal conduit, and (C) liquidtight flexible nonmetallic conduit.

*Reprinted with permission from NFPA 70-2014.

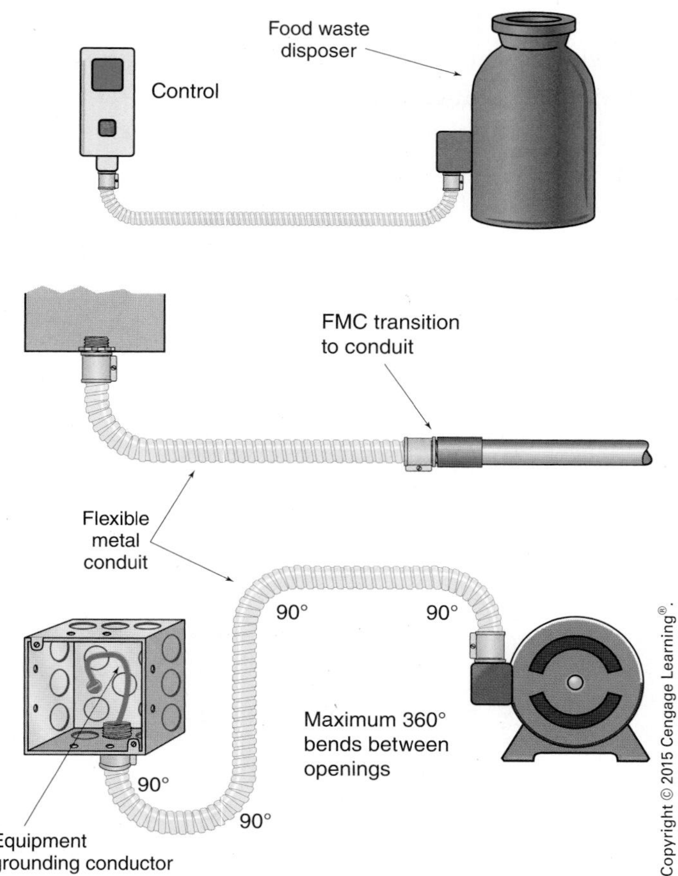

FIGURE 4-41 If FMC or LFMC are serving as an equipment ground return path, the total length shall not exceed 6 ft (1.8 m). Refer to *250.118(5)c*, and *250.118(6)d*.

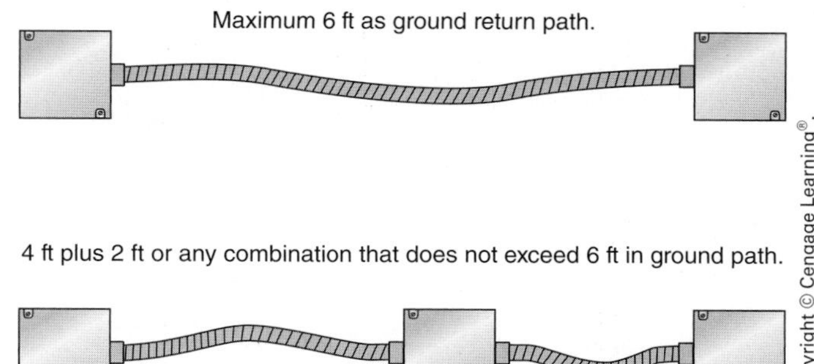

FIGURE 4-42 The "combined" length of flexible metal conduit and liquidtight flexible metal conduit shall not exceed 6 ft (1.8 m) to serve as a ground return path, *250.118(5)c*, and *250.118(6)d*.

as for feeders to panelboards where a separate neutral and equipment grounding conductor are required.

As indicated in the definition, Type USE cable is suitable for installation underground, but not in or on buildings. These cables are permitted to rise above ground for termination in meter bases or similar equipment on the outside of the building or structure. Type USE cable is also available with a combination of insulation types that makes it suitable for installation inside buildings. Sometimes this multiple insulation is referred to as "triple rating." The additional insulations are typically RHH and RHW. This results in a marking similar to USE/RHH/RHW. As shown in the Table 4-16, cable with this triple marking is permitted to be installed in any location indicated by the insulation type. For example, the USE insulation makes the cable suitable for direct burial applications; the RHH designation indicates the insulation is thermoset and rated

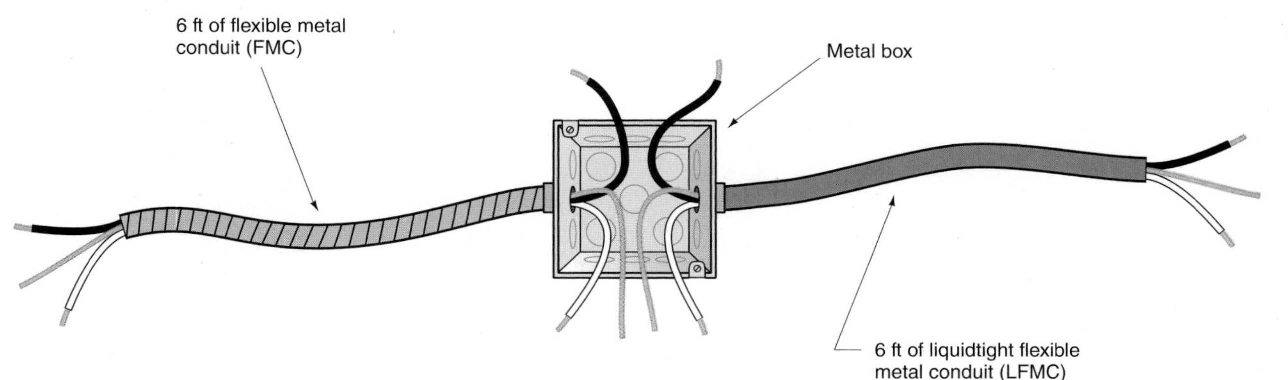

FIGURE 4-43 Use of flexible metal conduit or liquidtight flexible metal conduit.

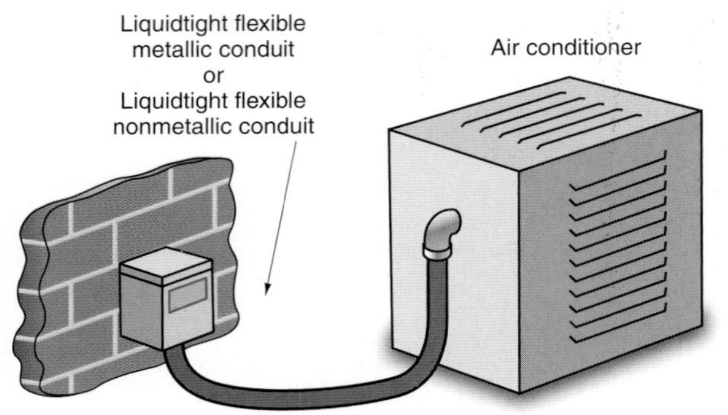

FIGURE 4-44 Common places where liquidtight flexible connections could be used.

90°C; the RHW designation indicates the insulation is thermoset, has a 75°C rating, and is suitable for wet locations.

Where permitted by local electrical codes, service-entrance cable is most often run from the utility's service drop (service point) to the meter base, then from the meter base to the main service panelboard.

It is also permitted as a branch circuit to connect major electrical appliances and as a feeder to a subpanel. Four-wire cable with three insulated conductors and a bare equipment grounding conductor is required for feeders to panelboards as well as to appliances such as electric ranges and dryers. For these situations, the choice of whether to use Type SE cable or nonmetallic-sheathed cable becomes a matter of choice, the major issue being the cost of the cable. Some Type SE cable with aluminum conductors may cost less than that for nonmetallic-sheathed cable with copper conductors. The labor for the installation of either type is the same.

There are two styles of Type SE cable: SER and SEU. Table 4-17 shows the *NEC* installation requirements and the key differences in construction of these cables. See Table 4-17 and Figure 4-47.

Type UF Cable (*Article 340*)

Type UF is an underground feeder and branch-circuit cable. Type UF cable is discussed in Chapter 16 of this text.

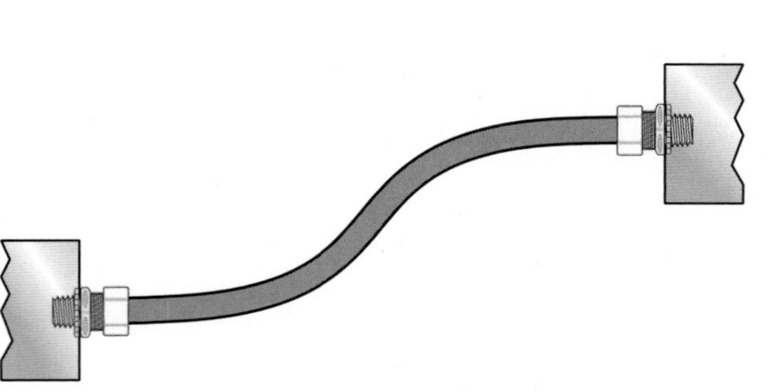

FIGURE 4-45 Grounding capabilities of liquidtight flexible metal conduit (LFMC). See *Article 350.*

Photos courtesy of Thomas and Betts Corp., www.TNB.com.

FIGURE 4-46 Color-coded ENT, a coupling, a threaded connector, and a snap-in connector. The blue color as applied to ENT is a registered trademark of Thomas & Betts International, Inc.

TABLE 4-17

Service-entrance cables Type SE, SER, SEU, and USE comparisons.

TYPE	COVERING	INSULATION	USES PERMITTED	BENDS	AMPACITY
SE Service Entrance Cable conforms to UL Standard 854: Service-Entrance Cables. See Figures 4-47 and 4-48.	Flame-retardant, moisture- and sunlight-resistant covering.	Type RHW, XHHW, or THWN insulation. Sizes 12 AWG and larger for copper. Sizes 10 AWG and larger for aluminum or copper-clad aluminum. If aluminum conductors, jacket will be marked "AL." If copper-clad aluminum conductors, jacket will be marked "AL (CU-CLAD)" or "Cu-Clad AL." If letter "H" on insulation, temperature rating 75°C, wet or dry locations.	600 volts or less. Above ground. Suitable for use where exposed to sun. Use "sill plate" where cable is bent to pass through a wall. See Figure 27-3. Permitted for interior wiring (feeder to a subpanel, ranges, dryers, cooktops, ovens) where all conductors are insulated. Must be installed according to the requirement in *Article 334* (nonmetallic-sheathed cable).	Internal radius of bend not to exceed 5 times the diameter of the cable.	See: *Table 310.15(B)(16)* and 230.42 for services. *215.2(A)(1)* for feeders. *310.15(B)(7)* for 120/240-volt, 3-wire single-phase dwelling services and feeders. *338.10(B)* for branch circuits and feeders. *338.10(B)(4)(a)* for interior installations. *338.10(B)(4)(b)* for exterior installations.

(*continues*)

TABLE 4-17 (*continued*)

TYPE	COVERING	INSULATION	USES PERMITTED	BENDS	AMPACITY
SE (continued)		If letter "HH" on insulation, temperature rating 75°C wet/90°C dry locations. If marked with a "–2", suitable for wet or dry locations as high as 90°C. May have one conductor uninsulated. May have reduced neutral conductor as permitted by *220.61* and *310.15(B)(7)*.	A bare, uninsulated conductor is permitted • as the neutral conductor when used as service-entrance cable. • as an equipment grounding conductor or when used for interior wiring. For existing installations only, may have a bare neutral conductor. Must originate at service panel, *338.10(B)(2), Exception*. Also see *250.140*. Not permitted for new work. Shall not be used where subject to physical damage unless protected. Shall not be used underground whether in or not in a raceway.		
SER Conforms to UL Standard 854: Service-Entrance Cables. Subject to the same rules as Type SE. See Figure 4-48.	Same as above.	Same as above. Available with two insulated phase conductors and one insulated neutral conductor. Also available with two insulated phase conductors, one insulated neutral conductor, and one bare equipment grounding conductor. Available with reduced neutral conductor as permitted by *220.61* and *310.15(B)(7)*. Type XHHW-2 conductors. Rated 90°C, wet or dry.	Same as above.	Same as above.	See: *Table 310.15(B)(16)* and 230.42 for services. *215.2(A)(1)* for feeders. *310.15(B)(7)* for 120/240-volt, 3-wire single-phase dwelling services and feeders. *338.10(B)* for branch circuits and feeders. *338.10(B)(4)(a)* for interior installations. *338.10(B)(4)(b)* for exterior installations.

(*continues*)

TABLE 4-17 *(continued)*

TYPE	COVERING	INSULATION	USES PERMITTED	BENDS	AMPACITY
SEU Conforms to UL Standard 854: Service-Entrance Cables. Subject to the same rules as Type SE.	Same as above.	Type XHHW-2 conductors. Rated 90°C, wet or dry. Type THHN/THWN Easily identified by the bare wraparound neutral conductor. Available with two and three insulated conductors plus the bare wraparound neutral conductor. Available with reduced neutral conductor as permitted by *220.61* and *310.15(B)(7)*.	Same as above.	Same as above.	See: *Table 310.15(B)(16)* and 230.42 for services. *215.2(A)(1)* for feeders. *310.15(B)(7)* for 120/240-volt, 3-wire single-phase dwelling services and feeders. *338.10(B)* for branch circuits and feeders. *338.10(B)(4)(a)* for interior installations. *338.10(B)(4)(b)* for exterior installations.
USE Service-Entrance Conductor. Individual conductors may be bundled in an assembly. Conforms to UL Standard 854: Service-Entrance Cables.	Moisture-resistant covering. Not required to have flame-retardant covering.	Insulation equivalent to Type RHW or XHHW. The temperature rating of the cable is the same as the temperature rating of the individual conductors. If this is not marked on the outer jacket, then the cable is rated 75°C. If Type USE-2, insulation equivalent to Type RHW-2 or XHHW-2. Rated 90°C, wet or dry. Cabled single conductor may have bare copper conductor with the assembly. May have reduced neutral conductor if multiple conductor assembly, as permitted by *220.61* and *310.15(B)(7)*.	Same as above except USE is also permitted for underground direct burial. Permitted to emerge above ground in meter bases and similar enclosures. Where cable emerges from the ground, protect according to the requirement found in *300.5(D)*. Not permitted for interior wiring.	In compliance with *NEC 312.6(B)(1)* where conductors do not enter or leave the enclosure opposite the terminal. In compliance with *NEC 312.6(B)(2)* where conductors enter or leave the enclosure opposite the terminal.	See: *Table 310.15(B)(16)* and 230.42 for services. *215.2(A)(1)* for feeders. *310.15(B)(7)* for 120/240-volt, 3-wire single-phase dwelling services and feeders. *338.10(B)* for branch circuits and feeders. *338.10(B)(4)(a)* for interior installations. *338.10(B)(4)(b)* for exterior installations.

(continues)

TABLE 4-17 *(continued)*

TYPE	COVERING	INSULATION	USES PERMITTED	BENDS	AMPACITY
USE/RHH/RHW Service-Entrance Conductor. Individual conductors that may be spiraled into an assembly. Complies with UL 44, Thermoset-Insulated Wire and UL 854, Service-Entrance Cable	Not required to have an overall outer covering. May have several conductors spiraled to create a multiconductor cable sometimes referred to as URD (underground rural distribution)	A single insulation system that is identified USE/RHH/RHW. It may also be identified as USE–2/RHH/RHW-2. May have a reduced neutral conductor if multiple conductor assembly as permitted by *220.61* and *310.15(B)(7)*.	As USE, for direct burial, including risers to meters or other enclosures inside or outside the building or structure. As RHH, for 90°C installations in dry or damp locations. As RHW, for 75°C installations in dry, damp, or wet locations.	Not less than 5 times the diameter of the cable. In compliance with *312.6(B)(1)* where conductors do not enter or leave the enclosure opposite the terminal. In compliance with *312.6(B)(2)* where conductors enter or leave the enclosure opposite the terminal.	See: *Table 310.15(B)(16)* 230.42 for Services; *215.2(A)(1)* for Feeders; *310.15(B)(7)* for 120/240-volt, 3-wire single-phase dwelling services and feeders.

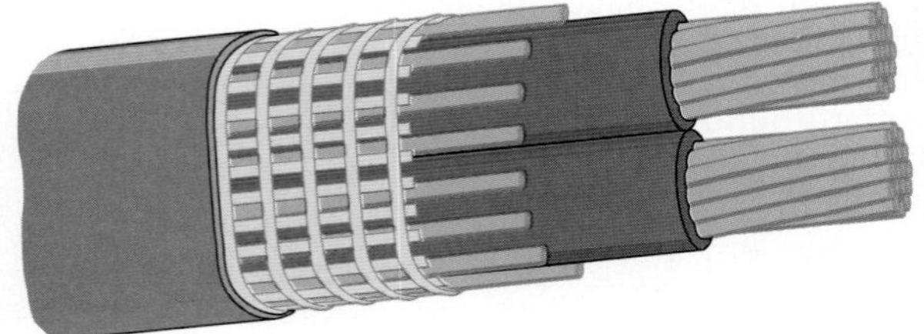

FIGURE 4-47 Type SEU service-entrance cable. The major characteristic is the uninsulated wraparound neutral conductor. See Table 4-17 for *NEC* requirements for the permitted uses of Type SE cables.

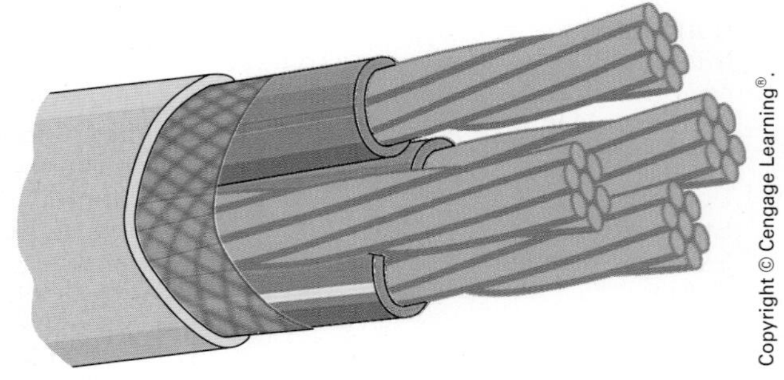

FIGURE 4-48 Type SER service-entrance cable. The major characteristics are the insulated neutral conductor and bare equipment grounding conductor. See Table 4-17 for *NEC* requirements for the permitted uses of Type SE cables.

REVIEW

Note: Refer to the *Code* or the plans where necessary.

1. The largest size solid conductor generally permitted to be installed in a raceway is (10 AWG) (8 AWG) (6 AWG). (Circle the correct answer.)
2. What is the minimum branch-circuit wire size that may be installed in a dwelling?

___

3. What exceptions, if any, are there to the answer for problem 2? ____________________

4. What determines the ampacity of a wire? ____________________

5. What unit of measurement is used for the diameter of wires? ____________________

6. What unit of measurement is used for the cross-sectional area of wires? ____________________

7. What is the voltage rating of the conductors in Type NMC cable? ____________________

8. Indicate the allowable ampacity of the following Type THHN (copper) conductors. Refer to *Table 310.15(B)(16)*.

   a. 14 AWG ____________ amperes
   b. 12 AWG ____________ amperes
   c. 10 AWG ____________ amperes
   d. 8 AWG ____________ amperes
   e. 6 AWG ____________ amperes
   f. 4 AWG ____________ amperes

9. What is the maximum operating temperature of the following conductors? Give the answer in degrees Celsius.

   a. Type XHHW ____________
   b. Type THWN ____________
   c. Type THHN ____________
   d. Type TW ____________

10. What are the colors of the conductors in nonmetallic-sheathed cable for

    a. 2-wire cable? ____________, ____________
    b. 3-wire cable? ____________, ____________, ____________

11. For nonmetallic-sheathed cable, may the uninsulated conductor be used for purposes other than grounding? ____________________

12. What size equipment grounding conductor is used with the following sizes of nonmetallic-sheathed cable?

    a. 14 AWG ____________
    b. 12 AWG ____________
    c. 10 AWG ____________
    d. 8 AWG ____________

13. Under what condition may nonmetallic-sheathed cable (Type NM) be fished in the hollow voids of masonry block walls? ____________________

14. a. What is the maximum distance permitted between straps on a cable installation?
____

 b. What is the maximum distance permitted between a box and the first strap in a cable installation? _______________

 c. Does the *NEC* permit 2-wire Romex stapled on edge? _______________

 d. When Type NM cable is run through holes in studs and joists, must additional support be provided? _______________

15. What is the difference between Type AC and Type ACT armored cables? _______________
____
____
____
____
____

16. Type ACT armored cable may be bent to a radius of not less than _______________ times the diameter of the cable.

17. When armored cable is used, what protection is provided at the cable ends?
____
____
____
____

18. What protection must be provided when installing a cable in a notched stud or joist, or when a cable is run through bored holes in a stud or joist where the distance is less than 1¼ in. (32 mm) from the edge of the framing member to the cable, or where the cable is run parallel to a stud or joist and the distance is less than 1¼ in. (32 mm) from the edge of the framing member to the cable? _______________
____
____
____
____

19. For installing directly in a concrete slab, (armored cable) (nonmetallic-sheathed cable) (conduit) may be used. (Circle the correct method of installation.)

20. Circle the correct answer to the following statements:

 a. Type SE service-entrance cable is for aboveground use. True False

 b. Type USE service-entrance cable is suitable for direct burial in the ground. True False

 c. Type SE cable with an uninsulated neutral conductor is permitted for hooking up an electric range. True False

 d. Type SE cable with an insulated neutral conductor is permitted for hooking up an electric range. True False

 e. Are Types SE and USE service-entrance cables used in your area? Yes No

21. When running Type NM cable through a bored hole in a stud, the nearest edge of the bored hole shall not be less than ______________ in. from the face of the stud unless the cable is protected by a 1⁄16 in. (1.6 mm) metal steel plate(s).

22. Where is the main service-entrance panel located in this residence? ______________

____

23. a. Is nonmetallic-sheathed cable permitted in your area for residential wiring?

____

b. From what source is this information obtained? ______________

____

24. Is it permitted to use flexible metal conduit over 6 ft (1.8 m) in length as a grounding means? (Yes) (No) (Circle the correct answer.)

25. Liquidtight flexible metal conduit may serve as a grounding means in sizes up to and including ______________ in. where used with listed fittings.

26. The allowable current-carrying capacity (ampacity) of aluminum wire, or the maximum overcurrent protection in the case of 14 AWG, 12 AWG, and 10 AWG conductors, is less than that of copper wire for a given size, insulation, and temperature of 30°C. Refer to *Table 310.15(B)(16)* and *240.4(D)* and complete the following table. Important: Where the ampacity of the conductor does not match the rating of a standard fuse or circuit breaker as listed in *240.6(A), 240.4(B)* permits the selection of the next "higher" standard rating of fuse or circuit breaker if the next higher standard rating does not exceed 800 amperes. Nonstandard ampere ratings are also permitted.

WIRE (AWG-kcmil)	COPPER		ALUMINUM	
	Ampacity	**Overcurrent Protection**	**Ampacity**	**Overcurrent Protection**
12 AWG THHN				
10 AWG THHN				
3 AWG THW				
4/0 THWN				
500 kcmil THWN				

27. It is permissible for an electrician to connect aluminum, copper, or copper-clad aluminum conductors together in the same connector. (True) (False) (Circle the correct answer.)

28. Terminals of switches and receptacles marked CO/ALR are suitable for use with ______________, ______________, and ______________ conductors.

29. Wire connectors marked AL/CU are suitable for use with ______________, ______________, and ______________ conductors.

30. A wire connector bearing no marking or reference to AL, CU, or ALR is suitable for use with (copper) (aluminum) conductors only. (Circle the correct answer.)

31. When Type NM cable is run through a floor, it must be protected by at least ______________ in. (______________ mm) of ______________

____

____

32. When nonmetallic-sheathed cables are bunched or bundled together for distances longer than 24 in. (600 mm), what happens to their current-carrying ability?

____

33. In diagrams A and B, nonmetallic-sheathed cable is run through the cold air return. Which diagram "Meets *Code*"? A __________ B __________
Check one.

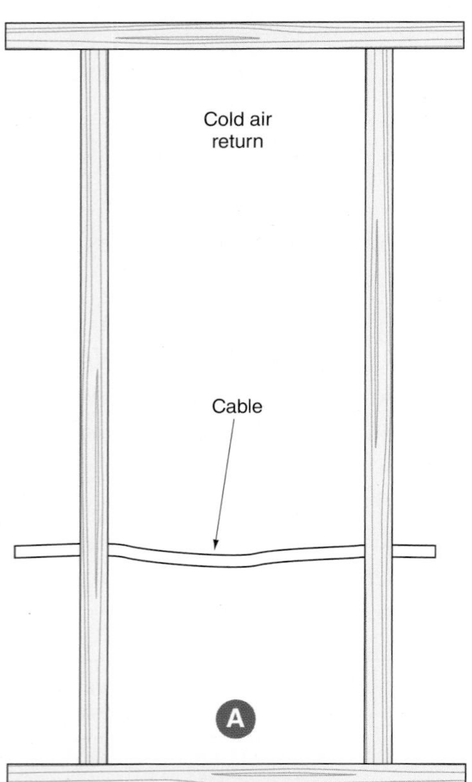

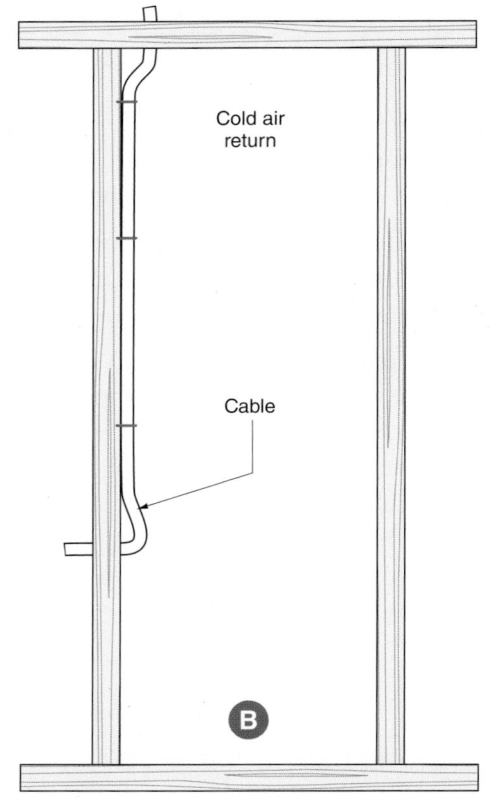

34. The marking on the outer jacket of a nonmetallic-sheathed cable indicates the letters NMC-B. What does the letter "B" signify? __________________________

____

____

35. A 120-volt branch circuit supplies a resistive heating load of 10 amperes. The distance from the panel to the heater is approximately 140 ft. Calculate the voltage drop using (a) 14 AWG, (b) 12 AWG, (c) 10 AWG, (d) 8 AWG copper conductors. See *210.19(A), Informational Note No. 4* and *215.2(A)(3), Informational Note No. 2*.

36. In problem 35, it is desired to keep the voltage drop to 3% maximum. What minimum size wire would be installed to accomplish this 3% maximum voltage drop? See *210.19(A), Informational Note No. 4* and *215.2(A)(3), Informational Note No. 2.*

37. *NEC 215.2*, *220.61*, *230.42*, and *310.15(B)(7)* state that the neutral conductor of a residential service or feeder can be smaller than the phase conductors, but only

38. The allowable ampacity of a 4 AWG THHN from *Table 310.15(B)(16)* is 95 amperes. What is this conductor's ampacity if connected to a terminal listed for use with 60°C wire? ________

39. If, because of some obstruction in a wall space, it is impossible to keep an Type NM cable at least 1¼ in. (32 mm) from the edge of the stud, then it shall be protected by a metal plate at least ________ in. thick.

40. The recessed fluorescent luminaires installed in the ceiling of the recreation room of this residence are connected with trade size ⅜ flexible metal conduit. These flexible connections are commonly referred to as fixture whips. Does the flexible metal conduit provide adequate grounding for the luminaires, or must a separate equipment grounding conductor be installed? ________

41. Flexible liquidtight metal conduit will be used to connect the air-conditioner unit. It will be trade size ¾. Must a separate equipment grounding conductor be installed in this FLMC to ground the air conditioner properly? ________

42. What size overcurrent device protects the air-conditioning unit? ________

43. May the 20-ampere small-appliance branch circuits in the kitchen of a residence also supply the lighting above the kitchen sink and under-cabinet lighting? Check the correct answer. Yes ________ No ________

44. A 30-ampere branch circuit is installed for an electric clothes dryer. What is the maximum ampere rating of a dryer permitted by the *NEC* to be cord-and-plug connected to this circuit? Check the correct answer.

    a. ______________ 30 amperes

    b. ______________ 24 amperes

    c. ______________ 15 amperes

45. In many areas, metal framing members are being used in residential construction. When using nonmetallic-sheathed cable, what must be used where the cable is run through the metal framing members? ______________

46. Are set screw–type connectors permitted to be used with armored cable that has an aluminum armor? ______________

47. Most armored cable today has 90°C conductors. What is the correct designation for this type of armored cable? ______________

48. If you saw two different types of SE cables, how would you distinguish them?

    ____

    ____

    ____

    ____

49. Circle the correct answer defining the type of location:

    a. A raceway installed underground (Dry) (Wet)

    b. A raceway installed in a concrete slab that is in direct contact with the earth (Dry) (Wet)

    c. A raceway installed in a concrete slab between the first and second floor (Dry) (Wet)

    d. A raceway installed on the outside of a building exposed to the weather (Dry) (Wet)

    e. The interior of a raceway installed on the outside of a building exposed to the weather (Dry) (Wet)

CHAPTER **5**

Conductor Identification, Switch Control of Lighting Circuits, Bonding/Grounding of Wiring Devices, and Induction Heating

OBJECTIVES

After studying this chapter, you should be able to

- identify the grounded and ungrounded conductors in cable or conduit (color coding).
- identify the various types of toggle switches for lighting circuit control.
- select a switch with the proper rating for the specific installation conditions.
- describe the operation that each type of toggle switch performs in typical lighting circuit installations.
- determine when a neutral conductor must be available for switch boxes.
- demonstrate the correct wiring connections for each type of switch per *Code* requirements.
- understand the various ways to bond wiring devices to the outlet box.
- understand how to design circuits to avoid heating by induction.

SAFETY ALERT

Improper wiring can result in fires, personal injuries, and electrocutions! All wiring must be done in conformance to the *NEC*. Anything less is unacceptable!

CONDUCTOR IDENTIFICATION [*NEC*® *ARTICLES 200* AND *210*]

Before making any electrical connections, an electrician must understand conductor color coding. The requirements are included in *200.6* for grounded (neutral) conductors, *210.5(C)* for branch circuit conductors, and *215.12(C)* for ungrounded ("hot") feeder conductors.

Ungrounded ("Hot") Conductor. An ungrounded conductor must have an outer finish that is a color other than green, white, gray, or three continuous white stripes. This conductor is commonly called the "hot" conductor. You will get a shock if this "hot" conductor and a grounded conductor or grounded surface such as a metal water pipe are touched at the same time.

Popular colors for cable wiring systems are black, red, and sometimes blue. Many other colors are available for conduit systems, including blue, yellow, brown, orange, and purple.

Grounded (Neutral) Conductor. For alternating-current circuits, the *NEC*® requires that the grounded (identified) conductor have an outer finish that is either continuous white or gray, or have an outer finish (not green) that has three continuous white stripes along the conductor's entire length. From an *NEC* definition standpoint, a *neutral conductor* connects to the *neutral point*. In multiwire circuits, the grounded conductor is also a *neutral* conductor.

Three Continuous White Stripes. In house wiring, you will probably never run across a conductor that has three continuous white stripes on other than green insulation. This method of identifying conductors is done by the cable manufacturer and would be found on larger size conductors and some that may be used by the electric utility. Some underground service conductors such as those used for a mobile home feeder may have two black conductors, one black conductor that has three white stripes, and one green conductor.

Grounded Neutral Conductor

For residential wiring, the grounded conductor with white insulation is commonly referred to as the *neutral*. However, this conductor is not always a neutral conductor, as you will learn.

For branch circuits and feeders, the grounded neutral conductor must be insulated, *310.106(D)*. For residential services, the grounded neutral conductor is permitted to be insulated or bare, *230.41*.

For residential wiring, the 120/240-volt electrical system is grounded by the electric utility at their transformer, and again by the electrician at the main service. This is in accordance with *250.20(B)(1)* requiring that the system be grounded if the maximum voltage to ground on the ungrounded conductors does not exceed 150 volts. *NEC 250.26(2)* requires that the neutral conductor be grounded for single-phase, 3-wire systems.

Electricians often use the term "neutral" whenever they refer to a white grounded circuit conductor. The correct definition is found in the *NEC*.

Neutral Conductor. The conductor connected to the neutral point of a system that is intended to carry current under normal conditions.* See Figure 5-1.

Neutral Point. The common point on a wye-connection in a polyphase system or midpoint on a single-phase, 3-wire system, or midpoint of a single-phase portion of a 3-phase delta system, or a midpoint of a 3-wire, direct current system.* See Figure 5-1.

By these definitions, the white grounded conductor in a 2-wire circuit is a neutral conductor if it is connected to the neutral point of the electrical system. In a multiwire 3-wire branch circuit, the white grounded conductor is a neutral conductor. See Figure 5-1.

---

*Reprinted with permission from NFPA 70-2014.

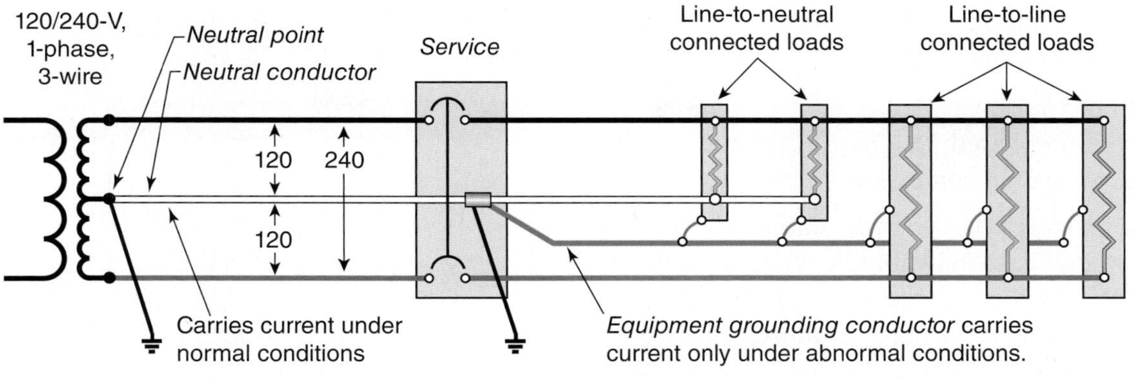

FIGURE 5-1 Definition of a neutral conductor and neutral point.

For those of you who will broaden your training into commercial and industrial wiring, you will find that there are electrical systems such as a 3-phase delta-connected corner grounded system where the grounded conductor is a phase conductor and is not a neutral conductor. This is covered in detail in the *Electrical Wiring—Commercial*, 15th Edition text.

A neutral conductor in residential wiring is always a grounded conductor—but as explained in the previous paragraph, a grounded conductor is not always a neutral conductor.

Although it may seem obvious to trained electricians, *NEC 200.2(B)* prohibits using a metal enclosure, a metal raceway, or metal cable armor to be part of a grounded conductor's path. Always connect the neutral conductor to the terminal bar for the neutral conductors rather than to an equipment grounding terminal bar.

Multiwire Branch-Circuit

The *Code* defines a *multiwire branch circuit* as A branch circuit consisting of two or more ungrounded conductors having a voltage between them, and a grounded conductor having equal voltage between it and each ungrounded conductor of the circuit and that is connected to the neutral or grounded conductor of the system.*

For additional information relative to the hazards involved with multiwire circuits, refer to Figures 17-5, 17-6, 17-7, 17-8, and 17-9 in Chapter 17 of this text.

*Reprinted with permission from NFPA 70-2014.

Color Coding (Cable Wiring)

The conductors in nonmetallic-sheathed cable (Romex) are color coded with insulation as follows:

2-wire: one black ("hot" phase conductor)
one white (grounded "identified" conductor)
one bare, covered, or green-insulated (equipment grounding conductor)

3-wire: one black ("hot" phase conductor)
one white (grounded "identified" conductor)
one red ("hot" phase conductor)
one bare, covered, or green-insulated (equipment grounding conductor)

4-wire: one black ("hot" phase conductor)
one white (grounded "identified" conductor)
one red ("hot" phase conductor)
one blue ("hot" phase conductor)
one bare, covered, or green-insulated (equipment grounding conductor)

Four-wire nonmetallic-sheathed cable is also available with two ungrounded ("hot") and two neutral conductors. This cable is designed for wiring two 120-volt branch circuits without using a common neutral. This avoids the requirement of installing a tie handle on the circuit breakers or installing a 2-pole circuit breaker. This cable has the following insulated conductors:

- one black ("hot" phase conductor)
- one white with a black stripe (grounded "identified" conductor)

- one red ("hot" phase conductor)
- one white with a red stripe ("hot" phase conductor)
- one bare, covered, or green-insulated (equipment grounding conductor)

Manufacturers of Type MC cable also make a "home run cable." This cable is available with six or eight 12 AWG conductors and with six, eight, twelve, and sixteen 10 AWG conductors. The insulation is THHN, so derating [required by *NEC 310.15(B)(3)*] is started in the 90°C column of *NEC Table 310.15(B)(16)*. Table 5-1 illustrates application of derating for the number of current-carrying conductors in the home run cable. Notice that the number of current-carrying conductors can change depending on how connections are made.

In every case, the allowable ampacity of the 12 AWG conductors after application of the adjustment factors permits a 20-ampere overcurrent device to be used. In some installations, home run cable can be used rather than installing a subpanel some distance away from the service equipment.

TABLE 5-1

Application of adjustment factors to home run cable.

	NO. OF CONDUCTORS	CONNECTED	NUMBER CURRENT-CARRYING CONDUCTORS	*TABLE 310.15(B)(16)*	FACTOR FROM *TABLE 310.15(B)(3)(A)*	ADJUSTED AMPACITY	PERMITTED OVERCURRENT PROTECTION
12 AWG	6	Three 2-wire circuits	6	30	0.8	24	20
		Two 3-wire circuits	4	30	0.8	24	20
	8	Four 2-wire circuits	8	30	0.8	24	20
		Two 3-wire circuits and one, 2-wire	6	30	0.8	24	20
10 AWG	6	Three 2-wire circuits	6	40	0.8	32	30
		Two 3-wire circuits	4	40	0.8	32	30
	8	Four 2-wire circuits	8	40	0.8	32	30
		Two 3-wire circuits and one, 2-wire	6	40	0.8	32	30
	12	Six 2-wire circuits	12	40	0.5	20	20
		Four 3-wire circuits	8	40	0.8	32	30
	16	Eight 2-wire circuits	16	40	0.5	20	20
		Five 3-wire circuits	15	40	0.5	20	20

Color Coding (Raceway Wiring)

When the wiring method is a raceway such as EMT, the choice of colors for the ungrounded "hot" conductors is virtually unlimited, with the following restrictions:

Green or green with yellow stripe(s): Reserved for use as an equipment grounding conductor only. Not permitted to be used as a grounded or ungrounded conductor. See *250.119* and *310.100(B)*.

White or gray: Reserved for use only as a grounded circuit conductor (neutral). See Table 5-2 for more information about identifying conductors. As explained elsewhere in this text, an exception will allow reidentifying a white conductor with marking tape and using it as a supply conductor in a switch loop or for supplying 240-volt loads such as water heaters.

What about Conductors with Gray Insulation?

The *NEC* recognizes a conductor with gray insulation to be used as a grounded conductor. See *200.6(A)* and *(B)*.

The *NEC* prohibits a conductor with gray insulation to be used as an ungrounded "hot" conductor. See *200.6(A)* and *(B)*.

TABLE 5-2

Changing colors of conductor insulation in a raceway or cable.

TO CHANGE THIS ...	TO THIS,	DO THIS
red, black, blue, etc.	an *equipment grounding conductor*	**Conductors 6 AWG or smaller:** No reidentification permitted for conductors 6 AWG and smaller. An equipment grounding conductor must be bare, covered, or insulated. If covered or insulated, the outer finish must be green or green with yellow stripes. The outer finish color must run the entire length of the conductor. **Conductors 4 AWG or larger:** Insulation on large conductors is usually black but may have other colors. Therefore, at time of installation, at both ends and at every place where the conductor is accessible except in a conduit body where there are no splices, do any one of the following: • Strip off the insulation for the entire exposed length, or • At the termination, encircle the insulation with green color, or • At the termination, encircle the insulation with a few wraps of green tape, green adhesive labels, or green heat shrink tubing See *210.5(B), 250.119,* and *310.110(B).*
red, black, blue, etc.	a *grounded* conductor	**Conductors 6 AWG or smaller:** For the conductor's entire length, the insulation must be white, gray, or have three continuous white stripes on other than green insulation. Reidentification in the field is not permitted when the wiring method is a raceway. **Conductors 4 AWG or larger:** For the conductor's entire length, the insulation can be white, gray, or have three continuous white stripes on other than green insulation. If not white, gray, etc., reidentification at terminations to be white or gray paint, tape, or heat shrink tubing. This marking must encircle the conductor or insulation. Do this marking at time of installation. Identification must encircle the conductor. See *200.2, 200.6(A)* and *(B); 200.7(A), (B),* and *(C); 210.5(A);* and *310.110(A).*
white, gray	an *ungrounded* ("hot") conductor	**Change permitted for cables but not for raceways:** Insulated conductors that are white, gray, or have three continuous white stripes are to be used only as grounded conductors in raceways. Conductors that are part of a cable assembly having white or gray insulation are permitted to be reidentified for use as an ungrounded conductor by marking tape, painting or other effective means. Reidentification required at each location the conductor is visible, accessible and at terminations. For switch loops, the reidentified conductor is to be the supply to the switch but not the return. See *200.7(A), (B),* and *(C);* and *310.110(C).*

For years, conductors with gray insulation were used in commercial and industrial installations as both grounded conductors and ungrounded conductors. Past editions of the *NEC* used the term *natural gray*, but no one really knew what it was. Was it "dirty white"? Was it "light gray"? Was it "dark gray"? Confusion was the end result. To bring an end to the confusion, all references to the term *natural gray* have been deleted in the *NEC*.

You might be saying to yourself, "No cable that I use for house wiring has a gray conductor in it." Right! But in commercial and industrial work, there are situations where two systems are installed in the same raceway, such as a 208Y/120-volt and a 480Y/277-volt circuit. This is an example of where you might want to use a white insulated conductor for the 208Y/120-volt system neutral conductor, and a gray insulated conductor for the 480Y/277-volt system neutral conductor.

Changing Colors When Conductors Are in a Cable

The basic rule in *200.6* is that grounded conductors must be white, gray, or have three continuous white stripes on other than green insulation.

For cable wiring such as nonmetallic-sheathed cable or armored cable, *200.7(C)(1)* permits the white or gray conductor to be used for single-pole, 3-way, or 4-way switch loops. These sections require that when used for a switch "loop," the conductor that is white, gray, or marked with three continuous white stripes is to be used for the supply to the switch, and not as the return conductor from the switch to the switched outlet. Also, the conductor must be permanently reidentified to indicate its use, by painting or other effective means, at its terminations and at each location where the conductor is visible and accessible. See Figure 5-2.

A requirement in *NEC 404.2(C)* for providing a neutral conductor at all switch boxes trumps the provisions in *200.7(C)(1)* for reidentifying the neutral conductor for switch loops. See the discussion later in this chapter as well as in Chapter 6.

How to Reidentify Conductor Insulation

Small conductors are available with insulation colors of black, blue, brown, gray, green, orange, pink, purple, red, tan, white, and yellow. There are a number of ways to reidentify a conductor's insulation. Plastic electrical tape, heat shrink tubing, permanent felt-tip marking pens, and fast-drying "touch-up" paint are generally considered acceptable ways to permanently reidentify conductor insulation. You might want to test the paint on a conductor to make sure that it does not have a harmful effect on the insulation, *NEC 310.10(G)*. Whichever method you choose to reidentify a conductor, be sure it encircles the insulation.

Increasingly, electricians are finding that permanent marking pens are a great way to reidentify conductor insulation. These pens are readily available with a wide felt tip (5⁄8 inch). Cut a notch or drill a small hole in the felt tip. Slide the notched marker up and down the conductor insulation, or slide the marker with the small hole over the conductor and slide it up and down. This makes the job of reidentifying a conductor very easy—and fast. Inspectors like this method.

Heat shrink tubing (UL recognized) with a shrink ratio of 2:1 is available at electrical distributors in packages of assorted colors and sizes, for example, 6 in. (150 mm) long. See Figure 5-2(B). This method is more labor intensive than using permanent marking pens or tape.

Some electrical inspectors have very strong feelings about what they will accept for reidentifying conductors. They might not consider plastic tape as being a "permanent" material. Check it out!

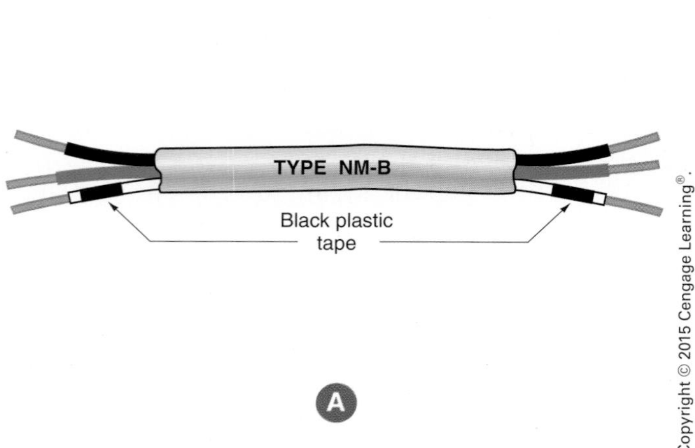

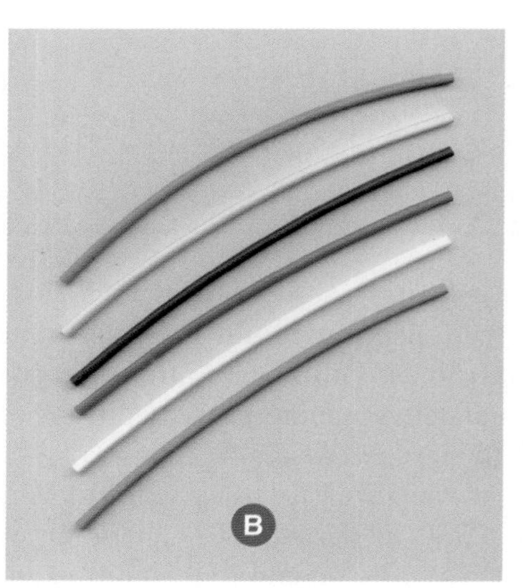

FIGURE 5-2 (A) Shows a white conductor reidentified with black plastic electrical tape. (B) Shows an assortment of easy-to-use, 6 in. long heat shrink tubing that is slid over the conductor to be reidentified, then shrunk tightly over the conductor by applying heat as specified in the instructions.

It is pretty obvious that the intent of the *Code* is to permanently reidentify a white conductor when it is used as a "hot" conductor. If not properly reidentified, someone someday will get a surprise. An accident waiting to happen—such "surprises" can prove to be fatal!

The wiring diagrams that follow are good examples of where permanent reidentification of the white conductor is necessary.

Changing Colors When Conductors Are in a Raceway

Because of the availability of so many insulation colors, it is rarely necessary to reidentify a conductor. If, however, it becomes necessary to change the actual color of a conductor's insulation, refer to Table 5-2.

CONNECTING WIRING DEVICES

Terminating and splicing conductors is a skill. The connections must be tight so as not to overheat under load.

NEC 110.3(B) requires us to follow the manufacturers' instructions regarding the listing and labeling of all electrical products. *NEC 110.14* provides clear requirements for making electrical connections.

Table 4-4 shows in detail the terminal identification markings found on wiring devices (switches and receptacles), indicating the types of conductors permitted to be connected to them.

Most commonly used wiring devices have screw terminals.

Some wiring devices have back-wiring holes only. For these, the conductor insulation is stripped off for the desired length, inserted into the hole, and held in place by the device's internal spring pressure mechanism. These wiring devices are referred to as having *push-in* terminals.

For convenience, some wiring devices have both screw terminals and back-wiring holes, as illustrated in Figures 5-3, and 5-4. These can be connected or hooked up in different ways. One way is to strip off the conductor insulation for the desired length and insert it into the back hole, and then properly tighten the screw terminal. In another way, the conductor insulation is stripped off for the desired length and terminated under the screw terminal, in which case the back-wiring holes are not used.

Using both the back-wiring terminals and the screw terminals, it would be possible to terminate as many as four conductors to the white terminal and four conductors to the brass terminal. This is a violation of *110.3(B)*. UL does not test receptacles with so many conductors connected to the device. A much better way is shown in Figure 5-5, where all of the necessary splicing is done independent of the receptacle. Short lengths of wire called *pigtails* are included in the splice, and these pigtails in turn connect to the receptacle. Removal of the receptacle does not have any effect on the splice.

Installing Switches and Receptacles

Always make sure that the yokes of switches and receptacles are rigidly held in place, so they do not twist. Not tightly securing wiring devices is unsafe as well as being an example of sloppy workmanship, *406.4(A)*.

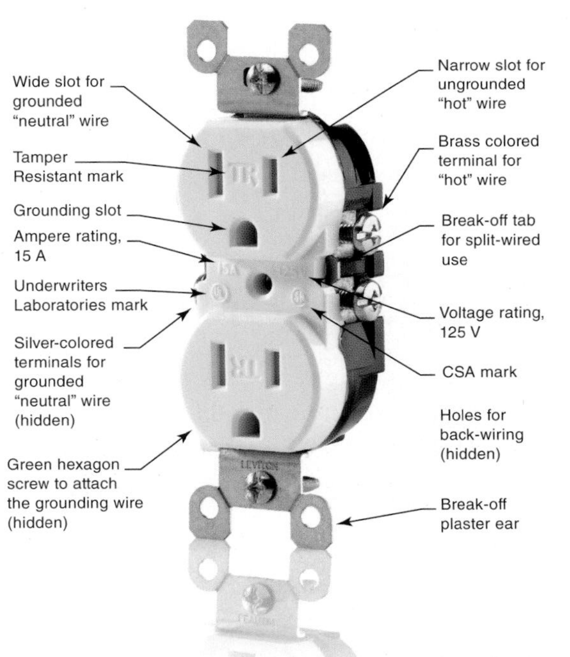

FIGURE 5-3 Grounding-type receptacle detailing various parts of the receptacle.

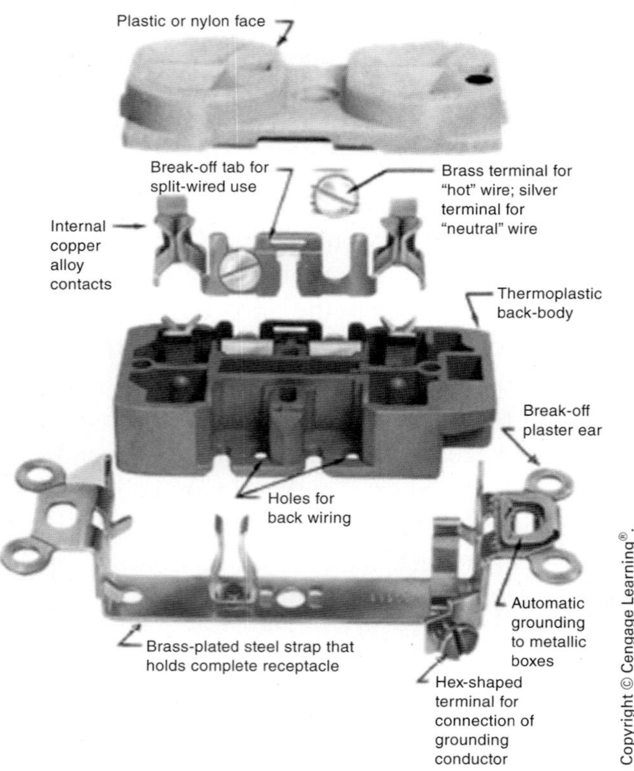

FIGURE 5-4 Exploded view of a grounding-type receptacle, showing all internal parts.

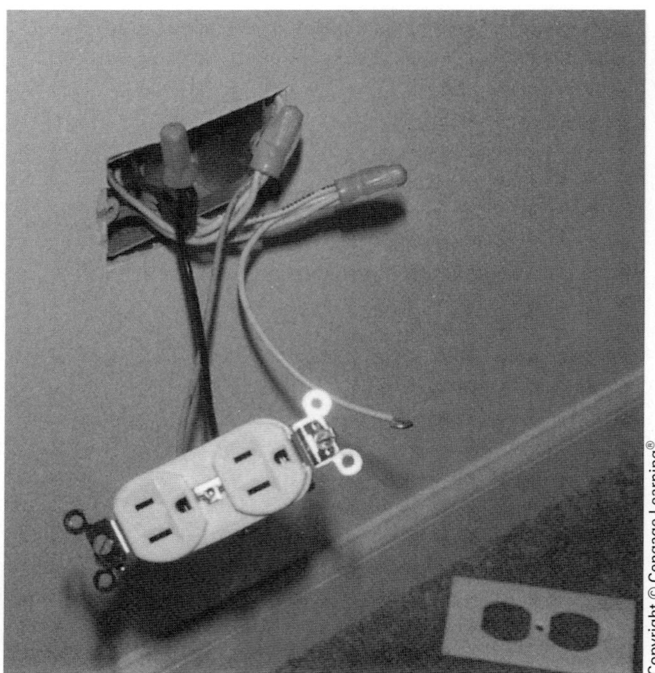

FIGURE 5-5 A receptacle connected with pigtails. Removal of the receptacle does not affect the circuit to the other receptacles. This particular installation has one receptacle switched and one receptacle on continuously. A 4 in. square outlet box trimmed with a single-gang plaster ring is used.

Receptacle Configuration

Figure 5-6 illustrates some of the common receptacle configurations used in homes in conformance to *Table 210.21(B)(3)*. Range and dryer receptacles are discussed in Chapter 20.

PUSH-IN TERMINATIONS

Be careful when using screwless push-in terminals.

Screwless push-in terminals on receptacles are listed by Underwriters Laboratories (UL) for use *only* with solid 14 AWG copper conductors. They *are not* to be used with

- aluminum or copper-clad aluminum conductors.
- stranded conductors.
- 12 AWG conductors. By design, the holes are large enough to take only a 14 AWG solid conductor.

Push-in terminals for 12 AWG solid copper conductors are still permitted on snap switches.

Figure 5-7 shows a typical tamper-resistant receptacle. These receptacles are required in most every location in a single-family dwelling. They were introduced into the *NEC* in an attempt to reduce the number of injuries caused by children who insert foreign objects into these receptacles. Electric shock and burn injuries commonly result when metallic foreign objects make contact with energized terminals in these receptacles.

The general rule in *406.12* is all nonlocking-type, 125-volt, 15- and 20-ampere receptacles are required to be tamper-resistant for new installations in dwelling units in all areas specified in *210.52*. If you take a look at that section, you will find it literally covers all locations in a dwelling including living areas, garages, unfinished basements, and

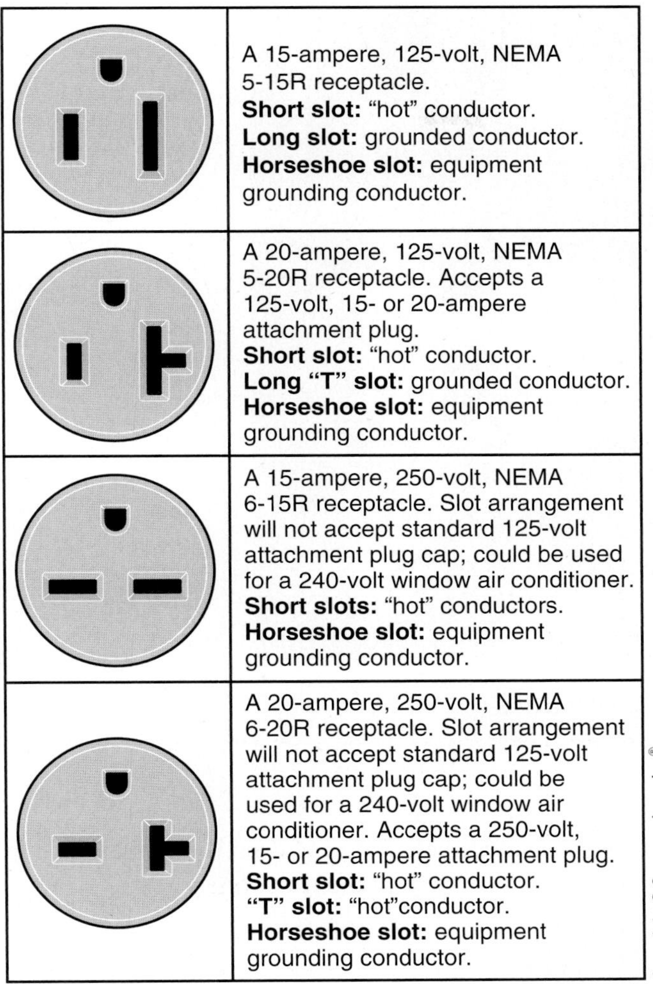

	A 15-ampere, 125-volt, NEMA 5-15R receptacle. **Short slot:** "hot" conductor. **Long slot:** grounded conductor. **Horseshoe slot:** equipment grounding conductor.
	A 20-ampere, 125-volt, NEMA 5-20R receptacle. Accepts a 125-volt, 15- or 20-ampere attachment plug. **Short slot:** "hot" conductor. **Long "T" slot:** grounded conductor. **Horseshoe slot:** equipment grounding conductor.
	A 15-ampere, 250-volt, NEMA 6-15R receptacle. Slot arrangement will not accept standard 125-volt attachment plug cap; could be used for a 240-volt window air conditioner. **Short slots:** "hot" conductors. **Horseshoe slot:** equipment grounding conductor.
	A 20-ampere, 250-volt, NEMA 6-20R receptacle. Slot arrangement will not accept standard 125-volt attachment plug cap; could be used for a 240-volt window air conditioner. Accepts a 250-volt, 15- or 20-ampere attachment plug. **Short slot:** "hot" conductor. **"T" slot:** "hot"conductor. **Horseshoe slot:** equipment grounding conductor.

FIGURE 5-6 Slot configuration of receptacles. Receptacles have the suffix "R" and plug caps have the suffix "P."

FIGURE 5-7 A typical duplex tamper-resistant receptacle.

accessory buildings. Scan through *210.52* and refresh your memory on the areas where tamper-resistant receptacles are required.

Several exceptions apply and exclude

1. receptacles located more than 5½ in. (1.7 m) above the floor.
2. receptacles that are part of a luminaire or appliance.
3. a single receptacle or a duplex receptacle for two appliances located within dedicated space for each appliance that in normal use is not easily moved from one place to another and that is cord-and-plug connected in accordance with *400.7(A)(6)*, *(A)(7)*, or *(A)(8)*.
4. Nongrounding receptacles used for replacements as permitted in *406.4(D)(2)(a)*.

Existing receptacles that are replaced in locations where tamper-resistant receptacles are now required must be replaced with tamper-resistant receptacles; see *406.4(D)(5)*.

TOGGLE SWITCHES (ARTICLE 404)

UL refers to them as "snap switches." Manufacturers and electricians refer to them as "toggle switches." They are one and the same.

The most frequently used switch in lighting circuits is the flush toggle switch, Figure 5-8. When mounted in a flush switch box, the switch is concealed in the wall, with only the insulated handle or toggle protruding through the cover plate.

Figure 5-9 shows one of the ways switches can be weatherproofed.

FIGURE 5-8 Toggle switches.

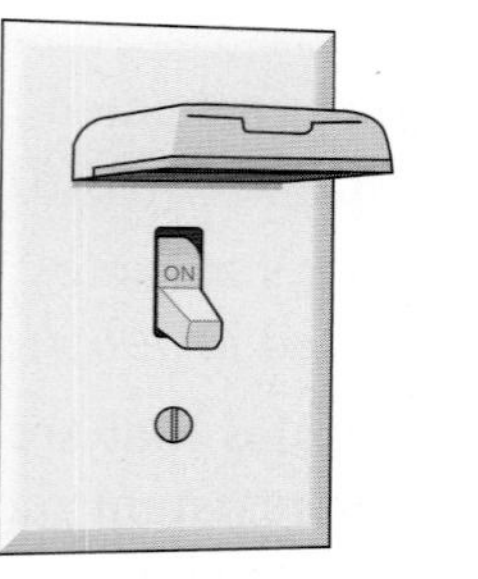

FIGURE 5-9 Switch is protected by a weatherproof cover.

Toggle Switch Ratings

UL lists toggle switches used for lighting circuits as *general-use snap switches*. The UL requirements are the same as *404.14(A)* and *(B)* of the *NEC*.

AC/DC General-Use Snap Switches *[404.14(B)]*

The requirements for general-use snap switches include the following:

- Alternating-current (ac) or direct-current (dc) circuits
- Resistive loads not to exceed the ampere rating of the switch at applied voltage
- Inductive loads not to exceed one-half the ampere rating of the switch at applied voltage
- Motor loads not to exceed the ampere rating of the switch at applied voltage only if the switch is marked in horsepower
- Tungsten filament lamp loads not to exceed the ampere rating of the switch at applied voltage when marked with the letter "T"
- For switches marked with a horsepower rating, a motor load not to exceed the rating of the switch at rated voltage

Why a "T" Rating?

A tungsten filament lamp draws a very high momentary inrush current at the instant the circuit is energized. (Recall that incandescent lamps have a tungsten filament.) This is because the *cold resistance* of tungsten is very low. For instance, the cold resistance of a typical 100-watt lamp is approximately 9.5 ohms. This same lamp has a *hot resistance* of 144 ohms when operating at 100% of its rated voltage.

Normal operating current would be

$$I = \frac{E}{R} = \frac{120}{144} = 0.83 \text{ ampere}$$

The instantaneous inrush current could be as high as

$$I = \frac{E}{R} = \frac{170 \text{ (peak voltage)}}{9.5} = 17.9 \text{ amperes}$$

This instantaneous inrush current drops off to normal operating current in about 6 cycles (0.10 second). The contacts of T-rated switches are designed to handle these momentary high inrush currents. See Chapter 13 for more information pertaining to inrush currents.

The ac/dc general-use snap switch normally is not marked ac/dc. However, it is always marked with the current and voltage rating, such as 10A-125V or 5A-250V-T.

AC General-Use Snap Switches *[404.14(A)]*

This is the type most commonly used for house wiring projects. Alternating-current general-use snap switches are marked "ac only," in addition to identifying their current and voltage ratings. A typical switch marking is 15A, 120–277V ac. The 277-volt rating is required on 277/480-volt systems. Other requirements include these:

- Alternating-current (ac) circuits only are allowed.
- Resistive and inductive loads are not to exceed the ampere rating of the switch at the voltage applied. This includes electric-discharge lamps that involve ballasts, such as fluorescent lamps.
- Tungsten-filament lamp loads are not to exceed the ampere rating of the switch at 120 volts.
- Motor loads are not to exceed 80% of the ampere rating of the switch at rated voltage. A UL requirement is that the load shall not exceed 2 horsepower.

What Are the Switching Conductors Called?

In *200.7(C)(1)*, we find a few unique words that have to be explained. There are also local and regional expressions for these words. It is always best to use the terms found in the *NEC*.

- *Switch loop:* The conductors that run between (to and from) a switched outlet and the switch controlling that outlet.
- *Supply:* The conductor that runs from the source to the switch. It might also be referred to as the *feed*, *hot*, *hot leg*, or *supply leg*.

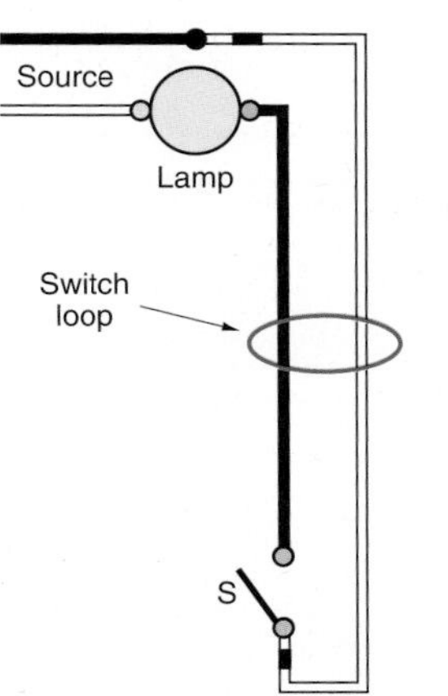

- *Return:* The conductor that runs from the switch back to the switched outlet. It might also be referred to as the *switch leg* or *switched leg*.

Conductor Color Coding for Switch Connections

The insulated conductors in nonmetallic-sheathed cable or armored cable that are commonly used for switching functions are typically black–white, black–white–red, or, in some cases, black–white–red–blue. We usually think of the black insulated conductor as the supply to a switch and a red insulated conductor as the return. However, when wiring with nonmetallic-sheathed cable or armored cable, we are "stuck" with the colors found in the cables. We must not use a white conductor as the return conductor from a switch, *200.7(C)(1)*. A color-coding scheme must be established for the supplies and returns. In the case of 3-way and 4-way switches, we also have "travelers"—the conductors that connect between 3-way and 4-way switches. The wiring diagrams in Figures 5-11 through 5-22 show the recommended color schemes when wiring with cable.

Do You Need the Grounded Circuit Conductor at the Switch? Maybe yes; maybe no.

We all know that conventional switches operate on a mechanical principle and are simply connected in series with the load. The contacts of the switch close (ON) and open (OFF). This type of switch does not require a grounded circuit conductor to operate correctly.

The world of electronic technology brings about change. Now we learn that many types of occupancy sensors require the grounded circuit conductor to operate

SAFETY ALERT

Untrained individuals will say, "Why not connect the occupancy sensor to the equipment grounding conductor as the grounded circuit conductor is not present at the switch?"

Wrong! It is a *Code* violation to use an equipment grounding conductor as a substitute for a grounded circuit conductor. *Never* use the equipment grounding conductor as a substitute for the circuit's grounded or neutral conductor. This could prove to be deadly. Equipment grounding conductors are intended to carry current *only* under ground fault conditions.

To address this dilemma, a new requirement was added to the *2011* edition of the *NEC* in *404.2(C)*. This new requirement brought with it a lot of controversy, confusion, and misapplication as to how it should be applied.

Many *Code* Proposals were submitted to revise this section for the *2014 NEC*. The Code Making Panel met, and a task force was set up to clarify the requirement. The final wording of *404.2(C)* was simplified and now reads:

▶ ***404.2(C): Switches Controlling Lighting Loads.*** The grounded circuit conductor for the controlled lighting circuit shall be provided at the location where switches control lighting loads that are supplied by a grounded general purpose branch circuit, for other than the following:

1. Where conductors enter the box enclosing the switch through a raceway, provided the raceway is large enough for all contained conductors, including a grounded conductor.
2. Where the box enclosing the switch is accessible for the installation of an additional or replacement cable without removing finish materials.
3. Snap switches with integral enclosures complying with 300.15(E).
4. Where a switch does not serve a habitable room or bathroom.
5. Where multiple switch locations control the same lighting load such that the entire floor area of that room or space is visible from the single or combined switch locations.
6. Where lighting in the area is controlled by automatic means.
7. A switch controlling a receptacle load◀*

Let's discuss the requirements of *404.2(C)*. A circuit layout needs a lot of thought beforehand. You think to yourself. You talk to yourself. You make mental comparisons of circuit possibilities, always keeping in mind to *let the* Code *decide.*

An important word, "habitable," is found in *404.2(C)(4)*. By definition, a habitable room is a space used for living, sleeping, eating, or cooking, or combinations thereof. Not considered to be a habitable room or space are bathrooms, toilet rooms, laundry rooms, closets, storage and utility rooms, equipment rooms, hallways, and garages.

Factors affecting cable layout include such considerations as how many feet of cable are needed; whether 2-wire, 3-wire, 4-wire, or 5-wire cable is needed; the number of conductors in the box, construction obstacles (beams, Lally columns, air-handling ducts, plumbing, etc.).

The cable layouts shown in this text are one of a number of possibilities. In some cases, it makes sense to run the grounded circuit conductor first to the switch location, then to the luminaire. See

*Reprinted with permission from NFPA 70-2014.

Figures 5-13 and 5-15. In other cases it makes sense to run the grounded circuit conductor first to the luminaire location then to the switch location. See Figure 5-14.

For 3-way and 4-way switch loops, the number of conductors in the cable and in the switch box varies according to which end of the switches the supply is connected to. See Figures 5-18, 5-19, 5-20, 5-22, and 5-23.

As stated earlier, electronic technology changes things. Electronic control devices such as dimmers, motion sensors, occupational sensors, photoelectric devices, timers, and fan speed controls are available that require a neutral conductor to operate correctly. It is a violation of the *NEC* and the listing requirements as well as the manufacturer's installation instructions to connect the neutral conductor to an equipment grounding conductor.

Some of these electronic switches, such as occupancy sensors, have internal electronic circuitry that allows a maximum of 0.5 milliampere (mA) leakage to ground. This tiny amount of leakage current is enough to power up the electronic circuitry in the device, yet does not pose a shock hazard. This is referred to as low-level standby current. The maximum leakage current of 0.5 mA is the same value of leakage current permitted for electric appliances. These electronic switches or controls may have an equipment grounding conductor rather than a neutral. Always follow the manufacturer's installation instructions.

Illuminated toggle switches, rocker switches, and illuminated faceplates do not require the grounded circuit conductor. These are merely in series with the load. Their operation is load ON, pilot lamp OFF; or load OFF, pilot lamp ON. See Figure 5-26.

True pilot lights require the grounded circuit conductor. Their operation is load ON, pilot lamp ON; or load OFF, pilot lamp OFF. See Figure 5-27.

Let's take a look at how the above rules apply to the lighting loads in this residence. See Table 5-3.

What Should an Electrician Do?

Most decisions on how switching circuitry is designed are made on an installed-cost basis, always keeping in mind the *NEC* requirements. The electrician or electrical contractor must look at the actual layout on the job to determine the most economical approach based on the cost of labor and materials (time and material).

What about the Color Coding of Conductors?

When the wiring method is a raceway or electrical metallic tubing (EMT), the choices of insulation colors are virtually unlimited.

When using cable, you have to be aware of the cables conductor colors and how to reindentify them where necessary.

Always connect a white wire to the white (silverish) terminal or to the white wire of a lampholder or receptacle, *200.9* and *200.10*.

Always connect the black switch-leg conductor (or red in some cases) to the black wire (or dark brass–colored terminal) of a lampholder or receptacle.

Never use a green-colored insulation for a grounded or ungrounded conductor. Green insulation is reserved for equipment grounding conductors. See *210.5* and *250.119*. See Table 5-2.

Induction Heating

Excessive heat damages conductor insulation. Care must be taken to avoid induction heating when running branch circuits, connecting switches, receptacles, and lighting outlets.

A conductor carrying an alternating current produces a magnetic field (flux) around the conductor. This magnetic field extends outside of the conductor. The greater the current, the stronger the magnetic field. In a 60 Hz (pronounced "60-hertz") circuit, the current and magnetic field reverses 120 times each second. If this single conductor is run through a steel raceway, a steel jacketed cable (i.e., Type AC or MC), a steel locknut, a steel bushing, or a knockout in a steel box, the alternating magnetic field "induces" heat in the steel. Heat in the steel, in turn, can harm the insulation on the conductor. When all conductors of the same circuit are run through the same raceway, the magnetic fields around the conductors are equal and opposite, thereby canceling one another out.

TABLE 5-3

Locations Where a Grounded Conductor Is Required.

ROOM/SPACE AREA	SWITCH LOCATION	IS GROUNDED CIRCUIT CONDUCTOR (GCC) REQUIRED AT SWITCH?	COMMENTS
1. Front Bedroom	Next to door	Yes	
2. Front Bedroom closet	Next to closet door	No	Not habitable
3. Master Bedroom	Next to door	Yes.	GCC required at only one of the two 3-way switches, *404.2(C)(5)*
4. Master Bedroom closets	Next to closet doors	No	Not habitable
5. Master Bedroom; luminaire outside kitchen sliding door.	Next to sliding door	No	Not habitable
6. Master Bedroom bath	Next to door	No	Not habitable
7. Hallway between bedrooms	Two locations	No	Not habitable
8. Bathroom off Hallway	Next to door	No	Not habitable
9. Front-Entry Porch	Inside front entrance	No	No habitable spaces/areas
10. Closet-Front Entry	In door jamb	No	Not habitable and not required by *404.2(C)(3)*
11. Kitchen lighting	Next to entry from living room and center hallway	Yes	GCC required at only one of the switch locations, *404.2(C)(5)*
12. Luminaire outside Kitchen sliding door	Next to sliding door	No	Not habitable
13. Living Room	At three entries to Living Room	No	GCC not required for switch controlled receptacles, *404.2(C)(7)*
14. Fireplace	Next to sliding door	No	Accent lighting, not general illumination
15. Track lighting	Next to entry to Living Room	No	Accent lighting, not general illumination
16. Study/Bedroom	Next to both entries to room	Yes	GCC required at only one of the switch locations, *404.2(C)(5)*
17. Study/Bedroom valance lighting	Next to entry from Hallway	No	Accent lighting, not general illumination
18. Laundry, Rear Entrance, Powder Room	In each room/area/space	No	Not habitable
19. Garage and outdoor lighting	At each entrance	No	Not habitable
20. Recreation Room	At foot of stairway	Yes	
21. Recreation Room closet and bar lighting	Next to closet and bar	No	GCC requirement already satisfied by 20
22. Workshop	Next to entry	No	Not habitable

To prevent induction heating and to keep the impedance of a circuit as low as possible where conductors are run in metal raceways, metal jacketed cables, and metal enclosures, the installation must follow certain *NEC* requirements.

NEC 300.3(B) requires that All conductors of the same circuit and, where used, the grounded conductor, all equipment grounding conductors, and bonding conductors shall be contained within the same raceway, auxiliary gutter, cable tray, trench, cable, or cord, unless otherwise permitted.* There are four exceptions: for paralleled installations, grounding conductors, nonferrous wiring methods, and certain enclosures.

NEC 300.20(A) requires that Where conductors carrying alternating current are installed in ferrous metal enclosures or metal raceways, they shall be so arranged as to avoid heating the surrounding metal by induction. To accomplish this, all phase conductors and, where used, the grounded conductor and all equipment grounding conductors shall be grouped together.*

Similar requirements are found in *300.3(B)* and *300.5*.

NEC 404.2(A) requires that Three-way and four-way switches shall be so wired that all switching is done only in the ungrounded circuit conductor. Where in metal raceways or metal-armored cables, wiring between switches and outlets shall be in accordance with 300.20(A).* Switch loops do not require a grounded conductor.

Figures 5-10 and Figure 5-11 illustrate the intent of the *NEC* to prevent induction heating.

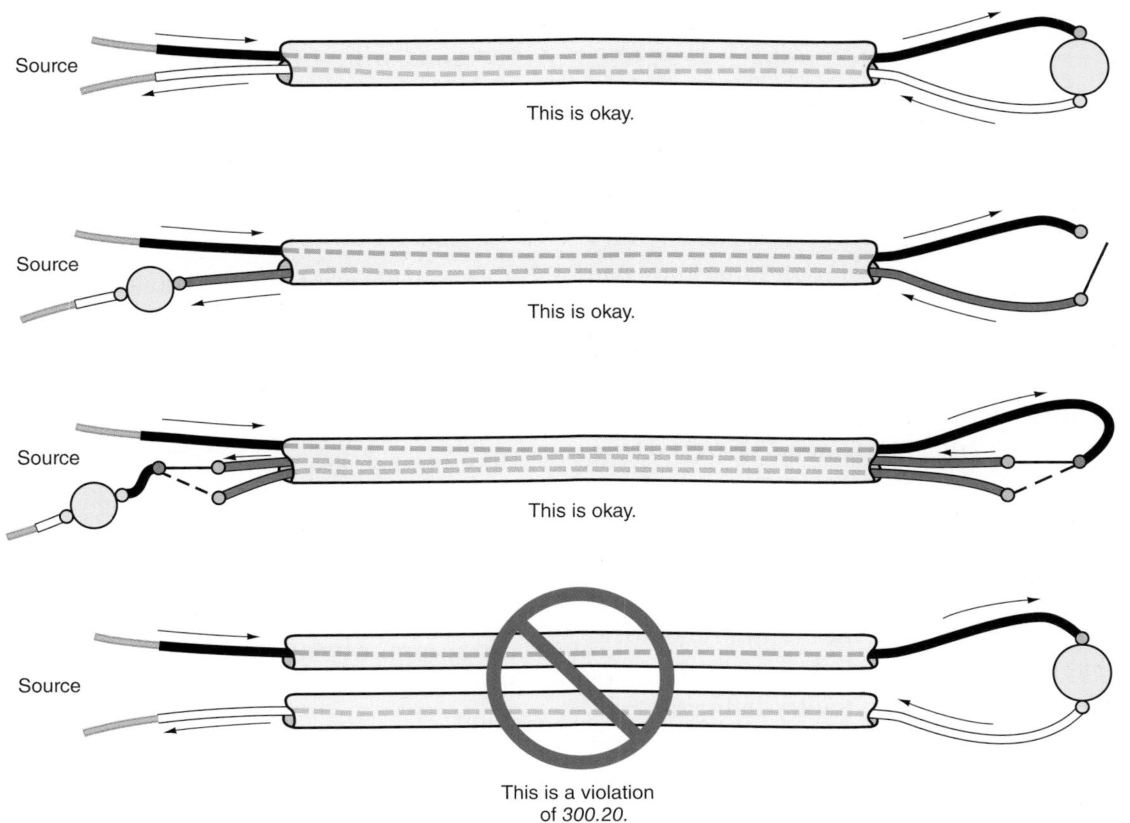

FIGURE 5-10 A few examples of arranging circuitry to avoid induction heating according to *300.20* when metal raceway or metal cables (Type MC) are used. Always ask yourself, "Is the same amount of current flowing in both directions in the metal raceway?" If the answer is *no*, there will be induction heating of the metal that can damage the insulation on the conductors.

*Reprinted with permission from NFPA 70-2014.

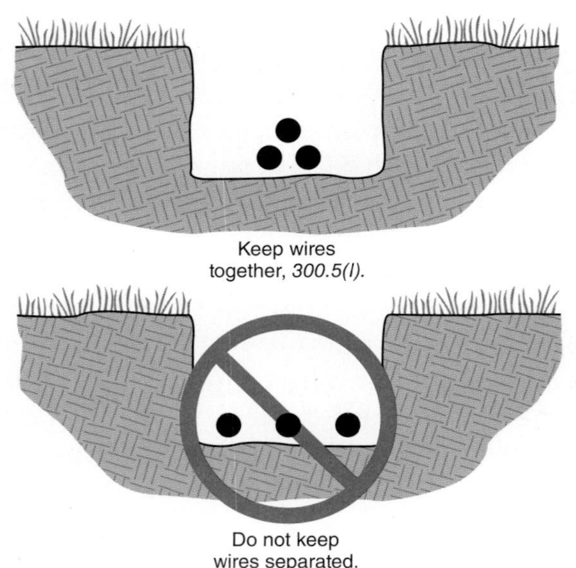

FIGURE 5-11 Arrangement of conductors in trenches. Refer to Chapter 16 for complete information pertaining to underground wiring.

Don't Do It!

Unusual switch connections may or may not conform to the *NEC*, depending on how the circuitry is run. Some electricians have found that they can save substantial lengths of 3-wire cable when hooking up 3-way and 4-way switches by using 2-wire cable.

Connections as shown in Figure 5-12 are in violation of *300.3(B)*, *300.20(A)*, and *404.3(C)* where metal boxes and/or steel-jacketed Types AC and MC are used. If nonmetallic-sheathed cable and plastic boxes are used, this connection continues to be *Code* violation because there is no grounded conductor at each switch. If the framing members are steel, we have a *Code* violation because of the induction heating issue. The circuitry in Figure 5-12 results in greater impedance and greater voltage drop than if wired in a conventional manner. High impedance will cause the tripping time of a breaker or opening time of a fuse to increase. Another possible problem would be that of picking up "noise" from the magnetic field surrounding the conductor. This could affect the operation of a computer or telephone located in the vicinity, particularly if non-twisted telephone cables are used. The circuitry is very confusing and is certainly an odd way to do a simple job.

Because of possible adverse interference created by magnetic fields, and confusion for someone who at some later date may be involved with that circuit for troubleshooting, repair, replacement, or remodel, the circuitry in Figure 5-12 is strongly discouraged. **Don't do it**.

Is It a Light? A Lamp? A Fixture?

Let's not get hung up on words! We usually say, "Turn on the light."

Light is "the radiant energy that is capable of exciting the retina and producing a visual sensation."

- A **lamp** is "a generic term for a manufactured source of light."
- A bulb is "a glass envelope—the glass component part used in a bulb assembly."
- A **luminaire** is "a complete lighting unit consisting of a light source such as a lamp or lamps, together with the parts designed to position the light source and connect it to the power supply." The *NEC* uses the term *luminaire* as the internationally used term for *lighting fixture*.

Use whatever term that you feel comfortable with. Most of us will continue to say, "Turn on the light."

Switch Types and Connections

Switches are available in single-pole, 3-way, 4-way, double-pole, dimmer, occupancy sensor, timer, and speed control types. They are available in snap (toggle), push, rocker, slider, rotary, electronic, and with or without pilot lights. Faceplates are commonly available in white, brown, and ivory plastic, stainless steel, brass, chrome, and wood.

In all of the following wiring diagrams, the equipment grounding conductors are not shown, as this is a separate issue covered elsewhere in this text.

Single-Pole Switches. A single-pole switch is marked with ON and OFF positions. Used when the load is to be controlled from one switching point, a single-pole switch is connected in series with the ungrounded ("hot") conductor.

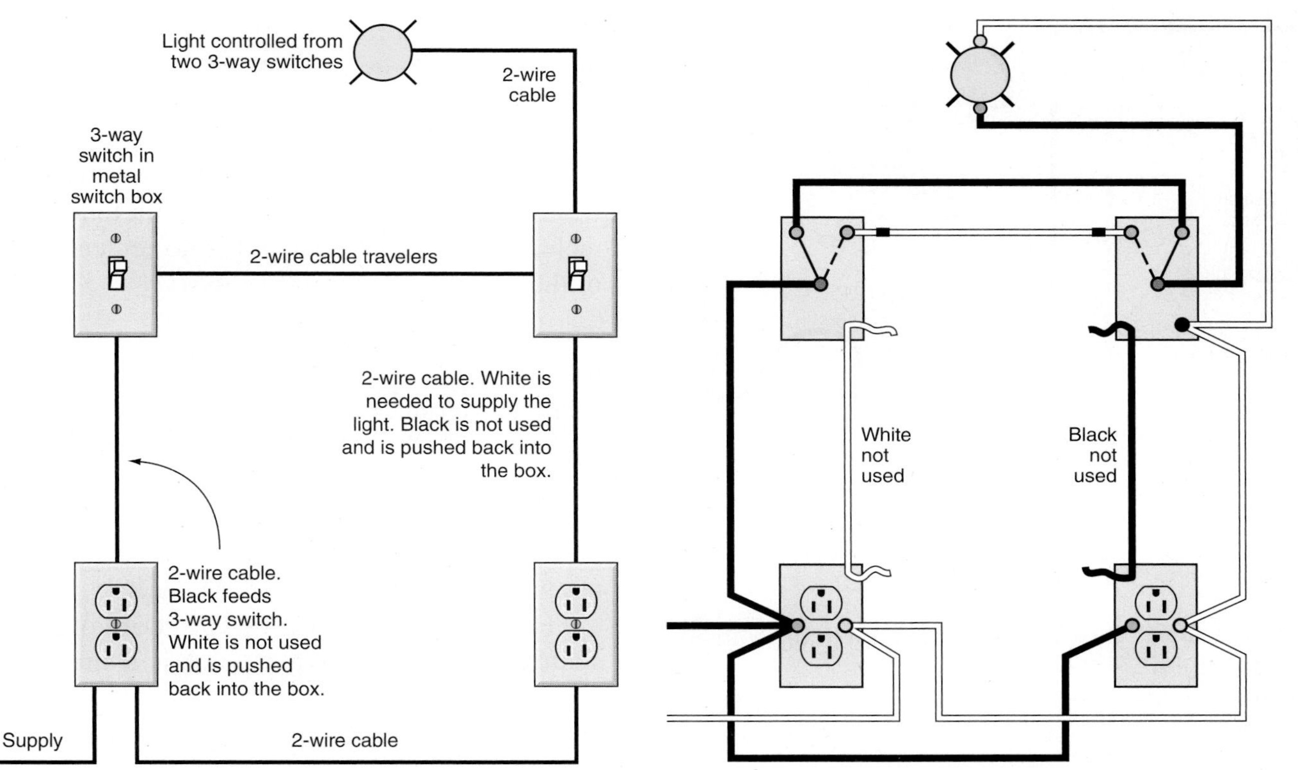

FIGURE 5-12 This connection is a violation of *300.3(B)*, *300.20(A)*, and *404.2(A)* where using steel-jacketed cables, steel raceways, or steel outlet and device boxes. See text for a full explanation. This is not a recommended way of connecting 3-way and 4-way switches.

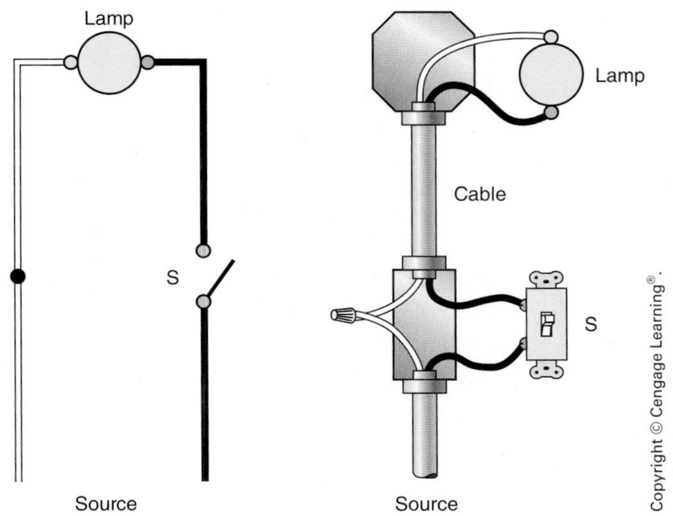

FIGURE 5-13 Single-pole switch in circuit with feed at switch.

Single-Pole Switch, Feed at Switch. In Figure 5-13, the 120-volt source enters the switch box where the white grounded conductor is spliced directly through to the lamp. The black conductor from the supply circuit connects to one terminal on the switch. The black conductor in the cable coming from the outlet box at the lamp is connected to the other terminal on the switch. Because a single-pole switch connection is simply a series circuit, it makes no difference which terminal is the supply and which terminal is the load.

This layout supplies a neutral conductor at the switch. That is why many electricians prefer to run the feed to the switch then up to the light.

Single-Pole Switch, Feed at Light. In Figure 5-14, the 120-volt source enters the outlet box where the luminaire will be connected. If a cable is used as the wiring method, a 3-wire cable is required. The conductor with black insulation is the supply to the switch and the conductor with red insulation is the return conductor. The conductor with white insulation is the neutral, so an electronic switch such as an occupancy sensor or timer can be connected properly.

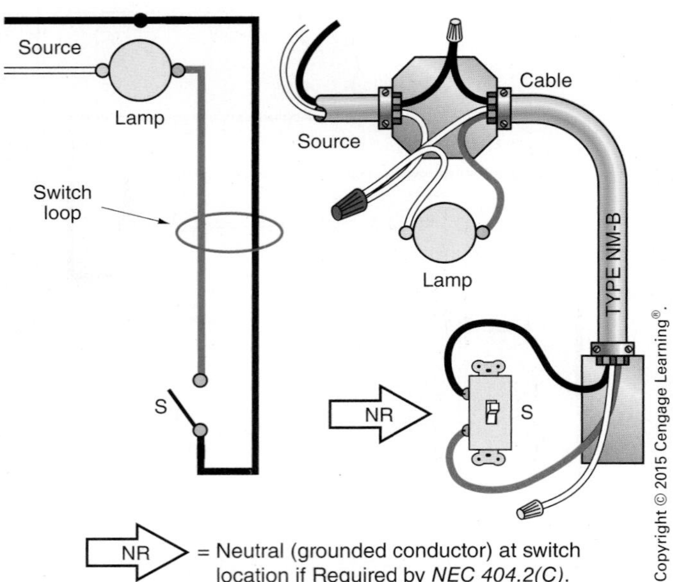

FIGURE 5-14 Single-pole switch in circuit with feed at light.

Single-Pole Switch, Feed at Switch, Receptacle ON Continuously. Figure 5-15 shows the 120-volt source entering the switch box where the switch is located. The switch controls the lamp. The receptacle remains "hot" at all times. The white conductor in the switch box is spliced straight through to the outlet box. The black conductor in the switch box is spliced through to the outlet box and also feeds the switch. The red conductor in the 3-wire cable is the switch leg connecting to the switch and to the lamp. A 2-wire cable from the outlet box carries the circuit to the receptacle. The circuitry is such that no reidentification of the white conductor is necessary.

Three-Way Switches. Three-way switches are used when a load is to be controlled from two locations. The term "3-way" is very misleading. Three-way switches have three terminals. Maybe they should have been named "3-terminal" switches. One terminal is called the "common" and is darker in color than the other two terminals. The other two terminals are called "traveler terminals."

Figure 5-16 shows the two positions of a 3-way switch. Actually, a 3-way switch is a single-pole, double-throw switch.

There is no ON/OFF marking on a 3-way switch, as can be seen in Figure 5-17.

Three-Way Switch Control, Feed at Switch. Figure 5-18 shows the 120-volt source entering at the first switch location. A 3-wire cable is run between the two switches. A 2-wire cable is run from the second switch to the lamp. The white conductors

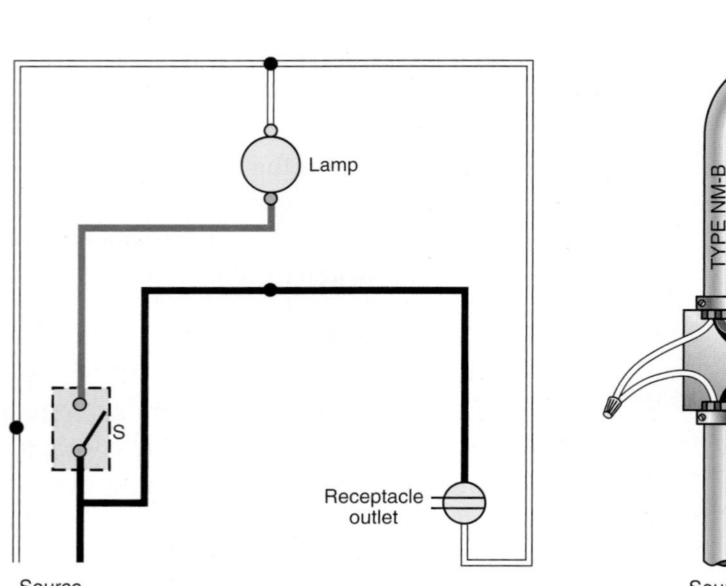

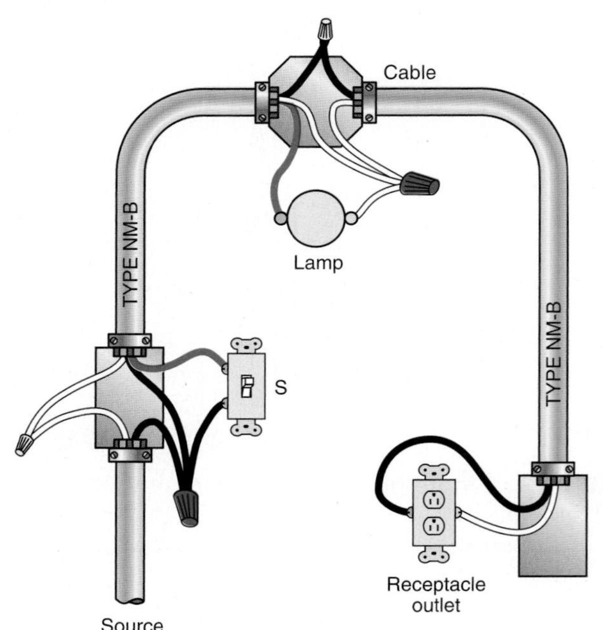

FIGURE 5-15 Lighting outlet controlled by single-pole switch with live receptacle outlet and feed at switch.

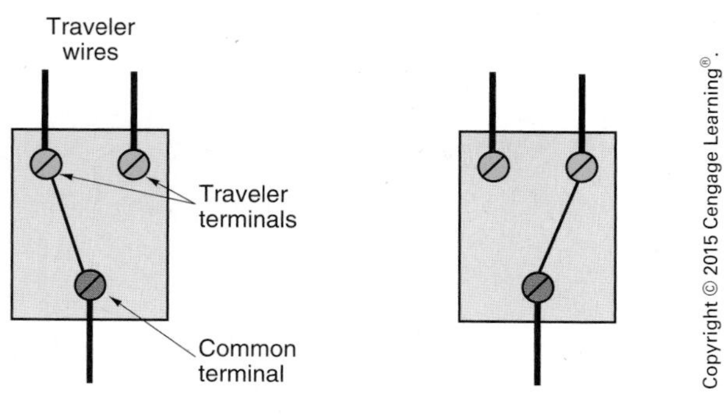

FIGURE 5-16 Two positions of a 3-way switch.

FIGURE 5-17 Toggle switch: 3-way snap switch.

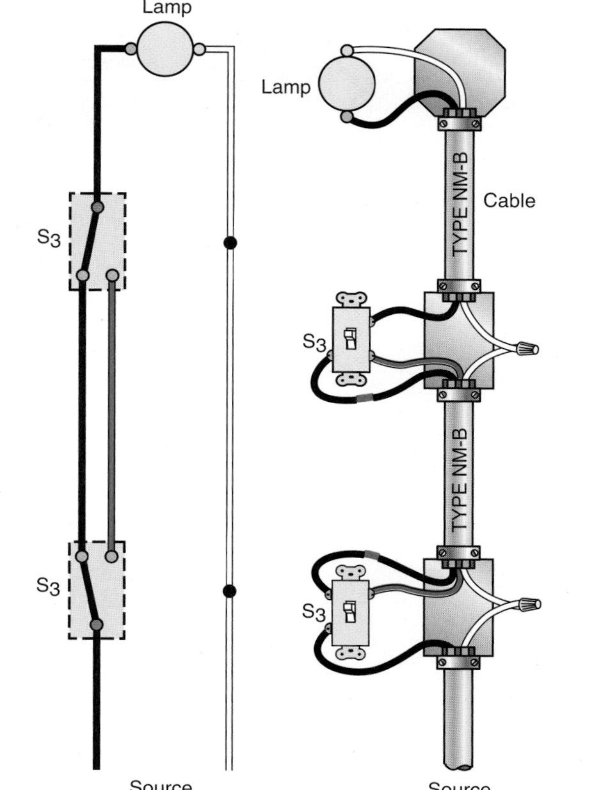

FIGURE 5-18 Circuit with 3-way switch control. The feed is at the first switch. The load is connected to the second switch. The black and red conductors are used for the travelers. The black conductor has been reidentified with red tape.

in both switch boxes are spliced directly through to the lamp. The red and black conductors in the 3-wire cable are the travelers. The black conductor in the 3-wire cable is reidentified with red tape, consistent with keeping both travelers red in color. The black conductor from the source is connected to the common terminal of the first switch. The black conductor in the cable coming from the outlet box at the lamp is connected to the common terminal of the second switch. The circuitry is such that no reidentification of the white conductor is necessary.

Three-Way Switch Control, Feed at Light. Figure 5-19 shows the 120-volt source feeding into the outlet box for the luminaire. A 3-wire cable is run from the outlet box to the first switch location. The conductor with black insulation is the supply to the switch and is connected to the black conductor in the 4-wire cable. A 4-wire cable is run from the first switch to the second switch. The neutral is carried through the first box and is capped with a wire connector in the second switch box.

The conductor with red insulation of the 3-wire cable is the return conductor to the lighting outlet and is connected to the common terminal of the first switch. The neutral conductor is capped off for future use. The red conductor and the blue conductor with red marking tape in the 4-wire cable are the travelers.

At the second switch location, the white conductor is capped off for future use. The black conductor of the 4-wire cable is connected to the common terminal of the 3-way switch. The blue

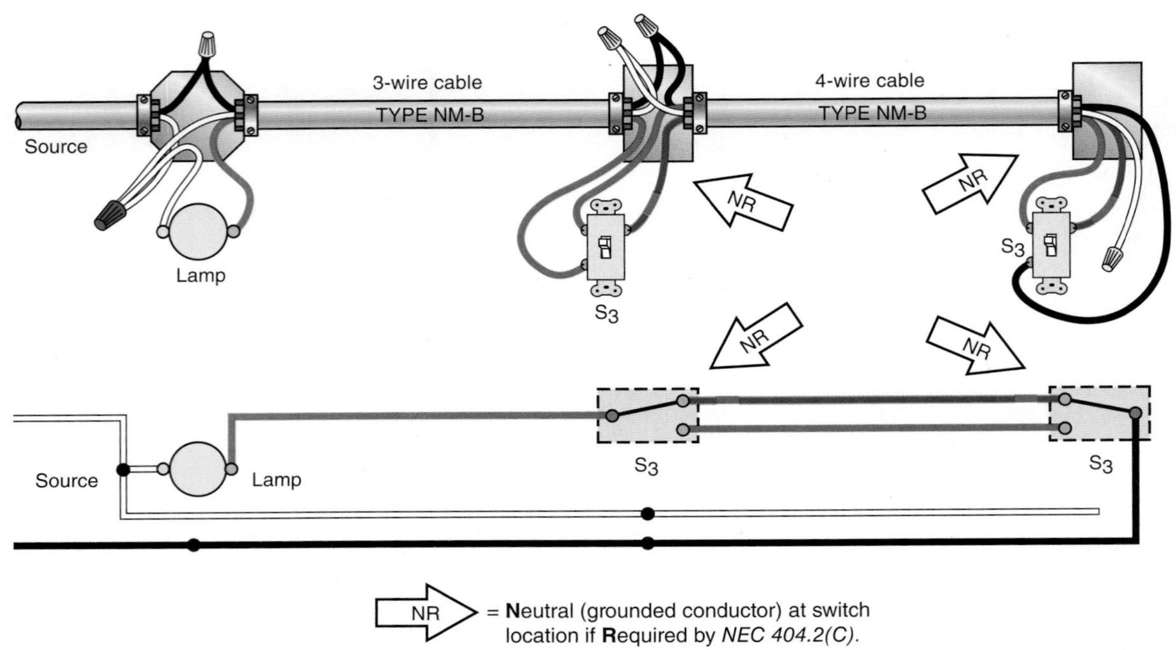

FIGURE 5-19 Circuit with 3-way switch control and feed at light.

conductor with red tape and the red conductor are the travelers.

Three-Way Switch Control, Feed at Light (Alternate Connection). Figure 5-20 shows the 120-volt source entering at the outlet box. The white conductor from the 120-volt source at the outlet box is connected to the neutral that is required at both switch boxes and to a jumper for the luminaire. A 4-wire cable is run to both switch boxes.

From the outlet box to the switch on the left, the black conductor in the 4-wire cable serves as the switch return to the lamp, and the red conductor and the blue conductor with the red marking tape are the travelers. From the outlet box to the switch on the right, the black conductor serves as the supply to the 3-way switch, and the red conductor and the blue conductor with red marking tape are the travelers. For uniformity, the conductor with red insulation and the conductor with blue insulation that has red marker tape used for travelers are identified and connected in the outlet box.

Four-Way Switches. Four-way switches are used where switch control is needed from more than two locations. The term "4-way" is misleading.

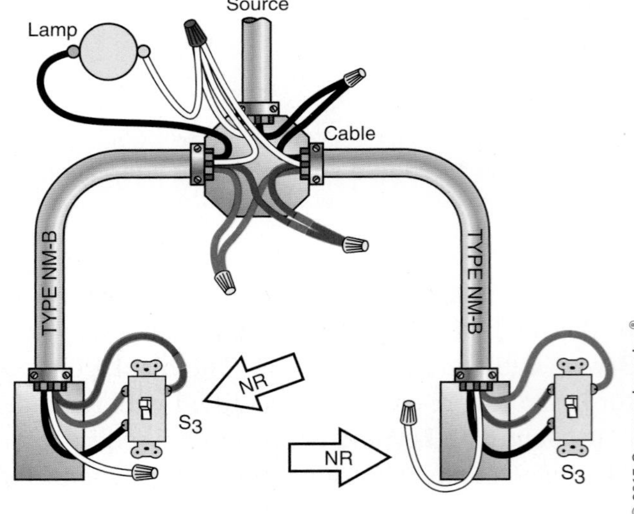

FIGURE 5-20 Circuit with a 3-way switch control with feed at the lamp. This is not the best way to wire multiple switches when wiring with cable. It is best to run the travelers between switches, not through the outlet box. Less confusion—fewer connections.

Four-way switches have four terminals. Maybe 4-way switches should have been called "four-terminal" switches. A 4-way switch does not have ON and OFF markings. Four-way switches are connected to the travelers *between* two 3-way switches. One 3-way switch is fed by the source; the other 3-way switch is connected to the load. Make sure that the travelers are connected to the proper pairs of terminals on a 4-way switch; otherwise, the switching will not work.

Figure 5-21 shows the internal switching of a 4-way switch.

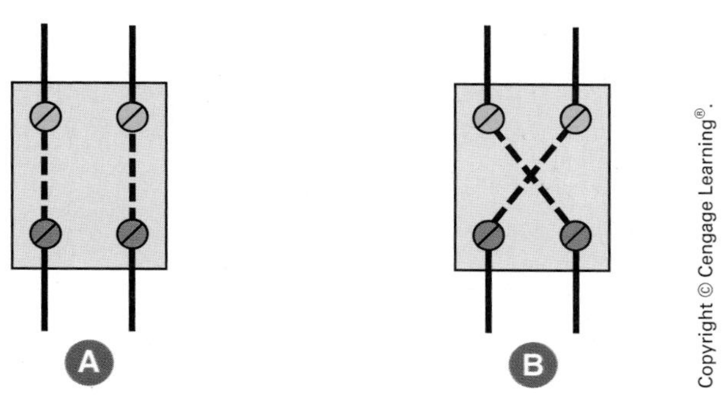

FIGURE 5-21 Two positions of a 4-way switch.

Figure 5-22 shows a luminaire controlled from three switching points. Because the white conductor from the source is connected straight through all of the switch boxes and on to the lamp, it satisfies the requirement for a neutral conductor to be present in all switch boxes.

The black conductor from the source is connected to the common terminal on the first 3-way switch. The black conductor at the third switch is connected to the common terminal on the 3-way switch and supplies the luminaire. The black and red conductors in the 3-wire cables are the travelers and are connected to the proper traveler terminals of the 3-way and 4-way switches. The black conductors in the 3-wire cables have been reidentified with red tape, consistent with keeping both travelers red in color. At the lamp, the white conductor of the 2-wire cable connects to the white conductor (or silverish terminal) of the lampholder, and the black conductor connects to the black conductor (or dark brassy terminal) of the lampholder.

Figure 5-23 shows a lamp controlled from three switching points with two 3-way switches and one 4-way switch. The supply comes into the outlet box

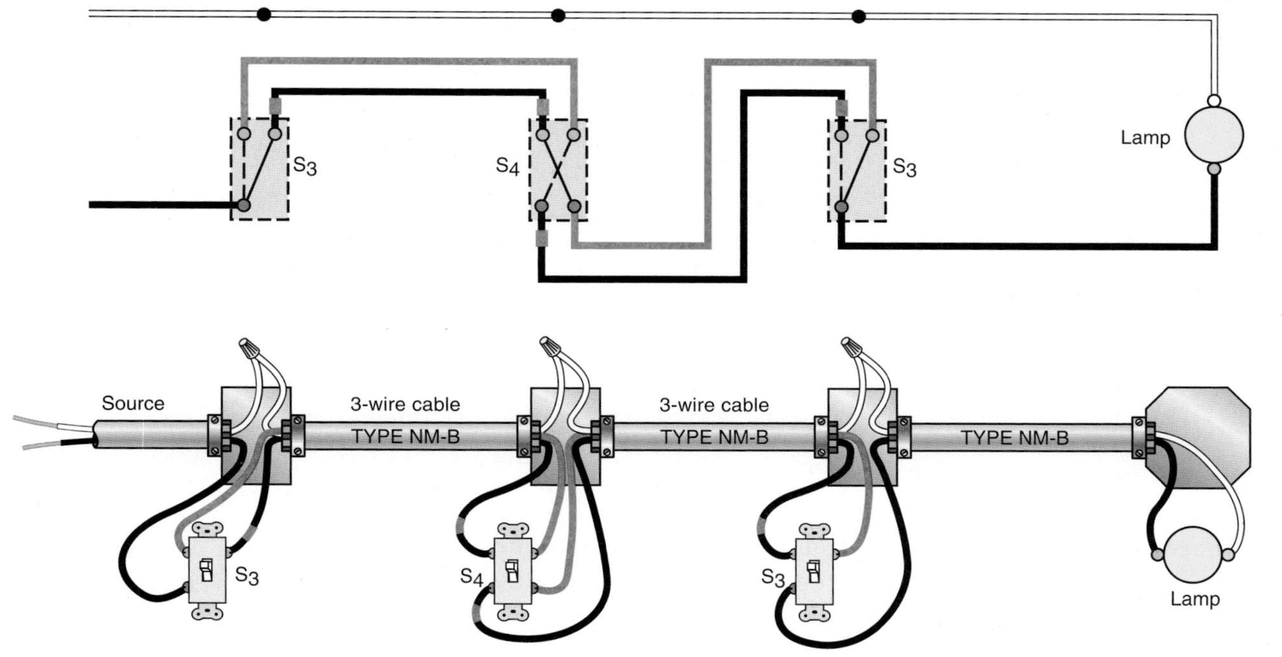

FIGURE 5-22 Circuit with switch control at three locations—feed at switch.

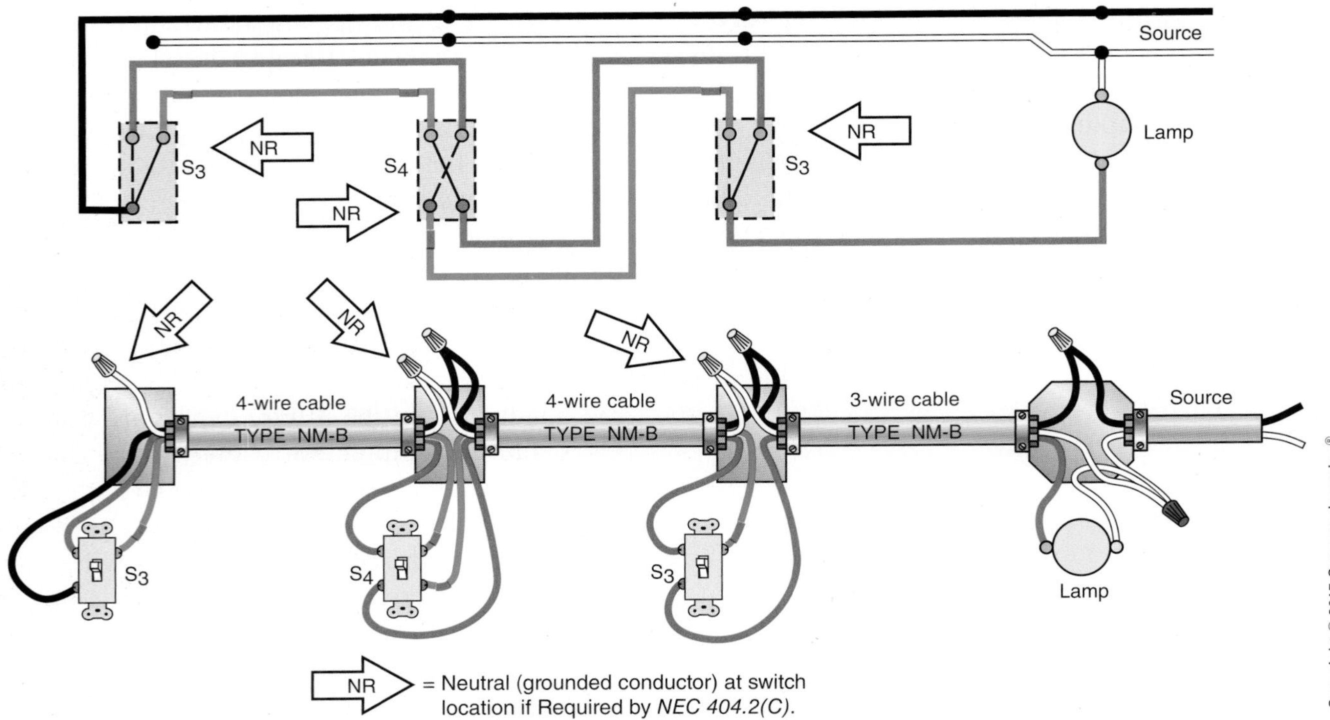

FIGURE 5-23 Circuit with switch control at three locations—feed at luminaire.

where the luminaire is connected. A 3-wire cable is run to the first switch box and 4-wire cables from the second to the third box, and from the third to the fourth box.

The conductor with black insulation is connected together at each switch box and is connected to the common terminal of the end 3-way switch. The conductor with white insulation is connected through each box and is available for future connection to an electronic switch or timer. The conductor with red insulation and the conductor with blue insulation are the travelers. Red marking tape is placed on the conductor with blue insulation for consistent identification of traveler conductors.

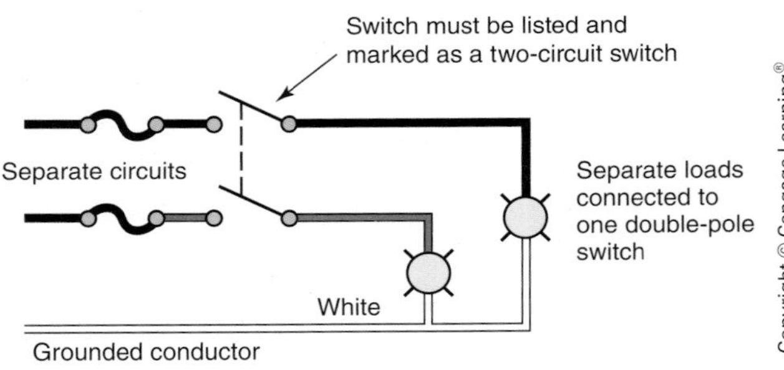

FIGURE 5-24 Application of a double-pole switch.

Double-Pole Switches. Double-pole switches are not common in residential work. They can be used when two separate circuits must be controlled with one switch, as in Figure 5-24. If used for this purpose, they must be marked "2-circuit." Otherwise, 2-pole switches must be used to switch two conductors of a single branch circuit.

They are commonly used to control 240-volt loads, such as electric heat, motors, electric clothes dryers, and similar 240-volt loads, as in Figure 5-25. When line-voltage thermostats for electric heat are double-pole, they switch both ungrounded "hot" conductors and are marked with ON and OFF positions.

Switches with Pilot Lights. There are instances when a pilot light is desired at the switch location. Figure 5-26 shows a "locator" type of switch where the lamp in the toggle glows when the switch is in the OFF position. This is sometimes referred to as a "glow" or "lit-handle" switch. A resistor inside the

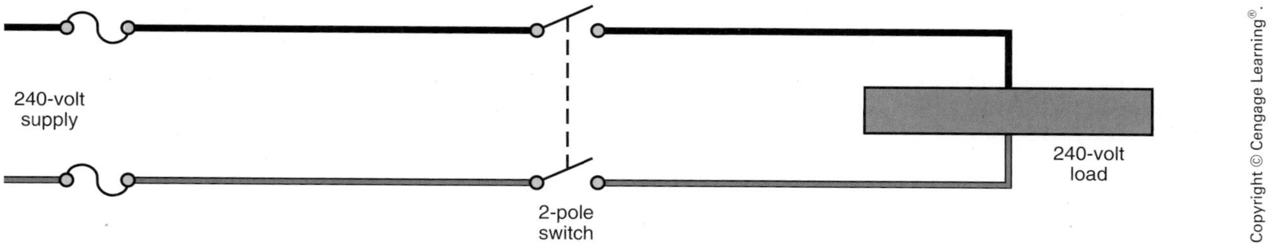

FIGURE 5-25 Double-pole (2-pole) disconnect switch.

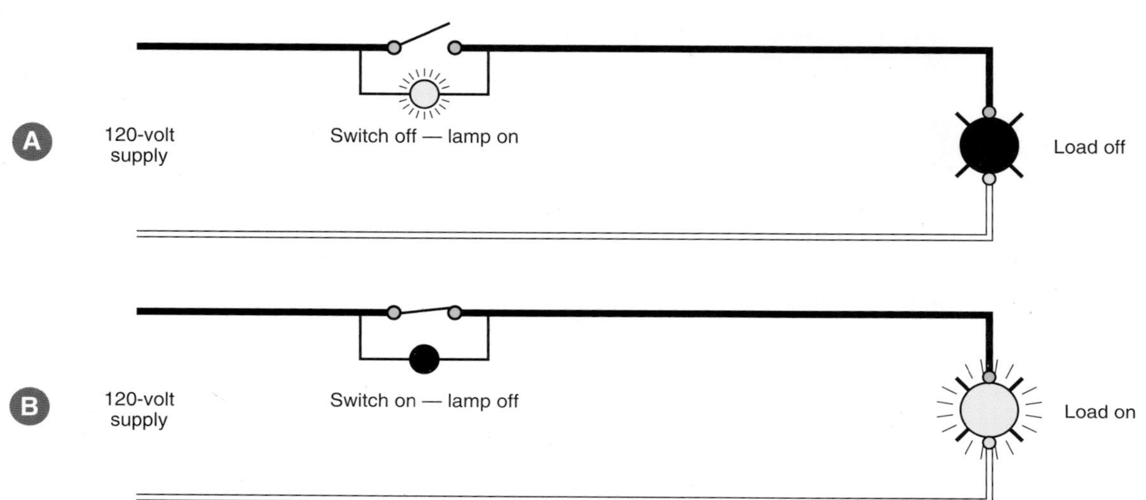

FIGURE 5-26 In (A) the switch is in the OFF position. The neon lamp in the handle of the switch glows and the load is off. This is a series circuit. The neon lamp has extremely high resistance. In a series circuit, voltage divides proportionately to the resistance. Therefore, the neon lamp "sees" line voltage (120 volts) for all practical purposes, and the load "sees" zero voltage. The neon lamp glows. In (B) the switch is in the ON position. The neon lamp is shunted out and therefore has zero voltage across it. Thus, the neon lamp does not glow. Full voltage is supplied to the load. This type of switch might be referred to as a *locator* because it glows when the load is off and does not glow when the load is on.

switch is connected in series with the pilot light. Be careful when working on this type of switch because when the switch is in the OFF position, the load and the pilot light are actually connected in series. Even though these switches are marked with an OFF position, they are not truly off. The circuit's path through the switch and connected load is still complete. There is a voltage drop (almost full open circuit voltage) across the pilot light/resistor and an infinitesimal voltage drop across the connected load. Although there is not enough voltage at the light to light it, there may be enough voltage and current present to cause a person to flinch, which might result in the person jerking back and falling off a ladder.

Figure 5-27 shows a true pilot light that is ON when the light is on.

COMBINATION WIRING DEVICES

In the space of one standard wiring device, it is possible to install combination wiring devices. Figure 5-28(A) shows three types of combination wiring devices. Other combinations are available. Figure 5-28(B) shows an interchangeable line of wiring devices, assembled on the metal yoke by the electrician. Switches, receptacles, and neon and incandescent pilot lights are available. Combination

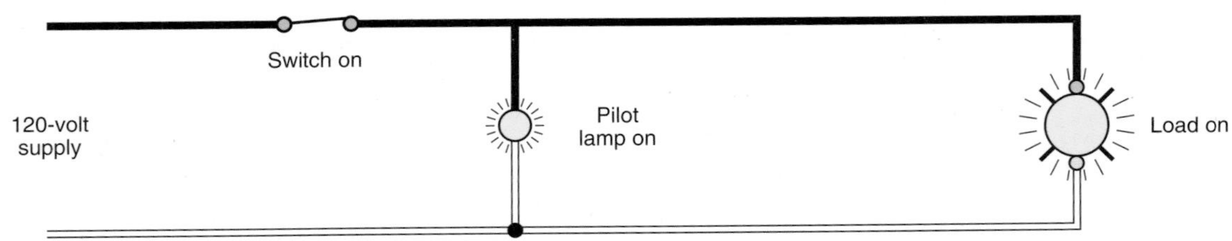

FIGURE 5-27 A true pilot light. The pilot lamp is an integral part of the switch. When the load is turned ON, the pilot light is also on. When the load is turned OFF, the pilot light is also off. This switch has three terminals because it requires a grounded circuit conductor.

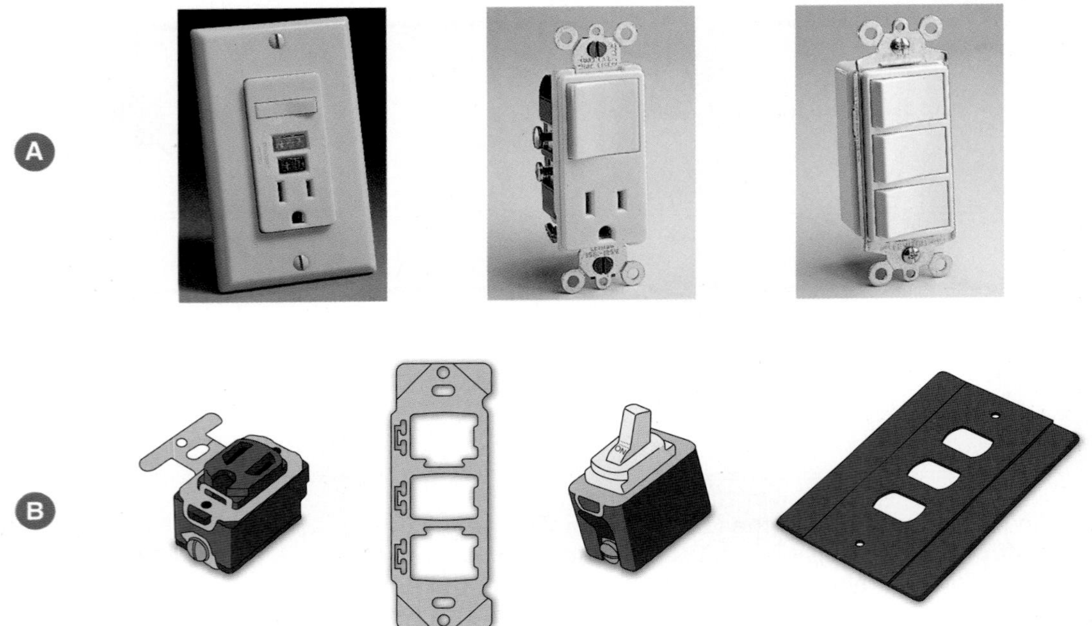

FIGURE 5-28 Combination wiring devices are shown in (A). Many combinations are available on a single yoke: slider, rocker, toggle, 3-way plus two single poles, four single-pole switches, switches and pilot lights, switches and receptacles. Interchangeable wiring devices are shown in (B).

wiring devices offer a neater appearance as opposed to having 2-gang or 3-gang faceplates at a given location. Another common use is where wall space is limited, such as between a door casing and a window casing where there is not enough space for a multigang box and faceplate. Be sure the wall box has plenty of room for the number of cable clamps, wiring devices, and conductors.

Miscellaneous Connections. Figures 5-29 and 5-30 are diagrams that somewhat combine parts of the previous wiring diagrams. These connections become rather complicated using cable because of the limitation on the available insulation colors. These connections are more suited to raceway wiring.

Does OFF Really Mean OFF?

The answer is *yes*. When a switching device has a marked OFF position, it must completely disconnect the ungrounded conductor(s) to the load it serves when turned to the OFF position. UL Standards require this. Also see *404.15(B)*.

Electronic controls, such as some motion (occupancy) sensors, some dimmers, some fan speed

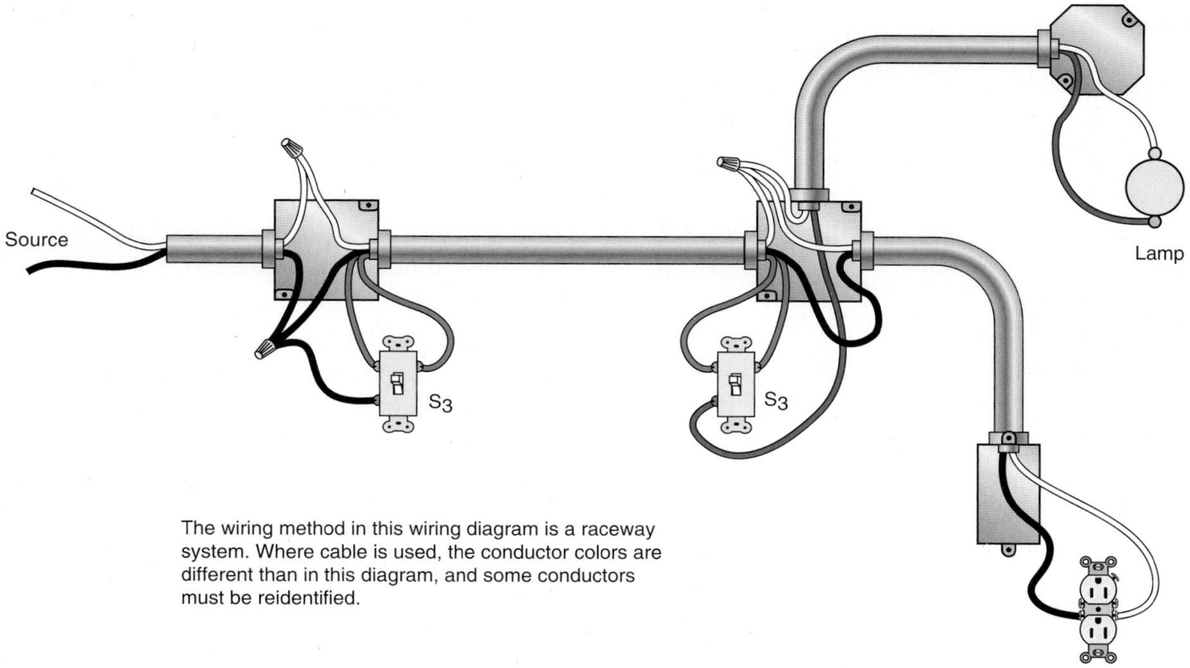

Source

Lamp

S_3

S_3

The wiring method in this wiring diagram is a raceway system. Where cable is used, the conductor colors are different than in this diagram, and some conductors must be reidentified.

FIGURE 5-29 A lighting outlet controlled by two 3-way switches. The receptacle is "live" at all times. The wiring method is a raceway. When cable is used, the conductor colors are different from those in this wiring diagram, and some conductors must be reidentified. A typical example of this might be between a house and a detached garage or other outbuilding.

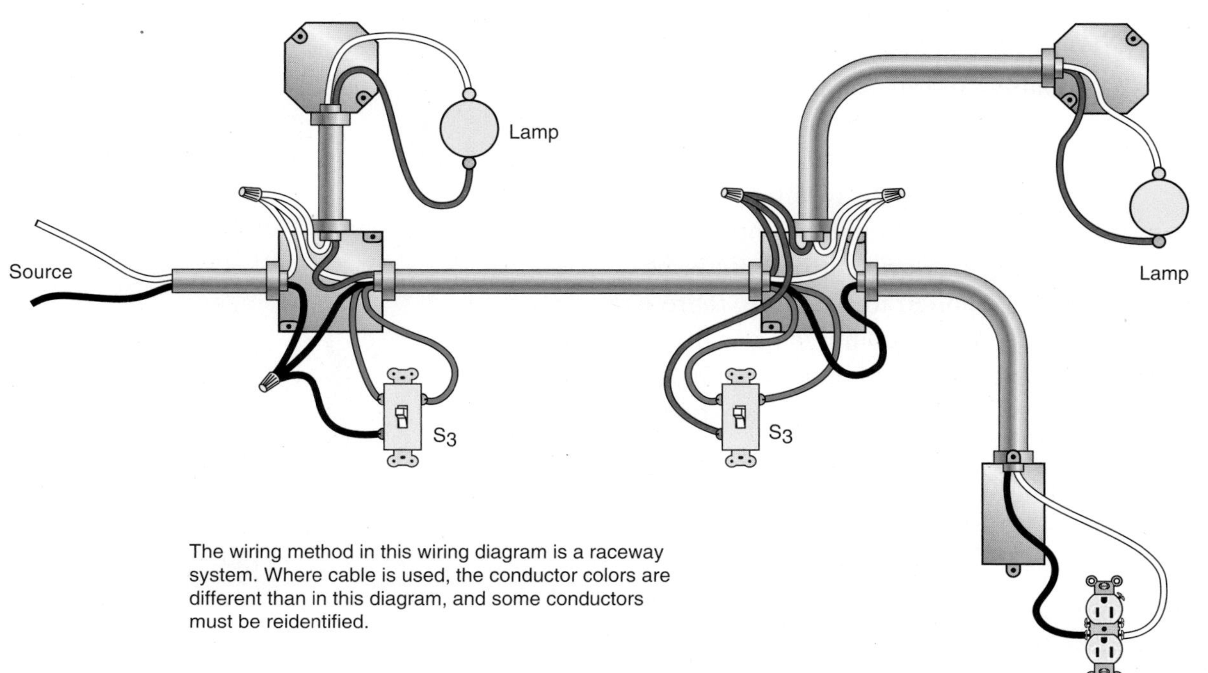

FIGURE 5-30 Two lighting outlets controlled by two 3-way switches. The receptacle is "live" at all times. An example of this hookup might be between a house and a detached garage or other outbuilding, where one lighting outlet is on the house and the other lighting outlet is on or in the garage or outbuilding.

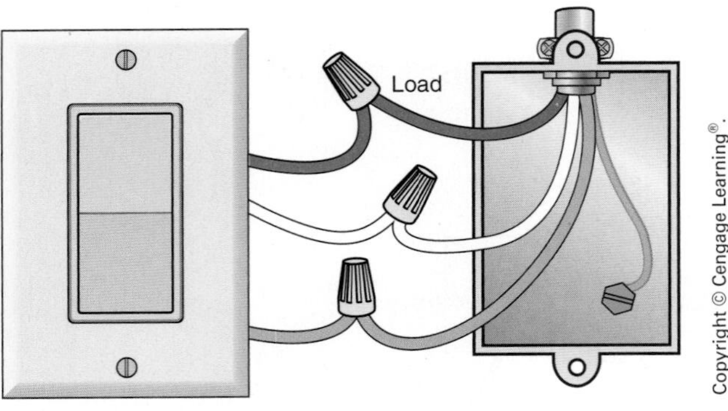

FIGURE 5-31 An X-10 switch. To operate properly, it requires the presence of the grounded circuit conductor. *Never* connect the white wire on the switch to an equipment grounding conductor in the box. This is hazardous and is a violation of *250.6(A)* and *250.24(A)(5)*.

controls, and X-10 transmitting and receiving devices (see Chapter 31), require a small voltage drop across the device to operate. When these devices are turned to what appears to be the OFF mode, there is still an ever-so-slight amount of voltage present at the load. The device is really not in a full OFF position. This can be hazardous to someone working on a luminaire, for example, thinking it is totally OFF. See Figure 5-31.

You will probably find instructions furnished with an electronic switch (control) that the load must be a minimum of X watts. This is to provide the proper voltage drop across the switch (control).

Compact fluorescent lamps will not operate on an electronic motion sensor or electronic timer because until voltage is applied to the lamp, there is no current draw. A fluorescent lamp is basically an open circuit until energized. The lamp cannot be energized because there will be no voltage across the electronic control to operate the electronic control.

Refer to Chapter 6 for additional information relating to the effects of an electrical shock.

Bonding and Grounding at Receptacles and Switches

A metal box is considered to be adequately grounded when the wiring method is armored cable, nonmetallic-sheathed cable with ground, or a metal raceway such as EMT, *250.118*. A separate equipment grounding conductor can also provide the required grounding, *250.134(B)*. Equipment grounding conductors may be bare or green.

Figure 5-32 illustrates a switch that has an equipment grounding conductor terminal. Using a switch of this type provides an easy means for providing the required grounding of a metal faceplate when using nonmetallic boxes.

Grounding and bonding of the equipment grounding conductor to a metal box, switch, or receptacle is important. Figures 5-32, 5-33, and 5-34 show how an equipment grounding conductor of a nonmetallic-sheathed cable can be attached to a metal box and to the equipment grounding screw of a receptacle. Most metal boxes have a No. 10-32 tapped hole for securing a green hexagon-shaped equipment grounding screw.

Other than for special "isolated ground" receptacles, the metal yoke of a receptacle is integrally bonded to the equipment grounding terminal of the receptacle.

Figures 5-4 and 5-5 show a receptacle that has a special listed "self-grounding" feature for the No. 6-32 mounting screws that results in effective grounding of the metal yoke and equipment grounding terminal without the use of a separate equipment

FIGURE 5-32 Single-pole toggle switch. Note that this switch has a terminal for the connection of an equipment ground conductor. Thus, when a metal face plate is installed, it will be grounded. See *404.9(A)* and *(B)*.

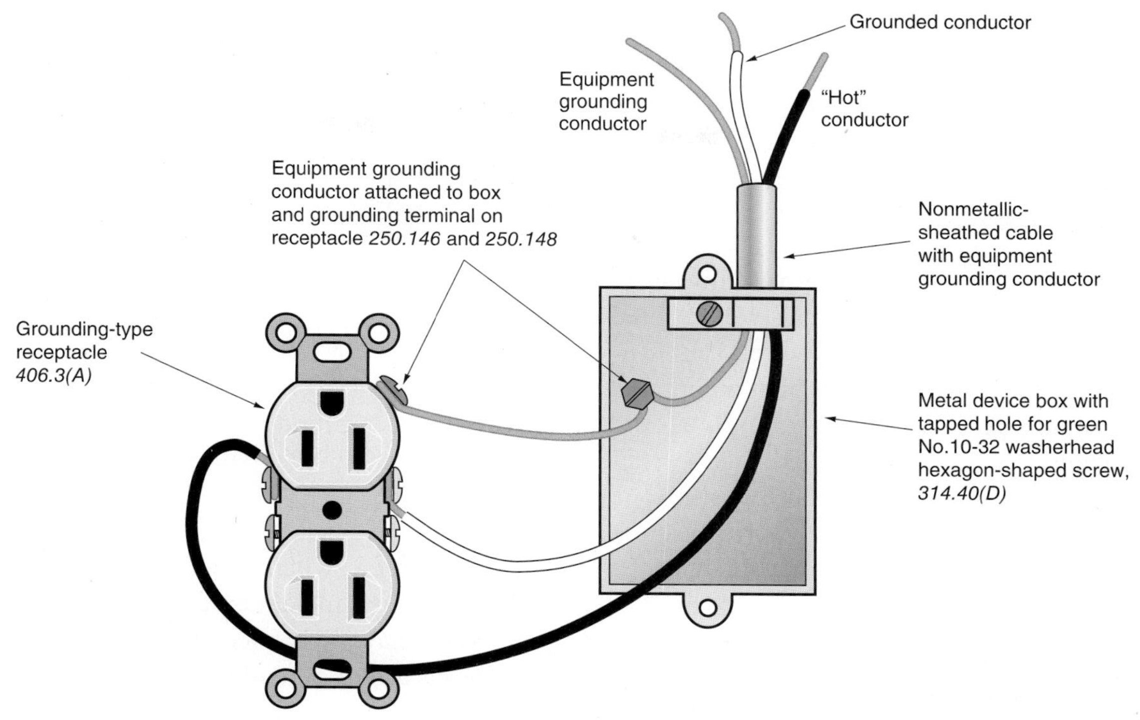

FIGURE 5-33 Connections for grounding-type receptacles.

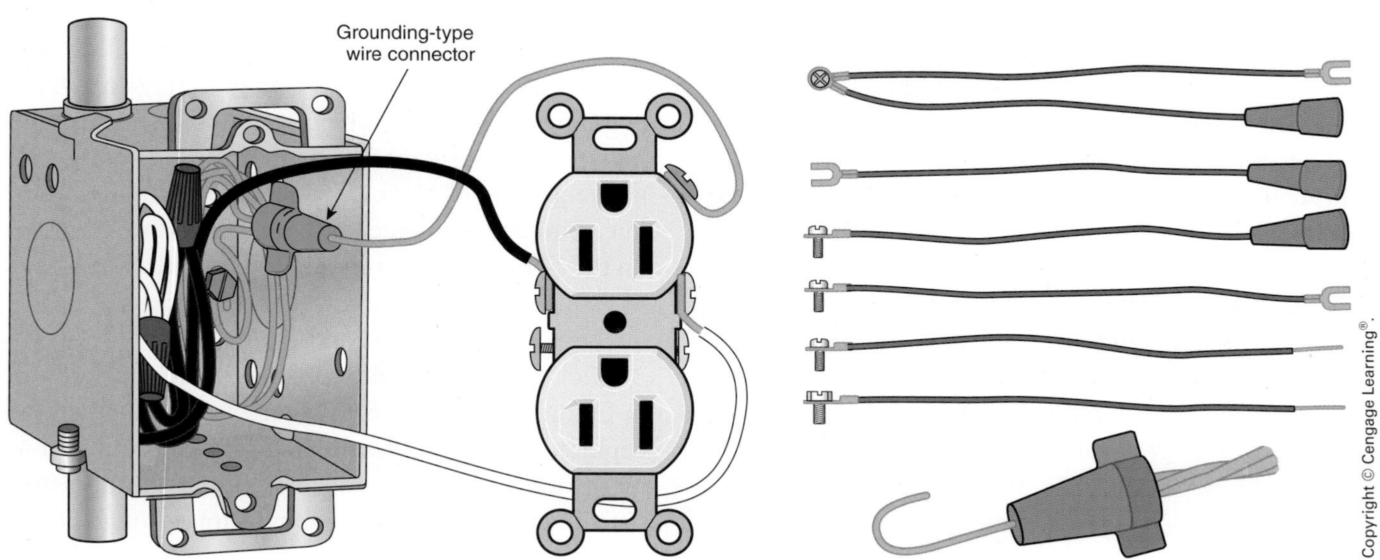

FIGURE 5-34 One method of connecting the grounding conductor using a special grounding-type wire connector. See *250.146* and *250.148*.

grounding conductor so long as a grounded metal box is used.

To ensure the continuity of the equipment grounding conductor path, *250.148* requires that where more than one equipment grounding conductor enters a box, they shall be spliced with devices "suitable for the use." Splices shall not depend on solder. The splicing must be done so that if a receptacle or other wiring device is removed, the continuity of the equipment grounding path will not be interrupted. This is clearly shown in Figure 5-33. Splicing must be done in accordance with *110.14(B)*.

Grounding Metal Faceplates and Switches

Assuming that an equipment grounding conductor has been carried to and properly connected at all metallic and nonmetallic boxes, we now need to be concerned with carrying that equipment grounding means to receptacles, switches, and metal faceplates.

Metal faceplates for receptacles shall be grounded, *406.6(B)*. This is accomplished through the No. 6-32 screw that secures the faceplate to the metal yoke on the receptacle.

A similar requirement for the grounding of metal faceplates is found in *314.25(A)*.

Snap switches and dimmer switches must provide a means to ground metal faceplates, *404.9(B)*. This is a requirement whether metallic or nonmetallic faceplates are installed. The actual grounding of the faceplate is through the No. 6-32 screws that secure the faceplate to the metal yoke on the switch.

In UL Standard 20, we find the requirement that flush-type switches intended for mounting in a flush-device box shall provide a grounding means. In most cases this is a green hexagon-shaped screw connected to the metal yoke of the switch. This can be seen on the switches illustrated in Figures 5-8, 5-17, and 5-32. The grounding means could also be a special self-grounding feature on the metal yoke, as illustrated in Figures 5-3 and 5-4, a bare copper wire, a copper wire with green insulation, or a copper wire with green insulation having one or more yellow stripes.

Connecting a bare equipment grounding conductor to the equipment grounding terminal on a switch complies with *404.9(B)* for the grounding of a metal faceplate. Where there are multiple switches, there is no need to attach an equipment grounding conductor on all of the switches. One connection is enough.

Figure 5-35 illustrates connecting equipment grounding conductors or jumpers to a metal box with a grounding clip.

If the equipment grounding path is accomplished through a self-grounding design of the metal yoke of a switch or receptacle, and the device is mounted in a metal device box that is grounded, you do not have to connect an equipment grounding conductor to the green hexagon-shaped grounding screw on the switch or receptacle.

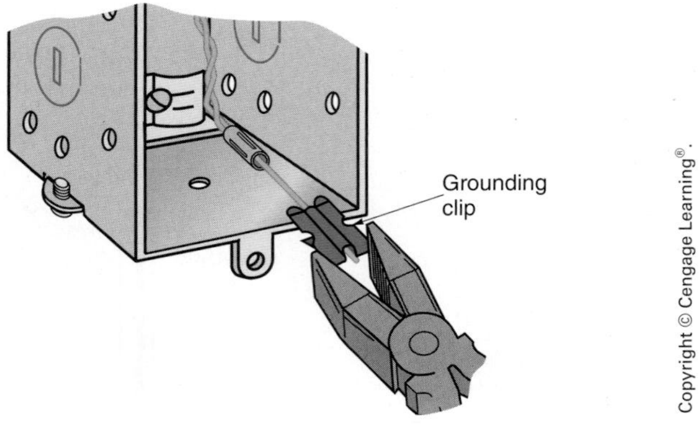

FIGURE 5-35 Method of attaching grounding clip to switch (device) box. See *250.146* and *250.148*.

Many wiring devices have small pieces of cardboard or plastic holding the No. 6-32 mounting screws from falling out of the yoke. Remove at least one of these small pieces of cardboard when direct metal-to-metal contact between the yoke of the wiring device and the box is needed, *250.146(A)*. An example of this would be a receptacle secured to a handy box for surface wiring. If there is no metal-to-metal contact, then an equipment grounding conductor must be connected to the grounding terminal on the wiring device, or the wiring device must have the special self-grounding feature discussed previously. See Figures 5-33 and 5-34.

For surface-mounted boxes of the type shown in Figures 18-8 and 18-9, where cover-mounted receptacles are used, the metal-to-metal requirement mentioned earlier is not acceptable. Some other listed means for providing satisfactory ground continuity must be used, *250.146(A)*. This section makes it clear that there must be direct metal-to-metal contact between the metal box and the device yoke to be considered an acceptable means of grounding the receptacle equipment grounding terminal and the metal cover.

To make sure that there is direct metal-to-metal contact between the metal yoke of a receptacle and a surface-mounted metal box, *250.146(A)* accepts the automatic self-grounding device on the yoke of the receptacle or the removal of at least one of the insulating washers. Automatic self-grounding devices on the yoke of a receptacle are clearly shown in Figures 5-4, and 5-5.

In general, where no effective grounding means exists in the switch box, it is acceptable to replace a snap switch with another snap switch that does not have a provision for grounding.

However, in existing installations where no effective grounding means is present and where the snap switch is located within reach of the earth, a conductive floor (i.e., concrete, tile), or other conductive surfaces (i.e., metal pipes, metal ducts), use nonmetallic faceplates or provide GFCI protection for the wiring to that particular snap switch. See *404.9(B), Exception*.

SAFETY ALERT

Never **make connections (jumpering) between the different colored terminals on receptacles. Lives have been lost because of this!**

- **Jumpering between the brass and silver terminals is a violation of *200.10* and *200.11*.**
- **Jumpering between the silver and green hexagon-shaped terminals is a violation of *250.24(A)(5)* and *250.142(B)*.**
- **Jumpering between the brass and green hexagon-shaped terminals is a violation of *250.126*.**

COMMON *CODE* VIOLATION TAPS

As defined in *240.2*, a "tap" is a conductor that has overcurrent protection ahead of its point of supply that exceeds the ampacity of the conductor. In general, conductors are required to be protected at their ampacities per *310.15*. Tap conductors are excluded from the basic rule, and this is found in *240.4(E)*. *Article 240* Part II generally covers feeder taps. Section *240.21(A)* refers to branch-circuit conductors meeting the requirements in *210.19* are permitted to have overcurrent protection as specified in *210.20*.

Several types of branch circuit taps are covered in *210.19(A)(3)* and *(4)*.

Some examples of taps are small luminaire wires. In some cases, taps can be cost-effective for hooking up electric ranges, counter-mounted cooktops, and wall-mounted ovens. This is discussed in Chapter 20.

Table 5-4 shows the maximum overcurrent protection for small conductors as permitted by *240.4(D)*. The overcurrent protection requirements or permissions for motors in *Article 430* act as an exception to this small conductor rule.

Figure 5-36 illustrates a *Code* violation. The switch-leg (loop) conductors are part of the branch-circuit wiring and are not to be considered a tap. Although the connected load may be well within the ampacity of the switch leg, the switch leg must also be capable of handling short circuits and ground faults. The switch leg shown in Figure 5-36 must be the same size as the branch-circuit conductors, which are 12 AWG 20-ampere conductors protected by a 20-ampere circuit breaker.

Connecting Receptacles to 20-Ampere Branch Circuits

Another very common *Code* violation is using short 14 AWG pigtails to connect receptacles to 20-ampere small-appliance branch-circuit conductors in a kitchen or dining room. These short

TABLE 5-4

Maximum overcurrent protection for small conductors.

CONDUCTOR SIZE	MAXIMUM OVERCURRENT PROTECTION
14 AWG copper	15 amperes
12 AWG copper	20 amperes
10 AWG copper	30 amperes

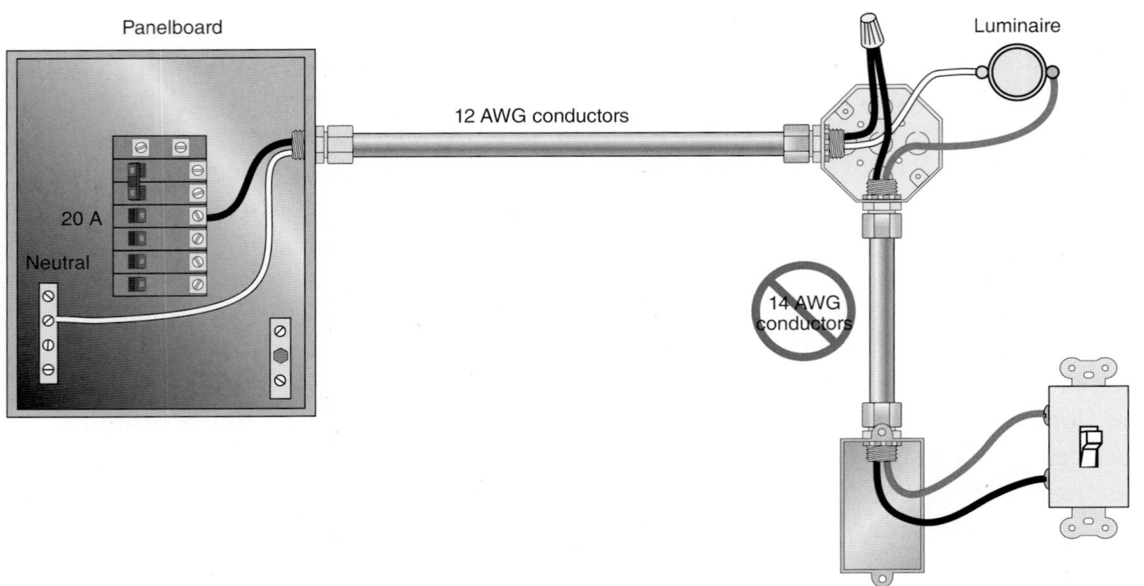

FIGURE 5-36 *Violation:* The electrician has run a 20-ampere branch circuit consisting of 12 AWG conductors to the outlet box where a luminaire will be installed. 14 AWG conductors are run as the switch legs between the lighting outlet and the switch. The switch legs are part of the branch circuit and must have an ampacity equal to that of the branch-circuit ampere rating, 20 amperes. Switch legs are not taps.

pigtails must be rated 20 amperes because they are an *extension* of the 20-ampere branch-circuit wiring—they are not taps. Figure 5-36 illustrates this type of connection.

Consider Using Larger Size Conductors for Home Runs

To minimize voltage drop, sometimes 12 AWG conductors are used for long "home runs" (50 ft or longer) even though the branch-circuit overcurrent device is rated 15 amperes. A home run is the branch-circuit wiring from the panel to the first outlet or junction box in that circuit. This is permitted by *210.3*, which states, Where conductors of higher ampacity are used for any reason, the ampere rating or setting of the specified overcurrent device shall determine the circuit rating.* In this case, 14 AWG conductors are permitted for the switch legs. But this leads to confusion later when someone removes the cover from the panel and wonders why 12 AWG branch-circuit conductors are connected to 15-ampere circuit breakers. For house wiring, it is best to use the same size conductor throughout a given branch circuit.

Specifications for commercial and industrial installations oftentimes will call for a home run of 50 ft or more (≥15.2 m) to be 10 AWG conductors, even though the branch-circuit rating is 20 amperes.

Table 210.24 is a summary of branch-circuit requirements. This table lists branch-circuit ratings, branch-circuit conductor sizes, tap conductor sizes, overcurrent protection, outlet (lampholder and receptacle) ratings, maximum loads, and permissible loads.

TIMERS

Timers are unique in that they provide automatic control of electrical loads. Timers are also referred to as time clocks.

Timers are used where a load is to be controlled for specific ON/OFF times of the day or night. The capabilities of timers are endless. Some timers are

*Reprinted with permission from NFPA 70-2014.

astronomical in that the location of the timer is entered when installed. After that, the timer turns lights on and off based on the dusk-to-dawn time rather than by a photocell.

Figure 5-37 shows two types of spring-loaded timers that are connected in series with the load—the same as a standard wall switch. A typical application is in a bathroom for the control of an exhaust fan or an electric heater. These timers install in a standard single-gang device box. Because they take up quite a bit of space, make sure the wall box is large enough to meet the box-fill requirements of *314.16*. Electronic timers that do not have any moving parts are available as well. These timers require a neutral conductor for proper operation.

Figure 5-38 is a 24-hour time switch commonly used to control security lighting, decorative lighting, or energy management. This particular time switch has two adjustable and removable "pins" that control one ON and one OFF operations (additional "pins" can be added). There is an override switch for manual control.

Figure 5-39 is an electronic, 7-day astronomical time switch. It can be programmed to skip certain days, and up to 14 events can be programmed. There is an override switch for manual control.

Courtesy of Intermatic, Inc.

FIGURE 5-38 A 24-hour time clock.

Courtesy of Intermatic, Inc.

FIGURE 5-37 Spring-loaded timers.

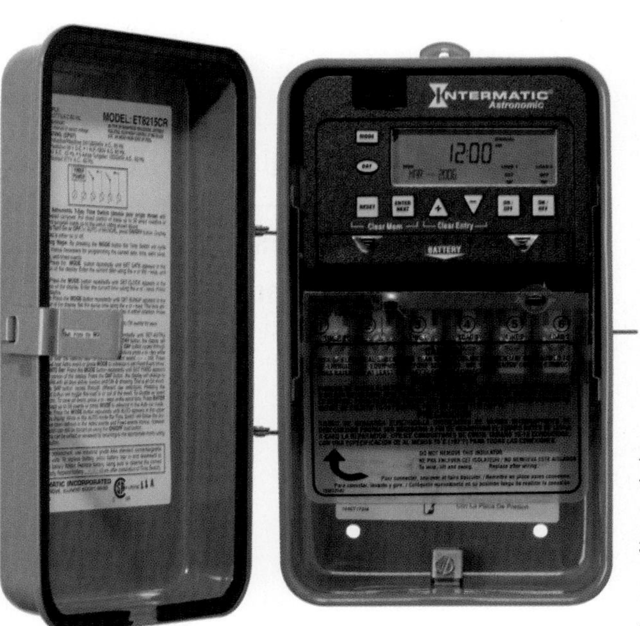

Courtesy of Intermatic, Inc.

FIGURE 5-39 An electronic time switch with astronomic feature.

REVIEW

Note: Refer to the *Code* or the plans where necessary.

1. The identified grounded circuit conductor must be ________________ or ________________ in color.
2. Explain how lighting switches are rated. ________________
3. A T-rated switch may be used to its ________________ current capacity when controlling an incandescent lighting load.
4. What switch type and rating is required to control five 300-watt tungsten filament lamps on a 120-volt circuit? Show calculations. ________________
5. List four types of lighting switches.
   a. ________________ c. ________________
   b. ________________ d. ________________
6. To control a lighting load from one control point, what type of switch would be used? ________________
7. Single-pole switches are always connected to the ________________ wire.
8. Complete the connections in the following arrangement so that both ceiling light outlets are controlled from the single-pole switch. Assume the installation is in cable.

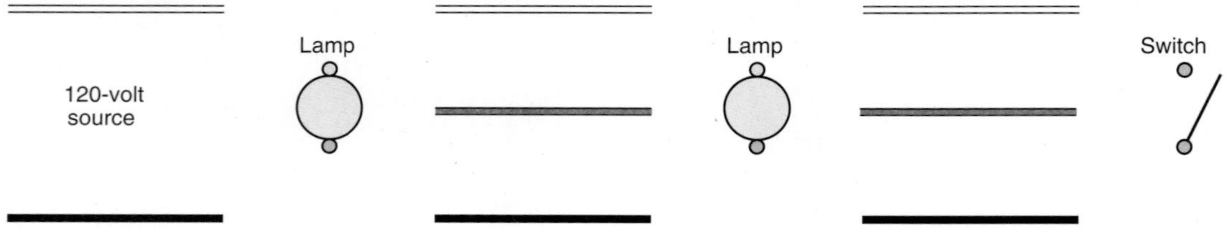

9. Complete the connections for the diagram. Installation is cable.

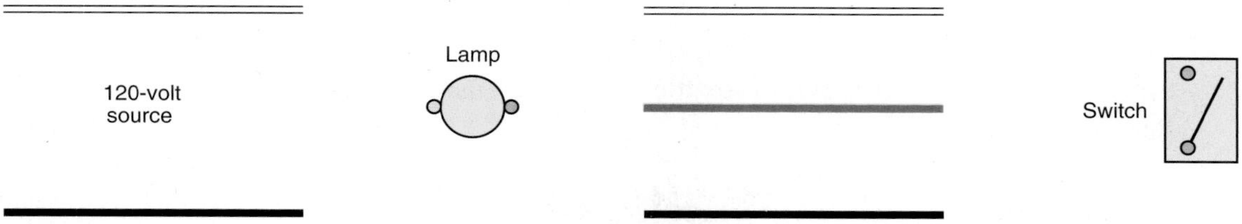

Lamp

120-volt source

Switch

10. A three-way switch may be compared to a ______________________ switch.
11. What type of switch is installed to control a luminaire from two different control points? How many switches are needed and what type are they? ______________________

____

12. Complete the connections in the following arrangement so that the lamp may be controlled from either 3-way switch.

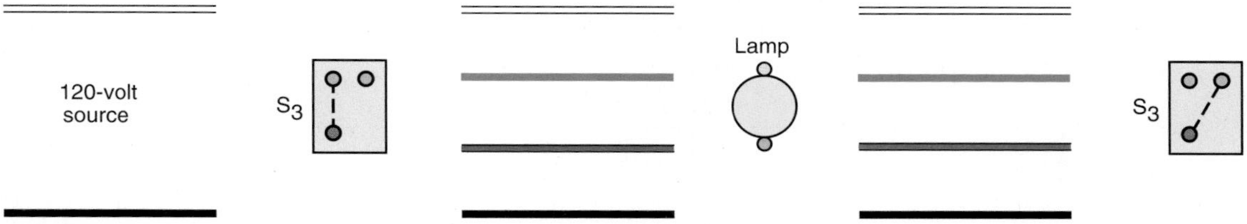

120-volt source

S_3

Lamp

S_3

13. When connecting 4-way switches, care must be taken to connect the travelers to the ______________________ terminals.
14. Show the connections for a ceiling outlet that is to be controlled from any one of three switch locations. The 120-volt feed is at the light. Use colored pens or pencils. Assume the installation is in cable.

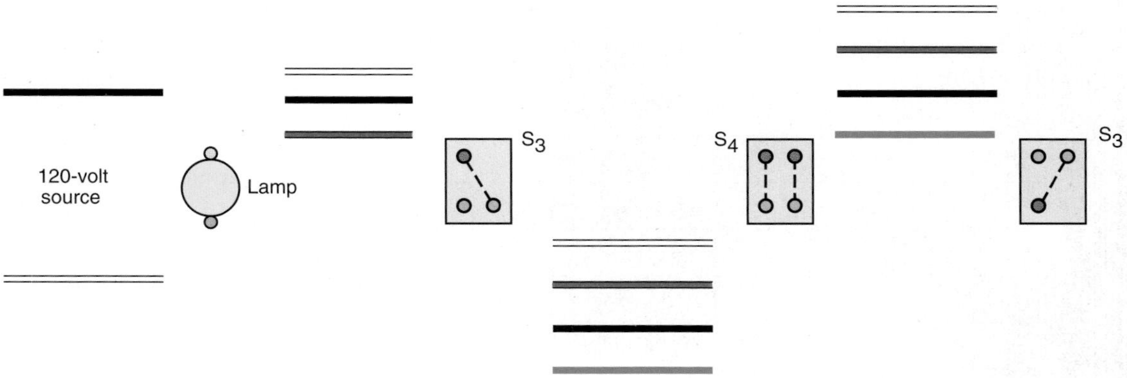

120-volt source

Lamp

S_3

S_4

S_3

15. Match the following switch types with the correct number of terminals for each.

Three-way switch	Two terminals
Single-pole switch	Four terminals
Four-way switch	Three terminals

16. When connecting single-pole, 3-way, and 4-way switches, they must be wired so that all switching is done in the ______________________ circuit conductor.

17. What section of the *Code* emphasizes the fact that all circuiting must be done so as to avoid the damaging effects of induction heating? ______

18. If you had to install an underground 3-wire feeder to a remote building using three individual conductors, which of the following installations "Meets *Code*"? Circle the correct installation.

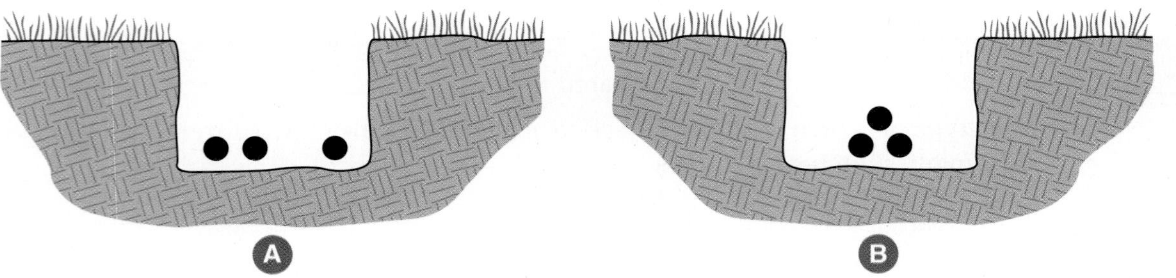

19. Is it always necessary to attach the bare equipment grounding conductor of a nonmetallic-sheathed cable to the green hexagon-shaped grounding screw on a receptacle? Explain.

20. List the methods by which an equipment grounding conductor is connected to a device box.

21. When two nonmetallic-sheathed cables (Romex) enter a box, is it permitted to bring both of the bare equipment grounding conductors directly to the grounding terminal of a receptacle, using the terminal as a splice point? ______

22. The installation in the accompanying drawing is wired with EMT. Receptacle outlet B is only a few feet from switch C. Receptacle outlet A is only a few feet from switch D. The electrician saved a considerable amount of wire by picking up the supply (feed) for switch C from receptacle B, and the switch leg (return) from switch D. What is wrong with this installation?

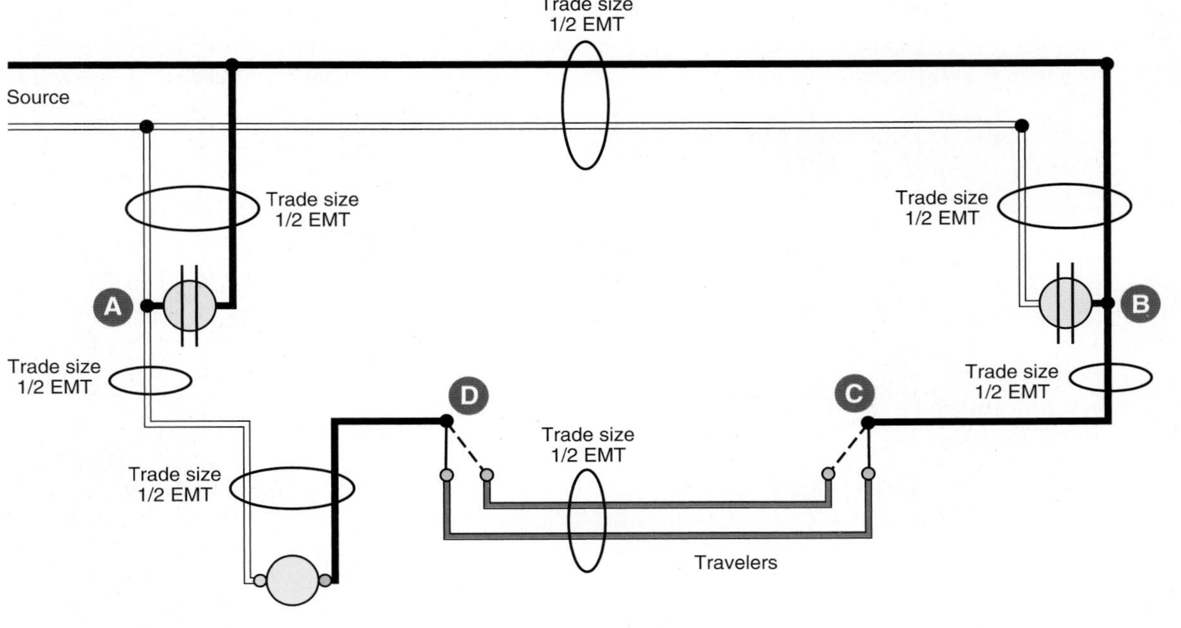

______

______

______

23. Define an *equipment grounding conductor*. ______

______

______

______

24. When metal toggle switchplates are used with nonmetallic boxes, the faceplate must be grounded according to *404.9(B)*. How is this accomplished? ______

______

______

25. Does the *Code* permit the ampacity of switch legs to be less than the ampere rating of the branch circuit? ______

26. Does the *Code* permit connecting receptacles to a 20-ampere branch circuit using short lengths of 14 AWG conductors as the pigtail between the receptacle terminals and the 12 AWG branch-circuit conductors? ______

27. Do receptacles that have back-wired push-in terminals accept 12 AWG conductors? Give a brief explanation for your answer. ______

______

______

______

CHAPTER **6**

Ground-Fault Circuit Interrupters, Arc-Fault Circuit Interrupters, Surge Protective Devices, Immersion Detection Circuit Interrupters, and Appliance Leakage Current Interrupters

OBJECTIVES

After studying this chapter, you should be able to

- understand how GFCIs, AFCIs, IDCIs, and ALCIs operate.
- understand *NEC* requirements of where and how GFCIs and AFCIs are to be installed and connected.
- understand why AFCIs and GFCIs should not be used on a shared neutral branch circuit unless it is listed as such.
- select and install other special-purpose devices, including tamper-resistant and weather-resistant receptacles.
- understand the important *NEC* requirements for replacing existing receptacles.
- know the rules for providing GFCI protection on construction sites.
- understand the basics of surge protective devices.

ELECTRICAL SHOCK HAZARDS

There can be no compromise where human life is involved!

Many injuries have occurred and many lives have been lost because of electrical shock. Coming in contact with live wires or with an appliance or other equipment that is "hot" spells danger. Problems arise in equipment when there is a breakdown of insulation because of wear and tear, defective construction, or misuse of the equipment. Insulation failure can result when the "hot" ungrounded conductor in the appliance comes in contact with the metal frame of the appliance. If the equipment is not properly grounded, the potential for an electrical shock is present.

The shock hazard exists whenever the user can touch both the defective equipment and grounded surfaces such as grounded equipment or appliances, water pipes, stainless steel sinks, metal faucets, grounded metal luminaires, earth, concrete in contact with the earth, or water.

A severe shock can cause considerably more damage to the human body than is visible. A person may suffer internal hemorrhages, destruction of tissues, nerves, and muscles. Further injury can result from a fall, cuts, burns, or broken bones.

The effect of an electric current passing through a human body varies, depending on circuit characteristics (i.e., current, frequency [60 Hz is the worst], voltage, body contact resistance [open wounds are extremely hazardous as compared to a callused hand]), internal body resistance, the path of current, duration of contact, and environmental conditions (e.g., humidity).

Experts generally agree on values of 800 to 1000 ohms as typical body and contact resistance, but the resistance can vary from a few hundred ohms to many thousand ohms. Let's see how important the role of resistance is for a person coming in contact with a 120-volt circuit.

Take a dry hand, for example:

$$I = \frac{E}{R} = \frac{120 \text{ volts}}{100{,}000 \text{ ohms}} = \begin{matrix}0.0012 \text{ ampere} \\ (1.2 \text{ milliamperes})\end{matrix}$$

This would probably be a little tingle.

Now take a wet hand and standing barefoot on the ground:

$$I = \frac{E}{R} = \frac{120 \text{ volts}}{1{,}000 \text{ ohms}} = \begin{matrix}0.12 \text{ ampere} \\ (120 \text{ milliamperes})\end{matrix}$$

This would probably be fatal.

Figure 6-1 shows a time/current curve that indicates the amount of current that a normal, healthy adult can stand for a certain time. Table 6-1 shows expected sensations at various current levels in milliamperes.

The path of current is very important. Where and how is contact to the live conductor made? What are you touching with the other hand? Where and in what are you standing? The current flow might be finger to finger, hand to hand, head to hand, head to foot, or any other combination such as a hand holding a "hot" wire, a hand holding a pair of pliers, a hand holding an electric drill, a hand grasping a grounded metal pipe, two hands grasping a grounded metal pipe, a hand in water, standing in water, hand to foot, foot to foot, and so on. Hand to foot can be fatal because the current is probably passing through the heart and lungs.

Generally, voltages of less than 50 volts to ground are considered safe. Voltages 50 volts and greater are considered lethal. All ac systems (with few exceptions) of 50 volts to 1000 volts that supply premises wiring are required to be grounded, *250.20(B)*. This is covered in Chapter 1.

Some circuits of less than 50 volts must be grounded according to *250.20(A)*

- if the circuit is supplied by a transformer where the primary exceeds 150 volts to ground.
- if the circuit is supplied by a transformer where the primary supply is ungrounded.
- if the circuit is overhead wiring outside of a building.

CODE REQUIREMENTS FOR GROUND-FAULT CIRCUIT INTERRUPTERS (*210.8*)

- To protect against electric shocks, *210.8(A)* requires that ground-fault circuit interrupter (GFCI) protection be provided for specific

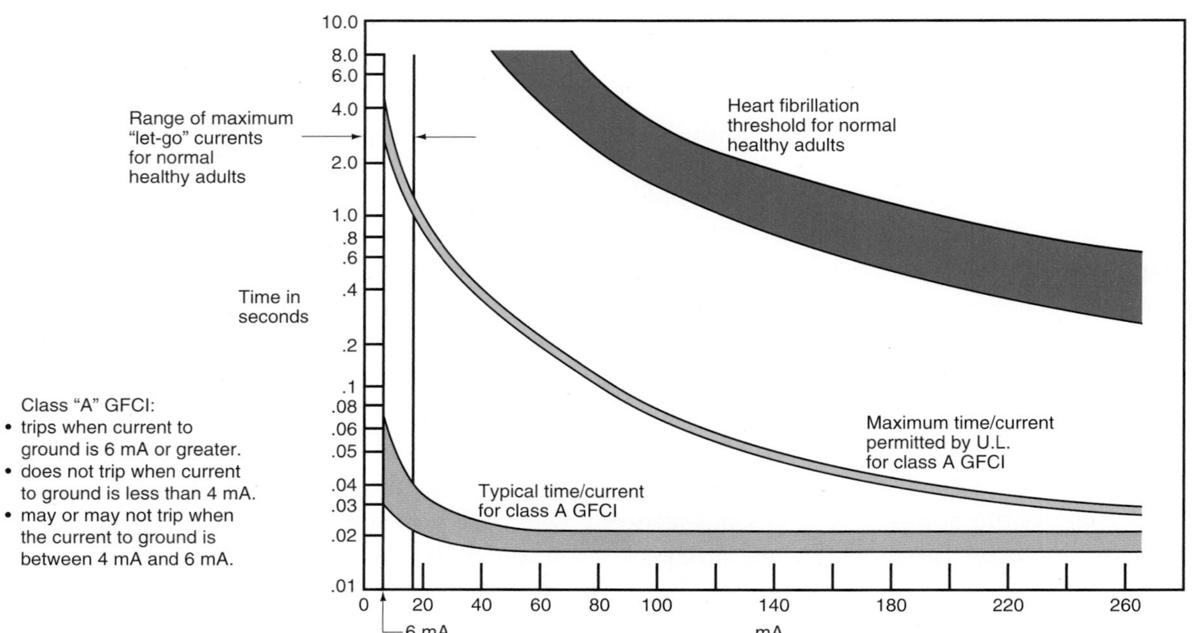

FIGURE 6-1 The time/current curve shows the tripping characteristics of a typical Class A GFCI. Note that if you follow the 6-mA line vertically to the crosshatched typical time/current curve, you will find that the GFCI will open in from approximately 0.035 second to just less than 0.1 second. One electrical cycle is 1⁄60 of a second (0.0167 second). An air bag in an automobile inflates in approximately 1⁄20 of a second (0.05 second).

TABLE 6-1

Effect of Electric Shock.

	CURRENT IN MILLIAMPERES @ 60 HERTZ	
	Men	**Women**
• Cannot be felt	0.4	0.3
• A little tingling—mild sensation	1.1	0.7
• Shock—not painful—can still let go	1.8	1.2
• Shock—painful—can still let go	9.0	6.0
• Shock—painful—just about to point where you can't let go—called "threshold"—may be thrown clear	16.0	10.5
• Shock—painful—severe—can't let go—muscles immobilize—breathing stops	23.0	15.0
• Ventricular fibrillation (usually fatal)		
• Length of time: 0.03 sec.	1000	1000
• Length of time: 3.0 sec.	100	100

single-phase, 125-volt, 15- and 20-ampere receptacles in dwellings. GFCI protection can be provided with GFCI circuit breakers or with GFCI receptacles. The basic requirements for GFCI protection of receptacles are as follows:

- **Bathrooms:** The word "bathroom" is defined in *NEC® Article 100* as An area, including a basin with one or more of the following: a toilet, a urinal, a tub, a shower, a bidet, or similar plumbing fixtures.* See Chapter 10 of this text for additional discussion of this subject.
- **Garages** (attached and detached) and in accessory buildings that are not habitable, such as sheds, workshops, storage buildings, or areas of similar use that have the floor at or below grade.
- **Outdoors:** An exception is that GFCI protection is not required for a receptacle not readily accessible and supplied by a dedicated branch circuit for snow melting or deicing equipment.

*Reprinted with permission from NFPA 70-2014.

Install these receptacles for snow melting or deicing equipment according to *426.28*.

- **Crawl spaces** that are at or below grade.
- **Unfinished basements:** An exception to this GFCI requirement is for a receptacle that serves a permanently installed fire alarm or burglar alarm system. See *760.41B* and *760.121(B)*. A receptacle on a porcelain or plastic lampholder must also be GFCI protected. Some individuals try to get around the requirement for GFCI protection in basements by calling the basement "finished," even though it is clearly evident that it is not "finished." If the basement is truly intended to be "finished," then the *Code* requirements for the spacing of receptacles, required lighting outlets, and switch controls must be followed.
- **Kitchens:** All receptacles that serve the countertop surfaces. This includes receptacles installed on islands and peninsulas, as in Figure 6-2. The key words here are *that serve the countertop surfaces*.

  GFCI protection is *not* required for receptacles that are obviously not intended to serve countertops, such as a receptacle installed

  – solely for a clock.

  – behind a range or refrigerator.
- ▶**Sinks:** Any receptacles within 6 ft (1.8 m) of the outside edge of a sink.◀ See Figure 6-2. This requirement is easy to understand. Just measure 6 ft (1.8 m) in any direction around a corner, even in another room or hallway. This requirement applies to all receptacles within a 6 ft dimension from the edge of a sink even those installed for a specific purpose such as for waste disposals, instant water heaters and microwave ovens.

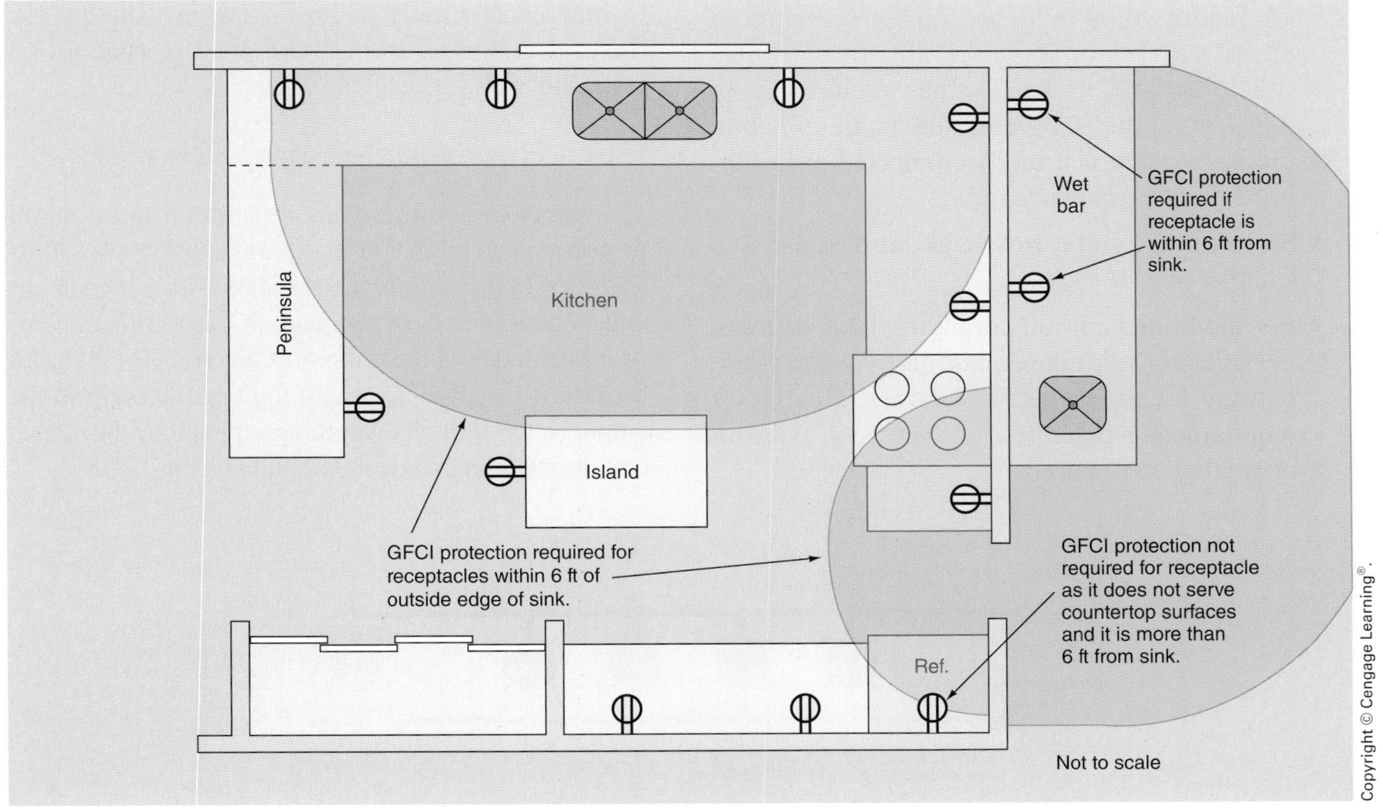

FIGURE 6-2 In kitchens, *all* 125-volt, single-phase, 15- and 20-ampere receptacles that serve countertop surfaces (walls, islands, or peninsulas) must be GFCI protected. All 125-volt, single-phase, 15- and 20-ampere receptacles within 6 ft (1.8 m) of a sink must be GFCI protected. See *210.8(A)(6)* and *210.8(A)(7)*. Drawing is not to scale.

- **Boathouses:** A boathouse for a boat is similar in concept to a garage for a car.
- ▶**Bathtub and Shower Stalls** Where receptacles are installed within 6 feet (1.8 m) of the outside edge of the bathtub or shower stall.◀* This is a new requirement for the 2014 *NEC*. The rule applies in addition to the requirement in *210.8(A)(1)* that GFCI protection be provided for receptacles in bathrooms. GFCI protection is now clearly required even where the bathtub or shower stall is not installed in a bathroom.
- ▶**Laundry Areas**◀* This is another new requirement for the 2014 *NEC*. The term "laundry area" is not defined in the *NEC*, but additional clarity might be obtained by referring to *NEC 210.52(F)*. There, a receptacle is required in the area designated for the installation of laundry equipment.
- ▶**Kitchen Dishwasher Branch Circuit**. GFCI protection shall be provided for outlets that supply dishwashers installed in dwelling unit locations. ◀* This is a new requirement in *210.8(D)* of the 2014 *NEC* and applies whether the dishwasher is directly connected or is supplied from a receptacle outlet.
- **Swimming pools, hot tubs, and spas:** See Chapter 30.
- For the branch circuit supplying heated floors in bathrooms, kitchens, and in hydromassage bathtub locations, see *424.44(G)*. This GFCI requirement is regardless of the type of flooring or heating cables used.

Readily Accessible Location

The circuit breaker or GFCI receptacle that provides the protection is required to be installed in a readily accessible location. See *NEC 210.8*. The term "readily accessible" is defined in *NEC Article 100* and means, Capable of being reached quickly for operation, renewal, or inspections without requiring those to whom ready access is requisite (required or needed) to climb over or remove obstacles or to resort to portable ladders, and so forth.*

This important requirement ensures that the device that provides the GFCI protection can be easily accessed to do the safety operation test that is required by the manufacturer. An operational test is required to be performed at least monthly. This is important, as we rely on the GFCI device functioning correctly when an actual ground fault occurs. A record of the test should be made and maintained.

The "push-to-test" button on the circuit breaker or receptacle should be pressed to test the device. Several manufacturers make plug-in type GFCI testers as well.

GFCI Receptacles and Breakers

The *Code* requirements for ground-fault circuit protection can be met in many ways. Figure 6-3 illustrates a GFCI circuit breaker installed on a branch circuit. A ground fault in the range of 4 to 6 milliamperes or more shuts off the entire circuit. If a GFCI circuit breaker is installed, a ground fault at any point in the circuit would shut off everything supplied by the branch circuit, rather than only the offending component.

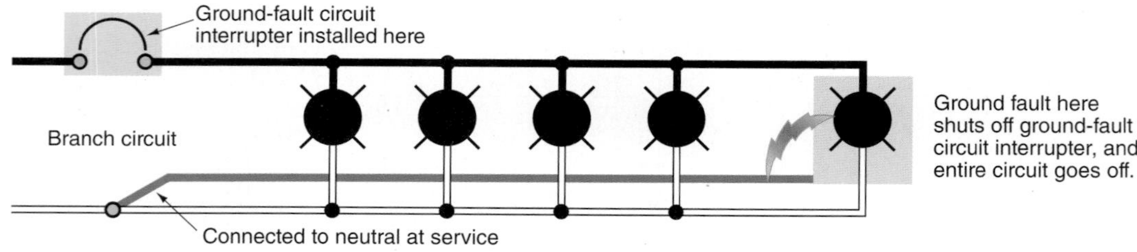

FIGURE 6-3 Ground-fault circuit interrupter as a part of the branch-circuit overcurrent device.

*Reprinted with permission from NFPA 70-2014.

When a GFCI receptacle is installed, then only that receptacle is shut off when a ground fault greater than 6 milliamperes occurs, as in Figure 6-4. GFCI receptacles can also be wired so they protect downstream receptacles or connected equipment.

GFCI receptacles break both the ungrounded ("hot") and grounded conductors.

GFCI circuit breakers break only the ungrounded ("hot") conductor.

Figure 6-5 shows the effect of a GFCI installed as part of a feeder supplying 15- and 20-ampere receptacle branch circuits. This is rarely used in residential wiring.

Figure 6-6 is a pictorial view of how a ground-fault circuit interrupter operates. The GFCI device in the upper drawing does not operate because the current through the torroidal coil is in balance. As shown in the lower drawing, the GFCI will trip very quickly if the current returning to its source is 6 mA or more and is not permitted to trip if the leakage current is less than 4 mA. This leakage current is simulated by the push-to-test circuit internal to the device. See the additional information later in this chapter.

What Kinds of GFCIs Are Available?

- 120-volt receptacles
- Single-pole, 120-volt breakers
- Single-pole, 120-volt dual-function, with both GFCI and AFCI features in one breaker
- Two-pole, 240-volt common trip breakers
- Two-pole, 120/240-volt common trip dual-function with both GFCI and AFCI features in one breaker
- Two-pole, 120/240-volt independent trip dual-function with both AFCI and GFCI features in one breaker. Suitable for use with a shared neutral on a multiwire branch circuit

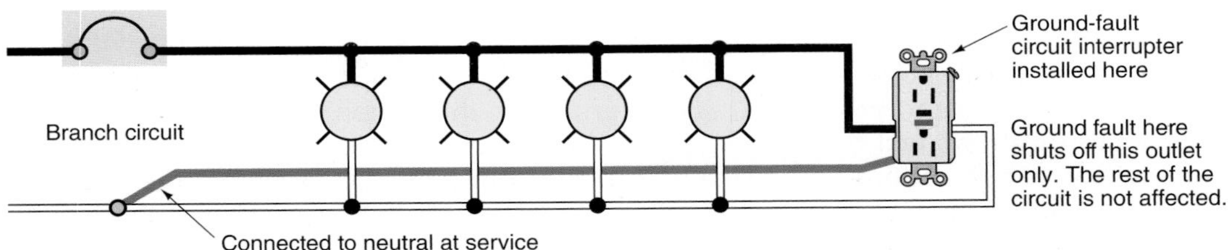

FIGURE 6-4 Ground-fault circuit interrupter as an integral part of a receptacle outlet.

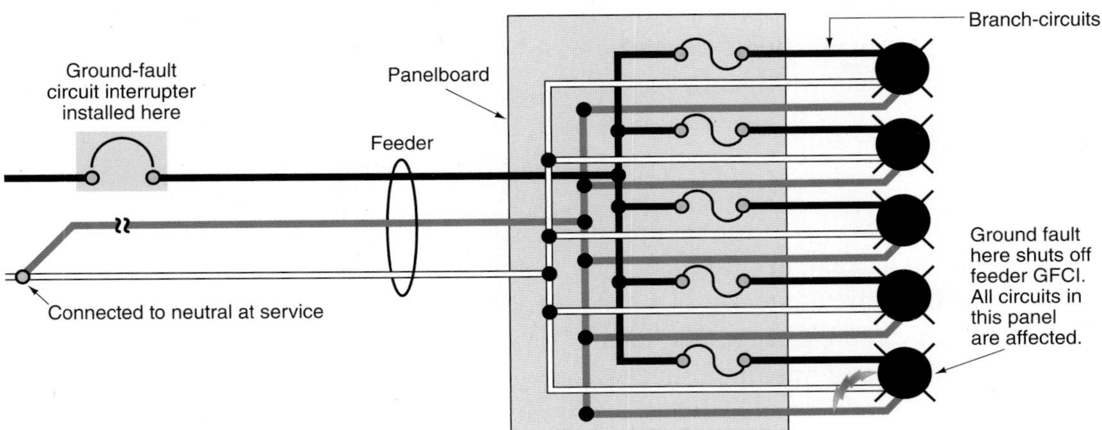

FIGURE 6-5 Ground-fault circuit interrupter as part of the feeder supplying a panel that serves a number of 15- and 20-ampere branch circuits. This is permitted by *NEC 215.9* but is rarely—if ever—used in residential wiring.

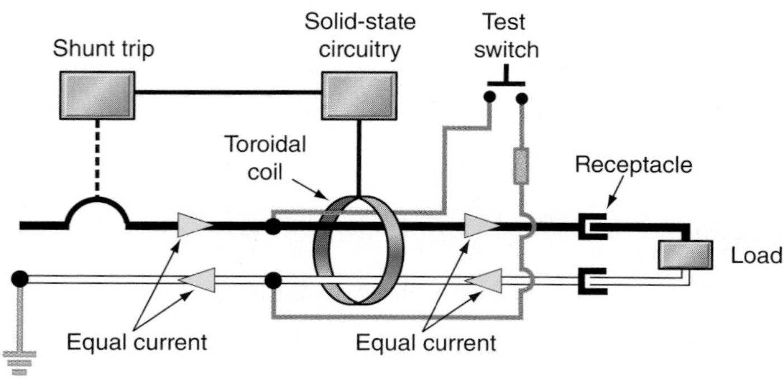

No current is induced in the toroidal coil because both circuit wires are carrying equal current. The contacts remain closed.

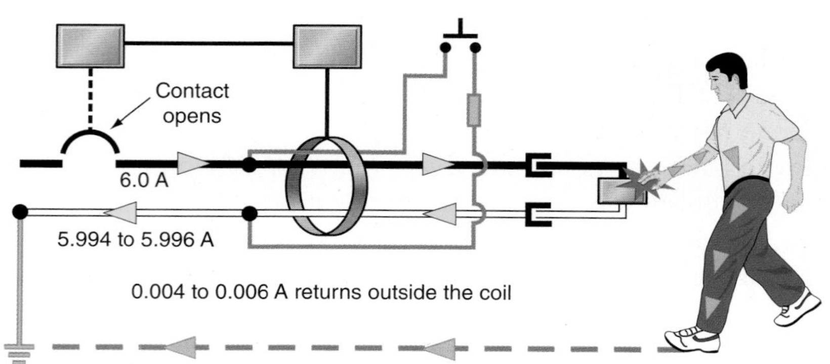

An imbalance of from 4 to 6 milliamperes in the coil will cause the contacts to open. The GFCI must open in approximately 25 milliseconds. Receptacle-type GFCIs have a switching contact in each circuit conductor.

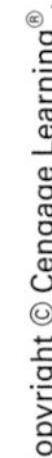

FIGURE 6-6 Basic principle of how a ground-fault circuit interrupter operates. Receptacle-type GFCIs switch both the phase ("hot") and grounded conductors. Note that when the test button is pushed, the test current passes through the test button, the sensor, then back around (bypasses, outside of) the sensor, then back to the opposite circuit conductor. This is how the "unbalance" is created, then monitored by the electronic circuitry to signal the GFCI's contacts to open. Note in the upper drawing that because the load currents passing through the sensor are equal, no unbalance is present.

- "Faceless." Strictly for GFCI protection, these have only the test and reset buttons, and do not have a receptacle. Mount in a single-device box. Commonly used to protect whirlpool tubs, etc.

Nuisance Tripping

Occasionally, a GFCI will trip for seemingly no reason. This could be leakage in extremely long runs of cable from a GFCI circuit breaker in a panel to the protected branch-circuit wiring. One manufacturer's literature specifies a maximum one-way length of 250 ft (76.2 m) for the branch circuit.

The allowable leakage current for listed cord-and-plug-connected appliances is so small that it should not cause nuisance tripping of GFCI devices. However, if nuisance tripping can be traced to an appliance, it might be a good idea to simply replace the GFCI circuit breaker or the GFCI receptacle before discarding the appliance. If the GFCI continues to nuisance trip, the appliance must be replaced.

Nuisance tripping occasionally can be traced to moisture somewhere in the circuit wiring, receptacles, or lighting outlets.

Some electricians are of the opinion that using nonmetallic staples (see Figure 4-15[C]) will reduce the possibility of nuisance tripping. No actual testing has been done to prove or disprove this opinion.

What a GFCI Does

A GFCI monitors the current balance between the ungrounded "hot" conductor and the grounded conductor. As soon as the current flowing through the "hot" conductor is in the range of 4 to 6 milliamperes more than the current flowing in the "return" grounded conductor, the GFCI senses this unbalance and trips (opens) the circuit off. The unbalance indicates that part of the current flowing in the circuit is being diverted to some path other than the normal return path along the grounded return conductor. If the "other" path is through a human body, as illustrated in Figure 6-6, the outcome could be fatal.

UL Standard No. 943 covers ground-fault circuit interrupters.

- Class A GFCI devices are the most common. They are designed to
  - trip when current to ground is 6 milliamperes (6⁄1000 of an ampere) or greater.
  - not trip when the current to ground is less than 4 milliamperes (4⁄1000 of an ampere).
  - may or may not trip when the current to ground is between 4 and 6 milliamperes.
  - open very quickly, in approximately 25 milliseconds.
- Class B GFCI devices are pretty much obsolete. They were designed to trip on ground faults of 20 milliamperes (20⁄1000 of an ampere) or more. They were used only for underwater swimming pool lighting installed before the adoption of the 1965 *NEC*. For this application, Class A devices were too sensitive and would nuisance trip.

What a GFCI Does *Not* Do

- It *does not* protect against electrical shock when a person touches both circuit conductors at the same time (two "hot" wires, or one "hot" wire and one grounded neutral conductor), because the current flowing in both conductors is the same. Thus, there is no unbalance of current for the GFCI to sense and trip.
- It *does not* limit the magnitude of ground-fault current. It *does* limit the length of time that a ground fault will flow. The GFCI should trip in about 25 milliseconds. In other words, you will still receive a severe shock during the time it takes the GFCI device to trip "off." See Figure 6-1.
- It *does not* sense solid short circuits between the "hot" conductor and the grounded "neutral" conductor. The branch-circuit fuse or circuit breaker provides this protection.
- It *does not* sense solid short circuits between two "hot" conductors. The branch-circuit fuse or circuit breaker provides this protection.
- It *does not* sense and protect against the damaging effects of arcing faults, such as would occur with frayed extension cords. This protection is provided by an arc-fault circuit interrupter (AFCI) discussed later in this chapter.
- It *does not* provide overload protection for the branch-circuit wiring. It provides *ground-fault protection only.*

GROUND-FAULT CIRCUIT INTERRUPTERS IN RESIDENCE CIRCUITS

In the residence covered in this text, GFCI protection for personnel must be provided for 125-volt, 15- and 20-ampere receptacle outlets installed in or near the following rooms or areas, as indicated:

- Bathrooms (all)
- Garages (all)
- Outdoors (all)
- Unfinished basements (specific receptacles)
- Kitchens (receptacles that serve countertop surfaces)
- ►Sinks (within 6 ft [1.8 m] of the sink)◄
- ►Laundry area (all)◄
- ►Dishwasher◄
- Waste disposal
- Hydromassage bathtub

GFCI protection can be provided by GFCI circuit breakers or by GFCI receptacles of the type shown in Figure 6-7. This is a design issue that confronts the electrician. The electrician must decide how to provide the GFCI personnel protection required by *210.8(A)*, (*C*) and (*D*) as well as other locations such as for hydromassage bathtubs. These receptacle outlets are connected to the circuits listed in Table 6-2.

Swimming pools also have special requirements for GFCI protection. These requirements are covered in Chapter 30.

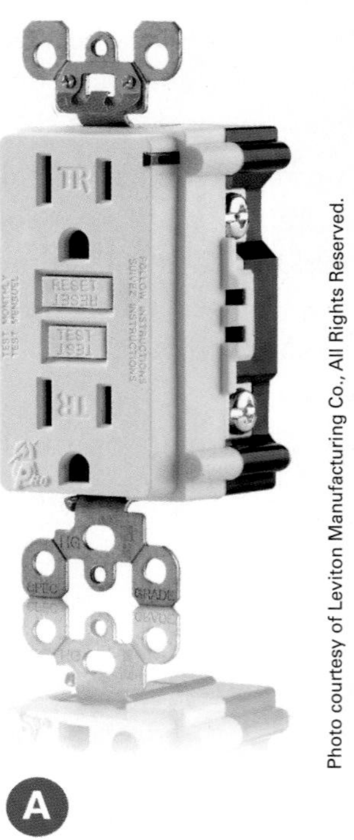

FIGURE 6-7 (A) A ground-fault circuit interrupter and tamper-resistant receptacle, and (B) a ground-fault circuit interrupter circuit breaker. The switching mechanism of a GFCI receptacle opens both the ungrounded "hot" conductor and the grounded conductor. The switching mechanism of a GFCI circuit breaker opens the ungrounded "hot" conductor only.

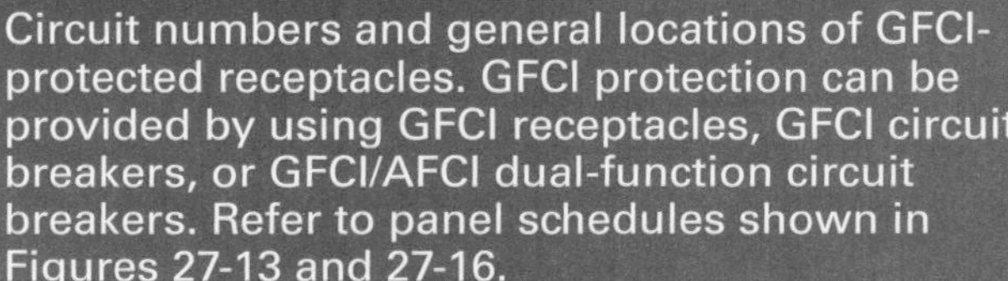

TABLE 6-2

Circuit numbers and general locations of GFCI-protected receptacles. GFCI protection can be provided by using GFCI receptacles, GFCI circuit breakers, or GFCI/AFCI dual-function circuit breakers. Refer to panel schedules shown in Figures 27-13 and 27-16.

CIRCUITS	GFCI RECEPTACLE LOCATIONS
A9	Master bath hydromassage
A15	Front-porch receptacle
A16	Outdoor receptacle on front of residence outside of Front Bedroom
A18	Workbench receptacles
A19	Outdoor receptacle on rear of residence outside of Master Bedroom
A20	Workshop receptacles on window wall
A22	Master Bath receptacle
A23	Bath (off hall) receptacle
B90	Wet-bar (Recreation Room) receptacles
B13	Kitchen receptacles
B14	Garage lighting and overhead door operator
B15	Kitchen receptacles
B16	Kitchen receptacles
B17	Outdoor receptacle outside of Living Room next to sliding door
B18	Laundry Room receptacle for clothes washer
B20	Laundry Room receptacles (2) and weatherproof receptacle (1) outside of laundry
B21	Powder Room receptacle
B23	Garage receptacles

GFCIs operate properly only on grounded electrical systems, as is the case in all residential, condo, apartment, commercial, and industrial wiring. The GFCI will operate on a 2-wire circuit even though an equipment grounding conductor is not included with the circuit conductor. In this residence, equipment grounding conductors are in the cables. In the case of a metal raceway (EMT) or armored cable (Type AC or Type MC), the equipment grounding conductor is provided by the metal raceway or armored cable that contains a bonding strip.

Never ground a system neutral conductor except at the service equipment. A GFCI would be inoperative or trip unnecessarily.

Never connect the neutral conductor of one circuit to the neutral conductor of another circuit.

When a GFCI feeds an isolation transformer (separate primary winding and separate secondary winding), as might be used for swimming pool underwater luminaires, the GFCI *will not* detect any ground faults on the secondary of the transformer.

Ground-fault circuit interrupters may be installed on other circuits, in other locations, and even when rewiring existing installations where the *Code* does not specifically call for GFCI protection.

Do not connect fire and smoke alarms to a circuit that is protected by a GFCI breaker. A nuisance tripping of the GFCI would render the fire alarm and smoke detectors inoperative. This requirement is found in NFPA 72, *The National Fire Alarm Code*.

FEED-THROUGH GROUND-FAULT CIRCUIT INTERRUPTER

With rare exception, all GFCI receptacles on the market today have the feed-through feature.

The decision to use more GFCIs rather than trying to protect many receptacles through one GFCI becomes one of economy and practicality. GFCI receptacles are more expensive than regular receptacles. Here is where knowledge of material and labor costs comes into play. The decision must be made separately for each installation, keeping in mind that GFCI protection is a safety issue recognized and clearly stated in the *Code*. However, the actual circuit layout is left up to the electrician.

Figure 6-8 illustrates how a feed-through GFCI receptacle supplies many other receptacles. Should a ground fault occur anywhere on this circuit, all 11 receptacles lose power—not a good circuit layout. Attempting to locate the ground-fault problem, unless it's obvious, can be very time-consuming. The use of more GFCI receptacles is generally the more practical approach.

Figure 6-9 shows how a feed-through receptacle is connected into a circuit. Figure 6-10 shows a feed-through GFCI receptacle connected midway into a circuit. In both diagrams, the feed-through GFCI receptacle and all downstream outlets are ground-fault protected.

In UL Standard 943, we find that a GFCI receptacle may have black and white line wire leads, red and gray load wire leads, and a green lead for the equipment grounding conductor, as illustrated in Figure 6-9. GFCI receptacles may have ordinary screw terminals: brass color for the "hot" conductors; silver color for the grounded conductors; and a green, hexagon-shaped screw for the equipment grounding conductor. The line and load connections are clearly marked.

Where the wiring method is a grounded method, such as electrical metallic tubing, flexible metal conduit, or armored cable, the grounding terminal of the GFCI receptacle is considered adequately

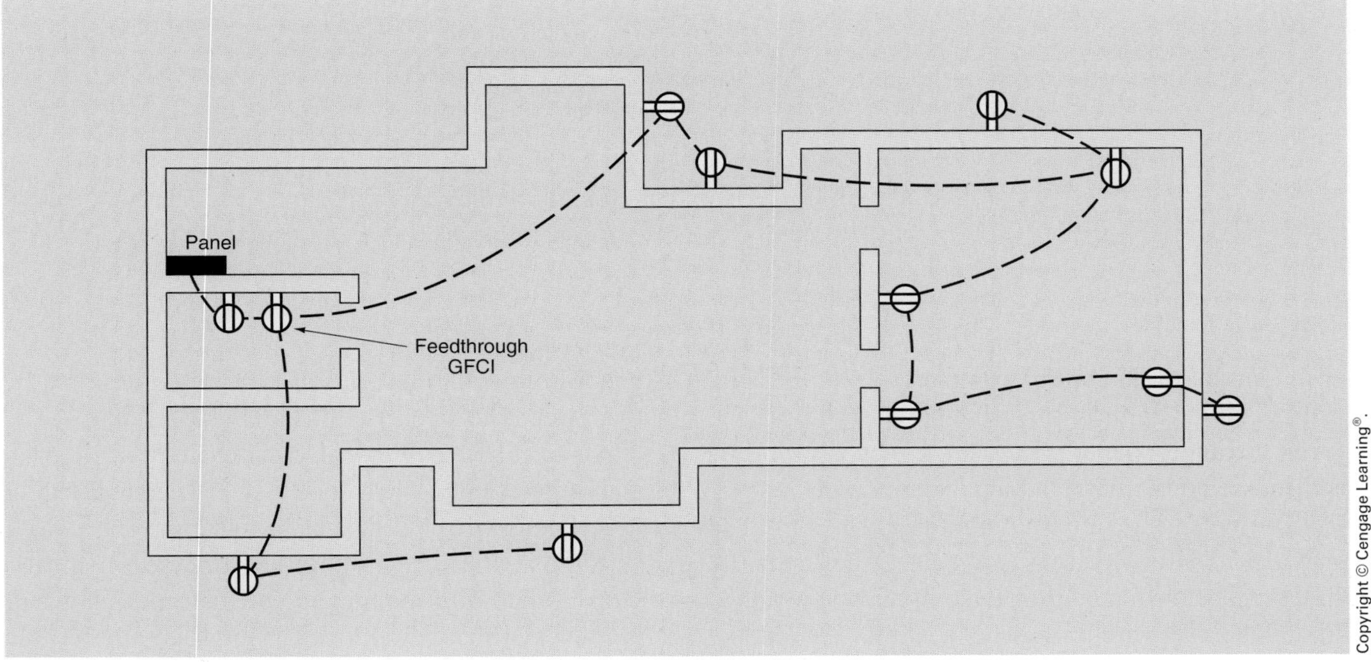

FIGURE 6-8 Illustration showing one GFCI feed-through receptacle protecting other receptacles. Although this might be an inexpensive way to protect many other downstream receptacles outlets that require GFCI protection, it could be considered as "putting too many eggs in one basket." The same situation would exist if a GFCI circuit breaker were to be installed in the panel. One major manufacturer of GFCI circuit breakers suggests that the maximum one-way length of a GFCI-protected branch circuit be limited to 250 ft (76.2 m).

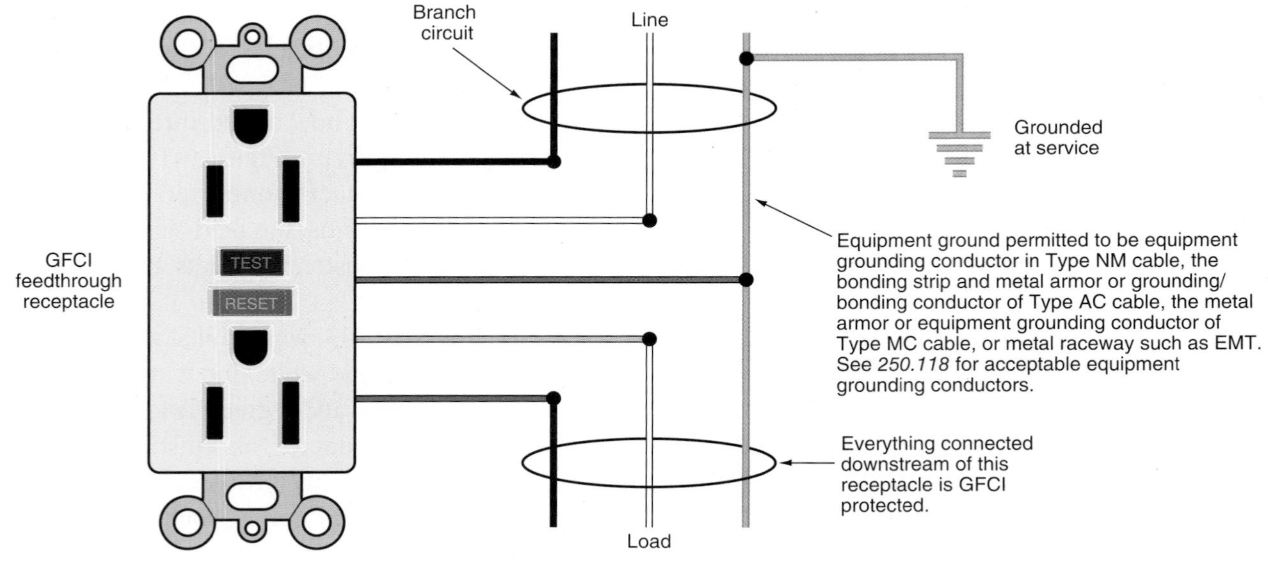

FIGURE 6-9 Connecting feed-through ground-fault circuit interrupter into circuit.

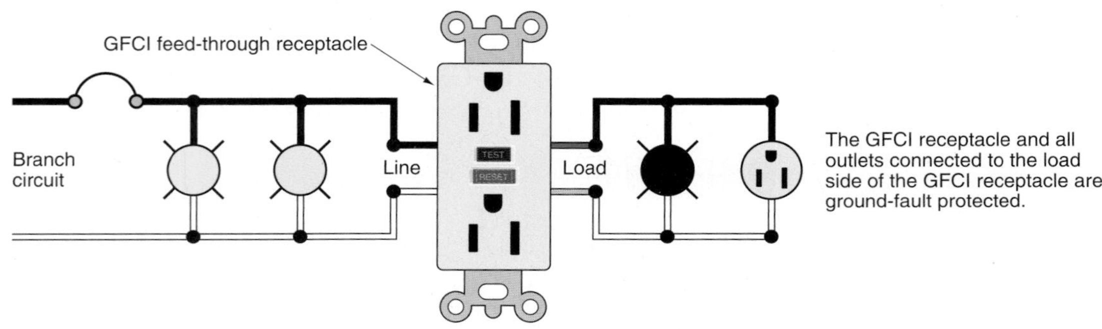

FIGURE 6-10 Feed-through GFCI receptacle installed midway in a branch circuit.

grounded by the metallic grounded wiring method, and a separate connection to the green lead or to the green hexagon-shaped screw is not necessary. This is discussed and illustrated in Chapter 5.

The wiring device manufacturer is required to provide a "safety yellow" adhesive label over the GFCI receptacle load terminals, or it may be wrapped around the load wire leads. The label must be marked with wording like this: *"Attention! The load terminals under this label are for feeding additional receptacles. Miswiring can leave this outlet without ground-fault protection. Read instructions prior to wiring."*

The most important factors to be considered are the continuity of electrical power and the economy of the installation. The decision must be made separately for each installation.

What about Multiwire Branch Circuits? It's plain and simple. Some AFCI and GFCI devices are listed for use on multiwire branch circuits—others are not.

Those that are not listed for use on multiwire branch circuits have clear instructions warning "Never share a neutral." Using two single-pole AFCIs or GFCIs on a multiwire branch circuit will not work because each needs to have the white grounded conductor from that branch circuit connected to that particular AFCI or GFCI.

As with any electrical equipment, follow the installation instructions (wiring diagrams).

SAFETY ALERT

Be very careful when hooking up GFCI receptacles. It is extremely important that the LINE and LOAD connections be done correctly. On older style GFCI receptacles, it was possible to reverse the line and load connections. Under a ground-fault condition, this sort of misconnection would result in the GFCI receptacle itself still being "live" even though the GFCI mechanism has tripped OFF. In the case of feed-through receptacles, a ground-fault condition shut off the circuit downstream, yet the feed-through GFCI receptacle itself was still energized.

It is easy to tell whether a GFCI receptacle has the line and load leads reversed. Push the test button. The GFCI will trip. If the GFCI receptacle is still "hot," it has been wired incorrectly.

The current UL Standard 943 requires that a GFCI receptacle must trip to the OFF position if the receptacle is miswired, such as in a reversal of the line/load connections. In other words, if misconnected, the receptacle simply will not work!

Figure 6-11(A) shows a 3-wire, multiwire circuit. This particular GFCI is not listed as suitable for sharing a neutral conductor in a multiwire branch circuit. Trace the current flow and note that the GFCI will sense a current unbalance of 5 amperes and trip OFF. In fact, as long as the unbalance exists, it will never be able to be turned ON.

Figures 6-11(B) and 6-11(C) show a 2-wire circuit. Figure 6-11(B) is an improper connection of a feed-through GFCI receptacle. Note how the GFCI detects a current unbalance of 1 ampere and will trip OFF. As long as the unbalance exists, the GFCI receptacle will continue to trip OFF until the misconnection is corrected. Figure 6-11(C) illustrates the proper connection of a GFCI receptacle.

Figure 6-11(B) and (C) also illustrate a possible violation of rules in *NEC 300.3(B)*. This section generally requires that all conductors of a circuit be in the same raceway or cable. *NEC 300.3(B)(3)* permits conductors in nonmetallic wiring methods to be run in different raceways or cables if the installation will not create the problem of heating metal enclosures by induction. This can be avoided by using all nonmetallic components, including cable and nonmetallic boxes.

Other Features for GFCI Receptacles

GFCIs have come a long way. All GFCI receptacles have a manual TEST button and a RESET button. In addition, look for GFCI receptacles that have:

- an LED that indicates that there is power to the receptacle.
- an LED that indicates that the device is not functioning properly.
- a built-in line/load reversal feature that prevents power to the receptacle face if the line/load connections are reversed.
- an End-of-Life feature that activates when the GFCI fails to respond to a manual test can be either:
  - a lockout feature so the GFCI cannot be reset if the internal GFCI circuit is not functioning properly, or
  - a failure light or alarm that indicates that the GFCI protection has failed yet the receptacle continues to supply power.

TESTING AND RECORDING OF TEST DATA FOR GFCI RECEPTACLES

Refer to Figure 6-7 and note that the GFCI receptacle has "T" (Test) and "R" (Reset) buttons. As shown in Figure 6-6, pushing the test button allows a small current to bypass the torroid coil in the

A **Improper** connection of single-pole GFCI circuit breaker on a multiwire branch circuit. The current flowing in and out of the GFCI circuit breaker is different (unbalanced) because the circuit is a multiwire circuit. This difference of current flow will cause the GFCI breaker to trip immediately.

Conventional single-pole circuit breaker.
5A
5A
5A
5A
N
5A
5A
10A
5A
5A
10A
10A
10A
GFCI or AFCI circuit breakers require a "neutral" connection.

B **Improper** connection: The current flowing in and out of the GFCI receptacle is different (unbalanced). The GFCI receptacle will trip immediately. The circuit may violate *NEC 300.3(B)* if metallic wiring methods or boxes are used.

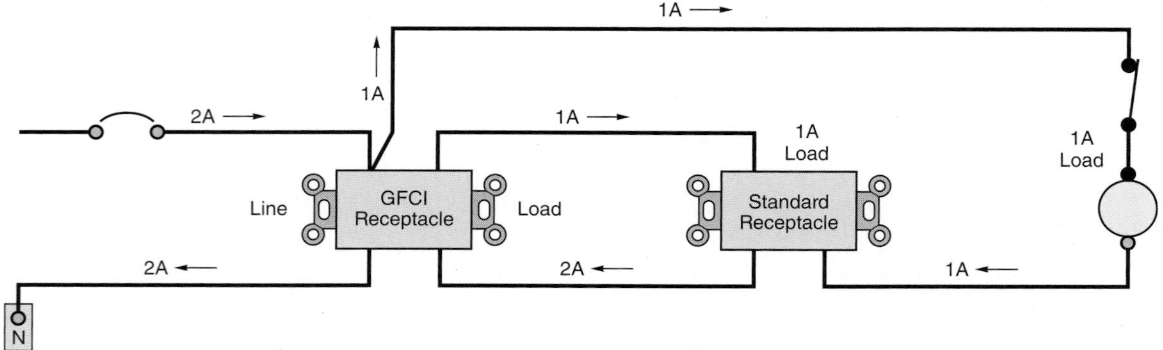

C **Proper** connection: The current flowing in and out of the GFCI receptacle is the same (balanced). The GFCI receptacle operates properly. The circuit may violate *NEC 300.3(B)* if metallic wiring methods or boxes are used.

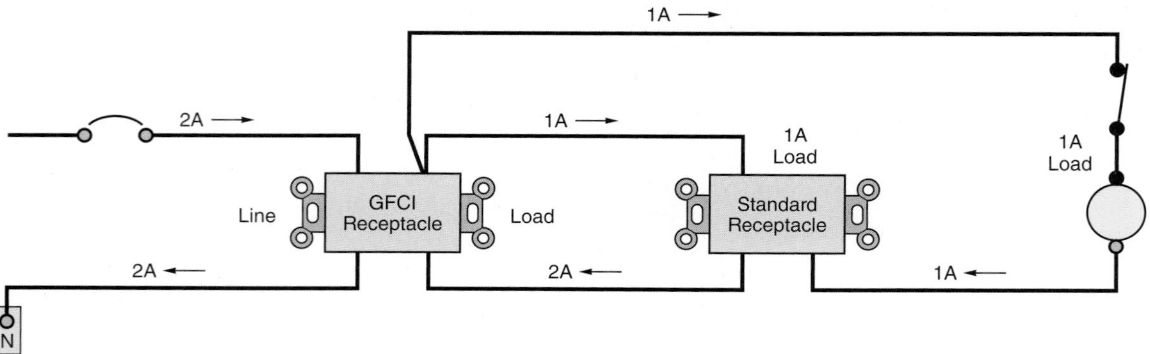

FIGURE 6-11 These diagrams show improper and proper connections of GFCI circuit breakers.

device. If operating properly, the GFCI receptacle should trip to the OFF position. Pushing the reset button will restore power. The operation is similar for the GFCI circuit breakers.

Because of the fragile nature of the electronic circuitry in GFCIs, test them often, particularly in high lightning strike areas of the country. GFCIs and other delicate electronic circuitry are susceptible to transient voltage surges.

"Things" are changing so fast in the electronic world, it is hard to keep up with the changes. At least one manufacturer recently announced the availability of a self-testing GFCI receptacle. Its circuitry automatically generates a simulated

OCCUPANT'S TEST RECORD

TO TEST, depress the TEST button. The RESET button should extend. Should the RESET button not extend, the GFCI will not protect against electrical shock. Call a qualified electrician.

TO RESET, depress the RESET button firmly into the GFCI unit until an audible click is heard. If reset properly, the RESET button will be flush with the surface of the receptacle.

This label should be retained and placed in a conspicuous location to remind the occupants that for maximum protection against electrical shock, each GFCI should be tested monthly.

Year	Jan	Feb	Mar	Apr	May	Jun	Jul	Aug	Sep	Oct	Nov	Dec

FIGURE 6-12 Homeowner's testing chart for recording GFCI testing dates. UL requires that the instructions furnished with GFCIs be marked and that testing be performed upon installation and at least once each month, as well as after severe lightning storms.

ground-fault leakage current every 60 seconds. Instead of relying on the homeowners to do the recommended monthly test of the GFCI receptacles in their homes, the test is performed automatically.

Underwriters Laboratories (UL) requires that detailed installation and testing instructions be included in the packaging for GFCI receptacles or circuit breakers. Instructions are not only for the electrician but also for homeowners, and must be left in a conspicuous place so homeowners can familiarize themselves with the receptacle, its operation, and its need for testing. Figure 6-12 shows a chart homeowners can use to record monthly GFCI testing.

ARC-FAULT CIRCUIT INTERRUPTERS (AFCIs)

Arc-fault circuit interrupters are "the new kids on the street" compared to GFCIs. First came ground-fault circuit interrupters (GFCIs); then came arc-fault circuit interrupters (AFCIs). As you will learn later in this chapter, dual-function circuit breakers are available that provide both AFCI and GFCI protection, making it very easy to "Meet *Code*." The definition of *Arc-Fault Circuit Interrupter* has moved to *Article 100* and reads:

A device intended to provide protection from the effects of arc faults by recognizing characteristics unique to arcing and by functioning to de-energize the circuit when an arc fault is detected.*

The requirements for providing AFCI protection for dwelling units have been extensively revised. The installer can choose between six methods for providing the AFCI protection. In addition, there are now more rooms or areas where AFCI protection is required. An exception has been added regarding the requirement for AFCI protection for existing branch circuits that are modified or extended. A new rule now requires AFCI protection in dormitory unit bedrooms, living rooms, hallways, closets, and similar rooms.

NEC 210.12(A) now reads, ▶All 120-volt, single phase, 15- and 20-ampere branch circuits supplying outlets *or devices* installed in dwelling unit *kitchens*, family rooms, dining rooms, living rooms, parlors, libraries, dens, bedrooms, sunrooms, recreation rooms, closets, hallways, *laundry areas*, or similar rooms or areas shall be protected by any of the means described in (1) through (6).◀* (The words added in processing the 2014 *NEC* are shown in italics for emphasis)

Note that the requirement now includes *devices* as well as *outlets*. Both of these terms are defined in *NEC Article 100*. We discuss the application of the requirement as well as how to comply with it in this chapter.

What Is an Outlet? This is an important term to understand for the correct application of the AFCI requirements. The word "outlet" is defined in *NEC Article 100* as A point on the wiring system at which current is taken to supply utilization equipment.* Several types of outlets that require AFCI protection are present in a dwelling unit. They include: lighting outlets, receptacle outlets, ceiling (paddle) fan outlets, and outlets for single station from fire alarms. AFCI protection is required for all of these outlets that are installed in any of the rooms mentioned. Note that a lighting switch is not an outlet. The box where the luminaire or ceiling (paddle) fan connects is the outlet.

What Is a Device?

The word "device" is defined in *NEC Article 100* as ▶A unit of an electrical system, other than a conductor, that carries or controls electric energy as its principal function.◀* This definition was revised during the processing of the 2014 *NEC* to add the phrase "other than a conductor" to clarify that it is not intended that a conductor be included in the definition of "device", although it's primary function is to carry electrical energy.

Common devices to which this rule on AFCI protection applies include light switches, occupancy sensors, timers, and receptacles. With this revised rule, these devices are required to be provided with AFCI protection if located in any of the rooms or areas specified in the rule, even though they may control a luminaire or other load located in a room or area where AFCI protection is not required. For example, a light switch may be located in the bedroom and control an outside luminaire. The luminaire on the outside of the building is not required to have AFCI protection. However, if the switch is located inside the dwelling in a bedroom or living room, for example, AFCI protection is required.

How to Provide AFCI Protection

The six different methods of providing AFCI protection that are permitted in *NEC 210.12(A)* are shown in Figure 6-13. Although the *Code* does not indicate a preference for one over the other, the cost of the installation, including both time and material, often influences the method chosen.

▶(1) A listed combination type arc-fault circuit interrupter, installed to provide protection of the entire branch circuit.◀* This method is perhaps the most simple. A listed combination type AFC I circuit breaker is installed in the panelboard where the branch circuit originates. The term "combination type" does not refer to a combination of GFCI and AFCI protection but that the device provides protection against both parallel and series type of arcing faults. The different types of AFC I devices are explained later in this chapter.

*Reprinted with permission from NFPA 70-2014.

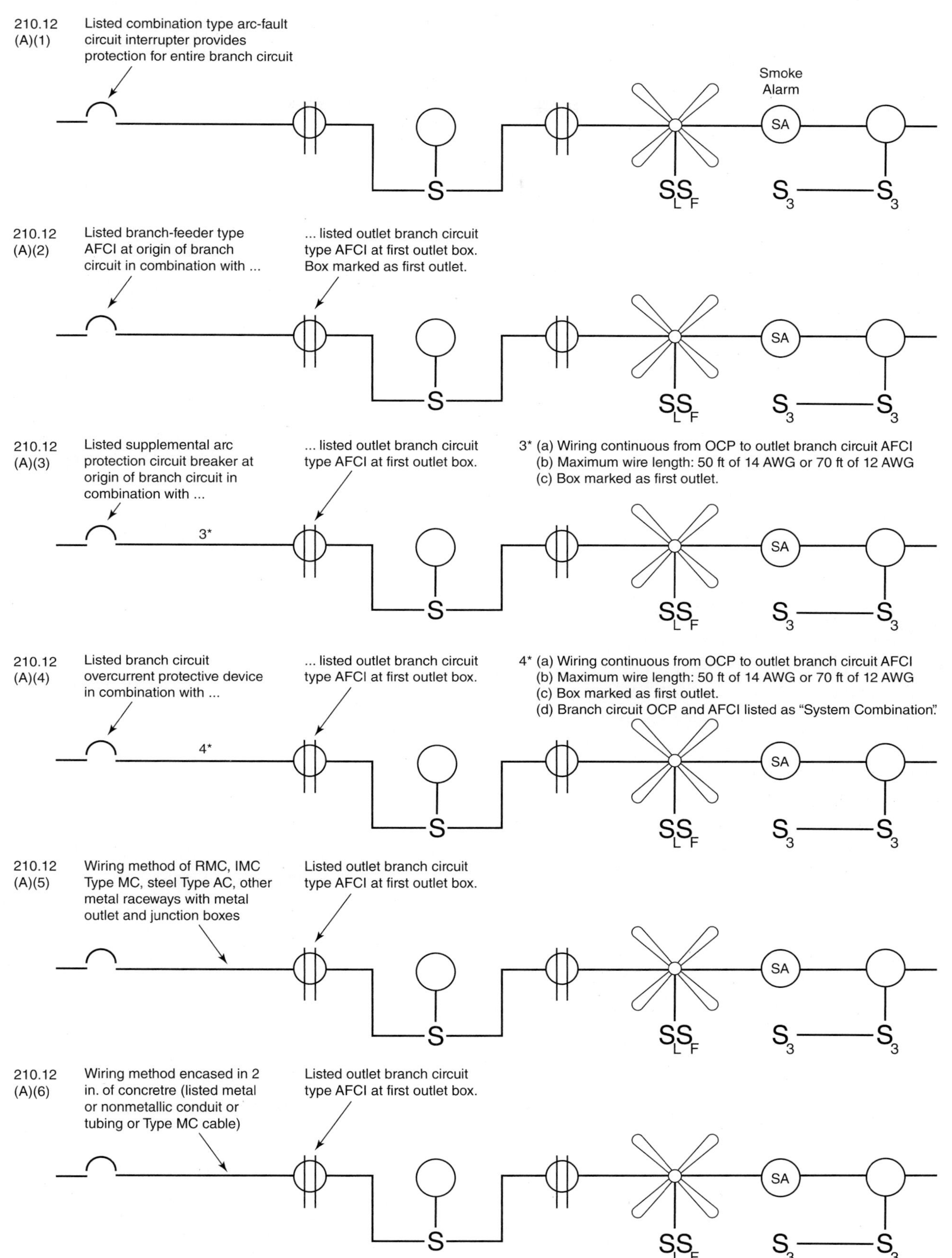

FIGURE 6-13 Methods permitted for providing the required AFCI protection for outlets or devices. The installer can choose any of the methods allowed.

▶(2) A listed branch/feeder type AFCI installed at the origin of the branch circuit in combination with a listed outlet branch circuit type arc-fault circuit interrupter installed at the first outlet box on the branch circuit. The first outlet box in the branch circuit is required to be marked to indicate that it is the first outlet of the circuit.◀*

This concept uses a branch/feeder-type AFCI installed in the circuit-breaker panelboard to provide protection against parallel arcing faults and an outlet branch-circuit type AFCI for protection of cord-connected appliances or luminaires. The outlet branch-circuit type AFCI will also provide protection for portions of the branch circuit on the load side of the device. The rule is silent on acceptable methods for identifying the first outlet box on the branch circuit. One method that it seems would comply with the rule is to use a label maker to place a label marked "First Outlet Box of Branch Circuit" on the finished surface below the device faceplate.

▶(3) A listed supplemental arc protection circuit breaker installed at the origin of the branch circuit in combination with a listed outlet branch circuit type arc-fault circuit interrupter installed at the first outlet box on the branch circuit is permitted if all of the following conditions are met:

(a) The branch circuit wiring is continuous from the branch circuit overcurrent device to the outlet branch circuit arc-fault circuit interrupter.

(b) The maximum length of the branch circuit wiring from the branch circuit overcurrent device to the first outlet does not exceed 50 ft (15.2 m) for a 14 AWG wire or 70 ft (21.3 m) for a 12 AWG wire.

(c) The first outlet box in the branch circuit is marked to indicate that it is the first outlet of the circuit.◀*

This method of providing AFCI protection is very conditional and very specific. The term "supplemental arc protection circuit breaker" refers to a type of protection that is in the developmental stage and not yet available in the marketplace. Conceptually, the supplemental arc protection circuit breaker will be required to provide arc-fault protection for the branch-circuit wiring between the overcurrent device and the listed outlet branch-circuit type arc-fault circuit interrupter installed at the first outlet box.

The additional requirements to use this installation method, shown in *(a)*, *(b)*, and *(c)*, are very straightforward.

▶(4) A listed outlet branch circuit type arc-fault circuit interrupter installed at the first outlet on the branch circuit in combination with a listed branch circuit overcurrent protective device where all of the following conditions are met:

(a) The branch circuit wiring is continuous from the branch circuit overcurrent device to the outlet branch circuit arc-fault circuit interrupter.

(b) The maximum length of the branch circuit wiring from the branch circuit overcurrent device to the first outlet is not permitted to exceed 50 ft (15.2 m) for a 14 AWG or 70 ft (21.3 m) for a 12 AWG conductor.

(c) The first outlet box in the branch circuit shall be marked to indicate that it is the first outlet of the circuit.

(d) The combination of the branch circuit overcurrent device and outlet branch circuit AFCI is identified as meeting the requirements for a "System Combination" type AFCI and is listed as such.◀*

This method of providing AFCI protection allows the use of a standard overcurrent protective device at the origination of the branch circuit. It could be either a circuit breaker or fuse rated for the overcurrent protection of the branch circuit. This overcurrent device provides overload, short-circuit, and ground-fault protection of the wires, devices, and equipment of the branch circuit.

The standard overcurrent device at the beginning of the branch circuit and the outlet branch circuit AFCI at the first outlet are required to be

*Reprinted with permission from NFPA 70-2014.

identified as meeting the requirements for a System Combination. The word "identified" is defined in *Article 100* as Identified (as applied to equipment). Recognizable as suitable for the specific purpose, function, use, environment, application, and so forth, where described in a particular *Code* requirement.

Informational Note: Some examples of ways to determine suitability of equipment for a specific purpose, environment, or application include investigations by a qualified testing laboratory (listing and labeling), an inspection agency, or other organizations concerned with product evaluation.*

As indicated in the definition, the identification required in the rule can be satisfied by the product being listed by an approved electrical products testing laboratory.

The additional requirements to use this installation method, as stated in *(a)*, *(b)*, and *(c)* of *210.12(A)(4)*, are very straightforward.

▶(5) If RMC, IMC, EMT, Type MC, or steel armored Type AC cables meeting the requirements of 250.118, metal wireways, metal auxiliary gutters and metal outlet and junction boxes are installed for the portion of the branch circuit between the branch-circuit overcurrent device and the first outlet, it is permitted to install a listed outlet branch-circuit type AFCI at the first outlet to provide protection for the remaining portion of the branch circuit.◀*

This method of providing AFCI protection also allows the use of a standard overcurrent protective device at the origination of the branch circuit. It could be either a circuit breaker or fuse rated for the overcurrent protection of the branch circuit. This overcurrent device provides overload, short-circuit, and ground-fault protection of the wires, devices, and equipment of the branch circuit.

However, this method of providing AFCI protection requires specific wiring methods for the portion of the branch circuit that does not have AFCI protection. It is assumed the Code Making Panel responsible for the rule concludes that the wiring methods that are permitted provide greater protection against physical damage than provided by the Type NM cable that is often used to wire dwelling units.

The listed outlet branch-circuit type AFCI at the first outlet provides AFCI protection for the remaining portion of the branch circuit. Note that there is no requirement to mark the first outlet of the branch circuit, but the AFCI device will be located at that point.

▶(6) Where a listed metal or nonmetallic conduit or tubing or Type MC Cable is encased in not less than 2 in. (50 mm) of concrete for the portion of the branch circuit between the branch circuit overcurrent device and the first outlet, it is permitted to install a listed outlet branch circuit type AFCI at the first outlet to provide protection for the remaining portion of the branch circuit.◀*

This method of providing AFCI protection also allows the use of a standard overcurrent protective device at the origination of the branch circuit. It could be either a circuit breaker or fuse rated for the overcurrent protection of the branch circuit. This overcurrent device provides overload, short-circuit, and ground-fault protection of the wires, devices, and equipment of the branch circuit.

However, this method of providing AFCI protection requires specific wiring methods for the portion of the branch circuit that does not have AFCI protection. The rule also requires that the wiring method be encased in not less than 2 in. (50 mm) of concrete. This concrete encasement provides protection against physical damage for the portion of the branch circuit between the overcurrent device and the outlet-type AFCI.

Although not specifically mentioned in *210.12(A)(6)*, it is presumed that one of the wiring methods permitted in *210.12(A)(5)* could be used for the portion of the branch circuit between a panel-board and a concrete floor and from a concrete floor to the outlet box where the AFCI device is installed. This would, for all practical purposes, be a mixture of the concepts in *210.12(A)(5)* and *(A)(6)*.

The listed outlet branch-circuit type AFCI at the first outlet provides AFCI protection for the remaining portion of the branch circuit. Note that there is no requirement to mark the first outlet of the branch circuit, but the AFCI device will be located at that point.

An exception is provided that appears to apply to all of *210.12(A)*. It applies to the installation of a branch circuit to a fire alarm system in the dwelling unit. It reads,

*Exception: Where an individual branch circuit to a fire alarm system installed in accordance with 760.41(B) or 760.121(B) is installed in RMC, IMC, EMT, or steel-sheathed cable, Type AC or Type MC, meeting the requirements of 250.118, with metal outlet and junction boxes, AFCI protection is permitted to be omitted.**

The requirements for wiring fire alarm systems in dwellings are contained in these two sections. The sections specifically prohibit the installation of AFCI and GFCI protection for the branch circuit that supplies a fire alarm system.

Note that most dwellings do not have a fire alarm system installed. Usually, single-station smoke alarms are installed in each room or area as required by the building code. Even though the smoke alarms are interconnected so that if one goes into alarm, all alarms sound, they are not considered to be a fire alarm system. Usually, a fire alarm system includes a fire alarm control cabinet as well as interconnection of detection and annunciation devices, and often reports an alarm off premises such as to a fire alarm monitoring station.

Informational Note No. 1 following *NEC 760.1* describes in detail the functioning of a typical fire alarm system. It reads,

Informational Note No. 1: Fire alarm systems include fire detection and alarm notification, guard's tour, sprinkler waterflow, and sprinkler supervisory systems. Circuits controlled and powered by the fire alarm system include circuits for the control of building systems safety functions, elevator capture, elevator shutdown, door release, smoke doors and damper control, fire doors and damper control and fan shutdown, but only where these circuits are powered by and controlled by the fire alarm system. For further information on the installation and monitoring for integrity requirements for fire alarm systems, refer to the NFPA 72-2010, National Fire Alarm and Signaling Code.*

A fire alarm system is not required by the *NEC* but, if installed, must be wired in accordance with *NEC Article 760*.

Certain rooms and areas are exempt from the AFCI requirement:

- Attics—AFCI protection not required, but no problem if you do provide AFCI protection.
- Bathrooms—The receptacle(s) require GFCI protection. You could put the lighting on an AFCI-protected circuit, but that is not required.
- Garages—most of the receptacle(s) require GFCI protection. You could put the lighting on an AFCI-protected circuit, but that is not required.
- Home Office—Careful here. If you take over a room that qualifies as a bedroom, the inspector is likely to require AFCI protection of the room.
- Outdoors—Outdoor receptacle(s) require GFCI protection, but not AFCI protection. Outdoor lighting is permitted to be supplied by a non-AFCI-protected branch circuit or an AFCI-protected branch circuit.
- Unfinished basements—Unfinished basement receptacle(s) require GFCI protection. You could put the lighting on an AFCI-protected circuit but are not required to do so.

A new *210.12(B)* was added to the 2011 *NEC*. It covers branch circuit extensions or modifications and reads, In any of the areas specified in 210.12(A), where branch circuit wiring is modified, replaced or extended, the branch circuit shall be protected by:

1. A listed combination AFCI located at the origin of the branch circuit; or
2. A listed outlet branch circuit AFCI located at the first receptacle outlet of the existing branch circuit.*

*Reprinted with permission from NFPA 70-2014.

This rule extends the AFCI requirements for areas of a dwelling where branch circuits are extended or modified, such as for room additions or remodels.

A new exception was added during the processing of the 2014 *NEC*. It reads as follows:

▶*Exception: AFCI protection shall not be required where the extension of the existing conductors is not more than 6 ft (1.8 m) and does not include any additional outlets or devices.*◀*

As can be seen, this exception has a very narrow application. It applies to an existing branch circuit that does not have AFCI protection. If more than 6 ft (1.8 m) of conductor is installed in the modification or extension of the existing branch circuit, AFCI protection *is* required. The AFCI protection can be provided in the form of an AFCI circuit breaker at the origination of the branch circuit or by an outlet-type AFCI receptacle. In addition, AFCI protection *is* required if an outlet or device is installed.

Don't Get Confused

Don't confuse an AFCI with a GFCI. These two devices protect against two different kinds of electrical hazards.

- An AFCI device is designed to sense and respond to an arcing fault that could develop into a fire. Some AFCI devices also provide personnel GFCI protection (6 mA). Some AFCI devices provide equipment ground-fault protection (GFP—30 mA). A typical application for equipment GFP in dwellings is for outdoor electric deicing and snow melting equipment, covered in *Article 426.*
- Always read the label and the instructions of an AFCI device to be sure you know where and how to connect it.
- A GFCI device is designed to sense and respond to a ground fault, protecting people from severe electrical shock that could lead to death.
- Dual-function AFCI/GFCI breakers provide normal overcurrent protection as well as arcing and ground-fault protection.

SAFETY ALERT

Be very careful when installing AFCI devices, whether circuit breaker or receptacle types, to follow the manufacturer's installation instructions. Some circuit breakers have pigtail neutral conductors that must be connected to the neutral bar in the panelboard and others have a clip that connects to the neutral bar when the circuit breaker is installed. These types of circuit breakers may not be interchangeable.

Receptacle-type AFCIs may be connected feed-through or standalone. If connected improperly, the vital safety protection required may not be functional.

Smoke Alarms and AFCI Protection

As you will learn in Chapter 26, smoke alarms are required to have secondary battery backup protection. Single-station smoke alarms require AFCI protection in the rooms or areas specified even if interconnected so that all alarms sound if one detector goes into alarm. This arrangement does not constitute a fire alarm system, as there is no fire alarm control panel. Permanently installed fire alarm systems are not permitted to be GFCI- or AFCI-protected, *760.41(B)* and *760.121(B).*

Why AFCI Protection?

Electrical arcing is an "early" event in the history of a typical electrical fire. Arcing is considered one of the leading causes of electrical fires in

*Reprinted with permission from NFPA 70-2014.

homes. Think about frayed cords, cords under carpets, cords pinched under legs of furniture, and nails being driven through plaster into cables in the wall. The Consumer Product Safety Commission recently cited 150,000 residential electrical fires annually in the United States, 850 deaths, 6000 injuries, and more than $1.5 billion in property loss. These statistics are pretty convincing of the need for greater electrical protection. The heat of an arc is extremely high, known to reach 10,000°F (5538°C) or more. Hot particles of metal expelled by the arc are enough to ignite most surrounding combustible materials.

We have all been guilty of blaming electrical fires on overloaded circuits, overloaded extension cords, and too many extension cords plugged into a receptacle. The heat of an overloaded cord by itself is generally not hot enough to ignite surrounding combustible materials. An overload condition can cause the breakdown of the insulation on the conductors, allowing the conductors to touch one another. This sets the stage for an arcing condition between the conductors, or between the "hot" conductor and ground. We now have the ingredients of an electrical fire.

For example, let's say that the cord on an electric iron becomes frayed and the conductor(s) break. The iron draws 10 amperes. The arcing of 10 amperes trying to bridge the gap in the broken conductor(s) develops the tremendous heat capable of igniting surrounding combustible material, yet 10 amperes flowing in the circuit will not trip a 15-ampere circuit breaker or open a 15-ampere fuse. This condition is not an overload and initially is not a short circuit. Conventional overcurrent devices "see" this as normal load current. They are not designed to detect and open under current levels below their trip setting. What is needed is a device that can detect the unique characteristics of an arcing condition—an AFCI.

Types of Arcs

Arcing is current flowing outside of the intended path. Arcing is a function of voltage and the amount of current flowing in the arc.

Taking the liberty of using familiar terms, we could say that arcs come in two varieties—*good* and *bad*.

Good arcs are not dangerous. For example, the slight arcing that occurs between the contacts of a switch when turned ON and OFF, or when a male attachment plug cap is inserted into a receptacle is not considered hazardous. This type of arcing creates a unique signature.

Bad arcs are dangerous. Arcs develop a tremendous amount of heat (I^2R) at the point of arcing (sputtering). Arcing could be line to line, line to neutral, or line to ground. This type of arcing also creates a signature.

To avoid nuisance tripping, the sophisticated electronic circuitry in an AFCI has the ability to recognize the signature, which is the shape of the alternating-current wave that is distorted from the normal sine wave and distinguish between *good* (harmless) and *bad* (dangerous) arcs. It makes a decision as to whether the arcing or glowing is dangerous or harmless.

Arcs are classified as follows:

- *Series Arcs:* When a conductor that is carrying current breaks or when a loose connection occurs, a "series" arcing fault is created. The current draw of the connected load is trying to jump across the opening created by the break or loose connection. Table 6-3 gives you a good idea of how fast an AFCI circuit breaker trips for various values of arcing current as specified in UL Standard 1699. Remember, there are 60 cycles in 1 second.
- *Parallel Arcs:* When arcing occurs between the black ("hot") and white grounded (neutral) circuit conductors, or between the black ("hot") conductor and ground, it is referred to as a "parallel" arc. These conditions might be caused by a nail being driven through a cable, a staple driven too tightly, or a clamp or connector squeezing the cable too tightly. The current is traveling outside of its intended path. A broken or frayed cord could result in a series or parallel arc, or both.

Table 6-4 shows UL Standard 1699 specified current values for parallel arc testing of an AFCI device.

TABLE 6-3

Clearing times for listed AFCI combination-type circuit breakers when subjected to various values of arcing-fault current.

	VARIOUS TIMES TO CLEAR ARCING CURRENT IN AMPERES			
Ampere Rating of AFCI Circuit Breaker	**5 Amperes**	**10 Amperes**	**Rated Current of the AFCI Circuit Breaker**	**150% of Rated Current of the AFCI Circuit Breaker**
15 amperes	1 second	0.40 second	0.28 second	0.16 second
20 amperes	1 second	0.40 second	0.20 second	0.11 second
30 amperes	1 second	0.40 second	0.14 second	0.10 second

TABLE 6-4

Arcing current values for parallel arcing tests.

ARCING CURRENT IN AMPERES					
75	100	150	200	300	500

For each of the above arcing current values, the total arcing time shall not exceed 8 half-cycles in 0.5 second. The 8 half-cycles of arcing could be consecutive or intermittent.

Types of AFCI Devices

The *NEC* tells us where to install AFCIs in homes. UL Standard 1699 specifies the sensing and tripping characteristics of AFCIs.

AFCIs are available in the following types:

1. *Combination AFCI:* This type of AFCI is mandatory in *210.12(A)*. It is installed in the panel where the branch circuit originates. It provides arc-fault protection as specified in the UL Standard for both *branch/feeder AFCIs* and *outlet AFCIs*.

   Note: Don't get confused. A *combination AFCI* means that the device has been tested and listed as meeting the requirements for both *branch/feeder AFCIs* and *outlet AFCIs*. When referring to a device that provides both AFCI and GFCI personnel protection, think of it as having *dual-function* characteristics. Don't use the word *combination* to describe a dual-function device!

2. *Branch/feeder AFCI:* This type of AFCI circuit breaker is installed in a panel. It provides arc-fault protection for the branch-circuit wiring. It is not quite as sensitive as an *outlet AFCI,* although it does provide limited protection for extension cords plugged into receptacles on the circuit.

3. *Outlet AFCI:* This is an AFCI receptacle. It provides protection for cord sets that are plugged into the receptacle. It is more sensitive than a *branch/feeder AFCI*. An outlet-type AFCI receptacle installed at the first receptacle is permitted by several of the methods in *210.12(A)*. Everything downstream from the feed-through AFCI receptacle is AFCI protected.

4. *Portable AFCI:* This is a plug-in device with one or more receptacles. It provides AFCI protection for cords that are plugged to it. This type of AFCI device is a must for workers using electrical tools on the job.

5. *Cord AFCI:* This is a plug-in device that provides AFCI protection to the power cord connected to it. It has no additional outlets. An example of this would be the AFCI device required to be an integral part of room air-conditioning equipment in accordance with *440.65*.

How Do You Know What Type It Is? The marking is found on the device. When installed in a panelboard, the marking must be visible after you remove the cover from the panelboard.

What Kinds of AFCIs Are Available?

- Single-pole, 120-volt breaker
- Single-pole, 120-volt dual-function with both AFCI and GFCI features in one breaker

- Two-pole, 240-volt, common trip breaker
- Two-pole, 120/240-volt breaker
- Two-pole, 120/240-volt common trip dual-function with both AFCI and GFCI features in one breaker
- Two-pole, 120/240-volt independent trip dual-function with both AFCI and GFCI features in one breaker; suitable for use with a shared neutral on a multiwire branch circuit
- 120-volt feed-through receptacles

UL Standards for AFCIs and GFCIs

Type	Conforms to Requirements of	If Circuit Breaker Type, Also Conforms to Requirements of
AFCI	UL 1699	UL489
GFCI	UL 943	UL 489
AFCI/GFCI combination	UL 1699 and 943	UL 489

What Types of AFCI Devices Do Electricians Install?

The most common AFCIs that electricians install today are the circuit-breaker type. Circuit-breaker AFCIs are the same width as a standard circuit breaker so they will fit into a standard electrical panel. See Figure 6-14.

How to Connect AFCIs

When installing a 125-volt, single-pole AFCI circuit breaker, you first "snap" the breaker into the panel. If the circuit breaker has an external wire with white insulation, it is connected to the neutral bar in the panel. Some manufacturers are producing AFCI circuit breakers that snap on a neutral bus that is spaced a short distance from the mounting bar for the breaker. These AFCI circuit breakers do not have an external wire pigtail. The black branch-circuit conductor is connected to the brass-colored load-side terminal on the breaker. The white branch-circuit conductor is connected to the silver-colored load-side terminal on the breaker. This is the same as for GFCI circuit breakers discussed earlier in this chapter.

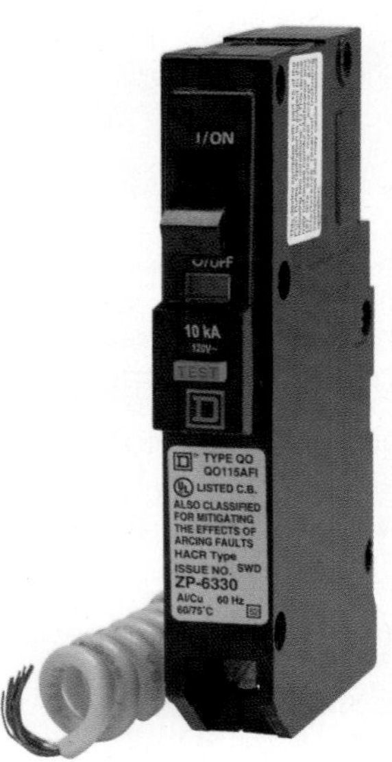

FIGURE 6-14 A listed combination-type arc-fault circuit-interrupter (AFCI) circuit breaker.

Never share the neutral conductor of a multiwire branch circuit when using GFCIs and AFCIs unless they are listed for sharing a neutral. See Figure 6-11. Two-pole AFCIs and GFCIs are available. Always follow the instructions furnished with these devices.

Is It Okay to Use AFCI and GFCI Devices in Series?

Yes! There should be no problem installing GFCI receptacles downstream from an AFCI circuit breaker in those locations where GFCI protection is required. It also is okay to use AFCI outlet-type devices on the load side of GFCI protection. The protection offered by the different devices does not conflict but is complementary.

On branch circuits that are required by *210.12(A)* to be AFCI-protected, you might want to pick up a receptacle(s) that requires GFCI protection. This occurs a few times in the residence discussed in this book. The Front Bedroom branch circuit A16 picks up an outdoor receptacle on the front of the house. The Master Bedroom branch circuit A19 picks up an outdoor receptacle on the back of the house. The Living Room branch circuit B17 picks up an outdoor receptacle on the back of the house. In addition, receptacles that serve countertops in the kitchen are required to have both GFCI and AFCI protection. Another example is receptacles that are in a laundry area are required to have AFCI protection. They are also required to have GFCI protection if they are within 6 ft of a sink.

You could also install a dual-function AFCI/GFCI circuit breaker in the panel, which would provide arc-fault and ground-fault protection for an entire branch circuit with one device. It is a matter of choice and availability of AFCI, GFCI, and AFCI/GFCI dual-function breakers.

Figure 6-15 illustrates various ways GFCI and AFCI protection might be combined to provide the protection required *by NEC 210.8(A)* or *210.12(A)*. As can be seen, several methods can be used. The order of protection is not covered in the *NEC* but is a matter of choice by the installer.

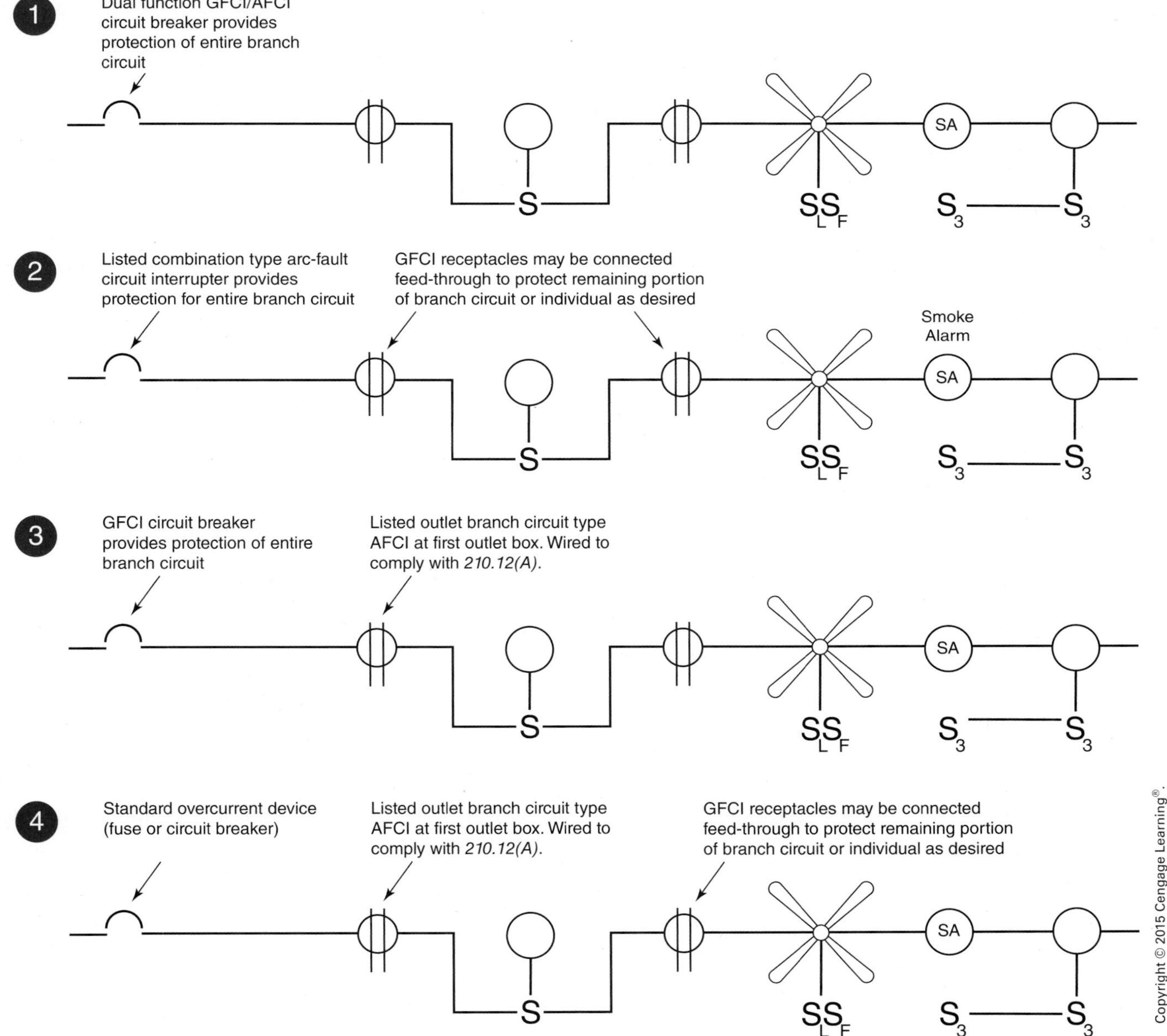

FIGURE 6-15 Four methods for providing both AFCI and GFCI protection where required. AFCI and GFCI protection technologies are compatible, complimentary and function in series.

As mentioned earlier, GFCI and AFCI technologies are compatible. There should not be a problem if the devices are connected in series.

One-line diagram (1) in Figure 6-15 shows a dual-function GFCI/AFCI circuit breaker installed in a panelboard at the origination of the branch circuit. This is perhaps the easiest and most straightforward method of providing both GFCI and AFCI protection. Verify whether this type of circuit breaker is available from your supplier for the brand of panelboard that is or that will be installed.

One-line diagram (2) in Figure 6-15 shows a listed combination-type AFCI circuit breaker installed in the panelboard. This device provides AFCI protection for the entire branch circuit. A GFCI-type receptacle can be connected for stand-alone protection or feed-through protection where required or desired. When connected for stand-alone protection, only the receptacles on the GFCI are protected. Downstream receptacles are not protected. When connected as feed-through, downstream receptacles, devices, and equipment are protected. The GFCI device can be connected feed-through to protect additional receptacles in the same room and stand-alone or feed-through for downstream receptacles. As you can see, there is a lot of flexibility on how connections are made to GFCI receptacles.

One-line diagram (3) in Figure 6-15 shows a GFCI-type circuit breaker installed at the origination of the branch circuit. This type circuit breaker provides GFCI protection for the entire branch circuit. A feed-through outlet-type AFCI receptacle is installed at the first outlet of the branch circuit to provide AFCI protection for cord-connected equipment as well as for the remainder of the branch circuit. Be cautious in providing this wiring arrangement. The wiring methods on the supply side of the AFCI device must comply with *NEC 210.12(A)(5)* (limited to metallic wiring methods and boxes) or *(A)(6)* (concrete-encased wiring).

One-line diagram (4) in Figure 6-15 shows a standard overcurrent device installed at the beginning of the branch circuit. This may be either a fuse or circuit breaker. A feed-through outlet-type AFCI receptacle is installed at the first outlet of the branch circuit to provide AFCI protection for cord-connected equipment as well as for the remainder of the branch circuit. As mentioned previously, the wiring methods must comply with *NEC 210.12(A)(5)* (limited to metallic wiring methods and boxes) or *(A)(6)* (concrete-encased wiring). A GFCI-type receptacle is installed where required or desired on the branch circuit. As discussed above, a GFCI-type receptacle can be connected for stand-alone protection or feed-through protection.

The intention of this discussion is to explain methods that can be used to comply with the *NEC* requirements. Be reminded that compliance with the *Code* is the minimum permitted for safety. Complete AFCI and GFCI protection can be provided for each of the applicable branch circuits even where not required.

AFCIs in Existing Homes

The mandatory requirement in *210.12(A)* for AFCI protection is for new work only. It is not retroactive. But it's not a bad idea to consider installing AFCIs on existing installations. It is easy to do. Just replace existing circuit breaker(s) with an AFCI circuit breaker(s). We covered earlier in this chapter the new requirement in *210.12(B)* for adding AFCI protection for branch circuit extensions or modifications.

In older homes, the insulation on the original wiring is rated 60°C. Over the years, the insulation becomes brittle. For example, in and above a luminaire, the heat from the luminaire literally bakes and chars the insulation on the conductors. AFCI protection could detect an arcing problem in its early stages. Today, conductors used in house wiring are rated 90°C, so the deterioration of conductor insulation because of heat has become a thing of the past.

NEC 406.4(D)(4) requires AFCI protection of receptacles that are replaced at a location where AFCI protection is required elsewhere in the *Code*. This requirement has an effective date of January 1, 2014. For example, a receptacle is replaced in a dwelling unit bedroom. AFCI protection is required for the branch circuit that supplies the receptacle by *210.12(A)*. As a result, the receptacle must have AFCI protection.

The AFCI protection can be provided in one of the following ways:

1. A listed outlet branch-circuit type arc-fault circuit interrupter receptacle
2. A receptacle protected by a listed outlet branch-circuit–type arc-fault circuit interrupter–type receptacle
3. A receptacle protected by a listed combination type arc-fault circuit interrupter–type circuit breaker

Checklist

A checklist is provided in the Appendix. It can be used to assist the installer in determining whether AFCI or GFCI protection is required or installed in any room or area in the dwelling. As you can see, the checklist can be customized to be used in any dwelling.

Testing AFCIs and GFCIs

AFCI/GFCI testers are used to test AFCI and GFCI receptacles and circuit breakers to make sure they are working properly. These devices simulate arc faults and ground faults. Figure 6-16 shows an arc-fault tester that also tests for proper polarity.

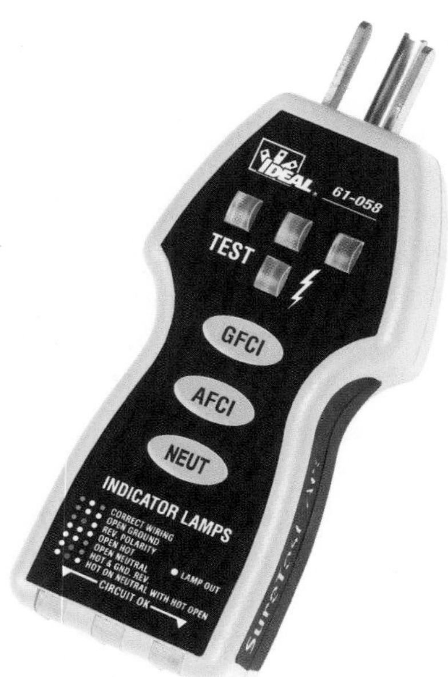

Courtesy Ideal Industries, Inc.

FIGURE 6-16 An arc-fault tester that also tests for proper polarity.

OTHER SPECIAL-PURPOSE RECEPTACLES

Tamper-Resistant Receptacles

A major step to reduce the number of injuries (burns), particularly to small children who always seem to find a way to insert things into the slots of a receptacle, was enacted for the 2008 *NEC* and was revised for the 2011 edition of the *Code*. The requirements for tamper-resistant receptacles in dwellings are found in *406.12(A)* and read as follows:

In all areas specified in *210.52*, all nonlocking type 125-volt, 15- and 20-ampere receptacles shall be listed tamper-resistant receptacles.

- *Exception 1: Receptacles located more than 5½ ft (1.7 m) above the floor are not required to be of the tamper-resistant type.*
- *Exception 2: Receptacles that are part of a luminaire or appliance are likewise not required to be of the tamper-resistant type.*
- *Exception 3: A single receptacle for one appliance or a duplex receptacle for two appliances that are located within dedicated space for each appliance that in normal use is not easily moved from one place to another and that is cord-and-plug connected in accordance with 400.7(A)(6), (A)(7), or (A)(8) do not require tamper-resistant receptacles.** (This exception recognizes that these receptacles would not be easily reached by children.)
- *Exception 4: Nongrounding receptacles used for replacements as permitted in 406.4(D)(2)(a) are not required to be of the tamper-resistant type.**

Tamper-resistant receptacles are identified by the words "Tamper Resistant" or the letters "TR" where they will be visible after installation, with the cover plate removed.

FIGURE 6-17 Two styles of tamper-resistant receptacles.

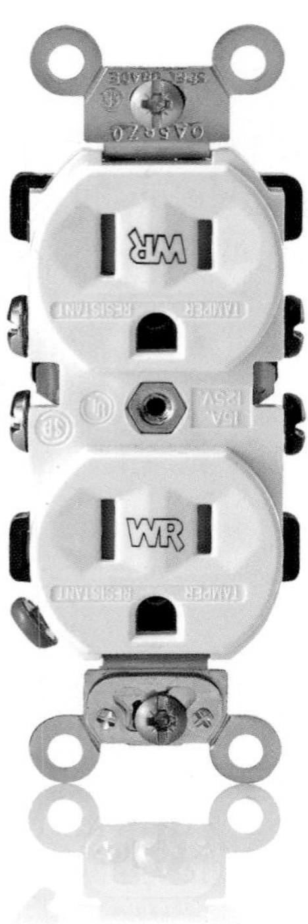

FIGURE 6-18 Weather-resistant receptacle that is also tamper-resistant.

Recall that *210.52* lists the rooms, areas, and other locations inside and outside of a dwelling where receptacles are required to be installed; the spacing for the receptacles is given as well as the number of receptacles required to be installed.

Tamper-resistant receptacles have a unique internal interlocking mechanism that energizes the blade contacts of the receptacle ONLY when a male attachment plug cap is inserted. If a paper clip or similar object is inserted into any one slot of the receptacle, no power is supplied to any of the blade contacts of the receptacle. Both slot shutters remain closed. Figure 6-17 is a typical tamper-resistant receptacle. Tamper-resistant receptacles have been required for many years in pediatric (children) areas of health care facilities, *517.18(C)*.

Weather Resistant. All 15- and 20-ampere, 125- and 250-volt nonlocking type receptacles installed in a damp or wet location are required to be listed weather-resistant type.

The construction of these receptacles makes them more durable if located where there is excessive moisture. These receptacles are required to be identified by the words "Weather Resistant" or the letters "WR" where they will be visible after installation, with a cover plate secured as intended. See Figure 6-18.

Note that the definition of *damp location* and *wet locations* are found in *NEC Article 100*. It is quite obvious that an outdoor location subject to rain is a wet location. Locations below the carport roof and protected from weather are no doubt damp locations. To avoid dispute with inspectors, it makes sense to use all weather-resistant receptacles outside of the dwelling.

Putting the Rules Together. A 15- or 20-ampere, 125-volt receptacle is installed outside the dwelling in a wet location. This receptacle is required to have GFCI protection or be of the GFCI-type, to be weather-resistant, and tamper-resistant. You're in luck! Manufacturers of GFCI receptacles produce a version that complies with all of these requirements. See Figure 6-19.

FIGURE 6-19 GFCI-type receptacle that is also weather resistant and tamper resistant.

REPLACING EXISTING RECEPTACLES

The *NEC* is very specific on the type of receptacle permitted to be used as a replacement for an existing receptacle. These *Code* rules are found in *406.4(D)*.

GFCI Protection. In *406.4(D)(3)* we find a retroactive rule requiring that when an existing receptacle is replaced in *any* location where the present *Code* requires GFCI protection (bathrooms, kitchens, outdoors, unfinished basements, garages, etc.), the replacement receptacle must be GFCI-protected. This could be a GFCI-type receptacle, or the branch circuit could be protected by a GFCI-type circuit breaker.

It is important to note that on existing wiring systems, such as "knob-and-tube," or older nonmetallic-sheathed cable wiring that did not have an equipment grounding conductor, a GFCI receptacle will still properly function. It does not need an equipment grounding conductor. Thus, protection against lethal shock is still provided. See Figure 6-6.

Arc-Fault Circuit-Interrupters. As detailed above, *406.4(D)(4)* requires AFCI protection where replacements are made at receptacle outlets that are required to be so protected elsewhere in the *Code*. A delayed effective date for this requirement of January 1, 2014, was included.

Tamper-Resistant Receptacles. Section *406.4(D)(5)* requires that listed tamper-resistant receptacles be provided where replacements are made at receptacle outlets that are required to be tamper-resistant elsewhere in the *Code*. *NEC 406.12* contains the requirements for tamper-resistant receptacles. See the discussion on tamper-resistant receptacles earlier in this chapter.

Weather-Resistant Receptacles. *NEC 406.4(D)(6)* requires that weather-resistant receptacles be provided where replacements are made at receptacle outlets that are required to be so protected elsewhere in the *Code*. See the discussion on weather-resistant receptacles earlier in this chapter.

Replacing Existing Two-Wire Receptacles Where Grounding Means Does Exist

When replacing an existing receptacle [Figure 6-20(A)] where the wall box is properly grounded [Figure 6-20(E)] or where the branch-circuit wiring contains an equipment grounding conductor [Figure 6-20(D)], at least two easy choices are possible for the replacement receptacle:

1. The replacement receptacle must be of the grounding type (Figure 6-20[B]) unless . . .
2. The replacement receptacle is of the GFCI type [Figure 6-20(C)].

If the an equipment grounding conductor is not present in the enclosure for the receptacle, *250.130(C)* permits an equipment grounding conductor to be run from the enclosure to be grounded, or from the green hexagon grounding screw of a grounding-type receptacle, to any one of these four locations:

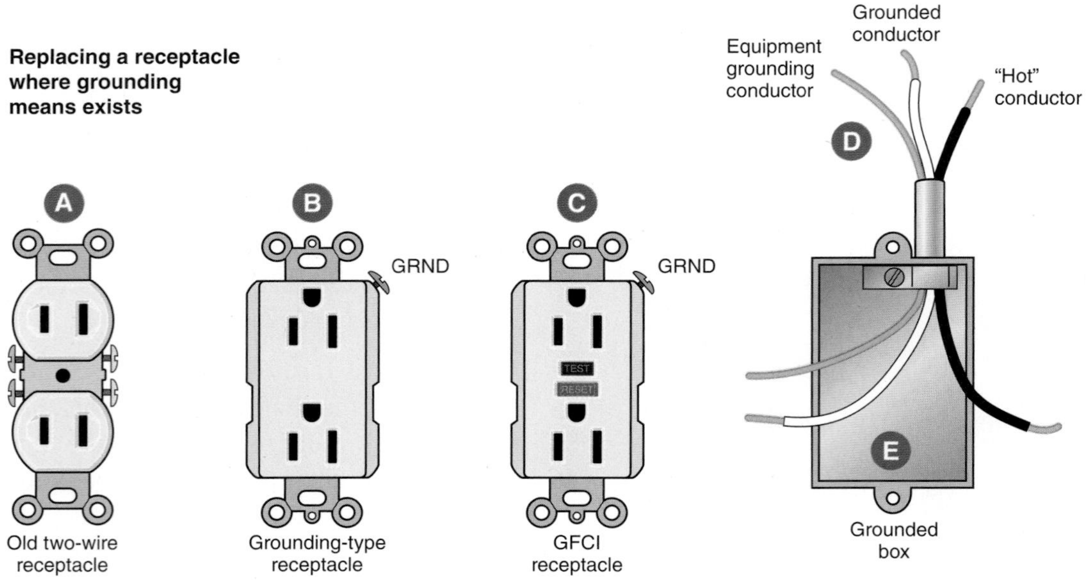

FIGURE 6-20 If a wall box (E) is properly grounded by any of the methods specified in *250.118* or if the branch-circuit wiring contains an equipment grounding conductor (D), an existing receptacle of the type shown in (A) must be replaced with a grounding-type receptacle (B) or with a GFCI receptacle (C). See *406.4(D)(1)*.

1. Any accessible point on the grounding electrode system
2. Any accessible point on the grounding electrode conductor
3. The equipment grounding terminal bar within the enclosure where the branch circuit for the receptacle originates
4. The grounded service conductor within the service-equipment enclosure

Note, however, that this permission is only for existing installations. See Figure 6-21.

Be sure that the equipment grounding conductor of the circuit is connected to the receptacle's green hexagon-shaped grounding terminal.

Do not connect the white grounded circuit conductor to the green hexagon-shaped grounding terminal of the receptacle.

Do not connect the white grounded circuit conductor to a metal box.

Replacing Existing Two-Wire Receptacles Where Grounding Means Does Not Exist

When replacing an existing 2-wire nongrounding-type receptacle [Figure 6-21(A)] where the box is not grounded [Figure 6-21(E)] or where an equipment grounding conductor has not been run with the circuit conductors [Figure 6-21(D)], four choices are possible for selecting the replacement receptacle:

1. The replacement receptacle may be a nongrounding type [Figure 6-21(A)].
2. The replacement receptacle may be a GFCI type [Figure 6-21(C)].
   - The GFCI replacement receptacle must be marked "No Equipment Ground," as in Figure 6-22.
   - The green hexagonal grounding terminal of the GFCI replacement receptacle does not have to be connected to any grounding means. It can be left "unconnected." The GFCI's trip mechanism will operate properly when ground faults occur anywhere on the load side of the GFCI replacement receptacle. Ground-fault protection is still there. Refer to Figure 6-6.
   - Do not connect an equipment grounding conductor from the green hexagonal grounding terminal of a replacement GFCI receptacle (see Figure 6-22) to any other downstream

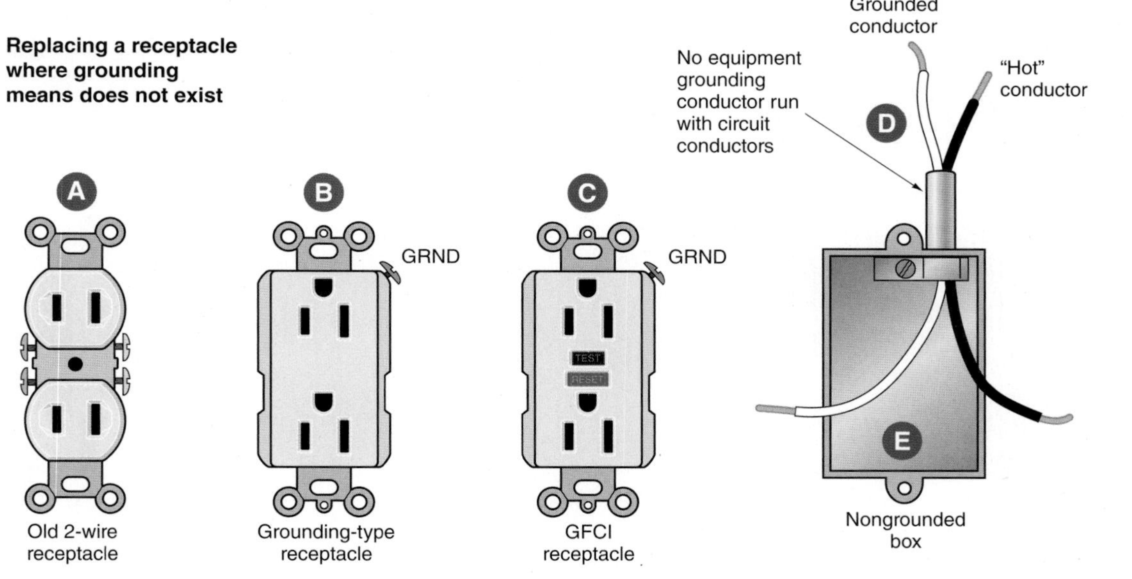

FIGURE 6-21 If a box (E) is *not* grounded or if in an equipment grounding conductor has not been run with the circuit conductors (D), an existing nongrounding-type receptacle (A) may be replaced with a nongrounding-type receptacle (A), a GFCI-type receptacle (C), a grounding-type receptacle (B) if supplied through a GFCI receptacle (C), or a grounding-type receptacle (B) if a separate equipment grounding conductor is run from the receptacle to any of the locations permitted by *250.130(C)*. See Figure 6-23.

receptacles that are fed through the replacement GFCI receptacle.

The reason this is not permitted is that if at a later date someone saw the conductor connected to the green hexagonal grounding terminal of the downstream receptacle, there would be an immediate assumption that the other end of that conductor had been properly connected to an acceptable grounding point of the electrical system. The fact is that the so-called equipment grounding conductor had been connected to the replacement GFCI receptacle's green hexagonal grounding terminal that was not grounded in the first place. We now have a false sense of security, a real shock hazard. The far better choice from a safety standpoint is to go to the time and expense of installing the equipment grounding conductor as permitted in *250.130(C)*.

3. The replacement receptacle may be a grounding type (Figure 6-21[B]) if it is supplied through a GFCI-type receptacle (Figure 6-21[C]).
   - The replacement receptacle must be marked "GFCI Protected" and "No Equipment Ground," as in Figure 6-22.
   - The green hexagonal equipment grounding terminal of the replacement grounding-type receptacle (Figure 6-21[B]) need not be connected to any grounding means. It may be left "unconnected." The upstream feed-through GFCI receptacle (Figure 6-21[C]) trip mechanism will work properly when ground faults occur anywhere on its load side. Ground-fault protection is still there. Refer to Figure 6-6.
   - Do not connect an equipment grounding conductor from the green hexagonal grounding terminal of a replacement GFCI receptacle (Figure 6-22) to any other downstream receptacles that are fed through the replacement GFCI receptacle.
4. The replacement receptacle may be a grounding type (Figure 6-21[B]) if
   - an equipment grounding conductor, sized per *Table 250.122*, is run from the replacement receptacle's green hexagonal grounding screw and properly connected to one of the four locations described earlier in this chapter. You are *not* permitted to make the equipment ground connection to the interior metal water piping anywhere beyond the first 5 ft (1.5 m).

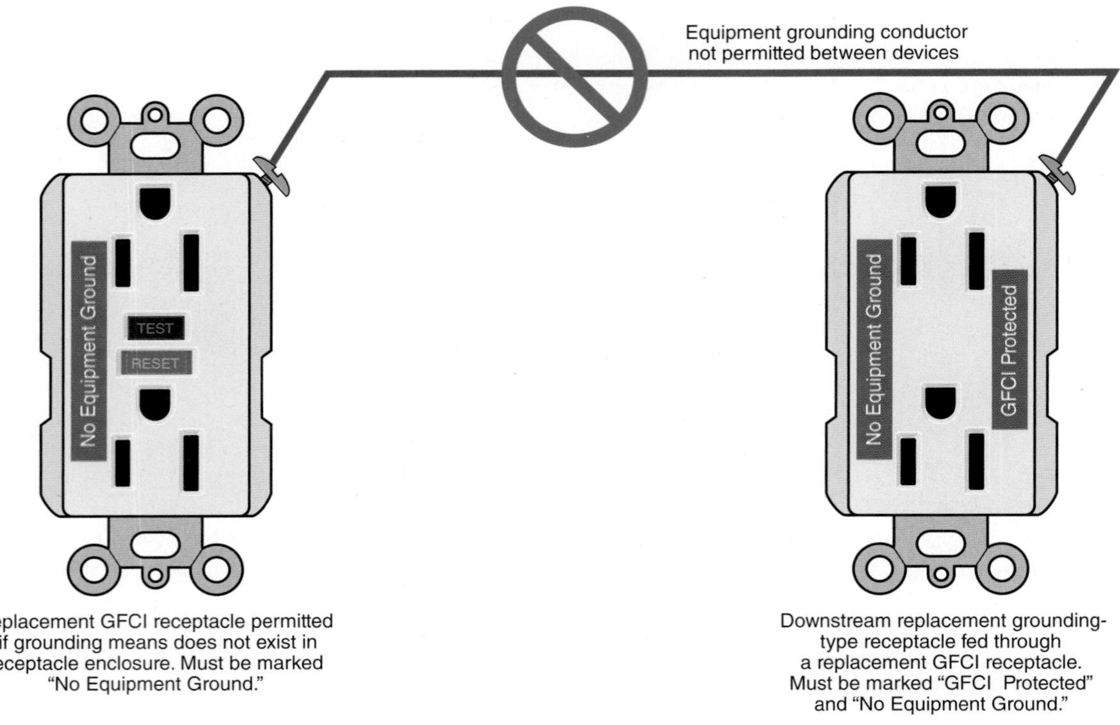

FIGURE 6-22 Where a grounding means does not exist, there are certain marking requirements, as presented in this diagram. *Do not* connect an equipment grounding conductor between these receptacles; see *406.3(D)(2)*.

See *250.130(C)* and the *Informational Note*. Also read *250.50* for the explanation of a "grounding electrode system." This permission is only for replacing a receptacle in existing installations. See Figure 6-23.

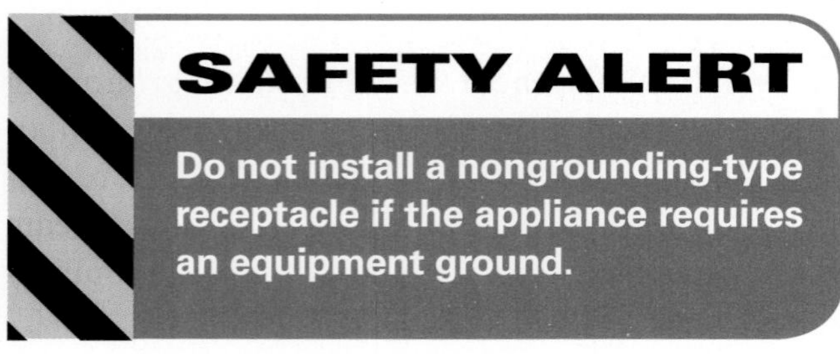

FIGURE 6-23 In old installations where an equipment grounding conductor does not exist in the box, one option is to install an equipment grounding conductor as permitted in *250.130(C)*. This is preferred over installing a GFCI receptacle without an equipment grounding conductor.

Many appliances are supplied by the manufacturer with a 3-wire cord that includes an equipment grounding conductor. These are often referred to as having a 3-wire cord. The installation instructions supplied by the manufacturer state the appliance is to be grounded or supplied from a receptacle that is grounded. It is a violation of the manufacturer's instructions as well as *NEC 110.3(B)* to supply these appliances from an ungrounded receptacle, even one that has been changed to a GFCI type.

NEC 250.114 contains requirements for cord-and-plug-connected equipment and appliances that are required to be connected to an equipment grounding conductor. Included are the following:

a. Refrigerators, freezers, and air conditioners
b. Clothes-washing, clothes-drying, dish-washing machines; ranges; kitchen waste disposers; information technology equipment; sump pumps and electrical aquarium equipment
c. Hand-held motor-operated tools, stationary and fixed motor-operated tools, and light industrial motor-operated tools
d. Motor-operated appliances of the following types: hedge clippers, lawn mowers, snow blowers, and wet scrubbers
e. Portable handlamps*

An exception to the requirement that the equipment be connected to an equipment grounding conductor is provided for tools and equipment that are double-insulated. This equipment is required to be distinctively marked.

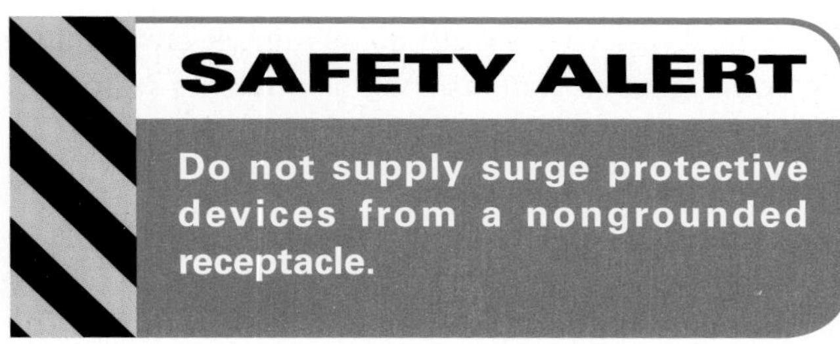

Surge protective devices (SPDs) are available in several different forms, including a receptacle outlet and plug strips. (Note that Underwriters Laboratories refers to plug strips as "relocatable power taps"; these are covered in detail in the *UL Guide Information for Electrical Equipment*, also referred to as the *White Book*.) See the discussion of surge protective devices later in this chapter. A photo of a receptacle-type SPD is shown in Figure 6-28 and a cord-type in Figure 6-29.

These surge protective devices rely on an equipment grounding conductor to function correctly. They cannot provide the safety protection from voltage spikes so prevalent from lightning events without having a properly connected equipment grounding conductor.

*Reprinted with permission from NFPA 70-2014.

PERSONNEL GROUND-FAULT PROTECTION FOR ALL TEMPORARY WIRING

Because of the nature of construction sites, there is a continual presence of shock hazard that can lead to serious personal injury or death through electrocution. Workers are standing in water, standing on damp or wet ground, or in contact with steel framing members. Electric cords and cables are lying on the ground, subject to severe mechanical abuse. All of these conditions spell danger.

All 125-volt, single-phase, 15-, 20-, and 30-ampere receptacle outlets that are not part of the permanent wiring of the building and that will be used by the workers on the construction site must be GFCI protected, *590.6(A)(1)*.

All 125-volt, single-phase, 15-, 20-, and 30-ampere receptacle outlets that are part of the actual permanent wiring of a building and are used by personnel for temporary power are also required to be GFCI protected *NEC 590.6(A)(2)*.

Receptacles on 15 kW or less Portable Generators. All 125-volt and 125/250-volt, single-phase, 15-, 20-, and 30-ampere receptacle outlets that are a part of a 15-kW or smaller portable generator shall have listed ground-fault circuit-interrupter protection for personnel. All 15- and 20-ampere, 125- and 250-volt receptacles, including those that are part of a portable generator, used in a damp or wet location shall comply with *406.9(A)* and *(B)*. Listed cord sets or devices incorporating listed ground-fault circuit-interrupter protection for personnel identified for portable use shall be permitted for use with 15 kW or less portable generators manufactured or remanufactured prior to January 1, 2011.

It should be clear to all users of the *NEC* that the Code Panel responsible for this article is serious about ensuring that people who use electrical power on construction sites have GFCI protection.

For receptacles *other* than 125-volt, single-phase, 15-, 20-, and 30-ampere receptacles, there are two options for providing personnel protection. In *590.6(B)(1)*, GFCI protection is one of the choices. In *590.6(B)(2)*, a written *Assured Equipment Grounding Conductor Program* is the second choice. This program is required to be continuously enforced on the construction site. A designated person must keep a

FIGURE 6-24 A temporary power box used on construction sites. This particular type is connected with a 50-ampere, 120/240-volt power cord. It has six 20-ampere, GFCI-protected, 125-volt receptacles plus additional 30-ampere and 50-ampere Twist-Lock receptacles.

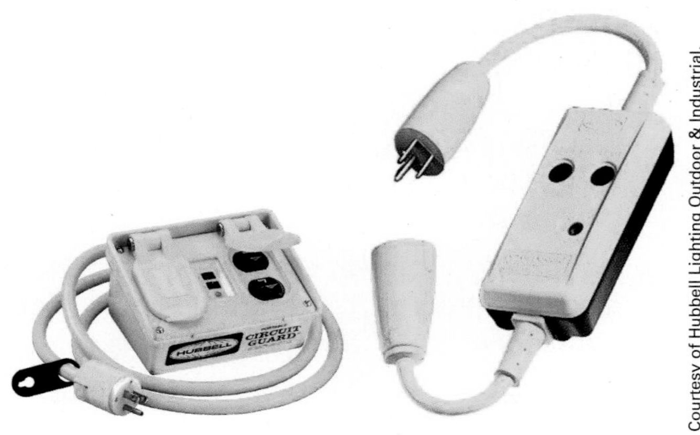

FIGURE 6-25 Cord-and-plug-connected portable GFCI devices are easy to carry around. They can be used anywhere when working with electrical tools and extension cords. These devices operate independently and are not dependent on whether the branch circuit is GFCI protected. These portable GFCI in-line cords are relatively inexpensive and should always be used when using portable electric tools.

written log, ensuring that all electrical equipment is properly installed and maintained according to the applicable requirements of *250.114*, *250.138*, *406.3(C)*, and *590.4(D)*. Because this option is difficult to enforce, many authorities having jurisdiction (AHJs) require GFCI protection, as stated in *590.6(B)(1)*.

Figure 6-24 is a listed manufactured power outlet for providing GFCI-protected temporary power on construction sites. It meets the requirements of *590.6(A)*.

Most electricians and other construction site workers carry their own listed portable ground-fault circuit-interrupter cord sets to be sure they are protected against shock hazard. These are shown in Figure 6-25 and also meet the requirements of *590.6(A)(2)* and *(A)(3)*.

Portable GFCI devices are easy to carry around. They have "open neutral" protection should the neutral conductor in the circuit supplying the GFCI open for whatever reason.

Portable GFCI devices are available with manual reset, which is advantageous should a power outage occur or if the GFCI is unplugged, so that equipment (drills, saws, etc.) will not start up again when the power is restored. This could cause injury to anyone using the equipment when the power is restored.

Portable devices are also available that reset automatically. These should be used on lighting, engine heaters, water (sump) pumps, and similar equipment where it would be advantageous to have the equipment start up as soon as power is restored.

Temporary Power for Construction Sites

In many parts of the country, it is common for the electrical contractor to furnish and install a panel for all trades to use at a construction site of a new home. Figure 6-26 shows a typical pedestal that might be secured to a pole or set in the ground. It contains a meter base, a main breaker, and GFCI branch breakers for 120-volt receptacles.

IMMERSION DETECTION CIRCUIT INTERRUPTERS (IDCIs) AND APPLIANCE LEAKAGE CURRENT INTERRUPTERS (ALCIs)

Major culprits for electrical shock in the home are personal grooming appliances. The worst offender was found to be handheld hair-drying appliances, most commonly used near a washbasin or, even worse, next to a water-filled bathtub with a person in it.

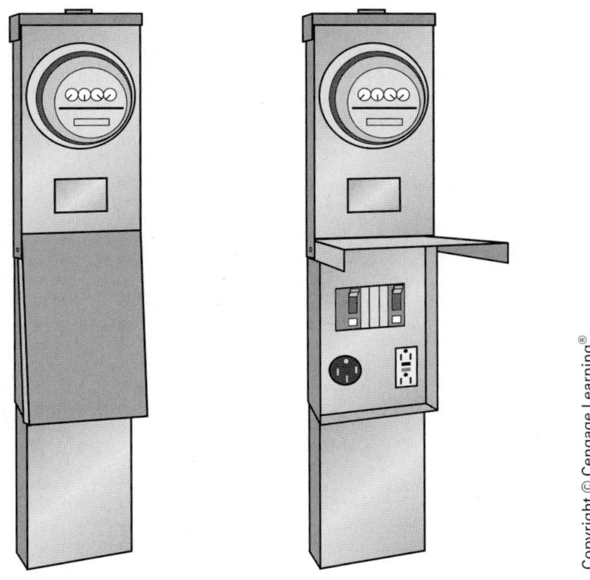

FIGURE 6-26 An example of a temporary construction site combination meter, breakers, and receptacles. The breakers and receptacles are under the lift-up cover.

Accidentally dropping or pulling the appliance into the tub can be deadly.

Because older homes probably do not have GFCI receptacles in bathrooms, *immersion detection circuit interrupters (IDCIs)* and *appliance leakage current interrupters (ALCIs)* were introduced to provide people with shock protection.

IDCI- and ALCI-protected appliances are easily recognized by the large attachment plug cap on the cord. IDCI and ALCI devices are an integral part of the attachment plug cap on handheld hair-drying appliances such as electric hair dryers, stylers, and heated air combs. These devices might be manually resettable or nonresettable.

The operation of an IDCI or ALCI device depends on a "third" wire probe in the cord, which is connected to a sensor inside the appliance. This probe detects leakage current in the range of 4 to 6 milliamperes when conductive liquid (e.g., water) enters the appliance and makes contact with any live part inside the appliance and the internal "sensor." IDCI and ALCI devices cut off power to the appliance regardless of whether the appliance switch is in the ON or OFF position.

UL requires that the plug cap be marked: *Warning: To Reduce the Risk of Electric Shock, Do Not Remove, Modify, or Immerse This Plug.*

Never replace the special attachment plug cap on these appliances with a standard plug cap. That would make the appliance unsafe!

IDCIs and ALCIs are a requirement of *NEC 422.41* and UL Standard 859.

SURGE PROTECTIVE DEVICES (SPDs)

Surge Protective Devices (SPDs) were previously referred to as Transient Voltage Surge Supressors (TVSSs). You may find this designation on older equipment.

The *Code* requirements for these devices that are permanently installed are found in *Article 285* in the *NEC*. Always install SPD devices that are listed under the requirements of UL Standard 1449. Carefully read and follow the instructions furnished with the SPD.

In today's homes, we find many electronic appliances (television, stereo, personal computer, fax equipment, word processor, VCR, VDR, CD player, digital stereo equipment, microwave oven), all of which contain many sensitive, delicate electronic components (printed circuit boards, chips, microprocessors, transistors, etc.).

Voltage transients, called *surges* or *spikes*, can stress, degrade, and/or destroy these components. They can cause loss of memory in the equipment or "lock up" the microprocessor.

The increased complexity of electronic integrated circuits makes this equipment an easy target for "dirty" power that can and will affect the performance of the equipment.

Voltage transients cause abnormal current to flow through the sensitive electronic components. This energy is measured in joules. A *joule* is the unit of energy when 1 ampere passes through a 1-ohm resistance for 1 second (like wattage, only on a much smaller scale).

Line surges can be line to neutral, line to ground, and line to line.

Transients

Transients are generally grouped into two categories:

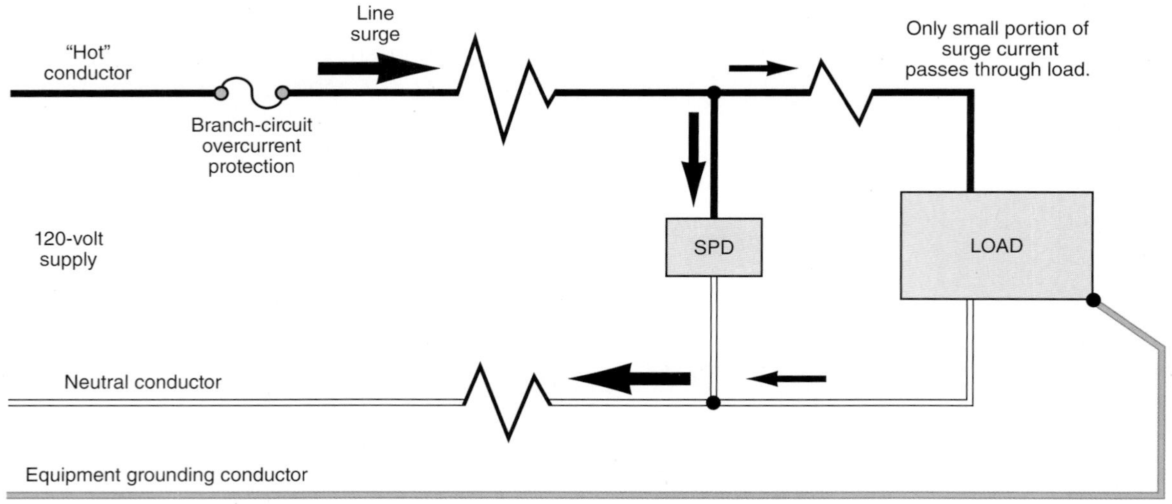

"Hot" conductor
Line surge
Only small portion of surge current passes through load.
Branch-circuit overcurrent protection
120-volt supply
SPD
LOAD
Neutral conductor
Equipment grounding conductor

FIGURE 6-27 A surge protective device absorbs and bypasses (shunts) surge currents around the load. The MOV dissipates the surge in the form of heat.

- **Ring wave:** These transients originate within the building and can be caused by copiers, computers, printers, HVAC cycling, spark igniters on furnaces, water heaters, dryers, gas ranges, ovens, motors, and other inductive loads.
- **Impulse:** These transients originate outside of the building and are caused by utility company switching, lightning, and so on.

To minimize the damaging results of these transient line surges, service equipment, panelboards, load centers, feeders, branch circuits, and individual receptacle outlets can be protected with a SPD, as shown in Figure 6-27.

A SPD contains one or more metal-oxide varistors (MOVs) that clamp the transients by absorbing the major portion of the energy (joules) created by the surge, allowing only a small, safe amount of energy to enter the actual connected load.

The MOV clamps the transient in times of less than 1 nanosecond, which is one billionth of a second, and keeps the voltage spike passed through to the connected load to a maximum range of 400 to 500 peak volts.

Typical SPD devices for homes are available as an integral part of receptacle outlets that mount in the same wall boxes as regular receptacles. They look the same as a normal receptacle and may have an audible alarm that sounds when an MOV has failed. The SPD device may also have a visual indication, such as a light-emitting diode (LED) that glows continuously until an MOV fails. Then the LED starts to flash on and off, as in Figure 6-28.

SPD devices are available in plug-in strips (Figure 6-29) and as part of desktop computer hardware power direction stands commonly placed under a monitor for plugging in a monitor, printer, scanner, and other peripheral equipment. These minimize the accumulation of the multitude of cords under and behind the equipment.

A SPD on a branch circuit provides surge suppression for all of the receptacles on the same

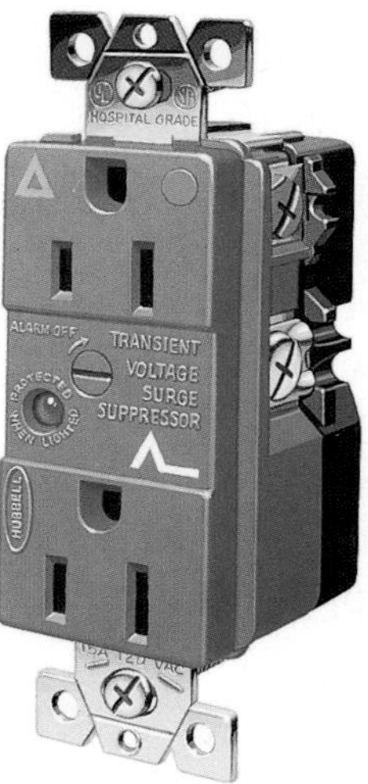

FIGURE 6-28 Surge protective device in the form of a receptacle.

FIGURE 6-29 Surge protective device included in a relocatable power tap (plug strip).

circuit. These surges are classified as Category A low-level surges.

Whole-House surge protectors are available that offer surge protection for the entire house. See Figure 6-30. These are generally available at electrical supply houses and are installed at the main electrical panel of a home.

When a Whole-House surge protector is installed, it is still a good idea to install surge protectors at the plug-in locations to more closely protect against low-level surges at the computer or other delicate electronic equipment locations.

Noise

Electromagnetic "noise," not to be confused with audible noise, is external interference that can be generated in varying degrees by all types of electrical and electronic equipment. You might recognize noise as snow on a TV screen or as humming/buzzing/static on a radio or telephone. Some typical sources of noise are fluorescent lamps and ballasts, motors running or being switched ON and OFF, loose electrical connections, or improper grounding. Even equipment running in the area such as X-ray equipment can be the culprit.

Although low-level noise is not physically damaging to electronic equipment, computers can lose memory, perform wrong calculations, have program malfunction, or lose Internet connection.

Noise comes from electromagnetic interference (EMI) and radio frequency interference (RFI).

EMI is usually caused by ground currents of very low values from such things as motors, utility switching loads, and lightning, and is transmitted through metal conduits.

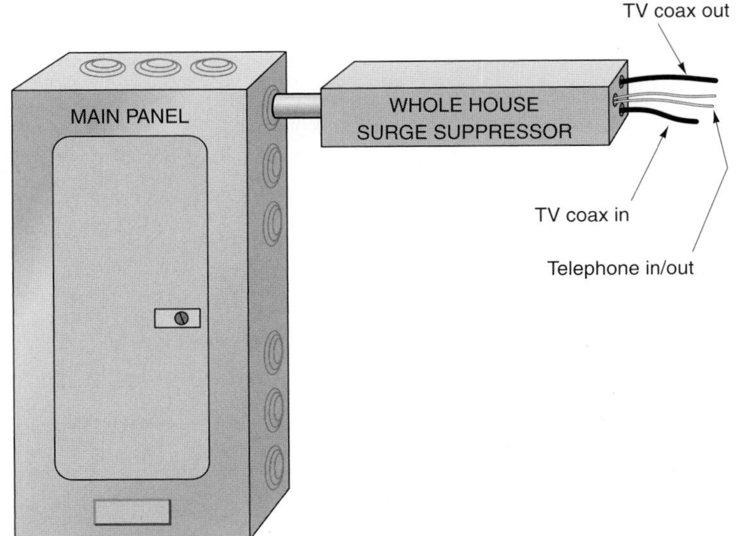

FIGURE 6-30 A Whole-House transient voltage surge suppressor installed next to the main service panel. *Article 285* in the *NEC* requires that these devices be listed and have their short-circuit rating marked on the device. Be sure to follow the manufacturer's installation and connection instructions.

RFI is noise, like the buzzing heard on a car radio when driving under a high-voltage transmission line. This interference "radiates" through the air from the source. Noise can be radiated or conducted from one residence to another. In commercial buildings, this radiated noise can be picked up by the grounding system of the building.

Consumer and commercial products today must meet Federal Communication Commission requirements for emission of electromagnetic and electronic noise so as not to interfere with communications equipment. Note: A cell phone has very limited amount of transmitting power.

In commercial work, isolated ground receptacles are installed in computer rooms to minimize noise.

In residential work, isolated ground receptacles are rarely, if ever, used.

Various types of electronic noise eliminators are available from electronic stores that filter out and virtually eliminate undesirable interference.

Here are some related UL Standards:

- UL 498 Attachment Plugs and Receptacles
- UL 943 Ground-Fault Circuit Interrupters
- UL 1283 Electromagnetic Interference Filters
- UL 1363 Relocatable Power Taps
- UL 1449 Transient Voltage Surge Suppressors
- UL 1644 Immersion Detection Circuit Interrupters
- UL 1699 Arc-Fault Circuit Interrupters

REVIEW

1. Explain the operation of a ground-fault circuit interrupter. Why are GFCI devices used? Where are GFCI receptacles required? ______

2. Residential GFCI devices are set to trip a ground-fault current above milliamperes. ______

3. Where must GFCI receptacles be installed in residential garages? ______

4. The *Code* requires GFCI protection for certain receptacles in the kitchen. Explain where these are required. ______

5. Is it a *Code* requirement to install GFCI receptacles in a fully carpeted, finished recreation room in the basement? ______

6. A homeowner calls in an electrical contractor to install a separate circuit in the basement (unfinished) for a freezer. Is a GFCI receptacle required? ____________________

7. GFCI protection is available as (a) branch-circuit breaker GFCI, (b) feeder circuit-breaker GFCI, (c) individual GFCI receptacles, and (d) feed-through GFCI receptacles. In your opinion, for residential use what type would you install? ________________

8. Extremely long circuit runs connected to a GFCI branch-circuit breaker might result in: (Circle the correct answer.)

   a. nuisance tripping of the GFCI.

   b. loss of protection.

   c. the need to reduce the load on the circuit.

9. If a person comes in contact with the "hot" and grounded conductors of a 2-wire branch circuit that is protected by a GFCI, will the GFCI trip OFF? Why? ____________

   ____

   ____

   ____

   ____

10. What might happen if the line and load connections of a feed-through GFCI receptacle were reversed? ____

    ____

    ____

    ____

    ____

11. May a GFCI receptacle be installed as a replacement in an old installation where the 2-wire circuit has no equipment grounding conductor? ____________________

12. What two types of receptacles may be used to replace a defective receptacle in an older home that is wired with knob-and-tube wiring where no equipment grounding means exists in the box? ____

    ____

13. You are asked to replace a receptacle. On checking the wiring, you find that the wiring method is conduit and that the wall box is properly grounded. The receptacle is of the older style 2-wire type that does not have a grounding terminal. You remove the old receptacle and replace it with: (Circle the correct answer.)

    a. the same type of receptacle as the type being removed.

    b. a receptacle that is of the grounding type.

    c. GFCI receptacle.

14. What color are the terminals of a standard grounding-type receptacle? ______________

    ____

    ____

15. What special shape and color are the grounding terminals of receptacle outlets and other devices? ____

16. Construction sites can be dangerous because of the manner in which extension cords, portable electrical tools, and other electrical equipment are used and abused. To reduce this hazard, *590.6(A)* of the *NEC* requires that all 125-volt, single-phase, 15-, 20-, and 30-ampere receptacle outlets used for temporary wiring during construction must be ____________________ protected.

17. In your own words, explain why the *Code* does not require certain receptacle outlets in kitchens, garages, and basements to be GFCI protected. ____________________

18. Circuit A1 supplies GFCI receptacles ① and ②. Circuit A2 supplies GFCI receptacles ③ and ④. Receptacles ① and ③ are feed-through type. Using colored pencils or marking pens, complete all connections.

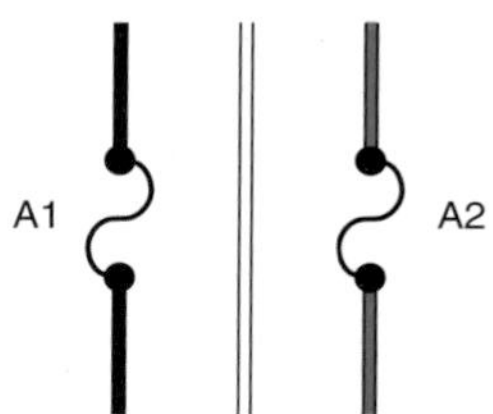

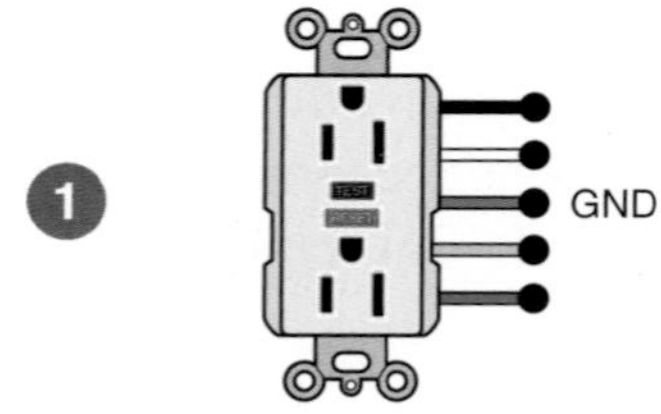

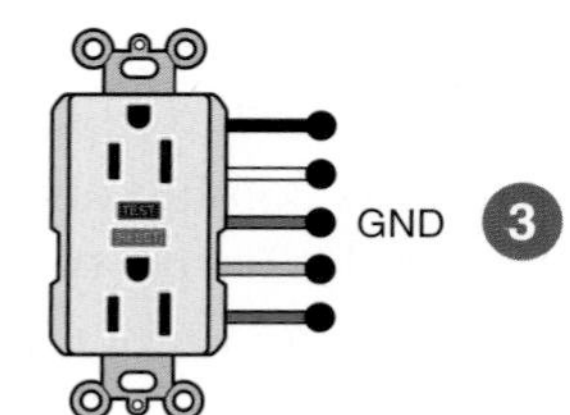

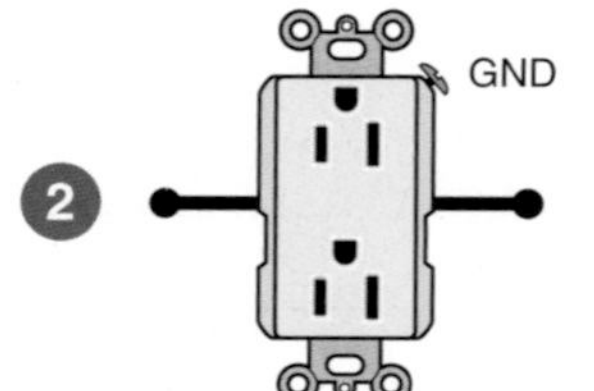

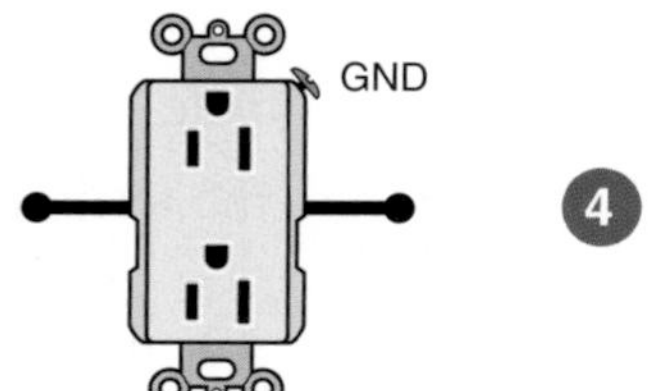

19. The term *SPD* is becoming quite common. What do the letters stand for? ________

20. Transients (surges) on a line can cause spikes or surges of energy that can damage delicate electronic components. A SPD device contains one or more ________ that bypass and absorb the energy of the transient.

21. Undesirable "noise" on a circuit can cause computers to lock up, lose their memory, and/or cause erratic performance of the computer. This noise does not damage the equipment. The two types of this noise are EMI and RFI. What do these letters mean?

22. Can SPD receptacles be installed in standard device boxes? ________

23. Some line transients are not damaging to electronic equipment but can cause the equipment to operate improperly. The effects of these transients can be minimized by installing ________.

24. Briefly explain the operation of an immersion detection circuit interrupter (IDCI).

25. What range of leakage current must trip an IDCI? ________

26. What amount of current flowing through a male human being will cause muscle contractions that will keep him from letting go of the live wire? ________ milliamperes

27. Other than for a few exceptions, the *NEC* requires that all systems having a voltage of over ________ volts shall be grounded.

28. In an old house, an existing nongrounding non-GFCI-type receptacle in a bathroom needs to be replaced. According to *406.4(D)(3)*, this replacement receptacle shall be: (Circle the correct answer.):

    a. a nongrounding-type receptacle.

    b. a GFCI-type receptacle or be GFCI protected in some other way, such as changing the breaker that supplies the receptacle to a GFCI type.

29. Do all receptacle outlets in kitchens have to be GFCI protected? Explain. ________

30. *NEC 210.12(A)* requires AFCI protection for the branch-circuit wiring that supplies all electrical outlets in specific rooms in dwellings. The branch circuits serving which areas, locations, or rooms in dwellings are not required to have AFCI protection?

____

____

____

31. When installing GFCI and AFCI circuit breakers in a panel, it is permissible to connect them to a multiwire branch circuit. (True) (False). (Circle the correct answer.)

32. The technologies of GFCI and AFCI protection are compatible and will function correctly if installed on the same branch circuit. (True) (False) (Circle the correct answer.)

33. Only the receptacles in the kitchen that are required to have GFCI protection such as those that serve counter surfaces are required to have AFCI protection. Other receptacles in the kitchen are not required to have AFCI protection. (True) (False) (Circle the correct answer.)

34. A switch that is located inside a bedroom for an outdoor light does not require AFCI protection since the switch is not an "outlet" as defined in *NEC Article 100*. (True) (False) (Circle the correct answer.)

35. Which of the following installations for a kitchen waste disposal that operates on a 15-ampere, 120-volt branch circuit requires AFCI protection? (Circle the correct answer.)

    a. The disposal operates by a switch that is installed above the kitchen counter.

    b. The disposal operates by a switch that is installed under the counter next to the disposal.

    c. Both (a) and (b)

    d. Neither (a) nor (b).

36. The receptacle outlets in the garage are required to have both GFCI and AFCI protection. (True) (False) (Circle the correct answer.)

37. Two 125-volt receptacles on a 20-ampere, 120-volt branch circuit are installed in the laundry room. One (1) is for the washer and is within 6 ft of the laundry sink and the other (2) is on a wall more than 6 ft from the sink. The receptacles are required to have: (Circle the correct answer.)

    a. GFCI and AFCI protection for both (1) and (2)

    b. GFCI and AFCI protection for (1) and only GFCI protection for (2)

    c. GFCI and AFCI protection for (2) and only AFCI protection for (1)

    d. GFCI and AFCI protection for (1) and only AFCI protection for (2)

38. In an old house that has two-wire branch circuits (without an equipment grounding conductor), it is okay to install a GFCI receptacle for a microwave oven in the kitchen without also installing an equipment grounding conductor. (True) (False) (Circle the correct answer.)

CHAPTER 7

Luminaires, Ballasts, and Lamps

OBJECTIVES

After studying this chapter, you should be able to

- understand luminaire terminology, such as Type IC and Type Non-IC.
- understand the *NEC* requirements for installing and connecting surface and recessed luminaires.
- realize that thermal insulation may have to be kept away from recessed luminaires.
- understand thermal protection requirements for recessed luminaires.
- know how to use "fixture whips."
- understand energy-saving ballasts and lamps.
- understand what a Class P ballast is.

WHAT IS A LUMINAIRE?

The *NEC*® defines a *luminaire* as A complete lighting unit consisting of a light source such as lamp or lamps, together with the parts designed to position the light source and connect it to the power supply.* *Luminaire* is the international term for "lighting fixture" and is used throughout the *NEC*.

TYPES OF LUMINAIRES

There are literally thousands of different types of luminaries from which to choose to satisfy certain needs, wants, desires, space requirements, and, last but not least, price considerations.

Note in Table 7-1 that whether the luminaire is incandescent or fluorescent, the basic categories are surface mounted, recessed mounted, and suspended ceiling mounted.

The *Code* Requirements

Article 410 sets forth the requirements for installing luminaires. The electrician must "Meet *Code*" with regard to mounting, supporting, grounding, live-parts exposure, insulation clearances, supply conductor types, maximum lamp wattages, and so forth.

Probably the two biggest contributing factors to fires caused by luminaries are installing lamp wattages that exceed that for which the luminaire has been designed, and burying recessed luminaires under thermal insulation if the luminaire has not been designed for such an installation.

TABLE 7-1

Mountings for basic categories of luminaires.

FLUORESCENT	INCANDESCENT
• Surface	• Surface
• Recessed	• Recessed
• Suspended Ceiling	• Suspended Ceiling

Nationally Recognized Testing Laboratories (NRTLs) test, list, and label luminaires that are in conformance with the applicable UL safety standards. Always install luminaires that bear the label from a qualified NRTL.

In addition to the *NEC*, the UL *Guide Information for Electrical Equipment* (*White Book*), and manufacturers' catalogs and literature are excellent sources of information about luminaires.

Read the Label

NEC 110.3(B) states that Listed or labeled equipment shall be installed and used in accordance with any instructions included in the listing or labeling.*

It is important to carefully read the label and any instructions furnished with a luminaire. Most *Code* requirements can be met by simply following this information.

Here are a few examples of label and instruction information:

- Maximum lamp wattage
- Type of lamp
- For supply connections, use wire rated for at least __________°C
- Type-IC
- Type Non-IC
- Suitable for damp, dry, or wet locations
- Thermally protected

Installing and Connecting Luminaires

The circuit conductors in a wall or ceiling box where luminaires are to be installed are usually

- white—the "identified" grounded conductor, and
- black—the ungrounded "hot" conductor. A "hot" switch leg might also be red or another color, but never white or green.

*Reprinted with permission from NFPA 70–2014.

Most surface-mounted luminaires will have a black and a white conductor in the canopy, making it easy to match these conductors to the circuit conductors in the box—white to white, black to black.

SAFETY ALERT

Be certain that the grounded (neutral) conductor is connected to the screw shell of lampholders. This protects persons from a shock hazard when installing or removing lamps in luminaires. A portion of the metal screw shell of lamps can be touched while inserting or removing a lamp in a luminaire. See *NEC 200.10(C)* for requirements for screw shells and *200.10(D)* for luminaires that have wire leads.

Chain-suspended luminaires usually come with a flexible, flat parallel conductor cord that weaves through the links of the chain. In a cord, it's a little more difficult to make a distinction between the "hot" conductor and the identified conductor.

To ensure proper polarity when making up the cord connections, generally connect as follows:

- The conductor with round insulation connects to the black "hot" circuit conductor.
- The conductor with grooved or raised insulation is the identified conductor that connects to the white circuit conductor.
- In some instances, the identified conductor will be tinned so it will have a silver color.
- The bare equipment grounding conductor (EGC) from the luminaire connects to the green hexagon-shaped screw in a metal electrical box or on the luminaire's mounting bar. Nonmetallic boxes have a bare EGC from the nonmetallic-sheathed cable(s) to which the luminaire's bare EGC is connected.

Surface-Mounted Luminaires

These are easy to install. It is simply a matter of following the manufacturers' instructions furnished

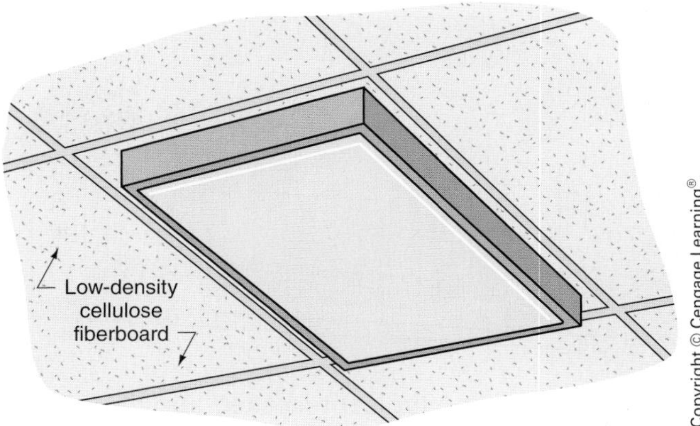

FIGURE 7-1 If they are to be mounted on *low-density cellulose fiberboard,* surface-mounted fluorescent luminaires must be marked "Suitable for Surface Mounting on Low-Density Cellulose Fiberboard."

with the luminaires. The luminaire is attached to a ceiling outlet box or wall outlet box using luminaire studs, hickeys, bar straps, or luminaire extensions. Do not exceed the maximum lamp wattage marked on the luminaire.

Special attention must be given when installing surface-mounted luminaires on low-density cellulose fiberboard, Figure 7-1. Because of the potential fire hazard, *410.136(B)* states that fluorescent luminaires that are surface mounted on this material must be spaced at least 1½ in. (38 mm) from the fiberboard surface unless the fixture is marked "Suitable for Surface Mounting on Low-Density Cellulose Fiberboard."

The *Informational Note* to *410.136(B)* explains combustible low-density cellulose fiberboard as sheets, panels, and tiles that have a density of 20 lb/ft^3 (320 kg/m^3) or less that are formed of bonded plant fiber material. It does not include fiberboard that has a density of over 20 lb/ft^3 (320 kg/m^3) or material that has been integrally treated with fire-retarding chemicals to meet specific standards. Solid or laminated wood does not come under the definition of *combustible low-density cellulose fiberboard.*

To obtain fire-resistance ratings, most acoustical ceiling panels today are made from mineral wool fibers or fiberglass.

FIGURE 7-2 Typical recessed luminaire.

Recessed Luminaires

Recessed luminaires, as shown in Figure 7-2, have an inherent heat problem. They must be suitable for the application and must be installed properly.

It is absolutely essential for the electrician to know early in the roughing-in stages of wiring a house what types of luminaires are to be installed. This is particularly true for recessed-type luminaires. To ensure that such factors as location, proper and adequate framing, and possible obstructions have been taken into consideration, the electrician must work closely with the general building contractor, carpenter, plumber, heating contractor, the thermal insulation installer, and the other building trades.

When working with recessed luminaires, you will come across unique terms such as

- ***Type Non-IC:*** See Figure 7-3(A). This type of luminaire is marked "Type Non-IC" and is for installations in noninsulated ceilings. Where installed in an insulated ceiling, thermal insulation must be kept at least 3 in. (75 mm) away from any part of the luminaire.
- ***Type IC:*** See Figure 7-3(B). This type of luminaire is marked "Type IC" and is permitted to be buried in thermal insulation.
- ***Inherently Protected:*** This type of luminaire is designed so that the outside surfaces of the luminaire do not exceed temperatures greater than 194°F (90°C)—even if buried in thermal insulation, if mislamped (a lamp type not specified on the product), or if overlamped (lamp wattage exceeds the maximum wattage rating marked on the product). It is marked "Inherently Protected."

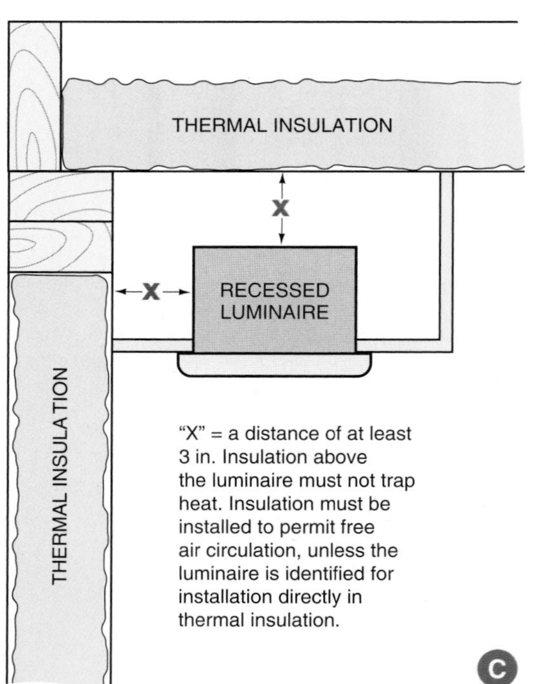

FIGURE 7-3 (A) is a Type Non-IC recessed luminaire that requires clearance between the luminaire and the thermal insulation. (B) is a Type IC recessed luminaire that may be completely covered with thermal insulation. (C) shows the required clearances from thermal insulation for a Type Non-IC luminaire. See *410.116*.

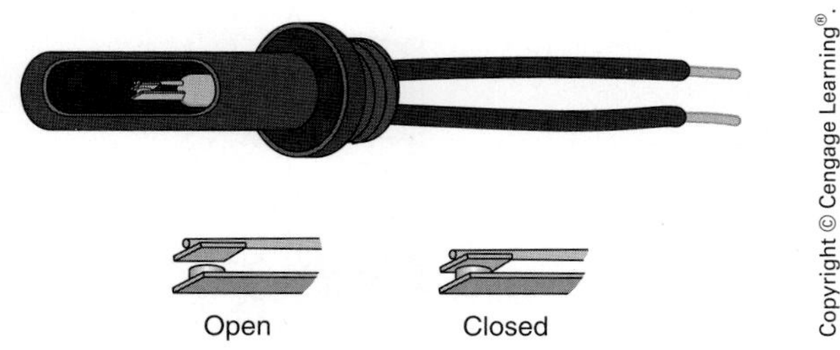

FIGURE 7-4 One type of recessed luminaire thermal protector.

When a Type Non-IC recessed luminaire gets overheated because of overlamping, mislamping, or being too close to thermal insulation, the thermal protector (Figure 7-4) trips off. When the luminaire cools down, it comes on again. This on–off cycling will repeat until the problem is corrected. These luminaires are marked "Blinking Light of This Thermally Protected Luminaire May Indicate Overheating."

In the trade, recessed luminaire housings are commonly referred to as "cans."

Installing recessed luminaire housings requires more attention than simple surface-mounted luminaires. The *Code* requirements for the installation of recessed luminaires are found in *410.110* through *410.122*. UL Standard 1571 covers recessed incandescent luminaires.

Figure 7-3(C) is a sketch that summarizes the minimum clearances for a Type Non-IC recessed luminaire that is not listed for direct burial in thermal insulation, *410.116(B)*.

Recessed luminaires are available for both new work and remodel work. Remodel work recessed luminaires can be installed from below an existing ceiling by cutting a hole in the ceiling, bringing power to the luminaire, making the electrical connections in the integral junction box on the luminaire, and then installing the housing through the hole.

Energy-Efficient Housings

Many states have residential energy-efficiency requirements, particularly for recessed luminaires. The most widely followed regulations are the State of California Title 24 requirements, and the State of Washington Restricted Air Flow Requirements.

Basically, these recessed "cans" have double-walled gasketed airtight housings that prevent heated or cooled air from escaping into attics and other unconditioned spaces in the house. They are listed for use in insulated ceilings and in direct contact with insulation.

Although energy-efficiency requirements generally are not applicable to new one- and two-family homes, they might have been adopted by your state or community. Check with your local electrical inspector and/or building official to see whether such laws are applicable in your area for one- and two-family dwellings.

Thermal Protection

To protect against the hazards of overheating, UL and the *NEC* in *410.115(C)* require that recessed luminaires be equipped with an integral thermal protector, as in Figure 7-4. These devices will cycle on and off repeatedly until the heat problem has been resolved. There are two exceptions to this requirement. Thermal protection is not required for recessed incandescent luminaires designed for installation directly in a concrete pour or for recessed incandescent luminaires that are constructed so as to not exceed "temperature performance characteristics" equivalent to thermally protected luminaires. These luminaires are so identified.

The Purpose of the Junction Box (Wiring Compartment) on a Recessed Luminaire Housing

Most recessed luminaires come equipped with a junction box (wiring compartment) that is an integral part of the luminaire housing. This is clearly shown in Figures 7-2 and 7-5. For most residential-type recessed luminaire installations, nonmetallic-sheathed cable or armored cable can be run directly into this junction box. These cables are required to have conductors rated 90°C. Check the marking on the luminaire for supply conductor temperature requirements.

Figure 7-6 shows the clearances normally needed between recessed luminaires and combustible framing members. This figure also summarizes the *NEC* requirements for connecting the supply

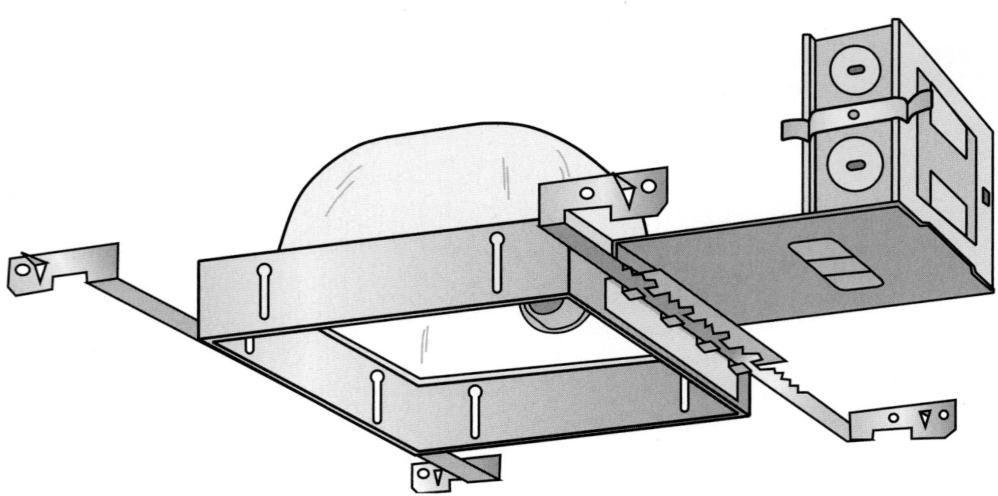

FIGURE 7-5 Roughing-in box of a recessed luminaire with mounting brackets and junction box (wiring compartment).

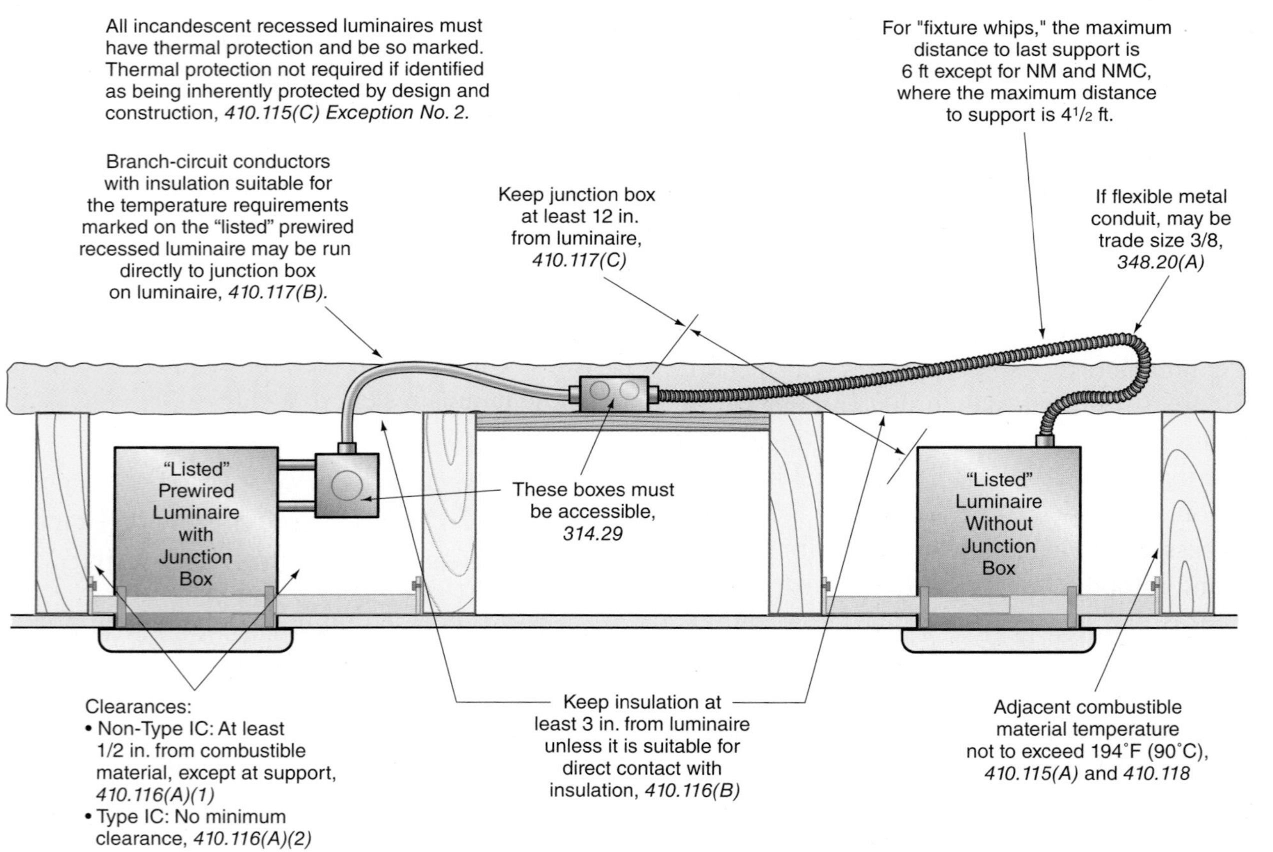

FIGURE 7-6 Requirements for installing recessed luminaires.

conductors to a recessed luminaire. Note in one case that the branch-circuit wiring is brought directly into the integral junction box on the luminaire. In the other case, a "fixture whip" is run from an adjacent junction box to the luminaire. In some instances, the luminaire comes with a flexible metal fixture whip, and in other situations the fixture whip is supplied by the electrician. Fixture whips are discussed later.

Maximum of ___ ___ AWG through branch-circuit conductors suitable for at least ___°C (___°F) permitted in a junction box (___ in ___ out).

FIGURE 7-7 A typical label found on a recessed luminaire specifying the maximum number and size of conductors permitted to be run through the wiring compartment on the luminaire.

In *410.21*, we find that Luminaires shall be of such construction or installed so that the conductors in outlet boxes shall not be subjected to temperatures greater than that for which the conductors are rated.*

NEC 410.21 further states that Branch-circuit wiring, other than 2-wire or multiwire branch circuits supplying power to luminaires connected together, shall not be passed through an outlet box that is an integral part of a luminaire unless the luminaire is identified for through-wiring.*

In *410.64*, we find that Luminaires shall not be used as a raceway for circuit conductors unless listed and marked for use as a raceway.*

Some recessed luminaires are marked "Identified for Through-Wiring," in which case branch-circuit conductors in addition to the conductors that supply the luminaire are permitted to be run through the outlet box or wiring compartment on the luminaire. A typical label will look like Figure 7-7.

The manufacturer of the luminaire, following UL requirements, will determine the maximum number, size, and temperature rating for the conductors entering and leaving the outlet box or wiring compartment, and will so mark the label.

Only those luminaires that have been tested for the extra heat generated by the additional branch-circuit conductors bear the label marking shown in Figure 7-7.

These fixtures have been tested for the added heat generated by the additional branch-circuit conductors.

Figure 7-8 shows three recessed luminaires, each with an integral outlet box. If these luminaires are marked "Identified for Through-Wiring," then it would be permitted to run conductors through the outlet boxes *in addition* to the conductors that supply the luminaires. If the "Identified for Through-Wiring" marking is not found on the luminaire, then it is a *Code* violation to run any conductors through the box other than the conductors that supply the luminaires that are "daisy chained."

Routing the Branch Circuit

Some electricians prefer to run the branch-circuit wiring to the junction box on the recessed luminaire, then drop the switch loop to the switch box location. Others prefer to run the branch circuit to the switch location, then to the junction box on the luminaire. Either way is acceptable.

Keep the requirement in *NEC 404.2(C)* in mind. This rule generally requires ▶The grounded circuit conductor for the controlled lighting circuit shall be provided at the location where switches control lighting loads that are supplied by a grounded general purpose branch circuit, for other than the following.◀* Several

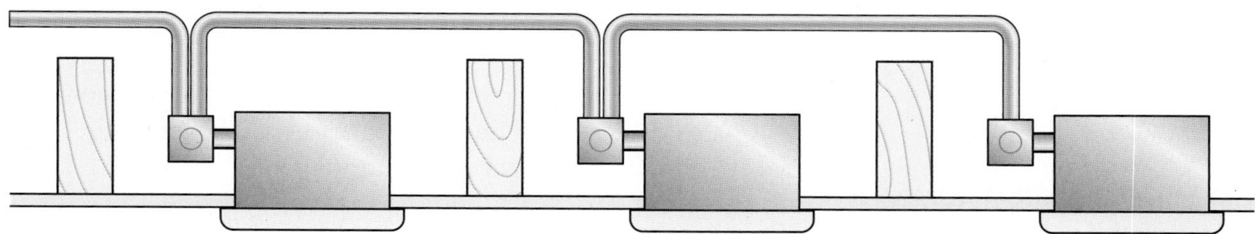

FIGURE 7-8 The only conductors permitted to run into or through the junction boxes on the recessed luminaires are those that supply the luminaires. Conductors other than those that supply the luminaires are permitted to run through the outlet boxes if the luminaire is marked "Identified for Through-Wiring." See *410.21* and *410.64*.

*Reprinted with permission from NFPA 70–2014.

exceptions are provided. This rule is extensively discussed in Chapter 5 of this text. As will be seen after reviewing the material, it will be simpler and save on the number of conductors required if the branch circuit wiring is routed through the switches rather than through the luminaire.

Recessed Luminaire Trims

There seems to be an endless choice of trims for recessed luminaires: eyeball, open, baffled, cone, pinhole, wall wash, and so on. Be very careful when choosing a trim for a recessed luminaire.

Recessed incandescent luminaires are listed by the NRTL with specific trims. Luminaire/trim combinations are marked on the label. Do not use trims that are not listed for use with the particular recessed luminaire. Installing a trim not listed for use with a specific recessed luminaire can result in overheating and possible on–off cycling of the luminaire's thermal protective device. Mismatching luminaires and trims is a violation of *110.3(B)* of the *NEC*, which states that Listed or labeled equipment shall be installed or used in accordance with any instructions included in the listing or labeling.*

Sloped Ceilings

Be very careful when selecting recessed luminaires. Conventional recessed luminaires are listed for installation in flat ceilings, not in sloped ceilings! Recessed luminaires intended for installation in a sloped ceiling are submitted to the NRTL specifically for testing in a sloped ceiling. Look for the marking "Sloped Ceiling." Instructions furnished with recessed luminaires provide the necessary information, such as being suitable for sloped ceilings having a 2/12, 6/12, or 12/12 pitch. Check the manufacturer's catalog and the NRTL listing.

Suspended Ceiling Lay-In Luminaires

Figure 7-9 illustrates luminaires installed in a typical suspended ceiling.

The recreation room of this residence has a "dropped" suspended acoustical paneled ceiling.

*Reprinted with permission from NFPA 70–2014.

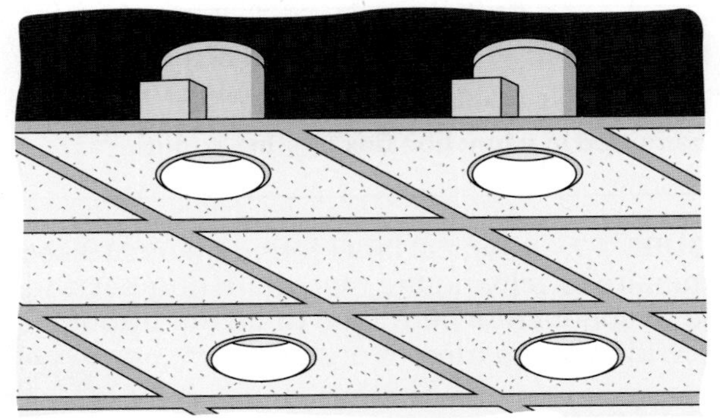

FIGURE 7-9 Suspended ceiling luminaires.

Luminaires installed in a dropped ceiling must bear a label stating "Suspended Ceiling Luminaire." These types of luminaires are not classified as recessed luminaires because in most cases there is a great deal of open space above and around them.

Support of Suspended Ceiling Luminaires

In *410.36(B)*, we find that when framing members of a suspended ceiling grid are used to support luminaires, all framing members must be securely fastened together and to the building structure. It is the responsibility of the installer of the ceiling grid to do this.

All lay-in suspended ceiling luminaires must be securely fastened to the ceiling grid members with bolts, screws, or rivets. Listed clips supplied by the manufacturer of the luminaire are also permitted to be used. Clips that meet this criteria are shown in Figure 17-1.

The logic behind all of the "securely fastening" requirements is that in the event of a major problem such as an earthquake or fire, the luminaires will not fall down and injure someone.

Connecting Suspended Ceiling Luminaires

The most common way to connect suspended ceiling lay-in luminaires is to complete all of the wiring above the ceiling using conventional wiring methods.

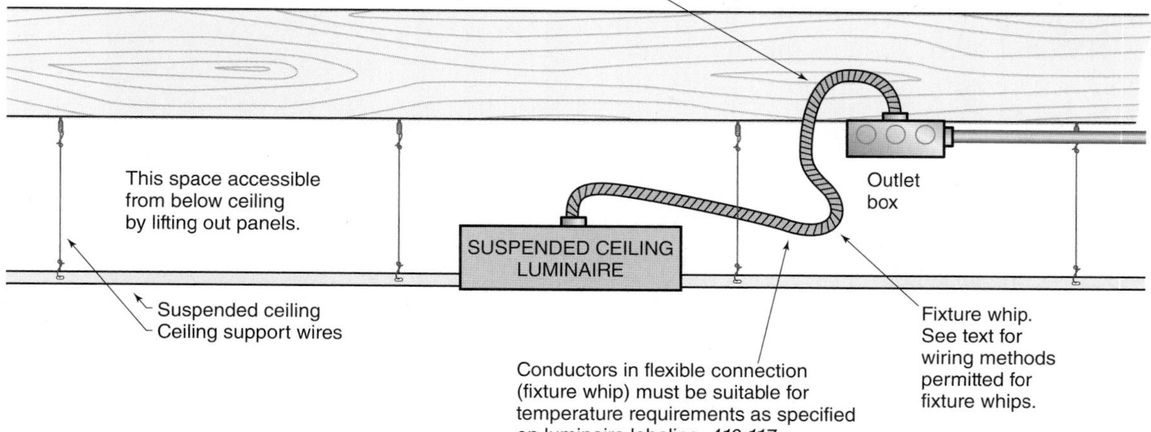

FIGURE 7-10 A suspended ceiling luminaire connected with a flexible fixture whip between an outlet box and the luminaire.

Then, from accessible outlet boxes strategically placed above the ceiling near the intended location of the lay-in luminaires, a flexible connection not more than 6 ft (1.8 m) long is made between the outlet boxes and the luminaires. See Figure 7-10. Lay-in ceiling tiles provides access to these outlet boxes.

In the electrical trade, these flexible connections are referred to as "fixture whips."

The *NEC* does not have a definition of a fixture whip.

Don't confuse a fixture whip with a tap. Most fixture whips in residential installations have conductors the same size as the branch-circuit conductors, so the whip is really an extension of the branch circuit, and by the *NEC* definition is not really a tap.

A tap is defined in *240.2* as a conductor that has overcurrent protection ahead of its point of supply that exceeds the maximum overcurrent protection for that conductor. This is quite common in commercial installations where the branch-circuit overcurrent protection is rated 20 amperes and the tap conductors are rated 15 amperes. Tap conductors must be in a suitable raceway or Type AC or MC cable of at least 18 in. (450 mm), but not more than 6 ft (1.8 m) in length.

A fixture whip might be

- armored cable (AC), *320.30(D)(3)*.
- metal-clad cable (MC), *330.30(D)(2)*.
- nonmetallic-sheathed cable (NM and NMC), *334.30(B)(2)*.
- flexible metal conduit (FMC): It is okay to use trade size ⅜, *348.30(A), Exceptions No. 3* and *No. 4*. If trade size ⅜ is used, length is not to exceed 6 ft (1.8 m), *348.20(A)(2)*.
- liquidtight flexible metallic conduit (LFMC): It is okay to use trade size ⅜, *350.30(A), Exceptions 3* and *4*. If trade size ⅜ is used, the length is not to exceed 6 ft (1.8 m), *350.20(A)*.
- liquidtight flexible nonmetallic conduit (LFNC): It is okay to use trade size ⅜, *356.30(2)*. If trade size ⅜ is used, the length is not to exceed 6 ft (1.8 m), *356.20(A)*. Where grounding is required, a separate equipment grounding conductor is required to be installed, *356.60*. Install according to *Article 250*. Size per *Table 250.122*.
- electrical nonmetallic-tubing (ENT), *362.30(A)*. Trade size smaller than ½ is not permitted, *362.20(A)*. Where grounding is required, a separate equipment grounding conductor is required

to be installed, *362.60*. Install according to *Article 250*. Size per *Table 250.122*.

Note: For the above wiring methods, the maximum distance to the last point of support is 6 ft (1.8 m) except for NM and NMC, where the maximum distance to the last point of support is 4½ feet (1.32 m).

Luminaire Grounding

Generally, all luminaires must be grounded, *410.42*. Assuming that the wiring is a metallic wiring method or nonmetallic-sheathed cable with an equipment grounding conductor, you have an acceptable equipment grounding connection at the lighting outlet. Grounding of a surface-mounted luminaire is easily accomplished by securing the luminaire to the properly grounded metal outlet box with the hardware provided with the luminaire. Metal outlet boxes have a tapped No. 10-32 hole in which to insert a green hexagonal grounding screw. The bare copper equipment grounding conductor in the nonmetallic-sheathed cable is usually terminated under this screw. Grounding and bonding is discussed in Chapter 5. If the outlet box is nonmetallic, the small bare equipment grounding conductor from the luminaire is connected to the equipment grounding conductor in the outlet box. For suspended ceiling luminaires, grounding of the luminaire is accomplished by using metallic fixtures whips or nonmetallic-sheathed cable with ground between the outlet box and the luminaire. Always follow the manufacturers' instructions for making the electrical connections.

What about replacing a luminaire in an older home? If there is no equipment grounding means in the outlet box, then a luminaire that is made of insulating material and has no exposed conductive parts must be used, *410.44 Exception No. 1*. *Exception No. 2* to *410.44(B)* allows a metallic replacement luminaire to be installed and connected to a system that does not have an equipment grounding conductor. In this case, a separate equipment grounding conductor must be installed in conformance with *250.130(C)*. This requirement is discussed in Chapter 6 of this text under the topic of replacing receptacles, so it is not repeated here. *Exception No. 3* to *410.44(B)* allows a metallic replacement luminaire to be installed and connected to a branch circuit that does not have an equipment grounding conductor if it is protected by GFCI device. Acceptable types of equipment grounding conductors are listed in *250.118*.

In Chapter 17, Figures 17-1 and 17-2 have additional information regarding recessed luminaire installations such as those in the recreation room of this residence.

Luminaires in Closets

There are special requirements for installing luminaires in closets. This is covered in Chapter 8.

Luminaires in Bathrooms

Because of the electrical shock hazards associated with electricity and water, there are special restrictions for luminaires installed in bathrooms. This is covered in Chapter 10.

FLUORESCENT BALLASTS AND LAMPS, INCANDESCENT LAMPS

The following is a brief introduction to a complex subject. For more information, check out lamp and ballast manufacturers' websites.

From an electrician's point of view, lamps and lightbulbs mean the same thing. Use any one of the words, and everyone will know what you mean.

Always read the label on a luminaire for the type and maximum wattage of the lamp to be installed in that luminaire.

Fluorescent Ballasts

An incandescent lamp contains a filament that has a specific "hot" resistance value when operating at rated voltage. When energized, the lamp provides the light output for which the lamp was designed.

A fluorescent lamp, on the other hand, cannot be connected directly to a circuit. It needs a ballast.

Fluorescent lamps do not have a filament running from end to end. Instead, they have a filament

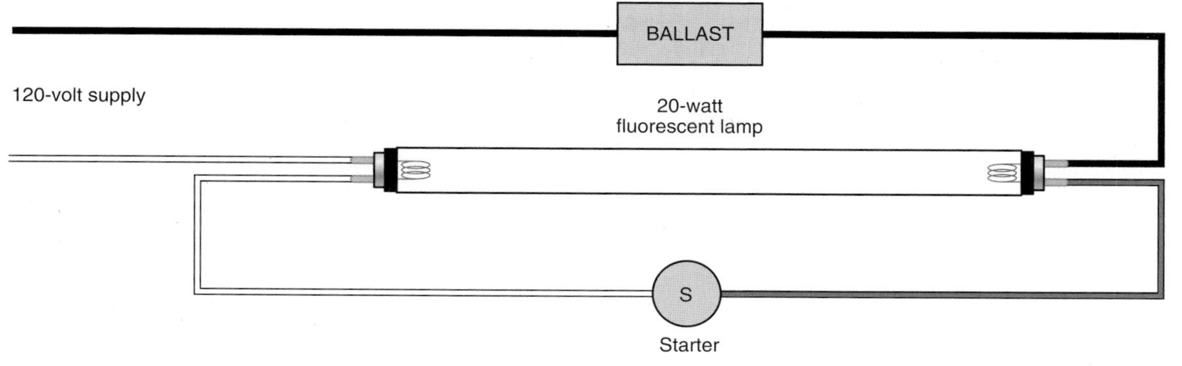

FIGURE 7-11 Simple series circuit.

at each end. The filaments are connected to a ballast, as in Figure 7-11. The ballast is needed to (1) control power, (2) control voltage to heat the filaments, (3) control voltage across the lamp to start the arc within the tube, and (4) limit the current flowing through the lamp. The arc is needed to ionize the gas and vaporize the droplets of mercury inside the lamp.

Without a ballast, the lamp would probably not start. But if it did start, it would "run away" with itself. The current flowing through the gases within the lamp would rise rapidly, destroying the lamp in a very short time.

Ballast Types

Preheat. Preheat ballasts are connected in a simple series circuit, Figure 7-11. They are easily identified because they have a starter. One type of starter is automatic and looks like a small roll of Lifesavers with two "buttons" on one end.

Another type of starter is a manual ON–OFF switch that has a momentary "make" position just beyond the ON position. When you push the switch on and hold it there for a few seconds, the lamp filaments glow. When the switch is released, the start contacts open, an arc is initiated within the lamp, and the lamp lights up.

Preheat lamps have two pins on each end.

Preheat lamps and ballasts are not used for dimming applications.

Rapid Start. Rapid-start ballasts are probably the most common type used today. Rapid-start ballasts/lamps do not require a starter. The lamps start in less than 1 second. For reliable starting, ballast manufacturers recommend that there be a grounded metal surface within ½ in. (12.7 mm) of the lamp and running the full length of the lamp, that the ballast be grounded, and that the supply circuit originates from a grounded system. T5 rapid-start lamps do not require a grounded surface for reliable starting.

Rapid-start lamps have two pins on each end.

Rapid-start lamps can be dimmed using a special dimming ballast. See Chapter 13.

Instant Start. Instant-start lamps do not require a starter. These ballasts provide a high-voltage "kick" to start the lamp instantly. They require special fluorescent lamps that do not require preheating of the lamp filaments. Because instant-start fluorescent lamps are started by brute force, they have a shorter life (as much as 40% less) than rapid-start lamps when older style magnetic ballasts are used. With electronic ballasts, satisfactory lamp life can be expected.

Instant-start lamps have one pin on each end.

Instant-start ballasts/lamps cannot be used for dimming applications.

Dimming Ballasts. Special dimming ballasts and dimmers are needed for controlling the light output of fluorescent lamps. Rapid-start lamps are used. Incandescent lamp dimmers cannot be used to control fluorescent lamps. An exception to this is that dimmers marked "Incandescent Only" can be used to dim compact fluorescent lamps. Dimming ballasts are discussed in Chapter 13.

Mismatching of Fluorescent Lamps and Ballasts

Be sure to use the proper lamp for a given ballast. Mismatching a lamp and ballast may result in poor starting and poor performance as well as shortened lamp and/or ballast life. The manufacturer's ballast and/or lamp warranty may be null and void.

T8 lamps are designed to be used interchangeably on magnetic or electronic rapid-start ballasts or electronic instant-start ballasts. Lamp life is reduced slightly when used with an instant-start ballast.

SAFETY ALERT

Do not insert spacers, washers, or shims between a ballast and the luminaire to make the ballast more quiet. This will cause the ballast to run much hotter and could result in shortened ballast life and possible fire hazard. Instead, replace the noisy ballast with a quiet, sound-rated one. Sometimes checking and tightening the many nuts, bolts, and screws of the luminaire will solve the problem.

Voltage

Operating a ballast at an over-voltage condition causes it to run hot and shortens its life.

Operating a ballast at an under-voltage situation can result in premature lamp failure and unreliable starting.

Most ballasts today will operate satisfactorily within a range of +5% to −7% of their rated voltage. The higher quality Certified Ballast Manufacturers Association (CBM) certified ballasts will operate satisfactorily within a range of ±10%.

Sound Rating

Most ballasts will hum, some more than others.

Ballasts are sound rated and are marked with letters "A" through "F." "A" is the quietest, and "F" is the noisiest. Look for an A or B sound rating for residential applications.

Magnetic ballasts (core and coil) hum when the metal laminations vibrate because of the alternating current reversals. This hum can be magnified by the luminaire itself, and/or the surface the luminaire is mounted on.

Electronic ballasts have little, if any, hum.

Energy Saving by Control

We have all heard the words "Shut off the lights when you leave the room." Sometimes this works—sometimes not.

This text is used in all 50 states in this country and in Canada. It is not the intent of this text to go into the detail of specific state and/or local energy requirements. Just be aware that new developments are happening relative to energy conservation, and you need to pay attention. As an example, the State of California recently put into place Title 24, Part 6. The intent is to reduce power consumption in new homes. The States of Washington and Wisconsin have stringent energy conservation laws. As time passes, other states will follow.

The Energy Policy Act of 2005 stipulates in Title XII the electricity that types of lamps are permitted based on lumens per watt (efficacy). This law in Section 1252 also calls for utilities to provide "time-based" metering for residential, commercial, and industrial customers. Customers then can vary their electrical demands based on the rates for different times of usage. You can learn more about this by visiting the government's website: http://energycommerce.house.gov/.

As you install residential lighting, you need to be familiar with the terms "efficacy" and "lumens per watt." Simply stated, efficacy is measured in lumens per watt.

This law simply requires that lamps in permanently installed luminaires

- have high-efficacy rating (high lumens per watt), or
- be controlled by occupancy sensors, or
- have a combination of lamp efficacy, occupancy sensor control, and/or dimmers.

Cord-connected floor or table lamps are not governed by current energy conservation laws.

Most fluorescent lamps with energy-saving ballasts, compact fluorescent lamps (CFLs), and halogen lamps meet this high-efficacy requirement. Conventional incandescent lamps do not. See Table 7-2 to view the comparison of lumens per watt for different types of lamps.

The following chart shows what is required to meet the lamp requirement.

LAMP WATTAGE	LAMP EFFICACY
Less than 15 watts	40 lumens per watt
15–40 watts	50 lumens per watt
Over 40 watts	60 lumens per watt

TABLE 7-2

Comparison of various lamps' characteristics. These characteristics are typical and vary by manufacturer. Always verify application with lamp and luminaire manufacturer's label, literature, and installation instructions.

TYPE OF LAMP	LUMEN PER WATT	DIMMING	COLOR AND APPLICATION	LIFE (APPROX. HOURS)	TYPICAL SHAPES
Incandescent	14–18	Yes	Warm and natural. Great for general lighting. Brand names have various trade names for their lamps, such as Reveal, Soft White, etc.	500, 750, 1000, 1500, 3000 hours. Depends on type of lamp. Lamp life typically is based on operating the lamp an average of 3 hours of operation per start.	Standard, spots, floods, decorative, flame, tubes, globes, PAR (similar to standard spots and floods but stronger). Use rough service bulbs where there is vibration, like garage door openers and ceiling fans. Base types: candelabra, intermediate, medium, mogul.
Halogen	16–22 Are more efficient than conventional incandescent lamps. More lumen output per watt.	Yes	Brilliant white. Excellent for accent and task lighting. Are filled with halogen gas and floods and have an inner lamp, allowing the filament to run hotter (whiter).	2000–4000 hours. Lamp life typically is based on operating the lamp an average of 3 hours of operation per start.	PAR spots and floods, flame, crystal, mini-reflector spots. Base types: candelabra, intermediate, and medium. Can replace most incandescent lamps.
Fluorescent	T12 82 T8 92 T5 104	Yes, but only 40-watt rapid-start lamps using special dimming ballast. See Chapter 13.	Warm and deluxe warm white, cool and deluxe cool white, plus many other shades of white. Great for general lighting, like the Recreation Room in this residence. The higher the K rating, the cooler (whiter) the color rendition.	6000 to 24,000 hours. Average life with lamps turned off and restarted once every 12 operating hours.	Straight, U-tube, circular. Single pin and double pin.

(*continues*)

TABLE 7-2 *(continued)*

TYPE OF LAMP	LUMEN PER WATT	DIMMING	COLOR AND APPLICATION	LIFE (APPROX. HOURS)	TYPICAL SHAPES
Compact Fluorescent	45/60 Because of their high lumens per watt, CFLs can save $20 to $50 over their lifetime when compared to equivalent incandescent lamps. Produces more lumens per watts (e.g., 27-watt CFL produces 1800 lumens vs. 1750 for a 100-watt incandescent.) Uses about 80% less energy, and generates 90% less heat than equivalent incan-descent lamp.	Yes, if so marked. CFLs have integral ballasts.	Soft warm white.	10,000–15,000 hours. Can last 9 to 13 times longer than incandescent. Lamp life typically is based on operating the lamp an average of 3 hours of operation per start. Read instructions carefully for restrictions such as *Not for use with dimmers, electronic timers, occupancy sensors, photocells, or lighted switches. Do not use in recessed or totally enclosed luminaires.*	Twisted (spiral) tubes, folded tubes, globe. R20, R30, R40, PAR38, "A" shape (like a standard incandescent). The number of tube loops are referred to as twin, double twin, triple twin, quad twin. Base types: medium. Can replace most incandescent lamps.
Light-Emitting Diodes (LEDs)	50/60+ An LED lamp contains a cluster (array) of many individual LEDs to produce this lumen output.	Yes if so marked. Check manu-facturer's instructions and warnings.	Bright white and soft white. Excellent color rendition. Applications include from flashlights to Christmas light strings to home, office, and parking lot area lighting.	60,000 to 100,000 hours.	Shapes: Standard "A," candle, spots. Base types: Medium and candelabra. Can replace incandescent lamps.

The Bottom Line

In a nutshell, this amounts to the mandatory use of high-efficiency luminaires (airtight), high-efficiency lamps, and various levels of control such as occupancy sensors and dimmers.

Today, occupancy sensor switches are available that install the same way as conventional toggle switches. They provide manual ON–OFF as well as automatic shut-off when no motion is detected after a preset time. As soon as motion is detected, they turn the lamps on. See Figure 7-12.

To learn more about the California energy conservation law, check out http://www.energy.ca.gov/title24 and http://www.haloltg.com. Browse for the summary of the California Title 24 law. Other lighting manufacturers' websites should also be checked out.

Energy-Saving Ballasts

The market for magnetic (core and coil) ballasts is shrinking!

The National Appliance Energy Conservation Amendment of 1988, Public Law 100-357 prohibited manufacturers from producing ballasts having a power factor of less than 90%. Ballasts that meet or exceed the federal standards for energy savings are marked with a letter "E" in a circle. Dimming ballasts and ballasts designed specifically for residential use were exempted.

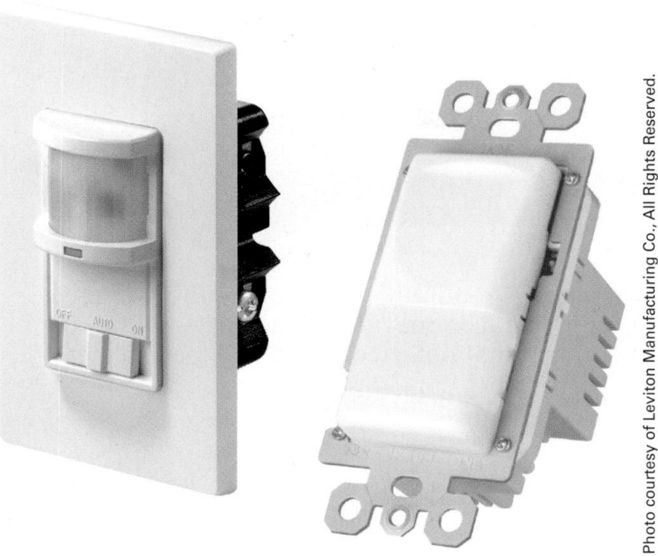

FIGURE 7-12 An occupancy (motion) sensor.

Today's electronic ballasts are much lighter in weight and considerably more energy efficient than older style magnetic ballasts (core and coil). Energy-saving ballasts might cost more initially, but the payback is in the energy consumption saving over time.

Old-style fluorescent ballasts get very warm and might consume 14 to 16 watts, whereas an electronic ballast might consume 8 to 10 watts. Combined with energy-saving fluorescent lamps that use 32 or 34 watts instead of 40 watts, energy savings are considerable. You are buying light, not heat.

When installing fluorescent luminaires, check the label on the ballast that shows the actual volt-amperes that the ballast and lamp will draw in combination. *Do not* attempt to use lamp wattage only when making load calculations because this could lead to an overloaded branch circuit. For example, a high-efficiency ballast might draw a total of 42 volt-amperes, whereas an old-style magnetic ballast might draw 102 volt-amperes. Refer to Table 7-3 for other comparisons.

The higher the power factor rating of a ballast, the more energy efficient. Look for a power factor rating in the mid to high 90s.

Energy-Saving Lamps

The Energy Policy Act of 1992 enacted restrictions on lamps. In October 1995, the common 4 ft, 40-watt T12 linear medium bipin fluorescent lamp was eliminated. This was replaced by an energy-efficient 34-watt T12 lamp, a direct replacement for the discontinued lamp.

The Energy Policy Act was amended drastically in 2005, particularly in the electricity section entitled Title XII.

Some lamps may be designated F40T12/ES, but the lamp draws 34 watts instead of 40 watts. The "ES" stands for "energy saving." ES is a generic designation. Manufacturers may use other designations such as "SS" for SuperSaver, "EW" for Econ-o-Watt, "WM" for Watt-Miser, and others.

The older, high-wattage incandescent R30, R40, and PAR38 lamps were also discontinued and

TABLE 7-3

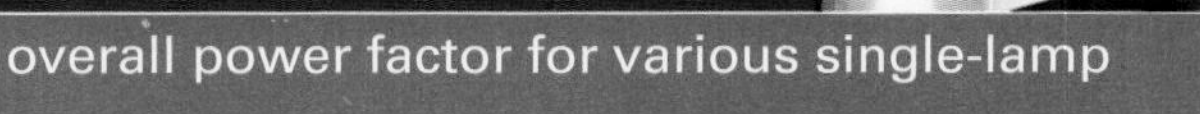

Various line currents, volt-amperes, wattages, and overall power factor for various single-lamp fluorescent ballasts.

BALLAST	LINE CURRENT	LINE VOLTAGE	LINE VOLT-AMPERES	LAMP WATTAGE	LINE POWER FACTOR
No. 1	0.35	120	42	40	0.95 (95%)
No. 2	0.45	120	54	40	0.74 (74%)
No. 3	0.55	120	66	40	0.61 (61%)
No. 4	0.85	120	102	40	0.39 (39%)
No. 5	0.22	120/277	26	30	0.99 (99%)

These values were taken from actual ballast manufacturers' data. For 2-lamp ballasts, double the values in this table. Note in line 5 how high the efficiency is of the latest type of electronic, high-performance ballast for T8 fluorescent lamps.

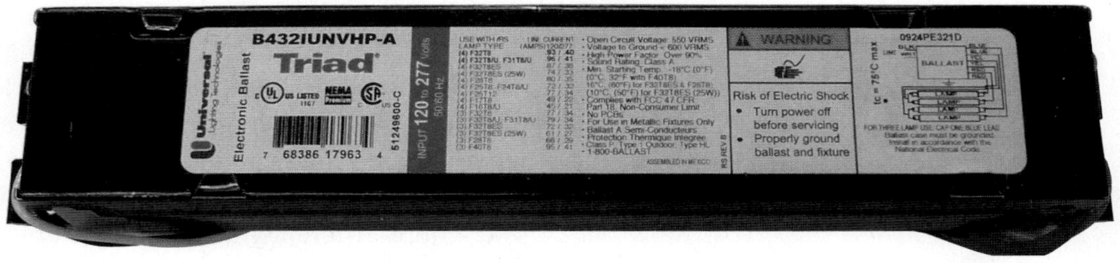

FIGURE 7-13 An electronic Class P ballast, thermally protected as required in *410.130(E)*. Electronic ballasts provide improved performance in fluorescent lighting installations. Electronic ballast lighting systems are 25% to 40% more energy efficient than conventional magnetic (core and coil) ballast fluorescent systems.

replaced with lower wattage lamps. See the sections on "Lamp Shapes" and "Lamp Diameters" in this chapter.

T12 lamps are still found in 4 ft shop lights and square luminaires that use U-tube lamps. Most newer square luminaires have U-tube T8 lamps. In new commercial installations, the T8 lamp has taken over from the T12 lamp.

Energy-saving fluorescent lamps use up to 80% less energy than incandescent lamps of similar brightness. Fluorescent lamps can last 13 times or more longer than incandescent lamps.

It has been estimated that the total electric bill savings across the country will exceed $250 billion over the next 15 years. Table 7-3 compares the power factor of various types of ballasts. Note the difference in line current for the different types of ballasts.

Class P Ballasts

In *410.130(E)*, we find that all fluorescent ballasts installed indoors (except simple reactance-type ballasts), both for new and replacement installations, must have thermal protection built into the ballast by the manufacturer of the ballast, as shown in Figure 7-13. Ballasts provided with built-in thermal protection are listed by UL as Class P ballasts. Under normal conditions, the Class P ballast has a case temperature not exceeding 194°F (90°C). The thermal protector must open within 2 hours when the case temperature reaches 230°F (110°C).

Some Class P ballasts also have a nonresetting fuse integral with the capacitor to protect against capacitor leakage and violent rupture. The Class P ballast's internal thermal protector will disconnect the ballast from the circuit in the event of excessive temperature. Excessive temperatures can be caused by abnormal voltage and improper installation, such as being covered with insulation.

The reason for thermal protection is to reduce the hazards of possible fire due to an overheated ballast when the ballast becomes shorted, grounded, covered with insulation, lacking in air circulation, and so on. Ballast failure has been a common cause of electrical fires.

See Figure 7-14 for some important requirements for fluorescent and incandescent luminaires.

Cold Temperature and Fluorescent Lamps

Conventional fluorescent lamps will operate satisfactorily in ambient temperatures of 60°F (16°C) or more.

In cold temperatures, standard fluorescent lamps will start poorly or possibly not even start. Special ballasts and cold-weather fluorescent lamps are needed. Refer to the manufacturer's literature for instructions relating to cold temperature.

Lamp Designations for Rapid-Start and Preheat Fluorescent Lamps

Example: F40T12/WWX/RS

F—fluorescent

40—wattage

T—tubular shape

12—diameter of lamp in eighths of an inch

WWX—color (warm white deluxe)

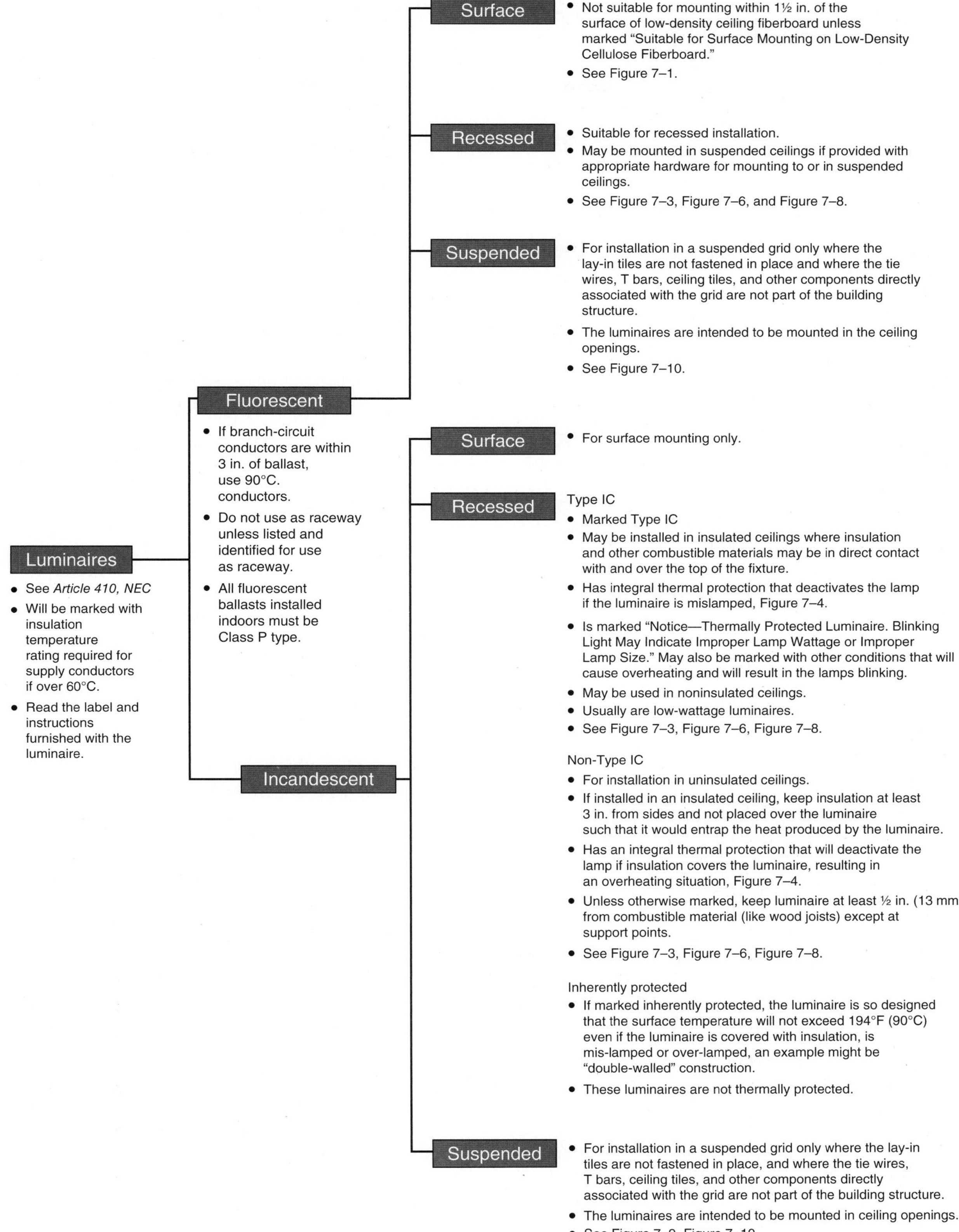

FIGURE 7-14 Some of the most important UL and *NEC* requirements for fluorescent and incandescent luminaires. Always refer to the UL Standards, the *NEC*, and the label and/or instructions furnished with the luminaire.

RS—rapid start

FC—shape (circular)

FB or FU—U-shaped, bent

Lamp Shapes

A—standard shape, general use

P or PS—pear shaped (150 watts and larger)

C—cone shape, like a night light or Christmas tree bulb

F—flame shape, decorative

G—globular

PAR—parabolic shape like a bowl, concentrates light

R—reflector type; might have reflective material near the base or at the bottom of the lamp; concentrates light

Burn Base Down. Some lamps will be designated as "Burn Base Down." Be sure to follow these words of warning.

Lamp Diameters

There is a simple way to determine the diameter of a lamp at its widest measurement. The industry uses an "eighths of an inch" rule. For example, the diameter of a T5 fluorescent lamp is $5 \div 8 = 0.625$ in. The diameter of a T8 fluorescent lamp is $8 \div 8 = 1$ in. The diameter of a T12 lamp is $12 \div 8 = 1\frac{1}{2}$ in.

Incandescent lamps follow the same system. The diameter of an A21 lamp is $21 \div 8 = 2.625$ in. The diameter of an R30 lamp is $30 \div 8 = 3\frac{3}{4}$ in.

Watts versus Volt-Amperes

It is very easy to overload a branch circuit that supplies fluorescent lighting. The most common mistake of making load calculations for fluorescent lighting loads is to use lamp wattage instead of the volt-amperes and total current draw as marked on the ballast's label.

The culprits are old-style, low-power-factor, low-efficiency ballasts!

Let's make a comparison of possibilities for the recreation room lighting where there are six recessed fluorescent luminaires, each containing two 2-lamp ballasts. The nameplate on each ballast indicates a line current rating of 0.70 ampere at 120 volts, which is 84 volt-amperes. The lamps are marked 40 watts.

The total current in amperes is

$$12 \times 0.70 = 8.4 \text{ amperes}$$

The total power in volt-amperes is

$$8.4 \times 120 = 1008 \text{ volt-amperes}$$

The total lamp wattage is

$$40 \times 24 = 960 \text{ watts}$$

If we had calculated the load for these luminaires based upon wattage, the result would have been

$$\frac{W}{E} = \frac{960}{120} = 8 \text{ amperes}$$

This result is not much different from the 8.4 amperes using data from the ballast nameplate.

Had we used low-cost, low-efficiency ballasts like No. 4 shown in Table 7-3, the result would be entirely different. The ballasts in Table 7-3 are single-lamp ballasts. We double the current values for a good approximation of a similar 2-lamp ballast. Using the low power factor ballast, we find that the total current in amperes is

$$12 \times 0.85 \times 2 = 20.4 \text{ amperes}$$

The total power in volt-amperes is

$$20.4 \times 120 = 2448 \text{ volt-amperes}$$

or

$$12 \times 102 \times 2 = 2448 \text{ volt-amperes}$$

The total lamp wattage is

$$24 \times 40 = 960 \text{ watts}$$

As previously shown, if we use lamp wattage to calculate the current draw:

$$I = \frac{W}{E} = \frac{960}{120} = 8 \text{ amperes}$$

But we really have a current draw of 20.4 amperes, which would overload the 15-ampere branch circuit B12. In fact, a load of 20.4 amperes is too much for a 20-ampere branch circuit. Two 15-ampere branch circuits would be needed to hook up the recreation room recessed fluorescent luminaires. Remember, *do not* load any circuit to more than 80% of the branch circuit's rating. That is 16 amperes for a 20-ampere branch circuit and 12 amperes for a 15-ampere branch circuit. And that is the maximum.

It is apparent that the *installed* cost using cheap luminaires is greater and more complicated than if high-quality luminaires using energy-efficient ballasts and lamps had been used. However, after the initial installation, the energy savings add up significantly.

Low power factor ballasts mean higher light bills! High-efficiency, high power-factor ballasts should always be used. More on the wiring of the lighting in the recreation room is covered in Chapter 17.

Other Considerations

- Because heat trapped by insulation around and on top of a luminaire can shorten the life of a ballast, always follow the manufacturer's installation requirements and the requirements found in *Article 410*. There is a nonscientific rule of thumb referred to as the "half-life rule." It means that for every 21°F (10°C) above electrical equipments' (motors, conductors, transformers, etc.) recommended maximum operating temperature, the expected life of that equipment is cut roughly in half. Raising the temperature another 21°F (10°C) will again cut the expected life in half. A 21°F (10°C) temperature rise (**Note:** This is a temperature difference, not an absolute temperature measurement), above the ballast's rated temperature of 90°C, which can reduce the ballast's life by one-half. This "half-life rule" is also true for conductors, motors, transformers, and other electrical equipment.
- Fluorescent lamps that are intensely blackened on both ends should be replaced. Operating a 2-lamp ballast with only one lamp working will cause the ballast to run hot and will shorten the life of the ballast. Severe blackening of one end of the lamp can also ruin the ballast. A flickering lamp should be replaced.
- Poor starting of a fluorescent lamp can be caused by poor contact in the lampholder, poor grounding, excessive moisture on the outside of the tube, or cold temperature (approximately 50°F [10°C]), as well as dirt, dust, and grime on the lamp.
- Most ballasts will operate satisfactorily within a range of +5% to −7% of their rated voltage. The higher quality CBM-certified ballasts will operate satisfactorily within a range of ±10%.
- Most T12 fluorescent lamps sold today are of the energy-efficient type. Generally, energy-efficient fluorescent lamps should not be used with old-style magnetic ballasts. To do so could result in poor starting, reduced lamp life, flickering, or spiraling. Home centers carry the most common T12 fluorescent lamps. Electrical supply houses usually carry different types of fluorescent lamps that can be matched with different types of ballasts. Check the markings on the ballast and the lamp carton, as well as the instructional literature available at the point-of-sale of the ballast or lamp, and/or in the manufacturer's catalogs.
- Lamp life generally is rated in X number of hours of operation based on 3 hours per start. Frequent switching results in shortened expected lamp life. Inversely, leaving the lamps on for long periods of time extends the expected lamp life. Vibration, rough handling, cleaning, and so on, shortens lamp life.
- Be careful of the type of conductors you use to connect fluorescent luminaires. Branch circuit conductors within 3 in. (75 mm) of a ballast must have an insulation temperature rating of not less than 90°C, *410.68*. Type THHN conductors, the conductors in nonmetallic-sheathed cable and in Type ACTHH armored cable are rated 90°C.

Voltage Limitations

The maximum voltage allowed for residential lighting is 120 volts between conductors, per *210.6(A)*.

In or on a home, lighting equipment that operates with an open-circuit voltage over 1000 volts is

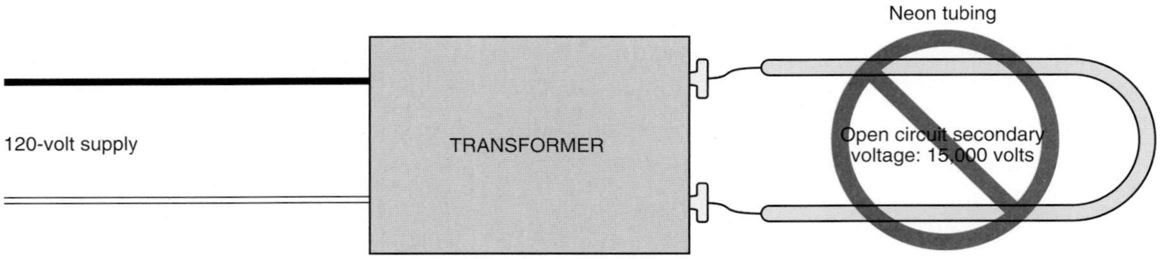

FIGURE 7-15 It is *not* permitted to install lighting in or on residences where the open-circuit voltage is over 1000 volts. This is a violation of *410.140(B)*.

not permitted, *410.140(B)*. This pretty much eliminates most neon lighting systems for decorative lighting purposes, as shown in Figure 7-15.

Incandescent Lamp Life at Different Voltages

Operating an incandescent lamp at other than rated voltage will result in longer—or shorter—lamp life. The following formula predicts the approximate expected lamp life at different voltages. For example, assume that a 120-volt incandescent lamp has a published lamp life of 1000 hours. The calculations show the expected lamp life of this lamp when operated at 130 volts and the expected lamp life when operated at 110 volts.

- At 130 volts:

$$\text{Expected Life} = \left(\frac{\text{Rated Volts}}{\text{Actual Volts}}\right)^{13} \times \text{Rated Life}$$
$$= \left(\frac{120}{130}\right)^{13} \times 1000$$
$$= (0.923)^{13} \times 1000$$
$$= 0.353 \times 1000$$
$$= 353 \text{ hours}$$

- At 110 volts:

$$\text{Expected Life} = \left(\frac{\text{Rated Volts}}{\text{Actual Volts}}\right)^{13} \times \text{Rated Life}$$
$$= \left(\frac{120}{110}\right)^{13} \times 1000$$
$$= (1.091)^{13} \times 1000$$
$$= 3.103 \times 1000$$
$$= 3103 \text{ hours}$$

Incandescent Lamp Lumen Output at Different Voltages

Operating an incandescent lamp at a voltage lower than the lamp's voltage rating will result in longer lamp life. A strong case might be made to install 130-volt lamps, particularly where they are hard to reach, such as flood lights located high up. However, the lumen output is reduced. For example, calculate the approximate lumen output of a 100-watt, 130-volt incandescent lamp that has an initial lumen output of 1750 lumens at rated voltage. The lamp is to be operated at 120 volts.

$$\text{Expected Lumens} = \left(\frac{\text{Actual Volts}}{\text{Rated Volts}}\right)^{3.4} \times \text{Rated Lumens}$$
$$= \left(\frac{120}{130}\right)^{3.4} \times 1750$$
$$= (0.923)^{3.4} \times 1750$$
$$= 0.762 \times 1750$$
$$= 1334 \text{ Lumens}$$

Formulae such as these are found in lamp manufacturer's catalogs. These formulae are useful in determining the effect of applied voltage on lamp wattage, line current, lumen output, lumens per watt, and lamp life. A calculator that has a y^x power function key is needed to solve these equations.

Exponents. Confused? If your calculator does not have a y^x function key, take the easy route. Just multiply the number again and again for the value of the exponents (power). Cumbersome . . . but it works. For example, the number 2^3 can be read "2 to the third power" or "2 raised to the third power."

$$2 \times 2 \times 2 = 8$$

Another example, the number 8^5 can be read "eight to the fifth power" or "eight raised to the fifth power." We simply multiply that number 8 by itself five times.

$$8 \times 8 \times 8 \times 8 \times 8 = 32{,}768$$

Some simple exponents can be read in a certain way; for example, a^2 is usually read as "a squared" and a^3 as "a cubed."

Figure 7-16 is a graph showing the affect of different voltages on lamp life and lumen output.

Keep It Simple!

The formulae for lamp life and lumen output are complicated, but here is a quick summary:

- An incandescent lamp operating at a voltage *less* than its rated voltage will last longer but will not burn as bright as it should.
- An incandescent lamp operating at a voltage *greater* than its rated voltage will not last as long but will burn brighter.

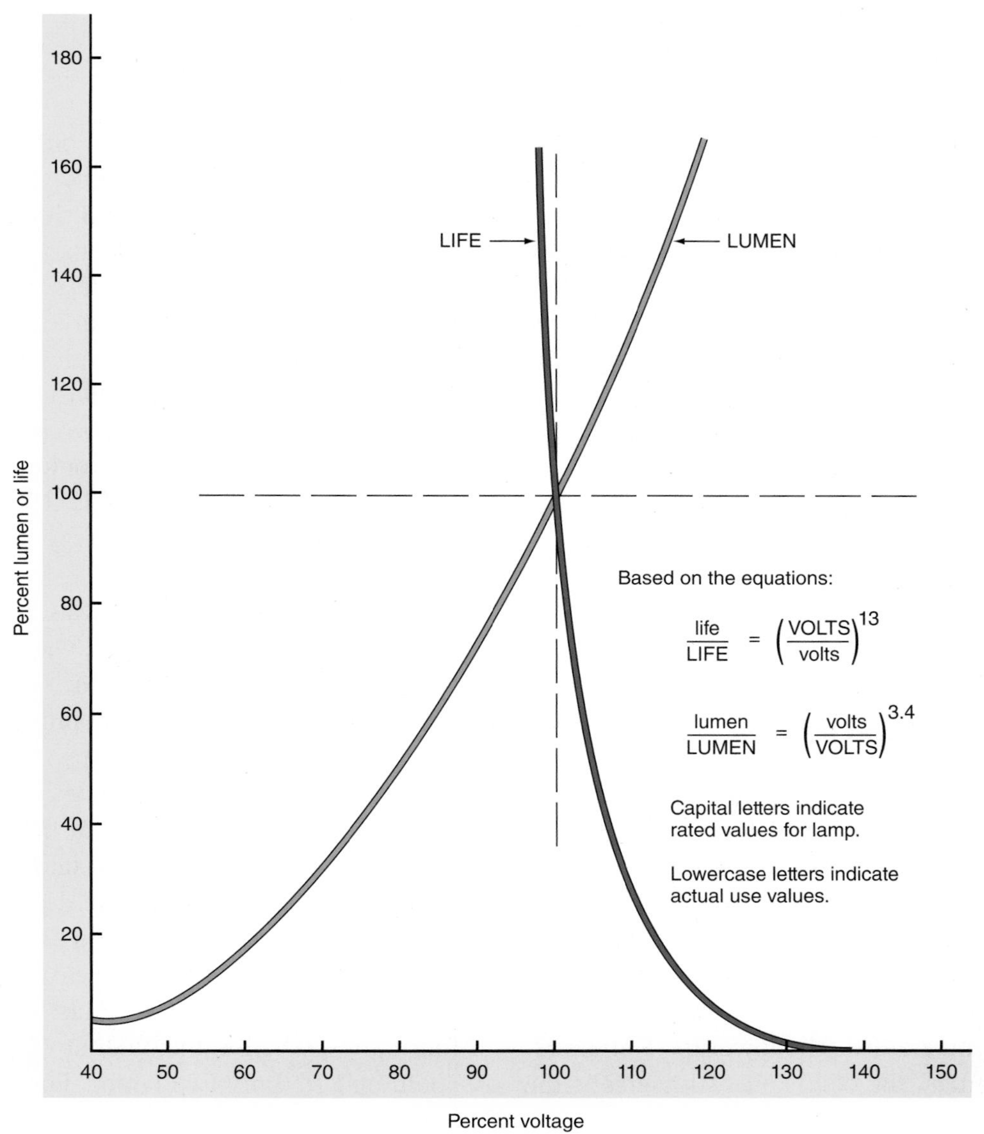

FIGURE 7-16 Typical operating characteristics of an incandescent lamp.

LAMP EFFICACY

A lumen is a measurement of visible light output from a lamp.

One lumen on 1 ft^2 of surface produces 1 *foot-candle*.

Another term you will hear in lighting is *efficacy*. Efficacy is the ratio of light output from a lamp to the electric power it consumes and is measured in *lumens per watt* (lm/w). In other words, efficacy is a measurement of input to output. Examples:

- Lamp #1: 26-watt compact fluorescent lamp (CFL) produces 1700 lumens at rated voltage, which equates to 65 lm/w.
- Lamp #2: 100-watt incandescent lamp produces 1200 lumens at rated voltage, which equates to 12 lm/w.
- Lamp #3: 100-watt lamp incandescent produces 1750 lumens at rated voltage, which equates to 17.5 lm/w.
- Lamp #4: An LED lamp equivalent to a 60 watt incandescent lamp is rated 800 lumens but consumes only 10.5 watts which equates to 76 lm/w.

The more lumens per watt, the greater the efficacy of the lamp. See Table 7-2.

It is obvious that typical incandescent lamps are very inefficient compared to compact fluorescent lamps. There is a dramatic trend toward the use of CFL lamps: more light for less energy, and less energy means lower electric bills!

LAMP COLOR TEMPERATURE

Lamp color temperature is rated in Kelvin degrees, and the term is used to describe the "whiteness" of the lamp light. In incandescent lamps, color temperature is related to the physical temperature of the filament. In fluorescent lamps where no hot filament is involved, color temperature is related to the light as though the fluorescent discharge were operating at a given color temperature. The lower the Kelvin degrees, the "warmer" the color tone. Conversely, the higher the Kelvin degrees, the "cooler" the color tone.

Incandescent lamps provide pleasant color tones, bringing out the warm red flesh tones similar to those of natural light. This is particularly true for the "soft" and "natural" white lamps.

Tungsten filament halogen lamps have a gas filling and an inner coating that reflects heat. This keeps the filament hot with less electricity. Their light output is "whiter." They are more expensive than the standard incandescent lamp.

Fluorescent lamps are available in a wide range of "coolness" to "warmth." Warm fluorescent lamps bring out the red tones. Cool fluorescent lamps tend to give a person's skin a pale appearance.

Fluorescent lamps might be marked daylight D (very cool), cool white CW (cool), white W (moderate), warm white WW (warm). These categories break down further into a deluxe X series (i.e., deluxe warm white—deluxe cool white), specification SP series, and specification deluxe SPX series.

Typical color temperature ratings for lamps are 2800K (incandescent), 3000K (halogen), 4100K (cool white fluorescent), and 5000K (fluorescent that simulates daylight). Note that a halogen lamp is "whiter" than a typical incandescent lamp.

Catalogs from lamp manufacturers provide detailed information about lamp characteristics.

Fluorescent lamps and ballasts are a moving target. In recent years, there have been dramatic improvements in both lamps and electronic ballast efficiency. First, the now-antiquated T12 fluorescent lamps (40 watts) were replaced by energy-saving T8 fluorescent lamps. These original T8 lamps are becoming a thing of the past. The latest T8 high-efficiency, energy-saving (25 watts vs. 32 watts) lamps have an expected 50% longer life than the original T8 lamps.

The newer T8 lamps use approximately 40% less energy than the older T12 lamps. At $0.06 per kWh, one manufacturer claims a savings of $27.00 per lamp over the life (30,000 hours) of the lamp. At $0.10 cents per kWh, the savings is said to be $45.00 per lamp over the life of the lamp. Using the newer T8 lamps on new installations and as replacements for existing installations makes the payback time pretty attractive. One electronic ballast can operate up to four lamps, whereas the older style magnetic ballast could operate only two lamps. For a three- or four-lamp luminaire, one ballast instead of two results in quite a saving. Some electronic ballasts can operate six lamps.

Hard to believe! You now can have reduced power consumption and increased light output using electronic ballasts.

Today's high-efficiency ballasts are available with efficiencies of from 98% to 99%.

The only way you can stay on top of these rapid improvements is to check out the websites of the various lamp and ballast manufacturers.

Today's magnetic and electronic ballasts handle most of the fluorescent lamp types sold, including standard and energy-saving preheat, rapid start, slimline, high output, and very high output. Again, check the label on the ballast.

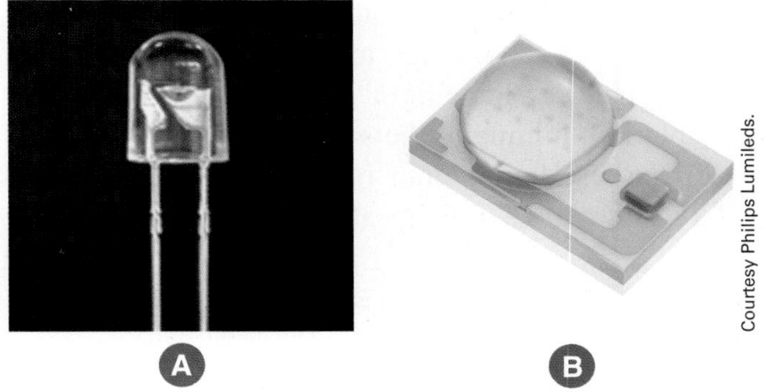

FIGURE 7-17 Some types of LEDs that luminaire manufacturers can use in their product. (A) is a typical single light-emitting diode (LED). (B) is a high light output LED.

LED Lighting

Light-emitting diode (LED) is pronounced "ell-eee-dee."

Reduce the electric bill! Save energy! Reduce energy consumption! Reduce the air-conditioning load!

It has been said that the incandescent bulb is from the dinosaur age, having been around since Thomas Edison applied for a patent on May 4, 1880.

Coming on strong is a new concept for lighting that uses light-emitting diodes (LEDs) as its source of light. It is called "solid-state lighting." LED lighting has very low power consumption. Electricians had better get ready for this new type of light source in luminaires.

LEDs are solid-state devices that have been around since 1962. When connected to a dc source, the electrons in the LED smash together, creating light. Think of an LED as a tiny lightbulb, but with no filament. Figure 7-17(A) is a typical LED that luminaire manufacturers can cluster in their luminaires to obtain the amount of light output they are looking for. Figure 7-17(B) is a high light output LED that lighting manufacturers can use in their luminaires. Figure 7-18 shows two LED-powered luminaires.

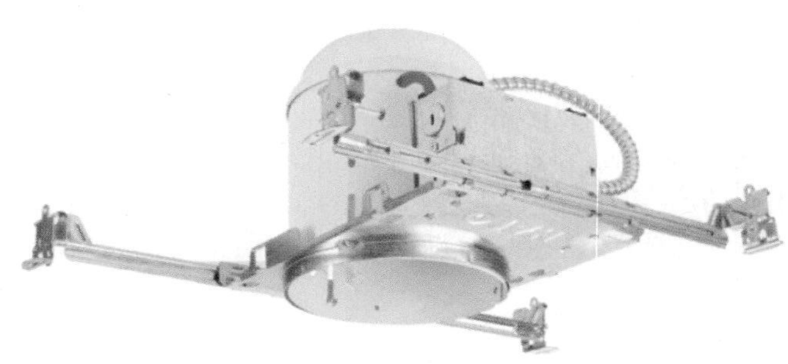

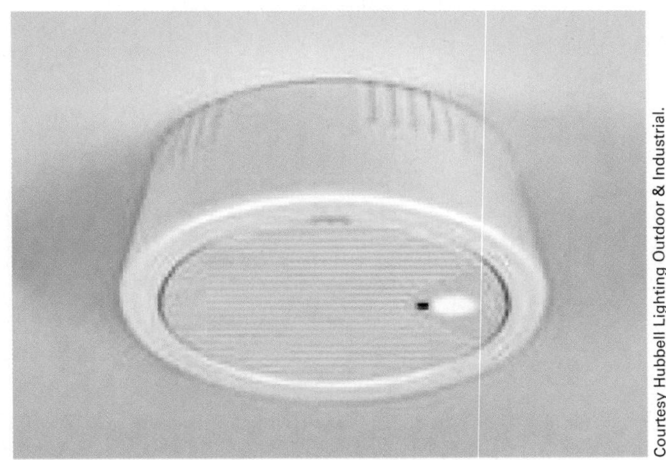

FIGURE 7-18 LED-powered luminaires.

Today, LEDs are all around us. They are commonly recognized by the tiny white, red, yellow, green, purple, orange, and blue lights found in the digital displays in TVs, radios, DVD and CD players, remotes, computers, printers, fax machines, telephones, answering machines, Christmas light strings, night lights, "locator" switches, traffic lights, digital clocks, meters, testers, tail lights on automobiles, strobe lights, occupation sensors, and other electronic devices, equipment, and appliances.

LEDs for lighting are a rather recent concept. Because the lumens per watt in LEDs are on the increase, it is now making sense to use LEDs in luminaires.

Individual LEDs are rather small. Putting a cluster (called an array) of LEDs together (i.e., 5, 20, 30, 60, 120) produces a lot of light. The result is an LED bulb, usable in a luminaire the same as a typical medium Edison-base lampholder incandescent bulb. Figure 7-18 shows three different styles of LED lamps.

LED Luminaires and the *NEC*

The *2008 NEC* recognized LED lighting for the first time in *Article 410*. LED lighting is now common in exit, recessed, surface, and under-cabinet luminaires; desk lamps; wall luminaires; down lights; and stop-and-go lights, to name a few. LEDs in flashlights have been around for quite a while.

Lamp manufacturers have come out with many different types of lamps for accent, task, conventional shape, flood, spots, and so on. They are available with the standard Edison base and candelabra base to replace existing incandescent lamps. The light is white, but with different LEDs other colors are available. See Figure 7-19 for an assortment of LED lamps.

FIGURE 7-19 An assortment of lightbulbs (lamps) powered by a number of individual light-emitting diodes (LEDs). These lamps screw into a standard medium Edison-base lampholder.

Today's LEDs last 60,000 to 100,000 hours. They have no filament to burn out, can withstand vibration and rough usage, contain no mercury, operate better when cool, lose life and lumen output at extremely high temperatures, and can operate at temperatures as low as −40°F (−40°C).

The lumen output of LEDs slowly declines over time. The decline varies. The industry seems to be settling on a lumen output rating of 70% after 50,000 hours of use.

LEDs use less electricity per lumen-output than incandescent or fluorescent lamps. LEDs are much more effecient in converting electrical energy to lighting energy than most other lighting sources. For the purpose of lighting is to have illumination—not heat. Heat is a waste.

Some LEDs may be dimmed, whereas others may not. Check with the manufacturer of the lamp (bulb) to verify the dimming or no dimming capability.

LEDs start instantly with no flickering. LEDs put out directional light as compared to the conventional incandescent lamp that shoots light out in almost all directions.

A study was recently made to compare an LED's predicted life of 60,000 hours (that's almost 7 years of continuous burning, or 21 years at 8 hours per day of usage) to a standard 60-watt incandescent lamp that has a rated life of 1000 hours. Over the 60,000 hour life:

- 60 standard incandescent lamps would be used compared to one LED bulb.
- the standard incandescent lamps would use 3600 kilowatt-hours, resulting in the cost of electricity at $360. The LED lamp would use 120 kilowatt-hours, resulting in a cost of electricity at $12.
- the total cost (lamps and cost of electricity) for the incandescent lamps was $400 compared to the cost of operating the LED lamp, which was $47.

Today, the lumen output of LED lamps is similar or somewhat greater when compared to CFLs, producing approximately 50+ lm/w. This is expected to improve to 150 lm/w in the future. Compare the LED lumen output to a typical incandescent lamp that has a lumen output of 14–18 lm/w. That's a significant increase in lumens per watt!

Today, many luminaires come complete with the proper LED lamps. The electrician must simply select the style or light output for the task or scene. Check out lighting showrooms or home centers.

Things Are Changing Fast

Recently announced is a line of LED lamps that are direct replacements for the conventional 40-watt fluorescent lamp. Nothing has to be done other than replace the existing fluorescent lamp with an LED lamp. They work on both magnetic and electronic ballasts. When compared to a conventional 40-watt fluorescent lamp, the LED replacement lamp typically has a 10-year life as opposed to a 2- or 3-year life—their power consumption is 20% less and their lumen output is comparable or slightly greater—and they can operate at 32°F (0°C). These LED lamps are currently available in warm white, cool white, daylight, neutral, and bright white.

LED lighting is accelerating at a rapid pace. The *NEC* Code-Making Panels will be seeing more and more proposals for changes in the *NEC* relating to LED-type luminaires. Keep an eye out for these changes. As with all electrical equipment, carefully read the label on the luminaire to be sure your installation "Meets *Code*."

For more information about LEDs and LED lighting, check out the website of the LED industry: http://www.ledsmagazine.com.

Outdoor Lighting

Before installing outdoor lighting, check with the local electrical inspector and/or building official to find out whether there are any restrictions regarding outdoor lighting.

A virtually unlimited array of outdoor luminaires is available that provide uplighting, downlighting, diffused lighting, moonlighting, shadow and texture lighting, accent lighting, silhouette lighting, and bounce lighting. After the luminaires are selected, the type and color of the lamp is then selected. Some luminaires have specific light "cutoff" data that are useful in determining whether the emitting light will spill over onto the neighbor's property. The method of control must also be considered. Switch control, timer control, dusk-to-dawn control, and motion sensors are ways of turning outdoor lighting on and off.

In recent years, more and more complaints are coming from neighbors claiming that they are being bothered by glare, brightness, and light spillover from their neighbor's outdoor luminaires. Security lighting, yard flood lights, driveway lighting, "moonlighting" in trees, and shrub lighting are examples of sources of light that might cross over the property line and be a "nuisance" to the next door neighbor.

Outdoor "lightscape" lighting considered by a homeowner to be aesthetically wonderful might be offensive to the neighbor. Nuisance lighting is also referred to as *light pollution*, *trespassing*, *intrusion*, *glare*, *spillover*, and *brightness*. Some quiet residential neighborhoods are beginning to look like commercial areas, used car lots, and airport runways because no restrictions govern outdoor lighting methods. This is not a safety issue and is not addressed in the *NEC*. However, the issue might be found in local building codes.

Many communities are being forced to legislate strict outdoor lighting laws, specifying various restrictions on the location, type, size, wattage, and/or footcandles for outdoor luminaires. Checking building codes in your area might reveal requirements such as these:

Light Source: The source of light (the lamp) must not be seen directly.

Glare: Glare, whether direct or reflected, such as from floodlights, and as differentiated from general illuminations, shall not be visible at any property line.

Exterior Lighting: Any lights used for exterior illuminations shall direct light away from adjoining properties.

The International Dark-Sky Association has a lot of information regarding light pollution issues on its website: http://www.darksky.org.

Replacing a Luminaire in an Older Home: Will There Be Problems?

Very possible!

Older homes were wired with conductors that were insulated with rubber (Type R) or thermoplastic (Type T). Both of these older style insulations were rated for maximum 60°C. The heat generated by the lamps over the years literally baked the conductor insulation to a hard, brittle material, particularly as the luminaires were over-lamped. If you aren't careful, pulling on or moving these conductors can easily break off the conductor insulation, which will result in a hazardous short-circuit or ground-fault situation.

Handle with care! Being very careful in handling the conductors while making up the splices and mounting the new luminaire—possibly sliding some readily available tubing-like electrical insulation—or taping over the conductor insulation with plastic electrical tape might be all that is needed.

Many modern luminaires are marked by the manufacturer with a requirement for wiring the luminaire with either 75°C or 90°C supply conductors. You may need to do some rewiring using the latest in nonmetallic-sheathed cables Type NM-B, Type NMC-B, or Type NMS-B. If the conductors are in a raceway, replacing the conductors with Type THHN might be a good idea. All of these conductors have a maximum temperature rating of 90°C.

Another common correction method is shown in Figure 7-20. Install a junction box approximately 24 inches from the location of the new luminaire. Connect a wiring method having conductors with an insulation system rated for the marked supply temperature from the junction box to the luminaire. It is best if the junction box is located in the attic or other accessible location. Otherwise, install a blank canopy over the junction box if it is exposed on the ceiling.

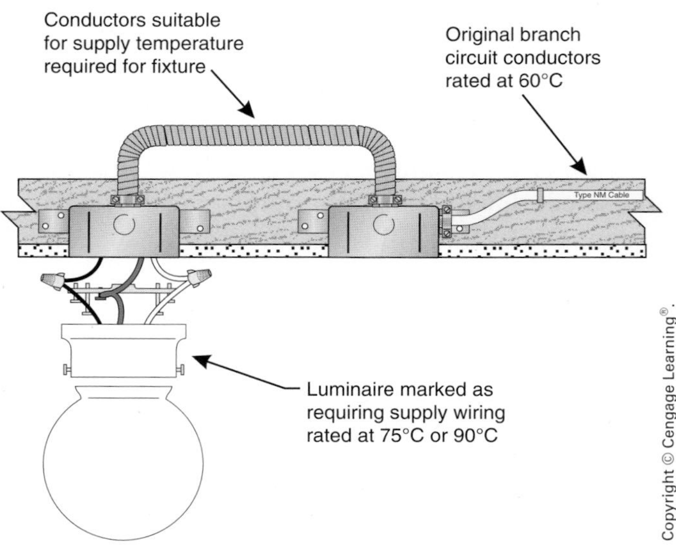

FIGURE 7-20 Installing luminaire requiring higher temperature supply conductors on a lower-rated older wiring system.

More on this subject is found in Chapter 4 in the nonmetallic-sheathed cable section.

REVIEW

1. Is it permissible to install a recessed luminaire directly against wood ceiling joists when the label on the luminaire does not indicate that the luminaire is suitable for insulation to be in direct contact with the luminaire? This is a Type Non-IC fixture.

___

2. If a recessed luminaire without an integral junction box is installed, what extra wiring must be provided? ___

___

___

___

___

___

3. Thermal insulation is not permitted to be installed within ______ in. (mm) of the top or ______ in. (mm) of the side of a recessed luminaire unless the luminaire is identified for use in direct contact with thermal insulation. This is a Type Non-IC fixture.
4. Recessed luminaires are available for installation in direct contact with thermal insulation. These luminaires bear the UL mark "Type ______."
5. Unless specifically designed, all recessed incandescent luminaires must be provided with factory-installed ______.
6. Plans require the installation of surface-mounted fluorescent luminaires on the ceiling of a recreation room that is finished with low-density ceiling fiberboard. What sort of mark would you look for on the label of the luminaire? ______
7. A recessed luminaire bears no marking indicating that it is "Identified for Through-Wiring." Is it permitted to run branch-circuit conductors *other* than the conductors that supply the luminaire through the integral junction box on the luminaire? ______
8. Fluorescent ballasts for all indoor applications must be ______ type. These ballasts contain internal ______ protection to protect against overheating.
9. Additional backup protection for ballasts can be provided by connecting a(n) ______ with the proper size fuse as recommended by the ballast manufacturer.
10. You are called upon to install a number of luminaires in a suspended ceiling. The ceiling will be dropped approximately 8 in. (200 mm) from the ceiling joists. Briefly explain how you might go about wiring these luminaires. ______
11. The *Code* places a maximum open-circuit voltage on lighting equipment in or on homes. This maximum voltage is (600) (750) (1000). (Circle the correct answer.) Where in the *NEC* is this voltage maximum referenced?
12. The letter "E" in a circle on a ballast nameplate indicates that the ballast ______

13. A 120-volt lamp fluorescent ballast for two 40-watt lamps is marked 85 volt-amperes. What is the power factor of the ballast? ______________________________

____

____

14. Can an incandescent lamp dimmer be used to control a fluorescent lamp load?

____

15. A good "rule-of-thumb" to estimate the expected life of a motor, a ballast, or other electrical equipment is that for every ______________ °C above rated temperature, the expected life will be cut in ______________.

16. A post light has a 120-volt, 60-watt lamp installed. The lamp has an expected lamp life of 1000 hours. The homeowner installed a dimmer ahead of the post light and leaves the dimmer set so the output voltage is 100 volts. The lamp burns slightly dimmer when operated at 100 volts, but this is not a problem. What is the expected lamp life when operated at 100 volts? ______________________________

____

____

____

17. Does your community have exterior outdoor lighting restrictions? ______________

If yes, what are they? ______________________________

____

____

18. Define the following terms.

    a. A *tap conductor* is ______________________________

    ____

    b. A *fixture whip* is ______________________________

    ____

19. Circle the correct answer for the following statements:

    a. Always match a fluorescent lamp's wattage and type designation to the type of ballast in the luminaire. (True) (False)

    b. It really makes no difference what type of fluorescent lamp is used, just so the wattage is the same as marked on the ballast. (True) (False)

20. Circle the correct answer for the following statements.

    a. When selecting a trim for a recessed incandescent luminaire, select any trim that physically fits and can be attached to the luminaire. (True) (False)

    b. When selecting a trim for a recessed incandescent luminaire, select a trim that the manufacturer indicates may be used with that luminaire. (True) (False)

21. Have you installed LED lighting? Add comments on your experience involving LED lighting. ______________________________

____

CHAPTER 8

Lighting Branch Circuit for the Front Bedroom

OBJECTIVES

After studying this chapter, you should be able to

- understand the meaning of general, accent, task, and security lighting.
- estimate loads for the outlets connected to a branch circuit.
- determine how many receptacles to connect to a branch circuit.
- determine how many branch circuits are needed.
- draw a cable layout and wiring diagram for a branch circuit.
- properly size outlet boxes based on the number of conductors and devices.
- understand the *NEC* requirements for luminaires in clothes closets.

INTRODUCTION

In the first seven chapters of this text, you studied the *National Electrical Code* (*NEC*®) requirements and the basics of house wiring. You will now apply all that you have learned to an actual house—room by room—circuit by circuit. You will use the text, the plans, the specifications that are in the back of this book, and the *NEC* to "bring it all together."

The home used in this text is a typical home—nothing prestigious or elegant. Other than the size of the service conductors and equipment, more branch circuits, and more gadgets, the *NEC* requirements are the same.

This might be a good time for you to refresh your memory about the arc-fault circuit-interrupter (AFCI) requirements for bedrooms. All 120-volt, single-phase, 15- and 20-ampere devices and outlets in bedrooms require AFCI protection. We covered the requirements extensively in Chapter 6 of this text.

RESIDENTIAL LIGHTING

Residential lighting is a personal thing. The homeowner, builder, and electrical contractor must meet to decide on what types of luminaires are to be installed in the residence. Many variables (cost, personal preference, construction obstacles, etc.) must be taken into consideration.

Residential lighting can be segmented into four groups; general lighting, accent lighting, task lighting, and security lighting.

General Lighting

General lighting often is referred to as ambient lighting.

General lighting provides overall illumination for a given area, such as the front hall, the bedroom hall, the workshop, or the garage lighting in this residence. General lighting can be very basic, which means just getting the job done by providing adequate lighting for the area involved. Or it can take the form of decorative lighting, such as a chandelier over a dining room table or other decorative luminaires that can be attached to ceiling paddle fans.

Lighting in hallways and stairways must not only be thought of as attractive general lighting, but also as safety lighting. These areas must have good lighting.

Accent Lighting

Accent lighting provides focus and attention to an object or area in the home. Examples are the recessed "spots" over the fireplace. Another example is the track lighting in the Living Room of this residence, which can accent a picture, photo, painting, or sculpture that might be hung on the wall. To be effective, accent lighting should be at least five times that of the surrounding general lighting.

Task Lighting

This is sometimes referred to as "activity" lighting. Task lighting provides proper lighting where tasks or activities are performed. The fluorescent luminaires above the workbench, the kitchen range hood, and the recessed luminaire over the kitchen sink are all examples of "task" lighting. To avoid too much contrast, task lighting should not exceed three times that of the surrounding general lighting.

Wall Washing. The term *wall washing* is used when luminaires are installed in a row, parallel to the wall that is to be "washed" with light. For example, if you wish to "wash" a 12 ft (3.7 m) wall, as illustrated in Figure 8-1, and you would like the luminaires to be 2 ft (600 mm) from the wall:

$$\frac{12}{2} = 6 \text{ luminaires recommended}$$

Security Lighting

Security lighting generally includes outdoor lighting, such as post lamps, wall luminaires, walkway lighting, and all other lighting that serves the purpose of providing lighting for security and safety reasons. Security lighting in many instances is provided by the normal types of luminaires found in the typical residence. In the residence discussed in this text, the outdoor bracket luminaires in front of the garage and next to the entry doors, as well as the post light, might be considered to be security lighting

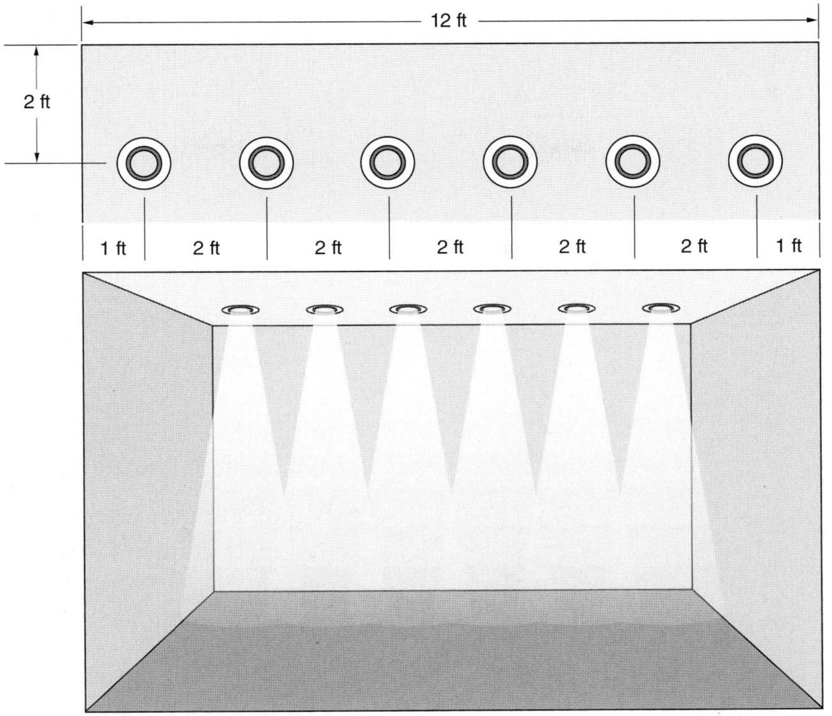

FIGURE 8-1 Illustration of how recessed luminaires can be installed in the ceiling close to a wall to provide beautiful "wall washing."

even though they add to the beauty of the residence. Security lighting can be controlled manually with regular wall switches, timers, light sensors, or motion detectors.

American Lighting Association

The American Lighting Association offers excellent residential lighting recommendations both online and hardcopy brochures. If you intend to become competent in residential lighting, a visit to its website at http://www.americanlightingassoc.com is a must!

This organization is comprised of many manufacturers of luminaires and electrical distributors that maintain extensive lighting showrooms across the country. A visit to one of these showrooms offers the prospective buyer an opportunity to select from thousands of luminaires. Most of the better-quality lighting showrooms are staffed with certified lighting consultants (CLCs), lighting specialists (LSs), and/or lighting associates (LAs) who are highly qualified to make recommendations to homeowners so that they will get the most value for their money.

Find copies of lighting manufacturers' catalogs. Visit their websites. They offer many lighting technique suggestions.

The lighting provided throughout this residence conforms to the residential lighting recommendations of the American Lighting Association. Certainly, many other variations are possible. For the purposes of studying an entire wiring installation for this residence, all of the load calculations, wiring diagrams, and so on, have been accomplished using the luminaire selection as indicated on the plans and in the specifications.

LAYING OUT GENERAL-PURPOSE LIGHTING AND RECEPTACLE CIRCUITS

In most residential installations, circuit layout is usually done by the electrician, always keeping in mind the *NEC* requirements. In larger, more costly residences, the circuit design might be specified by

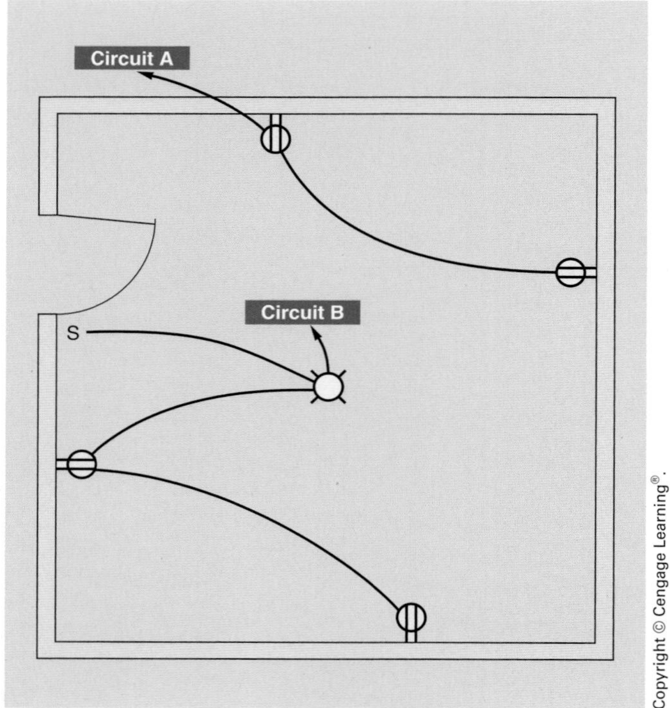

FIGURE 8-2 Wiring layout showing one room that is fed by two different circuits. If one of the circuits goes out, the other circuit will still provide electricity to the room.

the architect and included on the plans. Many circuit arrangements are possible.

Here are a few thoughts:

- Supply general-purpose receptacle and lighting outlets with more than one circuit. Mix it up. If one circuit has a problem, the second circuit continues to supply power to the other outlets in that room, as shown in Figure 8-2.
- Keep general-purpose receptacles on one circuit and lighting outlets on another branch circuit.
- Economize on wiring materials by connecting receptacle outlets back to back, as shown in Figure 8-3.
- Run the branch circuit first to the switch and then to the controlled lighting outlet. This keeps the outlet box at the lighting outlet or the wiring compartment on recessed cans less crowded. Having the grounded circuit conductor present at the switch location makes it easy to install switching devices that require a connection to the grounded circuit conductor. ▶Doing so satisfies the rule in *404.2(C)*.◀
- Install a receptacle below the room wall switch—an area never blocked by furniture.
- Think about where furniture might be placed. Try to locate receptacles to the side of these so as to be easily accessible after the furniture is in place. Don't let the 6 ft (1.8 m) receptacle requirement hinder your thinking.
- Although not a *Code* issue, it is poor practice to include outlets on different floors on the same circuit. Some local building codes limit this type of installation to lights at the head and foot of a stairway.

Wiring bedrooms became a little more complicated when AFCIs appeared on the scene, as required by *210.12*. The dilemma is whether to put all bedroom receptacles and lighting outlets on one AFCI-protected branch circuit, or to connect the receptacle and lighting outlets on different branch circuits, which means that more than one AFCI circuit breaker or receptacle will be needed.

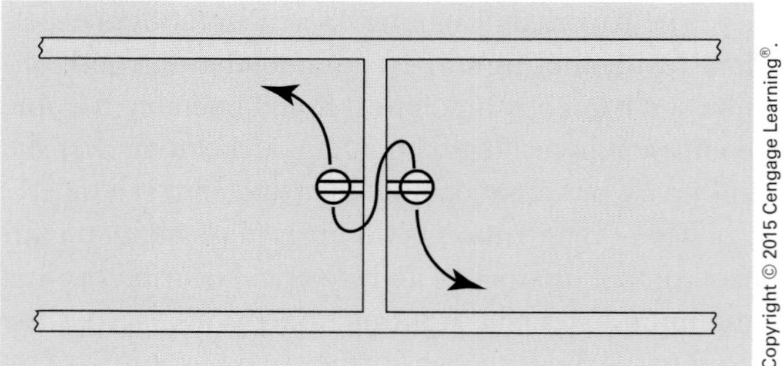

FIGURE 8-3 Receptacle outlets connected back to back. This can reduce the cost of the installation because of the short distance between the outlets. Do not do this in fire-rated walls unless special installation procedures are followed as illustrated in Figures 2-29, 2-30, and 2-31.

The number of receptacle and lighting outlets connected to one branch circuit is discussed a little later.

Cable Runs

A student studying the great number of wiring diagrams throughout this text will find many different ways to run the cables and make up the circuit connections. Sizing of boxes that will conform to the *Code* for the number of wires and devices is covered in great detail in this text. When a circuit is run "through" a recessed luminaire like the type shown in Figure 7-8, that recessed luminaire must be identified for "through-wiring." This subject is discussed in detail in Chapter 7. When in doubt as to whether the luminaire is suitable for running wires to and beyond it to another luminaire or part of the circuit, it is recommended that the circuitry be designed so as to end up at the recessed luminaire with only the two wires that will connect to that luminaire.

EXAMPLE

Calculate the wattage and volt-amperes of a 120-volt, 10-ampere resistive load.

Solutions:

a. $120 \times 10 \times 1 = 1200$ watts
b. $120 \times 10 = 1200$ volt-amperes

EXAMPLE

A 120-volt fluorescent ballast has an input current of 0.34 ampere and an input power rating of 22 watts. Calculate the power factor of the ballast.

Solution:

$$\text{Power Factor} = \frac{\text{Watts}}{\text{Volt-amperes}} = \frac{22}{120 \times 0.34} = 0.54 \text{ (54\% PF)}$$

ESTIMATING LOADS FOR OUTLETS

Although this text is not intended to be a basic electrical theory text, it should be mentioned that, when calculating loads,

$$\text{Volts} \times \text{Amperes} = \text{Volt-amperes}$$

Yet, many times we say that

$$\text{Volts} \times \text{Amperes} = \text{Watts}$$

What we really mean to say is that

$$\text{Volts} \times \text{Amperes} \times \text{Power factor} = \text{Watts}$$

In a pure resistive load such as a simple lightbulb, a toaster, an iron, or a resistance electric heating element, the power factor is 100%. Then

$$\text{Volts} \times \text{Amperes} \times 1 = \text{Watts}$$

With transformers, motors, ballasts, and other "inductive" loads, wattage is not necessarily the same as volt-amperes.

Therefore, to be sure that adequate ampacity is provided for in branch-circuit wiring, feeder sizing, and service-entrance calculations, the *Code* requires that we use the term *volt-amperes*. This allows us to ignore power factor and address the *true* current draw that will enable us to determine the correct ratings of electrical equipment.

However, in some instances the terms *watts* and *volt-amperes* can be used interchangeably without creating any problems. For instance, *220.55* recognizes that for electric ranges and other cooking equipment, the kVA and kW ratings shall be considered to be equivalent for the purpose of branch-circuit and feeder calculations.

A similar permission of equivalency of kilovolt-amperes (kVA) and kilowatts (kW) is given in *220.54* for electric clothes dryers.

The load calculation examples in *Annex D* of the *NEC* show kilowatt values in the list of connected loads, but the actual calculations are done using volt-ampere values. Throughout this text, *wattage* and *volt-amperes* are used when calculating and/or estimating loads.

For the residence in the plans, it is shown that six lighting circuits meet the minimum standards set by the *Code*. However, to provide sufficient capacity, 13 lighting circuits are to be installed in this residence.

How Many Outlets Are Permitted on One General-Purpose Lighting Branch Circuit?

This is a tough question, but it can be answered reasonably well using a few simple "rules of thumb" discussed next.

The *NEC* does not specify the maximum number of receptacle outlets or lighting outlets that may be connected to one 120-volt lighting or small-appliance branch circuit in a residence. It may not seem logical that 10 or 20 receptacle outlets and lighting outlets could be connected to one branch circuit and not be in violation of the *Code*. On our side is the fact that there is much load diversity in residential occupancies. We have no idea what will be plugged into the receptacles. We do know that rarely, if ever, will all receptacle outlets and lighting outlets be used at the same time. Having many "convenience" receptacles is safer because more receptacle outlets virtually eliminate or minimize the use of extension cords, one of the leading causes of electrical fires.

Recessed and surface-mounted incandescent luminaires are marked with their maximum lamp wattage. Fluorescent luminaires are marked with lamp wattage, and ballasts are marked in volt-amperes. In new construction, other than for recessed luminaires, we probably do not know what type of surface or hanging luminaires will be selected. Most luminaire manufacturers' catalogs provide excellent recommendations for lamp wattages for residential applications.

There are many items such as nightlights (1 to 7½ watts), smoke detectors (5 watts), clocks (5 watts), carbon monoxide detector (60 milliamperes), and clothes closet lights (60 to 75 watts) that use only a small portion of the branch circuit's capacity. It comes down to approximating, estimating, experience, and common sense! It is not an exact science.

Circuit Loading "Rules of Thumb"

Let's look at a few favorite "rules of thumb" that electricians use for determining how many outlets they connect to one general-purpose branch circuit.

Bear in mind that these "rules of thumb" are merely simple "ballpark" guidelines. They are not hard and fast *Code* requirements. Used together with common sense and good judgment as to how the circuits will be used will result in adequate circuiting.

Later on in this text, when performing actual load calculations, you will learn that the general lighting and receptacle load in dwellings is included in the 3 VA/ft^2. See *NEC Table 220.12* and *220.14(J)*.

It is important to note that *NEC 220.12* states that: For dwelling units, the calculated floor area shall not include open porches, garages, or unused or unfinished spaces not adaptable for future use.* All of this was discussed in detail in Chapter 3.

Outlets per Circuit Method. This is the most common method used by electricians across the country. It is very simple. No calculations are needed. Just count the outlets! Connect 10 outlets on a circuit where you anticipate fairly heavy-wattage lighting loads. Connect 15 outlets per circuit for circuits that have a lot of low-wattage lighting loads. Somewhere in-between might be the right number. This method is really estimating a load of 1 to 1½ amperes per outlet.

An example of this might be a bedroom with six or seven receptacles, one or two closet lights, and possibly a ceiling paddle fan with a light kit. This would probably be one 15-ampere lighting branch circuit with not much leeway for additional outlets.

Common sense tells us there can be more outlets when the circuit consists of low-wattage loads, and fewer outlets when the circuit consists of higher wattage loads. At best, it's a "guesstimate."

How Many Circuits Do You Need Using the Outlets per Circuit Method? If we were to wire a new house that has a count of 80 general-purpose lighting and receptacle outlets, how many 15-ampere lighting branch circuits do we need?

Where you anticipate somewhat high-wattage loads, figure on connecting 10 outlets per circuit.

$\frac{80}{10} = 8$ (a minimum of eight 15-ampere general-purpose lighting circuits)

Where you anticipate low-wattage loads, figure on connecting 15 outlets per circuit.

$\frac{80}{15} = 5.3$ (a minimum of six 15-ampere general-purpose lighting circuits)

Certainly more circuits than the results of the preceding calculations dictate would be better.

How Many Circuits Do You Need Using the per-Square-Foot Method? This method is more complicated because it involves calculations. In Chapter 3, we discussed how typical 15-ampere lighting branch circuits in homes could be figured at one 15-ampere branch circuit for every 600 ft^2 (55.8 m^2). Although not often used in residential wiring, if you choose to install 20-ampere lighting branch circuits, figure one 20-ampere branch circuit for every 800 ft^2 (74.4 m^2). This results in a good approximation for the minimum number of lighting branch circuits needed and is based on 3 volt-amperes per square foot (33 volt-amperes per square meter of floor area).

A 2400 ft^2 house would have a minimum of

$\frac{2400}{600} =$ four 15-ampere general-purpose lighting branch circuits

With so many electronic and electrical appliances in homes today, this method might prove to be inadequate. For this size house, you would probably want to install more than four 15-ampere general-purpose lighting branch circuits. It is better to be safe than sorry.

The 80% Rule. This method is also complicated because it involves calculations. Add up the wattages and/or the amperes of the loads you anticipate will be connected to a given branch circuit. Then limit the load to not more than 80% of the branch circuit's rating. Although this is a *Code* requirement for continuous loads and typically applies to commercial and industrial applications as found in *NEC 210.19(A)(1)(a)*, *210.20(A)*, *215.2(A)(1)(a)*, *215.3*, and *230.42(A)(1)*, it makes good sense to follow this guideline for most residential loads.

A 15-ampere, 120-volt lighting branch circuit would be calculated as

$15 \times 0.80 =$ 12 amperes maximum connected load

or

12 amperes $\times$ 120 volts $=$ 1440 watts (volt-amperes) maximum connected load

A 20-ampere, 120-volt lighting branch circuit would be calculated as

$20 \times 0.80 =$ 16 amperes (volt-amperes) maximum connected load

or

16 amperes $\times$ 120 volts $=$ 1920 watts (volt-amperes) maximum connected load

Wrap Up. In this residence, estimated loads for general-purpose outlets are figured at 120 volt-amperes (1 ampere). Receptacles in the garage and workshop are figured at 180 volt-amperes (1½ amperes) because these outlets will probably supply portable tools.

Obviously, the 10 to 15 outlets (1 to 1½ amperes per outlet) would not apply to the required small-appliance circuits in the Kitchen, the Laundry area, Workshop, or similar areas that supply a considerable number of plug-in tools and appliances. These circuits are separate issues and are discussed later on in this text.

Divide Loads Evenly

It is mandatory that loads be divided evenly among the various circuits, *210.11(B)*. The obvious reason is to avoid overload conditions on some circuits when other circuits might be lightly loaded. This is common sense. At the Main Service, don't connect the branch circuits and feeders to result in, for instance, 120 amperes on phase A and 40 amperes on phase B. Look at the probable and/or calculated loads to attain as close a balance as possible, such as 80 amperes on each A and B phase.

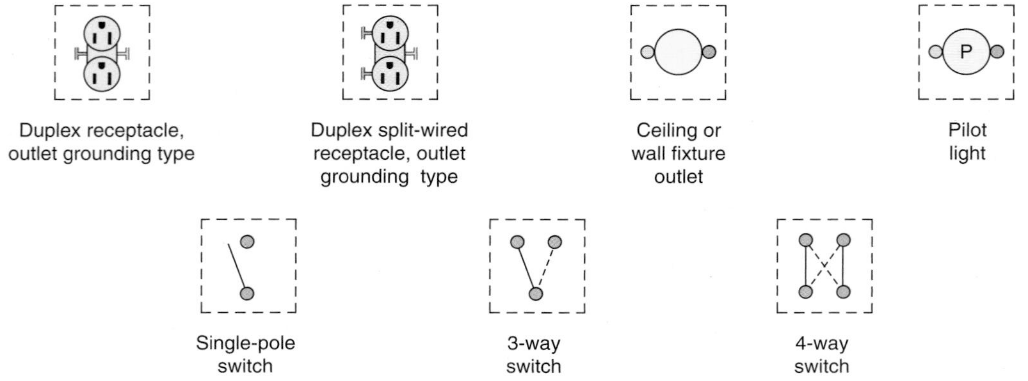

FIGURE 8-4 Pictorial illustrations used on wiring diagrams in this text.

SYMBOLS

The symbols used on the cable layouts in this text are the same as those found on the actual electrical plans for the residence. Refer to Figures 2-6, 2-7, 2-8, 2-9, and 2-10.

Pictorial illustrations are used on all wiring diagrams in this text to make it easy for the reader to complete the wiring diagrams, as in Figure 8-4.

DRAWING A CABLE LAYOUT AND WIRING DIAGRAM

Typical house plans show the electrical symbols on the construction floor plans. Rarely are separate electrical plans provided such as those included in the back of this text. Most residential plans do not provide much detail other than showing the location of receptacles, lighting, and appliances.

For the most part, the electrician gets on the job—studies the plans and then figures out how to make it work according to *Code* and how to do it in a cost-effective way. The electrician decides how to lay out the branch circuits, how many outlets to put on a circuit, how to run the cables, and from where the branch-circuit home runs will be taken.

Skilled electricians, after many years of experience, will generally not prepare detailed wiring diagrams, because they have ability to "think" the connections through mentally. They can visualize the size and type of boxes and cables needed.

Most electricians sketch a Cable Layout, bearing in mind that they are limited to working with 2-wire, 3-wire, and sometimes 4-wire cables. With the requirement in *404.2(C)* for having a neutral at every switch location, how the circuit is run will determine the number of conductors required in each cable. We discussed this extensively in Chapter 5.

The following simple steps will guide you through preparing a cable layout and wiring diagram. In later chapters, you will be asked to draw cable layouts and wiring diagrams. This is an exercise in how to make a color-coded cable layout and wiring diagram. Certainly, there are other ways to lay out the cable runs for this particular "typical" room.

Note that we are feeding the lighting portion of the branch circuit from the switch rather than from the lighting outlet. This allows us to comply with the rules in *404.2(C)* without using 4-conductor cable for a portion of the control circuit. That section requires a neutral conductor to be present in each switch box for possible connection of wiring devices such as timers and occupancy sensors.

DRAWING THE WIRING DIAGRAM OF A LIGHTING CIRCUIT

1. Refer to plans and make a cable layout of all lighting and receptacle outlets, as in Figure 8-5.
2. Draw a wiring diagram showing the traveler conductors for all 3-way switches, if any, as in Figure 8-6.
3. Draw a line between each switch and the outlet or outlets it controls. For 3-way

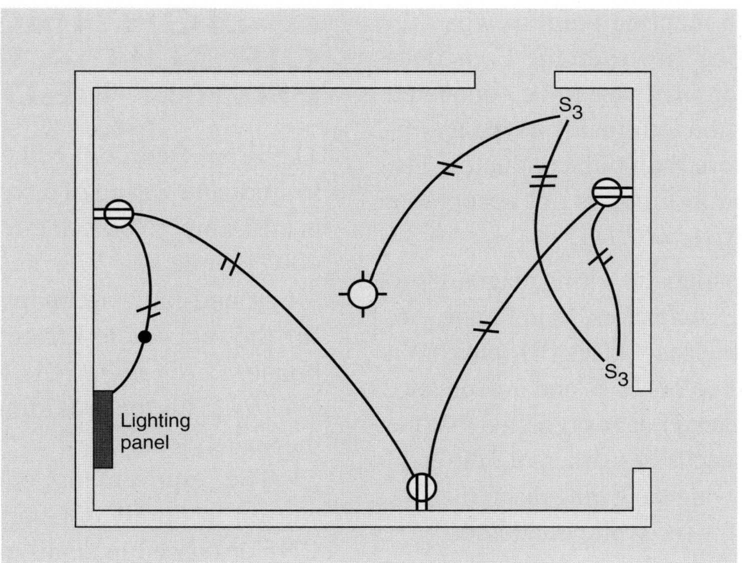

FIGURE 8-5 Typical cable layout.

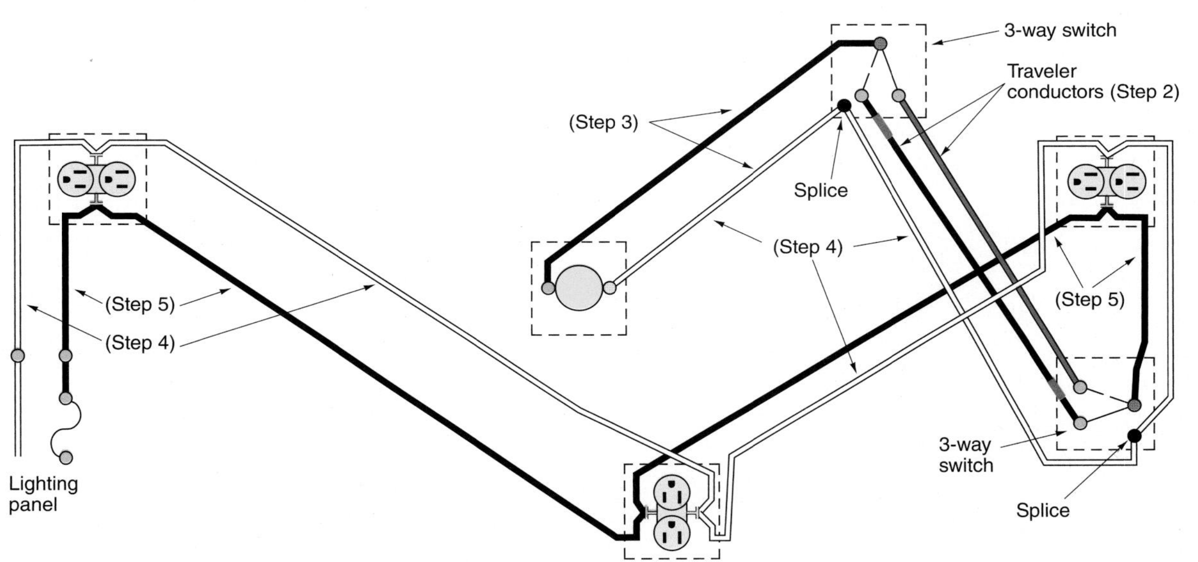

FIGURE 8-6 Wiring diagram of the circuit shown in Figure 8-5.

switches, do this for one switch only. This is the "switch leg."

4. Draw a line from the grounded terminal in the lighting panel to each current-consuming outlet. This line may pass through switch boxes but must not be connected to any switches. This is the white grounded circuit conductor, often called the "neutral" conductor.

 Note: An exception to step 4 may be made for double-pole switches. For these switches, all conductors of the circuit are opened simultaneously. They are rarely used in residential wiring.

5. Draw a line from the ungrounded "hot" terminal in the lighting panel to each switch and to each unswitched outlet. For 3-way switches, do this for one switch only. This is the "feed." The "switch leg" (switch loop, switch return) is covered in step 3.

6. Show splices as small dots where the various wires are to be connected together. In the wiring diagram, the terminal of a switch or outlet

may be used for the junction point of wires. In actual wiring practice, however, the *Code* does not permit more than one wire to be connected to a terminal unless the terminal is a type identified for use with more than one conductor. The standard screw-type terminal is not acceptable for more than one wire, *110.14*.

7. The final step in preparing the wiring diagram is to mark the color of the conductors, as in Figure 8-6. Note that the colors selected—black (B), white (W), and red (R)—are the colors of 2- and 3-conductor cables (refer to Chapter 5). It is suggested that the reader use colored pencils or markers for different conductors when drawing the wiring diagram.

   Use a fine line for the white conductor.

General comment: All cable layouts in this text represent only one way of laying out the circuits. Certainly, other layouts are possible. Some circuits were designed to give the student practice in hooking up similar circuits differently.

LIGHTING BRANCH CIRCUIT A16 FOR THE FRONT BEDROOM

The Front Bedroom is the beginning of our journey to study the wiring of a typical dwelling. Circuit A16 is a 15-ampere branch circuit. This branch circuit involves three key *NEC Code* issues: (1) *210.8(A)(3)* for ground-fault circuit interrupter (GFCI) protection for the receptacle located outside the front of the house, (2) ▶*210.12(A)* for AFCI protection, and (3) *404.2(C)* for providing the neutral conductor at the switch location.◀

The requirements of *NEC 404.2(C)* were discussed in detail in Chapter 5. GFCIs and AFCIs were discussed in detail in Chapter 6.

As we look at this circuit, we find five split-wired (half-switched) receptacle outlets. These will provide the general lighting through the use of table or other lamps that will be plugged into the switched receptacles, as shown in Figure 8-7.

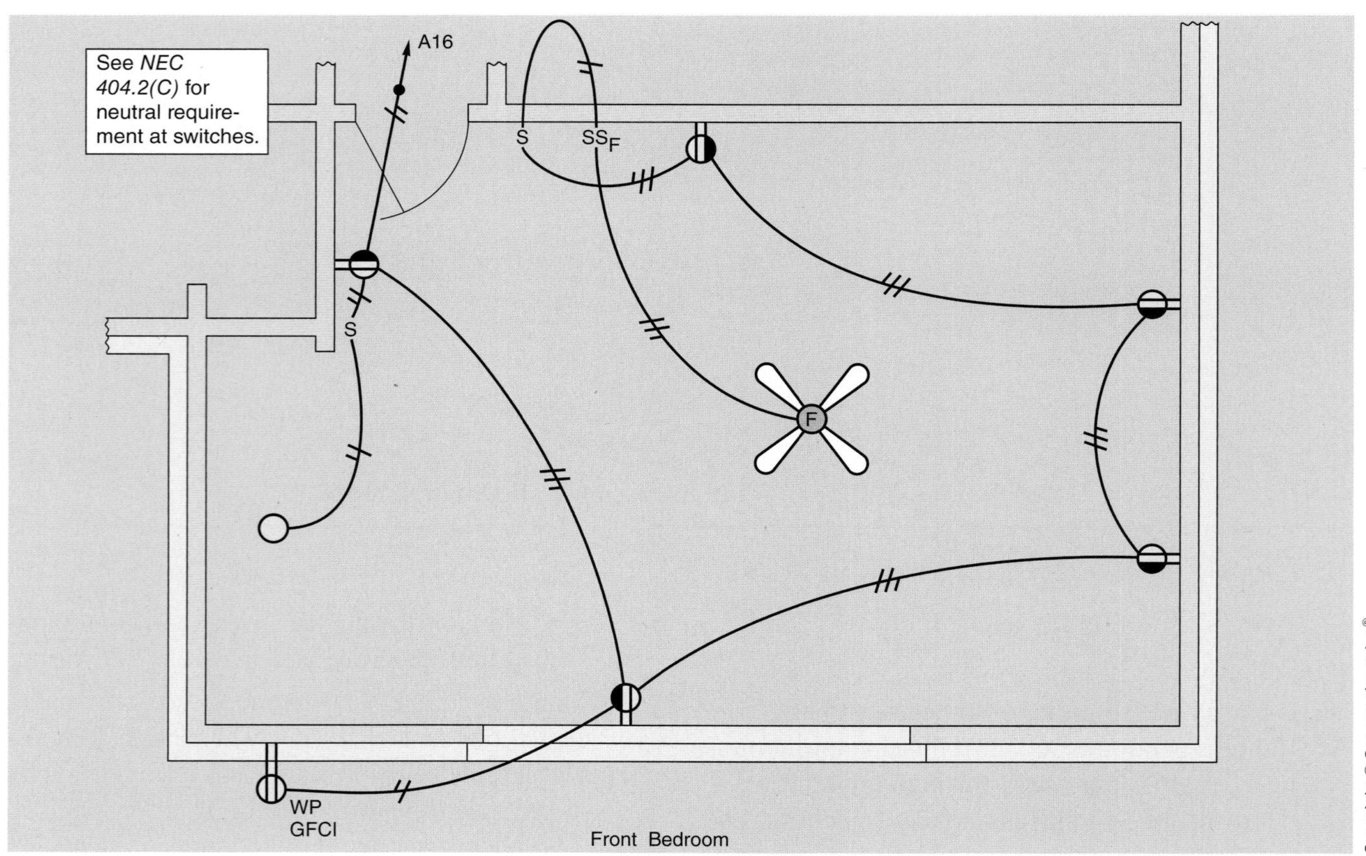

FIGURE 8-7 Cable layout for Front Bedroom circuit, A16.

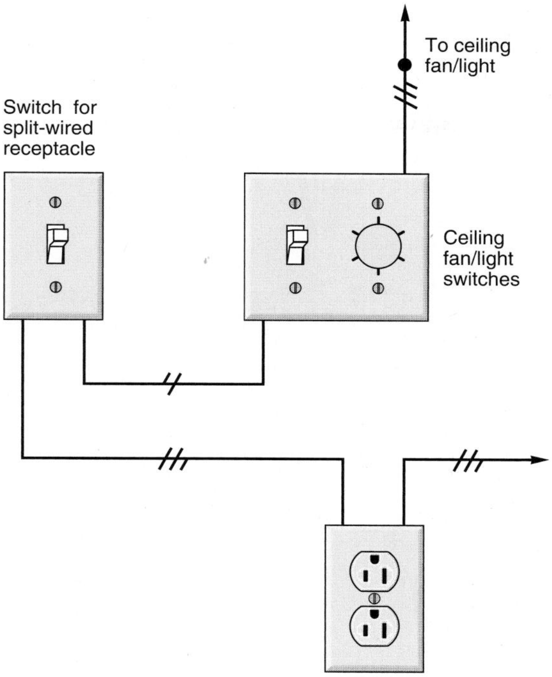

FIGURE 8-8 Conceptual view of how the Front Bedroom switch arrangement is to be accomplished.

Televisions, radios, clocks, computers, CD/DVD/DVR/VCR/VHS players and burners, scanners, printers, telephones, and other electrical and electronic items not intended to be controlled by a wall switch will be plugged into the "hot continuously" receptacle.

Next to the wall switch we find the control for the ceiling fan, which also has a luminaire as an integral part of the fan, as shown in Figure 8-8. A single-pole switch controls the ON–OFF for the light and a speed controller controls the speed of the fan motor.

The clothes closet light is controlled by a single-pole switch outside and to the right of the clothes closet door.

The outdoor weatherproof receptacle connected to this branch circuit meets the requirements of *210.52(E)*, covered in detail in Chapter 3. This receptacle is required to have GFCI protection, *210.8(A)(3)*, discussed in detail in Chapter 6.

As shown in Table 8-1, there are several ways to provide the required AFCI protection of the branch circuit and GFCI protection for receptacles. We discussed the requirements for AFCI and GFCI protection and methods of complying with the rules in Chapter 6 of this text.

Checking the actual electrical plans, we find one television outlet and one telephone outlet are to be installed in the Front Bedroom. Televisions and telephones are discussed in Chapter 25. Table 8-2 summarizes the outlets in the Front Bedroom and the estimated load.

TABLE 8-1

AFCI PROTECTION	GFCI PROTECTION
AFCI circuit breaker in panelboard	GFCI receptacle at outlet
AFCI receptacle at first receptacle outlet connected "feed-through." (Wiring method in compliance with *NEC 210.12(A)*)	GFCI breaker in panelboard
AFCI receptacle at first receptacle outlet connected "feed-through." (Wiring method in compliance with *NEC 210.12(A)*)	GFCI receptacle at outlet
Dual-function AFCI/GFCI circuit breaker in panelboard	

TABLE 8-2

Front Bedroom: outlet count and estimated load (Circuit A16).

DESCRIPTION	QUANTITY	WATTS	VOLT-AMPERES
Receptacles @ 120 watts each	5	600	600
Weatherproof receptacle	1	120	120
Clothes closet fixture	1	75	75
Ceiling fan/light	1		
Three 50-W lamps		150	150
Fan motor (0.75 A @ 120 V)		80	90
TOTALS	8	1025	1035

DETERMINING THE SIZE OF OUTLET BOXES, DEVICE BOXES, JUNCTION BOXES, AND CONDUIT BODIES

Box fill is discussed in detail in Chapter 2. Here is more on the subject. One of the most common problems found in residential wiring are wires jammed into electrical boxes. Factors that have to be considered are the size and number of conductors, the number of wiring devices, cable clamps, luminaire studs, hickeys, and splices. *NEC 314.16* provides the rules for sizing boxes.

Experience has shown that *minimum* sized boxes does not always make for easy working with the splices, conductors, wiring devices, and carefully putting them all into the box without crowding. Do not skimp on box sizing. The small additional cost of larger boxes is worth much more than the grief and added time it will take to trim out the box using *minimum* size boxes. You will be much happier with boxes that have more than the *minimum* size. When it comes to box sizing, think big—not small.

The following example shows how to calculate the *minimum* size box for a particular installation, as in Figure 8-9.

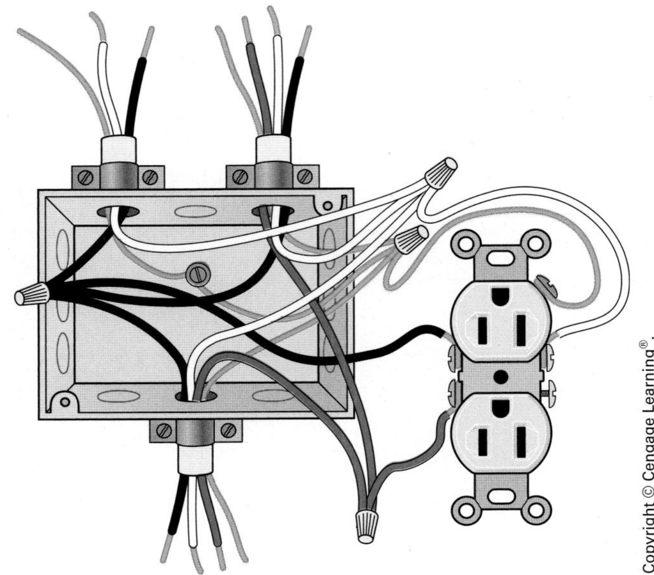

FIGURE 8-9 Determining size of box according to number of conductors.

1. Add the 14 AWG circuit conductors:
   2 + 3 + 3 = 8
2. Add equipment grounding wires (count one only) 1
3. Add two conductors for the receptacle 2

   Total 11 Conductors

Note that the "pigtails" connected to the receptacle in Figure 8-9 need not be counted when determining the correct box size. *NEC 314.16(B)(1)* states, A conductor, no part of which leaves the box, shall not be counted.*

Once the total number of conductors plus the volume count required for wiring devices, luminaire studs, hickeys, and clamps is known, refer to *Table 314.16(A)* and *(B)* and Figure 2-20 to select a box that has sufficient space. Also refer to Table 2-1 for a summary of box fill requirements. For example, a 4 × 2⅛ in. square box with a suitable plaster ring can be used.

The volume of the box plus the space provided by plaster rings, extension rings, and raised covers may be used to determine the total available volume. In addition, it is desirable to install boxes with external cable clamps. Remember that if the box contains one or more devices such as cable clamps, fixture studs, or hickeys, the number of conductors permitted in the box shall be one less than shown in *Table 314.16(A)* for *each type* of device contained in the box.

*Reprinted with permission from NFPA 70-2014.

GROUNDING OF WALL BOXES

The specifications for the residence and the *NEC* require that all metal boxes be grounded. Equipment grounding using nonmetallic-sheathed cable is accomplished by properly connecting the bare equipment grounding conductor in the cable to the metal box. An equipment grounding conductor is used *only* to ground the metal box and must *never* be used as a current-carrying conductor.

When using nonmetallic boxes, the equipment grounding conductor in nonmetallic-sheathed cable is connected to the green hexagon-shaped equipment grounding screw found on all switches and receptacles.

Improper connecting of the bare equipment grounding conductor has resulted in electrocutions!

Methods of connecting equipment grounding conductors are shown in many of the illustrations in Chapters 4 and 5 and in Figure 8-9.

Grounding-type receptacles must be installed on 15- and 20-ampere branch circuits, *406.3 (A)*.

POSITIONING OF SPLIT-WIRED RECEPTACLES

The receptacle outlets shown in the Front Bedroom are called split-wired or half-switched receptacles. The top portion of such a receptacle is "hot" at all times, and the bottom portion is controlled by the wall switch, as in Figure 8-10. It is recommended that the electrician wire the bottom half of the duplex receptacle as the switched section. As a result, when the attachment plug cap of a lamp is inserted into the bottom switched portion of the receptacle, the cord does not hang in front of the unswitched section.

When split-wired receptacles are horizontally mounted, which is common when 4 in. square boxes are used, because the plaster ring is easily fastened to the box in either a vertical or horizontal position, locate the switched portion to the right.

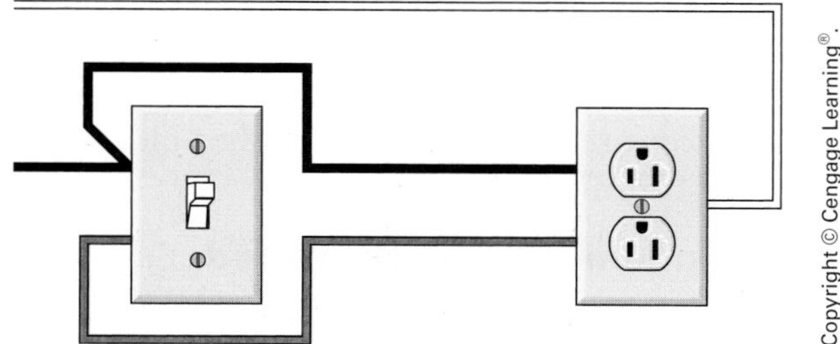

FIGURE 8-10 Split-wired or half-switched wiring for receptacles. The top receptacle is "hot" at all times. The bottom receptacle is controlled by the switch.

POSITIONING OF RECEPTACLES NEAR ELECTRIC BASEBOARD HEATING

Electric heating for a home can be accomplished with an electric furnace, wall-mounted, fan-forced units, electric baseboard heating units, heat pump, or resistance heating cables embedded in plastered ceilings or sandwiched between two layers of drywall sheets on the ceiling.

The important thing to remember at this point is that some baseboard electric heaters may not be permitted to be installed below a receptacle outlet, *210.52 Informational Note*.

According to *210.52*, receptacles that are part of an electric baseboard heating unit may be counted as one of the required number of receptacles in a given space, but only if the receptacles are not connected to the electric baseboard heating branch circuit. This is usually not a problem since the baseboard heater branch circuit is 240-volts with no neutral and the receptacle branch circuit is 120-volts with a neutral.

See Chapter 23 for a detailed discussion of installing electric baseboard heating units and their relative positions below receptacle outlets.

The Front Bedroom does have a ceiling fan/light on the ceiling. Ceiling fans are discussed in Chapter 9.

LUMINAIRES IN CLOTHES CLOSETS

The *NEC* defines a *clothes closet* as A non-habitable room or space intended primarily for storage of garments and apparel.* Many closets in homes are obviously "clothes closets." Others might be called storage closets, linen closets, broom closets, game closets, and laundry closets. These closets become a judgment call as to how to apply the *NEC* rules.

The *NEC* does not require luminaires in clothes closets, *210.70*. However, some local codes and some building codes do. When luminaires are installed in clothes closets, they must be of the correct type and must be installed properly.

*Reprinted with permission from NFPA 70-2014.

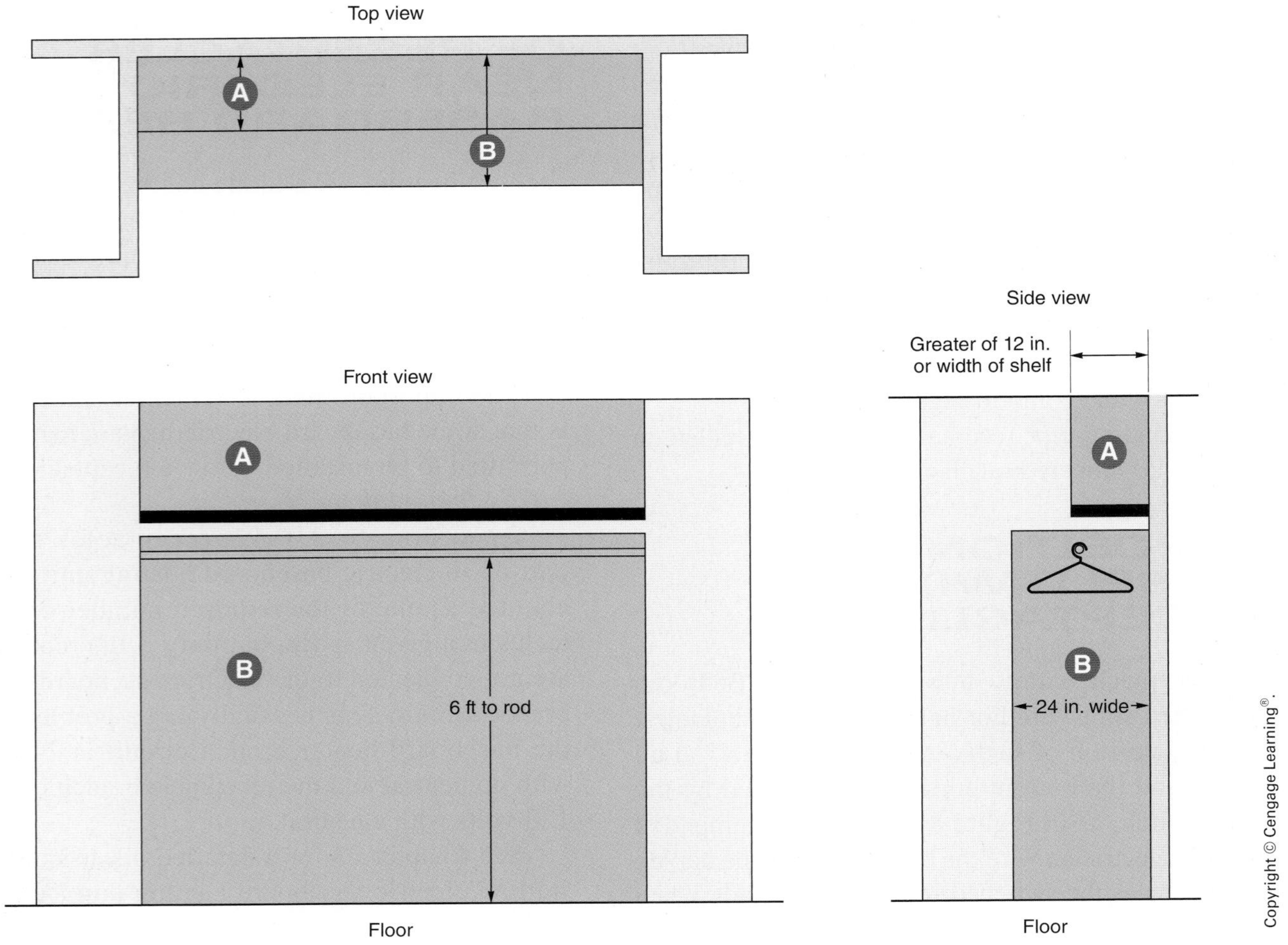

FIGURE 8-11 Typical clothes closet with one shelf and one rod. The shaded area defines "clothes closet storage space." Dimension A is width of shelf or 12 in. (300 mm) from wall, whichever is greater. Dimension B is below rod, 24 in. (600 mm) from wall; see *410.16.*

Clothing, boxes, and other material normally stored in clothes closets are a potential fire hazard. These items may ignite on contact with the hot surface of exposed lamps. Incandescent lamps have a hotter surface temperature than fluorescent lamps. In *NEC 410.16*, there are very specific rules for the location and types of luminaires permitted to be installed in clothes closets. Figures 8-11, 8-12, 8-13, and 8-14, illustrate the requirements of *NEC 410.16.*

In this residence, the closets in the bedrooms and front hall are 24 in. (600 mm) deep. The storage closet in the Recreation Room is 30 in. (750 mm) deep. The shelving is approximately 12 in. (300 mm) wide. Because of the required clearances between luminaires and the storage space, locating luminaires in closets can be difficult. In fact, some clothes closets are so small that luminaires of any kind are not permitted because of clearance requirements.

The *NEC* is very clear about which types of luminaires are permitted in a clothes closet. *NEC 410.16(A)* reads:

(A) Luminaire Types Permitted. Only luminaires of the following types shall be permitted in a closet:

(1) Surface-mounted or recessed incandescent or LED luminaires with completely enclosed light sources

(2) Surface-mounted or recessed fluorescent luminaires

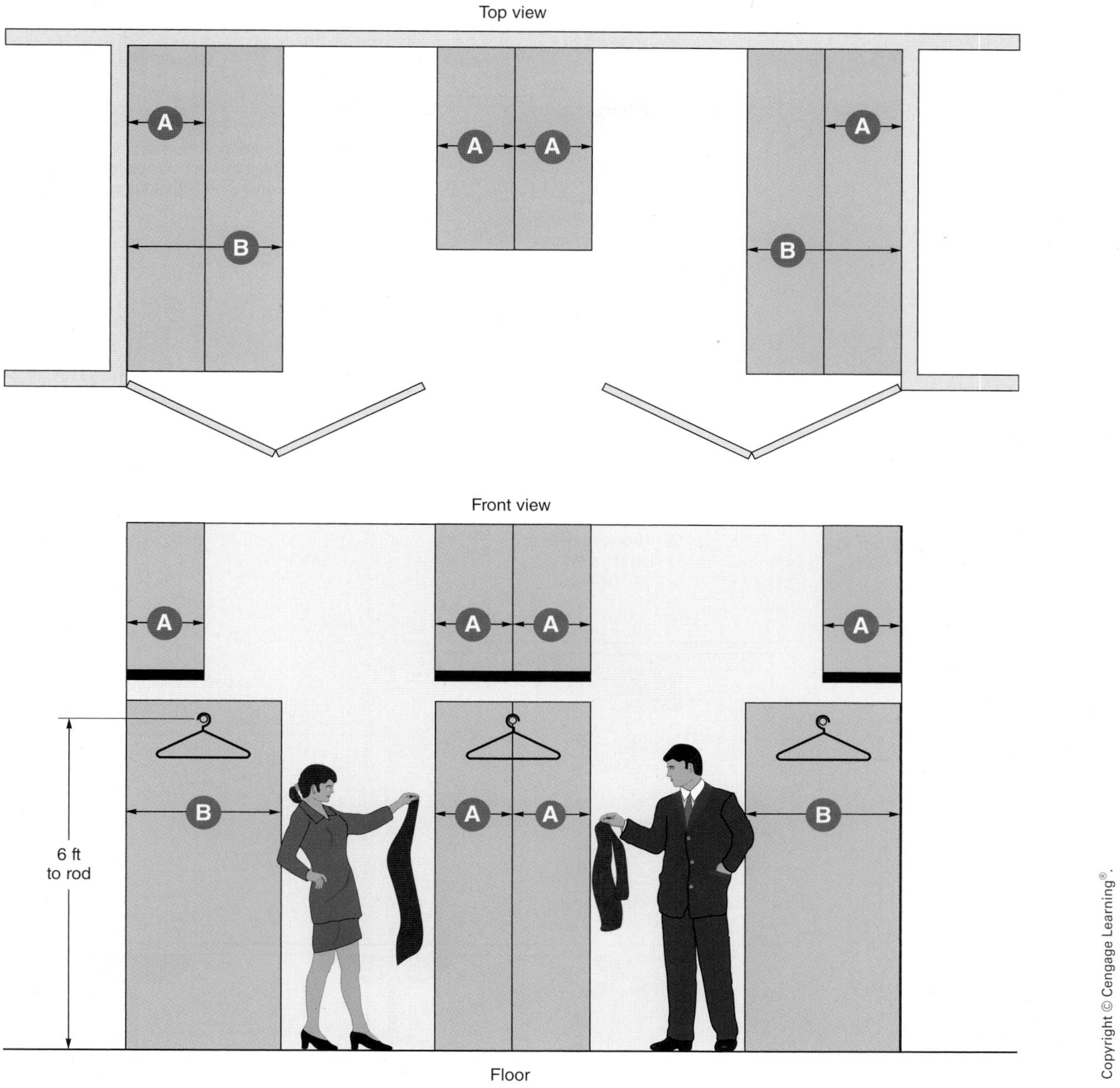

Top view

Front view

6 ft to rod

Floor

FIGURE 8-12 Large walk-in clothes closet where there is access to the center rod from both sides. The shaded area defines "clothes closet storage space." Dimension A is width of shelf or 12 in. (300 mm), whichever is greater. Dimension B is 24 in. (600 mm) from wall; see *410.16*.

(3) Surface-mounted fluorescent or LED luminaires identified as suitable for installation within the closet storage space*

These types are illustrated in Figure 8-14.

Whichever type is chosen, be sure that the minimum clearances are provided as required by *NEC 410.16(C)*.

Figure 8-15 shows typical bedroom-type luminaires.

Figure 8-16 shows unique luminaires referred to as *sconces*. They hug the wall and provide interesting uplighting. They are generally used to complement other lighting. They can be located just about anywhere—hallways, next to a fireplace, foyers, and similar places.

*Reprinted with permission from NFPA 70-2014.

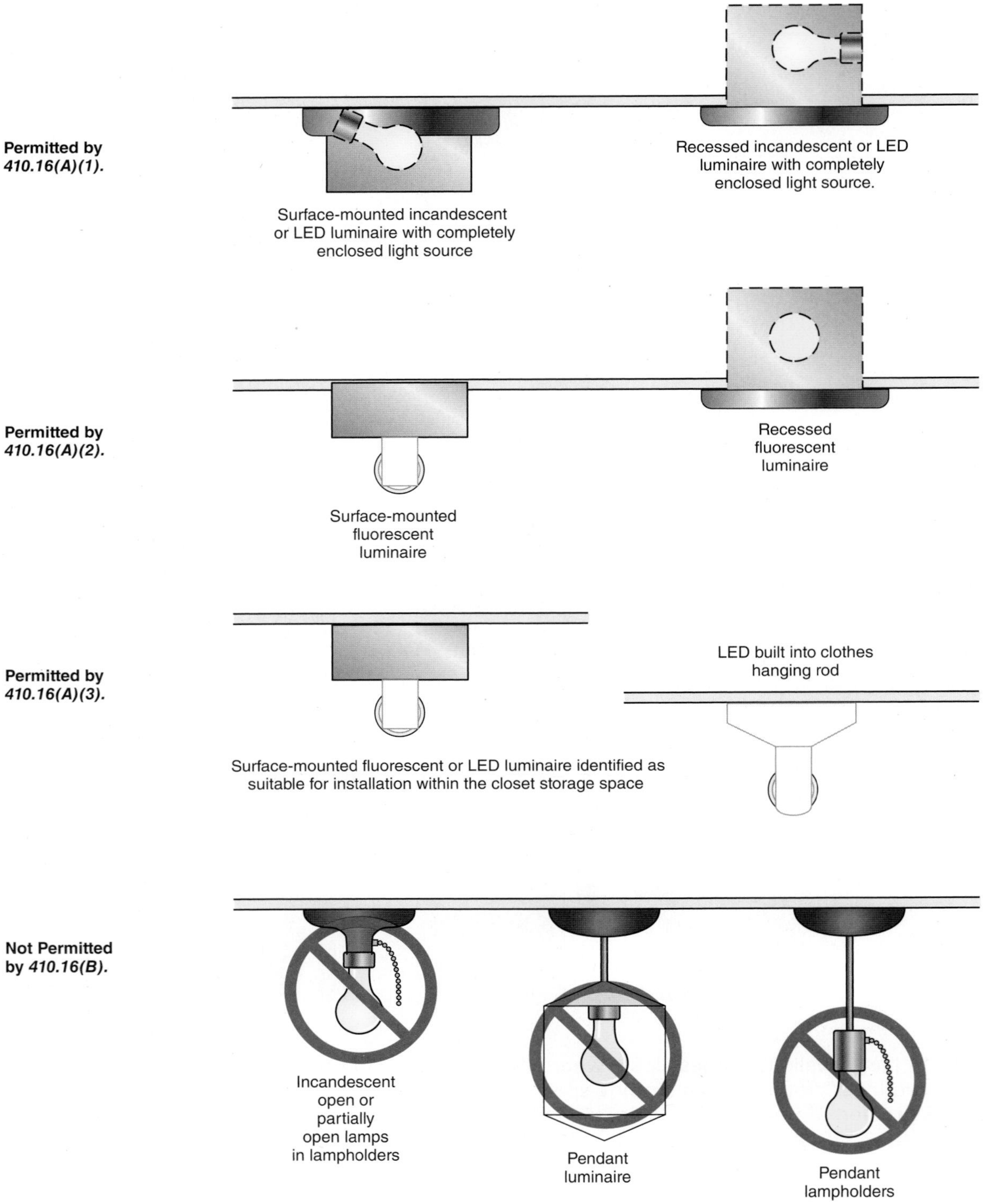

FIGURE 8-13 Illustrations of the types of luminaires permitted in clothes closets. The *Code* does not permit bare incandescent lamps, lampholders, pendant luminaires, or pendant lampholders to be installed in clothes closets.

Location of luminaires in clothes closets

410.16(C)(1):
Minimum clearances for surface-mounted incandescent or LED luminaires that have a completely enclosed light source

12" min.
12" min.
12" min.
STORAGE SPACE

410.16(C)(2):
Minimum clearances for surface-mounted fluorescent luminaires

6" min.
6" min.
6" min.
6" min.
STORAGE SPACE

410.16(C)(3) & (4):
Minimum clearances for recessed fluorescent luminaires and incandescent or LED luminaires that have a completely enclosed light source

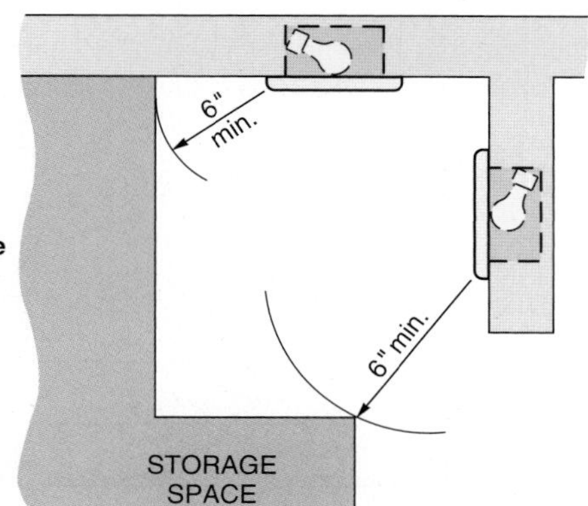

Note:
6 in. = 150 mm
12 in. = 300 mm

FIGURE 8-14 The above illustrations show the minimum clearances required between luminaires and the storage space in clothes closets. See Figure 8-11 and Figure 8-12 for a definition of storage space. *NEC 410.16(C)(3)* does permit surface-mounted fluorescent or LED luminaires to be installed within the storage space, but only if they are identified for this use.

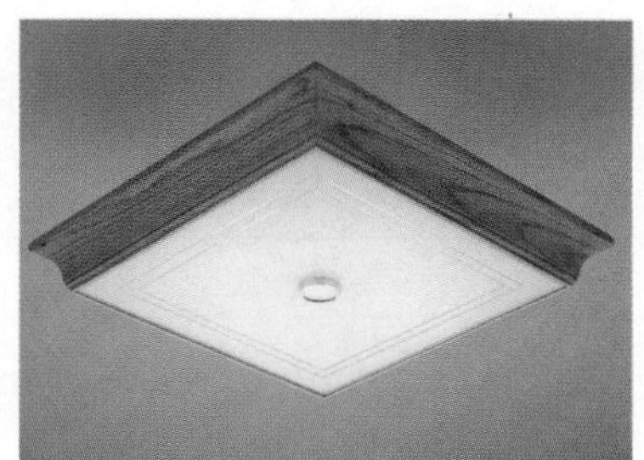

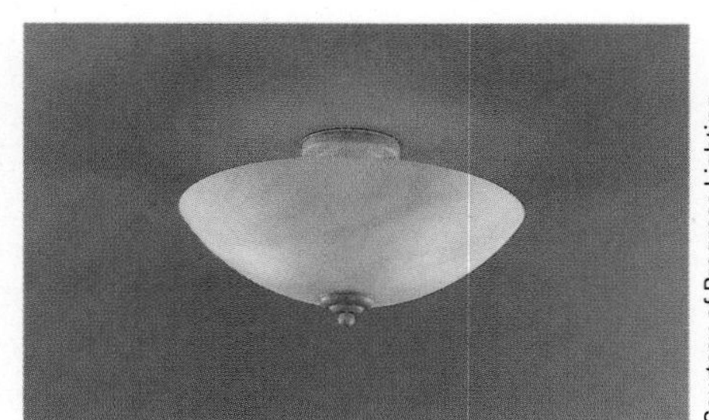

FIGURE 8-15 Typical bedroom-type luminaires.

Courtesy of Progress Lighting.

FIGURE 8-16 Sconces can provide attractive uplighting on a wall. Install approximately 66 in. (1.68 m) to the center from the floor.

REVIEW

Note: Refer to the *Code* or the plans where necessary.

1. Can the outlets in a circuit be arranged in different groupings to obtain the same result? Why? ____________________

2. Is it good practice to have outlets on different floors on the same circuit? Why? ____________________

3. A good rule to follow is to never load a circuit to more than ____________ % of the branch-circuit rating.

4. The *NEC* does not limit the number of lighting and receptacle outlets permitted on one branch circuit for residential installations. A guideline that is often used by electricians for house wiring is to allow ____________ to ____________ amperes per outlet.

   For a 15-ampere circuit, this results in ____________ to ____________ outlets on the branch circuit.

5. For residential wiring, not less than one 15-ampere lighting branch circuit should be provided for every ______ ft^2 of floor area. If 20-ampere lighting branch circuits are used, provide one 20-ampere lighting branch circuit for every ______ ft^2 of floor area.

6. For this residence, what are the estimated wattages used in determining the loading of branch circuit A16?

   Receptacles ______ watts (volt-amperes)

   Clothes closet fixture ______ watts (volt-amperes)

7. a. What is the ampere rating of circuit A16? ______

   b. What is the conductor size of circuit A16? ______

8. The *NEC* requires what type of unique protection for all 120-volt, 15- and 20-ampere branch-circuit outlets in residential homes? In which section of the *NEC* is this requirement found? ______

9. How many and what type of receptacles are connected to this circuit? ______

10. What main factors influence the choice of wall boxes? ______

11. How is a wall box grounded? ______

12. What is a split-wired receptacle? ______

13. Is the switched portion of an outlet mounted toward the top or the bottom? Why?

14. The following problems pertain to luminaires in clothes closets.
    a. Does the *Code* permit bare incandescent lamps to be installed in porcelain key-less or porcelain pull-chain lampholders?
    b. Does the *Code* permit bare fluorescent lamp luminaires to be installed?
    c. Does the *Code* permit pendant lampholders to be installed?
    d. What is the minimum clearance from the storage area to surface-mounted incandescent luminaires?
    e. What is the minimum clearance from the storage area to surface-mounted fluorescent luminaires?
    f. What is the minimum distance between recessed incandescent or recessed fluorescent luminaires and the storage area?
    g. Define the "storage area."
    h. If a clothes hanging rod is installed where there is access from both sides, such as might be found in a large walk-in clothes closet, define the storage area under that rod.

15. How many switches are in the bedroom circuit, what type are they, and what do they control?

16. The following is a layout of the lighting circuit for the Front Bedroom. Using the cable layout shown in Figure 8-7, make a complete wiring diagram of this circuit. Use colored pencils or pens.

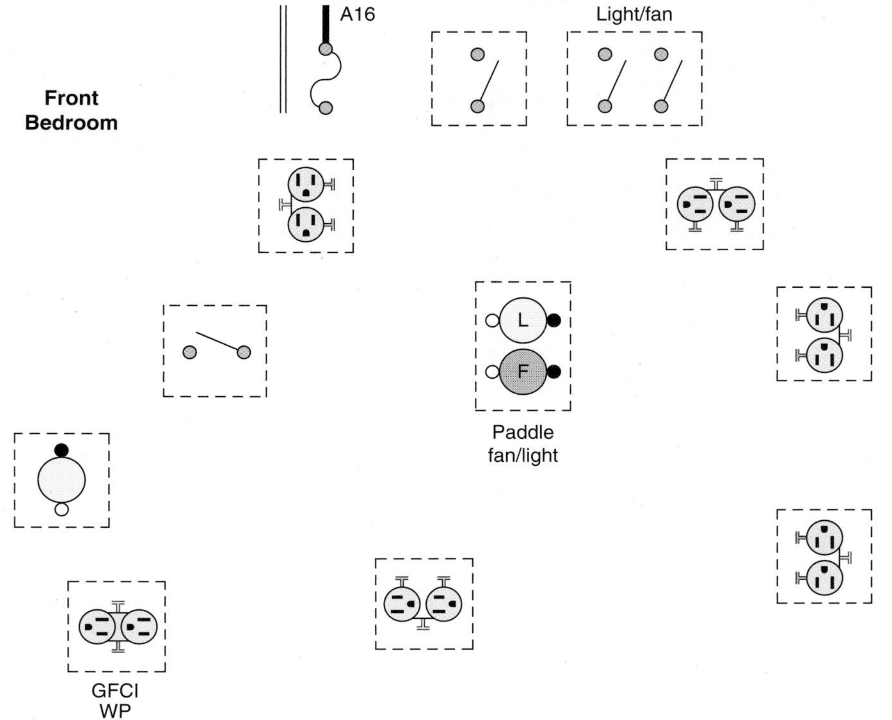

17. When planning circuits, what common practice is followed regarding the division of loads? ______________________________

____

18. The *Code* uses the terms *watts*, *volt-amperes*, *kW*, and *kVA*. Explain their significance in calculating loads. ______________________________

____

____

____

____

19. How many 14 AWG conductors are permitted in a device box that measures 3 in. × 2 in. × 2¾ in.? ______________________________

20. A 4 in. × 1½ in. octagon box has one cable clamp and one luminaire stud. How many 14 AWG conductors are permitted? ______________________________

____

____

____

____

CHAPTER **9**

Lighting Branch Circuit for the Master Bedroom

OBJECTIVES

After studying this chapter, you should be able to

- draw the wiring diagram of the cable layout for the Master Bedroom.
- understand that AFCIs are required for bedrooms.
- study *Code* requirements for the installation of ceiling suspended (paddle) fans.
- estimate the probable connected load for a room based on the number of luminaires and outlets included in the circuit supplying the room.
- gain more practice in determining box sizing based on the number of conductors, devices, and clamps in the box.

INTRODUCTION

The discussion in Chapter 3 and Chapter 8 about grouping outlets, estimating loads, selecting wall box sizes, and drawing wiring diagrams can also be applied to the circuit for the Master Bedroom.

The residence panel schedules show that the Master Bedroom is supplied by Circuit A19. Circuit A19 is a 15-ampere branch circuit. In accordance with *NEC® 210.12(A)*, overcurrent protection for this circuit is provided by a 15-ampere arc-fault circuit interrupter (AFCI) circuit breaker in Panel A.

In Chapter 8, we discussed how ground-fault circuit interrupter (GFCI) and AFCI protection could be achieved in the Front Bedroom. The same principle applies to the Master Bedroom.

AFCI and GFCI requirements are discussed in detail in Chapter 6.

Because Panel A is located in the basement below the wall switches next to the sliding doors, the home run for Circuit A19 is brought into the outdoor weatherproof receptacle. This results in six conductors in the outdoor receptacle box. The home run could have been brought into the corner receptacle outlet in the bedroom. Again, it is a matter of studying the circuit to determine the best choice for conservation of cable or conduit in these runs and to economically select the correct size of wall boxes.

LIGHTING BRANCH CIRCUIT A19 FOR THE MASTER BEDROOM

Figure 9-1, Table 9-1, and the electrical plans show that Circuit A19 has four split-wired receptacle outlets in this bedroom, one outdoor weatherproof GFCI receptacle outlet, two closet luminaires, each on a separate switch, plus a ceiling fan/luminaire, one telephone outlet, and one television outlet.

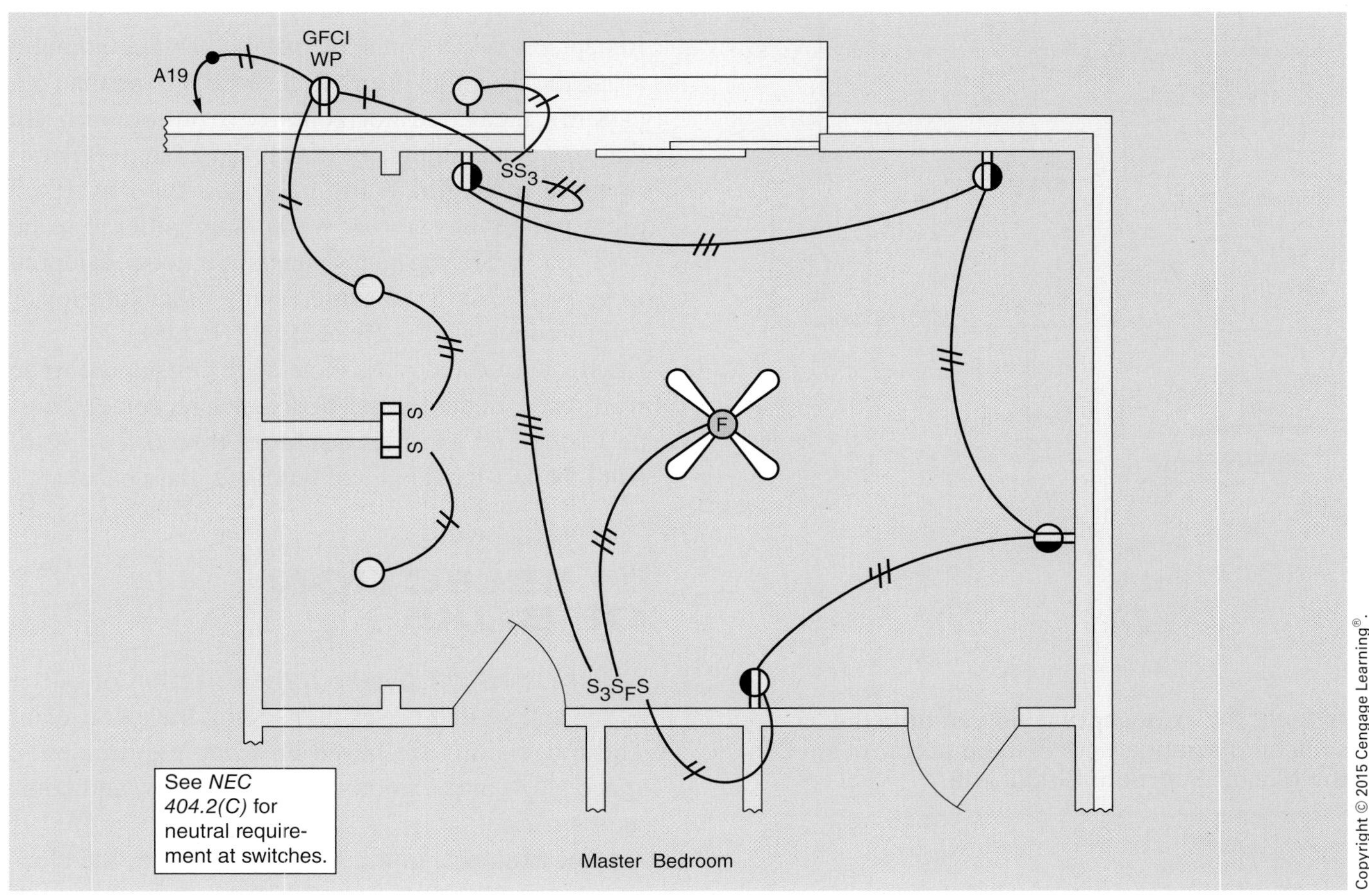

FIGURE 9-1 Cable layout for Master Bedroom.

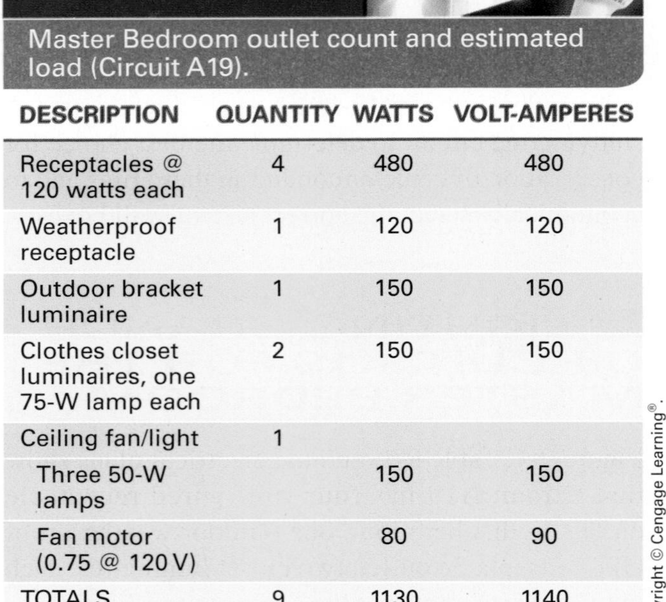

TABLE 9-1

Master Bedroom outlet count and estimated load (Circuit A19).

DESCRIPTION	QUANTITY	WATTS	VOLT-AMPERES
Receptacles @ 120 watts each	4	480	480
Weatherproof receptacle	1	120	120
Outdoor bracket luminaire	1	150	150
Clothes closet luminaires, one 75-W lamp each	2	150	150
Ceiling fan/light	1		
Three 50-W lamps		150	150
Fan motor (0.75 @ 120 V)		80	90
TOTALS	9	1130	1140

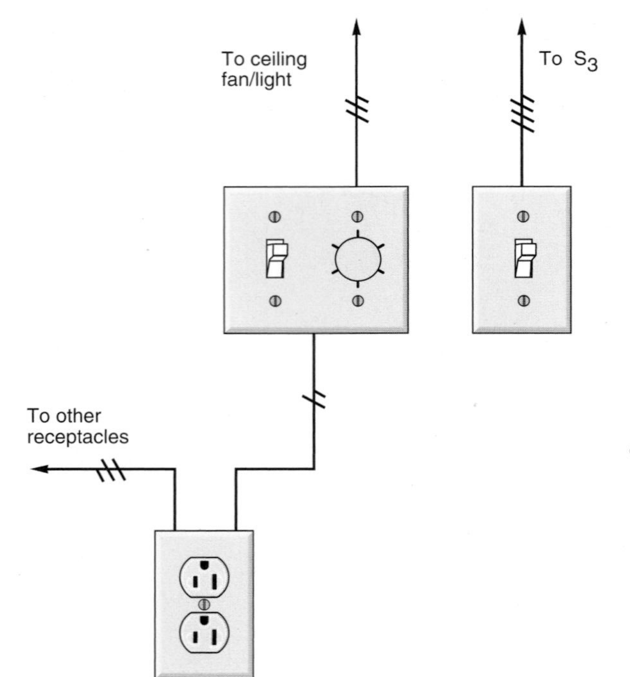

FIGURE 9-2 Conceptual view of how the switching arrangement is to be accomplished in the Master Bedroom Circuit A19.

In addition, an outdoor bracket luminaire is located adjacent to the sliding door and is controlled by a single-pole switch just inside the sliding door. Note that the single-pole switch (device) for the outdoor luminaire is located inside the bedroom. As a result, the wiring device is required to be protected by an AFCI. The AFCI circuit breaker at the origin of the branch circuit provides this protection.

The split-wired receptacle outlets are controlled by two 3-way switches. One is located next to the sliding door. As in the Front Bedroom, Living Room, and Study/Bedroom, the use of split-wired receptacles offers the advantage of having switch control of one of the receptacles at a given outlet, while the other receptacle remains "live" at all times. See Figures 3-3 and 12-15 for the definition of a receptacle.

Next to the bedroom door are a 3-way switch and the ceiling fan/light control, which is installed in a separate 2-gang box, as shown in Figure 9-2.

SLIDING GLASS DOORS AND FIXED GLASS PANELS

Receptacle outlets must be installed so that no point along the floor line in any wall space is more than 6 ft (1.8 m), measured horizontally, from an outlet in that space, including any wall space 24 in. (600 mm) or more in width. Wall space occupied by fixed glass panels in exterior walls is considered to be wall space. Sliding glass panels are not considered to be wall space, the same as any other interior or exterior door. This is covered in *210.52(A)(2)*.

Because fixed glass panels are considered to be wall space, in the Master Bedroom a receptacle must be located so as to be not more than 6 ft (1.8 m) from the left-hand edge of the fixed glass panel.

SELECTION OF BOXES

As discussed in Chapter 2, the selection of outlet boxes and switch boxes is made by the electrician. These decisions are based on *Code* requirements, space allowances, good wiring practices, and common sense.

For example, in the Master Bedroom, the electrician may decide to install a 4 in. square box with a 2-gang raised plaster cover at the box location next to the sliding door. Or two sectional switch boxes

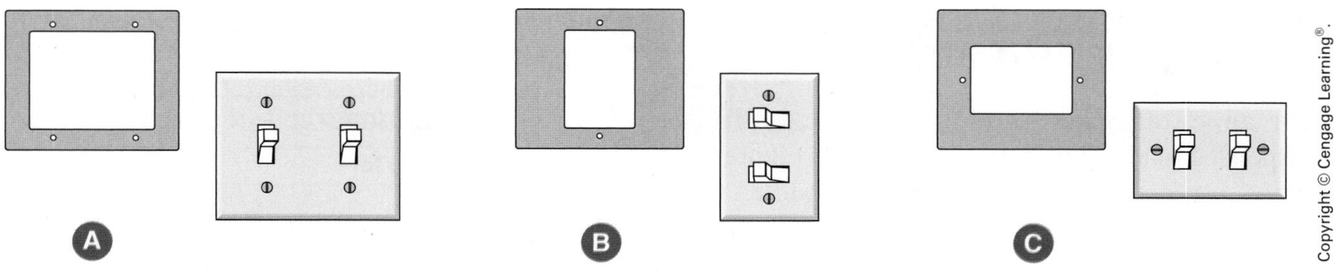

FIGURE 9-3 (A) shows two switches and a wall plate that attach to a 2-gang, 4 in. square raised plaster cover or to two device boxes that have been ganged together. (B) and (C) show two interchangeable-type switches and wall plate that attach to a 1-gang, 4 in. square raised plaster cover or to a single-device (switch) box. These types of wiring devices may be mounted vertically or horizontally.

TABLE 9-2

Height and width of standard wall plates.

NO. OF GANGS	HEIGHT	WIDTH
1	4½ in. (114.3 mm)	2¾ in. (69.85 mm)
2	4½ in. (114.3 mm)	4⁹⁄₁₆ in. (115.9 mm)
3	4½ in. (114.3 mm)	6⅜ in. (161.9 mm)
4	4½ in. (114.3 mm)	8³⁄₁₆ in. (207.9 mm)
5	4½ in. (114.3 mm)	10 in. (254 mm)
6	4½ in. (114.3 mm)	11¹³⁄₁₆ in. (300 mm)

ganged together may be installed at that location, as shown in Figure 9-3 and Table 9-2. Two-gang nonmetallic boxes are readily available and are particularly suitable for branch circuit wiring using nonmetallic cable.

The type of box to be installed depends on the number of conductors entering the box. The suggested cable layout, Figure 9-1, shows that four cables enter this box for a total of 14 conductors (ten 14 AWG circuit conductors and four equipment grounding conductors).

The requirements for calculating box fill are found in *314.16*. These requirements are summarized in Figure 2-35. Also refer to *Table 314.16(A)*, *314.16(B)*, and Table 2-1.

In this example, we have 10 circuit conductors, four equipment grounding conductors (counted as one conductor), four cable clamps (counted as one conductor), one single-pole switch (counted as two conductors), and one 3-way switch (counted as two conductors), for a total of 16 conductors.

1. Count the circuit conductors 2 + 2 + 3 + 3 =	10
2. Add one for one or more equipment grounding conductors	1
3. Add two for each switch (2 + 2)	4
4. Add one for one or more cable clamps	1
TOTAL	16

Now look at the Quik-Chek Box Selection Guide (Figure 2-35), and select a box or a combination of gangable device boxes that are permitted to hold 16 conductors.

Possibilities are:

1. Gang two 3 in. × 2 in. × 3½ in. device boxes together (9 × 2 = eighteen 14 AWG conductors allowed).
2. Install a 4 in. × 4 in. × 2⅛ in. square box (fifteen 14 AWG conductors allowed). Note that the raised plaster cover, if marked with its cubic-inch volume, can increase the maximum number of conductors permitted for the combined box and raised cover. See Figure 2-37.
3. Select a 2-gang nonmetallic box that has a marked volume of not less than 32 cubic inches. For nonmetallic boxes, *NEC Table 314.16(B)* requires not less than 2 $in.^3$ per volume allowance. The term *volume allowance* is used rather than *conductor* because a volume allowance is required for objects such as cable clamps and wiring devices.

CEILING-SUSPENDED (PADDLE) FANS

For appearance and added comfort, ceiling-suspended (paddle) fans are very popular. Figure 9-4 shows a typical ceiling-suspended (paddle) fan. Ceiling-suspended (paddle) fans rotate slowly in the range of approximately 60 r/min to 250 r/min for residential fans. Without forced air movement, warm air rises to the ceiling, and cool air drops to the floor. Air movement results in increased evaporation from our skin, and we feel cooler. Fans that hug the ceiling are not quite as efficient in moving air as those that are suspended a few inches or more from the ceiling.

Most ceiling-suspended (paddle) fans have a reversing switch, a HIGH/MED/LOW/OFF speed control pull chain switch for the fan, and an ON–OFF pull chain switch for the light accessory kit. Upscale models feature remote ON–OFF infinite speed and light control, eliminating the need for wiring of wall switch speed control. These remote controls are similar to those used with television, VCR, stereo, and cable channel selectors. Wiring for wall switches is discussed later in this chapter.

What Direction Should a Ceiling-Suspended (Paddle) Fan Rotate?

Major manufacturers of ceiling-suspended (paddle) fans recommend directing the air downward (counterclockwise) in the summer to benefit from the "wind chill" effect (see Figure 9-5[A]), and upward (clockwise) in the winter (see Figure 9-5[B]). Some people like air blowing on them; others do not. Upward or downward, the key is to get the stratified air moving. Try different speeds and direction, and do whatever feels best. Read the manufacturer's suggestions.

FIGURE 9-4 Typical home-type ceiling fan.

SUGGESTED FAN SIZES

For Rooms Up To:	Fan Size
144 ft²	42 in.
225 ft²	44–48 in.
400 ft²	52–54 in.
500 ft²	56 in.

Supporting Ceiling-Suspended (Paddle) Fans

This is a very important issue. There have been many reports of injuries (contusions, concussions, lacerations, fractures, head trauma, etc.) caused by the falling of a ceiling-suspended (paddle) fan. There is also the possibility of starting an electrical fire when the wires in the outlet box and fan canopy "short-circuit" as the fan falls from the ceiling. These problems are usually traced back to an improper installation.

The safe supporting of a ceiling-suspended (paddle) fan involves three issues:

1. the actual weight of the fan
2. the twisting and turning motion when the fan is started
3. vibration caused by unbalanced fan blades

Here's the scoop for installing ceiling-suspended (paddle) fans. These requirements are found in *NEC 314.27(C)* and *422.18*, and in the UL *White Book*.

The outlet box or outlet box system

- must be listed and marked as "Acceptable for fan support." Always look for this on the product.
- is permitted to be metallic or nonmetallic.
- must not support more than 70 lb (32 kg).
- if designed to support more than 35 lb (16 kg), but not more than 70 lb (32 kg), will be marked "Acceptable for fan support up to 70 lb (32 kg)."
- has a weight limitation that includes the fan and light kit.
- must be installed according to the manufacturer's instructions.

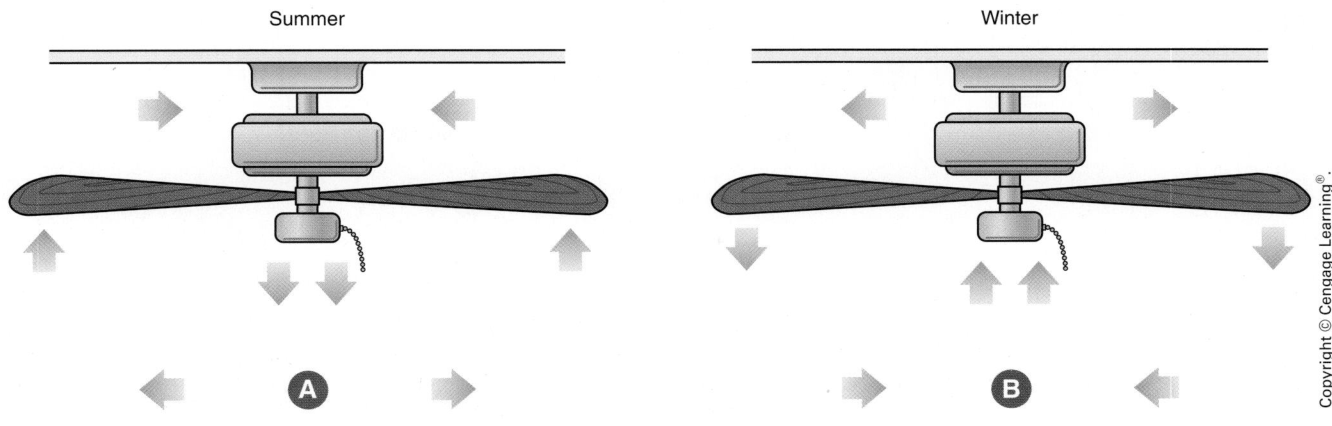

FIGURE 9-5 Suggested rotation of a ceiling-suspended (paddle) fan for winter and summer use.

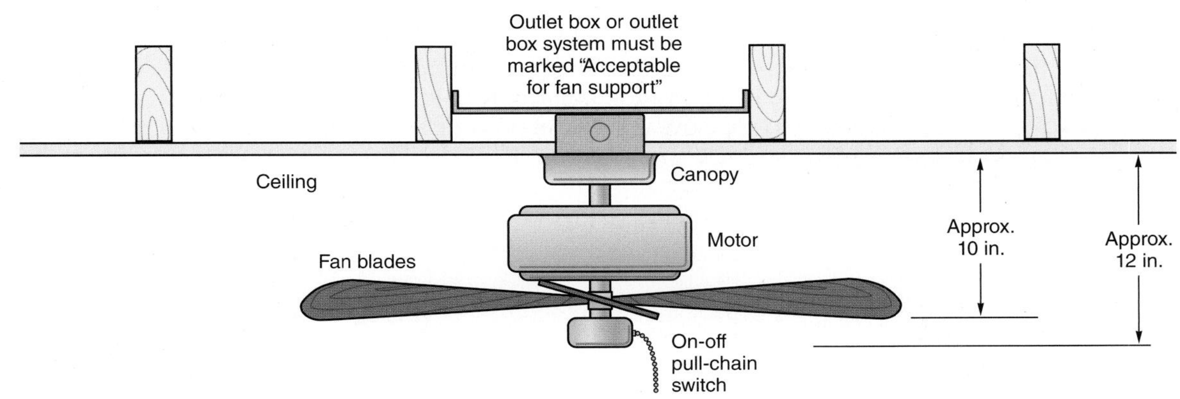

FIGURE 9-6 Typical mounting of a ceiling-suspended (paddle) fan.

- ▶Where spare, separately switched, ungrounded conductors are provided to a ceiling mounted outlet box, in a location acceptable for a ceiling-suspended (paddle) fan in single, two-family, or multi-family dwellings, the outlet box or outlet box system shall be listed for sole support of a ceiling-suspended (paddle) fan.◀*

What Is an Outlet Box System?

This includes such things as the outlet box, hanger, clamps, and other mounting hardware.

New Work. Install ceiling boxes marked "Acceptable for fan support" in all locations where it is likely that a ceiling-suspended (paddle) fan might be installed. This generally includes all habitable rooms.

Ceiling boxes installed close to a wall, in hallways, in closets, and similar nonhabitable rooms, and ceiling boxes installed for smoke alarms, fire alarms, and security devices do not require fan support boxes.

Some electricians are tired of trying to outguess what *is likely* and what the homeowner might do at some later date. They install suitable fan support outlet boxes for all ceiling outlets other than those that are clearly intended for smoke alarms, fire alarms, and security devices.

Old Work. When replacing an existing luminaire with a ceiling-suspended (paddle) fan, make sure that the existing outlet box is listed and marked as suitable for fan support. In older homes, it is unlikely that you will find this marking. You are taking a big chance if you secure a ceiling fan to an outlet box that is not listed. It might fall on someone's head! You may have to support the fan from the building structure. Carefully read and follow the manufacturer's installation instructions.

See Figure 9-6 for a typical mounting of a ceiling-suspended (paddle) fan.

*Reprinted with permission from NFPA 70-2014.

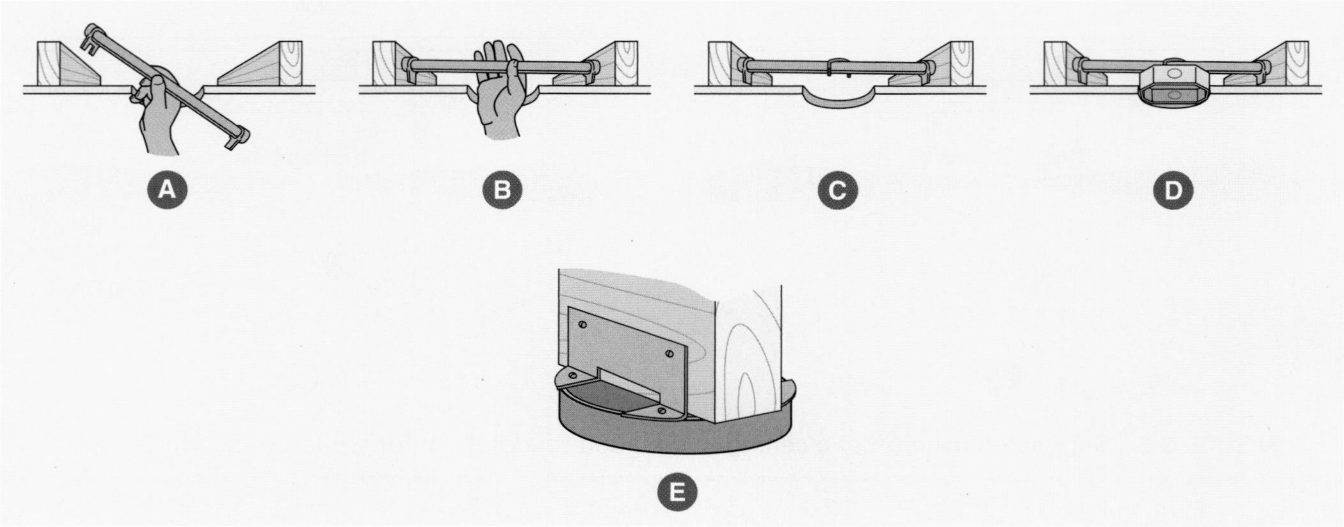

A B C D E

Courtesy of Legrand/Pass & Seymour.

FIGURE 9-7 (A), (B), (C), and (D) show how a listed ceiling fan hanger and box assembly is installed through a properly sized and carefully cut hole in the ceiling. This is done for existing installations. Similar hanger/box assemblies are used for new work. The hanger adjusts for 16 in. (400 mm) and 24 in. (600 mm) joist spacing but can be cut shorter if necessary. (E) shows a type of box listed and identified for the purpose where the fan is supported from the joist, independent of the box, as required by *422.18* and *314.27(C)* for fans that weigh more than 35 lb (16 kg).

You always have the choice of hanging a ceiling-suspended (paddle) fan independent of the outlet box, as illustrated in Figure 9-7(E).

Figure 9-7 shows one type of box/hanger assembly that can be used for new work or for existing installations.

Figure 9-8 shows an outlet box that is listed for both luminaire and fan support. This box has No. 8-32 holes for luminaire mounting and No. 10-32 holes for fan support.

Figure 9-9 shows the wiring for a fan/light combination with the supply at the fan/light unit. Figure 9-10 shows the fan/light combination with the supply at the switch. Figure 9-7 shows two types of ceiling fan hanger/box assemblies. Many other types are available.

Figure 9-11 illustrates a typical combination light switch and three-speed fan control. The electrician is required to install a deep 4 in. box with a 2-gang raised plaster ring for this light/fan control.

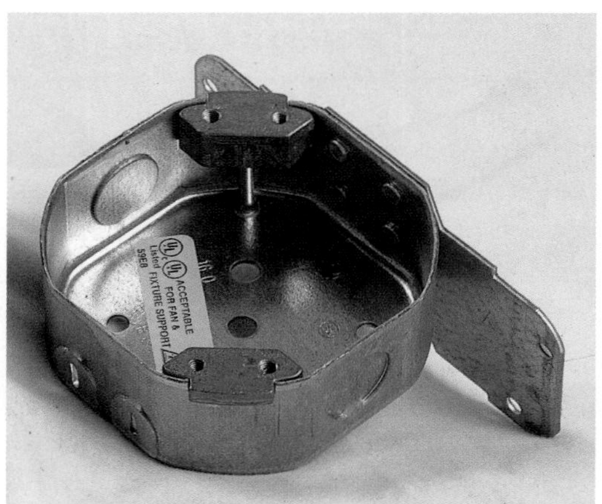

Courtesy of Legrand/Pass & Seymour.

FIGURE 9-8 A special outlet box for the support of a luminaire or a ceiling-suspended (paddle) fan. There are no ears on the box. Instead, special strong mounting brackets are provided. Note that the mounting holes on the brackets are of two sizes: one set of holes is tapped for No. 8-32 screws (for a luminaire), and the other set of holes is tapped for No. 10-32 screws (for a ceiling-suspended [paddle] fan).

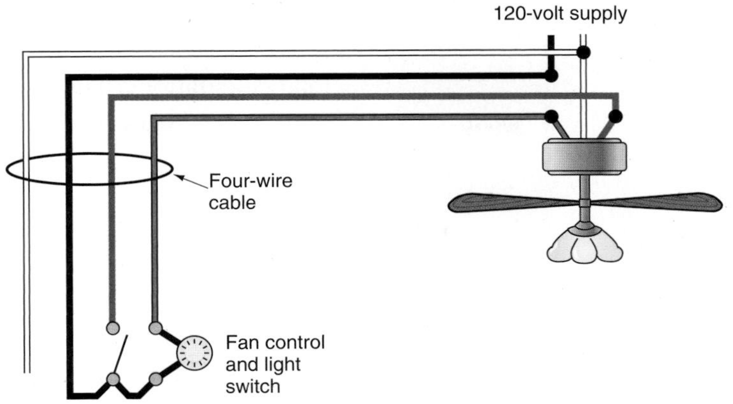

FIGURE 9-9 Fan/light combination with the supply at the fan/light unit.

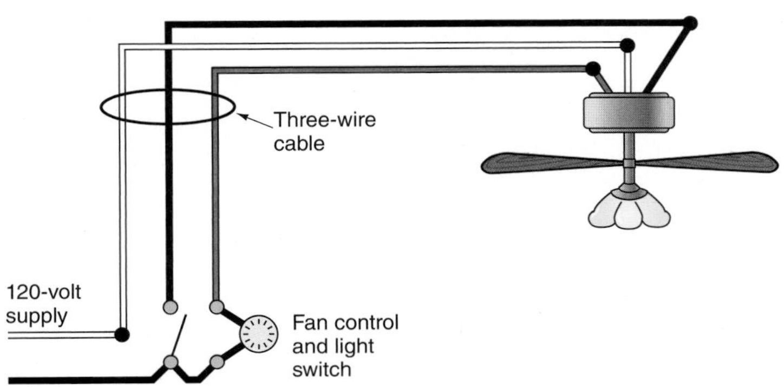

FIGURE 9-10 Fan/light combination with the supply at the switch.

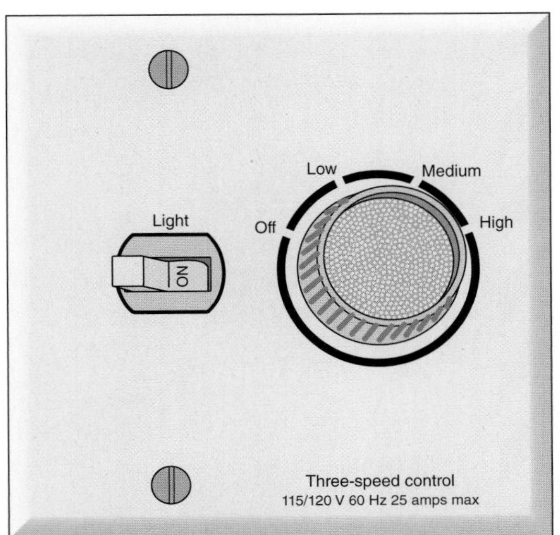

FIGURE 9-11 Combination light switch and three-speed fan switch.

Figure 9-12 illustrates a "slider" switch for speed control of a fan, and a dual control that provides 4-speed fan control and two-level HI/LO dimming for the incandescent lamps. Figure 9-13 shows two other styles of fan/light controls. Dual controls are needed where the ceiling fan also has a luminaire. These controls fit into a deep single-gang device box or, preferably, into a 4 in. square box with single-gang plaster cover. Be careful of "box fill." Make sure the wall box is large enough. Dimmer and speed control devices are quite large when compared to regular wall switches. Do not jam the control into the wall box.

A typical home-type ceiling fan motor draws 50 to 100 watts (50–100 volt-amperes). This is approximately 0.4 to slightly over 0.8 ampere.

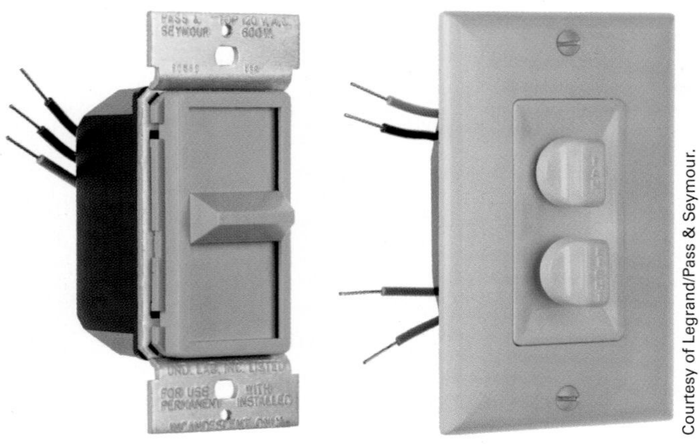

Courtesy of Legrand/Pass & Seymour.

FIGURE 9-12 The "slider" switch is for speed control of a fan. The dual-control switch provides 4-speed fan control and a two-level HI/LO dimming for an incandescent lamp.

Courtesy of Lutron Electronics Co., Inc.

FIGURE 9-13 Two types of fan/light controls. One has a toggle switch for turning a ceiling fan off and on, plus a "slider" that provides 3-speed control of the fan. The other has a toggle switch for turning a light off and on, plus two "sliders," one for 3-speed control of a ceiling fan and the other for full range dimming of lighting.

Ceiling fan/light combinations increase the load requirements somewhat. Some ceiling fan/light units have one lamp socket, whereas others have four or five lamp sockets.

For the ceiling fan/light unit in the Master Bedroom, 240 volt-amperes were included in the load calculations. The current draw is

$$I = \frac{VA}{V} = \frac{240}{120} = 2 \text{ amperes}$$

For most fans, switching on and off may be accomplished by the pull-switch, an integral part of the fan (HIGH/MED/LOW/OFF), or with a solid-state switch having an infinite number of speeds. Some of the more expensive models offer a remote control similar to that used with television, VCRs, stereos, and cable channel selectors.

Most ceiling fan motors have an integral reversing switch to blow air downward or upward.

Do not use a standard incandescent lamp dimmer to control fan motors, fluorescent ballasts, transformers, low-voltage lighting systems (they use transformers to obtain the desired low voltage), motor-operated appliances, or other "inductive" loads. Serious overheating and damage to the motor can result. Fan speed controllers have internal circuitry that is engineered specifically for use on inductive load circuits. Fan speed controllers virtually eliminate any fan motor "hum." Read the label and the instructions furnished with a dimmer to be sure it is suitable for the application.

SAFETY ALERT

Most fan manufacturers state that when using a wall speed control, the pull-chain speed control on the fan should be set at its highest speed setting. To use other pull-chain speed positions in combination with a wall speed control might damage the motor or the wall speed control. Several manufacturers of speed controls say that they do not test them at other than the high-speed setting on the fan. Others state in their instructions that their solid-state speed controls:

- **are to be used only with fans marked "Suitable for use with**

solid-state fan speed controls only."

- are not to be used in circuits protected by GFCIs.
- are not to be used to control lighting—just as a lighting dimmer control might be marked not to be used as a fan speed control.

NEC 110.3(B) requires that Listed or labeled equipment shall be installed and used in accordance with any instructions included in the listing or labeling.*

You can readily see the importance of carefully reading and following the manufacturer's installation instructions.

*Reprinted with permission from NFPA 70-2014.

REVIEW

Note: Refer to the *Code* or the plans where necessary.

1. What circuit supplies the Master Bedroom? ______________________
2. What unique type of electrical protection is required for all 120-volt, single-phase, 15- and 20-ampere branch circuits in certain habitable rooms and areas in new dwellings? Where in the *NEC* is this requirement found? ______________________
3. What type of receptacles are provided in this bedroom? How many receptacles are there? ______________________
4. How many ceiling outlets are included in this circuit? ______________________
5. What wattage for the clothes closet luminaires was used for calculating their contribution to the circuit? ______________________
6. What is the current draw for the ceiling fan/light? ______________________
7. What is the estimated load in volt-amperes for the circuit supplying the Master Bedroom? ______________________
8. Which is installed first, the switch and outlet boxes or the cable runs? ______________________
9. How many conductors enter the ceiling fan/light wall box? ______________________
10. What type and size of box may be used for the ceiling fan/light wall box? ______________________

11. What type of covers are used with 4 in. square outlet boxes? ______________________

____

____

12. a. Does the circuit for the Master Bedroom have a grounded circuit conductor?

____

b. Does it have an equipment grounding conductor? ______________________

c. Explain the difference between a "grounded" conductor and an "equipment grounding" conductor. ______________________

____

____

13. Approximately how many feet (meters) of 2-wire cable and 3-wire cable are needed to complete the circuit supplying the Master Bedroom? Two-wire cable: ____________ ft (____________ m). Three-wire cable: ____________ ft (____________ m).

14. If the cable is laid in notches in the corner studs, what protection for the cable must be provided? ______________________

____

15. How high are the receptacles mounted above the finished floor in this bedroom?

____

16. Approximately how far from the bedroom door is the first receptacle mounted? (See the plans.) ______________________

____

17. What is the distance from the finished floor to the center of the wall switches in this bedroom? ______________________

18. The Master Bedroom features a sliding glass door. For the purpose of providing the proper receptacle outlets, answer the following statements. (Circle the correct answer.)

- Sliding glass panels are considered to be wall space. (True) (False)
- Fixed panels of glass doors are considered to be wall space. (True) (False)

19. What type of receptacle will be installed outdoors, just outside of the Master Bedroom? ______________________

____

20. Show your calculations of how to select a proper wall box for the clothes closet luminaire switch. Keep in mind that the available space between the wood casings is small. See Figure 9-1. ______________________

____

____

____

____

21. When an outlet box is to be installed to support a ceiling fan, how must it be marked?

____

22. The following is a layout of the lighting circuit for the Master Bedroom. Using the cable layout shown in Figure 9-1, make a complete wiring diagram of this circuit. Use colored pencils or pens.

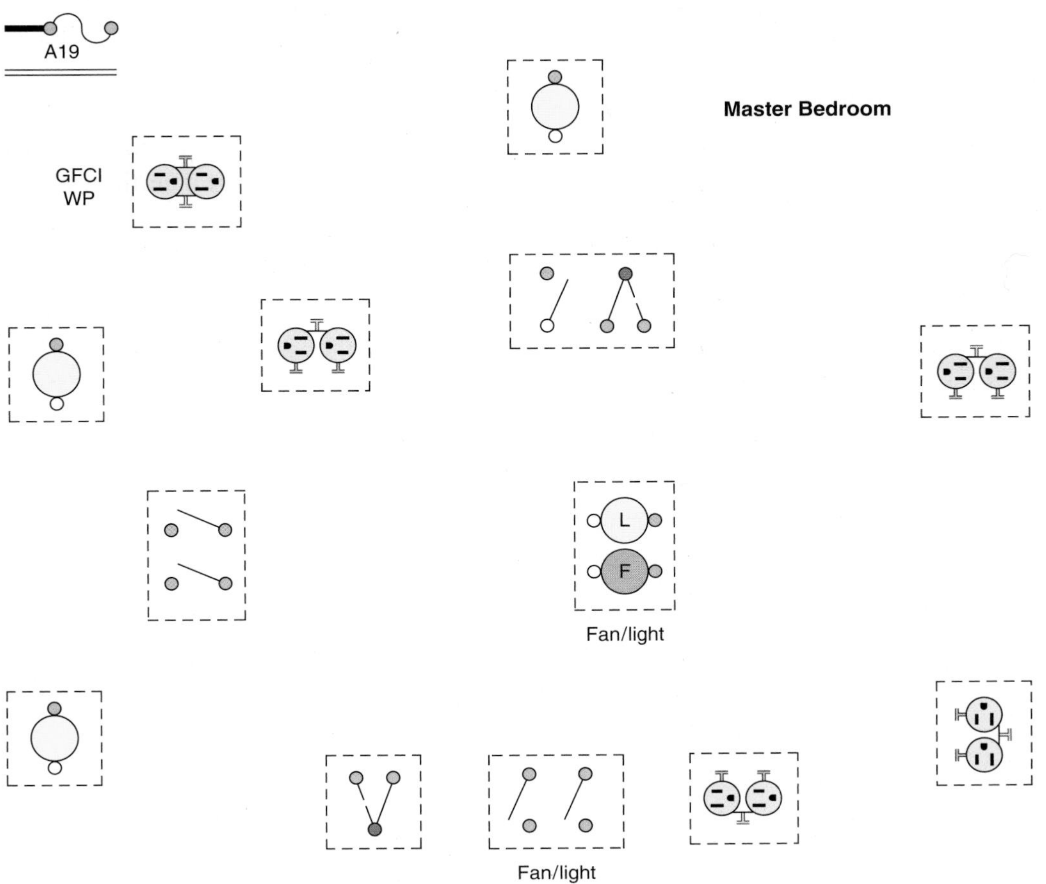

23. Connect the 4-wire cable, fan, light, fan speed control, and light switch. Refer to *200.7* to review the permitted use of the white conductor in the cable.

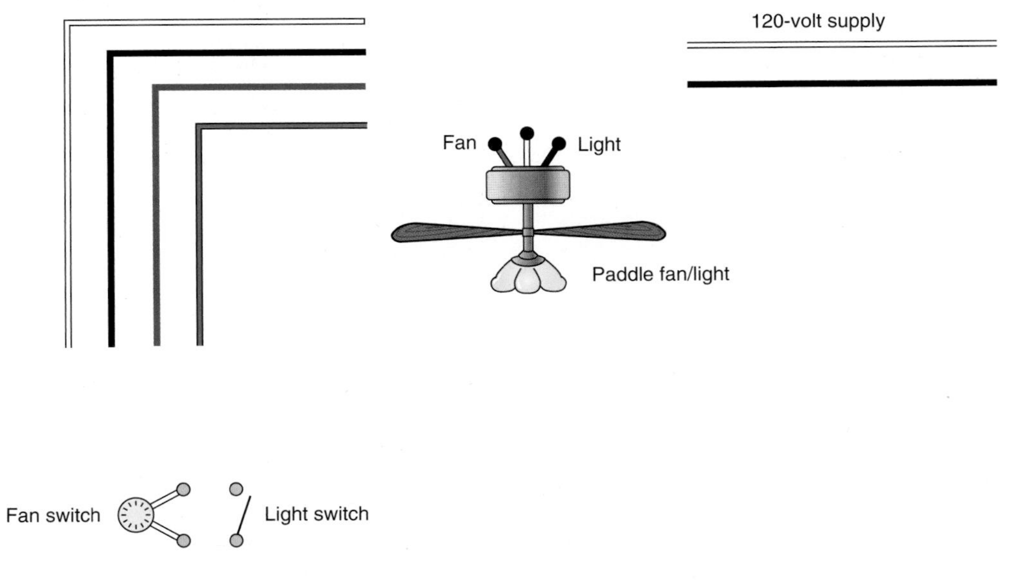

24. What is the width of a 2-gang wall plate? ______________________________

25. May a standard electronic dimmer be used to control the speed of a fan motor?

____

26. Outlet boxes that pass the tests at Underwriters Laboratories for the support of ceiling-suspended (paddle) fans are listed and marked "Acceptable for Fan Support." When roughing-in the electrical wiring for a new home, this type of outlet box must be installed wherever there is a likelihood of installing a ceiling-suspended (paddle) fan. Name those locations in homes that you consider likely to support a ceiling-suspended (paddle) fan. ______________________________

____

CHAPTER **10**

Lighting Branch Circuit—Bathrooms, Hallway

OBJECTIVES

After studying this chapter, you should be able to

- list equipment grounding requirements for bathroom installations.
- draw a wiring diagram for the bathroom and hallway.
- understand *Code* requirements for receptacles installed in bathrooms.
- understand *Code* requirements for receptacle outlets in hallways.
- discuss fundamentals of proper lighting for bathrooms.

INTRODUCTION

Circuit A14 is a 15-ampere branch circuit that supplies the hallway lighting, the hallway receptacle, and the vanity lighting in both bathrooms. Overcurrent protection and arc-fault protection for this circuit is provided by a 15-ampere AFCI circuit breaker in Panel A. The AFCI requirement is found in *210.12(A)*.

As previously discussed, *NEC® 404.2(C)* generally requires that the neutral conductor be brought to all switch locations. But there are a number of exceptions to this rule. Non-habitable rooms and bathrooms are exempt from the requirement. Now is a good time to review the detailed coverage on this *NEC 404.2(C)* requirement and what rooms, areas, or spaces are considered habitable and those that are considered non-habitable.

As required by *210.11(C)(3)*, the receptacles in the bathrooms are supplied by separate 20-ampere branch circuits A22 and A23. These receptacles are required to be GFCI protected, either by installing GFCI receptacles, or by installing GFCI circuit breakers in the panelboard. The GFCI requirement is found in *210.8(A)(1)*.

Table 10-1 summarizes outlets and estimated load for the bathrooms and bedroom hall.

Note that each bathroom shows a ceiling heater/light/fan that is connected to separate circuit A12 (Special-Purpose Ⓐ J) and separate circuit A11 (Special-Purpose Ⓐ K). These are discussed in detail in Chapter 22.

A hydromassage bathtub is located in the bathroom serving the master bedroom. It is connected to a separate circuit A9 (Special-Purpose Ⓐ A) and is also discussed in Chapter 22.

The attic exhaust fan in the hall is supplied by a separate circuit A10 (Special-Purpose Ⓐ L), also covered in Chapter 22.

TABLE 10-1

Bathrooms and bedroom hall: outlet count and estimated load (Circuit A14).

DESCRIPTION	QUANTITY	WATTS	VOLT-AMPERES
Receptacles @ 120 W each	1	120	120
Vanity luminaires. . . @ 200 W each	2	400	400
Hall luminaire	1	100	100
TOTALS	4	620	620

The receptacles in the master bathroom and hall bathroom are not included in this table as they are connected to separate 20-ampere branch circuits A22 and A23.

LIGHTING BRANCH CIRCUIT A14 FOR THE HALLWAY AND BATHROOMS

Figure 10-1 and the electrical plans for this area of the home show that each bathroom has a luminaire above the vanity mirror. Some typical luminaires are shown in Figure 10-2. Of course, the homeowner might decide to purchase a medicine cabinet complete with a self-contained luminaire. See Figure 10-3, which illustrates how to rough in the wiring for each type. These luminaires are controlled by single-pole switches at the doors.

You must remember that a bathroom (powder room) should have proper lighting for shaving, combing hair, grooming, and so on. Mirror lighting can accomplish this because a mirror will reflect what it "sees." If the face is poorly lit, with shadows on the face, that is precisely what will be reflected in the mirror.

A luminaire directly overhead will light the top of one's head but will cause shadows on the face. Mirror lighting and/or adequate lighting above and forward of the standing position at the vanity can provide excellent lighting in the bathroom. Figures 10-4, 10-5, and 10-6 show pictorial as well as section views of typical soffit lighting above a bathroom vanity.

Bathroom Receptacles

Figure 10-7 shows the definition of a bathroom according to the *NEC*. This definition was revised and expanded in the 2011 *NEC* to read, An area including a basin with one or more of the following: a toilet, a urinal, a tub, a shower, a bidet, or similar plumbing fixtures.* Note that the definition describes an *area*, not a *room*. The basin is the common denominator. As shown in Figure 10-8(D), the two rooms make up an area with the basin located in a space that is adjacent to the room with the tub and toilet.

*Reprinted with permission from NFPA 70-2014.

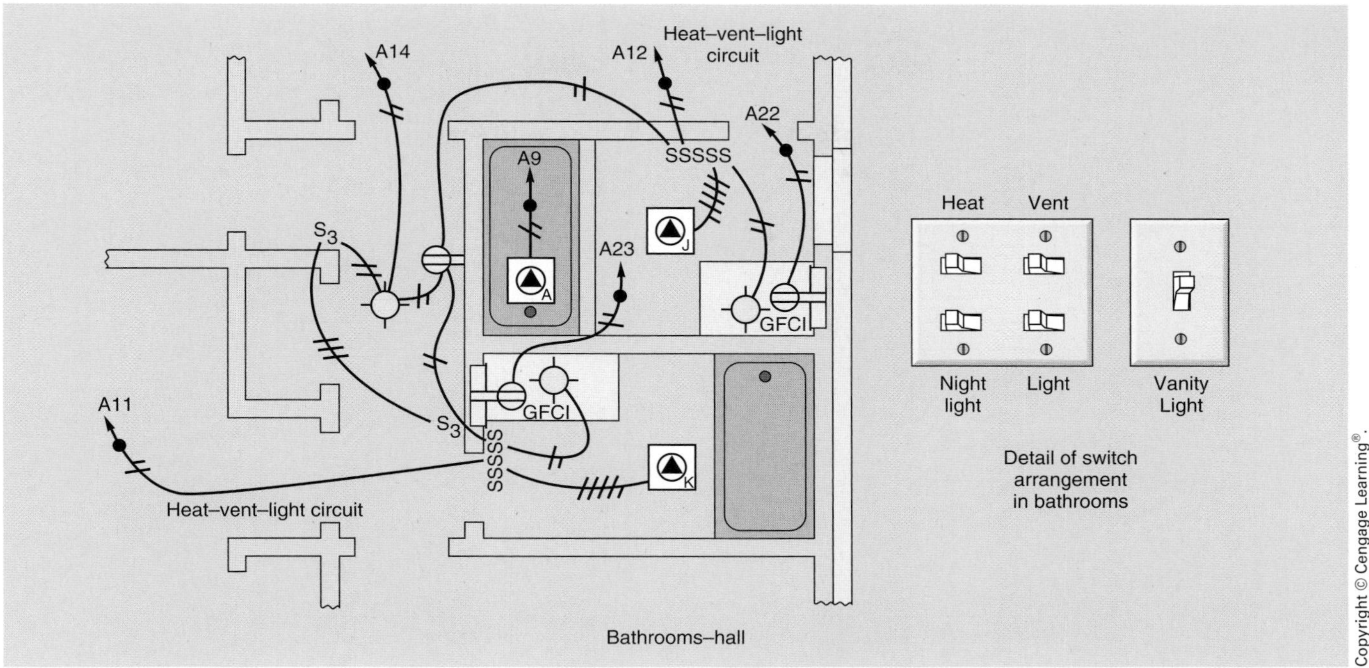

FIGURE 10-1 Cable layout for the master bathroom, hall bathroom, and hall. Layout includes lighting Circuit A14, two circuits (A11 and A12) for the heat/vent/lights, and two circuits (A22 and A23) for the receptacles in the bathrooms. The special-purpose outlets for the hydromassage tub, attic exhaust fan, and smoke detector are covered elsewhere in this text.

FIGURE 10-2 Typical vanity (bathroom) luminaires of the side bracket and strip types.

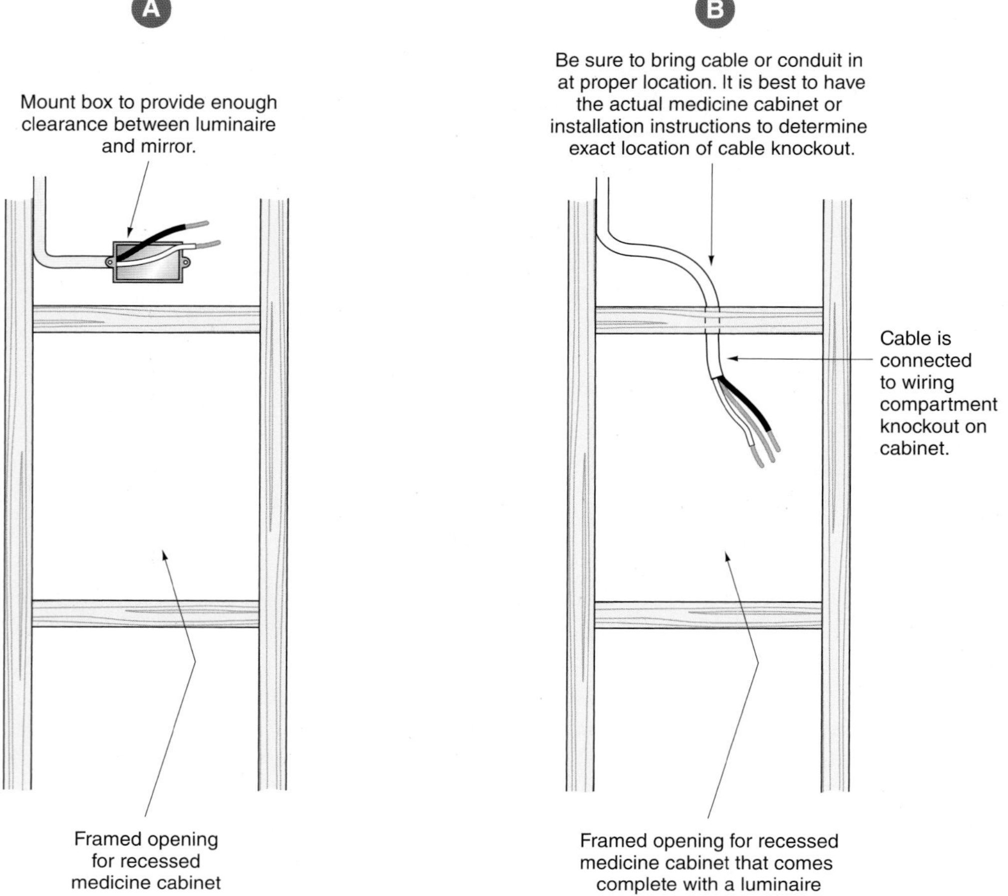

FIGURE 10-3 Two ways to rough in wiring for lighting above or on a vanity mirror or a medicine cabinet.

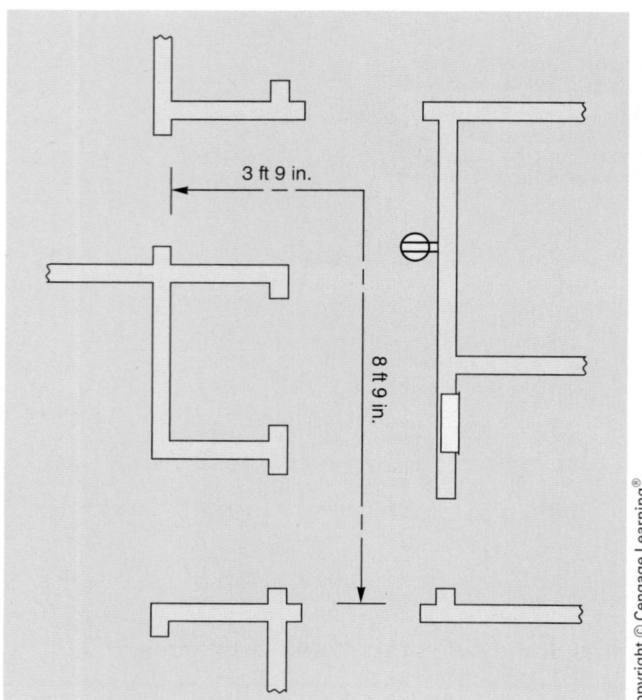

FIGURE 10-4 The centerline measurement of the bedroom hallway in this residence is 12 ft 6 in. (3.81 m), which requires at least one wall receptacle outlet, *210.52(H)*. This receptacle may be installed anywhere in the hall. The *Code* does not specify a location.

As required by *NEC 210.11(C)(3)*, the receptacles in the bathrooms are supplied by separate 20-ampere branch circuits A22 and A23. These receptacles are required to be GFCI protected, either by installing GFCI receptacles or GFCI circuit breakers in the panel. The GFCI requirement is found in *NEC 210.8(A)(1)*.

Although the *NEC* permits all bathroom receptacles to be on a single 20-ampere branch circuit so long as it doesn't supply other loads, we have chosen to run a separate 20-ampere branch circuit A22 to the receptacle in the Master Bedroom

FIGURE 10-5 Positioning of bathroom luminaires. Note the wrong way and the right way to achieve proper lighting.

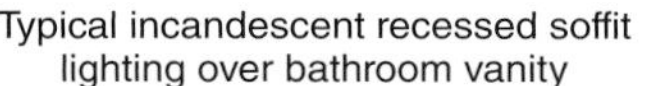
Typical incandescent recessed soffit lighting over bathroom vanity

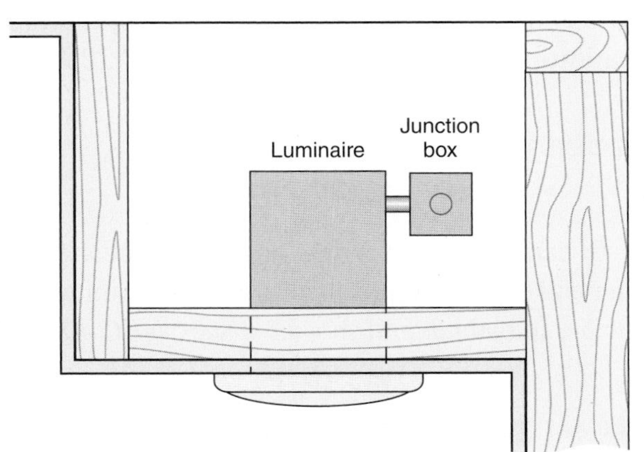

End cutaway of soffit showing recessed incandescent luminaires in typical soffit above bathroom vanity. Two or three luminaires generally installed to provide proper lighting.

FIGURE 10-6 Incandescent soffit lighting. Refer to Chapter 7 for minimum-clearance *Code* requirements for installation of recessed luminaires.

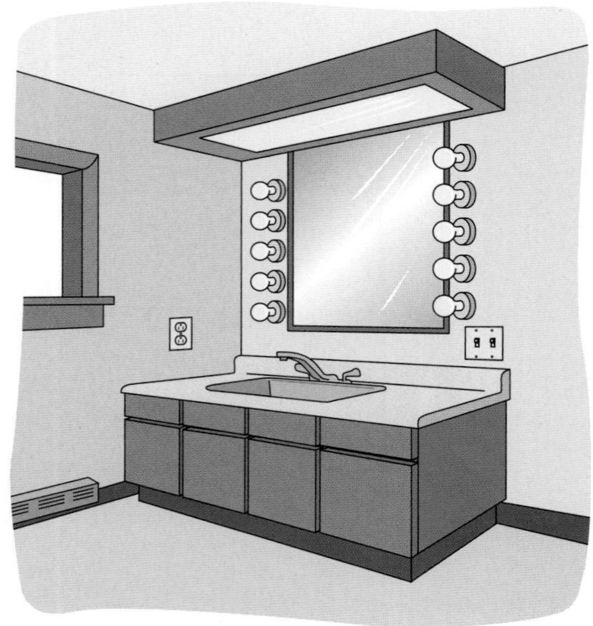

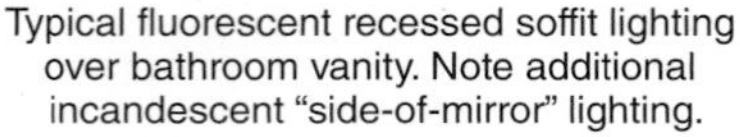

Typical fluorescent recessed soffit lighting over bathroom vanity. Note additional incandescent "side-of-mirror" lighting.

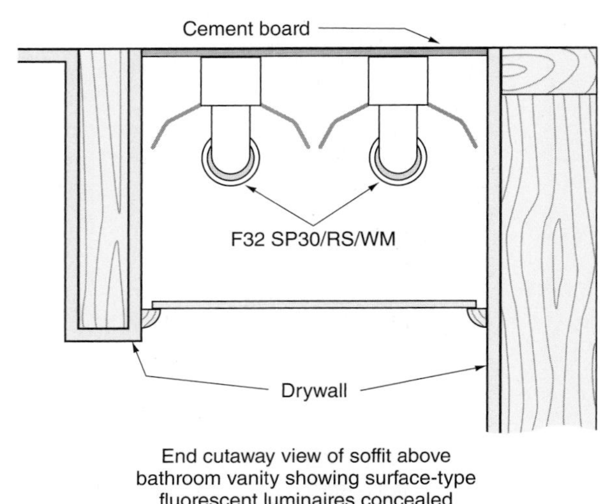

End cutaway view of soffit above bathroom vanity showing surface-type fluorescent luminaires concealed above translucent acrylic lens

FIGURE 10-7 Combination fluorescent and incandescent bathroom lighting.

bathroom, a separate 20-ampere branch circuit A23 to the receptacle in the hall bathroom, and another separate 20-ampere branch circuit B21 to the receptacle in the Powder Room located near the Laundry. These separate circuits are included in the general lighting load calculations; therefore, no additional load need be added, *NEC 220.14(J)*.

Receptacles in Bathtub and Shower Spaces

Because of the obvious hazards associated with water and electricity, receptacles are *not* permitted to be installed in bathtub and shower spaces, *NEC 406.9(C)*.

HANGING LUMINAIRES IN BATHROOMS

NEC 410.10(D) does not permit cord-connected luminaires; chain-, cable-, or cord-suspended luminaires; lighting track; pendants; or ceiling-suspended (paddle) fans to be located within a zone measured 3 ft (900 mm) horizontally and 8 ft (2.5 m) vertically from the top of a bathtub rim or top of a shower stall threshold. Figure 10-8 shows a top view of the 3 ft (900 mm) restriction. Figure 10-9 shows a side view of both permitted and not permitted installations.

Recessed or surface-mounted luminaires and recessed exhaust fans may be located within the restricted zone.

Luminaires Near Bathtub or Shower

Install luminaires listed for *wet* locations if subject to shower spray.

Install luminaires listed for *damp* or *wet* locations if located within the actual outside dimension of the bathtub or shower up to a height of 8 ft (2.5 m) vertically measured from the rim of the bathtub or shower threshold. The label on the luminaire will show the listing information.

HALLWAY LIGHTING

The hallway lighting is provided by one ceiling luminaire that is controlled with two 3-way switches located at either end of the hall. The home run to Main Panel A has been brought into this ceiling outlet box.

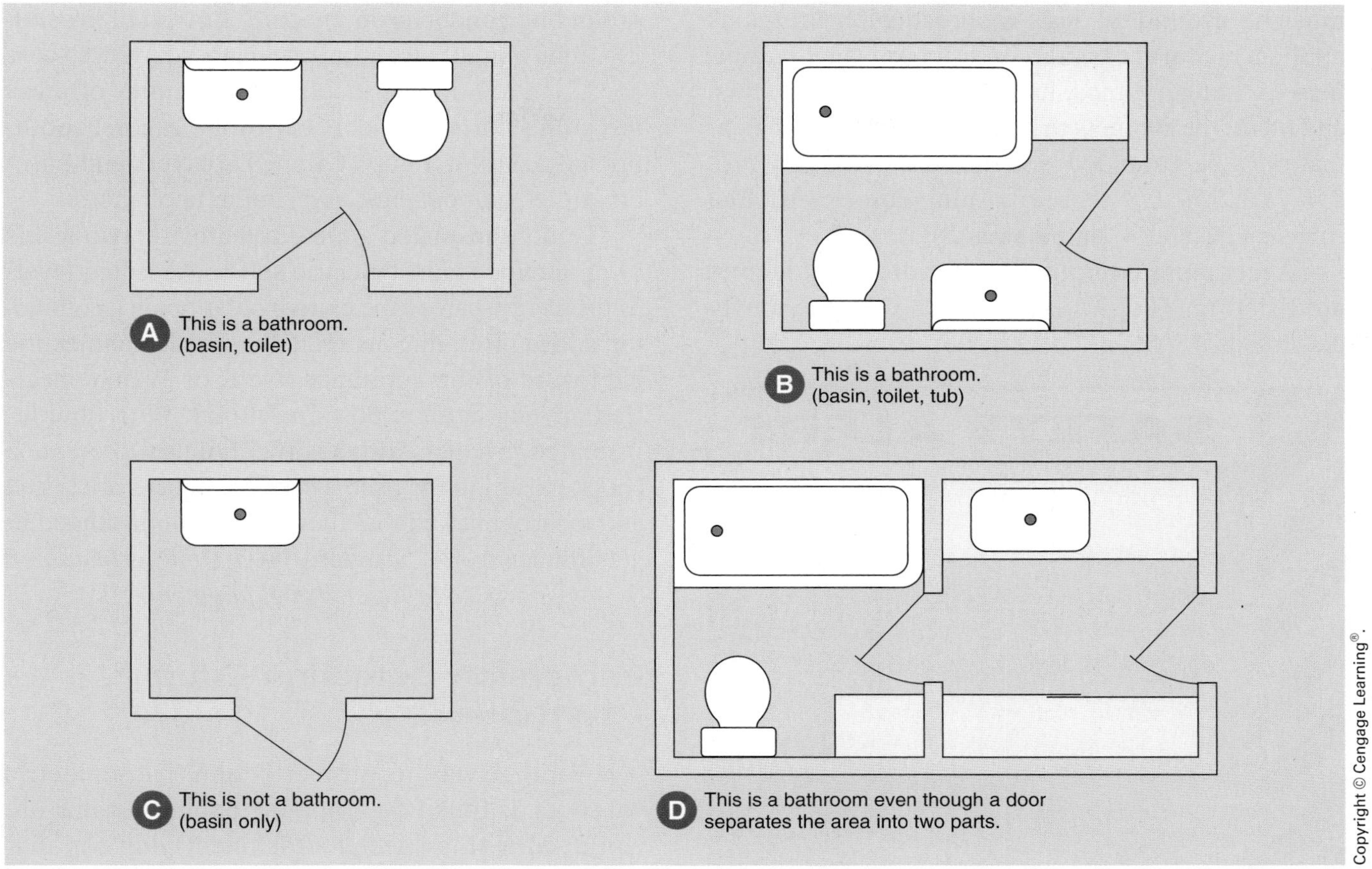

A This is a bathroom. (basin, toilet)

B This is a bathroom. (basin, toilet, tub)

C This is not a bathroom. (basin only)

D This is a bathroom even though a door separates the area into two parts.

FIGURE 10-8 The *Code* in *Article 100* defines a bathroom as An area including a basin with one or more of the following: a toilet, a urinal, a tub, a shower, a bidet, or similar plumbing fixtures.*

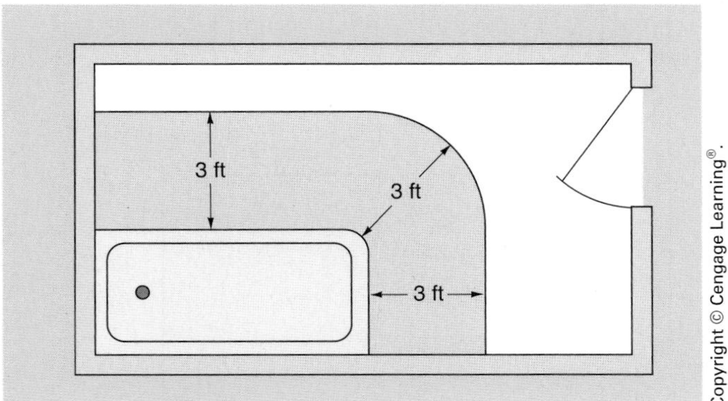

FIGURE 10-9 In a bathtub or shower area, no parts of cord-connected or cord-suspended luminaires or ceiling-suspended (paddle) fans are permitted in the shaded area. See text for details. Also see Figure 10-10.

RECEPTACLE OUTLETS IN HALLWAYS

One receptacle outlet has been provided in the hallway as required in *NEC 210.52(H)*, which states In dwelling units, hallways of 10 ft (3.0 m) or more in length shall have at least one receptacle outlet.*

For the purpose of determining the length of a hallway, the measurement is taken down the centerline of the hall, turning corners if necessary, but not passing through a doorway, Figure 10-4.

EQUIPMENT GROUNDING

With few exceptions, fixed and fastened-in-place electrical equipment with exposed non-current-carrying metal parts that could become energized

*Reprinted with permission from NFPA 70-2014.

must be grounded. This includes such things as appliances, furnaces, air conditioners, heat pumps, heat exchangers, fans, luminaires, electric heaters, and medicine cabinets that have lighting. Equipment that must be grounded is itemized in *NEC 250.110, 250.112, 250.114* (cord-and-plug connected), and *Article 410, Part V* (luminaires).

Acceptable equipment grounding conductors are listed in *NEC 250.118*, which was previously discussed.

SAFETY ALERT

When using nonmetallic-sheathed cable or armored cable, *never* use the bare equipment grounding conductor as a grounded circuit conductor (neutral), and *never* use the grounded circuit conductor as an equipment grounding conductor. These two conductors come together only at the main service panel, *never* in branch-circuit wiring. Refer to *250.142*.

Double Insulation in Lieu of Grounding

Appliances that are cord-and-plug connected with a 3-wire cord are grounded by the equipment grounding conductor in the cord. Instead of grounding, many small electrical appliances have a 2-wire cord, and are double insulated. Examples of these are some portable electrical tools, razors, toothbrushes, radios, TVs, VCRs, CD players and burners, stereo components, and similar appliances.

Double-insulated appliances provide two levels of protection against electric shock. Accepted levels of protection can be the extra insulation on a conductor, added insulating material, the plastic nonmetallic enclosure of the appliance itself, or even a specified air-gap at an exposed terminal. With double-insulated products, it takes two failures to create a shock hazard to the user. These appliances are tested and clearly marked that they are double insulated in accordance to UL Standard 1097 *Double-Insulated Systems for Use in Electrical Equipment.*

Immersion Detection Circuit Interrupters

Another way to protect people from electrical shock is found on grooming appliances that are equipped with a special attachment plug cap that provides the protection should the appliance be immersed in water.

Switches in Wet Locations

Do *not* install switches in wet locations, such as in bathtubs or showers, unless they are part of

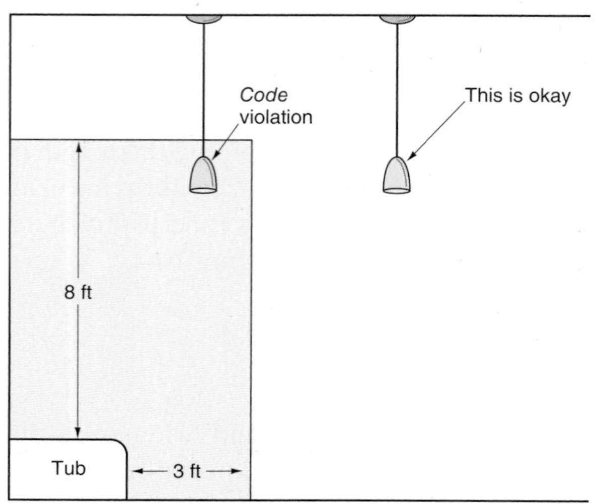

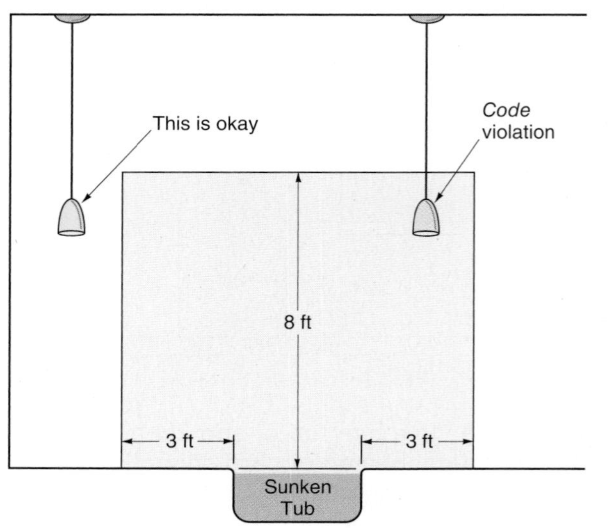

FIGURE 10-10 In a bathtub or shower area, no part of cord-connected or cord-suspended luminaires or ceiling-suspended (paddle) fans are permitted in the shaded area. See text for details. Also see Figure 10-9.

a listed tub or shower assembly, in which case the manufacturer has taken all of the proper precautions and submitted the assembly to a recognized testing laboratory to undergo exhaustive testing to establish the safety of the equipment; see *404.4*.

REVIEW

1. List the number and types of switches and receptacles used in Circuit A14.

2. There is a 3-way switch in the bedroom hallway leading into the living room. Show your calculation of how to determine the box size for this switch. The box contains cable clamps.

3. What wattage was used for each vanity luminaire to calculate the estimated load on Circuit A14? ______________________

4. What is the current draw for the answer given in problem 3?

5. Exposed non-current-carrying metallic parts of electrical equipment must be grounded if installed within __________ ft (__________ m) vertically or __________ ft (__________ m) horizontally of bathtubs, plumbing fixtures, pipes, or other grounded metal work or grounded surfaces.

6. What color are the faceplates in the bathrooms? Refer to the specifications.

   ____

7. Most appliances of the type commonly used in bathrooms, such as hair dryers, electric shavers, and curling irons, have 2-wire cords. These appliances are __________ insulated or __________ protected.

8. a. The *NEC* in Section __________ requires that all receptacles in bathrooms be __________ protected.

   b. The *NEC* in Section __________ requires that all receptacles in bathrooms be connected to one or more separate 20-ampere branch circuits that serve no other outlets.

c. The *NEC* in Section _____________ permits the additional required 20-ampere branch circuit for bathroom receptacles to supply more than one bathroom.

d. The *NEC* in Section _____________ permits other electrical equipment to be connected to the additional required 20-ampere branch circuit for bathroom receptacles, but only if the branch circuit supplies a single bathroom and the other equipment is located in that same bathroom.

e. The *NEC* in Section _____________ prohibits mounting receptacles in bathrooms face-up in the countertops and work surfaces near basins.

9. Hanging luminaires must be kept at least _____________ ft (_____________ m) from the edge of the tub as measured horizontally. In bathrooms with high ceilings, where the hanging luminaire is installed directly over the tub, it must be kept at least _____________ ft (_____________ m) above the edge of the tub.

10. The following is a layout of a lighting circuit for the bathroom and hallway. Using the cable layout shown in Figure 10-1, make a complete wiring diagram of this circuit. Use colored pencils or pens.

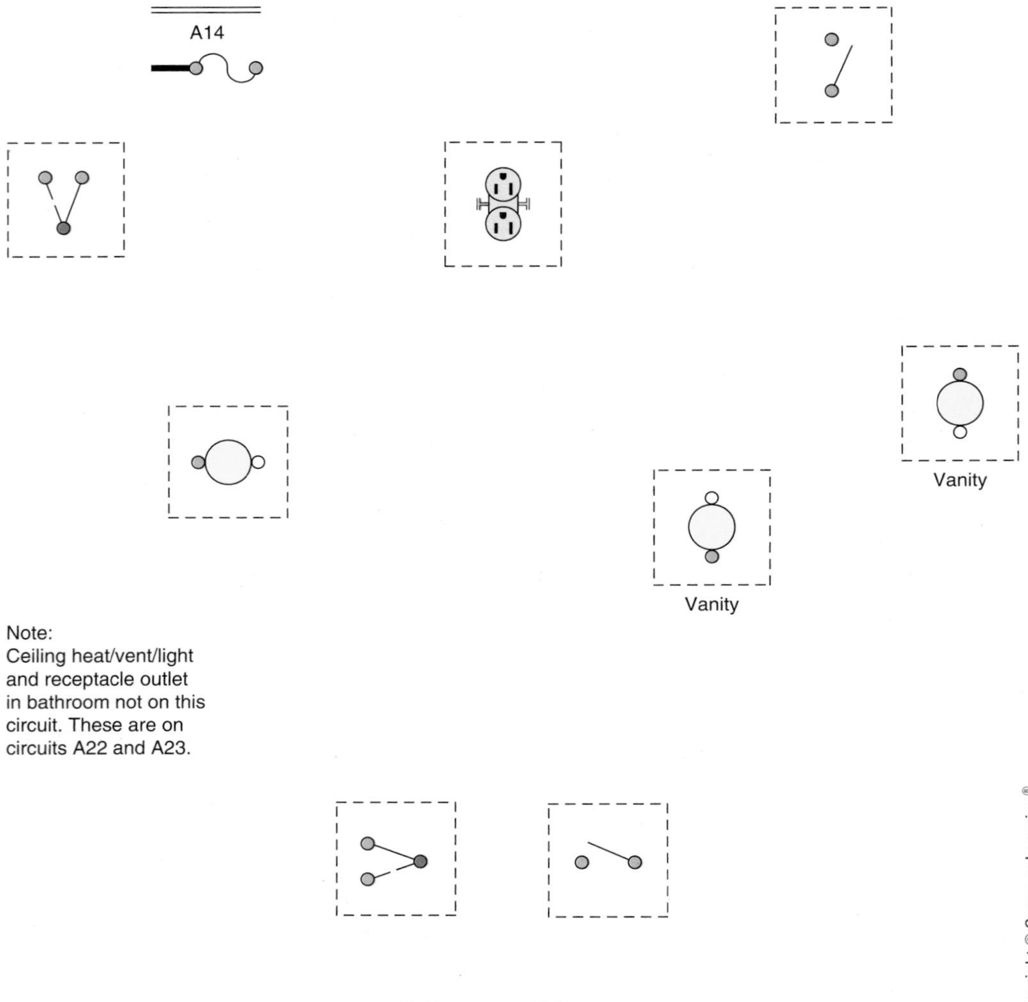

A14

Vanity

Vanity

Note:
Ceiling heat/vent/light and receptacle outlet in bathroom not on this circuit. These are on circuits A22 and A23.

Bathrooms—Hallway

11. Circle the correct answer as to whether a receptacle outlet is required in the following hallways.

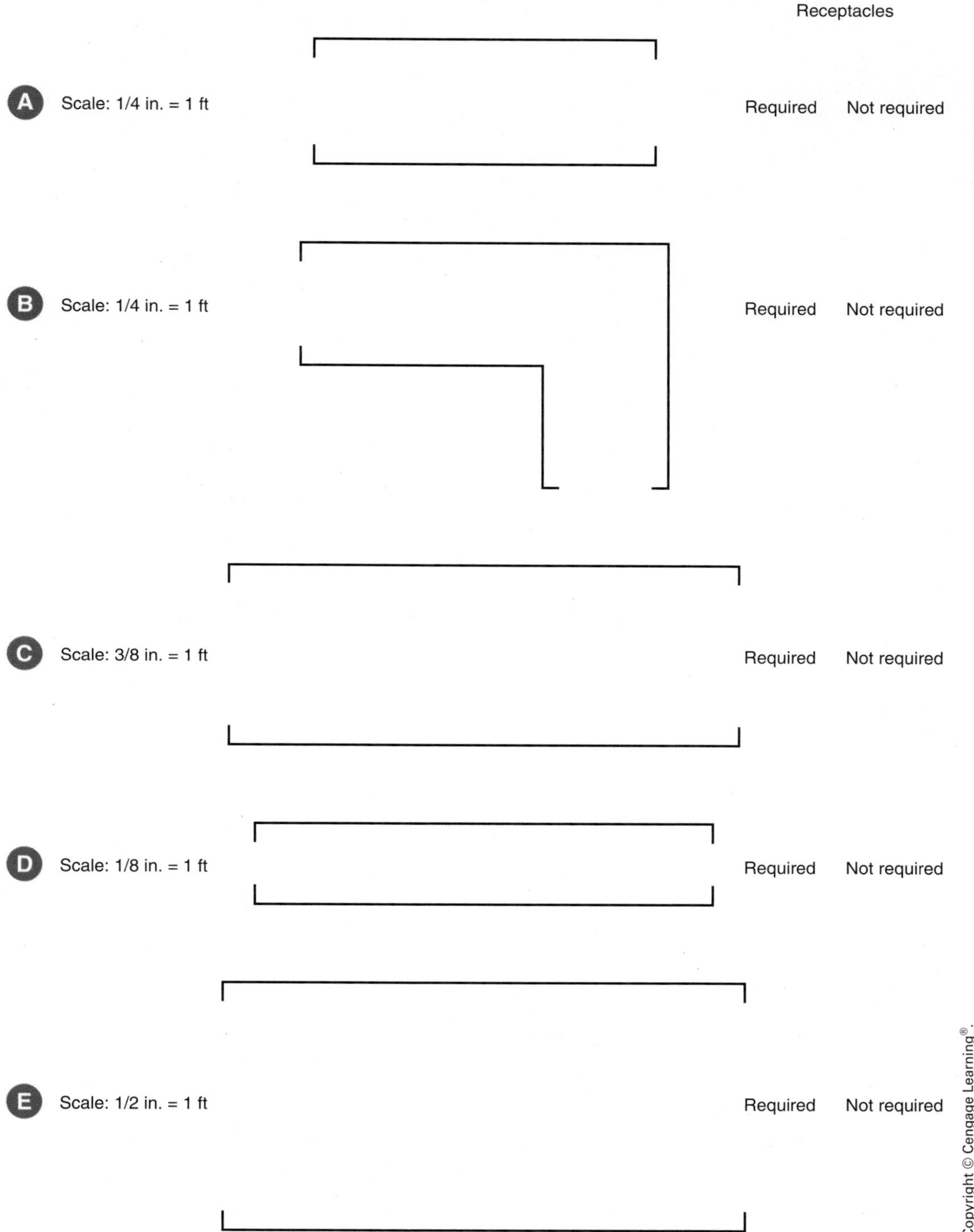

CHAPTER **11**

Lighting Branch Circuit—Front Entry, Porch, Post Light, Underground Wiring

OBJECTIVES

After studying this chapter, you should be able to

- understand how to install a switch in a doorjamb for automatic ON–OFF when the door is opened or closed.
- discuss types of luminaires recommended for porches and entries.
- complete the wiring diagram for the entry, porch, and post light circuit.
- discuss the advantages of switching outdoor receptacles from indoors.
- define wet and damp locations.
- understand the *NEC* requirements for underground cable and conduit wiring.
- become familiar with simple landscape lighting.

INTRODUCTION

The front entry area porch and post light are connected to Circuit A15. Overcurrent protection and arc-fault protection for this circuit is provided by a 15-ampere AFCI circuit breaker in Panel A. The AFCI requirement is found in *210.12(A)*. GFCI protection for the outdoor receptacle could be provided by a GFCI receptacle, or by an AFCI/GFCI dual-function circuit breaker in Panel A. GFCI protection is also required for the receptacle if it is installed in the post light.

The home run enters the ceiling box in the front entry. From this box, the circuit spreads out, feeding the closet light, the porch bracket luminaire, the two bracket luminaires on the front of the garage, one receptacle outlet in the entry, and one outdoor weatherproof GFCI receptacle on the porch, as shown in Figure 11-1. Table 11-1

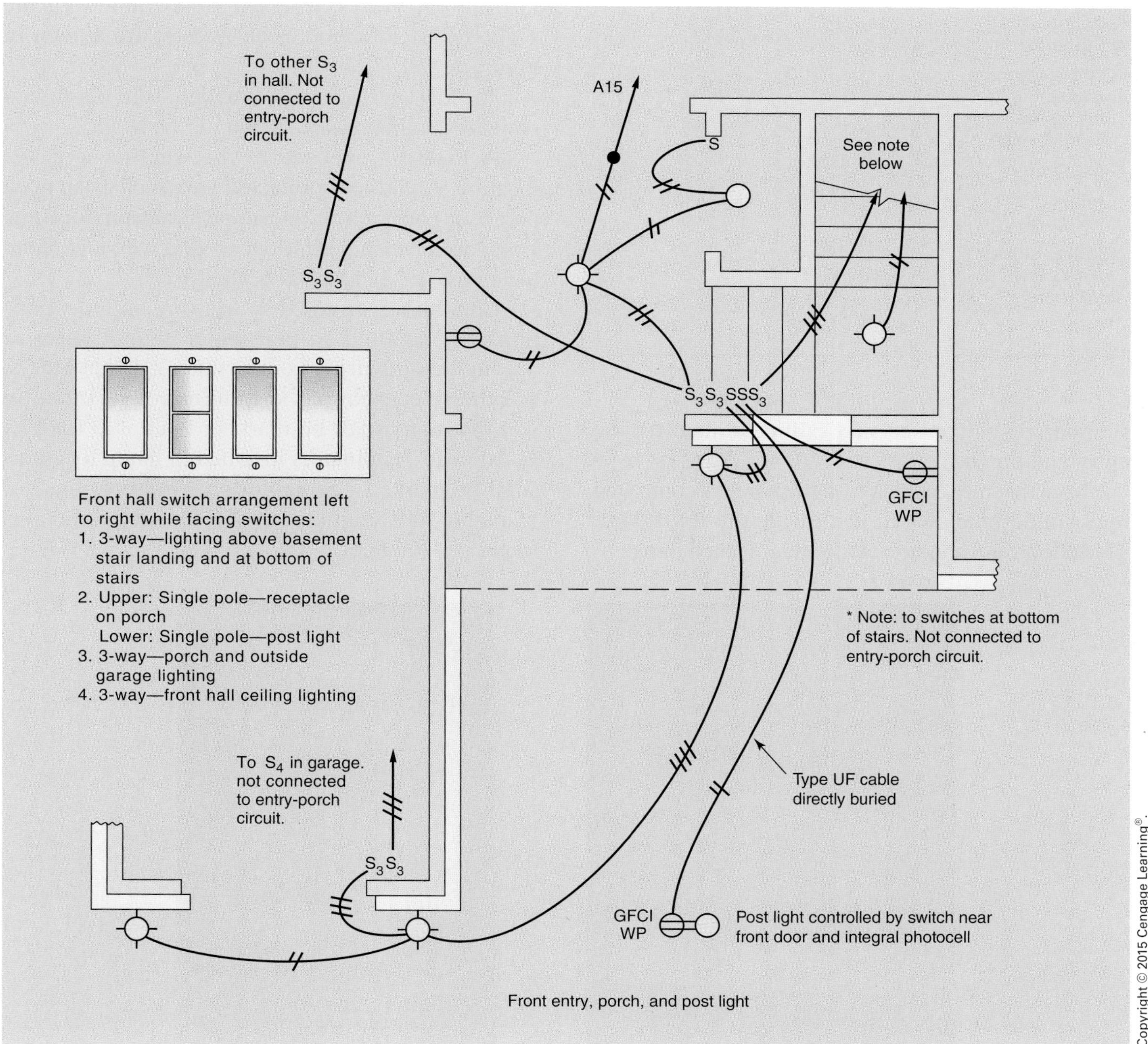

FIGURE 11-1 Cable layout of the front entry, porch, and post light Circuit A15. Note that cables from some of the other circuits are shown so that you can get a better idea of exactly how many wires will be found at the various locations.

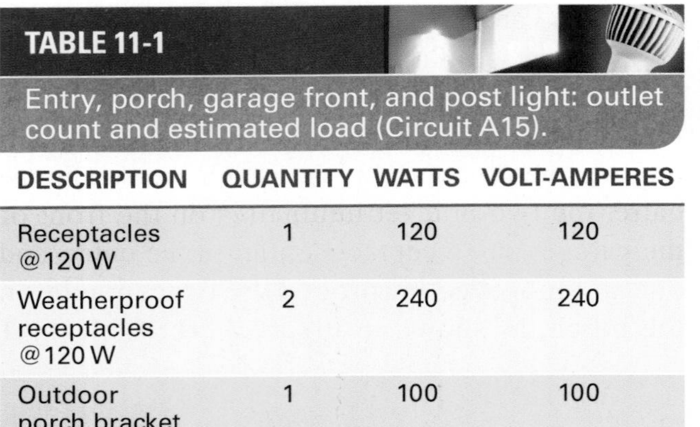

TABLE 11-1

Entry, porch, garage front, and post light: outlet count and estimated load (Circuit A15).

DESCRIPTION	QUANTITY	WATTS	VOLT-AMPERES
Receptacles @ 120 W	1	120	120
Weatherproof receptacles @ 120 W	2	240	240
Outdoor porch bracket luminaire	1	100	100
Outdoor garage bracket luminaires @ 100 W each	2	200	200
Post light	1	100	100
Ceiling luminaire	1	150	150
Clothes closet recessed luminaire	1	75	75
TOTALS	9	985	985

summarizes the outlets and estimated load for the entry and porch.

Note that the receptacle on the porch is controlled by a single-pole switch just inside the front door. This allows the homeowner to plug in such things as strings of ornamental holiday lighting or decorative lighting and have convenient switch control of that receptacle from inside the house, a nice feature.

Although it adds cost, consider an outdoor weatherproof receptacle on both sides of the front entry door. This makes for easy plugging in of outdoor decorations. Some electricians install a receptacle in the soffit. Switched or nonswitched? The decision is up to you.

Typical ceiling luminaires commonly installed in front entryways, where it is desirable to make a good first impression on guests, are shown in Figure 11-2.

Typical outdoor wall-bracket-style porch and entrance luminaires are shown in Figure 11-3.

A location exposed to the weather is a wet location. A partially protected area such as an open porch, or under a roof or canopy, is a damp location. For more precise definitions of dry, wet, and damp locations, see *Article 100* of the *NEC*®.

Luminaires installed in wet or damp locations must be installed so that water cannot enter or accumulate in wiring compartments, lampholders, or other electrical parts. Luminaires installed in wet locations shall be marked, "Suitable for Wet Locations." Luminaires installed in damp locations shall be marked, "Suitable for Wet Locations" or "Suitable for Damp Locations." This information is found in *410.10(A)*.

Courtesy of Progress Lighting.

FIGURE 11-2 Hall luminaires: ceiling mount and chain mount.

FIGURE 11-3 Outdoor porch and entrance luminaires: wall bracket styles.

Some inspectors consider the area within a 45° angle from a roof line or overhang to be a damp location, and outside the 45° angle to be a wet location. See Figure 11-4.

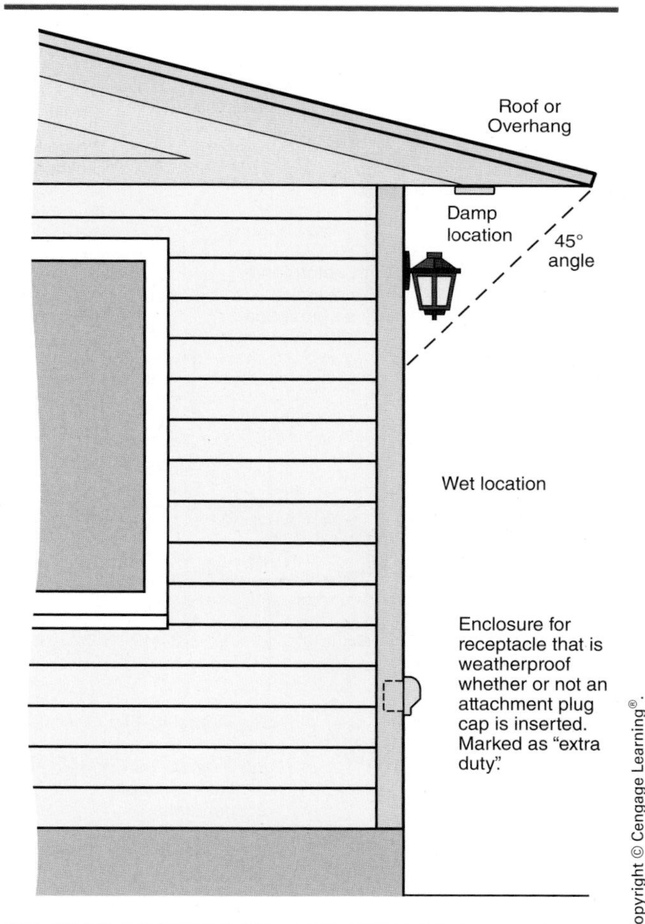

FIGURE 11-4 Outdoor areas considered to be either a *damp* or *wet* location.

Post Light Wiring

The post light shown in Figure 11-1 is shown to be supplied from a wall switch beside the front door. Although this location presents something of a problem for the electrician regarding space for wiring in a switch box as well as number of switches to be located in the box, these challenges are easily solved. Figure 11-1 shows the switch arrangement. The switches from left to right are:

1. a 3-way for the stairway luminaires.
2. two single-pole switches that are stacked by the manufacturer. The top switch controls the receptacle on the porch and the bottom switch controls the post light.
3. a 3-way switch for the luminaires on the porch and front of the garage.
4. a 3-way switch for the front hall luminaire.

Often, post light luminaires are controlled by a photocell that is installed on the post or in the luminaire. In addition, electronic timers are available that allow on and off times to be selected as desired.

Most post lights can be fitted with an optional 125-volt, 15- or 20-ampere receptacle. If installed, this receptacle is required to be provided with GFCI protection. The GFCI protection can be located as an integral part of the receptacle or on the supply side such as a dual-function AFCI/GFCI circuit breaker in the panelboard. As discussed in Chapter 16 of this text, the depth of cover for the Type UF cable that supplies the post light is reduced from 24 in. to 12 in. if the branch circuit has GFCI protection.

Figure 11-5 shows two methods of bringing the conduit into the basement: (1) It can be run below ground level and then can be brought through the basement wall, or (2) it can be run up the side of the building and through the basement wall at the ceiling joist level. When the conduit is run through the basement wall, the opening must be sealed to prevent moisture from seeping into the basement. The electrician must decide which of the two methods is more suitable for each installation. Additional requirements for underground wiring are found in Chapter 16 of this text. These requirements include the type of cable that is suitable for direct burial and the depth of wiring for protection against damage. This burial depth changes depending on whether the branch circuit has GFCI protection.

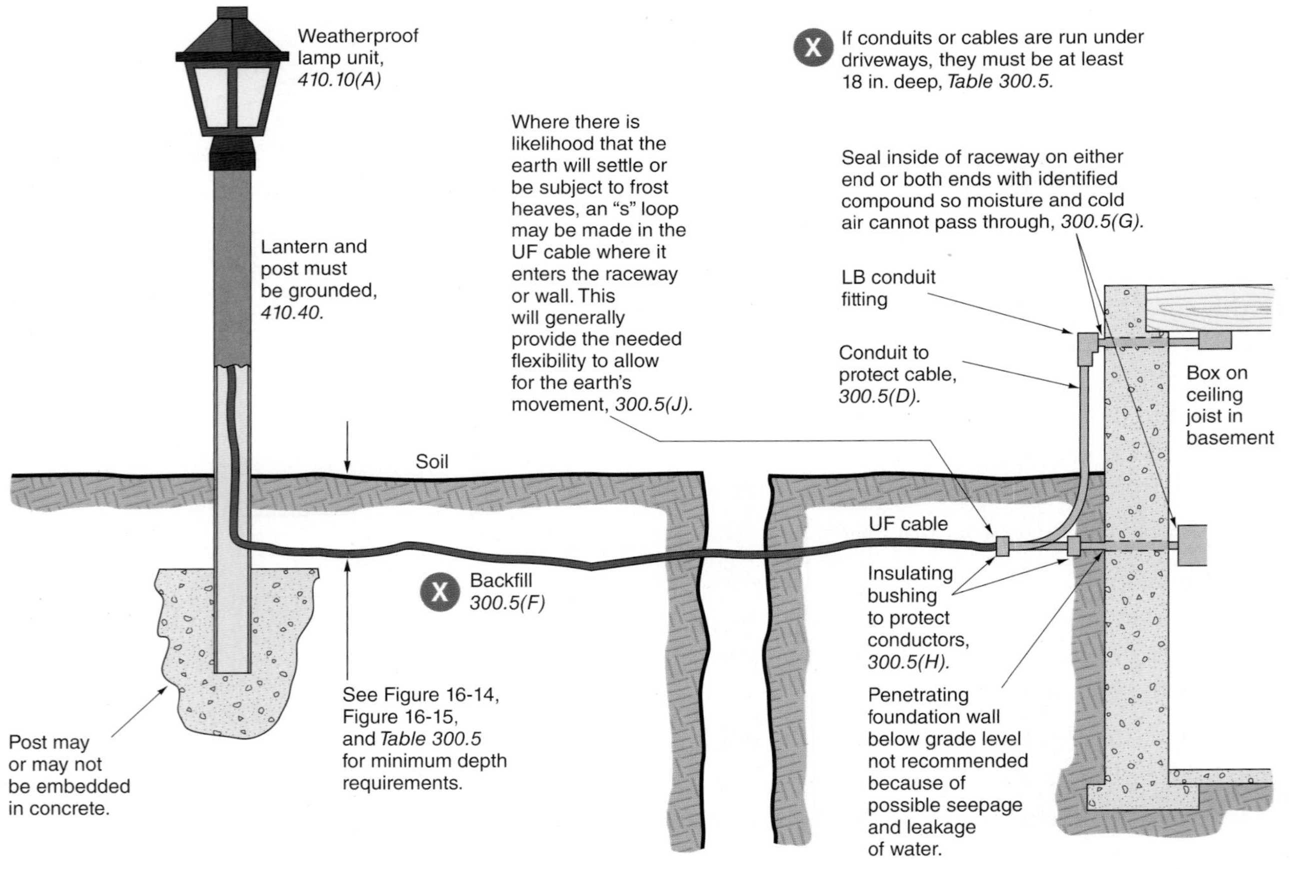

FIGURE 11-5 Methods of bringing cable and/or conduit through concrete wall and/or upward from a concrete wall into the hollow space within a framed wall.

According to the *Code*, any unprotected location exposed to the weather is considered a wet location.

Luminaires installed in wet locations are required to be marked "Suitable for Wet Locations." These luminaires are constructed so that water cannot enter or accumulate in lampholders, wiring compartments, or other electrical parts, *410.10(A)*.

Manufacturers' installation instructions may require the application of caulking to ensure protection against the entrance of water.

A damp location is defined as being protected from the weather but subject to moderate degrees of moisture. Partially protected areas, such as roofed open porches or areas under canopies, are damp locations. Luminaires in these locations must be marked "Suitable for Damp Locations."

A luminaire that is "Suitable for Wet Locations" is also suitable for damp locations. A luminaire that is "Suitable for Damp Locations" is not suitable for wet locations. Check the UL label on a luminaire to determine the suitability of the fixture for a wet or damp location.

The post in Figure 11-5 may or may not be embedded in concrete, depending on the consistency of the soil, the height of the post, and the size of the post lantern. Most electricians prefer to embed the base of the post in concrete to prevent rotting of wood posts and rusting of metal posts.

Figure 11-6 illustrates some typical post light luminaire heads.

Courtesy of Progress Lighting.

FIGURE 11-6 Post type lanterns.

CIRCUIT A15

Circuit A15 is a rather simple circuit that has two sets of 3-way switches: one set controlling the front entry ceiling luminaire, the other set controlling the porch bracket luminaire and the two bracket luminaires on the front of the garage. Also, one single-pole switch controls the weatherproof porch receptacle outlet, a second switch controls the post light, and another single-pole switch controls the clothes closet light. In the Review questions, you will be asked to follow the suggested cable layout for making up all of the circuit connections.

Probably the most difficult part of this circuit is planning and making up the connections at the 4-gang switch location just inside the front door. Refer to Figure 11-1 for details. Figure 11-7 shows a 4-gang nonmetallic box of the type that can be used for the front door switch location. Note that the 3-way switch for the luminaire over the stairway leading to the basement is part of the Recreation Room lighting branch circuit B12.

There are a lot of conductors, wiring devices, and cable clamps in this box. Let's figure out the proper wall size wall box.

1.	Add the circuit conductors $3 + 2 + 2 + 3 + 3 + 3 =$	16
2.	Add for equipment grounding wires (count 1 only)	1
3.	Add eight for the four switch yokes	8
4.	Add for cable clamps (count 1 only)	1
	Total	26

To select a box that has adequate cubic-inch capacity for this example, refer to *Table 314.16(A)* in the *NEC*, and Table 2-1 in this text. One possibility would be to use four 3 in. × 2 in. × 3½ in. device boxes ganged together, providing $4 \times 9 = 36$ conductor count based on 14 AWG conductors.

When using nonmetallic boxes, refer to *Table 314.16(B)*, and base the volume allowance on 2 in.^{3} for each 14 AWG conductor and for the volume allowance for the wiring devices, cable clamps, and equipment grounding conductors.

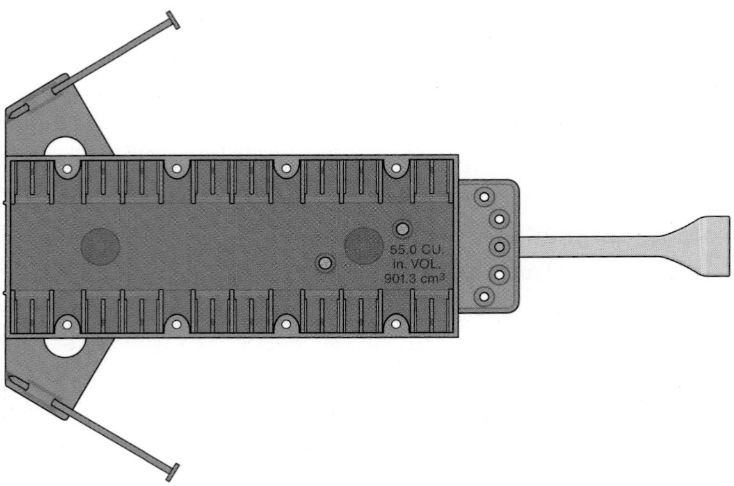

FIGURE 11-7 A 4-gang, nonmetallic box that has a marked volume of 55 in.^{3} (901.3 cm^3).

In the previous example, we are looking for a cubic-inch volume of $26 \times 2 = 52$ in.^{3}. See Figure 11-7.

Foyers. Section *210.52(I)* requires one or more receptacles be installed in foyers. That section reads, ▶Foyers that are not part of a hallway in accordance with 210.52(H) and that have an area that is greater than 60 ft^2 (5.6 m^2) shall have a receptacle(s) located in each wall space 3 ft (900 mm) or more in width. Doorways, door-side windows that extend to the floor, and similar openings shall not be considered wall space.◀* This results in one receptacle to be installed for the foyer.

DOORJAMB SWITCH

An interesting possibility presents itself for the front-entry clothes closet. Although the plans show that the clothes closet luminaire is turned on and off by a standard single-pole switch to the left as you face the closet, a doorjamb switch could have been installed.

A doorjamb switch comes as an assembly—the switch, the special box, and the cover.

*Reprinted with permission from NFPA 70-2014.

The instructions furnished with the doorjamb switch specify the maximum number and size of conductors the box can accommodate. These instructions also show the roughing-in dimensions of the box. The box furnished with a doorjamb switch is generally suitable only for the switch loop conductors. Because the wiring space is very limited, do not plan on using this box for splices or through wiring other than those necessary to make up the connections to the switch.

The grounded conductor (netural) is not required at the doorjumb switch because the switch does not control a luminaire in a habitable room. See *NEC 404.2(C)(4)*. A doorjamb switch is usually mounted about 6 ft (1.8 m) above the floor, on the inside of the 2×4 framing on the hinge side of the closet door. Run the switch loop cable to this point, and let it hang out until the carpenter cuts the proper size opening into the doorjamb for the box. After the finished woodwork is completed, the switch and cover plate can be installed. Figure 11-8 illustrates a door switch.

The plunger on the switch is pushed inward when the edge of the door pushes on it as the door is closed, shutting off the light. The plunger can be adjusted in or out to make the switch work properly.

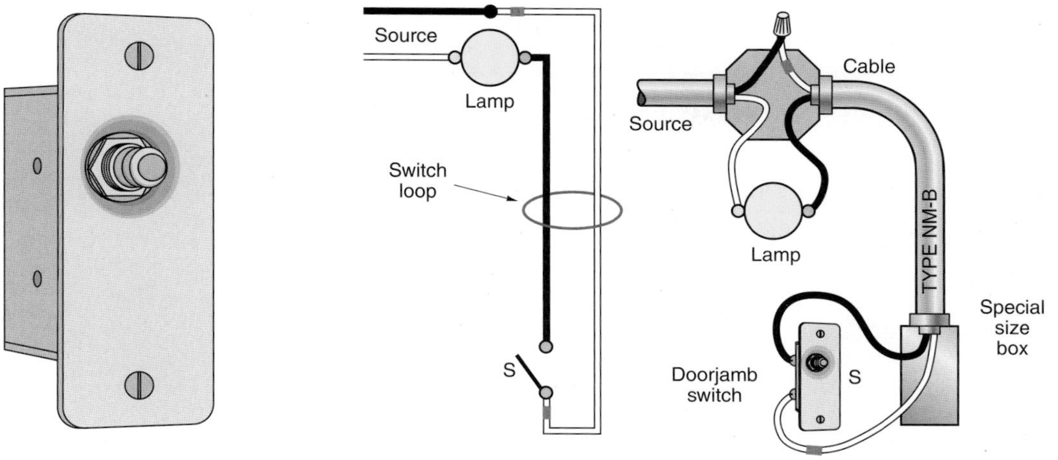

FIGURE 11-8 Doorjamb switch. Feed at light. Switch loop run from ceiling box to doorjamb box.

Here's another option. Install a low-voltage system. A simple, yet effective system can be created with a 24-volt transformer, a 120-volt lighting relay (with a 24-volt coil), and a 24-volt plunger-type switch. By using a Class 2 transformer, a box is not required for the switch. This can be accomplished quite simply by mounting a 4-square-deep metal box on the surface of the wall in the corner of the closet approximately 6 ft (1.8 m) above the floor. Pre-wire the 24-volt doorbell cable to the switch location and the power for the light through the location for the metal box. A cable or EMT is preinstalled to the outlet box for the closet luminaire. At the time the trim is installed, connect the transformer and relay along with the switch and you're finished.

REVIEW

1. a. How many circuit wires enter the entry ceiling box? ______________________

   b. How many equipment grounding conductors enter the entry ceiling box?________

2. Assuming that an outlet box for the entry ceiling has a fixture stud but has external cable clamps, what size box could be installed? Show calculations for both metal and nonmetallic boxes. ____

   ____

   ____

   ____

   ____

   ____

3. How many receptacle outlets and lighting outlets are supplied by Circuit A15? ______

____

4. Outdoor luminaires directly exposed to the weather must be marked as (Circle the correct answer.)

   a. suitable for damp locations.
   b. suitable for dry locations only.
   c. suitable for wet locations.

5. Make a material list of all types of switches and receptacles connected to Circuit A15.

____

____

____

____

____

____

6. From left to right, facing the switches, what do the switches next to the front door control?

____

____

____

____

____

____

____

7. Who is to select the entry ceiling luminaire? ______________________________

8. When installing the wiring for a doorjamb switch, the best choice is (Circle the correct answer.)

   a. run the supply conductors to the doorjamb box, then run the switched conductors to the luminaire.
   b. run the supply conductors to the outlet box where the luminaire is to be installed, then run the switch loop conductors to the doorjamb box.

9. The following layout is for Lighting Circuit A15, the entry, porch, post light, and front garage lights. Using the cable layout shown in Figure 11-1, make a complete wiring diagram of this circuit. Use colored pencils to indicate the color of the conductors' insulation.

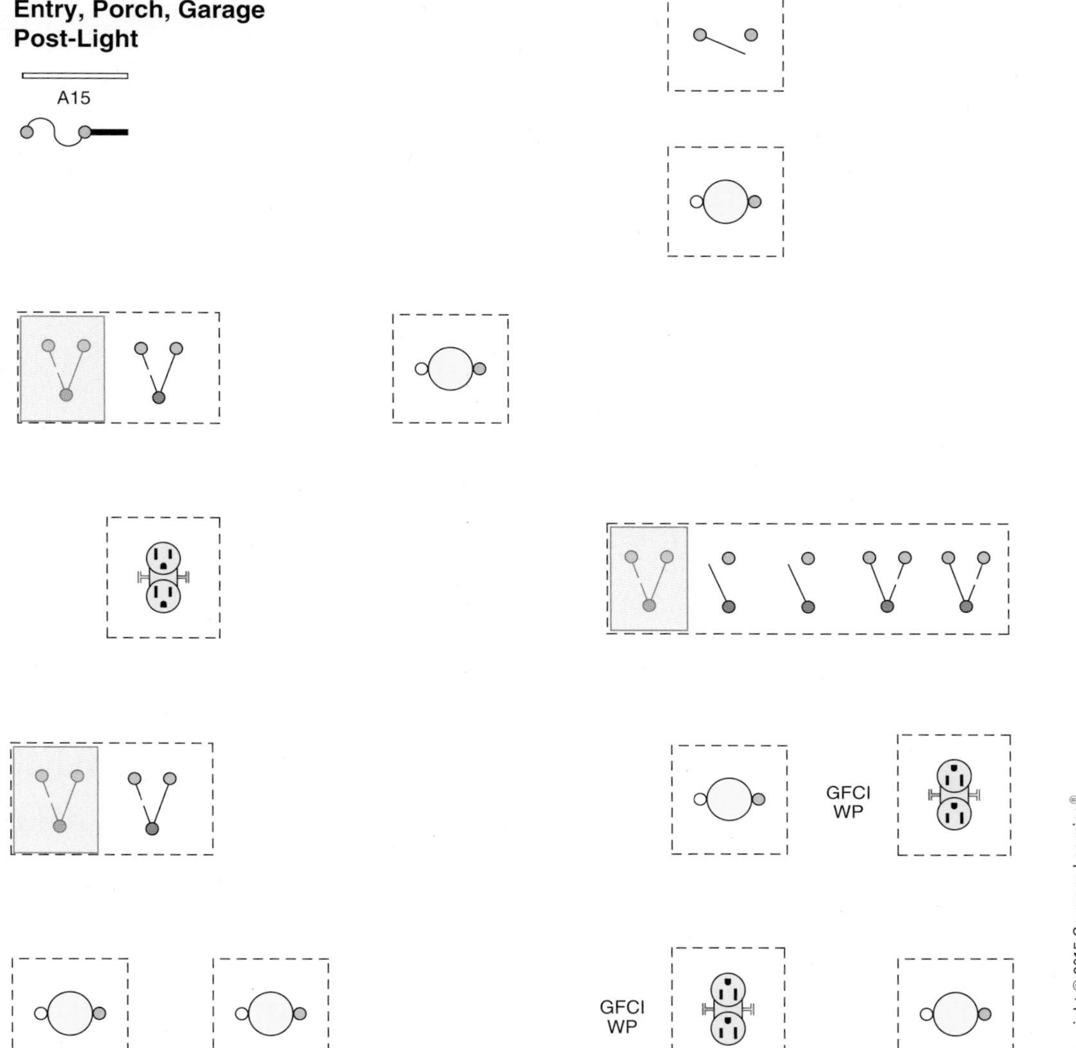

CHAPTER **12**

Lighting Branch Circuit and Small-Appliance Circuits for the Kitchen

OBJECTIVES

After studying this chapter, you should be able to

- understand lighting for a typical kitchen and dining room.
- understand the installation and operation of kitchen exhaust fans.
- know the *NEC* requirements for small-appliance branch circuits in kitchens.
- know the *NEC* requirements for GFCI protection for all receptacles that serve countertops in kitchens.
- decide whether an individual branch circuit should be installed for a refrigerator.
- decide where it might be desirable to install multiwire branch circuits and split-wired receptacles in areas where a high concentration of plug-in appliances might be used.

KITCHEN

The *NEC*® defines a *kitchen* as An area with a sink and permanent provisions for food preparation and cooking.* Although the *NEC* does not define *permanent*, it no doubt means *not temporary*. The concept is that permanent provisions are in place even though the appliance may not be permanent. For example, a dwelling is to have an electric range in the kitchen. The receptacle outlet for the range is permanent. However, the range can be placed in the dedicated space and the "pigtail" be plugged into the receptacle. The range can be moved and replaced with another appliance, yet the *provisions* for the supply to the range are permanent.

▶Changes in the 2014 *NEC* in *210.12(A)* now require that AFCI protection be provided for outlets and devices installed in kitchens.◀ We discussed this extensively in Chapter 6 of this text. As a result of this change, some receptacle outlets in the kitchen are required to have both AFCI and GFCI protection. As explained in Chapter 6, these technologies are compatible and complementary. In addition, branch circuits rated 125 V, single-phase, 15 or 20 amperes that supply outlets for appliances require AFCI protection. We'll remind you again of this requirement when we cover it in a later chapter. AFCI protection is not required for appliances that are supplied by branch circuits rated at 240 volts such as for electric ranges, cooktops, and built-in ovens.

LIGHTING CIRCUIT B7

The lighting circuit supplying the Kitchen originates at Panelboard B. Checking *NEC 210.12(A)*, the kitchen lighting branch circuit is required to be AFCI protected. Although a feed-through type receptacle may be used for this protection, perhaps a Type AFCI circuit breaker installed in the panelboard is more practical.

As stated previously, *NEC 404.2(C)* generally requires the grounded circuit conductor for the controlled lighting circuit to be provided at the location where switches control lighting loads that are supplied by a grounded general purpose branch circuit. Several exceptions are provided. One exception excludes rooms that are not considered habitable and bathrooms. There is no definition of habitable rooms or spaces in the *NEC*. It is suggested you consult with the AHJ regarding this issue as the building or residential code likely considers kitchens to be habitable spaces. As a result, we are providing a neutral conductor at switch locations.

The circuit number is B7. The cable layout for this circuit is shown in Figure 12-1. The home run is very short, leading from Panel B in the corner of the recreation room to the switch box to the right of the kitchen sink. From here, the circuit continues upward to the outlet box above one of the square surface-mounted luminaires on the ceiling, then across the ceiling to the outlet box above the eating area where a lighting track will be installed. From this outlet box, the circuit picks up the outdoor bracket luminaire. The range hood exhaust fan/light is fed from the outlet box above the other square luminaire on the ceiling.

As you study the cable layouts in this text, be assured that there are other ways to run the conductors. The thought process should be to sketch a few different cable-run possibilities. Then consider how many conductors enter the boxes, the ease or difficulty in making up the connections, and problems you might encounter because of obstructions (pipes and ducts) in the walls or ceilings. When installing recessed luminaires, be sure they are listed for "through wiring" if you intend to use the wiring compartment on the luminaire to continue on with the branch-circuit wiring. These luminaires are marked with the maximum number and size of conductors permitted in the wiring compartment.

Table 12-1 shows the outlets and estimated loads for the kitchen lighting Circuit B7.

KITCHEN LIGHTING

Kitchen lighting can become quite a challenge. Prior to installing the wiring, the electrician really has to consider the wishes of the homeowner. As stated earlier, lighting is a personal choice. Arrange lighting to avoid shadows in work areas.

The general lighting for the Kitchen in this residence is provided by two square surface-mounted luminaires, as shown in Figure 12-2. These luminaires might have four straight or two U-bent fluorescent lamps. These luminaires are controlled by a single-pole switch located next to the doorway leading to the living room.

*Reprinted with permission from NFPA 70-2014.

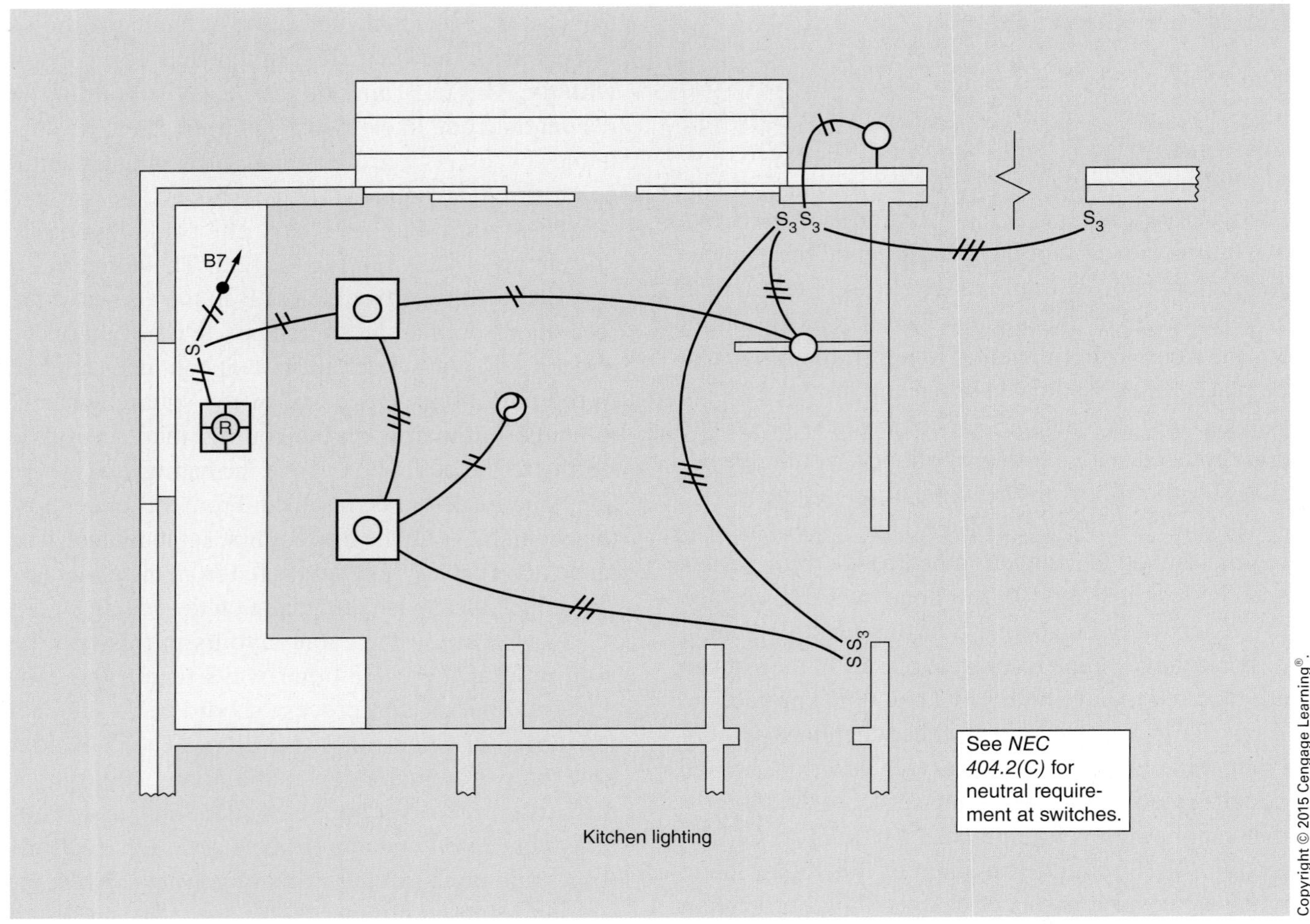

FIGURE 12-1 Cable layout for the kitchen lighting. This is Circuit B7. The appliance circuit receptacles are connected to 20-ampere circuits B13, B15, and B16 and are not shown on this lighting circuit layout.

Another consideration for kitchen lighting could be a series of recessed luminaires installed in the ceiling approximately 18 in. (450 mm) from the edge of the cabinets, spaced about 3 ft (0.9 m) to 4 ft (1.2 m) apart. This would offer pleasant lighting for the kitchen area, the countertops, the oven, the pantry, and the refrigerator.

Still another possibility would be a track lighting system on the ceiling approximately 18 in. (450 mm) from the edge of the cabinets in front of the kitchen cabinets above the sink wall (West), and on the ceiling in front of the refrigerator and pantry wall (South). A number of track lighting heads can be installed and adjusted to provide good lighting without shadows on work surfaces.

Above the breakfast area table is a short section of track lighting controlled by two 3-way switches: one located adjacent to the sliding door, the other at the doorway leading to the Living Room. Figure 12-3 illustrates just a few of the many types of luminaires that might be secured to the track. A one-light, stem-hung pendant luminaire or a few recessed luminaires could also be considered for this eating area.

Lighting for the electric range is in the range hood, with switching an integral part of the range hood.

Lighting over the sink is provided by the recessed luminaire above the sink, with switching to the right of the sink.

Figure 12-4 shows a variety of chain-suspended luminaires of the type commonly used in dining rooms.

Courtesy of Progress Lighting.

FIGURE 12-2 Typical ceiling luminaire of the type installed in the Kitchen of this residence. This type of luminaire might have four 20-watt fluorescent lamps, two 40-watt U-shaped fluorescent lamps, or four 60-watt incandescent lamps.

TABLE 12-1

Kitchen: outlet count and estimated load (Circuit B7).

DESCRIPTION	QUANTITY	WATTS	VOLT-AMPERES
Ceiling luminaires (eight 20-W FL lamps)	2	160	180
Recessed luminaire over sink	1	100	100
Lighting track	1	100	100
Outdoor rear bracket luminaire	1	100	100
Range hood fan/light (Two 60-W lamps)	1	120	120
Fan motor ($1\frac{1}{3}$ A @ 120 V)	1	160	180
TOTALS	7	740	780

Courtesy of Progress Lighting.

FIGURE 12-3 Fixtures that can be hung either singly or on a track over dinette tables. A lighting track is shown in Figure 12-1 over the breakfast eating area.

Undercabinet Lighting

In some instances, particularly in homes with little outdoor exposure, the architect might specify strip fluorescent or LED luminaires under the kitchen cabinets. Several methods may be used to install undercabinet luminaires. They can be fastened to the wall just under the upper cabinets (Figure 12-5[A]), installed in a recess that is part of the upper cabinets so that they are hidden from view (Figure 12-5[B]), or fastened under and to the front of the upper cabinets (Figure 12-5[C]). All three possibilities require close coordination with the cabinet installer to be sure that the wiring is brought out of the wall at the proper location to connect to the undercabinet luminaires.

Flat fluorescent and LED luminaires are available that mount on the underside of upper cabinets. Some of these luminaires are direct connected—others are cord-and-plug connected. These are pretty much out of sight but provide excellent lighting for countertop work surfaces.

Luminaire manufacturers also offer shallow undercabinet track lighting systems in both 120 volts and 12 volts. These systems include track, track lighting heads for small halogen lamps, end fittings, and transformer with cord and plug. Some have

FIGURE 12-4 Types of luminaires used over dining room tables.

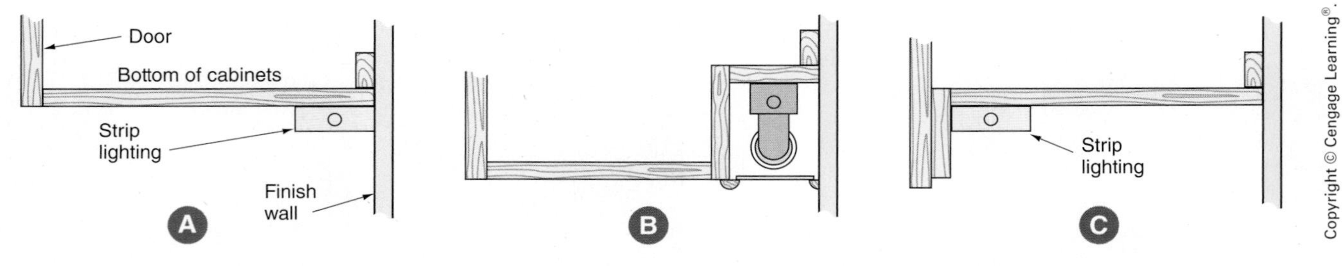

FIGURE 12-5 Methods of installing undercabinet fluorescent strip lighting.

two-level, high–low switching to vary the light intensity.

Visiting a lighting showroom, an electrical supply house, a home center, or a hardware store is an exciting experience. The choice of luminaires is almost endless.

Lamp Types

Good color rendition in the kitchen area is achieved by using incandescent lamps, or by using warm white deluxe (WWX), warm white (WW), or deluxe cool white (CWX) 3000 K lamps wherever fluorescent lamps are used. See Chapter 7 for details regarding energy-saving ballasts and lamps.

FAN OUTLET

Ducted fans exhaust air to the outside. They remove cooking odors, steam, and smoke. Ductless fans remove cooking odors but do not remove steam or smoke.

In this residence, the range hood exhaust fan above the electric range is connected to lighting branch circuit B7. The speed control, light, and light switch are integral parts of the fan. The electric supply is brought down from the ceiling to the wiring compartment on the fan. For exhaust fans mounted under cabinets along a wall, the electrical supply can be brought into the top or into the back of the fan housing wiring compartment.

FIGURE 12-6 Typical range hood exhaust fan. Speed control and light are integral parts of the unit.

In either case, a sufficient length of cable must be left during the roughing-in stage to be able to make the final connections in the wiring compartment. Figure 12-6 is a typical range hood exhaust fan.

NEC 422.16(B)(4) permits a range hood to be cord-and-plug connected. The receptacle would be installed inside the cabinetry above the range. To do this, all of the following conditions must be met:

- The flexible cord must be identified as suitable for use on a range hood in accordance to the manufacturer's installation instructions.
- The flexible cord must have a grounding-type attachment plug cap.
- The flexible cord shall not be less than 18 in. (450 mm) and not over 36 in. (900 mm).
- The receptacle shall be located to avoid physical damage to the flexible cord.
- The receptacle shall be accessible.
- The receptacle is supplied by an individual branch circuit.

The logic behind the individual branch-circuit requirement is that a range hood installed today might later be replaced with a combination microwave/range hood appliance that is generally cord-and-plug connected.

The *NEC* defines an *individual branch circuit as* A branch circuit that supplies only one utilization equipment.* To meet this requirement, either a single receptacle or a duplex receptacle that supplies no other load would have to be installed for the range hood. *NEC 210.21(B)(1)* requires that a single receptacle supplied by an individual branch circuit must have an ampere rating not less than the rating of the branch circuit. The range hood branch circuit could be a 15- or 20-ampere branch circuit. If you run a 15-ampere branch circuit, the single receptacle could be a 15- or a 20-ampere receptacle. If you run a 20-ampere branch circuit, the single receptacle shall be rated 20 amperes.

Instead of trying to guess what the current draw of a future microwave/range hood combination might be, it is best to install a 20-ampere individual branch circuit.

Fan Noise

The fan motor and air movement through an exhaust fan produce a certain amount of noise. Exhaust fan manufacturers specify a sound-level rating in their descriptive literature. This provides some idea of the noise level one might expect from the fan after it is installed. The sound-level rating unit used to define fan noise is a *sone* (rhymes with tone).

For simplicity, one sone is the noise level of an average refrigerator. The lower the sone rating, the quieter the fan. All manufacturers of exhaust fans provide this information in their descriptive literature.

CLOCK OUTLETS

Battery-operated wall and mantle clocks for the most part have replaced electric clocks. They can be put just about anywhere. Choices are virtually unlimited.

Clock hanger receptacles are still available. In a clock hanger receptacle, the receptacle is set back from the front edge of the faceplate, providing room for the typical small plug cap on the end of a cord. This type of receptacle can be hidden behind a painting or photograph that you want to light up.

Exception No. 1 to *NEC 210.52(B)(2)* allows a receptacle installed solely for the purpose of supplying an electric clock to be supplied from one of the required 20-ampere small-appliance branch circuits.

*Reprinted with permission from NFPA 70-2014.

SMALL-APPLIANCE BRANCH CIRCUITS FOR RECEPTACLES IN THE KITCHEN

The *Code* requirements for small-appliance circuits and the spacing of receptacles in kitchens are covered in *NEC 210.11(C)(1)*, *210.52(B)*, and *220.52(A)*. Important points follow:

- At least two 20-ampere small-appliance circuits are required to be installed, as shown in Figure 12-7.
- Either (or both) of the two circuits required in the kitchen is permitted to supply receptacle outlets in other rooms, such as a dining room, breakfast room, or pantry, Figure 12-8.
- These small-appliance circuits must be assigned a load of 1500 volt-amperes each when calculating feeders and service-entrance requirements.
- The receptacle for a refrigerator in a kitchen is permitted to be supplied by an individual 15- or 20-ampere branch circuit that is suitable for the load.
- A clock outlet is permitted to be connected to a small-appliance branch circuit when this outlet is to supply and support the clock only (see Figure 12-9).
- Countertop receptacles must be supplied by at least two 20-ampere small-appliance circuits, as in Figure 12-7 and 12-8.
- All 125-volt, single-phase, 15- and 20-ampere receptacles installed in kitchens to serve the countertops are required to be GFCI protected, *210.8(A)(6)*; the branch circuit is required to have AFCI protection, *210.12(A)*, Figures 12-7, 12-8, 12-10, 12-11 and 12-12. A person can receive a deadly electrical shock by accidentally touching a "hot" wire or defective appliance and a grounded surface at the same time. A grounded surface could be a stainless steel sink, a faucet, frames of a range, an oven, a cook-top, a refrigerator, a microwave oven, a countertop food processor, a luminaire, a metal water pipe, or metal hot/cold air registers.
- A receptacle may be installed for plugging in a gas appliance to serve the ignition and/or clock devices on the appliance. See Figure 12-7.
- The 20-ampere small-appliance circuits are provided to serve plug-in portable appliances. They are not permitted by the *Code* to serve lighting or appliances such as dishwashers, garbage disposals, or exhaust fans.
- A receptacle installed below the sink for easy plug-in connection of a food waste disposer

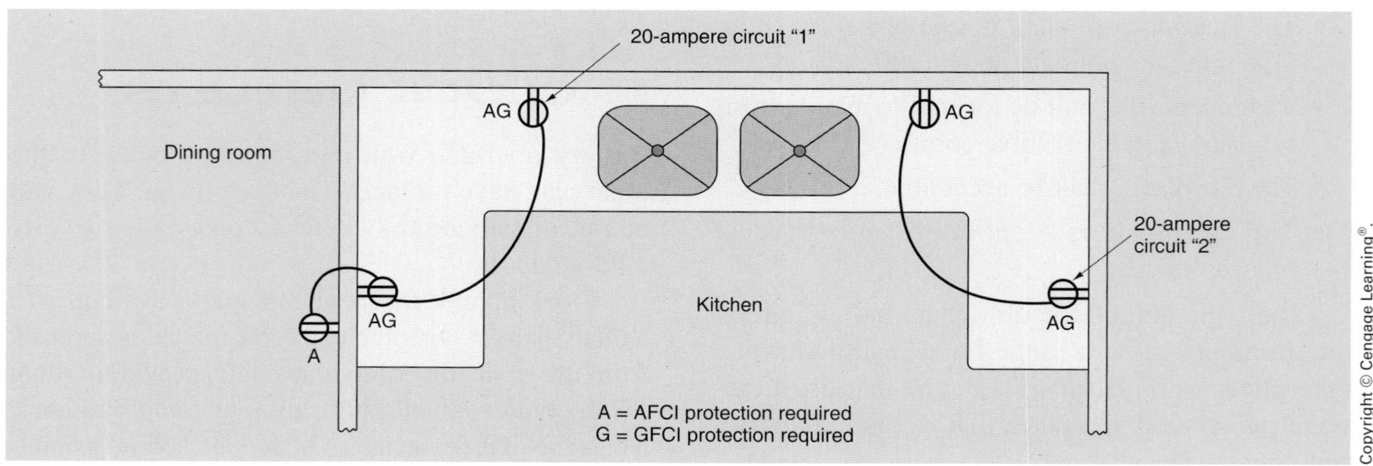

FIGURE 12-7 Countertop receptacles must be supplied by at least two 20-ampere small-appliance circuits, *210.52(B)(3)*. Lighting and receptacle outlets as well as devices in the kitchen are required to have AFCI protection in compliance with *NEC 210.12(A)*.

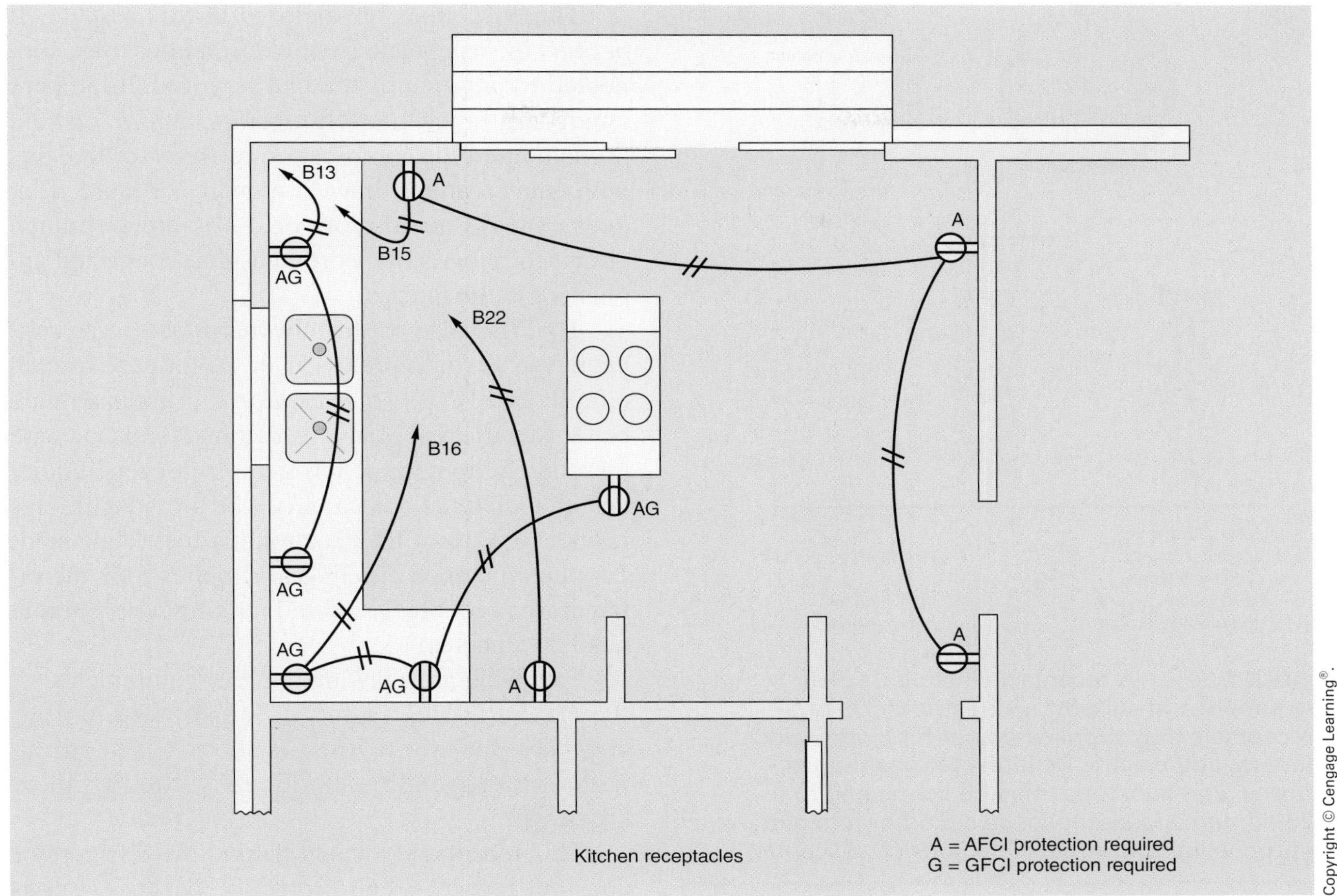

FIGURE 12-8 Cable layout for kitchen receptacles.

is *not* to be connected to the required small-appliance circuits, *210.52(B)(2)*, and it may be required to be GFCI protected, *210.8(A)(7)*. AFCI protection is required.

The 20-ampere small-appliance circuits prevent circuit overloading in the kitchen because of the heavy concentration and use of electrical appliances. For example, one type of cord-connected microwave oven is rated 1500 watts at 120 volts. The current required by this appliance alone is

$$I = \frac{W}{E} = \frac{1500}{120} = 12.5 \text{ amperes}$$

In *90.1(B)* we find a statement that *Code* requirements are considered necessary for safety. According to the *Code*, compliance with these rules will not necessarily result in an efficient, convenient, or adequate installation for good service or future expansion of electrical use.

Figure 12-8 shows four 20-ampere branch circuits feeding the receptacles in the kitchen area. These circuits are B13, B15, B16, and B22.

Figure 12-8 shows one way to arrange the 20-ampere branch circuits in the kitchen area. Obviously, there are other ways to arrange these branch circuits. The intent is to arrange the circuits so as to provide the best availability of electrical power for appliances that will be used in the heavily concentrated work areas in the Kitchen.

All receptacles serving countertops in kitchens must be GFCI protected, *210.8(A)(6)*. They must also have AFCI protection. See Chapter 6 for an explanation of methods that can be used to provide the required AFCI and GFCI protection.

In Figure 12-8, Circuits B13 and B16 are similar in that the circuit begins at an AFCI circuit breaker and then is routed to the line-side terminals of a feed-through GFCI receptacle. From the load-side

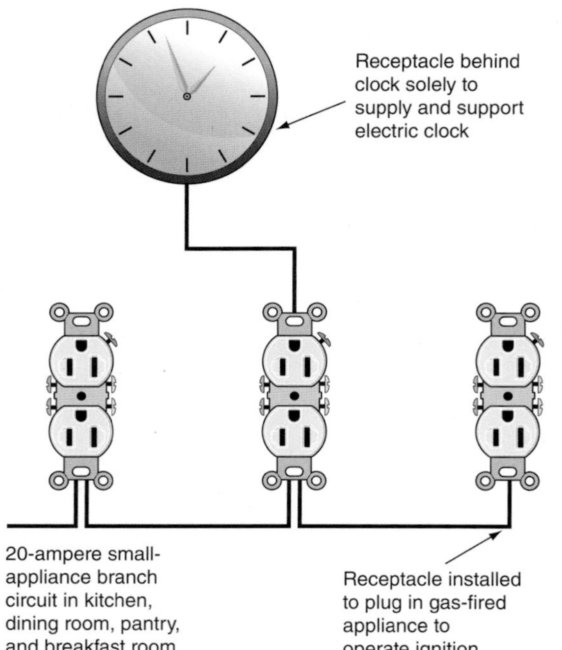

FIGURE 12-9 A receptacle installed solely to supply and support an electric clock, or a receptacle that supplies power for clocks, clock timers, and electric ignitions for gas ranges, ovens, and cooktops, may be connected to a 20-ampere small-appliance circuit or to a general-purpose lighting circuit, *210.52(B)(2), Exceptions.*

terminals of the feed-through GFCI receptacle, conductors are then run to the other grounding-type receptacle(s) on the circuit. Dual function AFCI/GFCI circuit breakers could have been installed on these circuits, in which case the receptacles would have been grounding-type receptacles. The choice is yours!

Circuits B15 and B22 do not require GFCI protection because the receptacles do not serve countertops. However, AFCI protection of the branch circuit or receptacle outlets is required.

Refrigerator Receptacle

So-called nuisance power outages in homes can be caused by using too many high-wattage small appliances in the kitchen at the same time—and then the refrigerator "kicks in."

The *NEC* does not help us in this respect. It *permits* the receptacle for the refrigerator to be connected to any one of the two required 20-ampere small-appliance branch circuits, *210.52(B)(1)*. Some home-type combination refrigerator/freezers with many features draw 11 amperes or more. That doesn't leave much room on a 20-ampere branch circuit for other small cord-and-plug-connected appliances in the kitchen.

The *NEC* also *permits* this receptacle to be connected to an individual 15- or 20-ampere branch circuit, *210.52(B)(1), Exception 2*. This is a much better way to do it. Many instructions furnished with these larger home-type refrigerator/freezers require that an individual branch circuit be provided. In this residence, Circuit B22 is an individual (dedicated) 20-ampere branch circuit that supplies only the refrigerator receptacle. This is a single receptacle rated 20 amperes.

Following are a few more *NEC* requirements:

110.3(B) states, Listed or labeled equipment shall be installed and used in accordance with any instructions included in the listing or labeling.*

This receptacle located behind the refrigerator does not require GFCI protection because it does not serve a countertop, *210.8(A)(6)*.

Branch Circuit, Individual. The *NEC* defines an *individual branch circuit* as A branch circuit that supplies only one utilization equipment.*

- When an individual branch circuit is run for the refrigerator, some inspectors require that a single receptacle be installed to "Meet *Code*." A single receptacle ensures that the receptacle will not supply other appliances, although this is highly unlikely because the receptacle is behind the refrigerator.
- Providing an individual branch circuit for the refrigerator diverts the refrigerator load from the receptacle branch circuits that serve the countertop areas. Because this individual branch circuit does not serve countertops, it is not considered a small-appliance branch circuit.

*Reprinted with permission from NFPA 70-2014.

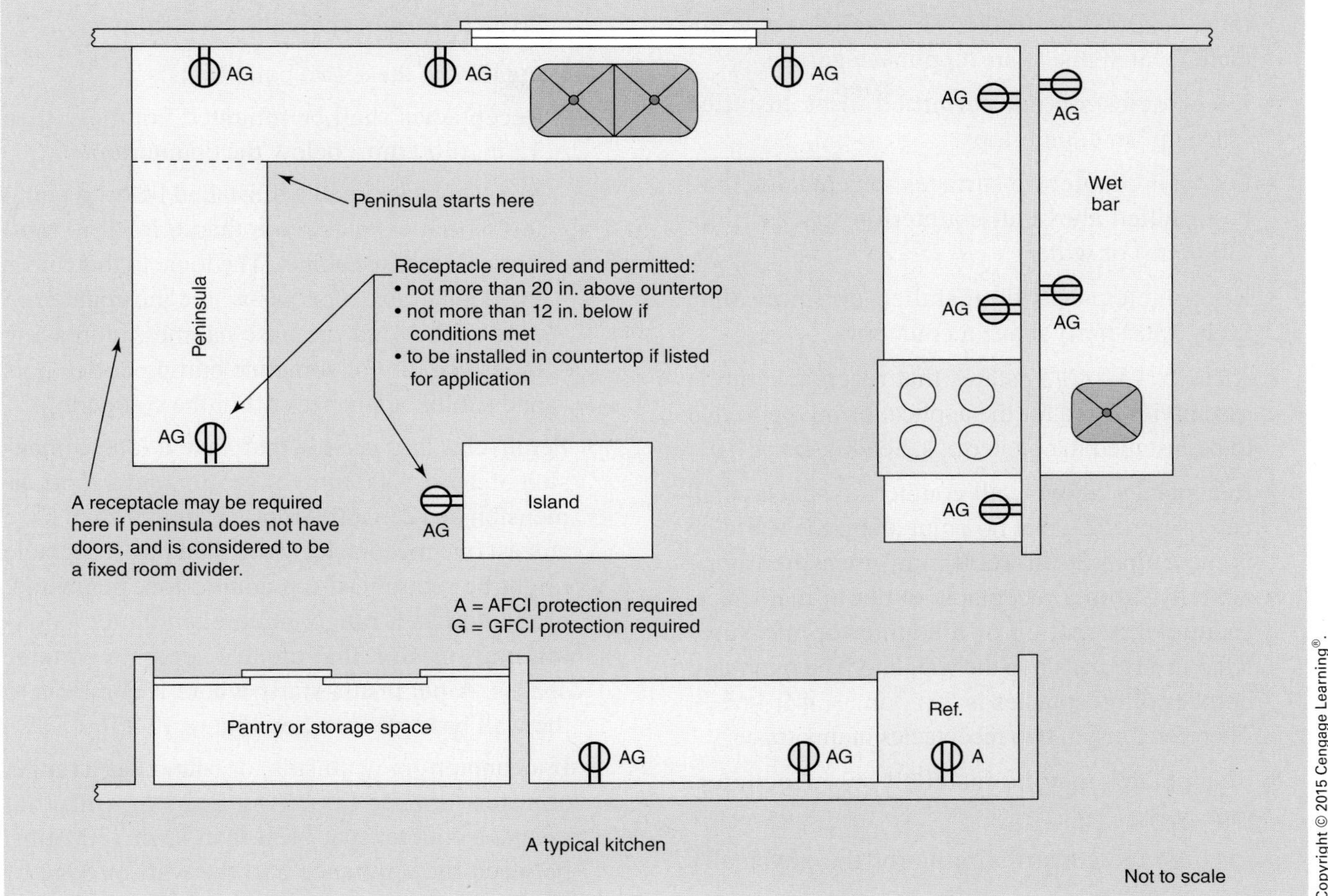

FIGURE 12-10 A receptacle outlet is required for each wall countertop space 12 in. (300 mm) or wider so no point is more than 24 in. (600 mm) from a receptacle. Not less than one receptacle is required for every island and peninsula cabinet. As shown for the wet bar area, GFCI protection is required for 125-volt, single phase, 15- and 20-ampere receptacles installed within 6 ft (1.8 m) of sinks that are not located in kitchens. See *210.8(A)(7)*.

The same logic would apply in other situations such as a receptacle located behind a gas range. It's up to the electrical inspector.

- A single receptacle must be rated not less than the branch-circuit rating, *210.21(B)(1)*. That's a 15-ampere receptacle on a 15-ampere branch circuit—a 20-ampere receptacle on a 20-ampere branch circuit.
- Cord-and-plug-connected appliances are not permitted to exceed 80% of the branch-circuit rating, *210.23(A)(1)*. That's 12 amperes on a 15-ampere circuit, and 16 amperes on a 20-ampere circuit.
- This individual branch circuit is not considered to be additional load for service-entrance calculations, *220.14(J)*. The refrigerator load would have been there no matter what circuit it was connected to. The advantage of doing this is that it diverts the refrigerator load from the small-appliance branch circuits in the kitchen.

Countertops in Kitchens

See Figure 12-10.

- The following requirements pertain to countertops in kitchens, pantries, breakfast rooms, dining rooms, and similar areas; see *210.52(C)*.

- Receptacles that serve countertops in kitchens must be GFCI protected. This includes a receptacle located inside an "appliance garage."
- Receptacles are not permitted to be installed "face up" in countertops.
- For wall countertop surfaces, receptacles shall be installed above all countertop spaces 12 in. (300 mm) or wider.
- Receptacles shall be installed not more than 20 in. (500 mm) above a countertop.
- *NEC 210.52(C)(5)* states that receptacle outlet assemblies listed for the application are permitted to be installed in countertops. See Figure 12-10.
- Receptacles above wall countertop spaces shall be positioned so that no point along the wall line is more than 24 in. (600 mm), measured horizontally, from a receptacle outlet in that space. Example: A section of a countertop measures 5 ft 3 in. (1.6 m) along the wall line. The minimum number of receptacles is 5 ft 3 in. ÷ 4 ft = 1+. Therefore, install two receptacles in this space.
- Receptacles may be installed below countertops only:
  – where the construction is for the physically impaired, or
  – for island and peninsula countertop spaces where the countertop is flat across its entire surface and there is no way to mount a receptacle within 20 in. (500 mm) above the countertop.

  For either of these two conditions:

  – Receptacles shall be mounted not more than 12 in. (300 mm) below the countertop.
  – No receptacles shall be installed below a countertop that extends more than 6 in. (150 mm) beyond its base cabinet. The logic to this rule is the popular use of stools where the countertop extends beyond the base cabinet, increasing the possibility of someone pulling coffee pots and similar appliances off of the countertop.
- Peninsulas and islands that have a long dimension of 24 in. (600 mm) or greater and a short dimension of 12 in. (300 mm) or greater shall have at least one receptacle. More than one receptacle might be required if the countertop, peninsula, or island has a range, counter-mounted cooking unit, or sink that creates separate counter spaces. A peninsula starts where it connects to the wall base cabinet. See Figure 12-11.
- If a countertop, peninsula, or island has a range, counter-mounted cooking unit, or sink that creates a counter space less than 12 in. (300 mm) between the appliance and the wall or edge of the countertop, separate countertop spaces on either side of these appliances have been created. Receptacles shall be installed to comply

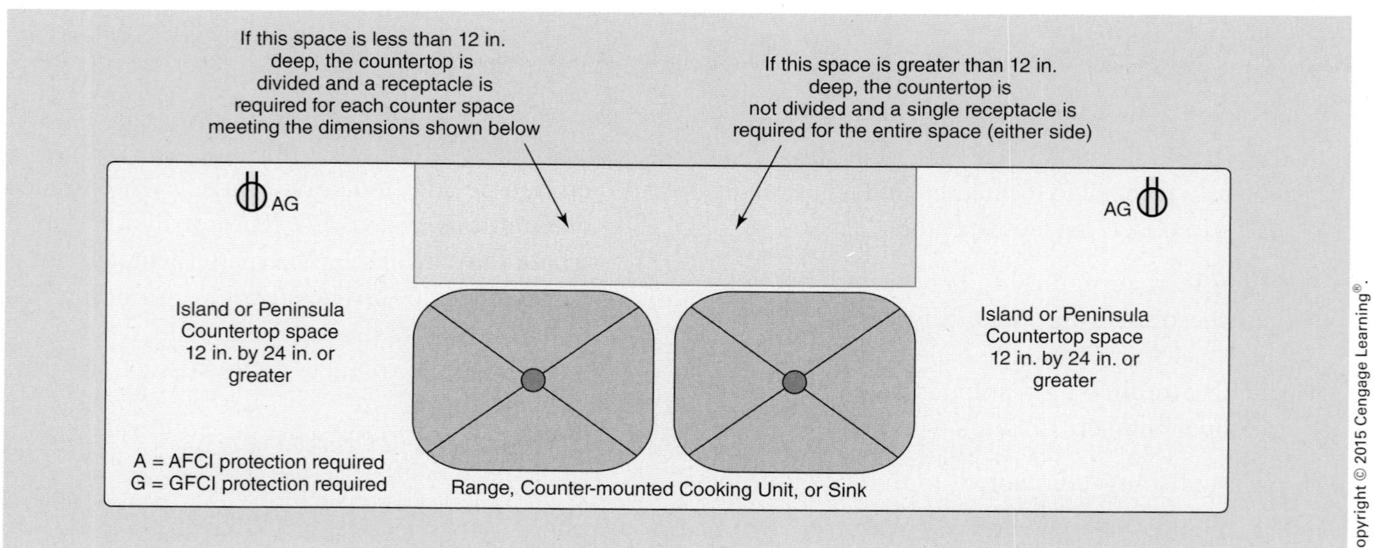

FIGURE 12-11 The depth of the space behind a range, counter-mounted cooking unit or sink determines whether a separate space is created.

with the required number of receptacles in each of these spaces.

- Receptacles that do not serve countertops, such as behind a refrigerator, under the sink for plugging in a food waste disposer or dishwasher, or behind or above an appliance that is fastened in place, are not considered part of the receptacle outlet requirement for serving countertops. These receptacles are not required to be GFCI protected.
- Receptacles inside an "appliance garage" are in addition to the receptacles required to serve countertops. They must be GFCI protected. Although permitted to be connected to one of the two required 20-ampere small-appliance circuits, consider running a separate 20-ampere branch circuit.
- A receptacle located inside an upper cabinet that serves a "hang below cabinet" microwave oven should be connected to an individual branch circuit. This is discussed in Chapter 20.
- Receptacles are generally not required on the wall behind a range, counter-mounted cooking unit, or sink. However, a receptacle is required
  - if the distance from the wall to the range, counter-mounted cooking unit, or sink is 12 in. (300 mm) or greater. In other words, if the space is 12 in. (300 mm) or greater, it is counter space; if less than 12 in. (300 mm), it is not counter space.
  - if the range, counter-mounted cooking unit, or sink is corner mounted on an angle, and the distance from the corner to the range, counter-mounted cooking unit, or sink is 18 in. (450 mm) or greater. In other words, if the space is 18 in. (450 mm) or greater, it is counter space; if less than 18 in. (450 mm), it is not counter space.

This is illustrated in *Figure 210.52* in the *NEC* and in Figures 12-12 and 12-13.

Sliding Glass Doors

See Chapters 3 and 9 for discussions on how to space receptacle outlets on walls where sliding glass doors are installed.

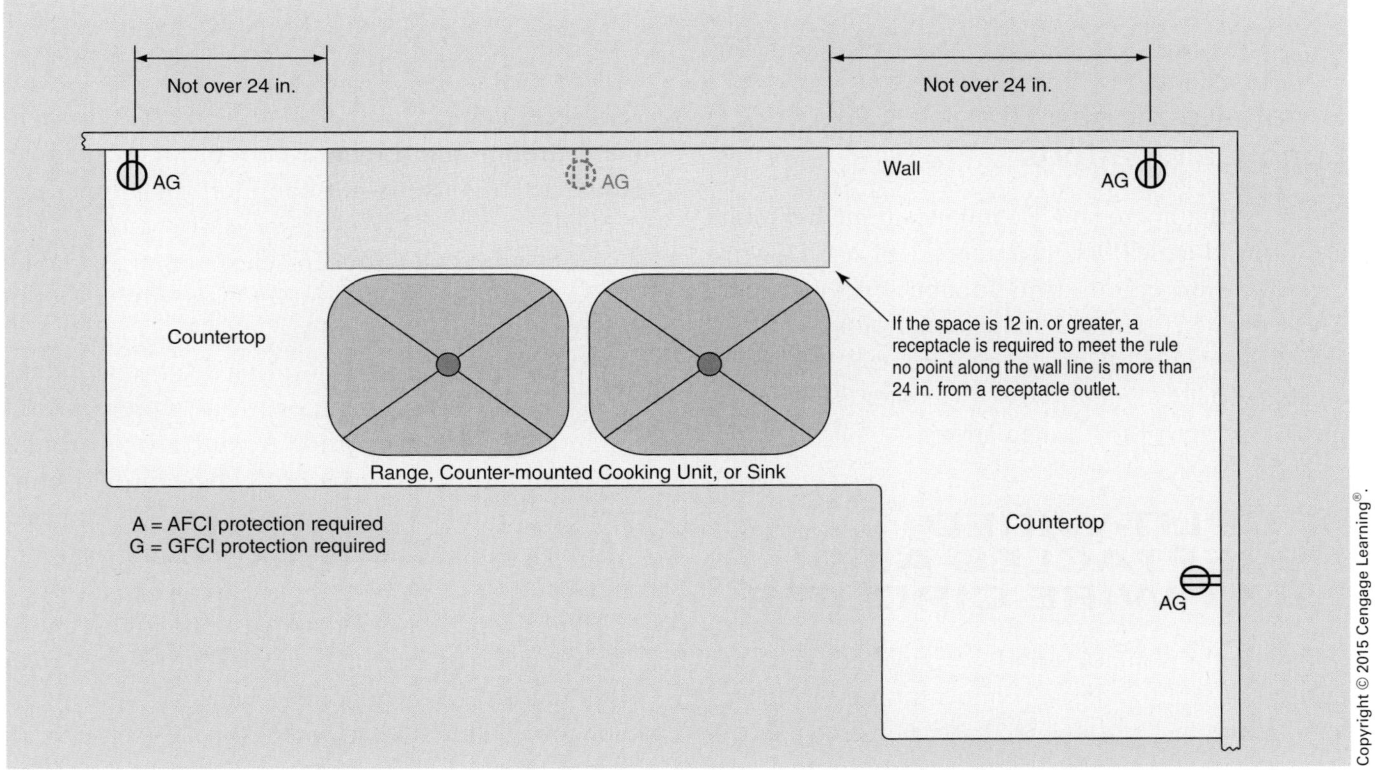

FIGURE 12-12 A receptacle is required in the space behind a range, counter-mounted cooking unit, or sink if the space behind the sink or range is 12 in. (300 mm) or greater. *210.52(C)(1).*

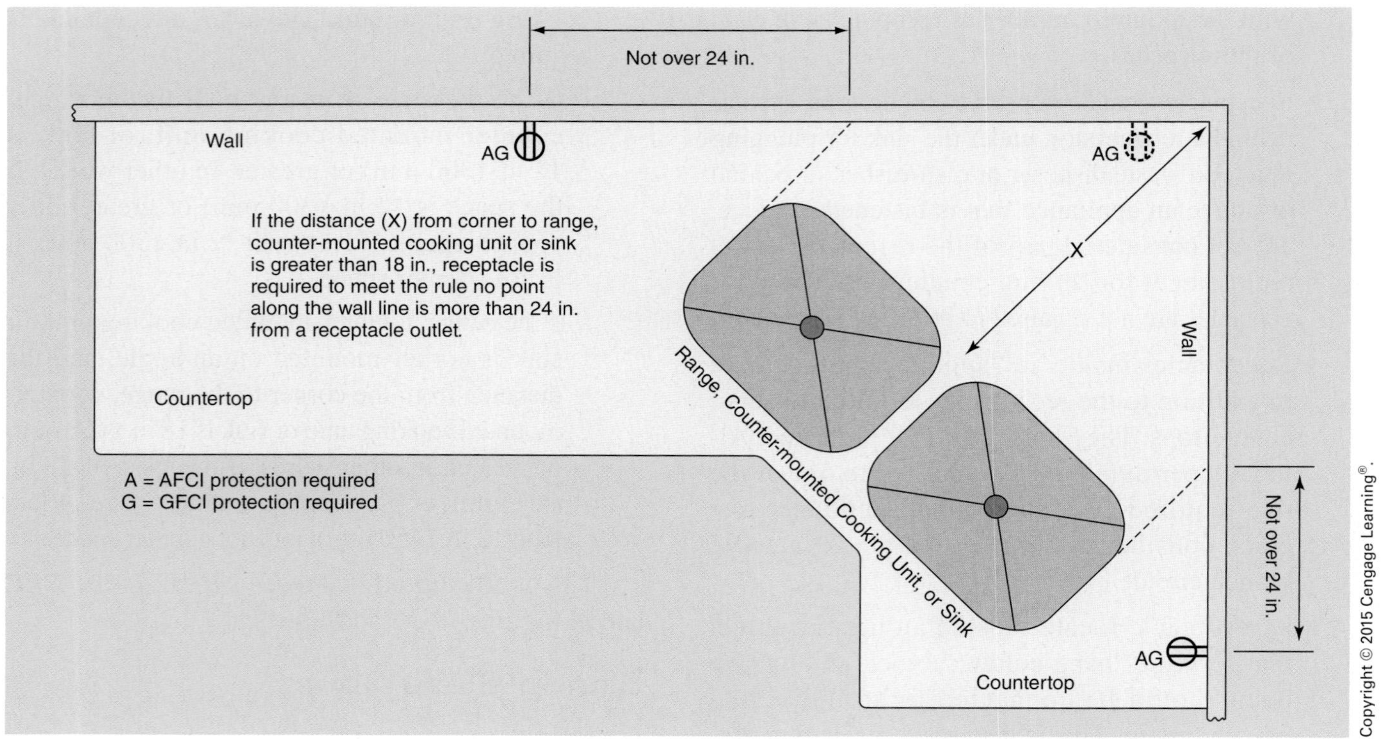

FIGURE 12-13 A receptacle is required in the space behind a range, counter-mounted cooking unit, or sink if the space behind the sink or range is 18 in. (300 mm) or greater, *210.52(C)(1)*.

Small-Appliance Branch Circuit Load Calculations

The 20-ampere small-appliance branch circuits are figured at 1500 volt-amperes for each 2-wire circuit when calculating feeders and services. The separate 20-ampere branch circuit for the refrigerator is excluded from this requirement, *220.14(J)* and *220.53(A), Exception*. See Chapter 29 for complete coverage of load calculations.

SPLIT-WIRED RECEPTACLES AND MULTIWIRE CIRCUITS

Split-wired receptacles may be installed wherever desired or where a heavy concentration of plug-in load is anticipated, in which case each receptacle is connected to a separate circuit, as in Figure 12-14 and 12-15.

A popular use of split-wired receptacles is to control one receptacle with a switch and leave one receptacle "hot" at all times, as is done in the bedrooms and Living Room of this residence. However, this use is not common for kitchen receptacles.

Split-wired GFCI receptacles are not available. Refer to Chapter 6, where GFCIs are discussed in great detail.

Therefore, the use of split-wired receptacles is rather difficult to incorporate where GFCI receptacles are required. A multiwire branch circuit could be run to a box where proper connections are made, with one circuit feeding one feed-through GFCI receptacle and its downstream receptacles.

This would be used when it becomes more economical to run one 3-wire circuit rather than installing two 2-wire circuits. Again, it is a matter of knowing what to do and when it makes sense to do it, as in Figure 12-16.

When multiwire circuits are installed, do not use the terminals of a receptacle or other wiring device to serve as the splice for the grounded

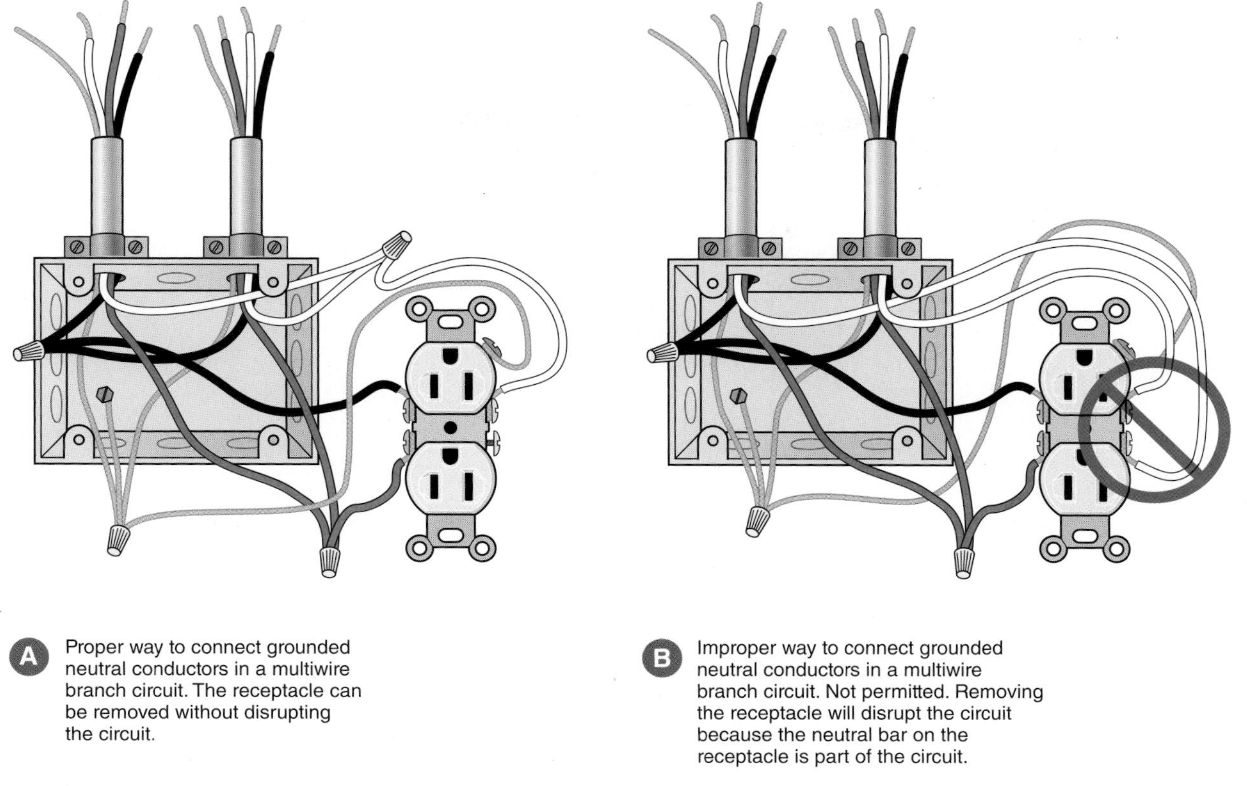

FIGURE 12-14 Connecting the grounded neutral conductors in a 3-wire (multiwire) branch circuit, *NEC 300.13(B)*.

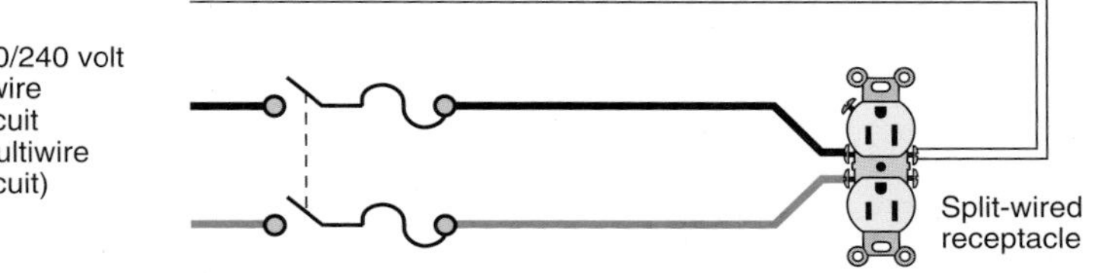

FIGURE 12-15 Split-wired receptacle connected to multiwire circuit.

neutral conductor. The *NEC* in *300.13(B)* states, In multiwire branch circuits, the continuity of a grounded conductor shall not be dependent upon device connections . . .* Any splicing of the grounded conductor in a multiwire circuit must be made independently of the receptacle or lampholder, or other wiring device. This requirement is illustrated in Figure 12-14. Note that the screw terminals of the receptacle *must* not be used to splice the neutral conductors. The hazards of an open neutral are discussed in Chapter 17. Figure 12-17 shows two types of grounding receptacles commonly used for installations of this type.

When multiwire branch circuits are used, a means must be provided to disconnect simultaneously (at the same time) all of the "hot" conductors at the panelboard where the branch circuit originates. Figure 12-15 shows a 120/240-volt, 3-wire (multiwire) branch circuit supplying a split-wired receptacle. Note that 240 volts are present on the wiring device, connected in such a way that 120 volts are connected to each receptacle on the wiring device. Multiwire branch circuits are also used to connect appliances such as electric clothes dryers and electric ranges. See *210.4(A)*, *210.4(B)*, and *210.7(B)*.

*Reprinted with permission from NFPA 70-2014.

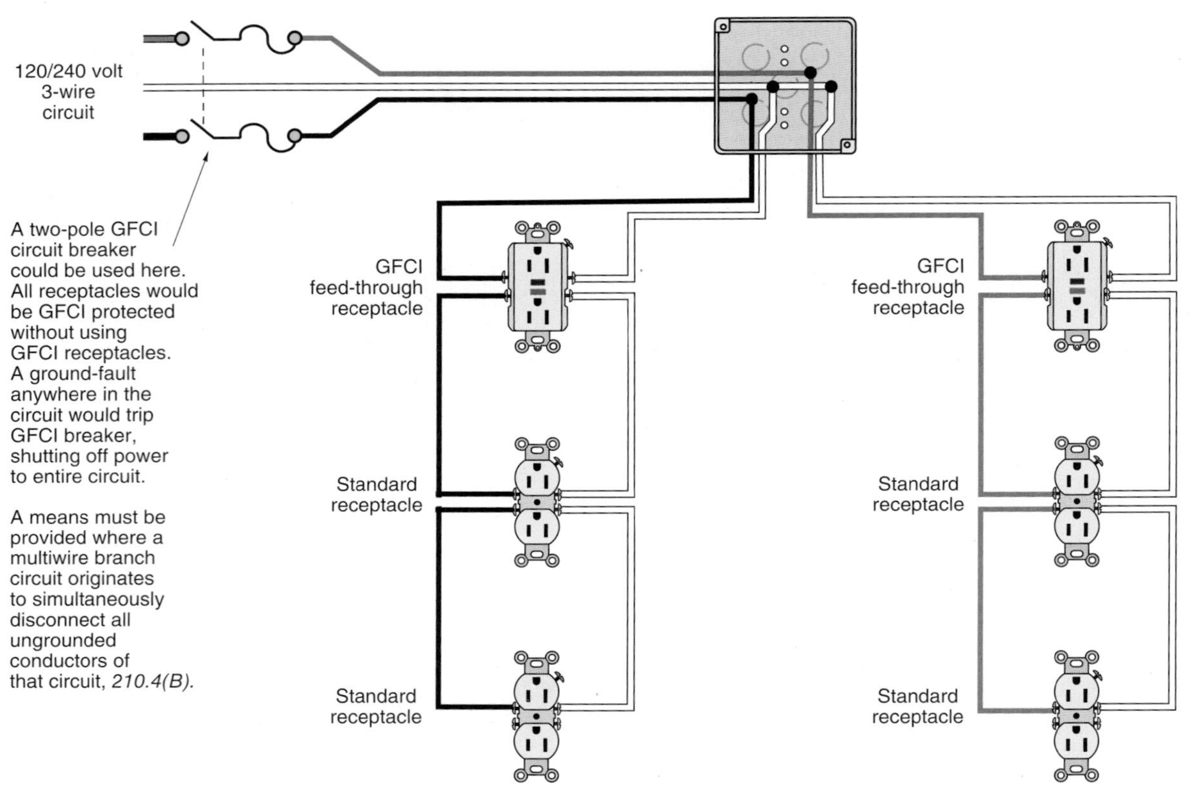

FIGURE 12-16 This diagram shows how a multiwire circuit can be used to carry two circuits to a box, then split the 3-wire circuit into two 2-wire circuits. This circuitry requires simultaneous disconnect of the ungrounded conductors as shown in Figure 12-15.

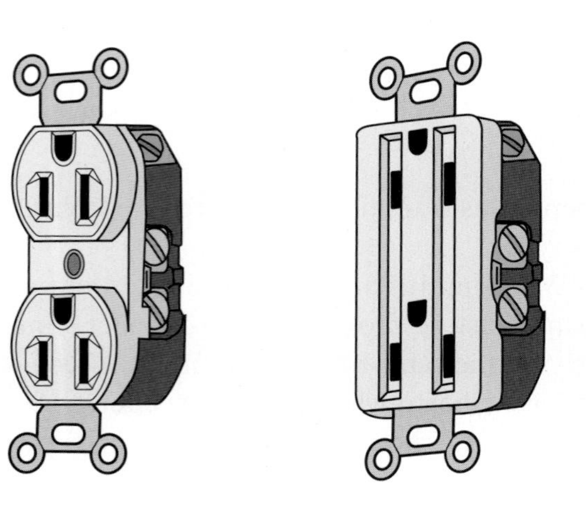

FIGURE 12-17 Grounding-type duplex receptacles, 15A-125V rating. Most duplex receptacles can be changed into "split-wired" receptacles by breaking off the small metal tab on the terminals. The figure clearly shows this feature.

SAFETY ALERT

Care must be taken when connecting a GFCI to a multiwire branch circuit. As discussed in detail in Chapter 6, some of these devices are listed for use on multiwire branch circuits where the neutral is shared; others are not listed for this application.

Important: "Handle ties" are discussed in detail in Chapter 28. Two very important safety issues must be considered. Handle ties between two single-pole circuit breakers

- *do* provide simultaneous disconnection of both circuit breakers when they are manually turned off.

- *do not* provide simultaneous tripping (common tripping) of both circuit breakers in the event of an overload, short-circuit, or ground fault on only one of the circuits of a multiwire branch circuit supplied by the two "handle-tied" circuit breakers.

RECEPTACLES AND OUTLETS

Article 100 defines a receptacle and a receptacle outlet, as illustrated in Figures 3-5, 12-18, and 12-19.

Receptacles are selected for circuits according to the following guidelines. A single receptacle connected to a circuit must have a rating not less than the rating of the circuit, *210.21(B)*. In a residence, typical examples for this requirement for single receptacles are the clothes dryer outlet (30 amperes), the range outlet (50 amperes), or the freezer outlet (15 amperes).

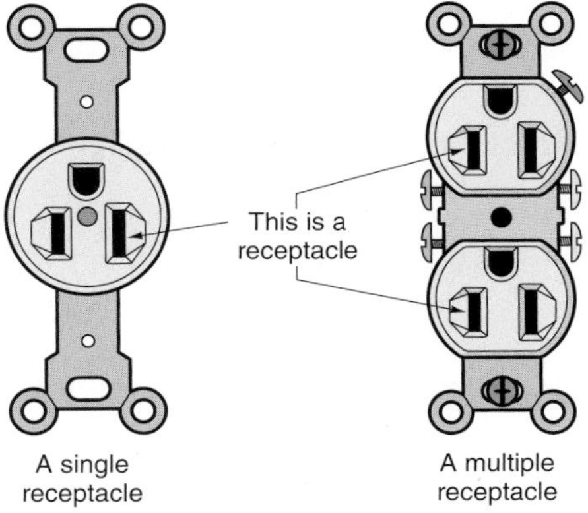

FIGURE 12-18 Receptacles: See *NEC* definitions, *Article 100.*

Circuits rated 15 amperes supplying two or more receptacles shall not contain receptacles rated over 15 amperes. For circuits rated 20 amperes supplying two or more receptacles, the receptacles connected to the circuit may be rated 15 or 20 amperes. See *Table 210.21(B)(3).*

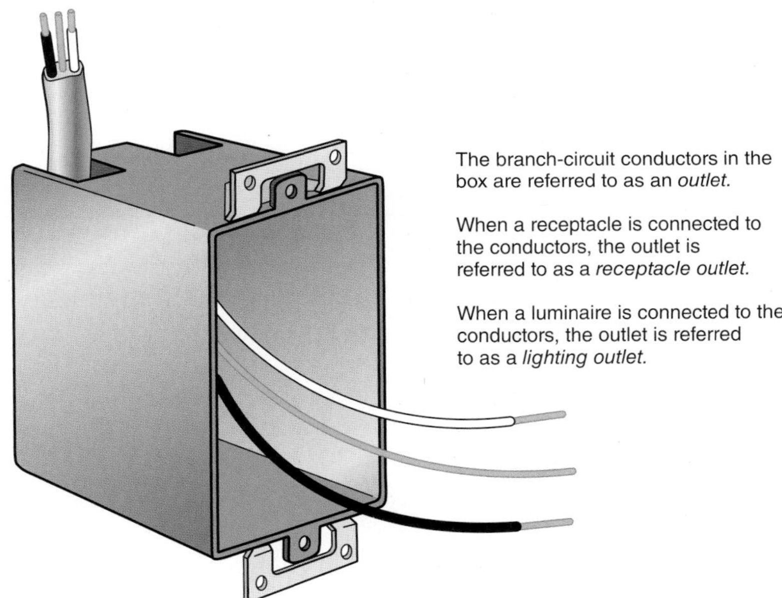

FIGURE 12-19 An *outlet* is defined in the *NEC* as A point on the wiring system at which current is taken to supply utilization equipment.*

*Reprinted with permission from NFPA 70-2014.

REVIEW

Note: Refer to the *NEC* or the plans in the back of this text if desired.

1. If everything on Circuit B7 were turned on, what would be the total current draw?
   ____

2. From what panel does the kitchen lighting circuit originate? What size conductors are used? ____________________
   ____

3. How many luminaires are connected to the kitchen lighting circuit?
   ____
   ____

4. What color fluorescent lamps are recommended for residential installations?
   ____
   ____

5. a. What is the minimum number of 20-ampere small-appliance circuits required for a kitchen according to the *Code*? ____________________
   b. How many are there in this kitchen? ____________________

6. How many receptacle outlets are provided in the kitchen? ____________________

7. What is meant by the term *split-wired receptacles*? ____________________
   ____
   ____
   ____
   ____

8. Duplex receptacles connected to the 20-ampere small-appliance branch circuits in kitchens and dining rooms (Circle the correct answer):
   a. may be rated 15 amperes. (True) (False)
   b. must be rated 15 amperes. (True) (False)
   c. may be rated 20 amperes. (True) (False)
   d. must be rated 20 amperes. (True) (False)

9. In kitchens, a receptacle must be installed at each counter space ____________ in. or wider.

10. A fundamental rule regarding the grounding of metal boxes, luminaires, and so on, is that they must be grounded when "in reach of ____________."

11. How many circuit conductors enter the box
    a. where the range hood will be installed? ____________________
    b. in the ceiling box over which the track will be installed? ____________________
    c. at the switch location to the right of the sliding door? ____________________

12. How much space is there between the countertop and upper cabinets? ____________

13. Where is the speed control for the range hood fan located? ____________

14. Who is to furnish the range hood? ____________

15. List the appliances in the kitchen that must be electrically connected. ____________

____

____

____

____

____

16. Complete the wiring diagram, connecting feed-through GFCI 2 to also protect receptacle 1, both to be supplied by Circuit A1. Connect feed-through GFCI 3 to also protect receptacle 4, both to be supplied by Circuit A2. Use colored pencils or markers to show proper color. Assume that the wiring method is EMT, where more freedom in the choice of insulation colors is possible.

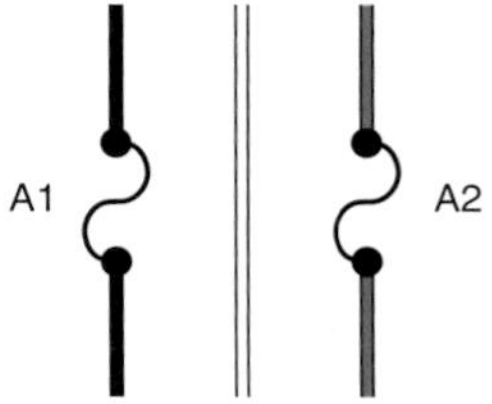

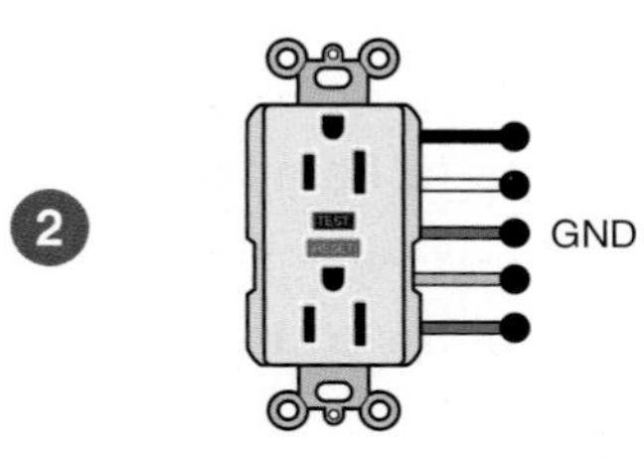

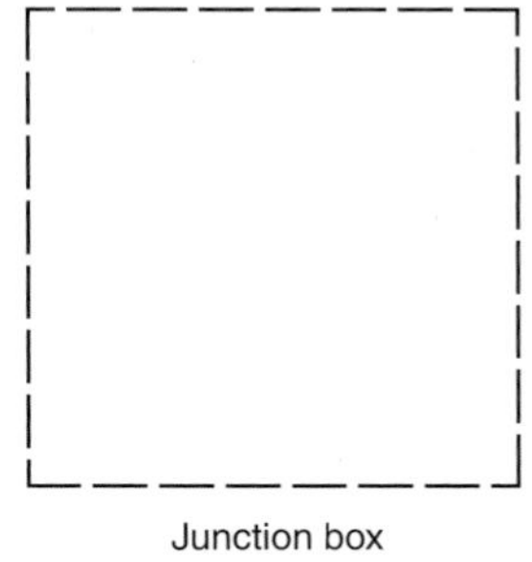

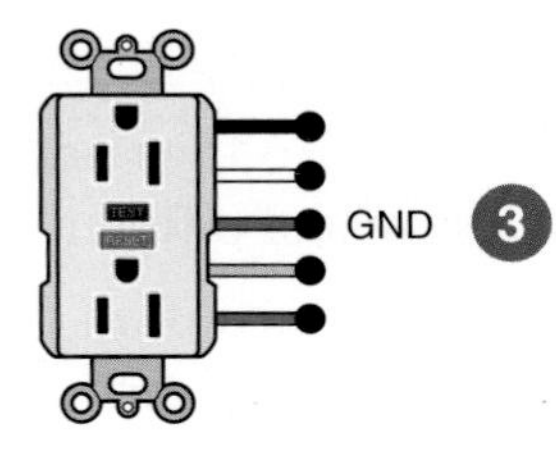

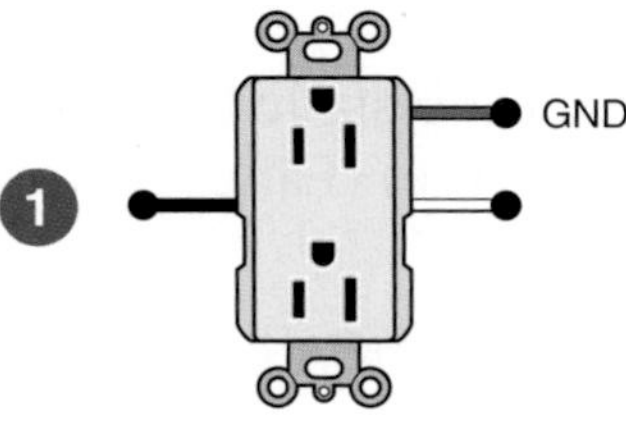

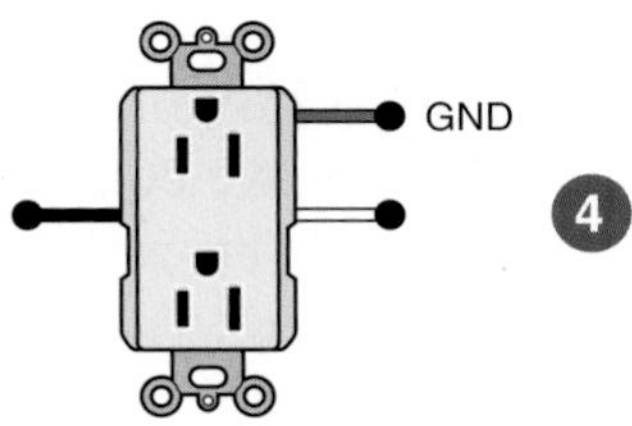

17. Each 20-ampere small-appliance branch-circuit load demand shall be determined at ____________. (Circle the correct answer.)

a. 2400 volt-amperes b. 1500 volt-amperes c. 1920 volt-amperes

18. a. Does the *NEC* permit the receptacle for the refrigerator to be supplied by one of the required 20-ampere small-appliance branch circuits? Yes ______________ No ______________ *Section* ______________.

    b. Does the *NEC* permit the receptacle for the refrigerator to be supplied by an individual branch circuit? Yes ______________ No ______________ *Section* ______________.

    c. What is the practice in your area?

       Run an individual branch circuit? ______________

       Connect it to one of the 20-ampere small-appliance branch circuits? ______________

19. a. The *Code* requires a minimum of two small-appliance circuits in a kitchen. Is it permitted to connect a receptacle in a dining room to one of the kitchen small-appliance circuits? ______________

       What *Code* section applies to this situation? ______________

    b. May outdoor weatherproof receptacles be connected to a 20-ampere small-appliance circuit? ______________

    c. Receptacles located above the countertops in the kitchen must be supplied by at least ______________ 20-ampere small-appliance circuits.

20. According to *210.52*, no point along the floor line shall be more than ______________ ft (______________ m) from a receptacle outlet. A receptacle must be installed in any wall space ______________ ft (______________ m) wide or greater.

21. The *Code* states that in multiwire circuits, the screw terminals of a receptacle must *not* be used to splice the neutral conductors. Why?

    ____

    ____

    ____

22. Electric fans produce a certain amount of noise. It is possible to compare the noise levels of different fans prior to installation by comparing their ______________ ratings.

23. Is it permitted to connect the white grounded circuit conductor to the equipment grounding terminal of a receptacle? ______________

24. What is unique about 120-volt appliances that are 2-wire cord-and-plug connected instead of being 3-wire cord-and-plug connected? ______________

    ____

25. a. All 125-volt, single-phase 15- and 20-ampere receptacles installed in a kitchen that serve countertop surfaces must be GFCI protected. (True) (False) (Circle the correct answer.)

    b. All 125-volt, single-phase 15- and 20-ampere receptacles installed within 6 ft (1.8 m) of the outside edge of a wet bar, laundry, utility, or similar sink must be GFCI protected. (True) (False) (Circle the correct answer.)

26. All receptacle outlets in the kitchen are required to be AFCI protected. (True) (False) (Circle the correct answer.)

27. The kitchen features a sliding glass door. For the purpose of providing the proper receptacle outlets, answer the following statements *true* or *false*.

    a. Sliding glass panels are considered to be wall space. ______________

    b. Fixed panels of glass doors are considered to be wall space. ______________

28. The following is a layout for the lighting circuit for the Kitchen. Complete the wiring diagram using colored pens or pencils to show the conductors' insulation color.

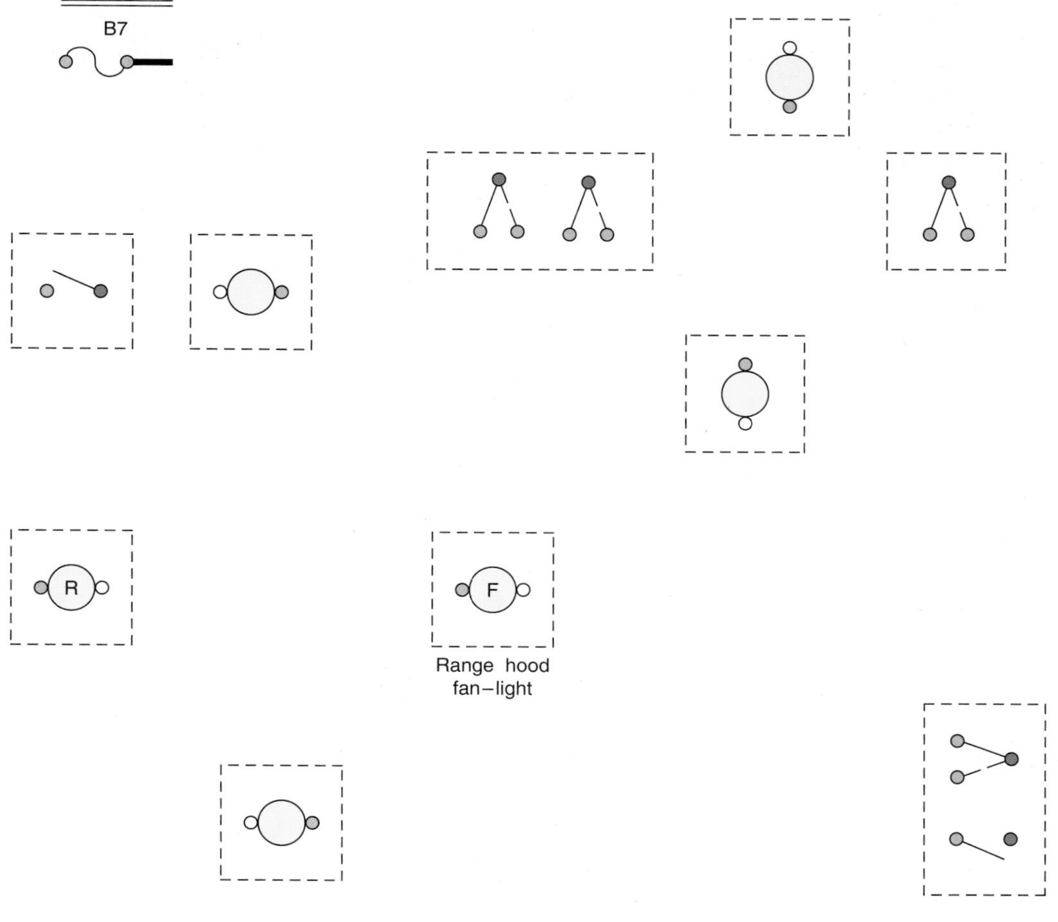

29. a. In your area, does the electrical inspector require that an individual branch circuit be installed for the refrigerator? (Yes) (No) (Circle the correct answer.)

    b. In your area, does the electrical inspector require that a single receptacle be installed for the refrigerator? (Yes) (No) (Circle the correct answer.)

CHAPTER **13**

Lighting Branch Circuit for the Living Room

OBJECTIVES

After studying this chapter, you should be able to

- lay out the wiring for a typical living room.
- understand the *NEC* requirements for track lighting.
- understand the basics for dimming incandescent and fluorescent lamps.

LIGHTING CIRCUIT B17 OVERVIEW

The feed for the living room lighting branch circuit is connected to Circuit B17, a 15-ampere circuit. Overcurrent protection and arc-fault protection for this circuit are provided by a 15-ampere AFCI circuit breaker in Panel B. The AFCI requirement is found in *NEC® 210.12(A)*. GFCI protection required by *210.8(A)(3)* for the outdoor receptacle could be provided by a GFCI receptacle or by an AFCI/GFCI dual-function circuit breaker in Panel B.

The home run is brought from Panel B to the weatherproof receptacle outside of the Living Room next to the sliding door. The circuit is then run to the split-wired receptacle just inside the sliding door, almost back to back with the outdoor receptacle, Figure 13-1.

A 3-wire cable is then carried around the Living Room, feeding in and out of the eight receptacles. This 3-wire cable carries the black and white circuit conductors plus the red wire, which is the switched conductor. Because these receptacles will provide most of the general lighting for the Living Room through the use of floor lamps and table lamps, it

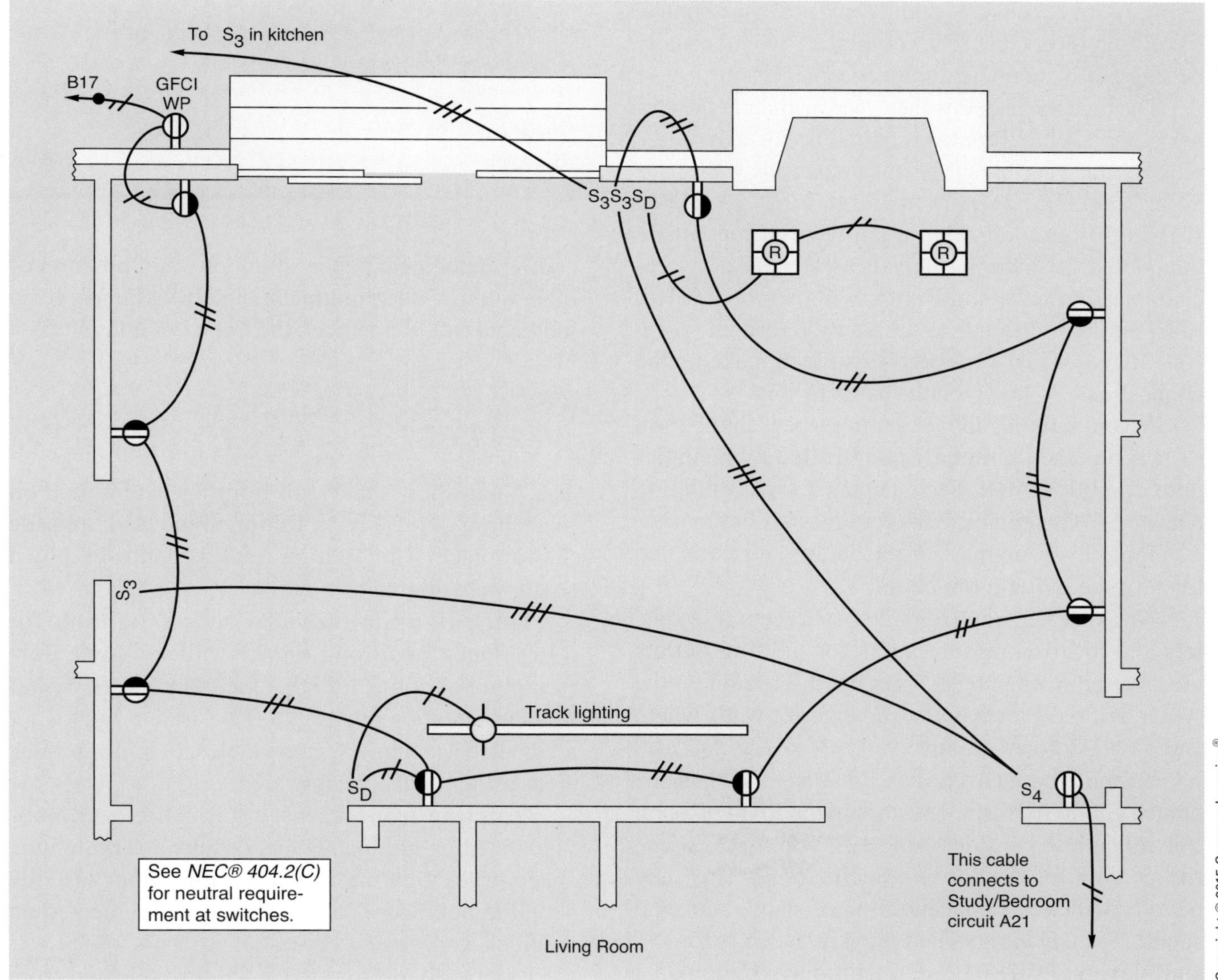

FIGURE 13-1 Cable layout for the Living Room. Note that the receptacle on the short wall where the 4-way switch is located is fed from the Study/Bedroom Circuit A21.

is advantageous to have some of these lamps controlled by the three switches—two 3-way switches and one 4-way switch.

A rule has been added to the 2014 *NEC* that prohibits the use of dimmers to control standard 15- or 20-ampere receptacles. The rule reads ▶**406.15 Dimmer-Controlled Receptacles.** A receptacle supplying lighting loads shall not be connected to a dimmer unless the plug/receptacle combination is a nonstandard configuration type that is specifically listed and identified for each such unique combination.◀*

This requirement has direct application to the wiring shown for the split-wired receptacles in the living room because 3- and 4-way switches control ½ of each receptacle. The receptacles would usually be standard 15-ampere duplex receptacles.

The receptacle below the 4-way switch in the corner of the Living Room is required because the *Code* requires that a receptacle be installed in any wall space 24 in. (600 mm) or more in width, *210.52(A)(2)(1)*. It is highly unlikely that a split-wired receptacle connected for switch control, as are the other receptacles in the Living Room, will ever be needed. Plus, the short distance to the Study/Bedroom receptacle makes it economically sensible to make up the connections as shown on the cable layout.

Accent lighting above the fireplace is in the form of two recessed luminaires controlled by a single-pole dimmer switch. See Chapter 7 for installation data and *Code* requirements for recessed luminaires.

Table 13-1 summarizes the outlets and estimated load for the living room circuit.

The spacing requirements for receptacle outlets and location requirements for lighting outlets are covered in Chapter 3, as is the discussion on the spacing of receptacle outlets on exterior walls where sliding glass doors are involved, *210.52(A)(2)(2)*.

You might want to consider running more than one branch circuit for the receptacles in the Living Room. You might also consider using 12 AWG for the home run or even for the entire branch circuit. This may be to provide for a concentration of stereo surround sound, TV, CD burners and players, video recorders and players, DVD recorders and players, personal computers, printers, fax machines, copy machines, adding machines, paper shredders, and all sorts of other audio/video equipment, and desk lamps. It's a judgment call. Remember, the *Code* is a minimum!

TABLE 13-1

Living Room outlet count and estimated load (Circuit B17).

DESCRIPTION	QUANTITY	WATTS	VOLT-AMPERES
Receptacles @ 120 W Note: Receptacle under S_4 is connected to Study/Bedroom circuit.	8	960	960
Weatherproof receptacle	1	120	120
Track light five lamps @ 40 W each	1	200	200
Fireplace recessed luminaires @ 75 W each	2	150	150
TOTALS	12	1430	1430

TRACK LIGHTING (ARTICLE 410, PART XV)

The plans show that track lighting is mounted on the ceiling of the Living Room on the wall opposite the fireplace. The lighting track is controlled by a single-pole dimmer switch.

Track lighting provides accent lighting for a fireplace, a painting on the wall, or some item (sculpture or collection) that the homeowner wishes to focus attention on, or it may be used to light work areas such as counters, game tables, or tables in such areas as the kitchen eating area.

Differing from recessed luminaires and individual ceiling luminaires that occupy a very definite space, track lighting offers flexibility because the actual lampholders can be moved and relocated on the track as desired.

Lampholders from track lighting selected from hundreds of styles are inserted into the extruded aluminum or PVC track at any point (the circuit conductors are in the track), and the plug-in connector

*Reprinted with permission from NFPA 70-2014.

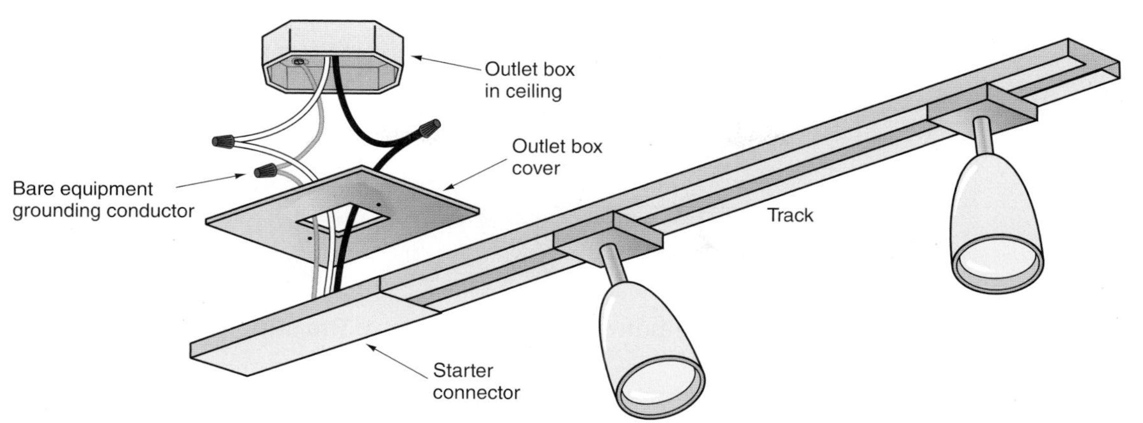

FIGURE 13-2 End-feed track light.

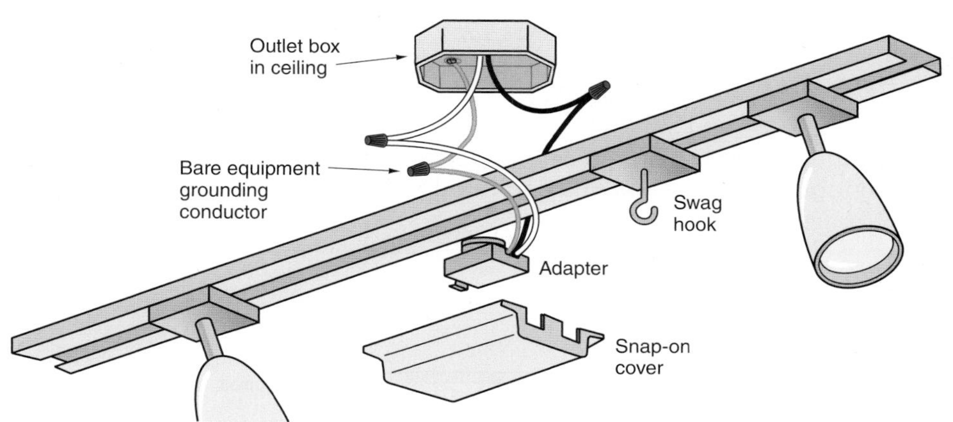

FIGURE 13-3 Center-feed track light.

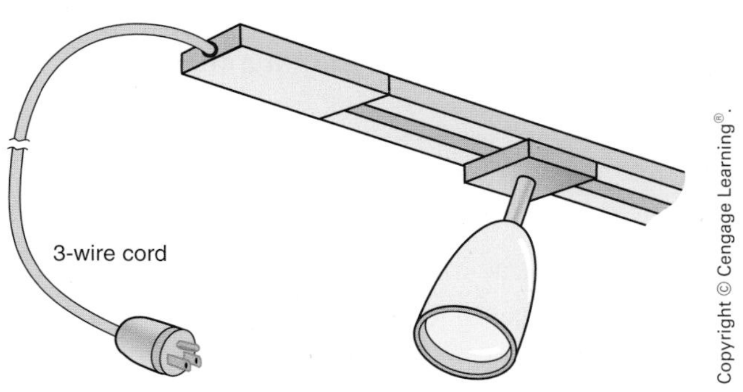

FIGURE 13-4 Plug-in track light.

on the lampholder completes the connection. In addition to being fastened to the outlet box, the track is generally fastened to the ceiling with toggle bolts or screws. Various track light installations are shown in Figures 13-2, 13-3, and 13-4.

Note on the plans for the residence that the living room ceiling has wood beams. This presents a challenge when installing track lighting that is longer than the space between the wood beams.

There are a number of things the electrician can do:

1. Install pendant kit assemblies that will allow the track to hang below the beam, as in Figure 13-5.
2. Install conduit conductor fittings on the ends of sections of the track, then drill a hole in the beam through which ½ in. EMT can be installed and connected to the conduit connector fittings, as in Figure 13-6.
3. Run the branch-circuit wiring concealed above the ceiling to outlet boxes located at the points

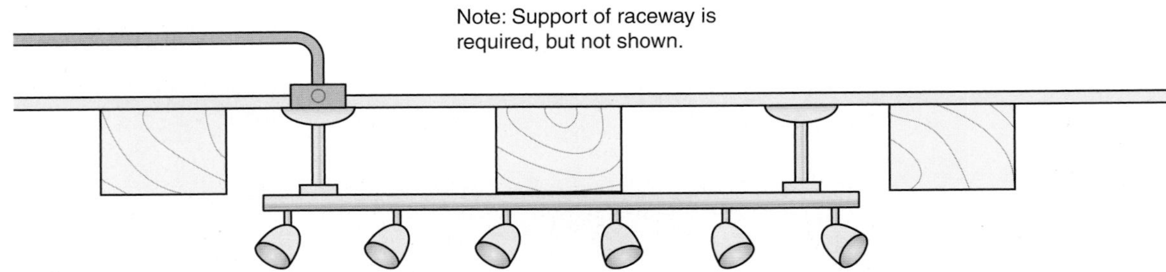

FIGURE 13-5 Track lighting mounted on the bottom of a wood beam. The supply wires to the track assembly are fed through the pendant.

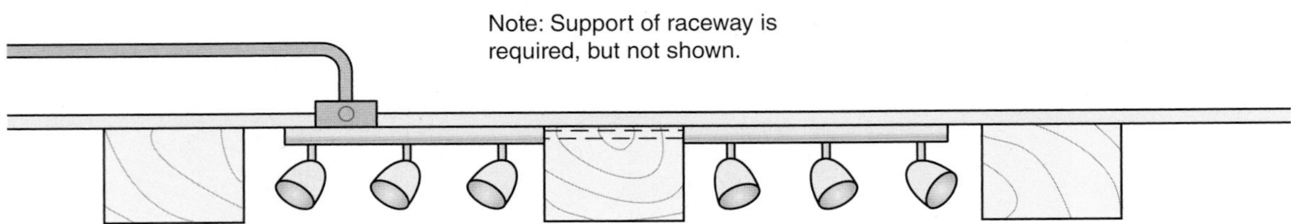

FIGURE 13-6 Track lighting mounted on the ceiling between the wood beams. The two track assembly sections are connected together by means of a conduit that has been installed through a hole drilled in the beam.

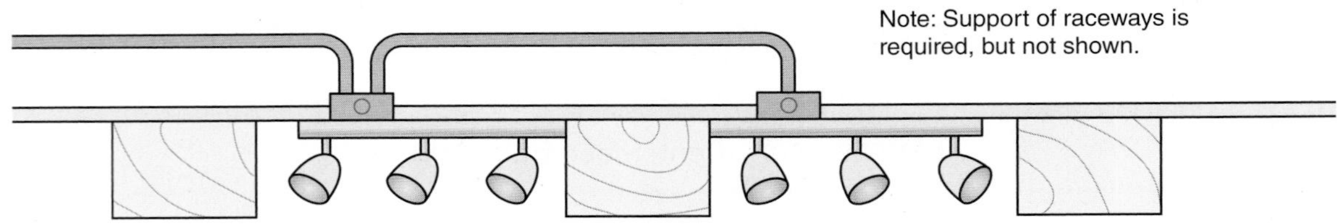

FIGURE 13-7 Track lighting mounted on the ceiling between the wood beams. Each of the two track assembly sections are connected to outlet boxes in the ceiling in the same manner that a regular luminaire is hung.

where the track is to be fed by the branch-circuit wiring, as in Figure 13-7.

Where to Mount Lighting Track

Lighting track manufacturers' catalogs and websites provide excellent recommendations for achieving the desired lighting results.

Lighting Track and the *Code*

Lighting track is listed under UL Standard 1574.

The *Code* rules for lighting track are found in *410.151* through *410.155*.

Lighting track in residential installations

- must be listed by an NRTL.
- are required to be installed according to the manufacturers' instructions, per *110.3(B)*.
- are not permitted to be installed
  - – where subject to physical damage.
  - – in damp or wet locations.
  - – concealed.
  - – through walls or partitions.
  - – less than 5 ft (1.5 m) above finished floor unless protected against physical damage, or if

the track is of the low-voltage type operating at less than 30 volts rms open-circuit voltage. See *Article 411* for low-voltage system requirements.

 – in the zone 3 ft (900 mm) horizontally and 8 ft (2.5 m) vertically from the top of a bathtub rim or shower stall threshold. This is discussed in Chapter 10.

- must be permanently installed and permanently connected to a branch circuit.
- are required to be supported twice if the track is not more than 4 ft (1.2 m), and shall have one additional support for each additional 4 ft (1.2 m) of track.
- must be grounded. Grounding of equipment is covered in previous chapters.
- are not permitted to be cut to length in the field unless permitted by the manufacturer. Instructions will specify where and how to make the cut, and how to close off the cut with a proper end cap.
- must have lampholders designed specifically for the particular track.
- must not have so many lampholders that the load would exceed the rating of the track.

Figure 13-8 shows cross-sectional views of a typical lighting track. Figure 13-9 shows typical lighting track lampholders.

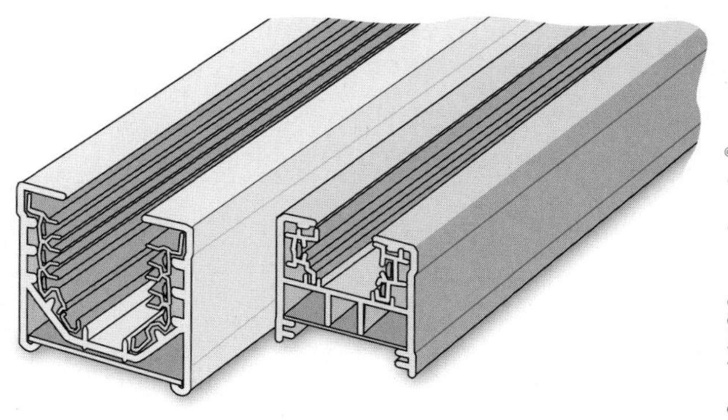

FIGURE 13-8 Tracks in cross-sectional views.

The track lighting in the Living Room of this residence is controlled by a single-pole dimmer switch.

Load Calculations for Track Lighting

For track lighting installed in homes, it is not necessary to add additional loads for the purpose of branch circuit, feeder, or service-entrance calculations. Track lighting is just another form of general lighting. Refer to *220.43(B)*.

In a commercial building, a loading factor of 150 VA for each 2 ft (600 mm) of lighting track must be added to the calculations for branch circuit, feeder, and service-entrance calculations, *220.43(B)*.

FIGURE 13-9 Typical lampholders that attach to track lighting systems.

For further information regarding track lighting, refer to *Part XV* of *Article 410*.

Portable Lighting Tracks

Portable lighting tracks, per UL Standard 153 must

- not be longer than 8 ft (2.4 m).
- have a cord not less than 5 ft (1.5 m) permanently attached to the track by the manufacturer.
- have overcurrent protection (fuse or circuit breaker) built into the plug cap. Overcurrent protection in the plug cap is not required if the conductors in the cord are 12 AWG, rated 20 amperes.
- not have its length altered in the field. To do so would void the UL listing.

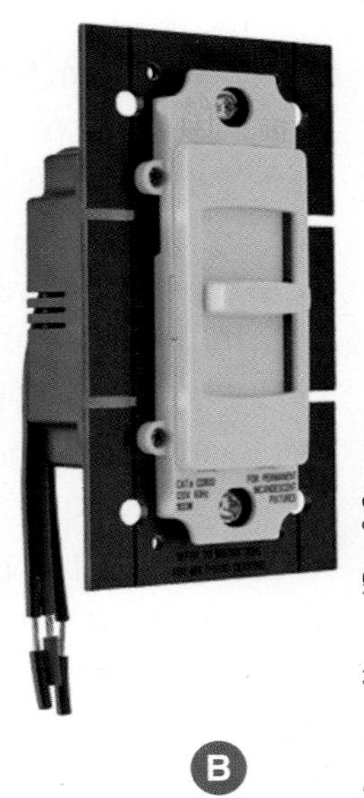

Courtesy of Legrand/Pass & Seymour.

FIGURE 13-10 Some typical electronic dimmers. Both perform the same function of varying the voltage to the connected incandescent lamp load: (A) rotating knob; (B) slider knob.

DIMMER CONTROLS FOR HOMES

Dimmers are used in homes to lower the level of light. There are two basic types: electronic and autotransformer.

Electronic Dimmers

Figure 13-10 shows two types (slide and rotate) of electronic dimmers. These are very popular. They fit in a standard switch (device) box. They can replace standard single-pole or 3-way toggle switches. See Figure 13-11 for the typical hookup of a single-pole and a 3-way dimmer.

The most common type of electronic dimmer for residential use is rated 600 watts, 125 VAC, 60 hertz. They are also available in 1000-watt ratings. For larger wattage requirements, consult manufacturers' literatures and websites. The connected load must not exceed the marked wattage rating of the dimmer.

Electronic dimmers offer many features such as LED nightlight, soft start when changing from OFF to the desired setting, and returning to the previous setting after a power outage.

Dimmer wiring devices and fan speed controls are physically larger than typical wiring devices (receptacles and switches). Think ahead to make sure you install a large enough wall box to accommodate the dimmer, conductors, and splices that are in the box. To jam the dimmer into the box may cause trouble, such as short circuits and ground faults.

Dimmers that install in flush device boxes are intended to control only permanently installed incandescent lighting unless listed for other applications, *404.14(E)*. The instructions will state something like "Do not use to control receptacle outlets." The reasons for this follow:

1. Serious overloading could result, because it may not be possible to determine or limit what other loads might be plugged in the receptacle.
2. Dimmers reduce voltage. Reduced voltage supplying a television, radio, stereo components, personal computer, vacuum cleaner, and so on, can result in costly damage to the appliance.

Electronic dimmers for use *only* to control incandescent lamps are marked with a *wattage* rating. Serious overheating and damage to the transformer and to the electronic dimmer will result if an "incandescent only" dimmer is used to control inductive loads.

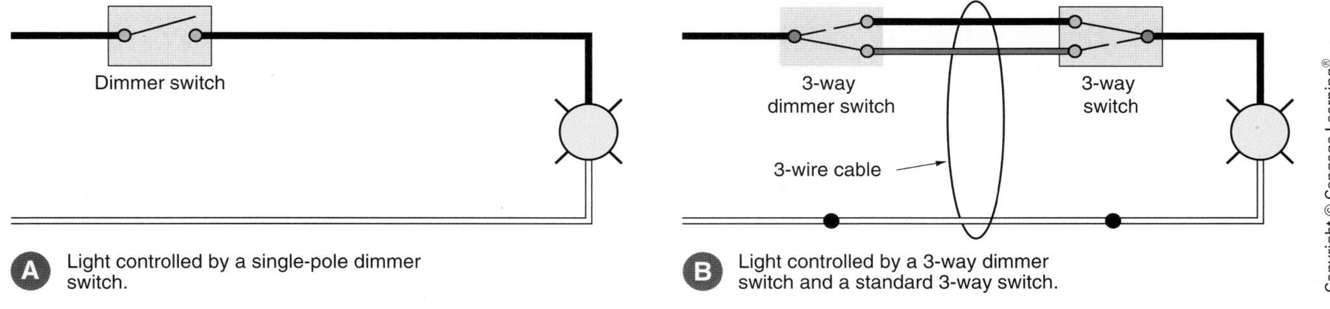

FIGURE 13-11 Use of single-pole and 3-way electronic dimmer controls in circuits.

For example, a dimmer marked 600 watts @120 volts is capable of safely carrying 5 amperes.

$$\text{Amperes} = \frac{\text{Watts}}{\text{Volts}} = \frac{600}{120} = 5 \text{ amperes}$$

Dimmers that are suitable for control of "inductive" loads, such as lighting systems that incorporate a transformer (as in low-voltage outdoor lighting and low-voltage track lighting), will be marked in *volt-amperes*. The instructions will clearly indicate that the dimmer can be used on inductive loads.

See Chapter 7 for a discussion of watts versus volt-amperes.

Do *not* hook up a dimmer with the circuit "hot." Not only is it dangerous because of the possibility of receiving an electrical shock, but the electronic dimmer can be destroyed when the splices are being made as a result of a number of "make and breaks" before the actual final connection is completed. The momentary inrush of current each time there is a "make and break" can cause the dimmer's internal electronic circuitry to heat up beyond its thermal capability. As one lead is being connected, the other lead could come in contact with "ground." Unless the dimmer has some sort of built-in short-circuit protection, the dimmer is destroyed. The manufacturer's warranty will probably be null and void if the dimmer is worked "hot." It is well worth the time to take a few minutes to de-energize the circuit.

Electronic dimmers use triacs to "chop" the ac sine wave, so some incandescent lamps might "hum" because the lamp filament vibrates. This hum can usually be eliminated or greatly reduced by turning the dimmer setting to a different brightness level; changing the lamp, installing a "rough service lamp"; or, as a last resort, changing the dimmer.

Autotransformer Dimmers

Figure 13-12 shows an autotransformer (one winding) type of dimmer. These are physically larger than the electronic dimmer and require a special wall box. They are not ordinarily used for residential applications unless the load to be

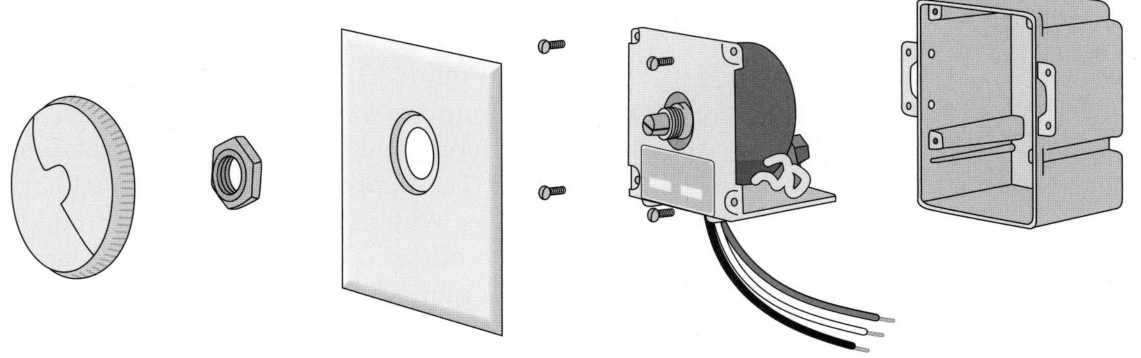

FIGURE 13-12 Dimmer control, autotransformer types.

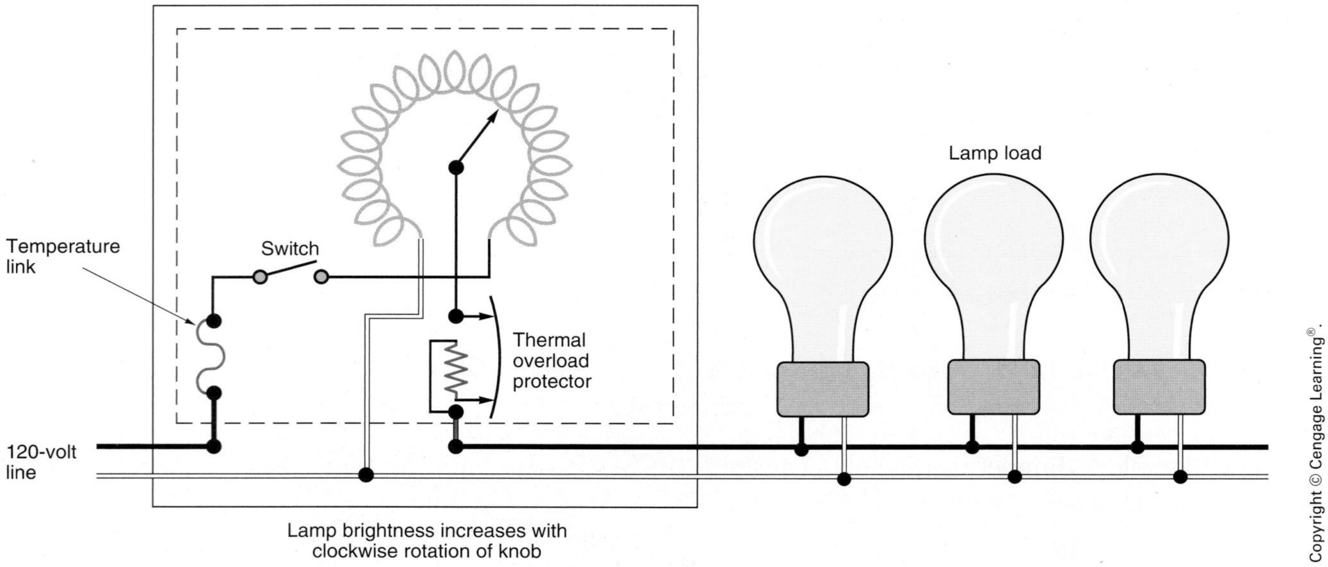

FIGURE 13-13 Dimmer control (autotransformer) wiring diagram for incandescent lamps.

controlled is very large. They are generally used in commercial applications for the control of large loads. The lamp intensity is controlled by varying the voltage to the lamp load. When the knob is rotated, a brush contact moves over a bared portion of the transformer winding. As with electronic dimmers, they will not work on direct current, but that is not a problem because all homes are supplied with alternating current. The maximum wattage to be controlled is marked on the nameplate of the dimmer. Overcurrent protection is built into the dimmer control to prevent burnout due to an overload.

Figure 13-13 illustrates the connections for an autotransformer dimmer that controls incandescent lamps.

Dimming Fluorescent Lamps

To dim incandescent lamps, the voltage to the lamp filament is varied (raised or lowered) by the dimmer control. The lower the voltage, the less intense the light. The dimmer is simply connected "in series" with the incandescent lamp load.

Fluorescent lamps are a different story. The connection is not a simple "series" circuit. To dim fluorescent lamps, special dimming ballasts and special dimmer switches are required. The dimmer control allows the dimming ballast to maintain a voltage to the cathodes that will maintain the cathodes' proper operating temperature, and also allows the dimming ballast to vary the current flowing in the arc. This in turn varies the intensity of light coming from the fluorescent lamp.

There is an exception to the need to use special fluorescent dimming ballasts and special dimmer controls. Some compact fluorescent lamps are available that have an integrated circuit built right into the lamp, allowing the lamp to be controlled by a standard incandescent wall dimmer. These dimmable compact fluorescent lamps can be used to replace standard incandescent lamps of the medium screw shell base type. No additional wiring is necessary. They are designed for use with dimmers, photocells, occupancy sensors, and electronic timers that are marked "Incandescent Only."

Figure 13-14 illustrates the connections for an autotransformer-type dimmer for the control of fluorescent lamps using a special dimming ballast. The fluorescent lamps are F40T12 rapid-start lamps. With the newer electronic dimming ballasts, 32-watt, rapid-start T8 lamps are generally used.

Figure 13-15 shows an electronic dimmer for the control of fluorescent lamps using a special dimming ballast.

Fluorescent dimming ballasts and controls are available in many types such as: simple ON–OFF,

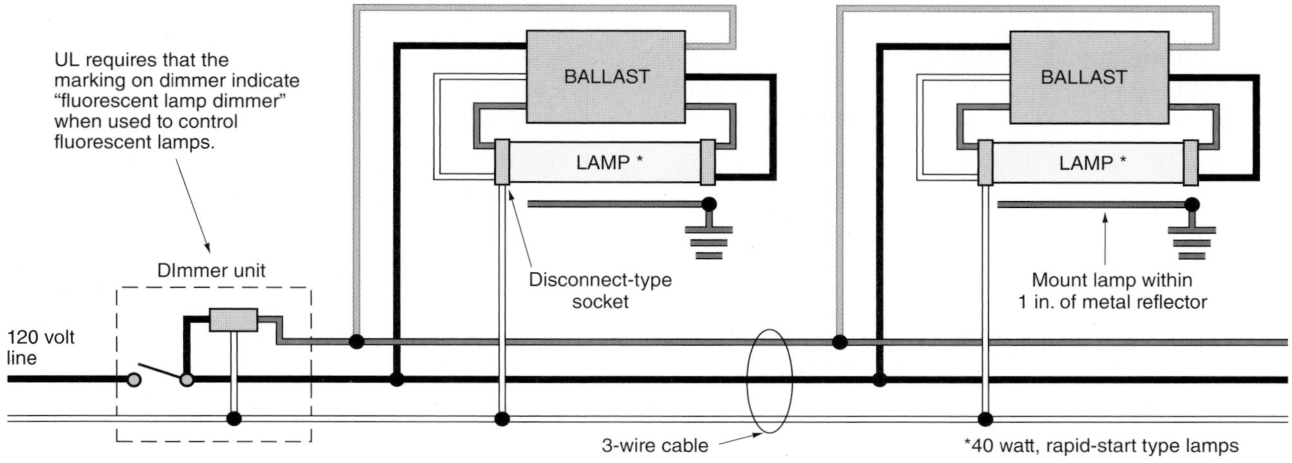

UL requires that the marking on dimmer indicate "fluorescent lamp dimmer" when used to control fluorescent lamps.

BALLAST

LAMP *

BALLAST

LAMP *

Dimmer unit

120 volt line

Disconnect-type socket

Mount lamp within 1 in. of metal reflector

3-wire cable

*40 watt, rapid-start type lamps

FIGURE 13-14 Dimmer control wiring diagrams for fluorescent lamps.

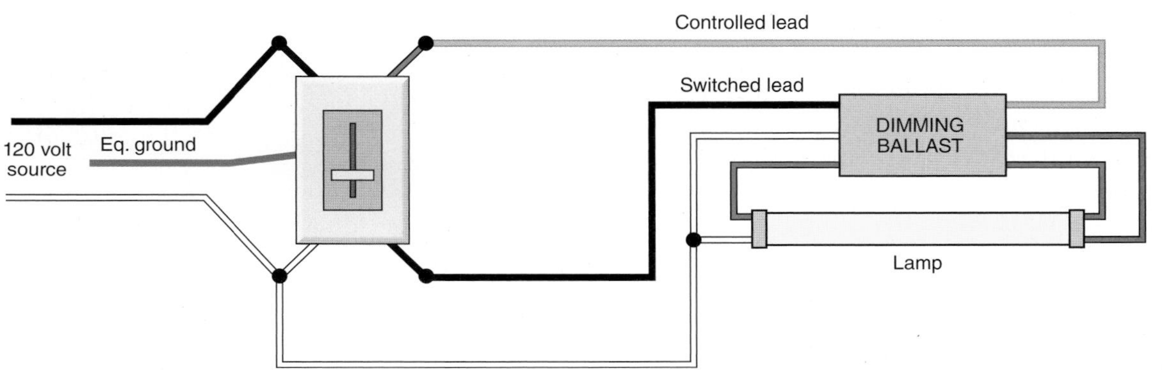

FIGURE 13-15 This wiring diagram shows the connections for an electronic dimmer designed for use with a special dimming ballast for rapid-start lamps. Always check the dimmer and ballast manufacturers' wiring diagrams, as the color coding of the leads and the electrical connections may be different than shown in this diagram. The dimmer shown is a "slider" type and offers wide-range (20% to 100%) adjustment of light output for the fluorescent lamp.

100% to 1% variable light output, 100% to 10% variable light output, and two-level 100% or 50% light output. Wireless remote control is also an option.

Here are some hints to ensure proper dimming performance of fluorescent lamps:

- Make sure that the lamp reflector and the ballast case are solidly grounded.
- Do not mix ballasts or lamps from different manufacturers on the same dimmer.
- Do not mix single- and 2-lamp ballasts.
- "Age" the lamps for approximately 100 hours before dimming. This allows the gas cathodes in the lamps to stabilize. Without "aging" (sometimes referred to as "seasoning"), it is possible to have unstable dimming, lamp striation (stripes), and/or unbalanced dimming characteristics between lamps.
- Because incandescent lamps and fluorescent lamps have different characteristics, they cannot be controlled simultaneously with a single dimmer control.
- Do not control electronic ballasts and magnetic ballasts simultaneously with a single dimmer control.

Always read and follow the manufacturer's instructions furnished with dimmers and ballasts.

Without special dimming fluorescent ballasts and controls, a simple way to have two light levels is to use 4-lamp fluorescent luminaires, such as those installed in the Recreation Room. Then, control the ballast that supplies the center two lamps with one wall switch, and control the ballast that supplies the outer two lamps with a second wall switch. This is sometimes referred to as an "inboard/outboard" connection. Lamps and ballasts are discussed in Chapter 7.

Dimming LED Lighting

The manufacturers of many LED lamps advertise that they are dimmable. Because there are so many variables to LED lighting installations, the electrician should carefully read the manufacturer's instructions to ensure proper installation. This includes both the lamp and dimmer. Be sure the lamp and dimmer are compatible.

If you want to get more information about LED lighting and LED dimming, you can find loads of technical information on the Internet. The manufacturers of LEDs, LED luminaires, and LED dimming circuitry have all of their installation instructions on their websites. If you know the manufacturer's name, do an Internet search for that manufacturer. If you are looking for a broad spectrum of data about LEDs, just enter the words "LED dimming" or "LED lighting." You will be amazed at how much information is available online.

SAFETY ALERT

Because LED lighting is becoming so popular, as always be sure to check the product labels to make sure that the product has been listed by a Nationally Recognized Testing Laboratory (NRTL) such as Underwriters Laboratories.

REVIEW

Note: Refer to the *NEC* or the plans in the back of this text if desired.

1. To what circuit is the living room connected? ______
2. How many receptacles are connected to the living room circuit? ______
3. a. How many wires enter the switch box at the 4-way switch location? ______
   b. What type and size of box may be installed at this location? ______

   ______

   ______

   ______
4. How many wires must be run between an incandescent lamp and its dimmer control?

   ______
5. Complete the wiring diagram for the dimmer and lamp.

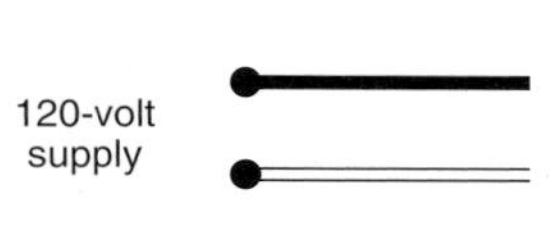

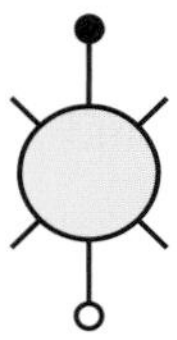

6. Is it possible to dim standard fluorescent ballasts? ______________________

7. a. How many wires must be run between a dimming-type fluorescent ballast and the dimmer control? ______________________

   b. Is a switch needed in addition to the dimmer control? ______________________

   ____

   ____

   ____

   ____

8. Explain why fluorescent lamps having the same wattage draw different current values.

   ____

   ____

   ____

   ____

   ____

9. What is the total current consumption of the track lighting and recessed luminaires above the fireplace? Show your calculations.

10. How many television outlets are provided in the Living Room? ______________________

11. Where is the telephone outlet located in the Living Room? ______________________

    ____

12. Electronic dimmers of the type sold for residential use (shall) (shall not) be used to control fluorescent luminaires. These dimmers (are) (are not) intended for speed control of small motors. (Circle the correct answers.)

13. When calculating a dwelling lighting load on the "volt-ampere per ft^2" basis, the receptacle load used for floor lamps, table lamps, and so forth (Circle the correct answer.),
    a. must be added to the general lighting load at 1½ amperes per receptacle.
    b. must be added to the general lighting load at 1500 volt-amperes for every 10 receptacles.
    c. is already included in the calculations as part of the volt-amperes per ft^2.
14. A layout of the outlets, switches, dimmers, track lighting, and recessed luminaires is shown in the following diagram. Using the cable layout of Figure 13-1, make a complete wiring diagram of this circuit. Use colored pencils or marking pens to indicate conductors.

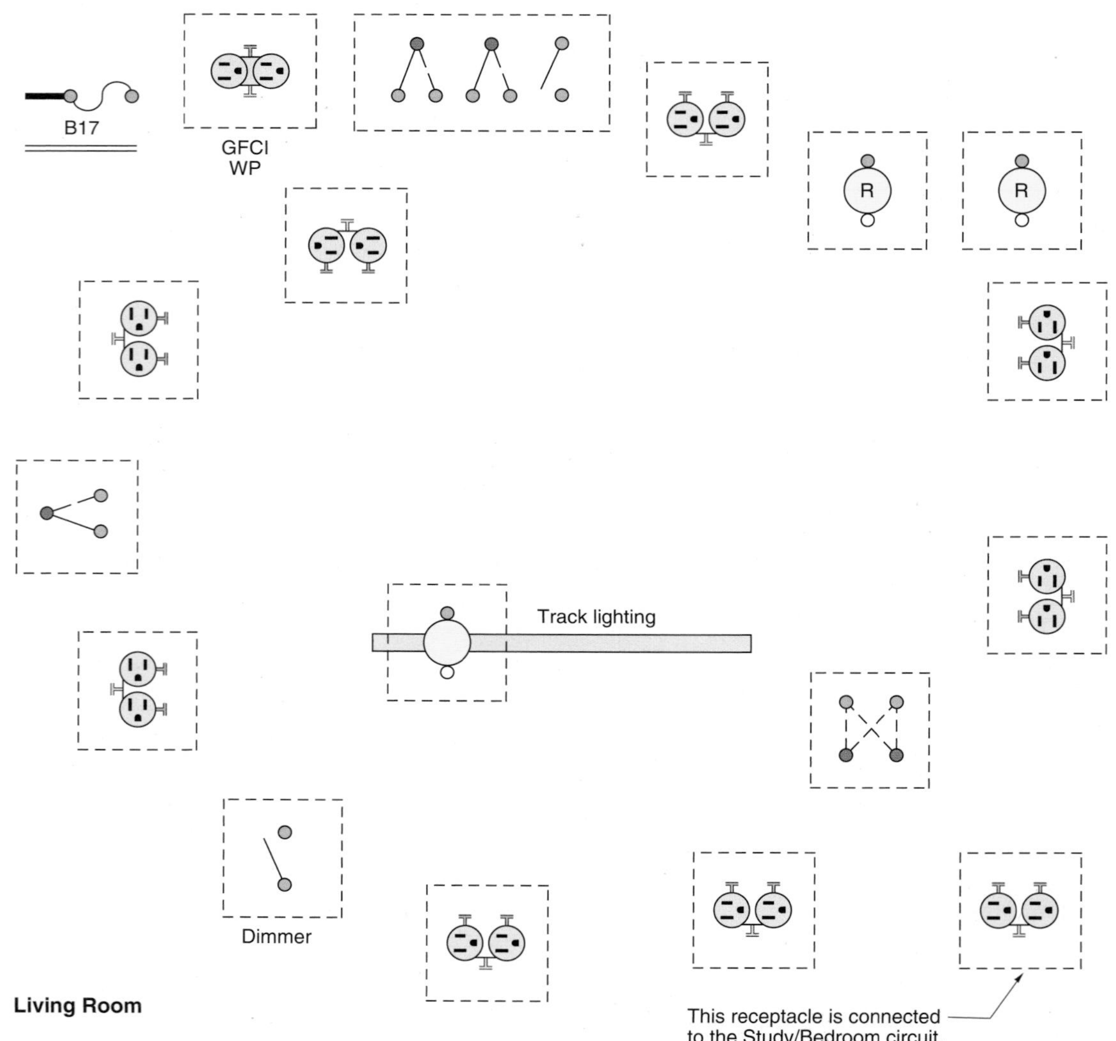

15. Prepare a list of television outlets, telephone outlets, and wiring devices shown on the plans and cable layout for the living room area. Include the number of each type present.

____

____

____

____

____

____

16. The Living Room features a sliding glass door. For the purpose of providing the proper receptacle outlets, answer the following statements *true* or *false*.
    a. Sliding glass panels are considered wall space. _______________
    b. Fixed panels of glass doors are considered wall space. _______________
17. a. May lengths of track lighting be added when the track is permanently connected?

    ____

    b. May lengths of track lighting be added when the track is cord connected?

    ____

18. Must track lighting always be fed (connected) at one end of the track? _______________

    ____

19. May a standard electronic dimmer be used to control low-voltage lighting that incorporates a transformer, or for controlling the speed of a ceiling fan motor?

    Yes _______________ No _______________
20. Does the *Code* permit cutting lengths of portable track lighting to change their length?

    Yes _______________ No _______________
21. Why should the power be turned off when hooking up a dimmer? _______________

    ____

    ____

CHAPTER 14

Lighting Branch Circuit for the Study/Bedroom

OBJECTIVES

After studying this chapter, you should be able to

- discuss valance lighting.
- make all connections in the Study/Bedroom for the receptacles, switches, fan, and lighting.
- discuss surge suppressors.

CIRCUIT A21 OVERVIEW

The Study/Bedroom circuit is so named because it can be used as a study for the present, providing excellent space for home office, personal computer, or den. Should it become necessary to have a third bedroom, the change is easily made.

Circuit A21 originates at Main Panel A. Circuit A21 is a 15-ampere branch circuit. Overcurrent protection and arc-fault protection for this circuit are provided by a 15-ampere AFCI circuit breaker in Panel A as required by *NEC® 210.12(A)*. AFCIs are discussed in detail in Chapter 6.

This circuit feeds the Study/Bedroom at the receptacle in the Living Room just outside of the Study. From this point, the circuit feeds the split-wired receptacle just inside the door of the Study. A 3-wire cable runs around the room to feed the other four split-wired receptacles. Figure 14-1 shows the cable layout for Circuit A21.

The black and white conductors carry the "live" circuit, and the red conductor is the switch return for the control of one portion of the split-wired receptacles.

There are four switched split-wired receptacles in the Study/Bedroom, controlled by two 3-way switches. A fifth receptacle on the short wall leading

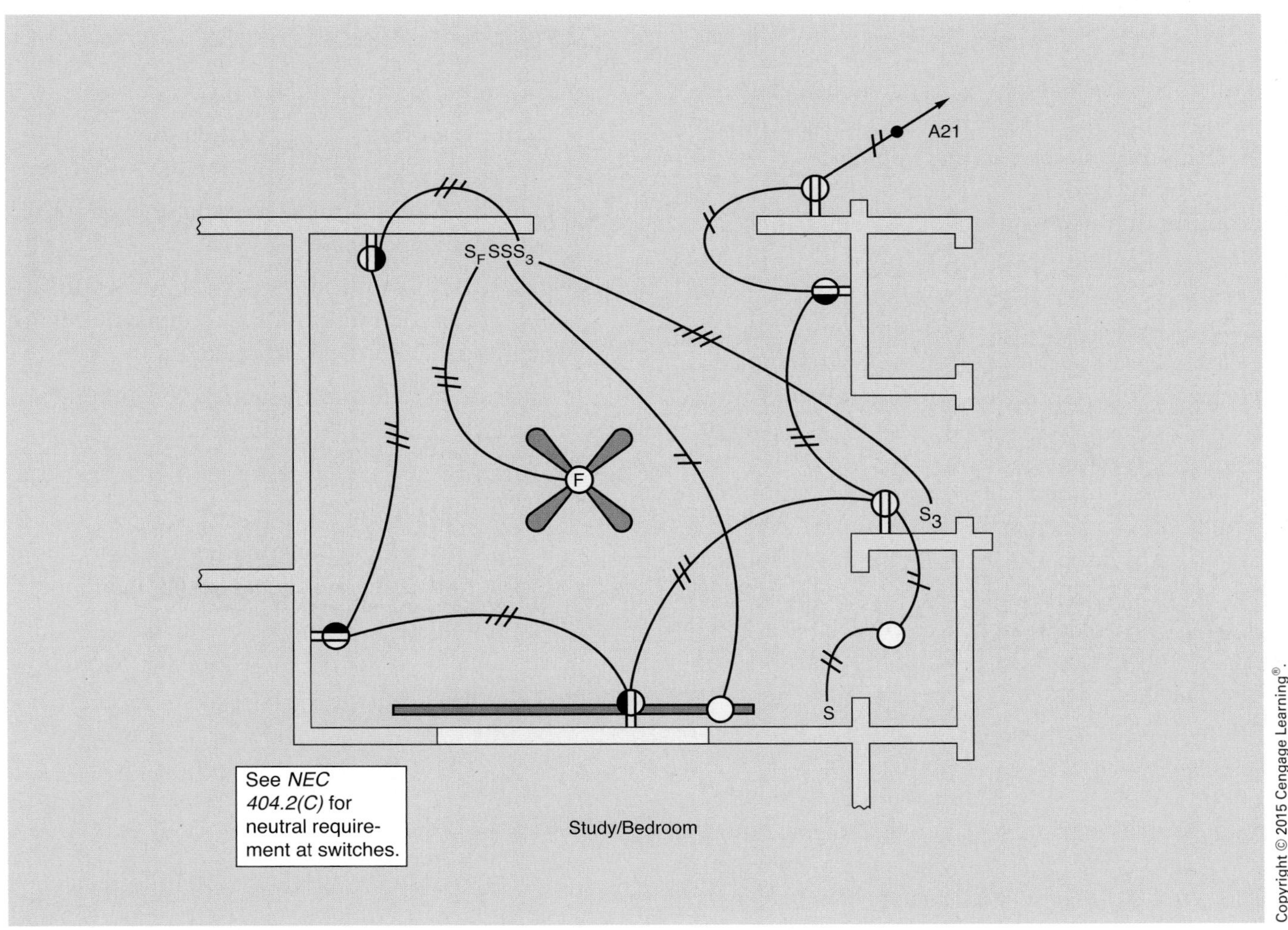

FIGURE 14-1 Cable layout for the Study/Bedroom.

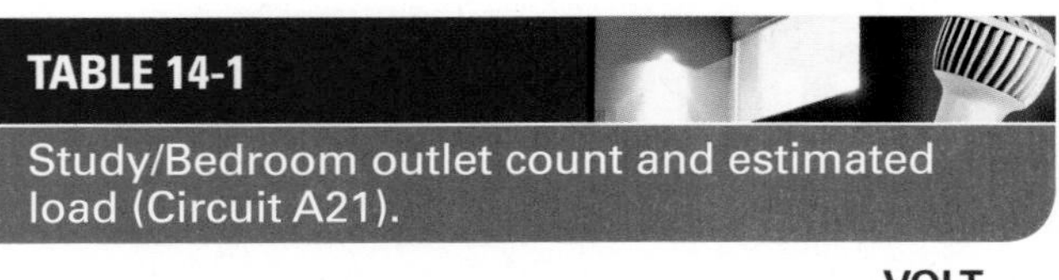

TABLE 14-1

Study/Bedroom outlet count and estimated load (Circuit A21).

DESCRIPTION	QUANTITY	WATTS	VOLT-AMPERES
Receptacles @ 120 W each	6	720	720
Closet luminaire	1	75	75
Paddle fan/light	1		
Three 50-W lamps		150	150
Fan motor (0.75 A 120 V)		80	90
Valance lighting	1		
Two 40-W fluorescent lamps		80	90
TOTALS	9	1105	1125

to the bedroom hallway is not switched. The ready access to this receptacle makes it ideal for plugging in vacuum cleaners or similar appliances.

A rule has been added to the 2014 *NEC* that prohibits the use of dimmers to control standard 15- or 20-ampere receptacles. The rule reads ▶**406.15 Dimmer-Controlled Receptacles.** A receptacle supplying lighting loads shall not be connected to a dimmer unless the plug/receptacle combination is a nonstandard configuration type that is specifically listed and identified for each such unique combination.◀*

This requirement has direct application to the wiring shown for the split-wired receptacles in the Study/Bedroom because 3-way switches control ½ of each receptacle. The receptacles would usually be standard 15-ampere duplex receptacles.

A ceiling fan/light is mounted on the ceiling. This installation is covered in Chapter 9. Table 14-1 summarizes the outlets and estimated load for the Study/Bedroom circuit.

As we all know, when someone decides to use any room, such as the Study/Bedroom, as a home office, a lot of electronic equipment will have to be plugged in. The current draw is very low, but the list for electronic equipment never ends! You will never have enough wall receptacles. Plug-in strips—probably of the surge protection type—are almost always needed.

*Reprinted with permission from NFPA 70-2014.

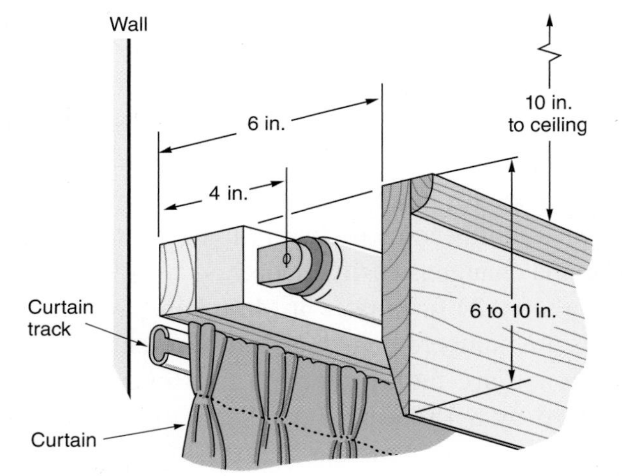

FIGURE 14-2 Fluorescent lighting behind the valance.

You could consider

- running more than one branch circuit for the receptacles and lighting in the Study/Bedroom.
- using 12 AWG for the home run or even for the entire branch circuit(s).
- that all 120-volt, single-phase, 15- and 20-ampere branch circuits that supply outlets in bedrooms require AFCI protection.
- wiring in quadplex receptacles instead of duplex receptacles.
- roughing in more wall receptacles than the *NEC* minimum.
- using only plug-in strips that are listed by a qualified electrical products testing laboratory.

As previously mentioned, the arrival of AFCI requirements might require new thinking about how to do the wiring for bedrooms.

VALANCE LIGHTING

Figure 14-2 illustrates an interesting indirect fluorescent lighting treatment above the windows and behind the valance board.

If the fluorescent valance lighting is to have a dimmer control, then special dimming ballasts would be required. See the discussion on fluorescent lamp dimming in Chapter 13.

The Study/Bedroom wiring is rather simple. Most of the concepts are covered in previous chapters in this text.

Figure 14-3 suggests one way to arrange the wiring for the ceiling fan/light control, the valance lighting control, and the receptacle outlet control.

SURGE SUPPRESSORS

Transient voltage surge suppressors, covered in Chapter 6, minimize the possibility of voltage surges damaging sensitive electronic equipment.

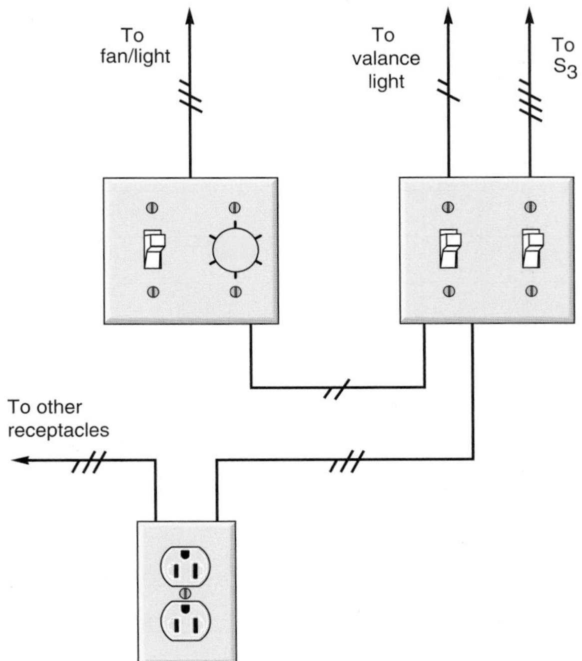

FIGURE 14-3 Conceptual view of how the Study/Bedroom switch arrangement is to be accomplished (Circuit A21).

REVIEW

Note: Refer to the *NEC* or the plans in the back of this text if desired.

1. Based on the total estimated load calculations, what is the current draw on the Study/Bedroom circuit? ______
2. a. The Study/Bedroom is connected to circuit ______.
   b. The conductor size for this circuit is ______.
3. a. What unique type of electrical protection is required for all 120-volt, single-phase, 15- and 20-ampere branch circuits for most habitable rooms and areas in new homes? ______
   b. What section of the *Code* covers this kind of protection? ______
4. Why is it necessary to install a receptacle in the wall space leading to the bedroom hallway? ______

5. Show the calculations needed to select a properly sized box for the receptacle outlet mentioned in problem 4. Nonmetallic-sheathed cable is the wiring method.

6. Show the calculations necessary to select a properly sized box for the receptacle outlet mentioned in problem 4. The wiring method is EMT, usually referred to as "thin-wall conduit."

7. Prepare a list of *all* wiring devices used in Circuit A21.

8. Using the cable layout shown in Figure 14-1, make a complete wiring diagram of Circuit A21. Use colored pencils or pens.

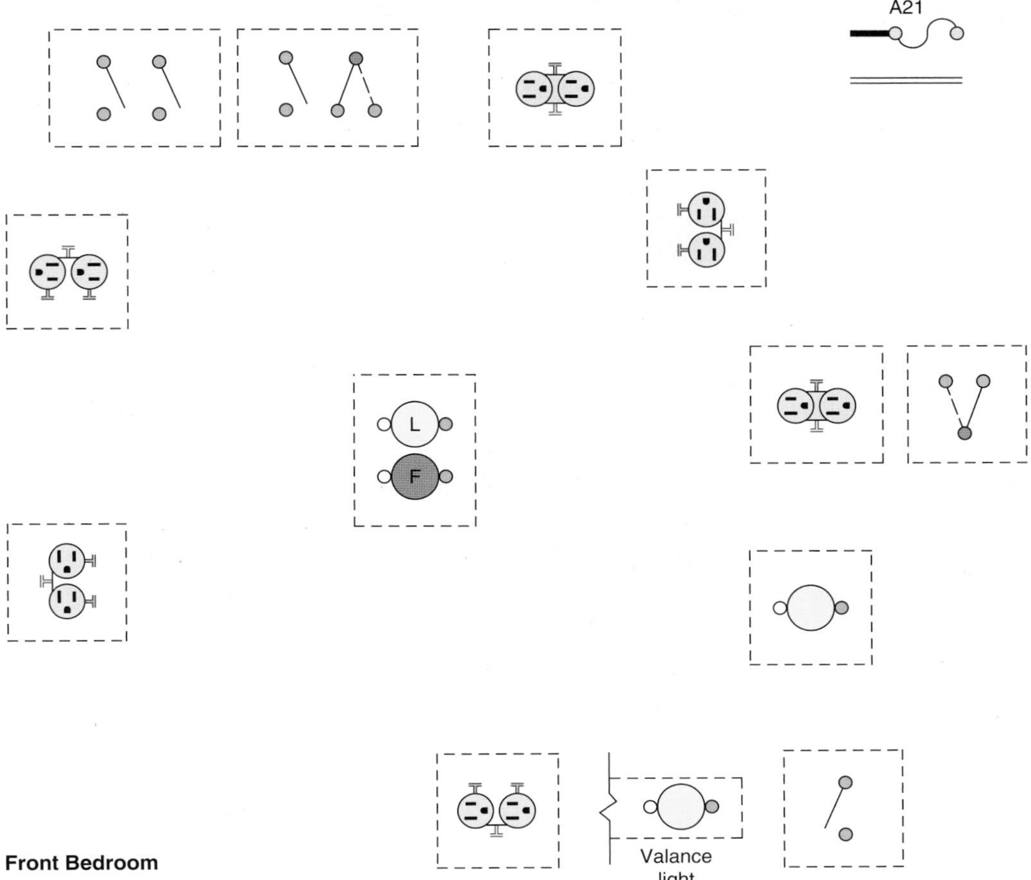

CHAPTER **15**

Dryer Outlet, Lighting and Receptacle Circuits for the Laundry, Powder Room, Rear Entry Hall, and Attic

OBJECTIVES

After studying this chapter, you should be able to

- understand the *Code* requirements for bathroom receptacles.
- understand the *Code* requirements for making load calculations and electrical connections for electric dryers.
- understand the *Code* requirements for receptacle outlets in laundry areas.
- understand the principles of exhaust fans.
- understand the *Code* requirements for attic wiring.

TABLE 15-1

Laundry, Rear Entrance, Powder Room, and Attic (Circuit B10).

DESCRIPTION	QUANTITY	WATTS	VOLT-AMPERES
Rear Entry Hall receptacle	1	120	120
Powder Room vanity luminaire	1	200	200
Rear Entry Hall ceiling luminaires @ 100 W each	2	200	200
Laundry Room fluorescent luminaire Four 32-W lamps	1	128	144
Attic luminaires @ 75 W each	4	300	300
Laundry exhaust fan	1	80	90
Powder Room exhaust fan	1	80	90
TOTALS	11	1108	1144

LIGHTING CIRCUIT B10

The estimated loads for the various receptacles and luminaires in the Laundry, Rear Entry, Powder Room, and Attic are shown in Table 15-1.

The lighting circuit is discussed a little later on in this chapter.

RECEPTACLE CIRCUIT B21

Receptacles in a residential bathroom require special treatment. This is discussed in detail in Chapter 3. Now would be a good time for you to review Chapter 3.

Circuit B21 is a separate 20-ampere branch circuit that supplies the receptacle in the Powder Room.

A bathroom is defined in *Article 100* as An area that has a basin, plus one or more of the following: a toilet, a urinal, a tub, a shower, a bidet, or similar plumbing fixtures.* Therefore, by definition, the *NEC*® considers the powder room to be a bathroom. As a result, special requirements apply to the receptacle outlet in the room.

*Reprinted with permission from NFPA 70-2014.

CLOTHES DRYER CIRCUIT Ⓐ$_D$

The electric clothes dryer requires a separate 120/240-volt, single-phase, 3-wire branch circuit. In this residence, the electric clothes dryer is indicated on the Electrical Plans in the laundry by the symbol Ⓐ$_D$. It is supplied by Circuit B(1–3) located in Panel B. Figure 15-1 shows the internal components and wiring of a typical electric clothes dryer.

Dryer Connection Methods

Electric clothes dryers can be connected in several ways.

One method is to run a 3-wire armored cable directly to a junction box on the dryer provided by the manufacturer, Figure 15-2. This flexible connection allows the dryer to be moved for servicing without disconnecting the wiring. The armor of the Type AC cable serves as the equipment grounding conductor. If Type AC cable is used for the branch circuit wiring, the disconnecting means must comply with *NEC 422.32*, as covered later in this chapter.

Another method is to run a conduit (EMT) to a point just behind the dryer. A combination coupling makes the transition from the EMT to a length of flexible metal conduit 2 ft (600 mm) to 3 ft (900 mm) long, as in Figure 15-3. This flexible connection also allows the dryer to be moved for servicing without disconnecting the wiring. A separate equipment grounding conductor must be installed in the EMT to meet the grounding requirements found in *NEC 250.118(5)*. If the wiring method described in this paragraph is used for the branch circuit wiring, the disconnecting means must comply with *NEC 422.32*, as covered later in this chapter.

Probably the most common method is to install a 4-wire, 30-ampere receptacle on the wall behind the dryer. A 4-wire, 30-ampere cord (pigtail) is then connected to the dryer. This arrangement is referred to as a "cord-and-plug connection" and is shown in Figure 15-4. In conformance to *NEC 210.23*, the connected load is not permitted to exceed the branch-circuit rating.

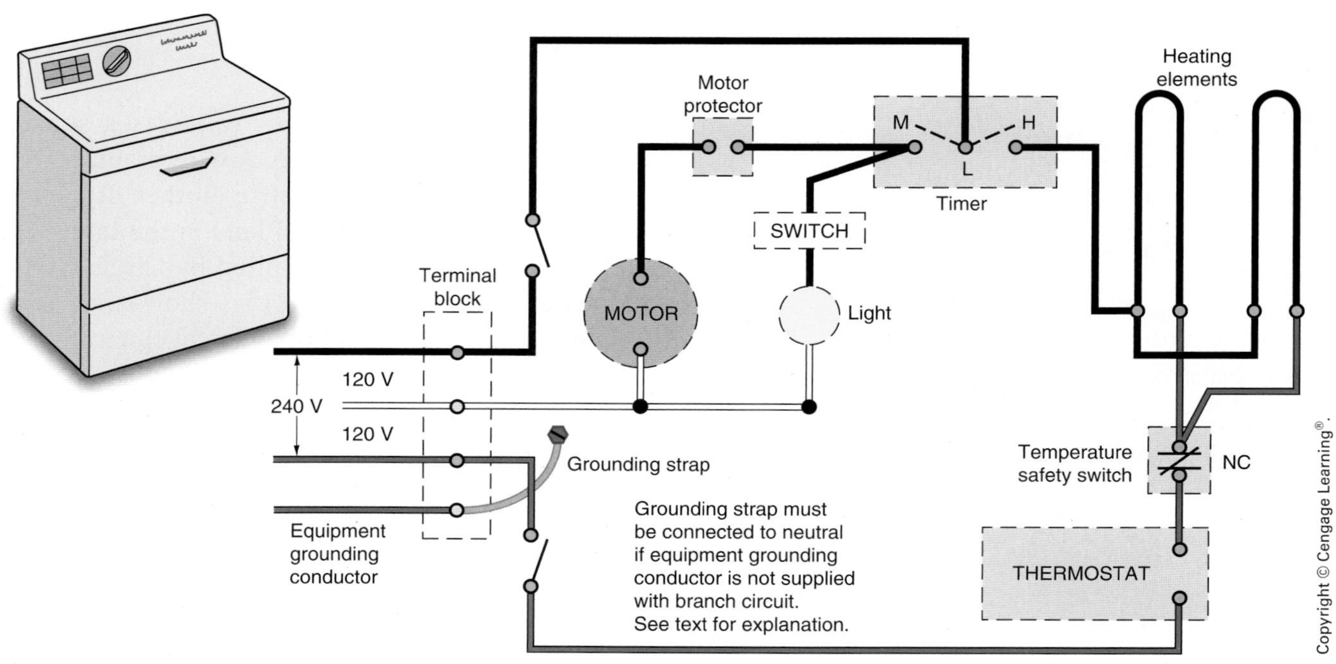

FIGURE 15-1 Clothes dryer: typical wiring and components.

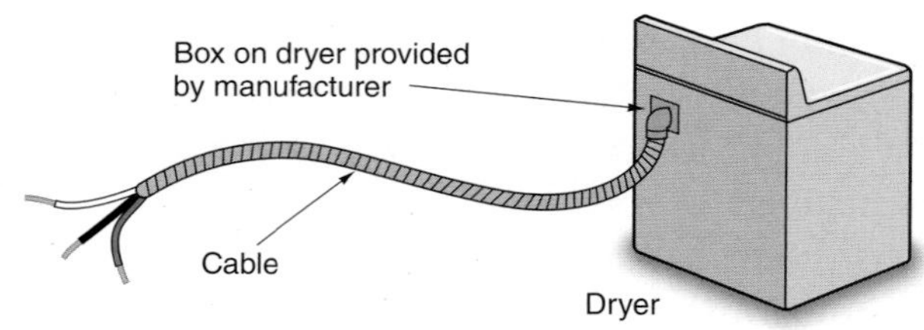

FIGURE 15-2 Dryer connected by armored cable.

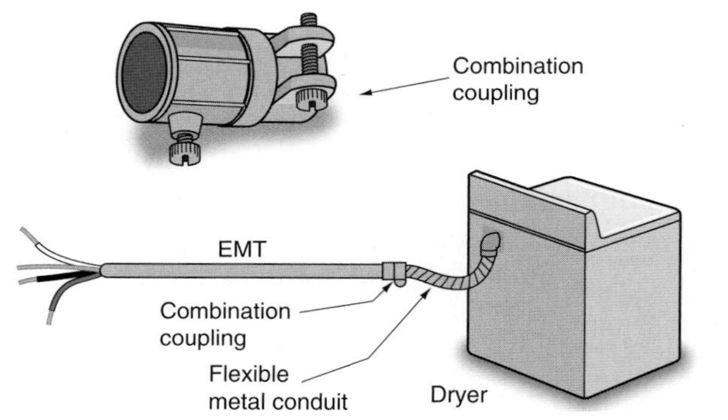

FIGURE 15-3 Dryer connected by EMT and flexible metal conduit.

Appliances

All new homes have appliances—some that operate entirely on electricity and some that require electricity as well as another energy source, such as a gas furnace.

Article 422 in the *NEC* is where to look for the majority of *Code* requirements for appliances. In *Article 422*, you will find quite a few references to other parts of the *NEC*, such as *Article 430* where motors are involved, and *Article 440* where hermetic-motor compressors (air conditioners and heat pumps) are involved.

As you continue studying this text, specific *Code* rules for appliances are discussed at the time it becomes necessary to mention the rule.

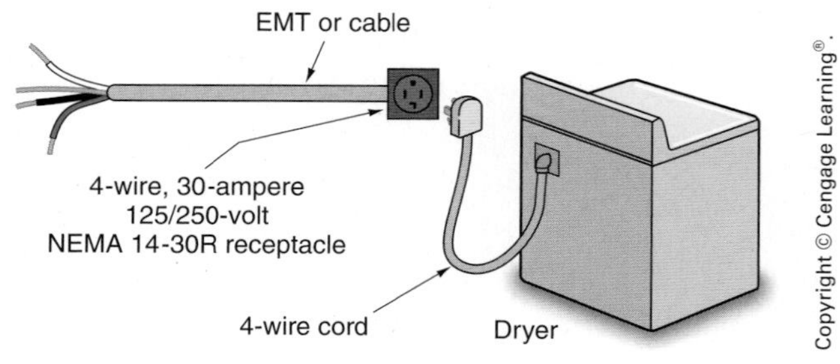

FIGURE 15-4 Dryer connection using a cord set.

Disconnecting Means

All electrical appliances are required to have a disconnecting means, *422.30*.

For permanently connected motor-driven appliances, *422.32* states that a disconnect switch or a circuit breaker in a panelboard may serve as the disconnecting means, provided the switch or circuit breaker is within sight of the appliance, or is capable of being locked in the OFF position. Breaker manufacturers produce "lock-off" devices that fit over the circuit breaker handle. Disconnect switches have provisions for locking the switch OFF.

The unit switch on an appliance is acceptable as a disconnect for an appliance, provided the unit switch has a marked OFF position and it disconnects all the conductors in the appliance; *422.34*. This does not normally apply to controllers of the type typically provided in dishwashers, as they do not shut off all the power to the appliance. If the appliance does have a qualifying unit switch, the service disconnect serving as the other disconnect does not have to be within sight of the appliance, *NEC 422.32, Exception*.

Cord-and-Plug Connection. For cord-and-plug-connected appliances, *NEC 422.33(A)* accepts the cord-and-plug arrangement as the disconnecting means.

Figure 15-4 shows a 30-ampere, 4-wire receptacle configuration for an electric clothes dryer installation in conformance to the latest *NEC*. Figure 15-5 shows (A) a surface-mount 4-wire receptacle and (B) a flush mount 4-wire receptacle.

For most residential electric clothes dryers, receptacles and cord sets are rated 30 amperes, 125/250 volts. Some large electric clothes dryers might require a 50-ampere, 125/250-volt receptacle and cord set.

Receptacles for dryers can be surface mounted or flush mounted. Flush-mounted receptacles are quite large and require a large wall box. A 4 in. square, 2⅛ in. deep outlet box with a suitable 2 gang plaster ring generally works fine. Correct box sizing is determined by referring to *Table 314.16(A)* or *Table 314.16(B)*, or to Figure 2-35 and Table 2-1.

To avoid duplication of text, refer to Chapter 20 for a complete description of ratings and blade and slot configurations for 3-wire and 4-wire dryer and range receptacles and their associated cord sets.

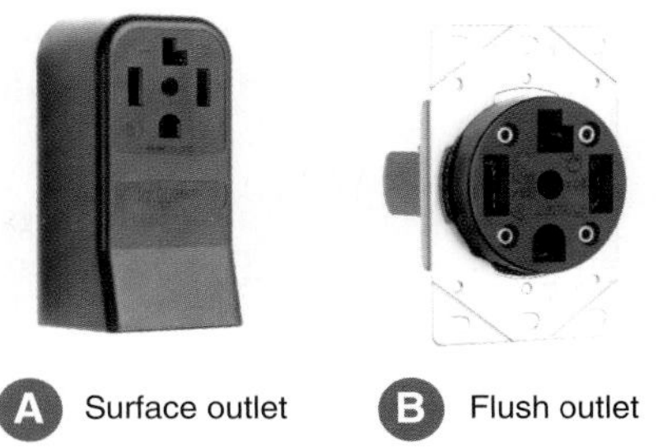

FIGURE 15-5 (A) shows a surface-mount 4-wire receptacle. (B) shows a flush-mount 4-wire receptacle. These are typically rated 30 amperes and are of the types required to supply electric dryers. Prior to the 1996 *NEC*, 3-wire receptacles and cords were permitted. This is no longer permitted. It is mandatory that a separate equipment grounding conductor be installed to ground the frames of electric clothes dryers and electric ranges. Also see Figure 20-2.

Courtesy of Leviton Manufacturing Company, Inc.

Load Calculations

In this residence, the dryer is installed in the Laundry Room. The Schedule of Special-Purpose Outlets in the specifications shows that the dryer is rated 5700 watts, 120/240 volts. The schematic wiring diagram in Figure 15-1 shows that the heating element is connected across 240 volts. The motor and lamp are connected to the 120-volt terminals.

It is impossible for the electrician to know the ampere rating of the motor, the heating elements, timers, lights, relays, and so on, and in what sequence they actually operate. This dilemma is answered in the *NEC* and UL listing requirements for all appliances.

The nameplate must show volts and amperes or volts and watts, *NEC 422.60(A)*. The nameplate must also show the minimum supply circuit conductor's ampacity and the maximum overcurrent protection, *NEC 422.62(B)(1)*.

A minimum load of 5000 watts (volt-amperes) or the nameplate rating of the dryer, whichever is larger, is used when calculating branch circuit, feeder, or service-entrance requirements, *NEC 220.54*.

The rating of an appliance branch circuit must not be less than the marked rating on the appliance, *NEC 422.10(A)*.

The electric clothes dryer in this residence has a nameplate rating of 5700 watts. This calculates out to be

$$I = \frac{W}{E} = \frac{5700}{240} = 23.75 \text{ amperes}$$

Because wiring devices used for dryer circuits are rated at 30 amperes and to comply with *NEC 210.3*, we have a minimum 30-ampere branch-circuit rating requirement.

Conductor Sizing

The minimum branch-circuit rating for the electric clothes dryer was found to be 30 amperes, and that is the minimum rating of the overcurrent protective device—fuses or circuit breaker. We need to find a conductor that has an ampacity of 30 amperes that will serve as the branch circuit for the dryer. The conductors will be protected by the 30-ampere overcurrent device.

According to *Table 310.15(B)(16)*, the allowable ampacity for a 10 AWG copper conductor is 30 amperes in the 60°C column. The maximum overcurrent protection for these conductors must not exceed 30 amperes, *NEC 240.4(D)*.

In this residence, Circuit B(1–3) is a 3-wire, 30-ampere, 240-volt branch circuit.

Nonmetallic-sheathed cable is run concealed in the walls from Panel B to a flush-mounted 4 in. square box behind the dryer location in the Laundry Room. A 2-gang nonmetallic deep box will work as well because dryer receptacles are designed to connect to 2-gang boxes. The nonmetallic-sheathed cable contains three 10 AWG conductors plus a 10 AWG equipment grounding conductor.

Where EMT is used as the wiring method, *Table C1, Annex C* of the *NEC* shows that three 10 AWG THHN conductors require trade size ½ EMT. The EMT serves as the equipment ground as provided in *NEC 250.118(4)*.

Overcurrent Protection for the Dryer

The motor of the dryer has integral thermal overload protection as required by *NEC 422.11(G)*. Note the motor protector in Figure 15-1. This protection is required by UL standards and prevents the motor from reaching dangerous temperatures as the result of an overload, bearing failure, or failure to start. Also note the temperature safety switch. This is a high-temperature cutoff that shuts off the heating element should the appliance's thermostat fail to operate. This high-temperature cutoff is also a UL requirement.

Overcurrent Protection for Branch Circuit

NEC 422.62 requires the appliance to be marked with the minimum supply circuit conductor ampacity and the maximum size overcurrent device.

The rating of the branch circuit must not be less than the rating marked on the appliance, *NEC 422.10(A)*.

In *NEC 422.11(A)*, we find that the overcurrent protection for the branch-circuit conductors must be sized according to *NEC 240.4*, which refers us right back to *NEC Article 422—Appliances*.

The maximum overcurrent protection for 10 AWG copper conductors is 30 amperes, *NEC 240.4(D)*.

Grounding Frames of Electric Dryers and Electric Ranges

For protection against electrical shock hazard, frames of electric clothes dryers and electric ranges must be grounded. Some key *Code* references are *NEC 250.134, 250.138, 250.140*, and *250.142(B)*.

There are now two methods for grounding the frames of electric dryers and electric ranges, namely *New Installations* and *Existing Installations*.

New Installations. One accepted grounding means is to connect the appliance with a metal raceway such as EMT, an equipment grounding conductor installed through flexible metal conduit (Greenfield), armored cable (BX), or other means listed in *NEC 250.118*. There are limitations on the use of flexible metal conduit as an equipment grounding conductor. See *NEC 250.118(5)* and *348.60*. This topic is covered in Chapter 4.

The separate equipment grounding conductor found in nonmetallic-sheathed cable is also an acceptable equipment grounding means. The equipment grounding conductor in nonmetallic-sheathed

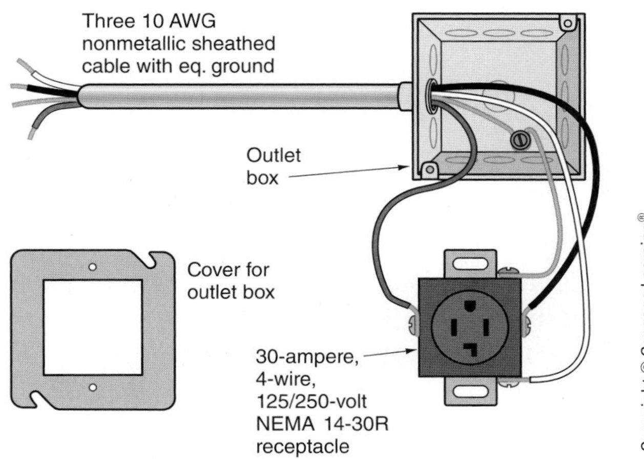

FIGURE 15-6 New installations of branch-circuit wiring for electric ranges, wall-mounted ovens, surface cooking units, and clothes dryers require that all equipment grounding be done according to *250.134, 250.138, 250.140,* and *250.142(B).* Receptacles and cord sets must be 4-wire (three circuit conductors *plus* equipment grounding conductor).

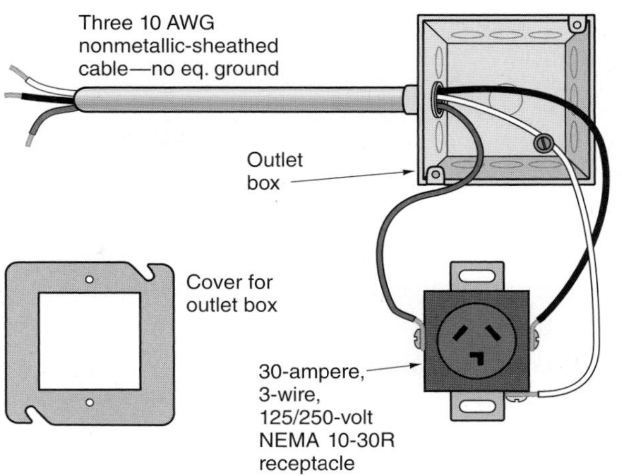

FIGURE 15-7 Prior to the 1996 *National Electrical Code,* it was permitted to ground the junction box to the neutral conductor only if the box was part of the circuit for electric ranges, wall-mounted ovens, surface cooking units, and clothes dryers. The receptacle was permitted to be a 3-wire type because the appliance grounding was accomplished by making a connection between the neutral conductor and the metal frame of the appliance. *This is no longer permitted.*

cable is sized according to *NEC Table 250.122*. See Figure 15-6.

Service-entrance cable is also permitted as a branch circuit to an electric dryer or electric range. The *Code* rules governing service-entrance cable are covered in Chapter 4.

When electric clothes dryers and electric ranges are cord-and-plug connected, the receptacle and the cord set must be 4-wire, the fourth wire being the separate equipment grounding conductor.

The small copper bonding strap (link) that is furnished with an electric clothes dryer or electric range *must not* be connected between the neutral terminal and the metal frame of the appliance.

Existing Installations. Prior to the 1996 edition of the *NEC 250.60* permitted the frames of electric ranges, wall-mounted ovens, surface cooking units, and clothes dryers to be grounded to the neutral conductor, as in Figure 15-7. By way of *Tentative Interim Amendment 53*, this special permission was put into effect in July of 1942 and was supposedly an effort to conserve raw materials during World War II. In effect, this special permission allowed the neutral conductor to serve a dual purpose: (1) the neutral conductor and (2) the equipment grounding conductor. This special permission remained in effect until the 1996 *NEC*.

Since then, grounding equipment such as ranges and dryers to a neutral conductor is not permitted!

Existing installations need not be changed if they were in conformance to the *NEC* at the time the branch circuit was installed.

If an electric clothes dryer or electric range is to be installed in a residence where the dryer branch-circuit wiring had been installed according to pre-1996 *NEC* rules, the wiring *does not* have to be changed. Merely use the grounding strap furnished by the dryer manufacturer to make a connection between the neutral conductor and the frame of the dryer.

RECEPTACLE OUTLETS—LAUNDRY

At least one receptacle outlet is required to be installed for the laundry equipment, *210.52(F)*.

NEC 210.12(A) requires that all outlets and devices in the laundry area be protected by an AFCI device. To comply with this requirement, a combination AFCI circuit breaker is installed in the panelboard for the branch circuits that serve the laundry area. Feed-through type GFCI receptacles can be installed at the first receptacle outlet in the room or area to provide the GFCI protection that is required. As an option, a dual function AFCI/GFCI circuit breaker can be installed in the panelboard.

For this laundry receptacle, a dedicated 20-ampere branch circuit is required to be provided to supply the laundry receptacle outlet or outlets. This circuit is not permitted to have other outlets, *NEC 210.11(C)(2)*. In this residence, the clothes washer receptacle is supplied by 20-ampere Circuit B18. This branch circuit is permitted to supply the other receptacle outlets in the room intended for laundry equipment but is not permitted to supply the outdoor weatherproof receptacle.

All 125-volt, single-phase, 15- and 20-ampere receptacles within 6 ft (1.8 m) of the outside edge of a sink are required to have GFCI protection, *NEC 210.8(A)(7)*. There are no exceptions to this rule. Because all three receptacles in the Laundry Room are within 6 ft (1.8 m) of the outside edge of the laundry sink, they are all required to be GFCI protected. As repeated a number of times in this text, GFCI protection can be provided by either a GFCI circuit breaker or a GFCI receptacle. This is covered in detail in Chapter 6.

Figure 15-8 shows the cable layout for the Laundry Room.

Circuit B20 is an additional 20-ampere branch circuit for the laundry area. It supplies the other two receptacles in the Laundry Room that can be used for plugging in an iron, sewing machine, TV, radio, or other plug-in appliance.

This circuit also serves the weatherproof receptacle on the outside wall of the laundry area. This branch circuit is *in addition* to the required separate 20-ampere circuit (B18) for the laundry equipment.

Circuits B18 and B20 are included in the calculations for service-entrance and feeder conductor sizing at a calculated load of 1500 volt-amperes per circuit, *NEC 220.52(B)*. This load is included with the general lighting load for the purpose of calculating the service-entrance and feeder conductor sizing

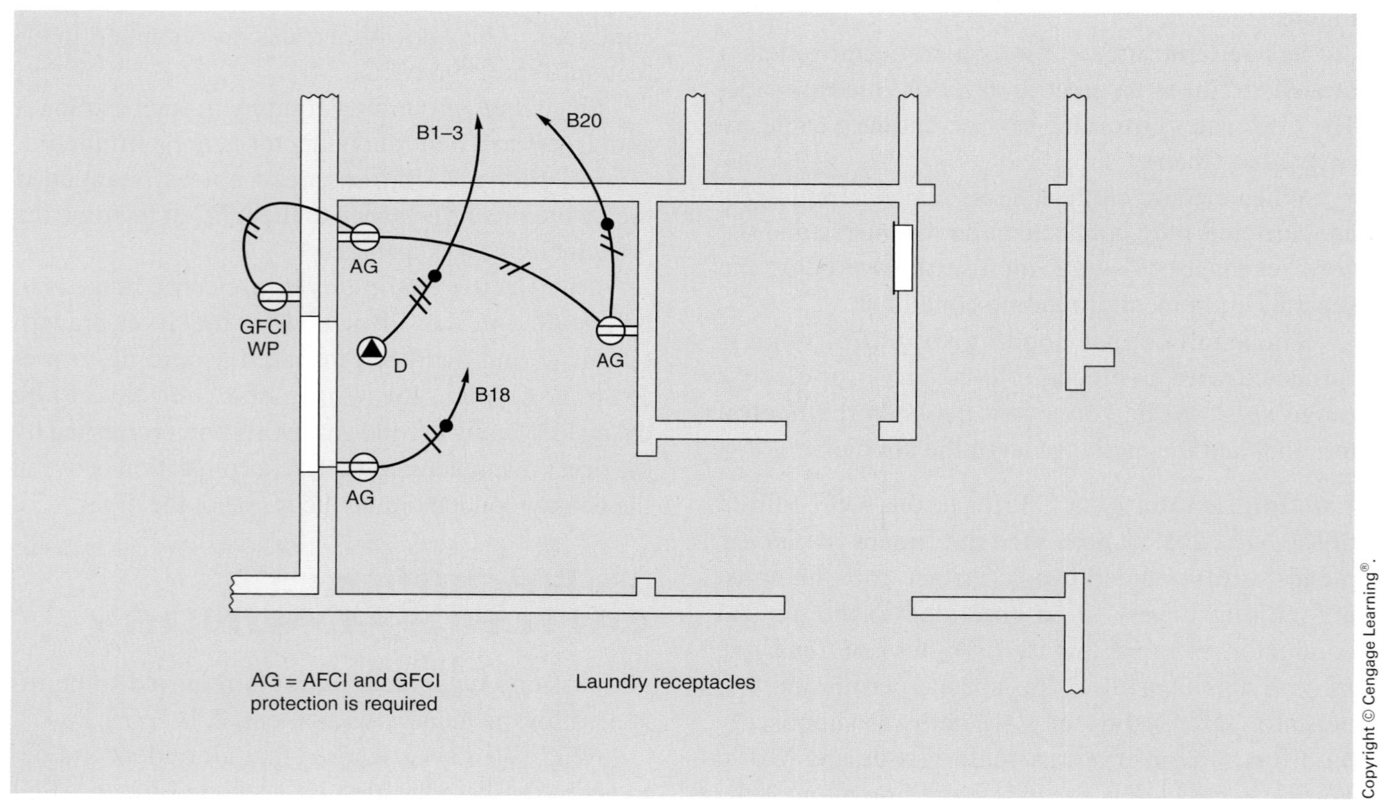

FIGURE 15-8 Cable layout for Laundry Room receptacles.

and is subject to the demand factors that are applicable to these calculations. See the complete calculations in Chapter 29.

Depth of Box—Watch Out!

Determining the cubic volume of a box is not the end of the story. Outlet and device boxes have to be deep enough so that conductors will not be damaged when installing the wiring device or equipment into the box. Merely calculating and providing the proper volume of a box is not always enough. The volume calculation might prove adequate, yet the size (depth) of the wiring device or equipment might be such that conductors behind it may possibly be damaged. See *NEC 314.24*.

Width of Box—Watch Out!

Large wiring devices, such as a 30-ampere, 3-pole, 4-wire dryer or range receptacle will not fit into a single gang box that is 2 in. (50.8 mm) wide. Dryer receptacles measure 2.10 in. (53.3 mm) in width. Likewise, a 50-ampere, 3-pole, 4-wire receptacle measures 2.75 in. (69.9 mm) in width. Consider using a 4 in. (101.6 mm) square box with a 2-gang plaster device ring for flush mounting or a 2-gang raised cover for surface mounting. Or, use a 2-gang device box. The center-to-center mounting holes of both the 30-ampere and the 50-ampere receptacles are 1.81 in. (46.0 mm) apart, exactly matching the center-to-center holes of a 2-gang plaster ring, raised cover, or gang device box. See *NEC 314.16(B)(4)*.

COMBINATION WASHER/DRYERS

Combination washer/dryers take up about half the floor space of a traditional washer and dryer pair. Some models have the washer/dryer units combined into one appliance, using the same drum for both wash and dry cycles. Others stack the dryer above the washer. Electric washer/dryer combinations generally require a 30-ampere branch circuit similar to that required for a typical electric clothes dryer. The receptacle would be a 30-ampere, 4-wire, 125/250-volt, NEMA 14-30R, Figure 15-6.

Gas washer/dryer combinations draw 10 or 12 amperes and plug into a standard grounding-type receptacle. If the receptacle is located within 6 ft (1.8 m) of the outside edge of a sink, then it must be GFCI protected, *210.8(A)(7)*. This receptacle and the branch circuit for the laundry appliances must conform to *210.52(F)* and *210.11(C)(2)*.

For installations that require a 30-ampere branch circuit and receptacle for a combination washer/dryer, a 120-volt receptacle and separate 20-ampere branch circuit must still be provided in conformance to *NEC 210.52(F)* and *210.11(C)(2)*.

LIGHTING CIRCUIT

The general lighting circuit for the Laundry, Powder Room, and hall area is supplied by Circuit B10. This circuit must be AFCI protected as required by *NEC 210.12(A)*.

Note on the cable layout, Figure 15-9, that in order to help balance loads evenly, the attic lights are also connected to Circuit B10.

The types of vanity and ceiling luminaires and wall receptacle outlets, as well as the circuitry and switching arrangements, are similar to types discussed in other chapters of this text and will not be repeated.

The receptacle in the Powder Room is supplied by Circuit B21, a separate 20-ampere branch circuit as required by *NEC 210.52(D)*, discussed in Chapter 10.

Exhaust Fans

Ceiling exhaust fans are installed in the Laundry and in the Powder Room. These exhaust fans are connected to Circuit B10.

The exhaust fan in the Laundry will remove excess moisture resulting from the use of the clothes washer, dryer, and ironing.

Exhaust fans may be installed in walls or ceilings, as in Figure 15-10. The wall-mounted fan can be adjusted to fit the thickness of the wall. If a ceiling-mounted fan is used, sheet-metal duct must be installed between the fan unit and the outside of the house. The fan unit terminates in a metal hood or grille on the exterior of the house. The fan has a shutter that opens as the fan starts up and

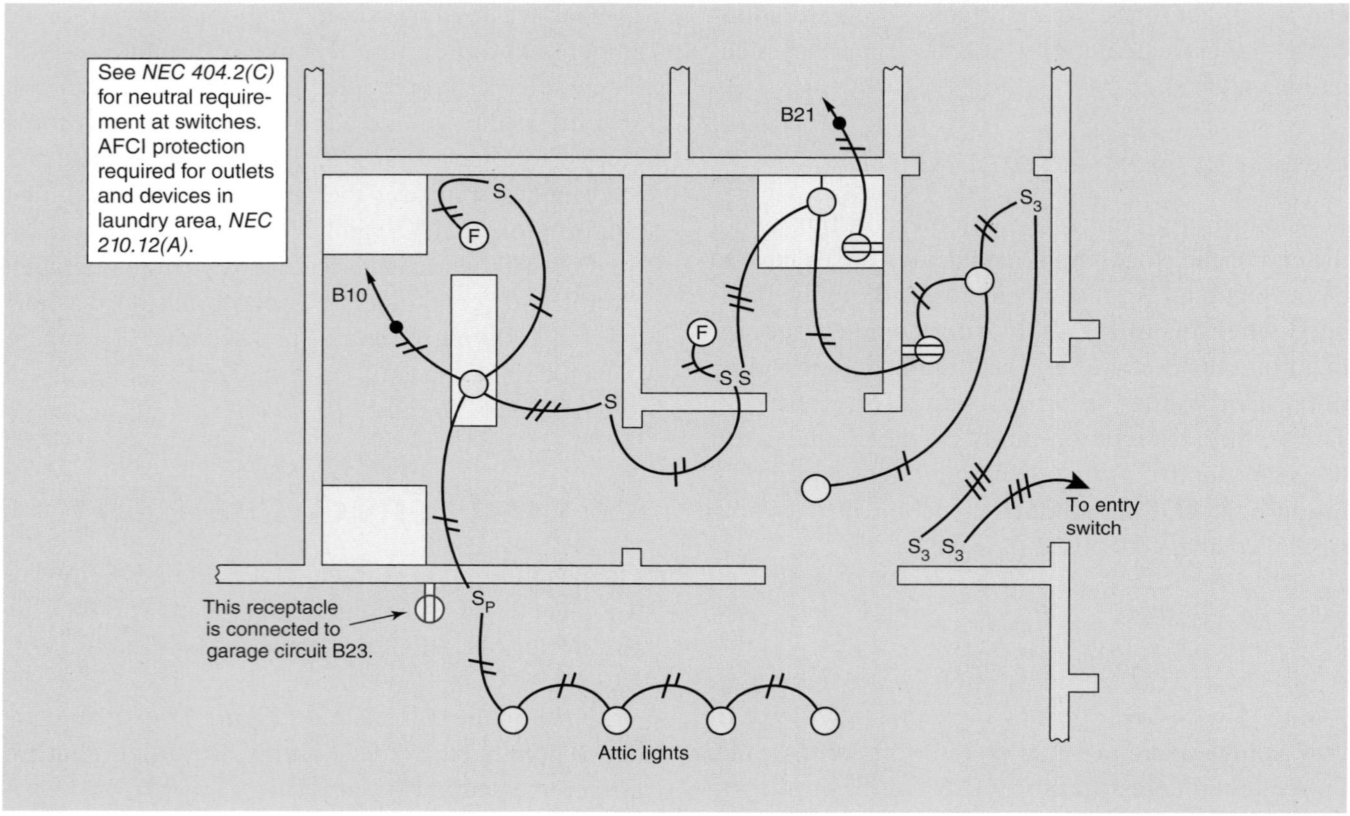

See *NEC 404.2(C)* for neutral requirement at switches. AFCI protection required for outlets and devices in laundry area, *NEC 210.12(A)*.

B10 B21 S S S S S_3 S_3 S_3 S_P F F

To entry switch

This receptacle is connected to garage circuit B23.

Attic lights

FIGURE 15-9 Cable layout for Laundry, Rear Entrance, Powder Room, and Attic (*Circuit B10*).

Courtesy of Broan-NuTone, LLC.

FIGURE 15-10 Exhaust fans.

closes as the fan stops. The fan may have an integral pull-chain switch for starting and stopping, or it may be used with a separate wall switch. In either case, single-speed or multispeed control is available. The fan in use has a very small power demand, 90 volt-amperes.

To provide better humidity control, both ceiling-mounted and wall-mounted fans may be controlled with a humidistat. This device starts the fan when the humidity reaches a certain value. When the humidity drops to a preset level, the humidistat turns the fan off.

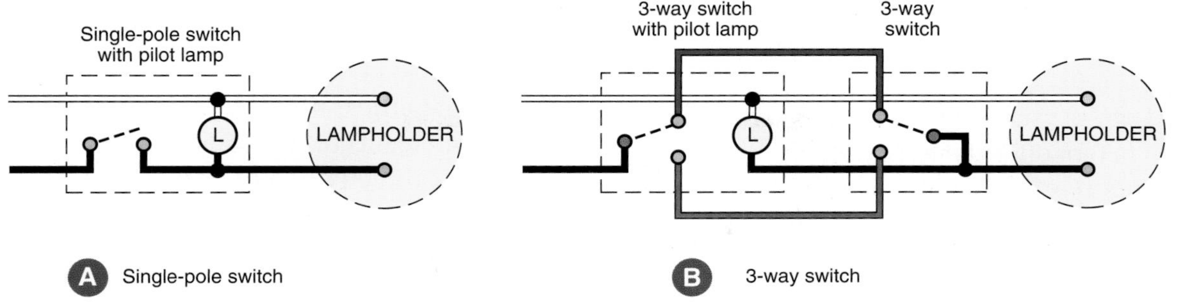

FIGURE 15-11 Pilot lamp connections.

Some of the more expensive exhaust fans have a built-in sensor that will automatically turn on the exhaust fan at a predetermined humidity level. The fan will automatically turn off when the humidity has been lowered to a preset level. This eliminates the need for a separate wall-mounted humidity control (humidistat).

ATTIC LIGHTING AND PILOT LIGHT SWITCHES

NEC 210.70(A)(3) requires that at least one lighting outlet must be installed in an attic if it is used for storage or has equipment that requires servicing. This lighting outlet must either contain a switch such as a pull-chain lampholder, or be controlled by a switch located near the entry to the attic. If the space contains equipment that might need servicing, such as air-conditioning equipment, *NEC 210.70(A)(3)* requires that the lighting outlet(s) be installed at or near this equipment. The lighting outlet must either contain a switch, such as a pull chain, or be controlled by a wall switch.

A 125-volt, single-phase, 15- or 20-ampere receptacle outlet is required to be installed in an accessible location within 25 ft (7.5 m) of air-conditioning or heating equipment in attics, in crawl spaces, or on the roof, *210.63*. Connecting this receptacle outlet to the load side of the equipment's disconnecting means is not permitted because this would mean that if you turned off the equipment, the power to the receptacle would also be off and thus would be useless for servicing the equipment.

The residence discussed in this text does not have air-conditioning, heating, or refrigeration equipment in the attic or on the roof.

Porcelain lampholders are available with a receptacle outlet for convenience in plugging in an extension cord. However, most electrical inspectors (authority having jurisdiction) would not accept the porcelain lampholder's receptacle in lieu of the required receptacle outlet, as stated in *NEC 210.63*.

The four porcelain lampholders in the attic are turned on and off by a single-pole switch on the garage wall, close to the attic storable ladder. Associated with this single-pole switch is a pilot light. The pilot light may be located in the handle of the switch, or it may be separately mounted. Figure 15-11 shows how pilot lamps are connected in circuits containing either single-pole or 3-way switches.

Because the attic in this residence is served by a folding storable ladder, switch control in the attic is not required. Where a permanent stairway of six or more risers is installed, *NEC 210.70(A)(2)(c)* requires switch control at both levels.

If a neon pilot lamp in the handle (toggle) of a switch does not have a separate grounded conductor connection, then it will glow only when the switch is in the OFF position, as the neon lamp will then be in series with the lamp load, Figure 15-12.

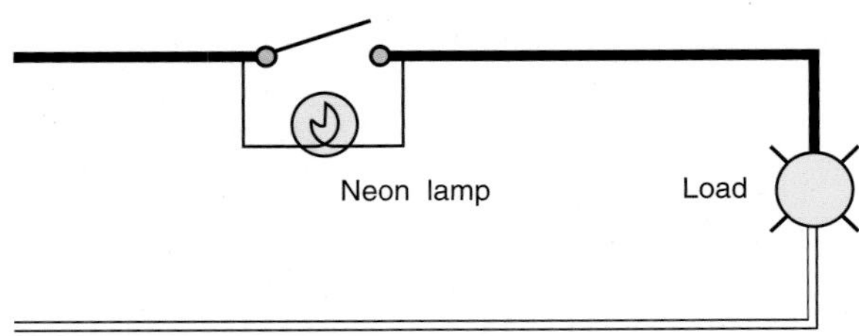

FIGURE 15-12 Example of neon pilot lamp in handle of toggle switch.

The voltage across the load lamp is virtually zero, so it does not burn, and the voltage across the neon lamp is 120 volts, allowing it to glow. When the switch is turned on, the neon lamp is bypassed (shunted), causing it to turn off and the lamp load to turn on.

Use this type of switch when it is desirable to have a switch glow in the dark, to make it easy to locate.

Installation of Cable in Attics

When nonmetallic-sheathed cable is installed in accessible attics, the installation must conform to the requirements of *NEC 334.23*. This section refers the reader directly to *NEC 320.23*, which describes how the cable is to be protected. See Figures 15-13 and 15-14.

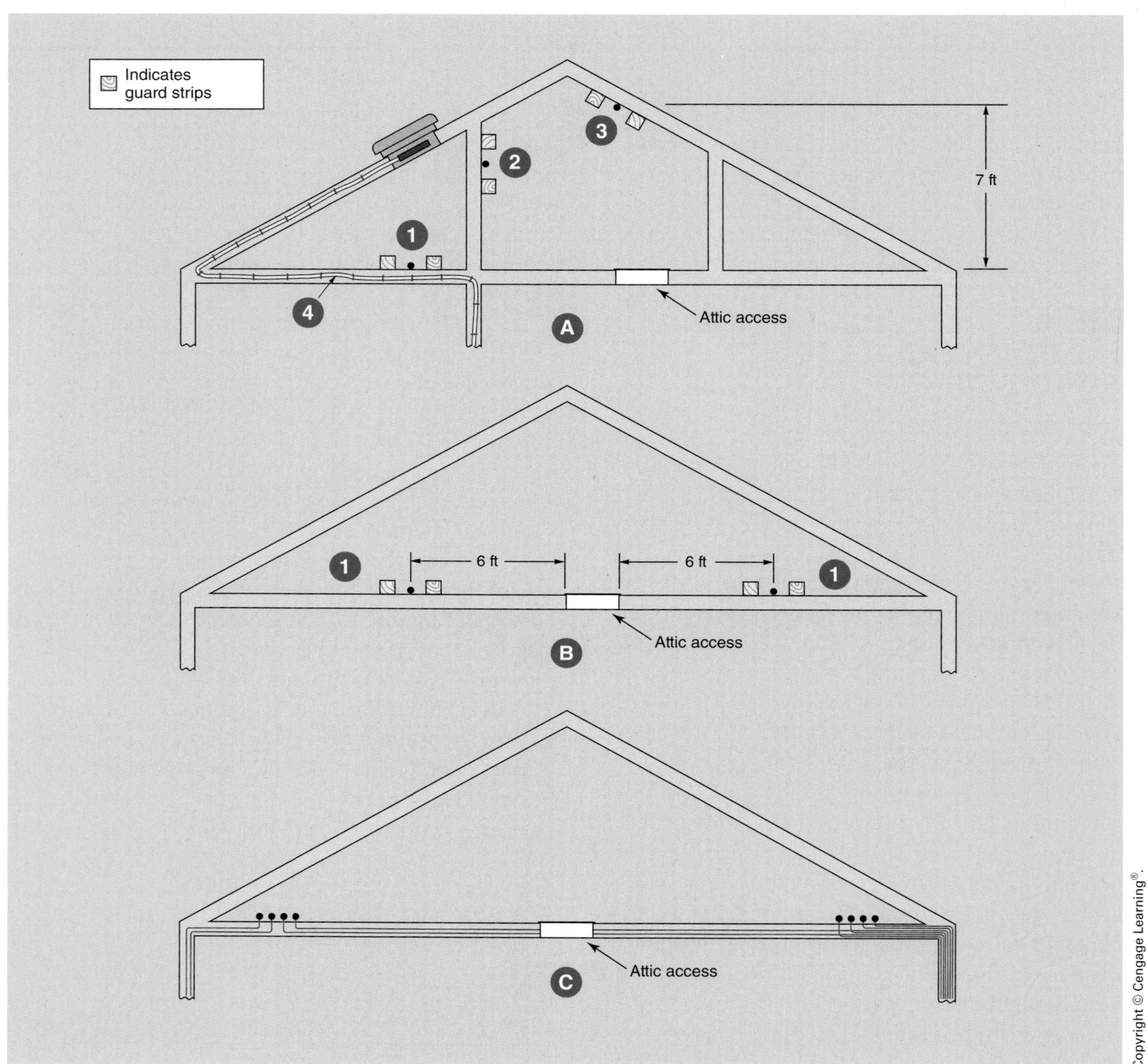

FIGURE 15-13 Protection of nonmetallic-sheathed cable or armored cable in an attic. Refer to *NEC 334.23* and *320.23*.

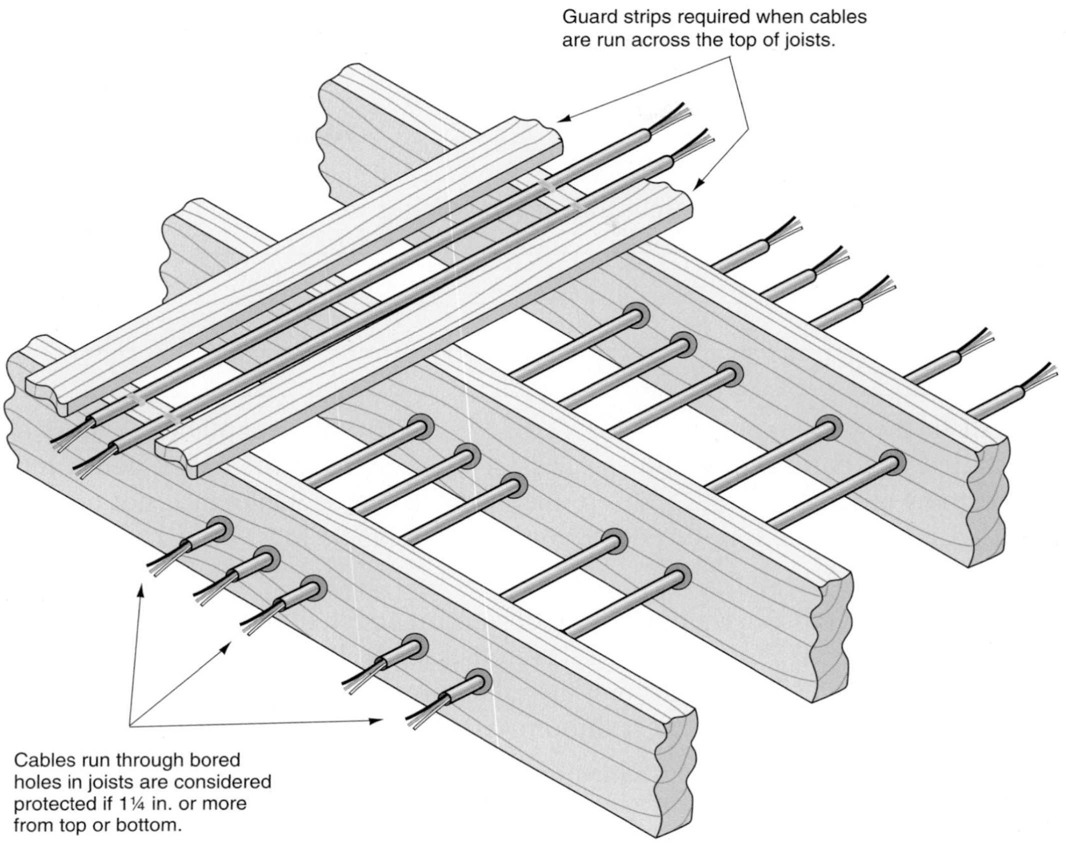

Guard strips required when cables are run across the top of joists.

Cables run through bored holes in joists are considered protected if 1¼ in. or more from top or bottom.

FIGURE 15-14 Methods of protecting cable installations in accessible attics. See *NEC 320.23* and *334.23*.

In accessible attics, cables must be protected by guard strips if

- they are run across the top of floor joists ①.
- they are run across the face of studs ② or rafters ③ within 7 ft (2.1 m) of the floor or floor joists.

Guard strips are *not* required if the cable is run along the sides of rafters, studs, or floor joists ④.

In attics not accessible by permanent stairs or ladders, as in Figure 15-13(B), guard strips are required only within 6 ft (1.8 m) of the nearest edge of the scuttle hole or entrance.

Figure 15-13(C) illustrates a cable installation that most electrical inspectors consider to be safe. Because the cables are installed close to the point where the ceiling joists and the roof rafters meet, they are protected from physical damage. It would be very difficult for a person to crawl into this space or store cartons in an area with such a low clearance. Although the plans for this residence show a 2 ft wide catwalk in the attic, the owner may decide to install additional flooring in the attic to obtain more storage space. Because of the large number of cables required to complete the circuits, it would interfere with flooring to install guard strips wherever the cables run across the tops of the joists, as in Figure 15-14. However, the cables can be run through holes bored in the joists and along the sides of the joists and rafters. In this way, the cables do not interfere with the flooring.

When running cables parallel to framing members or furring strips, be careful to maintain at least 1¼ in. (32 mm) between the cable and the edge of the framing member. This is a *Code* requirement referenced in *NEC 300.4(D)*, to minimize the possibility of driving nails into the cable. All of *300.4* is devoted to the subject of *Protection Against Physical Damage.*

See Chapter 4 in this text for more detailed discussion and illustrations relating to the physical protection of cables.

REVIEW

Note: Refer to the *Code* or plans where necessary.

1. List the switches, receptacles, and other wiring devices that are connected to Circuit B10. ____________________

   ____

   ____

   ____

2. a. Fill in the blank spaces. At least ________ 120-volt receptacle(s) must be installed in a bathroom within ________ in. of the basin, and it must be on the wall ________ to the basin. The receptacle must not be installed ________ on the countertop. The receptacle must be ________ protected.

   b. A separate (15) (20)-ampere branch circuit must supply the receptacle(s) in bathrooms and (shall) (shall not) serve other loads. If a separate 20-ampere branch circuit serves a single bathroom, then that circuit (is permitted) (is not permitted) to supply other loads in that bathroom. (Circle the correct answers.)

3. What special type of switch is controlling the attic lights? ____________________

   ____

   ____

4. When installing cables in an accessible attic along the top of the floor joists ________ must be installed to protect the cables.

5. The total estimated volt-amperes for Circuit B10 has been calculated to be 1144 volt-amperes. How many amperes is this at 120 volts? ________.

6. If an attic is accessible through a scuttle hole, guard strips are installed to protect cables run across the top of the joists only within (6 ft [1.8 m]) (12 ft [3.7 m]) of the scuttle hole. (Circle the correct answer.)

7. What section(s) of the *Code* refers (refer) to the receptacle required for the laundry equipment? State briefly the requirements.

   ____

   ____

   ____

   ____

   ____

   ____

   ____

   ____

8. What is the current draw of the exhaust fan in the Laundry? ______________________

9. The following is a layout of the lighting circuit for the Laundry, Powder Room, Rear Entry Hall, and Attic. Complete the wiring diagram using colored pens or pencils to indicate conductor insulation color. The GFCI receptacle in the Powder Room is not shown in the diagram because it is connected to a separate circuit, B21.

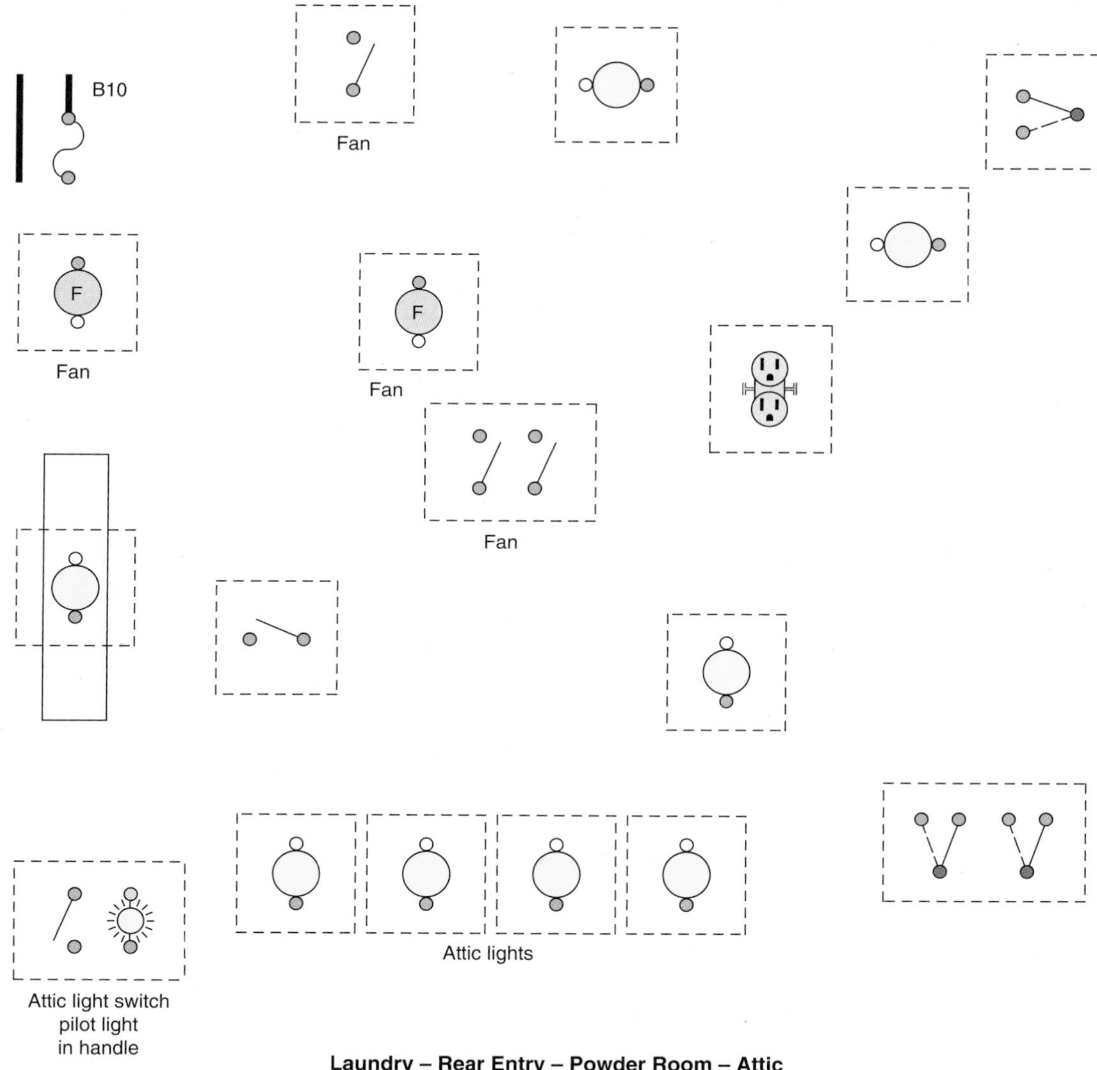

Laundry – Rear Entry – Powder Room – Attic

10. Laundry receptacle outlets are included in the residential load calculations at a value of ____________ volt-amperes per circuit. Choose one: 1000, 1500, 2000 volt-amperes.

11. List the various methods of connecting an electric clothes dryer. ____________________

____

12. a. What is the minimum power demand allowed by the *Code* for an electric dryer if no actual rating is available for the purpose of calculating feeder and service-entrance conductor sizing? ____________________

    b. What is the current draw? ____________________

13. What is the maximum permitted current rating of a portable appliance on a 30-ampere branch circuit? ____________________

14. What provides motor running overcurrent protection for the dryer? ____________________

    ____

15. a. Must the metal frame of an electric clothes dryer be grounded? ____________________

    b. Is it permitted to ground the metal frame of an electric clothes dryer to the neutral conductor? ____________________

    ____

16. An electric dryer is rated at 7.5 kW and 120/240 volts, 3-wire, single-phase. The terminals on the dryer and panelboard are marked 75°C.

    a. What is the wattage rating? ____________________

    b. What is the current rating? ____________________

    c. What minimum size type THHN copper conductors are required? ____________________

    ____

    ____

    d. If EMT is used, what minimum size is required? ____________________

    ____

    ____

    ____

    ____

    ____

    ____

17. When a metal junction box is installed as part of the cable wiring to a clothes dryer or electric range, may this box be grounded to the circuit neutral conductor?

18. A residential air-conditioning unit is installed in the attic.

    a. Is a lighting outlet required? ____________________

    b. What *Code* section applies? ____________________

    c. If a lighting outlet is required, how shall it be controlled? ____________________

    ____

    d. What *Code* section applies? ____________________

    e. Is a receptacle outlet required? ____________________

    f. What *Code* section applies? ____________________

19. The *Code* states that a means must be provided to disconnect an appliance. In the case of an electric clothes dryer, the disconnecting means can be (Circle the correct statemen(s).)
    a. a separate disconnect switch located within sight of the appliance.
    b. a separate disconnect switch located out of sight from the appliance. The switch is capable of being locked in the OFF position.
    c. a cord-and-plug-connected arrangement.
    d. a circuit breaker in a panel that is within sight of the dryer or capable of being locked in the OFF position.

CHAPTER **16**

Branch Circuits for the Garage and Outdoors

OBJECTIVES

After studying this chapter, you should be able to

- understand the basics for lighting a residential garage.
- understand the *NEC* requirements for receptacle outlets in a residential garage.
- become familiar with outdoor and simple landscape lighting.
- understand the *NEC* requirements for underground cable and conduit wiring.
- understand the operation and electrical connections for overhead door operators.
- understand the *NEC* article that contains requirements for electric vehicle charging systems.

LIGHTING BRANCH CIRCUIT B14

Circuit B14 is a 15-ampere circuit that originates at Panel B in the Recreation Room. This circuit is brought into the 2-gang switch box next to the service door on the front wall of the garage. From this point, the circuit is carried to the lighting outlets, switch locations, and overhead door operator.

Want More Than *Code* Minimum?

The garage cable layout in Figure 16-1 supplied by Circuits B14 and B23 "Meets *Code*" but could certainly be termed "minimum."

Some localities require an individual 20-ampere branch circuit for the overhead door operator. Residential overhead door operators and their UL requirements are covered later in this chapter.

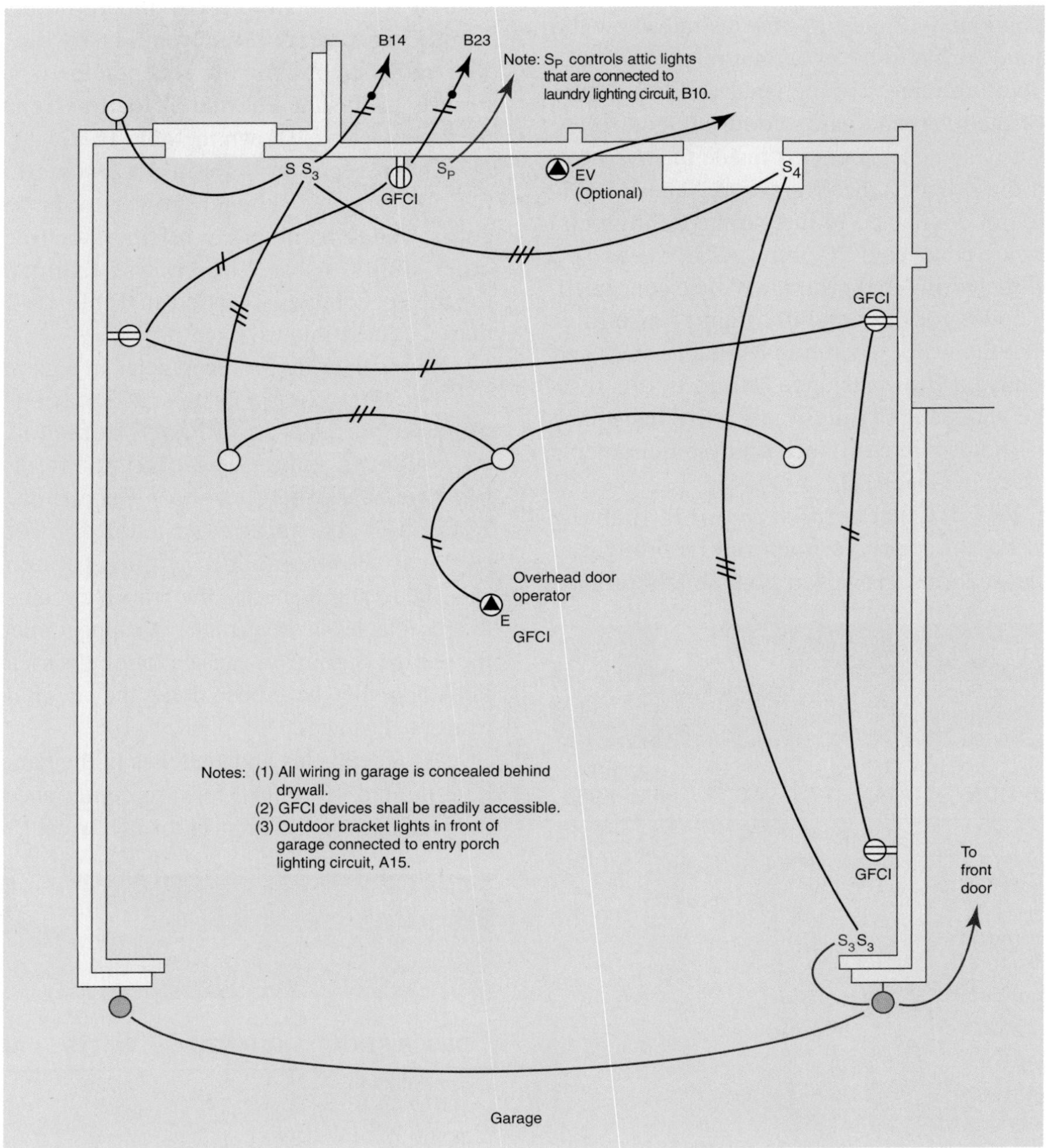

FIGURE 16-1 Cable layout for the garage circuit, B14.

Figure 16-1 shows the cable layout for Circuit B14. This branch circuit is not required to have AFCI protection by *NEC® 210.12(A)*.

The garage ceiling's porcelain lampholders are controlled from switches located at three entrances to the garage. This switching is accomplished by two 3-way switches and one 4-way switch.

The post light on the front lawn next to the sidewalk in the front of the house is controlled by a switch just inside the front door, as well as by a photocell that is an integral part of the post light. See the Notation on Plan 9. Figure 16-19 illustrates how the underground wiring to the post light might be done.

The two luminaires to the right and left on the outside of the overhead garage door are controlled by two 3-way switches, one just inside the overhead door and the other in the front entry. These luminaires are not connected to the garage lighting circuit. They are connected to Circuit A15.

All of the wiring in the garage will be concealed. Building codes for one- and two-family dwellings require that the walls separating habitable areas and a garage have a fire-resistance rating. In this residence, the walls and ceilings have a fire-resistance rating of 1 hour. Fire resistance rating requirements are discussed in Chapter 2.

NEC 404.2(C) generally requires that the grounded circuit conductor (neutral) be brought to all switch locations. However, nonhabitable rooms are exempted. By definition, a garage in not considered to be a habitable room. Therefore, the neutral conductor is not run to the 4-way switch and second 3-way switches that control the garage lighting.

Table 16-1 summarizes the outlets and the estimated load for the garage circuit B14.

TABLE 16-1

Garage outlet count and estimated load (Circuit B14).

DESCRIPTION	QUANTITY	WATTS	VOLT-AMPERES
Ceiling lampholders @ 100 W each	3	300	300
Outdoor garage bracket luminaire (rear)	1	100	100
Overhead door	1		
opener motor:			
4.8 A × 120 V			576
4.8 A × 120 V × 0.9 PF		518	
two 60-W lamps		120	120
TOTALS	5	1038	1096

RECEPTACLE BRANCH CIRCUIT B23

Circuit B23 is a simple 20-ampere branch circuit supplying the four receptacles in the garage. The cable layout for the receptacles is shown in Figure 16-1. The estimated load assignment for the receptacles is shown in Table 16-2.

These receptacles could very well be used for powering workbench power tools, extension cords, hedge trimmers, a refrigerator/freezer, and other similar uses. Who knows? Estimated loads for the receptacles as listed in Table 16-2 are just that . . . guesstimates!

Here's the rule for receptacles in a garage.

NEC 210.52(G)(1) states: ▶Garages. In each attached garage and in each detached garage with electric power. The branch circuit supplying this receptacle(s) shall not supply outlets outside of the garage. At least one receptacle outlet shall be installed for each car space.◀*

The requirement that the branch circuit supplying the receptacles in the garage not supply outlets outside the garage is going to cause a lot of electricians to rethink how they have been doing the circuit design for residential garages.

All receptacles and switches in the garage are to be mounted 46 in. (1.15 m) to center according to the Notation on the First Floor Electrical Plan.

TABLE 16-2

Garage receptacle count and estimated load (Circuit B23).

DESCRIPTION	QUANTITY	WATTS	VOLT-AMPERES
Receptacles @ 180 W each	4	720	720
TOTALS	4	720	720

*Reprinted with permission from NFPA 70-2014.

SAFETY ALERT

NEC 210.8(A)(2) **requires that all 125-volt, single-phase, 15- or 20-ampere receptacles installed in garages shall have ground-fault circuit-interrupter protection. GFCI requirements are discussed in detail in Chapter 6.**

To provide GFCI protection for the receptacles in the garage, there are a couple of possibilities. One is to install a GFCI circuit breaker for Circuit B23. Another is to install a GFCI feed-through receptacle at the first receptacle where Circuit B23 feeds the garage receptacles. See Figure 16-1.

GFCI devices are required to be readily accessible, *210.8*. This rule is intended to allow these safety devices to be tested at least monthly, as required in manufacturers' instructions. GFCI devices are not permitted behind appliances such as refrigerators or freezers or on the ceiling.

If a detached garage has no electricity provided, then obviously no rules are applicable; but just as soon as a detached garage is wired, then all pertinent *Code* rules become effective and must be followed. These would include rules relating to grounding, lighting outlets, receptacle outlets, GFCI protection, and so on.

ELECTRIC VEHICLE CHARGING SYSTEMS

Electric vehicle charging systems are becoming more and more popular. Many people have become aware of the environmental impact of using fossil fuels, such as gasoline for combustion engines of automobiles.

Electric vehicle charging systems require electrical power that is generated at power plants and then transmitted to the end user. Getting the electricity to the end user requires the generation, transmission, transformers and the service conductors. Many power plants are coal or nuclear powered. There are line losses, transformer losses, conductor losses, and so on. One must be aware that energy is being consumed somewhere in order to provide electrical power to the end use.

Figure 16-1 and Plan Sheet 9 for the residence associated with this text shows on the front wall of the garage a symbol marked EV. This is for an optional electrical outlet for a future electric vehicle charging system. If the branch circuit and receptacle are installed for the electric vehicle charging system, additional capacity must be provided in the feeder or service where the branch circuit originates. *NEC 220.14(A)* requires that the ampere rating of the appliance or load served be added to the load calculation. The rating of our optional outlet for an electric vehicle charging system is rated 40 A. If 40 A is added to the load calculation, a service larger than 200 A would be required. The electrical designer and installer must always be certain the electrical system that is installed is adequate for the calculated load on the branch circuit, feeder and service.

NEC Article 625 covers electric vehicle charging systems.

LIGHTING A TYPICAL RESIDENTIAL GARAGE

The recommended approach to provide adequate lighting in a residential garage is to install one 100-watt lamp or equivalent compact fluorescent or LED lamp on the ceiling above each side of an automobile, Figure 16-2. Figure 16-3 shows typical lampholders commonly mounted in garages.

- For a one-car garage, a minimum of two lampholders is recommended.
- For a two-car garage, a minimum of three lampholders is recommended.
- For a three-car garage, a minimum of four lampholders is recommended.

Lampholders arranged in this manner eliminate shadows between automobiles. Shadows are a hazard because they hide objects that can cause a person to trip or fall. It is highly recommended that these ceiling lampholders be mounted toward the front end of the automobile as it is

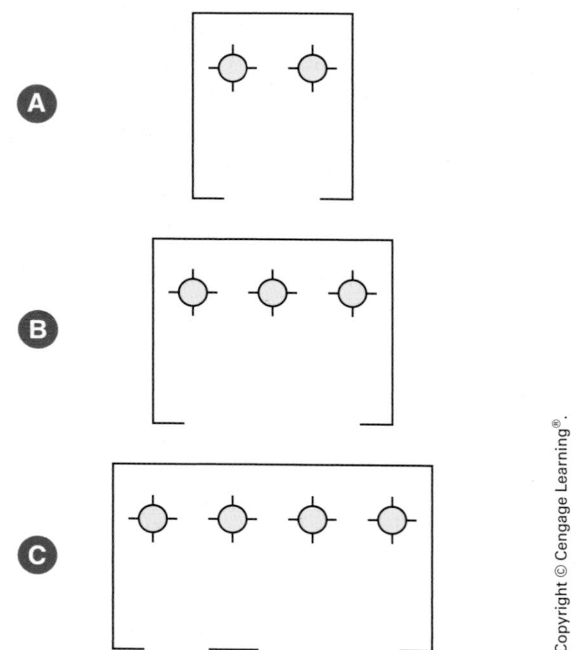

FIGURE 16-2 Positioning of lights in (A) a one-car garage, (B) a two-car garage, and (C) a three-car garage.

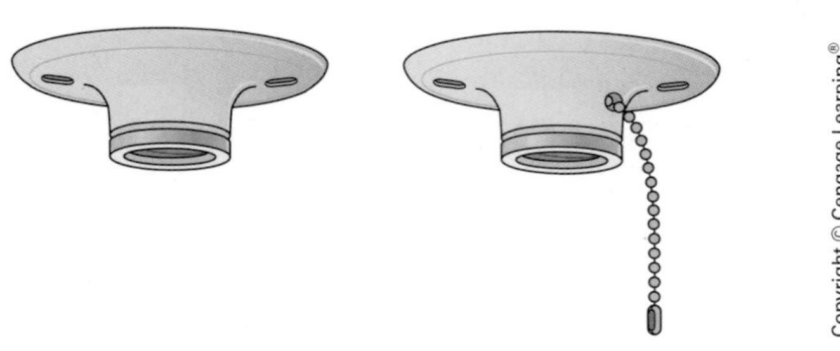

FIGURE 16-3 Typical porcelain or nonmetallic lampholders that would be installed in garages, attics, basements, crawl spaces, and similar locations. Note that one has a pull chain and one does not.

normally parked in the garage. This arrangement will provide better lighting where it is most needed when the owner is working under the hood. Do not place lights where they will be covered by the open overhead door.

Fluorescent luminaires are quite often used in the warmer parts (south and southwest) of the country. For example, instead of the three porcelain lampholders for conventional incandescent lamps indicated on the electrical plans on the ceiling of the garage, fluorescent strip lights or fluorescent "shop" lights could be installed.

Standard fluorescent lamps installed in a cold garage (northern parts of the country) would be difficult if not impossible to start. If they do start, the lamps will probably "flutter," and the light output will be significantly less than when operated at normal room temperatures. Special lamps and ballasts are required at temperatures below 50–60°F (10–16°C). It is generally cost-effective to use incandescent lamps in an unheated garage. LED lamps work extremely well in cold temperatures and should be considered for this application.

RECEPTACLE OUTLETS IN A GARAGE

At least one receptacle must be installed in an attached residential garage, *210.52(G)*. If a detached garage has no electricity provided, then obviously no rules are applicable, but just as soon as a detached garage is wired, then all pertinent *Code* rules become effective and must be followed. These would include rules relating to grounding, lighting outlets, receptacle outlets, GFCI protection, and so on.

NEC 210.8(A)(2) requires that all 125-volt, single-phase, 15- or 20-ampere receptacles installed in garages have ground-fault circuit-interrupter protection. GFCIs are discussed in detail in Chapter 6 of this text.

To provide GFCI protection for the receptacles in the garage, there are two possibilities. One is to install a GFCI circuit breaker for Circuit B14. Another simple possibility is to install a GFCI feed-through receptacle at the point where Circuit B14 feeds the garage. See Figure 16-1.

Note that all GFCI devices are required to be readily accessible, *NEC 210.8*. This rule is intended to allow these safety devices to be tested at least monthly, as required in manufacturers' instructions. GFCI devices are not permitted behind appliances such as refrigerators or freezers or on the ceiling near the garage door operator.

LANDSCAPE LIGHTING

One of the nice things about residential wiring is the availability of a virtually unlimited variety of sizes, types, and designs of outdoor luminaires for gardens, decks, patios, and for accent lighting. Luminaire showrooms at electrical distributors and the electrical departments of home centers offer a wide range of outdoor luminaires, both 120-volt and low-voltage types. Figure 16-4 and Figure 16-5 illustrate landscape lighting mounted on the ground. See also the following section on outdoor wiring.

UL Standard 1838 covers *Low-Voltage Landscape Lighting Systems* generally used outdoors in wet locations.

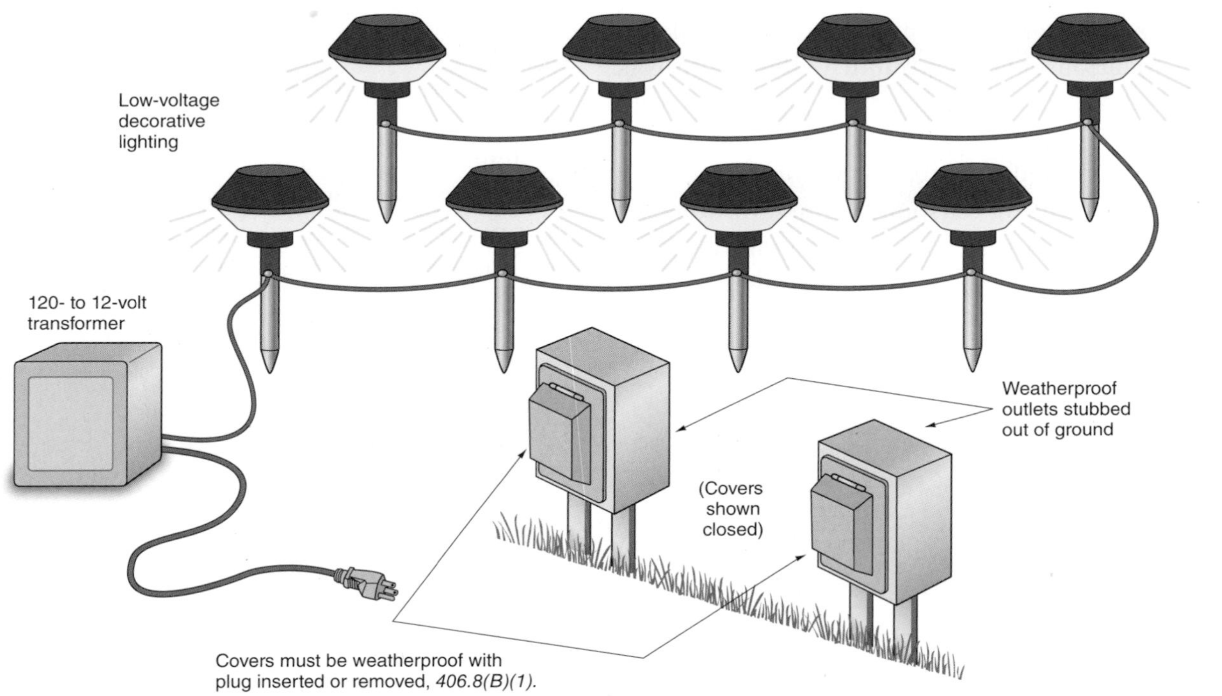

FIGURE 16-4 Weatherproof receptacle outlets stubbed out of the ground; either low-voltage decorative landscape lighting or 120-volt PAR luminaires may be plugged into these outlets.

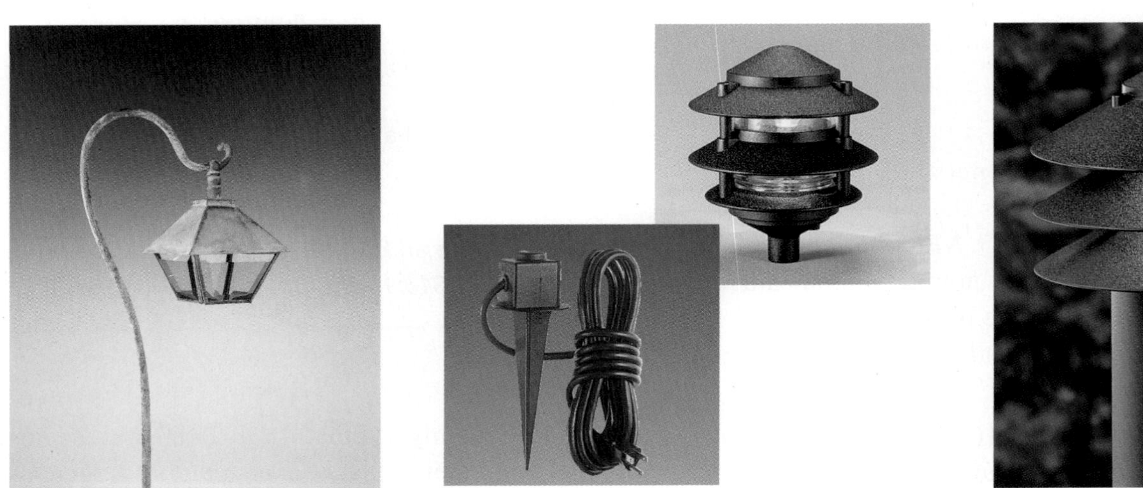

FIGURE 16-5 Luminaires used for decorative purposes outdoors, under shrubs and trees, in gardens, and for lighting paths and driveways.

UL Standard 2108 covers *Low-Voltage Lighting Systems* for use in dry locations.

Both of these UL standards require the manufacturer to include installation instructions with the product. It is extremely important to read these instructions to be sure the luminaires and associated wiring are installed and used in a safe manner.

Low-voltage lighting products will bear markings such as *Outdoor Use Only*, *Indoor Use Only*, or *Indoor/Outdoor Use*. If marked for use in damp or wet locations, the maximum voltage rating is 15 volts unless live parts are made inaccessible to contact during normal use. Low-voltage luminaires are intended for specific applications, such as surface mounting, suspended installations, or recessed installations, and are so marked. If of the recessed type, they must be installed according to the provisions of *410.116*, as discussed in Chapter 7.

More about Low-Voltage Lighting Systems

Low-voltage lighting systems are covered in *Article 411*. The basic requirements for these systems are that they shall

- operate at 30 volts rms (42.4 volts peak) or less.
- have one or more secondary circuits, each limited to 25 amperes maximum.
- be supplied through an isolating transformer. A standard step-down transformer does not meet this criterion. Look carefully at the nameplate on the transformer.
- be listed for the purpose.
- not have the wiring concealed or extended through a building unless the wiring meets the requirements of *Chapter 3, NEC*. An example of this is running Type NM nonmetallic-sheathed cable for the low-voltage portion of the wiring that is concealed or extended through the structure of the building.
- not be installed within 10 ft (3.0 m) of pools, spas, or water fountains unless specifically permitted in *Article 680*.
- not have the secondary of the isolating transformer grounded.
- not be supplied by a branch circuit that exceeds 20 amperes.

As stated previously, low-voltage lighting systems operate at 30 volts or less and are covered in *Article 411*. These systems are supplied through an isolating (two-winding) transformer that is connected to a 120-volt branch circuit of 20 amperes or less.

Careful consideration must be given to the length of the low-voltage wiring. When the low-voltage wiring is long, larger conductors are necessary to compensate for voltage drop. If voltage drop is ignored, the lamps will not get enough voltage to burn properly; the lamps will burn dim. Voltage drop calculations are discussed in Chapter 4.

You must also be concerned with the current draw of a low-voltage lighting system. For the same wattage, the current draw of a low-voltage lamp will be much greater than that of a 120-volt lamp. If not taken into consideration, the higher current can cause the conductors to run at dangerous temperatures; that could cause a fire!

EXAMPLE

A 480-watt load at 120 volts draws

$$480 \div 120 = 4 \text{ amperes}$$

This is easily carried by a 14 AWG conductor.

A 480-watt load at 12 volts draws

$$480 \div 12 = 40 \text{ amperes}$$

This requires an 8 AWG conductor, as determined by *Table 310.15(B)(16)*.

You can readily see that for this 480-watt example, the conductors must be considerably larger for the 12-volt application to safely carry the current.

Prewired low-voltage lighting assemblies that bear the UL label have the correct size and length of conductors. Adding conductors in the field must be done according to the manufacturers' instructions.

Voltage drop calculations are covered in Chapter 4.

OUTDOOR WIRING

If the homeowner requests that 120-volt receptacles be installed away from the building structure, the electrician can provide weatherproof receptacle outlets as illustrated in Figure 16-4.

Outdoor receptacles must have GFCI protection. See Chapter 6 for the complete discussion of GFCI protection.

These types of boxes have trade size ½ female threaded openings or hubs in which conduit fittings secure the conduit "stub-ups" to the box. Any unused openings are closed with trade size ½ plugs that are screwed tightly into the threads. Refer to Figure 16-6.

All receptacles installed in a wet location must have an enclosure that is weatherproof, whether or not an attachment plug cap is inserted, *406.8(B)(1)*. This requires a cover that is deep enough to shelter the attachment plug cap of the cord. See Figure 16-4, Figure 16-6, and Figure 16-7. These covers must be

Requirements for raceway-supported enclosures that do not contain wiring devices or support luminaires include:
- Enclosures are not larger than 100 in.^{3}
- Enclosures have threaded entries or identified hubs.
- Enclosures are supported by at least two conduits.
- Conduits must be secured within 18 in. of the enclosure if all conduit entries are on the same side.
- RMC or IMC are required for support. Extend conduit at least 18 in. into ground or pour concrete around conduits at grade.

Requirements for raceway-supported enclosures that do contain wiring devices or support luminaires include:
- The enclosure is not larger than 100 in.^{3}
- Enclosures have threaded entries or identified hubs.
- Enclosures are supported by at least two conduits.
- Conduits must be secured within 18 in. of the enclosure if all conduit entries are on the same side.
- RMC or IMC must be used for support. Extend conduit at least 18 in. into ground or pour concrete around conduits at grade.

Supporting an enclosure with one conduit is not permitted. The enclosure could easily twist, resulting in conductor insulation damage, and a poor connection between the conduit and the enclosure.
Do not use EMT or PVC to support enclosures. PVC conduit is permitted if the enclosure is supported independently by a post, strut, or other suitable methods. EMT not suitable for use underground.

RMC = Rigid metal conduit (*Article 344*)
IMC = Intermediate metal conduit (*Article 342*)
PVC = Rigid polyvinyl chloride conduit (*Article 352*)
EMT = Electrical metallic tubing (*Article 358*)

FIGURE 16-6 Raceway-supported enclosures with and without devices, *314.23(E)* and *(F)*.

FIGURE 16-7 Weatherproof receptacle covers are required to be marked "Extra Duty". They are available in metallic and nonmetallic types. Straight blade receptacles rated 15 and 20 amperes, 125 through 250 volts, installed in wet locations, must be listed as weather-resistant type.

Photos courtesy of Thomas and Betts Corp., www.TNB.com.

marked "Extra Duty" to identify they are suitable for the purpose.

What about the Receptacles?

Straight blade receptacles rated 15- and 20-amperes, 125- and 250-volts installed in wet locations are required to be listed as weather-resistant type, *406.9(B)(1)*.

Where flush-mounted weatherproof receptacles are desired for appearance purposes, a recessed weatherproof receptacle, as illustrated in Figure 16-7, may be used. This type of installation takes up more depth than a standard receptacle and requires special mounting, as indicated in the "exploded" view. Careful thought must be given during the "roughing-in" stages to ensure that the proper depth roughing-in box is installed.

Wiring with Type UF Cable (*Article 340*)

The plans show that Type UF underground cable, Figure 16-8, is used to connect the post light. The current-carrying capacity (ampacity) of Type UF cable is found in *Table 310.15(B)(16)*, using the 60°C column. According to *Article 340*, Type UF cable:

- is marked underground feeder cable.
- is available in sizes from 14 AWG through 4/0 AWG (for copper conductors) and from 12 AWG through 4/0 AWG (for aluminum conductors).
- may be used in direct exposure to the sun if the cable is of a type listed for sunlight resistance or is listed and marked for sunlight resistance.
- may be used with nonmetallic-sheathed cable fittings.
- is flame-retardant.
- is moisture-, fungus-, and corrosion-resistant.
- may be buried directly in the earth.
- may be used for branch-circuit and feeder wiring.
- may be used in interior wiring for wet, dry, or corrosive installations.
- is installed by the same methods as nonmetallic-sheathed cable (*Article 334*).
- must not be used as service-entrance cable.
- must not be embedded in concrete, cement, or aggregate.

FIGURE 16-8 Type UF underground cable.

Courtesy of Southwire Company.

- must be buried in the same trench where single-conductor cables are installed.
- shall be installed according to *300.5*.

You will also find underground feeder cable with the marking UF-B. This indicates that the conductors are rated 90°C dry and 60°C wet. The jacket is rated 75°C. The conductor ampacity is that of a 60°C conductor. The conductors meet the requirements of THWN but are not marked along the entire length of the conductors.

Equipment fed by Type UF cable is grounded by properly connecting the bare equipment grounding conductor found in the UF cable to the equipment to be grounded, Figure 16-9.

For underground wiring, it is significant to note that, whereas the *Code* does permit running up the side of a tree with conduit or cable (when protected from physical damage), as in Figure 16-10, the *Code* does *not* permit supporting conductors between trees, as in Figure 16-11. Figure 16-12 shows one type of luminaire that is permitted for the installation shown in Figure 16-10. Splices are permitted to be made directly in the ground, but only if the connectors are listed by a Nationally Recognized Testing Laboratory for such use, *110.14(B)* and *300.5(E)*.

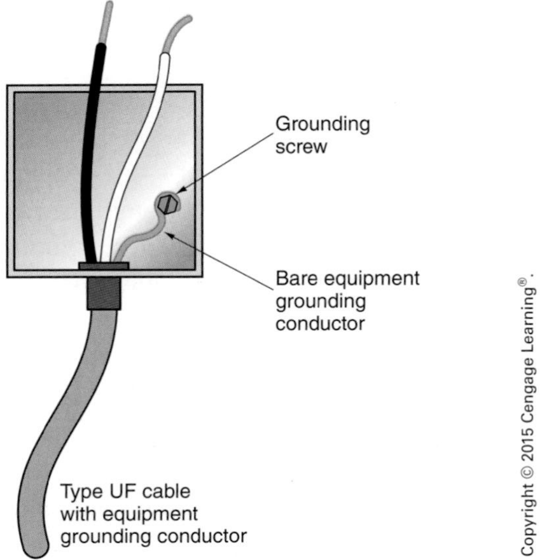

FIGURE 16-9 Illustration showing how the bare equipment grounding conductor of a Type UF cable is used to ground the metal box.

FIGURE 16-10 This installation conforms to the *Code*. *NEC 225.26* does not permit spans of conductors to be run between live or dead trees, but it does not prohibit installing decorative lighting as shown in this figure. *NEC 410.36(G)* permits outdoor luminaires and associated equipment to be supported by trees. Recognized wiring methods to carry the conductors up the tree must be used.

UNDERGROUND WIRING

Underground wiring is common in residential applications. Examples include wiring for decorative landscape lighting, post lamps, detached buildings such as a garage or tool shed, and low-voltage wiring for lawn sprinkling (irrigation) systems.

Grounding Equipment Supplied by Underground Wiring

For life safety, proper grounding is important! Figure 16-13 illustrates a lethal *Code* violation. The post lantern is grounded to the white grounded

FIGURE 16-11 *Violation:* Permanent overhead conductor spans are not permitted to be supported by live or dead vegetation, such as trees. Temporary (not over 90 days) holiday decorative lighting branch-circuit conductors or cables are permitted to be supported by trees. See *225.26, 230.10, 590.3(B),* and *590.4(J).*

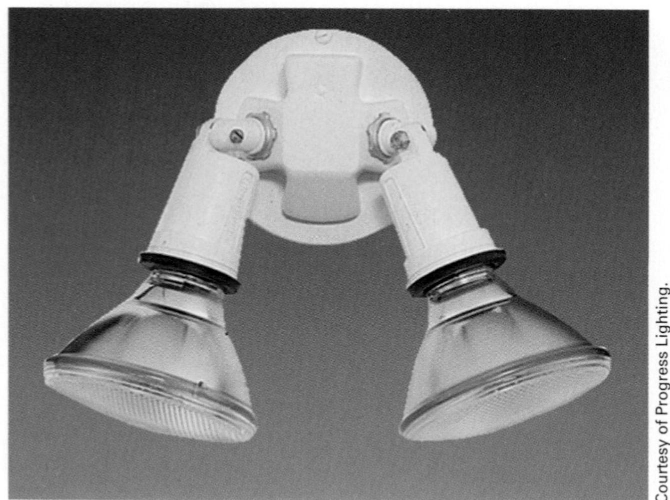

FIGURE 16-12 Outdoor luminaire with lamps.

circuit conductor. This violation results in a parallel return path for the current—one path through the ground and the second path through the person. That person might be a child playing around the post. This could result in a possible electrocution!

NEC 250.4(A)(5) is very clear about this scenario. It states that The earth shall not be considered as an effective ground-fault current path.*

Proper grounding of the metal post is achieved using the equipment grounding conductor in the UF cable. Read the instructions that come with the post lantern.

NEC 250.24(A)(5) and *250.142(B)* prohibit making any connections of equipment to the grounded

*Reprinted with permission from NFPA 70-2014.

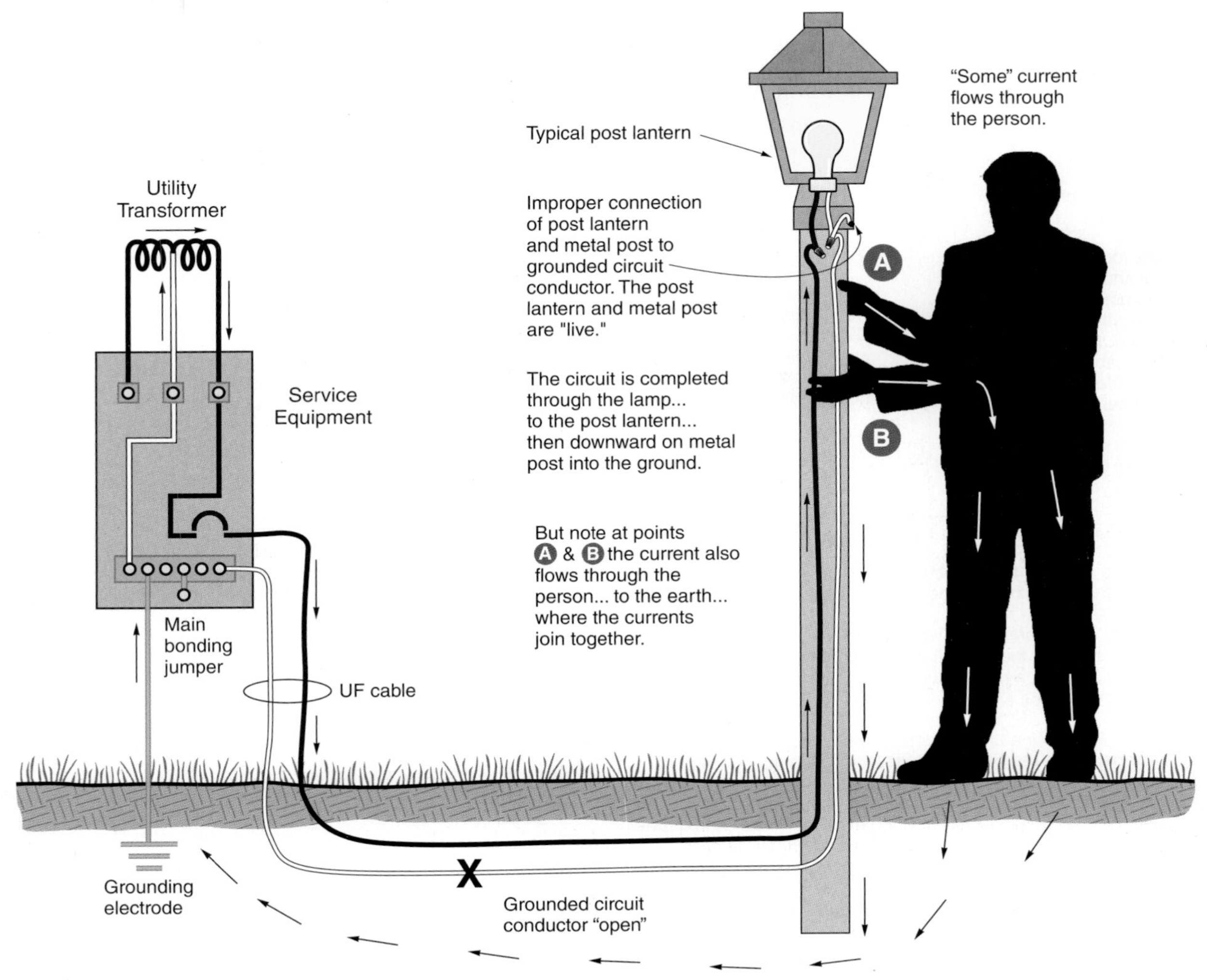

FIGURE 16-13 VIOLATION! This diagram clearly illustrates the electric shock hazard associated with the *Code* violation practice of grounding metal objects to the grounded circuit conductor. In this illustration, the post lantern and metal post have been improperly grounded (connected) to the grounded circuit conductor at (A). The grounded circuit conductor is "open." One path for the return current is through the metal post into the ground. The second path is through the person. This is a violation of *250.24(A)(5)* and *250.142(B)*.

circuit conductor anywhere on the load side of the service disconnect.

How Deep Should Underground Wiring Be?

Digging, aerating, rototilling, and similar yard work tasks can easily damage underground wiring. This is extremely hazardous from the electric shock standpoint. It can also result in costly repairs. The *NEC* addresses minimum burial depths of conduit and cable in great detail.

Table 300.5 shows the minimum depths required for the various types of wiring methods. Refer to Figures 16-14 and 16-15.

The wiring methods to which *Table 300.5* apply are shown across the top of the table. These wiring methods include direct burial cables or conductors,

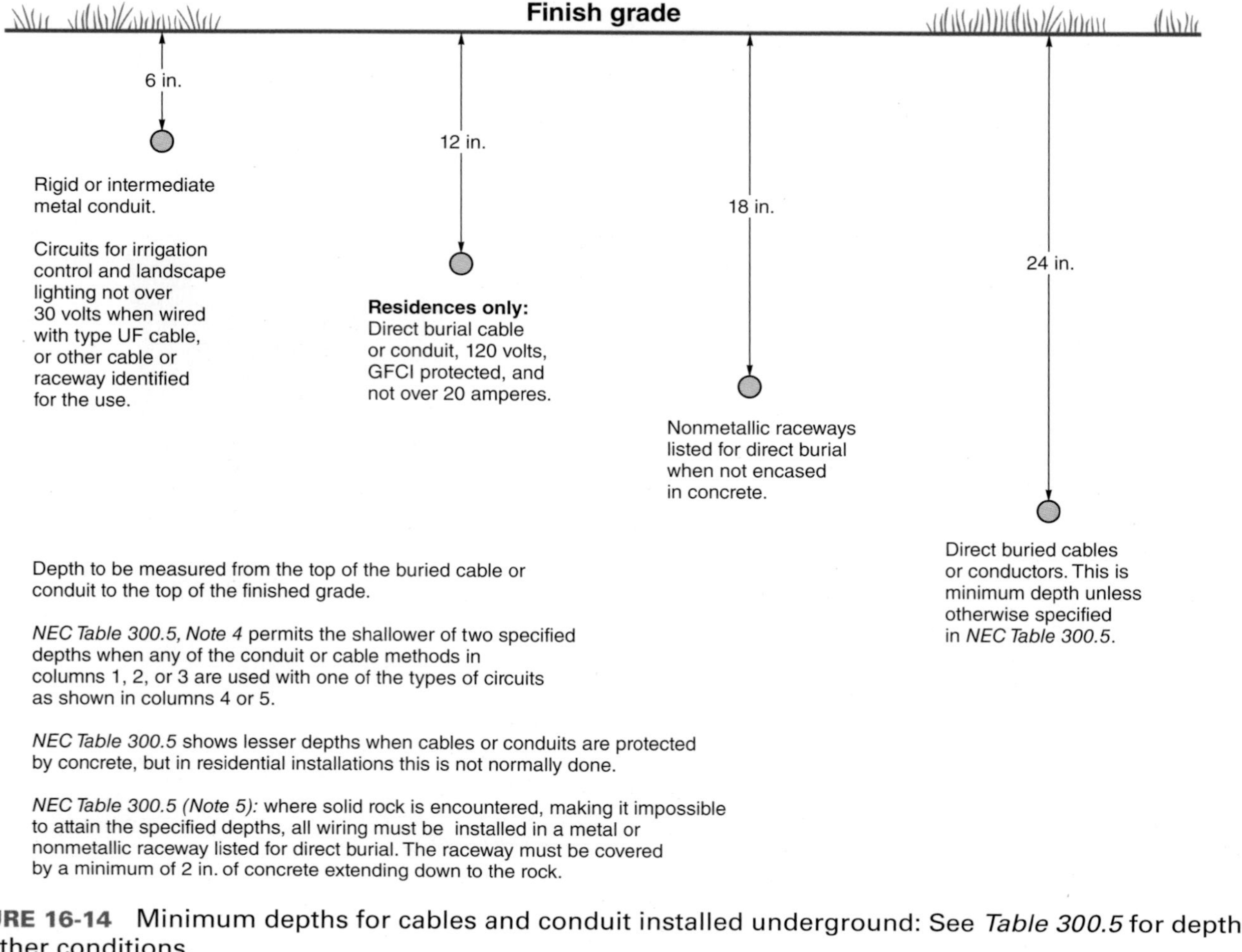

FIGURE 16-14 Minimum depths for cables and conduit installed underground: See *Table 300.5* for depths for other conditions.

rigid metal conduit, intermediate metal conduit, rigid PVC conduit that is approved for direct burial, special considerations for residential branch circuits rated 120 volts or less having GFCI protection and overcurrent protection not over 20 amperes, and low-voltage landscape lighting supplied with Type UF cable.

Down the left-hand side of the table we find the location of the wiring methods that are shown across the top of the table.

The notes to *Table 300.5* are very important. *Note 4* allows us to select the shallower of two depths when the wiring methods in Columns 1, 2, or 3 are combined with Columns 4 or 5. For example, the basic rule for direct buried cable is that it be covered by a minimum of 24 in. (600 mm). However, this minimum depth is reduced to 12 in. (300 mm) for residential installations when the branch circuit is not over 120 volts, is GFCI protected, and the overcurrent protection is not over 20 amperes. This is for residential installations only.

The measurement for the depth requirement is from the top of the raceway or cable to the top of the finished grade, concrete, or other similar cover.

Do not back-fill a trench with rocks, debris, or similar coarse material, *300.5(F)*.

Protection from Damage

Underground direct buried conductors and cables must be protected against physical damage according to the rules set forth in *300.5(D)*.

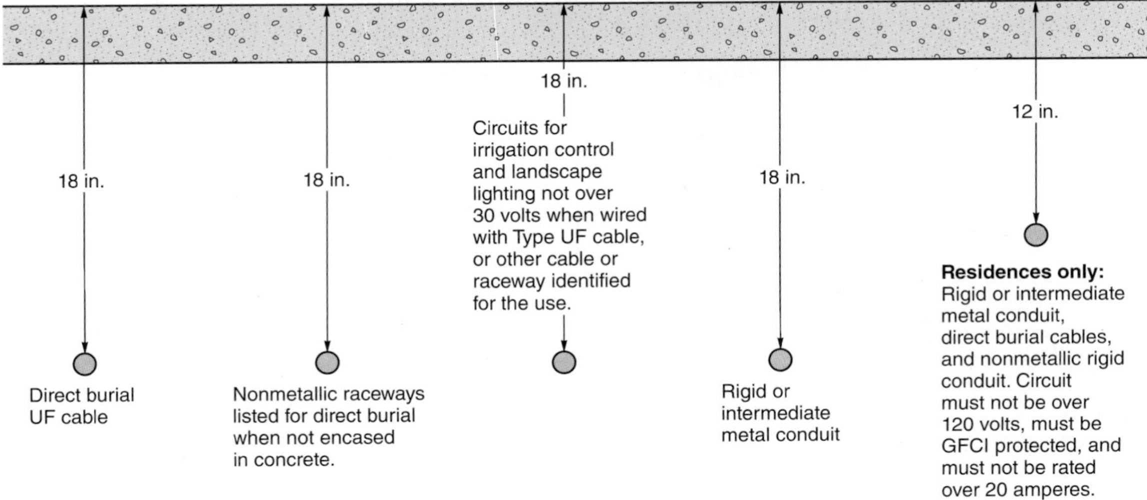

FIGURE 16-15 Depth requirements under 1- and two-family dwelling driveways and parking areas. These depths are permitted for one- and two-family residences. See *Table 300.5* for depths for other conditions.

Installation of Conduit Underground

By definition, an underground installation is a wet location. Cables and insulated conductors installed in underground raceways shall be listed for use in wet locations. Conductors suitable for use in wet locations will have the letter "W" in their type designation. For example RHW, TW, THW, THHW, THWN, and XHHW. See *300.5(B)* and *310.10(C)*.

All conduit installed underground must be protected against corrosion, *300.6(A)(3)*. The manufacturer of the conduit and UL Standards will furnish information as to whether the conduit is suitable for direct burial. Additional supplemental protection of metal conduit can be accomplished by "painting" the conduit with a nonmetallic coating. See *NEC Table 300.5*.

When metal conduit is used as the wiring method, the metal boxes, post lantern, and so forth that are properly connected to the metal raceways are considered to be grounded because the conduit serves as the equipment ground, *250.118* (Figure 16-16). Note that many electrical inspection authorities require an equipment grounding conductor of the wire type be installed inside all metal conduits that are installed underground and in conduit installed in concrete on grade. This is to address the problem of

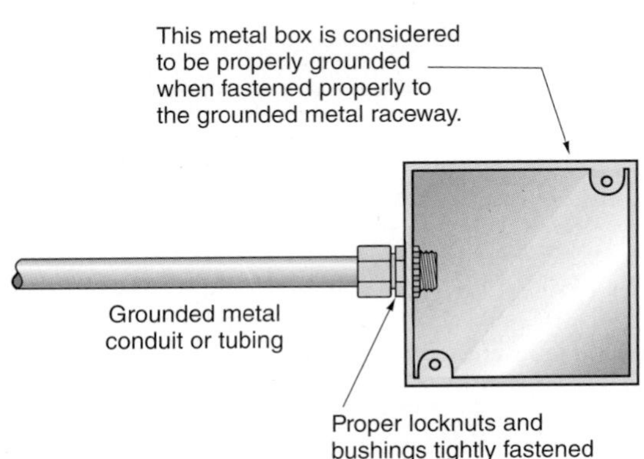

FIGURE 16-16 This illustration shows that a box is grounded when properly fastened to the grounded metal raceway. This is acceptable in *250.118*. More and more *Code*-enforcing authorities and consulting engineers are requiring that a separate equipment grounding conductor be installed in all raceways. This ensures effective grounding of the installation.

NEC TABLE 300.5

Minimum Cover Requirements, 0 to 1000 Volts, Nominal, Burial in Millimeters (Inches).

TYPE OF WIRING METHOD OR CIRCUIT

Location of Wiring Method or Circuit	Column 1 Direct Burial Cables or Conductors		Column 2 Rigid Metal Conduit or Intermediate Metal Conduit		Column 3 Nonmetallic Raceways Listed for Direct Burial Without Concrete Encasement or Other Approved Raceways		Column 4 Residential Branch Circuits Rated 120 Volts or Less with GFCI Protection and Maximum Overcurrent Protection of 20 Amperes		Column 5 Circuits for Control of Irrigation and Landscape Lighting Limited to Not More Than 30 Volts and Installed with Type UF or in Other Identified Cable or Raceway	
	mm	in.	mm	in.	mm	in.	mm	in.	mm	in.
All locations not specified below	600	24	150	6	450	18	300	12	150	6
In trench below 50 mm (2 in.) thick concrete or equivalent	450	18	150	6	300	12	150	6	150	6
Under a building	0 (in raceway or Type MC or Type MI cable identified for direct burial)	0	0	0	0	0	0 (in raceway or Type MC or Type MI cable identified for direct burial)	0	0 (in raceway or Type MC or Type MI cable identified for direct burial)	0
Under minimum of 102 mm (4 in.) thick concrete exterior slab with no vehicular traffic and the slab extending not less than 152 mm (6 in.) beyond the underground installation	450	18	100	4	100	4	150 (direct burial) 100 (in raceway)	6 4	150 (direct burial) 100 (in raceway)	6 4
Under streets, highways, roads, alleys, driveways, and parking lots	600	24	600	24	600	24	600	24	600	24
One- and two-family dwelling driveways and outdoor parking areas, and used only for dwelling-related purposes	450	18	450	18	450	18	300	12	450	18
In or under airport runways, including adjacent areas where trespassing prohibited	450	18	450	18	450	18	450	18	450	18

Notes:

1. Cover is defined as the shortest distance in millimeters (inches) measured between a point on the top surface of any direct-buried conductor, cable, conduit, or other raceway and the top surface of finished grade, concrete, or similar cover.
2. Raceways approved for burial only where concrete encased shall require concrete envelope not less than 50 mm (2 in.) thick.
3. Lesser depths shall be permitted where cables and conductors rise for terminations or splices or where access is otherwise required.
4. Where one of the wiring method types listed in Columns 1 through 3 is used for one of the circuit types in Columns 4 and 5, the shallowest depth of burial shall be permitted.
5. Where solid rock prevents compliance with the cover depths specified in this table, the wiring shall be installed in metal or nonmetallic raceway permitted for direct burial. The raceways shall be covered by a minimum of 50 mm (2 in.) of concrete extending down to rock.

Is supplemental corrosion protection required?

	IN CONCRETE ABOVE GRADE?	IN CONCRETE BELOW GRADE?	IN DIRECT CONTACT WITH SOIL?
Rigid Conduit[1]	No	No	No[2]
Intermediate Conduit[1]	No	No	No[2]
Electrical Metallic Tubing[1]	No	Yes	Yes[3]

[1]Severe corrosion can be expected where ferrous metal conduits come out of concrete and enter the soil. Some electrical inspectors and consulting engineers might specify the application of some sort of supplemental nonmetallic corrosion protection.

[2]Unless subject to severe corrosive effects. Different soils have different corrosive characteristics.

[3]In most instances, electrical metallic tubing is not permitted to be installed underground in direct contact with the soil because of corrosion problems.

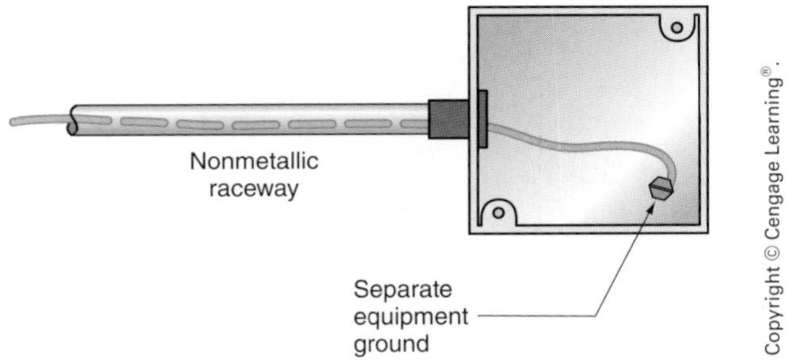

FIGURE 16-17 Grounding a metal box with a separate equipment grounding conductor.

conduit that may rust and fail due to corrosive conditions in the soil. See *Article 410, Part V*, for details on grounding requirements for luminaires.

When nonmetallic raceways are used, then a separate equipment grounding conductor, either green or bare, must be installed to accomplish the adequate grounding of the equipment served. See Figure 16-17. This equipment grounding conductor must be sized according to *Table 250.122, NEC*.

The last sentence of *250.4(A)(5)* states that The earth shall not be considered as an effective ground-fault current path.* See Figure 16-18.

According to the *Code*, any unprotected location exposed to the weather is considered a wet location.

Luminaires installed in wet locations shall be marked "Suitable for Wet Locations." These luminaires are constructed so that water cannot enter or accumulate in lampholders, wiring compartments, or other electrical parts, *410.10(A)*. Manufacturers' installation instructions may require the application of caulking to ensure protection against the entrance of water.

OVERHEAD GARAGE DOOR OPERATOR ⒺE

The receptacle for the garage overhead door operator is connected to Circuit B14. As you are about to learn, the maximum current draw for a typical overhead door operator is 5 amperes. Although not

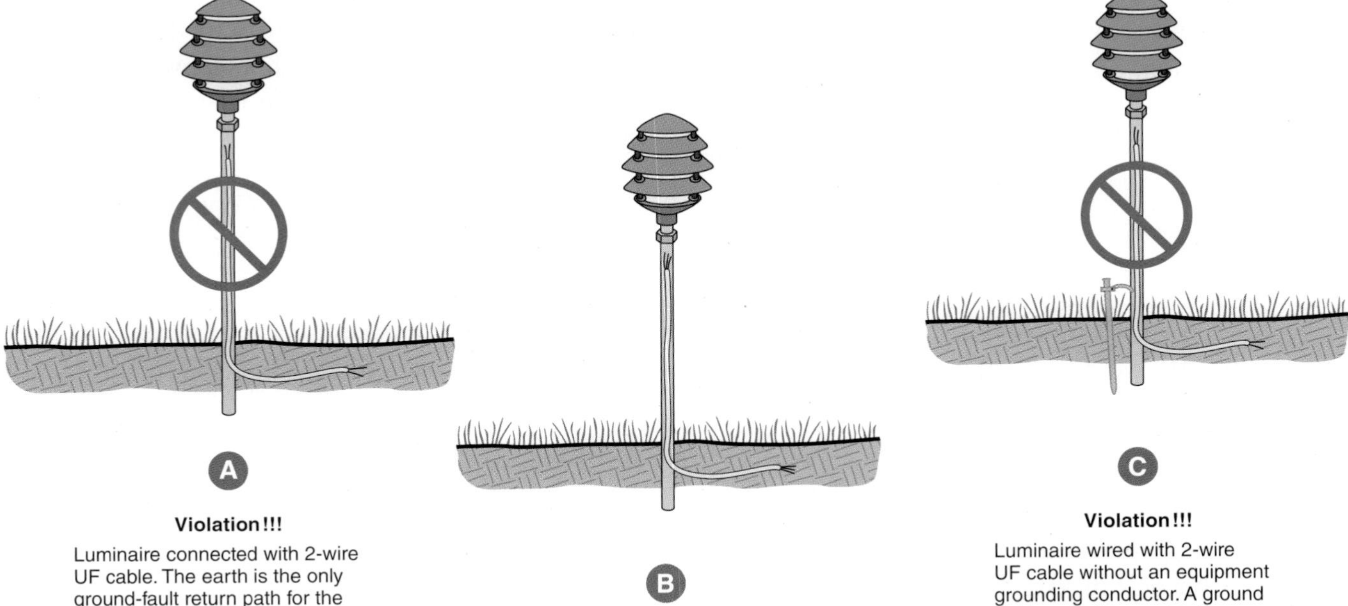

FIGURE 16-18 These drawings illustrate the intent of the last sentence of *250.4(A)(5)*, which states The earth shall not be considered as an effective ground-fault current path.*

*Reprinted with permission from NFPA 70-2014.

required by the *NEC*, some electrical inspectors insist that a separate branch circuit be run. This is an issue you will have to check out with your local electrical inspector.

Garage door operators are listed under UL Standard 325. This is the *Standard for Door, Drapery, Gate, Louver, and Window Operators and Systems.* Residential overhead garage door operators are required to have an inherent "entrapment protection system" that reverses the door in 2 seconds maximum when sensing an obstruction on its way down.

Standard 325 for residential overhead garage door operators also requires that under standard test conditions:

- The *maximum* current draw of the operator shall not exceed 5 amperes, excluding lamps or other devices.
- The *maximum* current of the motor in locked rotor condition plus the lamps and other devices such as a capacitor, microprocessor, or photoelectric sensors are not permitted to exceed 15 amperes.
- The maximum length of the power supply cord is 6 ft (1.8 m).
- The cord must have an equipment grounding conductor and have a grounding-type attachment plug cap.
- The photoelectric sensor shall be mounted not higher than 6 in. (150 mm) above the garage floor.
- Always follow the manufacturer's installation instructions.

Typical residential operators are advertised as ⅓ and ½ horsepower at 120 volts. The larger the horsepower rating, the more "lift" power. The manufacturer is permitted to show horsepower on the carton but *not* on the nameplate. Because the horsepower rating is a mechanical rating, the manufacturers' advertised horsepower rating cannot simply be converted to full-load amperes by referring to *NEC Table 430.248* as we do for conventional electric motors. The ampere rating marked on the nameplate is what counts. This nameplate ampere marking is based on standard test conditions and is the sum of the motor running current, the current draw of the lightbulbs, and any other small current draw of additional devices.

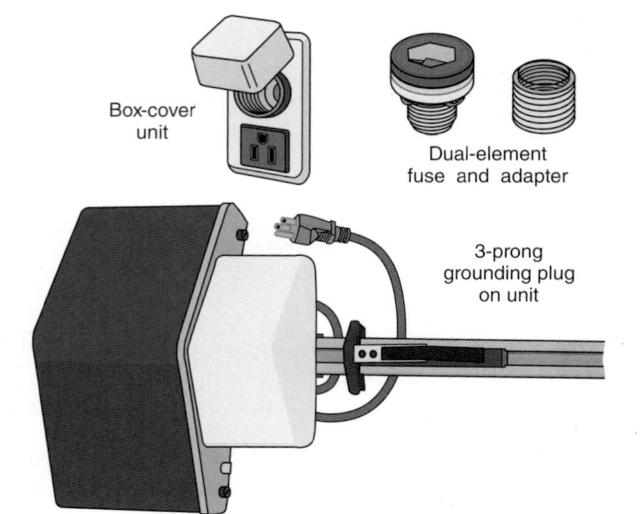

FIGURE 16-19 Overhead door operator with 3-wire cord connector.

The overhead door operator in this residence is marked 5.8 amperes, 120 volts, and is cord-and-plug connected as shown in Figure 16-19. The carton the unit came in was marked ⅓ horsepower. Overload protection for the motor is an integral part of the motor.

Where to Locate the Receptacle

For a typical residential overhead door operator, a receptacle is roughed-in on the ceiling approximately 3 ft (0.9144 m) greater than the height of the garage door header from the floor, in line with the center of the header. The approximate distance from the header to the receptacle would be

$$7\text{ ft} + 3\text{ ft} = 10\text{ ft}$$
$$(2.1336\text{ m} + 0.9144\text{ m} = 3.048\text{ m})$$
$$8\text{ ft} + 3\text{ ft} = 11\text{ ft}$$
$$(2.4384\text{ m} + 0.9144\text{ m} = 3.3528\text{ m})$$
$$9\text{ ft} + 3\text{ ft} = 12\text{ ft})$$
$$(2.7432\text{ m} + 0.9144\text{ m} = 3.6576\text{ m})$$
$$10\text{ ft} + 3\text{ ft} = 13\text{ ft}$$
$$(3.048\text{ m} + 0.9144\text{ m} = 3.9624\text{ m})$$

The cord supplied with the operator is long enough to reach this receptacle.

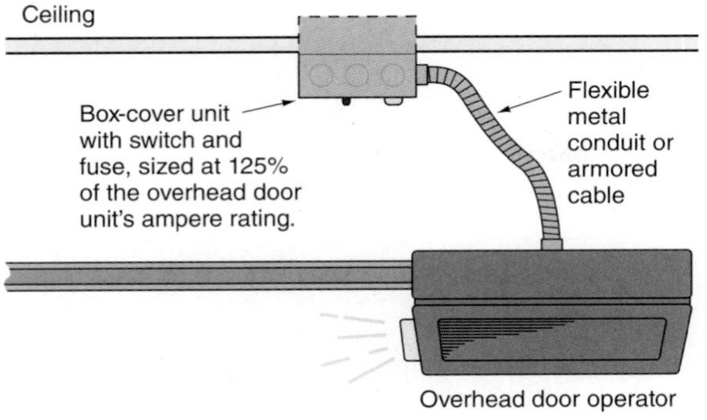

FIGURE 16-20 Permanent wiring for overhead door operator.

This receptacle is required to be GFCI protected. See *210.8(A)(2)*.

Although not commonly done, most residential overhead door operators can be "hard-wired" as shown in Figure 16-20.

Do Lamps in Overhead Door Opener Burn Out Often?

If you are experiencing too much lamp burnout in your garage door opener, the culprit is probably vibration. Garage door opener lamps are available. These are rough service bulbs. They have a more rugged filament than standard lamps and can withstand vibration for a longer period of time. They can also be used in portable work lights. Premature lamp burnout is often found where home laundry equipment is located above a luminaire, such as a washer and dryer located on the second floor.

Push Buttons

Push buttons are generally installed next to each door leading into the garage. Key-operated weatherproof switches or radio-transmitter type controls can also be installed outside next to the overhead door. The wiring between the overhead door operator and the push buttons is Class 2 wiring, which is similar to that used for chime and thermostat wiring. See Figure 16-21.

Push buttons should be installed at least 5 ft (1.5 m) or higher above the garage floor so as to be out of reach of small children.

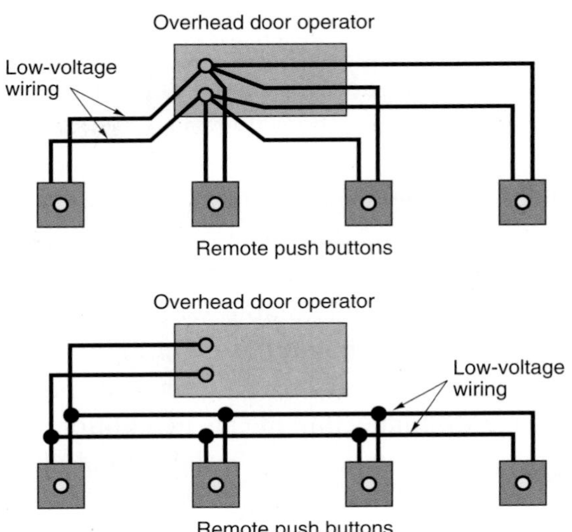

FIGURE 16-21 Push-button wiring for overhead door operator.

Quite often, electricians will install the low-voltage, Class 2 wiring during the rough-in stage so as to conceal this wiring between the push-button locations and the overhead door operator. A neat way to do this is to nail a single-gang, 4 in. square plaster ring to the stud at the desired push-button locations. When the house is "trimmed out," a single-gang faceplate with the push button is installed. See Figure 16-22.

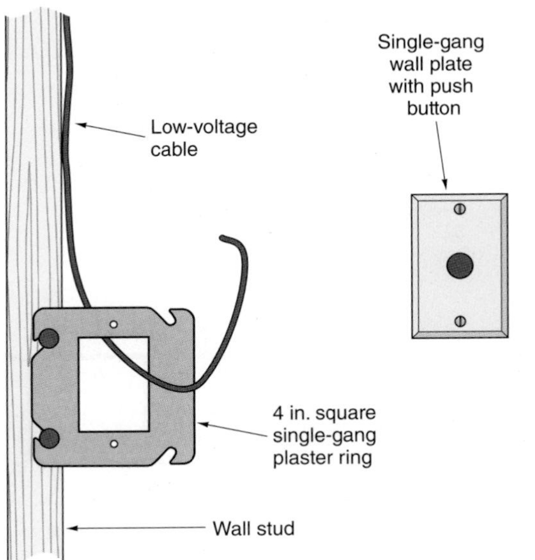

FIGURE 16-22 A sketch of how a 4 in. square cover can be used to provide an opening for low-voltage wiring.

Box-Cover Unit

The overhead door operator unit has overload protection built into it according to UL Standard 325. Additional short-circuit "backup" protection can be installed in the field. Here's how.

Instead of a standard receptacle on the ceiling near the operator, a box-cover unit can be installed. Box-cover units are available with *switch/fuse* or *receptacle/fuse* combinations, as in Figures 16-19, 16-20, 16-23, and 16-24. Box-cover units provide a convenient disconnecting means in addition to having a fuse holder in which to install a dual-element, time-delay Type S fuse.

Size the time-delay fuse in the range of *approximately* 125%–150% of the overhead door opener's marked ampere rating.

For example, the overhead door operator in this residence is marked 5.8 amperes. Then,

$$5.8 \times 1.25 = 7.25 \text{ amperes}$$
$$5.8 \times 1.5 = 8.7 \text{ amperes}$$

Therefore, installing an 8-ampere, time-delay, dual-element Type S fuse in the box-cover unit would be appropriate.

The objective is to have a lower ampere rating fuse than the 15- or 20-ampere branch-circuit breaker. Should a short circuit occur within the overhead door operator, the lower ampere rated fuse will open faster than the 15- or 20-ampere branch-circuit breaker. This takes the overhead door operator off the branch circuit. The power to the rest of the branch circuit stays on.

Courtesy of Eaton's Bussmann Business.

FIGURE 16-23 Box-cover unit that provides disconnecting means plus individual overcurrent protection.

NEC 210.8(A)(2) requires that all receptacles in a dwelling garage be GFCI protected. If you choose to install a receptacle switch box-cover unit on the ceiling for the overhead door operator(s), make sure that you lay out the garage circuitry so GFCI protection is provided for the receptacle/switch box-cover unit.

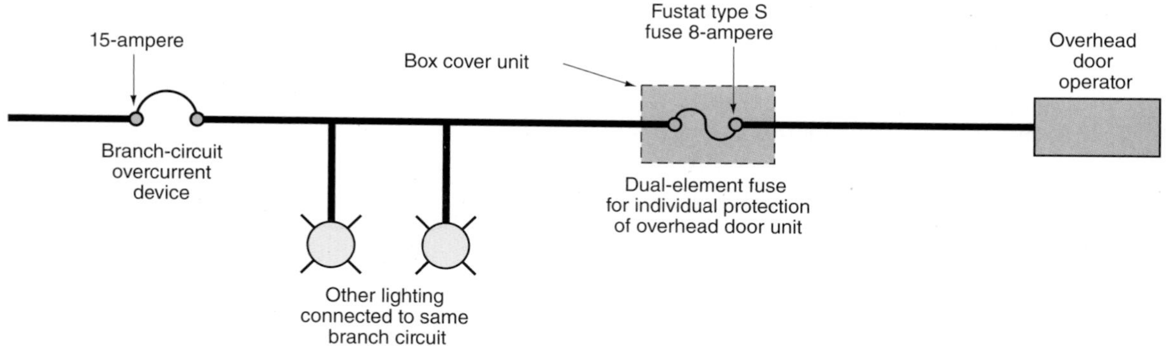

FIGURE 16-24 This diagram shows the benefits of installing a small-ampere-rated, dual-element, time-delay fuse that is sized to provide backup overload protection for the overhead door operator. If an electrical problem occurs at the overhead door operator, only the 8-ampere fuse will open. The 15-ampere branch circuit is not affected. All other loads connected to the 15-ampere circuit remain energized.

REVIEW

Note: Refer to the *Code* or plans where necessary.

1. What circuits supply the garage? ______________________________

2. What are the circuit ratings? ______________________________

3. How many GFCI-protected receptacle outlets are connected to the garage circuit? ______________________________

4. a. How many cables enter the wall switch box located next to the side garage door? ______________________________

   b. How many circuit conductors enter this box? ______________________________

   c. How many equipment grounding conductors enter this box? ______________________________

5. Show calculations on how to select a proper size box for problem 4. What kind of box would you use? ______________________________

6. a. How many lights are recommended for a one-car garage? __________ For a two-car garage? __________ For a three-car garage? __________

   b. Where are these lights to be located? ______________________________

7. From how many points in the garage of this residence are the ceiling lights controlled? ______________________________

8. GFCI breakers or GFCI receptacles are relatively expensive. How would *you* arrange the wiring of the garage circuit B14 to make as economical an installation as possible that complies with the *Code*? ______________________________

9. The total estimated volt-ampere load of the garage lighting branch circuit B14 draws how many amperes? Show your calculations. ______________________________

10. How high from the floor are the switches and receptacles to be mounted? ______

11. What type of cable feeds the post light? ______

12. All raceways, cables, and direct burial-type conductors require a "cover." Explain the term "cover." ______

13. In the spaces provided, fill in the cover (depth) for the following residential underground installations. See *NEC Table 300.5*.
    a. Type UF. No other protection. ______ in.
    b. Type UF below driveway. ______ in.
    c. Rigid metal conduit under lawn. ______ in.
    d. Rigid metal conduit under driveway. ______ in.
    e. Electrical metallic tubing between house and detached garage. ______ in.
    f. UF cable under lawn. Circuit is 120 volts, 20 amperes, GFCI protected. ______ in.
    g. Rigid PVC conduit approved for direct burial. No other protection. ______ in.
    h. Rigid PVC conduit. Circuit is 120 volts, 20 amperes, GFCI protected. ______ in.
    i. Rigid conduit passing over solid rock and covered by 2 in. (50 mm) of concrete extending down to rock. ______ in.

14. What section of the *Code* prohibits embedding Type UF cable in concrete? ______

15. Using the suggested cable layout of the garage lighting circuit B14, make a complete wiring diagram of this circuit, using colored pens or pencils to indicate conductor insulation colors. The receptacles in the garage are not shown in this diagram because they are connected to circuit B23.

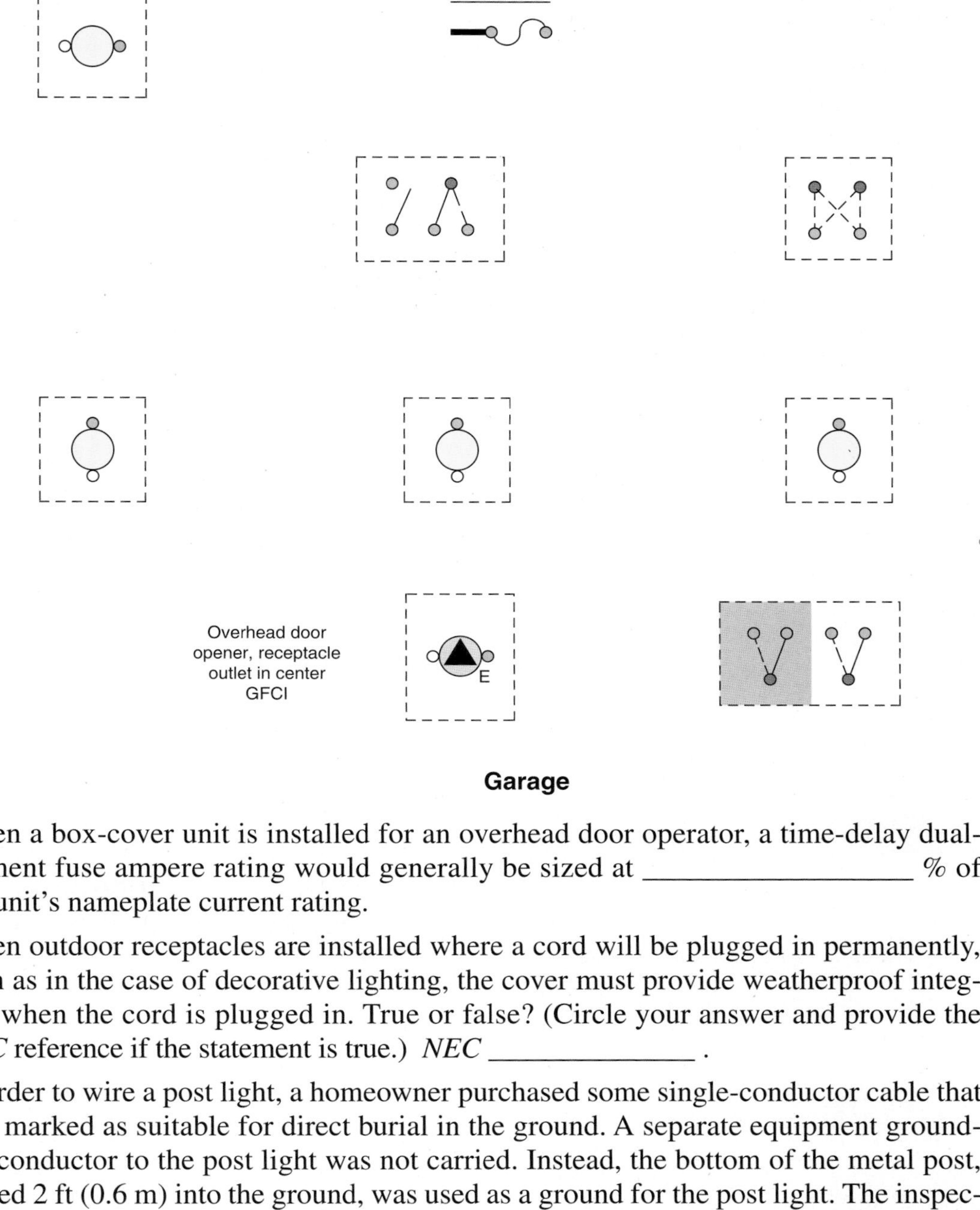

16. When a box-cover unit is installed for an overhead door operator, a time-delay dual-element fuse ampere rating would generally be sized at ________________ % of the unit's nameplate current rating.

17. When outdoor receptacles are installed where a cord will be plugged in permanently, such as in the case of decorative lighting, the cover must provide weatherproof integrity when the cord is plugged in. True or false? (Circle your answer and provide the *NEC* reference if the statement is true.) *NEC* ______________ .

18. In order to wire a post light, a homeowner purchased some single-conductor cable that was marked as suitable for direct burial in the ground. A separate equipment grounding conductor to the post light was not carried. Instead, the bottom of the metal post, buried 2 ft (0.6 m) into the ground, was used as a ground for the post light. The inspector "red tagged" (turned down) the installation. Who is right—the homeowner or the electrical inspector? Explain. ________________

____

____

19. What concerns are there when installing a low-voltage lighting system?

____

____

____

CHAPTER 17

Recreation Room

OBJECTIVES

After studying this chapter, you should be able to

- understand 3-wire (multiwire) branch circuits.
- understand how to install lay-in luminaires.
- calculate watts loss and voltage drop in 2-wire and 3-wire circuits.
- understand the term *fixture whips*.
- understand the advantages of installing multiwire branch circuits.
- understand problems that can be encountered on multiwire branch circuits as a result of open neutrals.

RECREATION ROOM LIGHTING (B9, 11, 12)

The Recreation Room is well lit through the use of six lay-in 2 ft × 4 ft fluorescent luminaires. These fixtures are exactly the same size as two 2 ft × 2 ft ceiling tiles. The luminaires rest on top of the ceiling tee bars, as in Figure 17-1.

Junction boxes are mounted above a dropped (suspended) ceiling usually within 12 to 24 in. (300 to 600 mm) of the intended luminaire location. A flexible conduit, referred to as a *fixture whip*, 4 to 6 ft (1.2 to 1.8 m) long, is installed between the junction box and the luminaire. These fixture whips contain the correct type and size of conductor suitable for the temperature ratings and load required by the *Code* for recessed luminaires. See Chapter 7 for a complete explanation of *Code* regulations for the installation of recessed luminaires, as in Figure 17-2.

Important: To prevent the luminaire from inadvertently falling, *410.36(B)* of the *Code* requires that (1) suspended ceiling framing members that support recessed luminaires must be securely fastened to each other and must be securely attached to the building structure at appropriate intervals, and (2) recessed luminaires must be securely fastened to the suspended ceiling framing members by bolts, screws, rivets, or special listed clips provided by the manufacturer of the luminaire for the purpose of attaching the luminaire to the framing member.

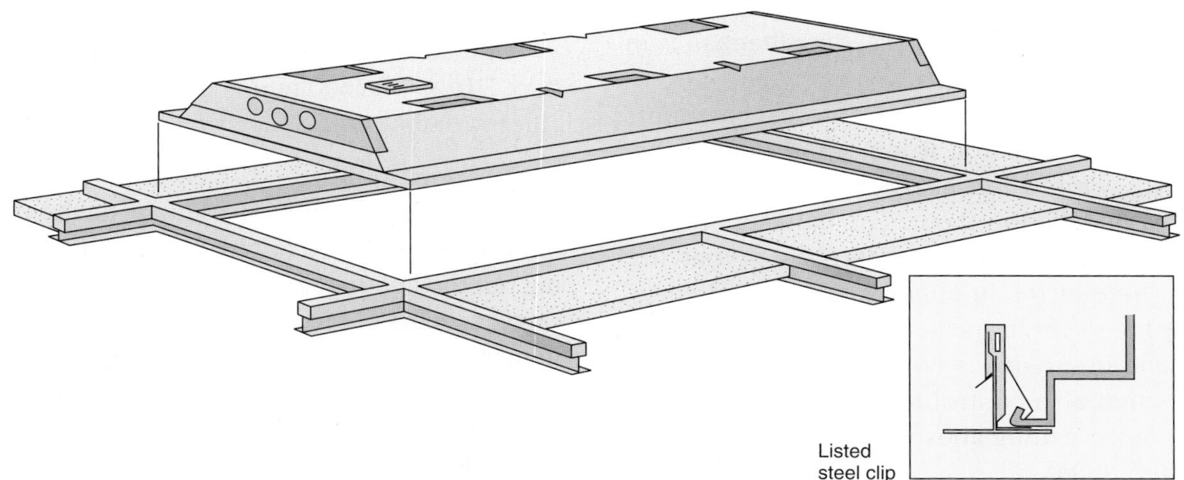

FIGURE 17-1 Typical lay-in fluorescent luminaire commonly used in conjunction with dropped suspended ceilings. See *410.36(B)*.

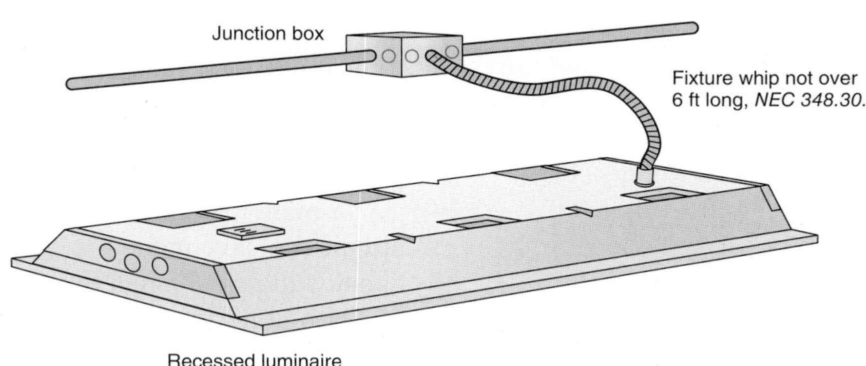

FIGURE 17-2 Typical suspended ceiling "lay-in" fluorescent luminaire showing flexible fixture whip connection. Conductors in whip must have temperature rating as required on label in luminaire; for example, "For Supply Use 90°C conductors."

Fluorescent luminaires of the type shown in Figure 17-1 might bear a label stating "Recessed Fluorescent Luminaire." The label might also state "Suitable for Use in Suspended Ceilings." Although they might look the same, there is a difference.

Recessed fluorescent luminaires are intended for installation in cavities in ceilings and walls and are to be wired according to *Article 410, Part XI*. These luminaires may also be installed in suspended ceilings if they have the necessary mounting hardware.

If marked "Suitable for Use in Suspended Ceilings," fluorescent luminaires are intended only for installation in suspended ceilings where the acoustical tiles, lay-in panels, and suspended grid are not part of the actual building structure.

Underwriters Laboratories (UL) Standard 1570 covers recessed and suspended ceiling luminaires in detail.

The six recessed "lay-in" fluorescent luminaires in the Recreation Room are 4-lamp luminaires with warm white, energy-saving F32SPX30/RS lamps installed in them. The ballasts in these luminaires are high-power factor, energy-saving ballasts. Five of the luminaires are controlled by a single-pole switch located at the bottom of the stairs. One luminaire is controlled by the 3-way switches that also control the stairwell luminaire hung from the ceiling above the stair landing midway up the stairs.

This was done so that the entire Recreation Room lighting would not be on continually whenever someone came down the stairs to go into the workshop. This is a practical energy-saving feature.

Circuit B12 feeds the fluorescent luminaires in the Recreation Room. See the cable layout in Figure 17-3. Table 17-1 summarizes the outlets and estimated load for the Recreation Room.

How to Connect Recessed and Lay-In Luminaires

Figures 17-1 and 17-2 show typical lay-in fluorescent luminaires. The installation and wiring connections for recessed luminaires and lay-in suspended-ceiling luminaires are covered in Chapter 7.

Recessed Luminaires between Ceiling Joists

Roughing-in recessed luminaires that will be installed between ceiling joists can be quite a challenge, particularly if you are trying to design spacing that makes sense. The conflict is between the 16 in. (400 mm) on center joists and the 24 in. (600 mm) square acoustic ceiling panels and their metal "T" bars and supports. Multiples of 16 in. (400 mm) and 24 in. (600 mm) don't agree except at 4 ft (1.2 m) intervals.

Dropped Ceiling. If there is a dropped ceiling that provides sufficient space between the finished ceiling and the bottom edge of the ceiling joists, you probably can space recessed incandescent luminaires easily to be in the exact centers of typical 2 ft × 2 ft (600 mm × 600 mm) acoustic panels.

In the case of recessed fluorescent luminaires, their 2 ft × 4 ft (600 mm × 1.2 m) size nicely coincides with the ceiling panels.

No Dropped Ceiling. If there is no drop, or if the drop is not as much as the height of the recessed luminaires, you have to give a lot of thought to laying out the lighting. You will find that the center-to-center spacing of recessed luminaires will have to be in increments of 4 ft (1.2 m). This spacing will find space between the ceiling joists that are 16 inches on center. With care, the recessed luminaires can be located to be exactly in the center of 2 ft × 2 ft (600 mm × 600 mm) acoustic ceiling panels.

RECEPTACLES AND WET BAR (B9–11)

The circuitry for the Recreation Room wall receptacles and the wet bar lighting area introduces a new type of circuit. This is termed a *multiwire branch circuit* or a *3-wire branch circuit*.

In many cases, the use of a multiwire branch circuit can save money in that one 3-wire branch circuit will do the job of two 2-wire branch circuits.

Also, if the loads are nearly balanced in a 3-wire branch circuit, the neutral conductor carries only the unbalanced current. This results in less voltage drop and watts loss for a 3-wire branch circuit as compared to similar loads connected to separate 2-wire branch circuits.

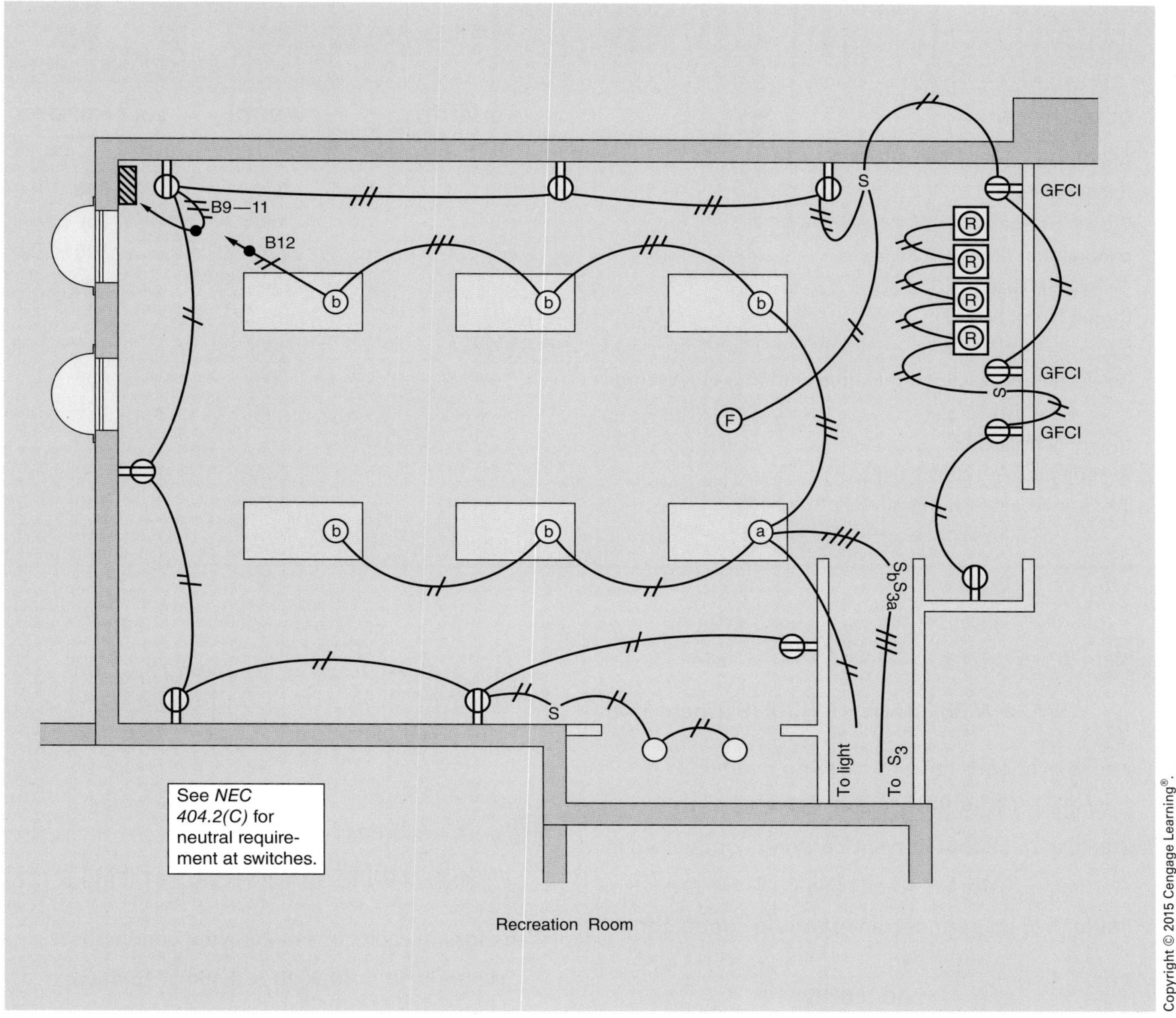

FIGURE 17-3 Cable layout for Recreation Room. The six fluorescent luminaires are connected to Circuit B12. The receptacles located around the room are connected to Circuit B11. The four recessed luminaires over the bar, the three receptacles located on the bar wall, and the receptacles located to the right as you come down the steps are connected to Circuit B9. Circuit B9 and Circuit B11 together make up a 3-wire circuit. Junction boxes are located above the dropped ceiling and to one side of each fluorescent luminaire. The electrical connections are made in these junction boxes. A flexible fixture whip connects between each junction box and each luminaire. See Figure 17-2.

Figures 17-4 and 17-5 illustrate the benefits of a multiwire branch circuit relative to watts loss and voltage drop. The example, for simplicity, is a purely resistive circuit, and loads are exactly equal.

The distance from the load to the source is 50 ft. The conductor size is 14 AWG solid, uncoated copper. From *Table 8* in *Chapter 9* in the *NEC®*, we find that the resistance of 1000 ft (300 m) of a 14 AWG uncoated copper conductor is 3.07 ohms (10.1 ohms/km). The resistance of 50 ft (15 m) is

$$\frac{3.07}{1000} \times 50 = 0.1535 \text{ ohms (round off to 0.154)}$$

TABLE 17-1

Recreation Room outlets and estimated load (three circuits).

DESCRIPTION	QUANTITY	WATTS	VOLT-AMPERES
Circuit Number B11			
Receptacles @ 120 W each	7	840	840
Closet luminaires @ 75 W each	2	150	150
Exhaust fan (2.5 A x 120V)	1	270	300
TOTALS	10	1260	1290
Circuit Number B12			
Luminaire on stairway	1	100	100
Recessed fluorescent luminaires with four 32-W lamps each	6	768	806
TOTALS	7	868	906
Circuit Number B9			
Receptacles @ 120 W each	4	480	480
Wet bar recessed luminaires @ 75 W each	4	300	300
TOTALS	8	780	780

EXAMPLE

TWO 2-WIRE BRANCH CIRCUITS (Figure 17-4)

Watts loss in each current-carrying conductor is

$$\text{Watts} = I^2R = 10 \times 10 \times 0.154 = 15.4 \text{ watts}$$

Watts loss in all four current-carrying conductors is

$$15.4 \times 4 = 61.6 \text{ watts}$$

Voltage drop in each current-carrying conductor is

$$E_d = IR$$
$$= 10 \times 0.154$$
$$= 1.54 \text{ volts}$$

E_d for both current-carrying conductors in each circuit:

$$2 \times 1.54 = 3.08 \text{ volts}$$

Voltage available at load:

$$120 - 3.08 = 116.92 \text{ volts}$$

Note in Figure 17-4 that the two 2-wire branch circuits result in four current-carrying conductors. If the wiring method is a raceway or cable, the ampacity of the conductors must be derated because there are more than three current-carrying conductors in the raceway. Adjustment and correction factors are discussed in Chapter 18 of this text. Refer to *NEC Table 310.15(B)(3)(c)*.

EXAMPLE

ONE 3-WIRE BRANCH CIRCUIT (Figure 17-5)

Watts loss in each current-carrying conductor is

$$\text{Watts} = I^2R = 10 \times 10 \times 0.154 = 15.4 \text{ watts}$$

Watts loss in both current-carrying conductors is

$$15.4 \times 2 = 30.8 \text{ watts}$$

Voltage drop in each current-carrying conductor is

$$E_d = IR$$
$$= 10 \times 0.154$$
$$= 1.54 \text{ volts}$$

E_d for both current-carrying conductors is

$$2 \times 1.54 = 3.08 \text{ volts}$$

Voltage available at loads:

$$240 - 3.08 = 236.92 \text{ volts}$$

Voltage available at each load:

$$\frac{236.92}{2} = 118.46 \text{ volts}$$

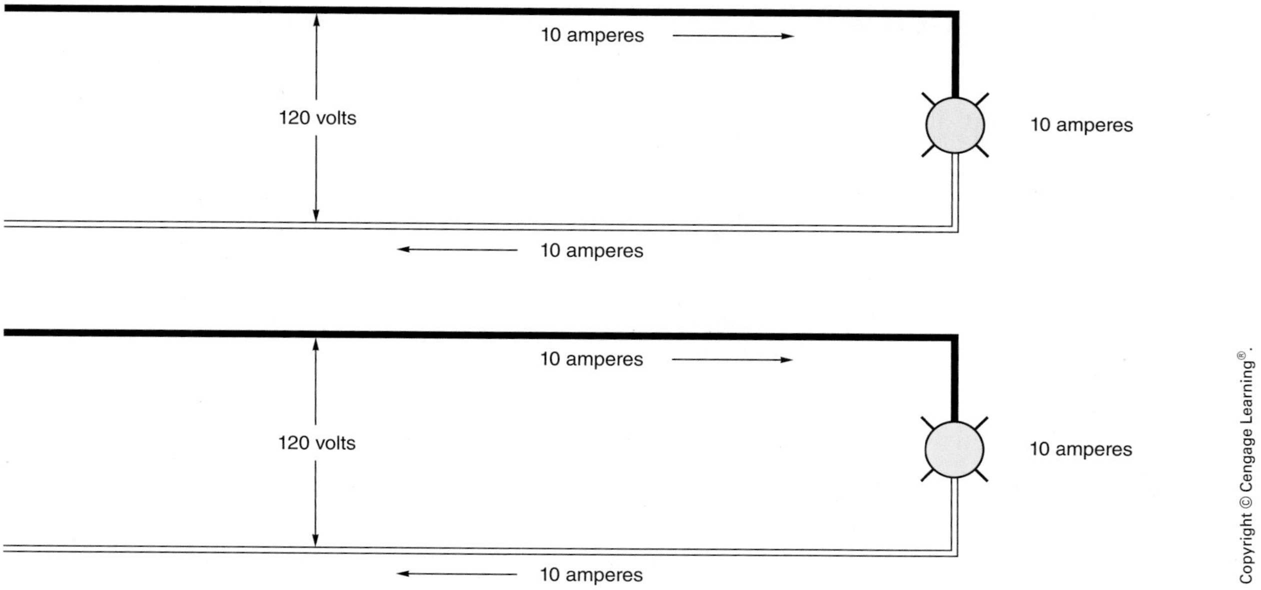

10 amperes
120 volts
10 amperes
10 amperes
10 amperes
120 volts
10 amperes
10 amperes
Copyright © Cengage Learning®.

FIGURE 17-4 The 2-wire branch circuits.

10 amperes
120 volts
10 amperes
240 volts
0 amperes
120 volts
10 amperes
10 amperes
Copyright © Cengage Learning®.

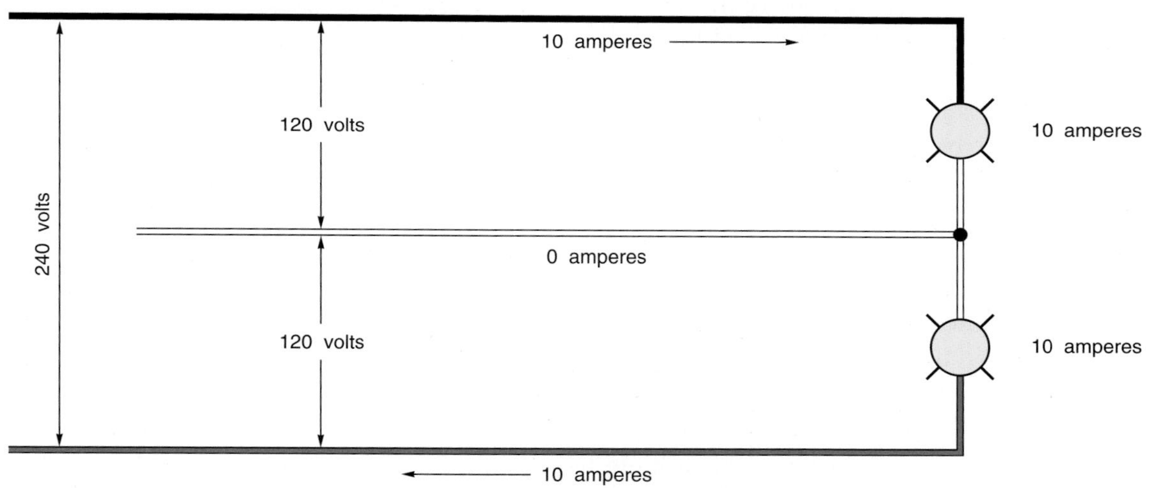

FIGURE 17-5 A 3-wire branch circuit.

Advantages of a 3-Wire Branch Circuit

1. Uses less wire for two circuits
2. There are three conductors instead of four. Might result in smaller raceway (if wiring in a raceway). Less need for correcting (derating) the conductor's ampacity due to having more than three current-carrying conductors in the raceway.
3. Less voltage drop in the conductors
4. Less watts loss in the conductors

These advantages are why there are 3-wire services and feeders, as shown in Panel B in this residence.

Disadvantages of a 3-Wire Branch Circuit

1. Possible burnout of equipment if the neutral conductor opens
2. Having to work with 240 volts present in outlet and device boxes
3. Possible unwanted power outage because *210.4(B)* requires all ungrounded conductors of a branch circuit to be disconnected simultaneously at the point the branch circuit originates
4. If you inadvertently open the neutral of a multiwire branch circuit and touch the open the conductors, you could be electrocuted. Remember the fundamental rule: line voltage appears across an open circuit.
5. Very important! When installing a 3-wire branch circuit where AFCIs and GFCIs are required, you must check the instructions furnished with these devices. Single-pole units are not designed to function on branch circuits where the neutral is shared. Double-pole units are designed to operate on shared-neutral branch circuits. This can be the key issue when choosing between 2-wire and 3-wire branch circuits.

The Recreation Room Wiring

At the beginning of this chapter, we pointed out that Circuit B12 supplies the ceiling lighting in the Recreation Room. Circuit B9 supplies the wet bar lighting and four receptacles. Circuit B11 powers the other receptacles in the Recreation Room, the ceiling exhaust fan, and the lighting in the storage closet.

Branch circuits B9, B11, and B12 present a challenge. The reason for this challenge is that these circuits are confronted with two distinct types of unique protection required by the *NEC.*

In accordance with *NEC 210.12(A)*, these branch circuits are required to have AFCI protection. AFCIs are discussed in Chapter 6.

In addition to the AFCI requirement, *NEC 210.8(A)* requires that receptacles *installed within 1.8 m (6 ft) of the outside edge of a sink be GFCI protected.* GFCIs are discussed in Chapter 6.

Depending on the availability of AFCI/GFCI products from different circuit breaker manufacturers, there are a couple of ways to "Meet *Code*" for connecting branch circuits B9, B11, and B12. As with all electrical devices, they shall be listed by a Nationally Recognized Testing Laboratory (NRTL).

Option 1:

1. For B12—install a single-pole AFCI circuit breaker.
2. For B9 and B11—install a 2-pole, independent-trip, dual-function AFCI/GFCI circuit breaker. The receptacles in the wet bar area could be conventional receptacles because GFCI protection for these receptacles is provided by the dual-function circuit breaker.

Option 2:

1. For B12—install a dual-function AFCI circuit breaker in Panel B.
2. For B9 and B11—install a 2-pole, independent-trip dual-function AFCI circuit breaker in Panel B.
3. Install GFCI receptacles where the receptacles are within 6 ft (1.8 m) of the outside edge of the wet bar sink.

Now let's take a look at the actual installation as suggested in the cable layout.

In the Recreation Room, a 3-wire cable carrying Circuit B9 and Circuit B11 feeds from Panel B to the receptacle closest to the panel. This 3-wire cable continues to each receptacle wall box along the same wall. At the third receptacle along the wall, the multiwire circuit splits, where the wet bar Circuit B9 continues on to feed into the receptacle outlet to the left of the wet bar.

The white conductor of the 3-wire cable is common to both the receptacle circuit and the wet bar circuit. The black conductor feeds the receptacles. The red conductor is spliced straight through the boxes and feeds the wet bar area.

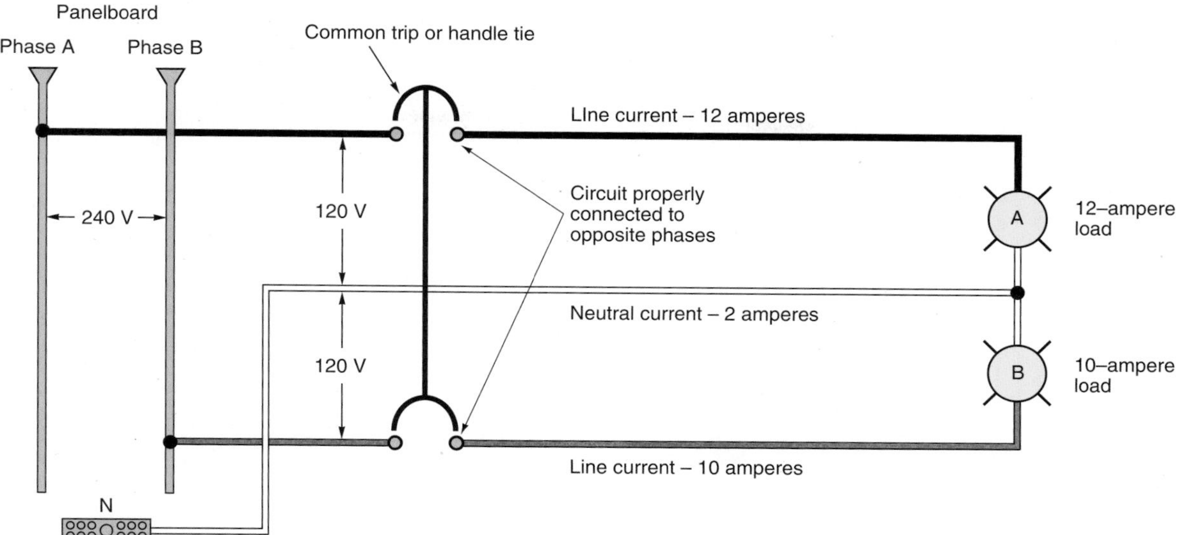

FIGURE 17-6 Correct wiring connections for a 3-wire (multiwire) branch circuit.

SAFETY ALERT

Care must be used when connecting a 3-wire circuit to the panel. The black and red conductors *must* be connected to the opposite phases in the panel to prevent heavy overloading of the neutral grounded (white) conductor. In addition, the circuit breaker must be either of the 2-pole, common trip type or have an identified tie-bar for two single-pole breakers installed. See *NEC 210.4(B)*. This is a safety issue and is intended to prevent an electric shock. A shock hazard exists when one circuit breaker of a multiwire branch circuit (shared neutral circuit) is opened and the neutral connection is broken. The neutral can be carrying current for the second circuit. You can get a severe electric shock that could result in death if your body completes the neutral circuit.

The neutral (grounded) conductor of the 3-wire cable carries the unbalanced current. This current is the difference between the current in the black wire and the current in the red wire. For example, if one load is 12 amperes and the other load is 10 amperes, the neutral current is the difference between these loads (2 amperes), Figure 17-6.

If the black and red conductors of the 3-wire cable are connected to the same phase in Panel A, Figure 17-7, the neutral conductor must carry the total current of both the red and black conductors rather than the unbalanced current. As a result, the neutral conductor will be overloaded. All single-phase, 120/240-volt panels are clearly marked to help prevent an error in phase wiring. The electrician must check all panels for the proper wiring diagrams.

Figure 17-7 shows how an improperly connected 3-wire branch circuit results in an overloaded neutral conductor.

If an open neutral occurs on a 3-wire branch circuit, some of the electrical appliances in operation may experience voltages higher than the rated voltage at the instant the neutral opens.

For example, Figure 17-8 shows that for an open neutral condition, the voltage across load A decreases and the voltage across load B increases. If the load on each circuit changes, the voltage on each circuit also changes. According to Ohm's law, the

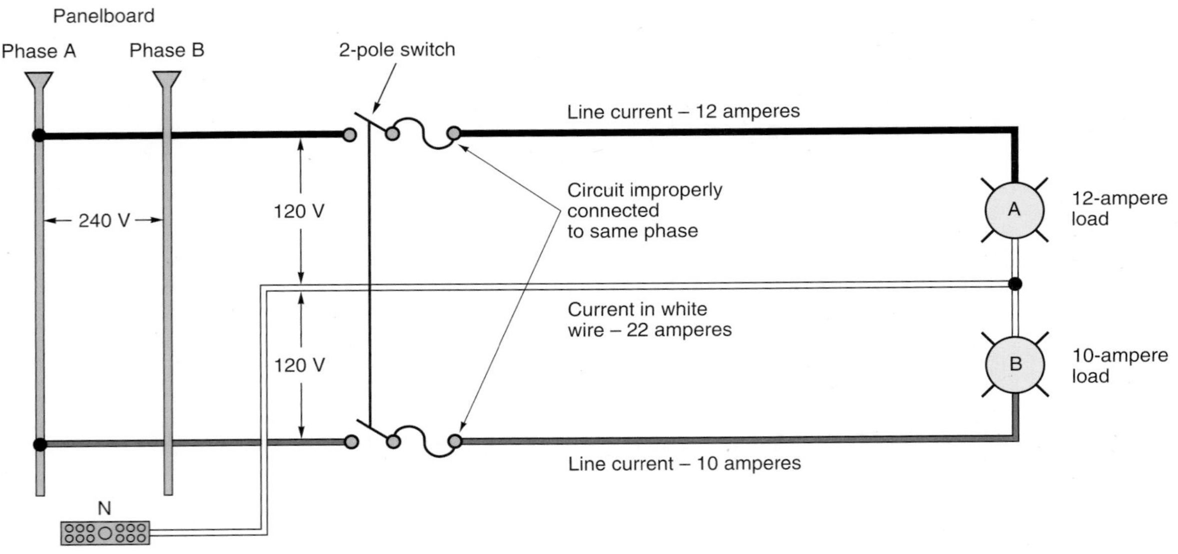

FIGURE 17-7 Improperly connected 3-wire (multiwire) branch circuit.

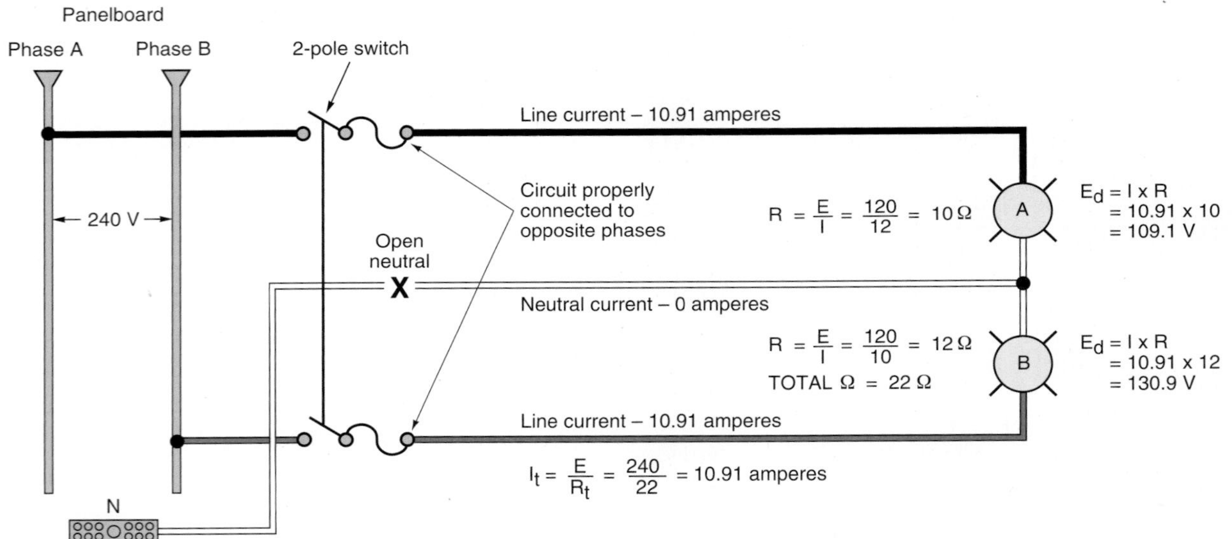

FIGURE 17-8 Example of an open neutral conductor.

voltage drop across any device in a series circuit is directly proportional to the resistance of that device. In other words, if load B has twice the resistance of load A, then load B will be subjected to twice the voltage of load A for an open neutral condition. To ensure the proper connection, care must be used when splicing the conductors.

An example of what can occur should the neutral conductor of a 3-wire multiwire branch circuit open is shown in Figure 17-9. Trace the flow of current from phase A through the television set, then through the toaster, then back to phase B, thus completing the circuit. The following simple calculations show why the television set (or stereo or home computer) can be expected to burn up.

$$R_t = 8.45 + 80 = 88.45 \text{ ohms}$$

$$I = \frac{E}{R} = \frac{240}{88.45} = 2.71 \text{ amperes}$$

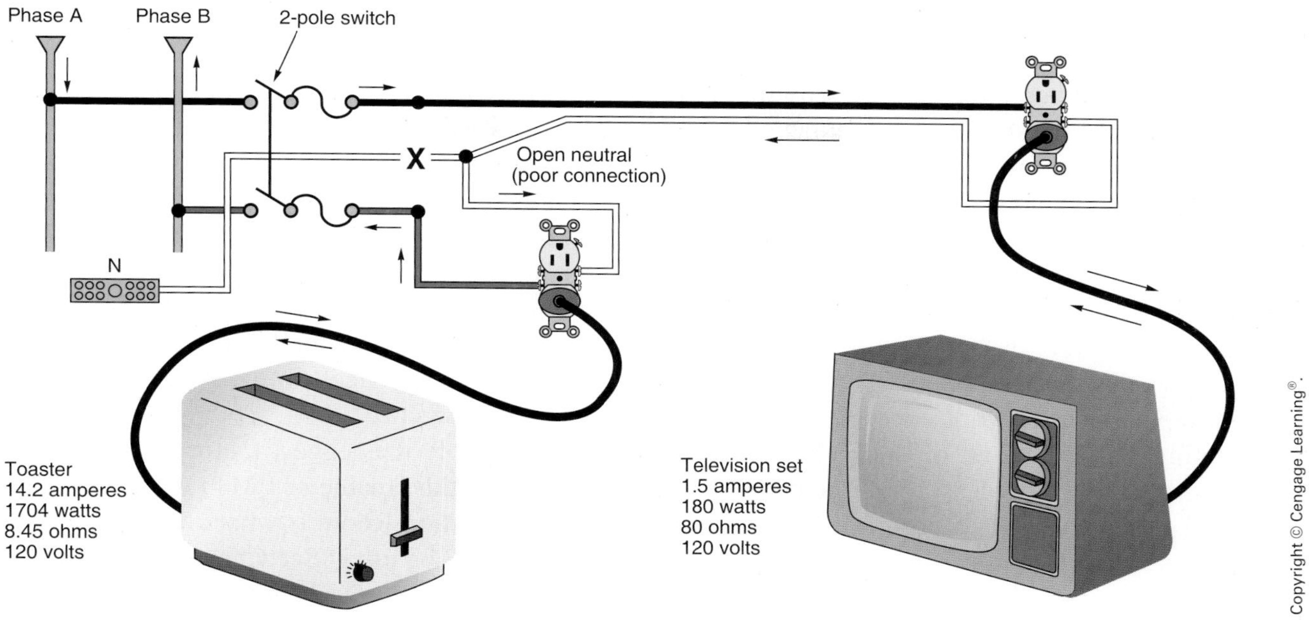

FIGURE 17-9 Problems that can occur with an open neutral on a 3-wire (multiwire) branch circuit.

Voltage appearing across the toaster:

$$IR = 2.71 \times 8.45 = 22.9 \text{ volts}$$

Voltage appearing across the television:

$$IR = 2.71 \times 80 = 216.8 \text{ volts}$$

This example illustrates the problems that can arise with an open neutral conductor on a 3-wire, 120/240-volt multiwire branch circuit.

The same problem can arise when the neutral of the utility company's incoming service-entrance conductors (underground or overhead) opens. The problem is minimized because the neutral point of the service is solidly grounded to the metal water piping system within the building. However, there are cases on record where poor service grounding has resulted in serious and expensive damage to appliances within the home because of an open neutral in the incoming service-entrance conductors.

For example, poor service grounding results when relying on a driven iron pipe (that rusts in a short period of time) as the only means of obtaining the service equipment ground. This was quite often done years ago when the electrical code was not as stringent as today's *NEC*. When working on older homes, check the service ground. It could very well be inadequate, unsafe, and in need of updating to the present *NEC*.

Keep in mind the purpose of the grounding electrode system is never to carry current for faulty circuits in the building or between the building and the utility transformer. In fact, the *NEC* states in *250.4(A)(5)* that the path through the earth is not an effective ground-fault return path.

Receptacles Near Sinks

All 125-volt, single-phase, 15- and 20-ampere receptacles within 6 ft (1.8 m) of the outside edge of a sink shall be GFCI protected. See *210.8(A)(7)*.

To meet this requirement for the receptacles located within 6 ft (1.8 m) of the wet bar sink, a feed-through GFCI receptacle could be installed at the first receptacle (upper right in Figure 17-3). With this arrangement, that receptacle and everything beyond it would be GFCI protected. The other choice would be to install a GFCI receptacle at each location as required to "Meet *Code*." Three GFCI receptacles are required to do this. As with all circuit designs, it becomes a cost issue (time and material) as well as keeping the wiring as simple as possible.

Single-pole GFCI and AFCI devices will not work properly when sharing a neutral, such as in the multiwire branch circuit for Circuits B9 and B11. A single-pole GFCI breaker in Panel B for Circuit B9 is not an option. Installing a feed-through GFCI receptacle for the first receptacle above the wet bar countertop will provide the required GFCI protection for the receptacles as well as the rest of the circuit. The confusion of how to "Meet *Code*" regarding a room that involves both GFCI and AFCI protection requirements is solved by using a 2-pole, independent-trip, AFCI/GFCI dual-function circuit breaker in the panel for Circuits B9 and B11.

Receptacles shall not be installed face up on the countertop surfaces or work surfaces. See *406.4(E)*.

A single-pole switch to the right of the wet bar controls the four recessed luminaires above it. See Chapter 7 for details pertaining to recessed luminaires.

An exhaust fan of a type similar to the one installed in the laundry is installed in the ceiling of the Recreation Room to exhaust stale, stagnant, and smoky air from the room.

The basic switch, receptacle, and luminaire connections are repetitive of most wiring situations discussed in previous chapters.

The wall boxes for the receptacles are selected based on the measurements of the furring strips on the walls. Two-by-two furring strips will require the use of 4 in. square boxes with suitable raised covers. If the walls are furred with 2 × 4s, then possibly sectional device boxes could be used. Select a box that can contain the number of conductors, devices, and clamps to "Meet *Code*." After all, the installation must be safe.

Probably the biggest factor in deciding which wiring method (cable or EMT) will be used to wire the Recreation Room is what size and type of wall furring will be used? See Chapter 4 for the in-depth discussion on the mechanical protection required for nonmetallic sheathed cables where the cables will be less than 1¼ in. (32 mm) from the edge of the framing members.

REVIEW

Note: Refer to the *Code* or plans where necessary.

1. What is the total current draw when all six fluorescent luminaires are turned on?

    ____

    ____

2. The junction box that will be installed above the dropped ceiling near the fluorescent luminaires closest to the stairway will have ____________ (number of) 14 AWG conductors.

3. Why is it important that the "hot" conductor in a 3-wire branch circuit be properly connected to opposite phases in a panel? ____________________

    ____

4. In the diagram, Load A is rated at 10 amperes, 120 volts. Load B is rated at 5 amperes, 120 volts.

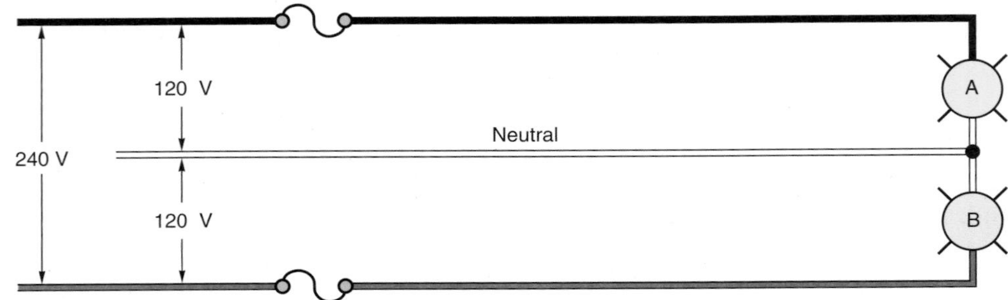

a. When connected to the 3-wire branch circuit as indicated, how much current will flow in the neutral conductor? ______

b. If the neutral conductor should open, to what voltage would each load be subjected, assuming both loads were operating at the time the neutral conductor opened? Show all calculations.

5. Calculate the watts loss and voltage drop in each conductor in the following circuit. Show calculations. Obtain resistance data from *Table 8, Chapter 9*, of the *NEC*.

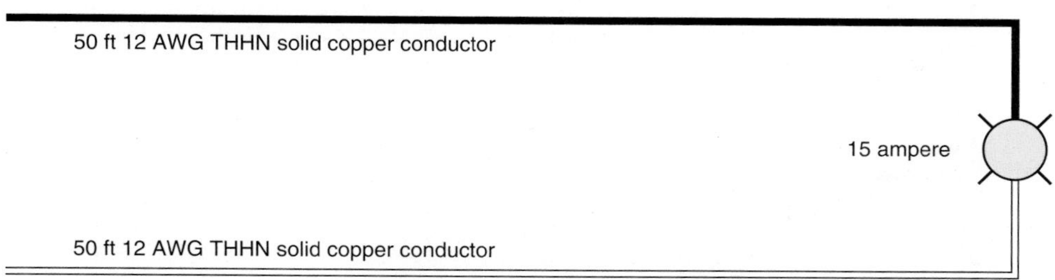

6. Unless specifically designed, all recessed incandescent luminaires must be provided with factory-installed ______ protection (see Chapter 7).

7. Where in the *NEC* would you look for the installation requirements regarding flush and recessed lighting luminaires?

______

8. What is the current draw of the recessed luminaires above the bar? ______

______

______

______

9. a. VA load for Circuit B9 is ______ volt-amperes.

   b. VA load for Circuit B11 is ______ volt-amperes.

   c. VA load for Circuit B12 is ______ volt-amperes.

10. Calculate the total current draw for Circuit B9, Circuit B11, and Circuit B12.

11. Complete the wiring diagram for the Recreation Room. Follow the suggested cable layout. Use colored pencils or pens to identify the various colors of the conductor insulation.

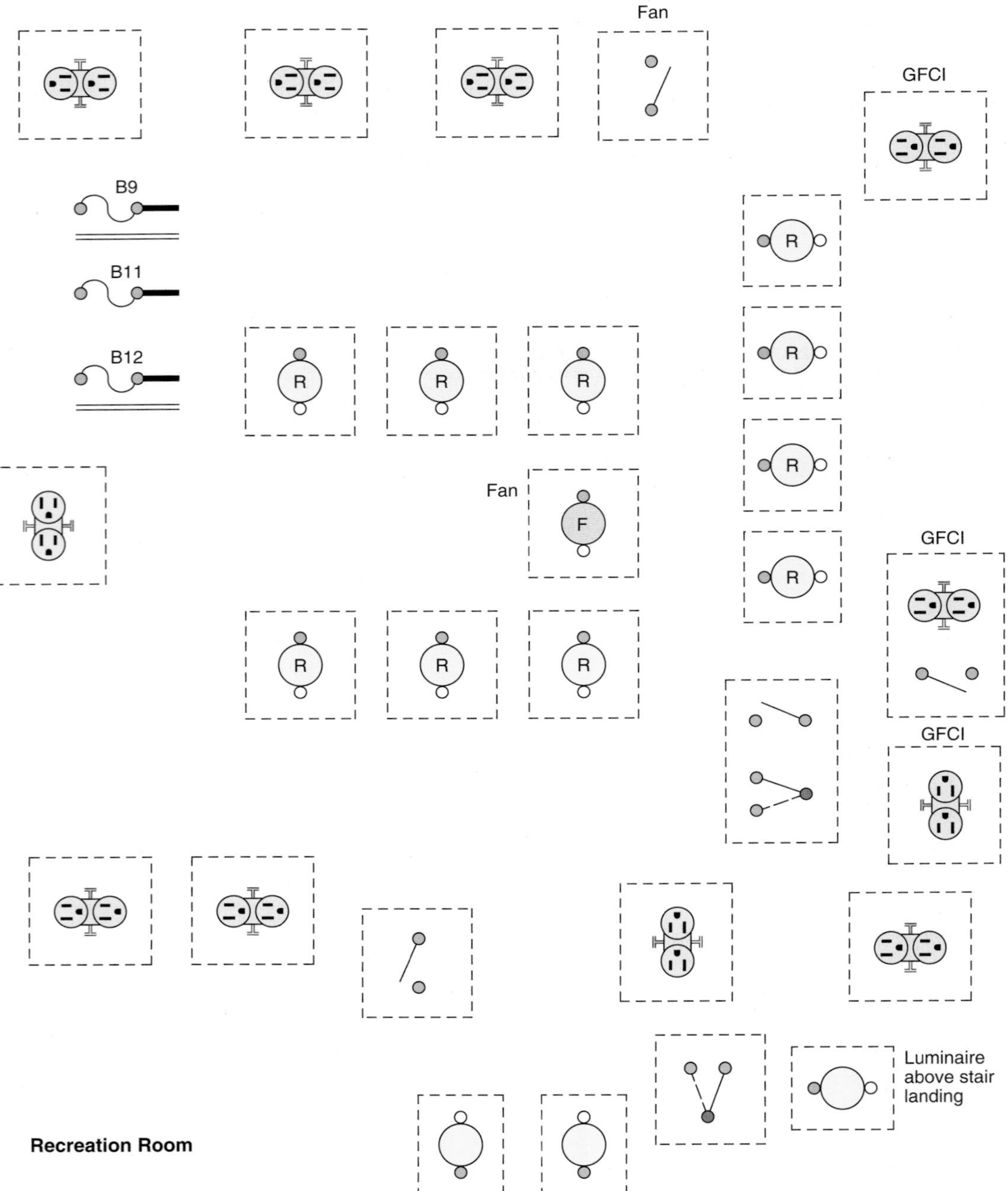

12. a. May a fluorescent luminaire that is marked "Recessed Fluorescent Luminaire" be installed in a suspended ceiling? ______________________

    b. May a fluorescent luminaire that is marked "Suspended Ceiling Fluorescent Luminaire" be installed in a recessed cavity of a ceiling? ______________________

13. The *NEC* requires GFCI protection for receptacles located within 6 ft (1.8 m) of the outside edge of a wet bar sink. This requirement is found in *NEC* ______________

14. Lay-in fluorescent luminaires that rest on the metal framing members of a suspended ceiling cannot just lay on the framing members. *NEC 410.36(B)* requires that (1) framing members used to support recessed luminaires must be securely fastened to each other and to the building structure at appropriate intervals, and (2) the luminaires must ______________________

    ____.

15. List the rooms and/or areas in a newly constructed single-family residence where arc-fault circuit-interrupter (AFCI) protection is required by the *NEC*.

    ____

    ____

    ____

CHAPTER **18**

Lighting Branch Circuit, Receptacle Circuits for Workshop

OBJECTIVES

After studying this chapter, you should be able to

- install cable and conduit wiring for a typical nonfinished basement.
- understand *NEC* location requirements for GFCI receptacles in basements.
- make conductor fill calculations for raceways.
- use derating and correction factors for determining conductor current-carrying capacity.
- understand maximum ratings of overcurrent protection for conductors.
- install multioutlet assemblies.

INTRODUCTION

The workshop area is supplied by more than one circuit:

A13 Separate branch circuit for freezer
A17 Lighting
A18 Plug-in strip (multioutlet assembly)
A20 Two receptacles on window wall

AFCI protection is not required for the unfinished basement area, *NEC® 210.12(A)*. GFCI protection is required for the receptacles located in an unfinished basement, *NEC 210.8(A)(5)*. As discussed in prior chapters, this can be accomplished by installing GFCI circuit breakers for Circuits A13, A18, and A20 in Panelboard A or by installing GFCI receptacles. However, keep in mind the GFCI devices are required to be installed in a readily accessible location. This means you have to be able to walk up to the device and check it for proper operation without using a ladder or moving objects such as a refrigerator or freezer. See *NEC 210.8.*

The wiring method in the workshop is electrical metallic tubing (EMT). Figure 18-1 shows the branch circuit layout for the workshop. An option if you prefer to use cable is to install a drop of EMT from the joist space to metal boxes that are secured to the wall. These boxes are often secured with masonry anchors. This is illustrated in Figure 18-2. This option is provided in *NEC 334.15(C)*. The installation is made by securing the box and EMT to the wall, installing a fitting or bushing at the top to protect the cable, and routing the cable to the box, leaving an adequate length to connect to the wiring device. The cable is required to be secured to within 12 in. (300 mm) of the point where the cable enters the EMT. The Type NM cable is not required to be secured to the EMT or to the box.

A single-pole switch at the entry controls all five ceiling porcelain lampholders. However, note on the plans that three of these lampholders have pull-chains, which would allow the homeowner to turn these lampholders on and off as needed (Figure 18-3). They would not have to be on all of the time, which would save energy.

In addition to supplying the lighting, Circuit A17 also feeds the chime transformer and ceiling exhaust fan. The current draw of the chime transformer is extremely small. Refer to Table 18-1. Refer to Table 18-2 for estimated loads on Circuit A13, Circuit A18, and Circuit A20. Note that these tables are used only for estimating the loading on the branch circuits. Actual load calculations for dwellings for lighting and receptacle loads used for general purposes are performed on a volt-amperes per square foot area. Detailed explanations of the load calculation are covered in Chapter 29 of this text.

WORKBENCH LIGHTING

Two 2-lamp, 32-watt fluorescent luminaires are mounted above the workbench to reduce shadows over the work area. The electrical plans show that these luminaires are controlled by a single-pole wall switch. A junction box is mounted immediately above or adjacent to the fluorescent luminaire so that the connections can be made readily.

When armored cable or flexible metal conduit is used as the wiring method between the luminaires and the junction box or boxes on the ceiling, either wiring method provides the necessary equipment grounding for the fluorescent luminaires, provided the metal outlet box is properly grounded. See *250.118* for the many acceptable means to accomplish equipment grounding.

Many workshop fluorescent luminaires are furnished with a 3-wire (black, white, green) flexible cord, suitable for the purpose, that has the attachment plug cap factory-attached to the cord. These luminaires can be plugged into a grounding-type receptacle mounted on the ceiling immediately above the fluorescent luminaires. This provides proper grounding of the luminaire.

RECEPTACLE OUTLETS

NEC 210.8(A)(4) and *210.8(A)(5)* require that all 125-volt, single-phase, 15- or 20-ampere receptacles installed in unfinished basements—such as workshop and storage areas that are not considered habitable, and crawl spaces that are at or below ground level—must be GFCI protected.

See Chapter 6 for a full discussion of GFCIs.

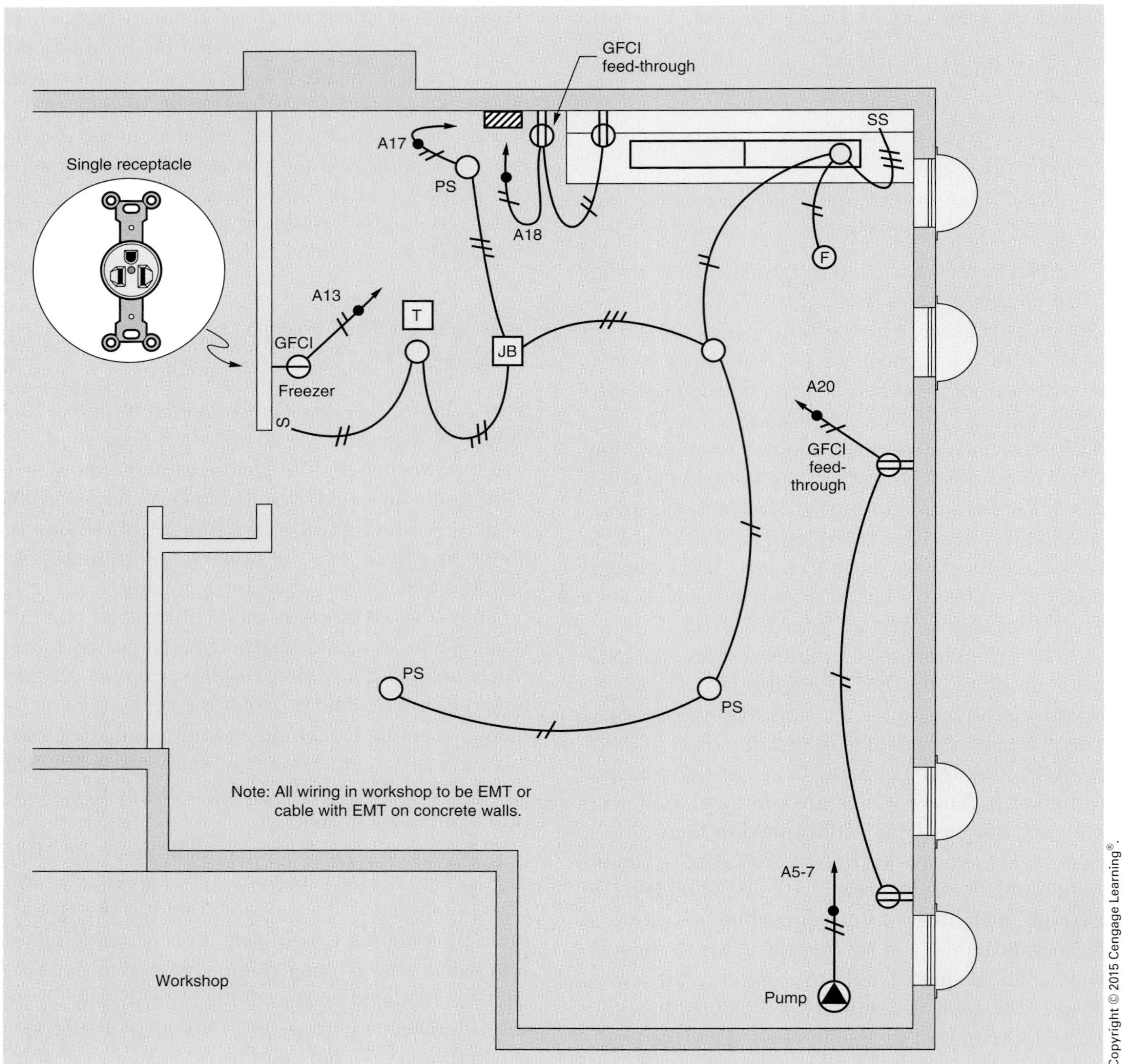

FIGURE 18-1 Conduit layout for workshop. Lighting is Circuit A17. Freezer single receptacle is Circuit A13. The two receptacles on the window wall are connected to Circuit A20. Plug-in strip receptacles and receptacle to the right of Main Panelboard are connected to Circuit A18. All receptacles are GFCI protected. A junction box mounted on the ceiling near the fluorescent luminaires above the workbench will provide a convenient place to make up the necessary electrical connections.

NEC 210.52(G) makes it mandatory to install at least one receptacle outlet in each unfinished basement area. If the laundry area is in the basement, then a receptacle outlet must *also* be installed for the laundry equipment, *210.52(F)*. This additional receptacle must be within 6 ft (1.8 m) of the intended location of the washer, *210.50(C)* and must be a 20-ampere circuit.

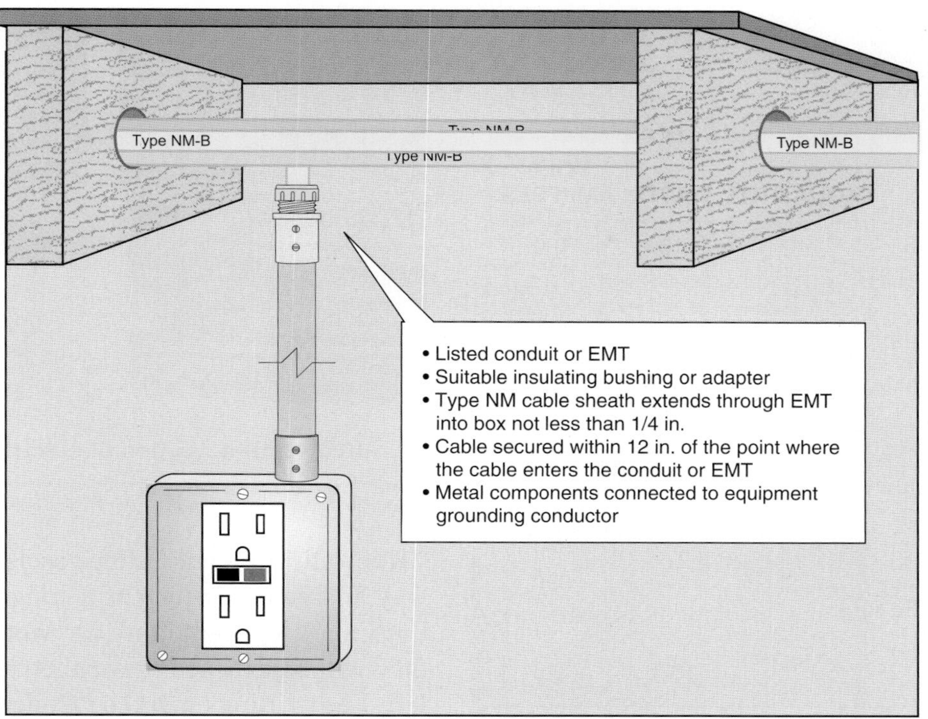

FIGURE 18-2 One method of installing Type NM cable on masonry walls in unfinished basement.

The circuit feeding the required laundry area receptacle(s) shall have no other outlets connected to it, *210.11(C)(2)*.

When a single receptacle is installed on an individual branch circuit, the receptacle rating must not be less than the rating of the branch circuit, *210.21(B)(1)*. So, if a single receptacle is installed on a 20-ampere branch circuit, as is shown for the freezer in Figure 18-1, the receptacle must have a rating not less than 20 amperes.

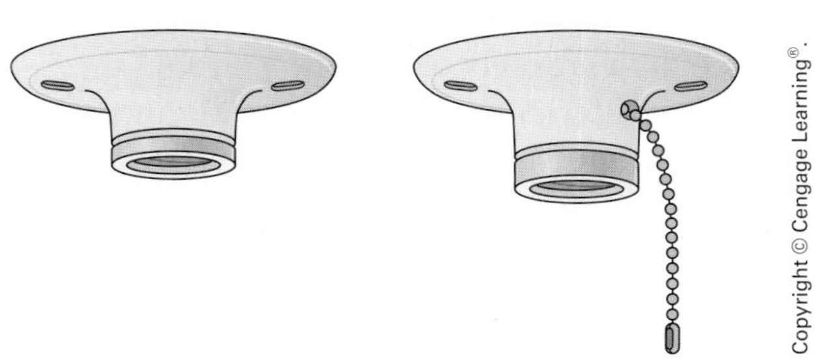

FIGURE 18-3 Porcelain and nonmetallic lampholders of both the keyless and pull-chain types. These lampholders are also available with a receptacle outlet, which must be GFCI protected when installed in locations where GFCI protection is required by the *Code*.

TABLE 18-1

Workshop: Outlet count and estimated load. Lighting load (Circuit A17).

DESCRIPTION	QUANTITY	WATTS	VOLT-AMPERES
Ceiling lights @ 100 W each	5	500	500
Fluorescent luminaires (Two 32-W lamps each)	2	128	160
Chime transformer	1	8	10
Exhaust fan	1	80	90
TOTALS	9	716	760

CABLE INSTALLATION IN BASEMENTS

Chapter 4 covers the installation of nonmetallic-sheathed cable (Type NM) and armored cable (Type AC).

Some additional rules apply when nonmetallic-sheathed cable and armored cable are run exposed, because of possible physical damage to the cables.

TABLE 18-2

Workshop: Outlet count and estimated load for three 20-A circuits A13, A18, A20.

DESCRIPTION	QUANTITY	WATTS	VOLT-AMPERES
20-A Circuit, Number A18 receptacles next to main panelboard	1	180	180
A six-receptacle, plug-in multioutlet assembly at 1½ A per outlet (180 VA)	1	1080	1080
TOTALS	2	1260	1260
20-A Circuit, Number A20 receptacles on window wall @ 180 W each	2	360	360
TOTALS	2	360	360
20-A Circuit, Number A13 single receptacle for freezer	1	—	696
TOTALS	1	1620	2316

In unfinished basements, nonmetallic-sheathed cables made up of conductors smaller than two 6 AWG or three 8 AWG must be run through holes bored in joists, or on running boards, *334.15(C)*. In unfinished basements, exposed runs of armored cable must closely follow the building surface or running boards to which they are fastened, *320.15*. *NEC 320.15* also permits armored cable (Type AC) to be run on the undersides of joists in basements, attics, or crawl spaces, where supported on each joist and so located as not to be subject to physical damage. This exception is not permitted when using nonmetallic-sheathed cable. The local inspection authority usually interprets the meaning of "subject to physical damage."

Review Chapter 4 and Chapter 15 for additional discussion regarding protection of cables installed in exposed areas of basements and attics.

CONDUIT INSTALLATION IN BASEMENTS

Where metal raceways are used in house wiring, in most cases the raceways will be EMT, as shown in Figure 18-4. EMT is also referred to in the trade as "Thinwall," which is defined as a metal tubing raceway and is not defined as a conduit. Use EMT fittings marked "Raintight" or "Wet Location" where they will be exposed to weather, rain, and/or water. Use EMT fittings marked "Concrete Tight" where used in poured concrete. The rules for EMT are found in *Article 358*.

FIGURE 18-4 Types of electrical raceways.

Rigid metal conduit (*Article 344*) is the most common conduit used for mast service; refer to Chapter 27. The following discussion only briefly touches on metal raceways, because most house wiring across the country is done with nonmetallic-sheathed cable, which is covered in detail in Chapter 4.

Local electrical codes may not permit cable wiring in unfinished basements other than where absolutely necessary. For example, a cable may be dropped into a basement from a switch at the head of the basement stairs. Transitions between cable and conduit also must be made. Usually this will be a junction box. The electrician must check the local codes before selecting conduit or cable for the installation to prevent costly wiring errors.

The electrician should never assume that the *NEC* is the recognized standard everywhere. Any city or state may pass electrical installation and licensing laws. In many cases these laws are more stringent than the *NEC*.

Outlet boxes may be fastened to masonry walls with lead or plastic anchors, shields, concrete nails, or power-actuated studs.

Conduit and EMT may be fastened to masonry or wood surfaces using straps or hangers, as illustrated in Figure 18-5. The advantage of using the

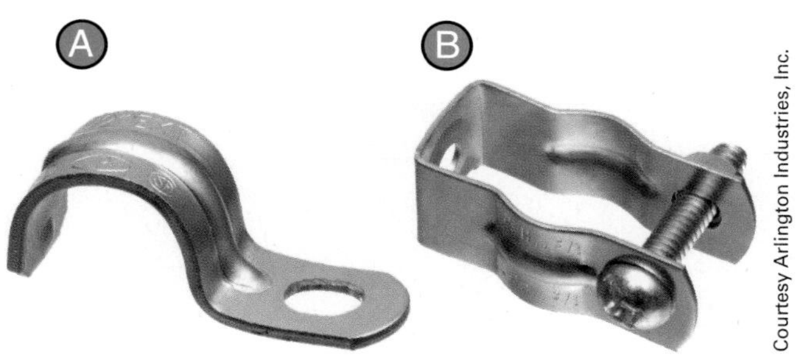

FIGURE 18-5 Methods of securing EMT on concrete wall. (A) a one-hole strap. (B) a conduit or EMT hanger.

hanger is that when fastened to the wall and with the EMT secured, the EMT will line up with the knockout in the metal box. This avoids having to bend an offset in the EMT.

The conduits on the workshop walls and on the ceiling are exposed to view. The electrician must install the exposed wiring in a neat and skillful manner while complying with the following practices:

- The conduit runs must be straight.
- Bends and offsets must be true.
- Vertical runs down the surfaces of the walls must be plumb.
- To ensure that the conductor insulation will not be injured, be sure to ream the cut ends of rigid conduit and electrical metallic tubing to remove rough edges and burrs.

All conduits and boxes on the workshop ceiling are fastened to the underside of the wood joists, as in Figure 18-6. Raceways are required to be installed as a complete system between pull points before pulling in the wires, as shown in Figure 18-7.

If the electrician can get to the residence before the basement concrete floor is poured, he might find it advantageous to install conduit runs in or under the concrete floor, from the main panelboard to the freezer outlet, the window wall receptacles, and the water pump disconnect switch location. This can be a cost-saving (labor and material) benefit.

Now might be a good time for you to review the *NEC* rules in *Article 358* for electrical metallic tubing.

Metal raceways and metal-clad cables (Type MC) are acceptable equipment grounding conductors, *250.118*. The runs would have to be continuous between other effectively grounded boxes, cabinets, and/or equipment. When a nonmetallic wiring method (NM, NMC, UF, etc.) is installed where short sections of metal raceway are provided for mechanical protection—or for appearance—these short sections of raceway need not be grounded, *250.86, Exception No. 2*.

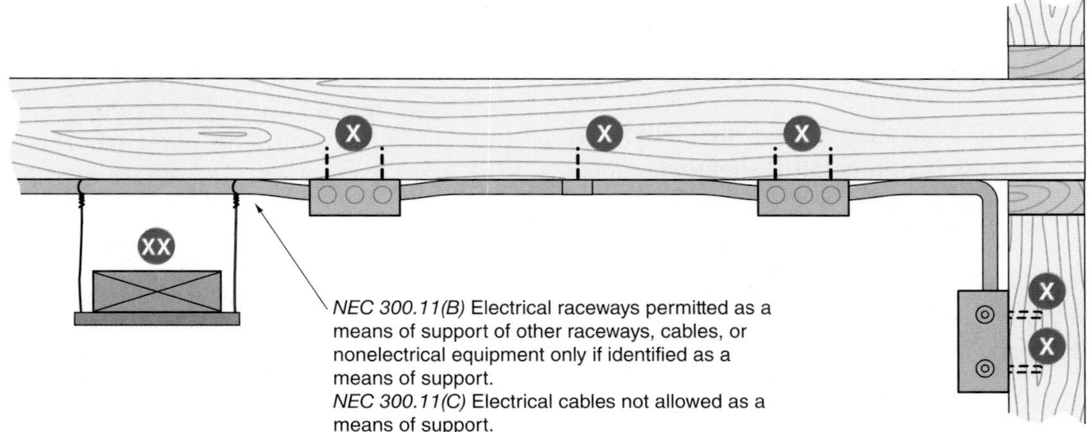

FIGURE 18-6 Raceways, boxes, fittings, and cabinets must be securely fastened in place (points X), *300.11(A)*. It is not permissible to hang other raceways, cables, or nonelectrical equipment from an electrical raceway or cable (point XX), *300.11(B)* and *(C)*. Refer to Figure 24-10 for an exception that allows low-voltage thermostat cables to be fastened to the conduit that feeds a furnace or other similar equipment. See Chapter 17 for a discussion of the use of suspended-ceiling support wire for supporting equipments.

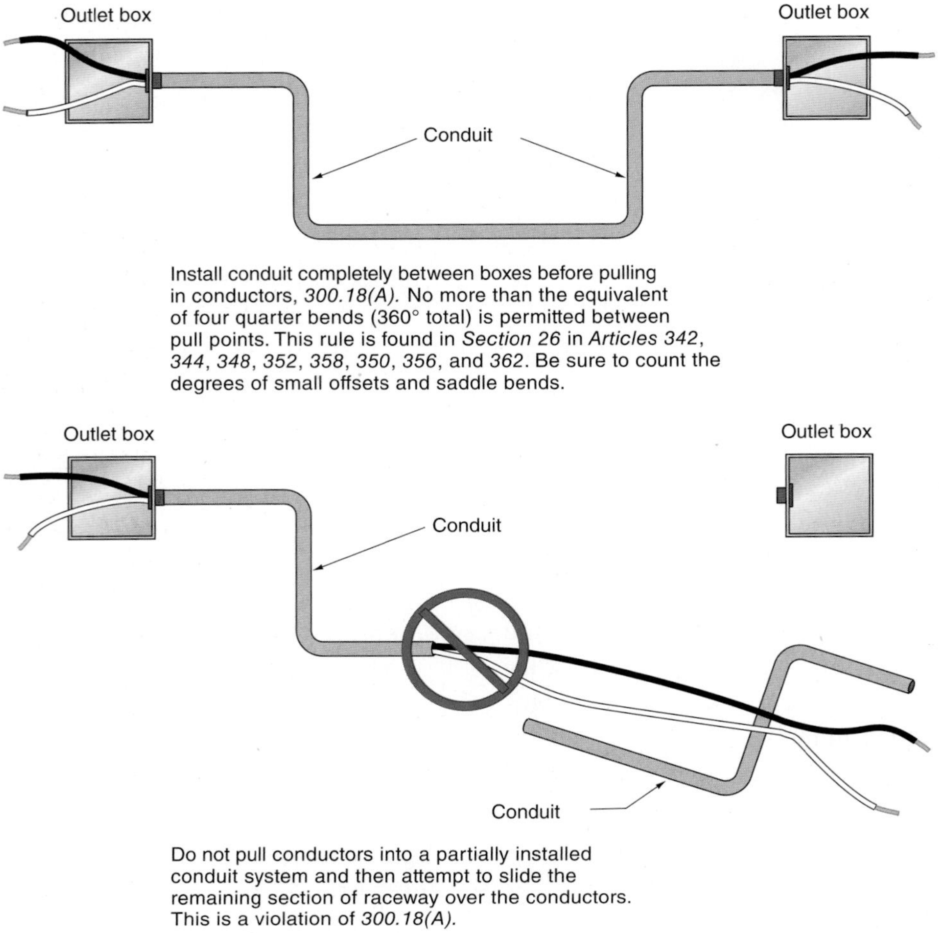

Outlet box

Outlet box

Conduit

Install conduit completely between boxes before pulling in conductors, *300.18(A)*. No more than the equivalent of four quarter bends (360° total) is permitted between pull points. This rule is found in *Section 26* in *Articles 342, 344, 348, 352, 358, 350, 356*, and *362*. Be sure to count the degrees of small offsets and saddle bends.

Outlet box

Outlet box

Conduit

Conduit

Do not pull conductors into a partially installed conduit system and then attempt to slide the remaining section of raceway over the conductors. This is a violation of *300.18(A)*.

FIGURE 18-7 Do not pull wires into the raceway until the conduit system is completely installed between pull points.

Cable in Ceilings—EMT on Walls

In many parts of the country, the wiring method in unfinished basements is nonmetallic-sheathed cable in the ceiling and EMT on the walls for receptacles and switches. This protects and provides a neat appearance for the cable. Figures 18-2 and 4-22 clearly illustrate the requirements for this type of installation. See *NEC 334.15(C)* and *300.18(A), Exception*.

Outlet Boxes for Use in Exposed Installations

Figure 18-8 illustrates outlet boxes and plaster rings that are used for concealed wiring, such as for the recreation room and first floor wiring in the residence. For surface wiring such as the workshop, various types and sizes of outlet boxes may be used. For example, a handy box, Figure 18-9, could be used for the freezer receptacle. The remaining workshop boxes could be 4 in. square outlet boxes with raised covers, Figure 18-10. The box size and type is determined by the maximum number of conductors contained in the box, *Article 314*.

NEC 406.5(C) prohibits mounting and supporting a receptacle to a cover (flat or raised) by only one screw unless the device assembly or box cover is listed and identified for securing by a single screw.

The listing and identifying for the purpose is a decision that is made by an NRTL, such as Underwriters Laboratories, Inc. In the past, securing a receptacle with a single screw often resulted in a loose receptacle when used over a long period

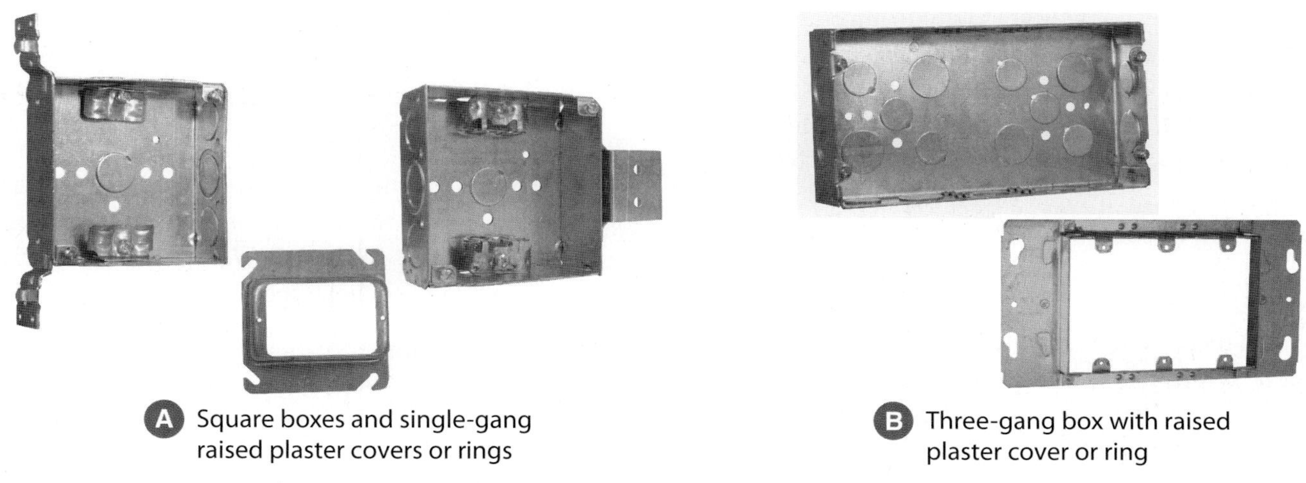

(A) Square boxes and single-gang raised plaster covers or rings

(B) Three-gang box with raised plaster cover or ring

FIGURE 18-8 Typical outlet boxes and raised covers.

Courtesy of Appleton Electric LLC, Emerson Industrial Automation.

FIGURE 18-9 Handy box and covers.

Courtesy of Appleton Electric LLC, Emerson Industrial Automation.

FIGURE 18-10 Four-inch square boxes and raised covers.

Courtesy of Appleton Electric LLC, Emerson Industrial Automation.

of time. Many short-circuits and ground-faults have been reported because of this.

For the typical raised cover, as illustrated in Figure 18-10, the duplex receptacle would be fastened to the raised cover by two small No. 6-32 bolts and nuts, usually furnished with the raised cover. The raised cover has two small "knockout" holes for these screws. The center No. 6-32 screw could also be used.

If a receptacle is installed on a surface-mounted box where there is direct metal-to-metal contact between the grounded metal box and the metal yoke of the receptacle, the receptacle is considered to be properly grounded without connecting an additional equipment grounding conductor to the equipment grounding terminal on the receptacle, *250.146*.

To make sure that there is direct metal-to-metal contact between the metal yoke of a receptacle and a surface-mounted metal box, *250.146(A)* accepts the automatic self-grounding device on the yoke of the receptacle or the removal of at least one of the insulating washers from one of the No. 6-32 screws on the receptacle. Automatic self-grounding devices on the yoke of a receptacle are clearly shown in Figures 5-3, 5-4, and 5-5.

Most commonly used outlet, device, and junction boxes have knockouts for trade size ½, ¾, and 1 connectors. The knockout size selected depends on the size of the conduits entering the box. The conduit size is determined by the maximum percentage of fill of the conduit's total cross-sectional area by the conductors pulled into the conduit.

Supporting equipment and/or raceways to the joists, walls, and so on, is accomplished with proper nails, screws, anchors, and straps. Be careful when trying to support other equipment from raceways as this is not always permitted. Refer to Figure 18-5.

Conduit Fill Calculations

For simplicity, the following conduit fill calculations show in. and in.^{2} To show calculations in in., in.^{2}, mm, and mm^2 would be very confusing.

An electrician must be able to select the proper size raceways based on the number, size, and type of conductors to be installed in a particular type of raceway. To do this, the electrician must become familiar with the use of certain tables found in *Chapter 9* and *Annex C* of the *NEC*.

Percent Fill. The *NEC* has established the following "percent fill" values for conduit and tubing. These percentages relate to the area of conduit and tubing as shown in *Table 4* of *NEC Chapter 9*. A separate table is provided for each type of circular raceway.

- one conductor 53% fill
- two conductors 31% fill
- three or more conductors 40% fill
- conduit or nipples not over 24 in. (600 mm) 60% fill

Derating of conductor ampacity is not required for the raceway that is not over 24 in. (600 mm) in length.

These percentages are taken from *Chapter 9, Table 1*, and the accompanying notes to the table.

All Conductors the Same Size and Type. When all conductors installed in the raceway are the same size and type:

- Determine how many conductors of one size and type are to be installed in the raceway.
- Refer to *Tables C1* through *C12A* in *Informative Annex C* of the *NEC* for the type of conduit, tubing, or raceway and for the specific size and type of conductors.
- Select the proper size raceway directly from the table.

These tables are not difficult to use. Here are a few examples of conductor fill for EMT. Look these up in the tables to confirm the accuracy of the examples.

- Not over five 10 AWG THHN conductors may be pulled into a trade size ½ EMT.
- Not over eight 12 AWG THW conductors may be pulled into a trade size ¾ EMT.
- Not over three 8 AWG THHN conductors may be pulled into a trade size ½ EMT.
- Not over six 8 AWG THW conductors may be pulled into a trade size 1 EMT.

Conductors of Different Sizes and/or Types. When conductors are different sizes or have different insulation types:

- Refer to *Chapter 9, Table 5*, of the *NEC*.
- Find the size and type of conductors used to determine the approximate cross-sectional area in in.^{2} of the conductors.
- Refer to *Chapter 9, Table 8*, for the in.^{2} area of bare conductors.
- Total the in.^{2} areas of all conductors.
- Refer to *Chapter 9, Table 4*, for the type and size of raceway used.
- Determine the minimum size raceway based on the total in.^{2} area of the conductors and the allowable in.^{2} area fill for the particular type of raceway used.
- The sum must not exceed the allowable percentage fill of the cross-sectional in.^{2} area of the raceway used.

For illustrative purposes, only the data for a few types of wires from *Table 5* are shown in Table 18-3. The dimensions and areas for some EMT from *Table 4* are shown in Table 18-4. You will need to refer to *Chapter 9* and *Annex C* of the *NEC* for the many other sizes and types of conductors and raceways.

TABLE 18-3

This sampling of typical conductors and their dimensions is taken from *Table 5, Chapter 9, NEC*. See *Table 5* for metric values.

TYPE	SIZE	APPROX, DIAM. IN.	APPROX. AREA IN.^{2}
THHN/THWN	14	0.111	0.0097
THHN/THWN	12	0.130	0.0133
THHN/THWN	10	0.164	0.0211
THHN/THWN	8	0.216	0.0366
THHN/THWN	6	0.254	0.0507
THHN/THWN	3	0.352	0.0973
THW	6	0.304	0.0726
THW	8	0.236	0.0437

EXAMPLE

If three 6 AWG THW conductors and two 8 AWG THW conductors are to be installed in one EMT, determine the proper size EMT.

Solution

1. Find the in.^{2} area of the conductors in *NEC Chapter 9, Table 5*.

Three 6 AWG THW conductors	0.0726 in.^{2}
	0.0726 in.^{2}
	0.0726 in.^{2}
Two 8 AWG THW conductors	0.0437 in.^{2}
	0.0437 in.^{2}
Total Area	0.3052 in.^{2}

2. In *NEC Chapter 9, Table 4,* look at the first chart entitled *Article 358*. Now look in the column marked "Over 2 Wires—40%." Note that a trade size ¾ EMT holds up to 0.213 in.^{2} of conductor fill, and a trade size 1 EMT holds up to 0.346 in.^{2} of conductor fill. Therefore, a trade size 1 EMT is the minimum size for the combination of conductors in the example.

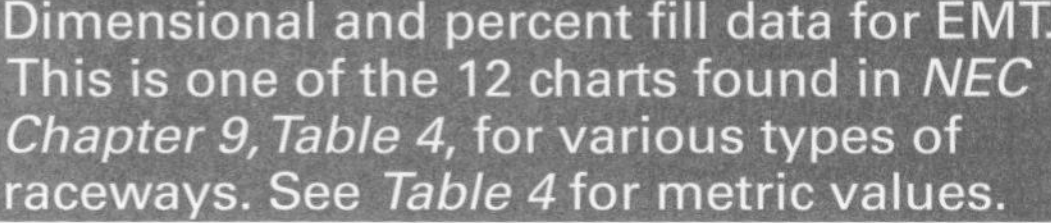

TABLE 18-4

Dimensional and percent fill data for EMT. This is one of the 12 charts found in *NEC Chapter 9, Table 4*, for various types of raceways. See *Table 4* for metric values.

ELECTRICAL METALLIC TUBING

TRADE SIZE (IN.)	OVER 2 WIRES 40% (IN.^{2})	1 WIRE 53% (IN.^{2})	2 WIRES 31% (IN.^{2})	INTERNAL DIAMETER (IN.)	TOTAL AREA 100% (IN.^{2})
½	0.122	0.161	0.094	0.622	0.304
¾	0.213	0.283	0.165	0.824	0.533
1	0.346	0.458	0.268	1.049	0.864
1¼	0.598	0.793	0.464	1.380	1.496
1½	0.814	1.079	0.631	1.610	2.036
2	1.342	1.778	1.040	2.067	3.356
2½	2.343	3.105	1.816	2.731	5.858
3	3.538	4.688	2.742	3.356	8.846
3½	4.618	6.119	3.579	3.834	11.545
4	5.901	7.819	4.573	4.334	14.753

Many electrical ordinances require that a separate equipment grounding conductor be installed in all raceways, whether rigid or flexible, whether metallic or nonmetallic. This requirement resulted from inspectors finding loose or missing set screws on set-screw type connectors and couplings. Corrosion of fittings also contributes to poor grounding. A separate equipment grounding conductor does take up space, and therefore must be counted when considering conduit fill.

EXAMPLE

A feeder to an electric furnace is fed with two 3 AWG THHN conductors plus one green insulated equipment grounding conductor. This feeder is installed in EMT. The feeder overcurrent protection is 100 amperes. What is the minimum size EMT to be installed?

Solution

According to *Table 250.122*, the minimum size equipment grounding conductor for a 100-ampere feeder is 8 AWG copper. Find the in.^{2} area of the conductors in *Table 5, Chapter 9, NEC.*

Two 3 AWG THHN	0.0973 in.^{2}
	0.0973 in.^{2}
One 8 AWG THHN	0.0366 in.^{2}
Total Area	0.2312 in.^{2}

From *Table 4, Chapter 9, NEC*:

40% fill for trade size ¾ EMT = 0.213 in.^{2}
40% fill for trade size 1 EMT = 0.346 in.^{2}
Therefore, install trade size 1 EMT.

Remember, *300.18(A)* states that a conduit system must be completely installed before pulling in the conductors. Refer to Figure 18-7.

Conduit Bodies [*314.16(C)*]

Conduit bodies are used with conduit installations to provide an easy means to turn corners where there is not adequate space to install an elbow with a standard radius, as well as to route the conduit or tubing around sharp corners and to terminate conduits. They are also used to provide access to conductors, to provide space for splicing (when permitted), and to provide a means for pulling conductors (Figure 18-11).

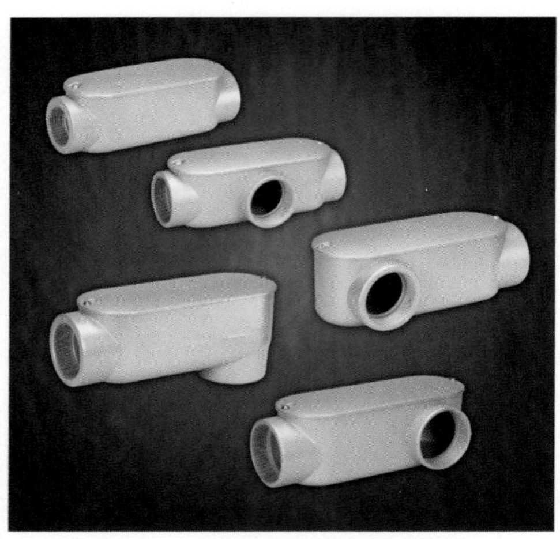

FIGURE 18-11 A common assortment of conduit bodies.

Photo courtesy of Thomas and Betts Corp., www.TNB.com.

A conduit body:

- must have a cross-sectional area not less than twice the cross-sectional area of the largest conduit to which it is attached, as shown in Figure 18-12(A). This is a requirement only when the conduit body contains 6 AWG conductors or smaller.
- may contain the same maximum number of conductors permitted for the size raceway attached to the conduit body.
- must not contain splices, taps, or devices unless the conduit body is marked with its in.^{3} capacity, as in Figure 18-12(B), and the conduit body must be supported in a rigid and secure manner.
- must have the conductor fill volume properly calculated according to *Table 314.16(B)*, which lists the free space that must be provided for each conductor within the conduit body. If the conduit body is to contain splices, taps, or devices, it must be supported rigidly and securely.
- must be sized according to *NEC 314.28* when it contains conductors 4 AWG or larger that are required to be insulated.

Conduit bodies are often referred to in the trade as Condulets and Unilets. The terms *Condulets* and

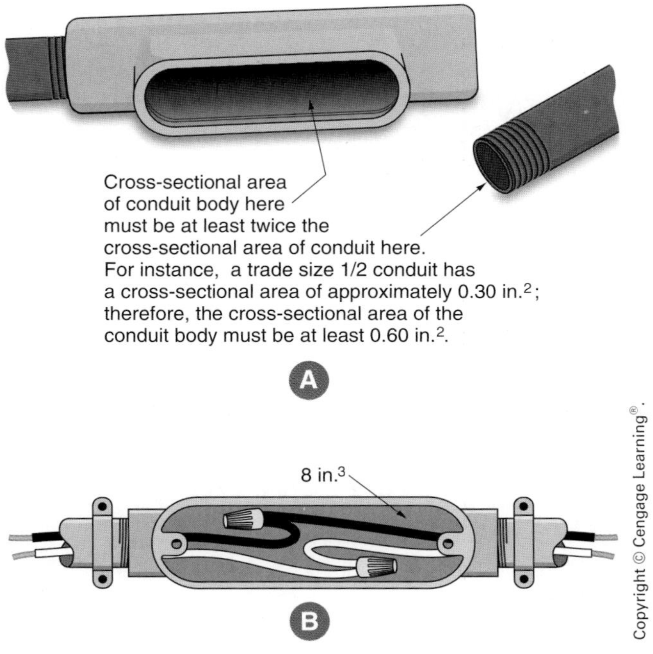

FIGURE 18-12 Requirements for conduit bodies.

Unilets are copyrighted trade names of Eaton's Crouse-Hinds Business and Appleton Electric, respectively, for their line of conduit bodies. Conduit bodies are excellent for turning corners and changing directions of a conduit run. They are available with threaded hubs for threaded conduit and threaded fittings, set screw hubs for direct insertion of EMT without needing an EMT connector, gasketed covers, built-in rollers to make it easy to pull wires, and in jumbo sizes to provide more room for pulling, splicing, and making taps. The chart that follows shows how typical conduit bodies are identified.

WHEN LOOKING AT THE FRONT OPENING, IF THE DIRECTION IS:	THE CONDUIT BODY IS CALLED:
to the back	LB
left	LL
right	LR
straight through	C
straight through in both directions	X
straight through and to the left or right	T
straight through and to the back	TB
one conduit entry only	E

ADJUSTMENT AND CORRECTION (DERATING) FACTORS FOR MORE THAN THREE CURRENT-CARRYING CONDUCTORS IN CONDUIT OR CABLE

These requirements apply equally if more than three current-carrying conductors are installed in a raceway or multiconductor cables are installed for a continuous length longer than 24 in. (600 mm) without maintaining spacing. For more than three current-carrying wires in conduit or cable, refer to Table 18-5.

NEC 310.15(B)(3)(a) states that for the conductors listed in the tables, the maximum allowable ampacity must be reduced when more than three current-carrying conductors are installed in a raceway or multiconductor cables are installed for a continuous length longer than 24 in. (600 mm) without maintaining spacing. The reduction factors based on the number of conductors are noted in Table 18-5. When the length of the grouped or bundled conductors does not exceed 24 in. (600 mm), the reduction factor as shown in Table 18-5 does not apply. See Figure 18-13.

According to *310.15(B)(5)(a)*, a neutral conductor of a multiwire branch circuit carrying only unbalanced currents shall not be counted in determining the number of current-carrying conductors, as shown in Table 18-5. See Figure 18-14.

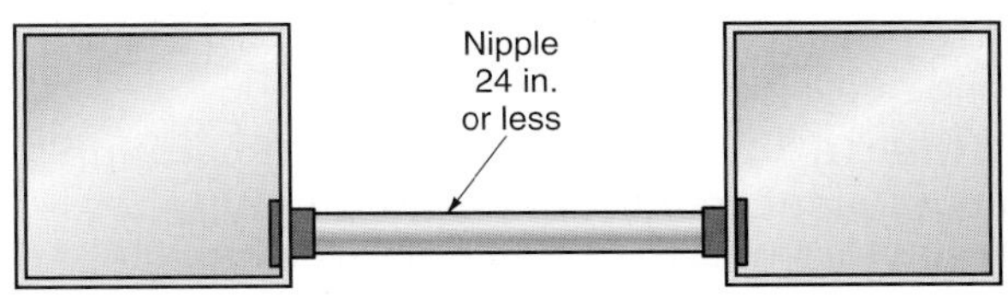

FIGURE 18-13 The derating factors in Table 18-5 are not required for short lengths of conduit that do not exceed 24 in. (600 mm), *310.15(B)(3)(a)(2)*. Conductor fill not to exceed 60% according to *Chapter 9, Table 1, Note 4*.

TABLE 18-5

Derating factors necessary to determine conductor ampacity when there are more than three current-carrying conductors in a raceway or cable. See *310.15(B)(3)(a)*.

NUMBER OF CURRENT-CARRYING CONDUCTORS	PERCENT OF VALUES IN *TABLE 310.15(B)(16)*. IF HIGH AMBIENT TEMPERATURES ARE PRESENT, ALSO APPLY CORRECTION FACTORS FOUND IN *TABLE 310.15(B)(2)(A)*.
4 through 6	80
7 through 9	70
10 through 20	50
21 through 30	45
31 through 40	40
41 and above	35

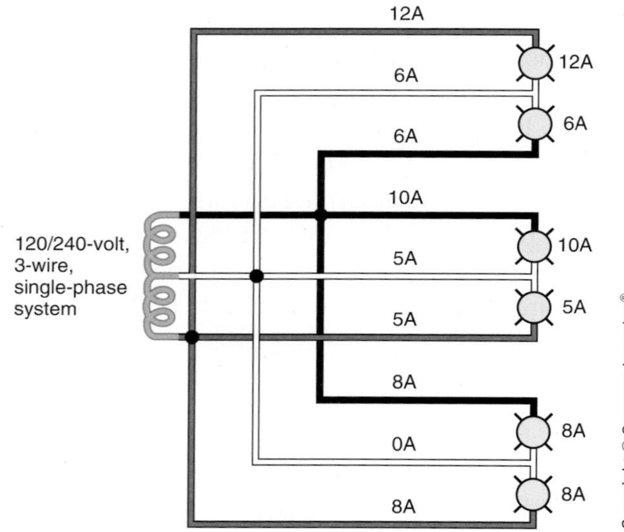

FIGURE 18-14 A neutral conductor carrying the unbalanced currents from the other conductors need not be counted when determining the ampacity of the conductor, according to *310.15(B)(5)(a)*. The *Code* would recognize this example as two current-carrying conductors. Thus, in the three multiwire circuits shown, the actual total is nine conductors, but only six current-carrying conductors are considered when the derating factor of Table 18-5 is applied.

EXAMPLE

What is the correct ampacity for size 6 AWG THHN/THWN when there are six current-carrying conductors in the conduit or cable and it is installed in a dry location? Note that we can begin the derating from the ampacity in the 90°C column of *Table 310.15(B)(16)* in a dry location because the THHN insulation has a 90°C rating.

$$75 \times 0.80 = 60 \text{ amperes}$$

CORRECTION FACTORS DUE TO HIGH TEMPERATURES

When conductors are installed in locations where the temperature is higher than 86°F (30°C), the correction factors noted in *Table 310.15(B)(2)(a)* must be applied.

For example, consider that four current-carrying 3 AWG THWN copper conductors are to be installed in one raceway or cable in an ambient temperature of approximately 90°F (32°C). "Ambient" temperature means "surrounding" temperature. This temperature is frequently exceeded in attics where the sun shines directly onto the roof.

The maximum ampacity for these conductors is determined as follows:

- The allowable ampacity of 3 AWG THWN copper conductors from *Table 310.15(B)(16)* = 100 amperes.
- Apply the correction factor for 90°F (32°C) found in *Table 310.15(B)(2)(a)*. The first column shows Celsius; the last column shows Fahrenheit.

$$100 \times 0.94 = 94 \text{ amperes}$$

- Apply the derating factor for four current-carrying conductors in one raceway.

$$94 \times 0.80 = 75.2 \text{ amperes}$$

- Therefore, 75.2 amperes is the new corrected ampacity for the example given.

Some AHJs in extremely hot areas of the country, such as the southwestern desert climates, will require that service-entrance conductors running up the side of a building or above a roof exposed to direct sunlight be "corrected" per the correction factors in *Table 310.15(B)(3)(c)*. Check this out before proceeding with an installation where this situation might be encountered.

EXAMPLE OF DERATING, CORRECTING, ADJUSTING, OVERCURRENT PROTECTION, AND CONDUCTOR SIZING

High ambient temperatures and/or more than three current-carrying conductors in the same raceway or cable call for the application of adjustment or correction factors relative to a conductor's ability to safely carry current.

Suppose that eight THHN/THWN copper conductors are installed in one raceway or cable(s)—the conditions are dry and the temperature is expected to reach 100°F (38°C). All conductors are considered to be current-carrying. The expected noncontinuous load on each conductor is 18 amperes. Let's analyze this.

- Consider 14 AWG THHN/THWN copper conductors:
  - The allowable ampacity in the 90°C column from *NEC Table 310.15(B)(16)* is 25 amperes.
  - The maximum overcurrent protection from Table 18-6 in this text or from *NEC 240.4(D)* is 15 amperes—not adequate to carry the 18-ampere load.
  - To find the corrected ampacity when eight current-carrying conductors are in one raceway or cable(s), apply the 0.70 adjustment factor (penalty) from Table 18-5 in this text or from *NEC Table 310.15(B)(3)(a)*.

  $$25 \times 0.70 = 17.5 \text{ amperes}$$

  - To find the corrected ampacity for the high ambient temperature, apply the 0.91 correction factor (penalty) from *NEC Table 310.15 (B)(2)(a)*.

  $$17.5 \times 0.91 = 15.9 \text{ amperes}$$

  - You can do this all in one step:

  $$25 \times 0.70 \times 0.91 = 15.9 \text{ amperes}$$

  - Analysis: 14 AWG THHN/THWN copper conductors are *too small* for the 18-ampere load. The 15-ampere maximum overcurrent protection *will not carry* the 18-ampere load.
- Consider 12 AWG THHN/THWN copper conductors:
  - The ampacity from the 90°C column of *NEC Table 310.15(B)(16)* is 30 amperes.
  - The maximum overcurrent protection from Table 18-6 in this text or from *NEC 240.4(D)* is 20 amperes.
  - To find the corrected ampacity when eight current-carrying conductors are in one raceway or cable(s), apply the 0.70 adjustment factor (penalty) from Table 18-5 in this text or from *NEC Table 310.15(B)(3)(A)*.
  - To find the corrected ampacity for the high ambient temperature, apply the 0.91 correction factor (penalty) from *NEC Table 310.15(B)(3)*.

  Let's do the calculation in one step:

  $$30 \times 0.70 \times 0.91 = 19.1 \text{ amperes}$$

  - Analysis: 12 AWG THHN copper conductors are all right for the 18-ampere load. The 20-ampere maximum overcurrent protection *will* carry the 18-ampere load. *NEC 240.4(B)* allows the next higher rated overcurrent device to be used. The next standard rating above 19.1 amperes is 20 amperes.

TABLE 18-6

Maximum size overcurrent protection for small-size conductors, *240.4(D)*.

CONDUCTOR SIZE (COPPER)	MAXIMUM AMPERE RATING OF OVERCURRENT DEVICE
14 AWG	15 amperes
12 AWG	20 amperes
10 AWG	30 amperes

Watch Out! The Overcurrent Device Depends on the Type of Load

The basic rule is that a conductor overcurrent protection shall not exceed its ampacity, *240.4*. There are exceptions.

Read *240.4(B)* closely! The permission to use the next higher standard size overcurrent device has tough restrictions. The key is whether the conductors are supplying a fixed load or supplying receptacles for cord-and-plug-connected portable loads. Let's take a closer look at the above example.

In the previous calculations, 14 AWG THHN copper conductors are obviously too small.

The 12 AWG THHN copper conductors are adequate for the 18-ampere load, and the 20-ampere overcurrent device is proper but *only* if the conductors serve a fixed load.

If the conductors supply receptacles for cord-and-plug-connected portable loads, we have no control over what loads might be plugged in. In this situation, the conductor's final allowable ampacity of 19.1 is adequate for the anticipated load of 18 amperes but is not permitted to be protected by a 20-ampere overcurrent device.

We would need to consider using 10 AWG THHN copper conductors.

From *Table 310.15(B)(16)*, using the 90° column, we find that the allowable ampacity of a 10 AWG THHN copper conductor is 40 amperes. We then apply the adjustment and correction factors. The new allowable ampacity becomes

$$40 \times 0.70 \times 0.91 = 25.5 \text{ amperes}$$

The conclusion here is that 10 AWG THHN conductors are more than adequate to serve the 18-ampere noncontinuous load and are properly protected by a 20-ampere fuse or circuit breaker.

As you can readily see, when derating, correcting, and adjusting the ampacity of a conductor, sizing the conductors and sizing the overcurrent protection become a little more complicated. On top of this, one must consider voltage drop where long runs are concerned.

OVERCURRENT PROTECTION FOR BRANCH CIRCUIT CONDUCTORS

The basic rule is that overcurrent protection for conductors shall not be more than the allowable ampacity of the conductor, *240.4*. Allowable ampacities for conductors are found in *Table 310.15(B)(16)*.

If the allowable ampacity of a conductor does not match a standard ampere rating of a fuse or circuit breaker, *240.4(B)* permits the use of the next higher rated fuse or circuit breaker, 800 amperes or less. For example, a conductor having an ampacity of 55 amperes could be protected by a 60-ampere overcurrent device. This permission does not apply if the conductors of the branch circuit supply more than one receptacle for cord-and-plug connection of a load, because it would be impossible to know just how much load might be plugged in.

Small-size branch-circuit conductors must be protected according to the values shown in Table 18-6. This table is based on *240.4(D)*.

An exception to this basic maximum size overcurrent protection rule is the protection of motor branch-circuit conductors. For motor branch circuits, the maximum size overcurrent protection values are found in *Table 430.52*.

If the allowable ampacity of a given conductor is *derated* and/or *corrected*, the overcurrent protection must be based on the new *derated* or *corrected* ampacity.

BASIC *CODE* CONSIDERATIONS FOR CONDUCTOR SIZING AND OVERCURRENT PROTECTION

The following is a quick review of *Code* requirements for conductor sizing and conductor overcurrent protection encountered in house wiring.

- *210.3:* The rating of a branch circuit is based on the rating of the overcurrent device.
- *210.19(A):* Branch-circuit conductors shall have an ampacity not less than the maximum load to be served. In addition, they are required to be sized not smaller than 100% of the noncontinuous load plus 125% of the continuous load. For dwellings, branch circuits for electric water heaters and electric space heating are considered continuous loads.
- *210.20:* An overcurrent device shall not be less than 125% for continuous loads plus 100% of noncontinuous loads on branch circuits.

- *210.20(B):* This section refers us to *240.4*, where we find that overcurrent protection for conductors shall not exceed the ampacity of a conductor. See *240.4* for exceptions such as "next standard size permitted" if the OCD and the conductor ampacity do not match.
- *210.23:* In no case shall the load exceed the branch-circuit ampere rating.
- *215.2(A):* Feeder conductors shall have an ampacity not less than required to supply the load. Like branch circuits, they are required to be sized not smaller than 100% of noncontinuous loads and 125% of continuous loads.
- *215.3:* An overcurrent device shall not be less than 100% of noncontinuous loads plus 125% for continuous loads on feeders. This section refers us to *Part 1* of *240.4*, where we find that overcurrent protection for feeders shall not exceed the ampacity of a conductor. Refer to *240.4* for specific requirements and exceptions to this basic rule.
- *220.4:* The total load shall not exceed the rating of the branch circuit.
- *240.4:* This section contains many of the overcurrent protection requirements for conductors.
- *230.42:* Service-entrance conductors are required to be sized not smaller than 100% of noncontinuous loads plus 125% of continuous loads for services.
- *422.13:* Storage-type water heaters. A fixed storage–type water heater that has a capacity of 120 gallons (450 L) or less is required to be considered a continuous load for the purpose of sizing branch circuits. As a result, the branch-circuit conductors and overcurrent device are required to be sized not smaller than 125% of the load.
- *424.3(B):* For fixed electric space-heating equipment such as an electric furnace, the branch circuit loads are considered to be continuous (the load is likely to continue for 3 hours or more); as a result, the branch-circuit conductors and the overcurrent protective devices shall not be less than 125% of the total load of the motor(s) and heater(s).
- *426.4:* Fixed outdoor electric deicing and snow-melting equipment, such as heating cables buried in the concrete of a driveway, are considered to be continuous loads. As such, the branch-circuit conductors and the overcurrent protective devices shall not be less than 125% of the total load of the heaters.
- *430.22:* Branch-circuit conductors for motors shall not be less than 125% of the motor's full-load current as shown in *NEC Table 430.248*. Motor branch circuit short-circuit and ground-fault protection is sized according to the percentages listed in *Table 430.52*.
- *Article 440, Part III:* Branch-circuit short-circuit and ground-fault protection for air-conditioning equipment is covered in Chapter 23 of this text.
- *Article 440, Part IV:* Conductor sizing for air-conditioning equipment is also covered in Chapter 23 of this text.

A few of the above *Code* sections make reference to *continuous loads*. The *Code* defines a *continuous load* as Where the maximum current is expected to continue for three hours or more.*

Loads in homes, such as certain sizes of electric water heaters, electric furnace, and snow-melting heating cables buried in the concrete of a driveway, are classified as continuous loads because they could be on for periods of 3 hours or more. The 125% factor for these loads must be applied to comply with *422.13*, *424.3(B)* and *426.4*.

Many loads in commercial and industrial installations can be considered to be continuous loads. For example, an electric range in a restaurant might be used continuously 24 hours a day, 7 days a week. An electric range in a home would not be used in this manner.

For continuous loads, the overcurrent protective device and the conductors must not be less than 125% of the continuous load. For example, a continuous load is 40 amperes. The minimum overcurrent device rating and the minimum conductor ampacity is $40 \times 1.25 = 50$ amperes.

---

*Reprinted with permission from NFPA 70-2014.

For combination continuous and noncontinuous loads, the overcurrent protective device and the conductors must not be less than 125% of the continuous load plus the noncontinuous load. For example, a feeder supplies a continuous load of 40 amperes and a noncontinuous load of 30 amperes. The minimum overcurrent device rating and the minimum conductor ampacity is $(40 \times 1.25) + 30 = 80$ amperes.

MULTIOUTLET ASSEMBLY

A multioutlet assembly (Wiremold) has been installed above the workbench. For safety reasons, as well as for adequate wiring, it is recommended that any workshop receptacles be connected to a separate circuit. In the event of a malfunction in any power tool commonly used in the home workshop (saw, planer, lathe, or drill), only receptacle Circuit A18 is affected. The lighting in the workshop is not affected by a power outage on the receptacle circuit.

As shown in Figure 18-15, the GFCI feed-through receptacle to the left of the workbench provides GFCI protection for the entire plug-in strip, *210.8(A)(5)*. GFCI protection could also have been provided by installing a GFCI circuit breaker in the circuit breaker panelboard.

The installation of multioutlet assemblies must conform to the requirements of *Article 380* of the *NEC*.

Load Considerations for Multioutlet Assemblies

NEC Table 220.12 gives the general lighting loads in volt-amperes per ft^2 (0.093 m^2) for types of occupancies. *NEC 220.14(J)* indicates that *all* general-use receptacle outlets in one, two, and multifamily dwellings are to be considered outlets for general lighting, and as such have been included in the volt-amperes per ft^2 calculations. Therefore, no additional load need be added.

Because the multioutlet assembly in Figure 18-15 has been provided in the workshop for the purpose of plugging in portable tools, a separate circuit, A18, is provided. To provide capacity above and beyond the minimum requirements of the *NEC*, this circuit has been shown in the service-entrance calculations at 1500 volt-amperes, similar to the load requirements for small-appliance circuits.

Many types of receptacles are available for various sizes of multioutlet assemblies. For example, duplex, single-circuit grounding, split-wired grounding, and duplex split-wired receptacles can be used with multioutlet assemblies.

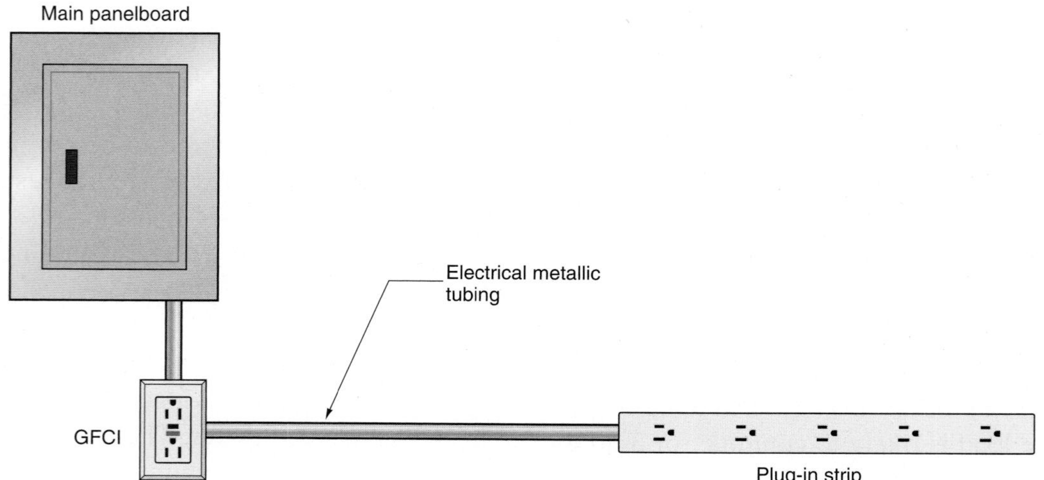

FIGURE 18-15 Detail of how the workbench receptacle plug-in strip is connected by feeding through a GFCI feed-through receptacle located below the main panelboard. This is Circuit A18.

Wiring a Multioutlet Assembly

A multioutlet assembly can be connected by bringing the supply into one end, as in Figure 18-16, or into the back of the assembly, as in Figure 18-17.

Receptacles for multioutlet assemblies can be spaced and wired on the job, in which case the covers are cut to the desired length. Receptacles can also be factory prewired, with six receptacles in a 3 ft (900 mm) long strip, and 10 receptacles in a 5 ft (1.5 m) long strip.

For other strip lengths, standard spacing of receptacles for precut assemblies are 6 ft (1.8 m), 9 in. (225 mm), 12 in. (300 mm), or 18 in. (450 mm). These measurements are center to center. Prewired multioutlet assemblies are generally wired with 12 AWG THHN conductors and have either 15- or 20-ampere receptacles, depending on how they are specified and ordered.

Multioutlet assemblies are not difficult to install. There is a wide variety of fittings, such as connectors, couplings, ground clamps, blank and end fittings, elbows for turning corners (flat, inside, outside, twist), flush plate adapters, starter boxes, raceway adapters, mounting clips, and so on, to fit just about any application.

Figure 18-18 shows a cord-connected multioutlet assembly of the type used in workshops or other locations where there is a need for the convenience of having a number of receptacles in close proximity. These prewired assemblies are available with six receptacles and a circuit breaker 40 in. long, or 10 receptacles and a circuit breaker, 52 in. long. These can also be used under kitchen cabinets. Figure 18-17 shows a

Courtesy Legrand/Wiremold.

FIGURE 18-16 A typical metal multioutlet assembly. Multioutlet assemblies are permitted to be run through (not within) a dry partition when no outlets are located within the partition, and all exposed portions of the assembly can have the cover removed. Nonmetallic multioutlet assemblies are also available.

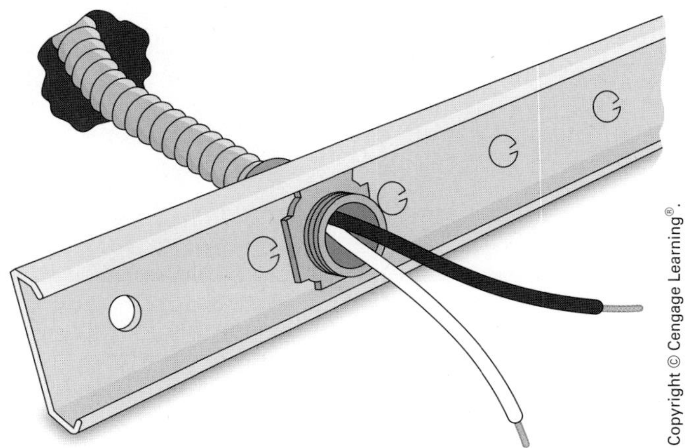

FIGURE 18-17 A multioutlet assembly back plate with the armored cable supply feeding into the back of the assembly.

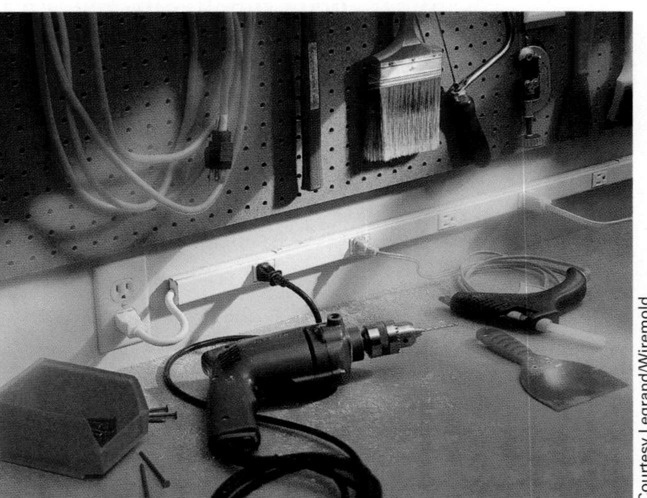

Courtesy Legrand/Wiremold.

FIGURE 18-18 A cord-connected, prewired multioutlet assembly installed above a workbench.

Courtesy Legrand/Wiremold.

FIGURE 18-19 A multioutlet assembly installed above the backsplash of a kitchen countertop.

multioutlet assembly back plate with the electrical supply feeding into the back of the assembly. Figure 18-19 shows a multioutlet assembly above the backsplash of a kitchen countertop.

EMPTY CONDUITS

Although not truly part of the workshop wiring, *Note 8* to the first-floor electrical plans indicates that two empty trade size ½ EMT raceways are to be installed, running from the workshop to the attic. This installation is unique but certainly welcomed if at some later date additional wiring for telephone, security, computer, structured wiring, home automation cables, or other similar future needs might be required. The empty raceways will provide the necessary route from the basement to the attic, thus eliminating the need to fish through the wall partitions.

REVIEW

1. a. What circuit supplies the workshop lighting? ______________________

   b. What circuit supplies the plug-in strip over the workbench? ______________________

   c. What circuit supplies the freezer receptacle? ______________________

2. Approximately how much trade size ½ EMT is used for connecting Circuit A13, Circuit A17, Circuit A18, and Circuit A20? ______________________

3. a. How is EMT fastened to masonry? ______________________

   ____

   b. The cut ends of rigid and electrical metallic tubing must be ______________________ to prevent the insulation on conductors from being damaged.

4. What type of luminaires are installed on the workshop ceiling?

   ____

5. A check of the plans indicates that two empty trade size ½ EMT conduits are installed between the basement and the attic. In your opinion, is this a good idea? Briefly explain your thoughts. ______________________

   ____

   ____

6. To what circuit are the smoke detectors connected?

____

7. Is it a *Code* requirement to connect smoke detectors to a GFCI-protected circuit?

____

8. When a freezer is plugged into a receptacle, be sure that the receptacle is protected by a GFCI. Is this statement true or false? Explain.

____

____

9. A sump pump is plugged into a single receptacle outlet. This receptacle shall (Circle the correct answer.)
   a. be GFCI protected.
   b. not required to be GFCI protected.
10. When a 120-volt receptacle outlet is provided for the laundry equipment in an unfinished basement, what is the *minimum* number of additional receptacle outlets required in that basement? What *Code* reference?____________________

____

11. Derating factors for conductor ampacities must always be taken into consideration when (a) ____________________ are contained in one raceway. Correction factors are applied when (b) ____________________
12. What current is flowing in the neutral conductor at
   a. ________________ amperes?
   b. ________________ amperes?
   c. ________________ amperes?
   d. ________________ amperes?

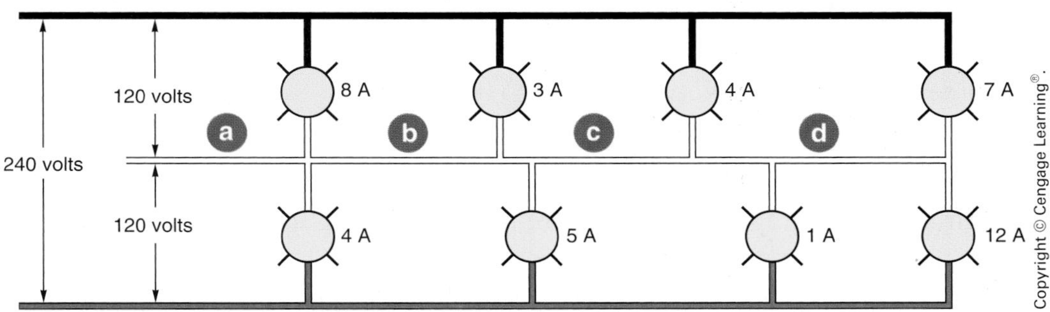

13. When a single receptacle is installed on an individual branch circuit, the receptacle must have a rating ________________ than the rating of the branch circuit.
14. Calculate the total current draw of Circuit A17 if all luminaires and the exhaust fan were turned on.
15. A conduit body must have a cross-sectional area not less than (two) (three) (four) times the cross-sectional area of the (largest) (smallest) conduit to which it is attached. (Circle the correct answers.)

16. A conduit body may contain splices if marked with (Circle the correct answer.)
    a. the UL logo.
    b. its in.^{3} area.
    c. the size of conduit entries.
17. What size box would you use where Circuit A17 enters the junction box on the ceiling? ______________________________

    ____
18. List the proper trade size of electrical metallic tubing for the following. Assume that THHN insulated conductors are used. See *Chapter 9, Table C1, NEC.*
    a. Three 14 AWG ______________________________
    b. Four 14 AWG ______________________________
    c. Five 14 AWG ______________________________
    d. Six 14 AWG ______________________________
    e. Three 12 AWG ______________________________
    f. Four 12 AWG ______________________________
    g. Five 12 AWG ______________________________
    h. Six 12 AWG ______________________________
    i. Six 10 AWG ______________________________
    j. Four 8 AWG ______________________________
19. According to the *Code*, what trade size EMT is required for each of the following combinations of conductors? Assume that these are new installations. Show all calculations. Refer to *Chapter 9, NEC, Table 4, Table 5,* and *Table 8.*
    a. Three 14 AWG THHN, four 12 AWG THHN

       ____
    b. Two 12 AWG THHN, three 8 AWG THHN

       ____
    c. Three 3/0 AWG THHN, two 8 AWG THHN

       ____
20. a. When more than three current-carrying conductors are installed in one raceway, their allowable ampacities must be reduced according to *NEC* ______________.
    b. The following is a project. We need four 15-ampere branch circuits. We will use two 3-wire multiwire branch circuits and 14 AWG THHN conductors. We will install the conductors in EMT. The room temperature will not exceed 86°F (30°C). The connected load is nonmotor and is noncontinuous.
       1. What is the minimum trade size EMT? ______________________________
       2. What is the ampacity of the conductors before derating? ______________________________

          ____
       3. What is the ampacity of the conductors after derating? ______________________________

          ____

4. What is the maximum overcurrent protection permitted for these conductors?

____

____

____

c. Let's increase the conductor size to 12 AWG THHN for the same installation to see what happens.

1. What is the minimum trade size EMT? ____

____

2. What is the ampacity of the conductors before derating? ____

____

3. What is the ampacity of the conductors after derating? ____

____

4. What is the maximum overcurrent protection permitted for these conductors?

____

____

d. It (is) (is not) necessary to count equipment grounding conductors, a neutral conductor that carries the unbalanced current from other conductors of the same circuit, and/or "travelers" when installing 3-way and 4-way switches for the purposes of derating the ampacity of these conductors. Circle the correct answer. Explain.

____

____

21. What is the minimum number of receptacles required by the *Code* for basements?

____

____

22. For laundry equipment in basements, what sort of electrical circuitry is required by the *Code*? ____

____

____

23. How many receptacles are required to complete the multioutlet assembly above the workbench? ____

24. A Type NM cable carries a branch circuit to the outside of a building, where it terminates in a metal weatherproof junction box. From this box, a short length of EMT is installed upward to another metal weatherproof junction box on which is attached a cluster of outdoor floodlights. The first box is located approximately 24 in. (600 mm) above the ground. To "Meet *Code*," do the metal boxes and metal raceway have to be grounded? Yes ______________ No ______________ *NEC* ______________.

25. When raised covers are used for receptacles, does the *Code* permit fastening the receptacle to the cover with the one center No. 6-32 screw? Yes ______________ No ______________ *NEC* ______________.

26. The following is a layout of the lighting circuits for the workshop. Using the conduit layout in Figure 18-1, make a complete wiring diagram. Use colored pencils or pens to indicate conductors.

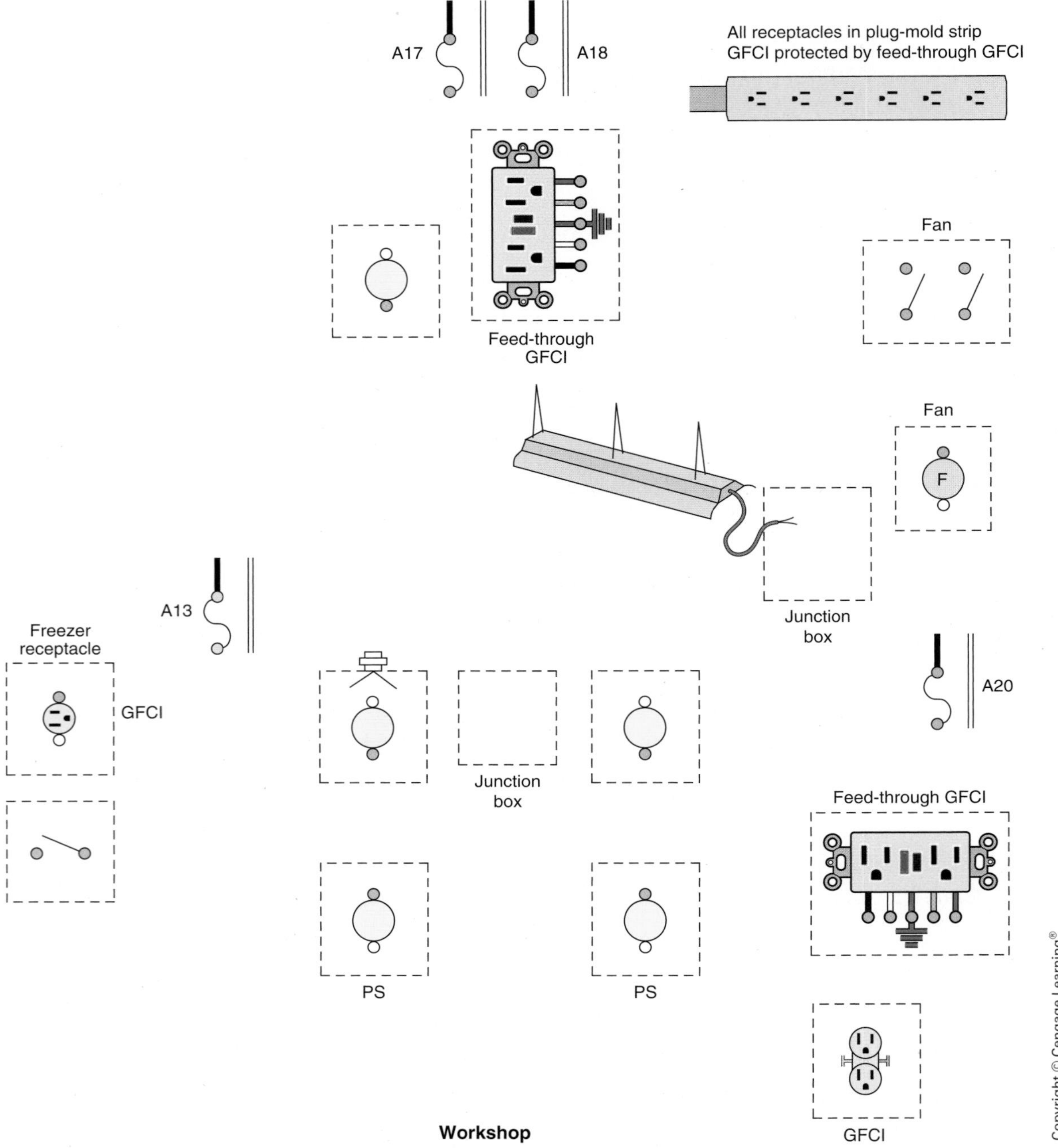

27. *NEC 210.8(A)(5)* requires that GFCI protection be provided for all receptacles in unfinished basements. Are there any exceptions?

____

28. Where loads are likely to be on continuously, the *calculated* load for branch circuits and feeders must be figured at (100%) (125%) (150%) of the continuous load, (plus) (minus) the noncontinuous load. This requirement generally (does) (does not) apply to residential wiring. (Circle the correct answers.)

CHAPTER **19**

Special-Purpose Outlets—Water Pump, Water Heater

OBJECTIVES

After studying this chapter, you should be able to

- understand the operation of jet pumps, submersible pumps, and their components.
- understand the operation of electric water heaters and their components.
- be familiar with the *NEC* requirements for designing the branch circuit, including conductors, cables, raceway, motor branch-circuit short-circuit ground-fault and overload protection, disconnecting means, and grounding for water pumps and electric water heaters.
- figure out various electrical connections for "Time-of-Use" metering.
- calculate the effect of voltage variation on heating elements and motors.
- understand the hazards of possible scalding.
- discuss heat pump water heaters.

WATER PUMP CIRCUIT ⒶB

In rural areas where there is no public water supply, dwellings need their own water supply. The Electrical Plans and Schedule of Special-Purpose Outlets for this residence show that the well pump is connected to Circuit A 5-7.

After a brief introduction to jet pumps and submersible pumps, we will discuss the key elements that make up a typical motor branch circuit. For the most part, submersible pumps dominate the residential market.

The manufacturer's installation instructions must always be followed.

JET PUMPS

Figure 19-1 shows the major components of a typical deep-well jet pump.

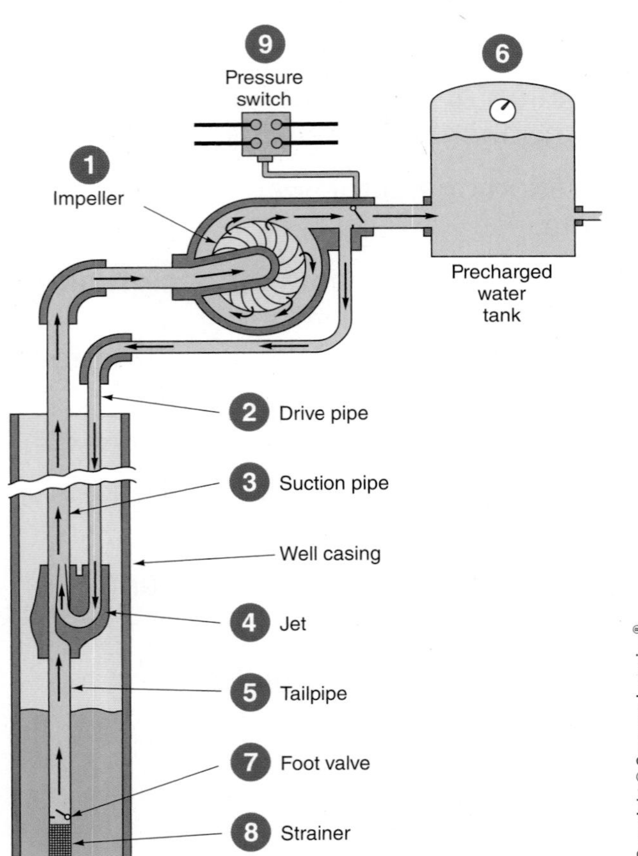

FIGURE 19-1 Components of a jet pump.

The pump impeller wheel, ①, forces water down a drive pipe, ②, at a high velocity and pressure to a point just above the water level in the well casing. Just above the water level, the drive pipe curves sharply upward and enters a larger vertical suction pipe, ③. The drive pipe terminates in a small nozzle or "jet," ④. The water emerges from the jet with great force and flows upward though the suction pipe.

Water rises in the suction pipe, drawn up through the tailpipe, ⑤, by the action of the jet. The water rises to the pump inlet and passes through the impeller wheel of the pump. Some of the water is forced down through the drive pipe again. The remaining water passes through a check valve and enters the storage tank, ⑥.

The foot valve, ⑦, prevents water in the pumping equipment from draining back into the well when the pump is not operating. The strainer, ⑧, keeps debris out of the system. The pressure switch, ⑨, starts and stops the pump motor.

Figure 19-2 shows the wiring of a typical water pump. Always follow the wiring instructions furnished by the specific pump manufacturer.

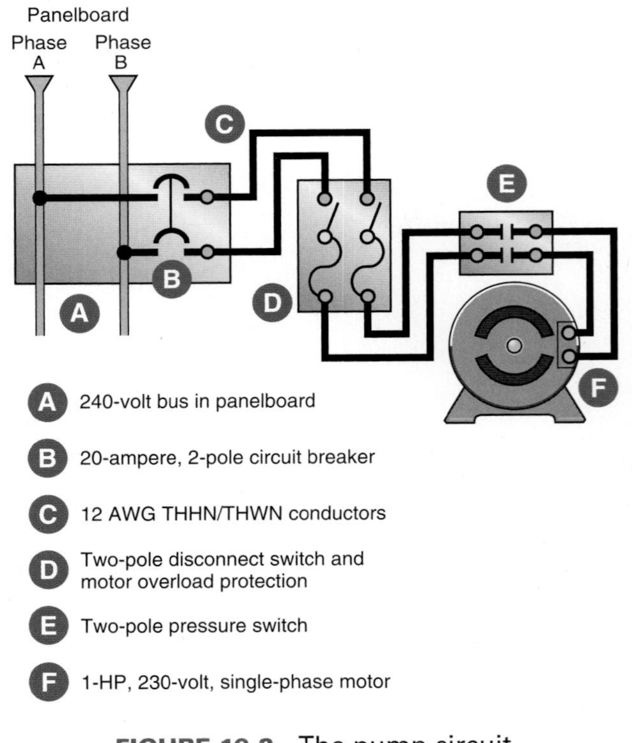

FIGURE 19-2 The pump circuit.

SUBMERSIBLE PUMPS

Figure 19-3 shows the major components of a typical submersible pump.

A submersible pump consists of a centrifugal pump, ①, driven by an electric motor, ②. The pump and the motor are contained in one housing, ③, submersed below the permanent water level, ④, within the well casing, ⑤. The pump housing is a cylinder 3–5 in. in diameter and 2–3 ft long. When running, the pump raises the water upward through the piping, ⑥, to the water tank, ⑦. Proper pressure is maintained in the system by a pressure switch, ⑧. The disconnect switch, ⑨, pressure switch, ⑧, limit switches, ⑩, and controller, ⑪, are installed in a logical and convenient location near the water tank.

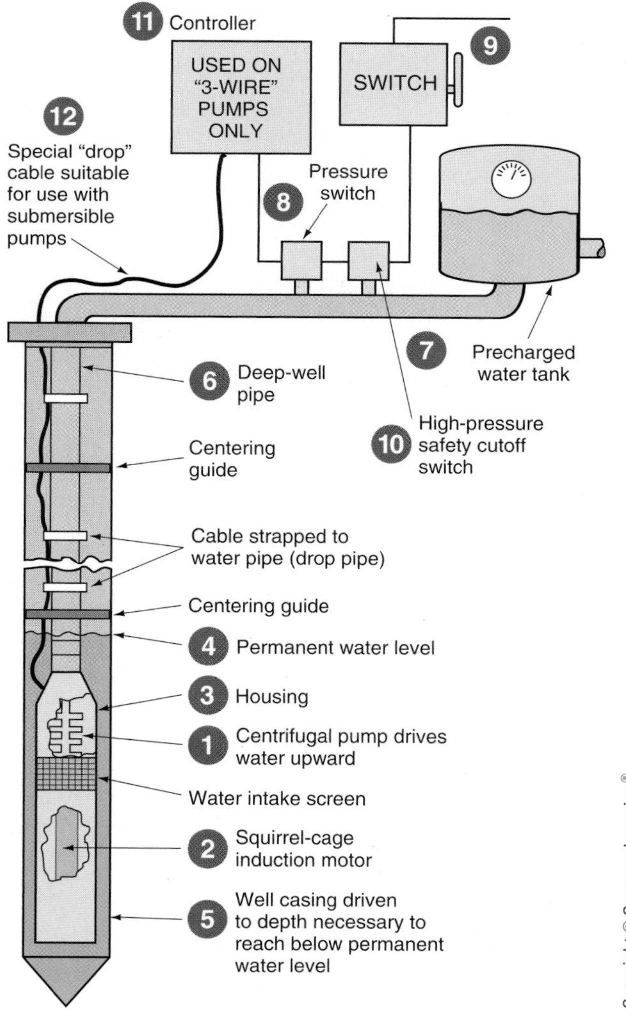

FIGURE 19-3 Submersible pump.

Pumps commonly called "2-wire pumps" contain all required starting and protection components and connect directly to the pressure switch—with no other aboveground controller. Generally, there are no moving electrical parts within the submersible pump, such as the centrifugal starting switch found in a typical single-phase, split-phase induction motor.

Other pumps, referred to as "3-wire pumps," require an aboveground controller that contains any required components not in the motor, such as a starting relay, overload protection, starting and running capacitors, lightning arrester, and terminals for making the necessary electrical connections.

Most water pressure tanks have a precharged air chamber in an elastic bag (bladder) that separates the air from the water. This ensures that air is not absorbed by the water. Water pressure tanks compress air to maintain delivery pressure in a usable range without cycling the pump at too fast a rate. When in direct contact with the water, air is gradually absorbed into the water, and the cycling rate of the pump will become too rapid for good pump life unless the air is periodically replaced. With the elastic bag, the initial air charge is always maintained, so recharging is unnecessary.

Submersible pumps are covered by UL Standard 778.

Submersible Pump Cable

Power to the motor is supplied by a "drop" cable, ⑫, especially designed for use with submersible pumps. This cable is marked "submersible pump cable," and it is generally supplied with the pump. This cable can also be purchased separately. The cable is cut to the proper length to reach between the pump and its controller, as shown in Figure 19-3, or between the pump and the well cap, Figure 19-4. When needed, the cable may be spliced according to the manufacturer's specifications.

Submersible water pump cable is "tag-marked" for use within the well casing for wiring deep-well water pumps, where the cable is not subject to repetitive handling caused by frequent servicing of the pump units.

A submersible pump cable is not designed for direct burial in the ground unless it is marked "Type USE" or "Type UF." Should it be required to run the

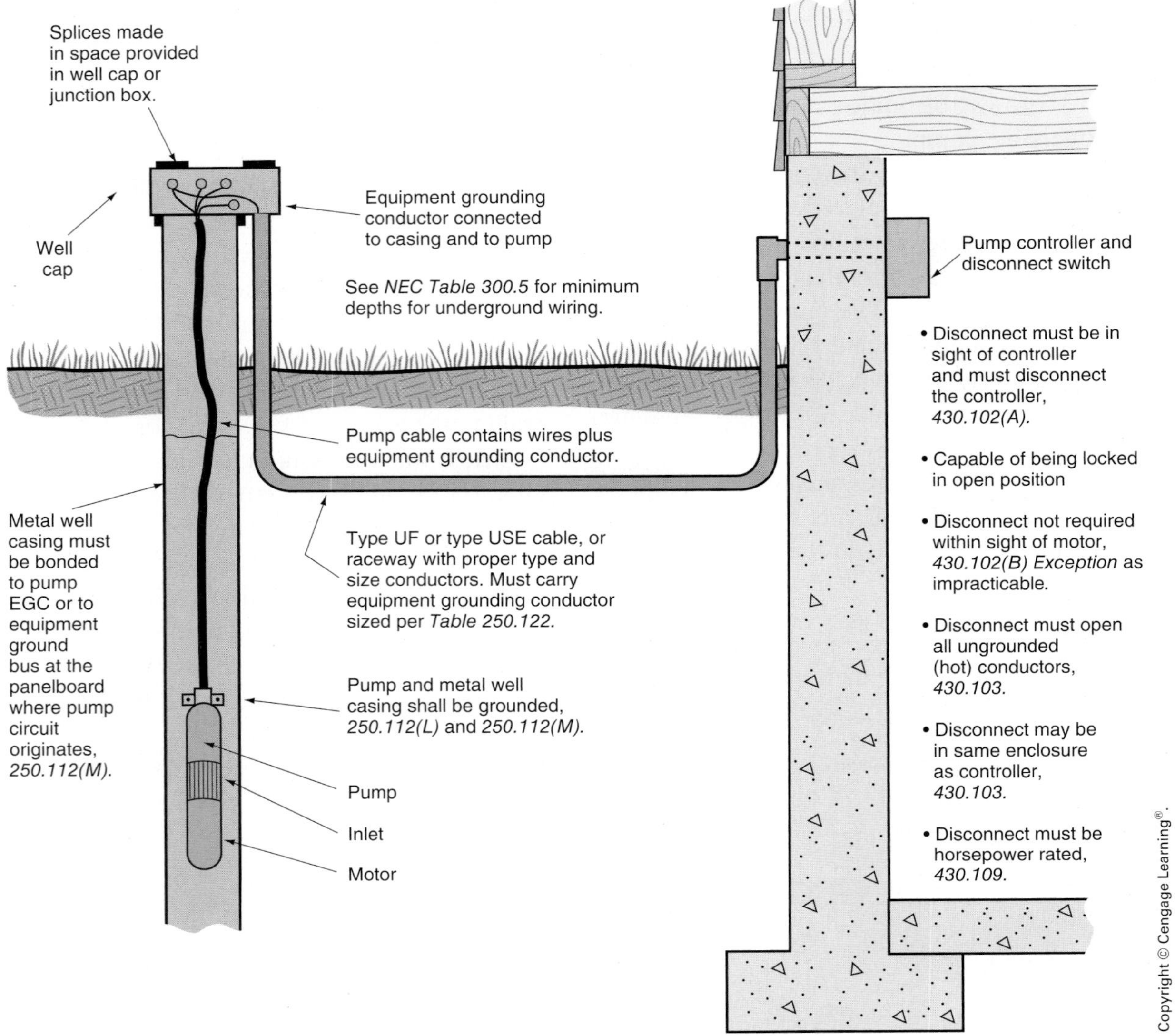

FIGURE 19-4 Grounding and disconnecting means requirements for submersible water pumps.

pump's circuit underground for any distance, it is necessary to install Type UF or Type USE cable or a raceway with suitable conductors, and then make up the necessary splices in a listed weatherproof junction box, as in Figure 19-4. Chapter 16 covers underground wiring in great detail.

MOTOR CIRCUIT DESIGN

We will now take a look at the *NEC*® requirements for all of the basics for a typical electrical motor branch circuit.

The pump circuit in this residence originates in Main Service Panelboard A, Circuit A5-7.

Current and Voltage

This information is found on the nameplate and instructions furnished with the pump.

The water pump in this residence is a 1-horsepower (hp), single-phase, 115/230-volt pump. Dual-voltage-rated pumps can be supplied by either 115 volts or 230 volts. When connected for the higher voltage, the current draw will be half of that when connected for the lower voltage.

For example, in *Table 430.248*, the full-load current rating of a 1-horsepower, 23-volt, single-phase motor is 8 amperes. The same motor connected to 115 volts will draw 16 amperes. Current is a big factor in determining conductor size, watts loss in the conductors, voltage drop, disconnect switch size, controller size, motor overload protection, and the size and type of branch-circuit overcurrent protection. Some pump controllers have a simple slide switch to change the connections from 115 volts to 230 volts. Pumps sized ½ horsepower are usually prewired for 115 volts. Pumps ¾ horsepower and larger are usually prewired for 230 volts.

For the purpose of our motor circuit design, we will connect this motor for 230 volts.

Table 430.248 for 115-volt motors covers the voltage range of 110–120 volts. For 23-volt motors, the voltage range is 220–240 volts. You will also find that some motors are marked 115/208–230 volts.

The NEMA Standard MG-1 *Information Guide for General Purpose Industrial AC Small and Medium Squirrel-Cage Induction Motor Standards* is a condensed version of the much larger NEMA Standard MG 1-2003 *Motors and Generators*. These standards show that most general-purpose NEMA-rated motors are designed to operate at ±10% of the motor's rated voltage.

Important: For determining the conductor size, switch size, and the branch-circuit short-circuit and ground-fault protection, use the current values from the tables in the *NEC*. See *430.6(A)(1)*.

For overload protection sizing, use the motor nameplate current value. See *430.6(A)(2)*. For the typical submersible pump motor, the running overload protection is almost always furnished by the manufacturer in the controller.

The nameplate rating of the pump motor is 8 amperes, 230 volts.

Conductor Size (*430.22*)

From *Table 430.248*, the full-load current rating of a 1-horsepower, 23-volt, single-phase motor is 8 amperes. To calculate the conductor size, use 125% of 8 amperes.

$$1.25 \times 8 = 10 \text{ amperes}$$

Use Type THHN/THWN conductors per the Specifications.

Checking *Table 310.15(B)(16)* of the *NEC*, we find that a 14 AWG Type THHN/THWN conductor has an ampacity in the 60°C column of 15 amperes and 20 amperes in the 75°C column, more than ample for the pump motor circuit. However, the Schedule of Special-Purpose Outlets found in this text, which is part of the Specifications for the residence, indicates that 12 AWG THHN/THWN conductors are to be used for the pump circuit. Installing larger size conductors can minimize voltage drop and might be suitable at a later date should the need arise to install a larger-capacity pump.

The two 12 AWG THHN/THWN conductors are connected to a 2-pole, 20-ampere branch-circuit breaker in Panelboard A, Circuit A(5–7).

Raceway Size

Two 12 AWG Type THHN/THWN conductors require trade size ½ EMT, *NEC Annex C, Table Cl.* This EMT will run across the ceiling from Panelboard A to the disconnecting means located in the southeast corner of the workshop. The controller and other equipment relating to the water pump are located in this corner of the workshop. Type UF cable or PVC conduit is buried underground to the well casing location. The cable or conduit is protected at both ends from the burial depth given in *Table 300.5* to the point of entry.

Motor Branch-Circuit Short-Circuit and Ground-Fault Protection (*430.52, Table 430.52*)

Fuses and circuit breakers are the most commonly used forms of motor branch-circuit short-circuit and ground-fault protection. Should the motor windings or branch-circuit conductors short-circuit or go to ground, the motor's branch-circuit short-circuit and ground-fault overcurrent device, either fuses or circuit breakers, will open.

Motor branch-circuit short-circuit and ground-fault protection does not protect the motor against overload burnout! That protection is provided in the motor controller for the submersible pump.

The various choices for motor branch-circuit short-circuit and ground-fault protection are found in *430.52* and *Table 430.52*. In almost all instances, the values found in *Table 430.52* will work just fine. In the event that the sizing calculation results in an unavailable ampere rating fuse or circuit breaker, select the next standard ampere rating, but do not exceed the absolute maximum sizing percentages shown in *430.52*. This will be discussed in more detail after we do the calculations.

From *Table 430.248*, the full-load current rating of a 1-horsepower, 23-volt, single-phase motor is 8 amperes. Here are the possibilities:

- Non-Time-Delay Fuses (300% desired—maximum size: 400%)

  8 × 3 = 24 amperes (use 25-ampere fuses)

  The disconnect switch would be a 30-ampere, 2-pole, 250-volt switch.

- Dual-Element, Time-Delay Fuses (175% desired—maximum size: 225%)

  8 × 1.75 = 14 amperes (use 15-ampere fuses)

  The disconnect switch would be a 30-ampere, 2-pole, 250-volt switch.

- Inverse Time Breakers (250%)

  8 × 2.5 = 20-ampere rating

  Maximum setting: 400% for FLA of 100 amperes or less.

- Instant Trip Breakers are never used in residential applications.

The motor branch-circuit overcurrent device must be capable of allowing the motor to start, *430.52(B)*. Where the values for branch-circuit protective devices (fuses or breakers) as determined by *Table 430.52* do not correspond to the standard sizes, ratings, or settings as listed in *240.6*, the next higher size, rating, or setting is permitted,*430.52(C)(1), Exception No. 1*.

If, after applying the values found in *Table 430.52* and after selecting the next higher standard size, rating, or setting of the branch-circuit overcurrent device, the motor still will not start, *430.52(C)(1), Exception No. 2*, permits using an even larger size, rating, or setting. The maximum values (percentages) are shown in this exception.

A nuisance opening of a fuse, or tripping of a breaker, can easily be predicted by referring to time/current curves for the intended fuse or breaker. Though a detailed discussion of this topic is beyond the scope of this book, a discussion of the operating characteristics of fuses and circuit breakers is covered in Chapter 28.

Motor Overload Protection (*430.31* through *430.44*)

Usually the manufacturer provides the necessary motor overload protection in the controller, and no additional protection is required.

Overload protection is needed to keep the motor from dangerous overheating due to overloads or failure of the motor to start. *NEC 430.31* through *430.44* cover virtually every type and size of motor, starting characteristics, and duty (continuous or noncontinuous) in use today. The overload protection for a motor might be thermal overloads (sometimes called *heaters*) in the controller, electronic sensing overload devices, built-in (inherent) thermal protection, or time-delay fuses.

We find the sizing requirements for motor overload protection in *430.32(A)(1)* and *430.32(B)(1)*. Although there are exceptions, most of the common installations require motor overload protection not to exceed 125% of the motor's full-load current draw, as indicated on the nameplate. If you need to provide this running overload protection, the running load must be obtained from the manufacturer's literature or from the nameplate on the motor. We will assume the nameplate current to be 7 amperes. For the pump motor this would be:

1.25 × 7 = 8.8 amperes

Therefore, for backup motor overload protection, we could install 10-ampere, time-delay, dual-element fuses in the disconnect switch. If we are interested only in motor branch-circuit short-circuit and ground-fault protection, we refer to *Table 430.52* and apply the multiplier of 175 percent.

7 × 1.75 = 12.3 amperes

We are permitted to go to the next standard size, which is a 15-ampere time-delay, dual-element fuse.

Instructions furnished with a particular motor or motor-operated appliance might indicate a maximum size overload protection and a maximum size branch-circuit protection. These values must be followed.

Disconnecting Means

The *NEC* requires that all motors be provided with a means to disconnect the motor from its electrical supply, *Article 430, Part IX*.

NEC 430.103 requires that a disconnecting means for a motor shall

- open all ungrounded conductors,
- be designed so all poles operate together (simultaneously), and
- be designed so that it cannot be closed automatically.

The disconnecting means for the water pump is a 30-ampere, two-pole, 250-volt switch mounted on the wall next to the pump controller. See Figures 19-3 and 19-4.

Follow these two simple rules for locating the disconnect switch.

Rule #1. *NEC 430.102(A)*: An individual disconnect must be in sight of the controller and must disconnect the controller. The *NEC* definition of "in sight" means that the controller must be visible and not more than 50 ft (15 m) from the disconnect.

Rule #2. *NEC 430.102(B)*: The disconnect must be in sight of the motor and driven machinery. If the disconnect, as required in *430.102(A)*, is in sight of the controller, the motor, and driven machinery, then that disconnect meets the requirements of both *430.102(A)* and *430.102(B)*.

How to Use the Exception to the "In-Sight" Rule

As with most good rules, there are exceptions! The *Exception* and the *Informational Notes* to *430.102(B)* tell us that the disconnect need not be in sight of the motor and driven machinery if the disconnect can be individually locked in the open OFF position. The locking provision must be of a permanent type installed on the switch or circuit breaker. Most disconnect switches have this LOCK-OFF feature. Listed circuit breaker lock-off devices that fit over the top of the circuit breaker handle also meet this requirement.

Anyone working on the pump needs to be assured that the power is off and stays off until he or she is ready to turn the power back on. This is particularly important in our installation because the motor is out of sight from the controller.

The disconnect switch for the submersible pump for this residence is located next to and in sight of the controller. It is not in sight of the motor and can't be, because the motor is located inside the well casing below the water level. Here's where the *Exception* comes into play. The *Exception* provides that disconnecting means is not required within sight of the motor if such a location is impracticable. The disconnecting means located on the line or supply side of the controller must be capable of being locked in the open position. See Figure 19-4.

GROUNDING

Figure 19-4 illustrates how to provide proper grounding of a submersible water pump and the well casing. Proper grounding of the well casing and submersible pump motor will minimize or eliminate stray voltage problems that could occur if the pump motor is not grounded.

Key *Code* sections that relate to grounding water-pumping equipment:

- *NEC 250.4(A)(5)*: The last sentence states that The earth shall not be considered as an effective ground-fault current path.*
- *NEC 250.86*: Metal enclosures for circuit conductors shall be grounded, other than short sections of metal enclosures that provide support or physical protection.
- *NEC 250.112(L)*: Motor-operated water pumps, including submersible pumps, must be grounded.
- *NEC 250.112(M)*: Bond metal well casing to the pump circuit's equipment grounding conductor.
- *NEC 250.134(B)*: Equipment fastened in place or connected by permanent wiring must have the equipment grounding conductor run with the circuit conductors.
- *NEC 430.241*: Grounding of motors—general.
- *NEC 430.242*: Grounding of motors—stationary motors.
- *NEC 430.244*: Grounding of controllers.
- *NEC 430.245*: Acceptable methods of grounding.

*Reprinted with permission from NFPA 70-2014.

Because so much nonmetallic (PVC) water piping is used today, some local electrical codes or interpretations require that a grounding electrode conductor be installed between the neutral bar of the main service disconnecting means and the metal well casing. The interpretation often related to the proximity of the well casing to the main building or structure. The *NEC* does not provide a distance beyond which a connection to the well casing is no longer required. Connecting to the metal well casing would satisfy the requirements of *250.104* (bonding), *250.50* (grounding electrode system), *250.52* (grounding electrode descriptions), and *250.53* (installation of grounding electrodes). If installed, the grounding electrode conductor would be sized in compliance with *Table 250.66* of the *Code*. At the metal casing, the grounding electrode conductor is attached to a lug termination or by means of exothermic (Cadweld) welding.

WATER HEATER CIRCUIT Ⓐc

Residential electric water heaters are listed under UL Standard 174, *Household Electric Storage Tank Water Heaters*.

Electrical contractors many times are also in the plumbing, heating, and appliance business. They need to know more than how to do the electrical hookup. The following text discusses electrical as well as other data about electric water heaters that will prove useful.

All homes require a supply of hot water. To meet this need, one or more automatic water heaters are generally installed as close as practical to the areas having the greatest need for hot water. Water piping carries the heated water from the water heater to the various plumbing fixtures and to appliances such as dishwashers and clothes washers.

For safety reasons, in addition to the regular temperature-control thermostat that can be adjusted by the homeowner, the *NEC* and UL Standards require that electric water heaters be equipped with a high-temperature cutoff. This high-temperature control limits the maximum water temperature to 190°F (88°C). This is 20°F (11°C) below the 210°F (99°C) temperature that causes the pressure/temperature relief valve to open. The high-temperature cutoff is factory preset and should never be tampered with or modified in the field. In conformance to *422.47* of the *NEC*, the high-temperature limit control must disconnect all ungrounded conductors.

In most residential electric water heaters, the high-temperature cutoff and upper thermostat are combined into one device, as in Figure 19-5.

Combination Pressure/ Temperature Safety Relief Valves

Look at Figure 19-5. A pressure/temperature relief valve is an important safety device installed into an opening in the water heater tank within 6 in. (150 mm) of the top. The opening is provided for and clearly marked by the manufacturer. Many water heater pressure/temperature relief valves are factory installed, although some models still require installation of the valve by the installing contractor. Proper discharge piping must be installed from the pressure/temperature relief valve downward to within 6 in. (150 mm) of the floor, preferably near a floor drain. Terminating the discharge pipe to within 6 in. (150 mm) of the floor is important to protect people from injury from scalding water or steam discharge should the relief valve activate for any reason. This also protects electrical equipment in and around the water heater from becoming soaked should the relief valve operate.

Pressure/temperature relief valves are installed for two reasons. Their pressure/temperature-sensing features are interrelated.

- *Pressure*: Water expands when heated. Failure of the normal temperature thermostat and the high-temperature-limiting device to operate could result in pressures that exceed the rated working pressure of the water heater tank and the plumbing system components. The relief valve, properly installed and properly rated to local codes, will open to relieve the excess pressure in the tank to the predetermined setting marked on the relief valve. The rating is stamped on the valve's identification plate.
- *Temperature*: Should runaway water heating conditions occur within the water heater, temperatures far exceeding the atmospheric boiling

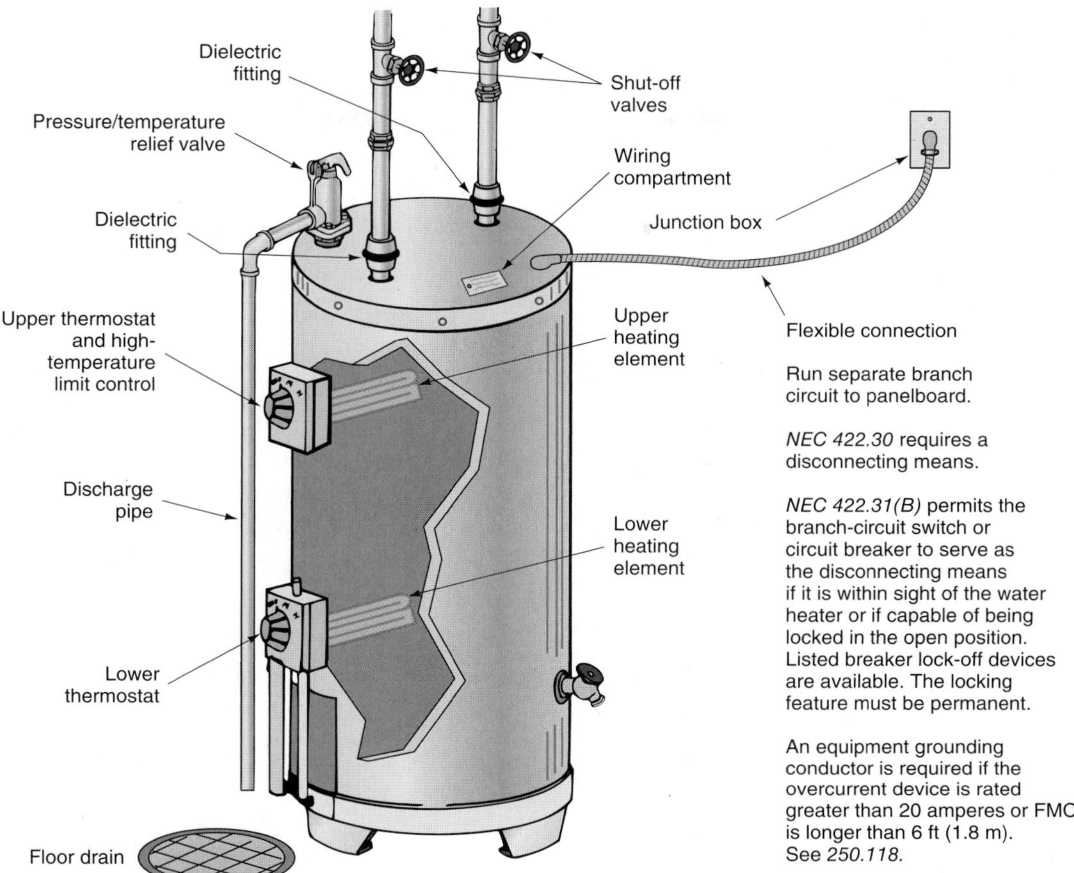

FIGURE 19-5 Typical electric water heater showing location of heating elements and thermostats and electrical connection.

point of water (212°F [100°C]) can take place. Under high pressures, water will stay liquid at extremely high temperatures. For a typical municipal water supply pressure of 45 psi, water will not boil until it reaches approximately 290°F (143°C). At 290°F (143°C), opening a faucet, a broken pipe, or a ruptured water heater tank would allow the pressure to instantly drop to normal atmospheric pressure. The superheated water would immediately flash into steam, having about 1600 times more volume than the liquid water occupied. This blast of steam can be catastrophic and could result in severe burns or death.

Thus, the function of the temperature portion of the pressure/temperature relief valve is to open when a temperature of 210°F (99°C) occurs within the water heater tank. Properly rated and maintained, the pressure/temperature relief valve will allow the water to flow from the water heater through the discharge pipe at a rate faster than the heating process, thereby keeping the water temperature below the boiling point. Should the cold water supply be shut off, the relief valve will allow the steam to escape without undue buildup of temperature and pressure in the tank.

Requirements for relief valves for hot water supply systems are found in the ANSI Standard Z21.22.

Sizes

Residential electric water heaters are available in many capacities, such as 6, 12, 15, 20, 30, 32, 40, 42, 50, 52, 66, 75, 80, 82, 100, and 120 gallons.

Corrosion of Tank

To reduce corrosion of the steel tank, most water heater tanks are glass lined. Glass lining is a combination of silica and other minerals (rasorite, rutile, zircon, cobalt, nickel oxide, and fluxes). This special mix is heated, melted, cooled, crushed, and then sprayed onto the interior steel surfaces and baked on at about 1600°F (871°C). To further reduce corrosion, aluminum or magnesium anode rods are installed in special openings, or as part of the hot water outlet fitting, to allow for the flow of a protective current *from* the rod(s) *to* any exposed steel of the tank. This is commonly referred to as *cathodic protection*. Corrosion problems within the water heater tank are minimized as long as the aluminum or magnesium rod(s) remain in an active state. Anodes should be inspected at regular intervals to determine when replacement is necessary, usually when the rod(s) is reduced to about one-third of its original diameter or when the rod's core wire is exposed. Further information can be obtained by contacting the manufacturer and consulting the installation manual.

Some water heater tanks are fiberglass or plastic lined.

Heating Elements

The wattage ratings of electric water heaters can vary greatly, depending on the capacity of the heater in gallons, the speed of recovery desired, local electric utility regulations, and codes. Typical wattage ratings are 1500, 2000, 2500, 3000, 3800, 4500, and 5500 watts.

A resistance heating element might be dual rated. For example, the element might be marked 5500 watts at 250 volts, and 3800 watts at 208 volts.

Most residential-type water heaters are connected to 240 volts except for the smaller 2-, 4-, and 6-gallon *point of use* sizes generally rated 1500 watts at 120 volts. Commercial electric water heaters can be rated single-phase or 3-phase, 208, 240, 277, or 480 volts.

UL requires that the power (wattage) input must not exceed 105% of the water heater's nameplate rating. All testing is done with a supply voltage equal to the heating element's *rated* voltage. Most heating elements may burn out prematurely if operated at voltages 5% higher than for which they are rated.

To help reduce the premature burnout of heating elements, some manufacturers will supply 250-volt heating elements, yet will mark the nameplate 240 volts, with its corresponding wattage at 240 volts. This allows a safety factor if slightly higher than normal voltages are experienced.

Heating Element Construction

Heating elements generally contain a nickel-chrome (nichrome) resistance wire embedded in compacted powdered magnesium oxide to ensure that no grounds occur between the wire and the sheath. The magnesium oxide is an excellent insulator of electricity, yet it effectively conducts the heat from the nichrome wire to the sheath.

The sheath can be made of tin-coated copper, stainless steel, or an iron-nickel-chromium alloy. The latter alloy comes under the trade name of Incoloy. The stainless steel and Incoloy types can withstand higher operating temperatures than the tin-coated copper elements. They are used in premium water heaters because of their resistance to deterioration, lime scaling, and dry firing burnout.

High watt density elements generally have a U-shaped tube and give off a great amount of heat per square inch of surface. Figure 19-6(A) illustrates a flange-mounted, high-density heating element. Figure 19-6(B) illustrates a high-density heating element that screws into a threaded opening in the tank. High-density elements are very susceptible to lime-scale burnout.

Low watt density elements, as in Figure 19-6 (C), generally have a double loop, like a U tube bent in half. They give off less heat per in.^{2}, have a longer life, and are less noisy than high-density elements.

An option offered by at least one manufacturer of electric water heaters is a dual-wattage heating element. A dual-wattage heating element actually contains two heating elements in one jacket. One element is rated 3800 watts, and one element is rated 1700 watts. Three leads are brought out to the terminal block. The 3800-watt heating element is connected initially, with the option of connecting the second 1700-watt heating element if the wiring to the water heater is capable of handling the higher

A B C

FIGURE 19-6 Heating elements.

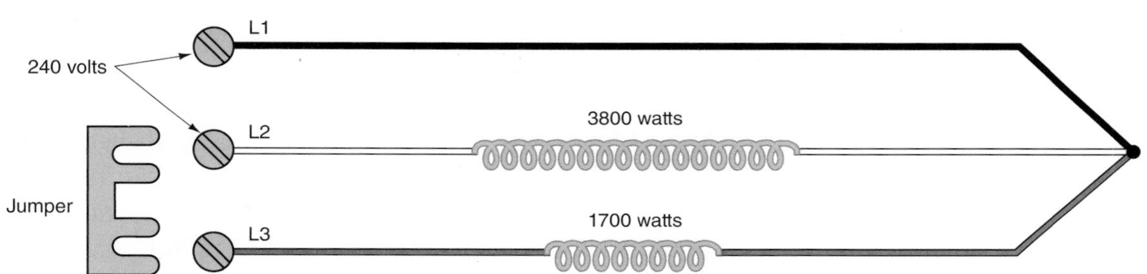

FIGURE 19-7 A dual-wattage heating element. The two wires of the incoming 240-volt circuit are connected to L1 and L2. These leads connect to the 3800-watt element. To connect the 1700-watt element, the jumper is connected between L2 and L3. With both heating elements connected, the total wattage is 5500 watts.

total combined wattage of 5500 watts. To attain the higher wattage, a jumper (provided by the manufacturer) is connected between two terminals on the terminal block, Figure 19-7.

Any electric heating element coated with lime scale may be noisy during the heating cycle. Limed-up heating elements should be cleaned or replaced, following the manufacturer's instructions.

Do not energize the electrical supply to a water heater unless you are sure it is full of water. Most water heater heating elements are designed to operate only when submersed in water.

Residential electric water heaters are available with one or two heating elements. Figure 19-5 shows two heating elements. Single-element water heaters will have the heating element located near the bottom of the tank. If the water heater has two heating elements, the upper element will be located about two-thirds of the way up the tank.

Speed of Recovery

Speed of recovery is the time required to bring the water temperature to satisfactory levels in a given length of time. The current accepted industry standard is based on a 90°F (32.22°C) rise. For example, referring to Table 19-1, we find that a 3500-watt element can raise the water temperature 90°F (32.22°C) of a little more than 16 gallons an hour. A 5500-watt element can raise water temperature 90°F (32.22°C) of about 25 gallons an hour. Speed of recovery is affected by the type and amount of insulation surrounding the tank, the supply voltage, and the temperature of the incoming cold water.

TABLE 19-1

Electric water heater recovery rate per gallon of water for various wattages.

HEATING ELEMENT WATTAGE	GALLONS PER HOUR RECOVERY FOR INDICATED TEMPERATURE RISE					
	60°F	70°F	80°F	90°F	100°F	Equivalent NET Btu Output
0750	5.2	4.4	3.9	3.5	3.1	02559
1000	6.9	5.9	5.2	4.6	4.1	03412
1250	8.6	7.4	6.5	5.7	5.2	04266
1500	10.3	8.9	7.8	6.9	6.2	05119
2000	13.8	11.8	10.3	9.2	8.3	06825
2250	15.5	13.3	11.6	10.3	9.3	07678
2500	17.2	14.8	12.9	11.5	10.3	08531
3000	20.7	17.7	15.5	13.8	12.4	10,238
3500	24.1	20.7	18.1	16.1	14.5	11,944
4000	27.6	23.6	20.7	18.4	16.6	13,650
4500	31.0	26.6	23.3	20.7	18.6	15,356
5000	34.5	29.6	25.9	23.0	20.7	17,063
5500	37.9	32.5	28.4	25.3	22.5	18,769
6000	41.4	35.5	31.0	27.6	24.6	20,475

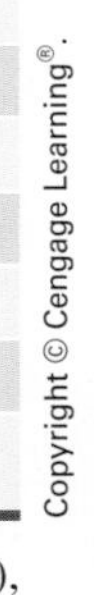

Notes: If the incoming water temperature is 40°F (Northern states), use the 80° column to raise the water temperature to 120°F.

If the incoming water temperature is 60°F (Southern states), use the 60° column to raise the water temperature to 120°F.

For example, approximately how long would it initially take to raise the water temperature in a 42-gallon electric water heater to 120°F where the incoming water temperature is 40°F? The water heater has a 2500-watt heating element.

Answer: 42 ÷ 12.9 = 3.26 hours (approximately 3 hours and 16 minutes).

To keep this table simple, noncluttered, and easy to read, only Fahrenheit temperatures are shown.

Table 19-2 shows the approximate time it takes to run out of hot water, using typical shower heads with various storage capacity residential electric water heaters. Current standards for new showerheads are typically 2.5 gallons per minute. When the flow restrictors are removed, the flow rate can increase to 6–18 gallons per minutes per showerhead. There are very low-flow-rate showerheads available in the range of 1–2 gallons per minute. All of this depends on the water pressure.

An easy way to determine a showerhead's flow rate is to turn on the water, hold a large bucket of known capacity under it, and time how long it takes to fill the bucket.

Scalding from Hot Water

The following are plumbing code issues but are discussed briefly because of their seriousness.

Temperature settings for water heaters are problematic!

Set the water heater thermostat high enough to kill bacteria on the dishes in the dishwasher and possibly get scalded in the shower. Or set the thermostat to a lower setting so as not to get scalded in the shower, but live with the possibility that bacteria will still be present on washed dishes.

The Consumer Product Safety Commission (CPSC) reports that 3800 injuries and 34 deaths occur each year due to scalding from excessively hot tap water. Unfortunately, the victims are the elderly and children under the age of 5. They recommend that the setting should not be higher than 120°F (49°C). Some state laws require that the thermostat be preset at no higher than 120°F (49°C). Residential water heaters are all preset by the manufacturer at 120°F (49°C).

SAFETY ALERT

Manufacturers post caution labels on their water heaters that read something like this:
SCALD HAZARD: *Water temperature over 120°f (49°c) can cause severe burns instantly or death from scalds. See instruction manual before changing temperature settings.*

TABLE 19-2

This chart shows the approximate time in minutes it takes to run out of hot water when using various shower heads, different incoming water temperatures, and different capacity water heaters.

SHOWER-HEAD(S) RATE IN GALLONS PER MINUTE	HOT WATER FLOW RATE IN GALLONS PER MINUTE BASED ON 120°F. STORED HOT WATER TO PROVIDE 105°F. MIXED WATER OUT OF SHOWERHEAD	HOT WATER FLOW RATE IN GALLONS PER MINUTE BASED ON 120°F. STORED HOT WATER TO PROVIDE 105°F. MIXED WATER OUT OF SHOWERHEAD.	LENGTH OF CONTINUOUS SHOWER TIME—IN MINUTES—FOR TYPICAL SIZES OF RESIDENTIAL ELECTRIC WATER HEATERS. CHART BASED ON 70% DRAW EFFICIENCY. RECOVERY RATE IS NOT FIGURED IN BECAUSE IT IS NORMALLY NOT A FACTOR IN CONTINUOUS DRAW OF WATER SITUATIONS.											
			100 GALLON		80 GALLON		66 GALLON		50 GALLON		40 GALLON		30 GALLON	
	40°F. Incoming water	60°F. Incoming water	40°F. Incom-ing water	60°F. Incom-ing water	40°F. Incom-ing water	60°F. Incom-ing water	40°F. Incom-ing water	60°F. Incom-ing water	40°F. Incom-ing water	60°F. Incom-ing water	40°F. Incom-ing water	60°F. Incom-ing water	40°F. Incom-ing water	60°F. Incom-ing water
02.0	1.6	1.5	43	47	34	37	28	31	22	23	17	19	13	14
02.5	2.0	1.9	34	38	28	30	23	25	17	19	14	15	10	11
03.0	2.4	2.3	29	31	23	25	19	21	14	16	11	12	09	09
04.0	3.3	3.0	22	23	17	19	14	15	11	12	09	09	06	07
05.0	4.0	3.8	17	19	14	15	11	12	09	09	07	07	05	06
06.0	4.9	4.5	14	16	11	12	09	10	07	08	06	06	04	05
08.0	6.5	6.0	11	12	09	09	07	08	05	06	04	05	03	04
12.0	9.8	9.0	07	08	06	06	05	05	04	04	03	03	02	02

Example #1: A 40-gallon electric water heater has a 40°F incoming water supply. How many minutes can a 3 GPM shower head be used to obtain 105°F hot water?

Answer: 11 minutes

Example #2: A residence has an 80-gallon electric water heater. The incoming water supply is approximately 60°F. How many minutes can two 2 GPM shower heads be used at the same time to obtain 105°F hot water? Hint: Two 2 GPM shower heads are the same as one 4 GPM shower head.

Answer: 19 minutes

To reduce the possibility of scalding, water heaters can be equipped with a thermostatic nonscald mixing valve or pressure-balancing valve that holds a selected water temperature to within one degree regardless of incoming water pressure changes. This prevents sudden unanticipated changes in water temperatures.

Table 19-3 shows time/temperature relationships relating to scalding.

What about Washing Dishes?

Older automatic dishwashers required incoming water temperature of 140°F (60°C) or greater to get the dishes clean. A water temperature of 120°F (49°C) is not hot enough to dissolve grease, activate power detergents, and kill bacteria. That is why commercial and restaurant dishwashers boost temperatures to as high as 180°F (82°C).

Newer residential dishwashers have their own built-in water heaters to boost the incoming water temperature in the dishwasher from 120°F (49°C) to 140°F–145°F (60°C–63°C). This heating element also provides the heat for the drying cycle. Today, a thermostat setting of 120°F (49°C) on the water heater should be satisfactory.

TABLE 19-3

Estimated time/temperature relationships related to scalding of an adult subjected to moving water. For children, the time to produce a serious burn is less than for adults.

TEMPERATURE	TIME TO PRODUCE SERIOUS BURN
120°F (48.89°C)	Approx. 9½ minutes
125°F (51.67°C)	Approx. 2 minutes
130°F (54.44°C)	Approx. 30 seconds
135°F (57.22°C)	Approx. 15 seconds
140°F (60.00°C)	Approx. 5 seconds
145°F (62.78°C)	Approx. 2½ seconds
150°F (65.76°C)	Approx. 18⁄10 seconds
155°F (68.33°C)	Approx. 1 second
160°F (71.11°C)	Approx. ½ second

Courtesy of Shriner's Hospitals for Children® - Cincinnati.

Courtesy of THERMO-DISC, Inc., subsidiary of Emerson Electric.

FIGURE 19-8 Typical electric water heater controls. The 7-terminal control combines water temperature control plus high-temperature limit control. This is the type commonly used to control the upper heating element as well as provide a high-temperature-limiting feature. The 2-terminal control is the type used to control water temperature for the lower heating element.

Thermostats/High-Temperature Limit Controls

All electric water heaters have a thermostat(s) and a high-temperature limit control(s).

Thermostats and high-temperature limit controls are available in many configurations. They are available as separate controls and as combination controls.

Figure 19-8 illustrates a combination thermostat/high-temperature limit control. This particular control is an interlocking type sometimes referred to as a snap-over type. It is used on 2-element water heaters and controls the upper heating element as well as provides the safety feature of limiting the water temperature to a factory-preset value. Another thermostat is used to control the lower heating element. When all of the water in the water heater is cold, the upper element heats the upper third of the tank. The lower thermostats are closed, calling for heat, but the interlocking characteristic of the upper thermostat/high limit will not allow the lower heating element to become energized. With this connection, both heating elements cannot be on at the same time. This is referred to as *nonsimultaneous* or *limited demand* operation.

When the water in the upper third of the tank reaches the temperature setting of the upper thermostat, the upper thermostat "snaps over," completing the circuit to the lower element. The lower element then begins to heat the lower two-thirds of the tank. Most of the time, it is the lower heating element that keeps the water at the desired temperature. Should a large amount of water be used in a short time, the upper element comes on. This provides fast recovery of the water in the upper portion of the tank, which is where the hot water is drawn from the tank.

AUTHOR'S TIP: Most residential water heater thermostats are of the single-pole type. They do not open both ungrounded conductors of the 240-volt supply. They open one conductor only, which is all that is needed to open the 240-volt circuit to the heating element. What this means is that a reading of 120 volts to ground will always be present at the terminals of the heating element on a 240-volt, single-phase system such as found in residential wiring. If the heating element becomes grounded to the metal sheath, the element can continue to heat even when the thermostat is in the OFF position because of the presence of the 120 volts. ●

The high-temperature limit control opens both ungrounded conductors.

Various Types of Electrical Connections

There are many variations for electric water heater hookups. Most electric utilities have their own unique time-of-use programs and specific electrical connection requirements for electric water heaters. It is absolutely essential that you contact your local electric utility for this information.

Because of the cumulative large power consumption of electric water heater loads, electric utilities across the country have been innovative in creating special lower rate structures (programs) for electric water heaters, air conditioners, and heat pumps. Electric utilities want to be able to shed some of this load during their peak periods. They will offer the homeowner choices of how the water heater, air conditioner, or heat pump is to be connected and, in some cases, will provide financial incentives (lower rates) to those who are willing to use these restricted "time-of-day" connections. "Time-of-day," "time-of use," and "off-peak" are terms used by different utilities. If homeowners have a large-storage-capacity electric water heater, they will have an adequate supply of hot water during those periods of time the power to the water heater is off. If they cannot live with this, they have other options. Check with the local power company for details on their particular programs.

The following text discusses some of the methods in use around the country for connecting electric water heaters, and in some cases air-conditioning and heat pump equipment.

Utility Controlled

The world of electronics has opened up many new ways for an electric utility to meter and control residential electric water heater, air-conditioning, and heat pump loads. Electronic watt-hour meters can be read remotely, programmed remotely, and can detect errors remotely. The options are endless.

The utility can install watt-hour meters that combine a watt-hour meter and a set of switching contacts. The contacts are switched on and off by the utility during peak hours, using a power line carrier signal. Older style versions of this type of watt-hour meter incorporated a watt-hour meter, time clock, and switching contacts. The time clock would be set by the utility for the desired times of the ON–OFF cycle. Today, the metering and the programming for timing and control of the ON–OFF cycling is accomplished through electronic devices. Electronic meters contain an optical assembly that scans the rotating disc and a microprocessor that records the number of turns and speed of the disc. These electronic digital meters can record the energy consumption during normal hours and during peak hours, allowing the electric utility to bill the homeowner at different rates for different periods of time. One type of residential single-phase watt-hour meter can register four different "time-of-use" rates. This type of meter is shown in Figure 29-3.

One major utility programs their residential meters for "peak periods" from 9 a.m. to 10 p.m.,

Monday through Friday—except for holidays. Their off-peak periods are weekend days and all other hours. Programming of electronic meters can extend out for more than 5 years. The possibilities for programming electronic meters are endless. Older style mechanical meters used gears to record these data and did not have the programming capabilities that electronic meters do as in Figures 19-9, 19-10, and 19-11.

The switching contacts in the meter are connected in *series* with the water heater, air conditioner, or heat pump load that is to be controlled for the special "time-of-day" rates, Figures 19-9, 19-10, and 19-11. The utility determines when they want to have the electric water heater on or off.

Figure 19-9 shows a separate combination meter/time clock for the electric water heater load. The utility sets the clock to be OFF during peak hours of high-power consumption. Thus the term *off-peak metering*. Off-peak metering is oftentimes called "time-of-use" or "time-of-day" metering. In exchange for lower electric rates, the homeowner runs the risk of being without hot water. Should homeowners run out of hot water during these off periods, they have no recourse but to wait until the power to the water heater comes on again. For this type of installation, large-storage-capacity water heaters are needed so as to be able to have hot water during these periods.

- **A** 120/240 volt service
- **B** Main watt-hour meter
- **C** Water heater watt-hour meter
- **D** Service conductors to main panelboard
- **E** Two-pole disconnect switch or circuit breaker
- **F** Combination high temperature limit and upper interlocking thermostat
- **G** Upper heating element
- **H** Lower thermostat
- **I** Lower heating element

Notations used in Figures 19-9, 19-10, 19-12, and 19-13.

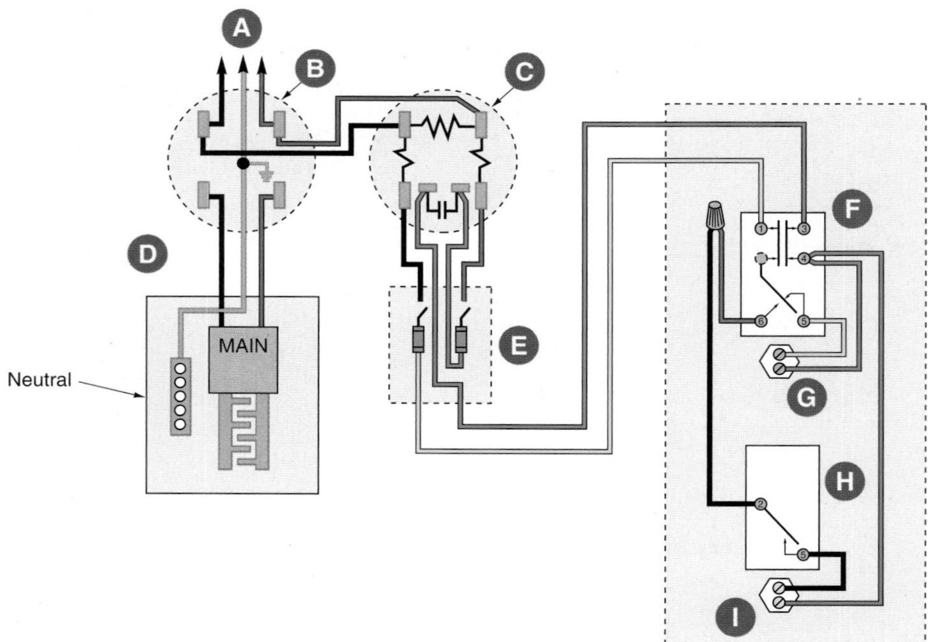

FIGURE 19-9 Wiring for a typical off-peak electric water heater circuit. A separate watt-hour meter/time clock controls the circuit to the water heater. The utility sets the time clock to turn the power off during certain peak periods of the day. This type of circuitry is called "time-of-day" programming.

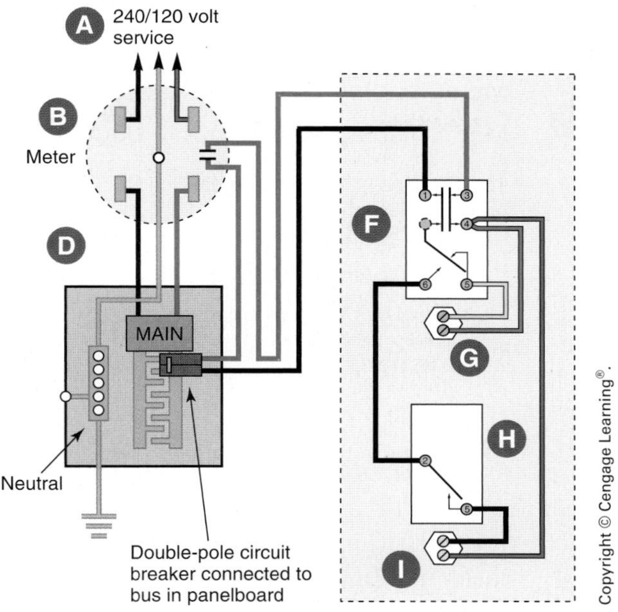

FIGURE 19-10 The meter contains a set of contacts that can be controlled by the utility. The water heater circuit is fed from the main panelboard, through the contacts, then to the water heater.

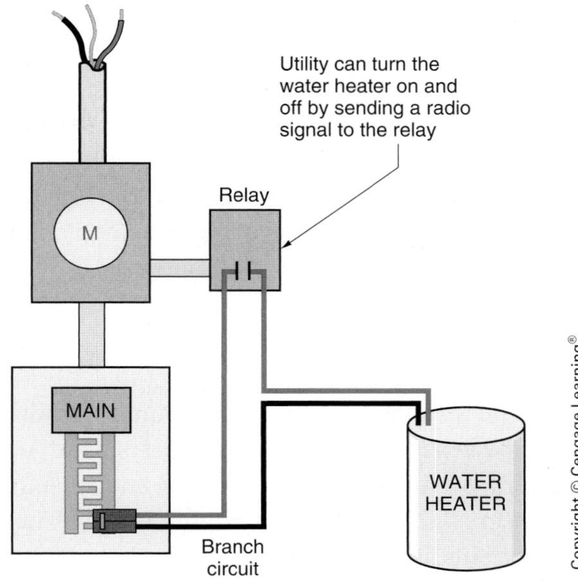

FIGURE 19-11 In this diagram, the utility sends a radio signal to the relay, which in turn can "shed" the load of a residential water heater, air conditioner, or heat pump. If the relay has two sets of contacts, four conductors would be required. Low-voltage relays are also used, in which case low-voltage conductors are run between the radio control unit and relay in the appliance. For the control of an air conditioner or heat pump, the radio control unit is connected "in series" with the room thermostat. This arrangement can be used with one watt-hour meter, using a sliding rate schedule with a "time-of-day" control. Contact the electric utility for technical data regarding how it wants the electrical connections to be made for its particular programs.

Figure 19-10 illustrates a type of meter that contains an integral set of contacts, controlled by the utility via signals sent over the power lines. These meters are available with either 30-ampere or 50-ampere contacts, the contact rating selected to switch the load to be controlled. The electric utility determines when to shed the load. The homeowner must be willing to live with the fact that the power to the water heater, air conditioner, or heat pump might be turned off during the peak periods.

In Figure 19-11, the utility mounts a radio-controlled relay to the side of the meter socket. One major utility using this method makes use of 10 different radio frequencies. When the utility needs to shed load, it can signal one or all of the relays to turn the water heater, air conditioner, or heat pump load off and on. This might only be a few times each year, as opposed to time-clock programs that operate daily. This method is a simple one. One watt-hour meter registers the total power consumption. The homeowner signs up for this program. The program offers a financial incentive, and the homeowner knows up front that the utility might occasionally turn off the load during crucial peak periods.

Customer Controlled

There are some rate schedules where customers can somewhat control their energy costs by doing their utmost to use electrical energy during off-peak hours. For example, they would not wash clothes, use the electric clothes dryer, take showers, or use the dishwasher during peak (premium) hours. The choice is up to homeowners. To accomplish this, watt-hour meters can be installed that have two sets of dials: one set showing the *total kilowatt hours*, and the second set showing *premium time kilowatt hours*. See Figure 29-2. With this type of meter, the utility will have the total kilowatt hours used as well as the premium time kilowatt hours used. With this information, the utility bills customers at one rate for the energy used during peak hours, and at another rate for the difference between total kilowatt hours and premium time kilowatt hours. The electric utility will determine and define *premium time*.

Some homeowners have installed time-clocks on their water-heater circuit to shut off during peak premium time hours to be sure they are not using energy during the periods with a high rate per kilowatt-hour.

Uncontrolled

The simplest and most common uncontrolled method is shown in Figure 19-12. There are no extra meters, switching contacts in the meter, extra disconnect switches, or controllable relays. The water heater circuit is connected to a 2-pole branch-circuit breaker in the main distribution panelboard. The electrical power to the water heater is on 24 hours a day. This is the connection for the water heater in the residence in this text. Instead of separate "low-rate" metering that requires an additional watt-hour meter, the utility offers a sliding scale of rate steps.

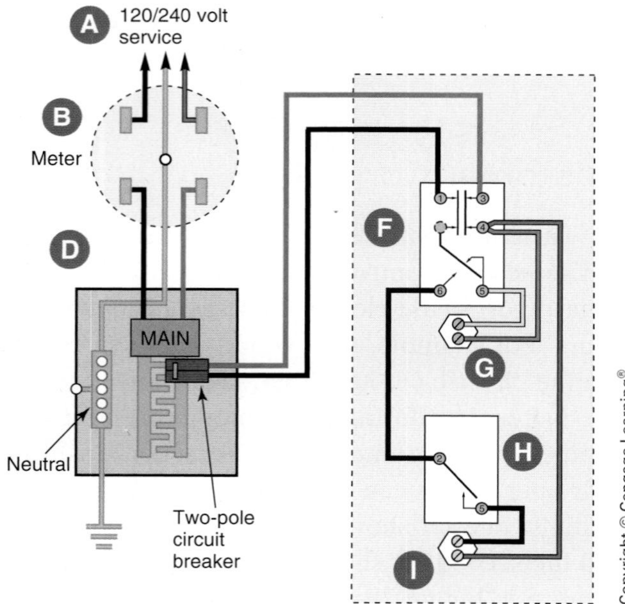

FIGURE 19-12 Water heater connected to 2-pole circuit in Main Panelboard.

EXAMPLE

Meter reading 06-01-20XX	00532	
Meter reading 05-01-20XX	00007	
Total kilowatt-hours used	00525	
Basic monthly energy charge		$11.24
1st 400 kWh × 0.10206		40.82
Next 125 kWh × 0.06690		8.36
Amount Due		$60.42

State, local, and applicable regulatory taxes would be added to the amount due.

Certain utilities will use the higher energy charge during the June, July, August, and September summer months.

Another uncontrolled method is to use two watt-hour meters, Figure 19-13. The water heater has power 24 hours per day. One meter will register the normal lighting load at the regular residential energy charge. The second meter will register the energy consumed by the water heater, air conditioner, or heat pump at some lower energy rate.

Are Water Heater Control Conductors Permitted in the Same Raceway as Service-Entrance Conductors?

Conductors other than service-entrance conductors are not permitted in the same raceway as the service-entrance conductors, *230.7*. However, *Exception No. 2* to this section does permit load management control conductors that have overcurrent protection to be installed in the service raceway.

ELECTRIC WATER HEATER BRANCH CIRCUITS

Branch-Circuit Rating

NEC 422.13 applies to all fixed storage electric water heaters having a capacity of 120 gallons (454.2 L) or less. This would include most typical residential electric water heaters. An electric water heater is a continuous load per *422.13*, which would

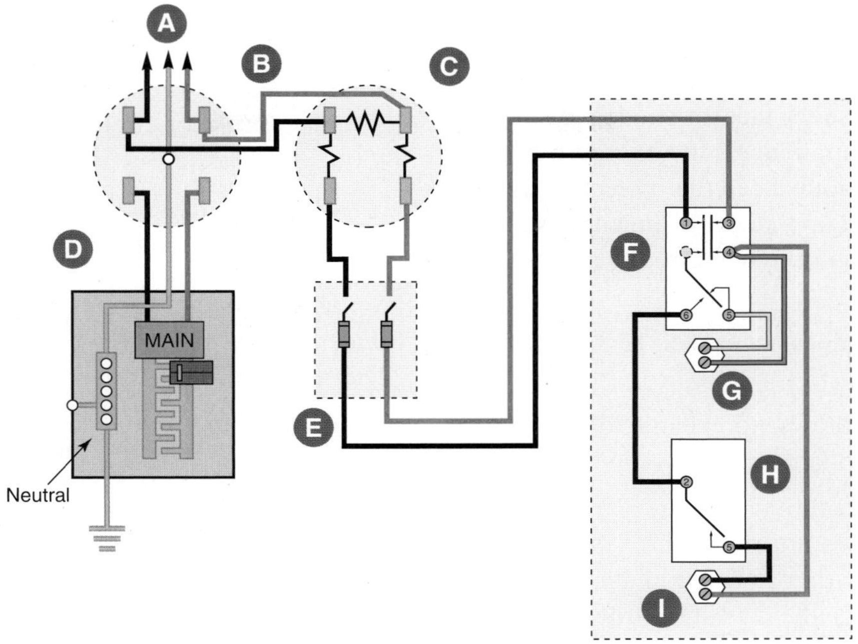

FIGURE 19-13 Twenty-four-hour water heater installation. The total energy demand for the water heater is fed through a separate watt-hour meter at a lower kWh rate than the regular meter.

require that the branch-circuit rating must not be less than 125% of the nameplate rating.

Disconnecting Means

For obvious safety reasons, *NEC 422.30* requires that an appliance have a means of disconnecting it from the power source. In most instances, this is a separate disconnect switch.

NEC 422.31(B) tells us that for permanently connected appliances rated greater than 300 volt-amperes or ⅛ horsepower, the disconnecting means is permitted to be the branch-circuit switch or circuit breaker for the appliance if the switch or breaker is within sight of the appliance *or* is capable of being locked in the OFF position.

The locking provision must be permanently installed on or at the switch or circuit breaker used as the disconnecting means. The locking provision must remain in place with or without the lock installed.

Workers have been seriously shocked when working on appliances that they thought were de-energized, because after turning off the power, someone came along wondering why the switch was off and turned the power back on.

Listed lock-off devices are available from the various manufacturers of circuit breakers. Disconnect switches have brackets with holes through which a padlock is installed. OSHA lock-off requirements are discussed in Chapter 1.

NEC 422.35 requires that switches and circuit breakers used as the disconnecting means be of the indicating type, meaning they must clearly show that they are in the ON or OFF position.

Overcurrent Protection and Conductor Size

There are two separate issues that must be considered.

- The overcurrent protective device ampere rating must be calculated.
- The conductor size must be determined.

Do both calculations. Then compare the results of both calculations to make sure that both overcurrent device and conductors are *Code* compliant.

Branch-Circuit Overcurrent Protection

NEC 422.11(E) states that for single non-motor-operated electrical appliances, the overcurrent protective device shall not exceed the protective device rating marked on the appliance. If the appliance has no such marking, then the overcurrent device is to be sized as follows:

- If the appliance does not exceed 13.3 amperes—20 amperes
- If the appliance draws more than 13.3 amperes—150% of the appliance rating

For a single non-motor-operated appliance, if the 150% sizing does not result in a standard size overcurrent device rating as listed in *240.6(A)*, then it is permitted to go to the next standard size.

EXAMPLE

A water heater nameplate indicates 4500 watts, 240 volts. What is the maximum size fuse permitted by the *Code*?

Solution

$$I = \frac{W}{E} = \frac{4500}{240} = 18.8 \text{ amperes}$$

An electric water heater is a continuous load per *422.13*. The branch-circuit rating must not be less than 125% of the water heater's nameplate rating. The minimum branch-circuit rating is

$$18.8 \times 1.25 = 23.5 \text{ amperes}$$

This would require a minimum 25-ampere fuse or circuit breaker, which is the next standard size larger than the calculated 23.5 amperes. See *240.6(A)* for standard ampere ratings for fuses and circuit breakers.

The maximum overcurrent device for the water heater is

$$18.8 \times 1.5 = 28.2 \text{ amperes}$$

This would be a 30-ampere fuse or circuit breaker, which is the next standard size larger than the calculated 28.2 amperes. But there is no reason to use a 30-ampere branch-circuit fuse or circuit breaker because we have already determined that a 25-ampere fuse or circuit breaker is suitable for the 4500-watt load.

Conductor Size

The conductors supplying the 4500-watt water heater will be protected by the 25-ampere breaker as determined above.

In *Table 310.15(B)(16)*, we find the allowable ampacity of conductors in the 60°C column as required by *110.14(C)*. In *240.4(D)*, we find the maximum overcurrent protection for small conductors. Putting this all together, we have the information shown in Table 19-4.

For the example, we selected a 25-ampere overcurrent device. A 30-ampere OCD would have been acceptable. Next, we need to find a conductor that is properly protected by a 25- or 30-ampere OCD. The conductor would have to be a minimum 10 AWG Type THHN, which has an allowable ampacity of 30 amperes, more than adequate for the minimum branch-circuit rating of 23.4 amperes. A 10 AWG conductor is properly protected by a maximum 30-ampere overcurrent device.

If a 12 AWG Type THHN had been selected, it would have been suitable for the load but would not have been properly protected by the 25-ampere OCD. The maximum OCD for a 12 AWG is 20 amperes unless the branch circuit is for a special application such as for motors.

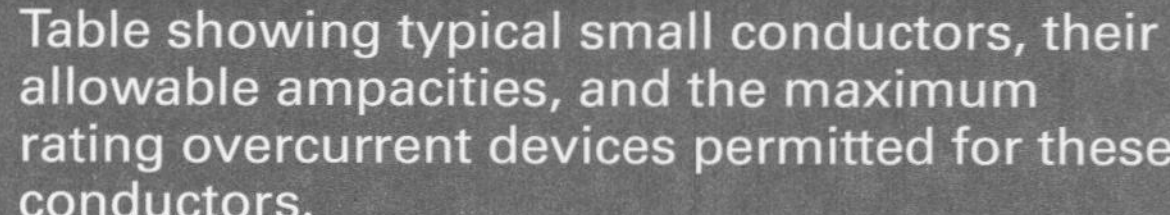

TABLE 19-4

Table showing typical small conductors, their allowable ampacities, and the maximum rating overcurrent devices permitted for these conductors.

CONDUCTOR SIZE	ALLOWABLE AMPACITY AMPERE FROM 60°C COLUMN OF *TABLE 310.15(B)(16)*	MAXIMUM RATING OF OVERCURRENT DEVICE
14 AWG	15 amperes	15 amperes
12 AWG	20 amperes	20 amperes
10 AWG	30 amperes	30 amperes

You might want to review Chapter 4 for a refresher as to why we use the 60°C column of *Table 310.15(B)(16)* for selecting the conductor size.

The Water Heater for This Residence

The water heater for the residence is connected to Circuit A6-8. This is a 30-ampere, 2-pole circuit breaker located in the main panelboard in the workshop. This circuit is a straight 240 volts and does not require a neutral conductor.

Checking the Schedule of Special-Purpose Outlets, we find that the electric water heater has two heating elements: a 4500-watt element upper and a 4500-watt lower element. Thus it is considered a quick-recovery unit. Because of the thermostats on the water heater, both elements cannot be energized at the same time. Therefore, the maximum load demand is

$$\text{Amperes} = \frac{\text{Watts}}{\text{Volts}} = \frac{4500}{240} = 18.75 \text{ or } 18.8 \text{ amperes}$$

For water heaters having two heating elements connected for nonsimultaneous operation, the nameplate on the water heater will be marked with the largest element's wattage at rated voltage. For water heaters having two heating elements connected for simultaneous operation, the nameplate on the water heater will be marked with the total wattage of both elements at rated voltage.

Conductor Size and Raceway Size for the Water Heater in This Residence

The conductor and overcurrent device is required to be not smaller than 125 percent of the current.

$$18.8 \text{ amperes} \times 1.25 = 23.5 \text{ amperes}$$

This results in a 10 AWG copper conductor and a 30 ampere overcurrent device because the small-conductor rule in *240.4(D)(4)* generally requires a 12 AWG conductor to have overcurrent protection not greater than 20 amperes.

Table C1 in *Annex C* of the *NEC* indicates that trade size ½ EMT will be okay for two 10 THHN/THWN conductors. For ease of installation, a short length of trade size ½ flexible metal conduit 18 to 24 in. (450 to 600 mm) long is attached to the EMT and is connected into the knockout provided for the purpose on the water heater. A 10 AWG equipment grounding conductor is required through the flexible metal conduit as required by the rules in *NEC 250.118(5)*.

Cord/Plug Connections Not Permitted for Water Heaters

NEC 400.7(A) and *422.16* list the uses where flexible cords are permitted. A key *Code* requirement, often violated, is that flexible cords shall be used only Where the fastening means and mechanical connections are specifically designed to permit ready removal for maintenance and repair, and the appliance is intended or identified for flexible cord connection.* Certainly the plumbing does not allow for the "ready removal" of the water heater. The conductors in flexible cords cannot handle the high temperatures encountered on the water heater terminals. Check the instruction manual for proper installation methods.

Flexible cords are not permitted to be

- used as a permanent wiring method.
- run through holes in walls, ceilings, and so on.
- run through doorways, windows, or similar openings.
- attached to building surfaces.
- concealed above ceilings, in walls, or under floors.
- installed in raceways.
- used where subject to physical damage.

Equipment Grounding

The electric water heater is grounded through the trade size ½ EMT and an internal 10 AWG equipment grounding conductor installed through the flexible metal conduit (FMC) or by the equipment grounding conductor contained within the Type NM cable if used as the wiring method. *Code* references are *250.110*, *250.118*, *250.134*, and

*Reprinted with permission from NFPA 70-2014.

348.60. The subject of equipment grounding is covered in Chapter 4.

EFFECT OF VOLTAGE VARIATION ON RESISTIVE HEATING ELEMENTS

The heating elements in the water heater in this residence are rated 240 volts. A resistive heating element will only produce rated wattage at rated voltage. It will operate at a lower wattage with a reduction in voltage. If connected to voltages above their rating, heating elements will have a very short life.

Ohm's law and the wattage formula show how the wattage and current depend on the applied voltage. Always use *rated* voltage to calculate the resistance, current draw, and wattage of a resistive heating element. Check manufacturers' specifications.

Nichrome wire, commonly used to make heating elements, has a "hot" resistance approximately 10% higher than its "cold" resistance.

SAFETY ALERT

When actually testing a heating element, never take an ohms reading on an energized circuit. Remove any other connected wires from the heating element terminal block on a water heater to prevent erratic ohms readings. Always check from the heating element leads to ground to confirm that there is no ground fault in the heating element.

Let's take a look at a 3000-watt, 240-volt heating element and calculate its resistance and current draw.

$$R = \frac{E^2}{W} = \frac{240 \times 240}{3000} = 19.2 \text{ ohms}$$

$$I = \frac{E}{W} = \frac{240}{19.2} = 12.5 \text{ amperes}$$

If a different voltage is substituted, the wattage and current values change accordingly.

At 220 volts:

$$W = \frac{E^2}{R} = \frac{220 \times 220}{19.2} = 2521 \text{ watts}$$

$$I = \frac{W}{E} = \frac{2521}{220} = 11.5 \text{ amperes}$$

Another way to calculate the effect of voltage variance is

$$\text{Correction Factor} = \frac{\text{Applied Voltage Squared}}{\text{Rated Voltage Squared}}$$

Using the previous example:

$$\text{Correction Factor} = \frac{220 \times 220}{240 \times 240} = \frac{48{,}400}{57{,}600} = 0.84$$

$$\text{then } 3000 \times 0.84 = 2521 \text{ watts}$$

In resistive circuits, current is directly proportional to voltage and can be simply calculated using a ratio and proportion formula. For example, if a 240-volt heating element draws 12.7 amperes at rated voltage, the current draw can be determined at any other applied voltage, say 208 volts.

$$\frac{208}{240} = \frac{X}{12.7}$$

$$\frac{208 \times 12.7}{240} = 11.0 \text{ amperes}$$

To calculate this example if the applied voltage is 120 volts:

$$\frac{120}{240} = \frac{X}{12.7}$$

$$\frac{120 \times 12.7}{240} = 6.4 \text{ amperes}$$

Also, in a resistive circuit, wattage varies as the square of the current. Therefore, when the voltage on a heating element is doubled, the current also doubles and the wattage increases four times. When the voltage is reduced to one-half, the current is halved and the wattage is reduced to one-fourth.

A 240-volt heating element connected to 208 volts will have approximately three-quarters of the wattage rating than at rated voltage.

TABLE 19-5

Table showing the approximate change in full-load current and starting-current for typical electric motors when operated at under voltage (90%) and over voltage (110%) conditions.

VOLTAGE VARIATION	FULL-LOAD CURRENT	STARTING CURRENT
110%	7% decrease	10–12% increase
90%	11% increase	10–12% decrease

EFFECT OF VOLTAGE VARIATION ON MOTORS

The above formulas work only for resistive circuits. Formulas for inductive circuits such as an electric motor are much more complex than this book is intended to cover. For typical electric motors, the following information will suffice (see Table 19-5).

As previously mentioned, NEMA rated electric motors are designed to operate at ±10% of their nameplate voltage.

HEAT PUMP WATER HEATERS

With so much attention given to energy savings, heat pump water heaters have entered the scene.

Residential heat pump water heaters can save 50% to 60% over the energy consumption of resistance-type electric water heaters, depending on energy rates. The heat energy transferred to the water is three to four times greater than the electrical energy used to operate the compressor, fan, and pump.

Heat pump water heaters work in conjunction with the normal resistance-type electric water heater. Located indoors near the electric water heater and properly interconnected, the heat pump water heater becomes the primary source for hot water. The resistance electric heating elements in the electric water heater are the backup source should it be needed.

A heat pump water heater removes low-grade heat from the surrounding air, but instead of transferring the heat outdoors as does a regular air-conditioning unit, it transfers the heat to the water. The unit shuts off when the water reaches 130°F (54.44°C). Moisture is removed from the air at approximately 1 pint per hour, reducing the need to run a dehumidifier. The exhausted cool air from the heat pump water heater is about 10° to 15°F (−12.22°C to −9.44°C) below room temperature. The cool air can be used for limited cooling purposes, or it can be ducted outdoors.

A typical residential heat pump water heater rating is 240 volts, 60 Hz, single-phase, 500 watts.

Heat pump water heaters operate with a hermetic refrigerant motor compressor. The electric hookup is similar to a typical 240-volt, single-phase, residential air-conditioning unit. *Article 440* of the *NEC* applies. This is covered in Chapter 23.

Electric utilities and numerous websites are good sources to learn more about the economics of installing heat pump water heaters to save energy.

REVIEW

Note: Refer to the *Code* or the plans where necessary.

WATER PUMP CIRCUIT (A)B

1. Does a jet pump have any electrical moving parts below the ground level? ________
2. For a jet pump, which is larger, the drive pipe or the suction pipe? ________
3. Where is the jet of the pump located? ________
4. What does the impeller wheel move? ________
5. Where does the water flow after leaving the impeller wheel?
   a. ________
   b. ________

6. What prevents water from draining back into the pump from the tank? ____________

____

7. What prevents water from draining back into the well from the equipment? ________

____

8. What is compressed in the water storage tank? ____________________

9. Explain the difference between a 2-wire submersible pump and a 3-wire submersible pump. ____________________

____

____

____

10. What is a common speed for jet pump motors? ____________________

11. Why is a 240-volt motor preferable to a 120-volt motor for use in this residence?

____

____

12. How many amperes does a 1-horsepower, 240-volt, single-phase motor draw? (See *Table 430.248.*) ____________________

13. What size are the conductors used for this circuit? ____________________

14. What is the branch-circuit protective device? ____________________

15. What provides the running overload protection for the pump motor? ____________

____

____

16. What is the maximum ampere setting permitted for running overload protection of the 1-horsepower, 240-volt pump motor? ____________________

17. Submersible water pumps operate with the electrical motor and actual pump located (Circle the correct answer.)

    a. above permanent water level.

    b. below permanent water level.

    c. half above and half below permanent water level.

18. Because the controller contains the motor starting relay and the running and starting capacitors, the motor itself contains ____________________.

19. What type of pump moves the water upward inside of the deep-well pipe? ________

____

20. Proper pressure of the submersible pump system is maintained by a ____________

____.

21. Fill in the data for a 16-ampere electric motor, single-phase, no *Code* letters.

    a. Branch-circuit protection, non-time-delay fuses:

    Normal size ____________ A   Maximum size ____________ A

    Switch size ____________ A   Switch size ____________ A

b. Branch-circuit protection dual-element, time-delay fuses:

Normal size ______________ A Maximum size ______________ A

Switch size ______________ A Switch size ______________ A

c. Branch-circuit protection instant-trip breaker:

Normal setting ______________ A

Maximum setting ______________ A

d. Branch-circuit protection—inverse time breaker:

Normal rating ______________ A

Maximum rating ______________ A

e. Branch-circuit conductor size, Type THHN ______________

Ampacity ______________ A

f. Motor overload protection using dual-element time-delay fuses:

Maximum size ______________ A

22. The *NEC* is very specific in its requirement that submersible electric water pump motors be grounded. Where is this specific requirement found in the *Code*?

____

____

23. Does the *NEC* allow submersible pump cable to be buried directly in the ground?

____

____

24. Must the disconnect switch for a submersible pump be located next to the well?

____

____

25. A metal well casing (shall) (shall not) be bonded to the pump's equipment grounding conductor or by grounding it with a separate equipment grounding conductor run all the way back to the same ground bus in the panelboard that supplies the pump circuit, *NEC*

____

WATER HEATER CIRCUIT ⓐC

1. According to *NEC 422.47* the high-temperature limit control must disconnect ______________ of the ungrounded conductors. The high-temperature control limits the maximum water temperature to ______________ °F (______________ °C).
2. A major hazard involved with water heaters is that they operate under whatever pressure the serving water utility supplies. Water stays liquid at temperatures higher than the normal boiling point of 212°F (100°C). If the high-temperature limit control failed to operate for whatever reason, should a pipe burst or a faucet be opened, the pressure would instantly drop to normal atmospheric pressure, causing the superheated water to turn into ______________ that could result in ______________. To prevent this from happening, water heaters are equipped with pressure/temperature relief valves.
3. Magnesium rods are installed inside the water tank to reduce ______________.

4. The heating elements in electric water heaters are generally classified into two categories. These are ____________ density and ____________ density elements.
5. An 80-gallon electric water heater is energized for the first time. Approximately how many hours would it take to raise the water temperature 80°F (26.67°C) (from 40°F to 120°F [4.44°C to 48.89°C])? The water heater has a 3000-watt heating element.

 ____

 ____

6. What term is used by utilities when the water heater power consumption is measured using different rates during different periods of the day? ____________
7. Explain how the electric utility in your area meters residential electric water heater loads.

 ____

 ____

 ____

8. For residential water heaters, the Consumer Product Safety Commission suggests a maximum temperature setting of ____________ °F (____________ °C). Some states have laws stating a maximum temperature setting of ____________ °F (____________ °C).
9. An 80-gallon electric water heater has 60°F (15.56°C) incoming water. How many minutes can a 3-gallon/minute showerhead be used to draw 105°F (40.56°C) water if the water heater thermostat is set at 120°F (48.89°C)?
10. Approximately how long would it take to produce serious burns to an adult with 140°F (60°C) water? ____________
11. Two thermostats are generally used in an electric water heater.
    a. What is the location of each thermostat? ____________

     ____

    b. What type of thermostat is used at each location? ____________

     ____

12. a. How many heating elements are provided in the heater in the residence discussed in this text? ____________
    b. Are these heating elements allowed to operate at the same time? ____________
13. When does the lower heating element operate? ____________

 ____

14. The *Code* states that water heaters having a capacity of 120 gallons (450 L) or less shall be considered ____________ duty and, as such, the circuit must have a rating of not less than ____________ percent of the rating of the water heater.
15. Why does the storage tank hold the heat so long? ____________
16. The electric water heater in this residence is connected for "limited demand" so that only one heating element can be on at one time
    a. What size wire is used to connect the water heater? ____________
    b. What size overcurrent device is used? ____________

17. a. If both elements of the water heater in this residence are energized at the same time, how much current will they draw? (Assume the elements are rated at 240 volts.)

    b. What size and Type THHN wire is required for the load of both elements? Show calculations.

18. a. How much power in watts would the two elements in problem 17 use if connected to 220 volts? Show calculations.

    b. What is the current draw at 220 volts? Show calculations.

19. A condominium owner complains of not getting enough hot water. The serviceman checks the voltage at the water heater and finds the voltage to be 208 volts. The serviceman checks further and finds the electrical service on the building to be a 120/208-volt, three-phase, 4-wire system. The main electrical panelboard for each condominium unit is fed with a 3-wire supply. The nameplate on the electric water heater is marked 4500 watts, 240 volts. What is the wattage output of the heating element in this water heater when connected to a 208-volt supply? ______

20. For a single, nonmotor-operated electrical appliance rated greater than 13.3 amperes that has no marking that would indicate the size of branch-circuit overcurrent device, what percentage of the nameplate rating is used to determine the maximum size branch-circuit overcurrent device? ______

21. A 7000-watt resistance-type heating appliance is rated 240 volts. What is the maximum size fuse permitted to protect the branch-circuit supplying this appliance?

    ______

22. The *Code* requirements for standard resistance electric water heaters are found in *Article 422* of the *Code*. Because heat pump water heaters contain a hermetic refrigerant motor compressor, the *Code* requirements are found in *Article* ______ of the *Code*.

23. In your area, is the branch-circuit breaker or switch for the electric water heater branch circuit acceptable as the disconnecting means, or are you required to install a separate disconnect within sight of the water heater? Provide a brief explanation.

    ______

    ______

CHAPTER **20**

Special-Purpose Outlets for Ranges, Counter-Mounted Cooking Unit $\triangle_G$, and Wall-Mounted Oven $\triangle_F$

OBJECTIVES

After studying this chapter, you should be able to

- understand the *NEC* requirements for installing and connecting free-standing ranges, counter-mounted cooking units, wall-mounted ovens, microwave ovens, and light energy ovens.
- make load calculations to determine proper size conductors, overcurrent protection, disconnecting means, and how to achieve proper grounding.
- be familiar with the different configurations for 30- and 50-ampere NEMA receptacles and cords.
- be aware of the many types of temperature controls found on cooking equipment.
- understand significant differences of color coding and terminal identification on foreign-made appliances.

BASIC CIRCUIT REQUIREMENTS FOR ELECTRIC RANGES, COUNTER-MOUNTED COOKING UNITS, AND WALL-MOUNTED OVENS

Electric ranges are covered in *Article 422* of the *NEC*® and by UL Standard 858, *Household Electric Ranges.*

UL Standard 1082 covers *Electric Household Cooking and Food Serving Appliances.*

Connection Methods

Most recognized wiring methods can be used to hook up appliances, such as nonmetallic-sheathed cable, armored cable, EMT, flexible metal or nonmetallic conduit, or service-entrance cable. Check local electrical codes to see whether there are any restrictions on the use of any of the wiring methods listed above.

Here are some of the more common methods for connecting electric ranges, wall-mounted ovens, and counter-mounted cooktops.

Direct Connection. One method is to run the branch-circuit wiring directly to the wiring compartment on the appliance. When "roughing-in" the branch-circuit wiring, sufficient length must be provided so as to allow flexibility to make up the connections and place the appliance into position, and for servicing.

The time and material required to run an individual branch circuit to each of the major electrical appliances might very well be the most economical way to do the hookups, instead of using junction boxes, tap conductors, and extra splices, as discussed next. Don't forget, "time" includes the time it takes to calculate the proper size tap conductors and junction boxes, plus the actual installation time.

Junction Box. Another method is to run the branch-circuit wiring to a junction box (an outlet box) behind, under, or adjacent to the range, oven, or cooktop location. By definition, this junction box is accessible because, should the need arise, the junction box can be reached by removing the appliance. The "whip" on the appliance is then connected to the junction box where the splicing of the tap conductors in the whip and the branch-circuit conductors are made. According to UL Standard No. 858, the flexible metal whip must not be less than 3 ft (900 mm) and no longer than 6 ft (1.8 m). The conductors in the whip are rated for the high temperatures encountered in the appliance.

Depending on the location of the appliances, it might be possible to connect more than one cooking appliance to a single branch circuit from a single junction box or from two junction boxes. This is illustrated in Figure 20-1.

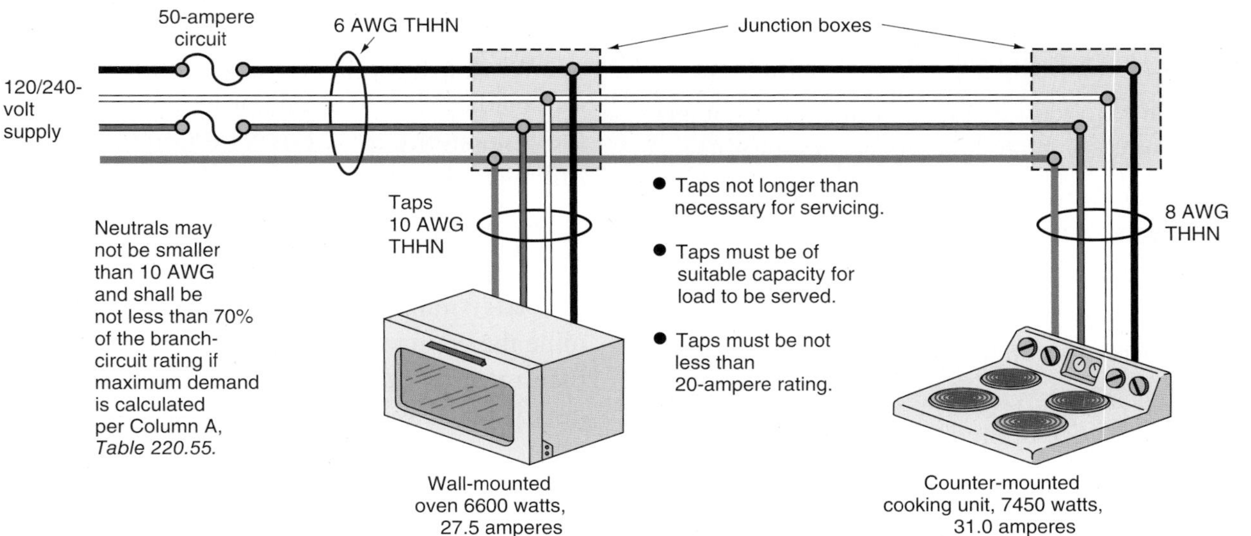

FIGURE 20-1 A counter-mounted cooktop unit and a wall-mounted oven connected to one branch circuit. Depending on the location of the appliances, it might be possible to make the installation with one junction box.

Cord-and-Plug Connection. Still another very common method is to use a cord-and-plug connection, *422.33(A)*. This is particularly true for freestanding ranges but is uncommon for wall-mounted ovens or counter-mounted cooktops. A cord-and-plug connection also serves as the disconnecting means for household cooking equipment. The range receptacle is permitted to be at the rear base of an electric range. Removing the bottom drawer makes this connection accessible, *422.33(B)*. The receptacle must have an ampere rating not less than the rating of the appliance, *422.33(C)*. For household cooking equipment, the demand factors of *Table 220.55* are also applicable to the receptacle, *422.33(C), Exception*. Range and dryer receptacles are similar but often have different ampere ratings; they are discussed later in this chapter.

Cord sets rated 30 amperes are referred to as *dryer cords*. Cord sets rated 40, 45, or 50 amperes are referred to as *range cords*.

Receptacles rated 30 amperes are referred to as *dryer receptacles*. Receptacles rated 50 amperes are referred to as *range receptacles*.

Nameplate on Appliance

NEC 422.60(A) requires that the nameplate must show the appliance's rating in volts and amperes or in volts and watts.

Load Calculations

In *210.19(A)(3)* and *Exceptions*, we find that

- the branch-circuit conductors must have an ampacity not less than the branch-circuit rating.
- the branch-circuit conductors' ampacity must not be less than the maximum load being served.
- the branch-circuit rating must not be less than 40 amperes for ranges of 8¾ kW or more.
- for ranges 8¾ kW or more, the neutral conductor of a 3-wire branch circuit may be 70% of the branch-circuit rating, but not smaller than 10 AWG.
- tap conductors connecting ranges, ovens, and cooking units to a 50-ampere branch circuit must be rated at least 20 amperes, must be adequate for the load, and must not be any longer than necessary for servicing the appliance.
- *NEC 220.55* permits the use of the values found in *Table 220.55* and its footnotes for the load calculation.
- *Table 220.55* shows the kW demand for different kW-rated household cooking appliances that are over 1¾ kW. Column C is used in most situations. Note that the maximum demand is considerably less than the actual appliance's nameplate rating. This is because rarely, if ever, will all of the heating elements be used at the same time. *Note 3* to the table tells us when we are permitted to use Columns A and B. This is discussed later in this chapter.
- *NEC 220.61* states that for calculating a feeder or service, the neutral conductor load can be figured at 70% of the load on the ungrounded conductors for ranges, ovens, and counter-mounted cooktops, using *Table 220.55* as the basis for the load calculation.

In *422.10(A)* we find that the branch-circuit rating shall not be less than the marked rating of the appliance. However, there is an exception for household cooking appliances. The last paragraph of this section states that Branch circuits for household cooking appliances shall be permitted to be in accordance with Table 220.55.*

Separate Branch Circuits

When a separate branch circuit is run to a wall-mounted oven or to a counter-mounted cooking unit, the branch-circuit size is based on the appliance's full nameplate rating, *Table 220.55, Note 4*.

Wire Size

After making the load calculations to determine the appliance's calculated load in amperes, we then refer to *Table 310.15(B)(16)* to select a conductor that has an allowable ampacity equal to or greater than the calculated load. As mentioned many times in this text, *110.14(C)(1)* as well as the UL Standards require that we use the 60°C column regardless of the conductor's insulation rating.

An exception to this basic rule is when the equipment is marked for 75°C. This is discussed in detail in Chapter 4 in the section entitled "The Weakest Link of the Chain."

Overcurrent Protection

NEC 422.11(A) states that the branch-circuit overcurrent protection shall not exceed the rating marked on the appliance.

For appliances that have no marked maximum overcurrent protection, the branch-circuit overcurrent protection is sized at the ampacity of the branch-circuit conductors supplying the appliance. See *240.4(B)*, *(D)*, and *(E)*.

Disconnecting Means

All appliances require a means of disconnecting them from the power source, *422.30*.

NEC 422.33(A) accepts a cord-and-plug arrangement as the required disconnecting means.

A branch-circuit switch or circuit breaker is acceptable as the disconnecting means if it is within sight of the appliance or is capable of being locked in the OFF position, *422.31(B)*. Breaker manufacturers supply "lock-off" devices that fit over the circuit breaker. Individual disconnect switches also have a lock-off provision. The lock-off provision must be permanently installed on the breaker or switch.

NEC 422.34 recognizes "unit switches" that have a marked OFF position to be an acceptable disconnect for the appliance. Controllers such as clocks, timers, temperature controls, and element controls do not qualify as "unit switches" as they do not disconnect all the ungrounded conductors in the appliance. A unit switch with a marked-off position disconnects all ungrounded conductors. This enables a person to work on a specific component of the appliance, knowing that the power is off to that particular component. Obviously, the line side terminals of the appliance are still "hot." Electric ranges, wall-mounted ovens, and counter-mounted cooktops are not normally equipped with unit switches, so another disconnecting means is required.

Using the branch-circuit switch or circuit breaker as the disconnecting means shuts off total power to the appliance and is much safer than relying on the unit switches. Be sure to use the required lock-off requirement.

GROUNDING FRAMES OF ELECTRIC RANGES, WALL-MOUNTED OVENS, AND COUNTER-MOUNTED COOKING UNITS

The frames of electric ranges, wall-mounted ovens, and counter-mounted cooking units must be grounded, *250.140*. Grounding is accomplished through the equipment grounding conductor in nonmetallic-sheathed cable, the metal armor of Type AC cable, or with metal raceways such as EMT and flexible metal conduit. See *250.118* and *250.134* for acceptable equipment grounding means.

The frames of electric ranges, wall-mounted ovens, counter-mounted cooking units, and clothes dryers *are not* permitted to be grounded to the neutral branch-circuit conductor, *250.140*.

Because of the ever-present confusion over using a separate equipment grounding conductor versus the old fashioned way of grounding the frame to the neutral conductor of the circuit, you had better review Chapter 15, where this subject is covered in detail. You might also want to review Chapter 4, where the grounding qualifications for various wiring methods are discussed.

Receptacles for Electric Ranges and Dryers

Because of their similarities, the following text applies to both range and dryer receptacles.

Residential electric ranges, electric dryers, wall-mounted ovens, and surface-mounted cooking units may be connected with cords designed specifically for this use. Depending on the appliance's wattage rating, the receptacles are usually rated 30 amperes or 50 amperes, 125/250 volts. Cord sets rated 40 amperes or 45 amperes contain 40- or 45-ampere conductors, yet the plug cap is rated 50 amperes. It becomes a cost issue as to whether to use a full 50-ampere-rated cord set in all cases or whether to match the cord ampere rating as closely as possible to that of the appliance being connected.

FIGURE 20-2 Illustrations of 30-ampere and 50-ampere receptacles used for cord-and-plug connection of electric ranges, ovens, counter-mounted cooking units, and electric clothes dryers. Four-wire receptacles and 4-wire cord sets are required for installations according to the *NEC*. Three-wire receptacles and 3-wire cord sets were permitted prior to the 1996 *NEC*. See *250.140*.

Range and dryer receptacles are available in both surface mount and flush mount.

Blade and Slot Configuration for Plug Caps and Receptacles

Figure 20-2 shows the different configurations for receptacles commonly used for electric range and dryer hookups. NEMA uses the letter "R" to indicate a receptacle, and the letter "P" for plug cap.

- **30-ampere, 3-wire:** The L-shaped slot on the receptacle (NEMA 10-30R) and the L-shaped blade on the plug cap (NEMA 10-30P) are for the white neutral conductor.
- **30-ampere, 4-wire:** The L-shaped slot on the receptacle (NEMA 14-30R) and the L-shaped blade on the plug cap (NEMA 14-30P) are for the white neutral conductor. The horseshoe-shaped slot on the receptacle and the round or horseshoe-shaped blade on the plug cap are for the equipment ground.
- **50-ampere, 3-wire:** The wide flat slot on the receptacle (NEMA 10-50R) and the matching wide blade on the plug cap (NEMA 10-50P) are for the white neutral conductor.
- **50-ampere, 4-wire:** The wide flat slot on the receptacle (NEMA 14-50R) and the matching wide blade on the plug cap (NEMA 14-50P) are for the white neutral conductor. The horseshoe-shaped slot on the receptacle and the round or horseshoe-shaped blade on the plug cap are for the equipment ground.

Cord Sets for Electric Ranges and Dryers

A 30-ampere cord set contains:

- **3-wire:** three 10 AWG conductors. The attachment plug cap (NEMA 10-30P) is rated 30 amperes.
- **4-wire:** four 10 AWG conductors. The attachment plug cap (NEMA 14-30P) is rated 30 amperes.

A 40-ampere cord set contains:

- **3-wire:** three 8 AWG conductors. The attachment plug cap (NEMA 10-50P) is rated 50 amperes.
- **4-wire:** three 8 AWG conductors and one 10 AWG conductor. The attachment plug cap (NEMA 14-50P) is rated 50 amperes.

A 45-ampere cord set contains:

- **3-wire:** two 6 AWG conductors and one 8 AWG conductor. The attachment plug cap (NEMA 10-50P) is rated 50 amperes.
- **4-wire:** three 6 AWG conductors and one 8 AWG conductor. The attachment plug cap (NEMA 14-50P) is rated 50 amperes.

A 50-ampere cord set contains:

- **3-wire:** two 6 AWG conductors and one 8 AWG conductor. The attachment plug cap (NEMA 10-50P) is rated 50 amperes.
- **4-wire:** three 6 AWG conductors and one 8 AWG conductor. The attachment plug cap (NEMA 14-50P) is rated 50 amperes.

Terminal Identification for Receptacles and Cords

Receptacle and cord terminals are marked as follows:

- "X" and "Y" for the ungrounded conductors.
- "W" for the white grounded conductor. The "W" terminals will generally be whitish or silver (tinned) in color, in accordance with *200.9* and *200.10* of the *NEC*.
- "G" for the equipment grounding conductor. This terminal is green colored, hexagon shaped, and is marked "G," "GR," "GRN," or "GRND" in accordance with *250.126* and *406.9(B)* of the *NEC*. The grounding blade on the plug cap must be longer than the other blades so that its connection is made before the ungrounded conductors make connection, in accordance with *406.9(D)* of the *NEC*.

WALL-MOUNTED OVEN CIRCUIT ⒶF

The wall-mounted oven in this residence is located to the right of the kitchen sink.

The branch-circuit wiring, using nonmetallic-sheathed cable, is run from Panelboard B to the oven location, coming into the cabinet space where the oven will be installed from behind or from the bottom. Earlier in this chapter we discussed the various methods for hooking up appliances. The choices are direct connection, junction box, or cord-and-plug connection. In most cases, a built-in wall-mounted oven will have a flexible metal conduit attached to it, which is run to a junction box where the electrical connections are completed.

Circuit Number. The circuit number is B(6-8). This information is found in the Schedule of Special-Purpose Outlets contained in the Specifications at the rear of this text, and on the Directory for Panelboard B found in Chapter 27.

Nameplate Data. This appliance is rated 6.6 kW (6600 watts) @ 120/240 volts.

Load Calculations. Use nameplate rating per *220.55, Note 4*.

$$I = \frac{W}{E} = \frac{6600}{240} = 27.5 \text{ amperes}$$

Wire Size. The conductors in nonmetallic-sheathed cable are rated for 90°C. However, in conformance to *334.80*, the allowable ampacity is determined using the 60°C column of *Table 310.15(B)(16)*. Here we find that a 10 AWG conductor has an allowable ampacity of 30 amperes—more than adequate for the appliance's calculated load of 27.5 amperes. This will be a 10 AWG 3-conductor plus equipment grounding conductor nonmetallic-sheathed cable.

Overcurrent Protection. A 2-pole, 30-ampere circuit breaker located in Panelboard B.

Disconnecting Means. The disconnecting means is the 30-ampere, 2-pole circuit breaker in Panelboard B. If a cord-and-plug connection were made, this also would serve as a disconnecting means.

Grounding. The equipment grounding conductor contained in the nonmetallic-sheathed cable provides the means for grounding the frame of the oven. According to *Table 250.122*, the equipment grounding conductor for a circuit protected by a 30-ampere overcurrent device is a 10 AWG copper conductor. That is the size EGC contained in 10 AWG nonmetallic-sheathed cable.

COUNTER-MOUNTED COOKING UNIT CIRCUIT ⒶG

The cooktop in this residence is located in the kitchen island.

The branch-circuit wiring is run from Panelboard B to the island. The nonmetallic-sheathed cable is concealed above the recreation room ceiling, upward into the island cabinetry, ending in a junction box below the cooking unit. Earlier in this chapter, we discussed the various methods for hooking up appliances. The choices are direct connection, junction box, or cord-and-plug connection. In most cases, the cooktop will have a flexible metal conduit attached to it, which also runs to the junction box where the electrical connections are completed.

Note on the plans that a receptacle outlet is to be installed on the side of the island. This receptacle outlet is supplied by small-appliance branch circuit B16. This receptacle is required by *210.52(C)(2)*.

Circuit Number. The circuit number is B(2-4). This information is found in the Schedule of Special-Purpose Outlets contained in the Specifications at the rear of this text, and on the Directory for Panelboard B found in Chapter 27.

Nameplate Data. This appliance is rated 7450 watts @ 120/240 volts.

Load Calculations. Use nameplate rating per *Table 220.55, Note 4*.

$$I = \frac{W}{E} = \frac{7{,}450}{240} = 31 \text{ amperes}$$

Wire Size. The conductors in nonmetallic-sheathed cable are rated for 90°C. However, according to *334.80*, the allowable ampacity is determined using the 60°C column of *Table 310.15(B)(16)*. Here we find that an 8 AWG conductor has an allowable ampacity of 40 amperes—more than adequate for the appliance's calculated load of 31 amperes. This will be an 8 AWG 3-conductor plus equipment grounding conductor nonmetallic-sheathed cable.

Overcurrent Protection. A 2-pole, 40-ampere circuit breaker located in Panelboard B.

Disconnecting Means. The disconnecting means is the 40-ampere, 2-pole circuit breaker in Panelboard B. A cord-and-plug connection would also be a disconnecting means.

Grounding. The equipment grounding conductor contained in the nonmetallic-sheathed cable provides the means for grounding the frame of the appliance. According to *Table 250.122*, the equipment grounding conductor for a circuit protected by a 40-ampere overcurrent device is a 10 AWG copper conductor. That is the size EGC contained in an 8 AWG nonmetallic-sheathed cable.

FREE-STANDING RANGE

Load Calculations

The load calculations for a free-standing electric range are simple.

According to *Table 220.55*, the demand is permitted to be figured at 8 kW for an electric range that is rated 12 kW or less. For example, if an electric range is marked 11.4 kW, the maximum demand can be figured at 8 kW because, under normal situations, the surface heating elements, broiler element, baking element, and lights would not all be on at the same time.

Checking *Table 310.15(B)(16)*, we find that an 8 AWG 3-conductor plus equipment grounding conductor nonmetallic-sheathed cable with an ampacity of 40 amperes is more than adequate.

Larger Ranges

Load calculations for double-oven ranges are a little more complex than for typical one-oven electric ranges.

Larger double-oven free-standing ranges are probably rated more than 12 kW. Let us consider a range that has a rating of 14,050 watts. This is the same wattage as the previously discussed wall-mounted oven and counter-mounted cooking unit.

The total connected load of this free-standing electric range is

$$I = \frac{W}{E} = \frac{14{,}050}{240} = 58.5 \text{ amperes}$$

However, *Table 220.55* permits the calculated load to be considerably less than the connected load.

Table 220.55 and the footnotes provide the information we need to make the load calculation. The column C load of 8 kW is permitted to be used with 5% added for each kW or major fraction thereof by which the rating of the cooking appliances exceed 12 kW.

1. 14,050 − 12,000 watts = 2050 watts (2 kW)
2. According to *Note 1* of *Table 220.55*, a 5% increase for each kW and major fraction in

excess of 12 kW must be added to the load demand in Column C. 2 kW (the kW over 12 kW) × 5% per kW = 10%.

3. 8 kW (from Column C) × 0.10 = 0.8 kW
4. The calculated load is 8 kW + 0.8 kW = 8.8 kW.
5. In amperes, the calculated load is

$$I = \frac{W}{E} = \frac{8{,}800}{240} = 36.7 \text{ amperes}$$

Wire Size. The conductors in nonmetallic-sheathed cable are rated for 90°C. However, in conformance to *334.80*, the allowable ampacity is determined using the 60°C column of *Table 310.15(B)(16)*. Here we find that an 8 AWG conductor has an allowable ampacity of 40 amperes—more than adequate for the appliance's calculated load of 36.7 amperes. This will be an 8 AWG 3-conductor plus equipment grounding conductor nonmetallic-sheathed cable.

NEC 210.19(A)(3), Exception No. 2, states that for ranges of 8¾ kW or more, where the maximum demand has been calculated using Column C of *Table 220.55*, the neutral conductor may be reduced to not less than 70% of the branch-circuit rating but never smaller than 10 AWG.

$$40 \times 0.70 = 28 \text{ amperes}$$

Checking *Table 310.15(B)(16)*, we find that a 10 AWG conductor has an allowable ampacity of 30 amperes—more than adequate to carry the calculated neutral load. If the wiring method were a raceway method, this electric range could be served by two 8 AWG phase conductors and one 10 AWG neutral conductor.

Using the 70% multiplier is a moot point when installing nonmetallic-sheathed cable or armored cable, because the insulated phase conductors in these cables are all the same size.

Why the Reduced Neutral?

The heating elements are connected across the 240-volt source. Therefore, no neutral current from these elements is flowing in the 240/120-volt supply to the range.

Some loads in a range are 120-volt loads. These might be controls, timers, convection fans, rotisseries, and similar items. These loads do result in a current flow in the neutral conductor of the 240/120-volt supply to the range.

The permission to reduce the neutral conductor supplying an electric range to 70% is more than adequate to carry the above mentioned "line-to-neutral" connected loads within the range.

Some manufacturers supply a small 240/120-volt transformer inside the range that is connected across the two "hot" 240-volt conductors. The 120-volt secondary supplies the 120-volt loads within the range. With this arrangement, there is no neutral current flowing in the 240/120-volt supply to the range. In fact, the branch circuit to this type of electric range really only needs the two "hot" ungrounded conductors. Because you have no way of knowing what type of electric range will be installed, you must install a 4-wire branch circuit consisting of two "hot" ungrounded conductors, one grounded neutral conductor, and one equipment grounding conductor.

When using cable, reducing the neutral conductor is a moot point because all conductors in the cable are the same size, other than the equipment grounding conductor. When using a raceway wiring method, it is possible to take advantage of reducing the size of the neutral conductor.

Overcurrent Protection. A 2-pole, 40-ampere circuit breaker.

Disconnecting Means. The disconnecting means would be the 2-pole, 40-ampere circuit breaker. A cord-and-plug connection behind the base of a freestanding electric range would also serve as a disconnecting means, *422.33(B)*.

Grounding. The equipment grounding conductor contained in the nonmetallic-sheathed cable provides the means for grounding the frame of the appliance. According to *Table 250.122*, the equipment grounding conductor for a circuit protected by a 40-ampere overcurrent device is a 10 AWG copper conductor. That is the size EGC contained in an 8 AWG nonmetallic-sheathed cable.

CALCULATIONS WHEN MORE THAN ONE WALL-MOUNTED OVEN AND COUNTER-MOUNTED COOKING UNIT ARE SUPPLIED BY ONE BRANCH CIRCUIT

The following discusses how to make the load calculations when more than one wall-mounted oven and counter-mounted cooking unit that are supplied by one branch circuit.

Load Calculations

Note 4 to *Table 220.55* states that when a single branch circuit supplies a counter-mounted cooking unit and not more than two wall-mounted ovens, all located in the same room, the calculation is made by adding up the nameplate ratings of the individual appliances, then treating this total as if it were one electric range.

Figure 20-1 shows a wall-mounted oven and a counter-mounted cooking unit supplied by one 50-ampere branch circuit.

The total connected load of these two appliances is

$$I = \frac{W}{E} = \frac{14{,}050}{240} = 58.5 \text{ amperes}$$

However, because we are permitted to treat the two individual cooking appliances as one, we use the same calculations as we did for the free-standing range previously discussed.

The resulting conductor size, overcurrent protection, disconnecting means, and grounding are the same as for the equivalent free-standing range.

The cost of installing one 50-ampere branch circuit to supply both appliances might be higher than the cost of installing separate branch circuits to each appliance. There will be additional junction boxes, different size cables, cable connectors, conduit fittings, and splices. In addition, there are restrictions in the *NEC* on how to size the smaller tap conductors.

Factory-installed flexible metal conduit furnished and connected to these built-in cooking appliances will have the proper size and temperature-rated conductors for the particular appliance.

If you provide the flexible connections (whips) in the field, the minimum conductor sizing would be calculated as discussed above for each appliance. After the minimum conductor size is determined, we then must make sure that the conductors meet the "tap rule." This is important because the taps will have overcurrent protection rated greater than the ampacity of the tap conductors.

NEC 210.23(C) states that fixed cooking appliances that are fastened in place may be connected to 40- or 50-ampere branch circuits. If taps are to be made to a 40- or 50-ampere branch circuit, the taps must be able to carry the load, and in no case may the taps be less than 20 amperes, *210.19(A)(3)*.

For example, the 50-ampere circuit shown in Figure 20-1 has 10 AWG THHN tap conductors for the oven, and 8 AWG THHN tap conductors for the counter-mounted cooking unit. Each of these taps has an ampacity of greater than 20 amperes.

According to *Table 310.15(B)(16)*, a 6 AWG THHN conductor has an allowable ampacity of 55 amperes, using the 60°C column of the table.

USING A LOAD CENTER

The terms **load center** and **panelboard** are used interchangeably in the electrical trade. They are really one and the same thing. *Load center* is pretty much a manufacturers' term for panelboard. Panelboards might contain additional features that a typical residential load center might not have. Refer to Key Terms in the Appendix in this text for a more complete definition of "load center" and "panelboard."

In this residence, the built-in oven and range are very close to Panelboard B, in which case it is economical to install a separate branch circuit to each appliance. However, there are situations where the panelboard and the built-in cooking appliances are far apart.

The decision becomes one of installation cost (labor and material) as to whether to run separate branch circuits or to run one larger branch circuit and use the tap rules.

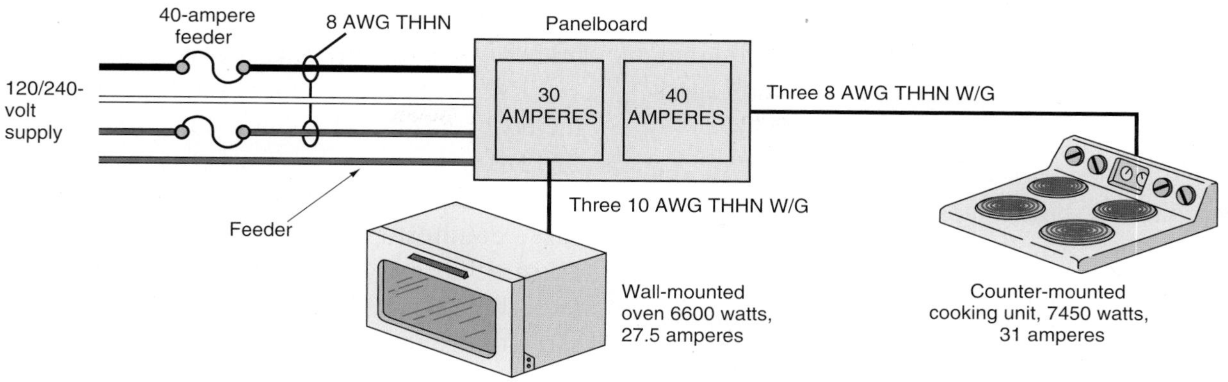

FIGURE 20-3 Appliances connected to a panelboard.

The tap rules can be ignored if a smaller panelboard is installed, as illustrated in Figure 20-3. There will be no taps—just branch circuits to each appliance sized for that particular appliance.

The calculations for the feeder from the main panelboard to another smaller panelboard are based on the two cooking appliances being treated as one. The conductors from the smaller panelboard are calculated for each appliance as previously discussed.

In homes with a basement, the panelboard could be installed below the kitchen where the appliances are installed.

CALCULATIONS WHEN MORE THAN ONE ELECTRIC RANGE, WALL-MOUNTED OVEN, OR COUNTER-MOUNTED COOKING UNIT IS SUPPLIED BY A FEEDER OR SERVICE

Although generally not an issue for wiring one- and two-family dwellings, the calculations for services and feeders of larger multifamily dwellings is more complex where a number of household cooking units are installed.

We need to refer to *Table 220.55*. This table has three columns. Column C is almost always used, with one exception. *Exception No. 3* shows that for household cooking units having a nameplate rating of more than 1¾ kW through 8¾ kW, Columns A and B are permitted to be used.

Column A is for household cooking units rated more than 1¾ kW but less than 3½ kW. Column B is for household cooking units rated 3½ kW through 8¾ kW.

When using Columns A and B, apply the demand factors for the number of household cooking units whose kW ratings fall within that column. Then add the results of each column together.

EXAMPLE

Calculate the demand for a feeder that has four 3-kW wall-mounted ovens and four 6-kW counter-mounted cooktops connected to it.

Step 1: 4 ovens × 3 kW = 12 kW
12 kW × 6% = 7.92 kW

Step 2: 4 cooktops × 6 kW = 24 kW
24 kW × 50% = 12 kW

Step 3: 7.92 kW + 12 kW = 19.92 kW

Therefore, 19.92 kW is the demand load for the four ovens and four cooktops, even though the actual connected load is 36 kW. Diversity plays a large factor in arriving at the calculated demand load.

Review of Choices for Hooking Up Electric Cooking Appliances

- Run a separate branch circuit from the main panelboard to each appliance.
- Run one large branch circuit to a junction box(es); then make taps to serve each appliance. See Figure 20-1.

- Run one large feeder from the main panelboard to a panelboard located near the appliances, then run a separate branch circuit to each appliance. See Figure 20-3.

When one considers the complicated calculations, additional splices, taps, working with different size conductors, extra time and material, the best and simplest choice is probably to run a separate branch circuit from the main panelboard to each appliance.

MICROWAVE OVENS

Microwave ovens are available in countertop, over-the-range, and over-the-counter models and as part of a free-standing electric range.

The electrician's primary concern is to install the proper size and type of branch circuit and to provide proper equipment grounding for the appliance.

A free-standing electric range that includes a microwave oven is connected according to the methods previously discussed in this chapter.

Countertop microwave ovens are plugged into the most convenient receptacle. If the receptacles that serve the countertops have been wired according to the *NEC*, they would be supplied by at least two 20-ampere small-appliance branch circuits, *210.11(C)(1)* and *210.52(B)(1)*.

Operating a microwave oven and another appliance at the same time might be too much load for a 20-ampere small-appliance branch circuit. The branch-circuit breaker might trip, or the branch-circuit fuse might open. That is why microwave cooking appliance manufacturers suggest a dedicated 15- or 20-ampere branch circuit for the appliance.

Microwave ovens fastened to the underside of a cabinet above a countertop or range generally are cord-and-plug connected. They come with a rather short power supply cord. A receptacle outlet must be provided inside the upper cabinet. Instructions furnished with these microwave ovens describe where to position this receptacle. A receptacle installed inside the cabinet is *in addition* to the receptacles required to serve countertop surfaces, *210.52*. This receptacle must not be connected to any of the 20-ampere small-appliance branch circuits that serve the countertop areas, *210.52(B)(2)*.

If a receptacle is installed inside the upper cabinet, the best choice is to connect the receptacle to an individual, dedicated, separate 120-volt, 20-ampere branch circuit for the microwave oven.

This receptacle is not required to be GFCI protected because it does not serve the countertop, *210.8(A)(6)*.

NEC 210.52(C)(5) states that receptacles that serve countertops shall not be located more than 20 in. (500 mm) above the countertop. A receptacle located inside the upper cabinets does not serve the countertop and therefore is not subject to *210.52(C)(5)*.

A cord-and-plug-connected appliance must not exceed 80% of the branch-circuit rating, *210.23(A)(1)*. A 20-ampere branch circuit may supply a maximum cord connected load of $20 \times 0.80 = 16$ amperes.

LIGHTWAVE ENERGY OVENS

A recent innovation is an oven that combines and cycles on and off lightwave energy from high-wattage halogen lamps and microwave energy. Cooking time is about 4 to 8 times faster than a conventional oven. These ovens are mounted to the underside of an upper cabinet or in a wall cabinet. Some require a dedicated 3-wire, 30-ampere, 120/240-volt circuit. A 4-wire, 30-ampere cord and plug is furnished with these ovens. A 30-ampere, 4-wire, NEMA 14-30R receptacle must be installed in the cabinet above the oven. Lower wattage units require a dedicated conventional 20-ampere, 120-volt branch circuit.

UL Standard 923 covers microwave and lightwave energy cooking appliances.

SURFACE HEATING ELEMENTS

Figures 20-4 and 20-5 illustrate typical electric range surface units.

Electric heating elements generate a large amount of radiant heat. Heat is measured in British Thermal Units (Btus). A Btu is defined as the amount of heat required to raise the temperature of one pound of water by 1 degree Fahrenheit. A 1000-watt heating element generates approximately 3412 Btu of heat per hour.

Heating elements in surface-type cooking units typically consist of spiral-wound nichrome

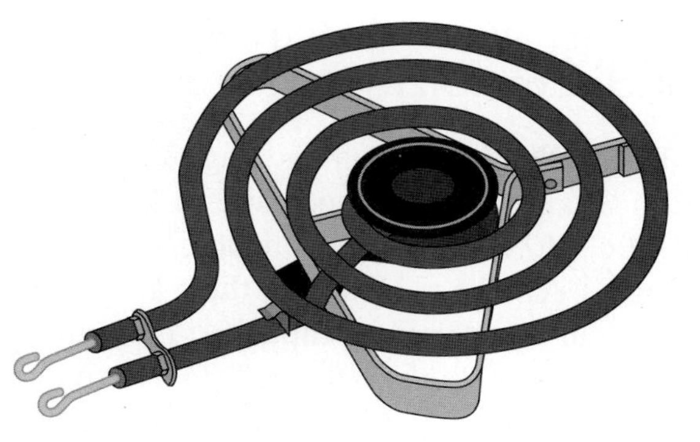

FIGURE 20-4 Typical 240-volt electric range surface heating element used with infinite heat controls.

resistance wire carefully imbedded in magnesium oxide—a white, chalklike powder. The wire/magnesium oxide is then encased in a nickel-steel alloy sheath, flattened under very high pressure, and then formed into coils. In the world of electric ranges, we use the term *surface heating elements*. In the world of gas ranges, the term *burners* is used.

Surface heating elements with one-coil (one heating element) are used with infinite heat controls. A one-coil element is shown in Figure 20-6. Two-coil (two heating elements) surface unit heating elements are used with fixed-heat 3-, 5-, and 7-position controls. A 2-coil element is shown in Figure 20-7. Three-coil (three heating elements) surface unit heating elements are used with 3-position switches.

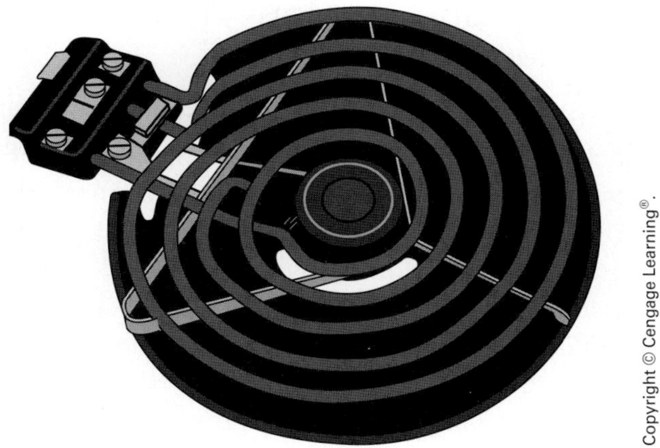

FIGURE 20-5 Typical electric range surface unit of the type used with 7-position controls.

TEMPERATURE CONTROLS

Different types of controls are used to adjust the heating elements on electric ranges.

Infinite-Position Controls

Most electric ranges are equipped with infinite-position temperature controls. These controls contain electrical contacts that open and close constantly, "pulsing" the power supply to the heating element, thus maintaining the desired temperature of the surface heating unit.

Figure 20-6 shows the inner workings of this type of control.

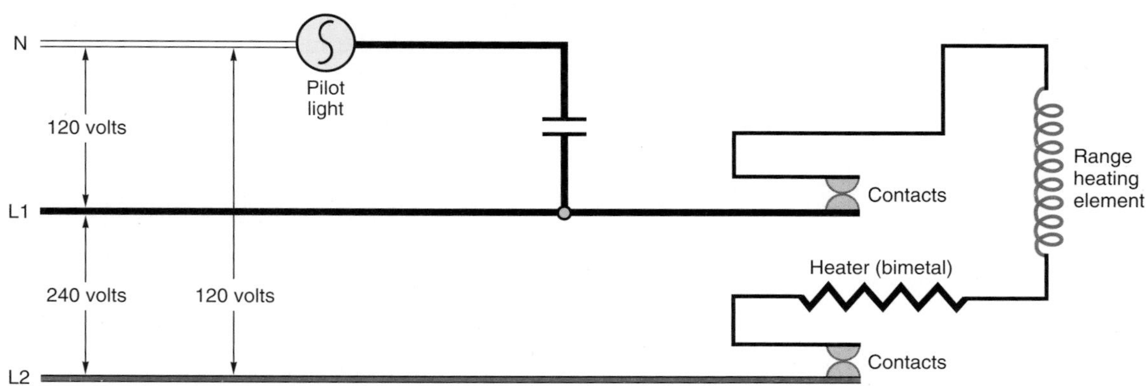

FIGURE 20-6 Typical internal wiring of an infinite heat surface element control.

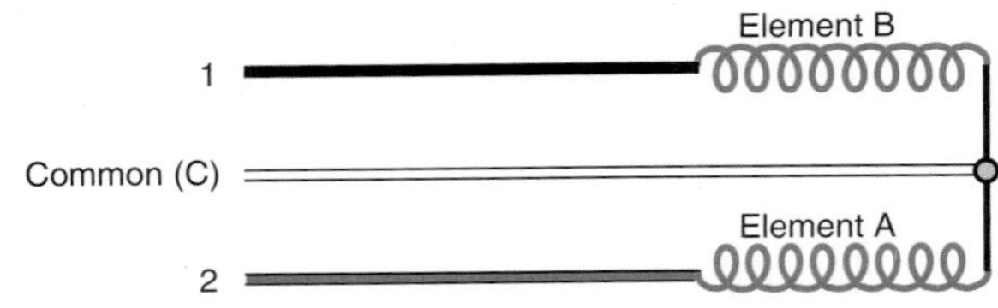

FIGURE 20-7 Typical surface unit wiring.

Fixed-Heat Controls

Still in use are rotary controls that have specific "indents" that lock the switch contacts into place as the knob is rotated to the various heat positions. Similar switching arrangements are found in pushbutton-type controls. These switches vary the connections to the heating elements individually, in series, in parallel, and in series/parallel to attain the various heat levels (high-low-medium, medium-high, etc.). These switches are commonly available in three, five, and seven positions.

Coil Size Selector Controls

Coil size selector controls are used with 3-coil heating elements. In position one, the center heating element comes on. In position two, the center and middle heating elements come on. In the third position, the center, middle, and outer heating elements come on. The fourth position is OFF.

Automatic Sensor Controls

This type of surface heating element has a sensor in the center of the unit. A special 12-volt control circuit is provided through a small transformer. When the control knob is set at a specific temperature, a "responder" pulses the heating element off and on to maintain the set temperature.

Caution When Selecting Conductor Size and Type

Recall from previous coverage in this chapter and in Chapter 4 that, according to UL Standards and *NEC 110.14(C)*, when conductor sizes 14 AWG through 1 AWG are installed, their allowable ampacity (current-carrying capacity) is found in the 60°C column of *Table 310.15(B)(16)*. If you are using nonmetallic-sheathed cable, the ampacity shall be in accordance with the 60°C conductor temperature rating, *334.80*.

That's why it is so important to read the appliance manufacturer's installation data carefully for recommended branch-circuit rating, conductor size, and conductor type.

Connecting Foreign-Made Appliances

Be careful when connecting foreign-made appliances. It is quite possible that the color coding of the terminals and conductors is different from what you are accustomed to. Throughout this text, we have discussed color coding as white for the grounded circuit conductor, black and red for the ungrounded "hot" conductor, and green or bare for the equipment grounding conductor. In this country, we use the term *ground*. In Europe, "ground" is referred to as "earth." Instructions might indicate something like this:

Green, green/yellow, or bare	= earth
Blue	= neutral
Brown	= live

The equipment grounding terminal might be marked with the letter "E." Just be aware that there could be differences in the identification of the terminals and conductors.

REVIEW

Note: Refer to the *Code* or the plans where necessary.

COUNTER-MOUNTED COOKING UNIT CIRCUIT ⓐG

1. a. What circuit supplies the counter-mounted cooking unit in this residence?

   ____

   b. What is the rating of this circuit? ______________________________

2. What three methods may be used to connect counter-mounted cooking units?

   ____

   ____

3. Is it permissible to use standard 60°C insulated conductors to connect all counter-mounted cooking units? Why? ______________________________

   ____

4. What is the maximum operating temperature (in degrees Celsius) for a

   a. Type TW conductor? ______________________________

   b. Type THW conductor? ______________________________

   c. Type THHN conductor? ______________________________

5. The *NEC* permits the following wiring methods to supply household built-in ovens and ranges. True or false? (Circle the correct answer.)

   a. Nonmetallic-sheathed cable (True) (False)

   b. Armored cable (True) (False)

   c. Flexible metal conduit (True) (False)

6. For electric ranges that have a calculated demand of 8¾ kW or more, the neutral conductor can be reduced to ______________ % of the branch-circuit rating but shall not be smaller than ______________ AWG.

7. One kilowatt equals ______________ Btu per hour.

8. Older style electric ranges had heat control knobs that had "indents" that could be felt when rotating the knob to adjust temperature to three, five, or seven different heat positions. Most new electric ranges have control knobs that do not have "indents." This type of heat control is known as a(n) ______________ heat control.

9. When a separate circuit supplies a counter-mounted cooking unit, what *Code* reference tells us that it is "against *Code*" to apply the demand factors of *Table 220.55* and requires us to calculate the load based on the appliance's actual nameplate rating?

   ____

WALL-MOUNTED OVEN CIRCUIT $\textcircled{\blacktriangle}_F$ AND FREE-STANDING RANGE

1. To what circuit is the wall-mounted oven connected? ______
2. An oven is rated at 7.5 kW. This is equal to
   a. ______ watts.
   b. ______ amperes at 240 volts.
3. a. What section of the *Code* governs the grounding of a wall-mounted oven? ______
   b. By what methods may wall-mounted ovens be grounded? ______
4. What is the type and ampere rating of the overcurrent device protecting the wall-mounted oven in this residence? *Hint*: Refer to the Schedule of Special-Purpose Outlets in the Specifications. ______
5. Approximately how many ft (m) of cable are required to connect the oven in the residence? ______
6. When connecting a wall-mounted oven and a counter-mounted cooking unit to one feeder, how long are the taps to the individual appliances? ______
7. The branch-circuit load for a single wall-mounted self-cleaning oven or counter-mounted cooking unit shall be the ______ rating of the appliance.
8. A 6 kW counter-mounted cooking unit and a 4 kW wall-mounted oven are to be installed in a residence. Calculate the maximum demand according to *Column C, Table 220.55*. Show all calculations. Both appliances will be connected to the same branch circuit.

9. The size of the neutral conductor supplying an electric range may be based on _______________ % of the branch-circuit rating.

10. A free-standing electric range is rated 11.8 kW, 120/240 volts. The wiring method is nonmetallic-sheathed cable. Use the 60°C column of *Table 310.15(B)(16)* in conformance to *110.14(C)*. Answer the following questions:

    a. According to Column C, *Table 220.55*, what is the maximum demand? _______ kW

    b. What is the minimum size for the ungrounded copper conductors? _________ AWG

    c. What is the minimum size neutral copper conductor? ___________________ AWG

    d. What is the correct rating for the branch-circuit overcurrent protection? _____ amperes

11. A double-oven electric range is rated at 18 kW, 120/240 volts. Calculate the maximum demand according to *Table 220.55*. Show all calculations.

12. For the range discussed in problem 11:
    a. What size 60°C ungrounded conductors are required? ______________
    b. What size 60°C neutral conductor is required? ______________
    c. What size 75°C ungrounded conductors are required? The terminal block and lugs on the range, the branch-circuit breaker, and the panelboard are marked as suitable for 75°C wire. ______________
    d. What size 75°C neutral conductor is required? Terminations are the same as in part (c). ______________
13. For ranges of 8¾ kW or higher rating, the minimum branch-circuit rating is ______________ amperes.
14. A nonmetallic-sheathed cable is used to connect a wall-mounted oven. The insulated conductors are 10 AWG. What is the size of the equipment grounding conductor in this cable? ______________
15. Match the following statements with the correct letter. Some letters may be used more than once.

____ A 30-ampere, 3-wire receptacle	a. for the equipment grounding conductor
____ A 50-ampere, 4-wire receptacle	b. for the white neutral conductor
____ The L-shaped blade on a 30-ampere plug cap	c. for the "hot" ungrounded conductors
____ The horseshoe-shaped slot on a 50-ampere receptacle	d. NEMA 10-30R
____ A 50-ampere, 4-wire plug cap	e. NEMA 14-50R
____ Terminals X and Y	f. NEMA 14-50R
____ Terminal W	
____ Terminal G	

16. A receptacle is mounted inside of a kitchen cabinet for the purpose of plugging in a microwave oven. Circle the correct answer.
    a. The receptacle may be connected to the same circuit that supplies the other receptacles that serve the countertop areas.
    b. The receptacle shall not be connected to the same circuit that supplies the other receptacles that serve the countertop areas.

CHAPTER 21

Special-Purpose Outlets—Food Waste Disposer $\triangle_H$, Dishwasher $\triangle_I$

OBJECTIVES

After studying this chapter, you should be able to

- install circuits for a typical food waste disposer and dishwasher.
- know the meaning of "continuous feed" and "batch feed" food waste disposers.
- understand direct connections (hard wired) and cord-and-plug connections.
- understand the meaning of branch-circuit protection and overload protection.
- be aware of the requirements for providing a means to disconnect appliances.
- understand the various acceptable methods for grounding appliances.
- understand the significance of water temperature for dishwashers.

INTRODUCTION

This chapter discusses the circuit requirements for food waste disposers (also called kitchen waste disposer, garbage disposer, garbage disposal, waste disposer, In-Sink-Erator®, Disposall®) and dishwashers of the type normally installed in homes. *Code* requirements for appliances are found in *Article 422* of the *NEC*®.

These 120-volt major appliances are *not* permitted to be connected to the 20-ampere small-appliance branch circuits that serve the receptacles above the countertops in the kitchen.

Note that *NEC 210.12(A)* requires that 15 and 20 A, 125-volt branch circuits that supply outlets or devices in dwelling unit kitchens have AFCI protection. The most practical location for this protection is an AFCI type circuit breaker installed at the origination of the branch circuit in panelboard B. See the discussion in Chapter 6 of this text for additional information on other methods of providing AFCI protection.

FOOD WASTE DISPOSER Ⓐ$_H$

Types of electrical connections are shown in Figures 21-1 and 21-2. The food waste disposer outlet is shown on the plans by the symbol Ⓐ$_H$. It is rated 120 volts, 7.2 amperes, and it is connected to Circuit B19, a 20-ampere, 120-volt branch circuit that originates in Panelboard B. See the discussion in Chapter 6 of this text for additional information on providing the required AFCI protection.

Controls

Continuous feed food waste disposers are usually controlled by a single-pole wall switch conveniently located above the countertop, as shown in Figure 21-2. The branch-circuit wiring is run first to the switch box. A second cable is then run from the switch box to the junction box on the disposer.

"Batch-feed" food waste disposers are equipped with an integral switch that starts and stops the

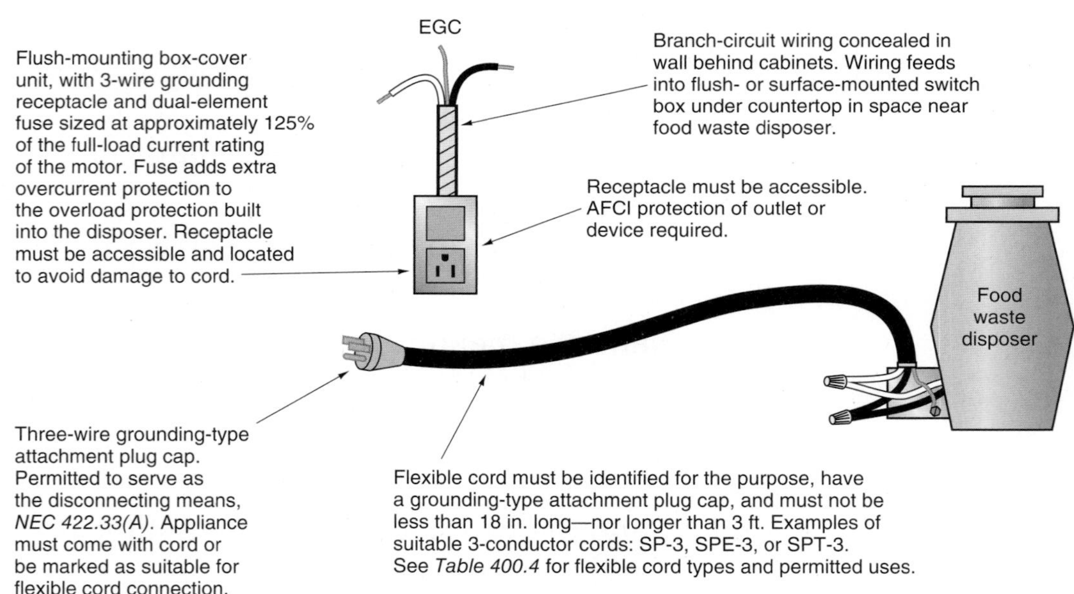

FIGURE 21-1 A cord-and-plug-connected food waste disposer, *422.16(A)* and *(B)(1)*. The same rules apply to built-in dishwashers and trash compactors, except the cord length must be 3 to 4 ft (900 mm to 1.2 m). The receptacle/fuse box-cover unit or a conventional receptacle could be used. ▶A receptacle under the kitchen sink is required to be GFCI protected if it is within 6 ft (1.8 m) of the sink,◀ *210.8(A)(7)*.

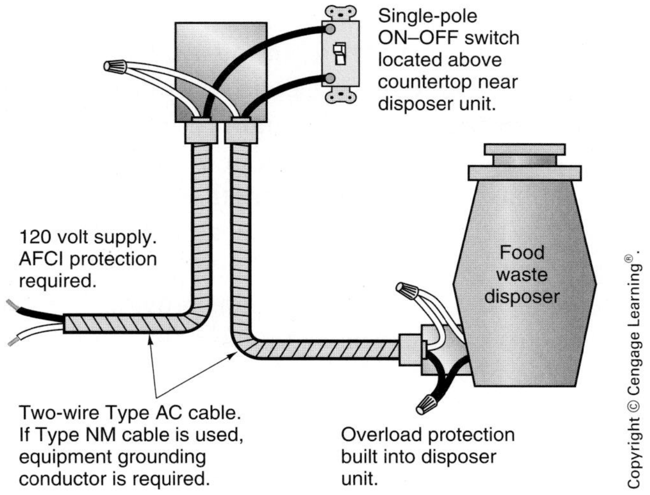

FIGURE 21-2 Wiring for a food waste disposer operated by a separate switch located above the countertop near the sink. Many local codes require this wall switch for all installations to be sure that there is a safe means of easily disconnecting the appliance.

disposer when the user twists the drain cover into place after filling the disposer with waste food. A wall switch is not required. See Figure 21-3.

Although rarely used in residential installations, a plumber could install a flow switch in the cold water line under the sink to ensure that the food waste disposer is not run without water, Figure 21-4. Generally, water should be left running for at least 15 seconds after shutting off the disposer to flush the drain clear of food waste particles.

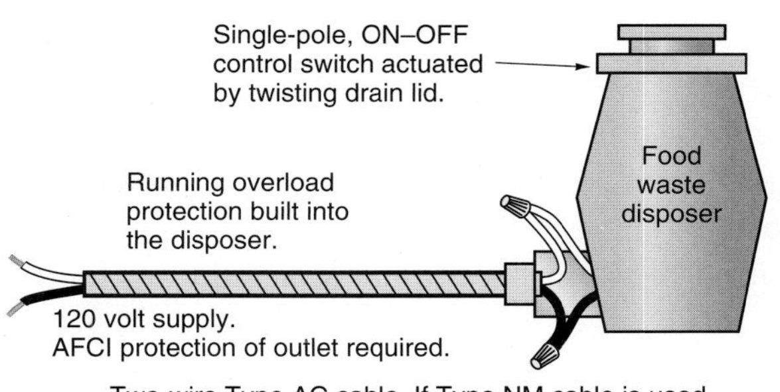

FIGURE 21-3 Wiring for a "batch-feed" food waste disposer with an integral ON–OFF switch.

DISHWASHER Ⓐ$_{\text{I}}$

The dishwasher outlet is shown on the plans by the symbol Ⓐ$_{\text{I}}$. The dishwasher is connected to Circuit B5, a 20-ampere, 120-volt branch circuit that originates in Panelboard B. ▶This circuit is required to be provided with both AFCI and GFCI protection,◀ see *210.8(D)* and *210.12(A)*. See the discussion in Chapter 6 of this text for the various methods of providing the required protection. A typical dishwasher contains a motor; an electric heating element; and, in some models, a small fan to assist in circulating the hot air during the drying cycle. For example, one manufacturer's dishwasher total connected load as marked on the nameplate is 9.2 amperes. This includes the motor, heating element, controls, relays, and an 875-watt heating element.

In most residential dishwashers, the motor and the booster heater operate at the same time.

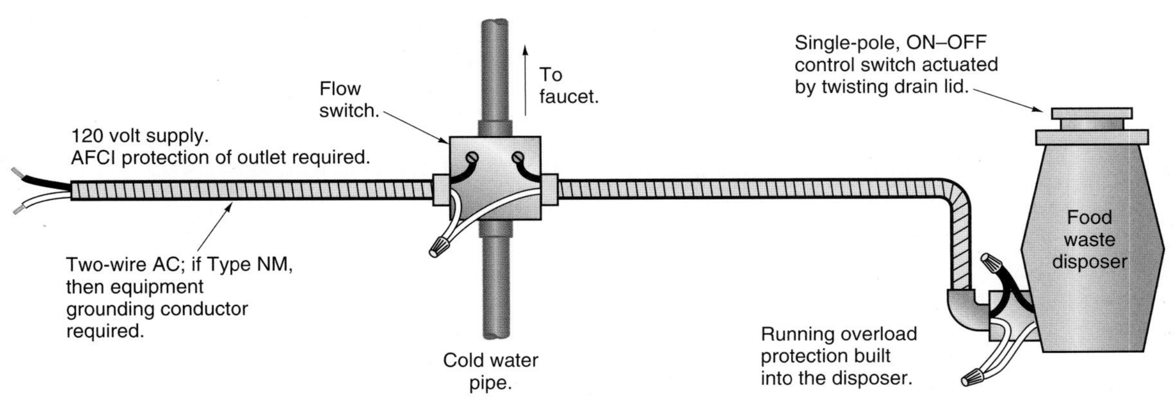

FIGURE 21-4 "Batch-feed" disposer with a flow switch in the cold water line.

The heating element boosts the incoming hot water from a typical water heater setting of 120°F (49°C) to approximately 140°F (60°C) or higher so as to better clean, dissolve detergents, and somewhat sanitize the dishes. During the dry cycle, only the heating element is on, plus a circulating fan if so equipped.

CODE RULES COMMON TO THE FOOD WASTE DISPOSER AND THE DISHWASHER

Branch-Circuit Rating

The manufacturer will specify the recommended rating of the branch circuit. Although it is possible to connect both appliances to one branch circuit, much depends on the current draw of the appliances. The preferred way to connect these appliances is to supply each appliance with a separate 20-ampere branch circuit. This provides a little spare capacity if the homeowner should decide to add a booster water heater at the sink location later on.

Do not connect the food waste disposer or dishwasher to any of the required 20-ampere small-appliance branch circuits in the kitchen. This is a violation of *210.52(B)(2)*. Small-appliance circuits are intended to serve countertops for cord-connected *portable* appliances only. See Chapter 12.

Conductors

Per the Specifications in the back of this text, the branch-circuit conductors are 12 AWG. The wiring method might be nonmetallic-sheathed cable, armored cable, or EMT, depending on the wiring method required by the code in your locality.

Direct Connection

See Figures 21-2, 21-3, and 21-4. This connection is sometimes referred to as being "hard wired." These appliances have a junction box (wiring compartment) for the electrical hookup. The branch-circuit supply (cable or flex conduit) is run directly to this box. During the rough-in stage of new wiring, the electrician must make sure that the supply cable or flex conduit is brought in at the proper location. This might be under the sink, brought up from underneath, or brought in behind the appliance. A sufficient length of wiring is necessary to be able to make the electrical connections and to position the appliance into place. The manufacturer's installation instructions provide all of the necessary information as to where to bring in the power supply.

GFCI and AFCI Protection

▶Both GFCI and AFCI protection is required as applicable. GFCI protection of the waste disposal is required if it is cord-and-plug-connected and the receptacle is within 6 ft (1.8 m) of the kitchen sink. *NEC 210.8(A)(7)*. GFCI protection of the dishwasher in required by *NEC 210.8(D)*.

AFCI protection of the waste disposal and dishwasher is required since the words "kitchens" and "devices" was added to *NEC 210.12(A)*.◀ The word "outlets" existed in *NEC 210.12(A)*. So, when you look at these rules, waste disposals and dishwashers are connected to "outlets" in the kitchen and the waste disposal is controlled by a device.

Cord-and-Plug Connection

See Figure 21-1. Cord-and-plug connection is permitted by *422.16(B)(2)*.

If 3-wire cord-and-plug connection is desired for easy disconnecting and servicing, the appliance manufacturers make available a power supply cord kit for this purpose. The appliances would have to be listed as being intended or identified for flexible cord connection, *NEC 400.7* and *422.16(A)* and *(B)*.

UL lists these appliances for connection with and without a cord-and-plug assembly.

Cord lengths are restricted by the *NEC* and UL Standards:

- Food waster disposers: no shorter than 18 in. (450 mm) and no longer than 3 ft (900 mm)
- Dishwashers: no shorter than 3 ft (0.9 m) and no longer than 4 ft (1.2 m)

A duplex grounding-type receptacle is usually installed in an accessible location under the sink for plugging in the dishwasher and food waste disposer. ▶This receptacle is required to be GFCI protected if it is located within 6 ft of the sink, *210.8(A)(7)*.◀

Remember that *210.23(A)(2)* limits cord-connected equipment to 80% of the branch-circuit ampere rating.

Overload Protection

Overload protection is an integral part of the appliance. Overload protection prevents the motor from burning out if it stalls or overheats for any reason. This protection is required in the UL Standard.

Branch-Circuit Protection

Branch-circuit short-circuit and ground-fault protection is provided by the branch-circuit fuse or circuit breaker, sized according to the manufacturers' recommendations. In our case, each appliance is fed by a separate 20-ampere branch circuit.

Disconnecting Means

All appliances must be provided with some means of disconnecting the appliance, *Article 422, Part III*. The branch-circuit breaker can serve as the disconnecting means if it is within sight of the appliance or is capable of being locked in the OFF position. Rarely, if ever, are these appliances within sight of the breaker panel. Lock-off provisions must be permanently installed on the breaker or switch, *422.31(B)*. Some local codes actually require that these appliances be cord-and-plug connected so the appliances can easily be disconnected for servicing by pulling out the plug. This requires that a receptacle be installed—usually under the sink in an accessible location.

Figure 21-1 shows a cord-and-plug connection that serves as the disconnecting means, *422.33(A)*.

Figure 21-2 shows a wall switch that serves as the disconnecting means.

If you do install a receptacle(s) below the sink or behind the appliance for cord-and-plug connection of a dishwasher or food waste disposer, the receptacle(s) installed do not have to be GFCI protected because they are not there to serve countertops.

Grounding

- *Article 250, Part VI*, covers equipment grounding and equipment grounding conductors. In *Part VI*, we find:
  - *250.114*: Grounding requirements for cord-and-plug-connected appliances.
  - *250.118*: Listing of all the acceptable methods of grounding equipment, such as the equipment grounding conductor in nonmetallic-sheathed cable, armored cable, or flexible metal conduit.
  - *250.119*: Requirements for the equipment grounding conductor to be bare or have green insulation, or have green insulation with one or more yellow stripes.
  - *250.122*: The minimum size for equipment grounding conductors.
  - *250.126*: Requirement for the equipment grounding conductor terminal to be bare or green and hexagon shaped.
- *Article 250, Part VII* covers methods of equipment grounding. In *Part VI*, we find:
  - *250.134*: Discussion of grounding of equipment fastened in place, or "hard-wired"; referral back to *250.118*.
  - *250.138*: Acceptance of the equipment grounding conductor in flexible cords for cord-and-plug-connected appliances.
  - *250.142(B)*: Grounding appliances to the grounded circuit conductor is not permitted.

Supplemental Overcurrent Protection and Disconnecting Means

Instead of installing a duplex grounding-type receptacle under the sink, electricians will sometimes install a box-cover switch/fuse unit under the sink of a type similar to that shown in Figure 16-23. For a cord-and-plug connection, the box-cover unit would be a receptacle/fuse type, as shown in Figure 21-1. This receptacle/fuse unit does not have to be GFCI protected because it does not serve the countertop. However, AFCI protection of the branch circuit that supplies the outlet is required. Box-cover units provide both supplemental overload and short-circuit protection. They also serve as a convenient accessible disconnecting means, as required by *422.30*, making it easy for the technician to be assured that the circuit is off when working on the dishwasher. Time-delay fuses in this box-cover unit would be sized at approximately 125% of the nameplate rating of the dishwasher.

Read the Installation Instructions

Always read the installation instructions, the owner's manual, and the nameplate for specific electrical requirements.

PORTABLE DISHWASHERS

In addition to built-in dishwashers, portable models are available. Most portable models are convertible to become built-in units. Portable dishwashers have one hose that connects to the water faucet and a water drainage hose that hangs in the sink. The dishwasher probably will be plugged into the receptacle nearest the sink.

Portable dishwashers are supplied with a 3-wire cord and plug. If the 3-wire plug cap is plugged into a properly connected 3-wire grounding-type receptacle, the dishwasher is adequately grounded.

In older homes that do not have grounding-type receptacles, the use of a portable cord-and-plug-connected dishwasher will require the installation of a grounding-type receptacle. Assuming the appliance will be used in the kitchen, the receptacle will also have to be GFCI protected, *210.8(A)(6)*, as well as AFCI protection as required in *210.12 (A)*. These topics are covered in Chapter 6 of this text.

Cord-and-plug-connected portable dishwashers are generally rated 10 amperes or less at 120 volts.

WATER TEMPERATURE

Electric water heaters are shipped with the thermostat set as low as possible. When the water heater is installed, the thermostat should be reset to recommended settings such as 120°F (49°C). In fact, because of the problem of people being scalded, many municipalities and states have implemented code requirements that water heater thermostats be set at not over 120°F (49°C). Although this temperature is acceptable for baths and showers, 120°F (49°C) is not hot enough to properly clean dishes and glasses. Some manufacturers of dishwashers provide a feature that turns on the heating element for the final rinse cycle. This raises the water temperature from 120°F (49°C) to the desired 140°F (60°C)–145°F (63°C).

See Chapter 19 for a discussion on water heaters and the dangers of scalding with hot water of varying temperatures.

REVIEW

Note: Refer to the *Code* or plans where necessary.

FOOD WASTE DISPOSER CIRCUIT $Ⓐ_H$

1. How many amperes does the food waste disposer draw? ____________________
2. a. To what circuit is the food waste disposer connected?

   ____

   b. What size wire is used to connect the food waste disposer?

   ____

3. Means must be provided to disconnect the food waste disposer. The homeowner need not be involved in electrical connections when servicing the disposer if the disconnecting means is ____

____

4. How is motor overload protection provided in most food waste disposer units?

____

5. When running overcurrent protection is not provided by the manufacturer or if additional backup overcurrent protection is wanted, dual-element, time-delay fuses may be installed in a separate box-cover unit. These fuses are sized at not over ____________% of the full-load rating of the motor.

6. Why are flow switches sometimes installed on food waste disposers? ____________

____

7. Where in the *NEC* would you look for basic *Code* rules relating to grounding appliances?

____

8. Do the plans show a wall switch for controlling the food waste disposer?

____

9. A separate circuit supplies the food waste disposer in this residence. Approximately how many ft (m) of cable will be required to connect the disposer?

____

10. If a receptacle is installed underneath the sink for the purpose of plugging in a cord-connected food waste disposer, must the receptacle be GFCI protected?

____

11. Is it required that the branch circuit for the waste disposer have AFCI protection? (Yes) (No) (Circle your answer.)

DISHWASHER CIRCUIT ⒽI

1. a. To what circuit is the dishwasher in this residence connected? ____________
   b. What size wire is used to connect the dishwasher? ____________
   c. Is it required that the branch circuit for the dishwasher have AFCI protection? (Yes) (No) (Circle your answer.)
2. The dishwasher in this residence is rated (Circle the correct answer.)
   a. 5 A.
   b. 9.2 A.
   c. 15 A.
3. The heating element is rated at (Circle the correct answer.)
   a. 875 watts.
   b. 1000 watts.
   c. 1250 watts.

4. How many amperes at 120 volts do the following heating elements draw?
   a. 750 watts ____________________
   b. 1000 watts ____________________
   c. 1250 watts ____________________
5. How is the dishwasher in this residence grounded? ____________________
   ____
   ____
   ____
   ____
6. What type of cord is used on most portable dishwashers?
   ____
7. How is a portable dishwasher grounded? ____________________
   ____
   ____
8. Who is to furnish the dishwasher?
   ____
   ____
9. What article of the *Code* specifically addresses electrical appliances?
   ____

CHAPTER 22

Special-Purpose Outlets for the Bathroom Ceiling Heat/Vent/Lights ◬K ◬J, the Attic Fan ◬L, and the Hydromassage Tub ◬A

OBJECTIVES

After studying this chapter, you should be able to

- explain the operation and control of heat/vent/lights and make electrical connections in conformance to the *NEC*.
- understand the operation and control of attic exhaust fans and make electrical connections in conformance to the *NEC*.
- understand humidity problems, solutions, and humidistats.
- understand *NEC* requirements for hydromassage bathtub branch circuits.

BATHROOM CEILING HEATER CIRCUITS Ⓐ$_K$ Ⓐ$_J$

Both bathrooms contain a combination heater, light, and exhaust fan installed in the ceiling. The heat/vent/light is shown in Figure 22-1. The symbols Ⓐ$_K$ and Ⓐ$_J$ represent the outlets for these units.

Each heat/vent/light contains a heating element similar to the surface burners of electric ranges but usually has fins attached that are similar in purpose to the fins on a radiator for a car or truck. The appliances also have a single-shaft motor with a blower wheel, a lamp with a diffusing lens, and a means of discharging air to the outside of the dwelling.

The heat/vent/lights specified for this residence are rated 1500 watts at 120 volts. The unit in the Master Bedroom bathroom Ⓐ$_J$ is connected to Circuit A12, a 20-ampere,120-volt circuit. In the Front Bedroom bathroom, the unit is connected to Circuit A11, also a 20-ampere,120-volt circuit.

The current rating of the 1500-watt unit is

$$I = \frac{W}{E} = \frac{1500}{120} = 12.5 \text{ amperes}$$

Article 424 of the *NEC*® covers fixed electric space heating. *NEC 424.3(B)* considers fixed electric space heating to be a continuous load. Simply stated, this means that the 125% multiplier must be applied for the branch-circuit conductors, *210.19(A)*, and the branch-circuit overcurrent device, *210.20(A)*.

Therefore,

$$12.5 \text{ amperes} \times 1.25 = 15.625 \text{ amperes.}$$

We have chosen to supply each of the bathroom ceiling heaters with a 20-ampere,120-volt branch circuit. The 12 AWG conductors and the 20-ampere circuit breakers "Meet *Code*."

Wiring

At first, the connections look rather complicated. Let's walk through it!

In Figure 22-2, we see the wiring diagram for the heat/vent/light. The 120-volt line is run from Panel A to the wall box. A raceway (flexible or electrical metallic tubing) is run between the wall box where the switches are located to the heat/vent/light unit. In this raceway, five wires are needed—one common white grounded circuit conductor (neutral), and one conductor each for the heating element, fan motor, 100-watt lamp, and the 7-watt night light. The four ungrounded switch leg wires may be any color except white, gray, or green. Color coding of conductors is discussed in detail in Chapter 5. Because the circuit rating is 20 amperes, 12 AWG conductors are installed. See the Schedule of Special-Purpose Outlets contained in the Specifications for this text.

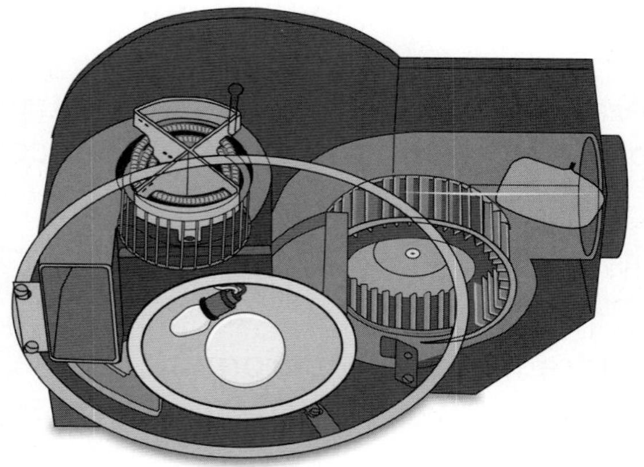

Courtesy of Broan-Nu Tone, LLC.

FIGURE 22-1 Heat/vent/light.

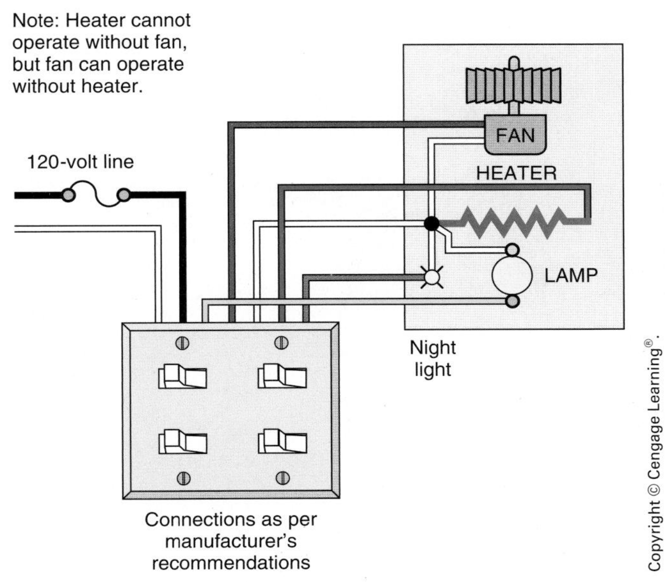

FIGURE 22-2 Wiring for Heat/vent/light.

Nontrained electricians often violate the *Code* by running one 3-wire and one 2-wire cable between the wall box and heat/vent/light. *NEC 300.3(B)* states that All conductors of the same circuit and, where used, the neutral and all equipment grounding conductors shall be contained within the same raceway, cable tray, trench, cable, or cord.*

The switch location requires a 2-gang opening, with the wall box being large enough to accommodate seven 12 AWG conductors, four switches, and cable clamps. For this installation, a 4 in. square, 2⅛ in. deep box with a 2-gang raised plaster ring is used. This provides ample room for the conductors and switches. Box sizing is covered in Chapter 2 and Chapter 8.

Operation of the Heat/Vent/Light

The heat/vent/light is controlled by four switches, as shown in Figure 22-2:

one switch for the heater
one switch for the light
one switch for the exhaust fan
one switch for the night light

When the heater switch is turned on, the heating element begins to give off heat. The heat activates a bimetallic coil attached to a damper section in the housing of the unit. The heat-sensitive coil expands until the damper closes the discharge opening of the exhaust fan. Air is taken in through the outer grille of the unit. The air is blown downward over the heating element. The blower wheel circulates the heated air back into the room area.

If the heater is turned off so that only the exhaust fan is on, the air is pulled into the unit and is exhausted to the outside of the house by the blower wheel.

Exhaust fans and luminaires are available with integral automatic moisture sensors and motion detectors. Step into a dark bathroom, and the light goes on. Step into and run the shower, and the exhaust fan turns on at a predetermined moisture level.

Grounding the Heat/Vent/Light

The heat/vent/light must be properly grounded!

Proper grounding is accomplished by using any of the accepted equipment grounding methods found in *250.118*. For residential wiring, the equipment grounding conductor in nonmetallic-sheathed cable, the armor and bonding wire in armored cable, or flexible metal conduit provides the necessary equipment ground. As stated many times before, NEVER ground an appliance to the white grounded circuit conductor, *250.142(B)*.

ATTIC EXHAUST FAN CIRCUIT Ⓐ L

Read the Installation Instructions!

Before beginning any installation of electrical equipment, always refer to manufacturers' literature for specific installation instructions.

The attic exhaust fan in this residence has a 2-speed, ¼-horsepower, direct-drive motor. "Direct-drive" means that the fan blade is attached directly to the shaft of the motor without the use of pulleys and belts.

*Reprinted with permission from NFPA 70-2014.

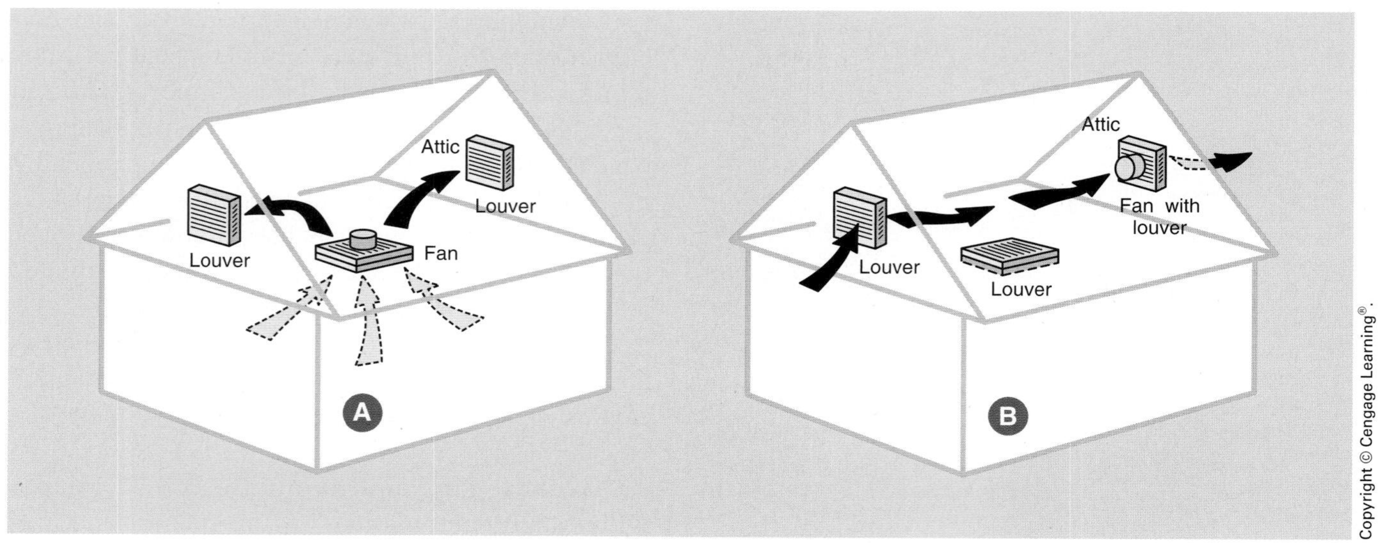

FIGURE 22-3 Exhaust fan installation in an attic.

The exhaust fan is mounted in the hall ceiling between the Master Bedroom and the Front Bedroom, as shown in Figure 22-3(A). Exhaust fans can also be installed in a gable of a house, as shown in Figure 22-3(B), or in the roof.

When the fan is not running, the louvers on the fan and house gable remain in the CLOSED position.

When running, the louvers on the fan and house gable OPEN, removing hot, stagnant, humid, or smoky air from the house by drawing in fresh air through open windows and doors, in turn exhausting the air through the louvers.

An exhaust fan can lower the indoor temperature of the house as much as 10°F to 20°F (−6°C to −11°C). On hot days, the air in an attic can reach superheated temperatures of 150°F (66°C) or more. The heat from the attic will radiate and conduct through the ceiling into the living areas. This results in increased loading of the air-conditioning system. These problems are minimized by properly vented exhaust fans. Humidity issues are discussed later on.

Bathroom exhaust fans, vents for clothes dryers, and similar high-humidity sources are *never* to be vented directly into the attic or other dead spaces. They are to be vented to the outside.

Sizing a Home-Type Exhaust Fan

Exhaust fan ventilating capability is rated in cubic feet of air per minute (CFM).

To determine the minimum amount of CFM needed, multiply the square footage of the house by 3. For example, an 1800 ft^2 house would need an exhaust fan rated $180 \times 3 = 5400$ CFM.

To determine the minimum amount of attic exhaust louver square footage area, divide the CFM by 750. In this example, $5400 \div 750 = 7.2$ ft^2.

Never exhaust air into spaces within walls, ceilings, attics, crawl spaces, or garages. The humidity may damage the structure and insulation.

Figure 22-4 shows a typical ceiling-mounted whole-house ventilator exhaust fan.

FIGURE 22-4 Ceiling exhaust fan as viewed from the attic.

Branch Circuit

The exhaust fan in this residence is rated 115 volts nominal, 5.8 amperes (696 VA). It is connected to Circuit A10, a 120-volt,15-ampere circuit breaker in the main panel. This branch circuit supplies the attic exhaust fan only. It supplies no other loads.

Conductors (*430.22*)

The *NEC* requires that the minimum size conductor is to be not less than 125% of the motor's full-load current rating.

Therefore,

$$5.8 \times 1.25 = 7.25 \text{ amperes}$$

Table 310.106(A) shows a minimum size conductor for general wiring to be 14 AWG copper.

Table 310.15(B)(16) shows that a 14 AWG THHN copper has an allowable ampacity of 15 amperes in the 60°C column.

In conformance to these requirements, the Schedule of Special-Purpose Outlets tells us that the branch-circuit conductors are 14 AWG conductors. The branch-circuit cable is run to a 4 in. square deep outlet box installed by the electrician near the exhaust fan. A switch/fuse box-cover unit is to be mounted on this outlet box. From this box, a flexible connection like armored cable, flexible metal conduit, or nonmetallic-sheathed cable is made to the factory-installed junction box on the frame of the fan. Depending on the size of junction box on the fan assembly, it might be possible to run the branch-circuit cable directly into that junction box.

Overload Protection (*430.31* through *430.44*)

The exhaust fan motor is provided with integral (built into the motor) overload protection. Overload protection prevents the motor from burning out if it stalls or overheats for any reason. This protection is required in the UL Standard.

NEC 430.32(C) states that for automatically started motors, the overload protection must not exceed 125% of the full-load running current of the motor. *Table 430.248* shows that a ¼-horsepower motor has a full-load current draw of 5.8 amperes. Thus, the overload device rating is $5.8 \times 1.25 = 7.25$ amperes. The motor's built-in overload protection meets this requirement.

Motor Branch-Circuit Short-Circuit and Ground-Fault Protection (*430.52, Table 430.52*)

Motor branch-circuit short-circuit and ground-fault protection is covered in Chapter 19. Here is a brief review.

For a circuit breaker, *Table 430.52* indicates a maximum rating of 250% of the motor's full-load current rating. Therefore, $5.8 \times 2.5 = 14.5$ amperes. The next standard rating is permitted by *430.52(C)(1)*. This is a 15–ampere circuit breaker according to *240.6(A)*.

For a dual-element, time-delay fuse, *Table 430.52* indicates a maximum rating of 175% of the motor's full-load current rating. Therefore, $5.8 \times 1.75 = 11+$. The next standard rating is permitted by *430.52(C)(1)*. This is a 15-ampere, dual-element, time-delay fuse according to *240.6(A)*.

There are exceptions to the above rules that permit even higher-rated fuses or circuit breakers under certain conditions as stated in *430.52(C)*.

Disconnecting Means (*Article 430, Part IX*)

A switch/fuse box-cover unit is mounted on the 4 in. square outlet box. This meets the requirement that the disconnecting means shall be in sight of the motor, *430.102(B)*. Anyone working on the exhaust fan can easily turn off the power. The 15-ampere circuit breaker in the main panel also serves as a disconnecting means since it can be locked in the OFF position.

Grounding (*Article 250, Part VI*)

The exhaust fan must be properly grounded!

Proper grounding is accomplished by using any of the accepted equipment grounding methods found in *250.118*. For residential wiring, the equipment grounding conductor in nonmetallic-sheathed cable, the armor and bonding wire in armored cable, or flexible metal conduit provides the necessary equipment ground. DO NOT ground the exhaust fan metal housing to the white grounded circuit conductor (neutral), *250.142(B)*.

Supplemental Overcurrent Protection

Integral overload protection in the motor is the first line of defense against motor burnout and potential fire. A dual-element, time-delay fuse installed in the switch/fuse box-cover unit adjacent to the exhaust fan per the specifications provides backup overload protection as the second line of defense against motor burnout. Dual-element, time-delay fuses are available in many ampere ratings, such as 7½, 8, 9, 10, and 12 amperes.

Should the motor fail to start, short-circuit, or go to ground, the fuse in the box-cover switch/fuse unit will open without tripping the branch-circuit breaker back in Panel A. This makes it easy for servicing the unit.

Refer to Table 28-1 in this text for much more information relating to fuses and circuit breakers.

Fan Control

Typical residential fan controls are illustrated in Figures 22-5 and 22-6. These may be mounted adjacent to or above the other wall-mounted lighting switches.

Figure 22-7 shows many additional options available for controlling exhaust fans.

A. A simple ON–OFF switch.

B. A speed control switch that allows multiple and/or an infinite number of speeds. (The exhaust fan in this residence is controlled by this type of control.) See also Figure 22-5.

C. A timer switch (S_T) that allows the user to select how long the exhaust fan is to run, up to 12 hours. See also Figure 22-6. Figure 22-6(A) shows a spring-wound timer and Figure 22-6(B) an electronic timer. The electronic timer requires a connection to the branch circuit neutral.

D. A humidity control (H) switch that senses moisture buildups. See "Humidity Control" later in this chapter for further details on this type of switch.

E. Exhaust fans mounted into end gables or roofs of residences are available with an adjustable temperature control on the frame of the fan. This thermostat generally has a start range of 70°F to 130°F (21°C to 54°C), and will automatically stop at a temperature 10°F (−12°C) below the start setting, thus providing totally automatic ON–OFF control of the exhaust fan.

F. A high temperature automatic heat sensor that will shut off the fan motor when the temperature reaches 200°F (93°C). This is a safety feature so that the fan will not spread a fire. Connect in series with other control switches.

Courtesy of Broan-Nu Tone, LLC.

FIGURE 22-5 A typical 2-speed fan control. Also available in 3-speed, 4-speed, and totally variable speed electronic types.

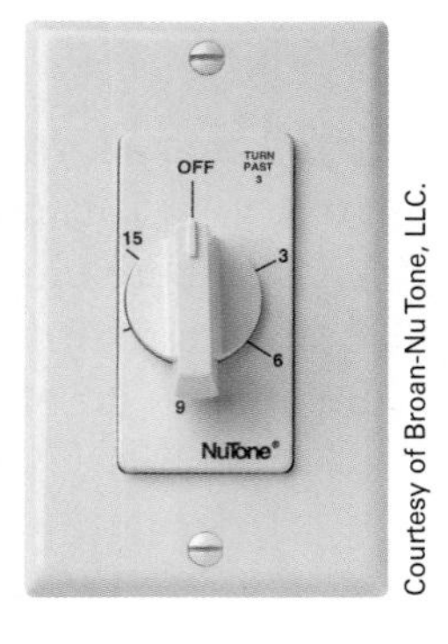

Courtesy of Broan-Nu Tone, LLC.

Courtesy of Leviton Manufacturing Company. All Rights Reserved.

FIGURE 22-6 (A) A manual timer fan control. some timer controls have a continuous ON position. (B) An electronic timer fan control.

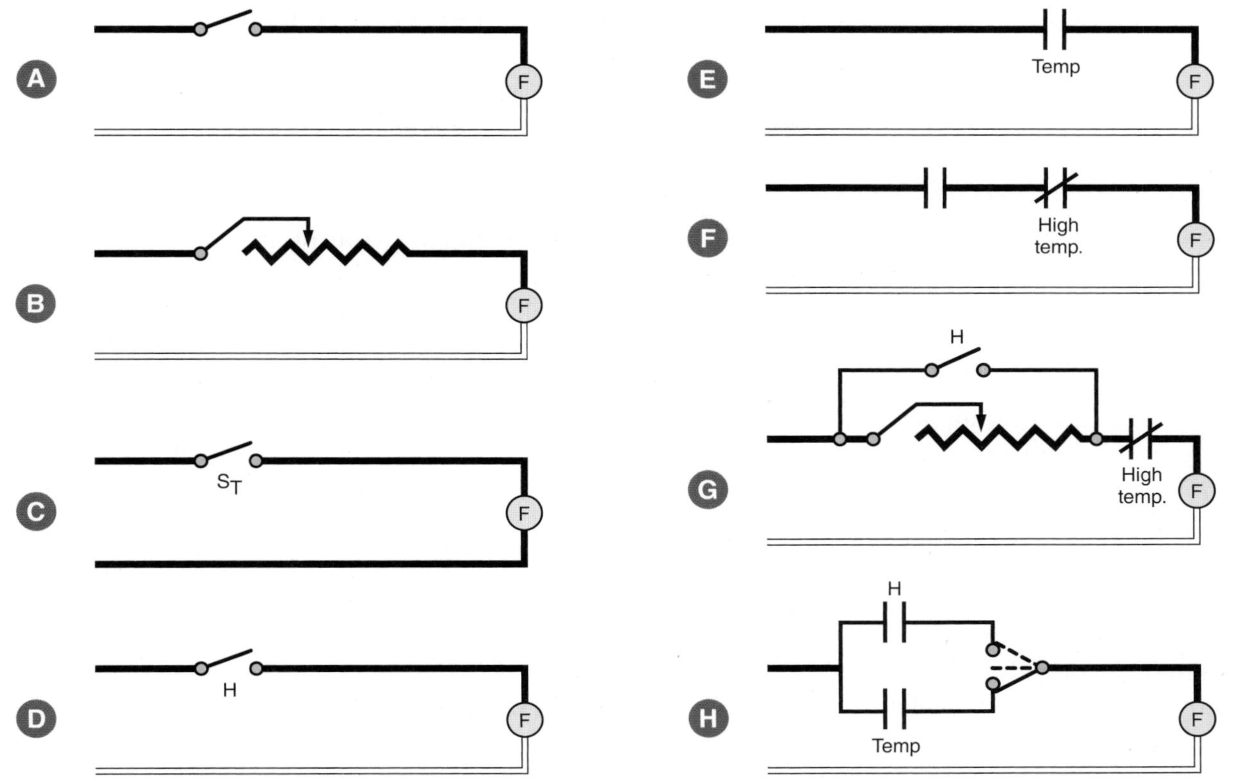

FIGURE 22-7 Different options for the control of exhaust fans.

G. A combination of controls. This circuit shows an exhaust fan controlled by an infinite-speed switch, a humidity control (H), and a high-temperature heat sensor. Note that the speed control and the humidity control are connected in parallel so that either can start the fan. The high-temperature heat sensor is in series so that it will shut off the power to the fan when it senses 200°F (93°C), even if the speed control or humidity control is in the ON position.

H. This circuitry combines a humidistat (D) and a thermostat (E) in such a way that the home-owner can select between controlling humidity and temperature in the attic. A selector switch is installed in a readily accessible place for ease of use by the homeowner. The switch has a CENTER OFF position. The humidity and temperature sensors are mounted in the attic.

Any of the above variations for controlling an exhaust fan could also include a pilot light to give clear visual indication that the fan is running. This pilot light could be mounted in the same switch box as the speed control or timer control, or it could be mounted next to these switches.

HUMIDITY

Humidity can be a big problem in homes.

Well-insulated homes can experience problems with excess humidity due to the "tightness" of the house. Thermal insulation and vapor barriers that keep moisture out of the insulation must be properly installed. Insulation must be kept dry or its efficiency decreases. Whole-house vinyl wraps (Tyvek®) stop airflow through wall cavities, as well as preventing wind-driven rain from getting behind wood, vinyl, brick, and stucco siding.

High humidity is uncomfortable, but a bigger problem is that high humidity promotes the growth of mold and the deterioration of fabrics and floor coverings. Framing members, wall panels, plaster, or drywall may also deteriorate because of the humidity.

Instead of manual ON–OFF control of an exhaust fan, a desired humidity level can be maintained automatically with a device called a humidistat. Humidistats have a humidity-sensitive element such as human hair (moisture causes hair to contract, which is why hair frizzes when exposed to high humidity), nylon strips as in Figure 22-8 (nylon expands and contracts with a change in humidity), or electronic components (that change resistance as the humidity changes) for the more sophisticated applications. When the relative humidity reaches a certain set level, the humidity-sensing component reacts, in turn activating the actual switching mechanism, such as a small microswitch, which turns the motor on and off. A bimetallic element cannot be used because it reacts to temperature changes only.

Adjustable settings are from 5% to 95% relative humidity. A comfort level of about 50% relative humidity is considered acceptable.

Humidistats having proper voltage rating (120 or 240 volts) and ampere rating are used to automatically switch the motor directly. See Figure 22-9(A). They are available in styles similar to typical speed and timer controls, with a knob and faceplate to match the other switch and receptacle faceplates in a home. The wiring is as simple as wiring in a switch leg for switching lighting outlets.

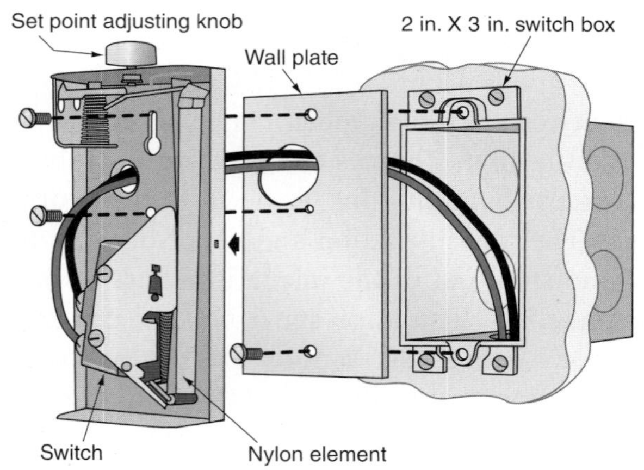

FIGURE 22-8 Details of a humidity control used with an exhaust fan.

Where the 120-volt power wiring becomes very lengthy because of the distance between the controller and the fan, or for large motors having a high current draw, a relay can be used. See Figure 22-9(B). This relay is mounted near the whole-house exhaust fan. The branch-circuit wiring is run from the main panelboard to a 4 in. square, 2⅛ in. deep outlet box conveniently located near the fan. A box-cover switch/fuse unit on this outlet box serves as the disconnecting means within sight of the motor, as required by *422.30* and *430.102*, and has a dual-element, time-delay fuse sized at approximately 125% of the motor's full-load current rating for motor overload backup protection. Typical home-type whole-house exhaust fans do not require relays.

Dehumidifiers used in basements and other damp locations and humidifiers on furnaces use a humidistat to turn the power to the motor on and off.

APPLIANCE DISCONNECTING MEANS

To clean, adjust, maintain, or repair an appliance, it must be disconnected to prevent personal injury. *NEC 422.30* through *422.35* outline the basic methods for disconnecting fixed, portable, and stationary electrical appliances. Note that each appliance in the dwelling conforms to one or more of the disconnecting methods listed. The more important *Code* rules for appliance disconnects are as follows:

- Each appliance must have a disconnecting means.
- The disconnecting means may be a separate disconnect switch if the appliance is permanently connected.
- The disconnecting means may be the branch-circuit switch or circuit breaker if the appliance is permanently connected and not over ⅛ horsepower or 300 volt-amperes.
- The disconnecting means may be an attachment plug cap if the appliance is cord connected.
- The disconnecting means may be the unit switch on the appliance only if other means for disconnection are also provided. In a single-family residence, the service disconnect serves as the other means.

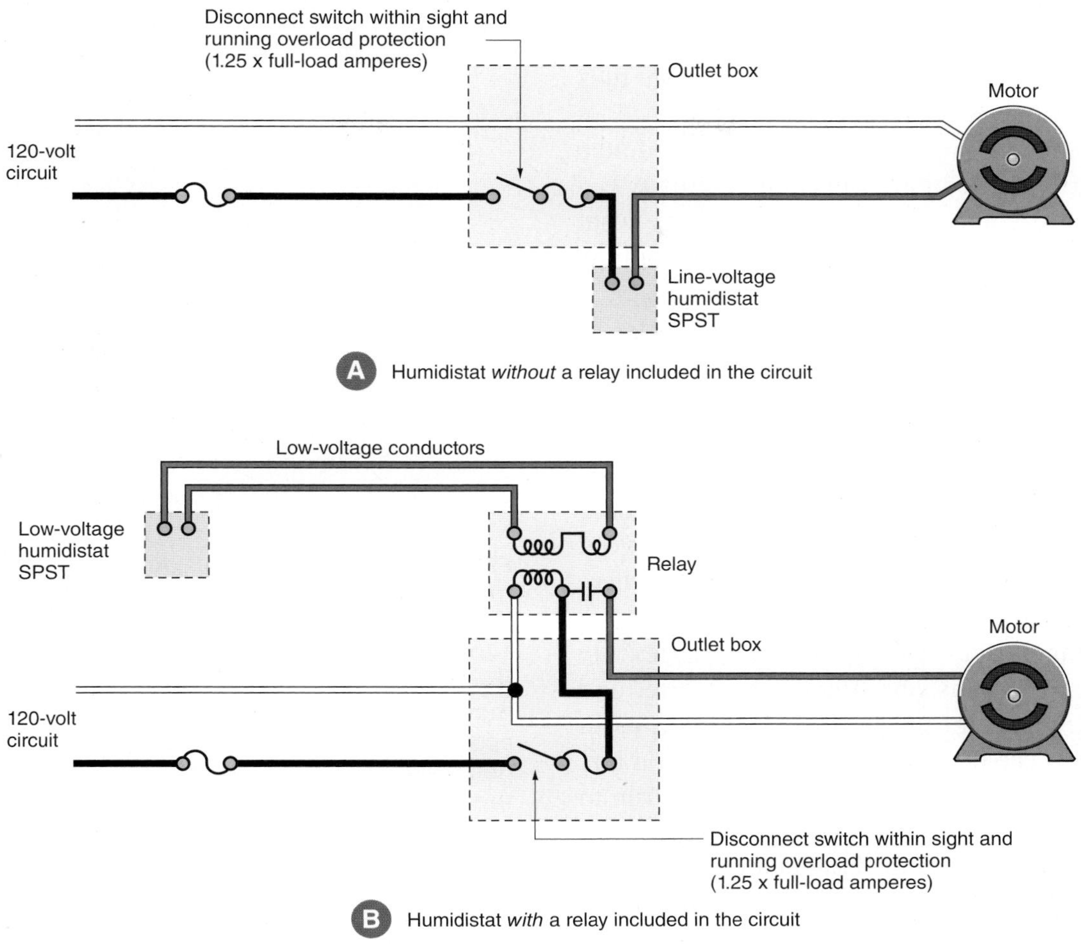

FIGURE 22-9 Wiring for the humidistat control.

- The disconnect must have a positive ON–OFF position.
- The disconnect must be capable of being locked in the OFF position or be within sight of a motor-driven appliance where the motor is more than ⅛ horsepower. The lock-off provision must be permanent.

Read *422.30* through *422.35* for the details of appliance disconnect requirements.

HYDROMASSAGE BATHTUB CIRCUIT Ⓐ A

The Master Bedroom is equipped with a hydromassage bathtub. A hydromassage bathtub is sometimes referred to as a whirlpool bath.

NEC 680.2 defines a *hydromassage bathtub* as A permanently installed bathtub equipped with a recirculating piping system, pump and associated equipment. It is designed so that it can accept, circulate, and discharge water upon each use.*

In other words, fill—use—drain.

The significant difference between a hydromassage bathtub and a regular bathtub is the recirculating piping system and the electric pump that circulates the water. Both types of tubs are drained completely after each use.

Spas and hot tubs are intended to be filled, then used. They are *not* drained after each use, because they have a filtering and heating system.

*Reprinted with permission from NFPA 70-2014.

Electrical Connections

The *NEC* requirements for hydromassage tubs are found in Chapter 30 of this text.

The hydromassage bathtub in this residence is supplied by a dedicated branch circuit A9, a 15-ampere,120-volt circuit. If you are not sure what the actual current draw is, run a 20-ampere branch circuit.

The Schedule of Special-Purpose Outlets indicates that the hydromassage bathtub in this residence has a ½-horsepower motor that draws 10 amperes.

The hydromassage bathtub electrical control is prewired by the manufacturer. The electrician needs only to install an individual 15-ampere,120-volt, GFCI-protected circuit to the end of the tub, where the pump and control are located.

Some manufacturers supply a short flexible conduit that contains one black, one white, and one green equipment grounding conductor, as in Figure 22-10. The electrician installs a junction box near the pump. A dedicated branch circuit is run to this box, where the branch-circuit wiring is connected to the hydromassage conductors. This junction box must be accessible. The pump and power panel may also need servicing. Access may be from underneath or the end, whichever is convenient for the installation, as in Figure 22-11.

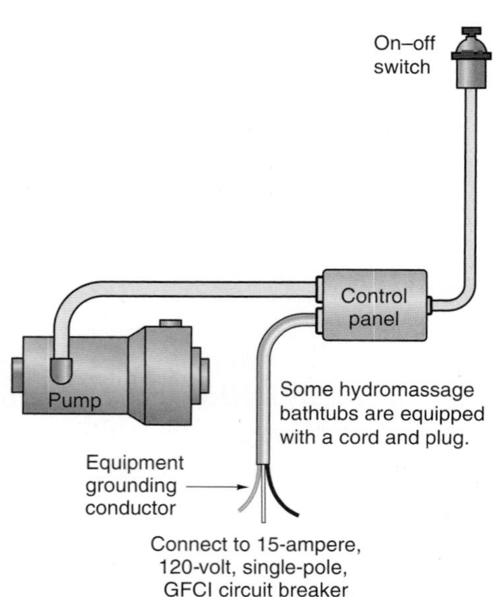

FIGURE 22-10 Typical wiring diagram of a hydromassage tub showing the pump motor, control panel, and electrical supply leads.

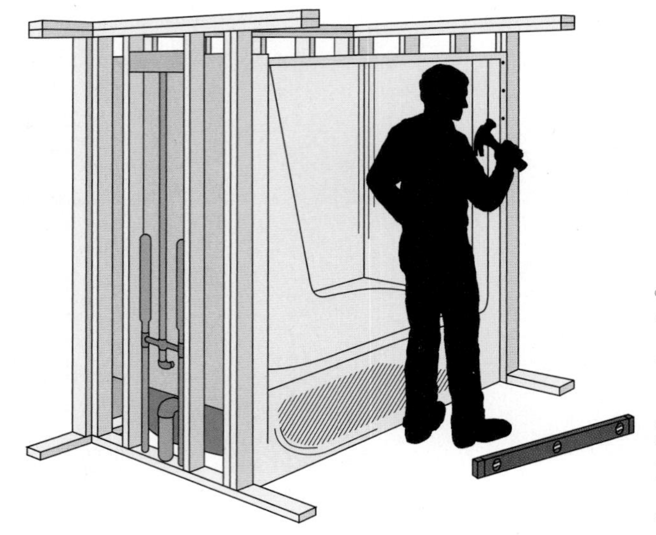

FIGURE 22-11 The basic roughing-in of a hydromassage bathtub is similar to that of a regular bathtub. The electrician runs a separate 15-ampere, 120-volt, GFCI-protected circuit to the area where the pump and control are located. Check the manufacturer's specifications for these data. An access panel from the end or from below is necessary to service the wiring, the pump, and the control panel.

Make sure that the equipment grounding conductor of the circuit is properly connected to the green equipment grounding conductor of the hydromassage tub. *Never connect the green equipment grounding conductor to the grounded (white) branch-circuit conductor.*

Some hydromassage tubs come equipped with a cord and plug, in which case the electrician installs a receptacle within 24 in. (600 mm) of the pump. Usually the branch circuit is protected by a GFCI circuit breaker, although a GFCI receptacle could be used. If the hydromassage bathtub is cord- and plug-connected with the supply receptacle accessible only through a service access opening, the receptacle is required to be installed so its face is within direct view and not more than 1 ft (300 mm) of the opening, *680.73*.

Read and follow the manufacturer's installation instructions.

Figure 22-12 is a photograph of a typical hydromassage tub.

FIGURE 22-12 A typical hydromassage bathtub, sometimes referred to as a whirlpool.

REVIEW

Note: Refer to the *Code* or plans where necessary.

BATHROOM CEILING HEATER CIRCUITS ▲K ▲J

1. What is the wattage rating of the heat/vent/light? ______
2. To what circuits are the heat/vent/lights connected? ______
3. Why is a 4 in. square, 2⅛ in. deep box with a 2-gang raised plaster ring or similar deep box used for the switch assembly for the heat/vent/light? ______
4. a. How many wires are required to connect the wall switches and the heat/vent/light? ______
   b. What size wires are used? ______
5. Can the heating element be energized when the fan is not operating? ______
6. Can the fan be turned on without the heating element? ______
7. What device can be used to provide automatic control of the heating element and the fan of the heat/vent/light? ______

8. Where does the air enter the heat/vent/light? ______
9. Where does the air leave this unit? ______
10. Who is to furnish the heat/vent/light? ______
11. For a ceiling heater rated 1200 watts at 120 volts, what is the current draw? ______
12. An electrician wired up a heat/vent/light as described in this chapter. The wiring diagram showed that the electrician needed to run five conductors between the switches and the heat/vent/light. The electrician used one 12/3 nonmetallic-sheathed cable and one 12/2 nonmetallic-sheathed cable. Both cables had equipment grounding conductors. During the rough-in inspection, the electrical inspector cited "Noncompliance with *NEC 300.3(B)*." What was the problem? ______

ATTIC EXHAUST FAN CIRCUIT Ⓐ$_L$

1. What is the purpose of the attic exhaust fan? ______
2. At what voltage does the fan operate? ______
3. What is the horsepower rating of the fan motor? ______
4. Is the fan direct- or belt-driven? ______
5. How is the fan controlled? ______
6. What is the rating of the dual-element, time-delay fuse in the switch/fuse box-cover unit? ______
7. What is the rating of the running overcurrent protection if the motor is rated at 10 amperes? ______
8. What is the basic difference between a thermostat and a humidistat? ______
9. What size conductors are to be used for this circuit? ______
10. How many ft (m) of cable are required to complete the wiring for the attic exhaust fan circuit? ______
11. May the metal frame of the fan be grounded to the grounded circuit conductor? ______
12. Which section of the *Code* prohibits grounding equipment to a grounded circuit conductor? ______

HYDROMASSAGE BATHTUB CIRCUIT Ⓐ A

1. Which circuit supplies the hydromassage bathtub? ____________________
2. Which of the following statements "Meets *Code*"? (Circle the correct answer.)
   a. Hydromassage tubs and all of their electrical components shall be GFCI protected.
   b. Hydromassage tubs and all of their electrical components shall not be GFCI protected.
3. Which conductor size feeds the hydromassage tub in this residence? ____________________

   ____
4. What is the fundamental difference between a hydromassage bathtub and a spa? ______

   ____

   ____
5. Which sections of the *Code* reference hydromassage bathtubs? ____________________

   ____
6. Must the metal parts of the pump and power panel of the hydromassage tub be grounded? ____________________

   ____

   ____

   ____
7. Is it permissible to connect the hydromassage tub's green equipment grounding conductor to the branch circuit's grounded (white) conductor?

   ____
8. All 120–volt, single-phase receptacles within (6 ft [1.8 m]) (10 ft [3.0 m]) (15 ft [4.5 m]) of a hydromassage tub must be GFCI protected. (Circle the correct answer.)

Special-Purpose Outlets—Electric Heating ◬M, Air Conditioning ◬N

OBJECTIVES

After studying this chapter, you should be able to

- understand the *NEC* requirements for embedded resistance heating cable, electric furnaces, electric baseboards, and heat pumps.
- understand how the heating output of resistive heating elements is affected by voltage variation.
- understand the data found on the nameplate of HVAC equipment, and determine electrical installation requirements in conformance to the *NEC*.
- realize that a 120-volt receptacle must be installed near HVAC equipment.
- understand energy rating terminology.

There are many types of electric heat available for heating homes (e.g., heating cable, unit heaters, boilers, electric furnaces, duct heaters, baseboard heaters, heat pumps, and radiant heating panels). This chapter discusses many of these types. The residence discussed in this text is heated by an electric furnace located in the workshop.

Detailed *Code* requirements for fixed electric space-heating equipment are found in *Article 424*.

This text cannot cover in detail the methods used to calculate heat loss and the wattage required to provide a comfortable level of heat in the building. For this residence, the total estimated wattage is 13,000 watts. Depending on the location of the residence (in the Northeast, Midwest, or South, for example), the heating load will vary.

Electric heating has gained wide acceptance when compared with other types of heating systems. It has a number of advantages. Electric heating is flexible when baseboard heating or wall insert, fan-operated heaters are used, because each room can have its own thermostat. Thus, one room can be kept cool while an adjoining room is warm. This type of zone control for an electric, gas- or oil-fired central heating system is more complex and more expensive.

Electric heating is safer than heating with fuels. The system does not require storage space, tanks, or chimneys. Electric heating is quiet. Electric heat does not add or remove anything from the air. As a result, electric heat is cleaner. This type of heating is considered to be healthier than fuel heating systems that remove oxygen from the air. The only moving part of an electric baseboard heating system is the thermostat. This means that there is a minimum of maintenance.

As with all heating systems, adequate insulation must be provided. Proper insulation can keep electric bills to a minimum. Insulation also helps to keep the residence cool during the hot summer months. The cost of extra insulation is offset through the years by the decreased burden on the heating and air-conditioning equipment.

RESISTANCE HEATING CABLES

Resistance heating cables can be embedded in the plaster or between two layers of drywall. Installation requirements are found in *424.34* through *424.45* of the *NEC*®. Because premises wiring (nonmetallic-sheathed cable, etc.) will be located above these ceilings and will be subjected to the heat created by the heating cables, the *Code* requires the following:

- Keep the wiring not less than 2 in. (50 mm) above the heated ceiling, and *also* reduce the ampacity of the wiring to the 122°F (50°C) correction factors found in *Table 310.15(B)(2)(a)*, or
- Keep the wiring above the thermal insulation that is at least 2 in. (50 mm) thick—in which case the correction factors do not have to be applied.

For ease in identification, *424.35* requires that the leads on heating cables be color coded as follows:

- 120 volts yellow
- 208 volts blue
- 240 volts red
- 277 volts brown
- 480 volts orange

ELECTRIC FURNACES

In an electric furnace, the source of heat is the resistance heating elements. The blower motor assembly, filter section, condensate coil, refrigerant line connections, fan motor speed control, high-temperature limit controls, relays, and similar components are much the same as those found in gas furnaces.

As with all appliances, many of the *NEC* requirements have already been met by virtue of the product being listed by a recognized testing laboratory. Your job is to make sure the premises wiring is installed according to the *NEC*. The recognized testing laboratory takes care of the safety requirements of the appliance.

Summary of Important *NEC* Rules

Many of the following *NEC* requirements are emphasized in Figure 23-1.

- *NEC 220.51*: When making a calculation for sizing a service or feeder, include 100% of the rating of the total connected fixed space-heating load. This is shown in Chapter 29.

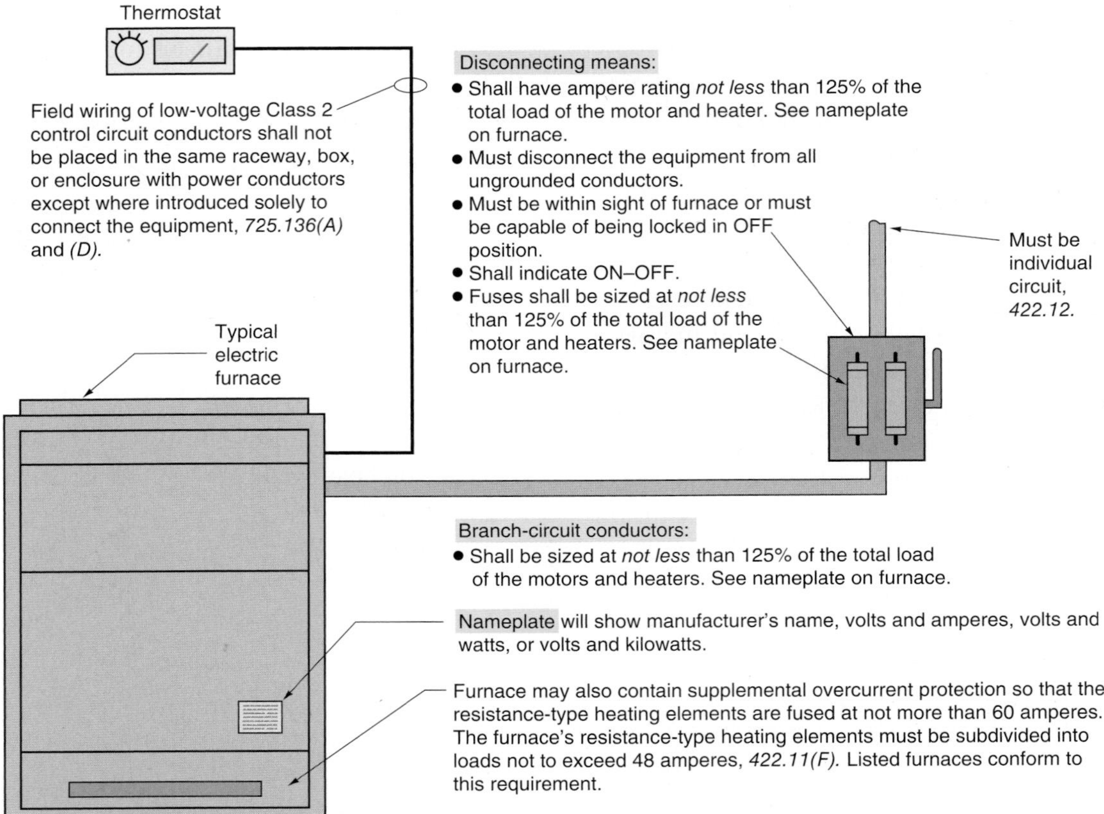

FIGURE 23-1 An electric furnace is considered to be a *continuous load*, subject to sizing the conductors, disconnecting means, and overcurrent devices as 125% of the load, *NEC 424.3(B)*. Central electric heating equipment must be supplied by an individual branch-circuit, *NEC 422.12*.

- *Article 250, Part VII*: Contains most of the requirements for equipment and appliances grounding. Grounding is covered in Chapters 4, 15, 20, 21, and 22.
- *NEC 250.118*: Lists the acceptable methods for equipment and appliance grounding. This is generally the equipment grounding conductor in nonmetallic-sheathed cable, the metal armor and bonding strip in armored cable, or a metal raceway such as EMT or FMC. Severe limitations are imposed on FMC in *250.118(5)*.
- *NEC 422.10(A)*: The branch-circuit rating for an appliance must not be less than the marked rating.
- *NEC 422.11*: The branch-circuit overcurrent protection must be sized according to the value indicated on the nameplate. The electrician need do nothing more than size the conductors and overcurrent protection according to the nameplate data and instructions furnished with the furnace. Overcurrent protection for internal components is provided by the manufacturer.
- *NEC 422.12*: Central heating equipment must be supplied by an individual branch circuit. *Associated* equipment such as humidifiers and electrostatic air cleaners is permitted to be connected to the same branch circuit. The logic behind this "individual circuit" requirement is that if the heating system were to be connected to some other branch circuit, such as a lighting branch circuit, a fault on that circuit would shut off the power to the heating system. We find this same requirement in *8.6.4* of NFPA 54, the *National Fuel Gas Code*.
- *NEC 422.30*: A disconnecting means must be provided.

- *NEC 422.31(B)*: The disconnecting means must be within sight of the furnace. For electric furnaces, the disconnect might be mounted on the side of the furnace, on an adjacent wall, or the circuit breaker in the panelboard, if the panelboard is within sight of the furnace. For gas furnaces, the disconnect could simply be a snap switch, switch/fuse mounted in a handy box on the side of the furnace, or the circuit breaker in the panelboard, if the panelboard is in sight of the furnace. See *430.102(A)* and *(B)*.
- *NEC 422.32*: This is another reminder that the disconnect for motor-driven appliances having a motor greater than ⅛ horsepower must be within sight of the appliance. See *430.102(A)* and *(B)*.
- *NEC 422.62(B)(1)*: The nameplate shall specify the minimum size supply circuit conductor's ampacity and the maximum branch-circuit overcurrent protection.
- *NEC 424.3(B)*: Fixed electric space heating is considered to be a continuous load. The branch-circuit conductors and the rating or setting of overcurrent protective devices supplying fixed electric space-heating equipment consisting of resistance elements with or without a motor shall not be less than 125% of the total load of the motors and the heaters. If applying the 125% rule results in a nonstandard rating of overcurrent protective device, then *240.4(B)* permits the use of the next higher rating. See Figure 23-1 for the 125% calculations.
- *NEC 430.60(B)*: The nameplate must show volts and watts or volts and amperes.
- *NEC 430.102(A)*: The disconnect shall be within sight of the controller and shall disconnect the controller. With a residential furnace, the controls are an integral part of the furnace.
- *NEC 430.102(B)*: The disconnect shall be located within sight of the motor and driven machinery. There is an exception to the "in sight" requirement, but the exception does not pertain to residential installations.

Figure 23-1 illustrates the key *NEC* requirements for hooking up an electric furnace.

EXAMPLE

What size copper conductors (THHN), fuses, and disconnect switch are required for a furnace marked 79 amperes, 240 volts, single phase, 60 cycles? Terminals on furnace and switch marked 75°C. To select the proper ampacity of the conductors in accordance with *110.14(C)*, be sure to use the 75°C ampacity column in *Table 310.15(B)(16)*.

Answer:

Conductor size: $79 \times 1.25 = 98.8$ amperes
From *Table 310.15(B)(16)*, select 3 AWG THHN (100 amperes at 75°C).

Fuse size: $79 \times 1.25 = 98.8$ amperes
Install 100-ampere fuses.

Switch: 100-ampere switch

Supply Voltage

Because proper supply voltage is critical to ensure full output of the furnace's heating elements, voltage must be maintained at not less than 98% of the furnace's rated voltage. The effect of voltage drop is discussed in Chapters 4 and 19. Mathematically, for every 1% drop in voltage, there will be a 2% drop in wattage output of a resistance heating element. Manufacturers' installation instructions usually have tables that show the wattage or kilowatt output at different voltages. Table 23-1 is a typical table.

A simple way to calculate the effect of voltage variance on a resistive heating element is

$$\text{Correction Factor} = \frac{\text{Applied Voltage Squared}}{\text{Rated Voltage Squared}}$$

TABLE 23-1

Kilowatt rating.

@ 240 VOLTS	@ 208 VOLTS
12.0	9.0
15.0	11.3
20.0	15.0
25.0	18.8
30.0	22.5

EXAMPLE

The rating of the heating elements in an electric furnace is 30 kW at 240 volts. What is the kW output if the supply voltage is 208 volts?

Step 1: Correction Factor $= \dfrac{208 \times 208}{240 \times 240}$

$= \dfrac{43{,}264}{57{,}600} = 0.75$

Step 2: 30 kW × 0.75 = 22.5 kW

Therefore, the actual voltage at the supply terminals of an electric furnace or electric baseboard heating unit will determine the true wattage output of the heating elements.

Cord-and-Plug Connection Not Permitted

Flexible cords are not a recognized wiring method.

NEC 400.7(A)(8) and *422.16* describe the permitted uses for flexible cords. A key requirement, oftentimes violated, is that flexible cords shall be used only Where the fastening means and mechanical connections are specifically designed to permit ready removal for maintenance and repair, and the appliance is intended or identified for flexible cord connection.* Certainly the gas piping and the size of a gas or electric furnace do not allow for "ready removal" of the furnace. Furthermore, the conductors in flexible cords cannot withstand the high temperature encountered on the terminals of a furnace.

ANSI Standard Z21.47 does not allow cord-and-plug connection for gas furnaces.

CONTROL OF ELECTRIC BASEBOARD HEATING UNITS

Line-voltage and low-voltage thermostats can be used to control electric baseboard heating units, as shown in Figure 23-2. When the current rating or ampere rating of the load exceeds the rating of a line-voltage thermostat, then it is necessary to utilize a relay in conjunction with the low-voltage thermostat. This is illustrated in Figure 23-3. Color-coding requirements are explained later in this chapter, and are covered in detail in Chapter 5.

Most line-voltage thermostats are ampere rated for noninductive resistive loads. You may need to convert watts to amperes to select the right size thermostat, for example, if you are controlling a 2500-watt, 120-volt resistive heating element or a 5000-watt, 240-volt resistive heating element:

$$I = \frac{W}{E} = \frac{2500}{120} = 20.8 \text{ amperes}$$

—or—

$$I = \frac{W}{E} = \frac{5000}{240} = 20.8 \text{ amperes}$$

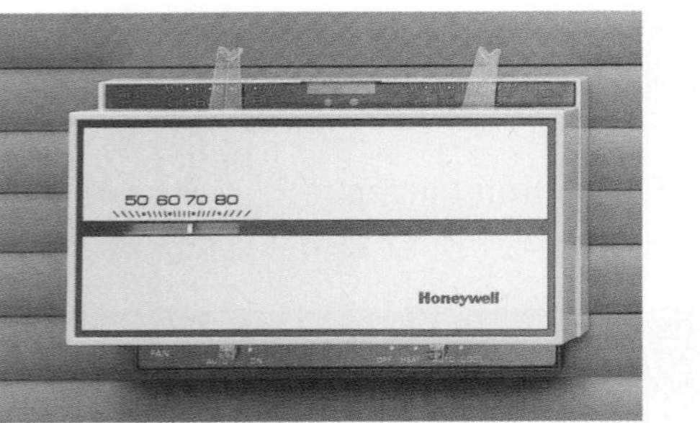

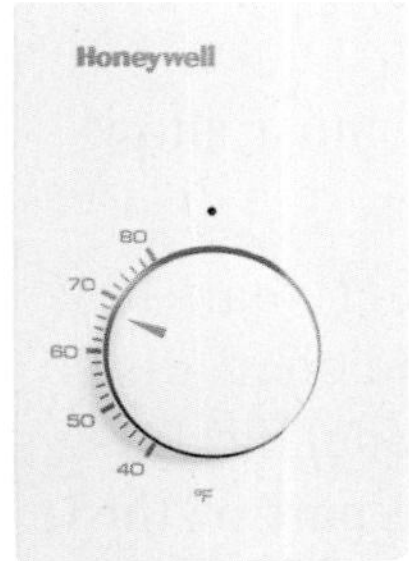

FIGURE 23-2 Thermostats for electric heating systems. A thermostat or other switching device that is marked with an OFF position must open all ungrounded conductors of the circuit when turned to the OFF position. Typical mounting height is 52 in. (1.3 m) to center.

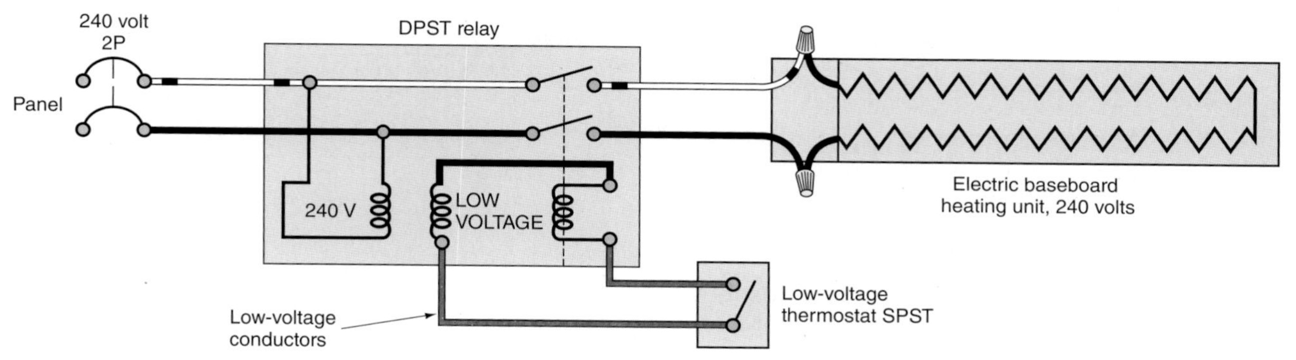

FIGURE 23-3 Wiring for an electric baseboard heating unit having a low-voltage thermostat and relay. *NEC 200.7* requires that a white wire in cable systems used as an ungrounded conductor must be permanently reidentified at points where connections are made or where it is visible and accessible. This becomes necessary when nonmetallic-sheathed cable or armored cable is the wiring method. Reidentification can be accomplished with black paint, black tape, or heat-shrink tubing. Color coding and reidentification of conductors is discussed in Chapter 5 and later in this chapter.

The total connected load for this thermostat must not exceed 20.8 amperes.

Disconnection of Baseboard Heaters. Like other fixed electric space-heating equipment, baseboard heaters are required to be provided with disconnecting means. The line-voltage thermostat is permitted to serve as both the controller and the disconnecting means provided it meets all the following conditions:

1. The thermostat has a marked OFF position.
2. It directly opens all ungrounded conductors when manually placed in the OFF position.
3. It is designed so the circuit cannot be energized automatically after the device has been manually placed in the OFF position.
4. It is located within sight of the heaters it controls and disconnects.

As can be seen, a 2-pole thermostat is required to serve as a disconnecting means for baseboard heaters. A single-pole thermostat does not have a marked OFF position and will not satisfy these requirements to serve as a disconnecting means. See *424.20*.

If 2-pole thermostats are not used to satisfy the disconnecting means requirements, the installation must comply with *424.19(B)*. This permits the circuit breaker to serve as the disconnecting means if it's within sight from the heater or is capable of being locked in the open position. If the circuit breaker is located out of sight from the baseboard heaters, it must be provided with a means of being locked open.

MARKING THE CONDUCTORS OF CABLES

Two-wire cable contains one white and one black conductor. Because 240-volt circuits are permitted to supply electric baseboard heaters, it would appear that the use of 2-wire cable is in violation of the *Code*.

According to *200.7(C)*, 2-wire cable is permitted for the 240-volt heaters provided the white conductor is permanently reidentified by paint, colored tape, or other effective means. This is necessary because it might be assumed that an unmarked white conductor is a grounded conductor having no voltage to ground. Actually, the white wire is connected to a "hot" phase in the panel and has 120 volts to ground on it. A person could be subjected to a lethal shock by touching this conductor and the grounded baseboard heater (or any other grounded object) at the same time. The white conductor must be permanently reidentified at the electric heater terminals and at the panels where these cables originate. Figures 23-3, 23-4, and 23-5 show this reidentification. Review the section on "Color Coding" in Chapter 5.

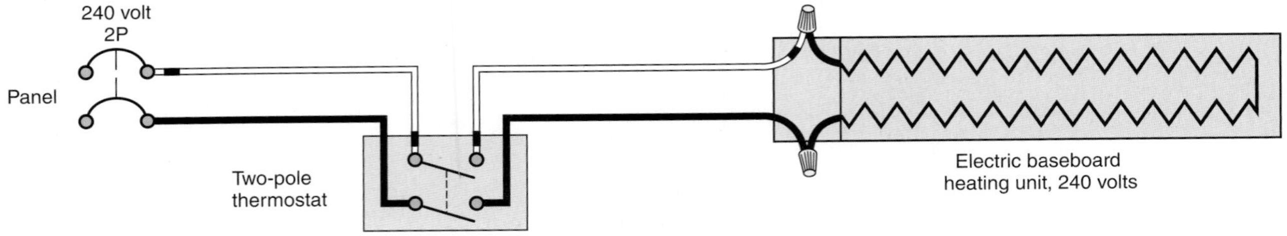

FIGURE 23-4 Wiring for a single electric baseboard heating unit. The thermostat is a line-voltage thermostat. See Figure 23-3 caption for color-coding requirements.

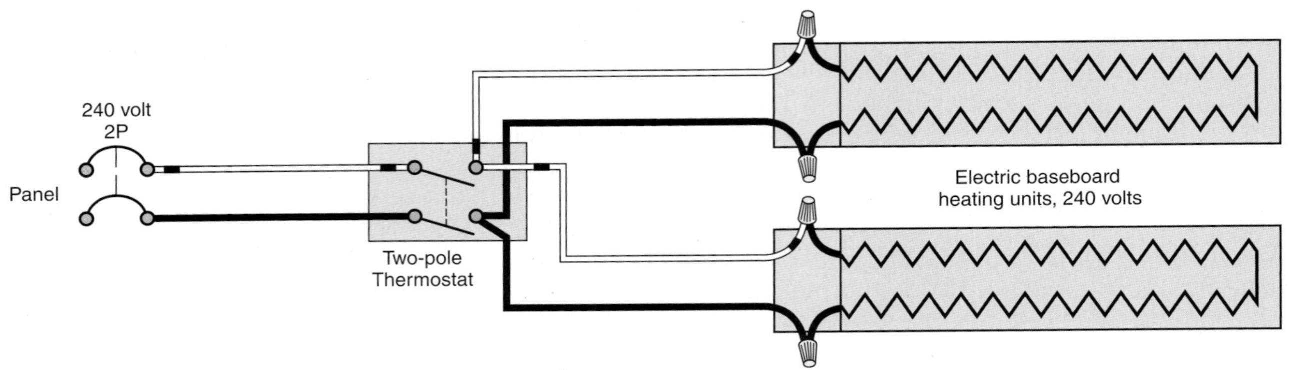

FIGURE 23-5 Wiring for two electric baseboard heating units. Note the circuit connections at the thermostat location. The thermostat is a line-voltage thermostat. See Figure 23-3 caption for color-coding requirements.

CIRCUIT REQUIREMENTS FOR ELECTRIC BASEBOARD HEATING UNITS

Figure 23-3 shows an electric baseboard heater controlled by a low-voltage thermostat. The low-voltage contacts of the relay are connected to the thermostat. The line-voltage contacts of the relay are used to switch the actual heater load. Relays are used whenever the load exceeds the ampere or wattage limitations of a line-voltage thermostat. Color coding for conductors is explained in Chapter 5 and is also discussed later in this chapter.

Figure 23-6 shows typical electric baseboard heating units. The heater to the right has an integral thermostat on it. Electric baseboard heating units are available in both 240-volt and 120-volt ratings and in a variety of wattage ratings.

The wiring for an individual baseboard heating unit or group of units is shown in Figures 23-4 and 23-5. A 2-wire cable (armored cable or nonmetallic-sheathed cable with ground) would be run from a 240-volt, 2-pole circuit in the main panel to the outlet box or switch box installed at the thermostat location. A second 2-wire cable runs from the thermostat to the junction box on the heater unit. The proper connections are made in this junction box. Most heating unit manufacturers provide knockouts at the rear and on the bottom of the junction box. The supply conductors can be run through these knockouts. Most baseboard units also have a channel or wiring space running the full length of the unit, usually at the bottom. When two or more heating units are joined together, the conductors are run in this wiring channel. Most manufacturers indicate the type of wire required for these units because of conductor temperature limitations.

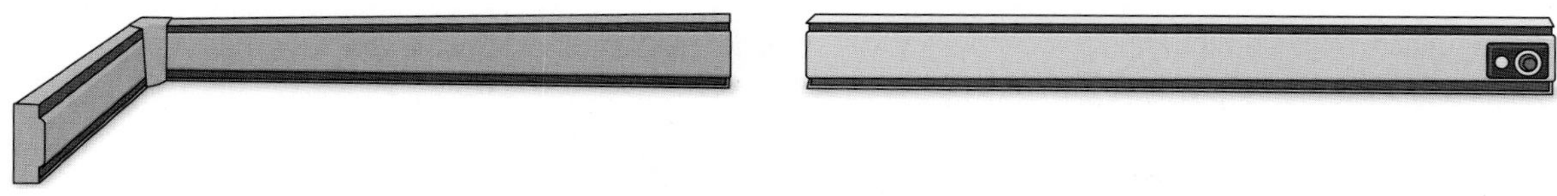

FIGURE 23-6 Electric baseboard heating systems.

Supply conductors generally must be rated 90°C. These would be Type THHN conductors, or nonmetallic-sheathed cable or armored cable that has 90°C conductors. The nameplate and/or instructions furnished with the electric baseboard heating unit provides this information.

Most wall and baseboard heating units are available with built-in thermostats, as shown in Figure 23-6. These thermostats are also available as an accessory that can be added in the field. The branch-circuit supply cable for such a unit runs from the main panelboard to the junction box on the unit.

Some heaters have 120-volt, 15- or 20-ampere receptacle outlets. UL states that when receptacle sections are included with the other components of baseboard heating systems, they must be supplied separately using conventional wiring methods. See Figure 23-7.

A variety of fittings such as internal and external elbows (for turning corners) and blank sections are available from the manufacturer.

Manufacturers of electric baseboard heaters list "blank" sections ranging in length from 2 to 10 ft (600 mm to 3.0 m). These blank sections give the installer the flexibility needed to spread out the heater sections, and to install these blanks where wall receptacle outlets are encountered, as in Figure 23-8.

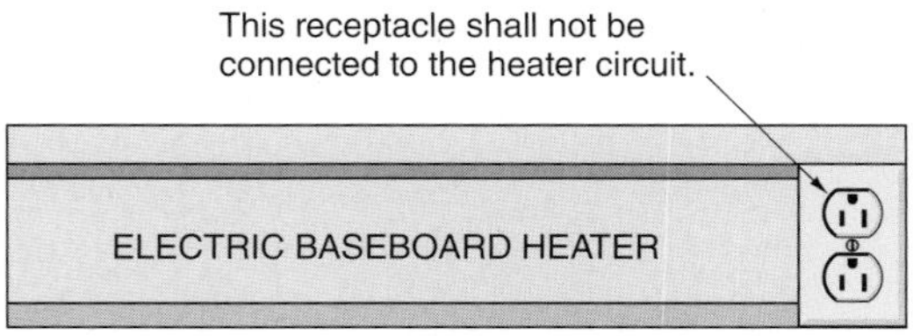

This receptacle shall not be connected to the heater circuit.

ELECTRIC BASEBOARD HEATER

FIGURE 23-7 Factory-installed receptacle outlets or receptacle outlet assemblies provided by the manufacturer for use with its electric baseboard heaters may be counted as the required receptacle outlet for the space occupied by a permanently installed heater. See the second paragraph of *210.52*.

LOCATION OF ELECTRIC BASEBOARD HEATERS IN RELATION TO RECEPTACLE OUTLETS

Listed electric baseboard heaters shall not be installed below wall receptacle outlets unless the instructions furnished with the baseboard heaters indicate that they may be installed below receptacle outlets. See the *Informational Note* to *210.52*. The reasoning for generally not allowing electric baseboard heaters to be installed below receptacle outlets is the possible fire and shock hazard resulting when a cord hangs over and touches the heated electric baseboard unit. The insulation on the cord can melt from the heat. See Figures 23-9 and 23-10.

The UL Guide Card KLDR makes the following statement: "Electrical cords, drapes, and other furnishings should be kept away from baseboard

FIGURE 23-8 Use of blank baseboard heating sections.

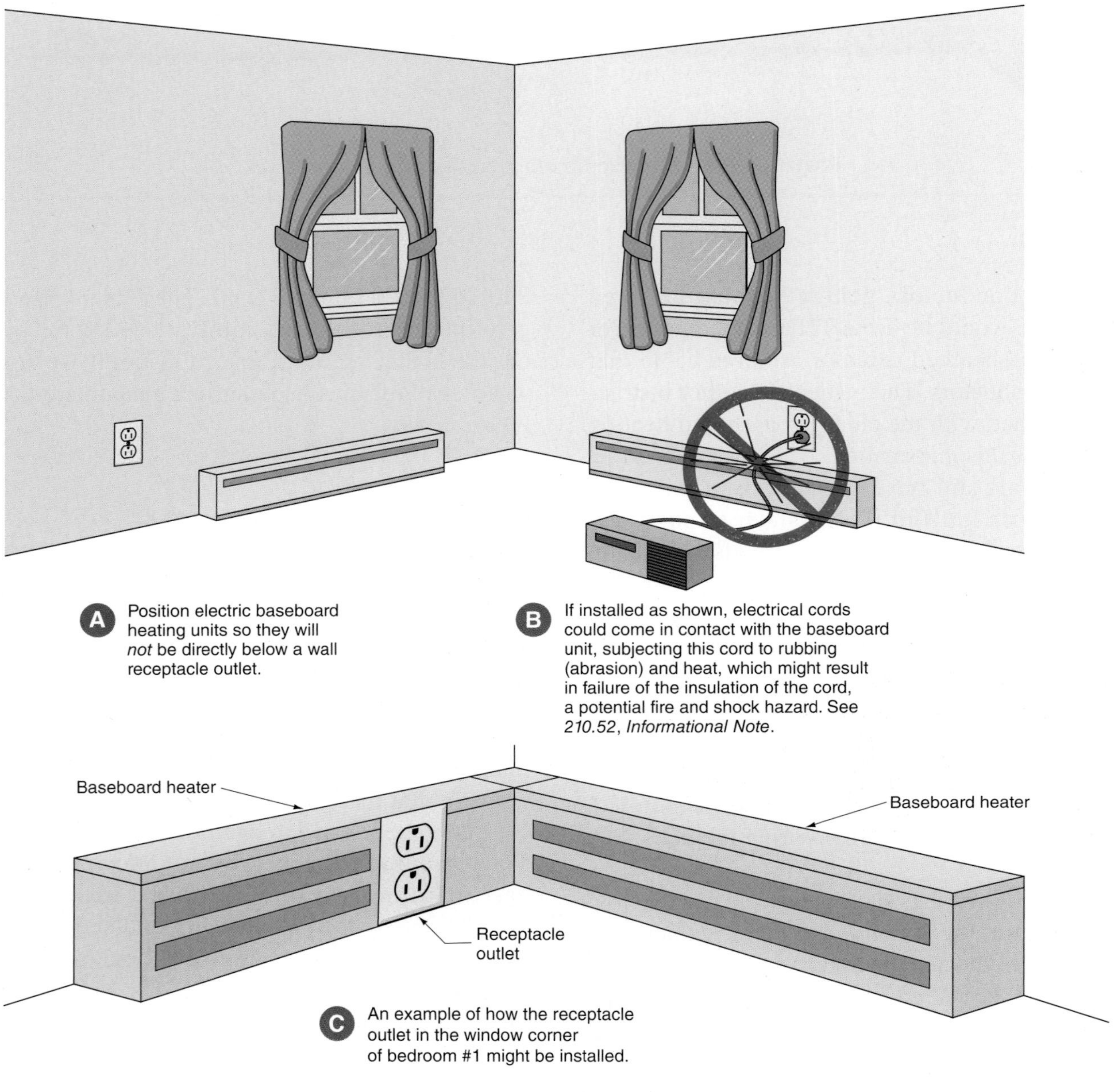

A Position electric baseboard heating units so they will *not* be directly below a wall receptacle outlet.

B If installed as shown, electrical cords could come in contact with the baseboard unit, subjecting this cord to rubbing (abrasion) and heat, which might result in failure of the insulation of the cord, a potential fire and shock hazard. See *210.52, Informational Note*.

Baseboard heater

Baseboard heater

Receptacle outlet

C An example of how the receptacle outlet in the window corner of bedroom #1 might be installed.

FIGURE 23-9 Position of electric baseboard heaters.

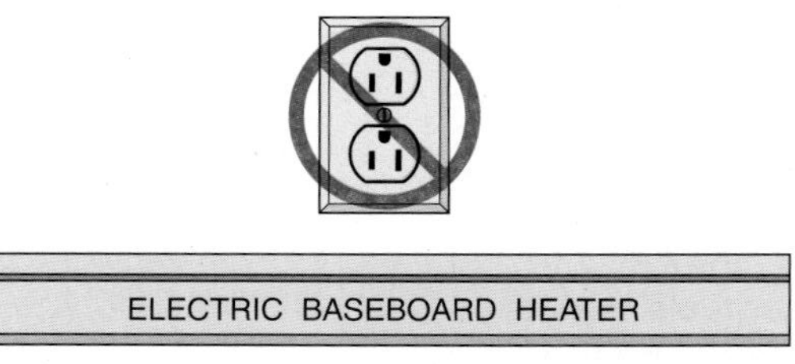

FIGURE 23-10 *Code violation.* Unless the instructions furnished with the electric baseboard heater specifically state that the unit is listed for installation below a receptacle outlet, then this installation is a violation of *110.3(B)*. In this type of installation, cords attached to the receptacle outlet would hang over the heater, creating a fire hazard and possible shock hazard should the insulation on the cord melt. See *Informational Note* to *210.52*.

Courtesy of Cadet Manufacturing.

FIGURE 23-11 A toe-kick heater designed to be located near the bottom of a floor-mounted cabinet.

heaters. To reduce the likehood of cords contacting the heater, the heater is not to be located beneath electrical receptacles."

Electricians must pay close attention to the location of receptacle outlets and the electric baseboard heaters. They must study the plans and specifications carefully. They may have to fine-tune the location of the receptacle outlets and the electric baseboard heaters, and/or install blank spacer sections, keeping in mind the *Code* requirements for spacing receptacle outlets and just how much (wattage and length) baseboard heating is required (Figure 23-7 and Figure 23-10).

The *Code* does permit factory-installed or factory-furnished receptacle outlet assemblies as part of permanently installed electric baseboard heaters. These receptacle outlets shall not be connected to the heater circuit (see second paragraph of *210.52*, Figure 23-7 and Figure 23-9).

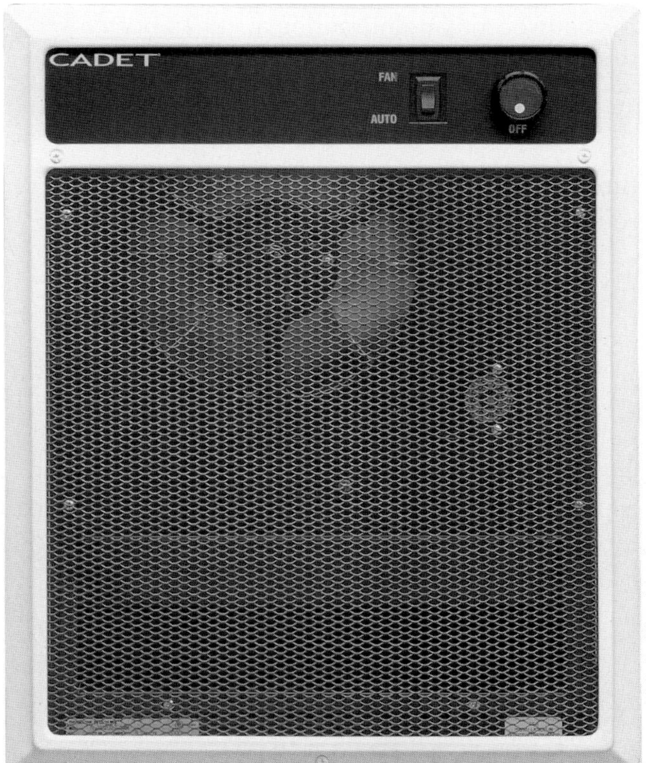

Courtesy of Cadet Manufacturing.

FIGURE 23-12 Wall-mounted space heaters are designed for heating from small to large rooms or areas.

WALL-MOUNTED HEATERS

Electric space heaters are available in many sizes or heating capacities and designed for a variety of locations. For example, heaters in bathrooms can be installed in the wall or in the toe-kick space under the cabinet. Figure 23-11 shows a heater designed for installation in the toe-kick space near the floor. These heaters are designed for bathrooms or kitchens or other areas having floor-mounted cabinets. Figure 23-12 shows a wall heater designed for larger rooms or areas. The heater shown in Figure 23-13 is particularly well suited for bathrooms or nurseries as it is provided with a timer as well as a thermostat.

These heaters typically have a can or enclosure that is installed at the rough-in stage of the project. The branch circuit is routed to the unit either

Courtesy of Cadet Manufacturing.

FIGURE 23-13 A wall-mounted heater well suited for bathrooms and nurseries as it has both a thermostat and timer.

through the thermostat location or directly to the rough-in can.

Wall-mounted heaters should be located carefully to avoid conflict with furniture layout. At times, this can best be accomplished if the heater is installed near the entry door to the room.

Wall-mounted heaters are available with or without a built-in thermostat and in a variety of heating capacities to match the heat loss of the room or area. Smaller capacity heaters are available in 120-volt models. Heaters designed to operate on 240 volts are available in ratings from 750 to 4000 watt capacities. Heater manufacturers provide online calculators to assist in proper sizing of heaters.

HEAT PUMPS

A heat pump is dual-purpose equipment that functions as a heating unit and an air conditioner operating in reverse. A "reversing valve" inside the unit changes the direction of the flow of the system's refrigerant.

A "split" heat pump system consists of an outdoor unit with a hermetic motor-compressor, a fan, and a coil. The indoor unit consists of a refrigerant coil and usually makeup resistance heating inside of the air handler. The movement of air inside the home is accomplished by the blower fan on the air handler. The "split system" is the most common type for residential use.

When all of the components are housed in one outdoor unit, the system is referred to as a "packaged system."

The hermetic motor-compressor is the heart of the system. The refrigerant circulates between the indoor and outdoor units and through the coils, tubing, and compressor, absorbing and releasing heat as it travels through the system.

In the winter, the heat pump (the outdoor coil) extracts heat from the outdoor air and distributes it through the warm air ducts in the home. In extremely cold climates, a heat pump is supplemented by electric heating elements in the air handler.

In the summer, the process reverses. The heat pump (indoor coil) absorbs heat and condenses humidity from the indoor air. The heat is transferred to the outside by the refrigerant, and the humidity (condensation) is removed by piping it to a suitable drain.

Heat pump water heaters are discussed in Chapter 19.

Because a heat pump contains a hermetic motor-compressor, the *Code* requirements found in *Article 440* apply. These are discussed later in this chapter.

GROUNDING

Grounding of electrical equipment has already been discussed and will not be repeated here. Grounding is covered in *Article 250* of the *NEC.* To repeat a caution, *never* ground electrical equipment to the grounded (white) circuit conductor of the circuit.

CIRCUIT REQUIREMENTS FOR ROOM AIR CONDITIONERS

For homes that do not have central air conditioning, window or through-the-wall air conditioners may be installed. These types of room air conditioners are available in both 120-volt and 240-volt ratings. Because room air conditioners are cord-and-plug connected, the receptacle outlet and the circuit capacity must be selected and installed according to applicable *Code* regulations. The *Code* rules for air conditioning are found in *Article 440*. The *Code* requirements for room air-conditioning units are found in *440.60* through *440.65* of the *NEC*.

The basic *Code* rules for installing these cord-and-plug-connected units and their receptacle outlets are as follows:

- The air conditioners must be grounded.
- The air-conditioner rating may not exceed 40 amperes at 250 volts, single phase.
- The rating of the branch-circuit overcurrent device must not exceed the branch-circuit conductor rating or the receptacle rating, whichever is less.
- The air-conditioner load shall not exceed 80% of the branch-circuit ampacity if no other loads are served.
- The air-conditioner load shall not exceed 50% of the branch-circuit ampacity if other loads are served.
- The attachment plug cap may serve as the disconnecting means.
- The maximum cord length is 10 ft (3.0 m) for 120-volt units, and 6 ft (1.8 m) for 208- or 240-volt units. This is also a UL requirement.
- There is an ever-present potential arcing-fault hazard related to the power supply cords on room air conditioners that could cause a fire or create an electrical shock hazard. In addition to requiring that the air-conditioner unit be properly grounded, *NEC 440.65* requires that factory-installed leakage current detector interrupter (LCDI) or arc-fault circuit interrupter (AFCI) devices be provided on single-phase cord-and-plug-connected room air conditioners. These devices are permitted to be an integral part of the attachment plug cap, or may be located in the power supply cord within 12 in. (300 mm) of the attachment plug cap.

LCDIs are Leakage Current Detection and Interruption Protection devices. They can sense leakage current flowing between or from the cord conductors and interrupt the circuit at a predetermined level of leakage current.

AFCIs are discussed in Chapter 6.

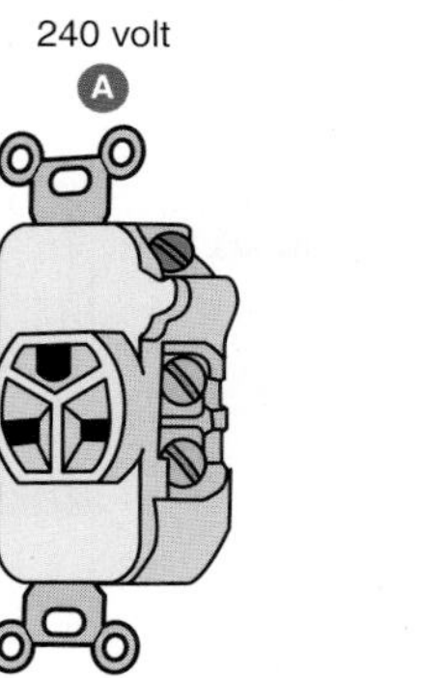

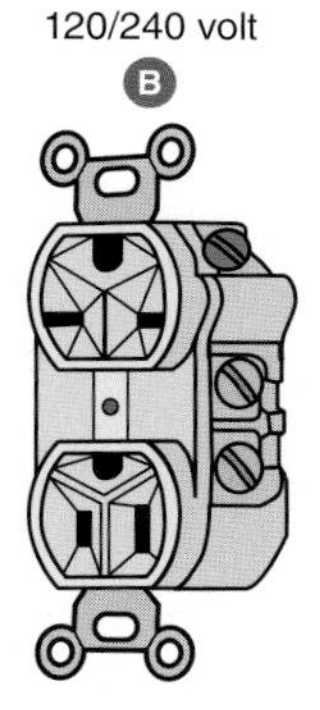

FIGURE 23-14 Types of receptacles.

Receptacles for Room Air Conditioners

Figure 23-14(A) is a straight 240-volt receptacle. Figure 23-14(B) is a combination receptacle. The lower portion is for 120-volt use, and the upper portion is for 240-volt use. Note the different slot configurations that meet the noninterchangeable requirements of *406.3(F)*.

CENTRAL HEATING AND AIR CONDITIONING

The residence discussed throughout this text has central electric heating and air conditioning consisting of an electric furnace and a central air-conditioning unit.

The wiring for central heating and cooling systems is shown in Figure 23-15. Basic circuit requirements are shown in Figure 23-16. Note that one branch circuit runs to the electric furnace and another branch circuit runs to the air conditioner or heat pump outside the dwelling. Low-voltage wiring is used between the inside and outside units to provide control of the systems. The low-voltage Class 2 circuit wiring shall not be run in the same raceway as the power conductors, *725.136(A)*.

Refer to Chapter 25 for more information about low-voltage circuitry.

"Time-of-Use" Lower Energy Rates

Many utilities offer lower energy rates for air-conditioner and/or heat pump loads through special

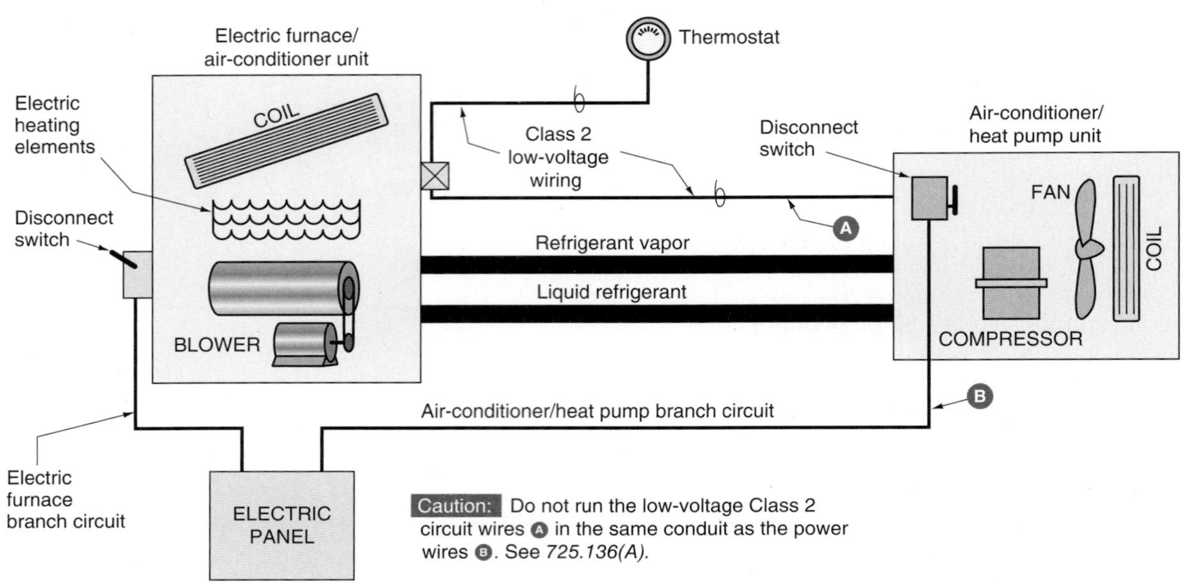

FIGURE 23-15 Connection diagram showing typical electric furnace and air-conditioner/heat pump installation.

programs, such as "time-of-use" usage and sliding scale energy rates. This is covered in Chapter 19.

UNDERSTANDING THE DATA FOUND ON AN HVAC NAMEPLATE

The letters HVAC stand for "heating, ventilating, and air-conditioning."

Article 440 applies when HVAC equipment employs a hermetic refrigerant motor-compressor(s). *Article 440* is supplemental to the other articles of the *NEC* and is needed because of the unique characteristics of hermetic refrigerant motor-compressors. The most common examples are air conditioners, heat pumps, and refrigeration equipment.

Hermetic refrigerant motor-compressors

- combine the motor and compressor into one unit.
- have no external shaft.
- operate in the refrigerant.
- do not have horsepower and full-load current (often referred to as full-load amperes or FLA) ratings like standard electric motors. This is because as the compressor builds up pressure, the current increases, causing the windings to get hot. At the same time, the refrigerant gets colder, passes over the windings, and cools them. Because of this, a hermetically sealed motor can be "worked" much harder than a conventional electric motor of the same size.

Special terminology is needed to understand how to properly make the electrical installation. These special terms are found in *440.2* and in UL Standard 1995. An air-conditioning nameplate from Shiver Manufacturing Company helps explain these special terms.

- **Rated-Load Current (RLC):** Rated-load current is established by the hermetic motor-compressor manufacturer under actual operation at rated refrigerant pressure and temperature, rated voltage, and rated frequency. RLC is marked on the nameplate. In most instances, the marked RLC is at least equal to 64.1% of the hermetic refrigerant motor-compressor's maximum continuous current (MCC). In our example, the RLC is 17.8 amperes. See *440.2*.
- **Branch-Circuit Selection Current (BCSC):** The manufacturer of the HVAC end-use equipment might design better cooling and better heat dissipation into the equipment, in which case the

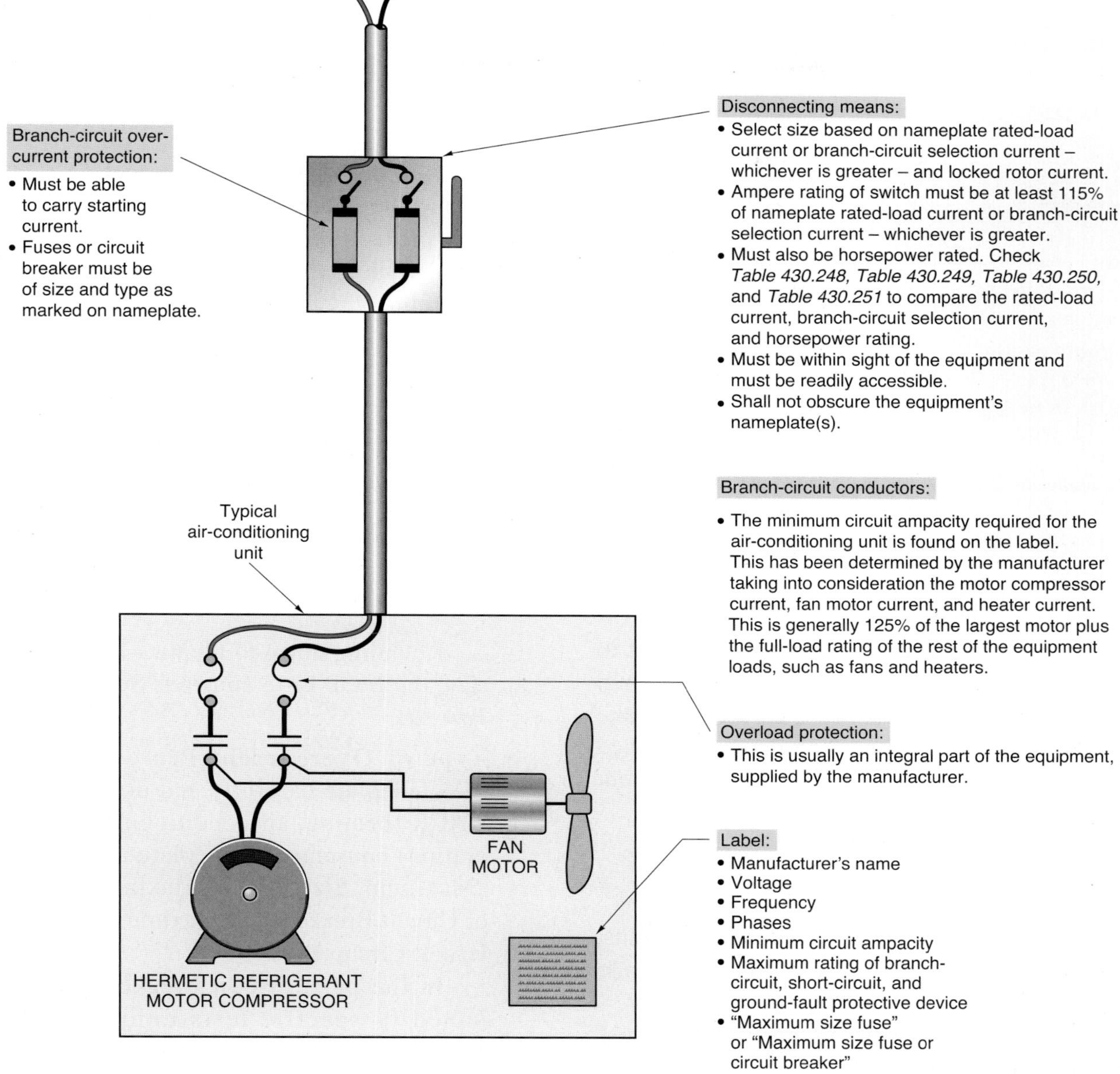

FIGURE 23-16 Basic circuit requirements for a typical residential-type air conditioner or heat pump. Reading the label is important in that the manufacturer has determined the minimum circuit ampacity and the maximum size and type of fuse or circuit breakers.

hermetic refrigerant motor-compressor might be capable of being continuously "worked" harder than other equipment not so designed. "Working" the hermetic refrigerant motor-compressor harder will result in a higher current draw. This is safe insofar as the motor-compressor is concerned, but the conductors, disconnect switch, and branch-circuit overcurrent devices must also be capable of safely carrying this higher current draw. This higher value of current is marked on the nameplate as "BCSC."

Because the BCSC ampere value is always equal to or greater than the RLC ampere value, this BCSC value must be used instead of the RLC value when selecting conductors, disconnects, overcurrent devices, and other associated electrical equipment. Table 23-2 shows that the BCSC is 19.9 amperes. See *440.2* and *440.4(C)*.

TABLE 23-2

Shiver Manufacturing Company.

MODEL NO. XYZ – ELECTRICAL RATINGS

	VAC	HZ	PH	RLC	LRC	FLA
Compressor	230	60	1	17.8	107	—
Outdoor fan motor ¼ HP	230	60	1	—	—	1.5
Branch-circuit selection current	19.9 amperes					
Minimum circuit ampacity	26.4 amperes					
Maximum fuse or HACR type breaker	45.0 amperes					
Operating voltage range:	197 min. 253 max.					

- **Minimum Circuit Ampacity (MCA):** MCA is the minimum circuit ampacity requirement to determine the conductor size and switch rating. MCA data are always marked on the nameplate of the end-use equipment and are determined by the manufacturer of the end-use equipment as follows:

$$\text{MCA} = (\text{RLC or BCSC} \times 1.25) + \text{other loads}$$

Other loads would include condensing fans, electric heaters, coils, and so forth, that operate concurrently (at the same time).

In our example, the marked MCA is 26.4 amperes, calculated by the manufacturer of the end-use equipment as follows:

$$(19.9 \times 1.25) + 1.5 = 26.375 \text{ amperes (round off to 26.4)}$$

Referring to *Table 310.15(B)(16)*, using the 60°C column, 10 AWG copper conductors would be adequate for this air-conditioning unit.

See *Article 440, Part IV.*

A mistake that electricians often make is to multiply the MCA by 1.25 again. Certainly there is no hazard in doing this, but it might result in an unnecessary higher installation cost. For long runs, larger conductors would keep voltage drop to a minimum.

Outdoor installations of air-conditioning equipment might be in very hot parts of the country. You might want to review Chapter 18 for the *NEC* correction factor requirements where high ambient temperatures are encountered.

- **Maximum Overcurrent Protection (MOP):** The MOP value is the maximum size overcurrent protection, fuse, or circuit breaker. The MOP value is marked on the nameplate and is determined by the manufacturer of the end-use equipment as follows:

$$(\text{RLC or BCSC}) \times 2.25 + \text{other loads}$$

Other loads include condensing fans, electric heaters, coils, and so on, that operate at the same time. In our example, the marked MOP is 45 amperes, determined by the manufacturer as follows:

$$(19.9 \times 2.25) + 1.5 = 46.3 \text{ amperes}$$

Rounding down to the next lower standard size, the MOP is 45 amperes. See *Article 440, Part III.*

- **Type of Overcurrent Protection:** Check the nameplate carefully and do what it says! HACR (heating, air conditioning, and refrigeration) equipment nameplate might indicate "Maximum Size Fuse," "Maximum Size Fuse or Circuit Breaker," or "Maximum Size Fuse or HACR Circuit Breaker."

In the past, circuit breakers were subjected to specific tests unique to HVAC equipment and were marked with the letters "HACR." Today, no additional tests are made. All currently listed circuit breakers are now suitable for HVAC application. These circuit breakers and HVAC equipment may or may not show the letters "HACR." Because of existing inventories, it will take many years for the marking "HACR" to disappear from the scene. In the meantime, read and follow the information found on the nameplate of the equipment.

See Figures 23-17, 23-18, 23-19, and 23-20.

- **Disconnecting Means to Be in Sight:** *NEC 440.14* of the *Code* requires that the disconnecting means must be within sight of the unit, and it

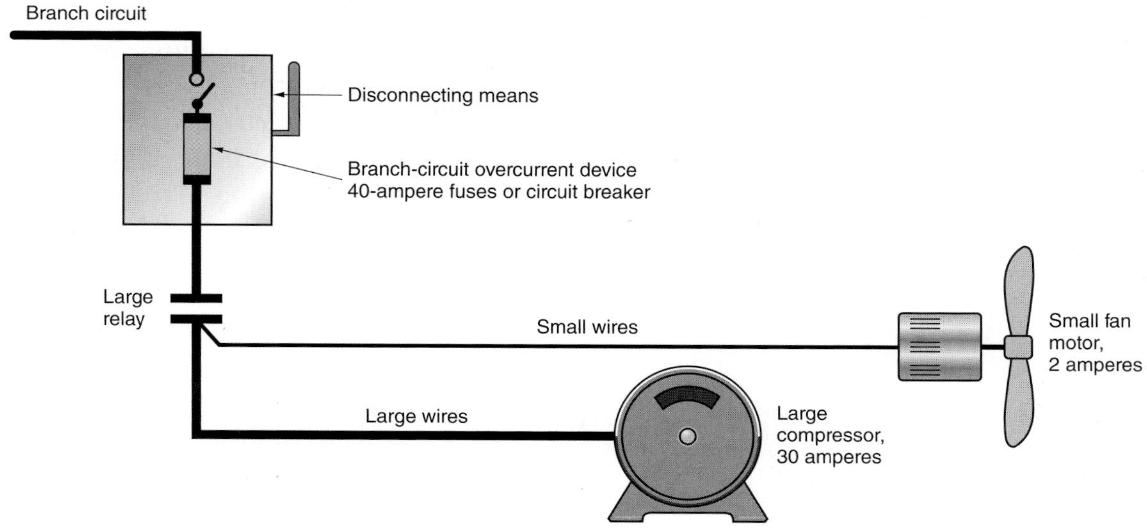

FIGURE 23-17 In a typical air-conditioner unit, the branch-circuit overcurrent device must protect the large components (large wires, large relay, hermetic motor-compressor) as well as the small components (small wires, fan motor, crankcase heater) under short-circuit and/or ground-fault situations. It is extremely important to install the proper size and type of overcurrent device. Read the nameplate on the equipment and the instructions furnished with the equipment.

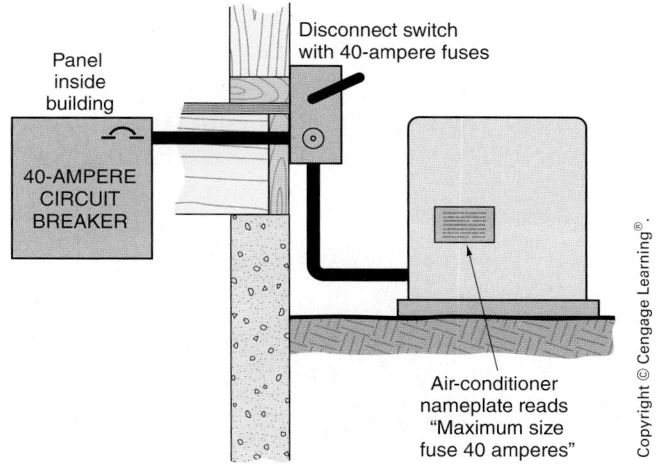

FIGURE 23-18 This installation "Meets *Code*." The disconnect is within sight of the unit. The disconnect contains fuses as specified on the air-conditioner nameplate. Refer to *NEC 110.3(B)*.

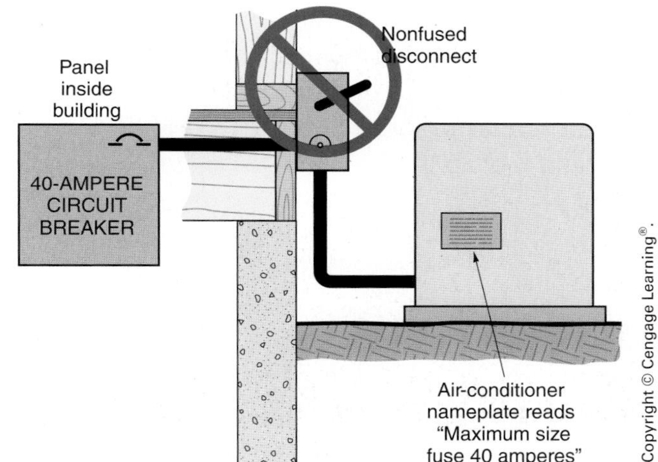

FIGURE 23-19 This installation does not "Meet *Code*" because the overcurrent device inside the building is not of the type specified on the air-conditioner nameplate. Refer to *NEC 110.3(B)*.

must be *readily accessible*. See Figures 23-18, 23-19, and 23-20. Mounting the disconnect on the side of the house instead of on the air conditioner itself allows for easy replacement of the air-conditioning unit, should that become necessary. Mounting the disconnect behind the air conditioner or heat pump would not be considered "readily accessible" by most electrical inspectors. Generally, there is just not enough space to work safely on the disconnect. Squeezing behind or leaning across the top of the air conditioner to gain access to

FIGURE 23-20 In the past, air-conditioning equipment and circuit breakers were marked with the letters "HACR." As time passes and current inventory is used up, this marking will disappear from the nameplates of equipment and on the circuit-breaker label. In the meantime, follow the manufacturer's installation instructions and the data found on the nameplate. Refer to *NEC 110.3(B)*.

the disconnect is certainly not a safe practice. Mounting the disconnect on the outside wall of the house, to one side of the AC unit, is the accepted practice. See *Article 440, Part II,* and *110.26*.

NEC 440.14 states that if you mount the disconnect on the equipment, do not obscure the equipment's nameplate(s). It is hard to believe that some installers mount the disconnect switch on top of the equipment's nameplate, making it impossible to read the nameplate.

- **Disconnecting Means Rating:** Figure 23-21 shows a fusible and a nonfusible "pull-out" disconnect commonly used for residential air-conditioning and/or heat pump installations. The horsepower rating of the disconnecting means must be at least equal to the sum of all of the individual loads within the end-use equipment, at rated load conditions, and at locked rotor conditions, *440.12(B)(1)*.

  The ampere rating of the disconnecting means must be at least 115% of the sum of all of the individual loads within the end-use equipment, at rated load conditions, *440.12(B)(2)*. Installing a disconnect switch that equals or exceeds the MOP value will in almost all cases be the correct choice. The MOP in our example is 40 amperes. Thus, a 60-ampere disconnect switch is the correct size. In cases where the MOP value is close to the ampere rating of the disconnect switch (i.e., 30, 60, 100, 200, etc.), *locked-rotor current* values must be considered.

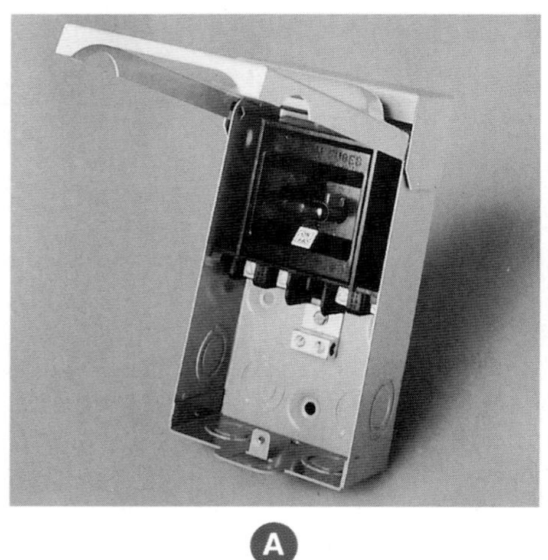

A

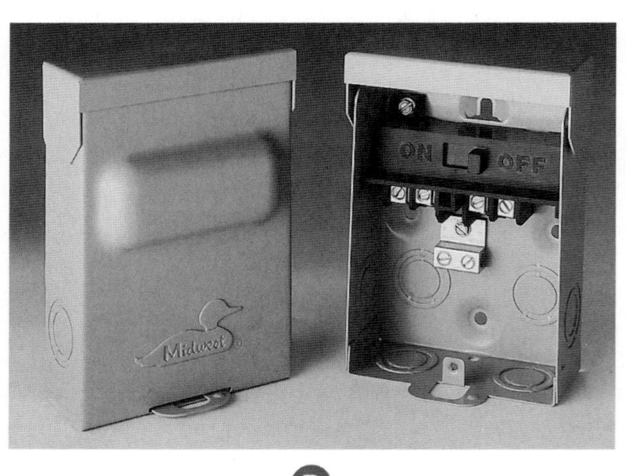

B

FIGURE 23-21 (A) is a pull-out fusible disconnect. Fuses are inserted into fuse clips on the pull-out device. Insert the pull-out device for ON, and remove the pull-out device for OFF. (B) is a nonfusible disconnect, available in 30- and 60-ampere ratings at 240 volts. Both have padlock provisions to prevent unauthorized tampering. Air-conditioner disconnects are available that have a GFCI receptacle as part of the disconnect enclosure. The GFCI receptacle must be wired to a separate 20-ampere, 120-volt branch circuit.

- **Locked-Rotor Current (LRC):** This is the maximum current draw when the motor is in a LOCKED position. When the rotor is locked, it is not turning. The disconnect switch and controller (if used) must be capable of safely interrupting locked-rotor current. See *440.12* and *440.41*. In our example, the nameplate indicates that the hermetic refrigerant motor-compressor has an LRC of 107 amperes. The fan motor has an FLA of 1.5 amperes and would have an LRC approximately 6 times higher. According to *440.12(B)(1)(b)*, all locked-rotor currents and other loads are added together when combined loads are involved. Therefore, the total locked-rotor current that the disconnect switch in our example would be called on to interrupt is

$(1.5 \times 6) + 107 = 116$ locked-rotor amperes

Checking *Table 430.251(A)* for the conversion of locked-rotor current to horsepower, we find that the locked-rotor current for a 230-volt, single-phase, 3-horsepower motor is 102 amperes, and the locked-rotor current for a 230-volt, single-phase, 5-horsepower motor is 168 amperes. Selecting a disconnect switch on the basis of the locked-rotor current in our example, the air-conditioning unit is considered to be a 5-horsepower unit, because 116 falls between 102 and 168. Manufacturers' technical literature indicates that a 30-ampere, 240-volt, heavy-duty, single-phase disconnect switch has a 3-horsepower rating, and a 60-ampere, 240-volt heavy-duty, single-phase disconnect switch has a 10-horsepower rating. Thus, a 60-ampere disconnect is selected. Note that this is the same size disconnect selected previously on the basis of MOP.

ENERGY RATINGS

Energy ratings indicate the efficiency of heating and cooling equipment. Basically, these ratings are a comparison of output (heating or cooling) to input (electricity, gas, or oil).

Energy Efficiency Rating (EER): Cooling efficiency rating for room air conditioners. The ratio of the rated cooling capacity in Btu per hour divided by the amount of electrical power used in kilowatt-hours. The higher the EER number, the greater the efficiency.

Seasonal Energy Efficiency Rating (SEER): Cooling efficiency rating for central air conditioners and heat pumps. SEER is determined by the total cooling of an air conditioner or heat pump in Btu during its normal usage period for cooling divided by the total electrical energy input in kilowatt-hours during the same period. The higher the SEER number, the greater the efficiency.

Annual Fuel Utilization Efficiency (AFUE): Tells you how efficiently a furnace converts fuel (gas or oil) to heat. For example, an AFUE of 85% means that 85% of the fuel is used to heat your home, and the other 15% goes up the chimney. The higher the efficiency, the lower the operating cost. Old furnaces might have an AFUE rating as low as 60%. Mid-efficiency ratings are approximately 80%. High-efficiency ratings are 90% or higher. Maximum furnace efficiency available is approximately 96.6%. The higher the AFUE number, the greater the efficiency.

Heating Seasonal Performance Factor (HSPF): Heating efficiency of a heat pump. HSPF is determined by the total heating of a heat pump in Btu during its normal usage period for heating divided by the total electrical energy input in kilowatt-hours during the same period. The higher the HSPF rating, the greater the efficiency.

NONCOINCIDENT LOADS

Loads such as heating and air conditioning are not likely to operate at the same time. The *NEC* recognizes this diversity in *220.60*. Therefore, when calculating a feeder that supplies both types of loads, or when sizing service equipment, only the larger of the two loads need be considered. This is discussed in Chapter 29 where service-entrance calculations are presented. Of course, the branch circuit supplying the heating load is sized for that particular load, and the branch circuit supplying the air-conditioning load is sized for that particular load.

RECEPTACLE NEEDED FOR SERVICING HVAC EQUIPMENT

For servicing HVAC equipment, *NEC 210.63* requires that a 125-volt, single-phase, 15- or 20-ampere-rated receptacle be installed

- at an accessible location,
- on the same level as the HVAC equipment,
- within 25 ft (7.5 m) of the equipment; and
- the receptacle shall not be connected to the load side of the equipment disconnecting means.

The outdoor receptacles required by *210.52(E)* (one in front and one in back) *might* or *might not* meet the requirement of *210.63*. If the HVAC equipment is located on a roof, in an attic, or in a similar location, a receptacle must be installed in that location so as to be on the same level as the equipment.

A receptacle is not required if the equipment served is an evaporative cooler in one- and two-family dwellings. This type of equipment requires very little servicing.

Some air-conditioner disconnects have a GFCI receptacle as an integral part of the disconnect enclosure, Figure 23-22. The GFCI receptacle must be supplied by a separate 15- or 20-ampere, 120-volt branch circuit. Installing this type of air-conditioner disconnect eliminates the need for installing a receptacle as part of the premises wiring. But this could also become a nightmare, trying to figure out how to wire it. When wiring with this type of disconnect, both the AC branch circuit *and* the 20-ampere, 120-volt branch circuit for the GFCI receptacle must somehow be run to the disconnect. This might require more time and material than wiring the required receptacle for the AC as required by *210.63* in the customary way. The choice is yours.

Courtesy of Midwest Electric Products, Inc.

FIGURE 23-22 Photo of an AC disconnect that has an integral GFCI receptacle.

Adjustment factors might have to be applied when more than three current-carrying conductors are installed in the same raceway or cable, *310.15(B)(2)*. All of this depends on the wiring method: Is it NM cable or is it a raceway?

GAS EXPLOSION HAZARD

Although it might seem a bit out of place in an electrical book to talk about gas explosions, we must talk about it.

Often overlooked is a requirement in 2.7.2(c) of NFPA 54, *National Fuel Gas Code*, that gas meters be located at least 3 ft (900 mm) from sources of ignition. An electric meter or a disconnect switch are possible sources of ignition. Some utility regulations require a minimum of 3 ft (900 mm) clearance between electric metering equipment and gas meters and gas-regulating equipment. It is better to be safe than sorry! Check this issue out with the local electrical inspector and/or the local electric utility before installing the air-conditioner or heat-pump disconnect on the outside of the house. Also consider the location of the service watt-hour meter.

REVIEW

Note: Refer to the *Code* or plans where necessary.

ELECTRIC HEAT

1. a. What is the allowance in watts made for electric heat in this residence? ______
   b. What is the value in amperes of this load? ______

2. What are some of the advantages of electric heating? ______

3. List the different types of electric heating system installations. ______

4. There are two basic voltage classifications for thermostats. What are they? ______

5. What device is required when the total connected load exceeds the maximum rating of a thermostat? ______

6. The electric heat in this residence is provided by what type of equipment? ______

7. At what voltage does the electric furnace operate? ______

8. A certain type of control connects electric heating units to a 120-volt supply or a 240-volt supply, depending on the amount of the temperature drop in a room. These controls are supplied from a 120/240-volt, 3-wire, single-phase source. Assuming that this type of device controls a 240-volt, 2000-watt heating unit, what is the wattage produced when the control supplies 120 volts to the heating unit? Show all calculations.

9. What advantages does a 240-volt heating unit have over a 120-volt heating unit?

10. The white wire of a cable may be used to connect to a "hot" ungrounded circuit conductor only if ______

11. Receptacle outlets furnished as part of a permanently installed electric baseboard heater, when not connected to the heater's branch circuit, (may) (may not) be counted as the required receptacle outlet for the space occupied by the baseboard heater. Electric baseboard heaters (shall) (shall not) be installed beneath wall receptacle outlets unless the instructions furnished with the heaters indicate that it is acceptable to install the heater below a receptacle outlet. (Circle the correct answers.)

12. The branch circuit supplying a fixed electric space heater must be sized to at least __________ percent of the heater's rating according to *NEC* __________.

13. Calculate the current draw of the following electric furnaces. The furnaces are all rated 240 volts.

    a. 7.5 kW _____ amperes b. 15 kW _____ amperes c. 20 kW _____ amperes

14. For ballpark calculations, the wattage output of a 240-volt electric furnace connected to a 208-volt supply will be approximately 75% of the wattage output had the furnace been connected to a 240-volt supply. In problem 13, calculate the wattage output of (a), (b), and (c) if connected to 208 volts. __________

    ____

    ____

    ____

15. A central electric furnace heating system is installed in a home. The circuit supplying this furnace

    a. is limited to a branch circuit rated not greater than 30 amperes.

    b. is limited to not more than one branch circuit.

    c. is considered to be a continuous load.

16. What section of the *Code* provides the correct answer to problem 15? __________

17. Electric heating cable embedded in plaster, or sandwiched between layers of dry wall, creates heat. Therefore, any nonmetallic-sheathed cable run above these ceilings and buried in insulation must have its current-carrying capacity derated according to the (40°C) (50°C) (60°C) correction factors found in *Table 310.15(B)(2)(a)* in the *NEC*. (Circle the correct answer.)

AIR CONDITIONING

1. a. When calculating air-conditioner load requirements and electric heating load requirements, is it necessary to add the two loads together to determine the combined load on the system? __________

   b. Explain the answer to part (a). __________

   ____

   ____

2. The total load of a cord-connected room air conditioner shall not exceed what percentage of a separate branch circuit? (Circle the correct answer.)

   a. 75% b. 80% c. 125%

3. The total load of a cord-connected air conditioner shall not exceed what percentage of a branch circuit that also supplies lighting? (Circle the correct answer.)

   a. 50% b. 75% c. 80%

4. a. Must an air conditioner installed in a window opening be grounded if a person on the ground outside the building can touch the air conditioner? ______________

   b. What *Code* section governs the answer to part (a)? ______________

5. A 120-volt air conditioner draws 13 amperes. What size is the circuit to which the air conditioner will be connected? ______________

6. What is the *Code* requirement for receptacles connected to circuits of different voltages and installed in one building? ______________

7. When a central air-conditioning unit is installed and the label states "Maximum Size Fuse 50 Amperes," is it permissible to connect the unit to a 50-ampere circuit breaker? ______________

8. When the nameplate on an air-conditioning unit states "Maximum Size Fuse or HACR Circuit Breaker," what type of circuit breaker must be used? ______________

   ____

9. What section of the *NEC* prohibits the installation of Class 2 control circuit conductors in the same raceway as the power conductors? ______________

   ____

10. Match the following terms with the statement that most closely defines the term. Enter the letters in the blank space.

    RLC ________ a. The maximum ampere rating of the overcurrent device

    MCA ________ b. The current value to be used instead of the rated-load ampere

    BCSC ________ c. The value used to determine the minimum ampere rating for the branch circuit

    MOP ________ d. A current value for the hermetic motor-compressor that was determined by operating it at rated temperature, voltage, and frequency

11. The disconnect for an air conditioner or heat pump must be installed ______________ of the unit.

CHAPTER **24**

Gas and Oil Central Heating Systems

OBJECTIVES

After studying this chapter, you should be able to

- understand the basics of typical home warm air and hot water (hydronic) heating systems.
- understand and apply the *NEC* requirements for branch-circuit wiring for central heating systems.
- define all of the major components of typical heating systems.
- understand and apply the *NEC* requirements for Class 2 control circuit wiring.

INTRODUCTION

In Chapter 23, we discussed various types of electric heating systems. In this chapter, we will take a look at gas- and oil-fired systems.

FORCED–WARM AIR FURNACES

Gas- and oil-fired heating systems provide the heat source for forced-warm air furnaces and hot water systems. Gas-fired, forced-warm air furnaces are the most common. Forced-warm air systems move hot air from the furnace through supply air ducts that in turn connect to supply air registers strategically located throughout the home. The room air is then pulled into return air registers, returning to the furnace through return air ducts, where the air is again filtered, heated, and forced out through the supply air ducts.

Air-conditioning, humidity control, and fresh air intake are but a few of the things that can be added to a typical forced-warm air heating system.

HOT WATER SYSTEMS

Hot water systems, also known as hydronic systems, move hot water through pipes to radiators (i.e., baseboards) and, in some instances, through tubing embedded in the ceiling or in the floor or on the underside of the floor to heat the floors from below. The water is returned to the boiler through a return pipe. Hot water systems are very adaptable to zone control by using a combination of one or more circulating pumps and/or valves for the different zones. Hot water heating systems are referred to as *hydronic* (wet) systems.

PRINCIPLE OF OPERATION

Most residential heating systems operate as follows. A room thermostat is connected to the proper electrical terminals on the furnace or boiler. Inside the equipment, the various controls and valves are interconnected in a manner that will provide safe and adequate operation of the furnace.

Most residential heating systems are "packaged units" in which the burner, heating component igniter or hot spark igniter, safety controls, valves, fan controls, sensors, high-temperature-limit switches, fan blower motor, blower fan speed control, draft inducer motor, primary and secondary heat exchanger, and so on, are preassembled, prewired parts of the furnace or boiler. Several of the control system components are included in an integrated printed circuit board module, which is the "brain" of the system. This module provides proper sequencing and safety features.

Older furnaces have gas pilot lights for gas burner ignition and thermocouples that enable the gas valve to function.

Similar electronic circuitry is used in oil burners where the ignition, the oil burner motor, blower fan motor, high-temperature limit controls, hot water circulating pump in the case of hot water systems, and the other devices previously listed are all interconnected to provide a packaged oil burner unit.

Evaporator coils for air-conditioning purposes might be an integral part of the unit, or they might be mounted in the ductwork above or below the furnace.

Wiring a Residential Central Heating System

Gas- and oil-fired heating systems are usually "packaged units," which means that all of the internal wiring of the components has been done by the manufacturer of the unit. They are considered "appliances" in the *NEC*® and are required to comply with *Article 422*.

In Chapter 23, we discussed the major *NEC* requirements for wiring an electric furnace.

For the electrician, the "field wiring" for a typical residential furnace or boiler consists of the following:

1. Installing and connecting the low-voltage wiring between the furnace and the thermostat. This might be a cable consisting of two, three, four, or five small 18 AWG or 20 AWG conductors. Small knockouts are provided on the furnace through which the low-voltage Class 2 wires are brought into the furnace's wiring compartment (see Figure 24-1). Class 2 wiring must not be run through the same raceway or cable as power wiring, *725.136(A)*.
2. Installing and connecting the branch-circuit power supply to the furnace. Knockouts are provided on the furnace through which the line voltage power supply is brought into the wiring compartment of the furnace. Depending on local codes, the line voltage power supply wiring might be in EMT, armored cable, or other accepted wiring methods (Figure 24-1).

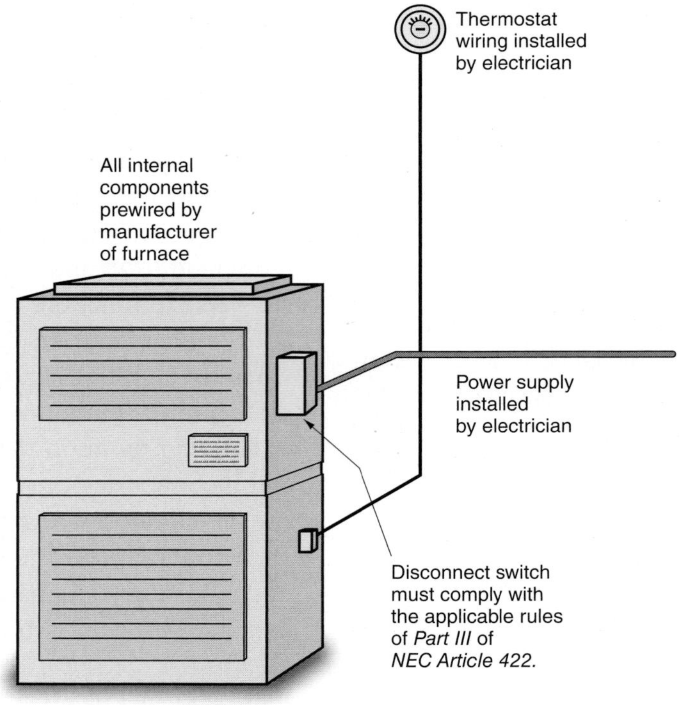

FIGURE 24-1 Diagram showing thermostat wiring, power supply wiring, and disconnect switch. ANSI Standard Z21.47 does not allow cord-and-plug connections for a gas furnace.

Disconnecting Means

The disconnecting means for the motor-operated appliances such as the fuel-fired furnace discussed in this chapter is required to comply with *NEC 422.31(C)*.

Motor-Operated Appliances Rated over 1⁄8 Horsepower. ►The disconnecting means shall comply with 430.109 and 430.110. For permanently connected motor-operated appliances with motors rated over 1⁄8 hp, the disconnecting means shall meet 422.31(C)(1) or (2).

(1) The branch-circuit switch or circuit breaker shall be permitted to serve as the disconnecting means where the switch or circuit breaker is within sight from the appliance.

(2) The disconnecting means shall be installed within sight of the appliance.◄*

An exception relating to unit switches is provided. This does not apply to fuel-fired furnaces such as used in the dwelling in this text.

The reference to *NEC 430.109* ensures that the proper type of device is used as the disconnecting means. Several types are permitted. Often a snap switch with the proper rating or an enclosed circuit breaker is installed. The requirements in *NEC 430.110* cover the rating of the disconnecting means. The ampere rating is generally required to be not less than 115% of the full-load current rating of the motor. This information is usually provided on the appliance nameplate required in *NEC 422.60*.

As can be seen in the rules in *NEC 422(C)(1)* and *(C)(2)*, there is no reference to *NEC 110.25* for locking of disconnecting means that are located out of sight of the appliance. This is due to the requirement that the disconnecting means be located within sight of the appliance. The *NEC* defines "within sight" as being visible and not more than 50 ft (15 m) from the equipment the disconnect controls.

Some furnaces come with a disconnect switch. If not, a toggle switch can be mounted on a box on the side of the furnace. This is simple for gas-fired furnaces. Electric furnaces require a larger ampere-rated disconnect. Some electric furnaces come complete with circuit breakers that provide the overcurrent protection for the heating elements as required by UL Standard, and to serve as the disconnecting means required by the *NEC*.

*Reprinted with permission from NFPA 70-2014.

Individual Branch Circuit Required

NEC 422.12 requires that central heating equipment be supplied by an individual branch circuit. By definition in *Article 100* of the *NEC*, an individual branch circuit is A branch circuit that supplies only one utilization equipment.*

The size of the branch circuit depends on the requirements of the particular furnace. A typical gas-forced warm air furnace might require only a 120-volt, 15-ampere branch circuit, whereas an electric furnace requires a 240-volt branch circuit and much larger conductors. Instructions furnished with a given furnace will indicate the branch-circuit requirements.

Section *422.10(A)* of the *NEC* requires The branch-circuit rating for an appliance that is a continuous load, other than a motor-operated appliance, shall not be less than 125 percent of the marked rating, or not less than 100 percent of the marked rating if the branch-circuit device and its assembly are listed for continuous loading at 100 percent of its rating.* Typically, this rule applies to branch circuits for water heaters and fixed electric space heating equipment, but not to gas- or oil-fired furnaces.

Some typical wiring diagrams are shown in Figures 24-2, 24-3, 24-4(A), and 24-4(B). These wiring diagrams show the individual electrical components of a typical residential heating system. In older systems, many of these components were wired by the electrician. In newer systems, most components other than the thermostat are prewired as part of a packaged furnace or boiler. The manufacturer's literature includes detailed wiring diagrams for a given model.

MAJOR COMPONENTS

Gas- and oil-fired warm air systems and hot water boilers contain many individual components. Here is a brief description of these components.

Aquastat. An aquastat is a direct immersion water temperature thermostat that regulates boiler or tank temperature in hydronic heating systems, ensuring that the circulating water maintains proper, satisfactory temperature.

Cad Cell. A cad cell is a cadmium sulfide sensor for sensing flame. Cad cells have pretty much replaced the older style stack-mounted bimetal switches.

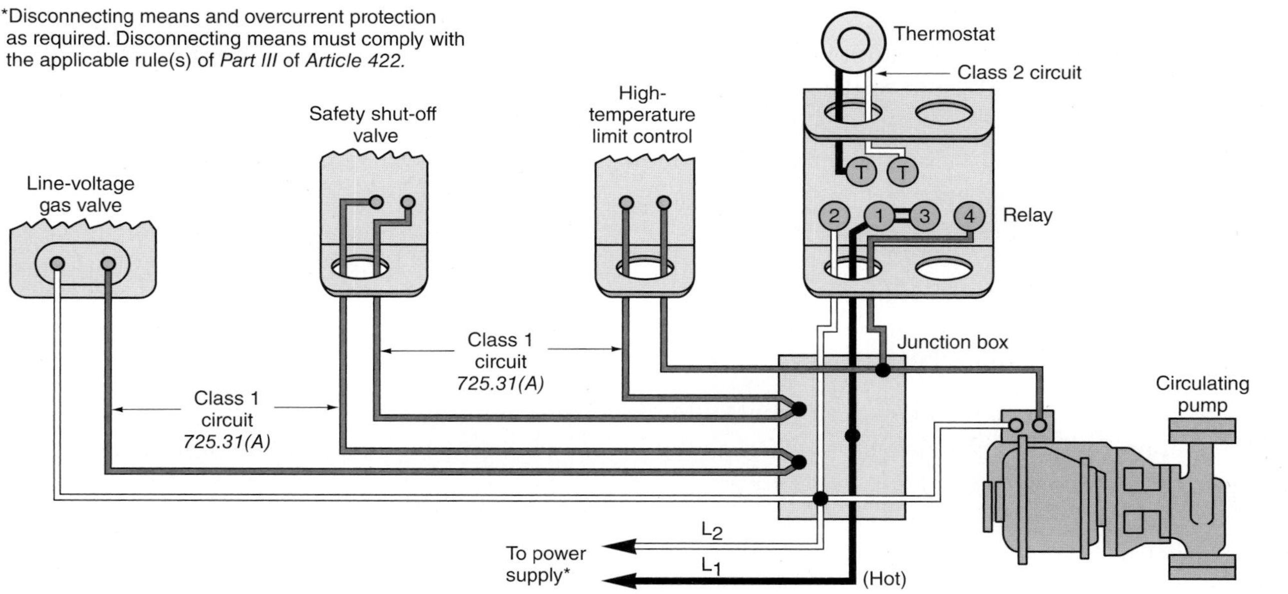

FIGURE 24-2 Typical wiring diagram for a gas burner, forced hot water system. Always consult the manufacturer's wiring diagram relating to a particular model heating unit.

*Reprinted with permission from NFPA 70-2014.

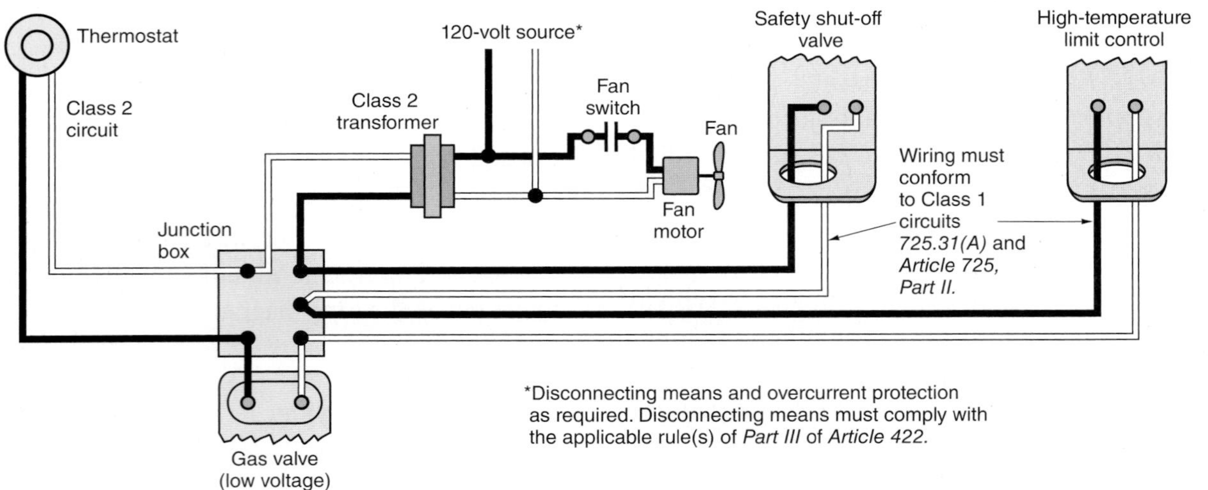

FIGURE 24-3 Typical wiring diagram for a gas burner, forced-warm air system. Always consult the manufacturer's wiring diagrams relating to a particular model heating unit.

Circulating Pump. A circulating pump circulates hot water in a hydronic central heating system.

Control Circuit. A control circuit is any electric circuit that controls any other circuit through a relay or an equivalent device; also referred to as a remote-control circuit.

Combustion Chamber. The combustion chamber surrounds the flame and radiates heat back into the flame to aid in combustion.

Combustion Head. A combustion head creates a specific pattern of air at the end of the air tube. The air is directed in such a way as to force oxygen into the oil spray so the oil can burn. A combustion head might also be referred to as the turbulator, fire ring, retention ring, or end cone.

Draft Regulator. A draft regulator is a counter-weighted swinging door that opens and closes to help maintain a constant level of draft over the fire.

Fan Control. A fan control is used to control the blower fan on forced-warm air systems. On newer furnaces, the fan control is an electronic timer that starts the fan motor a given time in seconds after the main burner has come on. This timing ensures that the blower will blow warm air. Older type adjustable temperature fan controls are still around.

Fan Motor. The fan motor is the electric motor that forces warm air through the heating ducts. In typical residential heating systems, the motor is provided with integral overload protection. The motor may be single, multiple, or variable speed.

Flue. A flue is a channel in a chimney or a pipe for conveying flame and smoke to the outer air.

Heat Exchanger. A heat exchanger transfers the heat energy from the combustion gases to the air in the furnace or to the water in a boiler.

High-Temperature Limit Control. A high-temperature limit control is a safety device that limits the temperature in the plenum of the furnace to some predetermined safe value as determined by the manufacturer. When this predetermined temperature is reached, the high-temperature limit control shuts off the power to the burner. In newer furnaces, the high-temperature limit control is part of an integrated circuit board and is communicated to by a sensor strategically located in the plenum. Older furnaces might have individual high-temperature limit controls and a separate fan control, or they might have combination high-temperature limit/fan controls.

Hot Surface Igniter. In furnaces that do not have a pilot light that is lit continuously, the hot surface igniter heats to a cherry red/orange color. When it reaches the proper temperature, the main gas valve is allowed to open. If the hot surface igniter does not come on, the main gas valve will not open. Most energy-efficient furnaces use this concept for ignition.

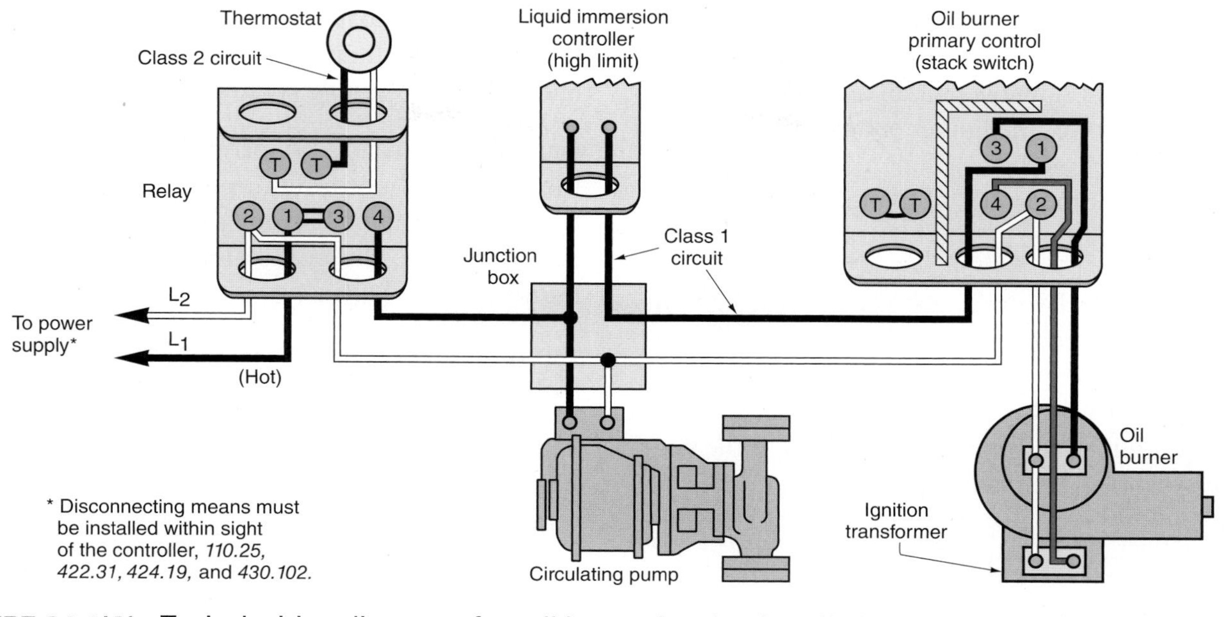

FIGURE 24-4(A) Typical wiring diagram of an oil burner heating installation. Always consult the manufacturer's wiring diagram relating to a particular model heating unit.

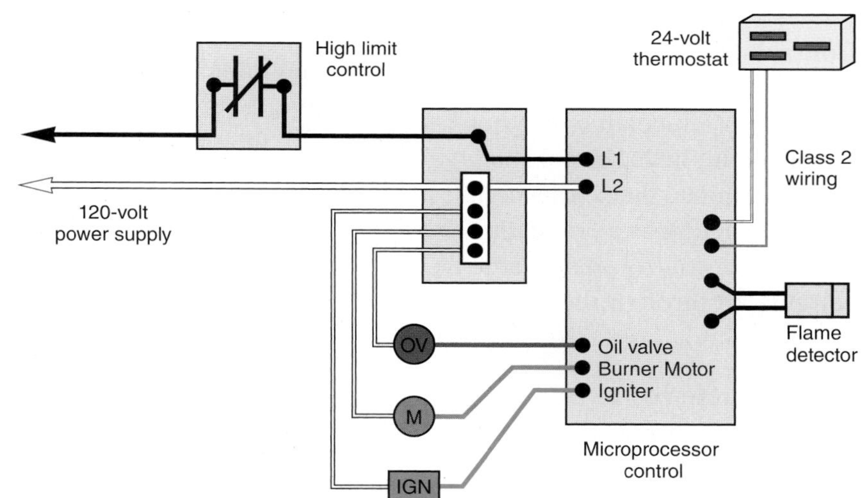

FIGURE 24-4(B) An oil burner featuring an electronic microprocessor that controls all facets of the burner operation. Most of the components are an integral part of the unit, thus keeping field wiring to a minimum. All manufacturers provide detailed wiring diagrams for their products. Always consult the manufacturer's wiring diagrams relating to a particular model heating unit.

Hydronic System. The hydronic system is a system of heating or cooling that involves transfer of heat by circulating fluid such as water in a closed system of pipes.

Ignition.

- **Constant:** The igniter is designed to stay on continuously.
- **Intermittent Duty:** Defined by UL 296 as "ignition by an energy source that is continuously maintained throughout the time the burner is firing." In other words, the igniter is on the entire time the burner is firing.
- **Interrupted Duty:** Defined by UL 296 as "an ignition system that is energized each time the

main burner is to be fired and de-energized at the end of a timed trial for ignition period or after the main flame is proven to be established." In other words, the igniter comes on to light the flame; then, after the flame is established, the igniter is turned off and the main flame keeps burning.

Induced Draft Blower. When the thermostat calls for heat, the induced draft blower starts first to expel any gases remaining in the combustion chamber from a previous burning cycle. It continues to run, pulling hot combustion gases through the heat exchangers, then vents the gases to the outdoors.

Integrated Control. An integrated control is a printed circuit board containing many electronic components. When the thermostat calls for heat, the integrated circuit board takes over to manage the sequence of events that allow the burner to operate safely.

Liquid Immersion Control. This is also referred to as an aquastat. A liquid immersion control controls high water and circulating water temperature in a hot water system. A high-temperature limit control shuts off the burner when a predetermined dangerous high temperature is reached. When acting as a circulating water temperature control, this control makes sure that the water being circulated through the piping system is at the desired temperature—not cold.

Low-Water Control. Also referred to as a low-water cutoff, this device senses low water levels in a boiler. When a low-water situation occurs, the low-water control shuts off the electrical power to the burner.

Main Gas Valve. The main gas valve is the valve that allows the gas to flow to the main burner of a gas-fired furnace. It may also function as a pressure regulator, a safety shut-off, and as the pilot and main gas valve. Should there be a "flame-out," if the pilot light fails to operate, or in the event of a power failure, this valve will shut off the flow of gas to the main burner.

Nozzle. A nozzle produces the desired spray pattern for the particular appliance in which the burner is used.

Oil Burner. The function of an oil burner is to break fuel oil into small droplets, mix the droplets with air, and ignite the resulting spray to form a flame.

Primary Control. A primary control controls the oil burner motor, ignition, and oil valve in response to commands from a thermostat. If the oil fails to ignite, the controller shuts down the oil burner.

Pump and Zone Controls. Pump and zone controls regulate the flow of water or steam in boiler systems to specific zones in the building.

Safety Controls. Safety controls such as pressure relief valves, high-temperature limit controls, low-water cutoffs, and burner primary controls protect against appliance malfunction.

Solenoid. A solenoid is an electrically operated device; when the valve coil is energized, a magnetic field is developed, causing a spring-loaded steel valve piston to overcome the resistance of the spring and immediately pull the piston into the stem. The valve is now in the open position. When de-energized, the solenoid coil magnetic field instantly dissipates, and the spring-loaded valve piston snaps closed, stopping oil flow or gas flow to the nozzle. The flame is extinguished, allowing fuel to flow. Usually a spring returns the valve back to the closed position.

Spark Ignition. In spark ignition systems, the spark comes on while pilot gas flows to the pilot orifice. Once the pilot flame is "proven" through an electronic sensing circuit, the main gas valve opens. Once ignition takes place, the sensor will monitor and prove existence of a main flame. If for some reason the main flame goes off, the furnace shuts down. Energy-efficient furnaces may use this concept for ignition.

Switching Relay. A switching relay is used between the low-voltage wiring and line-voltage wiring. When a low-voltage thermostat calls for heat, the relay "pulls in," closing the line-voltage contacts on the relay, which turns on the main power to the furnace or boiler. Usually relays have a built-in transformer that provides the low-voltage Class 2 control circuit.

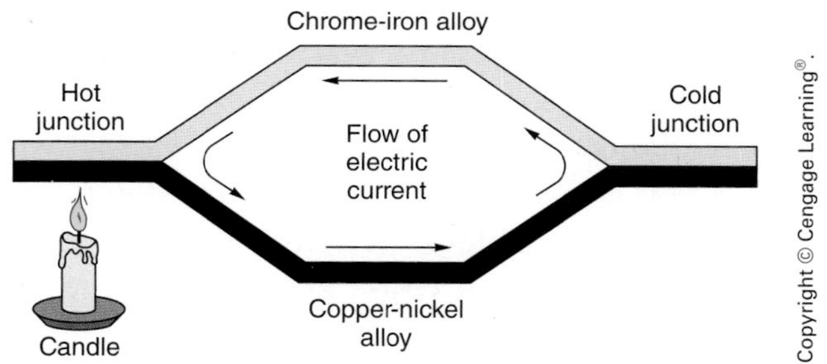

FIGURE 24-5 Principle of a thermocouple.

Thermocouple. A thermocouple is used in systems having a gas pilot light to hold open a gas valve pilot solenoid magnet. A thermocouple consists of two dissimilar metals connected together to form a circuit. The metals might be iron and copper, copper and iron-constantan, copper-nickel alloy and chrome-iron alloy, platinum and platinum-rhodium alloy, or chromel and alumel. The types of metals used depend on the temperatures involved. When one of the junctions is heated, an electrical current flows (Figure 24-5). To operate, there must be a temperature difference between the metal junctions. In a gas burner, the source of heat for the thermocouple is the pilot light. The cold junction of the thermocouple remains open and is connected to the pilot safety shut-off gas valve circuit. A single thermocouple develops a voltage of about 25 to 30 dc millivolts. Because of this extremely low voltage, circuit resistance is kept to a very low value.

Thermopile. More than one thermocouple connected in series is a thermopile. See Figure 24-6. The power output of a thermopile is greater than that of a single thermocouple—typically either 250 or 750 dc millivolts. For example, 10 thermocouples (25 dc millivolts each) connected in series result in a 250 dc millivolt thermopile. Twenty-six thermocouples connected in series result in a 750 dc millivolt thermopile.

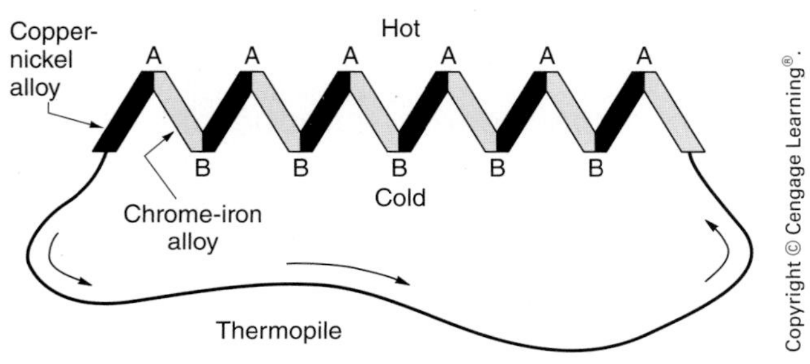

FIGURE 24-6 Principle of a thermopile.

Thermostat. A thermostat turns the heating/cooling system on and off. Programmable thermostats might have day/night setback settings, a clock, a digital thermometer, humidity control, ventilation, and filtration control. Thermostats are usually mounted approximately 52 in. (1.3 m) above the finished floor. Do not mount thermostats where influenced by drafts or air currents from hot or cold air registers, near fireplaces or concealed hot or cold water pipes or ducts, in the direct rays of the sun, or on outside walls. See Figure 24-7. Some thermostats contain a small vial of mercury that "tips," so the mercury closes the circuit by bridging the gap between the electrical contacts inside the vial. These thermostats must be kept perfectly level to ensure accuracy.

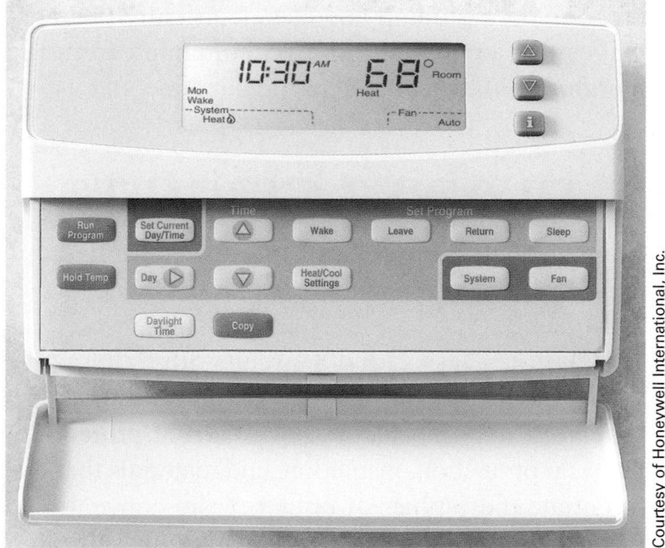

FIGURE 24-7 Digital electronic programmable thermostat that provides proper cycling of the heating system, resulting in comfort as well as energy savings. This type of thermostat, depending on the model, can be programmed to control temperature, time of operation, humidity, ventilation, filtration, circulation of air, and zone control. Installation and operational instructions are furnished with the thermostat.

SAFETY ALERT

Do not throw mercury thermostats into the trash. The mercury is considered a hazardous waste material. Mercury thermostats must be returned to a waste management organization for proper disposal.

Transformer. The transformer converts (transforms) the branch-circuit voltage to low voltage. For residential applications these are Class 2 transformers, 120 volts to 24 volts. The low-voltage wiring on the secondary of this transformer is Class 2 wiring.

Water Circulating Pump. In hydronic (wet) systems, a circulating pump circulates hot water through the piping system. Multizone systems have more than one pump, each controlled by a thermostat for a particular zone. In multizone systems, a high-temperature control (aquastat) might be set to maintain the boiler water temperature to a specific temperature. Some multizone systems have one circulating pump and use electrically operated valves that open and close on command from the thermostat(s) located in a particular zone(s).

CLASS 2 CIRCUITS

Remote-control, signaling, and power-limited circuits fall into three categories—Class 1, Class 2, and Class 3.

Class 1, Class 2, and 3 circuits offer alternative wiring methods regarding voltage, power limitations, wire size, derating factors, overcurrent protection, physical protection, insulation, and materials that differentiate these types of circuits from conventional wiring methods, as covered in *Chapter 1* through *Chapter 4* in the *NEC*. Figure 24-8 defines remote control, signaling, and power-limited terminology.

Low-voltage wiring in homes is generally Class 2 wiring. Class 2 wiring is very tolerant. We will limit our discussion to Class 2 wiring. Class 1 and Class 3 circuits are more restrictive than Class 2 circuits and are more commonly found in commercial and industrial applications. You can learn about Class 1 and Class 3 circuits in *Article 725* in the *NEC*.

Wiring for Class 2 Circuits

Electricians describe wire and cable used for low-voltage wiring *bell wire* or *thermostat wire*. What you are really looking for are listed Class 2 conductors and cables, referred to as CL2 or CL2X cables. Examples of Class 2 wiring are the low-voltage conductors and cables used for heating and air conditioning, thermostats, security systems, remote control and signal wiring (chimes), intercom wiring, and other low-voltage applications.

Class 2 conductors are generally made of copper, have a thermoplastic insulation, and have a voltage rating of not less than 150 volts as required by *725.179(G)*. Because the current required for Class 2 circuits is rather small, 18 AWG and 20 AWG conductors are most commonly used. Conductors as small as 24 AWG also are available; however, small conductors and long runs can result in voltage drop problems.

Multiconductor Class 2 cables consist of two to as many as 12 single conductors. These cables are available with or without a protective PVC outer jacket. The type with the outer jacket is preferred because there is less chance of damage to individual conductors, and it gives a neat appearance.

The conductors within these cables are color coded to make circuit identification easy. Table 24-1 is a table showing the color coding for the conductors in Class 2 cables.

Here is a summary of the *NEC* requirements for installing of Class 2 wiring:

- Class 2 circuits are "power-limited" and are considered to be safe from fire hazard and electric shock hazard.

TABLE 24-1

Table showing the color coding of Class 2 conductors and cables.

NUMBER OF CONDUCTORS	COLOR	NUMBER OF CONDUCTORS	COLOR
1	Red	7	Orange
2	White	8	Black
3	Green	9	Pink
4	Blue	10	Gray
5	Yellow	11	Tan
6	Brown	12	Purple

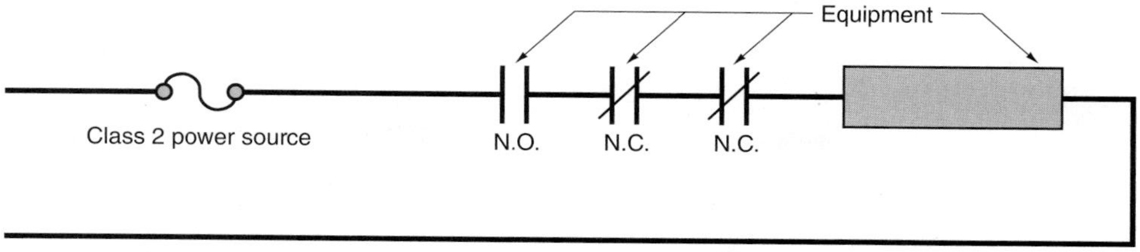

A A Class 2 circuit is The portion of the wiring system between the load side of a Class 2 power source and the connected equipment,* *725.2.*

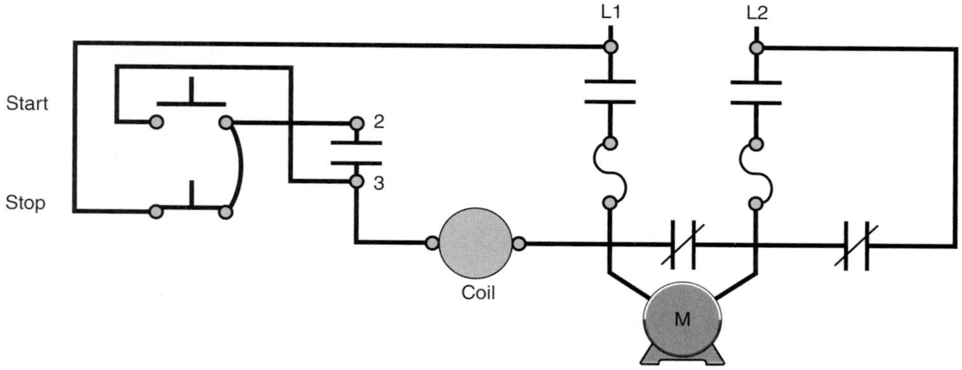

B A remote-control circuit is Any electrical circuit that controls any other circuit through a relay or equivalent device,* *Article 100, Definitions.*

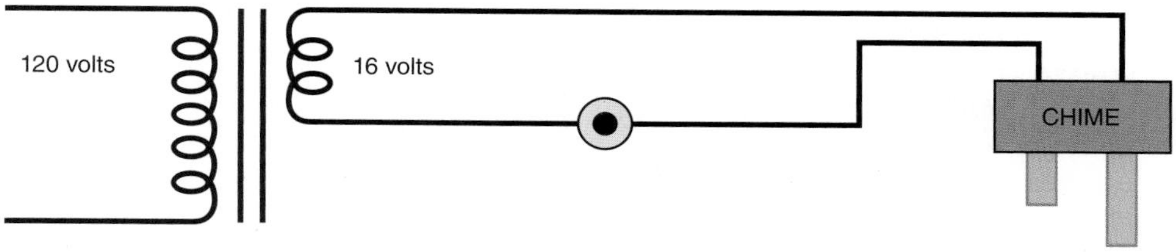

C A signaling circuit is Any electrical circuit that energizes signaling equipment,* *Article 100, Definitions.*

FIGURE 24-8 The above diagrams explain a remote-control circuit and a signaling circuit.

- Class 2 wiring does not have to be in a raceway. If a failure in the wiring should occur, such as a short circuit or an open circuit, there would not be a direct fire or shock hazard.
- Class 2 wiring is not permitted in the same raceway, cable, compartment, outlet box, or fitting with light and power conductors, *725.136(A)*. Even if the Class 2 conductors have the same 600-volt insulation as the power conductors, this is not permitted. See Figures 24-9 and 24-10. Manufacturers of heating and cooling equipment provide separate openings for bringing in the Class 2 wiring and separate openings for bringing in the power supply. *NEC 725.136(D)* addresses this issue.
- Keep Class 2 conductors at least 2 in. (50 mm) away from light or power wiring. This really pertains to old, open knob-and-tube wiring. Today, however, where light and power wiring is installed in Type NMC, Type AC, or in conduit, there is no problem, *725.136(I)*.
- Class 2 wiring shall not enter the same enclosure unless there is a barrier in the enclosure that separates the Class 2 wiring from the light and power conductors, *725.136(B)*.

*Reprinted with permission from NFPA 70-2014.

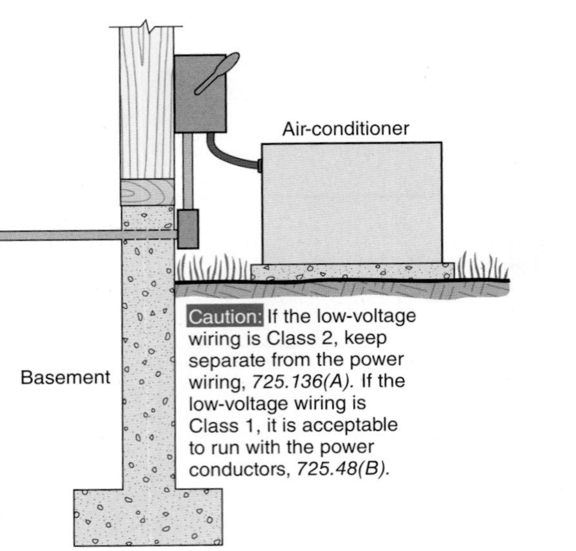

FIGURE 24-9 Most residential air-conditioning units, furnaces, and heat pumps have their low-voltage circuitry classified as Class 2. *NEC 725.136(A)* prohibits installing low-voltage Class 2 conductors in the same raceway as the power conductors, even if the Class 2 conductors have 600-volt insulation. Therefore, the conduit running out of the basement wall in the above diagram would not be permitted to contain both the 240-volt power conductors and the 24-volt low-voltage Class 2 conductors. Always read the instructions furnished with these types of appliances to be sure your wiring is in compliance with these instructions, per *110.3(B)* and the *NEC.* Review Chapter 4 for grounding methods.

- Class 2 circuits are inherently current limiting and do not require separate overcurrent protection. A Class 2 transformer is an example of this.
- Transformers that are intended to supply Class 2 circuits are listed and marked "Class 2 Transformer." These transformers have built-in overcurrent protection to prevent overheating should a short circuit occur somewhere in the secondary winding or in the secondary circuit wiring. These transformers may be connected to branch circuits having overcurrent protection not more than 20 amperes, *725.127.*
- Class 2 cables shall have a voltage rating of not less than 150 volts, *725.179(G).*

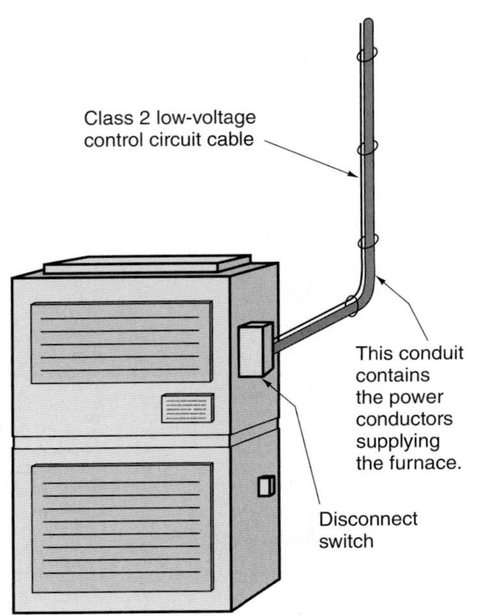

FIGURE 24-10 The *Code* in *300.11(B)(2)* and *725.143* allows Class 2 control circuit conductors (cables) to be supported by the raceway that contains the power conductors supplying electrical equipment, such as the furnace shown.

Securing Class 2 Cables

- Class 2 cables must be installed in a neat and workmanlike manner and shall be adequately supported by the building structure in such a manner that they will not be damaged by normal building use, *725.24.*
- Class 2 cables shall be attached to structural components by straps, staples, hangers, or similar fittings designed and installed so as not to damage the cable, *725.24.*
- Class 2 cables may be secured directly to surfaces with insulated staples or may be installed in raceways.
- The insulation and jacket of Class 2 cables and conductors are rather thin. Be careful during installation not to pierce or crush the wires. Secure these cables and conductors with care, using the proper type of staples. Staple "guns" that use rounded staples are available. They straddle the cable nicely instead of flattening it, which might result in shorted-out wires.

- Do not support Class 2 cables from other raceways, cables, or nonelectric equipment. Taping, strapping, tie wrapping, hanging, or securing Class 2 cables to electrical raceways, piping, or ducts is generally prohibited, *725.143*.
- *NEC 725.143* refers back to *300.11(B)(2)*, which allows Class 2 conductors to be supported by the raceway that contains the power supply conductors for the same equipment that the Class 2 conductors are connected to. See Figure 24-10.
- Do not run Class 2 conductors or cables through the same holes in studs and joists, holes that contain nonmetallic-sheathed cable, armored cable, raceways, conduits, or other pipes. There is too much chance of physical damage to the Class 2 cables.

Class 2 Power Source Requirements

Class 2 power source voltage and current limitations, and overcurrent protection requirements, are found in *Chapter 9, Table 11(A)* and *Table 11(B)* of the *NEC*. You should refer to the nameplate data on Class 2 transformers, power supplies, and other equipment to verify that they are listed by a Nationally Recognized Testing Laboratory (NRTL) for a particular class.

REVIEW

1. The residence in this text is heated with (Circle the correct answer.)
   a. gas.
   b. electricity.
   c. oil.
2. The *NEC* requires that the disconnecting means for central heating equipment, other than fixed electric space heating equipment, must comply with __________ of __________ __________.
3. The *NEC* in __________ requires that a central heating system be supplied by __________ (a separate) (an individual) branch circuit. (Circle the correct answer.)
4. In a home, the low-voltage wiring between the thermostat and furnace is (Circle the correct answer.)
   a. Class 1.
   b. Class 2.
   c. Class 3.
5. Class 2 circuits are __________-limited and are considered to be __________ from __________ hazard and from __________.
6. The nameplate on a transformer will indicate whether or not it is a Class 2 transformer.
   (True) (False) (Circle the correct answer.)
7. If Class 2 conductors are insulated for 150 volts, and the power conductors are insulated for 600 volts, does the *Code* permit pulling these conductors through the same raceway?
   (Yes) (No) (Circle the correct answer.) Give the *NEC* section number. __________

8. If you use 600-volt rated conductors for Class 2 wiring, does the *Code* permit pulling these conductors through the same raceway as the power conductors? (Yes) (No) (Circle the correct answer.) Give the *NEC* section number. ______________________

9. The *NEC* is very strict about securing anything to electrical raceways. Name the *Code* sections that prohibit this practice. *NEC* ______________________

____

10. Does the *Code* permit attaching the low-voltage thermostat cable to the EMT that brings power to a furnace? (Yes) (No) (Circle the correct answer.) Give the *NEC* section numbers. ______________________

11. Explain what a thermocouple is and how it operates. ______________________

____

____

____

____

CHAPTER **25**

Television, Telephone, and Low-Voltage Signal Systems

OBJECTIVES

After studying this chapter, you should be able to

- install residential telephone and television wiring, antennas, and CATV cables, in conformance to *NEC* requirements.
- describe the basic operation of satellite antennas.
- install typical low-voltage wiring for chimes.

INSTALLING THE WIRING FOR HOME TELEVISION

This chapter discusses the basics of installing outlets, cables, receivers, antennas, amplifiers, and multiset couplers, as well as some of the *NEC*® rules for home television. Television is a highly technical and complex field. To ensure a trouble-free installation of a home system, a competent television technician should do the work. These individuals receive training and certification through the Electronic Technicians Association in much the same way that electricians receive their training through apprenticeship and journeyman training programs.

Some cities and states require that when more than three television outlets are to be installed in a residence, the television technician making the installation be licensed and certified. Just as electricians can experience problems because of poor connections, terminations, and splices, poor reception problems can arise as a result of poor terminations. In most cases, poor crimping is the culprit. Electrical shock hazards are present when components are improperly grounded and bonded.

According to the plans for this residence, television outlets are installed in the following rooms:

Room	Outlets
Front Bedroom	2
Kitchen	1
Laundry	1
Living Room	3
Master Bedroom	2
Recreation Room	3
Study/Bedroom	2
Workshop	1
Total	15

Because TV is a low-voltage system, metallic or nonmetallic standard single-gang device boxes, 4 in. square outlet boxes with a plaster ring, plaster rings only, or special mounting brackets can be installed during the rough-in stage at each likely location of a television set (more on this later). See Figure 25-1. For new construction, shielded coaxial cables are installed concealed in the walls. For remodel work, cables can be concealed in the walls by fishing the cables through the walls and installing mounting brackets that are inserted and snapped into place through a hole cut into the wall.

Shielded 75-ohm RG-6 coaxial cable is most often used to hook up television sets to minimize interference and keep the color signal strong. There are different kinds of shielded coaxial cable. Double-shielded cable that has a 100% foil shield covered with a 40% or greater woven braid is recommended. This coaxial cable has a PVC outer jacket. The older style, flat 300-ohm twin-lead cable, can still be found on existing installations, but the reception might be poor. To improve reception, 300-ohm cables can be replaced with shielded 75-ohm coaxial cables.

For this residence, 15 television outlets are to be installed. A shielded 75-ohm RG-6 coaxial cable will be run from each outlet location back to one central point, as in Figure 25-2.

This central point could be where the incoming Community Antenna Television (CATV) cable comes in, where the cable from an antenna comes in, or where the TV output of a satellite receiver is carried. For rooftop or similar antenna mounting, the cable is run down the outside of the house, then into the basement, garage, or other convenient location where the proper connections are made. CATV companies generally run their incoming

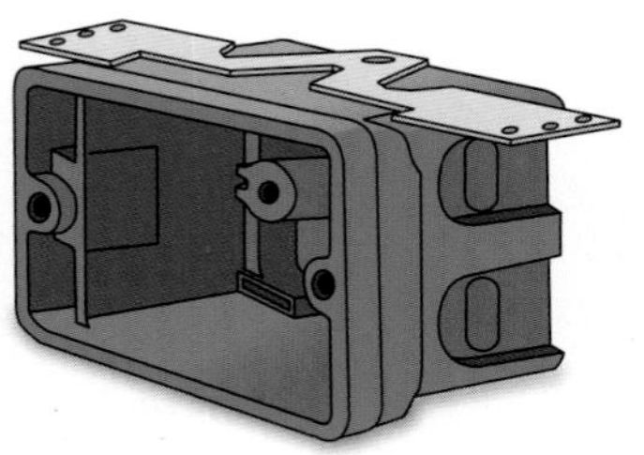
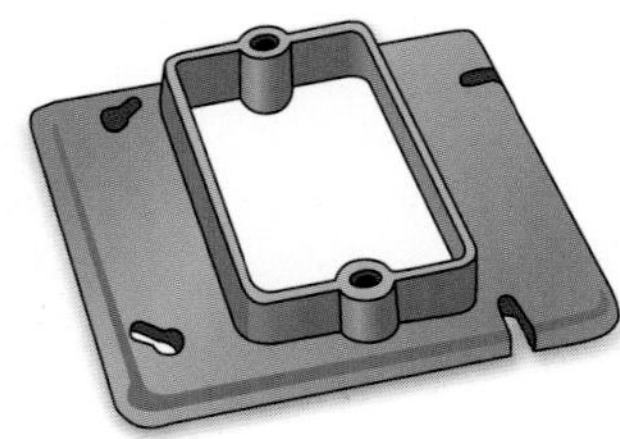
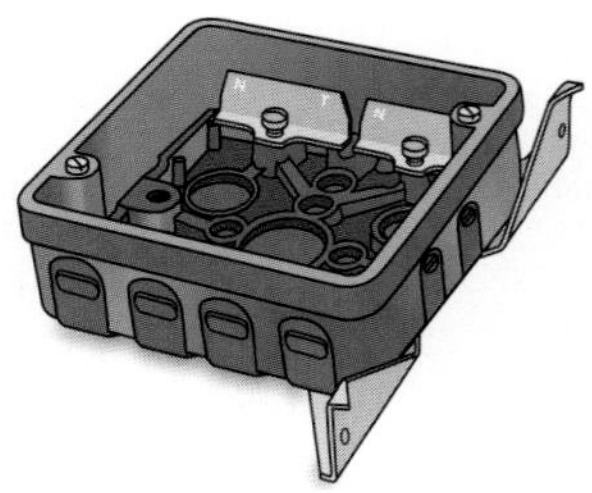

FIGURE 25-1 Nonmetallic boxes and a nonmetallic plaster ring.

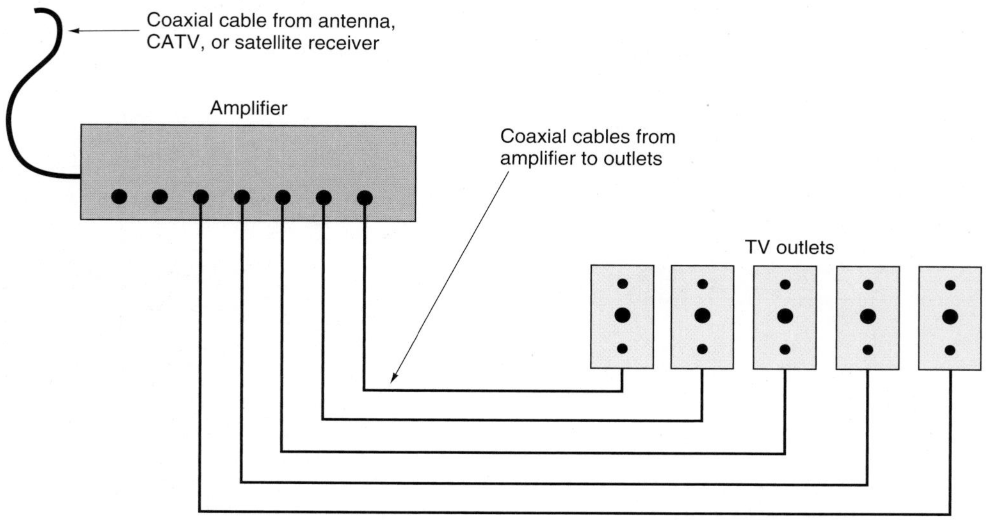

FIGURE 25-2 A television master amplifier distribution system may be needed where many TV outlets are to be installed. This will minimize the signal loss. In simple installations, a multiset coupler can be used, as shown in Figure 25-3.

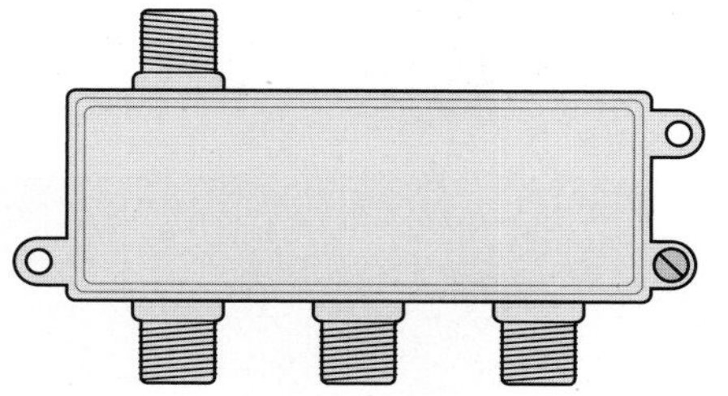

FIGURE 25-3 A 3-way CATV cable splitter: one in, three out. Two-way and 4-way splitters are also commonly available.

coaxial cables underground to some convenient point just inside of the house. Here, the technician hooks up all of the coaxial cables coming from the TV outlets to an amplifier that boosts the signal and improves reception, Figure 25-2, or to a multiset splitter, Figure 25-3. An amplifier may not be necessary if *only* those coaxial cables that will be used are hooked up.

Cable television does not require an antenna on the house because generally the cable company has the proper antenna to receive signals from satellites. The cable company then distributes these signals throughout the community they have contractually agreed to serve. These contracts usually require the cable company to run their coaxial cable to a point just inside the house. Inside the house, they will complete the installation by furnishing cable boxes, controls, and the necessary wiring. In geographical areas where cable television is not available, antennas, as shown in Figure 25-4 and Figure 25-5, are needed.

Faceplates for the TV outlets are available in styles to match most types of electrical faceplates in the home.

Hazards of Mixing Different Voltages

For safety reasons, voice/data/video (VDV) wiring must be separated from the 120-volt wiring. If both 120-volt and low-voltage circuits are run into the same box, then a permanent barrier must be provided in the box to separate the two systems. A better choice is to keep the two systems totally separate using two wall boxes.

Another popular choice is where VDV cables are run to a location right next to a 120-volt receptacle. Mounting brackets that are commercially available can be installed around the electrical device box during the rough-in stage. This lets you trim out a 120-volt receptacle and VDV

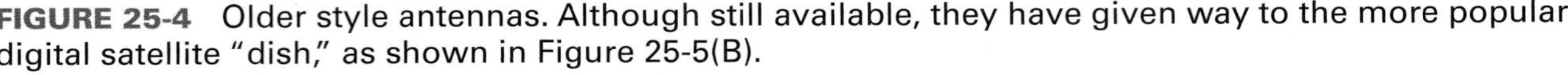

FIGURE 25-4 Older style antennas. Although still available, they have given way to the more popular digital satellite "dish," as shown in Figure 25-5(B).

FIGURE 25-5 (A) A digital satellite system receiver, and a remote control. (B) An 18 in. (450 mm) dish-type satellite antenna.

jacks with one faceplate, as shown in Figure 25-6. These mounting brackets keep the center-to-center measurements of the device mounting holes for the electrical box and the VDV wiring precisely in alignment. With this method, a wall box is not provided for the VDV wiring. The VDV cables merely come out of the drywall next to the electrical box, and the faceplate takes care of trimming out the receptacle and the VDV wiring jacks.

FIGURE 25-6 A special mounting bracket fastened to the wood stud. The bracket fits nicely over the electrical wall box to accommodate a 120-volt receptacle, two coaxial outlets, and one telephone outlet. A single 2-gang faceplate is used for this installation. Many other combinations are possible.

When wiring a new house, it certainly makes sense to run at least one coaxial cable and one Category 5 cable to a single wall box wherever TV, telephone, and/or computers might be used. This could be a single box located close to a 120-volt receptacle outlet, or it could be a mounting bracket as shown in Figure 25-6. Run a separate coaxial cable and a separate Category 5 cable from each location to a common point in the house. You will use a lot of cable, but this will enable you to interconnect the cables as required, such as for a local area network (LAN) to serve more than one computer located in different rooms, or other home automation systems. This is the beginning of structured wiring, as discussed in Chapter 31.

For adding VDV wiring to an existing home, many types of nonmetallic brackets are available that snap into a hole cut in the drywall and lock in place.

More *NEC* rules are discussed later in this chapter.

Chapter 31 discusses structured wiring using Category 5 unshielded twisted pair (UTP) cable and coaxial shielded cable.

Although shielded coaxial cable minimizes interference, it is good practice to keep the coaxial cables on one side of the stud space and keep the light and power cables on the other side of the stud space.

Intersystem Bonding Termination

An intersystem bonding termination is required to be installed at or near the service equipment and at or near the building disconnecting means if remote buildings are supplied with electric power, Figure 25-7. The intersystem bonding termination is intended to provide a convenient location where systems such as telephone, television, and antenna systems can be bonded together. This ensures that all the electrical systems that supply the building or structure are bonded (connected together). This helps reduce dangerous flashover should overvoltages

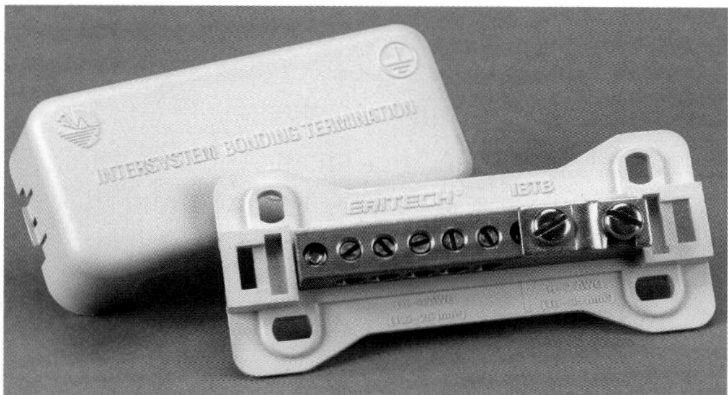

FIGURE 25-7 Intersystem bonding termination equipment.

be imposed on the electrical systems. These overvoltages originate from lightning events or from problems related to the electric utility system.

The intersystem bonding termination equipment is required to

1. be accessible for connection and inspection.
2. consist of a set of terminals with the capacity for connection of not less than three intersystem bonding conductors.
3. not interfere with opening the enclosure for a service, building, or structure disconnecting means, or metering equipment.
4. at the service equipment, be securely mounted and electrically connected to an enclosure for the service equipment, to the meter enclosure, or to an exposed nonflexible metallic service raceway, or be mounted at one of these enclosures and be connected to the enclosure or to the grounding electrode conductor with a minimum 6 AWG copper conductor.
5. at the disconnecting means for a building or structure, be securely mounted and electrically connected to the metallic enclosure for the building or structure disconnecting means, or be mounted at the disconnecting means and be connected to the metallic enclosure or to the grounding electrode conductor with a minimum 6 AWG copper conductor.
6. be listed as grounding and bonding equipment.

Code Rules for Cable Television (CATV) (*Article 820*)

The letters CATV stand for "Community Antenna Television," often simply referred to as "Cable TV."

CATV systems are installed both overhead and underground in a community. Then, to supply an individual customer, the CATV company runs coaxial cables through the wall of a residence at some convenient point. Up to this point of entry *(Article 820, Part II)* and inside the building *(Article 820, Part V)*, the cable company must conform to the requirements of *Article 820,* plus local codes if applicable.

Coaxial cable generally used for one- and two-family dwellings is Type CATV and CATVX. The cable shall be listed. See *820.154*, *820.179(C)*, and *820.179(D)*. Other types of acceptable cables are shown in *Table 820.154(b)*.

Here are some of the key rules to follow when making a coaxial cable installation:

1. The outer conductive shield of the coaxial cables must be grounded as close to the point of entry as possible, *820.93*.
2. Coaxial cables shall not be run in the same conduits or box with electric light and power conductors, *820.133(A)(1)(c)*.
3. Do not support coaxial cables from raceways that contain electrical light and power conductors, *820.133(B)*.
4. Keep the coaxial cable at least 2 in. (50 mm) from light and power conductors unless the conductors are in a raceway, nonmetallic-sheathed cable, armored cable, or UF cable, *820.133(A)(2)*. This clearance requirement really pertains to old knob-and-tube wiring. The 2 in. (50 mm) clearance is not required if the coaxial cable is in a raceway.
5. Where underground coaxial cables are run, they must be separated by at least 12 in. (300 mm) from underground light and power conductors, unless the underground conductors are in a raceway, Type UF cable, or Type USE cable, *820.47(B)*. The 12 in. (300 mm) clearance is not required if the coaxial cable has a metal cable armor.
6. The bonding and grounding conductor *(Article 820, Part III* and *Part IV)*:
   a. outer conductive shield shall be grounded as close as possible to the coaxial cable entrance or attachment to the building, *820.93(A)*.
   b. shall not be smaller than a 14 AWG copper or other corrosion-resistant conductive material. It need not be larger than 6 AWG copper. The bonding or grounding electrode conductor is permitted to be insulated, covered or bare. It shall have a current-carrying capacity not less than that of the coaxial cable's outer metal shield. See *NEC 810.100(A)(1), (A)(2)* and *(A)(3)*.

c. length shall be as short as possible, but not longer than 20 ft (6.0 m). If impossible to keep the grounding conductor to this maximum permitted length, then the *Exception* to *820.100(A)(4)* permits installing a separate grounding electrode. These two electrodes shall be bonded together with a bonding conductor not smaller than 6 AWG copper.

d. may be solid or stranded, *820.100(A)(2)*.

e. shall be run in a line as straight as practical, *820.100(A)(5)*.

f. is required to be protected against physical damage. If run in a metal raceway, the metal raceway shall be bonded to the bonding or grounding electrode conductor at both ends, *820.100(A)(6)*.

g. shall be connected to the nearest accessible location on one of the following, *820.100(B)*:

- bonded to the intersystem bonding termination if one exists, *820.100(B)(1)*.
- If the building or structure has no intersystem bonding means such as a terminal bar, the bonding conductor or grounding electrode conductor is to be connected to one of the following:
    1. The building grounding electrode system
    2. The grounded interior metal water piping pipe, within 5 ft (1.5 m) of where the pipe enters the building; refer to *250.52*
    3. The power service accessible means external to the enclosure
    4. The metallic power service raceway
    5. The service equipment enclosure
    6. The grounding electrode conductor or its metal enclosure
    7. The grounding electrode conductor or grounding electrode of a disconnecting means that is grounded to an electrode according to *250.32*. This pertains to a second building on the same property served by a feeder or branch circuit from the main electrical service in the first building.

h. If *none* of the options in part (g) is available, then ground to any of the electrodes, per *250.52*, such as metal underground water pipes, the metal frame of the building, a concrete encased electrode, or a ground ring. Watch out for the maximum length permitted for the bonding or grounding electrode conductor, as mentioned above.

When installing CATV wiring and equipment, do it in a neat and workmanlike manner, not subject to physical damage when run on building surfaces. Secure cables with listed hardware, including straps, cable ties, and so on, *820.24*. The requirements found in *300.4(D)* (*Cables and Raceways Parallel to Framing Members and Furring Strips*) and *300.11* (*Securing and Supporting*) also apply to CATV cables.

SAFETY ALERT

DO NOT simply drive a separate ground rod to ground the metal shielding tape on the coax cable. Difference of potential between the coax cable shield and the electrical system ground during lightning strikes could result in a shock hazard as well as damage to the electronic equipment. If for any reason one grounding electrode is installed for the CATV shield grounding and another grounding electrode for the electrical system, bonding these two electrodes together with no smaller than a 6 AWG copper conductor will minimize the possibility that a difference of potential voltage might exist between the two electrodes.

SATELLITE ANTENNAS

A satellite antenna is often referred to as a "dish." See Figure 25-5. A satellite dish has a parabolic shape that concentrates and reflects the signal beamed down from one of the many stationary satellites orbiting 22,245 miles (35,800 kilometers) above the equator. Stationary means that the satellite is traveling at the same speed that the Earth rotates. The orbit in which satellites travel is called the "Clarke Belt," as shown in Figure 25-8.

The latest satellite technology is the digital satellite system (DSS). A digital system allows the use of small antennas, approximately 18 in. (450 mm) in diameter, which can easily be mounted on a roof, a chimney, the side of the house, a pipe, or a pedestal, using the proper mounting hardware.

Although rarely used today for residential TV reception, large dish satellite antennas are still permitted to be used. Because of their huge size and weight, proper installation is important.

The basic operation of a satellite system is for a transmitter on Earth to beam an "uplink" signal to a satellite in space. Electronic devices on the satellite reamplify and convert this signal to a "downlink" signal, then retransmit the downlink signal back to Earth, as shown in Figure 25-9.

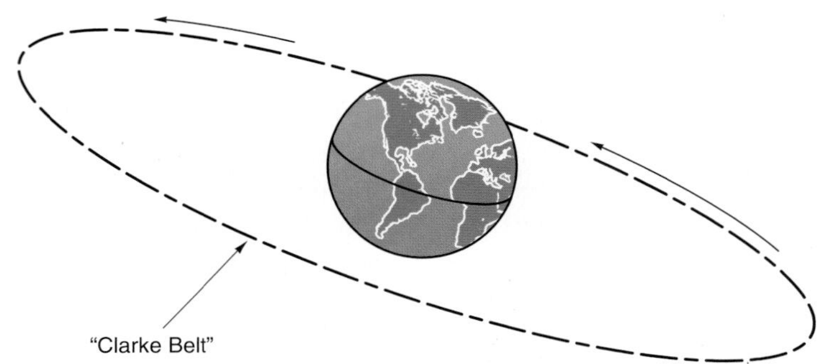

FIGURE 25-8 All television and communications satellites travel around the Earth in the same orbit called the "Clarke Belt." They appear to be stationary in space because they are rotating at the same speed at which the Earth rotates. This is called "geosynchronous orbit (geostationary orbit)." The satellite receives the uplink signal from Earth, amplifies the signal, and transmits it back to Earth. The downlink signal is picked up by the satellite antenna.

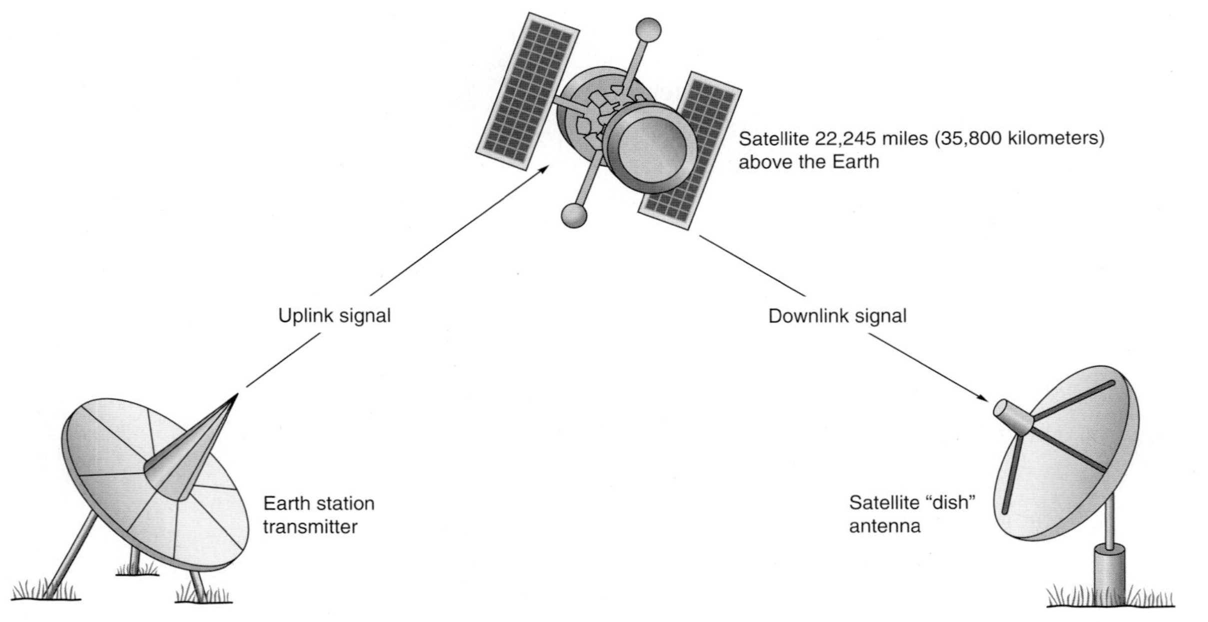

FIGURE 25-9 The Earth station transmitter beams the signal up to the satellite, where the signal is amplified, converted, and beamed back to Earth.

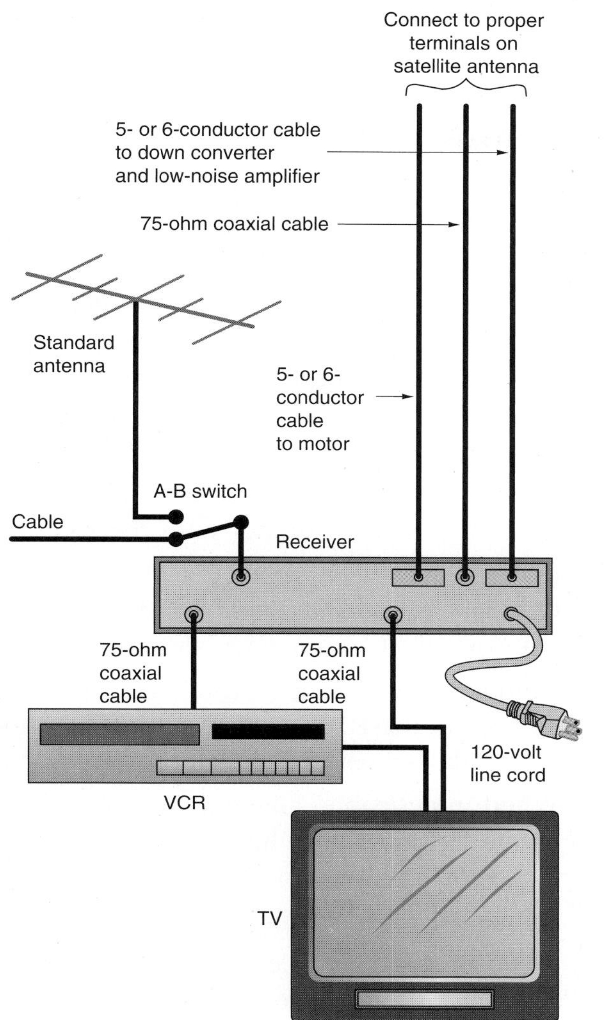

FIGURE 25-10 Typical satellite antenna/cable/standard antenna wiring. Connection to a standard antenna cable, CATV cable, or to a satellite cable enables the homeowner to watch one channel while recording another channel. The installation manual for the receiver usually has a number of different hookup diagrams. In this diagram, the A–B switch allows a choice of connecting to a standard antenna or to the CATV cable. This hookup can be desirable should the CATV or satellite reception fail.

Regardless of the type of antenna used, the line of sight from the antenna to the satellite must not be obstructed by trees, buildings, utility poles, or other structures. Instructions furnished with an antenna provide the necessary data for direction and up angle based on ZIP codes.

Before installing an antenna, check with your local inspection department. Local codes might have restrictions and requirements in addition to those in the *NEC*. Some residential communities or subdivisions may have covenents that restrict or regulate the types or locations of outdoor TV antennas. It is wise to become aware of all regulations that apply before beginning an installation.

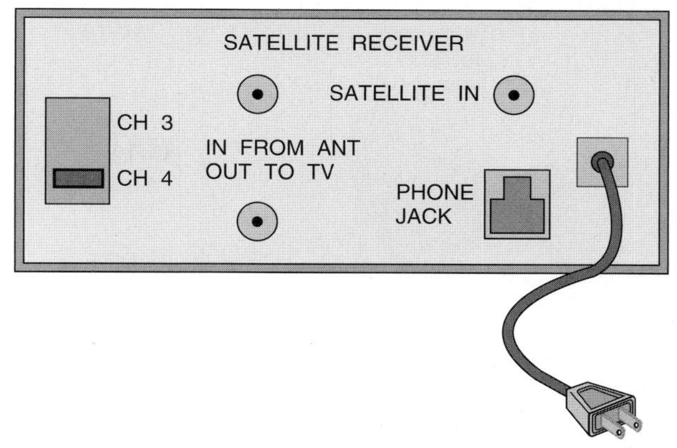

FIGURE 25-11 The terminals on the back of a digital satellite receiver. The connections are similar to those in Figure 25-10. Note the difference in that the digital satellite receiver has a modular telephone plug, a feature that uses a toll-free number to update the access card inside the receiver to ensure continuous program service. The telephone can also handle program billing.

Figures 25-10 and 25-11 show typical connections between a television set, receiver, VCR, and antenna. The manufacturer's installation instructions are always to be followed.

Figure 25-12 illustrates one method of installing a large antenna post in the ground in accordance to a manufacturer's instruction.

CODE RULES FOR THE INSTALLATION OF ANTENNAS AND LEAD-IN WIRES (*ARTICLE 810*)

Home television and AM and FM radios generally come complete with built-in antennas. For those locations in outlying areas on the fringe or out of reach of strong signals, it is quite common to install a separate antenna system or a satellite antenna system.

You will want to read *NEC 250.94* to learn more about the requirements for intersystem bonding. We covered these requirements earlier in this chapter.

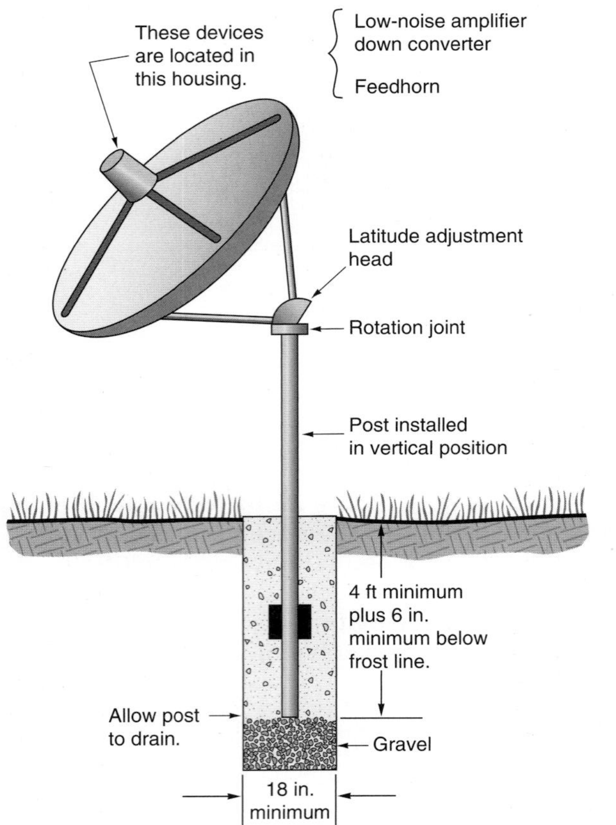

FIGURE 25-12 Satellite antenna solidly installed in the ground according to the manufacturer's instructions. The popularity of these large antennas has given way to the small digital satellite system antenna, as shown in Figure 25-5.

Indoor or outdoor antennas may be used with televisions. The front of an outdoor antenna is aimed at the television transmitting station. When there is more than one transmitting station and they are located in different directions, a rotor is installed. A rotor turns the antenna on its mast so that it can face in the direction of each transmitter. The rotor is controlled from inside the building. The rotor controller's cord is plugged into a regular 120-volt receptacle to obtain power. A 3-wire or 4-conductor cable is usually installed between the rotor motor and the control unit. The wiring for a rotor may be installed during the roughing-in stage of construction, running the rotor's cable into a device wall box, allowing 5 to 6 ft (1.5 to 1.8 m) of extra cable, then installing a regular single-gang switchplate when finishing.

Article 810 covers radio and television equipment. Although instructions are supplied with antennas, the following key points of the *Code* regarding the installation of antennas and lead-in wires should be followed:

1. Antennas and lead-in conductors shall be securely supported, *810.12*.
2. Antennas and lead-in conductors shall not be attached to the electric service mast, *810.12*.
3. Antennas and lead-in conductors shall not be attached to any poles that carry light and power wires over 250 volts between conductors, *810.12*.
4. Lead-in conductors shall be securely attached to the antenna, *810.12*.
5. Antennas and lead-in conductors shall be kept away from all light and power conductors to avoid accidental contact with the light and power conductors, *810.13*.
6. Outdoor antennas and lead-in conductors shall not cross over light and power conductors, *810.13*.
7. Outdoor antennas and lead-in conductors shall be kept at least 24 in. (600 mm) away from open light and power conductors, *810.13*.
8. Where practicable, antenna conductors shall not be run under open light and power conductors, *810.13*.
9. On the outside of a building:
   a. Position and fasten lead-in conductors so they cannot swing closer than 24 in. (600 mm) to light and power conductors having *not* over 250 volts between conductors; 10 ft (3.0 m) if *over* 250 volts between conductors, *810.18(A)*.
   b. Keep lead-in conductors at least 6 ft (1.8 m) away from a lightning rod system, *810.18(A)*, or bonded together according to *250.60*.
   c. Underground lead-in radio and television conductors and cables shall be separated by at least 12 in. (300 mm) from underground light and power conductors.

   Note: The clearances in *a*, *b*, and *c* are not required if the light and power conductors or the lead-in conductors are in a metal raceway or metal cable armor.
10. On the inside of a building:
    a. Keep the antenna and lead-in conductors at least 2 in. (50 mm) from other open wiring (as in old houses) unless the other wiring is in a metal raceway or cable, *810.18(B)*.

b. Keep lead-in conductors out of electric boxes unless there is an effective, permanently installed barrier to separate the light and power wires from the lead-in wire, *810.18(C)*.

11. Grounding:

a. All metal masts and metal structures that support antennas shall be grounded, *810.21*. See Figure 25-12.

b. The grounding conductor must be copper, aluminum, copper-clad steel, bronze, or a similar corrosion-resistant material, *810.21(A)*.

c. The bonding or grounding electrode conductor need not be insulated. It must be securely fastened in place, may be attached directly to a surface without the need for insulating supports, shall be protected from physical damage or be large enough to compensate for lack of protection, and shall be run in as straight a line as is practicable, *810.21(B)*, *(C)*, *(D)*, and *(E)*.

d. In buildings with an intersystem bonding termination, the bonding conductor shall be connected to the intersystem bonding termination. See Figure 25-13.

e. If the building or structure does not have an intersystem bonding means, the bonding conductor or grounding electrode conductor is required to be connected to the nearest accessible location on one of the following:

- the building or structure grounding electrode system. Refer to *250.50* for more details.
- the grounded interior metal water pipe, within 5 ft (1.52 m) of where the water pipe enters the building. Refer to *250.53* and *250.68(C)* for more details.
- the metallic power service raceway.
- the service equipment enclosure.
- the grounding electrode conductor or the grounding electrode conductor metal enclosure of the power service.

f. If neither *d* nor *e* is available, then connect a grounding electrode conductor to any one of the grounding electrodes, per *250.52*, such as metal underground water pipe, metal frame of building, concrete-encased electrode, or ground ring, *810.21(F)(3)*.

g. The grounding conductor may be run inside or outside of the building, *810.21(G)*.

h. The grounding conductor shall not be smaller than 10 AWG copper or 8 AWG aluminum, *810.21(H)*.

i. Protect the grounding conductor from physical damage. If run in a metal raceway, the metal raceway shall be bonded to the grounding conductor at both ends, *800.21(D)*.

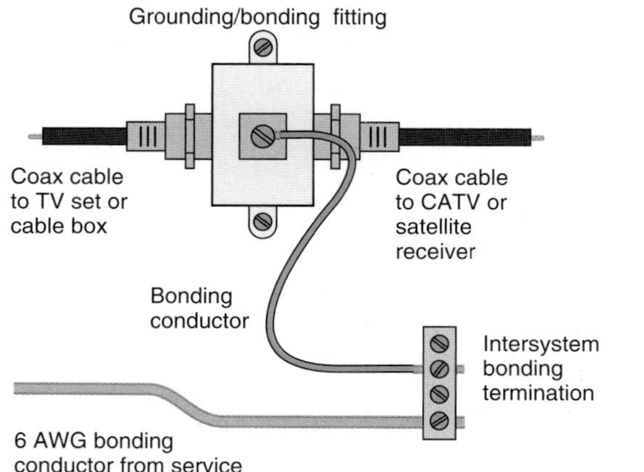

FIGURE 25-13 Typical connection for the required grounding of the metal shield on coaxial cable to the intersystem bonding termination.

SAFETY ALERT

DO NOT simply drive a separate ground rod to ground the metal mast, structure, or antenna. Difference of potential between the metal mast, structure, or antenna and the electrical system ground during a lightning strike could result in a shock hazard as well as damage to the electronic equipment. If any reason permitted by the *Code* results in one grounding electrode for the antenna and another grounding electrode for the electrical system, bond the two electrodes together with a bonding jumper not smaller than 6 AWG copper or equivalent, *810.21(J)*.

The objective of grounding and bonding to the same grounding electrode as the main service ground is to reduce the possibility of having a difference of voltage potential between the two systems.

TELEPHONE WIRING (*ARTICLE 800*)

Since the deregulation of telephone companies, residential "do-it-yourself" telephone wiring has become quite common. The following is an overview of residential telephone wiring.

The *NEC* in *800.156* requires that For new construction, a minimum of one communications outlet shall be installed within the dwelling and cabled to the service provider demarcation point.* Prior to the *2008 NEC*, telephone outlets were not required, but if installed, *Article 800* spelled out the installation requirements.

The telephone company will install the service line to a residence and terminate at a protector device, as in Figure 25-14. The protector protects the system from hazardous voltages. The protector may be mounted either outside or inside the home. Different telephone companies have different rules. Always check with the phone company before starting your installation.

The point where the telephone company ends and the homeowner's interior wiring begins is called the *demarcation point*, as shown in Figure 25-14. The preferred demarcation point is outside of the home, generally near the electric meter, where proper grounding and bonding can be done. A *network interface* might be of the type depicted in Figure 25-14. This is a combination unit that provides the compartment for the utility with controlled access and an owner's compartment. The utility makes a connection from the grounding/bonding terminal in the network interface unit to the intersystem bonding termination that is installed near the service equipment. The utility provides the dial tone to the customer's compartment. The customer can make connections of individual or multipair cables to the terminals provided.

*Reprinted with permission from NFPA 70-2014.

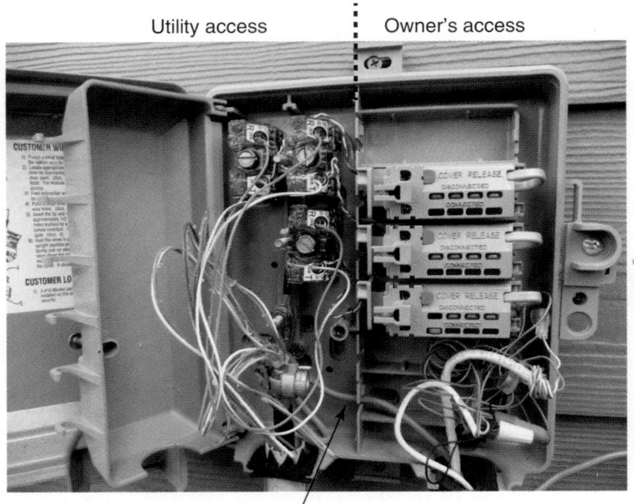

FIGURE 25-14 Typical residential telephone network interface unit (NIU) installation. The telephone company installs the NIU and connects the underground cable to the protector in the NIU. The utility installs the bonding conductor to the intersystem bonding terminal bar that is installed near the service equipment. The customer installs the cable to the telephone outlet(s) and makes the connections in the customer section of the NIU.

In the past, interior telephone wiring in homes was installed by the telephone company. Today, the responsibility for roughing in the interior wiring is usually left up to the electrician. Telephone wiring is *not* to be run in the same raceway with electrical wiring. In addition, it cannot be installed in the same outlet and junction boxes as electrical wiring. Whatever the case, the rules found in *Article 800* apply. Refer to Figures 25-14, 25-15, and 25-16.

For a typical residence, the electrician will rough in wall boxes wherever a telephone outlet is wanted. The boxes can be of the single-gang types illustrated in Figures 25-1, 2-12, 2-18, 2-20, and 2-32. Concealed telephone wiring in new homes is generally accomplished using multipair cables that are run from each telephone outlet to the customer compartment of the network interface unit. The cable can also take the form of a "loop system," as in

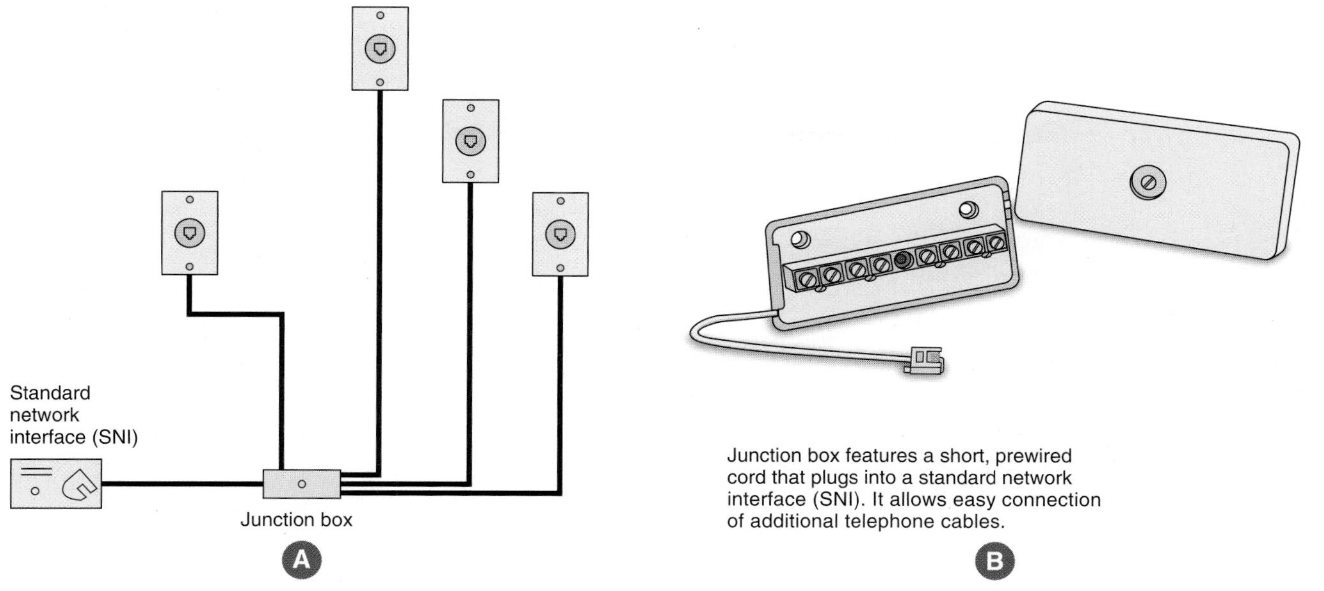

FIGURE 25-15 (A) Individual telephone cables run to each telephone outlet from the common connection point. This is sometimes referred to as a star-wired system. (B) A junction box for easy connection of multiple telephone cables.

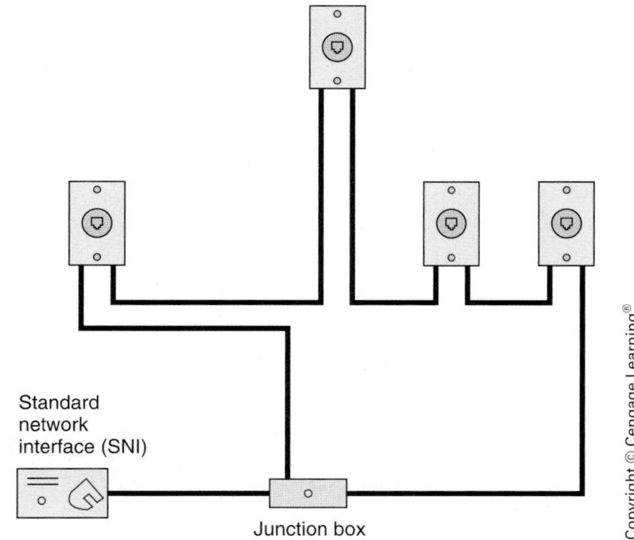

FIGURE 25-16 A complete "loop system" (daisy-chain) of the telephone cable. If something happens to one section of the cable, the circuit can be fed from the other direction.

Figure 25-16. A 3-pair, 6-conductor twisted cable is shown in Figure 25-17. There are some localities that require that these cables be run in a metal raceway, such as EMT.

The outer jacket is usually of a thermoplastic material. This outer jacket most often is a neutral color, such as white, light-gray, or beige, that blends in with decorator colors in the home. This is particularly important in existing homes where the telephone cable may have to be exposed. Cables, mounting boxes, junction boxes, terminal blocks, jacks, adaptors, faceplates, cords, hardware, plugs, and so on, are all available through electrical distributors, builders' supply outlets, telephone stores, electronic stores, hardware stores, and similar wholesale and retail outlets. The selection of types of the preceding components is endless.

Telephone Conductors

The color coding of telephone cables is shown in Figure 25-17. Figure 25-18 shows some of the many types of telephone cords available.

Cables are available with various numbers of conductors. Here are a few:

2 pair . . . . 4 conductors
3 pair . . . . 6 conductors
4 pair . . . . 8 conductors
6 pair . . . 12 conductors
12 pair . . . 24 conductors

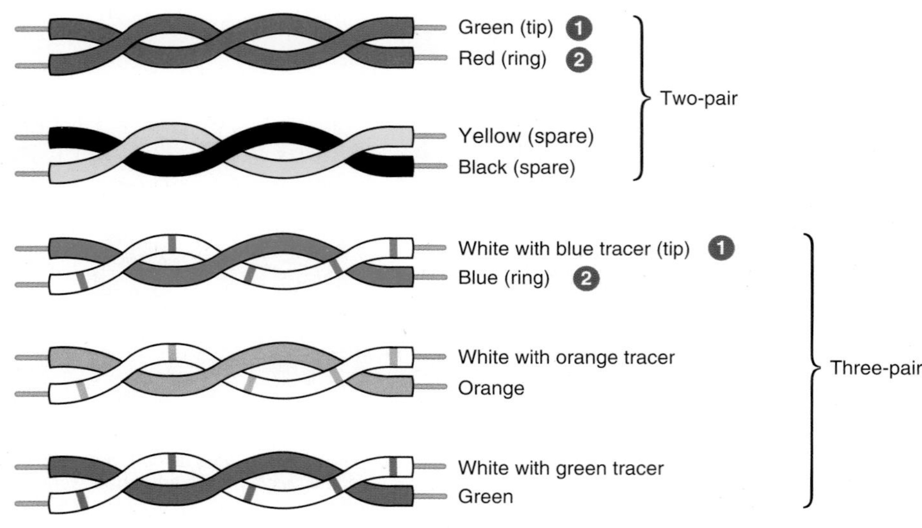

FIGURE 25-17 Color coding for a 2-pair and a 3-pair, 6-conductor twisted telephone cable. The color coding for 2-pair telephone cable stands alone. In multipair cables, the additional pairs can be used for more telephones, fax machines, security reporting, speakerphones, dialers, background music, and so on.

Telephone circuits require a separate pair of conductors from the telephone all the way back to the phone company's central switching center. To keep to a minimum interference that could come from other electrical equipment, such as electric motors and fluorescent fixtures, each pair of telephone wires (two wires) is twisted.

For three or more pair cables, the color coding becomes white/blue, white/orange, white/green, white/brown, white/slate, red/blue, red/orange, red/green, red/brown, red/slate, black/blue, black/orange, black/green, black/brown, black/slate, yellow/blue, yellow/orange, yellow/green, yellow/brown, yellow/slate, violet/blue, violet/orange, violet/green, violet/brown, violet/slate.

In multipair cables, the additional pairs can be used for more telephones, fax machines, security reporting, speaker phones, dialers, background music, and so forth.

Installing Communication Wiring

When installing communication wiring and equipment, do it in a neat and workmanlike manner, not subject to physical damage when run on building surfaces. Secure cables with listed hardware, including straps, cable ties, and so on, *800.24*. The requirements found in *300.4(D)* (*Cables and Raceways Parallel to Framing Members and Furring Strips*) and *300.11* (*Securing and Supporting*) also apply to communication cables.

Cross-Talk Problems

A common telephone interference problem in homes is "cross-talk." This is apparent if you hear someone faintly talking in the background when you are using the telephone. Another problem is losing an online connection to your computer. The signals traveling in one pair of conductors are picked up by the adjacent pair of conductors in the cord or cable. Quite often, these problems can be traced to older style flat (untwisted) 2-line telephone lines. Older style cables could carry audio signals quite well, but they are unsatisfactory for data transmission. These problems are virtually eliminated by using cables and cords that have conductors twisted at proper intervals, such as Category 5 cables. In Figure 25-17,

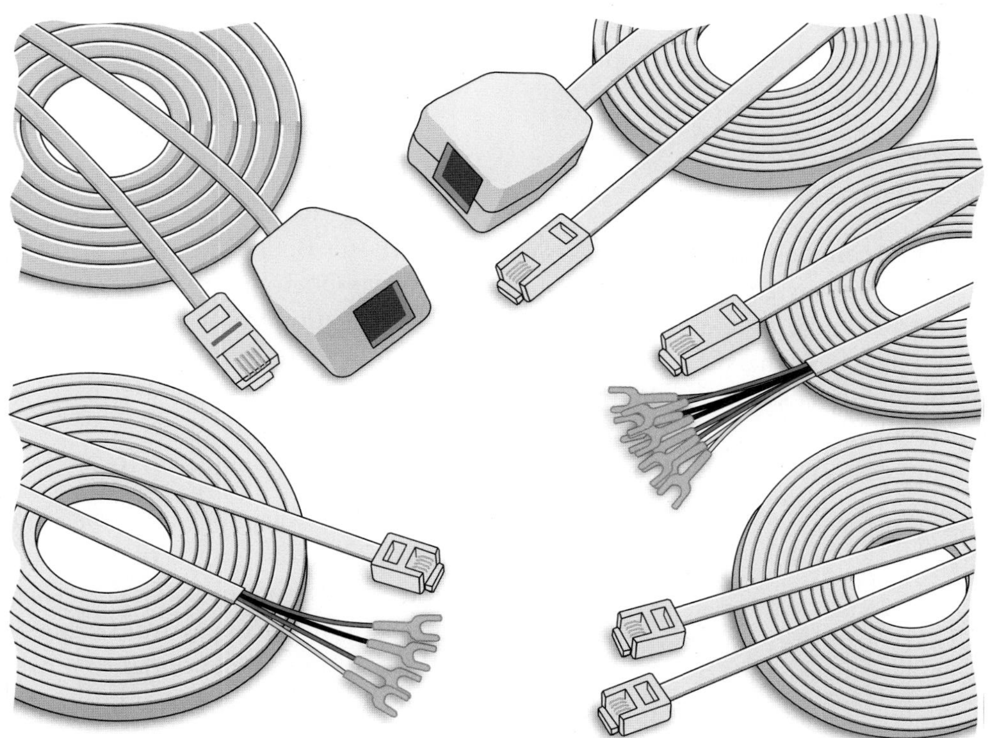

FIGURE 25-18 Some of the many types of telephone cable available are illustrated. Note that for ease of installation, the modular plugs and terminals have been attached by the manufacturer of the cables. These cords are available in round and flat configurations, depending upon the number of conductors in the cable.

each pair of conductors is twisted, and then all of the pairs are twisted again. Generally, you should avoid flat 2-line cords.

The recommended circuit lengths are as follows:

24 AWG gauge—not over 200 ft (60 m)

22 AWG gauge—not over 250 ft (75 m)

Ringer Equivalence Number (REN)

Telephones, fax machines, answering machines, ringers, flashers, and so on, that have a sounder (ringer) are assigned an REN number. Generally, the maximum number of RENs on any single telephone line is 5. This maximum number of RENs ensures that all telephones will ring. When too many telephones are connected, the ring signal can become unreliable, resulting in devices failing to respond.

Older style electromechanical ringers were considered to be 1 REN. Electronic ringers have a minimal REN value. Mixing electromechanical and electronic ringers could present a problem, as the electromechanical ringers tend to hog the current. It all depends on your telephone line. Try it and see what happens.

Pay close attention to the total RENs connected in your home. REN values are found on the label of the device.

EXAMPLE

Four electronic telephones:

$4 \times 0.3 = 1.2$ RENs

One fax machine:	1.5 RENs
One desk top telephone/ Answering machine:	0.7 RENs
One desk top telephone:	0.4 RENs
Total	3.8 RENs

Call your telephone company if you think you have a problem because of too many telephones connected to one telephone line.

Plug-in or Permanent Connection?

Most residential telephones plug into a jack of the type illustrated in Figures 25-19 and 25-20. It might be a good idea to have at least one permanently connected telephone, such as a wall phone in the kitchen. If all of the phones were of the plug-in type, it is possible (but highly unlikely) that all of them could be unplugged. As a result, there would be no audible signal.

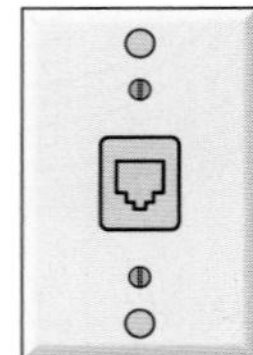

FIGURE 25-19 Telephone modular jack.

FIGURE 25-20 Three styles of wall plates for modular telephone jacks: rectangular stainless steel, weatherproof for outdoor use, and circular.

The electrical plans show that nine telephone outlets are provided as follows:

Front Bedroom	1
Kitchen	1
Laundry	1
Living Room	1
Master Bedroom	1
Rear Outdoor Patio	1
Recreation Room	1
Study/Bedroom	1
Workshop	1

Wireless Telephones

Wireless telephones have become all the rage! A master telephone unit is plugged into the telephone outlet somewhere in the house. The unit usually requires a power-supply transformer that plugs into a 120-volt receptacle outlet to power the telephone master unit. The master unit acts as a transmitter/receiver for the other remote telephones located at desirable locations. The master unit often provides services for the entire system, including answering machine, telephone number list, and intercom. Each of the remote telephones sits on a charging cradle and communicates to the wired telephone network by high-frequency radio signal.

Though very flexible, these telephones rely on power from the electric utility for operation. You should always have a telephone set that will function without reliance on 120-volt power, for use during power outages, storms, and emergencies.

Installation of Telephone Cables (*Article 800, Part V*)

Telephone cables

1. are generally Type CM or Type CMX (for one- and two-family dwellings only), listed for telephone installations as being resistant to the spread of fire. See *800.154*, *800.179(D)*, and *800.179(E)*. Also refer to *Table 800.154(b)* and *Table 800.179* for other acceptable types.
2. shall be separated by at least 2 in. (50 mm) from light and power conductors unless the light and power conductors are in a raceway, or in nonmetallic-sheathed cable, Type AC cable, or Type UF cable, *800.133(A)(2)*.

3. shall not be placed in any conduit or boxes with electric light and power conductors unless the conductors are separated by a partition, *800.133(A)(1)(c)*.
4. do not support telephone cables from raceways that contain light and power conductors, *800.133(B)*.
5. should not share the same bored holes as electrical wiring, plumbing, gas pipes, and so on.
6. may be terminated in either metallic or nonmetallic boxes or plaster rings.
7. should be secured using rounded or depth-stop plastic staples. Do not use metal staples.
8. should not be installed using nail guns. Do not crush the cable.
9. should be kept away from hot water pipes, hot air ducts, and other heat sources that might harm the insulation.
10. should not be run in the same stud space as electrical branch-circuit wiring and should be kept at least 12 in. (300 mm) from power wiring where the cables are run parallel to the power wiring. These cautions were written before twisted pairs and category-rated cables entered the scene. With the advent of *properly* installed twisted and shielded cables, these "old wives' tale" recommendations from the past are probably not an issue.

Follow the specifications and instructions for the cable you are installing.

Raceway

In some communities, electricians prefer to install regular device boxes at a telephone outlet location, then "stub" a trade size ½ EMT to the basement or attic from this box. Then, at a later date, the telephone cable can be fished through the raceway.

Grounding (*Article 800, Part IV*)

The telephone company will provide the proper grounding of their incoming cable sheath and primary protector, generally using an insulated conductor not smaller than 14 AWG copper or other corrosion-resistant material, and not longer than 20 ft (6.0 m).

For one- and two-family dwellings, where it might be impractical to keep the grounding conductor 20 ft (6.0 m) or less, an additional communications grounding electrode must be installed. This additional ground rod must be bonded to the power grounding electrode system with a bonding conductor not smaller than a 6 AWG copper. In just about all instances, all grounding and bonding of telephone equipment is done by the telephone company personnel.

It is very important that an intersystem bonding termination device, as shown in Figure 25-7, be installed for terminating the bonding conductors from the communication system protector. The installation requirements for the intersystem bonding termination device as required in *NEC 250.94* was covered earlier in this chapter.

Safety

Open-circuit voltage between conductors of an idle pair of telephone conductors is approximately 48 volts dc. The superimposed ringing voltage can reach 90 volts ac. Therefore, always work carefully with insulated tools, and stay clear of bare terminals and grounded surfaces. Disconnect the interior telephone wiring if work must be done on the circuit, or take the phone off the hook, in which case the dc voltage level will drop to approximately 7 to 9 volts dc, and there should be no ac ringing voltage delivered.

Wiring for Computers and Internet Access

Wiring for telephones, high-speed Internet access, computers, television, printers, modems, security systems, intercoms, and similar home automation equipment could be thought of as one big project. Planning ahead is critical. Deciding on where this equipment is likely to be located will help you to design an entire voice/data/video system.

You will want to install Category 5e cables between each telephone outlet and a central distribution point so as to have the flexibility to connect the voice/data/video equipment in any configuration.

Recently, there has been a trend to install Category 7 cables for higher performance by fully shielding each twisted pairs in the cable.

This virtually eliminates crosstalk and improves resistance to noise. Category 7 cable is better suited for long distances than Category 5, Category 5e, or Category 6.

Something generally not taken into consideration is that the estimated life cycle of Category 5, Category 5e, or Category 6 cables is 10 years, whereas the estimated life cycle of Category 7 cable is 15 years.

For the complex installations, the selection and installation of these types of cables should be left up to a skilled professional trained in that technology.

In addition to the previously mentioned equipment, there very well might be a scanner, an answering machine, desk lamps, floor lamps, a clock, an adding machine, a calculator, a television, a radio, a printer, CD/DVD/DVR/VCR/VHS players and burners, ZIP drives, and others. They draw little current, but all need to be plugged into a 120-volt receptacle! Managing and plugging in that many power cords is a major problem. Give consideration to installing two or three duplex or quadplex 120-volt receptacles at these locations. Multioutlet plug-in strips with surge protection will most likely be needed.

Chapter 31 contains much information about home automation and structured wiring.

Bringing Technology into Your Home

In most areas, the telephone or CATV company provides the entire "package" for homes. This might include digital local and long-distance telephone service, Internet access, cable modems for personal computers, cable television, digital subscriber lines (DSL) (through existing copper telephone lines), and whatever other great new things may come next.

Wi-Fi (also spelled *Wifi* or *WiFi*) is a popular technology that allows an electronic device to exchange data wirelessly (using radio waves) over a computer network, including high-speed Internet connections. The Wi-Fi Alliance defines Wi-Fi as any "wireless local area network (WLAN) products that are based on the Institute of Electrical and Electronics Engineers' (IEEE) 802.11 standards." However, because most modern WLANs are based on these standards, the term "Wi-Fi" is used in general English as a synonym for WLAN. Only Wi-Fi products that complete Wi-Fi Alliance interoperability certification testing successfully may use the "WI-FI CERTIFIED" trademark.

A device that can use Wi-Fi (such as a personal computer, video-game console, smartphone, digital camera, tablet, or digital audio player) can connect to a network resource such as the Internet via a wireless network access point. Such an access point (or hotspot) has a range of about 65 feet (20 meters) indoors and a greater range outdoors. Hotspot coverage can comprise an area as small as a single room with walls that block radio waves or as large as many square miles; this is achieved by using multiple overlapping access points.

Wi-Fi can be less secure than wired connections (such as Ethernet) because an intruder does not need a physical connection. Web pages that use SSL are secure, but intruders can easily detect unencrypted Internet access. Because of this, Wi-Fi has adopted various encryption technologies. The early encryption WEP, proved easy to break. Higher quality protocols (WPA, WPA2) were added later. An optional feature added in 2007, called Wi-Fi Protected Setup (WPS), had a serious flaw that allowed an attacker to recover the router's password. The Wi-Fi Alliance has since updated its test plan and certification program to ensure all newly certified devices resist attacks.

SIGNAL SYSTEM (CHIMES)

A signaling circuit is described in the *NEC* as Any electric circuit that energizes signaling equipment.* Signaling equipment includes such devices as chimes, doorbells, buzzers, code-calling systems, and signal lights.

Door Chimes (Symbol CH)

Present-day dwellings use chimes rather than bells or buzzers to announce that someone is at a door.

*Reprinted with permission from NFPA 70-2014.

Courtesy of Broan-NuTone, LLC.

FIGURE 25-21 Push buttons for door chimes.

A musical tone is sounded rather than a harsh ringing or buzzing sound. Chimes are available in single-note, 2-note, 8-note (4-tube), and repeater tone styles. In a repeater tone chime, both notes sound as long as the push button is depressed. In an 8-note chime, contacts on a motor-driven cam are arranged in sequence to sound the notes of a simple melody when the chime button is pushed. This type of chime is usually installed in dwellings having three entrances. The chime can be connected so that the 8-note melody sounds for the front door, two notes sound for the side door, and a single note sounds for the rear door. Chimes are also available with clocks and lights.

Electronic chimes may relay their chime tones through the various speakers of an intercom system. When any chime is installed, the manufacturer's instructions must be followed.

The plans show that two chimes are installed in the residence. Two-note chimes are used. Each chime has two solenoids and two iron plungers. When one solenoid is energized, the iron plunger is drawn into the opening of the solenoid. A plastic peg in the end of the plunger strikes one chime tone bar. When the solenoid is de-energized, spring action returns the plunger, where it comes to rest against a soft felt pad so that it does not strike the other chime tone bar. Thus, a single chime tone sounds. As the second solenoid is energized, one chime tone bar is struck. When the second solenoid is de-energized, the plunger returns and strikes the second tone bar. A 2-tone signal is produced. The plunger then comes to rest between the two tone bars. Generally, two notes indicate front door signaling, and one note indicates rear or side door signaling.

Figure 25-21 shows various push-button styles used for chimes. Many other styles are available. Figure 25-22 shows several typical residential-type wall-mounted chimes. The symbols used to indicate push buttons and audible signals on the plans are shown in Figure 25-23.

Figure 25-24 shows how to provide proper backing for chimes. This is done during the rough-in stages of the electrical installation.

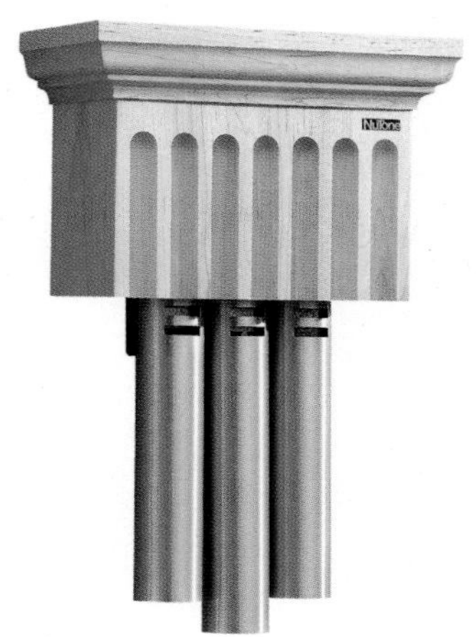

Courtesy of Broan-NuTone, LLC.

FIGURE 25-22 Typical residential door chimes.

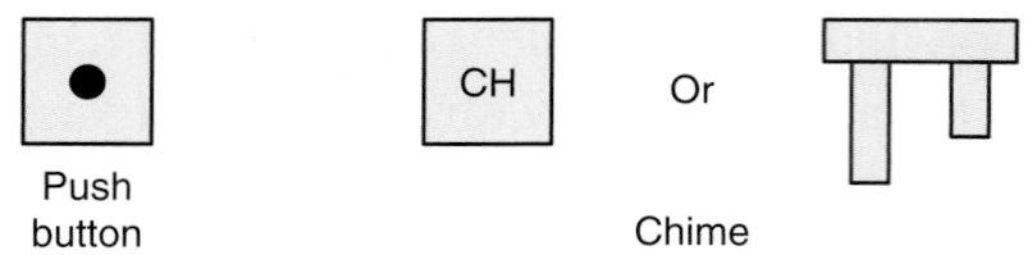

FIGURE 25-23 Symbols for chimes and push buttons.

Chimes for the Hearing Impaired

Chimes for the hearing impaired are available with accessory devices that can turn on a dedicated lamp at the same time the chime is sounded. Figure 25-25 shows the devices needed to accomplish this. Figure 25-26 shows the wiring details.

Chime Transformers

Chime transformers are covered under UL Standard 1585. Because of their low-voltage rating and power limitation, they are listed as Class 2 transformers. The low-voltage wiring on the secondary side of the transformer is Class 2 wiring. Class 2 circuits are discussed in Chapter 24 and in *Article 725*.

Figure 25-27 shows a chime transformer. The top view shows the 120-volt leads. Note the small set screw for ease of mounting into a conduit knockout in an outlet box. The bottom view shows the low-voltage terminals.

Chime transformers used in dwellings are generally rated 16 volts. They have built-in thermal overload protection. Should a short circuit occur in the low-voltage wiring, the overload device opens and closes repeatedly until the short circuit is cleared.

The UL standard requires that all metal parts of a transformer be properly grounded. To accomplish this, some chime transformers have a bare copper equipment grounding conductor in addition to the black and white supply conductors. Internal to the transformer, this equipment grounding conductor is connected to the metal parts (laminations) of the transformer. This bare equipment grounding conductor must be connected to the equipment grounding conductor or grounding screw in the outlet box. Do not connect this equipment grounding conductor to the branch-circuit grounded (white) conductor.

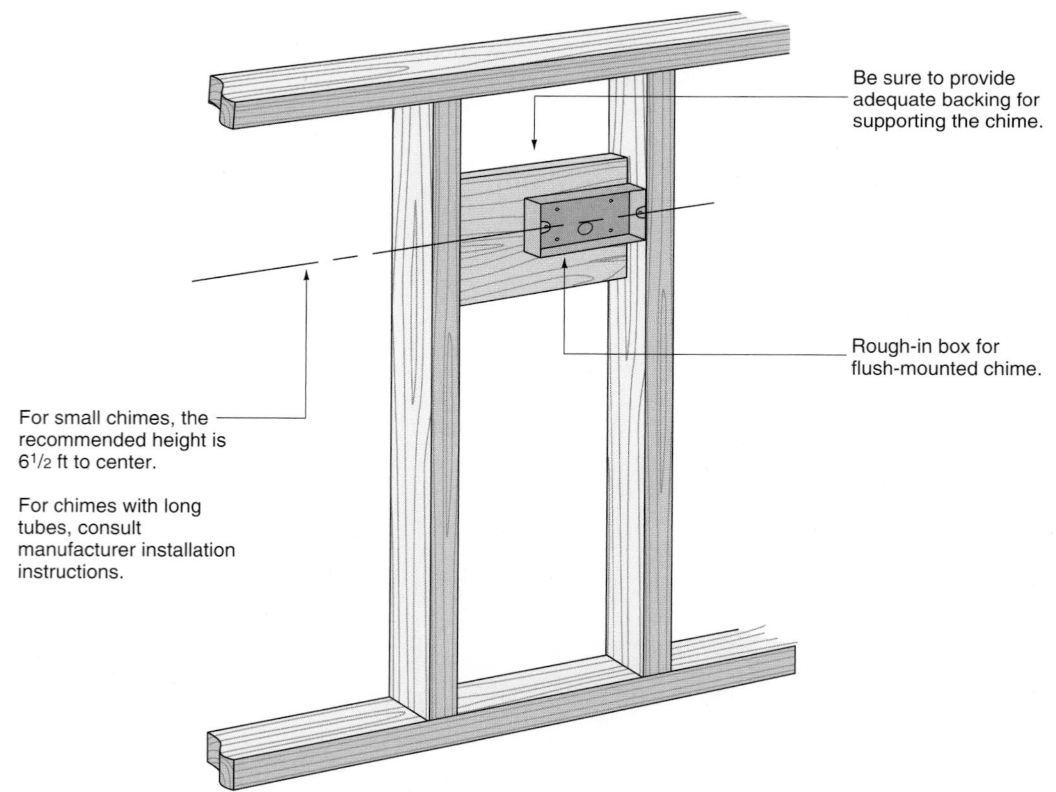

FIGURE 25-24 Roughing-in for a flush-mounted chime.

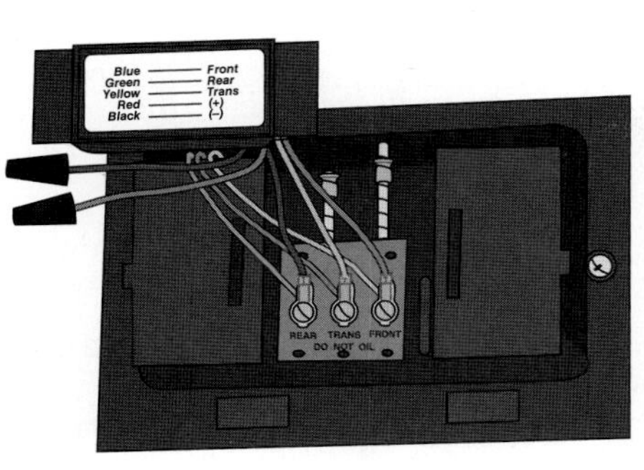

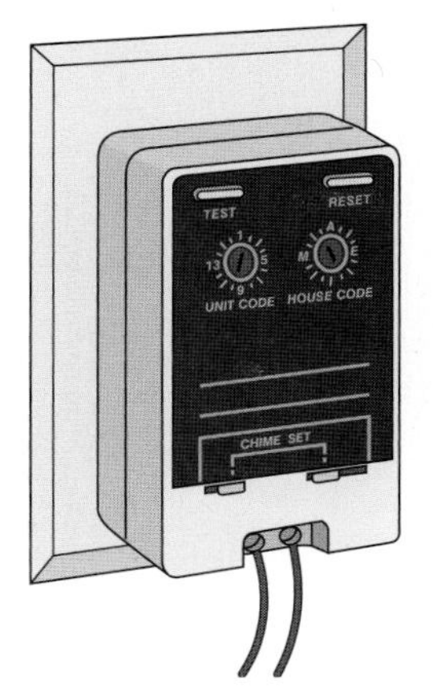

FIGURE 25-25 Accessory devices can be attached to new or existing chimes to provide visual signals when the chime is sounded. The first illustration shows the module that is connected and mounted inside the chime. The second illustration shows a transmitter that is plugged into a nearby receptacle. The third illustration shows a receiver into which a lamp is plugged.

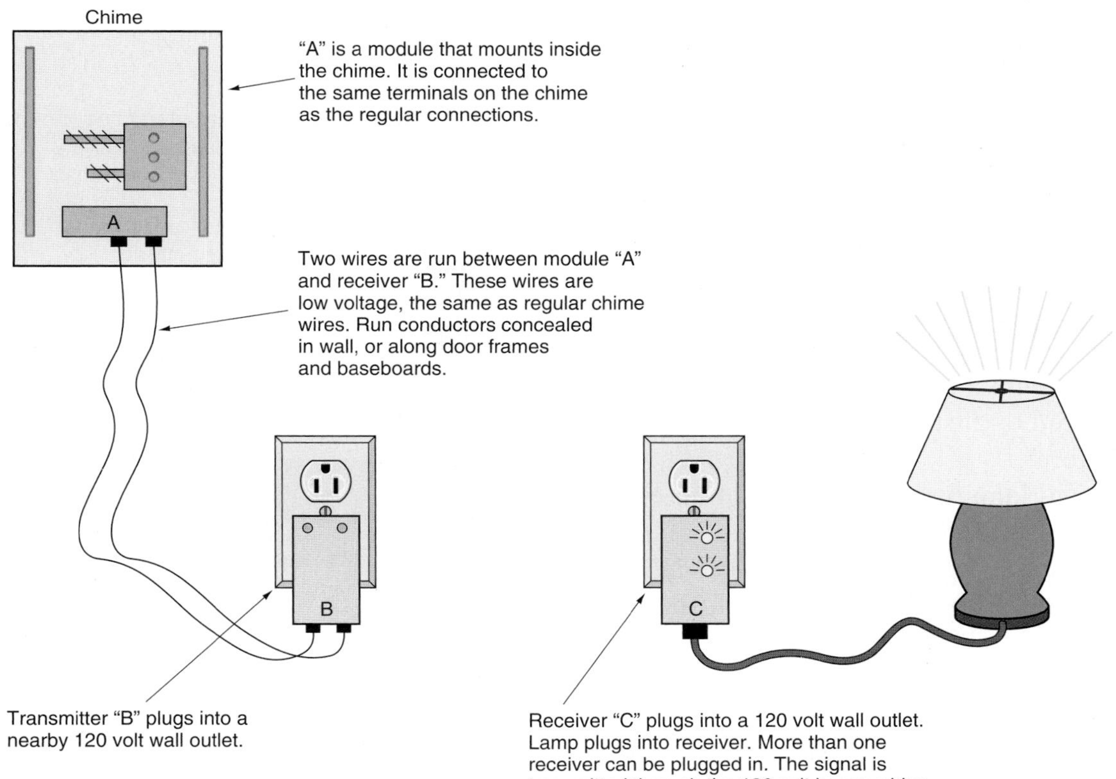

FIGURE 25-26 Installation diagram for auxiliary devices for use with chimes that can help the hearing impaired. When the chime sounds, the signal is transmitted to the receiver, turning on the lamp that is plugged into the receiver. The ON time is adjustable at "C." The lamp should be a "dedicated" lamp that lights up only when the chime sounds.

FIGURE 25-27 Chime transformer.

Some chime transformers are marked "Install in Metal Box Only."

Additional Chimes

To extend a chime system to cover a larger area, a second or third chime may be added. In the residence, two chimes are used: one chime is mounted in the front hall, and a second (extension) chime is mounted in the recreation room. The extension chime is wired in parallel to the first chime. The wires are run from one chime terminal board to the terminal board of the other chime. The terminals are connected as follows: transformer to transformer, front to front, and rear to rear, Figure 25-28.

When more than one chime is installed, it will probably be necessary to install a transformer with a higher volt-ampere (wattage) rating. Do not install a transformer with a higher voltage rating. Check the manufacturers' instructions furnished with the chime transformer.

Do not add another transformer to an existing transformer to solve the problem of multiple chimes not responding properly. The *NEC* in *725.121(B)* prohibits connecting transformers in parallel unless specifically listed for interconnection. Residential chime transformers are not listed for interconnection.

If a buzzer (or bell) and a chime are connected to a single transformer and are used at the same time, the transformer will put out a fluctuating voltage. This condition does not allow either the buzzer or the chime to operate properly. The use of a transformer with a larger rating may solve this problem.

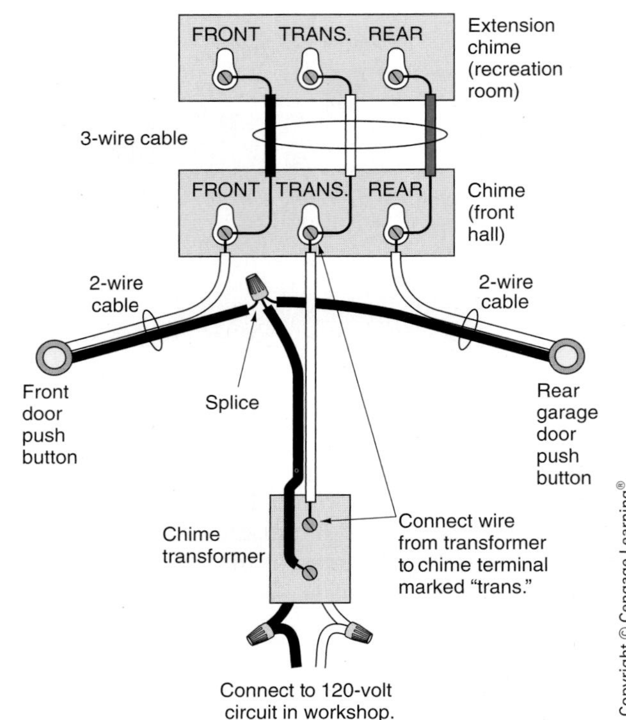

FIGURE 25-28 Circuit for chime installation.

The wattage consumption of chimes varies with the manufacturer. Typical ratings are shown in Table 25-1.

Transformers with ratings of 5, 10, 15 20, and 30 watts (VA) are available. For a multiple chime installation, wattage ratings for the individual chimes are added. The total value is the minimum transformer rating needed to do the job properly.

TABLE 25-1

Typical chimes with their power consumption.

TYPE OF CHIME	POWER CONSUMPTION
Standard 2-note	10 watts
Repeating chime	10 watts
Internally lighted, two lamps	10 watts
Internally lighted, four lamps	15 watts
Combination chime and clock	15 watts
Motor-driven chime	15 watts
Electronic chime	15 watts
Programmable musical	15 watts

Low-Voltage Wiring

The low-voltage wiring for a chime(s) is classified as Class 2 wiring, as discussed in Chapter 24.

Wiring the Chime

The circuit shown in Figure 25-28 is recommended for this chime installation because it provides a "hot" low-voltage circuit at the front hall location. However, it is not the only way in which these chimes may be connected.

Figure 25-28 shows that a 2-wire cable runs from the transformer in the utility room to the front hall chime. A 2-wire cable then runs from the chime to both the front and rear door push buttons. A 3-wire cable also runs between the front hall chime and the recreation room chime. Because of the "hot" low-voltage circuit at the front hall location, a chime with a built-in clock can be used. A 4-conductor cable may be run to the extension chime so that the owner may install a clock-chime at this location also.

The plans show that the chime transformer is mounted into the knockout on one of the ceiling boxes in the utility room. The line-voltage connections are easy to make here. Some electricians prefer to mount the chime transformer on the top or side of the distribution panelboards. The electrician decides where to mount the transformer after considering the factors of convenience, economy, and good wiring practice.

Don't Conceal Boxes!

It may seem obvious, but the issue of accessibility bears repeating. Electrical equipment junction boxes and conduit bodies must be accessible, *314.29*. It is permitted to have certain electrical equipment such as a chime transformer, junction boxes, and conduit bodies above a dropped lay-in ceiling because these are accessible by dropping out a ceiling panel. Never install electrical equipment, junction boxes, or conduit boxes above a permanently closed-in ceiling or within a wall. There may come a time when the electrical equipment and junction boxes need to be accessed. This includes chime transformers.

REVIEW

Note: Refer to the *Code* or plans where necessary.

TELEVISION CIRCUIT

1. How many television outlets are installed in this residence? ______
2. Which type of television cable is commonly used and recommended? ______
______
3. What determines the design of the faceplates used? ______
______
4. What must be provided when installing a television outlet and receptacle outlet in one wall box? ______
______
______
______
______
______

5. From a cost standpoint, which system is more economical to install: a master amplifier distribution system or a multiset coupler? Explain the basic differences between these two systems. ______

6. How many wires are in the cable used between a rotor and its controller?

7. Digital satellite systems use an antenna that is approximately (18 in. [450 mm]) (36 in. [900 mm]) (72 in. [1.8 mm]) in diameter. (Circle the correct answer.)

8. List the requirements for cable television inside the house. ______

9. Which article of the *Code* references the requirements for cable community television installation? ______

10. It is generally understood that grounding and bonding together all metal parts of an electrical system and the metal shield of the cable television cable to the same grounding reference point in a residence will keep both systems at the same voltage level should a surge, such as lightning, occur. Therefore, if the incoming cable that has been installed by the CATV cable company installer has the metal shield grounded to a driven ground rod, does this installation conform to the *NEC*? ______

11. All television satellites rotate above the Earth in (the same orbit) (different orbits). (Circle the correct answer.)

12. Television satellites are set in orbit (10,000) (18,000) (22,245) miles above the Earth, which results in their rotating around the Earth at (precisely the same) (different) rotational speed as the Earth rotates. This is done so that the satellite "dish" can be focused on a specific satellite (once) (one time each month) (whenever the television set is used). (Circle the correct answers.)

13. Which section of the *Code* prohibits supporting coaxial cables from raceways that contain light or power conductors? *NEC* ______

14. When hooking up one CATV cable and another cable from an outdoor antenna to a receiver that has only one antenna input terminal, a(n) ______ switch is usually installed.

15. In your own words, explain the purpose of an intersystem bonding termination device.

16. For a new home, an intersystem bonding termination devices is required (a) at the service equipment, (b) at the subpanel. (Circle the correct answer.)

TELEPHONE SYSTEM

1. How many locations are provided for telephones in the residence? ________
2. At what height are the telephone outlets in this residence mounted? Give measurement to center.
____
____
____
____
3. Sketch the symbol for a telephone outlet.
4. Is the telephone system regulated by the *NEC*? ________
5. a. Who is to furnish the outlet boxes required at each telephone outlet? ________
   b. Who is to furnish the faceplates? ________
6. Who is to furnish the telephones? ________
7. Who does the actual installation of the telephone equipment? ________
8. How are the telephone cables concealed in this residence? ________
____
____
____
9. The point where the telephone company's cable ends and the interior telephone wiring meets is called the ________ point. The device installed at this point is called a(n) ________.
10. What are the colors contained in a four-conductor telephone cable assembly, and what are they used for? ________
____
____
____
____
____
____
11. Itemize the *Code* rules for the installation of telephone cables in a residence. ________
____
____
____
12. If finger contact were made between the red conductor and green conductor at the instant a "ring" occurs, what shock voltage would be felt? ________
13. What section of the *Code* prohibits supporting telephone wires from raceways that contain light or power conductors? *NEC* ________

14. The term "cross-talk" is used to define the hearing of the faint sound of voices in the background when you are using the telephone. Cross-talk can be reduced significantly by running (flat cables) (twisted pair cables) to the telephones. (Circle the correct answer.)
15. Which section of the *Code* prohibits telephone cables from being installed in the same box or enclosure or from being pulled into the same raceway as light and power circuits? ______

SIGNAL SYSTEM

1. What is a signal circuit? ______

   ______
2. What style of chime is used in this residence? ______
3. a. How many solenoids are contained in a 2-tone chime? ______

   b. What closes the circuit to the solenoid of a chime? ______
4. Explain briefly how two notes are sounded by depressing one push button (when two solenoids are provided). ______

   ______

   ______

   ______

   ______

   ______

   ______

   ______
5. a. Sketch the symbol for a push button. ______

   b. Sketch the symbol for a chime. ______
6. a. At what voltage do residence chimes generally operate? ______

   b. How is this voltage obtained? ______
7. What is the maximum volt-ampere rating of transformers supplying Class 2 systems?

   ______
8. What two types of chime transformers for Class 2 systems are listed by UL?

   ______

   ______
9. Is the extension chime connected with the front hall chime in series or in parallel?

   ______
10. How many bell wires terminate at

    a. the transformer? ______

    b. the front hall chime? ______

    c. the extension chime? ______

    d. each push button? ______

11. a. What change in equipment may be necessary when more than one chime is connected to sound at the same time on one circuit? ______

 b. Why? ______

12. What type of insulation is usually found on low-voltage wires? ______
13. What size wire is installed for signal systems of the type in this residence? ______
14. a. How many wires are run between the front hall chime and the extension chime in the recreation room? ______
 b. How many wires are required to provide a "hot" low-voltage circuit at the extension chime? ______
15. Why is it recommended that the low-voltage secondary of the transformer be run to the front hall chime location and separate 2-wire cables be installed to each push button? ______

16. a. Is it permissible to install low-voltage Class 2 systems in the same raceway or enclosure with light and proper wiring? ______
 b. Which *Code* section covers this? ______
17. a. Where is the transformer in the residence mounted? ______
 b. To which circuit is the transformer connected? ______
18. a. How many feet of 2-conductor bell wire cable are required? ______
 b. How many feet of 3-conductor bell wire cable are required? ______
19. How many insulated staples are needed for the bell wire if it is stapled every 24 in. (600 mm)? ______
20. Should low-voltage bell wire be pulled through the same holes in studs and joists that also contain nonmetallic-sheathed cable? ______
21. What section of the *Code* prohibits supporting fire alarm conductors from raceways that contain light or power conductors? *NEC* ______.

22. Complete this typical chime wiring diagram using colored pencils. The 2-wire, low-voltage cables contain one red and one white conductor. Use yellow to represent the white conductors.

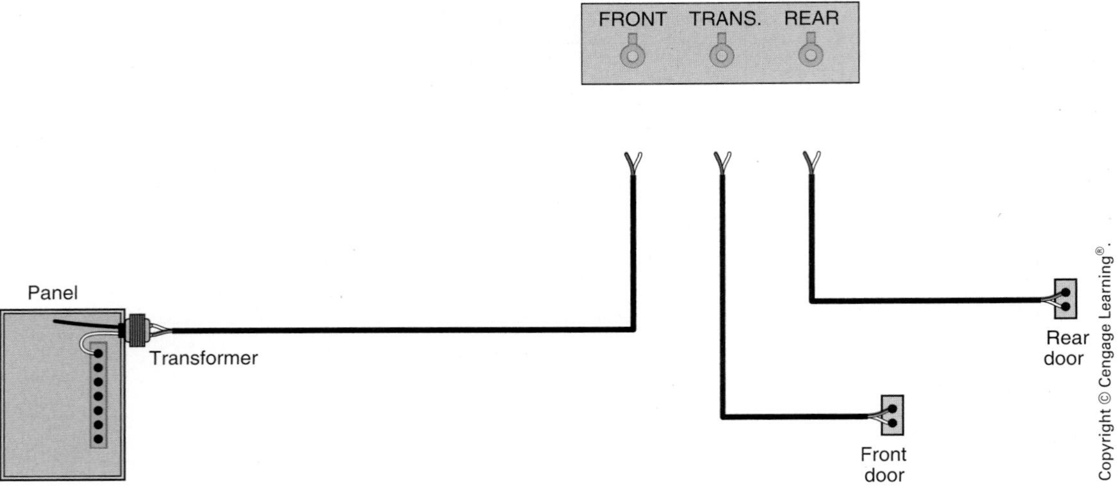

23. What does the term "Wi-Fi" stand for? ____________________

24. A Wi-Fi system is (required) (optional). (Circle the correct answer.)

CHAPTER **26**

Smoke, Heat, Carbon Monoxide Alarms, and Security Systems

OBJECTIVES

After studying this chapter, you should be able to

- understand the basics of the *National Fire Alarm Code NFPA 72 (2013)*, the *Standard for the Installation of Carbon Monoxide (CO) Detection and Warning Equipment NFPA 720 (2012)*, and the *National Electrical Code NFPA 70 (2014)*.
- understand the basics of smoke, heat, and carbon monoxide alarms.
- understand the location requirements for the installation of smoke, heat, and carbon monoxide alarms for *minimum* acceptable levels of protection.
- understand the location requirements for the installation of smoke and heat alarms that *exceed* the minimum acceptable levels of protection.
- discuss general requirements for the installation of security systems.
- be aware of important UL Standards covering fire warning equipment.

NATIONAL FIRE ALARM CODE (*NFPA 72*)

NFPA 72 is the *National Fire Alarm Code*.

The importance of installing smoke and fire alarms in homes is supported by results from exhaustive investigations of home fires indicating that measurable quantities of smoke come before detectable quantities of heat. In other words, smoke generally comes before fire. Because this is true, smoke alarms are considered the primary means of protection against fire in homes.

Following are some interesting facts about home fires.

Fires in homes were the third leading cause of deaths in homes—more deaths from asphyxiation than from burns.

Half of fires in homes occurred between 10 p.m. and 8 a.m.. That's sleep time.

Chapter 29 of the *National Fire Alarm Code* specifically covers *Single- and Multiple-Station Alarms* and *Household Fire Alarm Systems*. It contains the minimum requirements for the proper selection, installation, operation, and maintenance of fire warning equipment that will provide *reasonable* fire safety. *Chapter 29* also covers small residential board and care occupancies, which are defined as, "A building or portion thereof that is used for lodging and boarding of four or more residents, not related by blood or marriage to the owners or operators, for the purpose of providing personal care services."**

In *NFPA 72*, *Section 29.2*, we find that the primary purpose of fire warning equipment is, "To provide a reliable means to notify the occupants of the presence of a threatening fire and the need to escape to a place of safety before such escape might be impeded by untenable conditions in the normal path egress."** *NFPA 72* is not intended to provide protection to property, *29.1.5*.

This chapter deals with the types of smoke and heat alarms typically installed in one- and two-family dwellings. These smoke and heat alarms are not classified as fire alarm systems.

This chapter does not involve the more complex household fire alarm systems, but rather the devices and a fire alarm control unit(s) that receive, monitor, and process signals from detection devices. These types of fire alarm systems communicate via telephone or wireless with a central station that in turn calls the fire department. Fire alarm systems are required to meet stricter rules regarding the sound level and time of audible signals. There are also tougher rules regarding sleeping area requirements. Fire alarm systems require rechargeable batteries from the ac power source supplying the control unit(s). Typical home-type smoke and heat alarms are permitted to have replaceable batteries.

Chapter 29 of the *National Fire Alarm Code* tells us where to install smoke and heat detectors and alarms in homes.

Article 760 in the *National Electrical Code®* (*NEC®*) tells us how to install the wiring for fire alarm systems.

Section 29.3.2 of the *National Fire Alarm Code* states that fire warning equipment is required to be installed according to the listing and manufacturer's published instructions.

There is nothing in the *NEC* that deals with carbon monoxide. To learn about carbon monoxide detectors and alarms in homes, we refer to *NFPA 720*, the *Standard for the Installation of Carbon Monoxide (CO) Detection and Warning Equipment*.

Because carbon monoxide is said to be the number one killer, many communities have passed legislation adopting all or parts of *NFPA 720*. *NFPA 720* requires carbon monoxide detectors and alarms for all dwellings. Homeowners and landlords had better pay close attention to this! How often have you heard on the news or read in the newspaper that the carbon monoxide detectors did not work! The batteries were either dead or had been removed.

Equally important is that there is nothing in the *National Fire Alarm Code* that will prevent injury or death if proper escape routes have not been planned, *29.4.2*.

Fire warning devices commonly used in a residence are heat detectors, Figure 26-1, and combination smoke alarms/detectors, Figure 26-2.

The more elaborate systems connect to a central monitoring customer service center through a telephone line. These systems offer instant contact with police or fire departments, a panic button for emergency medical problems, low-temperature detection,

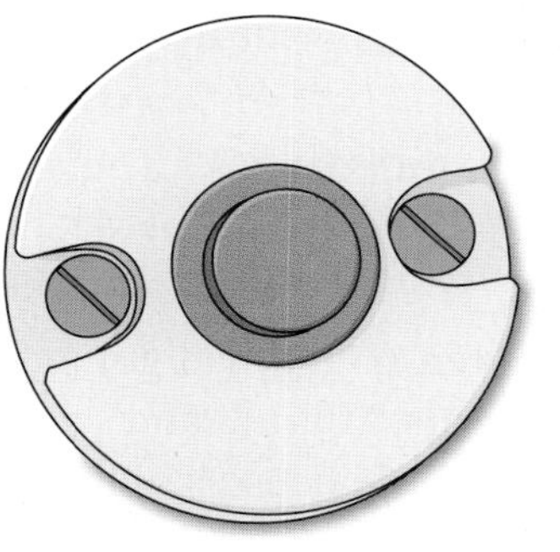

FIGURE 26-1 Heat detector.

FIGURE 26-2 Combination smoke alarm/detector.

flood (high-water) detection, perimeter protection, interior motion detection, and other features.

All detections are transmitted first to the company's customer service center, where personnel on duty monitor the system 24 hours a day, 7 days a week. They will verify that the signal is valid. After verification, they will contact the police department, the fire department, and individuals whose names appear on a previously agreed-on list that the homeowner prepared and submitted to the company's customer service center.

NFPA Standards are not law until adopted by a city, state, or other governmental body. Some NFPA Standards are adopted totally, whereas only portions of others are adopted. Installers must be aware of all requirements in the locality in which they are doing work. Most communities require a special permit for the installation of fire warning equipment and security systems in homes, and require registration of the system with the police and/or fire departments. In most instances, fire protection requirements are found in the building codes of a community and are not necessarily spelled out in the electrical code.

Are Smoke and Heat Alarms Required?

Pay particular attention to the words found in *NFPA 72 29.1.2*, which states, "Smoke and heat alarms shall be installed in all occupancies where required by governing laws, codes, or standards."** Similar wording is found in *29.5.1.1*. This puts the responsibility on the local building department. It might seem confusing to have so many codes. Without question, you must become familiar with your local applicable laws, codes, or standards.

Most communities adopt building codes published by the International Code Council (ICC). These codes contain requirements for the installation of smoke alarms. Your local building department officials can explain what is required for new work and remodel work (alterations, repairs, and additions) where a permit is required. Whereas new work requires interconnected, hard-wired, battery backup smoke alarms, it might be acceptable on remodel jobs to install battery-powered-only smoke alarms for existing areas where interior walls or ceiling finishes are not removed to expose structural framing members. To install, interconnect, and hard-wire smoke alarms in these existing areas could result in damage to walls and ceilings, requiring patching, repainting, or other repairs—an uncalled for and tremendous expense. If there is access from an attic, crawl space, or basement, the inspector might require interconnected, hard-wired, battery backup smoke alarms if you can do the job without having to remove interior finishes. Check this out with your local inspector!

DEFINITIONS

Here are some very important definitions from *National Fire Alarm and Signaling Code NFPA-72* and *National Electrical Code NFPA-70*:

An **alarm** is "A warning of danger."** *NFPA-72 3.3.11*

A **heat alarm** is "A single- or multiple station alarm responsive to heat."** *NFPA-72 3.3.123*

A **smoke alarm** is "A single- or multiple station alarm responsive to smoke."** *NFPA-72 3.3.268*

**Reprinted with permission from NFPA 72-2013.

A **detector** is "A device suitable for connection to a circuit that has a sensor that responds to a physical stimulus such as heat or smoke."** *NFPA-72 3.3.66*

A **heat detector** is "A fire detector that detects either abnormally high temperature or rate of temperature rise, or both."** *NFPA-72 3.3.66.10*

A **smoke detector** is "A device that detects visible or invisible particles of combustion."** *NFPA-72 3.3.66.20*

A **household fire alarm system** is "A system of devices that uses a fire alarm control unit to produce an alarm signal in the household for the purpose of notifying the occupants of the presence of a fire so that they will evacuate the premises."** *NFPA-72 3.3.105.2*

Shall "Mandatory rules of this *Code* are those that identify actions that are specifically required or prohibited and are characterized by the use of the terms shall or shall not."* *NFPA-70 90.5(A)*

A **combination detector** is "A device that either responds to more than one of the fire phenomenon or employs more than one operating principle to sense one of these phenomenon. Typical examples are a combination of a heat detector with a smoke detector or a combination rate-of-rise and fixed-temperature heat detector. This device has listings for each sensing method employed."** *NFPA-72 3.3.66.4*

A **single station alarm** is "A detector comprising an assembly that incorporates a sensor, control components, and an alarm notification appliance in one unit operated from a power source either located in the unit or obtained at the point of installation."** *NFPA-72 3.3.262*

A **multiple station alarm** is "A single station alarm capable of being interconnected to one or more additional alarms so that the actuation of one causes the appropriate alarm signal to operate in all interconnected alarms."** *NFPA-72 3.3.161*

A **dwelling unit** is A single unit, providing complete and independent living facilities for one or more persons, including permanent provisions for living, sleeping, cooking, and sanitation.* *NFPA-70 Article 100*

A **living area** is "Any normally occupiable space in a residential occupancy, other than sleeping rooms or rooms that are intended for combination sleeping/living, bathrooms, toilet compartments, kitchens, closets, halls, storage or utility spaces, and similar areas." ** *NFPA-72 3.3.143*

A **separate sleeping area** is "The area of a dwelling unit where the bedrooms or sleeping rooms are located."** *NFPA-72 3.3.252*

In this chapter, rather than going back and forth between the words "alarm" and "detector," we refer to fire warning devices as alarms, knowing full well that before an alarm can sound, the cause must be detected. Most home-type fire warning devices combine the detector and alarm in one device. The location requirements for smoke and heat alarms are pretty much the same.

SMOKE, HEAT, AND CARBON MONOXIDE ALARMS

This chapter covers the basic requirements for protection in homes against the hazards of fire, heat, smoke, and carbon monoxide. Fire alarm systems for commercial installations are much more complicated than the typical household system.

Fire is the third leading cause of accidental death. Home fires account for the biggest share of these fatalities, most of which occur at night during sleeping hours. Rapidly developing high-heat fires and slow smoldering fires are the culprits. Both produce smoke and deadly gases.

Smoke, heat, and carbon monoxide alarms are installed in a residence to give the occupants *early warning* of the presence of fire or toxic fumes. Fires produce smoke and toxic gases that can overcome the occupants while they sleep. Most fatalities result from the inhalation of smoke and toxic gases rather than from burns. Heavy smoke reduces visibility.

In nearly all home fires, detectable smoke precedes detectable levels of heat. People sleeping are less likely to smell smoke than people who are awake. The smell of smoke probably will not awaken a sleeping person. Therefore, smoke alarms are considered to be the primary devices for protecting lives.

---

*Reprinted with permission from NFPA 70-2014.
**Reprinted with permission from NFPA 72-2013.

Heat alarms DO NOT take the place of smoke alarms. Heat alarms are installed *in addition* to smoke detectors.

Another lifesaving device is the carbon monoxide (CO) alarm, which senses dangerous carbon monoxide emitted from a malfunctioning furnace or other source. *NFPA 720* is the *Standard for the Installation of Carbon Monoxide (CO) Detection and Warning Equipment.*

Home-type smoke, heat, and carbon monoxide devices as a rule have both the detector and the alarm in one device. There also are alarms that combine more than one type of sensor into one unit, such as smoke/heat, smoke/carbon monoxide, or carbon monoxide/explosive gases.

In larger, more complex commercial installations, detectors and alarms are generally separate devices that, when triggered, send a signal to a central control panel, which in turn sounds the general alarm.

As with all electrical installations, always install fire warning equipment that has been listed by a nationally recognized testing laboratory (NRTL).

DETECTOR TYPES

AC/Battery: Required in new construction. These conform to the *NFPA 72, Section 29.6,* requirement that alarms have two power supplies—regular household 120-volt ac for the primary source of power and 9-volt battery for the secondary source. The alarm "chirps" when the battery gets low. See Figure 26-3.

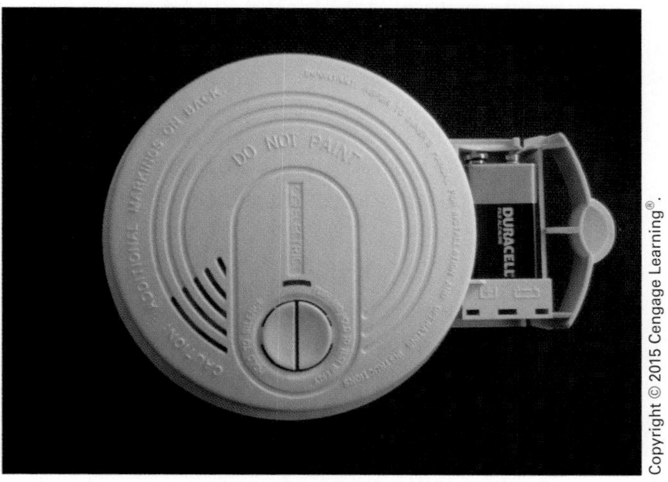

FIGURE 26-3 An ac-dc smoke alarm that operates on 120 volts ac as the primary source of power and a 9-volt battery as the secondary source of power. In the event of a power outage, the alarm continues to operate on the battery.

Interconnected: Required in new construction. Should any smoke, heat, or carbon monoxide alarm trigger, that alarm will sound and simultaneously send a signal to set off all other interconnected alarms. See *NFPA-72 29.5.2.* Check the manufacturer's instructions to determine how many units may be interconnected. Never connect more units than the number specified by the manufacturer. Never "mix" different manufacturer's devices. Testing laboratories do not perform tests using more than one manufacturer's devices in the test. The alarm "chirps" when the battery gets low. These units usually have a small 3-wire power connector—the wires connect to the field wiring, and the other end has a connector that plugs into the back of the unit. See Figure 26-4. The pigtail may have a red- or orange-colored "interconnect" wire. The field wiring between alarms is 3-wire nonmetallic-sheathed cable that contains a red insulated conductor that is used as the "interconnect" wire.

Plug-in: Not permitted in new construction. These alarms plug into a wall outlet that is not controlled by a wall switch. They operate on 120 volts as the primary source of power and on a 9-volt battery as the secondary source. The alarm "chirps" when the battery gets low. If of the plug-in type, they must have some sort of restraining means so as to reduce the possibility of being unplugged.

Battery-Operated Only: Not permitted in new construction. These alarms operate on a 9-volt battery only. The alarm "chirps" when the battery gets low. These alarms are commonly used in existing homes.

For the Hearing Impaired: These alarms have a bright strobe light that provides a visual alarm for the hearing impaired. This complies with the requirements found in the Americans with Disabilities Act (ADA).

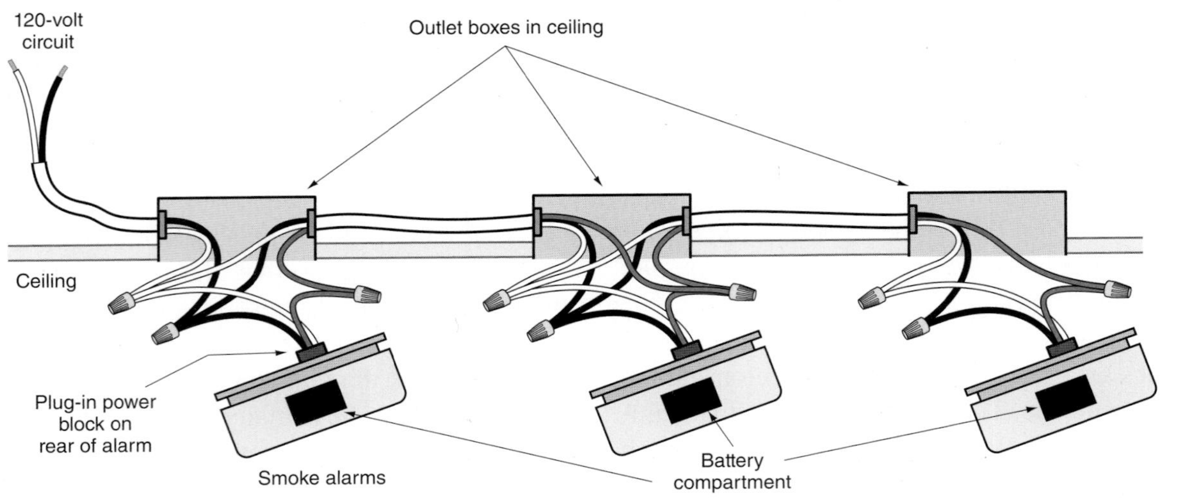

FIGURE 26-4 Interconnected smoke or heat alarms. The 2-wire branch circuit is run to the first alarm; then 3-wire nonmetallic-sheathed cable interconnects the other alarms. If one alarm in the series is triggered, all other alarms will sound off. See *NFPA 72, 29.5.2.1.1.*

WIRELESS SYSTEMS (LOW-POWER RADIO)

Wireless technology has come a long way in recent years. Here is an overview of what wireless means as related to smoke and fire alarms.

When one alarm sounds off, all other alarms in the house are required to sound off at the same time. Stated another way, the alarms must be interconnected so as to "communicate" with one another. In new installations, this is generally done with hard-wiring, as indicated in Figure 26-4.

Hard-wiring smoke alarm systems—or adding more smoke alarms to an existing installation and trying to fish wires through walls and ceilings—can be messy, time-consuming, and costly.

Today, wireless systems that are listed in conformance to UL Standard 217, the standard for single- and multiple-station smoke alarms, are available. Both 120-volt ac- and battery-powered devices are used. Wireless systems might be the answer to installing a smoke alarm system economically and should be considered.

120-volt ac Powered: For existing 120-volt ac-powered systems, one interconnected smoke alarm is replaced with a 120-volt ac-powered smoke alarm that is designed specifically to send a wireless signal to other wireless smoke alarms of the same manufacturer. This triggers all of the other wireless smoke alarms to sound off. Additional smoke alarms can be installed anywhere in the house. That's all there's to it!

Battery Powered: For existing homes that do not have a smoke alarm system, the world of wireless technology can be quite advantageous. Battery-powered wireless smoke alarms can be installed. When one smoke alarm senses smoke, it triggers and sends a signal (wireless) to all other wireless battery-powered smoke alarms in the home. They are "interconnected" via wireless technology. No hard-wiring is required.

The maximum number of smoke alarms that can be interconnected is specified in the instructions furnished by the manufacturer, based on the testing and listing of the devices.

IMPORTANT: Local codes may or may not accept wireless smoke alarm systems. Check this out with your local building officials.

IMPORTANT: Whether the system is hard-wired or wireless, never mix different manufacturers' components.

IMPORTANT: Check the system regularly in accordance with manufacturer's instructions.

IMPORTANT: Be sure to install equipment "listed for the purpose," *NFPA-72, 29.3.1.*

See *NFPA 72, 6.17*, for a complete list of requirements for wireless systems. Also see *NFPA 72, 29.7.7.*

TYPES OF SMOKE ALARMS

Two common types of smoke alarms are the **photoelectric type** (sometimes called photoelectronic) and the **ionization type**. They usually contain an indication light to show that the unit is functioning properly. They also may have a test button that simulates smoke, so that when the button is pushed, the detector's smoke-detecting ability as well as its circuitry and alarm are tested.

Some alarms are tested with a magnet. Others are tested with a listed spray from an aerosol can.

Smoke alarms generally do not sense heat, flame, or gas. However, some smoke alarms can be set off by acetylene and propane gas.

Some alarms have a "hush button" that can temporarily silence a nuisance alarm for about 15 minutes.

Photoelectric Type

The photoelectric type of smoke alarm has a light sensor that measures the amount of light in a chamber. When smoke is present, an alarm sounds, indicating a reduction in light due to the obstruction of the smoke. This type of sensor detects smoke from burning materials that produce large quantities of smoke, such as furniture, mattresses, and rags. The photoelectric type of alarm is less effective for gasoline and alcohol fires, which do not produce heavy smoke. This type of alarm can become more sensitive to smoke as it gets older.

Photoelectric sensors are a good choice for areas subject to steam, such as in or outside bathrooms, utility rooms, and kitchens.

There are two types of photoelectric devices: light obscuration and light scattering.

Ionization Type

The ionization type of alarm contains a low-level radioactive source (less than used in luminescent watch and clock dials), which supplies particles that ionize the air in the detector's smoke chamber. Plates in this chamber are oppositely charged. Because the air is ionized, an extremely small amount of current (millionths of an ampere) flows between the plates. Smoke entering the chamber impedes the movement of the ions, reducing the current flow, which triggers the alarm. This type often sets off a nuisance alarm because cooking routinely gives off small "invisible" smoke particles. This type of alarm can become more sensitive to smoke as it get older.

The ionization type of alarm is effective for detecting small amounts of smoke, as in gasoline and alcohol fires, which are fast flaming with little or no smoke.

Some smoke alarms are available with both ionization and photoelectric sensors combined. Other smoke alarms are available with smoke and carbon monoxide detection capabilities or smoke and heat detection capabilities in one unit.

TYPES OF HEAT ALARMS

Heat alarms respond to heat—not smoke!

Three types of heat alarms are described as follows.

- Fixed temperature heat detectors sense a specific fixed temperature, such as 135°F (57°C) or 200°F (93°C). They are required to have a temperature rating of at least 25°F (14°C) above the normal temperature expected, but not to exceed 50°F (28°C) above the expected temperature. Fixed temperature detectors are sometimes combined with smoke alarms and carbon monoxide alarms.
- Rate-of-rise heat detectors sense rapid changes in temperature (12°F to 15°F per minute) such as those caused by flash fires.
- Fixed/rate-of-rise temperature detectors are available as a combination unit.
- Fixed/rate-of-rise/smoke combination detectors are available in one unit.

INSTALLATION REQUIREMENTS

The following information includes specific recommendations for installing smoke alarms and heat alarms in homes. Remember that smoke alarms are the primary fire warning devices, and heat detectors are installed in addition to the required smoke

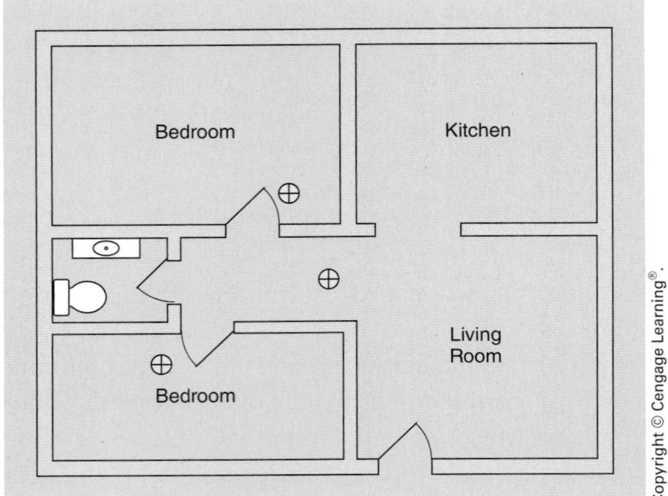

FIGURE 26-5 Smoke alarms are required in all sleeping areas, outside all sleeping areas, and on every level. Acceptable locations are on the ceiling or on the wall space above the door. See Figures 26-7, 26-8, and 26-9 for location restrictions.

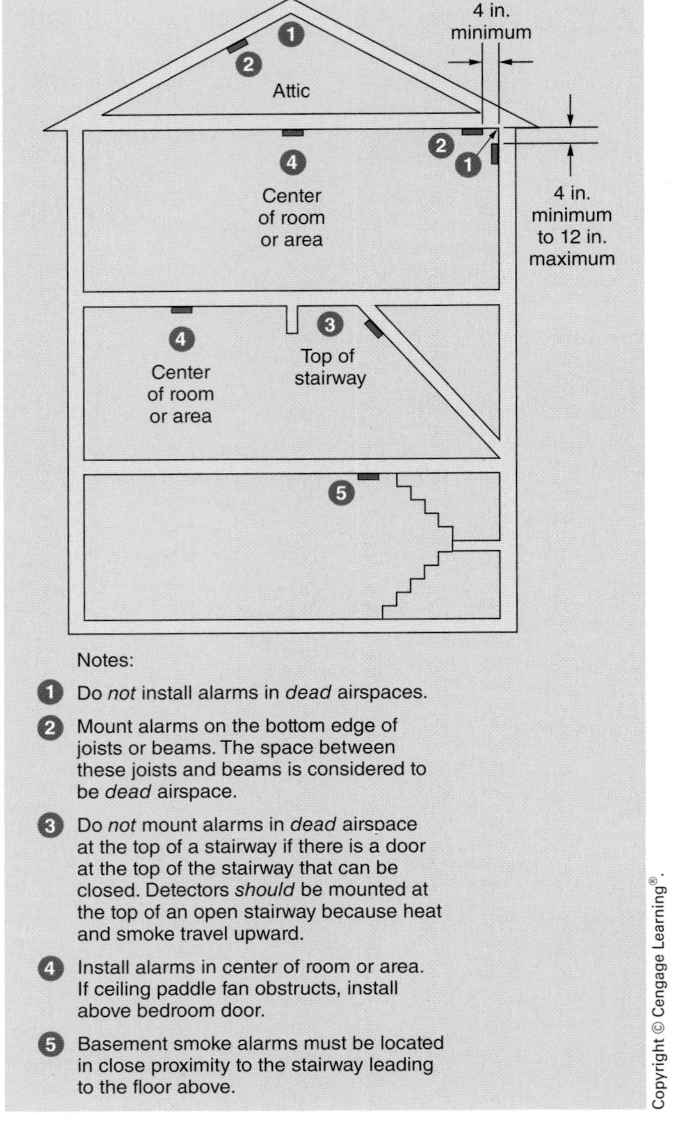

FIGURE 26-6 Recommendations for the installation of heat and smoke alarms. Check manufacturers' installation instructions for their recommendations about installing smoke and heat alarms in attics.

alarms. As you continue reading, study Figures 26-5, 26-6, 26-7, 26-8, and 26-9. Complete data are found in NFPA Standard 72, Chapter 29, and in the instructions furnished by the manufacturer of the equipment.

The Absolute Minimum Level of Protection

Here is a brief summary of the absolute minimum level of protection for smoke alarms for new residential construction in conformance to *NFPA 72*, Chapter 29. Some communities have adopted portions of *NFPA 72* for existing dwellings. You will have to check with your local authority having jurisdiction (AHJ). Because life safety is so important, attending an NFPA seminar on *NFPA 72* is highly recommended.

- Install multiple station smoke alarms
  a. in all sleeping rooms and guest rooms.
  b. outside each separate sleeping area, within 21 ft (6.4 m) of any door to the sleeping area. This distance is measured along the path of travel.
  c. on every level of a dwelling unit. This includes basements but excludes crawl spaces and unfinished attics.
  d. in guest bedrooms and suites: more elaborate homes may perhaps have a guest bedroom(s) or suite, completely self-contained with bedroom(s), bathroom(s), and closet(s).

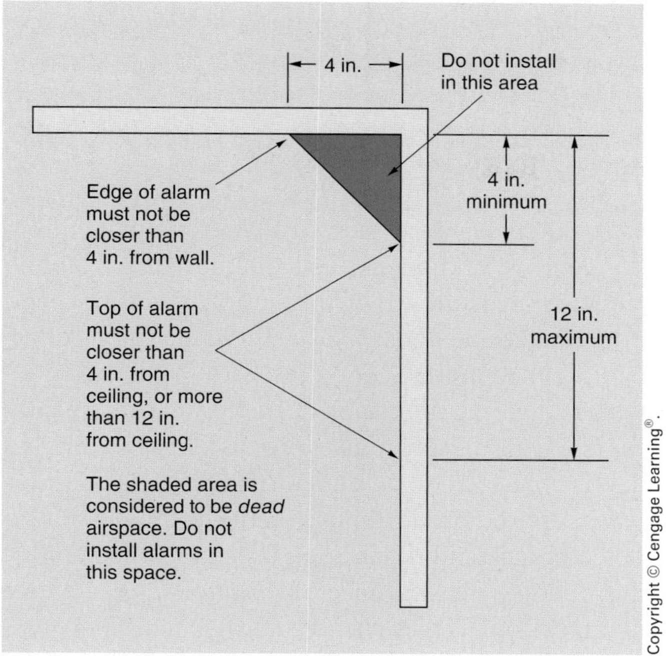

FIGURE 26-7 Do not mount smoke or heat alarms in the *dead* airspace where the ceiling meets the wall.

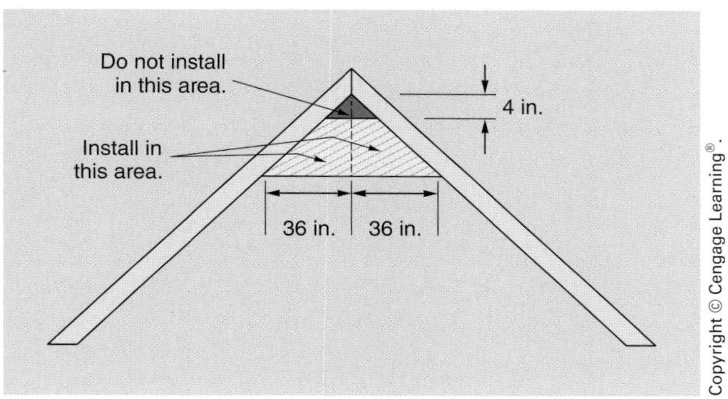

FIGURE 26-8 Installing smoke and heat detectors/ alarms in peaked (cathedral) ceilings.

If these are separated from the rest of the house by a door, a smoke alarm shall be installed in the guest bedroom or suite, plus an additional smoke alarm shall be installed on the living area side of the door. In instances where a hallway is outside the sleeping area, a smoke alarm shall be installed in that hallway. If that hallway is closed off by a door,

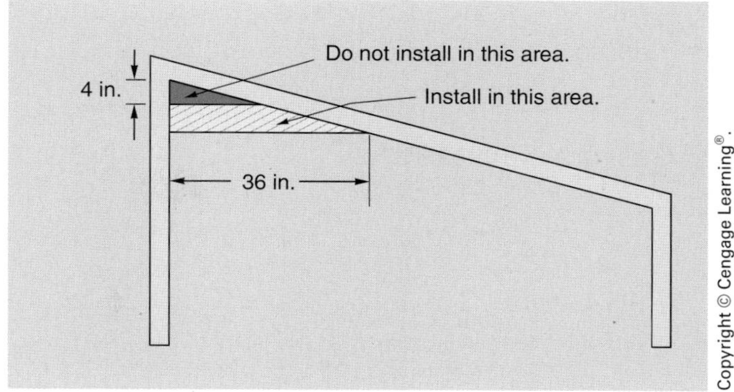

FIGURE 26-9 Installing smoke and heat detectors/ alarms in sloped ceilings.

an additional smoke alarm shall be installed on the living area side of the door. The bottom line is that three smoke alarms might be needed.

e. After installing smoke alarms required in (a), (b), (c), and (d), check one more thing for large homes. It might be necessary to install additional smoke alarms for a given floor area 1000 ft^2 (93 m^2) or greater. Don't include the garage area. To conform, either of the following is acceptable:

   1. Check to verify that no point on the ceiling of the given floor area is more than 30 ft (9.1 m) from a smoke alarm.
   2. Install a smoke alarm for every 500 ft^2 (46.5 m^2) in the given floor area.

f. for cathedral and/or vaulted ceilings extended over more than one floor. Meeting requirements (a), (b), (c), and (d) for the upper level will meet requirement (e).

g. For existing dwellings, listed battery-powered smoke alarms that are not interconnected are generally permitted. This provides a reasonable level of protection as opposed to having no smoke alarms at all. However, your local code might require ac/dc dual-powered interconnected alarms.

- Smoke alarms must have at least two independent sources of power. The primary source is the 120-volt ac circuit, and the secondary (standby) source is the integral battery power supply.

- The ac source of power is permitted to be a dedicated branch circuit or an unswitched part of a conventional branch circuit.

Where to Install

In *bedrooms* and *halls outside of sleeping areas.* See Figure 26-5.

In *other rooms and areas.* See Figure 26-6.

On *walls*—not closer than 4 in. (100 mm), but not farther than 12 in. (300 mm) from the adjoining ceiling. Do not install in dead airspace. See Figures 26-6 and 26-7.

On *flat ceilings*—not closer than 4 in. (100 mm) from the adjoining wall. See Figures 26-6 and 26-7. A level ceiling is one that is level or has a slope of not more than 1 ft in 8 ft (1 m in 8 m).

On *peaked ceilings*—locate with 36 in. (900 mm) horizontally from the peak, but not closer than 4 in. (100 mm) vertically from the peak. A peaked ceiling might also be called a cathedral ceiling. Do not install in dead airspace. See Figure 26-8.

On *sloped ceilings*—a sloped ceiling is defined as having a slope greater than 1 ft in 8 ft (1 m in 8 m). Locate within 36 in. (900 mm) of the high side but not closer than 4 in. (100 mm) from the adjoining wall. Do not install in dead airspace. See Figure 26-9. A level ceiling is one that is level or has a slope of not more than 1 ft in 8 ft (1 m in 8 m).

How to Wire Smoke and Heat Alarms

In new construction, smoke and heat alarms require both primary (120-volt) and secondary (battery) sources of power. The *National Fire Alarm Code NFPA 72* does not specify where to pick up the 120-volt supply.

Smoke and Fire Alarms and AFCI Protected Branch Circuits

An often asked question is: Are home-type smoke and fire alarms permitted to be connected to an AFCI protected branch circuit? As it now stands, the answer is yes. Read the following words carefully.

NEC 210.12(A) states that in dwelling units, ▶All 120-volt, single phase, 15- and 20-ampere branch circuits supplying outlets or devices installed in dwelling unit kitchens, family rooms, dining rooms, living rooms, parlors, libraries, dens, bedrooms, sunrooms, recreation rooms, closets, hallways, laundry areas, or similar rooms or areas shall be protected by any of the means described in (1) through (6):◀*

The changes for the 2014 *NEC* include adding "or devices," "kitchens," and "laundry areas" to the rule. As a result, AFCI protection is required for both outlets and devices installed in the rooms or areas included in the Section.

A single-station smoke detector/alarm typically installed in these rooms or areas is both a device and is connected to an outlet. As a result, AFCI protection is required. Even though the single station detectors are interconnected so that if one sounds, all sound, the interconnected detectors are not a fire alarm system as there is no control panel.

It should be noted that the requirements in *NEC 760.41(B)* and *121(B)* that fire alarm systems not be supplied by a GFCI- or AFCI-protected branch circuit does not apply to the interconnected single-station smoke alarms we have been speaking about because this installation does not qualify as a fire alarm system.

Probably the simplest way to hard-wire and interconnect smoke alarms when wiring new homes is to pick up one smoke alarm from one of the bedroom branch circuits, then run the interconnecting wiring to all of the other required and optional smoke alarms. The electrical plans for this residence indicate the location of the required and optional smoke and heat alarms.

Although smoke alarms are required to have battery backup power, some local codes do not permit smoke alarms to be supplied by the same branch circuit as the bedroom lighting. They require that some other branch circuit supply the smoke alarms. Check this out with your electrical inspector.

Wireless systems are discussed earlier in this chapter.

Combination Smoke and Carbon Monoxide Detectors/Alarms

Combination smoke and carbon monoxide detectors/alarms are readily available. Some communities are requiring that this type be installed in

*Reprinted with permission from NFPA 70-2014.

new construction. Check with your local electrical inspector.

Some inspectors require a lock-off device over the handle of the circuit breaker that serves the alarms so that it will not unintentionally be turned off. This lock-off device is really not necessary. Should the breaker supplying the alarms get turned off intentionally or unintentionally, or if it trips due to trouble in the circuit, the battery backup would power the alarms. When the batteries get low, the alarms start "chirping."

The usual way to connect alarms is to run the 2-wire branch circuit to the first alarm, then run 3-wire nonmetallic-sheathed cable from the first alarm to all other alarms. The third conductor provides the interconnect feature that will sound all alarms should any one of the alarms activate. See Figure 26-4.

Summary of Dos and Do Nots

When installing detectors and alarms, ask yourself, "Where will the smoke and greatly heated air travel?" Because smoke and heated air rise, detectors and alarms must be in the path of the smoke and greatly heated air.

Figures 26-5, 26-6, 26-7, 26-8, and 26-9 illustrate some of the more important issues regarding smoke and heat alarms.

Here is a list of the *Dos* and *Do nots* in no particular order of importance.

The Dos

1. *Do* install listed smoke alarms, and install them according to the manufacturer's instructions.
2. *Do* install smoke alarms on the ceiling, as close as possible to the center of the room or hallway. Note: Because ceiling paddle fans are often installed in bedrooms, install the alarms on the wall space above the bedroom door, not closer than 4 in (102 mm) nor more than 12 in. (305 mm) from the ceiling.
3. *Do* make sure that the path of rising smoke will reach the alarm when the alarm is installed in a stairwell. This is usually at the top of the stairway. The alarm must not be located in a dead airspace created by a closed door at the head of the stairway.
4. *Do* space according to the instructions furnished with the devices.
5. *Do* install the type that has a "hush" button if located within 20 ft (6.1 m) of a cooking appliance, or install a photoelectric type.
6. *Do* install smoke alarms at the high end of a room that has a sloped, gabled, or peaked ceiling where the rise is greater than 1 ft (305 mm) per 8 ft (2.6 m). Do not install in the dead airspace.
7. *Do* mount on the bottom of open joists or beams.
8. *Do* install a smoke alarm on the basement ceiling close to the stairway to the first floor.
9. *Do* install smoke alarms in split-level homes. A smoke alarm installed on the ceiling of an upper level can suffice for the protection of an adjacent lower level if the two levels are not separated by a door. Better protection is to install alarms for each level.
10. *Do* install in new construction dual-powered smoke alarms that are hard-wired directly to a 120-volt ac source and also have a battery. In existing homes, battery-powered alarms are the most common, but 120-volt ac alarms or dual-powered alarms can be installed. Remember, battery-powered alarms will not operate with dead batteries. Ac-powered alarms will not operate when the power supply is off.
11. *Do* install interconnected smoke alarms in new construction so that the operation of any alarm will cause all other alarms that are interconnected to sound.
12. *Do* install smoke alarms so they are not the only load on the branch circuit. Connect smoke alarms to a branch circuit that supplies other lighting outlets in habitable spaces. Because smoke alarms draw such a miniscule amount of current, they can easily be connected to a general lighting branch circuit, such as a bedroom branch circuit.
13. *Do* always consider the fact that doors, beams, joists, walls, partitions, and similar obstructions

will interfere with the flow of smoke and heat, and, in most cases, create new areas needing additional smoke alarms and heat alarms.

14. *Do* make sure that the gap around the ceiling outlet box is sealed to prevent dust from entering the smoke chamber. This is a big problem for ceiling-mounted alarms.
15. *Do* clean smoke alarms according to the manufacturers' instructions.
16. *Do* consider that the maximum distance between heat alarms mounted on flat ceilings is 50 ft (15 m) and 25 ft (7.5 m) from the detector to the wall. This information is explained in detail by *NFPA 72*. Where obstructions such as beams and joists will interfere with the flow of heat, the 50 ft distance is reduced to 25 ft (7.5 m), and the 25 ft distance is reduced to 12½ ft (3.75 m).
17. *Do* install alarms after construction clean-up of all trades is complete and final.
18. *Do* install interconnected units of the same manufacturer. Different manufacturers' units may or may not be compatible.
19. *Do* connect all interconnected units to the same branch circuit. Different circuits cannot be shared.

The Do Nots

1. *Do not* install smoke alarms in the dead airspace at the top of a stairway that can be closed off by a door.
2. *Do not* install within 36 in. (914 mm) of a door to a kitchen or to a bathroom containing a shower or tub.
3. *Do not* locate where the smoke alarm will be subject to temperature and/or humidity that exceeds the limitations stated by the manufacturer.
4. *Do not* install within a 36 in. (914 mm) horizontal path from a supply hot air register. Install outside the direct flow of air from these registers.
5. *Do not* install within a horizontal path 36 in. (914 mm) from the tip of the blade of a ceiling-suspended paddle fan.
6. *Do not* install where smoke rising in a stairway could be blocked by a closed door or an obstruction.
7. *Do not* place the edge of a ceiling-mounted smoke alarm or heat alarm closer than 4 in. (102 mm) from the wall.
8. *Do not* place the top edge of a wall-mounted smoke alarm or heat alarm closer than 4 in. (102 mm) from the ceiling and farther than 12 in. (305 mm) down from the ceiling. Some manufacturers recommend placement not farther down than 6 in. (153 mm).
9. *Do not* install smoke alarms or heat detectors on an outside wall that is not insulated or one that is poorly insulated. Instead mount the alarms on an inside wall.
10. *Do not* install smoke alarms or heat alarms on a ceiling where the ceiling will be excessively cold or hot. Smoke and heat will have difficulty reaching the alarms. This could be the case in older homes that are not insulated or are poorly insulated. Instead mount the alarms on an inside wall.
11. *Do not* install smoke alarms or heat alarms where the ceiling meets the wall, because this is considered dead airspace where smoke and heat may not reach the detector.
12. *Do not* connect smoke alarms or heat alarms to wiring that is controlled by a wall switch.
13. *Do not* install smoke alarms or heat alarms where the relative humidity exceeds 85%, such as in bathrooms with showers, laundry areas, or other areas where large amounts of visible water vapor collect. Check the manufacturer's instructions.
14. *Do not* install smoke alarms or heat alarms in front of air ducts, air conditioners, or any high-draft areas where the moving air will keep the smoke or heat from entering the detector.
15. *Do not* install smoke alarms or heat alarms in kitchens where the accumulation of household smoke can result in setting off an alarm, even though there is no real hazard. The person in the kitchen will know why the alarm sounded, but other people in the house may panic. This problem exists in multifamily dwelling units, where

unwanted triggering of the alarm in one dwelling unit might cause people in the other units to panic. The photoelectric type may be installed in kitchens but must not be installed directly over the range or cooking appliance. A better choice in a kitchen is to install a heat alarm.

16. *Do not* install smoke alarms where the temperature can fall below 32°F (0°C) or rise above 120°F (49°C) unless the detector is specifically identified for this application.
17. *Do not* install smoke alarms in garages where vehicle exhaust might set off the detector. Instead of a smoke alarm, install a heat detector.
18. *Do not* install smoke alarms in airstreams that will pass air originating at the kitchen cooking appliances across the alarm. False alarms will result.
19. *Do not* install smoke alarms or heat detectors on ceilings that employ radiant heating.
20. *Do not* install smoke alarms or heat alarms in a recessed location.
21. *Do not* connect smoke alarms or heat alarms to a switched circuit or a circuit controlled by a dimmer.
22. *Do not* install smoke detectors in attics because of possible nuisance triggering. Check the manufacturers' instructions.

MAINTENANCE AND TESTING

Once installed, smoke, heat, and carbon monoxide alarms must be tested periodically and maintained (blow out accumulated dust) to make sure they are operating properly. This is a requirement of *NFPA 72*, which states, "Homeowners shall inspect and test smoke alarms and all connected appliances in accordance with the manufacturer's instructions at least monthly."**

Failure rate. Field studies indicate a probable failure rate of:

3% in the first year
30% in the first 10 years
50% in the first 15 years
Nearly 100% in 30 years

**Reprinted with permission from NFPA 72-2013.

The need for periodic testing and replacement is obvious.

When to replace. In conformance to 10.4.7 and 29.8.1.4(5) in *NFPA 72*, smoke alarms "Shall be replaced when they fail to respond to tests and shall not remain in service longer than 10 years from the date of manufacture."**

Note: The 10-year mandatory replacement requirement is often overlooked. This requirement is based on studies of tens of thousands of smoke alarms in operation, to determine the acceptable number of failures over many years. Code committees and manufacturers analyzed the data and agreed that 10 years in service provided a reasonable lifetime. The 10-year replacement program, along with regular testing per the manufacturer's instruction, results in very few homes going unprotected for any extended period of time. Hard-wired and battery-operated alarms are equally affected by age.

The expected life of lithium batteries used in some alarms is said to be 10 years.

At least one major manufacturer of smoke alarms has incorporated a feature that

- sets off a chirping sound after the alarm has been in service for 10 years. The chirping sound repeats every 30 seconds.
- sets off a chirping sound when the battery is low. The chirping sound repeats every minute.

Suggestion. It is difficult to know when a smoke, heat, or carbon monoxide alarm was installed. The instructions are usually filed away or were discarded or otherwise separated from these devices. It might not be a bad idea when installing the alarm to mark the installation date on the device.

After You Install Smoke and Heat Alarms, Then What?

After installing smoke and heat alarms, there is still more to do. *NFPA 72* requires that the installer of the fire warning system provide the homeowner with the following information:

1. An instruction booklet illustrating typical installation layouts
2. Instruction charts describing the operation, method, and frequency of testing and maintenance of fire warning equipment

3. Printed information for establishing an emergency evacuation plan
4. Printed information to inform owners where they can obtain repair or replacement service, and where and how parts requiring regular replacement, such as batteries or bulbs, can be obtained within 2 weeks
5. Information stating that unless otherwise recommended by the manufacturer, smoke alarms shall be replaced when they fail to respond to tests
6. Instruction that smoke alarms shall not remain in service longer than 10 years from the date of installation

This information is usually part of the instructions furnished by the manufacturer of the alarms.

Exceeding Minimum Levels of Protection

The following are recommendations for attaining levels of fire warning protection that exceed the minimum level stated previously and include guidelines for installing smoke and heat detectors in a home.

- Install smoke alarms in *all* rooms, basements, hallways, heated attached garages, and storage areas. Installing alarms in these locations will increase escape time, particularly if the room or area is separated by a door(s). In some instances, smoke alarms are installed in attics and crawl spaces.
- Install heat alarms in kitchens because conventional smoke alarms can nuisance sound an alarm.
- Install heat alarms in garages because gasoline fires give off little smoke.
- Consider special alarms that light up or vibrate for occupants who are hard of hearing.
- Consider smoke alarms that have an "escape" light.
- Consider low temperature alarms (e.g., 45°F [7°C]) that can detect low temperatures should the heating system fail. The damage caused by frozen water pipes bursting can be extremely costly.

Building Codes

In most instances the various building codes published by the International Code Council (ICC) (formerly BOCA, ICBO, and SBCCI) adopt *NFPA 72* and *NFPA 720* by reference, but it would be wise to check with your local AHJ to find out if there are any differences between these codes that might affect your installation. See Chapter 1 for information about these building code organizations.

Many companies specialize in installing complete fire and security systems. They also offer system monitoring at a central office and can notify the police or fire department when the system gives the alarm. How elaborate the system should be is up to the homeowner.

The ICC Code requires that smoke alarms be connected to a circuit that also supplies lighting outlets in habitable spaces. The logic is that should the alarms be connected to a separate circuit, that circuit could unknowingly be off.

Manufacturers' Requirements

Here are a few of the responsibilities of the manufacturers of smoke alarms and heat alarms. Complete data are found in *NFPA 72*.

- The power supply must be capable of operating the signal for at least 4 minutes continuously.
- Battery-powered units must be capable of a low-battery warning "chirp" of at least one chirp per minute for 7 consecutive days.
- Direct-connected 120-volt ac alarms must have a visible indicator that shows "power on."
- Alarms must not signal when a power loss occurs or when power is restored.

  Note: In *4.3.1* and *2.3.1* in *NFPA 72*, we find that we must

- always install fire alarm equipment that is "listed for the purpose."
- always install fire alarm equipment in accordance with the manufacturers' installation and maintenance instructions.

In this residence, the smoke alarms and heat detectors are connected to Circuit A17 located in Panel A in the Workshop. Run two conductors (the circuit) to the first alarm, and then interconnect the other alarms by running three conductors between all other detectors.

According to the plans for this residence, smoke, heat, and carbon monoxide alarms (detectors) are located as follows:

Smoke: Bedrooms	3
Hall between bedrooms	1
Entry	1
Recreation Room	1
Rear Hall	1
Total	7
Heat: Kitchen	1
Laundry Room	1
Workshop	1
Garage	1
Total	4
Carbon monoxide:	
Recreation Room	1
Hall between bedrooms	1
Total	2

CARBON MONOXIDE ALARMS

Carbon monoxide alarms are installed in addition to smoke and heat alarms!

Carbon monoxide (CO) is referred to as the "silent killer"! It is responsible for more deaths than any other single poison.

When a heat or smoke alarm is triggered, it is generally not difficult to determine the source of the problem. But when a carbon monoxide alarm goes off, it may be difficult to determine the source.

Carbon monoxide is odorless, colorless, and tasteless—undetectable by any of a person's five senses, taste, smell, sight, touch, hearing—but is highly poisonous. Carbon monoxide replaces oxygen in the bloodstream, resulting in brain damage or total suffocation.

Carbon monoxide is produced by the incomplete burning of fuels. Common sources of carbon monoxide in a home are a malfunctioning furnace, gas appliances, kerosene heaters, automobile or other gas engines, charcoal grills, fireplaces, or a clogged chimney.

Symptoms of carbon monoxide poisoning are similar to those of flulike illnesses, such as dizziness, fatigue, headaches, nausea, diarrhea, stomach pains, irregular breathing, and erratic behavior.

Carbon monoxide has about the same characteristics as air. It moves about just like air.

Carbon monoxide alarms detect carbon monoxide from any source of combustion. They are not designed to detect smoke, heat, or gas. The Consumer Product Safety Commission (CPSC) recommends that at least one carbon monoxide alarm be installed in each home, and preferably one on each floor of a multilevel house. Some cities and states require this. If only one carbon monoxide alarm is installed, it should be in the area just outside individual bedrooms.

Carbon monoxide alarms monitor the air in the house and sound a loud alarm when carbon monoxide above a predetermined level is detected. They provide early warning before deadly gases build up to dangerous levels.

For residential applications, carbon monoxide alarms are available for hard-wiring (direct connection to the branch-circuit wiring), plug-in (the male attachment plug is built into the back of the detector for plugging directly into a 120-volt wall receptacle outlet), and combination 120-volt and battery-operated units. They are available with the interconnect feature for interconnecting with smoke and heat alarms. Some carbon monoxide alarms also have an explosive gas sensor.

NFPA 720 contains the installation recommendations for carbon monoxide alarms. Here is a summary of these recommendations.

Installation "Dos" for Carbon Monoxide Alarms

1. *Do* install carbon monoxide alarms that are tested and listed in conformance to UL Standard 2034.
2. *Do* install carbon monoxide alarms according to the manufacturer's instructions.
3. *Do* use household electricity as the primary source of power. In existing homes, monitored battery units are permitted.
4. *Do* install carbon monoxide alarms in or near bedrooms and living areas.
5. *Do* install carbon monoxide alarms in locations where smoke alarms are installed.

6. *Do* test carbon monoxide alarms once each month or as recommended by the manufacturer.
7. *Do* remove carbon monoxide alarms before painting, stripping, wallpapering, or using aerosol sprays. Store in a plastic bag until the project is completed.
8. *Do* carefully vacuum carbon monoxide alarms once each month or as recommended by the manufacturer.
9. *Do* interconnect alarms where multiple alarms are installed.

Installation "Do Nots" for Carbon Monoxide Alarms

1. *Do not* connect carbon monoxide alarms to a switched circuit or a circuit controlled by a dimmer.
2. *Do not* install carbon monoxide alarms closer than 6 in. (150 mm) from a ceiling. This is considered dead airspace.
3. *Do not* install carbon monoxide alarms in garages, kitchens, or furnace rooms. This could lead to nuisance alarms, may subject the detector to substances that can damage or contaminate the alarm, or the alarm might not be heard.
4. *Do not* install carbon monoxide alarms within 15 ft (4.5 mm) of a cooking or heating gas appliance.
5. *Do not* install carbon monoxide alarms in dusty, dirty, or greasy areas. The sensor inside of the alarm can become coated or contaminated.
6. *Do not* install carbon monoxide alarms where the air will be blocked from reaching the alarm.
7. *Do not* install carbon monoxide alarms in dead airspaces such as in the peak of a vaulted ceiling or where a ceiling and wall meet.
8. *Do not* install carbon monoxide alarms where the detector will be in the direct airstream of a fan.
9. *Do not* install carbon monoxide alarms where temperatures are expected to drop below 40°F (4°C) or get hotter than 100°F (38°C). Their sensitivity will be affected.
10. *Do not* install carbon monoxide alarms in damp or wet locations such as showers and steamy bathrooms. Their sensitivity will be affected.
11. *Do not* mount carbon monoxide alarms directly above or near a diaper pail. The high methane gas will cause the detector to register, and this is not carbon monoxide.
12. *Do not* clean carbon monoxide alarms with detergents or solvents.
13. *Do not* spray carbon monoxide alarms with hair spray, paint, air fresheners, or other aerosol sprays.

FIRE ALARM SYSTEMS

These are the more comprehensive installations involving a central control panel, alarms, sensors, detectors as covered in the *National Fire Alarm Code, NFPA 72, Chapters 1* through *10*.

The wiring of fire alarm systems is covered in *Article 760* of the *NEC*.

NEC 760.2 defines a fire alarm circuit as The portion of the wiring system between the load side of the overcurrent device or the power-limited supply and the connected equipment of all circuits powered and controlled by the fire alarm system. Fire alarm circuits are classified as either non power-limited or power-limited.*

Fire-protective signaling systems installed in homes have power-limited fire alarm (PLFA) circuits where the power output is limited by the listed power supply for the system. Here are the key requirements in *Article 760, Part III*:

- The power supply must have power output capabilities limited to the values specified in *Chapter 9, Table 12(A)* and *Table 12(B)*. Refer to *760.121*. Residential fire-protective signaling systems are generally "power-limited."
- Do not connect the equipment to a GFCI- or AFCI-protected branch circuit, *760.41(B)* and *760.121(B)*.

*Reprinted with permission from NFPA 70-2014.

- Wiring on the supply side of the equipment is installed according to the conventional wiring methods found in *Chapter 1* through *Chapter 4* of the *NEC*. See *760.127*.
- The branch-circuit overcurrent protection supplying the fire alarm system shall not exceed 20 amperes, *760.127*.
- Cables may be run on the surface or be concealed and shall be adequately supported and protected against physical damage, *760.130(B)(1)*.
- When run exposed within 7 ft (2.1 m) of the floor, the cable shall be securely fastened at intervals not over 18 in. (450 mm), *760.130(B)(1)*.
- Cables shall not be run in the same raceway, cable, outlet box, or device box, or be in the same enclosure as light and power conductors, unless separated by a barrier or introduced solely to connect that particular piece of equipment. Refer to *760.136(A)* and *(B)*. This is also a ruling in *725.136(A)* and *(B)*.
- Cables shall not be supported by taping, strapping, or attaching by any means to any electrical conduits, or other raceways, *760.143*.
- Cables shall be supported by structural components, *760.24*.
- Wiring on the load side shall be insulated solid or stranded copper conductors of the types listed in *Table 760.179(A)*.
- *Table 760.179(I)* shows different types of cable that can be substituted for one another when necessary.
- If installed in a duct or plenum, the cable must be plenum rated, *760.179(D)*. The cable will be marked FPLP.
- Type FPL cable is permitted in one- and two-family dwellings, *760.179(F)*.
  - Conductors in cable shall not be smaller than 26 AWG. Single conductors shall not be smaller than 18 AWG. Refer to *760.179(B)*.
  - Cable minimum voltage rating is 300 volts, *760.179(C)*.
- Remove abandoned conductors and cables. They add fuel to a fire.

It makes a lot of sense to contact your local inspection authority before installing a fire alarm system to be sure the system you intend to install meets the requirements of the local codes! Contractors who specialize in this kind of work are well aware of the *Code* requirements and will install the proper types of cables.

SECURITY SYSTEMS

It is beyond the scope of this text to cover all types of residential security systems. What follows is a typical system.

Professionally installed and homeowner-installed security systems can range from a simple system to a very complex one. Features and options include master control panels, remote controls, perimeter sensors for doors and windows, motion sensors, passive infrared sensors, wireless devices, interior and exterior sirens, bells, electronic buzzers, strobe lights to provide audio (sound) as well as visual detection, heat sensors, carbon monoxide alarms, smoke alarms, glass break sensors, flood sensors, low temperature sensors, and lighting modules. Figure 26-10 shows some of these devices. Figure 26-11 shows a schematic one-line diagram for a typical security system. Some suppliers of Internet services provide a means to monitor and control security and alarm systems remotely, including by "smart" cellular telephones.

Monitored security systems are connected by telephone lines to a central station, a 24-hour monitoring facility. If the alarm sounds, security professionals will contact the end user to verify the alarm before sending the appropriate authorities (fire, police, or ambulance) to the home.

The decision about how detailed an individual residential security system should be generally begins with a meeting between the homeowner and the security consultant or licensed electrical contractor before the actual installation. Most security system installers, consultants, and licensed electrical contractors are familiar with a particular manufacturer's system.

The systems are usually explained and even demonstrated during an in-home security presentation. They are then professionally installed by

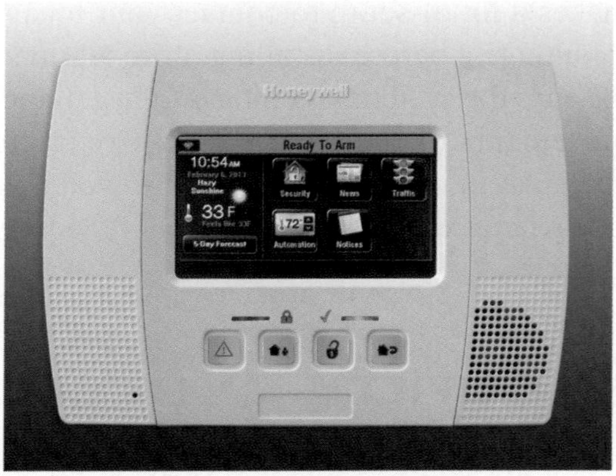

Wireless self-contained home/business control system allows the end-user to control their security system, thermostats, garage doors, lighting, locks and more.

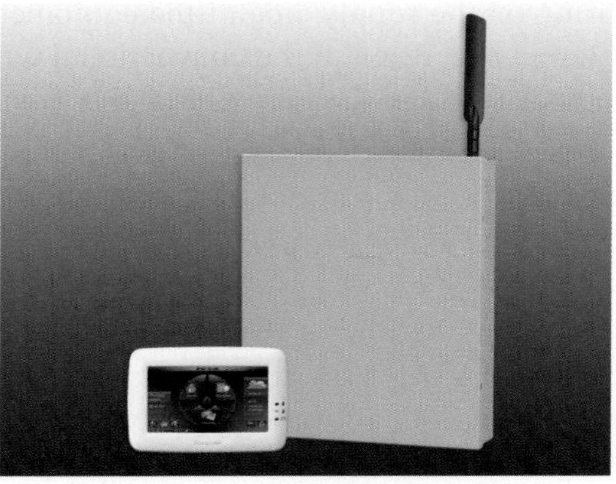

Hybrid control panels support both wired and hardwired sensors and can control security, manage home and building automation, provide energy savings and more.

Wireless keys allow the end-user to arm and disarm their security system with the press of a button. They can also be used to operate lights and appliances.

Professionally monitored smoke/heat detectors are on 24/7, even if the burglary protection is turned off.

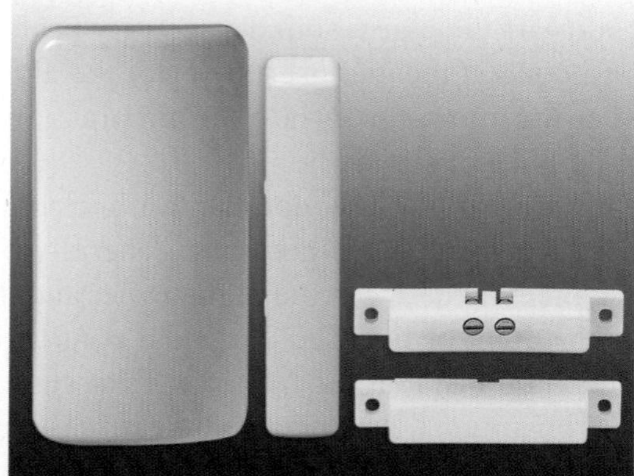

Contacts are often located at vulnerable entry points such as doors and windows.

Sirens and sounders alert users to emergencies.

Courtesy of Honeywell International, Inc.

FIGURE 26-10 Types of security systems and devices. (*continued*)

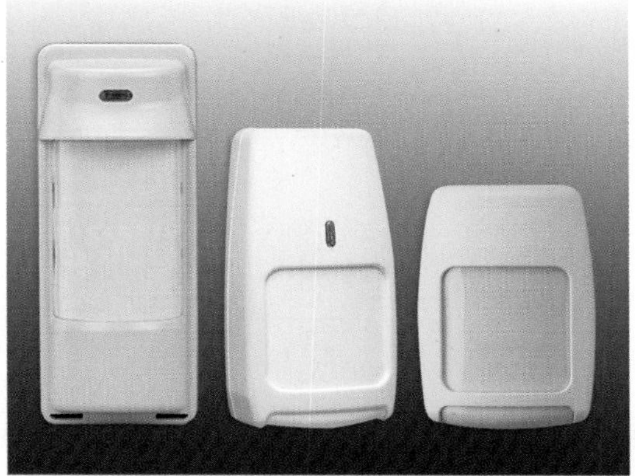

Motion detectors are security devices that sense motion inside or outside the home or business.

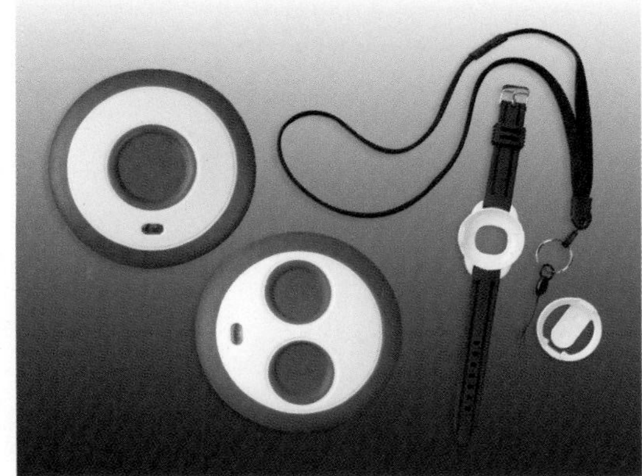

Wireless personal panic transmitters provide end-users with an extra level of convenience, comfort and peace of mind.

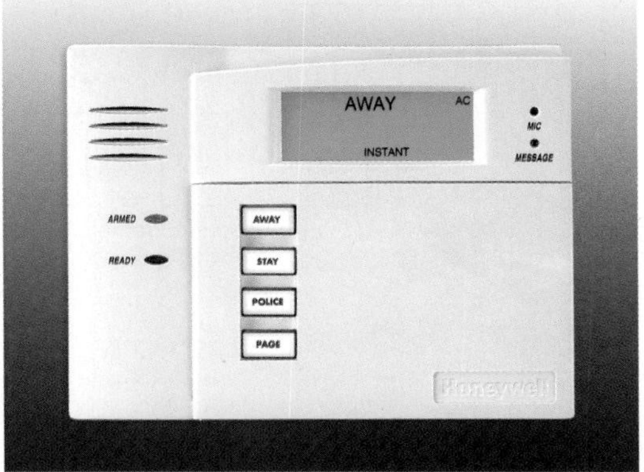

Designed to be mounted on walls or placed on any tabletop, wireless keypads are versatile and provide easy-to-use portable protection.

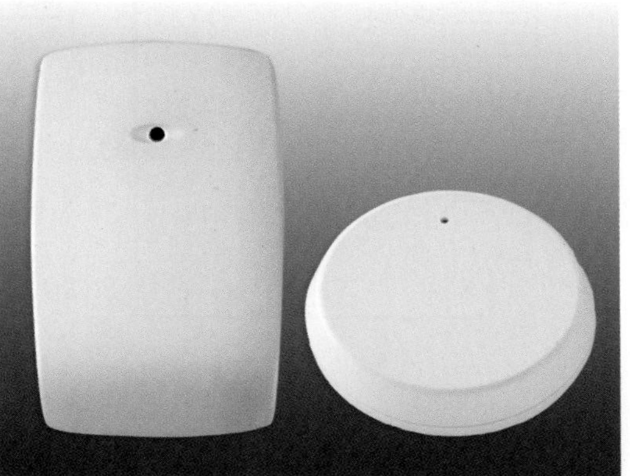

Glassbreak detectors will sound the alarm when they detect the sound of breaking glass. They provide perimeter protection and are considered the first line of defense in the protection of a home or business.

FIGURE 26-10 Types of security systems and devices.

the security system company or licensed electrical contractor.

Do-it-yourself kits that contain many of the features provided by the professionally installed systems can be found at home centers, hardware stores, electrical distributors, and electronics stores. Depending on the system, it may be possible to connect to a central monitoring facility.

The wiring of a security system consists of small, easy-to-install low-voltage, multiconductor cables made up of 18 AWG conductors. These conductors should be installed after the regular house wiring is completed to prevent damage to these smaller cables. Usually the wiring can be done at the same time as the chime wiring is being installed.

Security system wiring comes under the scope of *Article 725*.

When wiring detectors such as door entry, glass break, floor mat, and window foil detectors, circuits are electrically connected in series so that if any part of the circuit is opened, the security system will detect the open circuit. These circuits are generally referred to as "closed" or "closed loop." Alarms, horns, and other signaling devices are connected in parallel, because they will all signal at the same time when the security system is set off. Heat detectors and smoke detectors are generally connected in parallel because any of these devices will "close" the circuit to the security master control unit, setting off the alarms.

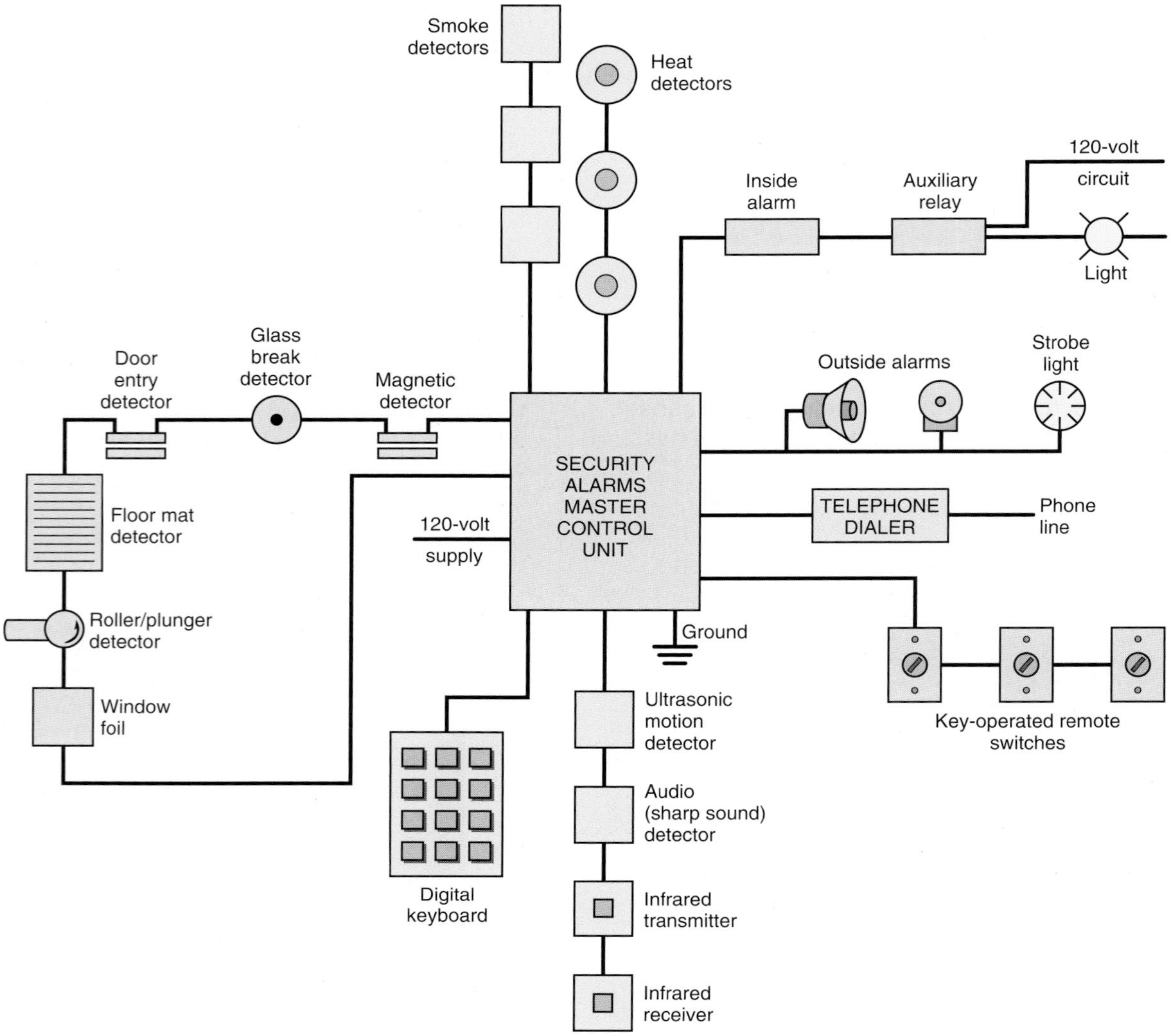

FIGURE 26-11 Diagram of typical residential security system showing some of the devices available. Complete wiring and installation instructions are included with these systems. Check local code requirements in addition to following the detailed instructions furnished with the system. Most of the interconnecting conductors are 18 AWG.

The instructions furnished with all security systems cover the installation requirements in detail, alerting the installer to *Code* regulations, clearances, suggested locations, and mounting heights of the systems' components.

Always check with the local electrical inspector to determine if there are any special requirements in your locality relative to the installation of security systems.

Sprinkler Systems Code

Home fire protection sprinkler systems are covered by NFPA 13D, which is the *Standard for the Installation of Sprinkler Systems in One- and Two-Family Dwellings and Manufactured Homes*.

The 2012 *International Residential Code* (*IRC*) in section *R313.2* requires, as of January 1, 2011, that new one- and two-family dwellings have an automatic residential fire sprinkler system.

An exception provides that the automatic residential fire sprinkler system is not required for *additions* or *alterations* to existing buildings that are not already provided with an automatic residential sprinkler system.

Automatic residential fire sprinkler systems are required to be designed and installed in accordance with *IRC* Section *P2904* or NFPA 13D.

This rule will apply where the *International Residential Code* is adopted and enforced. A parallel requirement is in the 2012 *International Building Code*.

Standards Relating to Fire, Smoke, Carbon Monoxide, and Security Devices

- UL 217—Standard for Single- and Multiple-Station Alarms
- UL 268—Standard for Smoke Detectors for Fire Protective Signaling Systems
- UL 365—Standard for Police Station Connected Burglar Alarm Units
- UL 521—Standards for Heat Detectors for Fire Protection Signaling Systems
- UL 539—Standard for Single- and Multiple-Station Heat Detectors
- UL 609—Standard for Local Burglar Alarm Units and Systems
- UL 827—Standard for Central Station Alarm Services
- UL 1023—Standard for Household Burglar Alarm System Units
- UL 1610—Standard for Central Station Burglar Alarm Units
- UL 1641—Standard for Installation and Classification of Residential Burglar Alarm Systems
- UL 2034—Standard for Single- and Multiple-Station Carbon Monoxide Alarms

REVIEW

1. In your own words, explain the terms *alarm* and *detector*.

____

____

____

____

____

2. The requirements for household fire alarm systems are found in what *NFPA Code* and in what chapter of this *Code*?

____

3. Do heat alarms and carbon monoxide alarms take the place of smoke alarms? Explain.

____

____

4. Name two basic ways that smoke alarms are powered.

____

____

5. Name the two types of smoke alarms.

____

____

6. List the absolute minimum level of smoke alarm protection in a new one-family dwelling.

____

____

7. Circle the correct answer from the following statements:
   a. Smoke alarms installed in new one- and two-family homes shall be battery operated only, so as not to be affected by a power outage.
   b. Smoke alarms installed in new one- and two-family homes shall be dual-powered by a 120-volt circuit and a battery.
8. Circle the correct answer from the following statements:
   a. Smoke alarms installed in new one- and two-family homes shall be interconnected so that if any one of them is triggered, all other alarms will also sound off.
   b. Smoke alarms installed in new one- and two-family homes shall not be interconnected so that each alarm will operate independently of all other alarms in the home.
9. Circle the correct answer from the following statements:
   a. Always install smoke and fire alarms in dead airspaces.
   b. Never install smoke and fire alarms in dead airspaces.
10. Circle the correct answer from the following statements:
    a. Mount wall-mounted smoke alarms so that the top edge is not closer than 4 in. (102 mm) from the ceiling and not more than 12 in. (305 mm) from the ceiling.
    b. Mount wall-mounted smoke alarms anywhere on the wall, but not lower than 36 in. (914 mm) from the ceiling.
11. Cooking and baking in the kitchen can produce quite a bit of smoke. The best choice for a smoke alarm in the kitchen is the photoelectric type. The better choice would be to install a (heat alarm) (carbon monoxide alarm). Circle the correct answer.
12. An important but often overlooked requirement in the *National Fire Alarm Code, NFPA 72* is that alarms must be replaced after being in service for (circle the correct answer)
    a. not more than 5 years.
    b. not more than 10 years.
    c. not more than 15 years.
13. Carbon monoxide is ______________, and is ______________, ______________, and ______________.
14. Circle the correct answer from the following statements:
    a. Carbon monoxide is heavier than air.
    b. Carbon monoxide will always rise to the ceiling.
    c. Carbon monoxide will drop to the floor.
    d. Carbon monoxide is about the same weight as air.
15. The wiring of more complex household fire alarm systems is covered in what article of the *National Electrical Code*? (Circle the correct answer.)
    a. *Article 725*
    b. *Article 310*
    c. *Article 760*

16. Circle the correct answer from the following statements:
    a. Always connect a fire alarm circuit or system to a power supply that has ground-fault circuit interrupter (GFCI) protection.
    b. Never connect a fire alarm circuit or system to a power supply that has ground-fault circuit interrupter (GFCI) protection.
    c. It makes no difference whether the branch circuit is GFCI or AFCI protected because the alarms are required to have battery backup power. In the event of a loss of power for whatever reason, the alarms would still be operative.
17. Circle the correct answer from the following statements:
    a. Conductors for fire alarm systems shall not be installed in the same raceways, cables, or electrical boxes as the light and power conductors.
    b. It is permissible to install fire alarm conductors in the same raceways, cables, or electrical boxes as the light and power conductors because the conductors for fire alarm systems covered in *Article 760* are small and would easily fit into other electrical raceways and boxes.
18. Fire alarm cable generally used in more complex residential fire alarm and security systems is marked (FPL) (low-voltage bell wire) (THHN). (Circle the correct answer.)
19. When installing a fire alarm system or a complete security system package in a home, always follow the installation instructions from (Circle the correct answer.)
    a. the manufacturer of the product and applicable codes.
    b. your neighbor, because he knows a lot about codes and standards.
    c. the man at the home center who sold you the product.
20. Residential sprinkler fire protection systems are required to be installed in compliance with the following (Circle the best answer.):
    a. NFPA 13D
    b. *IRC P2904*
    c. either *NFPA 13D* or *IRC P2904*
    d. neither *NFPA 13D* nor *IRC P2904*

CHAPTER **27**

Service-Entrance Equipment

OBJECTIVES

After studying this chapter, you should be able to

- understand the *NEC* terminology and requirements for installing all types of residential electrical services.
- calculate the proper size of residential service conductors and raceways.
- understand and install the required grounding and bonding for residential electrical services.
- understand the meaning of the term *UFER* ground.
- know the *NEC* requirements for grounding electrical equipment in a separate building.
- calculate the cost of using electricity.
- understand the meaning of conductor *withstand rating*.

INTRODUCTION

This chapter covers a lot of ground. Mastering the material in this chapter is quite an accomplishment. Be patient, understanding, and thorough as you tackle it. Congratulate yourself when you have completed this chapter.

An electric service is required for all buildings containing an electrical system and receiving electrical energy from a utility company. The *NEC*® defines the term service as, The conductors and equipment for delivering electric energy from the serving utility to the wiring system of the premises served.* The point where the utility's supply ends and the customer's premises wiring begins is called the *service point*. The utility company generally must be contacted to determine where they want the meter to be located.

NEC 230.79(C) requires a minimum 100-ampere, 3-wire disconnect for a one-family dwelling. Let's continue with the full details regarding residential electrical services.

IMPORTANT DEFINITIONS

Several definitions that are very important to planning and installing the service equipment for dwellings are located in *NEC Article 100*. Many of these clarify the portion of the service supply that is the responsibility of the electric utility and the portion that is the responsibility of the property owner or installer.

Service drop, The overhead conductors between the utility distribution system and the Service Point.* These conductors will be installed, owned, and maintained by the electric utility.

Service Conductors, Overhead. The overhead conductors between the Service Point and the first point of connection to the service-entrance conductors at the building or other structure.* These conductors will be installed by the electrician rather than by the electric utility. However, this rarely happens in practice. The electric utility almost always installs a *service drop* to the building or structure.

The definition of *Service Point* in *NEC Article 100* is important to our discussion of these issues. It reads, The point of connection between the facilities of the serving utility and the premises wiring.* This *service point* establishes the point or line of demarcation between the electric utility and its customer and is most often established by the electric utility. The electric utility is responsible for the conductors and equipment on the supply side of the *service point*, and the customer is responsible for the electrical conductors and equipment on the load side of the *service point*. See Figure 27-1 for an illustration of these definitions.

Furthermore, *NEC 90.2(B)(5)* states that the electric utility's service drop is not covered by the *NEC*. As a result, the title of *NEC 230.24* was revised to cover *Overhead Service Conductors* rather than *Service Drops*. This section clearly does not apply to the electric utility's service drop, but rather to the customer's overhead service conductors (which are rarely installed). However, *NEC 230.26* continues to require the point of attachment for the utility service drop conductors be located so as to provide for the requirements in *NEC 230.9* for building openings, and *NEC 230.24* for clearances above roofs and above the ground. The bottom line is the point of attachment for the utility's service drop must provide for the *NEC*-required clearances of overhead service conductors.

OVERHEAD SERVICE

An overhead service includes the service raceways, fittings, meter, meter socket, the main service disconnecting means, and the service conductors between the main service disconnecting means and the point of attachment to the utility's service-drop conductors. Usually, residential-type watt-hour meters are located on the exterior of a building. In some cases, the entire service-entrance equipment may be mounted outside the building. This includes the watt-hour meter and the disconnecting means. Figure 27-2 illustrates the *Code* terms for the various components of a service entrance.

Overhead service-drop conductors might be attached to a through-the-roof mast-type service or to the side of the house. In either case, proper

*Reprinted with permission from NFPA 70-2014.

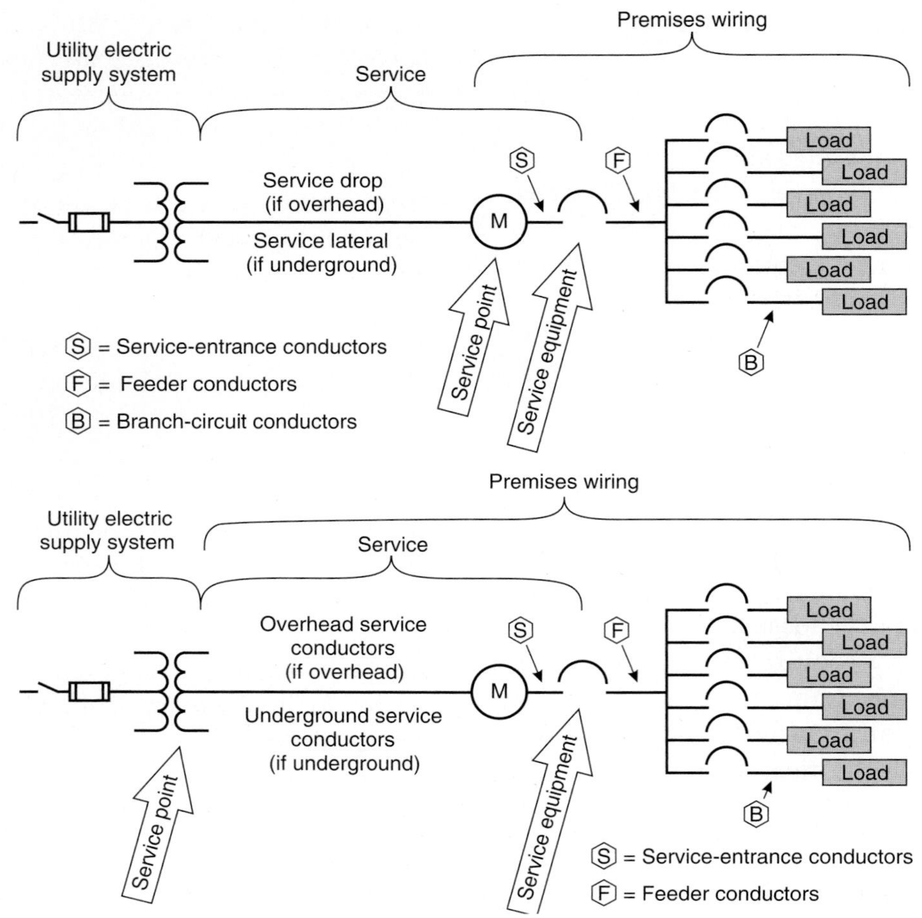

FIGURE 27-1 Definitions from *NEC* relative to overhead- and underground-supplied services.

clearance is required above the ground, roof, fences, windows, sidewalks, decks, and so on. Locate the mast insulator or screw-in knob so when the electric utility installs the service drop, the clearances above the ground or building features required in *NEC 230.24* are met.

MAST-TYPE SERVICE

The mast service, Figures 27-3, and 27-4, is a commonly used method of installing an overhead service entrance. The overhead mast service is most commonly used in areas where the electric utility distribution system has not been installed underground.

The service raceway is run through the roof, as shown in several figures including 27-2, 27-3, and 27-4, using a roof flashing and neoprene seal fitting, as illustrated in Figures 27-3(A) and 27-5. This fitting keeps water from seeping in and damaging the structure where the raceway penetrates the roof. The conduit is securely fastened to the building to comply with *Code* and electric utility requirements. Many types of fittings are readily available at electrical supply houses for securely fastening raceways and other electrical equipment to wood, brick, masonry, and other surfaces. Figures 27-3(B) shows typical methods of securing the conduit mast to

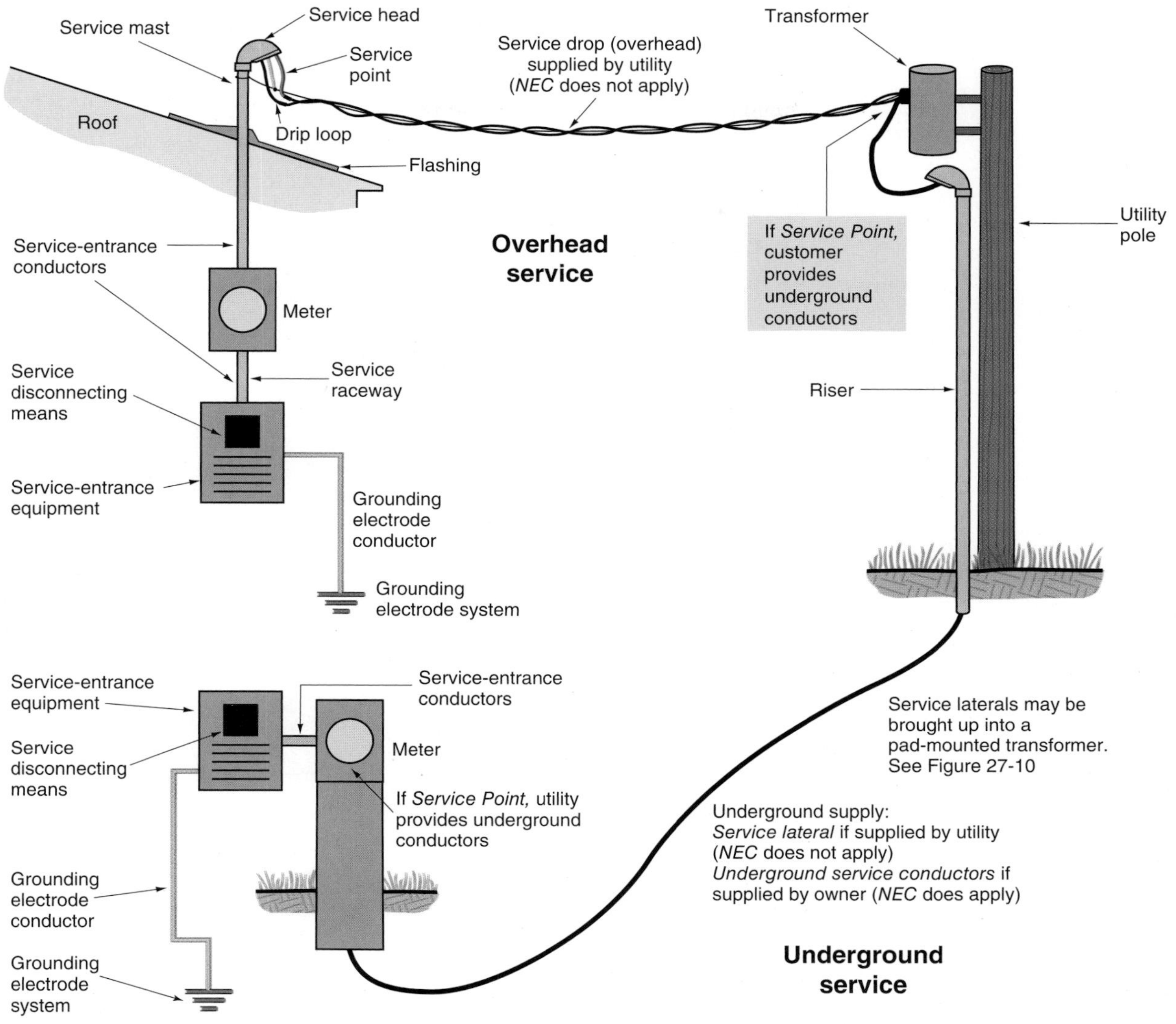

FIGURE 27-2 *Code* terms for services.

ensure it will safely carry the load of the service drop. Several conduit mast fittings are shown in Figure 27-5.

Service Mast as Support *(230.28)*

The bending force on the conduit increases with an increase in the distance between the roof support and the point where the service-drop conductors are attached. The pulling force of a service drop on a mast service conduit increases as the length of the service drop increases. As the length of the service drop decreases, the pulling force on the mast service conduit decreases.

If extra support is not to be provided, the mast service conduit must be not smaller than trade size 2 (metric designator 53). This size prevents the conduit from bending due to the strain of the service-drop conductors.

If extra support is needed, it is usually provided in the form of guy wires attached to a mast fitting and to the roof in a "Y" configuration. For fairly flat roofs, "stiff-legs" made out of galvanized pipe or conduit are often installed from the mast to the roof. Figure 27-3 shows several methods for securing conduit masts that are used to support the utility service drop. Some of the hardware that may be used for terminating the service drop is shown

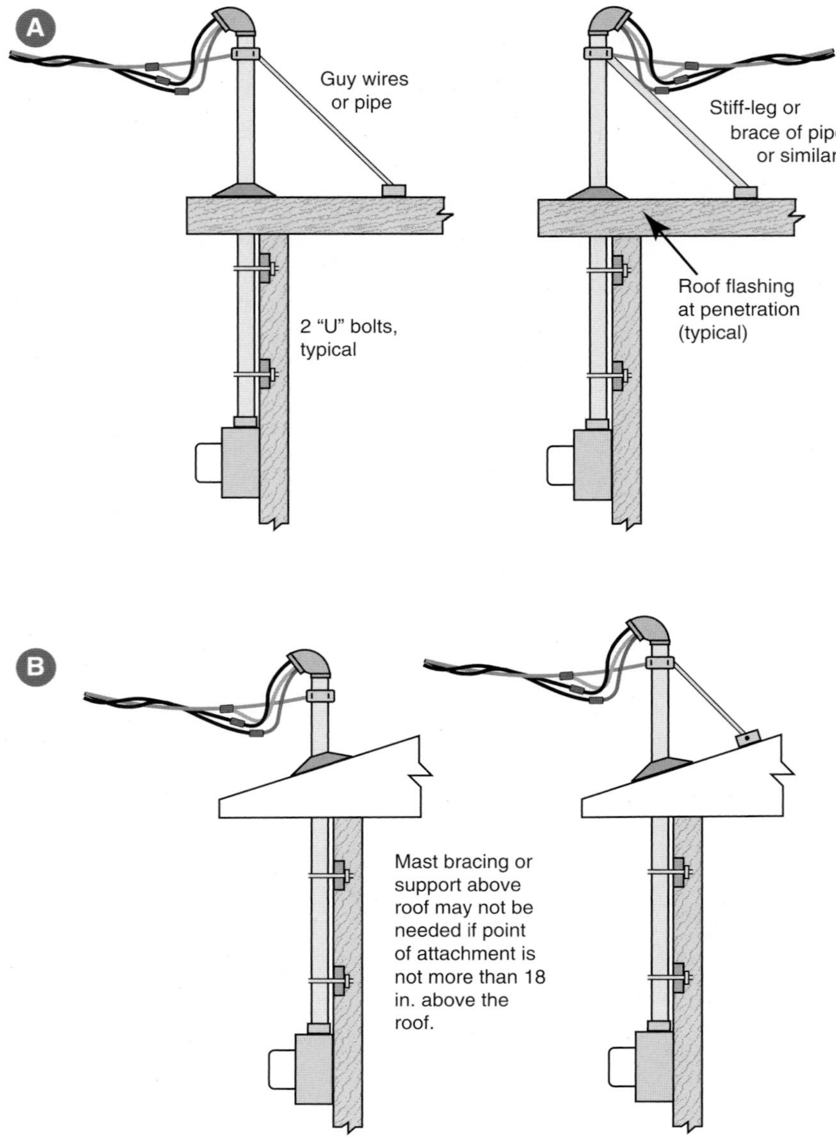

FIGURE 27-3 Typical methods of securing and supporting a mast-type service. Mast clamp positioned to provide for clearances in *NEC 230.24(A)* and *(B)*. Follow electric utility service requirements for all clearances.

in Figure 27-5(B), 27-5(C), 27-5(E). Again, many support fittings and devices are available at electrical supply houses.

When used as a through-the-roof service mast, rigid metal conduit or intermediate metal conduit is not required to be supported within 3 ft (900 mm) of the service head, *344.30(A)* and *342.30(A)*. Verify the application of this provision with the local electrical inspector and electrical utility, as many utilities have specific installation requirements for securing masts to which their service drop will be attached. The roof mast kit provides adequate support when properly installed.

Clearance Requirements for Service Drop and Overhead Service Conductors for Mast Installations

Several factors determine the maximum length of conduit that can be installed between the roof support and the point where the service-drop

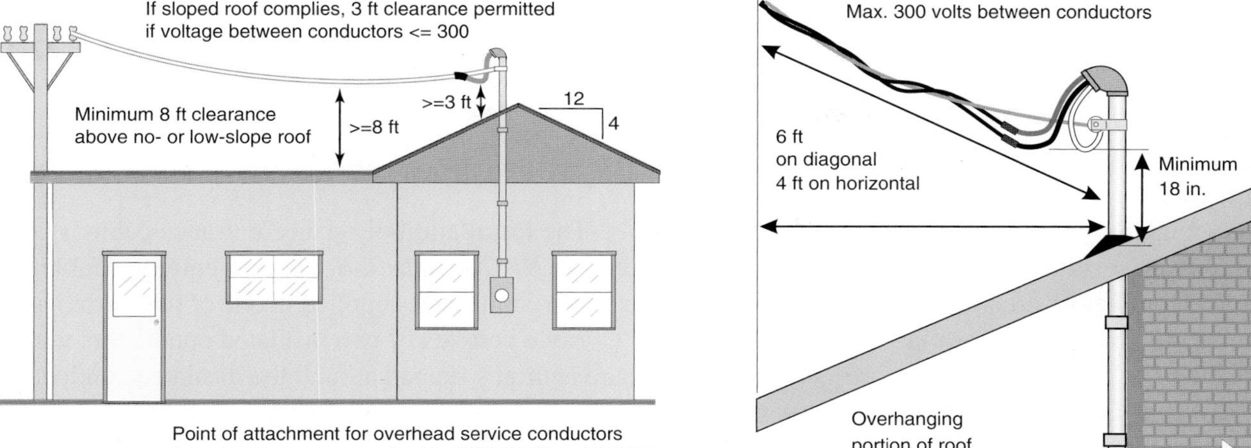

FIGURE 27-4 Clearances of overhead service conductors over roofs.

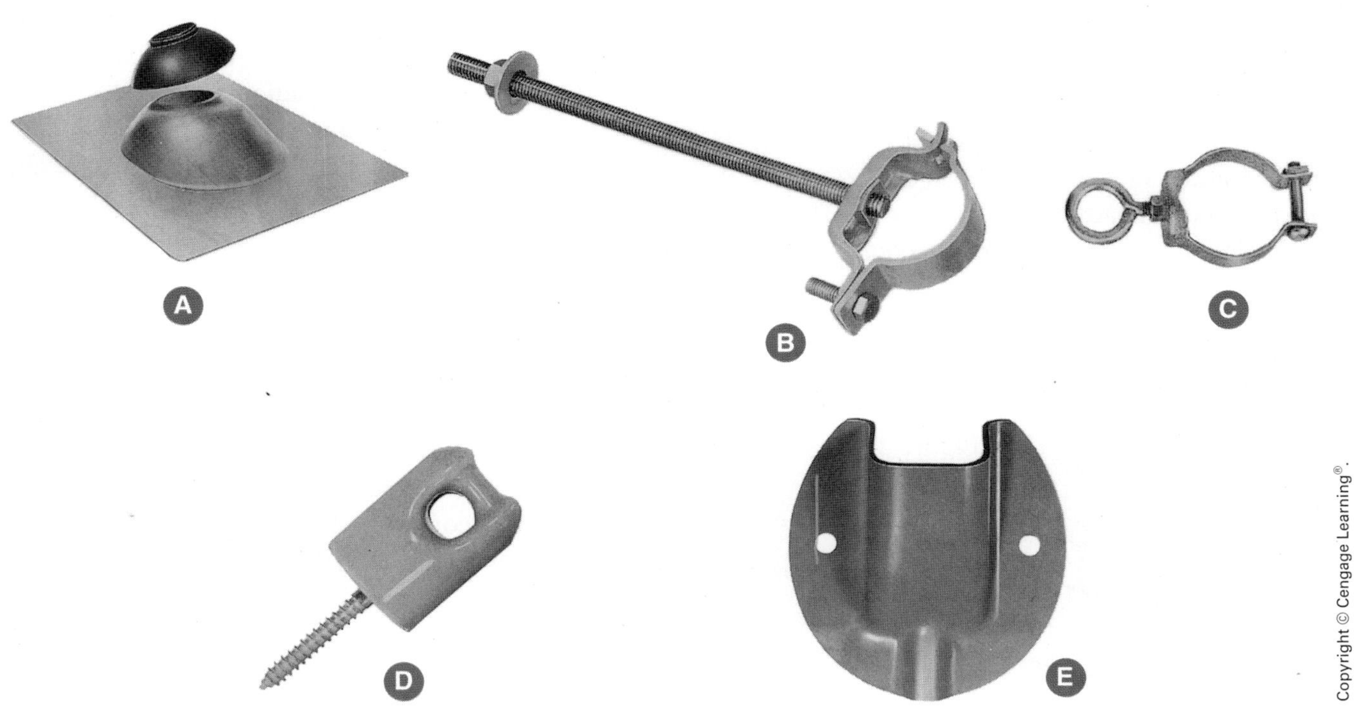

FIGURE 27-5 Fittings for conduit mast include (A) neoprene seal with metal flashing to use where the service raceway comes through the roof; (B) through-the-wall support for securing the service raceway; (C) clamp that fastens to the service raceway to which a guy wire can be attached for support of the service riser; (D) insulator that can be screwed into a barge rafter or at the gable end of the building; (E) an insulating clamp that is fastened to a conduit mast for terminating the service drop.

conductors are attached. Verify these factors with the local electric serving utility. Typical requirements are shown in Figures 27-3 and 27-4.

The *NEC* rules for insulation and clearances apply to the service-drop and service-entrance conductors. For example, the service conductors must be insulated except where the voltage to ground does not exceed 300 volts. In this case, the grounded neutral conductors are not required to have insulation.

Clearance above Roofs. Consult the electric utility that will serve the residence. *NEC 230.26* requires that the point of attachment is to provide for the clearances in *NEC 230.24(A)* for overhead service conductors. The utility service drop is treated as an equivalency to the customer's overhead service conductors. See Figure 27-4.

Only power service-drop conductors or overhead service conductors are permitted to be attached to and supported by the service mast, *230.28*. Refer to Figure 27-6.

All fittings used with raceway-type service masts must be identified for use with service masts, *230.28(A)*.

Consult the utility company and electrical inspection authority for information relating to their specific requirements for clearances, support, and so on, for service masts.

Clearance from Ground. Once again, consult the electric utility that will serve the residence. *NEC 230.26* requires the point of attachment to provide for the clearances in *NEC 230.9* and *230.24(B)* for the customer's overhead service conductors. These rules give the required clearances for overhead service conductors above ground. See Figure 27-7 for clearances.

Clearance from Windows. *NEC 230.24(C)* refers back to 230.9 to determine the required clearances for service conductors from windows and other building openings. See Figure 27-7.

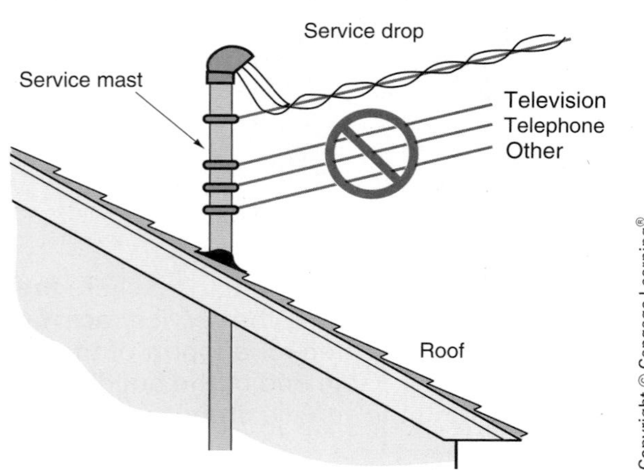

FIGURE 27-6 Only power service-drop or overhead service conductors are permitted to be attached to and supported by the service mast. *Do not* attach or support television cables, telephone cables, or anything else to the mast, *230.28*.

The installation requirements for a typical service entrance are shown in Figures 27-2, 27-3, 27-4, 27-7, and 27-8.

Installing Service-Entrance Cable

The installation of service-entrance cable is covered in *NEC Article 338*. Service-entrance cable for aboveground use is supplied in one of two varieties. A "U" style consists of two insulated conductors with a bare neutral wrapped around the insulated conductors in a spiral configuration. At terminations, the spiral neutral conductor is unwound and twisted together to form the third or neutral conductor. A 4-wire variety of Type SE cable is manufactured and is suitable for installation where a separate neutral and equipment grounding conductor are required, such as for wiring ranges, dryers, and some feeder panelboards.

As shown in Figure 27-9, fittings are specifically designed to facilitate installation of service-entrance cable outside the building. These fittings include a weatherhead and raintight and non-raintight cable clamps. Sill plates are available to protect the cable and weatherproof the installation where the cable is routed inside the building.

The cable must be protected against physical damage in accordance with *NEC 230.50*.

Be sure to check with the local electrical inspector, as some jurisdictions limit the length of service-entrance cables inside a building. It is considered more fragile than a metal raceway such as EMT, IMC, or rigid steel conduit and requires protection against physical damage in accordance with *NEC 300.4*.

Verify Which Overhead Clearances Apply

Electricians must abide by the *NEC*. Utilities must go along with the *National Electrical Safety Code*. These codes may have different clearance requirements for service-drop conductors and drip loops. When involved with installing an overhead service, check with your electrical inspector to determine which code is enforced in your area.

The installation requirements of the serving electric utility must always be complied with so long as they do not reduce the requirements of the *NEC*.

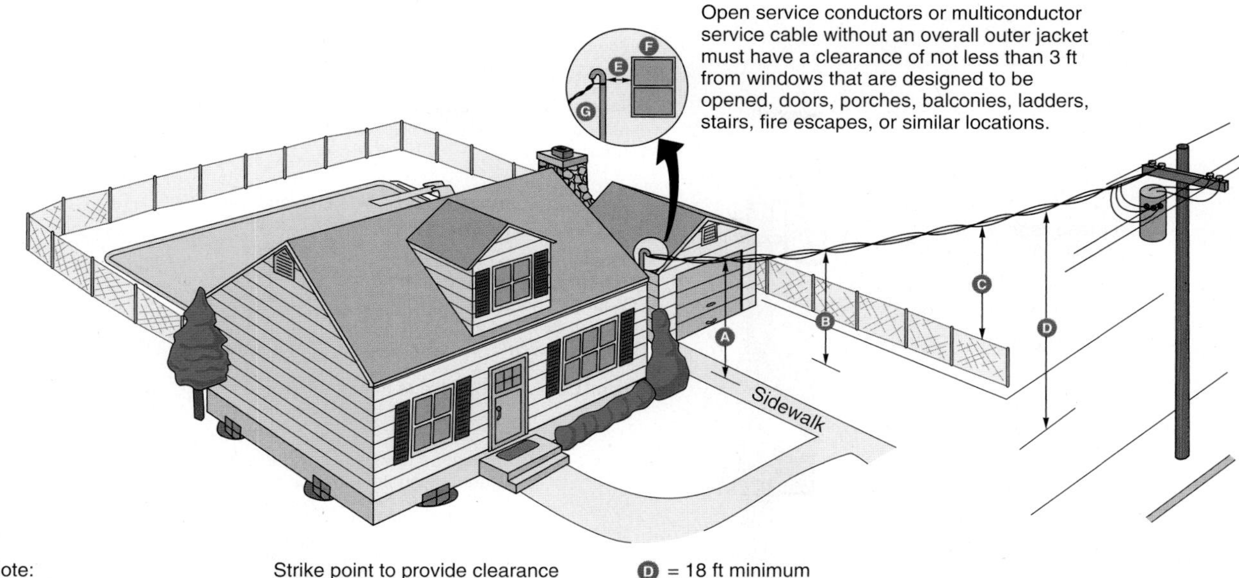

Note:
Clearances are designated by letters A through G

Strike point to provide clearance for overhead service conductors 120/240-volt, single-phase:
A = 10 ft minimum
B = 12 ft minimum
C = 10 ft minimum
D = 18 ft minimum
E = 3 ft minimum
F = Conductors run above top level of window are considered out of reach from that window. 3 ft clearance not required.
G = 10 ft minimum from drip loop to finish grade

Note: Electric utilities follow the *National Electrical Safety Code (NESC)*. The clearance requirements in the *NESC* may differ from those in the *National Electrical Code* (*NEC*). The deciding factor: Is the installation installed, owned, and maintained by the customer, or is the installation installed, owned, and maintained by the utility?

FIGURE 27-7 Clearances from ground for a residential service in accordance with typical utility requirements. Clearances (A) and (C) of 10 ft (3.0 m) permitted only if the service-entrance drop conductors are supported on and cabled together with a grounded bare messenger wire where the voltage to ground *does not* exceed 150 volts. This 10 ft (3.0 m) clearance must also be maintained from the lowest point of the drip loop. Clearance (B) of 12 ft (3.7 m) is permitted over residential property such as lawns and driveways where the voltage to ground *does not* exceed 300 volts. The circled insert shows clearance from building openings.

UNDERGROUND SERVICE

Underground service means conduit with conductors or direct-burial cable is installed underground from the point of connection at the service equipment to the system provided by the utility company.

New residential developments often include underground installations of the high-voltage electrical systems. The conductors in these distribution systems connect to pad-mounted transformers, Figure 27-10. These transformers are often placed at the rear lot line or in other inconspicuous locations in the development. The transformer and primary high-voltage conductors are generally installed by, and are the responsibility of, the utility company.

If installed by the electric utility, the conductors installed between the pad-mounted transformer and the meter are called *service lateral conductors*. Normally, the electric utility furnishes and installs them. In many areas, the utility will install both the service lateral conductors and the communication cables in the same trench. In some areas, but not too commonly, service laterals, communication cables, and gas lines are run in the same trench. Figure 27-10 shows a typical underground installation.

The wiring from the external meter to the main service equipment is the same as the wiring for a service connected from overhead lines, as in

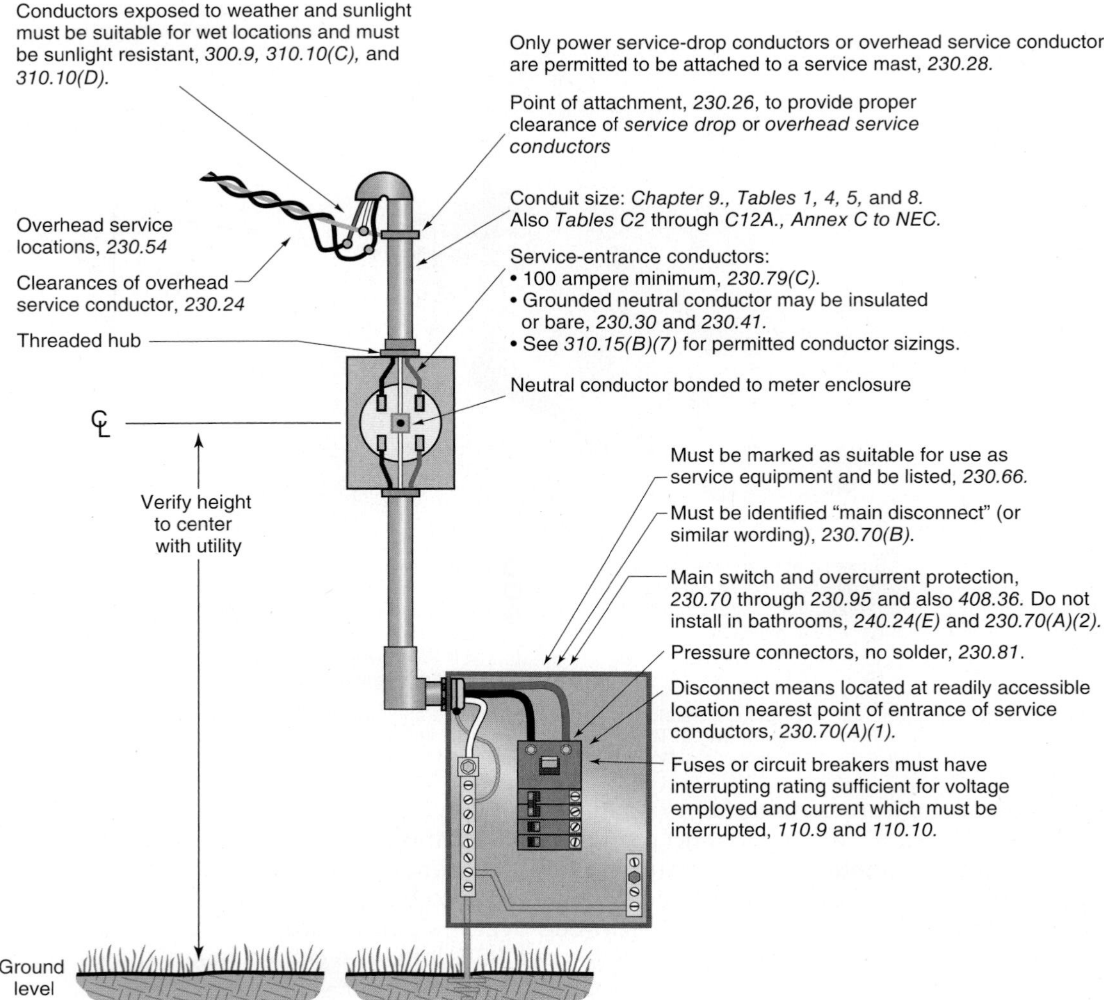

FIGURE 27-8 The wiring of a typical service-entrance installation. Today, most services are supplied underground, as shown in Figure 27-10.

Figure 27-8. Some local codes may require conduit to be installed underground from the pole to the service-entrance equipment. *NEC* requirements for underground services are given in *Article 230, Part III*. If installed by the owner or electrician, the underground conductors must be suitable for direct burial in the earth, Type USE single-conductor or Type USE cable containing more than one conductor, *NEC Article 338*.

If the electric utility installs the underground service conductors, the work must comply with the rules established by the utility. Utilities follow the *National Electrical Safety Code (NESC)*. These rules may not be the same as those given in *NEC 90.2(B)(5)*.

As stated in *NEC 90.2(B)(5)*, the *NEC* does not cover installations under the exclusive control of an electric utility where such installations:

a. consist of service drops or service laterals and associated metering, or
b. are on property owned or leased by the electric utility for the purpose of communications, metering, generation, control, transformation,

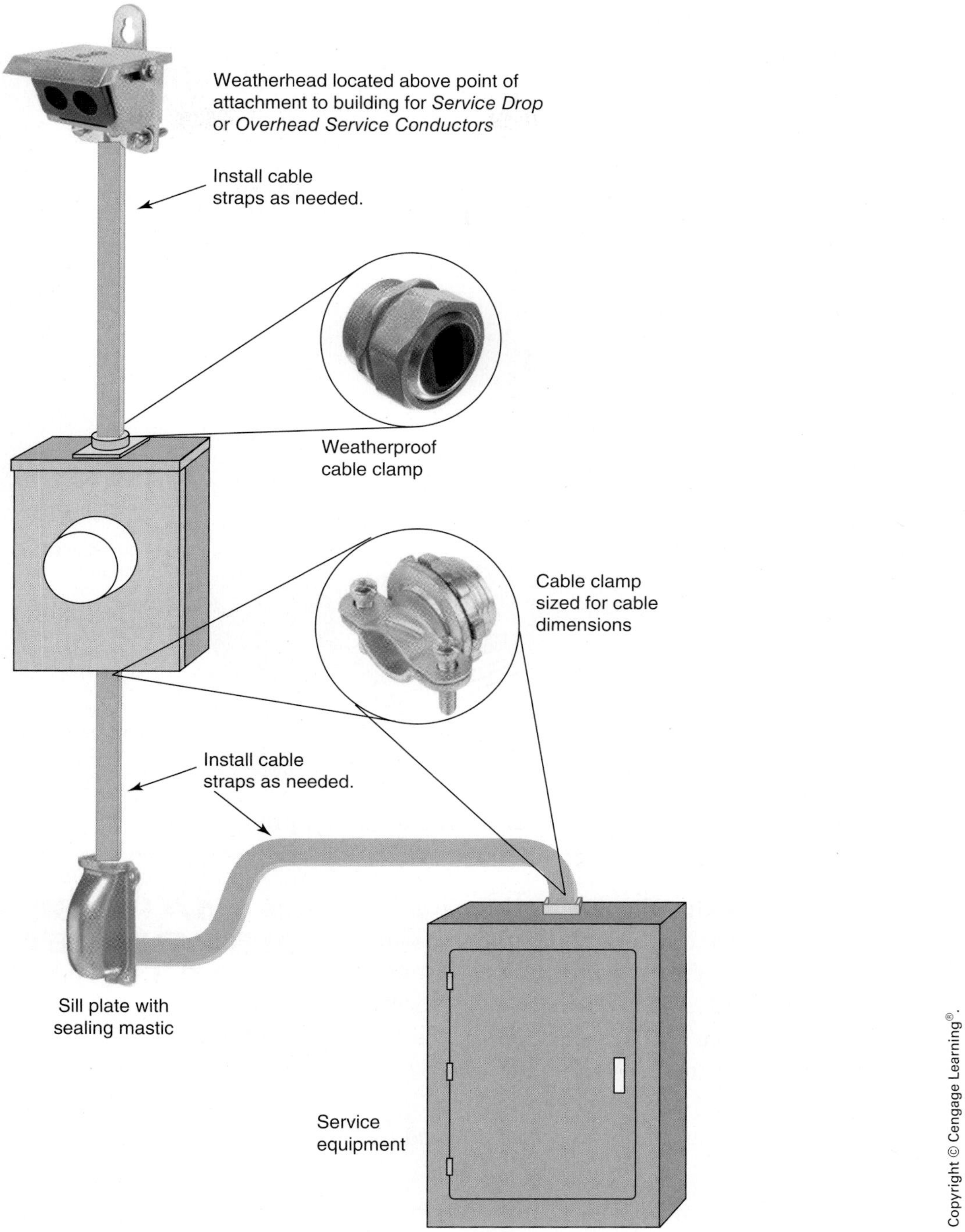

FIGURE 27-9 Typical hardware used for installation of service-entrance cable for services.

transmission, or distribution of electric energy, or

c. are located in legally established easements, rights-of-way, or

d. are located by other written agreements either designated by or recognized by public service commissions, utility commissions, or other regulatory agencies having jurisdiction for such installations. These written agreements are limited to installations for the purpose of communications, metering, generation, control, transformation, transmission, or

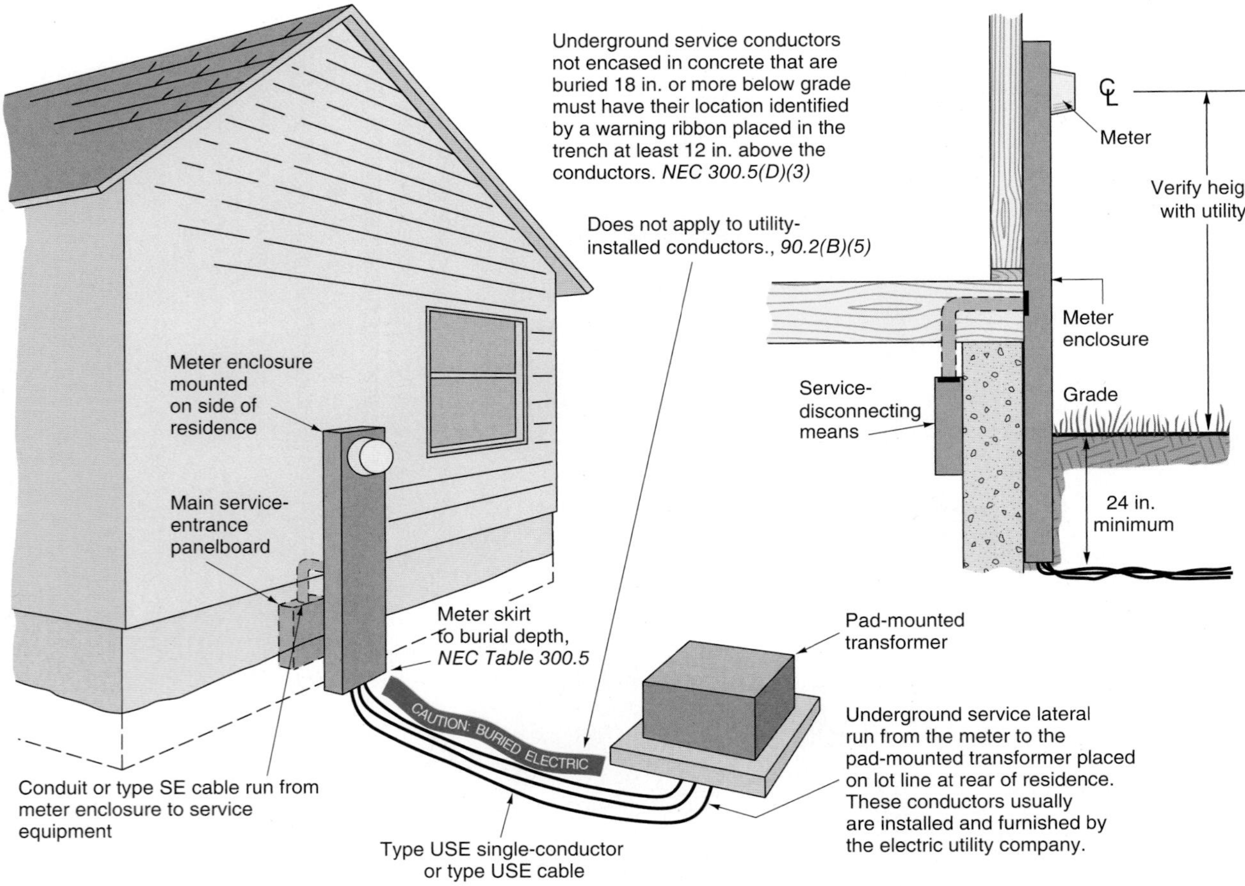

FIGURE 27-10 Underground service.

distribution of electric energy where legally established easements or rights-of-way cannot be obtained. These installations are limited to Federal Lands, Native American Reservations through the U.S. Department of the Interior Bureau of Indian Affairs, military bases, lands controlled by port authorities and State agencies and departments, and lands owned by railroads.

When underground service conductors or service-entrance cables are installed by the electrician, *NEC 230.50* applies. This section deals with the protection of service-entrance conductors and service-entrance cables against physical damage. *NEC 230.50* also refers to *300.5*, which covers all situations involving underground wiring, such as the sealing of raceways where they enter a building. Sealing raceways is covered in "Meter/Meter Base Location" later on in this chapter.

MAIN SERVICE DISCONNECT LOCATION

In conformance to *230.70(A)(1)* and *230.70(A)(2)*, the main service disconnecting means shall

- be located inside the building at a readily accessible location nearest the point of entrance of the service conductors, Figure 27-11, or
- be located outside the building at a readily accessible location, and
- not be installed in a bathroom.

If for some reason it is necessary to locate service-entrance equipment some distance inside the building, *NEC 230.6* considers service-entrance conductors to be outside of a building if the service-entrance conductors are one of the following:

- installed under not less than 2 in. (50 mm) of concrete beneath the building, or

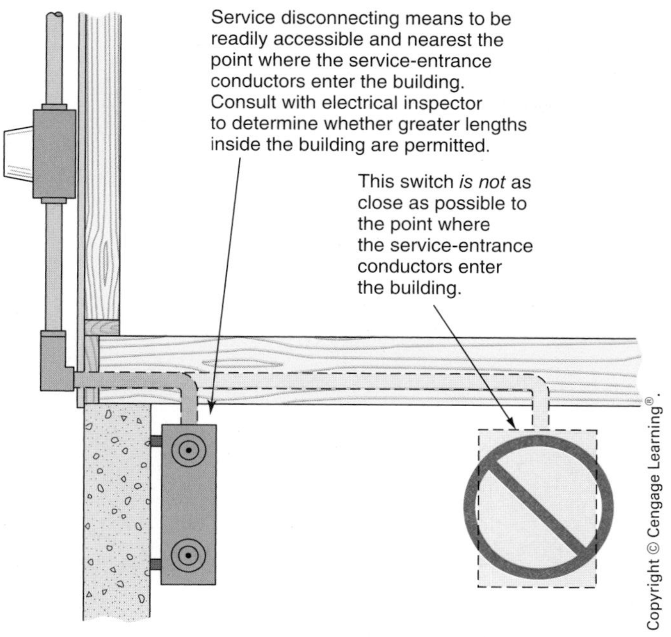

FIGURE 27-11 Main service disconnect location, *230.70*. Do not install in a bathroom. Refer to *230.70(A)(2)* and *240.24(E)*.

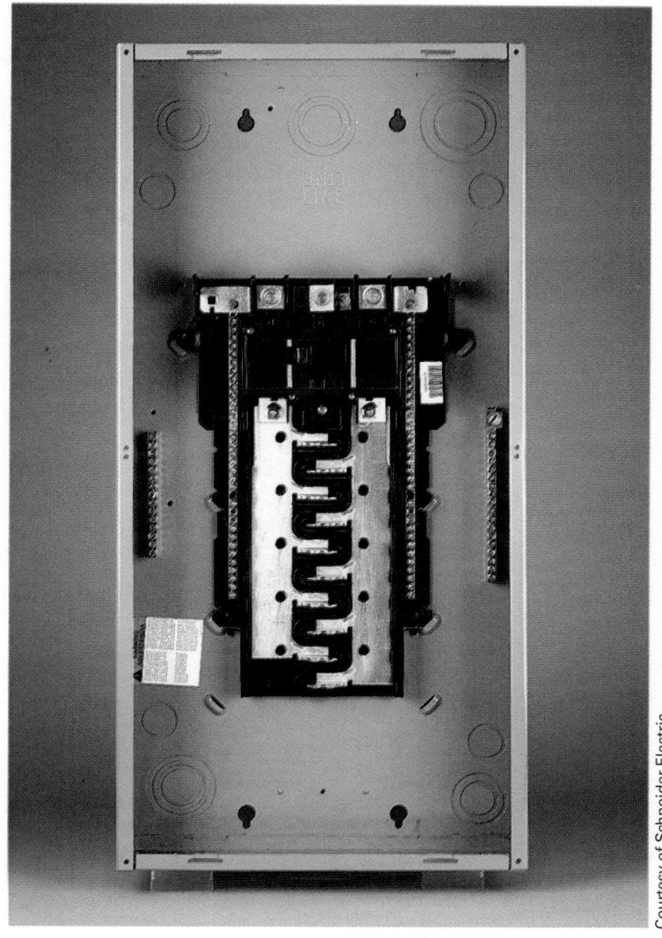

FIGURE 27-12 A typical 120/240-volt, single-phase, 3-wire load center. This load center has a main circuit breaker. Note the neutral terminal bars for termination of the white branch-circuit grounded conductors, and the equipment grounding terminal bars for termination of the equipment grounding conductors of nonmetallic-sheathed cable. See *230.66*.

- installed within the building in a raceway encased in at least 2 in. (50 mm) of concrete or brick, or
- installed in conduit, and covered by at least 18 in. (450 mm) of earth beneath the building.

Circuit Directory/Circuit Identification

NEC 408.4 is very clear as to its requirements on the subject. It requires that Every circuit and circuit modification shall be legibly identified as to its clear, evident, and specific purpose or use. The identification shall include sufficient detail to allow each circuit to be distinguished from all others. Spare positions that contain unused overcurrent devices or switches shall be described accordingly. The identification shall be included in a circuit directory that is located on the face or inside of the panel door in the case of a panel board, and located at each switch on a switchboard. No circuit shall be described in a manner that depends on transient conditions of occupancy.*

Figure 27-12 shows a typical 240/120-volt, single-phase, 3-wire panelboard.

Figure 27-13 shows the circuit identification for Main Service Panel A. Figure 27-14 shows the circuit identification for sub-Panel B.

NEC 110.22 also requires that disconnect switches be "legibly marked," describing what the disconnect is for.

*Reprinted with permission from NFPA 70-2014.

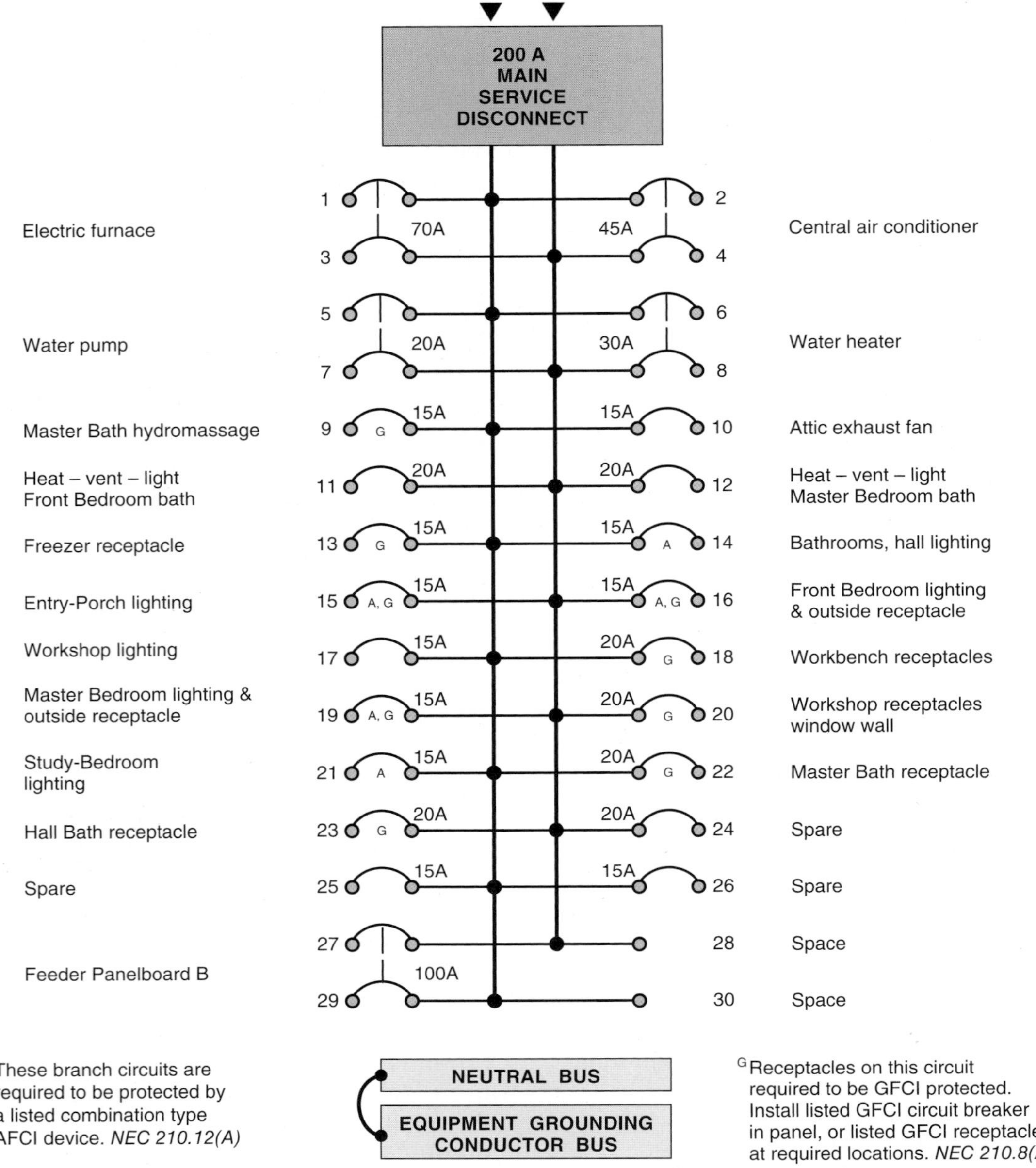

FIGURE 27-13 Circuit schedule of Main Service Panel A.

Service Disconnecting Means (Panel A)

It is extremely important that a disconnect or panelboard used as service equipment be marked "Suitable for Use as Service Equipment." This is a requirement of *230.66*. Some disconnects and panelboards are so marked—others are not. Such marking ensures that proper bonding together of the neutral bus, equipment ground bus, and metal enclosure is provided, as well as a termination (lug) for the grounding electrode conductor. This section also requires that service equipment be listed. This means the equipment must bear the listing mark or label from a qualified electrical products testing laboratory.

More requirements for service disconnect means are found in *230.70* through *230.82*. *NEC 230.71(A)* requires that the service disconnecting means consist of not more than six switches or six circuit breakers mounted in a single enclosure, in a group of

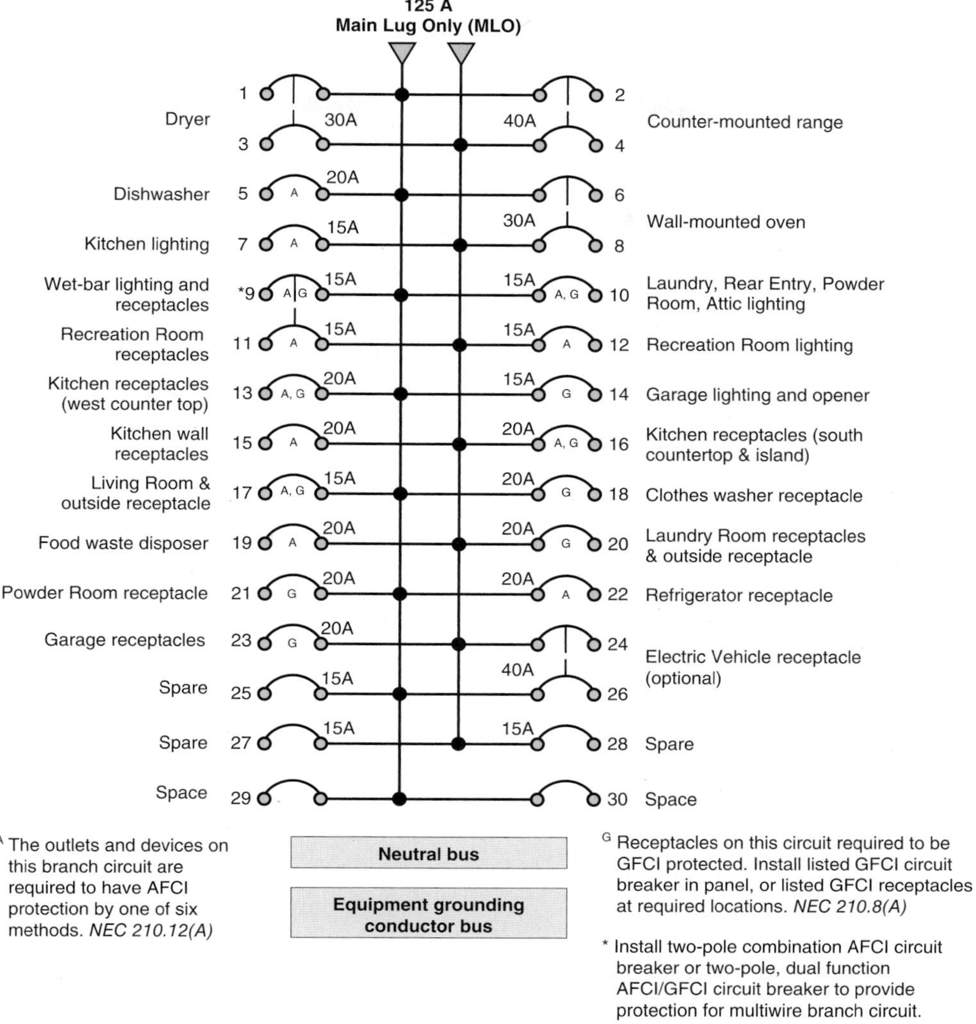

FIGURE 27-14 Circuit schedule of Panelboard B.

separate enclosures in the same location, or in or on a switchboard. This permits the disconnection of all electrical equipment in the house with not more than six hand operations. Some local codes take exception to the six-disconnect rule and require a single main disconnect. Some localities require the main disconnecting means to be located outside. If located outside, *230.70(A)(1)* requires that the service disconnect be installed at a readily accessible location.

You need to check this out with the local authority that has jurisdiction in your area.

In damp and wet locations, to prevent the accumulation of moisture that could lead to rusting of the enclosure and damage to the equipment inside the enclosure, *312.2* specifies that there be at least a ¼ in. (6 mm) airspace between the wall and a surface-mounted enclosure. Most disconnect switches, panelboards, meter sockets, and similar equipment have raised mounting holes that provide the necessary clearance. This is clearly visible in the open-meter socket in Figure 27-15 of the "Meter/Meter Base Location" section in this chapter. The *NEC* (in Definitions) considers masonry in direct contact with the earth to be a wet location. The *Exception* to *312.2* allows nonmetallic enclosures to be mounted without an airspace on concrete, masonry, tile, and similar surfaces.

Panelboards are required to be installed in accordance with the listing of the panelboard. A panelboard used as service equipment that has multiple overcurrent devices is not required to have overcurrent

Courtesy of Milbank Manufacturing Company.

FIGURE 27-15 Typical self-contained meter socket. Combination meter socket/main breaker in one enclosure is also available.

protection on the line side of the panelboard, *408.36, Exception No. 1*. When used as other than service equipment, overcurrent protection for the panelboard is not permitted to exceed the rating of the panelboard.

NEC 230.79 tells us that the service disconnect must have a rating not less than the calculated load to be carried, determined in accordance with *Parts III, IV*, or *V* of *Article 220*. We show how these calculations are made in Chapter 29 of this text.

For a one-family residence, the minimum size main service disconnecting means is 100 amperes, 3-wire, *NEC 230.79(C)*.

NEC 230.70(B) requires that the main service disconnect be permanently marked to identify it as service disconnect.

When nonmetallic-sheathed cable is used, a panelboard must have a terminal bar for attaching the equipment grounding conductors, *408.20*. See Figure 27-16.

For this residence, panelboard A serves as the main disconnecting means for the service. This panelboard consists of a main circuit breaker and many branch-circuit breakers. Figure 27-12 is a photo of this type of panelboard. Figure 27-13 is the schedule of the circuits connected to panelboard A. Panelboard A is located in the Workshop.

Working Space. To provide for safe working conditions around electrical equipment, *110.26* contains a number of rules that must be followed. For residential installations, there are some guidelines for working space.

- A working space not less than 30 in. (762 mm) wide and 3 ft (914 mm) deep must be provided in front of electrical equipment, such as panelboards A and B in this residence.
- This working space must be kept clear and not used for storage.
- Do not install electrical panelboards inside of cabinets, nor above shelving, washers, dryers, freezers, work benches, and so forth.
- Do not install electrical panels in bathrooms.
- Do not install electrical panels above or close to sump pump holes.
- The hinged cover (door) on the panel shall be able to be opened to a full 90°.
- A "zone" equal to the width and depth of the electrical equipment, from the floor to a height of 6 ft (1.8 m) above the equipment or to the structural ceiling, whichever is lower, shall be dedicated to the electrical installation. No piping, ducts, or equipment foreign to the electrical installation shall be located in this zone. This zone is dedicated space intended for electrical equipment only! See *110.26(E)*.

Mounting Height. Disconnects and panelboards must be mounted so the center of the main disconnect handle in its highest position is not higher than 6 ft 7 in. (2.0 m) above the floor, *404.8*.

Headroom. To provide for safe adequate working space and easy access to service equipment and electrical panels, the *NEC* in *110.26(A)(3)* requires that there be at least 6½ ft (2.0 m) of headroom. If you are installing service equipment in an existing

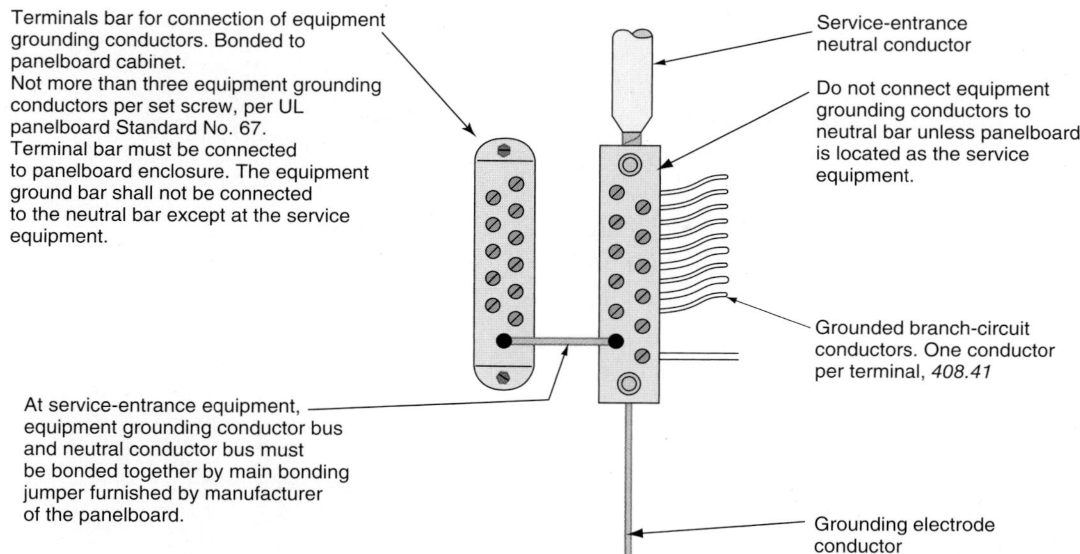

FIGURE 27-16 Connections of service neutral conductor, branch-circuit neutral conductors, and equipment grounding conductors at panelboards, *408.40* and *408.41*.

dwelling, services of 200 amperes or less are permitted to be located in areas where the headroom is less than 6½ ft (2.0 m).

Lighting Required for Service Equipment and Panels. Working on electrical equipment with inadequate lighting can result in injury or death. In *110.26(D)*, there is a requirement that illumination be provided for service equipment and panelboards. But the *Code* does not spell out how much illumination is required. It becomes a judgment call on the part of the electrician and/or electrical inspector. Note that the electrical plans for the workshop show a porcelain pull-chain lampholder to the front of and off to one side of the main service panelboard A. In many cases, adjacent lighting, such as the fluorescent luminaires in the recreation room, is considered to be the required lighting for equipment such as panelboard B.

Panelboard Setback in Walls

Panelboards are sometimes installed within a wall for flush mounting. The setback and repair opening rules for panelboards are found in *NEC 312.3* and *312.4*:

- In plaster, drywall, or plasterboard: repair wall so no gap or opening is greater than ⅛ in. (3 mm).
- In combustible walls: flush.
- In wall of concrete, tile, or other noncombustible material: set back not more than ¼ in. (6 mm).

Close Openings in Panelboards

Inside a panelboard, there are live parts! There is a real fire and shock hazard when there are openings (knockouts) on the sides, top, or bottom of a panelboard or when circuit breaker openings in the cover are inadvertently broken out.

NEC 408.7 requires that Unused opening for circuit breakers and switches shall be closed using identified closures, or other approved means that provide protection substantially equivalent to the wall of the enclosure.*

Identified means Recognizable as suitable for the specific purpose, function, use, environment, application, and so forth, where described in a particular Code requirement.*

Approved means Acceptable to the authority having jurisdiction.*

*Reprinted with permission from NFPA 70-2014.

Panelboard B

When a second panelboard is installed that is supplied by a feeder, it is oftentimes called a subpanel. It would no doubt be more correctly referred to as a feeder panelboard.

For installation where the main service panelboard is a long distance from areas that have many circuits and/or heavy load concentration, as in the case of the kitchen and laundry of this residence, it is recommended that at least one additional panel be installed near these loads. This results in the branch-circuit wiring being short. Line losses (voltage and wattage) are considerably less than if the branch circuits had been run all the way back to the main panelboard.

The cost of material and labor to install an extra panelboard should be compared to the cost of running all of the branch-circuit home runs back to the main service panelboard.

Panelboard B in this residence is located in the Recreation Room. It is fed by three 3 AWG THHN or THWN conductors run in a trade size 1 EMT originating at the service (panelboard A). This feeder is protected by a 100-ampere, 2-pole circuit breaker located in panelboard A.

Do not install panelboards in clothes closets, *240.24(D)*. There is a fire hazard because of the presence of ignitable materials.

Do not install panelboards in bathrooms, *240.24(E)*. Damaging moisture problems may result from the presence of water, and shock hazard may result from both the presence of water and the close proximity of metal faucets, and other plumbing fixtures.

Panelboard B is shown in Figure 27-17. The circuit schedule for panelboard B is found in Figure 27-14.

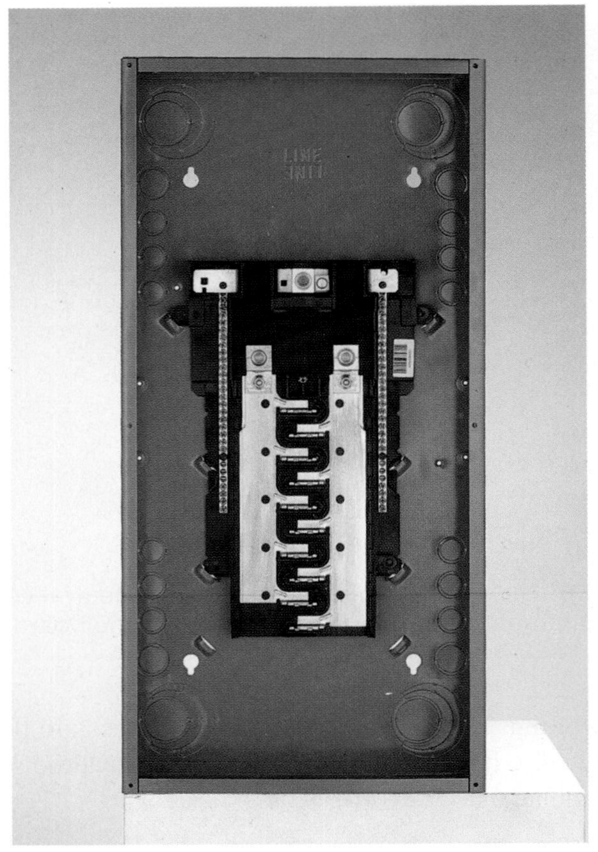

Courtesy of Schneider Electric.

FIGURE 27-17 Typical 120/240-volt, single-phase, 3-wire panelboard, sometimes called a load center. This panelboard does not have a main disconnect and is referred to as a *main lug only (MLO)* panelboard. The branch-circuit breakers are not shown. This is the type of panelboard that could be used as panelboard B in the residence discussed in this text. Most panelboards have a separate terminal bar to terminate the equipment grounding conductors of nonmetallic-sheathed cable, as in Figure 27-12.

SERVICE-ENTRANCE CONDUCTOR SIZING

NEC 230.42(A) states that the minimum size service-entrance conductors must be of sufficient size to carry the load as calculated according to *Article 220*.

NEC 230.79(C) calls for a minimum service of 100 amperes for a one-family residence.

The standard and optional methods for calculating the minimum size of service-entrance conductors are discussed in detail in Chapter 29 of this text.

RUNNING CABLES INTO TOP OF SERVICE PANEL

See Figure 4-21 for running nonmetallic-sheathed cables into the top of a panelboard. Note that *312.5(C)* generally requires that where cable is installed, each cable be secured to the cabinet. This directly applies to our installation, as panelboards are installed in cabinets.

SERVICE-ENTRANCE OVERCURRENT PROTECTION

Overcurrent protection (fuses and circuit breakers) is discussed in Chapter 28 of this text.

SERVICE-ENTRANCE RACEWAY SIZING

This residence is served with an underground service. The conduit running from the meter to Main Panelboard A must be sized correctly for the conductors it will contain. In the case of a through-the-roof mast service, the conduit might be sized for both mechanical strength reasons and conductor fill. All utilities have requirements for size and guying of mast services. Table 27-1 shows the calculation for sizing the service-entrance conduit.

METER/METER BASE LOCATION

The electric utility must be contacted before the installation of the service equipment begins. The utility will determine when, where, and how the connection will be made to the homeowner's service conductors. Because of the ever-increasing number of wood decks and concrete patios being added to homes, many utilities prefer that the meter be mounted on the side of the house rather than in the rear. If there is a raised deck, the meter could be only a short distance above the deck, making it both subject to physical damage and hard to read. If concrete patios are poured over underground utility cables, it is very difficult to repair or replace these cables should there be problems. Utilities furnish manuals and brochures detailing their requirements for services that are not found in the *NEC*.

The electric utility furnishes and installs the watt-hour meter. The meter base, also referred to as a meter socket, is usually furnished and installed by the electrical contractor, although some utilities furnish the meter base to the electrical contractor for installation.

Figure 27-7 shows a service supplied from an underground service. Figure 27-11 discusses the requirement that the main disconnect must be as close as possible to the point of entrance of the service-entrance conductors. Figure 27-15 shows a typical meter socket for an overhead service, usually mounted at eye level on the outside wall of the house. The service raceway or service-entrance cable is connected to the threaded hub (boss) on top of the meter socket. Figure 27-18 shows how to seal the raceway where it enters the building, to keep out moisture.

A combination meter socket/main breaker in one enclosure is also available.

Raintight and Draining

Outdoor locations are generally wet locations. However, under an overhang may be considered a damp rather than a wet location. If you are using EMT outdoors in a wet location, the fittings

TABLE 27-1

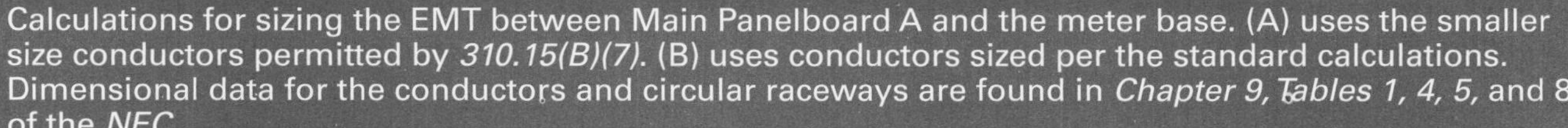

Calculations for sizing the EMT between Main Panelboard A and the meter base. (A) uses the smaller size conductors permitted by *310.15(B)(7)*. (B) uses conductors sized per the standard calculations. Dimensional data for the conductors and circular raceways are found in *Chapter 9, Tables 1, 4, 5,* and 8 of the *NEC*.

(A) CONDUCTOR SIZE BASED ON AMPACITY PER *310.15(B)(7)*			**(B) CONDUCTOR SIZE BASED ON AMPACITY PER *TABLE 310.15(B)(16)***		
Two 2/0 AWG THWN Copper		0.2223 in.^{2}	Two 3/0 AWG THWN Copper		0.2679 in.^{2}
		0.2223 in.^{2}			0.2679 in.^{2}
One 1 AWG Bare Copper		0.0870 in.^{2}	One 1 AWG Bare Copper		0.0870 in.^{2}
	TOTAL	0.5316 in.^{2}		TOTAL	0.6228 in.^{2}
EMT		Trade Size 1¼	EMT		Trade Size 1½

NEC 300.5(G) and *300.7* state that when raceways pass through areas having great temperature differences, some means must be provided to prevent passage of air back and forth through the raceway. Note that outside air is drawn in through the conduit whenever a door opens. Cold outside air meeting warm inside air causes the condensation of moisture. This can result in rusting and corrosion of vital electrical components. Equipment having moving parts, such as circuit breakers, switches, and controllers, is especially affected by moisture. The sluggish action of the moving parts in this equipment is undesirable.

Sealant shall be identified for such use, and shall not have an adverse effect on the conductor insulation, *230.8*.

NEC 230.8 requires seals where service raceways enter from an underground distribution system.

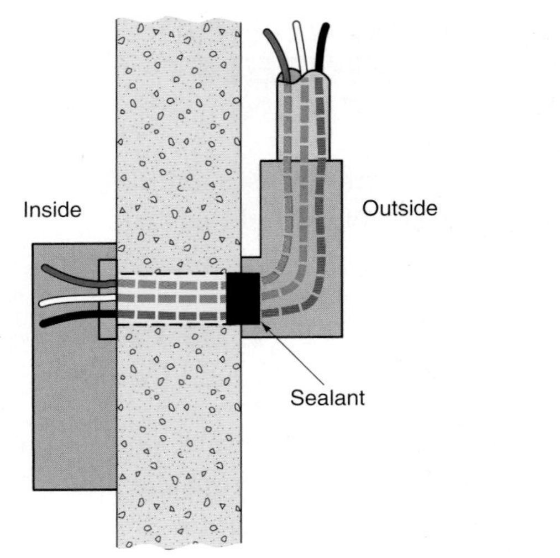

FIGURE 27-18 Installation of conduit through a basement wall.

(couplings and connectors) must be listed as "Raintight." If the fittings and the container they come in are not marked "Raintight," then the fittings have not been listed for raintight applications. Raintight fittings are generally of the compression type that has a special sealing ring.

NEC 225.22 states that Raceways on exteriors of buildings or other structures shall be arranged to drain and shall be raintight in wet locations.*

NEC 230.53 states that Where exposed to the weather, raceways enclosing service-entrance conductors shall be raintight and arranged to drain. Where embedded in masonry, raceways shall be arranged to drain.*

Pedestals

Meter "pedestals," as shown in Figure 27-19, are sometimes used for residential services. Usually mounted on the outside wall of the house, a meter "pedestal" could also be installed on the lot line between residential properties. Contact the electric utility on this issue.

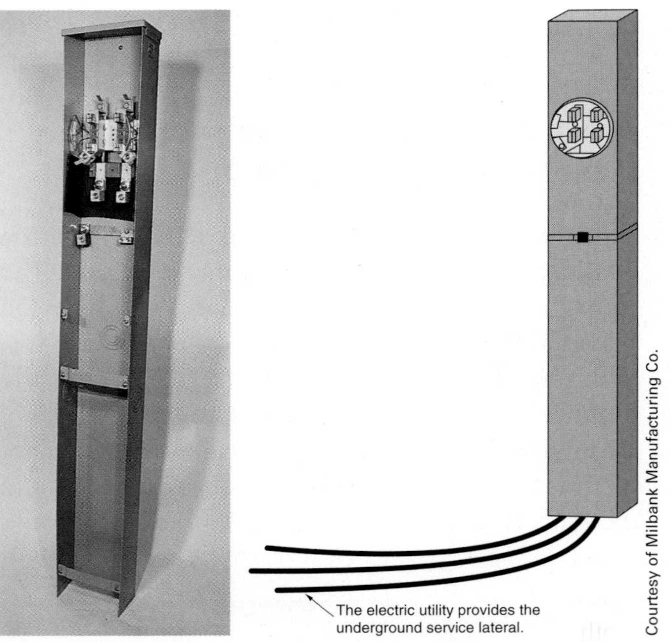

FIGURE 27-19 Installing a metal "pedestal" allows for ease of installation of the underground service-entrance conductors by the utility.

For underground services, the utility generally runs the underground lateral service-entrance conductors in a trench from a pad-mount transformer to the line-side terminals of the meter base. The electrician then takes over to complete the service installation by running service-entrance conductors from the load side of the meter to the line side of the main service disconnect.

Quite often, the electric utility and the telephone company have agreements whereby the electric utility will lay both power cables and telephone cables

*Reprinted with permission from NFPA 70-2014.

in the same trench. The electric utility carries both types of cables on its underground installation vehicles. This is a much more cost-effective way to do this than having each utility dig and then backfill its own trenches.

Often overlooked is a requirement in *NFPA Standard No. 54 (National Fuel Gas Code), Section 2.7.2(c)*, requiring that gas meters be located at least 3 ft (900 mm) from sources of ignition. Examples of sources of ignition might be an air-conditioner disconnect where an arc can be produced when the disconnect is opened under load. Furthermore, some electric utilities require a minimum of 3 ft (900 mm) clearance between electric metering equipment and gas meters and gas regulating equipment. It is better to be safe than sorry. Check this issue with the local electrical inspector if there might be a problem with the installation you are working on.

COST OF USING ELECTRICAL ENERGY

A watt-hour meter is always connected into some part of the service-entrance equipment to record the amount of energy used. Billing by the utility is generally done on a monthly basis. The meter might be mounted on the side of the house or on a pedestal somewhere on the premise on the lot line. The utility makes this decision. The residence discussed in this text has one meter, mounted on the back of the house near the sliding doors of the Master Bedroom.

The kilowatt (kW) is a convenient unit of electrical power. One thousand watts (w) is equal to one kilowatt. The watt-hour meter measures and records both wattage and time.

For residential metering, most utilities have rate schedules based on "cents per kilowatt-hour." Stated another way: How much wattage is being used and for how long?

Burning a 100-watt lightbulb for 10 hours is the same as using a 1000-watt electric heater for 1 hour. Both equal 1 kilowatt-hour.

$$\text{kWh} = \frac{\text{watts} \times \text{hours}}{1000} = \frac{100 \times 10}{1000} = 1 \text{ kWh}$$

$$\text{kWh} = \frac{\text{watts} \times \text{hours}}{1000} = \frac{1000 \times 1}{1000} = 1 \text{ kWh}$$

In these examples, if the electric rate is \$0.08 cents per kilowatt-hour, the cost to operate the 100-watt lightbulb for 10 hours and the cost to operate the electric heater for 1 hour are the same—\$0.08. Both loads use 1 kilowatt-hour of electricity.

Simple Formula for Calculating the Cost of Using Electricity

The cost of electrical energy used can be calculated as follows:

$$\text{Cost} = \frac{\text{watts} \times \text{hours used} \times \text{cost per kWh}}{1000}$$

EXAMPLE

Find the cost of operating a color television set for 8 hours. The label on the back of the television set indicates 175 watts. The electric rate is \$0.10494 per kilowatt-hour.

$$\text{Cost} = \frac{175 \times 8 \times \$0.10494}{1000} = \begin{matrix}\$0.1469 \\ \text{(approx. 15¢)}\end{matrix}$$

EXAMPLE

Find the approximate cost per day of operating a central air conditioner that on average runs 50% of the time during a 24-hour period on a typical hot summer day. The unit's nameplate is marked 240 volts, single-phase, 23 amperes. The electric rate is \$0.09 per kilowatt-hour. The steps are as follows:

1. The time the air conditioner operates each day is: $24 \times 0.50 = 12$ hours.
2. Convert the nameplate data to use in the calculations:

$$\begin{aligned}\text{Watts} &= \text{volts} \times \text{amperes} = 240 \times 23 \\ &= 5520 \text{ watts}\end{aligned}$$

3. $\text{Cost} = \dfrac{\text{watts} \times \text{hours used} \times \text{cost per kWh}}{1000}$

$$= \frac{5520 \times 12 \times \$0.09}{1000} = \$5.96$$

Note: The answers to these examples are approximate because they were based on watts. Power factor and efficiency factors were not included, as they generally are unknown when making rough estimates of an electric bill. For all practical purposes, the answers are acceptable.

Table 27-2 is an example of what a typical monthly electric bill might look like.

Some utilities increase their rates during the hot summer months, when the air-conditioning load is high. Other utilities provide a second meter for specific loads such as electric water heaters, air conditioners, heat pumps, or total electric heat. Some utilities use electronic watt-hour meters that have the capability of registering kilowatt-hour consumption during a specific "time of use." These electronic watt-hour meters might have up to four different "time-of-use" periods, each period having a different cents-per-kilowatt-hour rate.

Other charges that might appear on a "light bill" might be a fuel adjustment charge based on a per kWh basis. Such charges enable a utility to recover from the consumer extra expenses it might incur for fuel costs used in generating electricity. Fuel charges can vary with each monthly bill, without the utility having to apply to the regulatory agency for a rate change.

GROUNDING/ BONDING (*ARTICLE 250*)

All residential electrical systems are grounded systems. Within the system, some things are *grounded* and some things are *bonded*. The terms *grounding* and *bonding* are used throughout the *NEC*. There is a distinct difference between grounding and bonding. Each serves a different purpose. Let's take a look at these two very important terms.

What Does Grounded (Grounding) Mean?

Grounded (grounding) is defined in *NEC Article 100* as Connected to ground or to a conductive body that extends the ground connection.* See Figure 27-20. As shown, the electrical system is grounded by the electric utility at the transformer and again at the service equipment.

What Does Bonded (Bonding) Mean?

Bonded (bonding) is defined in *NEC Article 100* as Connected to establish electrical continuity and conductivity.* See Figure 27-21. In its simplest form, bonding means "connected together."

TABLE 27-2

A typical monthly electric bill.

GENERIC ELECTRIC COMPANY ANYPLACE, USA

DAYS OF SERVICE	FROM 06-01-20XX	TO 06-28-20XX	DUE DATE 07-25-20XX
Present reading			84,980
Previous reading			83,655
Kilowatt-hours used			1,325
Rate/kWh 1st 400 kWh @ $0.10494			$41.98
Remaining 925 kWh @ $0.06168			57.05
Energy charge			99.03
Basic service charge (single-dwelling)			8.91
State tax			4.24
Total current charges			$112.18
Total amount due by 07-25-20XX			$112.18
Total amount due after 07-25-20XX			$117.79

*Reprinted with permission from NFPA 70-2014.

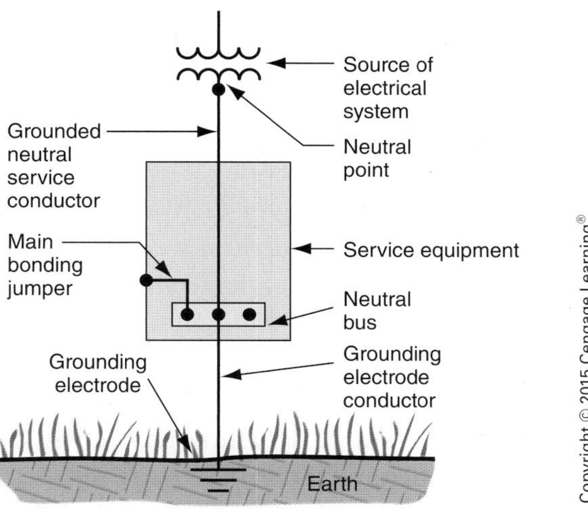

FIGURE 27-20 Grounding of an electrical system at the utility transformer and at the service connects the electrical system to earth.

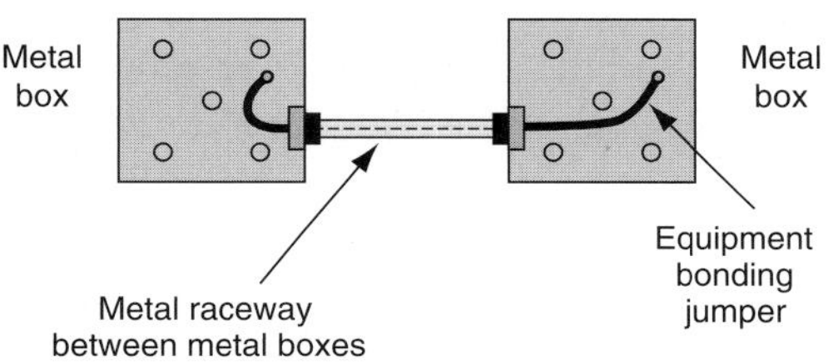

FIGURE 27-21 Two methods for bonding the metal boxes or other metal parts together. One method uses the metal raceway. The second method is to install a separate equipment grounding conductor between the two metal boxes, as would be required if a nonmetallic raceway had been used.

Now that we have talked about the meaning of the words "grounding" and "bonding," here are more detailed definitions you should be familiar with.

GROUNDING

The following definitions related to grounding are found in *NEC Article 100*, *250.2*, and *250.4(A)*.

Ground: The Earth.* Throughout this country and, for that matter, throughout the world, the earth is the common reference point for all electrical systems.

Grounded Conductor: A system or circuit conductor that is intentionally grounded.* In residential wiring, this is the white conductor. It is also referred to as the neutral conductor.

Grounding Conductor, Equipment (EGC): The conductive path(s) that provides a ground-fault current path and connects normally non-current-carrying metal parts of equipment together and to the system grounded conductor or to the grounding electrode conductor, or both.* The definition speaks of connecting equipment together. When you think about it, it is logical to conclude that the equipment grounding conductor also functions as a bonding conductor, because it serves to connect conductive parts together. In residential wiring, the equipment grounding conductor is the bare conductor in nonmetallic-sheathed cable, the metal jacket plus the bonding strip in armored cable, metal raceways, green insulated conductors, and so on. Different acceptable EGCs are listed in *250.118*. Size the EGC per *Table 250.122*.

EGCs are illustrated in Figures 5-30, 5-32, 5-33, 5-34, 27-16, and 30-4.

Ground Fault: An unintentional, electrically conducting connection between a normally current-carrying conductor of an electrical circuit, and the normally non-current-carrying conductors, metallic enclosures, metallic raceways, metallic equipment, or earth.* Ground faults occur when an ungrounded "hot" conductor comes in contact with a grounded surface or grounded conductor. This could be a result of insulation failure or an ungrounded conductor connection coming loose.

Ground-Fault Current Path: An electrically conductive path from the point of a ground fault on a wiring system through normally non-current-carrying conductors, equipment, or the earth to the electrical supply source.* This is the path that the flow of ground-fault current will take. Whatever fault current flows, that fault current must return to its source. What goes out must come back! The current might return through connectors, couplings, bonding jumpers, grounding conductors, ground clamps, and other components that make up the ground-fault return path.

Ground-Fault Current Path, Effective: As defined in *250.2*, An intentionally constructed, reliable, low-impedance electrically conductive path designed and intended to carry current under ground-fault conditions from the point

*Reprinted with permission from NFPA 70-2014.

of a ground fault on a wiring system to the electrical supply source and that facilitates the operation of the overcurrent protective device or ground fault detectors on high-impedance grounded systems.* If and when a ground fault occurs, we want the overcurrent device ahead of the ground fault to open as fast as possible to clear the fault. To accomplish this, the integrity of the ground-fault current path must be unquestionable.

We sometimes think that electricity follows the path of least resistance. That is not totally correct! Electricity follows all paths. As stated earlier, ground-fault current might return through connectors, couplings, bonding jumpers, grounding conductors, ground clamps, and any other components that make up the ground-fault return path. Some of the fault current might flow on sheet metal ductwork and on metal water piping or metal gas piping. That is why the *NEC* is so strict about keeping the impedance of the ground-fault current path as low as possible. We want the ground-fault current to return on the electrical system—not on other nonelectrical parts in the building. Think about it this way—the lower the impedance, the higher the current flow. The higher the current flow, the faster the overcurrent device will clear the fault.

As required by *250.4(A)(5)*, Electrical equipment and wiring and other electrically conductive material likely to become energized shall be installed in a manner that creates a low-impedance circuit facilitating the operation of the overcurrent device or ground detector for high-impedance grounded systems. It shall be capable of safely carrying the maximum ground-fault current likely to be imposed on it from any point on the wiring system where a ground fault may occur to the electrical supply source. The Earth shall not be considered as an effective ground-fault current path.*

Grounding Electrical Systems: Electrical systems that are grounded shall be connected to earth in a manner that will limit the voltage imposed by lightning, line surges, or unintentional contact with higher voltage lines and that will stabilize the voltage to earth during normal operation.* See Figure 27-20.

Grounding of Electrical Equipment: Normally non-current-carrying conductive materials enclosing electrical conductors or equipment, or forming part of such equipment, shall be connected to earth so as to limit the voltage to ground on these materials.*

Grounding Electrode: A conducting object through which a direct connection to earth is established.* In residential wiring, the grounding electrode might be an underground metal water piping supply, a concrete-encased reinforcing steel bar or bare copper conductor (UFER ground), or it might be a ground rod(s). Acceptable grounding electrodes are listed in *250.52(A)(1)* through *(7)*. See *250.52*.

Grounding Electrode Conductor: A conductor used to connect the system grounded conductor or the equipment to a grounding electrode or to a point on the grounding electrode system.* See *250.62*. For typical residential wiring, the grounding electrode conductor is run between the main service panel neutral/ground terminal bar and the grounding electrode system.

Grounding electrode conductors

- are permitted to be aluminum or copper [see the limitation on the installation of aluminum conductors in *250.64(A)*].
- shall be protected against physical damage, *250.64(B)*.
- If required, splices or connections shall be made as permitted in parts (1) through (4) of *250.64(C)*.

1. Splicing of the wire-type grounding electrode conductor shall be permitted only by irreversible compression-type connectors listed as grounding and bonding equipment or by the exothermic welding process.
2. Sections of busbars shall be permitted to be connected together to form a grounding electrode conductor.
3. Bolted, riveted, or welded connections of structural metal frames of buildings or structures.
4. Threaded, welded, brazed, soldered or bolted-flange connections of metal water piping are sized according to Table 250.66.*

*Reprinted with permission from NFPA 70-2014.

In this residence, a 4 AWG armored copper grounding electrode conductor (GEC) runs from Main Panel A, across the workshop ceiling to the water pump area. The GEC is terminated with a listed ground clamp to the metal water piping where the piping comes through the basement wall. Depending on the size of the metal water pipe, ground clamps of the types illustrated in Figures 27-22, 27-23, 27-24, and 27-25 are used. As previously mentioned, because of the ever-increasing use of nonmetallic water pipe, there is a corresponding decrease in the use of a water pipe grounding electrode and more use of the concrete-encased grounding electrode, referred to as the UFER ground. To comply with the requirements in *NEC 250.50* to create a grounding electrode system, a 4 AWG copper bonding conductor was connected to a 20 ft (6.0 m) section of the reinforcing steel bar in the foundation footing before the concrete was poured. The end of the bonding conductor was coiled and left in the area of the water pipe, so the two grounding electrodes could be bonded together.

Reference to grounding electrode conductors is found in several figures in this chapter.

Ungrounded: Not connected to ground or a conductive body that extends the ground connection.*

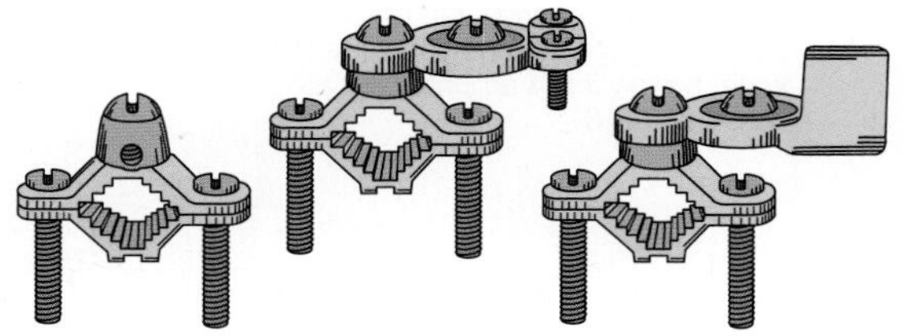

FIGURE 27-22 Typical ground clamps used in residential systems.

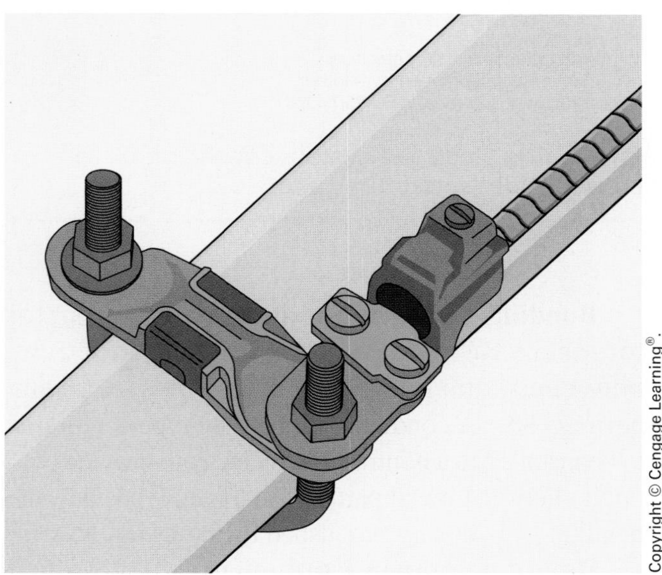

FIGURE 27-23 Armored grounding electrode conductor connected with ground clamp to water pipe.

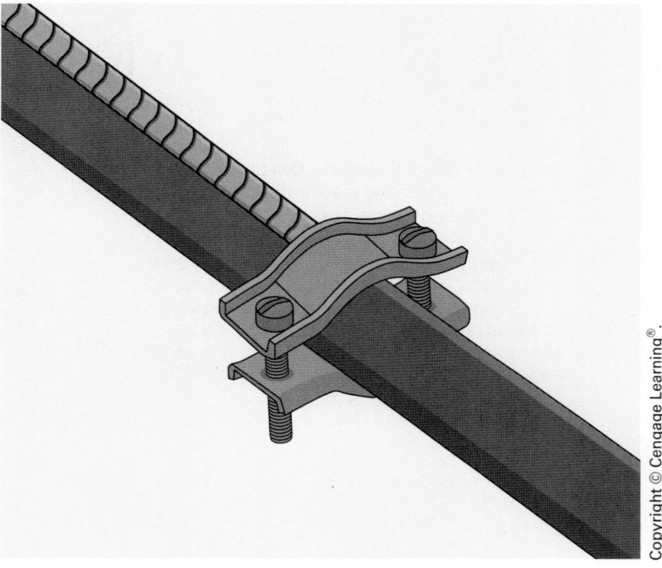

FIGURE 27-24 Ground clamp of the type used to bond (jumper) around water meter.

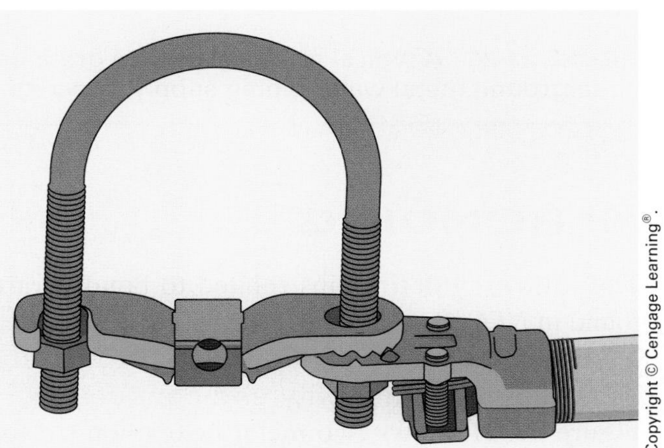

FIGURE 27-25 Ground clamp of the type used to attach ground wire to well casings.

*Reprinted with permission from NFPA 70-2014.

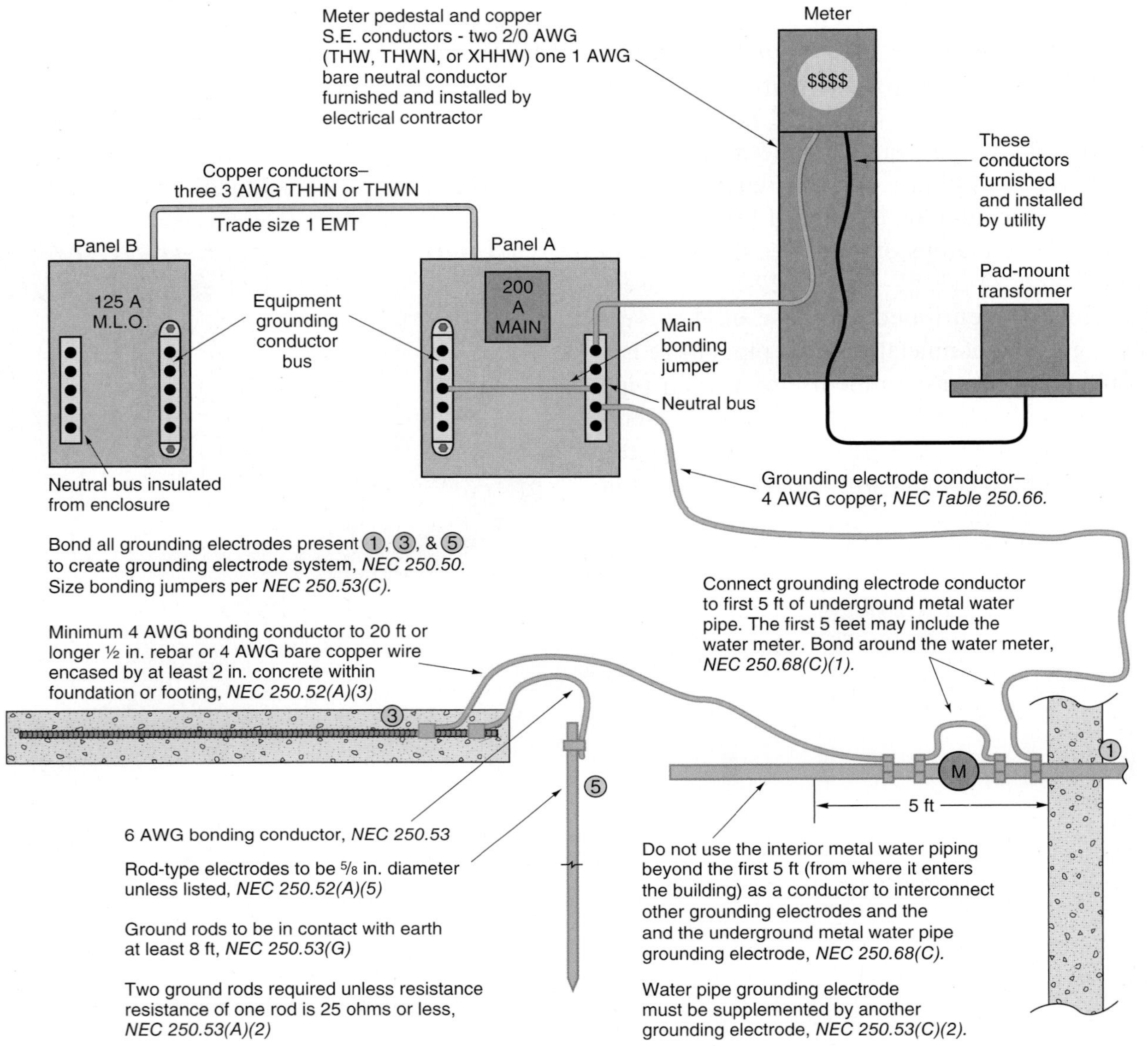

FIGURE 27-26 A typical electrical service grounded to a grounding electrode system consisting of an underground metal water piping supply, concrete-encased electrode, and ground rod.

BONDING

The following definitions related to bonding are found in *NEC Article 100.*

Bonded (Bonding): Connected to establish electrical continuity and conductivity.* Figure 27-21 shows two metal boxes bonded together with the metal raceway installed between the two boxes. The bonding jumper could also have been a conductor. Requirements for the size of a conductor of the wire type are found at various locations in *Article 250.*

Bonding Conductor or Jumper: A conductor to ensure the required electrical conductivity between metal parts required to be electrically connected.* A conductor of the wire type is required to be installed in a nonmetallic raceway to provide continuity between metal parts. With a metal raceway, the bonding would be accomplished by the metal raceway.

Bonding Jumper, Equipment: The connection between two or more portions of the equipment grounding conductor.* Bonding equipment

*Reprinted with permission from NFPA 70-2014.

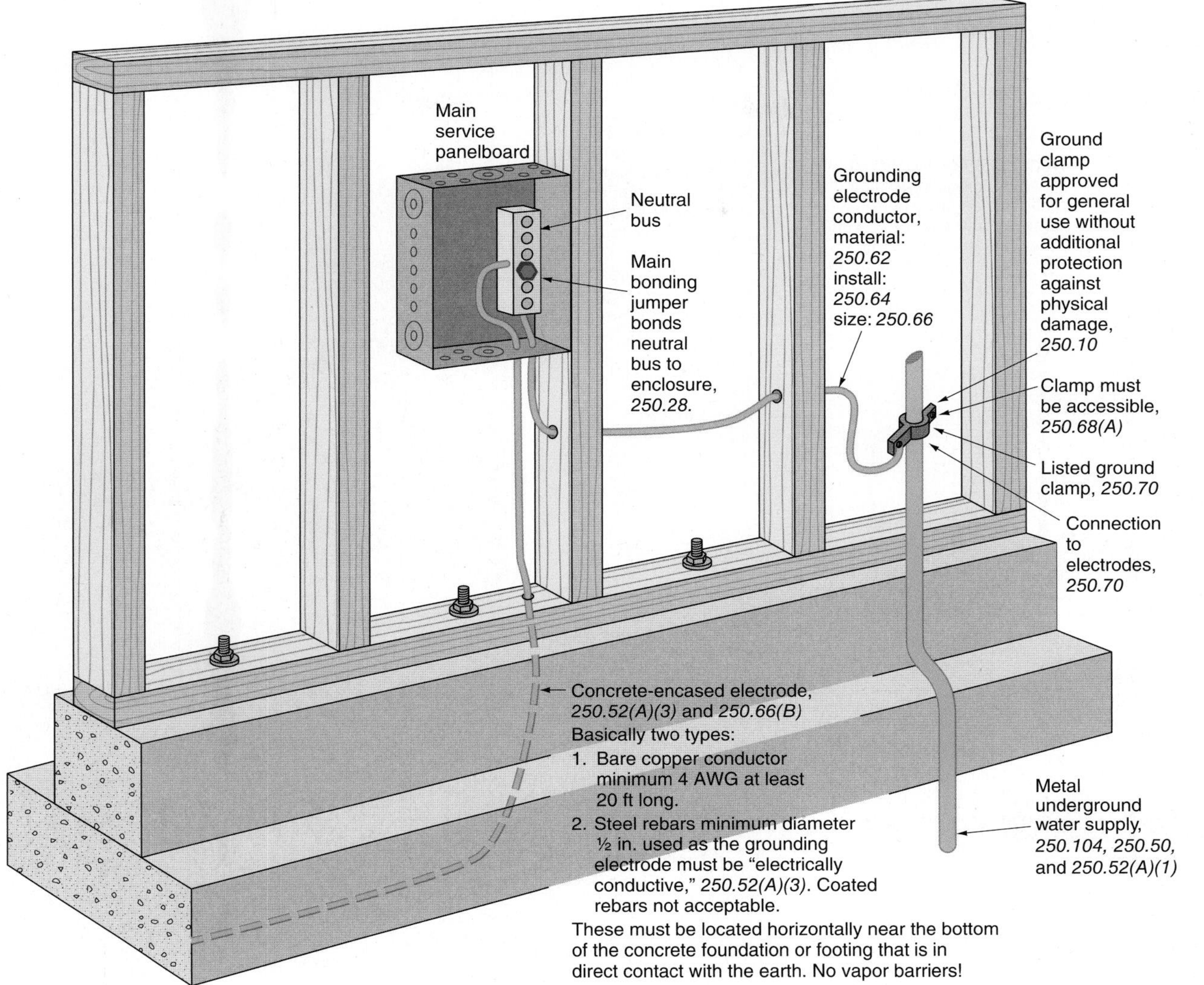

FIGURE 27-27 Connection of concrete-encased grounding electrode and underground metal water pipe.

together does not mean that the equipment is grounded. However, typical electrical systems in residential occupancies are grounded. The equipment grounding conductor then serves a triple function of extending the earth or ground connection, bonding metal enclosures together, and providing a ground-fault return path back to the source.

Bonding Jumper, Main: The connection between the grounded circuit conductor and the equipment grounding conductor at the service.* See *250.28*. Main bonding jumpers are clearly illustrated in Figures 27-8, 27-16, 27-20, 27-26, and 27-28.

Bonding of Electrically Conductive Materials and Other Equipment: Normally non-current-carrying electrically conductive materials that are likely to become energized shall be connected together and to the electrical supply source in a manner that establishes an effective ground-fault current path. *NEC 250.4(A)(4)*.*

Bonding of Electrical Equipment: Normally non-current-carrying conductive materials enclosing electrical conductors or equipment, or forming part of such equipment, shall be

*Reprinted with permission from NFPA 70-2014.

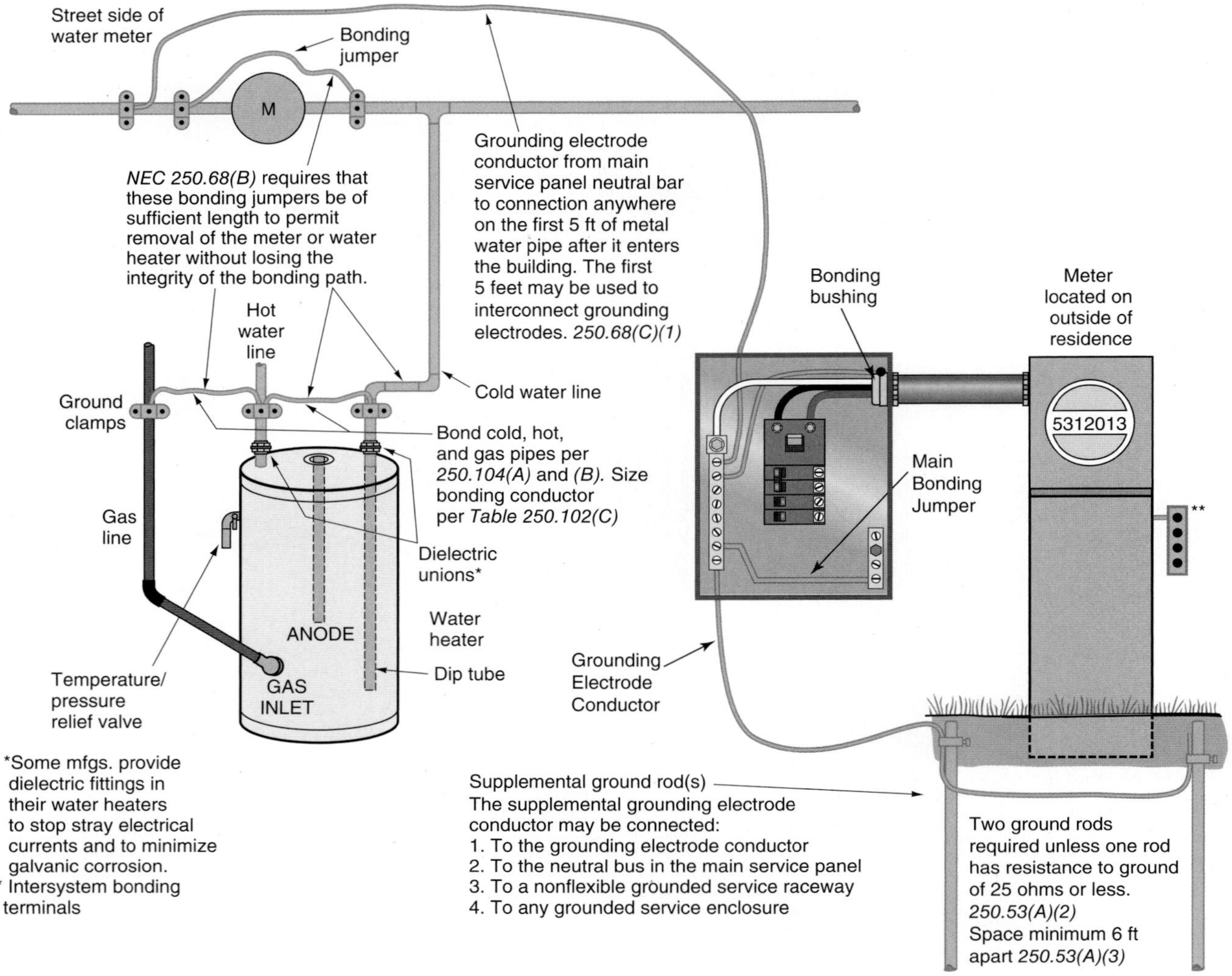

Street side of water meter

Bonding jumper

M

NEC 250.68(B) requires that these bonding jumpers be of sufficient length to permit removal of the meter or water heater without losing the integrity of the bonding path.

Grounding electrode conductor from main service panel neutral bar to connection anywhere on the first 5 ft of metal water pipe after it enters the building. The first 5 feet may be used to interconnect grounding electrodes. *250.68(C)(1)*

Hot water line

Cold water line

Ground clamps

Bond cold, hot, and gas pipes per *250.104(A)* and *(B)*. Size bonding conductor per *Table 250.102(C)*

Gas line

Dielectric unions*

ANODE

Water heater

Dip tube

Temperature/ pressure relief valve

GAS INLET

Bonding bushing

Meter located on outside of residence

5312013

Main Bonding Jumper

**

Grounding Electrode Conductor

*Some mfgs. provide dielectric fittings in their water heaters to stop stray electrical currents and to minimize galvanic corrosion.

** Intersystem bonding terminals

Supplemental ground rod(s)

The supplemental grounding electrode conductor may be connected:

1. To the grounding electrode conductor
2. To the neutral bus in the main service panel
3. To a nonflexible grounded service raceway
4. To any grounded service enclosure

Two ground rods required unless one rod has resistance to ground of 25 ohms or less. *250.53(A)(2)*

Space minimum 6 ft apart *250.53(A)(3)*

FIGURE 27-28 One method that may be used to provide proper electrical system grounding of service-entrance equipment, and bonding of the cold and hot water piping and gas piping. **Provide a listed intersystem bonding terminal strip at the meter enclosure, service equipment enclosure, or on the grounding electrode conductor.

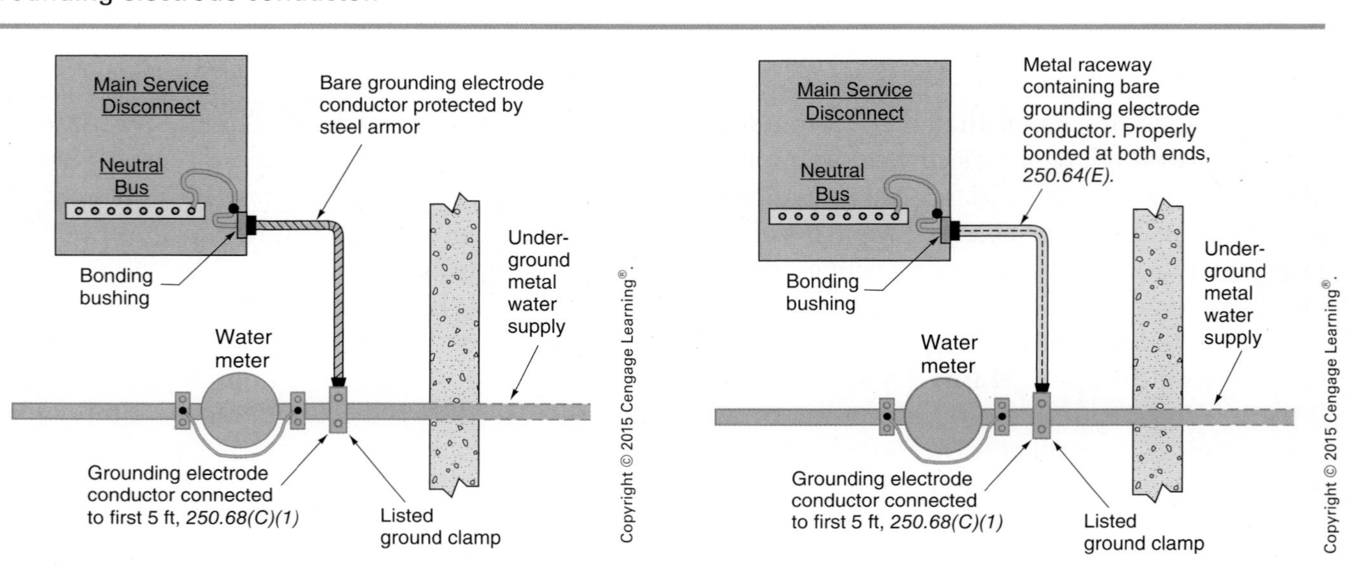

FIGURE 27-29 This installation "Meets *Code*."

FIGURE 27-30 This installation "Meets *Code*."

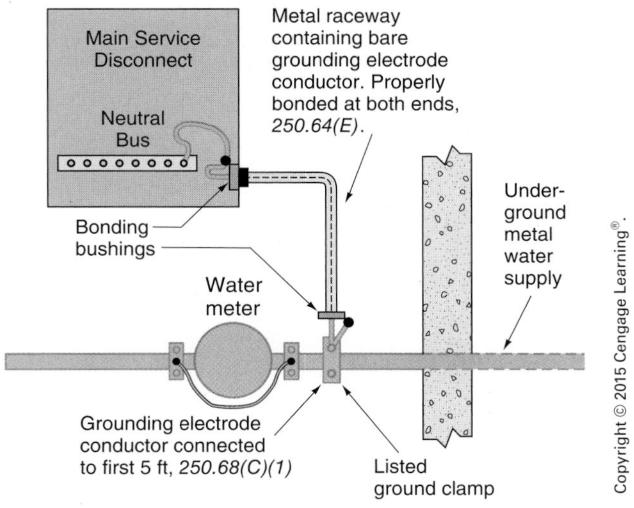

FIGURE 27-31 This installation "Meets *Code*."

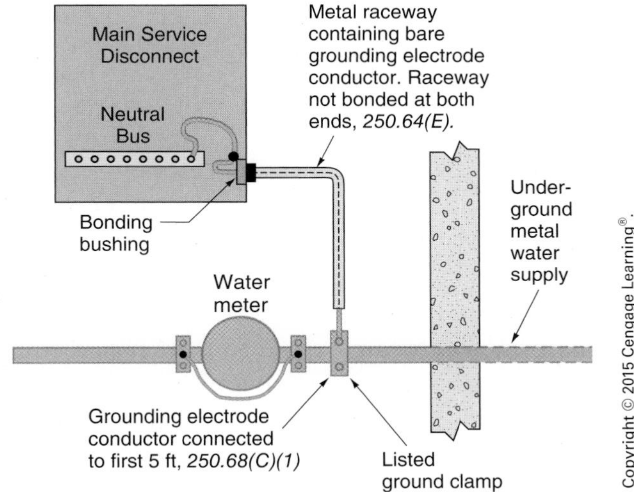

FIGURE 27-32 This installation *does not* "Meet *Code*."

connected together and to the electrical supply source in a manner that establishes an effective ground-fault current path. *NEC 250.4(A)(3).**

These are a few other key *Code* sections relating to system grounding:

250.20(B)(1): This section requires that an electrical system be grounded where the maximum voltage to ground on the ungrounded "hot" conductors does not exceed 150 volts. The serving electric utility grounds the midpoint of their transformer. This midpoint becomes the system *neutral point*.

The electrical system in a typical home is single-phase, 3-wire, 120/240 volts. The voltage between

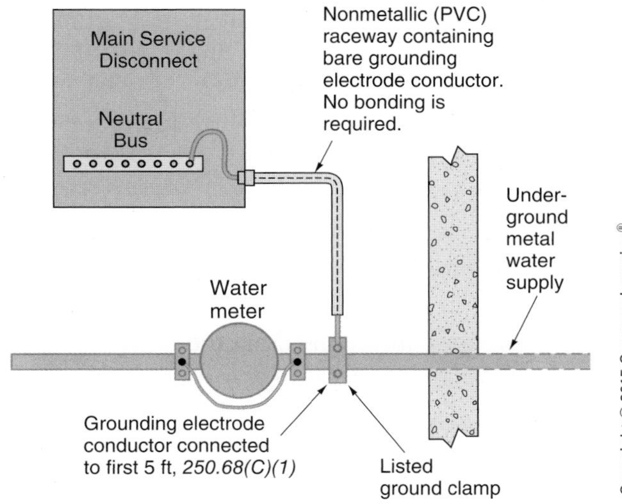

FIGURE 27-33 This Installation "Meets *Code*."

the ungrounded "hot" phase conductors is 240 volts, and the voltage between the ungrounded "hot" phase conductors and the neutral (the grounded conductor) is 120 volts. The grounded neutral service conductor is again grounded at the main service.

The electrical system in some large multifamily buildings (condos, apartments, etc.) is derived from a 3-phase, 4-wire, 208Y/120-volt system. Here, the voltage between the ungrounded "hot" phase conductors is 208 volts, and the voltage between the ungrounded "hot" phase conductors and the neutral (the grounded conductor) is 120 volts. The supply to each dwelling unit is referred to as a single-phase, 3-wire, 120/208-volt system.

250.26(2): This section requires that the *neutral conductor* be grounded for single-phase, 3-wire systems. As mentioned earlier, the transformer midpoint (neutral point) is grounded at the utility transformer and again at the main service.

Providing a proper ground-fault current path helps ensure that overcurrent protective devices will operate fast when responding to ground faults. One essential component of an effective ground-fault current path consists of a low-impedance (ac-resistance) ground path. Ohm's law verifies that, in a given circuit, "the lower the impedance, the higher the value of current." As the value of ground-fault current increases, there is an increase in the speed with which a fuse will open or a circuit breaker will trip. This is called an *inverse time* relationship.

*Reprinted with permission from NFPA 70-2014.

Clearing Ground Faults

Clearing a ground fault or short circuit is important because arcing damage to electrical equipment, as well as conductor insulation damage, is closely related to a value called *ampere-squared-seconds* (I^2t), where

I = the current in amperes flowing phase to ground, or phase to phase.
t = the time in seconds that the current is flowing.

Thus, there will be less equipment and/or conductor damage when the fault current is kept to a low value and when the time that the fault current is allowed to flow is kept to a minimum. The impedance of the circuit determines the *amount* of fault current that will flow. The speed of operation of a fuse or circuit breaker determines the amount of *time* the fault current will flow.

Grounding electrode conductors and equipment grounding conductors carry an insignificant amount of current under normal conditions. However, when a ground fault occurs, the equipment grounding conductor as well as the ungrounded ("hot") circuit conductors must be capable of carrying whatever value of fault current might flow (*how much?*) for the time (*how long?*) it takes for the overcurrent protective device to clear the fault and reduce the fault current to zero.

This potential hazard is recognized in the *Note* below *Table 250.122*, which states that Where necessary to comply with 250.4(A)(5) or 250.4(B)(4), the equipment grounding conductor shall be sized larger than given in this table.* This note calls attention to the fact that equipment grounding conductors may have to be sized larger than indicated in the table when high-level ground-fault currents are possible.

What is a high-level ground fault? To determine the available short-circuit and ground-fault current, a short-circuit calculation is necessary. It is also necessary to know the time/current characteristic of the overcurrent protective device. This is discussed in *Electrical Wiring—Commercial* (Copyright © Cengage Learning®).

This is referred to as the conductor's "withstand rating."

Table 250.66 and *Table 250.122* are based on the fact that copper conductors and their bolted connections can withstand

- one ampere
- for 5 seconds
- for every 30 circular mils.

Thermoplastic insulation on a copper conductor rated 75°C can safely withstand

- one ampere
- for 5 seconds
- for every 42.25 circular mils.

To exceed these values will result in damage to the conductors, with possible burn-off and loss of the grounding or bonding path.

When bare equipment grounding conductors are in the same raceway or cable as insulated conductors, always apply the insulated conductor withstand rating limitation.

Properly selected fuses and circuit breakers for normal residential installations generally will protect conductors and other electrical equipment against the types of ground faults and short circuits to be expected in homes. Fault currents can be quite high in single-family homes where the pad-mounted transformer is located close to the service equipment, and in multifamily dwellings (apartments, condos, town houses), making it necessary to take equipment short-circuit ratings and conductor withstand rating into consideration.

The subject of conductor and equipment withstand rating is covered in greater depth in *Electrical Wiring—Commercial* (Copyright © Cengage Learning®).

Grounding Electrode System

Article 250, Part III, covers the requirements for establishing a grounding electrode system.

NEC 250.50 requires that metal underground water piping, the metal frame of a building, a concrete-encased electrode, a ground ring, and rod-pipe-plate electrodes be bonded together to create a grounding electrode system if any or all of the grounding electrodes *are present* in a new installation.

Metal gas piping shall never be used as a grounding electrode, *250.52(B)(1)*. Past experience has shown that because of galvanic action, gas pipes

*Reprinted with permission from NFPA 70-2014.

have deteriorated, resulting in serious incidents. However, metal gas piping is required to be bonded. Normally, the equipment grounding conductor for the branch circuit supplying a gas furnace serves as the required bond. *NEC 250.104(B)* covers the rules for bonding metal gas piping.

Should any one of these become disconnected, the integrity of the grounding system is maintained through the other paths. Here are a few key points:

- *250.90* states that Bonding shall be provided where necessary to ensure electrical continuity and the capacity to conduct safely any fault current likely to be imposed.*
- *250.92(A)* explains what parts of a service must be bonded together.
- *250.94* explains what is acceptable as a bonding means.
- *250.96(A)* states in part that bonding of metal raceways, cable armor, enclosures, frames, fittings, and so on, that serve as the grounding path Shall be effectively bonded where necessary to ensure electrical continuity and the capacity to conduct safely any fault current likely to be imposed on them.*

By having all metal parts bonded together for a grounding electrode system, potential differences between non-current-carrying metal parts are virtually eliminated. In addition, the grounding electrode system serves as a means to bleed off lightning, stabilize the system voltage, and ensure that the overcurrent protective devices will operate.

Hazard of Improper Bonding

Figure 27-34 and the following steps illustrate what can happen if the electrical system is not properly grounded and bonded:

1. A "live" wire contacts the gas pipe. The bonding jumper A has not been installed.
2. The gas pipe now has 120 volts on it. The pipe is energized. It is "hot."
3. The insulating joint in the gas pipe at the gas meter results in no current flow as the circuit is open.
4. The 20-ampere overcurrent device does not open, but the gas pipe remains energized.
5. If a person touches the "live" gas pipe and the water pipe at the same time, current flows through the person's body. If the hand-to-hand body resistance is 1100 ohms, the current is

$$I = \frac{E}{R} = \frac{120}{1,100} = 0.11 \text{ ampere}$$

This amount of current passing through a human body can cause death. See Chapter 6 for a discussion regarding electric shock.

6. The overcurrent device does not open.
7. If the bonding jumper A had been installed, it would have kept the voltage difference between the water pipe and the gas pipe at or near zero. The overcurrent device would have opened. Checking *Table 8, Chapter 9, NEC*, we find the dc resistance of 1000 ft (305 m) of an uncoated 4 AWG copper wire is 0.308 ohm.

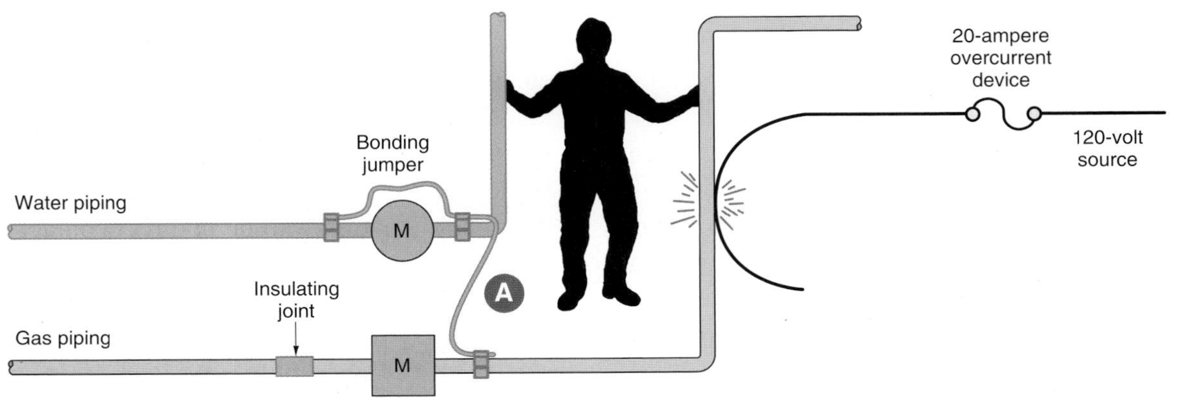

FIGURE 27-34 Proper bonding of gas piping system eliminates shock hazard.

*Reprinted with permission from NFPA 70-2014.

The resistance of 100 ft (30 m) is 0.0308 ohm. The resistance of 10 ft (3.0 m) is 0.00308 ohm. The current would be

$$I = \frac{E}{R} = \frac{120}{0.00308} = 38{,}961 \text{ amperes}$$

In an actual installation, the total impedance of *all* parts of the circuit would perhaps be much higher than these simple calculations. A lower current would result. The value of current would be enough to cause the overcurrent device to open.

Where to Connect the Grounding Electrode Conductor

NEC 250.24(A) tells us that the grounding electrode conductor shall be connected to the grounded (neutral) service conductor. See Figure 27-26. *NEC 250.24(A)(1)* goes on to tell us that the connection is permitted to be made at any accessible point from the load end of the service drop or service lateral up to, and including, the terminal or bus to which the grounded (neutral) service conductor is connected at the service disconnecting means.

All residential panelboards have

- a neutral bus for the white grounded circuit conductors, and
- an equipment grounding bus for the bare equipment grounding conductors when nonmetallic-sheathed cable is used as the wiring method.

The neutral conductors and equipment grounding conductors are often connected to the same terminal bar in the service equipment as permitted by *NEC 408.40*. Only the equipment grounding conductors are connected to the equipment grounding conductor terminal bar in subpanels.

For just about all residential services, the connection of the grounding electrode conductor is made to the neutral bus. *NEC 250.24(A)(1)*. Many residential panelboards have a green hexagon-shaped No.10-32 screw that becomes the main bonding jumper between the neutral bus and the enclosure when properly installed.

Figures 27-16, 27-20, 27-29 through 27-33, and 27-41 show the grounding electrode conductor connected to the neutral bus in the main service panelboard.

Grounding and Bonding the Electrical System in a Typical Residence

A metal underground water piping system 10 ft (3.0 m) or longer, in direct contact with the earth, is acceptable as a grounding electrode, *250.52(A)(1)*. The connection of the grounding electrode conductor must be made on the first 5 ft (1.52 m) of where the metal underground water piping enters the building, *250.68(C)(1)*.

When metal underground water piping is used as the only grounding electrode, it must be supplemented by at least one additional grounding electrode, *250.53(D)(2)*. In this residence, the concrete-encased electrode and a driven ground rod are all connected together to create the grounding electrode system and automatically satisfy the requirement for the supplemental grounding electrode, as permitted by *NEC 250.50* and *250.52(A)(5)*.

As shown in Figure 27-26, the 20 ft (6.0 m) length of steel reinforcing bar that is encased in 2 in. (50 mm) of concrete is recognized in *NEC 250.52(A)(3)* as a grounding electrode. A 4 AWG bare copper conductor at least 20 ft (6.0 m) long installed identically to the steel reinforcing rod is also permitted as a concrete-encased grounding electrode. This is often referred to as a UFER ground, named after Herbert G. Ufer, who worked for Underwriters Laboratories. Connecting the grounding electrode conductor to a concrete-encased electrode (a minimum 4 AWG bare copper conductor or reinforcing bars (rebars) are required by *250.66(B)*. A concrete encased electrode does not require a supplemental electrode as does underground metal water piping, *250.53(D)(2)*.

A concrete-encased electrode could also be installed vertically in a foundation wall. The key is that the foundation wall be in direct contact with the earth, *250.52(A)(3)*.

Watch out when using ground rods, pipes, or plates as grounding electrodes! *NEC 250.53(A)(2)* states that two rods, pipes, or plates be installed unless it can be shown that a single grounding electrode has a resistance to ground of 25 ohms or less. There are a number of manufacturers of testers for measuring ground–earth resistance.

What's So Good about a Concrete-Encased Electrode?

Unquestioned reliability!

The increasing use of nonmetallic water mains brought about the need to use concrete-encased electrodes. In addition, a concrete-encased electrode is recognized as a grounding electrode that provides a good connection to the earth.

When used as a grounding electrode, a concrete-encased electrode does not have to be supplemented as does a metal underground water piping system. There is no need to check for the maximum 25-ohm requirement as there is for ground rods. Many communities have mandated using a concrete-encased electrode as the primary electric service electrode because of its proven performance record of providing an excellent connection to earth.

The permanent moisture under a concrete foundation or footing ensures a low-impedance direct connection to earth. When using a concrete-encased grounding electrode, be sure that the footing or foundation is in direct contact with the earth. Make sure that there is no vapor barrier underneath the footing or foundation.

The electrician must work closely with the concrete and rebar contractor. It is necessary to bring one end (stub-up) of a reinforcing bar (called "rebar") or the bare 4 AWG copper conductor upward out of the concrete slab or footing at a location near the likely location of the electrical main service. This makes for a rather easy connection point for the grounding electrode conductor. See *250.52(A)(3)*. If the rebar is brought up close to the metal water service, the bonding together of the rebar and the metal water service is easily accomplished.

If a grounding electrode conductor connection is made underground or is imbedded in concrete, the connectors must be listed for direct burial. See Figure 27-26.

Coated rebars are not acceptable.

More Ground Rod Rules for Residential Application

- The most common ground rods are copper-clad steel. Copper makes for an excellent connection between the rod and the ground clamp. The steel gives it strength to withstand being driven into the ground. Galvanized and stainless steel rods are also available. Aluminum rods are not permitted.
- Ground rods must be at least ⅝ in. (15.87 mm) in diameter unless listed by a qualified electrical testing laboratory.
- They must be installed below the permanent moisture level if possible.
- They must not be less than 8 ft (2.5 m) in length.
- They must be driven to a depth so that at least 8 ft (2.5 m) is in contact with the soil.
- If solid rock is encountered so the rod cannot be driven vertically to the proper depth, the rod must be driven at an angle not greater than 45° from vertical.
- If the ground rod cannot be driven at an oblique angle not greater than 45° from the vertical, the rod is permitted to be installed in a trench that is at least 2½ ft (750 mm) deep.
- The rod should be driven so the upper end is flush with or just below ground level. If the upper end is exposed, the ground rod, the ground clamp, and the grounding electrode conductor must be protected from physical damage.
- If more than one rod is needed, they must be kept at least 6 ft (1.8 m) apart. Driving the rods close together reduces their effectiveness because their sphere of influence overlaps. Actually, it is better to space multiple rods twice the length of the longest rod. For example, when driving two 8-ft (2.5 m) ground rods, space them 16 ft (4.9 m) apart.

Table 27-3 shows accurate multipliers for multiple ground rods spaced one rod-length apart. To use this chart, divide the resistance value of one rod by the number of rods used, then apply the multiplier.

TABLE 27-3

Multipliers for multiple ground rods.

NUMBER OF RODS	MULTIPLIER
2	1.16
3	1.29
4	1.36

EXAMPLE

The ground–earth resistance reading of one 8 ft (2.5 m) ground rod is 30 ohms. This exceeds the *NEC* maximum of 25 ohms. A second ground rod is driven and connected in parallel to the first ground rod. What is the approximate ground–earth resistance of these two ground rods when spaced 8 ft (2.5 m) apart?

$$\frac{30}{2} \times 1.16 = 17.4 \text{ ohms}$$

- ▶Where the grounding electrode conductor is connected to a single or multiple rod, pipe, or plate electrode, or any combination thereof as permitted in *250.52(5)* or *(A)(7)*, that portion of the conductor that is the sole connection to the grounding electrode(s) is not be required to be larger than 6 AWG copper wire or 4 AWG aluminum wire, *NEC 250.66(A)*.◀*

In some parts of the country, a water pipe ground is considered unreliable. The concern is the prolific use of insulating fittings and nonmetallic water piping services. These parts of the country have found that a concrete-encased electrode provides excellent grounding capabilities. Refer to *250.52(A)(3)*.

Other Acceptable Grounding Electrodes

- The metal frame of a building (this residence is constructed of wood), *250.50(A)(2)*.
- A ground ring that encircles the building consisting of not less than 20 ft (6.0 m) of bare copper wire, minimum size 2 AWG, buried directly in the earth at least 2½ ft (750 mm) deep. See *250.52(A)(4)* and *250.52(G)*.
- Ground plates must be at least 2 ft^2 (0.186 m^2), *250.52(A)(7)*. Two ground plates are required unless the resistance of one plate is found to be 25 ohms or less, *250.53(A)(2)*. The ground plates must be buried at least 2½ ft (750 mm) deep.
- Other Listed Electrodes,* *NEC 250.52(A)(6)*. This "catch-all" title was added specifically to accommodate chemical ground rods. These consist of copper tubing that is filled with a natural earth salt chemical. Because they are quite expensive, it is uncommon to use them for other than commercial installations.
- *250.53(D)(2)* tells us that a metal underground water piping system must have a supplemental electrode installed. Ground rods, ground plates, and underground metal water piping systems are required to have a supplemental grounding electrode. Other types of grounding electrodes, such as structural steel electrodes, concrete-encased electrodes, and ground rings, do not need a supplemental electrode.

Bond All Grounding Electrodes Together

NEC 250.50 requires that if any or all of the following are present in a new installation, they must be bonded together: metal underground water pipe, metal frame of the building, concrete-encased electrode, ground ring, rod and pipe electrodes, and plate electrodes. This creates the grounding electrode system.

Some Issues Regarding Bonding of Metal Pipes Inside or Attached to the Outside of the Home

Metal water piping and metal gas piping must be properly bonded, *250.104(A)* and *(B)*.

Over the years, underground natural gas services installed from the street to the home might have been black iron pipe, copper tubing, or plastic piping designed for gas underground installations. You might even find lead water and gas pipes in homes built prior to 1963, but not likely after 1963. Today, metal gas piping installed underground has a factory-applied corrosion protection coating and/or wrapping as required by *NFPA 54, Natural Fuel Gas Code*. Most underground gas pipes today are plastic.

Homes have been blown off their foundations as a result of ignition and explosion of leaking gas from underground piping. The culprit causing the leak usually can be traced to corrosion. Where gas is found

*Reprinted with permission from NFPA 70-2014.

to be leaking from an underground copper pipe, the gas company workers pull the copper pipe out of the ground, pulling in its place an approved plastic pipe. A 14 AWG yellow tracer wire is pulled in (buried) with the plastic pipe so as to be able to locate the underground nonmetallic piping at some later date.

Underground gas line piping is not a major concern for electricians! What is of concern to the electrician is that the metal water and gas piping within or on the home must be bonded together.

Here are the *NEC* requirements that deal with bonding metal water and gas piping:

- *250.52(B)(1):* Metal underground gas piping is not permitted to be used as a grounding electrode.

  Similarly, the *National Fuel Gas Code, NFPA 54, Section 7.13.3* states that "Gas piping shall not be used as a grounding conductor or electrode."

  To prevent galvanic corrosion of underground metal gas piping, gas utilities install dielectric (insulating) fittings at the line side port of the gas meter. A dielectric fitting "isolates" the underground metal gas piping from the interior metal piping and prevents the corrosion problem. Today, gas companies generally install plastic piping underground. This solves the corrosion problem present when metal piping is installed underground.

- *250.104(A):* This requires that interior metal water piping be bonded, and that the bonding conductor be sized according to *Table 250.66*. Figure 27-28 clearly shows this bond.

  To control galvanic corrosion of the water heater tank and any directly connected steel, galvanized, and/or copper piping, you will usually find dielectric fittings (unions) in the water heater's cold and hot water lines. Always install the bonding jumper *above* the dielectric fittings.

- *250.104(B):* All metal piping, including gas piping, shall be bonded. Figure 27-28 clearly shows this bond.

- *250.104(B):* The bonding conductor for metal gas piping shall be sized according to *Table 250.122* for the ampere rating of the circuit that might energize the metal piping.

  Similarly, the *National Fuel Gas Code, NFPA 54*, in *Section 7.13.1* requires that

  "Each above ground portion of gas piping other than Corrugated Stainless Steel Tubing (CSST) that is likely to become energized shall be electrically continuous and bonded to an effective ground-fault current path. Gas piping, other than CSST, shall be considered to be bonded when it is connected to gas utilization equipment that is connected to the appliance grounding conductor of the circuit supplying the appliance."**

  In *NFPA 54 7.13.2,* CSST gas piping systems are required to be bonded to the electrical service grounding electrode system. The bonding jumper is required to be connected to a metallic pipe or fitting between the point of delivery and the first downstream CSST fitting. The bonding jumper is not permitted to be smaller than 6 AWG copper wire or equivalent. Gas piping systems that contain one or more segments of CSST are required to be bonded in accordance with these rules.

  The logic to this requirement is that if the "hot" conductor of the electrical circuit supplying a gas utilization equipment such as a gas furnace comes in contact with the metal frame of the appliance, the ground-fault current path through the equipment grounding conductor will cause the branch-circuit overcurrent device to clear the fault. In the meantime, bonding maintains an equal voltage potential between the metal frame of the gas appliance and its metal gas supply piping. Equal voltage potential is also maintained between the faulted appliance and other nearby equipment.

  If there is no likelihood of the gas pipe becoming energized (no gas appliance that is also served by electricity), then no bonding of the gas piping is required.

- *250.104(B), Informational Note:* Bonding all piping and metal air ducts within the premises will provide additional safety.* *Informational Notes* are not mandatory but certainly provide excellent recommendations to improve safety.

In a typical home, there is usually a large number of bonds between the hot and cold water pipes.

---

*Reprinted with permission from NFPA 70-2014.

**Reprinted with permission from NFPA 54-2012.

There are interconnections at metallic mixer faucets, water heaters, clothes washers, and similar plumbing connections. Many plumbers use short lengths of copper tubing to bridge a stud space or joist space to support and keep water lines in place. They solder (tack) the water lines to these cross pieces. This bonds the hot and cold water pipes in a number of places inside the walls.

Figure 27-28 illustrates the bonding together of the cold and hot water metal pipes and the metal gas piping. This bonding jumper is sized per *Table 250.66*. This installation is easy and eliminates the questionable bonding/grounding of the many interconnections of the metal water and gas piping through various removable gas appliances. This bonding above the water heater *does not* mean that the gas pipe is serving as a grounding electrode.

Some electrical inspectors will consider the hot/cold mixing faucets and the water heater as an acceptable means of bonding the hot and cold water metal pipes together. Most, however, will require a bonding jumper as illustrated in the figures, consistent with the thinking that electrical grounding and bonding shall not be dependent on the plumbing trade.

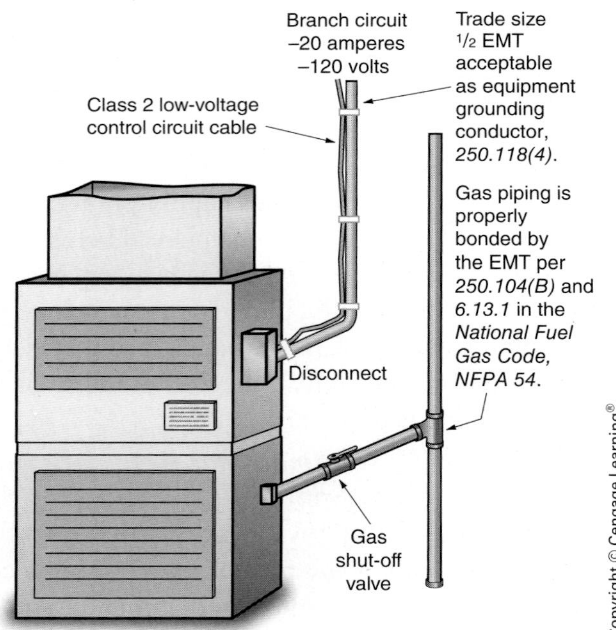

FIGURE 27-35 Gas piping is considered to be properly bonded by the appliance's branch-circuit equipment grounding conductor. The equipment grounding conductor in this illustration is the EMT, which provides an excellent ground-fault current return path.

NEC TABLE 250.66

Grounding Electrode Conductor for Alternating-Current Systems

Size of Largest Ungrounded Service-Entrance Conductor or Equivalent Area for Parallel Conductors[a] (AWG/kcmil)		Size of Grounding Electrode Conductor (AWG/kcmil)	
Copper	**Aluminum or Copper-Clad Aluminum**	**Copper**	**Aluminum or Copper-Clad Aluminum[b]**
2 or smaller	1/0 or smaller	8	6
1 or 1/0	2/0 or 3/0	6	4
2/0 or 3/0	4/0 or 250	4	2
Over 3/0 through 350	Over 250 through 500	2	1/0
Over 350 through 600	Over 500 through 900	1/0	3/0
Over 600 through 1100	Over 900 through 1750	2/0	4/0
Over 1100	Over 1750	3/0	250

Notes:
1. If multiple sets of service-entrance conductors connect directly to a service drop, set of overhead service conductors, set of underground service conductors, or service lateral, the equivalent size of the largest service-entrance conductor shall be determined by the largest sum of the areas of the corresponding conductors of each set.
2. Where there are no service-entrance conductors, the grounding electrode conductor size shall be determined by the equivalent size of the largest service-entrance conductor required for the load to be served.

[a]This table also applies to the derived conductors of separately derived ac systems.
[b]See installation restrictions in 250.64(A).

Don't depend on the plumber to do your job!

Do it right! Follow Figure 27-28 for reliably bonding together all metal water piping and metal gas piping.

The ground-fault current return path in Figure 27-35 is the EMT. Where nonmetallic-sheathed cable is used, the equipment grounding conductor in the cable serves as the ground-fault current return path. Where armored cable is used, the armor plus the bonding strip is the ground-fault current return path. See *NEC 250.118* for other accepted equipment grounding conductors.

NEC 300.6 addresses corrosion that an electrician needs to be concerned with when working in wet locations, underground, in concrete, and similar installations.

Ground Clamps

Ground clamps used for bonding and grounding must be listed for the purpose. The use of solder to make up bonding and grounding connections is not acceptable because under high levels of fault current, the solder would probably melt, resulting in loss of the integrity of the bonding and/or grounding path. See *250.8*.

Various types of ground clamps are shown in Figures 27-22, 27-23, 27-24, and 27-25. These clamps and their attachment to the grounding electrode must conform to *250.70*.

Connection of Equipment Grounding and Grounded Conductors in Main Panel and Subpanel

According to the second paragraph of *408.20*, equipment *grounding* conductors (the bare copper equipment grounding conductors found in nonmetallic-sheathed cable) shall not be connected to the *grounded* conductor (the neutral conductor) terminal bar (neutral bar) unless the bar is identified for the purpose, and is located where the *grounded* conductor is connected to the *grounding* electrode conductor. In this residence, this occurs in Main Panel A. The green main bonding jumper screw furnished with Panel A is installed in Panel A. This bonds the neutral bar, the ground bus, the grounded neutral conductor, the grounding electrode conductor, and the panel enclosure together, as in Figure 27-16. *Grounding* conductors and *grounded* conductors are not to be connected together anywhere on the load side of the main service disconnect, *250.24(5)* and *250.142(B)*. The green bonding screw furnished with Panel B will *not* be installed in Panel B. More on this a little later in this chapter.

Sheet Metal Screws Not Permitted for Connecting Grounding Conductors

Sheet metal screws are not permitted to be used to connect grounding conductors or connection devices to enclosures, *250.8*. Sheet metal screws do not have the same fine thread that No. 10-32 machine screws have, which match the tapped No. 10-32 threaded holes in outlet boxes, device boxes, and other enclosures. Sheet metal screws "force" themselves into a hole instead of nicely threading themselves into a pretapped matching hole. Sheet metal screws have *not* been tested for their ability to safely carry ground-fault currents as required by *250.4(A)(5)*.

A typical bonding jumper (equipment) is illustrated in Figure 27-36.

Bonding Service Equipment

At the main service-entrance equipment, the grounded neutral conductor must be bonded to the metal enclosure. For most residential panels, this main bonding jumper is a bonding screw that is furnished with the panel. This bonding screw is inserted through the neutral bar into a threaded hole in the back of the panel.

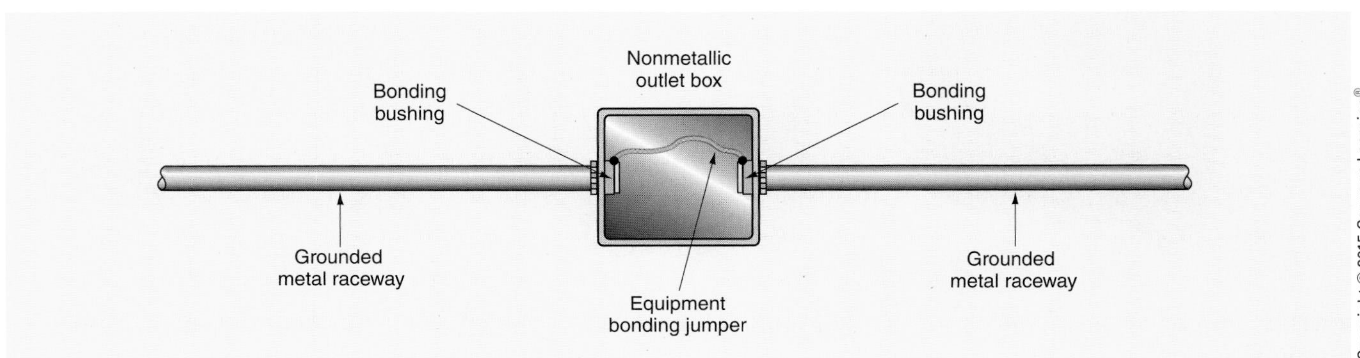

FIGURE 27-36 Bonding jumper (Equipment).

This bonding screw must be green and must be clearly visible after it is in place, *250.28(B)*.

The required main bonding jumpers for services must not be smaller than the grounding electrode conductor, *250.28(D)*. The lugs on bonding bushings are based on the ampacity of the conductors that would normally be installed in that particular size of raceway. The lugs become larger as the trade size of the bushing increases. However, there is no need to calculate the adequacy of the device furnished by the panelboard manufacturer that is intended to function as the main bonding jumper so long as the panelboard is listed by a qualified electrical testing laboratory such as UL.

Bonding at service-entrance equipment is very important because service-entrance conductors do not have overcurrent protection at their line side, other than the electric utility company's primary transformer fuses. Overload protection for service-entrance conductors is at their load end. High available fault currents can result in severe arcing, which is a fire hazard. For all practical purposes, available short-circuit current is limited only by (1) the kVA rating and impedance of the transformer that supplies the service equipment, and (2) the size, type, and length of the service conductors between the transformer and service equipment. Fault currents can easily reach 20,000 amperes or more at the main service in residential installations. Much higher fault current is available in multifamily dwellings such as apartments and condominiums.

Fault current calculations are presented later in this chapter. The text *Electrical Wiring—Commercial* (Copyright © Cengage Learning®) covers fault current calculations in greater depth.

NEC 250.92(A) and *(B)* lists the parts of the service-entrance equipment that must be bonded. These include the meter base, service raceways, service cable armor, and main disconnect. *NEC 250.94* lists the methods acceptable for bonding all of the previous equipment.

Grounding/Bonding Bushings

Grounding bushings have a means (lug) for connecting a grounding or bonding conductor to it. A grounding bushing might also have a means (one or more set screws) that ensure reliable bonding of the bushing to a metal raceway, in which case the bushing serves as a grounding and bonding bushing. These bushings are available with an insulated throat. See Figure 27-37. These are the most commonly used for bonding residential electrical services.

Why Are Grounding/Bonding/Insulating Bushings Installed?

Grounding/bonding bushings, bonding wedges, grounding conductors, and bonding jumpers are installed to ensure a low-impedance path if a fault occurs. Bonding jumpers are required where there are concentric or eccentric knockouts, or where reducing washers are installed at service equipment. *NEC 300.4(F)* states that if 4 AWG or larger ungrounded circuit conductors enter a cabinet, box enclosure or raceway, an insulating bushing or equivalent must be used, shown in Figures 27-37,

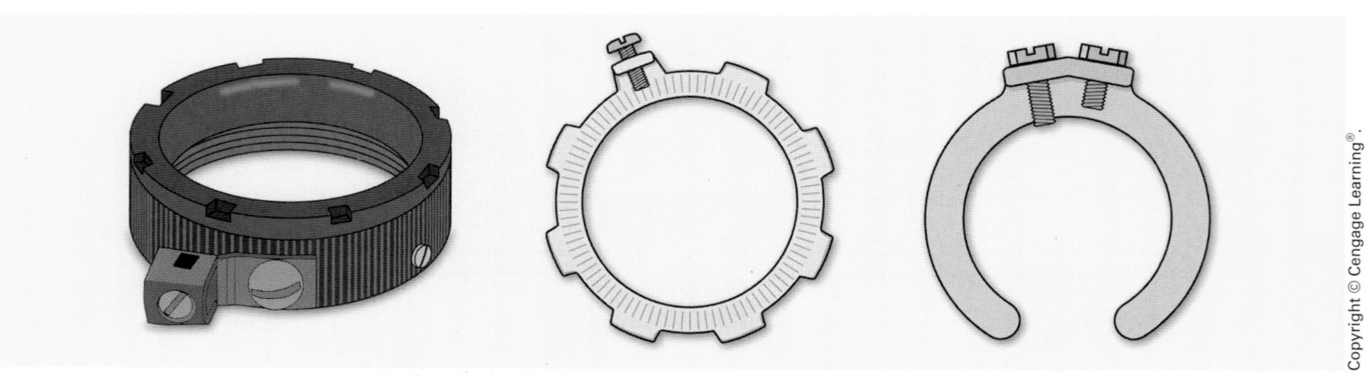

FIGURE 27-37 An insulating grounding/bonding bushing with grounding lug, a bonding locknut, and a bonding wedge.

FIGURE 27-38 Insulating bushings.

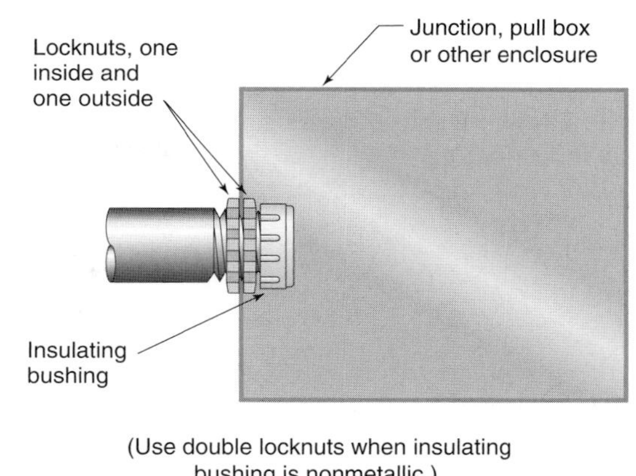

FIGURE 27-39 Insulating bushing in enclosure.

27-38, and 27-39. Similar insulating bushing requirements are found in *312.6(C)*, *314.17(D)*, *314.28(A)*, *354.46*, and *362.46*. Insulating bushings protect the conductors from abrasion where they pass through the bushing. Combination metal/insulating bushings can be used. Some EMT connectors have an insulated throat. Insulating sleeves are also available that slide into a bushing between the conductors and the throat of the bushing.

If the bushing is made of insulating material only, as in Figure 27-38, then two locknuts must be used, as shown in Figure 27-39.

Installing a Grounding Electrode Conductor from Main Service Disconnect to Grounding Electrode

The *Code* specifies in *NEC 250.64* how a grounding electrode conductor is to be installed, as shown in Table 27-4. This table shows the major requirements for installing grounding electrode conductors.

Specific requirements are provided to ensure that the conductor is suitable for the environment, is protected from physical damage, and is installed in a continuous length, unless a splice is necessary.

TABLE 27-4

NEC 250.64: Installing a Grounding Electrode Conductor.

(A)	Aluminum or copper-clad aluminum conductors. This section places restrictions on the use of aluminum or copper-clad aluminum conductors as grounding electrode conductors. If installed outside these conductors are not permitted to be terminated within 18 in. (450 mm) of the earth.		
(B)	Securing and protection against physical damage. If installed exposed a grounding electrode conductor or its enclosure is required to be securely fastened to the surface on which it is carried. Grounding electrode conductors are permitted to be installed on or through framing members.		
	4 AWG or larger	6 AWG	Smaller than 6 AWG
	May be attached to the surface on which it is carried. Does not need additional mechanical protection unless exposed to physical damage.	If not subject to physical damage, may be run along the surface of the building structure without additional protection. If subject to physical damage, install in RMC, EMT, IMC, or Schedule 80 PVC, or cable armor.	Must be installed in rigid metal conduit, electrical metallic tubing, intermediate metal conduit, rigid PVC conduit, or cable armor.
(C)	Continuous. (1) Generally, grounding electrode conductors are required to be installed in one continuous length without a splice or joint. If it is necessary to splice grounding electrode conductors of the wire type, irreversible compression connectors listed as grounding and bonding equipment or the exothermic welding process must be used. (2) Sections of busbars are permitted to be connected together to create a grounding electrode conductor of the desired length. (3) Metal frames of buildings or structures are permitted to have connections that are bolted, riveted or welded to create a grounding electrode conductor. (4) For metal water piping, connections that are threaded, welded, brazed, soldered or of the bolted-flange type are permitted to create a grounding electrode conductor.		
(E)	Enclosures for grounding electrode conductors. A ferrous metal raceway such as conduit or EMT is required to be bonded to the grounding electrode conductor at both ends. This is illustrated in Figures 27–30 and 27–31.		

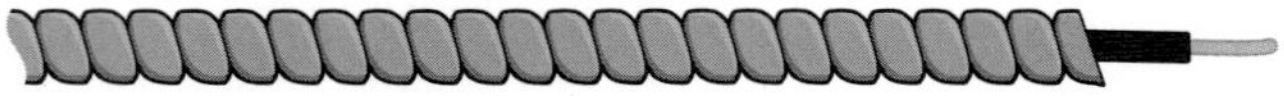

FIGURE 27-40 A typical armored grounding electrode conductor.

NEC 250.64(E) requires that the metal raceway that encloses a grounding electrode conductor be bonded to the conductor at both ends. At first thought, it might appear that simply installing a grounding electrode conductor in a metal raceway makes for a neat and workmanlike installation. However, unless the enclosing metal raceway is properly bonded on both ends, the ground path will be a high-impedance path. Bonding the enclosing metal raceway at one end only will result in a ground path impedance approximately twice that of bonding the metal raceway on both ends. Remember—the higher the impedance, the lower the current flow.

A Typical Armored Grounding Electrode Conductor

Figure 27-40 shows a typical grounding electrode conductor protected by a spiraled metal tape. It is important that this metal tape be electrically connected to the grounding electrode conductor at both ends. This can be accomplished at the service equipment by installing a cable clamp that is designed to connect both the armor and the grounding electrode conductor to the enclosure. At the grounding electrode, a ground clamp must be used that is designed to connect both the armor and the grounding electrode conductor together. Proper bonding at both ends of the armored grounding electrode conductor is necessary to ensure a low impedance path.

Figures 27-29, 27-30, 27-31, 27-32, and 27-33 illustrate accepted methods of installing a grounding electrode conductor between the main panel and the ground clamp. Additional information concerning grounding electrode conductors can be found in Chapter 4.

GROUNDING ELECTRICAL EQUIPMENT AT A SECOND BUILDING

Detached Garage—Grounding

A detached garage or other second building on the same residential property is almost always served by the same electrical service as the house.

For dwellings, a detached garage is not required to have any electrical supply. However, if you do run electric power to a detached garage, specific *Code* requirements must be complied with.

We discussed the receptacle and lighting requirements for garages in Chapter 3. We will now discuss the grounding requirements for a second building, *250.32*. See Figure 27-41.

Branch Circuit to Detached Garage

If a single or multiwire branch circuit is run to a detached garage or other second building, the simplest way to provide proper and adequate grounding for the equipment in that detached garage is to install an underground cable Type UF that contains an equipment grounding conductor. A grounded metal raceway between the house and the outbuilding may also serve as the equipment grounding conductor.

A separate grounding electrode at the second building is not required if only a single branch circuit supplies the building, *250.32(A), Exception*.

Feeder to Detached Garage

When a feeder supplies a panelboard in a detached garage or other second building that has more than one branch circuit, proper grounding at the second building is accomplished by:

- installing a metal raceway, which is an acceptable equipment grounding conductor in *NEC 250.118*, between the main building service equipment or panelboard and the panelboard in the second building, or
- installing a separate equipment grounding conductor of the wire type (in a cable or in a

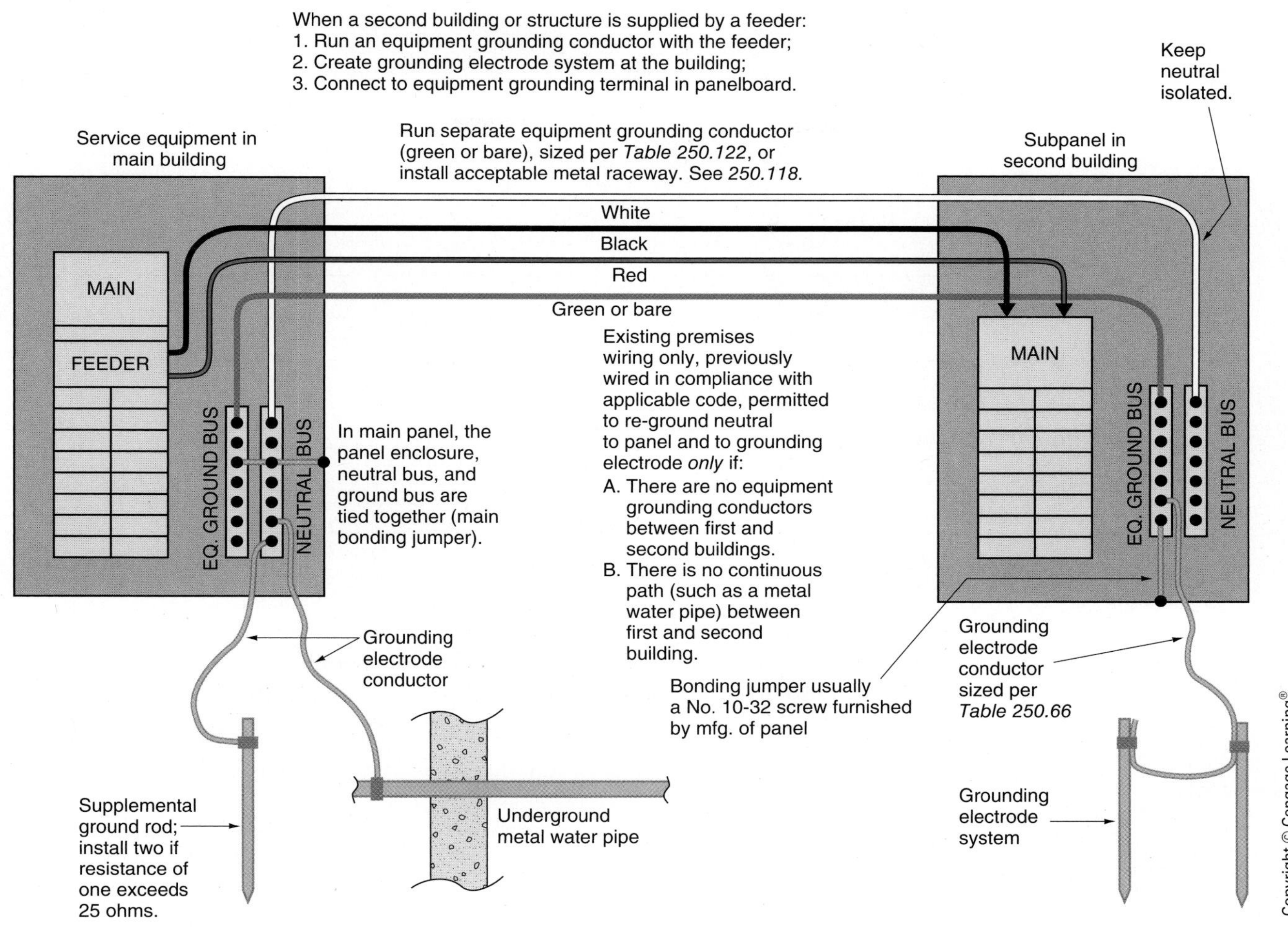

FIGURE 27-41 Grounding at a second building or structure, *250.32.*

nonmetallic raceway) between the main building service equipment or panelboard and the panelboard in the second building.

In either case, a grounding electrode system is required at the second building. A grounding electrode conductor is run between the grounding electrode and the equipment grounding bus in the panelboard in the second building.

The grounding electrode conductor is sized according to *Table 250.66*, but it does not have to be larger than the largest ungrounded supply conductor, *250.32(E)*.

The neutral bus and the equipment grounding bus in the second building are not connected together. If these two buses were tied together in the second building, there would be more than one path (a parallel path) for neutral current to flow back on. Take a look at Figure 27-41. Imagine a tie between the neutral bus and the equipment grounding bus in the second building. Now imagine that there is some value of neutral current. You can readily see that this neutral current would have many ways to return to the source at the main building panel. Some current would return on the neutral conductor, some on the equipment grounding conductor, some on the grounding electrode conductor—to the grounding electrode—and back through the earth, and some through a metal water pipe if one is present. These multiple paths of neutral return current are not permitted.

All of the above requirements are found in *NEC 230.32*, and are illustrated in Figure 27-33.

REVIEW

Note: Refer to the *NEC,* the text, and/or the plans where necessary.

1. Define the *Service Point.* ______________________________

2. Define *service-drop conductors.* ______________________________

3. Who is responsible for determining the service location? ______________

4. a. The service head must be located (above) (below) the point where the service-drop conductors are spliced to the service-entrance conductors. (Circle the correct answer.)
   b. What *Code* section provides the answer to part (a)? ______________

5. What is a mast-type service entrance? ______________________________

6. When a conduit is extended through a roof, must it be guyed? ______________

7. a. What size and type of conductors are installed for the service in this text? ______________
   b. What size conduit is installed? ______________
   c. What size grounding electrode conductor is installed? (not neutral) ______________
   d. Is the grounding electrode conductor insulated, armored, or bare? ______________

8. How and where is the grounding electrode conductor attached to the water pipe?
____

9. What are the minimum distances or clearances for the following?
   a. Overhead service conductor clearance over private driveway ______________
   b. Overhead service conductor clearance over private sidewalks ______________
   c. Overhead service conductor clearance over alleys ______________
   d. Overhead service conductor clearance over a roof having a roof pitch of not less than 4/12. (Voltage between conductors does not exceed 300 volts.) ______________
   e. Overhead service conductor horizontal clearance from a porch ______________
   f. Overhead service conductor clearance from a fence that can be climbed on ______

10. What are the minimum size ungrounded conductors using Type THW copper for the following residential electrical services? The terminals in the panelboard are marked 75°C.

AMPERE RATING OF SERVICE	SERVICE-ENTRANCE CONDUCTORS SIZED PER *TABLE 310.15(B)(16)*	SERVICE-ENTRANCE CONDUCTORS SIZED PER *310.15(B)(7)*
100		
200		
400		

11. What is the minimum size copper grounding electrode conductor for each of the following residential electrical services? Refer to *Table 250.66*. The ungrounded conductors are Type THW copper.

AMPERE RATING OF SERVICE	GROUNDING ELECTRODE CONDUCTOR WHEN THE SERVICE-ENTRANCE CONDUCTORS ARE SIZED PER *TABLE 310.15(B)(16)*	GROUNDING ELECTRODE CONDUCTOR WHEN THE SERVICE-ENTRANCE CONDUCTORS ARE SIZED PER *310.15(B)(7)*
100		
200		
400		

310.15(B)(7) is only for 120/240-volt, single-phase residential services and feeders. This Section does not apply to services and feeders other than residential.

12. What is the recommended height of a meter socket from the ground? ____________

13. a. May the bare grounded neutral conductor of a service be buried directly in the ground? ____________

b. What section of the *Code* covers this? ____________

14. How far must mechanical protection be provided when underground service conductors are carried up a pole? ____________

15. a. A service disconnect may consist of not more than how many switches or circuit breakers? ____________

b. Must these devices be in one enclosure? ____________

c. What type of main disconnect is provided in this residence? ____________

d. Does your city permit more than one service disconnect? ____________

16. Complete the following table by filling in the columns with the appropriate information.

	CIRCUIT NUMBER	AMPERE RATING	POLES	VOLTS	CONDUCTOR SIZE
A. Living room receptacle outlets					
B. Workbench receptacle outlets					
C. Water pump					
D. Attic exhaust fan					
E. Kitchen lighting					
F. Hydromassage tub					
G. Attic lighting					
H. Counter-mounted cooking unit					
I. Electric furnace					

17. a. What size conductors supply Panel B? ______
    b. What size raceway? ______
    c. Is this raceway run in the form of electrical metallic tubing or rigid conduit?
    d. What size overcurrent device protects the feeders to Panel B? ______
18. How many electric meters are provided for this residence? ______
19. a. According to the *NEC*, is it permissible to ground a residential rural electrical service to driven ground rods only when a metallic water system is available?

    ______

    b. What sections of the *Code* apply? ______
20. What table in the *NEC* lists the sizes of grounding electrode conductors required for electrical services of various sizes? *Table* ______
21. *NEC 250.53(D)(2)* requires that a supplemental ground be provided if the available grounding electrode is: (Circle the correct answer.)
    a. metal underground water pipe.
    b. building steel.
    c. concrete encased ground.
22. Do the following conductors require mechanical protection?
    a. 8 AWG grounding electrode conductor ______
    b. 6 AWG grounding electrode conductor ______
    c. 4 AWG grounding electrode conductor ______
23. Why is bonding of service-entrance equipment necessary? ______

    ______

    ______

24. What special types of bushings are required on service entrances? ______

    ______

25. When insulated conductors 4 AWG or larger that are required to be insulated are installed in conduit, what additional provision is required on the conduit ends? ______

    ______

26. What minimum size copper bonding jumpers must be installed to bond properly the electrical service for the residence discussed in this text? ______

    ______

27. a. Where is Panel A located? ______
    b. On what type of wall is Panel A fastened? ______
    c. Where is Panel B located? ______
    d. On what type of wall is Panel B fastened? ______
28. When conduits pass through the wall from outside to inside, the conduit must be ______ to prevent air circulation through the conduit.

29. Briefly explain why electrical systems and equipment are grounded. ____________

____

____

____

____

30. In general, systems are required to be grounded if the maximum voltage to ground does not exceed: (Circle the correct response.)
   a. 120 volts.
   b. 150 volts.
   c. 300 volts.

31. To ensure a complete grounding electrode system: (Circle the correct response.)
   a. everything must be bonded together.
   b. all metal pipes and conduits must be isolated from one another.
   c. the service neutral is grounded to the water pipe only.

32. An electric clothes dryer is rated at 5700 watts. The electric rate is 10.091 cents per kilowatt-hour. The dryer is used continuously for 3 hours. Find the cost of operation, assuming the heating element is on continuously. $____________

33. A heating cable rated at 750 watts is used continuously for 72 hours to prevent snow from freezing in the gutters of the house. The electric rate is 8.907 cents per kilowatt-hour. Find the cost of operation. $____________

34. When used as service equipment, a panelboard (load center) must be ____________ that is suitable for use as service equipment, *230.66*. If the panelboard (load center) contains the main service disconnect, it must be clearly marked ____________ ____________, *230.70(B)*.

35. Here are five commonly used terms in the electrical industry. Enter the letter of the term that corresponds to its definition.
   a. Grounding electrode conductor
   b. Main bonding jumper
   c. Grounded circuit conductor
   d. Equipment grounding conductor
   e. Underground service conductors (Service Point at transformer)

   ______ The neutral conductor.

   ______ The term used to define underground service-entrance conductors that run between the meter and the utilities connection.

   ______ The conductor (sometimes a large threaded screw) that connects the neutral bar in the service equipment to the service-entrance enclosure.

   ______ The conductor that runs between the neutral bar in the main service equipment to the grounding electrode (water pipe, ground rod, etc.).

   ______ The bare copper conductor found in nonmetallic-sheathed cable.

36. The electrician mounted the disconnect switch for a central air-conditioning unit 8 ft (2.5 m) above the ground. This was easy to do because all he had to do was run the conduit across the ceiling joists in the basement, through the outside wall, and directly into the back of the disconnect switch. The electrical inspector turned the job down, citing *NEC* ________________ that requires that the disconnecting operating handle must not be higher than ________________ above the ground.

37. a. According to the *NEC* definition of a wet location, a basement cement wall that is in direct contact with the earth is considered to be a wet location. A panelboard mounted on this wall must have at least [¼ in. (6 mm)], [½ in. (13 mm)] [1 in. (25 mm)] space between the wall and the panel. (Circle the correct answer.)

    b. How is this required spacing accomplished? ________________

    ____

38. When using rebars as the concrete-encased electrode, the rebars must be (a) insulated with plastic material so they will not rust, or (b) bonded together by the usual steel tie wires or other effective means. Underline the correct answer.

39. The circuit directory in a panelboard (may) (shall) be filled out according to *NEC* ________________. Circle the correct answer, and enter the correct *NEC* section number.

40. A main service panel is located in a dark corner of a basement, far from the basement light. In your opinion, does this installation meet the requirements of *110.26(D)*? Explain your answer. ________________

    ____

41. Does the electric utility in your area allow the location of the meter to be on the back side of a residence? ________________

42. The term used when the utility charges different rates during different periods during the day is ________________.

43. A ground rod is driven below the meter outside of the house. A grounding electrode conductor connects between the meter base and the ground rod. The ground clamp is buried under the surface of the soil. This is permitted if the ground clamp is ________________ for direct burial according to *NEC* ________________.

44. Copper-coated steel ground rods are the most commonly used grounding electrodes. These rods shall:

    a. if not listed, be at least ______ in. (______ mm) in diameter, *250.52(A)(5)*.

    b. be driven to a depth of at least ______ ft (______ m) unless solid rock is encountered, in which case the rod may be driven at a ______-° angle, or it may be laid in a trench that is at least ______ ft (______ mm) deep, *250.53(G)*.

    c. be separated by at least ______ ft (______ m) when more than one rod is driven, *250.53(A)(3)*.

    d. have a ground resistance of not over ______ ohms for one rod, *250.53(A)(2) Exception*.

45. What section of the *Code* prohibits the use of sheet metal screws as a means of attaching grounding conductors to enclosures? *NEC* ____________________

46. Which section of the *NEC* prohibits using the space below and in front of an electrical panel as storage space? *NEC* ____________________

47. What section of the *NEC* prohibits using an underground metal gas pipe as the grounding electrode for an electrical service? ____________________

48. When wiring a gas furnace, what additional steps, if any, are necessary in order to make sure that the gas piping supplying the furnace is adequately bonded? Explain.

____

____

____

CHAPTER **28**

Overcurrent Protection–Fuses and Circuit Breakers

OBJECTIVES

After studying this chapter, you should be able to

- understand the important *NEC* requirements for fuses and circuit breakers.
- discuss the five possible circuit conditions.
- understand the various types and operation of fuses and circuit breakers.
- know when to use single-pole and 2-pole circuit breakers.
- understand the term *interrupting rating* for fuses and circuit breakers.
- calculate available short-circuit current using a simple formula.
- understand *series-rated* panelboards.
- understand the meaning of *selective coordination* and *nonselective coordination*.

THE BASICS

The *NEC*® covers overcurrent protection of conductors in *Article 240*. Overcurrent protection for residential services, branch circuits, and feeders is provided by circuit breakers or fuses. These are the "safety valves" of an electrical circuit.

Fuses and circuit breakers are sized by matching their ampere ratings to conductor ampacities and connected load currents. They sense overloads, short circuits, and ground faults, and protect the wiring and equipment from reaching dangerous temperatures.

A fuse will function only one time. It does its job of protecting the circuit. When a fuse opens, find out what caused it to open, fix the problem, and replace it with the size and type that will provide proper overcurrent protection. Do not keep replacing a fuse without finding out why the fuse blew in the first place.

When a circuit breaker trips, find out what caused it to trip, fix the problem, then reset it. Do not repeatedly reset the breaker again and again into a fault. This is looking for trouble.

KEY *NEC* REQUIREMENTS FOR OVERCURRENT PROTECTION

- *NEC Table 210.24:* This table shows the maximum overcurrent protection for branch-circuit conductors.
- *NEC 240.4:* Overcurrent protection is sized according to the ampacity of a conductor.
- *NEC 240.4(B):* Where the standard ampere rating of fuses or circuit breaker does not match the ampacity of the conductor, the next higher ampere rating is permitted. This is permitted if the branch circuit does not supply more than one receptacle for cord-and-plug-connected portable loads and for overcurrent devices not exceeding 800 amperes.
- *NEC 240.4(D):* The maximum overcurrent protection of small conductors is

CONDUCTOR SIZE	MAXIMUM OVERCURRENT PROTECTION
14 AWG copper	15 amperes
12 AWG copper	20 amperes
10 AWG copper	30 amperes

- *NEC 240.6(A):* Lists the standard ampere ratings for fuses and circuit breakers.
- *NEC 240.20(A):* Overcurrent protection is required to be provided for each ungrounded ("hot") conductor.
- *NEC 240.21:* Overcurrent protection is generally required at the point where a conductor receives its supply. The tap rules act like exceptions to the general requirement even though they are not written in typical style for exceptions.
- *NEC 240.22:* Overcurrent devices are generally not permitted in the grounded conductor. Exceptions are: (1) if the overcurrent device opens *all* conductors of the circuit at the same time, and (2) where the overcurrent devices are used for motor overload protection.
- *NEC 240.24(A):* Overcurrent protective devices are required to be accessible.
- *NEC 240.24(D):* Overcurrent protective devices are not permitted near easily ignitable material, such as in clothes closets.
- *NEC 240.24(E):* Other than for supplementary overcurrent protection, overcurrent devices are not permitted to be located in bathrooms.
- *NEC 240.24(F):* Overcurrent devices are not permitted to be located over steps of a stairway.
- *NEC 230.70(A)(2):* Service disconnect(s) are not permitted to be located in bathrooms.
- *NEC 230.79(C):* 100 amperes is the minimum-size service for a one-family dwelling.
- *NEC 230.90(A):* Overload protection is required to be provided for each ungrounded "hot" service-entrance conductor. This is accomplished by the overcurrent protective device(s) in the service-entrance main panel.
- *NEC 230.91:* The main service overcurrent device is usually an integral part of the service disconnecting means. For large services, the service disconnecting means could be a separate disconnect switch.
- *NEC 310.15(B)(7):* This section allows the ampacity of service-entrance conductors and specific feeders for 3-wire, single-phase, 120/240-volt supplies for dwellings to be rated 83% of the overcurrent device. This rule applies to services and specific feeders rated 100 through 400 amperes.

FIVE CIRCUIT CONDITIONS

For the following five conditions, each conductor is 5 feet (1.52 m) long and is solid copper. From *Table 8* in *Chapter 9* in the *NEC*, we find the resistance of a 14 AWG solid copper conductor to be 3.07 ohms per 1000 feet (304.8 m). This equates to 0.307 ohm per 100 feet (30.48 m), 0.0307 ohm per 10 feet (3.05 m), and 0.01535 ohm per 5 feet (1.52 m). To keep the calculation simple, we rounded off the resistance values for each 5-foot (1.52-m) conductor length to 0.015 ohm. We are not trying to be rocket scientists. We are merely pointing out the fundamentals of different circuit conditions.

Normal: Normal loading of a circuit is when the current flowing is within the capability of the circuit and/or the connected equipment. In Figure 28-1, we have a 15-ampere circuit carrying approximately 10 amperes.

An *overload* is a condition where the current flowing is more than the circuit and/or connected equipment is designed to safely carry. Figure 28-2 shows a 15-ampere circuit carrying 20 amperes. A momentary overload is harmless, but a continuous overload condition will cause the conductors and/or equipment to overheat—a potential cause for fire. The current flows through the "intended path"—the conductors and/or equipment.

A *short circuit* is a condition when two or more normally insulated circuit conductors come in contact with one another, resulting in a current flow that bypasses the connected load. A short circuit might be two "hot" conductors coming together, or it might be a "hot" conductor and a "grounded" conductor coming together. In either case, the current flows outside of the "intended path." The only resistance (impedance) is that of the conductors, the source, and the arc. This low resistance results in high levels of short-circuit current. The heat generated at the point of the arc can result in a fire, as shown in Figure 28-3.

A *ground fault* is a condition when a "hot" or ungrounded conductor comes in contact with a grounded surface, such as a grounded metal raceway, metal water pipe, sheet metal, and so on, as shown in Figure 28-4. The current flows outside of the intended path. A ground fault can result in a flow of current greater than the circuit rating, in which case the overcurrent device will open. A ground fault can also result in a current flow less than the

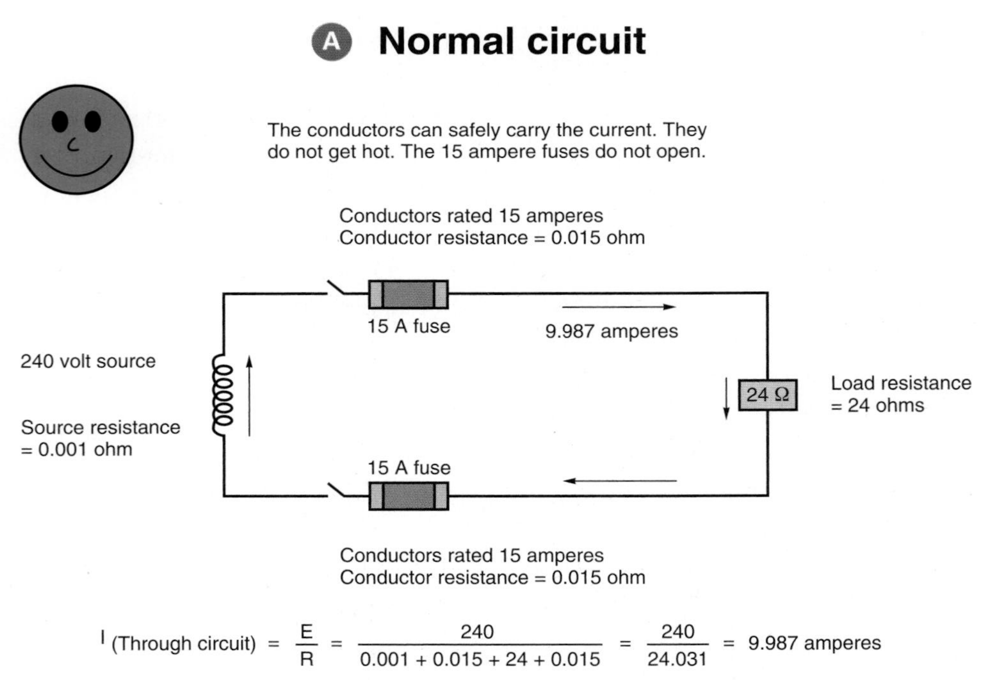

FIGURE 28-1 A normally loaded circuit.

B Overloaded circuit

The conductors begin to get hot, but the 15 ampere fuses will open before the conductors are damaged.

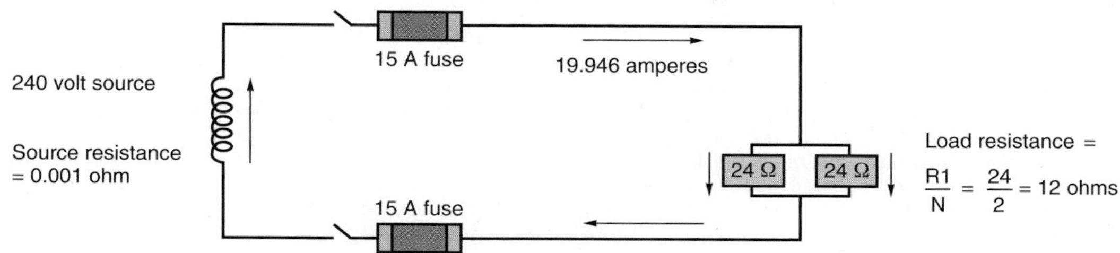

$$I_{\text{(Through circuit)}} = \frac{E}{R} = \frac{240}{0.001 + 0.015 + 12 + 0.001} = \frac{240}{12.031} = 19.946 \text{ amperes}$$

FIGURE 28-2 An overloaded circuit.

C Short circuit

The conductors get extremely hot. The insulation will melt off, and the conductors will melt unless the fuses open in a very short (fast) period of time. Current-limiting overcurrent devices will limit the amount of "let-through" current by opening so fast (fraction of a cycle) that the full value of fault current will not be reached.

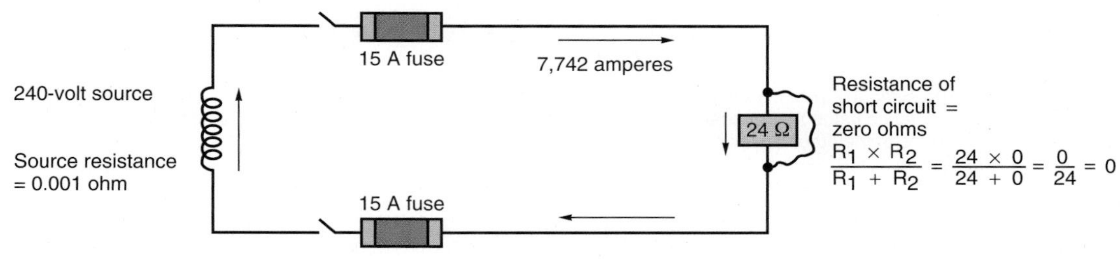

$$I_{\text{(Through circuit)}} = \frac{E}{R} = \frac{240}{0.001 + 0.015 + 0.015} = \frac{240}{0.031} = 7{,}742 \text{ amperes}$$

FIGURE 28-3 Note that the connected load is short-circuited.

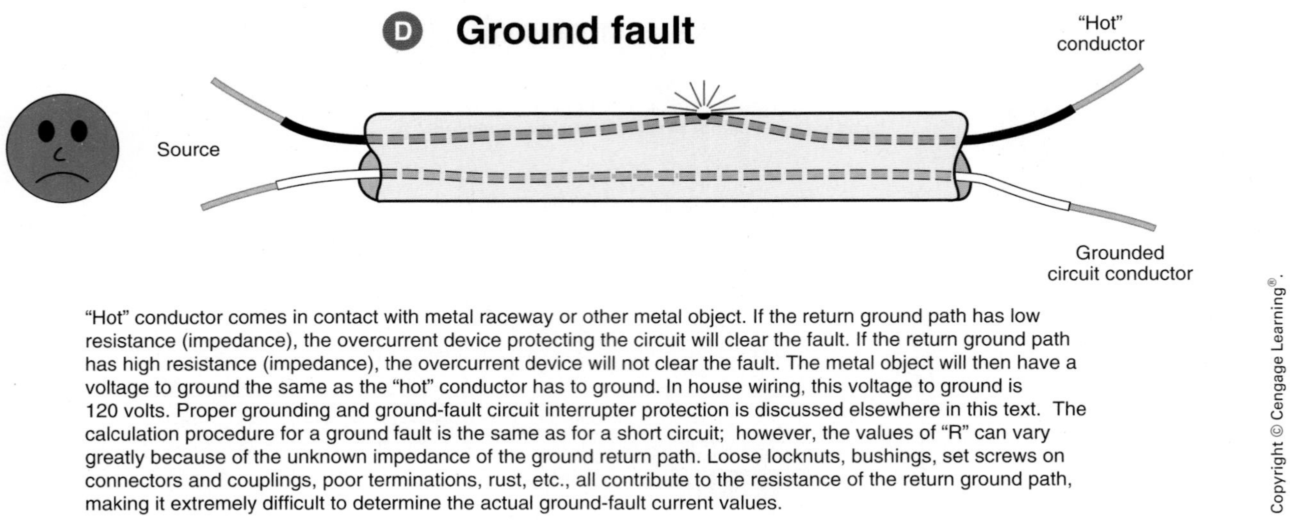

D Ground fault
"Hot" conductor
Source
Grounded circuit conductor
"Hot" conductor comes in contact with metal raceway or other metal object. If the return ground path has low resistance (impedance), the overcurrent device protecting the circuit will clear the fault. If the return ground path has high resistance (impedance), the overcurrent device will not clear the fault. The metal object will then have a voltage to ground the same as the "hot" conductor has to ground. In house wiring, this voltage to ground is 120 volts. Proper grounding and ground-fault circuit interrupter protection is discussed elsewhere in this text. The calculation procedure for a ground fault is the same as for a short circuit; however, the values of "R" can vary greatly because of the unknown impedance of the ground return path. Loose locknuts, bushings, set screws on connectors and couplings, poor terminations, rust, etc., all contribute to the resistance of the return ground path, making it extremely difficult to determine the actual ground-fault current values.
Copyright © Cengage Learning®.

FIGURE 28-4 The insulation on the "hot" conductor has come in contact with the metal conduit. This is termed a "ground fault."

E Open circuit
The circuit has an "open." No current will flow in the circuit.
Conductors rated 15 amperes
Conductor resistance = 0.015 ohm
15 A fuse
X
"Open"
240 volt source
Source resistance = 0.001 ohm
24 Ω
Load resistance = 24 ohms
15 A fuse
Zero amperes
Conductors rated 15 amperes
Conductor resistance = 0.015 ohm
I (Through circuit) = $\frac{E}{R} = \frac{240}{\infty}$ = Zero amperes*
* The symbol ∞ represents an immeasurably high value. An open circuit has such a high resistance value that ordinary ohmmeters cannot read it. We can ignore the resistance values of the other components of the circuit and use infinity for the "R" value.
Caution: Line voltage (in this case 240 volts) appears across the "open."
Copyright © 2015 Cengage Learning®.

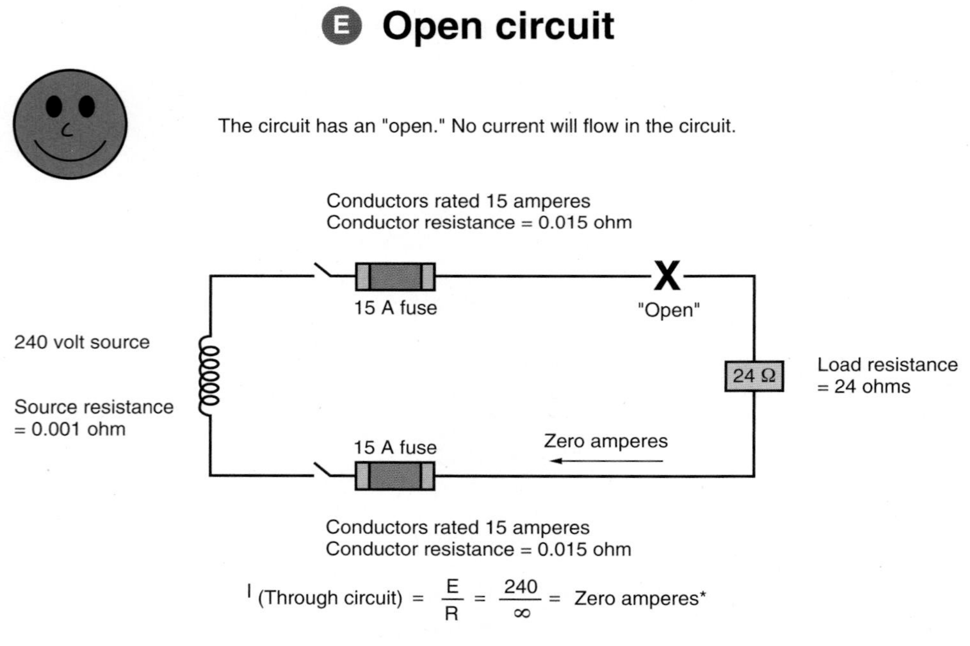

FIGURE 28-5 The circuit is "open" where marked "X." No current flows in the circuit so no heat is created.

circuit rating, in which case the overcurrent device will not open.

A *GFCI* is an example of a device that protects against very low levels of ground-fault current passing through the human body when a person touches a "live" wire and a grounded surface. See Chapter 6 in this text for additional information on GFCI protection.

An *open circuit* is a condition where the circuit is not closed somewhere in the circuit, as in Figure 28-5, and no current can flow.

FUSES

Fuses are a reliable and economical form of overcurrent protection. *NEC 240.50* through *240.54* provides the requirements for plug fuses, fuseholders, and adapters. Here is a brief recap:

Edison-Base Plug Fuses

- Plug fuses may be used (1) in circuits that do not exceed 125 volts between conductors, and (2) in circuits having a grounded neutral where no conductor operates at over 150 volts to ground. In a residential 120/240-volt electrical system, the voltage between the two "hot" conductors is 240 volts, and the voltage from either "hot" conductor to ground is 120 volts.
- The fuses shall have ampere ratings of 0 through 30 amperes.
- Plug fuses shall have a hexagonal configuration somewhere on the fuse when rated at 15 amperes or less.
- The screw shell of the fuseholder must be connected to the load side of the circuit.
- Edison-base plug fuses may be used only to replace fuses in existing installations where there is no sign of overfusing or tampering.
- Typical applications for plug fuses are using box-cover units where you need to provide close overcurrent protection, such as for an attic fan, a gas furnace, an appliance, and similar loads. This already has been discussed elsewhere in this text.

Type S Fuses

All new plug fuse installations are required to be Type S fuses, *240.52*. That is because the ampere ratings of conventional Edison-base plug fuses are all interchangeable—any ampere rating up to and including 30 amperes.

- Type S fuses are classified at 0 through 15 amperes, 16 through 20 amperes, and 21 through 30 amperes. The reason for this classification is given in the following paragraph. In the 0- to 15-ampere range, there are many ampere ratings to choose from, excellent for protecting motors. See Table 28-1. Type S fuses are used with a corresponding size adapter.

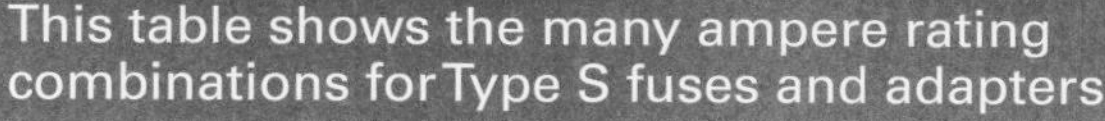

TABLE 28-1

This table shows the many ampere rating combinations for Type S fuses and adapters.

TYPE S FUSE INFORMATION

Type S Fuse Ampere Ratings	Type S Adapter Ratings	Type S Fuse Ampere Ratings That Fit into This Adapter
3/10, 1/2, 8/10, 4/10, 6/10, 1	1	1 ampere & smaller
1 1/8, 1 1/4	1 1/4	All smaller
1 4/10, 1 6/10	1 6/10	All smaller
1 8/10, 2	2	1 8/10, 2
2 1/4, 3 1/2	2 1/2	1 8/10, 2, 2 1/4, 2 1/2
2 8/10, 3 8/10	3 2/10	1 8/10, 2, 2 1/4, 2 1/2, 2 8/10, 3 2/10
3 1/2, 4	4	3 1/2, 4
4 1/2, 5	5	3 1/2, 4, 4 1/2, 5
5 6/10, 6 1/4	6 1/4	3 1/2, 4, 4 1/2, 5, 5 6/10, 6 1/4
7, 8	8	7, 8
9, 10	10	7, 8, 9, 10
12, 14	14	7, 8, 9, 10, 12, 14
15	15	15
20	20	20
25	30	20, 25, 30
30	30	20, 25, 30

When the electrician installs fusible equipment, the ampere rating of the circuit(s) must be determined. After determining the ampere rating of the Type S fuse to be used, an adapter of the proper size is inserted into the Edison-base fuseholder. The proper Type S fuse is then screwed into the adapter. Because of the adapter, the fuseholder is nontamperable and noninterchangeable. For example, assume that a 15-ampere adapter is inserted for a 15-ampere branch circuits; it is impossible to substitute a Type S fuse with a higher ampere rating without removing the 15-ampere adapter.

A Type S fuse and adapter is shown in Figure 28-6.

Courtesy of Eaton's Bussmann Business.

FIGURE 28-6 Type S fuse and adapter.

Fuse Characteristics

Fuses have different time-current characteristics. The term *time-current* refers to how long it will take a fuse or circuit breaker to open under different current values.

- Nontime-delay: This type of fuse has one link (fusible element). One part of the link is "necked down" so when excessive current flows, it will open in the weakest part of the link—the necked down portion. Nontime-delay fuses are not the best choice for motor circuits because of the high starting inrush current of motors. See the "W" fuse, Figure 28-7.

Courtesy of Eaton's Bussmann Business.

FIGURE 28-7 Three types of plug fuses: single-element, nontime-delay (W); dual-element, time-delay (T); loaded link, time-delay (TL).

- Time-delay, loaded link: This type of fuse has one link (fusible element) that is "loaded" with a heat sink next to the "necked down" portion of the link. This "load" acts as a heat sink that absorbs a considerable amount of heat before the "necked down" portion of the link melts open. This heat sink provides the fuse with time delay. See the "TL" fuse, Figure 28-7.
- Time-delay, dual-element: One fuse element opens quickly when a short circuit, heavy overload, or ground fault occurs. The other element in series with the first element opens slowly on overload conditions. Dual-element, time-delay fuses are an excellent choice for motor circuits because they will not open needlessly on momentary overloads. See the "T" fuse, Figure 28-7.

Cartridge Fuses

Cartridge fuses are covered in *NEC 240.60* and *240.61*.

Cartridge fuses are available with the same three basic types of time-current characteristics as plug fuses.

The most common type of dual-element cartridge fuse is shown in Figure 28-8. They are available in 250-volt and 600-volt ratings with ampere ratings from 0 through 600 amperes.

Large ampere rating cartridge fuses may have more than one fuse element.

Time-delay, dual-element fuses provide a time-delay of about 10 seconds at a current of 500% of the fuse rating before it opens. This provides accurate protection for prolonged overloads.

Dual-element fuses are used on motor and appliance circuits where the long time-delay characteristic is required. Single-element fuses are more suitable for circuits that are not expected to have high inrush currents.

Cartridge fuses are available in Class H and Class R types. Refer to Table 28-2.

General-Purpose Cartridge Fuses

These are single-element cartridge fuses that have no intentional time-delay designed into them, as are dual-element, time-delay fuses. These are the type of fuses used in conventional disconnect switches.

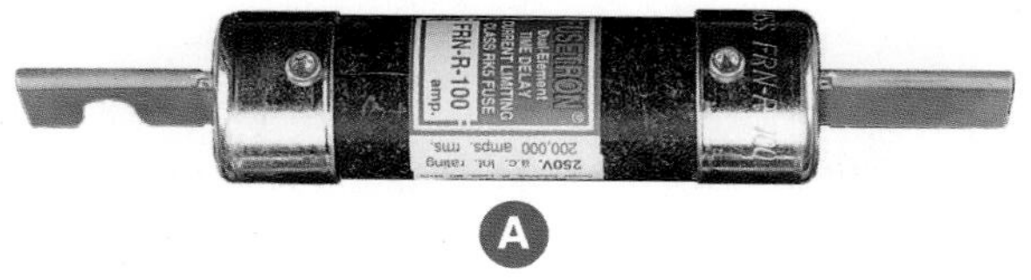

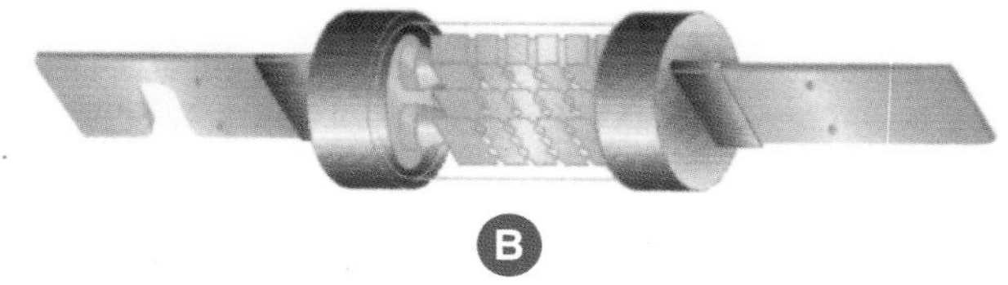

FIGURE 28-8 Cartridge-type dual-element fuse (A) is a 250-volt, 100-ampere fuse. The cutaway view in (B) shows the internal parts of the fuse. When this type of fuse has a rejection slot in the blades of 70–600-ampere sizes, or a rejection ring in the end ferrules of the 0–60-ampere sizes, it is a UL Class RK1 or RK5 fuse.

They are available in ampere ratings from 1/8 through 600 amperes, in both 250-volt and 600-volt ratings.

Those rated through 60 amperes are classified as K5 and have an interrupting rating of 50,000 amperes. Sizes 0 through 60 have a brass ferrule on each end. See Figure 28-9.

Those rated 65–600 are classified as class H and have an interrupting rating of 10,000 amperes. These sizes have a blade coming out of each end and are therefore referred to as "knife blade" fuses. See Figure 28-9.

Other Classes of Fuses

Class CC, Class G, and Class T fuses are special-purpose fuses that are found in original equipment, or in commercial and industrial panelboards, multimetering equipment, and motor controllers. They are used in original equipment because of their small physical size. If you need more information about these types of fuses, consult the manufacturer's technical literature and/or websites. These fuses are rarely used in residential applications but are mentioned here to make you aware of them.

TABLE 28-2

The interrupting ratings of the more common fuses and circuit breakers.

TYPE OF OVERCURRENT DEVICE	INTERRUPTING RATING
Circuit breakers	Are marked with their interrupting rating if other than 5000 amperes.
Plug fuses (Edison base and Type S)	10,000 amperes ac RMS symmetrical. Plug fuses 15 amperes and less will have a hexagon shape in the window—or the fuse body of the fuse will have the hexagon shape.
Class H cartridge fuses	10,000 amperes ac RMS symmetrical. This is the most common type of low-cost cartridge fuse. Renewable link Class H cartridge fuses are permitted only on existing installations where there is no evidence of overfusing or tampering. Rarely, if ever, will these be found in residential electrical systems. The interrupting rating is marked when other than 10,000 amperes.
Class R cartridge fuses	The letter R stands for "Rejection." Knife blade types have a notch in the blade. Ferrule types have an annular ring in one of the ferrules. This means that if the switch or fuseholder is designed for Class R fuses, then ordinary Class H fuses will not fit. These are further broken down into RK1 and RK5 categories. RK fuses are available with interrupting ratings of 50,000, 100,000, and 200,000 amperes ac RMS symmetrical. The interrupting rating is marked on the fuse.

Note: Using a fuse or circuit breaker having a high interrupting rating does not necessarily constitute a safe installation. The entire assembly (panelboard, switch, controller) is marked with its short-circuit rating when used with a specific type of overcurrent device. That is why it is so important to "read the label."

Courtesy of Eaton's Bussmann Business.

FIGURE 28-9 A 60-ampere and a 100-ampere general purpose, one-time fuse.

Sometimes poor connections at the terminals or loose fuse clips can be found because the fuse tubing will appear to be charred. This problem can be dangerous and should be corrected as soon as possible. Quite often, the heat resulting from poor connections in the fuse clips or fuseholder will cause the brass or copper fuse clips or fuseholders to turn from a bright, shiny appearance to that of antique brass or copper.

Fuse Ampere Ratings

The standard ampere ratings for fuses are given in *240.6(A)*. In addition to the standard ratings listed, there are many other "in-between" ampere ratings available to match specific load requirements. Check a fuse manufacturer's catalog for other available ampere ratings.

CIRCUIT BREAKERS

Installations in dwellings normally use thermal-magnetic circuit breakers. On a continuous overload, a bimetallic element in such a breaker moves until it unlatches the inner tripping mechanism of the breaker. Momentary small overloads do not

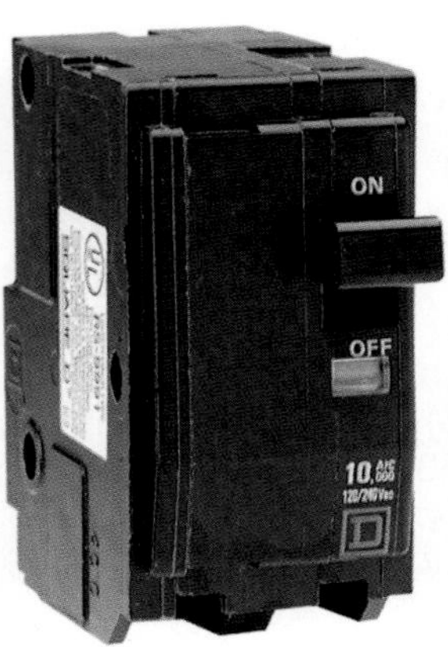

Courtesy of Schneider Electric.

FIGURE 28-10 Typical molded-case circuit breakers.

cause the element to trip the breaker. If the overload is heavy or if there is a short circuit, a magnetic coil in the breaker causes it to interrupt the branch-circuit almost instantly. Figure 28-10 illustrates a typical single-pole and a 2-pole circuit breaker.

NEC 240.80 through *240.86* give the requirements for circuit breakers. The following points are taken from these sections.

- Circuit breakers are required to be trip-free so that even if the handle is held in the ON position, the internal mechanism will trip to the OFF position.
- Breakers are required to indicate clearly whether they are on or off.
- A breaker is required to be nontamperable so that it cannot be readjusted (trip point changed) without dismantling the breaker or breaking the seal.
- The rating is required to be durably marked on the breaker. For small breakers rated at 100 amperes or less and 600 volts or less, the rating must be molded, stamped, or etched on the handle or on another part of the breaker that will be visible after the cover of the panel is installed.
- Every breaker with an interrupting rating other than 5000 amperes is required to have this rating marked on the breaker. Circuit breakers used in dwellings typically are rated to interrupt 10,000 amperes during a short circuit or a ground-fault condition.
- Circuit breakers rated at 120 volts and 277 volts and used for fluorescent loads are not permitted to be used as switches unless marked "SWD."

Thermal-magnetic circuit breakers are temperature sensitive. Some circuit breakers are ambient (surrounding temperature) compensated. This means that the tripping characteristic of the breaker is partially or completely neutralized by the effect of ambient temperature. Underwriters Laboratories (UL) Standard 489 specifies calibration testing at various loads and ambient temperatures. The continuous test is run at 104°F (40°C). One manufacturer states that no rerating is necessary when the circuit breaker is installed in an ambient temperature in the range of 14°F to 140°F (−10°C to 60°C). If you are installing circuit breakers in extremely hot or extremely cold temperatures, consult the manufacturer's literature.

It is a good practice to turn the breaker on and off periodically to "exercise" its moving parts.

NEC 240.6(A) gives the standard ampere ratings of circuit breakers.

How Much Load Can a Circuit Breaker Safely Carry?

The *NEC* permits 100% continuous loading on an overcurrent device *only* if that overcurrent device is listed for 100% loading. See *NEC 210.20* and *215.3*. At the time of this writing, UL has no listing of residential-type molded-case circuit breakers that are suitable for 100% continuous loading. A *maximum* loading of 80% of the continuous is good practice—less is even better!

One 2-Pole or Two Single-Pole Circuit Breakers

Use 2-pole circuit breakers on 3-wire 120/240-volt branch circuits (two "hots" and one neutral) for loads such as electric ranges, electric ovens, and electric clothes dryers.

Use 2-pole circuit breakers for straight 240-volt branch circuits (two "hots") for such loads as electric water heaters, electric furnaces, electric baseboard heaters, air conditioners, and heat pumps.

The use of handle ties, which connect the handles of two single-pole circuit breakers together so both poles trip together, is permitted for a few very specific applications.

Handle ties *do* allow two single-pole circuit breakers to be switched simultaneously to the OFF or ON positions.

Handle ties generally *do not* simultaneously trip both handle-tied, single-pole circuit breakers on overloads, short circuits, or ground faults on one of the circuit breakers. This could result in a false sense of security for someone working on an appliance where two single-pole circuit breakers with a handle tie has tripped. This person may think that the branch circuit is totally in the OFF position, when in fact it is quite possible that one of the circuit breakers may still be ON.

Identified Handle Ties and the *Code*

First appearing in the 2005 *NEC* is that handle ties for circuit breakers must be *identified*. *Identified* in the *NEC* means Recognizable as suitable for the specific purpose, function, use, environment, application, and so forth, where described in a particular Code requirement.* In addition, they may have been tested and listed by a Nationally Recognized Testing Laboratory (NRTL) for use on a specific manufacturer's circuit breaker.

- *210.4(B):* Where multiwire branch circuits are used, a means must be provided to simultaneously disconnect all ungrounded conductors. Identified handle ties on a circuit breaker meet this requirement. A disconnect switch also meets this requirement. We discussed the pros and cons of multiwire branch circuits earlier in this text.
- *210.4(C):* A multiwire branch circuit is generally only permitted to serve line-to-neutral loads. There are two exceptions to this rule:
    1. where the branch-circuit supplies only one piece of equipment such as an electric range, in which case the branch circuit is actually supplying both line-to-neutral and line-to-line loads, or
    2. where the overcurrent device simultaneously disconnects all ungrounded conductors of the multiwire branch circuit.
- *225.33(B):* Identified handle ties are permitted to disconnect all conductors of an outside

*Reprinted with permission from NFPA 70-2014.

multiwire branch circuit or feeder with no more than six operations of the hand.

- *230.71(B):* Identified handle ties are permitted for circuit breakers to disconnect all conductors of a service as long as it takes no more than six operations of the hand to shut everything off.
- *240.15:* This section requires that a circuit breaker open all ungrounded conductors of a circuit.
- *240.15(B)(1):* Individual single-pole circuit breakers, with or without handle ties, are permitted to protect the ungrounded conductors of a multiwire branch circuit only when serving single-phase, line-to-neutral loads. See *210.4(B).*
- *240.15(B)(2):* Individual single-pole circuit breakers with identified handle ties are permitted to protect the ungrounded conductors for line-to-line loads for single-phase circuits.

With all of the above restrictions on the use of handle ties, they should only be used as the last resort. If you have to use handle ties, use only those that are *identified.*

INTERRUPTING RATINGS FOR FUSES AND CIRCUIT BREAKERS

Here are some very important *NEC* sections regarding interrupting ratings.

- *NEC 110.9:* **Interrupting Rating.** Equipment intended to interrupt current at fault levels shall have an interrupting rating not less than the nominal circuit voltage and the current that is available at the line terminals of the equipment. Equipment intended to interrupt current at other than fault levels shall have an interrupting rating at nominal circuit voltage not less than the current that must be interrupted.*
- *NEC 110.10:* **Circuit Impedance, Short-Circuit Current Ratings, and Other Characteristics.** The overcurrent protective devices, the total impedance, the component short-circuit current ratings, and other characteristics of the circuit to be protected shall be selected and coordinated to permit the circuit protective devices used to clear a fault to do so without extensive damage to the electrical components of the circuit. This fault shall be assumed to be either between two or more of the circuit conductors, or between any circuit conductor and the grounding conductor or enclosing metal raceway. Listed products applied in accordance with their listing shall be considered to meet the requirements of this section.*
- UL Standard No. 67 requires that panelboards be marked with their short-circuit current rating.

The overcurrent protective device must be able to interrupt the current that may flow under any condition (overload or short circuit). Such interruption must be made with complete safety to personnel and without damage to the panel or switch in which the overcurrent device is installed.

SAFETY ALERT

Overcurrent devices with inadequate interrupting ratings are, in effect, bombs waiting for a short circuit to trigger them into an explosion. Personal injury may result and serious damage will be done to the electrical equipment.

Series Rated versus Fully Rated Panelboards

Panelboards are available in two types: fully rated and series rated. Figures 28-11, 28-12, 28-13, and 28-14 explain the meaning of these terms.

When you hear and see the term *series rated*, the term means exactly what it says. Two circuit breakers are rated in series as provided in *NEC 240.86.* The main or upstream overcurrent device and the downstream or branch-circuit overcurrent devices are connected in series.

*Reprinted with permission from NFPA 70-2014.

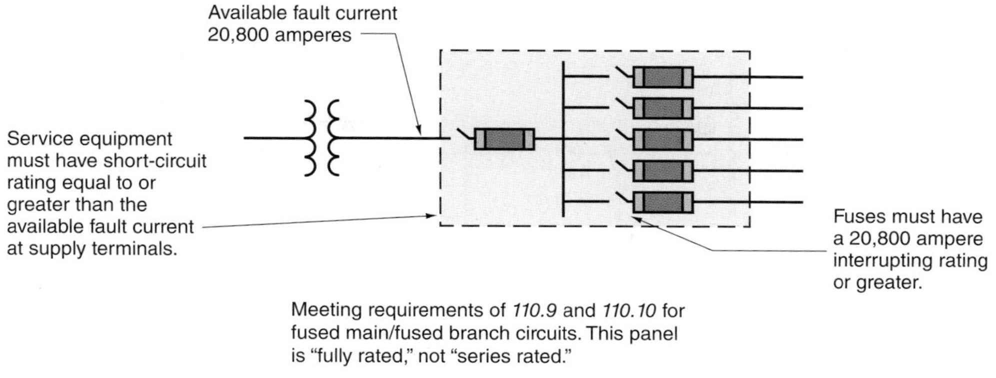

FIGURE 28-11 A *fully rated* system using main fuses and branch-circuit fuses. The entire assembly (fuses and panelboard) is tested, listed, and marked with a maximum short-circuit rating. Look for the marking on the panelboard. This type of combination can be found in commercial and industrial facilities.

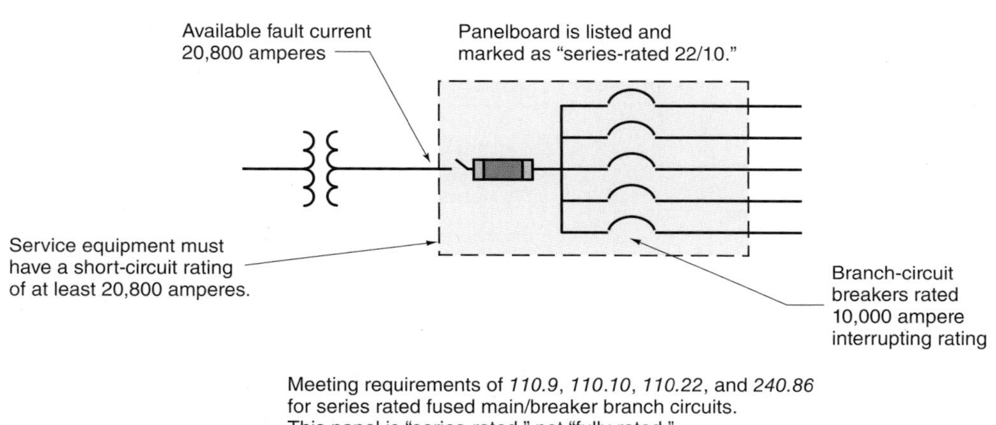

FIGURE 28-12 A *series-rated* system using main fuses and branch-circuit breakers. The entire assembly (main fuses, branch-circuit breakers, and panelboard) is tested, listed, and marked with a maximum short-circuit rating. Look for the marking on the panelboard. The lower interrupting rating of the branch-circuit breakers is acceptable because the combination has been tested and listed as a series-rated system. This type of combination can be found in light commercial installations such as multimetering service equipment.

The unique characteristic of a series-rated combination is that the upstream circuit breaker has a high interrupting rating and the downstream circuit breakers have a low interrupting rating. Series-rated systems are less costly than fully rated systems. It is safe to say that high interrupting rated circuit breakers cost more than low interrupting rated circuit breakers. What is compromised is selectivity, discussed a little later.

The upstream high interrupting rated overcurrent device is permitted to be in the same enclosure as the lower interrupting rated devices—or is permitted to be remote, such as at a main switchboard panelboard—and the lower interrupting rated circuit breakers are located in a subpanel.

Interrupting Ratings

Interrupting ratings of overcurrent devices, as shown in Table 28-2, are given in root-mean-square (RMS) values.

The interrupting rating of a fuse or circuit breaker does not indicate the short-circuit rating of a panelboard. These are stand-alone ratings. The short-circuit rating of a panelboard is derived by testing the panel as an assembly with breakers and/or fuses installed in the panel. The short-circuit rating of a panelboard is marked on its label.

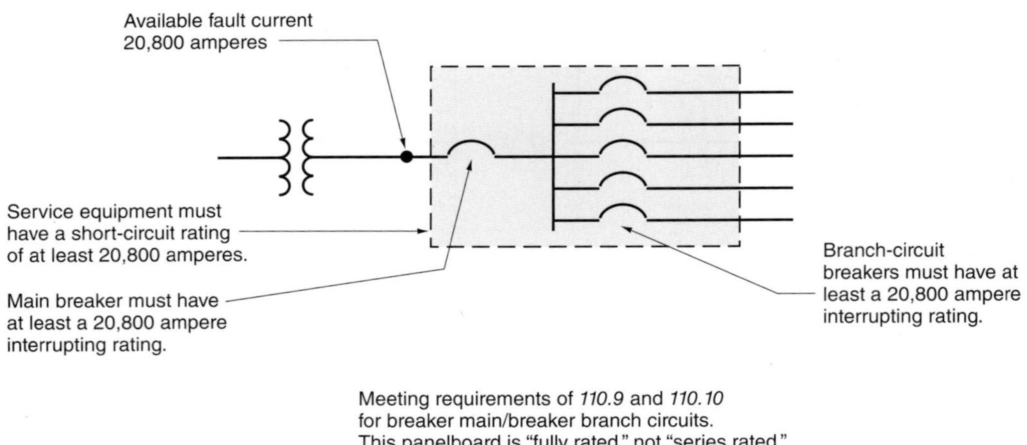

FIGURE 28-13 A *fully rated* system using a main circuit breaker circuit and branch-circuit breakers. The entire assembly (main breaker, branch-circuit breakers, and panelboard) is tested, listed, and marked with a maximum short-circuit rating. Look for the marking on the panelboard. This type of combination can be found in commercial and industrial installations.

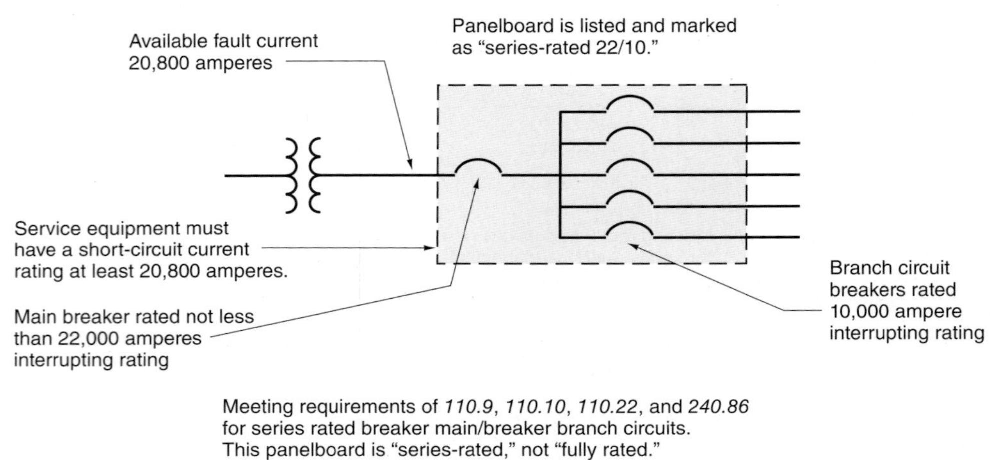

FIGURE 28-14 A *series-rated* system using a main circuit breaker and branch-circuit breakers. The entire assembly (main breaker, branch-circuit breakers, and panelboard) is tested, listed, and marked with a maximum short-circuit rating. Look for the marking on the panelboard. The lower interrupting rating of the branch-circuit breakers is acceptable because the assembly has been listed as a series–rated system. This type of combination is most commonly found in typical residential and light commercial applications.

SHORT-CIRCUIT CURRENTS

This text does not cover in detail the methods of calculating short-circuit currents. For a detailed discussion, see *Electrical Wiring—Commercial* (Copyright © Cengage Learning®). The ratings required to determine the maximum available short-circuit current delivered by a transformer are the kVA and impedance values of the transformer. The size and length of wire installed between the transformer and the overcurrent device must be considered as well.

The transformers used in modern electrical installations are efficient and have very low impedance values. A low-impedance transformer having a given kVA rating delivers more short-circuit current than a transformer with the same kVA rating and a higher impedance. When an electrical service is connected to a low-impedance transformer, the problem of available short-circuit current is very serious.

The examples given in Figures 28-11, 28-12, 28-13, and 28-14 show that the available short-circuit current at the transformer secondary terminals is 20,800 amperes.

HOW TO CALCULATE SHORT-CIRCUIT CURRENT

Short-circuit current is also referred to as fault current.

Calculating fault currents is just as important as calculating load currents. Overloads will cause conductors and equipment to run hot, will shorten their life, and will eventually destroy them. Fault currents that exceed the interrupting rating of the overcurrent protective devices and panel rating can cause violent electrical explosions of equipment, with the potential of serious injury to people standing near it. Fire hazard is also present.

The local electric utility and the electrical inspector are good sources of information when short-circuit current is to be determined.

A simplified method is given as follows to determine the approximate available short-circuit current at the terminals of a transformer. This method assumes that there is an infinite amount of primary short-circuit current available (infinite bus), and a bolted fault at the secondary transformer terminals.

1. Determine the normal full-load secondary current delivered by the transformer.

   For single-phase transformers:

   $$I = \frac{\text{kVA} \times 1000}{E}$$

   For 3-phase transformers:

   $$I = \frac{\text{kVA} \times 1000}{E \times 1.73}$$

   where

   I = current, in amperes
   kVA = kilovolt-amperes (transformer nameplate rating)
   E = secondary line-to-line voltage (transformer nameplate rating)

2. Using the impedance value given on the transformer nameplate, find the multiplier to determine the short-circuit current.

   $$\text{multiplier} = \frac{100}{\text{percent impedance}}$$

3. The short-circuit current = normal full-load secondary current × multiplier.

EXAMPLE

A transformer is rated at 100 kVA and 120/240 volts. It is a single-phase transformer with an impedance of 1% (from the transformer nameplate). Find the short-circuit current.

For a single-phase transformer:

$$I = \frac{\text{kVA} \times 1000}{E}$$

$$I = \frac{100 \times 1000}{240}$$

= 416 amperes, full-load current

The multiplier for a transformer impedance of 1% is

$$\text{multiplier} = \frac{100}{\text{percent impedance}}$$

$$= \frac{100}{1}$$

$$\text{multiplier} = 100$$

The short-circuit current = I × multiplier
= 416 amperes × 100
= 41,600 amperes

Thus, the available short-circuit current at the terminals of the transformer is 41,600 amperes.

This value decreases as the distance from the transformer increases as a result of the impedance added by the conductors.

If the transformer impedance is 1.5%, the multiplier is

$$\text{multiplier} = \frac{100}{1.5} = 66.6$$

The short-circuit current = 416 × 66.6
= 27,706 amperes

If the transformer impedance is 2%,

$$\text{multiplier} = \frac{100}{2} = 50$$

The short-circuit current = 416 × 50
= 20,800 amperes

(**Note:** A short-circuit current of 20,800 amperes is used in Figures 28-11, 28-12, 28-13, and 28-14.)

AUTHOR'S TIP: The line-to-neutral short-circuit current at the secondary of a single-phase transformer is approximately 1½ times greater than the line-to-line short-circuit current. ●

For the previous example,

Line-to-line short-circuit current = 20,800 amperes
Line-to-neutral short-circuit current
= 20,800 × 1.5 = 31,200 amperes

It is significant to note that UL Standard 1561 allows the marked impedance on the nameplate of a transformer to vary plus or minus (±) 10% from the actual impedance value of the transformer. Thus, for a transformer marked "2%Z," the actual impedance could be as low as 1.8%Z to as much as 2.2%Z. Therefore, the available short-circuit current at the secondary of a transformer as calculated previously should be increased by 10% if one wishes to be on the safe side when considering the interrupting rating of the fuses and breakers to be installed.

For an easy and fun computer program for making short-circuit calculations, visit the Bussmann website at www.bussmann.com. Here you will find access to their point-to-point short-circuit calculation method that can be performed on your computer. You will also find the trail to downloading the short-circuit app program to your smartphone so you can make short-circuit calculations wherever you might be. You will also find *Code* compliance programs relating to arc-flash calculations, arc-flash labeling, coordination analysis, and time-current curve analysis.

PANELBOARDS . . . WHAT ARE THEY?

The terms *panel*, *panelboard*, and *load center* are all used, and for all practical purposes mean the same thing. The *NEC* and the UL Standards use the term *panelboard*. Neither the *NEC* nor the UL Standards contain the term *load center*. A load center is really a panelboard with less gutter (wiring) space and less depth and width, and generally does not have features such as integral relays, remote-controlled circuit breakers, load-shedding devices, and other similar optional features. However, some load centers are available for home automation applications that do have certain remote control features. Circuit breakers used in load centers are usually of the "plug-on" type as opposed to the "bolt-on" type found in commercial and industrial panelboards. Load centers generally are for use in a shallow wall (2 × 4 studs) such as found in residential and light commercial construction. They are available with a main circuit breaker or main fuses, or with main lugs only (MLO).

Panelboards are tested and listed under UL Standard 67.

The most popular residential panelboard installed nowadays are *series rated*, as shown in Figure 28-14. The most common rating is the 22/10 panelboard. The main circuit breaker has an interrupting rating of 22,000 amperes. The branch-circuit breakers have an interrupting rating of 10,000 amperes. The series rating of 22/10 is adequate for most typical residential main panelboards. If you are installing a large electrical service where a large kVA rated transformer is involved, don't take a chance! You should have real concern for services over 200 amperes. Obtain the available fault current on the line side of the service equipment from the local electric utility; then select panels, fuses, and circuit breakers suitable for the available fault current.

Table 28-3 compares the requirements for different types of panelboards.

Selective Coordination/ Nonselective Coordination

These are terms used by the real pros!

Selective Coordination: Under overload, short-circuit, or ground-fault conditions, only the overcurrent device nearest the fault opens. The main fuses or circuit breaker remain closed. Refer to *240.2*. See the definition in *NEC Article 100*.

Nonselective Coordination: Under overload, short-circuit, or ground-fault conditions, the branch breaker or fuses might open, the main breaker or fuses might open, or both might open. For example, a short circuit in a branch circuit could also cause the main breaker to trip, resulting in a total power outage. Series-rated systems are susceptible to nonselectivity when available fault currents are high.

TABLE 28-3

Comparison of panelboard types.

	Fused Main/ Fused Branches (Fully Rated) Figure 28-11	Fused Main/ Fused Branches (Series Rated) Figure 28-12	Breaker Main/ Breaker Branches (Fully Rated) Figure 28-13	Breaker Main/ Breaker Branches (Series Rated) Figure 28-14
Must be listed and marked as suitable for use as service equipment when used as service equipment. Does not need this marking when used as a subpanel.	Yes	Yes	Yes	Yes
Must have short-circuit current rating adequate for the available fault current at the line side of the panelboard.	Yes	Yes	Yes	Yes
Must be marked with its short-circuit current rating.*	Yes	Yes	Yes	Yes
Must be marked "Series Rated."	No	Yes Marking: Caution—Series Combination System Rated ______ Amperes. Identified Replacement Components Required.	No	Yes Marking: Caution—Series Combination System Rated ______ Amperes. Identified Replacement Components Required.
Main fuses or circuit breaker must have interrupting rating adequate for the available fault current on the line side of the panelboard.	Yes	Yes	Yes	Yes
Branch fuses or breakers must have interrupting rating adequate for the available fault current on the line side of the panelboard.	Yes	No Main fuses might have 200,000-ampere interrupting rating. Branch breakers might have 10,000-ampere interrupting rating. Other combinations available. Check manufacturers' data.	Yes	No Main breaker might have 22,000-ampere interrupting rating. Branch breakers might have 10,000-ampere interrupting rating. Other combinations available. Check manufacturers' data.
Main fuses are current limiting. See *240.2.*	Yes	Yes	Not applicable	Not applicable
Must be sized as determined by service-entrance calculations and/or local codes.	Yes	Yes	Yes	Yes
Check manufacturer's data for time-current characteristic curves for fuses and breakers, and for "unlatching times" of breakers.	Always a good idea.	Always a good idea.	Always a good idea.	Always a good idea.

*If a panelboard has no marked short-circuit rating, then its short-circuit rating is based on the lowest interrupting-rated device installed.

The theory behind selective and nonselective systems is covered in detail in *Electrical Wiring—Commercial* (Copyright © Cengage Learning®). Time-current curves for fuses and circuit breakers, circuit-breaker "unlatching time" data, and fuse melting/total clearing data are discussed.

REVIEW

1. What is the minimum size service for a one-family dwelling? ____________________
2. a. What is a Type S fuse? ____________________
   b. Where must Type S fuses be installed? ____________________
3. a. What is the maximum voltage permitted between conductors when using plug fuses? ____________________
   b. May plug fuses (Type S) be installed in a switch that disconnects a 120/240-volt clothes dryer? ____________________
   c. Give a reason for the answer to (b). ____________________
4. Will a 20-ampere, Type S fuse fit properly into a 15-ampere adapter? ____________________
5. A time-delay, dual-element fuse will hold 500% of its rating for approximately (5 seconds) (10 seconds) (20 seconds). (Circle the correct answer.)
6. What part of a circuit breaker causes the breaker to trip
   a. on an overload? ____________ b. on a short circuit? ____________
7. What is meant by an ambient-compensated circuit breaker? ____________________
8. List the standard sizes of fuses and circuit breakers up to and including 100 amperes. ____________________
9. When hooking up a 240-volt electric baseboard heater, you should use (Circle the correct answer.)
   a. a 2-pole circuit breaker.
   b. two single-pole breakers and a handle tie.
10. Using the method shown in this unit, what is the approximate short-circuit current available at the terminals of a 50-kVA single-phase transformer rated 120/240 volts? The transformer impedance is 1%.
    a. Line-to-line
    b. Line-to-neutral

11. State four possible combinations of service equipment that meet the requirements of *110.9* and *110.10* of the *Code*.

    a. ______________________ c. ______________________

    b. ______________________ d. ______________________

12. Which *Code* section states that all overcurrent devices must have adequate interrupting ratings for the available fault current to be interrupted? ______________________

    ____

13. All electrical components have some sort of "withstand rating." The *NEC* and UL Standards refer to this as the component's short-circuit current rating. This rating indicates the ability of the component to withstand a specified amount of fault current for the time it takes the overcurrent device upstream of the component to open the circuit. *NEC* ______________________ refers to short-circuit current rating.

14. Arc-fault damage is closely related to the value of ______________________.

15. The utility company has provided a letter to the contractor stating that the available fault current at the line side of the main service-entrance equipment in a residence is 17,000 amperes RMS symmetrical, line to line. In the space following each statement, write in "Meets *Code*" or "Violation" of *110.9* and *110.10* of the *Code*.

    a. Main breaker has a 10,000-ampere interrupting rating; branch breakers have a 10,000-ampere interrupting rating. ______________________

    b. Main current-limiting fuse has a 200,000-ampere interrupting rating; branch breakers have a 10,000-ampere interrupting rating. The panel is marked "Series-Rated."

       ____

    c. Main breaker has a 22,000-ampere interrupting rating; branch breakers have a 10,000-ampere interrupting rating. The panel is marked "Series-Connected."

       ____

16. While working on the main panel with the panel energized, the electrician inadvertently causes a direct short circuit (line to ground) on one of the branch-circuit breakers. The available fault current at the main service equipment is rather high. The panel is labeled "Series-Connected." The main breaker is rated 100 amperes. Both branch-circuit breaker and main breaker trip OFF. This condition is referred to as being (selective) (nonselective). (Circle the correct answer.)

17. Repeat problem 13 for a main service panel that consists of 100-ampere main current-limiting fuses and breakers for the branch circuits. The branch circuit trips; the 100-ampere fuses do not open. This condition is referred to as being (selective) (nonselective). (Circle the correct answer.)

18. *NEC* ______________________ prohibits the connecting of fuses and/or circuit breakers into the grounded circuit conductor.

19. In general, overcurrent protective devices are inserted where the conductor receives its ______________________.

20. a. Does the *NEC* permit panelboards to be installed in clothes closets? ____________

    b. Does the *NEC* permit panelboards to be installed in bathrooms? ____________

21. Overcurrent devices must be accessible. (True) (False). (Circle the correct response.)

CHAPTER 29

Service-Entrance Calculations

OBJECTIVES

After studying this chapter, you should be able to

- calculate the size of the service entrance, including the size of the neutral conductors, using conventional and optional methods.
- understand why the neutral conductor is permitted to be smaller than the ungrounded conductors.
- understand special *Code* rules that permit smaller size service-entrance conductors for single-family dwellings.
- realize that derating service-entrance conductors may be necessary if the installation is located in a hot climate.
- understand how to read a watt-hour meter.

INTRODUCTION

The various load values determined in earlier chapters of this text are now used to illustrate the proper method of determining the size of the service-entrance conductors for the residence. The calculations are based on *NEC*® requirements. The student must check local and state electrical codes for any variations in requirements that may take precedence over the *NEC*.

SIZING OF SERVICE-ENTRANCE CONDUCTORS AND SERVICE DISCONNECTING MEANS

NEC 230.42 tells us that service-entrance conductors must be sufficient to carry the load as calculated according to *Article 220* and shall have an ampacity not less than the rating of the service disconnect.

For a one-family residence, the minimum-size main service disconnecting means is 100 amperes, 3 wire, *230.79(C)*.

There are two methods for calculating service size for homes:

1. The conventional method is found in *Article 220, Part III*.
2. The optional method is found in *Article 220, Part IV*.

Method 1 (*Article 220, Part III*)

All of the volt-ampere values in the following calculations are taken from previous chapters of this text.

Single-Family Dwelling Service-Entrance Calculations

1. General Lighting Load, *220.12*

   3232 ft^2 @ 3 VA per ft^2 = 9696 VA

Note: Included in this floor area calculation are all lighting outlets and general-use receptacles. Do not include open porches, garages, or unused or unfinished spaces not adaptable for future use. See *NEC 220.12*, *Table 220.12*, and *220.14(J)*.

2. Minimum Number of 15-Ampere Lighting Branch Circuits

   9696 VA ÷ 120 volts = 80.8 amperes

   $$\text{then, } \frac{\text{amperes}}{15} = \frac{80.8}{15} = 5.387$$

   (round up to 6) 15-ampere branch circuits

3. Small-Appliance Load, *210.11(C)(1)*, *210.52(B)*, and *220.52(A)*

   (minimum of two 20-ampere branch circuits)

   2 branch circuits @ 1500 VA each = 3000 VA

4. Laundry Branch Circuit, *210.11(C)(2)*, *210.52(F)*, and *220.52(B)*

   (Minimum of one 20-ampere branch circuit)

   1 branch circuit(s) @ 1500 VA = 1500 VA

   Total = 1500 VA

5. Total General Lighting, Small-Appliance, and Laundry Load

   Lines 1 + 3 + 4 = 14,196 VA

6. Net Calculated General Lighting, Small Appliance, and Laundry Loads (less ranges, ovens, and "fastened-in-place" appliances). Apply demand factors from *Table 220.42*.

   a. First 3000 VA @ 100% = 3000 VA

   b. Line 5 14,196 − 3000 = 11,196 @ 35% = 3919 VA

   Total a + b = 6919 VA

7. Electric Range, Wall-Mounted Ovens, Counter-Mounted Cooking Units *(Table 220.55)*

Wall-mounted oven	7450 VA
Counter-mounted cooking unit	6600 VA
Total	14,050 VA (14 kW)

   14 kW exceeds 12 kW by 2 kW

   2 kW × 5% = 10% increase; therefore:

   8 kW + 0.8 kW = 8.8 kW = 8,800 VA

8. Electric Dryer, *220.54* = 5700 VA

9. Electric Furnace, *220.51*, Air Conditioner, Heat Pump, *Article 440*

   Air conditioner: 26.4 × 240 = 6336 VA

   Electric furnace: 13,000 VA

   (Enter largest value, *220.60*) = 13,000 VA

10. Net Calculated General Lighting, Small Appliance, Laundry, Ranges, Ovens, Cooktop Units, HVAC

    Lines 6 + 7 + 8 + 9 = 34,419 VA

11. List "Fastened-in-Place" Appliances *in addition* to Electric Ranges, Air Conditioners, Clothes Dryers, Space Heaters

Appliance		**VA Load**
Water Heater	=	4500 VA
Dishwasher 9.2 × 120	=	1104 VA
Garage Door Opener 5.8 × 120	=	696 VA
Food Waste Disposer 7.2 × 120	=	864 VA
Water Pump 8 × 240	=	1920 VA
Hydromassage Tub 10 × 120	=	1200 VA
Attic Exhaust Fan 5.8 × 120	=	696 VA
Heat/Vent/Lights 1500 × 2	=	3000 VA
Freezer 5.8 × 120	=	696 VA
	Total	14,676 VA

12. Apply 75% demand factor, *220.53*, if four or more "Fastened-in-Place" appliances. If less than four, figure at 100%. Do not include electric ranges, clothes dryers, electric space heaters, or air-conditioning equipment.

    Line 11 14,676 × 0.75 = 11,007 VA

13. Total Calculated Load (Lighting, Small-Appliance, Ranges, Dryers, HVAC, "Fastened-in-Place" Appliances)

    Line 10 35,469
    + Line 12 11,007 = 45,426 VA

14. Add 25% of Largest Motor (*220.50* and *430.24*)

    This is the water pump motor:
    8 × 240 × 0.25 = 480 VA

Note: The largest motor can be difficult to determine because nothing is in place when service-entrance load calculations are made. It might be an air-conditioning unit or a heat pump. If the dwelling is cooled by an evaporative cooler, the largest motor might be a water pump motor, a large attic exhaust fan, a large food waste disposer, or a sump pump. For simplicity in this example, the water pump motor was chosen for this example calculation.

Had we used the air-conditioning unit as the largest motor, we would use the Branch-Circuit Selection Current value of 19.9 amperes. See Chapter 23 for complete explanation of how air-conditioning current ratings are derived. This would result in 19.9 × 240 = 4,776 VA. Next: 4,776 × 0.25 = 1,194 VA. Next: 46,401 + 1,194 = 46,620 VA. Next: 46,620 ÷ 240 = 194 amperes. The service-entrance size would be 200 amperes.

15. Total of Line 13 + Line 14 = 45,906 VA

16. Minimum Ampacity for Ungrounded Service-Entrance Conductors

$$\text{Amperes} = \frac{\text{Line 15}}{240} = \frac{45{,}906}{240}$$

$$= 191 \text{ amperes}$$

17. Ungrounded Conductor Size (copper)
    2/0 AWG THWN

Note: This could be a 3/0 AWG THW, THHW, THWN, XHHW, or THHN per *Table 310.15(B)(16)*, or a AWG 2/0 (same types) using *310.15(B)(7)*. *Section 310.15(B)(7)* may only be used for 120/240-volt, 3-wire, residential, single-phase service-entrance conductors, underground service conductors, and feeder conductors that serve as the main power feeder to a dwelling unit. This rule *cannot* be applied to the feeder to Panel B because that feeder carries only part of the load in the residence. We have assumed that the terminals in the equipment are marked 75°C. We read the ampacity from the 75°C column in *Table 310.15(B)(16)* as stated in *110.14(C)*.

18. Minimum Ampacity for Neutral Service-Entrance Conductor, *220.61*, and *310.15(B)(7)*. Do not include loads connected only phase to phase such as the water heater (240-volt loads).

    a. Line 6 = 6919 VA

    b. Line 7, Oven and Cooktop 8800 × 0.70 = 6160 VA

    c. Line 8 Dryer 5700 × 0.70 = 3990 VA

    d. Line 11 (include only 120-volt loads)

Freezer	=	696 VA
Food Waste Disposer	=	864 VA
Garage Door Opener	=	696 VA
Heat/Vent/Light	=	3000 VA
Attic Fan	=	696 VA
Hydromassage Tub	=	1200 VA
Dishwasher	=	1104 VA
	Total	8256 VA

e. Line d Total @ 75% Demand Factor per *220.53*: 8256 × 0.75 = 6192 VA

f. Add 25% of Largest 120-volt Motor. This is the hydromassage pump motor: 10 × 120 × 0.25 = 300 VA

g. Total a + b + c + e + f = 23,561 VA

$$\text{Amperes} = \frac{\text{volt-amperes}}{\text{volts}} = \frac{23{,}561}{240}$$

$$= 98 \text{ amperes}$$

19. Neutral Conductor Size (copper), *220.61*, 3 AWG (75°C rated insulation)

    Note: The calculations indicate that a 3 AWG copper neutral conductor would be adequate based on the unbalanced load calculations. *NEC 310.15(B)(7)* states that the neutral conductor is permitted to be smaller than the ungrounded "hot" conductors if the requirements of *215.2*, *220.61*, and *230.42* are met. *NEC 220.61* states that a feeder or service neutral load shall be the maximum unbalance of the load determined by *Article 220*. The specifications for this residence specify that a 1 AWG neutral conductor be installed. The neutral conductor shall not be smaller than the grounding electrode conductor, *250.24(C)(1)*. Where bare conductors are used with insulated conductors, their ampacity is based on the ampacity of the other insulated conductors in the raceway, *310.15(B)(4)*.

20. Grounding Electrode Conductor Size (copper), *Table 250.66*, 4 AWG

21. Conduit (see Table 27-1 for calculations) Trade Size 1¼

    Note: A one-page short version of the above service-entrance calculation form has been added to the Appendix of this text to encourage the student to perform additional residential load calculations. Practice makes perfect!

Section 310.15(B)(7) provides special consideration for 120/240-volt, single-phase service and feeder conductors for a dwelling unit.

NEC Section 310.15(B)(7) has been revised for the 2014 edition, and the previous *Table 310.15(B)(7)* has been deleted. The rule continues to apply to one-family dwellings and the individual dwelling units of two-family and multifamily dwellings that are supplied by 120/240-volt, single-phase services or feeders. The service or feeder conductors to which this rule applies are required to supply the entire load associated with the individual dwelling unit. See Figure 29-1 The service or feeder conductors that comply with this rule are permitted to have an allowable ampacity not less than 83% of the service or feeder rating.

This section now reads, ▶120/240 Volt, Single-Phase Dwelling Services and Feeders. For one-family dwellings and the individual dwelling units of two-family and multifamily dwellings, service and feeder conductors supplied by a single phase, 120/240-volt system shall be permitted be sized in accordance with 310.15(B)(7)(a) through (d).

(a) For a service rated 100 through 400 amperes, the service conductors supplying the entire load associated with a one-family dwelling or the service conductors supplying the entire load associated with an individual dwelling unit in a two-family or multifamily dwelling shall be permitted to have an ampacity not less than 83% of the service rating.

(b) For a feeder rated 100 through 400 amperes, the feeder conductors supplying the entire load associated with a one-family dwelling or the feeder conductors supplying the entire load associated with an individual dwelling unit in a two-family or multifamily dwelling shall be permitted to have an ampacity not less than 83% of the feeder rating.

(c) In no case shall a feeder for an individual dwelling unit be required to have an ampacity greater than that of its 310.15(B)(7)(a) or (b) conductors.

(d) Grounded conductors shall be permitted to be sized smaller than the ungrounded conductors provided the requirements of 220.61 and 230.42 for service conductors or the requirements of 215.2 and 220.61 for feeder conductors are met. ◀*

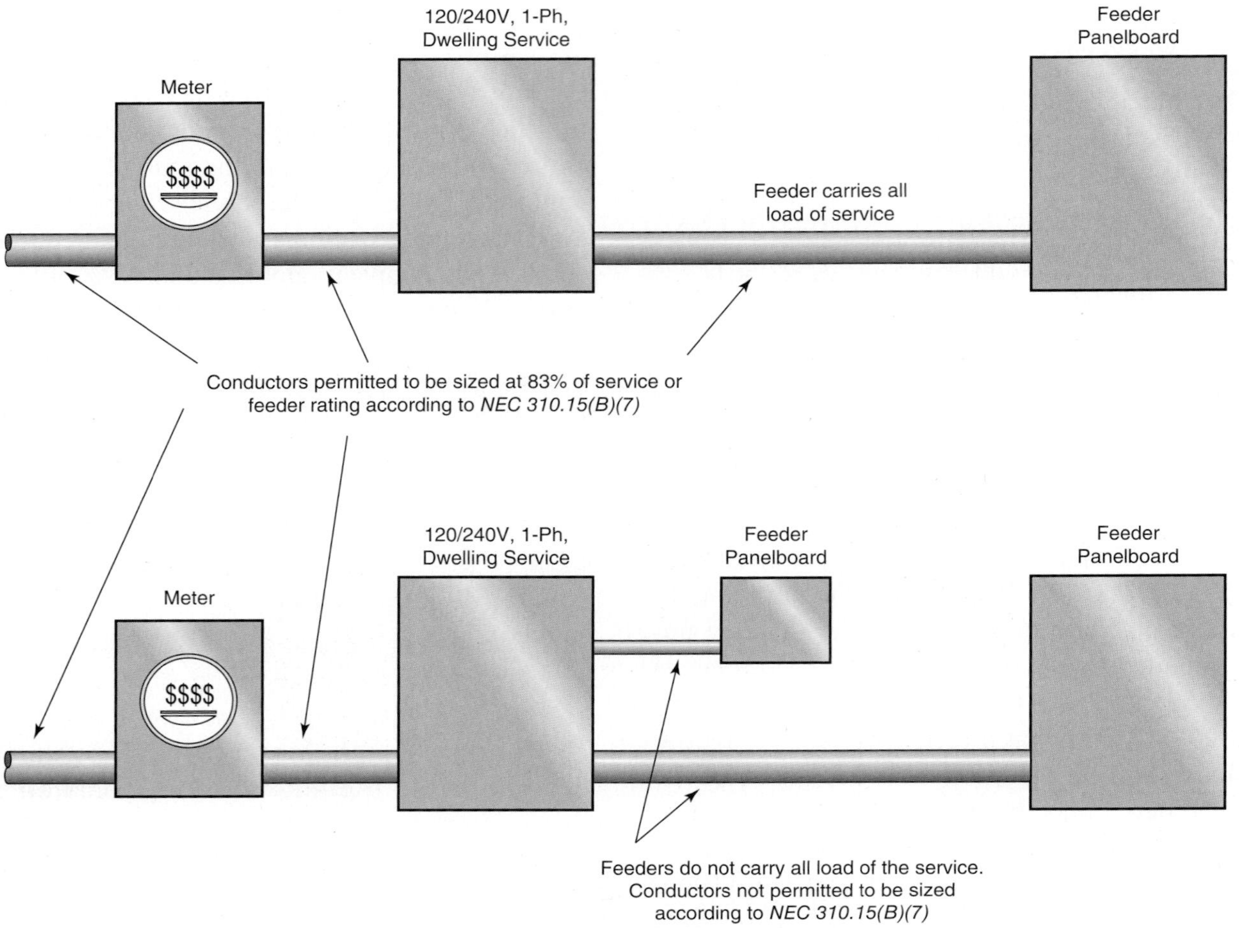

FIGURE 29-1 The application of *NEC 310.15(B)(7)* for services and feeders.

Note that the Code-Making Panel chose to use the words "service conductors" rather than designating the type of service conductors such as service-entrance conductors or service-lateral conductors. Also note that because this article does not apply to installations under the jurisdiction of an electric utility, it cannot be used for sizing conductors provided or installed by an electric utility.

Table 29-1 shows the application of this 83% rule to the sizing of service or feeder conductors for both copper and aluminum. The conductor ampacity in the 75°C columns of *NEC Table 310.15(B)(16)* has been used to calculate the minimum conductor size application of this rule.

Informational Note 1 indicates it is possible that the conductor ampacity will require other correction or adjustment factors applicable to the conductor installation. This indicates that if the service or feeder is installed where the ambient temperature exceeds 86°F (30°C), a correction factor from *Table 310.15(B)(2)(a)* must be applied. If more than three current-carrying conductors are installed in a raceway or cable longer than 24 inches (610 mm), an adjustment factor must be applied from *Table 310.15(B)(3)(a)*. See Chapter 18 of this text for examples of applying correction and adjustment factors for high temperatures or excessive numbers of current-carrying conductors.

This 83% rule is based on the load diversity found in homes. Typically, the load actually experienced is far lower than the load calculation indicates. These conductor sizes are not permitted to be used for feeder conductors that do not carry the entire load of the service. An example of this would be the feeder from the Main Panel A to Panel B in the residence discussed in this text, shown in Figure 29-1.

Section 310.15(B)(7) cannot be used to determine the feeder size to Panel B in the dwelling in this text because this feeder does not carry the entire electrical load supplied by the service to the home.

TABLE 29-1

Minimum conductor sizes with 75°C insulation for 120/240-volt, single-phase dwelling services and feeders.

SERVICE OR FEEDER RATING (AMPERES)	83% OF SERVICE OR FEEDER RATING	MINIMUM COPPER WIRE WITH 75°C INSULATION	MINIMUM ALUMINUM WIRE WITH 75°C INSULATION
100	83	4 AWG	2 AWG
110	91	3 AWG	1 AWG
125	104	2 AWG	1/0 AWG
150	125	1 AWG	2/0 AWG
175	145	1/0 AWG	3/0 AWG
200	166	2/0 AWG	4/0 AWG
225	187	3/0 AWG	250 kcmil
250	208	4/0 AWG	300 kcmil
300	249	250 kcmil	350 kcmil
350	291	350 kcmil	500 kcmil
400	332	400 kcmil	600 kcmil

Note: Installations exposed to temperatures in excess of 86°F (30°C) will require that correction factors from *Table 310.15(B)(2)(a)* be applied.
Installations of more than three current-carrying conductors in a raceway or cable longer than 24 in. (600 mm) will require that adjustment factors from *Table 310.15(B)(3)(a)* be applied.

Be careful! Underground wiring, raceways in direct contact with the earth, installations subject to saturation of water or other liquids, and unprotected locations exposed to the weather are wet locations per the definition in the *NEC*. Insulated conductors and cables in these locations are required to be listed for wet locations, *300.5(B)*.

A similar wet location requirement found in *NEC 300.9* requires, Where raceways are installed in wet locations above grade, the interior of these raceways shall be considered to be a wet location. Insulated conductors and cables installed in raceways in wet locations above grade shall comply with 310.8(C).* Note the importance of the words "interior of these raceways."

Suffice it to say that insulated conductors and cables in wet locations must have a "W" designation in their type marking. See *310.10(C)*.

*Reprinted with permission from NFPA 70-2014.

High Temperatures

In certain parts of the country, such as the southwestern desert climates where extremely hot temperatures are common, the AHJ will probably enforce *310.10(D)* and *310.15(A)(3)* regarding temperature limitations of insulated conductors. This means that conductors in raceways or cables exposed to direct sunlight (on a roof or side of a building) be corrected (derated) according to the "Correction Factors" found in *Table 310.15(B)(2)(a)*. See Figure 29-2.

EXAMPLE

If the U.S. Weather Bureau lists the average summer temperature as 113°F (45°C), then a correction factor of 0.82 must be applied. For instance, a 3/0 XHHW copper conductor per *Table 310.15(B)(16)* is 200 amperes at 86°F (30°C). At 113°F (45°C), the conductor's ampacity is

$$200 \times 0.82 = 164 \text{ amperes}$$

A properly sized conductor capable of carrying 200 amperes safely requires an ampacity before application of the temperature correction factor of

$$\frac{200}{0.82} = 243.9 \text{ amperes}$$

Therefore, according to *Table 310.15(B)(16)* and the applied correction factor, a 250 kcmil conductor with Type XHHW insulation is required.

Should you wish to learn more about the derating of conductors installed in raceways in direct exposure to sunlight on or above a roof, *NEC 310.15(B)(3)(c)* is covered in detail in the *Electrical Wiring—Commercial* text, a companion to *Electrical Wiring—Residential*.

Neutral Conductor

Why is the neutral conductor permitted to be smaller than the "hot" ungrounded conductors?

The neutral conductor must be able to carry the calculated maximum neutral current, *NEC 220.61*. Loads connected only phase to phase or at 240 volts are not included in the calculation of neutral conductors, as no load from these circuits is placed on the neutral. The calculations resulted in a calculated load of approximately 191 amperes for the ungrounded conductors and approximately 98 amperes for the

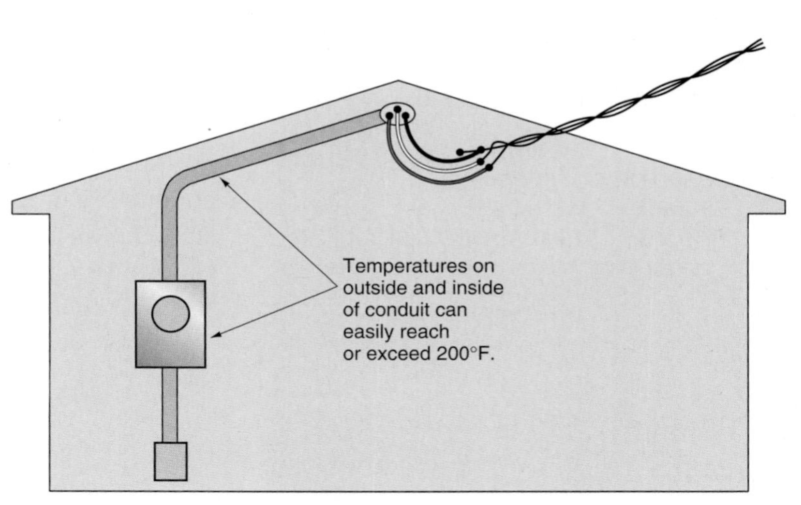

FIGURE 29-2 Example of a high-temperature location.

neutral conductor. Referring to *Table 310.15(B)(16)*, this could be two 3/0 AWG THW, THHW, THWN, XHHW, or THHN copper conductors and one 2 AWG neutral conductor. The neutral conductor may be insulated or bare per *230.30* and *230.41*.

NEC 310.15(B)(7) also permits the neutral conductor to be smaller than the "hot" ungrounded conductors, but only if it can be verified by calculations that the requirements of *215.2*, *220.61*, and *230.42* are met. We verified this in our calculations for the residence discussed in this text.

In general, service-entrance conductors must be insulated, *230.30*, *230.41*, and *310.2*. However, the exceptions to these rules permit the use of bare neutral conductors.

Where bare conductors are used with insulated conductors in raceways or cables, their allowable ampacity is limited to the allowable ampacity for the adjacent insulated conductors, *310.15(B)(4)*.

For example, for a service consisting of two 3/0 AWG Type THWN conductors and one 1/0 AWG bare neutral conductor, the allowable ampacity of the bare neutral conductor is found in *Table 310.15(B)(16)* in the THWN column.

3/0 AWG THWN phase conductors	200 amperes
1/0 AWG bare neutral conductor	150 amperes

The specifications for this residence call for two 2/0 AWG ungrounded "hot" conductors and one 1 AWG neutral conductor.

The service neutral conductor serves another purpose. In addition to carrying the unbalanced current, the neutral conductor must also be capable of safely carrying any fault current that it might be called on to carry, such as a line-to-ground fault in the service equipment back to the utility transformer. Sizing the neutral conductor only for the normal neutral unbalance current could result in a neutral conductor too small to safely carry fault currents. The neutral conductor might burn off, causing serious voltage problems, damage to equipment, and the creation of fire and shock hazards.

In sizing neutral conductors for dwellings, this generally is not a problem. But installing reduced size neutrals on commercial and industrial services can present a major problem. The grounded conductor of *any* service must never be smaller than the required grounding electrode conductor for that particular service. Refer to *230.23*, *230.31*, *230.42(C)*, *250.2*, *250.4*, and *250.24(B)*.

Subpanel B Calculations. We will next calculate the load on the feeder to panelboard B. This will determine the minimum ampacity of the feeder conductors and the rating of the overcurrent device protecting the feeder. We will include in the calculation only the loads that will be supplied from this panelboard and will apply appropriate demand factors. Note that some demand factors are not permitted, because a reduced number of appliances are supplied. The concept for application of demand factors is that all connected loads will not be operating at the same time, and reduced loading of the service or feeder occurs.

1. General Lighting Load (*220.12, 220.43*)

   1420 sq. ft @ 3 volt-amperes per ft^2 = 4260 VA

   (Kitchen, Living Room, Laundry, Rear Entry Hall, Powder Room, Recreation Room)

2. Small-Appliance Loads [*220.52(A)*]

   Kitchen 2
   Laundry 1 [*220.52(B)*]
   Total 3 @ 1500 VA per circuit = 4500 VA

3. Total General Lighting and Small-Appliance Load = 8,760 VA

4. Application of Demand Factors (*Table 220.42*)
   3000 volt-amperes @100% = 3000 VA
   8,760 − 3000 = 5760 @35% = 2016 VA

5. Net Calculated Load (Less Range and "Fastened-in-Place" Appliances) 5016 VA

6. Wall-Mounted Oven and Counter-Mounted Range:
   Table 220.55, Note 4:

Wall-mounted oven	7450 VA
Counter-mounted range	6600 VA
Total	14,050 VA

   14 kW exceeds 12 kW by 2 kW:
   2 kW × 5% = 10% increase
   8 kW × 0.10 = 0.8 kW
   (*Column C, Table 220.55*)
   Cooking appliances
   8 + 0.8 = 8.8 kW = 8800 VA

7. Dryer (*220.54*) 5700 VA

8. Net Calculated Load = Lines 5 + 6 + 7 19,516 VA

9. List "Fastened-in-Place" Appliances *in addition* to Electric Ranges, Air Conditioners, Clothes Dryers, Space Heaters

Appliance		**VA Load**
Dishwasher Motor: 9.2 × 120	=	1104 VA
Food waste disposer 7.2 × 120	=	864 VA
Garage door opener 5.8 × 120	=	696 VA
Total		2664 VA

   Note that a 75% demand factor cannot be applied as fewer than four of these appliances are supplied by the feeder.

10. Calculated load (lines 8 and 9) = 22,180 VA

11. Add 25% of Largest Motor
    7.2 × 120 × 0.25 = 216 VA

12. Net Calculated Load (Lighting, Small Appliance Circuits, Ranges, Dryer, Dishwasher, Food Waste Disposer, Garage Door Opener) 22,396 VA

    $$\text{Amperes} = \frac{\text{volt-amperes}}{\text{volts}} = \frac{22{,}396}{240} = 93 \text{ amperes}$$

13. The feeder conductors supplying Panel B could be a 3 AWG copper conductor with 75°C insulation per *NEC Table 310.15(B)(16)*. We have not shown the calculations for the neutral conductor to Panel B. The specifications and also Figure 27-26 require that three 3 AWG THHN feeder conductors supply Panel B.

Method 2 (Optional Calculations) (*Article 220, Part IV*)

A second method for determining the load for a one-family dwelling is given in *Article 220,*

Part IV. This method simplifies the calculations. But remember, the minimum-size service for a one-family dwelling is 100 amperes, *230.42(B)* and *230.79(C)*.

Let's take a look at *Part IV* of *Article 220*. This is an alternative method of calculating service loads and feeder loads. It is referred to as the *optional method*.

NEC 220.80 permits us to calculate service-entrance conductors and feeder conductors (both phase conductors and the neutral conductor) using an optional method. *NEC 220.82* addresses single-family dwellings. Note that several of the loads require nameplate information. For example, you are not permitted to use the general loads for electric ranges or dryers from *Part III* of *Article 220*. If the nameplate information is not available at the time the load calculation is made, a standard, rather than optional, load calculation must be performed. *NEC 220.84* addresses multifamily dwellings.

NEC 220.82(B) tells us to

1. include 1500 volt-amperes for each 20-ampere small-appliance and laundry branch circuit.
2. include 3 volt-amperes per ft^2 for lighting and general-use receptacles.
3. include the nameplate rating of appliances
   - that are fastened in place, or
   - that are permanently connected, or
   - that are located to be connected to a specific circuit, such as ranges, wall-mounted ovens, counter-mounted cooking units, clothes dryers, and water heaters.
4. include nameplate ampere rating or kVA for motors and all low-power-factor loads. The intent of the reference to low-power factor is to address such loads as low-cost, low-power-factor fluorescent ballasts. It is always recommended that high-power-factor ballasts be installed.

If there is heating and air conditioning, *NEC 220.82(C)* tells us to make the load calculation by selecting the largest load in kVA from a list of six types of loads:

1. 100% of nameplate rating of air conditioning and cooling.
2. 100% of nameplate rating of heat pump that does not have supplemental electric heating.
3. 100% of nameplate rating of heat pump compressor and 65% of supplemental electric heat. The supplemental electric heating load does not have to be added in if it cannot come on at the same time as the heat pump compressor.
4. 65% of electric space heating if less than four separately controlled units.
5. 40% of electric space heating if four or more separately controlled units.
6. 100% of electric thermal storage heating if expected to be continuous at full nameplate value.

So let's begin our optional calculation for the residence discussed in this text. The residence has an air conditioner and an electric furnace.

Air conditioner

$26.4 \times 240 = 6336$ volt-amperes

Electric furnace

$13{,}000 \times 0.65 = 8450$ volt-amperes

Therefore, we will select the electric furnace load for our calculations because it is the largest load. It is also a noncoincidental load, as defined in *220.60*. We can omit the air-conditioner load from our calculations from here on.

We now add up all of the other loads:

General lighting load 3232 ft^2 @ 3 volt-amperes per ft^2 =	9696 volt-amperes
Small-appliance and laundry circuits (3) @ 1500 volt-amperes each	4500 volt-amperes
Wall-mounted oven (nameplate rating)	6600 volt-amperes
Counter-mounted cooking unit (nameplate rating)	7450 volt-amperes
Water heater (nameplate rating)	4500 volt-amperes
Clothes dryer (nameplate rating)	5700 volt-amperes
Dishwasher 9.2×120	1104 volt-amperes
Food waste disposer 7.2×120	864 volt-amperes
Water pump 8×240	1920 volt-amperes
Garage door opener 5.8×120	696 volt-amperes
Heat-vent-lights (2) 1500×2	3000 volt-amperes
Attic exhaust fan 5.8×120	696 volt-amperes
Hydromassage tub 10×20	1200 volt-amperes
Total other loads	47,926 volt-amperes

We can now complete our optional calculation:
Enter electric furnace load

LOADS	VOLT-AMPERES (VA)	FACTOR	VOLT-AMPERES
Electric furnace	13,000	0.65	8,550
Other loads	10,000	1.00	10,000
Remainder of other loads	47,926 − 10,000 = 37,926	0.40	15,170
		Total	33,720
Amperes = volt-amperes ÷ volts = 33,720 VA ÷ 240 = 140 amperes			

Using the optional calculation method, and referring to *310.15(B)(7)*, we find that for the 140-ampere calculated load, a 1 AWG copper conductor is permitted to be used for a service with a rating of 150 amperes and could be installed for this service entrance. This seems a bit small, and that is why some local electrical codes do not permit the use of this optional calculation with the smaller conductor sizes permitted in *310.15(B)(7)*.

The specifications for the residence discussed in this text call for a full 200-ampere service consisting of two AWG 2/0 THW, THHN, THWN, or XHHW copper phase conductors and one 1 AWG bare copper neutral conductor.

ELECTRIC VEHICLE CHARGING

An outlet is shown in Figure 16-1 for electric vehicle charging equipment. We also show a circuit in the directory for panelboard B in Figure 27-14 and indicate that it is an optional 40-ampere branch circuit.

There are many types of charging equipment for electric vehicles. These range from fast chargers to trickle chargers. The electrician needs to know what type of charging system is installed so the correct branch circuit can be installed. Typically, this information is provided on the blueprints as well as in the specifications for the project.

In addition to providing information for the branch circuit to be installed, the electrician needs to be furnished with the load for the charging equipment so adequate capacity can be assured in the panelboards, feeders, and service for the building. If a 40-ampere branch circuit were to be installed in the project included in this text, there is adequate capacity in the service, particularly if the optional load calculation procedure is used.

Electric vehicle charging systems are covered extensively in *NEC Article 625*.

Existing Homes

When adding new loads in an existing home, *220.83* provides an optional calculation method that may be used to confirm that the existing service is large enough to handle the additional load. The procedure is a mirror image of the optional calculation method we discussed above.

Service-Entrance Conductor Size Table

Table 29-2 has been taken from one major city that prefers to show minimum service-entrance conductor size requirements rather than having electrical contractors make calculations each and every time they install a service.

It is within the realm of local authorities to publish code requirements specifically for their communities.

Main Disconnect

The main service disconnecting means in this residence is rated 200 amperes, discussed in Chapter 27.

Grounding Electrode Conductor

Grounding electrode conductors are sized according to *Table 250.66* of the *Code*.

As an example, this residence is supplied by 2/0 AWG copper service-entrance conductors. Checking *Table 250.66*, we find that the minimum grounding electrode conductor must be a 4 AWG copper conductor.

The grounding electrode conductor for the service in this residence is a 4 AWG copper armored ground cable, discussed in Chapter 27.

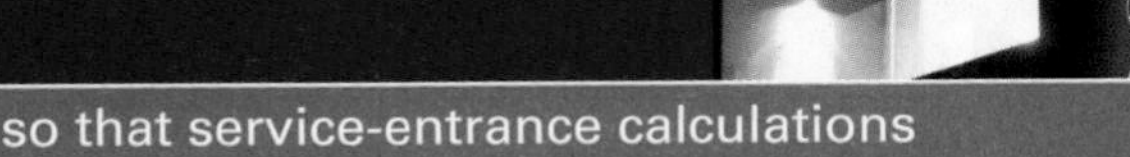

TABLE 29-2

Service-entrance conductor sizing used in some cities so that service-entrance calculations do not have to be made for each service-entrance installation.

SERVICE-ENTRANCE CONDUCTOR SIZE

	COPPER			ALUMINUM		
Size (amperes)	**Phase (Hot) Conductors**	**Neutral Conductors**	**Conduit Trade Size**	**Phase (Hot) Conductors**	**Neutral Conductors**	**Conduit Trade Size**
100	2 AWG	4 AWG	1½	1 AWG	2 AWG	1½
125	1 AWG	4 AWG	1½	2/0 AWG	1 AWG	1½
150	1/0 AWG	4 AWG	1½	3/0 AWG	1/0 AWG	2
175	2/0 AWG	2 AWG	2	4/0 AWG	2/0 AWG	2
200	3/0 AWG	2 AWG	2	250 kcmil	3/0 AWG	2

NEC TABLE 250.66

Grounding Electrode Conductor for Alternating-Current Systems

Size of Largest Ungrounded Service-Entrance Conductor or Equivalent Area for Parallel Conductors[a] (AWG/kcmil)		Size of Grounding Electrode Conductor (AWG/kcmil)	
Copper	**Aluminum or Copper-Clad Aluminum**	**Copper**	**Aluminum or Copper-Clad Aluminum[b]**
2 or smaller	1/0 or smaller	8	6
1 or 1/0	2/0 or 3/0	6	4
2/0 or 3/0	4/0 or 250	4	2
Over 3/0 through 350	Over 250 through 500	2	1/0
Over 350 through 600	Over 500 through 900	1/0	3/0
Over 600 through 1100	Over 900 through 1750	2/0	4/0
Over 1100	Over 1750	3/0	250

Notes:

1. If multiple sets of service-entrance conductors connect directly to a service drop, set of overhead service conductors, set of underground service conductors, or service lateral, the equivalent size of the largest service-entrance conductor shall be determined by the largest sum of the areas of the corresponding conductors of each set.
2. Where there are no service-entrance conductors, the grounding electrode conductor size shall be determined by the equivalent size of the largest service-entrance conductor required for the load to be served.

[a]This table also applies to the derived conductors of separately derived ac systems.
[b]See installation restrictions in 250.64(A).

TYPES OF WATT-HOUR METERS

Figure 29-3 shows different styles of watt-hour meters. The first, (A), is a typical electronic residential kilowatt-hour-only meter. The second, (B), is a single-phase watt-hour meter that has two sets of registers: the electromechanical register records the total kilowatt-hours, and the electronic register records and displays kilowatt-hour consumption for up to four different rate schedules. The third, (C), is a solid-state programmable polyphase watt-hour meter that can register kilowatt-hour consumption for up to five different rate schedules, demand, reactive measurements, and load profile if so equipped. Utilities that offer special "time-of-use" rates install meters of the second and third types.

Wiring diagrams for the metering and control of electric water heaters, air conditioners, and heat pumps are discussed in Chapter 19. Electric utilities across the country have different requirements for hooking up these loads. Always consult your local utility for their latest information regarding electric service.

Time-Based Metering

A rather recent innovation for residential metering is "time-based" metering, also referred to as "real-time" or "smart" metering. These electronic solid-state meters are capable of reading and calculating

A

B

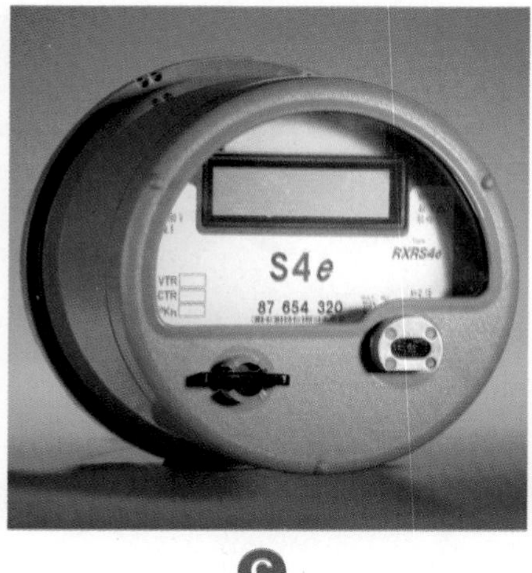

C

Courtesy of Landis+Gyr.

FIGURE 29-3 Photo A shows an electronic watt-hour meter. Photo B shows a single-phase watt-hour meter that has two sets of registers: one registers the total kilowatt-hours, the second records and displays kilowatt-hour consumption for up to four different rate schedules. Photo C shows an electronic programmable polyphase watt-hour meter that is capable of registering five different time-of-use rates, demand, reactive measurement, and load profile if so equipped.

electric usage by the hour. Utilities offering this type of time-based metering have rate pricing that changes by the hour. You can plan ahead and use high-energy-consumption appliances (dishwasher, washer, clothes dryer, clothes iron, ranges, ovens, cooktops, air conditioning, etc.) when the rate is low.

Low rates generally are early in the morning and late at night. Typically, midnight is the lowest rate, which slowly increases until noon; then a steeper increase occurs in the afternoon, peaking between 3 and 6 p.m., and then into the night rates drop off.

Some utilities have a website where you can check their rates for all times of the day. You can keep track of your actual usage. This puts the burden on the consumer to choose when to do household tasks to reduce the electric bill.

The world of electronics is amazing. The transition from off-peak metering to time-based metering will no doubt become very popular.

Commercial and industrial customers have had this type of "time-based" metering for years.

- Does your local electric utility offer this type of metering?
- Does it charge a monthly premium for this type of metering?
- It just might pay to check it out!

READING WATT-HOUR METERS

Reading a digital watt-hour meter is easy. Just read the numbers.

For dial-type meters, starting with the first dial, record the last number the pointer has just passed. Continue doing this with each dial until all dial readings have been recorded. The reading on the five-dial watt-hour meter in Figure 29-4 is 18,672 kilowatt-hours.

If the meter reads 18,975 one month later, as indicated in Figure 29-5, by subtracting the previous reading of 18,672, it is found that 303 kilowatt-hours of electricity were used during the month. The 303 kilowatt-hours of electricity used is multiplied by the rate per kilowatt-hour, and the power company bills the consumer for the energy used. The utility may also add a fuel adjustment charge or other charges.

FIGURE 29-4 The reading of this five-dial meter is 18,672 kilowatt-hour.

FIGURE 29-5 One month later, the meter reads 18,975 kilowatt-hours, indicating that 303 kilowatt-hours were used during the month.

REVIEW

Note: Refer to the *Code* or the plans where necessary.

1. According to *NEC* ________________, the minimum-size service required for a one-family dwelling is ________________ amperes.
2. a. What is the unit load per ft^2 for the general lighting load of a residence? ________

   ____

   b. What are the demand factors for the general lighting load in dwellings? ________

   ____

   ____

   ____

3. a. What is the ampere rating of the circuits that are provided for the small-appliance loads? ____

   b. What is the minimum number of small-appliance circuits permitted by the *Code*?

   ____

   c. How many small-appliance circuits are included in this residence? ____________

4. Why is the air-conditioning load for this residence omitted in the service calculations?

5. What demand factor may be applied when four or more fixed appliances are connected to a service, in addition to an electric range, air conditioner, clothes dryer, or space-heating equipment? *(220.53)*

6. What load may be used for an electric range rated at not over 12 kW? *(Table 220.55)*

7. What is the calculated load for an electric range rated at 16 kW? *(Table 220.55)* Show calculations.

8. What is the calculated load when fixed electric heating is used in a residence? *(220.51)*

9. On what basis is the neutral conductor of a service entrance determined?

10. Why is it permissible to omit an electric space heater, water heater, and certain other 240-volt equipment when calculating the neutral service-entrance conductor for a residence?

11. a. What section of the *Code* contains an optional method for determining residential service-entrance loads?

    b. Is this section applicable to a two-family residence?

12. Calculate the minimum size of copper service-entrance conductors and grounding electrode conductor required for a residence containing the following: floor area is 24 ft × 38 ft (7.3 m × 11.6 m). The dwelling will have:
    - a 12-kW, 120/240-volt electric range.
    - a 5-kW, 120/240-volt electric clothes dryer consisting of a 4-kW, 240-volt heating element, a 120-volt motor, and 120-volt lamp that have a combined load of 1 kW.
    - a 2200-watt, 120-volt sauna heater.

- six individually controlled 2-kW, 240-volt baseboard electric heaters.
- a 12-ampere, 240-volt air conditioner.
- a 3-kW, 240-volt electric water heater.
- a 1000-watt, 120-volt dishwasher.
- a 7.2-ampere, 120-volt food waste disposer.
- a 5.8-ampere, 120-volt attic exhaust fan.

Determine the sizes of the ungrounded "hot" conductors and the neutral conductor. Use Type THWN. Use *310.15(B)(7)*. Also determine the correct size grounding electrode conductor.

Do not forget to include the required two 20-ampere small-appliance circuits and the laundry circuit.

The panelboard is marked for 75°C terminations.

You may use the blank Single-Family Dwelling Service-Entrance Calculation form in the Appendix of this text, or use the form as a guide to make your calculations in the proper steps.

Two ________________ AWG THWN ungrounded "hot" conductors

One ________________ AWG THWN or bare neutral conductor

One ________________ AWG grounding electrode conductor

STUDENT CALCULATIONS

13. Read the meter shown. Last month's reading was 22,796. How many kilowatt-hours of electricity were used for the current month?

14. After studying this chapter, we realize that electric utilities across the country have many ways to meter residential customers. They may have lower rates during certain hours of the day. A common term used to describe these programs is ________________.

15. a. Does your electric utility offer time-based metering or some form of it?

    ____

    b. Do you use it to reduce your electric bills? ________________________

CHAPTER **30**

Swimming Pools, Spas, Hot Tubs, and Hydromassage Baths

OBJECTIVES

After studying this chapter, you should be able to

- recognize the importance of proper swimming pool wiring with regard to human safety.
- discuss the hazards of electrical shock associated with faulty wiring in, on, or near pools.
- understand and apply the basic *NEC* requirements for the wiring of swimming pools, spas, hot tubs, and hydromassage bathtubs.

INTRODUCTION

The risk of electric shock and electrocution increase significantly when people, water, and electricity are combined.

To protect people against the hazards of electric shock associated with swimming pools, spas, hot tubs, and hydromassage bathtubs, special *Code* requirements are necessary.

A picture is worth a thousand words. A detailed drawing of the *NEC*® requirements for swimming pools appears on Sheet 10 of 10 in the Plans found in the back of this text. Refer to this drawing often as you study this chapter.

ELECTRICAL HAZARDS

A person can suffer an immobilizing or lethal shock in a residential-type pool in either of two ways.

1. *Direct contact:* An electrical shock can be deadly to someone in a pool who touches a "live" wire or touches an object such as a metal ladder, metal fence, metal parts of a luminaire, metal enclosures of electrical equipment, or appliances that have become live by an ungrounded conductor coming in contact with the metal enclosure or exposed metal parts of an appliance. Figure 30-1 shows a person in the water touching a faulty appliance.
2. *Indirect contact:* An electrical shock can be lethal to someone in a pool if the person is merely in the water when voltage gradients in the water are present. This is illustrated in Figure 30-2.

As shown in Figure 30-2, "rings" of voltage radiate outward from the radio to the pool walls and bottom of the pool. These rings can be likened to the rings that form when a rock is thrown into the water. The voltage rings, or gradients, range from 120 volts at the radio to zero volts at the pool walls and bottom. The pool walls and bottom are assumed to be at ground, or zero, potential. The gradients, in varying degrees, are found throughout the body of water. Figure 30-2 shows voltage gradients in the pool of 90 volts and 60 volts. This figure is a simplification of the actual situation in which there are many voltage gradients. In this case, the voltage differential, 30 volts, is an extremely hazardous value. The person in the pool immersed in these voltage gradients is subject to severe shock, immobilization (which can result in drowning), or actual electrocution. Tests conducted over the years have shown that a voltage gradient of 1½ volts per foot can cause paralysis.

The shock hazard to a person in a pool is quite different from that of the normal "touch" shock hazard to a person not submersed in water. The water makes contact with the entire skin surface of the body rather than just at one "touch" point. Skin wounds, such as cuts and scratches, reduce the body's resistance to shock to a much lower value than that of the skin alone. Body openings such as

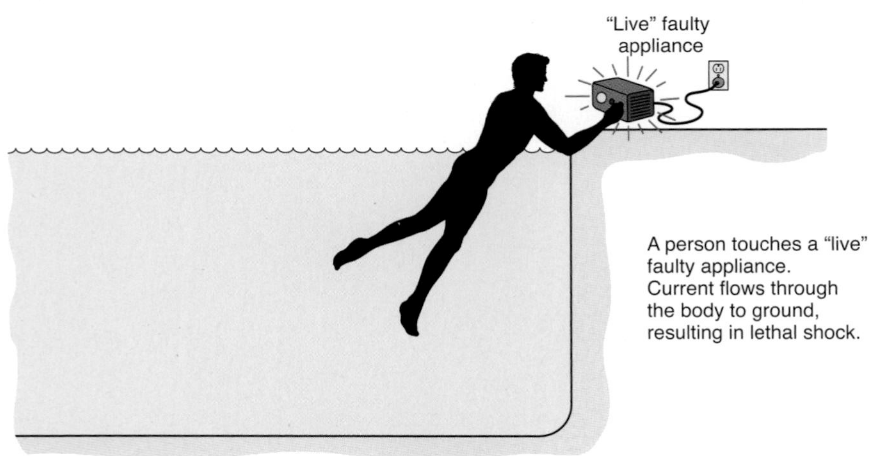

FIGURE 30-1 Touching a "live" faulty appliance can cause lethal shock.

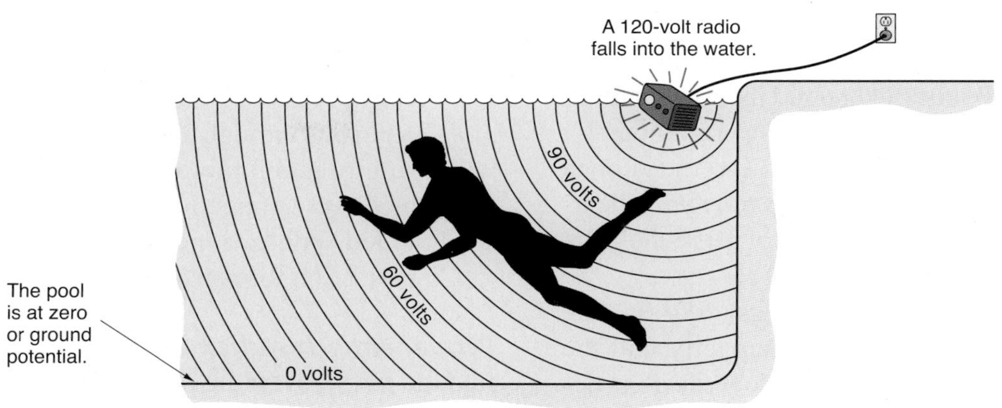

FIGURE 30-2 Voltage gradients surrounding a person in the pool can cause severe shock, drowning, or electrocution.

ears, nose, and mouth further reduce the body resistance. As Ohm's law states, for a given voltage, the lower the resistance, the higher the current.

KEY *NEC* REQUIREMENTS—WIRING FOR SWIMMING POOLS

Article 680 covers swimming pools, wading pools, hydromassage bathtubs, fountains, therapeutic and decorative pools, hot tubs, and spas.

Here are some of the general *Code* rules. Comprehensive details for swimming pool requirements are found on Sheet 10 of 10 of the Plans at the back of this text.

General, Part I. Contains general requirements that apply to all types of swimming pools, spas, hot tubs, hydromassage tubs, and similar water related equipment. *Parts II, III, IV, V, VI,* and *VII* discuss the wiring requirements for specific equipment.

Scope, *680.1:* The scope tells us exactly what is covered by *Article 680.*

Definitions, *680.2:* There are many one-of-a-kind terms listed in *680.2*. These terms might seem like a foreign language. It makes sense to read these definitions carefully to make it easier to understand the rest of *Article 680.*

Other Articles, *680.3:* Many of the basic requirements of *Chapter 1* through *Chapter 4* in the *NEC* are modified by special rules in *Article 680* that are unique to swimming pools, wading pools, hydromassage bathtubs, therapeutic pools, decorative pools, hot tubs, and spas. This section of the *NEC* contains a convenient table that lists other articles and chapters that apply.

Ground-Fault Circuit Interrupters, *680.5:* GFCIs are required to be listed circuit breakers or receptacles, or other listed types. GFCIs are discussed in Chapter 6. Throughout *Article 680*, you will find the need for GFCIs.

Be careful when connecting underwater luminaires. Once you run conductors beyond (on the load side of) a GFCI device protecting that wiring, you are not permitted to have other conductors in the same raceway, box, or enclosure. The exceptions to this rule are when (1) the other conductors are protected by GFCIs; (2) the other conductors are equipment grounding conductors; (3) the other conductors are supply conductors to a feed-through type GFCI; or (4) the GFCI is installed in a panelboard, where obviously there will be a "mix" of other circuit conductors that are not GFCI protected. This is covered in *680.23(F)(3).*

Grounding, *680.6:* In general, grounding is required to be done as required in *Article 250.* But there are modifications that specifically address swimming pools.

Cord-and-Plug-Connected Equipment, *680.7:* Here we find permission to use a flexible cord-and-plug connection for fixed or stationary equipment associated with a permanently installed pool. Underwater luminaires are not included in this permission.

Overhead Conductors, ***680.8:*** Overhead open conductors must be kept out of reach. Clearances are shown in Plan 10 of 10 in the back of this text.

Electric Pool Water Heaters, ***680.9:*** The branch-circuit conductor ampacity and the rating of the branch-circuit overcurrent devices must not be less than 125% of the total nameplate rated-load ampere rating of the electric heater. Figure 30-3 shows the requirements for unit heaters, radiant heaters, and embedded heating cables.

Underground Wiring Location, ***680.10:*** Underground wiring is not permitted under pools or within 5 ft (1.5 m) horizontally from the inside wall of the pool. There are a few exceptions.

Maintenance Disconnecting Means, ***680.12:*** A means must be provided to simultaneously disconnect all ungrounded conductors of a circuit supplying pool-associated equipment. The disconnecting means is required to be readily accessible, within sight, and located at least 5 ft (1.5 m) horizontally from the inside walls of the pool, spa, or hot tub. Pool lighting is not covered by *680.12*.

▶**GFCI Protection for Pump Motors,** ***680.21(C):*** Outlets supplied by branch circuits protected by an overcurrent device rated 120 volts through 240 volts, single phase, that supply pump motors whether by receptacle or direct connection are required to be provided with ground-fault circuit interrupter protection for personnel.◀

Permanently Installed Pools, Part II. Here we find the rules governing permanently installed pools such as for motors; clearances for overhead lighting; underwater lighting; existing installations; receptacles; requirements for switching; equipment associated with the pool; bonding, grounding, and audio equipment; and electric heaters. Electric deck heater *Code* requirements are shown in Figure 30-3.

Grounding and bonding together all metal parts in and around pools ensures that these metal parts are at the same voltage potential to ground, greatly reducing the shock hazard. Stray voltage and voltage gradient (see Figure 30-2) problems are kept to a minimum. Proper grounding and bonding also facilitates the opening of the overcurrent protection devices should a fault occur in the circuit(s).

You will see on Sheet 10 of 10 of the Plans that a metal conduit by itself is not considered an adequate equipment grounding means for equipment in and around pools.

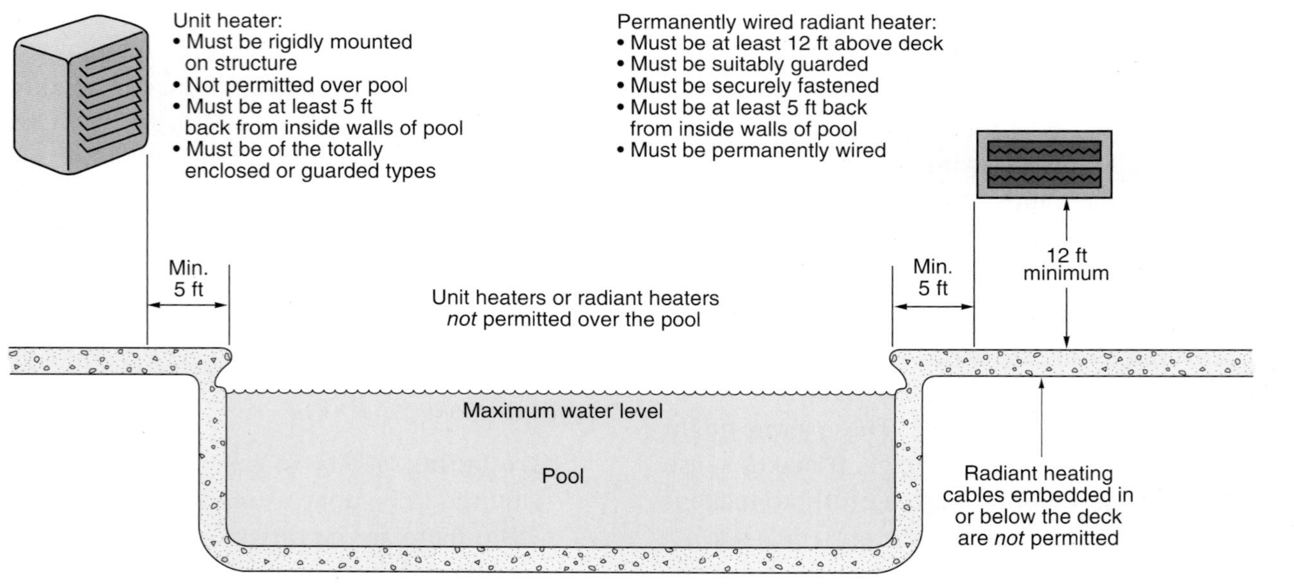

FIGURE 30-3 Deck-area electric heating within 20 ft (6.0 m) from inside edge of swimming pools, *680.27(C)*.

Figure 30-4 is a detailed illustration showing the grounding of important metal parts of a permanently installed pool. As can be seen, the equipment grounding conductors also bond or connect the enclosures together, as stated in the *Informational Note* following the definition of Equipment Grounding Conductor in *NEC Article 100*.

Figure 30-5 shows an equipotential bonding grid that consists of the conductive reinforcing steel bars of a permanently installed swimming pool.

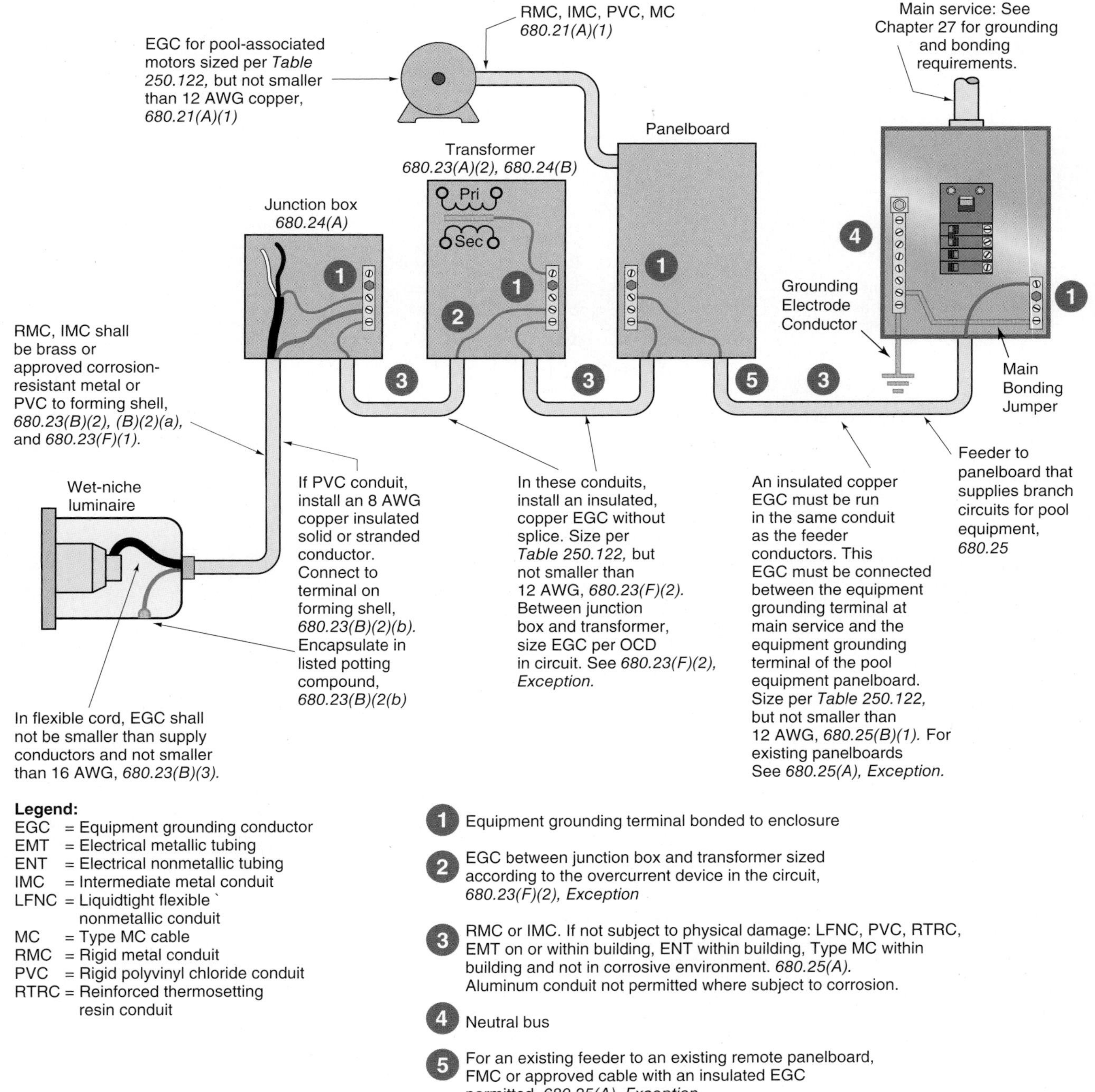

FIGURE 30-4 A "trail" of the grounding and bonding requirements from a wet-niche luminaire back to the main service panelboard.

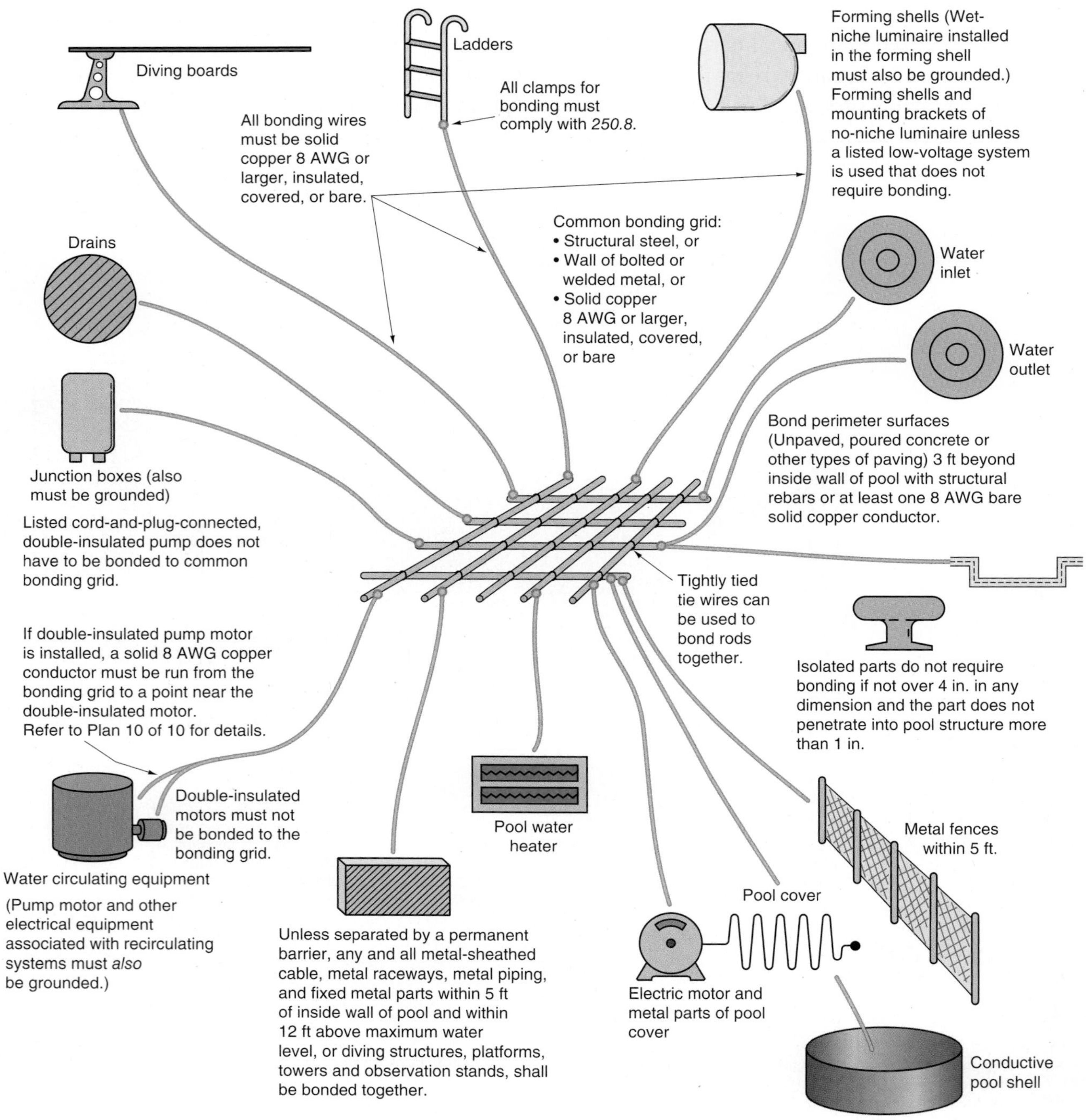

Diving boards

Ladders

All clamps for bonding must comply with *250.8*.

All bonding wires must be solid copper 8 AWG or larger, insulated, covered, or bare.

Forming shells (Wet-niche luminaire installed in the forming shell must also be grounded.) Forming shells and mounting brackets of no-niche luminaire unless a listed low-voltage system is used that does not require bonding.

Drains

Common bonding grid:
- Structural steel, or
- Wall of bolted or welded metal, or
- Solid copper 8 AWG or larger, insulated, covered, or bare

Water inlet

Water outlet

Junction boxes (also must be grounded)

Bond perimeter surfaces (Unpaved, poured concrete or other types of paving) 3 ft beyond inside wall of pool with structural rebars or at least one 8 AWG bare solid copper conductor.

Listed cord-and-plug-connected, double-insulated pump does not have to be bonded to common bonding grid.

Tightly tied tie wires can be used to bond rods together.

If double-insulated pump motor is installed, a solid 8 AWG copper conductor must be run from the bonding grid to a point near the double-insulated motor. Refer to Plan 10 of 10 for details.

Isolated parts do not require bonding if not over 4 in. in any dimension and the part does not penetrate into pool structure more than 1 in.

Double-insulated motors must not be bonded to the bonding grid.

Water circulating equipment

(Pump motor and other electrical equipment associated with recirculating systems must *also* be grounded.)

Pool water heater

Metal fences within 5 ft.

Pool cover

Unless separated by a permanent barrier, any and all metal-sheathed cable, metal raceways, metal piping, and fixed metal parts within 5 ft of inside wall of pool and within 12 ft above maximum water level, or diving structures, platforms, towers and observation stands, shall be bonded together.

Electric motor and metal parts of pool cover

Conductive pool shell

FIGURE 30-5 Equipotential Bonding Grid: The purpose of bonding together all metal parts and other conductive surfaces that are part of, in, and around a permanently installed swimming pool is to reduce voltage gradients in the pool and adjacent areas around the pool. Bonding is done with solid 8 AWG copper conductors (insulated, covered, or bare) or with rigid metal conduit made of brass or other identified corrosion-resistant metal. See the list of items required to be included in the equipment bonding grid in *NEC 680.26* and Plan 10 of 10. In installations where the structural steel is encapsulated with nonconductive material, you will have to install a copper conductor grid. See *680.26(B)(1)(a)* and *(b)*.

The conductor that connects various components together must not be smaller than 8 AWG and must be a solid copper conductor. If the pool shell does not have conductive steel reinforcing bars, a copper bonding grid is required. See *NEC 680.26(B)* for details about the following. Components that must be bonded include these:

- Conductive pool shells
- Perimeter surfaces within 3 ft (1 m)
- Metallic components
- Underwater lighting
- Metal fittings
- Electrical equipment associated with the water circulating system
- Fixed metal parts within 5 ft (1.5 m).

Note that the bonding conductor does not have to be connected to any service equipment or any grounding electrode. It merely ties all metal parts together.

Pool Water. Pool water contains dissolved chemicals that are conductive to some extent. Some pools are made of fiberglass (nonconductive) and some have insulating liners (nonconductive). Where there are no metal parts in contact with the pool water, there still is a need to bond the water to the pool deck to eliminate voltage gradients between the pool water and the deck. For these situations, the *NEC* covers this in *680.26(C)* which reads:

> **(C) Pool Water.** ▶Where none of the bonded parts is in direct connection with the pool water, the pool water shall be in direct contact with an approved corrosion-resistant conductive surface that exposes not less than 9 in.^{2} (5800 mm^2) of surface area to the pool water at all times. The conductive surface shall be located where it is not exposed to physical damage or dislodgement during usual pool activities, and it shall be bonded in accordance with 680.26(B).◀*

Storable Pools, Part III. Storable pools are just that: they are used on or above the ground and can be dismantled and stored.

*Reprinted with permission from NFPA 70-2014.

The wiring for storable pools must follow the basic requirements found in *Article 680, Part 1*, plus the additional requirements found in *Part III*. The additional requirements for storable pools are as follows:

- Cord-connected pool filter pumps are double-insulated, and have provisions for grounding only internal and nonaccessible noncurrent-carrying metal parts. This is the equipment grounding conductor in the cord that has a grounding-type attachment plug cap.
- All electrical equipment used with storable pools are required to be GFCI protected. For cord-connected pool filter pumps, the GFCI device is required to be part of the attachment plug cap or located in the power supply cord not more than 12 in. (300 mm) from the attachment plug cap, *680.31*.
- All 125-volt, 15- and 20-ampere receptacles within 20 ft (6.0 m) of inside walls of a storable pool must be GFCI protected, *680.32*.
- Receptacles are not permitted less than 6 ft (1.83 m) from the inside walls of a storable pool, *680.22(A)(1)*.
- Receptacles rated 15 or 20 amperes, 125 through 250 volts, that supply pool pump motors are required to be GFCI protected.

Spas and Hot Tubs, Part IV. What's the difference? In reality, there is no difference. Many years ago, hot tubs were constructed of wood such as redwood, teak, cypress, or oak. Along came acrylics, plastics, fiberglass, and other synthetic products. The name "spa" was created. Today, both names are used interchangeably depending on where you live. They have therapeutic jets that create whirling currents of hot water, a filtering system, and some have mood lighting, sound systems, and waterfalls. Spas and hot tubs are not designed or intended to have the water drained or discharged after each use. Here are the *Code* rules for spas and hot tubs. (See Figure 30-6.)

***Outdoor Installations, 680.42*:** Spa and hot-tubs in outdoor locations must be wired according to *Parts I* and *II of Article 680*, except for the following:

a. Flexible connections using flexible conduit and cord-and-plug connections are permitted.

Courtesy of Hot Spring Spas.

FIGURE 30-6 A typical spa.

b. Different metal components where there is metal-to-metal mounting of the same metal frame are acceptable as the required bond.

c. For one-family dwellings, the interior wiring to a self-contained spa or hot tub or a packaged spa or hot tub assembly is permitted to be conventional wiring methods that contain a copper equipment grounding conductor not smaller than 12 AWG. Conventional wiring methods are found in *Chapter 3* of the *NEC*. However, outdoor wiring to the spa or hot tub must be those wiring methods required in *Parts I* and *II* of *Article 680* for an in-ground swimming pool.

d. ►Equipotential bonding of perimeter surfaces in accordance with 680.26(B)(2) shall not be required to be provided for spas and hot tubs where all of the following conditions apply:

(1) The spa or hot tub shall be listed as a self-contained spa for aboveground use.

(2) The spa or hot tub shall not be identified as suitable only for indoor use.

(3) The installation shall be in accordance with the manufacturer's instructions and shall be located on or above grade.

(4) The top rim of the spa or hot tub shall be at least 28 in. (710 mm) above all perimeter surfaces that are within 30 in. (760 mm), measured horizontally from the spa or hot tub. The height of nonconductive external steps for entry to or exit from the self-contained spa shall not be used to reduce or increase this rim height measurement.◄*

e. Specific requirements or provisions are made for wiring methods that are located in the interior of a dwelling unit or another building or structure that is associated that supply a spa or hot tub that is located outdoors. *NEC 680.42(C)* reads as follows:

►**Interior Wiring to Outdoor Installations.** In the interior of a dwelling unit or in the interior of another building or structure associated with a dwelling unit, any of the wiring methods recognized or permitted in Chapter 3 of this *Code* that contains a copper equipment grounding conductor that is insulated or enclosed within the outer sheath of the wiring method and not smaller than 12 AWG shall be permitted to be used for the connection to motor, heating, and control loads that are part of a self-contained spa or hot tub or a packaged spa or hot tub equipment assembly. Wiring to an underwater luminaire shall comply with 680.23 or 680.33.◄*

This provision allows the use of a cable type wiring method including Type AC, Type MC, and Type NM rather than a typical raceway. This provision acts like an exception to the general requirement for a raceway such as EMT. Note that this rule is

*Reprinted with permission from NFPA 70-2014.

applicable only inside a dwelling unit or an associated building or structure. Since the spa or hot tub has a control or distribution where overcurrent protection is supplied, the supply conductors are considered a feeder rather than a branch circuit. As a result, the wiring method for outdoor and underground wiring must comply with *680.25(A)*.

Indoor Installations, *680.43* states that indoor installations are required to be wired according to *Parts I* and *II* of *Article 680* and is required to be connected using any of the wiring methods of *NEC Chapter 3*, except for the following:

a. Wiring Methods: General requirements for wiring methods are provided in several sections of *Part I* and *Part II* of *Article 680*. Several of these requirements mandate wiring in conduit, EMT, or Type MC cables. Three exceptions are provided that offer relief for grounding and bonding and wiring methods. The exceptions read:

   Exception No. 1: Listed spa and hot tub packaged units rated 20 amperes or less shall be permitted to be cord-and-plug-connected to facilitate the removal or disconnection of the unit for maintenance and repair.

   Exception No. 2: The equipotential bonding requirements for perimeter surfaces in 680.26(B)(2) shall not apply to a listed self-contained spa or hot tub installed above a finished floor.

   ▶*Exception No. 3: For a dwelling unit(s) only, where a listed spa or hot tub is installed indoors, the wiring method requirements of 680.42(C) shall also apply.*◀*

   The reference to *680.42(C)* permits wiring methods such as Type NM cable to be used as the wiring method for spas and hot tubs that are located inside dwellings.

b. Receptacles:

   - At least one 125-volt, 15- or 20-ampere receptacle connected to a general-purpose branch circuit must be installed at least 6 ft (1.83 m), but not more than 10 ft (3.0 m) from the inside wall of the spa or hot tub.
   - All other 125-volt, 15- or 20-ampere receptacles must be located at least 6 ft (1.83 m), measured horizontally, from the inside wall of the spa or hot tub.
   - All 125-volt receptacles rated 30 amperes or less located within 10 ft (3.0 m) of the inside walls of a spa or hot tub must be GFCI protected.
   - A receptacle that supplies power to a spa or hot tub must be GFCI protected.

c. Luminaires, lighting outlets, and ceiling-suspended paddle fans above spa or hot tub and within 5 ft (1.5 m) from the inside walls of the spa or hot tub:

   - are to be not less than 12 ft (3.7 m) above if not GFCI protected.
   - are to be not less than 7 ft 6 in. (2.3 m) if GFCI protected.
   - if less than 7 ft 6 in. (2.3 m), must be GFCI protected and be suitable for damp locations and be:
     - recessed luminaires with a glass or plastic lens and nonmetallic or isolated metal trim.
     - surface-mounted luminaires with a glass or plastic globe, a nonmetallic body, or a metallic body isolated from contact.

   If underwater lighting is installed, the rules for regular swimming pools apply, in which case refer to Sheet 10 of 10 of the Plans.

d. Switches:

   - wall switches must be kept at least 5 ft (1.5 m), measured horizontally, from the inside edge of the spa or hot tub.

e. Bonding: Bonding requirements are similar to those of conventional swimming pools. Bonding keeps everything at the same potential. Excluded are small conductive surfaces not likely to become energized such as drains and air and water jets. Bond together:

   - all metal fittings within or attached to the spa or hot tub structure.
   - metal parts of associated electrical equipment including pump motors.

*Reprinted with permission from NFPA 70-2014.

- metal raceways, metal piping, and other metal surfaces within 5 ft (1.5 m) of the inside walls of the spa or hot tub, unless separated by a permanent barrier, such as a wall or building.
- electrical devices and controls not associated with the pool or hot tub and located not less than 5 ft (1.5 m) from the spa or hot tub. Bonding to the spa or hot tub system also is acceptable.

f. Bonding all of the above things together can be accomplished by means of threaded piping and fittings, by metal-to-metal mounting on a common base or frame, or by a solid copper 8 AWG or larger solid, insulated, covered or bare bonding jumper.

g. Bond all electrical equipment within 5 ft (1.5 m) of the inside wall of a spa or hot tub. Also bond all electrical equipment associated with the circulating system of the spa or hot tub.

Protection, 680.44: The word *protection* refers to GFCI protection. Outlets for spas and hot tubs are required to be GFCI protected unless the unit has integral GFCI protection. A few other exceptions are mentioned in *680.44*.

Fountains, Part V. The residence discussed in this text does not have a fountain. Permanently installed fountains come under the requirements of *Part V* of *Article 680*. For fountains that share water with a regular swimming pool, the wiring must conform to the requirements found in *Part II* of *Article 680*, which covers permanently installed swimming pools.

Small self-contained portable fountains that have no dimension over 5 ft (1.5 m) do not come under the requirements of *Part V*.

Pools and Tubs for Therapeutic Use, Part VI. The residence discussed in this text does not have any therapeutic equipment.

Hydromassage Bathtubs, Part VII. A hydromassage tub is intended to be filled, used, then drained after each use. These are also known as whirlpool bathtubs. Figure 22-12 illustrates one manufacturer's hydromassage bathtub.

- Protection, *680.71*: The word *protection* refers to GFCI protection. Hydromassage bathtubs, together with their associated electrical components, must be protected by a readily accessible GFCI device. Be sure to review the definition of "readily accessible" in *NEC Article 100*. This requirement will not permit a GFCI receptacle behind an access panel that requires tools to be removed. This GFCI protection could be provided by a GFCI circuit breaker or by a GFCI receptacle at the hydromassage bathtub location so long as it meets the definition of "readily accessible." The hydromassage bathtub must also be supplied by an individual branch circuit. This means the branch circuit supplies only the hydromassage bathtub.
- Protection, *680.71*: All 125-volt, single-phase receptacles that are located within 6 ft (1.83 m), measured horizontally, from the inside walls of the hydromassage bathtub must be GFCI protected.
- Other Electrical Equipment, *680.72*: A hydromassage bathtub does not constitute any more of a shock hazard than a regular bathtub. All wiring (fixtures, switches, receptacles, and other equipment) in the same room but not directly associated with the hydromassage bathtub is installed according to all the normal *Code* requirements covering installation of that equipment in bathrooms.
- Accessibility, *680.73*: Electrical equipment associated with a hydromassage bathtub must be accessible without damaging the building structure or finish.

  If the hydromassage bathtub is cord-and-plug-connected with the supply receptacle accessible only through a service access opening, the receptacle is required to be installed so that its face is within direct view and not more than 1 ft (300 mm) of the opening.
- Bonding, *680.74*: All metal piping systems and all grounded metal parts in contact with the circulating water are required to be bonded together using a copper 8 AWG or larger solid, insulated, covered, or bare. The water piping systems, both supply and drainage, for most dwellings is nonmetallic. As a result, it is not too common to see a bonding conductor installed.

The 8 AWG or larger solid copper bonding jumper, if installed, is required to be long enough to terminate on a replacement non-double-insulated pump motor and is required to be terminated to the equipment grounding conductor of the branch circuit of the motor when a double-insulated circulating pump motor is used.

This residence has a hydromassage tub in the master bathroom. The wiring of it is covered in Chapter 22 of this text.

GETTING TRAPPED UNDER WATER

There is another real hazard, not truly electrical in nature, but solvable electrically.

Drownings have occurred when a person is trapped under water by the tremendous suction at an unprotected drain. The grate over the drain may have been removed or broken. One report indicated that three adult males could not pull a basketball from a drain because of the suction!

To address this hazard, *680.41* requires that a clearly labeled emergency shut-off switch for the control of the recirculation system and jet system for spas and hot tubs shall be installed not less than 5 ft (1.5 m) away, adjacent to, and within sight of the spa or hot tub. Although this requirement does not apply to single-family dwellings, some states require this disconnect for single-family dwelling pools.

Another hazard associated with spas and hot tubs is prolonged immersion in hot water. If the water is too hot and/or the immersion too long, lethargy (drowsiness, hyperthermia) can set in. This can increase the risk of drowning. UL Standard 1563 establishes the maximum water temperature at 104°F (40°C). The suggested maximum time of immersion is generally 15 minutes. Instructions furnished with spas and hot tubs specify the maximum temperature and time permitted for using the spa or hot tub. Hot water, in combination with drugs and alcohol, presents a real hazard to life. Caution must be observed at all times.

Although not an electrical issue, it is interesting to note that Underwriters' Laboratories has a category referred to as *Suction Fittings for Swimming Pools, Wading Pools, Spas, and Hot Tubs.* For listing these fittings, UL uses the American Society of Mechanical Engineers Standard ASME A112.19.8-2007 that addresses the safety issues regarding hair, body, finger, and limb entrapment in suction fittings that are designed to be totally submerged in swimming pools, wading pools, spas, and hot tubs.

UNDERWRITERS LABORATORIES STANDARDS

UL standards of interest are these:

UL 676	Underwater Lighting Fixtures
UL 943	Ground-Fault Circuit Interrupters
UL 1081	Swimming Pool Pumps, Filters, and Chlorinators
UL 1241	Junction Boxes for Underwater Pool Luminaires
UL 1563	Electric Spas, Equipment Assemblies, and Associated Equipment
UL 1795	Hydromassage Bathtubs

REVIEW

Note: Refer to the *Code* or the plans where necessary.

1. Most of the requirements for the wiring of swimming pools are covered in *NEC Article* ______________________.
2. Name the two ways in which a person may sustain an electrical shock when in a pool.
   a. ____
   b. ____

3. Using the text, the Plans in the back of this text, and the *NEC*, determine whether the following items must be grounded. Place an "X" in the correct space.

	True	False
a. Wet-and-dry niche luminaires	________	________
b. Electrical equipment located within 5 ft (1.52 m) of inside walls of pool	________	________
c. Electrical equipment located within 10 ft (3.0 m) of inside walls of pool	________	________
d. Recirculating equipment and pumps	________	________
e. Luminaires and ceiling fans installed more than 15 ft (4.5 m) from inside walls of pool	________	________
f. Junction boxes, transformers, and GFCI enclosures that contain conductors that supply wet-niche luminaires	________	________
g. Panelboards that supply the electrical equipment for the pool	________	________
h. Panelboards 20 ft (6.0 m) from the pool that do not supply the electrical equipment for the pool	________	________

4. Is it permitted to run other conductors in the same conduit as the GFCI-protected conductors that run to underwater luminaires? ____________________

5. *NEC* ____________________ prohibits running electrical conduits under a swimming pool.

6. What is the purpose of grounding and bonding? ____________________

____

7. Does the *NEC* require that, after all of the metal parts of a swimming pool are bonded together to form a common grounding grid, all of this grid must be connected to a grounding electrode? ____________________

8. May conductors be run above the pool? Explain. ____________________

____

9. What is the closest distance from the inside wall of a pool that a receptacle may be installed? ____________________

10. Receptacles located within 20 ft (6.0 m) from the inside wall of a pool must be ____________________ protected.

11. Luminaires installed above a new outdoor swimming pool must be mounted not less than (10 ft. [3.0 m]) (12 ft. [3.7 m]) above the maximum water level. (Circle the correct answer.)

12. Junction boxes and enclosures for transformers and GFCIs have one thing in common. They all (are made of brass) (have threaded hubs) (must be mounted at least 8 in. [200 mm] above the deck). (Circle the correct answer.)

13. For indoor spas and hot tubs, the following statements are either true or false. Check one. Refer to *680.43(A)*, *(B)*, and *(C)*.

	True	False
a. Receptacles may be installed within 5 ft (1.5 m) from the edge of the spa or hot tub.	________	________
b. All receptacles within 10 ft (3.0 m) of the spa or hot tub must be GFCI protected.	________	________
c. Any receptacles that supply power to pool equipment must be GFCI protected.	________	________
d. Wall switches must be located at least 5 ft (1.5 m) from the pool.	________	________
e. In general, luminaires above a spa shall not be less than 7 ft 6 in. (2.3 m) above the maximum water level if the circuit is GFCI protected.	________	________

14. Bonding and grounding of electrical equipment in and around spas and hot tubs (is required by the *NEC*) (is decided by the electrician). (Circle the correct answer.)

15. Where a spa or hot tub is installed in an existing bathroom or other suitable location, an existing receptacle outlet within 5 ft (1.5 m) of the tub is permitted to remain, but only if the circuit feeding the receptacle has ____________________ protection.

16. Underwater pool luminaires shall be installed so that the top of the lens is not less than ____________________ in. below the normal water line unless they are listed for a lesser depth.

17. a. Hydromassage bathtubs and their associated electrical components (shall) (shall not) be protected by a GFCI. (Circle the correct answer.)

    b. All receptacles located within 6 ft (1.83 m) of a hydromassage tub (shall) (shall not) be protected by a GFCI. (Circle the correct answer.)

CHAPTER **31**

Wiring for the Future: Home Automation Systems

OBJECTIVES

After studying this chapter, you should be able to

- understand the basics of Insteon, ZigBee, Z-Wave, and X10 systems.
- understand the basics of wireless and structured wiring systems.
- understand some of the terminology used in the home automation industry.
- understand the types of category-rated cables.

ORGANIZATIONS

The mostly highly recognized organizations regarding standards and installation requirements for communications systems are listed here.

American National Standards Institute (ANSI)
Electronic Industries Alliance (EIA)
International Electrotechnical Commission (IEC)
Institute of Electrical and Electronics Engineers (IEEE)
International Organization for Standardization (ISO)
National Electrical Manufacturers Association (NEMA)
National Fire Protection Association (NFPA)
Telecommunications Industry Association (TIA)
Underwriters Laboratories, Inc. (UL)

Visit these organizations' websites.

LET'S GET STARTED

Before the ink on this page is dry, companies and products for home automation technology will have come and gone. Some systems are wireless, some systems are hard-wired, and some systems are hybrid, using both wireless and hard-wiring. Without question, home automation is a moving target. It is almost impossible to stay on top of this fast-changing technology.

Wiring for the future comes under many names: *home automation*, *advanced home automation*, *structured wiring*, *voice-data-video (VDV)*, and *smart house*, to name a few.

Will home automation (wireless, hard-wired, or hybrid) become the norm? Will conventional premises wiring become a thing of the past? Will homeowners really want to have elaborate automated control of things? Will everyone want a home theater? The answer to these and similar questions could be yes, no, or maybe. Only time will tell.

There is an increasing trend in upper-end, upscale residential wiring in the installation of advanced home automation systems. These systems are generally installed in larger, more expensive homes but are also suitable for the typical home on a limited basis.

This chapter is but a mere introduction to home automation wiring. It is beyond the scope of this book to cover the subject in detail because of the many manufacturers' products available today. Some of these products can be interchanged; others cannot. There are texts, videos, and websites devoted specifically to advanced home automation. Visit the manufacturers' websites and download specific information that fits your needs. Check the Cengage Learning catalog of available texts on this subject.

We hear about, read about, and might even attend seminars about home automation systems. The list of companies in manufacturing and installing home automation systems is endless! Some of the names associated with home automation systems, past and present, are ActiveHome, Avaya, BlueTooth, CEBus, Cisco, Decora Home Controls (Leviton), Eaton, Echelon, Elan, Enerlogic, Greyfox, Honeywell, IBM, IES Technologies, Insteon, Intermatic, Logitech, LightTouch, LonWorks, Lutron, Motorola Home Sight, On-Q/Legrand, Perceptive Automation, Radio Shack's "Plug 'N Power," PowerMark, SmartHome, SmartHouse, SmartLinc, Lightminder, Wi-Fi, X10 Powerhouse, ZigBee, and Z-Wave. Manufacturers develop and design their devices around one of these technologies. In most cases, one does not mix different manufacturers' products.

Home automation components are available at electronic retail stores and home centers as well as online from many sources.

Everyone is familiar with individual control devices. Examples are wall switches, thermostats, motion detectors, occupancy sensors, timers, TV remote controllers, and similar devices. They control one "thing."

Home automation provides the ability to control lighting (on, off, and dimming), heating, cooling, ventilation, appliances, home theater, audio/video entertainment, phone systems, irrigation control, and security in a home. Control of single or multiple loads from any number of locations is possible. Modules are available that can be controlled through a personal computer. Quite a few components merely plug in (plug-and-play). Others need to be wired:

- Those devices using infrared function when the receiver is in sight of the sender.
- Those devices using radio frequency (RF) function through walls.

Today, there are different types of advanced home automation systems for homes.

Many new devices have been produced that allow control of your home automation system or other activities through your smart phone, tablet, or computer. These include

- checking your home status at any time, including lighting, temperature, and security.
- allowing access to anyone at any time.
- managing lights when you're on the road.
- remotely changing temperature settings of thermostats.
- monitoring video cameras inside or outside.
- getting a text alert when anyone enters the house or a lock is accessed.
- receiving an alert when a new recorded video is available.
- observing live streaming videos.
- and other new features that are being added almost daily!

THE X10 SYSTEM

The X10 home automation system has been around the longest. X10 technology was invented in Scotland in 1975. The original X10 patent expired in 1997, so it is now an open standard for any manufacturer to apply to their products.

X10 technology is simple and is the least costly home automation system, especially for retrofit wiring in existing homes. For the most part, special wiring is not required for basic functions. For many applications, all that is necessary is to plug in the various components as needed.

This system transmits carrier wave (120 kilohertz [kHz]) signals that are superimposed on the regular 120/240-volt, 60-cycle branch-circuit wiring in the home. Components communicate with each other over the existing electrical wiring without a central controller.

Transmitting devices take a small amount of power from the 120-volt power line, modulate or change it to a higher frequency, and then superimpose this high-frequency signal back onto the ac circuit. At the other end, a receiving device responds to the high-frequency burst. X10 is basically a one-way communications system. Some devices "talk" and others "listen."

The control of appliances, audio/video equipment, outdoor and indoor lighting, and receptacles, as well as the arming of a security system, is possible from just about any place in the home. You might call it a "plug-and-play" system. Components are easily added as needed.

An X10 system might include programmers; receiving modules (Figure 31-1); transmitters; keypad wall or tabletop controller (Figure 31-2); dimmers; single-pole, 3-way, 4-way, and double-pole switches; on/off/dim lamp modules; on/off appliance modules; receptacle modules; thermostats; timers; surge protection devices; burglar alarm devices; motion sensors (turn on lights, sound an alarm); wireless controllers; handheld remotes; photocells for light/darkness sensing; drapery controllers; telephone responders; tie-in with a personal computer; voice activation; and so on. A typical system of this type will have the capability of 256 easily adjusted different "addresses," more than adequate for

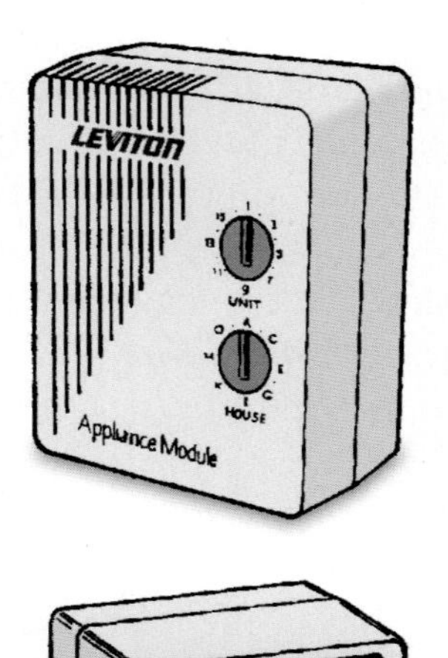

Courtesy of Leviton Manufacturing Company, Inc.

FIGURE 31-1 Various types of typical receiver modules used with X10 systems.

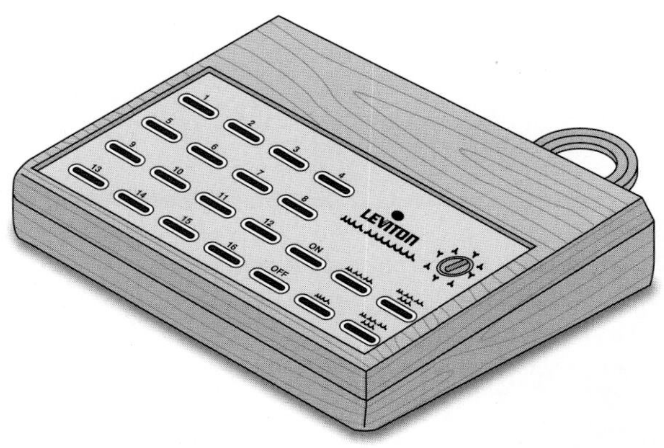

FIGURE 31-2 A typical X10 tabletop keypad controller makes it easy to transmit preprogrammed signals to receiver modules.

Courtesy of Leviton Manufacturing Company, Inc.

just about any size home. Receivers and switches generally mount in standard wall boxes. X10 wall switches and receptacles can replace conventional switches and receptacles. Make sure that the wall box has ample space for the larger X10 devices, so the wires in the wall box will not be jammed.

To ensure trouble-free operation, a phase-coupler (Figure 31-3) connected at the main panel is used to provide proper strength command signals to both sides of the ac wiring.

Limitations of X10 Technology

X10 modules are susceptible to interference from an ac power line. Many things can generate "noise" on 120-volt power lines, including fluorescent luminaire ballasts, appliances, and wireless intercoms. And because standard home wiring (most often Type NM cable) is unshielded, it can pick up unintended induced signals in the same way that an antenna does, even from devices that are not attached to the same circuit.

In fact, where two or more houses are connected to the secondary of the same utility transformer, signals could be transmitted from one house to another. X10 filters are available that install at the service panel to block interference from outside the house.

X10 signals are relatively slow. The maximum data rate is about 60 bits per second (bps) as compared to systems that handle 10,000 bps. This means there can be a noticeable delay between the time you activate a control and the time a controlled action takes place.

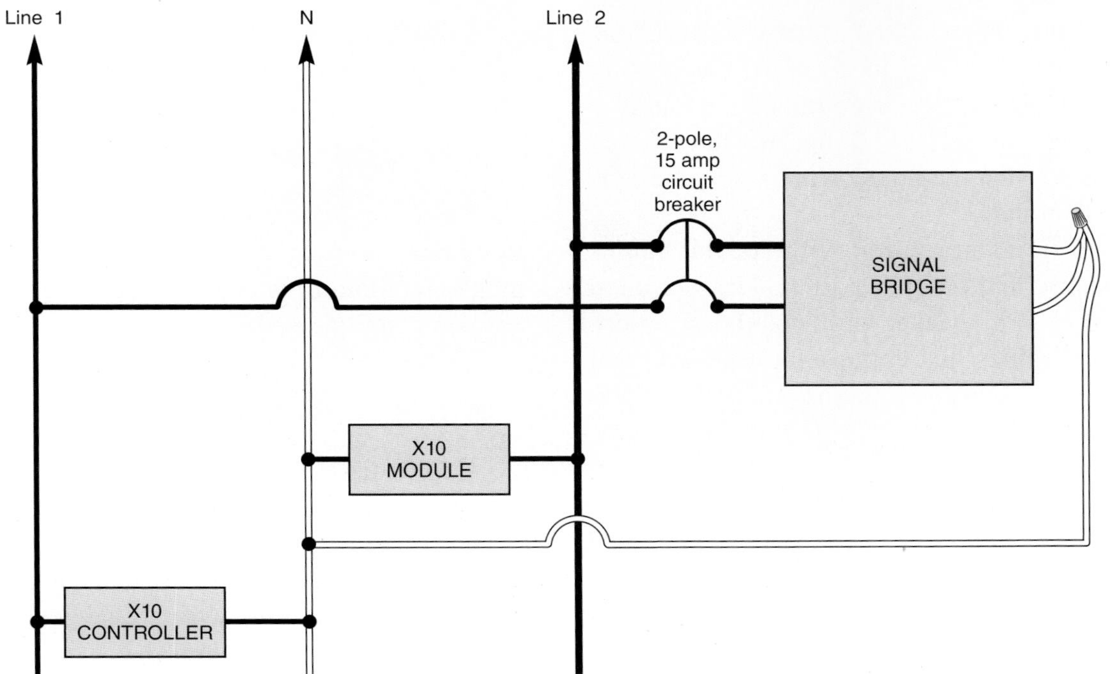

FIGURE 31-3 A signal bridge phase coupler connected to the main panel to ensure proper signals on both sides of an ac line.

INSTEON

Insteon systems use both wireless and hard-wired technology. The system is compatible with X10 devices. As with most typical home automation systems, the Insteon system incorporates remote control (on/off, dimming) lighting, security sensors and alarms, heating, cooling, humidity, door locking, and control of appliances, as well as audio and video control.

Insteon devices have the ability to send and receive X10 commands. Thus, they are said to be compatible with X10 devices. If X10 devices are already in use in a home, it will not be necessary to discard the X10 devices if newer Insteon devices are used. Insteon's power line frequency is 131.65 kilohertz (kHz). Its wireless radio frequency is 902 to 924 megahertz (MHz). Insteon systems work through walls.

STRUCTURED RESIDENTIAL WIRING SYSTEMS

Another type of system is referred to as "structured wiring" or "infrastructure wiring" and is completely separate from the conventional branch-circuit wiring in the home. It is nicely suited for new construction. Some electricians refer to this concept as "future proofing," which means wiring the new home during the original construction (rough-in) stage for all the applications that the homeowner may want now or at some later date.

These systems are wired with special cables such as unshielded twisted pair (UTP), shielded twisted pair (STP), speaker wire, coaxial, fiber optic (FO), and cables that contain multiple types of conductors, as shown in Figure 31-4. The UL Standard refers to FO cables as *optical fiber cables*. A single pair of FO cables can handle the same amount of voice traffic as 1400 pairs of copper conductors.

Structured premises wiring systems, sometimes called "home LANs" (for local area networks), are intended to allow distribution of high-speed data, audio, and video signals throughout the home. They offer such advantages as faster Internet access for home computers and convenient networking of data devices (computers, printers, and fax machines) in home offices.

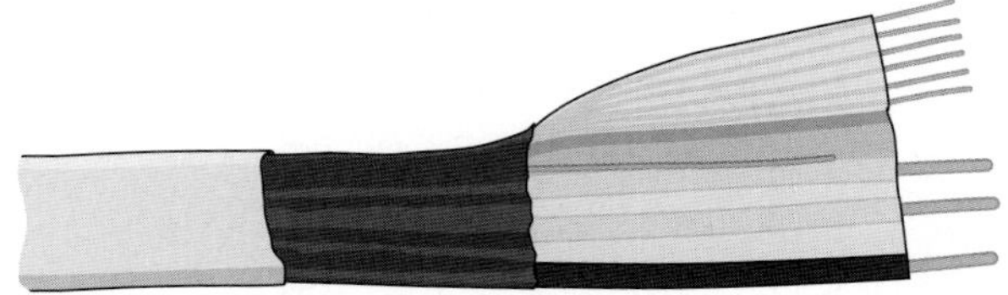

FIGURE 31-4 A typical cable consisting of ac power, telephone, coaxial cable, and low-voltage control conductors.

Structured premises wiring systems get their names from ANSI/TIA/EIA standards 568, 568A, 568B, 570, 570A, 570B, and other standards that describe "structured," or standardized, cabling systems for businesses and homes (see box on Home Wiring Standards).

Standard 570B primarily addresses residential installations.

Although different systems may not be interchangeable, they all provide the same wiring infrastructure and similar product interfaces such as cables, FO cables, modules, receptacles, switches, outlets, connectors, jacks, keypads, and so forth.

Here are a few of the common devices used in a "structured" wiring installation.

Combination Outlets

Telecommunications, coaxial, and sometimes FO connectors can be combined in the same wall plates, with multiple cables to each outlet location, as shown in Figure 31-5.

Central Distribution

Cables from room outlets are run to a central cabinet, sometimes referred to as a "service-center." Except for outlets, most system components are located at this cabinet, which also functions as a patch panel for making most wiring changes at a central location, as shown in Figure 31-6.

You will need a 120-volt receptacle near the location where the central distribution panel is located.

Courtesy of Leviton Manufacturing, Inc.

FIGURE 31-5 A single-gang combination wall outlet with two telecommunications outlets and two coaxial cable outlets. Many other combinations are available.

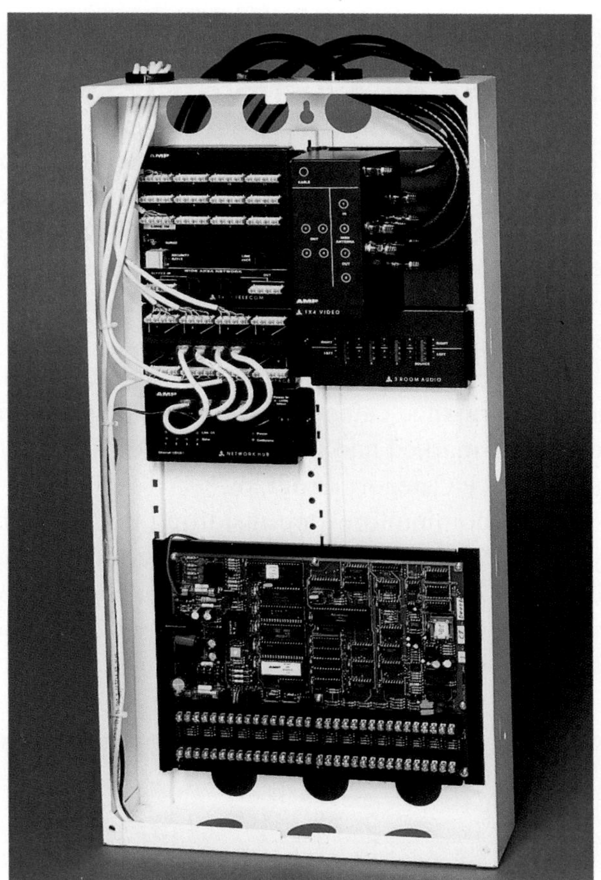

Courtesy of Leviton Manufacturing, Inc.

FIGURE 31-6 A "service-center" is the hub of a structured wiring system with telecommunications, audio, video, and home automation controllers installed.

STANDARDS

Home Wiring Standards

The EIA/TIA 570-A *Telecommunications Cabling Standard* is available from Global Engineering, 15 Inverness Way East, Englewood, CO 80112-5776. This standard was jointly developed by the Electronic Industries Association (EIA) and the Telecommunications Industry Association (TIA). It describes a wiring system that uses twisted 4-pair telecommunications cabling, modular jack outlets, and star wiring (home runs) to each room from the distribution panel.

The NEMA Standard WC 63.1-2005 entitled *Performance Standard for Twisted Pair Premise Voice and Data Communications Cable* contains the minimum electrical performance as well as material and mechanical specifications, definitions, and test methods for these types of cables. Check out www.nema.org.

CABLE TYPES AND INSTALLATION RECOMMENDATIONS

Communication cable is tested and listed in accordance with UL Standard 444. Optical fiber cable is tested and listed in accordance to UL Standard 1651. These standards are mirror images of the EIA and TIA standards. Cables that have passed the UL testing will bear a "listing" marking.

If the cable has also passed the data transmission performance category program, the cable will be surface-marked and tag-marked "Verified to UL Performance Category Program."

Some communities are installing underground hybrid FO/coaxial cable systems that combine telephone, CATV, and community intranet capabilities. In other areas, the telephone utility provides both voice and Internet services through its distribution system. In more remote locations, Internet services may be available only from satellite signals.

The features or functions of category-rated cables are shown in Table 31-1.

Proper selection and installation of any type of audio/video/data cable is extremely important. The final performance will be that of the weakest link in the system. Improper installation, connection hardware, outlets, and other components can reduce the performance to that of an old-fashioned phone system.

Although you may not experience problems with audio transmission, using the proper cable for data transmission is critical. Losing your online Internet connection for no apparent reason could very well be pinpointed to older style untwisted-pair connecting cables that are unable to handle high-speed data transmission. Using high-quality cables is extremely important.

TABLE 31-1

Types of category-rated cables.

Category 1**	Four-pair nontwisted cable. Referred to as *quad wire*. Not suitable for modern audio, data, and imaging applications. This is the old-fashioned Plain Old Telephone Service (POTS) cable. If you have any, get rid of it!
Category 2**	Usually 4-pair with slight twist to each pair. Not suitable for modern audio, data, and imaging applications. If you have any, get rid of it!
Category 3	UTP.* Data networks up to 16 MHz. Still available, but better choices are Categories 5e, 6, 6A, and 7.
Category 4*	UTP.* Data networks up to 20 MHz. Not suitable for modern audio, data, and imaging applications. If you have any, get rid of it!
Category 5*	UTP.* Data networks up to 100 MHz. Four pairs with 24 AWG copper conductors, solid (for structured wiring) and stranded (for patch cords). Cable has an outer PVC jacket. Was the most popular prior to advent of Categories 5e, 6, and 7.
Category 5e	UTP.* Basically an enhanced Cat 5 manufactured to tighter tolerances for higher performance. Data networks up to 350 MHz. Audio/video, phones, communication. This is detailed in the new TIA/EIA 568A-5 Standard. Preferred over Cat 5 for new installations.
Category 6	UTP.* Four twisted pairs. Data networks up to 250 MHz. Audio/video, phones, communication. Has a spline between the four pairs to minimize interference between the pairs.
Category 6A	Augmented cable. Unshielded. Excellent for data center environment.
Category 7	Best choice where high performance is required. Four twisted pairs . . . each pair fully foil shielded. Entire cable has added layer of insulation. Has an outer polyurethane jacket. Eliminates crosstalk. Improves resistance to electromagnetic interference (noise). Good choice for long runs. Estimated 15-year life cycle compared to 10 years for other category cables. Data networks up to 800 MHz. Excellent for commercial installations.
Category 7a	Enhanced Category 7. Excellent choice for long runs. Data networks up to 1000 MHz. Excellent for commercial installations.
Category 8	Newly adopted in 2013 by ISO/IEC. Has Class II performance. Excellent for commercial installations.

*UTP means unshielded twisted pair.

**Category 1, 2, 4, and 5 no longer supported by EIA/TIA. These cables are no longer manufactured and are listed for reference only.

Courtesy of Honeywell International, Inc.

FIGURE 31-7 Typical bundled cables showing both coaxial and Category 5E cables.

For residential applications, a minimum of one Category 5e, 6, or 6e unshielded twisted pair (UTP) cable with four twisted pairs of 24 AWG copper conductors or two Category 5e UTP cables with two twisted pairs of 24 AWG copper conductors should be installed for the home runs between each outlet location and a central distribution panel where all of the audio/video/data cables are run. Also install one or two RG-6 quad-shielded coax cables to each outlet location. For overall superior performance, install Category 7 or Category 7a cables. Figure 31-7 illustrates a typical bundled cable with two coaxial cables, two Category 5E cables, and a Category 7 cable.

There are still widespread problems with existing telephone lines outside of the home causing havoc with online Internet connections. There is not much you can do about this, other than complain to the telephone company.

Installation Recommendations

Because it is so easy to damage the cable, thus affecting its transmission properties, here are some cautions for installing any audio/video/data cables:

- Keep the runs as short as possible.
- Run the cables in one continuous length from Point A to Point B without splices.
- Do not to make sharp bends in the cable. Make slow sweeping bends: radius 4D for Category 5 cable, radius 10D for coaxial cable.
- Do not pull so hard as to damage the conductor insulation and the cable jacket (25 lb is the maximum it can stand).
- Do not kink or knot the cable.
- Use rounded or depth-stop plastic staples.
- Where a number of cables are bundled, use tie-wraps loosely around the bundle; then nail the tie-wrap to the framing member.
- Do not use nail guns or metal staples. This can only lead to trouble. Use cable-tie guns.
- Always untwist the least amount, about ½ in. (13 mm), of the conductors at point of termination. The tighter the twist in a cable, the less distortion and interference.
- Do not run these cables through the same holes as power wiring, conduit, pipes, and ducts. There is too much possibility of damage to the cables.
- Keep cables away from hot water pipes, hot air ducts, and other sources of high heat.
- Keep cables at least 12 in. (300 mm) from power wiring where the cables are run parallel to the power wiring. This caution was written before twisted pairs and category-rated cables entered the scene. With the advent of properly installed twisted and shielded Category 5e and 6 cables, this recommendation from the past is probably not an issue. Category 5e and 6 cables are highly immune to interference.

- Run a dedicated 15-ampere, 120-volt branch circuit to the central distribution panel that will be hub of the audio/video/data system.
- Always check the specifications and recommendations of the cable you are installing.
- Consider installing two or three trade size 2 PVC empty conduits from the basement to the attic. Additional cables can be installed at a later date without time-consuming fishing of cables or damaging walls and ceilings.

Many types of bundled cables are available that contain a multitude of cables. Examples are one RG-6 (75-ohm) coaxial cable and one Cat 5e cable; two RG-6 (75-ohm) coaxial cables, two Cat 5e cables, and one ground; two FO cables, two RG-6 (75-ohm) coaxial cables, two Cat 5e cables, and one ground; or two RG-6 (75-ohm) coaxial cables, and one Cat 5e cable. The combinations are endless, and you need to check out manufacturers' catalogs to find the cable that meets your needs.

TERMINOLOGY

The letters RG stand for *Radio Grade*.

The jacks (connectors) for Cat 5 cables are generally the RJ-11 6-pin jack and the RJ-45 8-pin jack. The letters RJ stand for *Registered Jack*.

Connectors for coaxial cable are referred to as *F* connectors. The threaded type is preferred over the push-on type.

The letter M means *mega*, which is 1 million (for short: *meg*).

The letter G means *giga*, which is 1 billion (for short: *gig*).

The letters bps mean *bits per second*.

The letters Hz mean *hertz*, which are cycles per second.

The letters MHz mean *megahertz*, which are millions of cycles per second.

A *bit* is the smallest unit of measurement in the binary system. The expression is derived from the term *binary digit*.

It takes 8 bits to make 1 byte. A *byte* is the amount of data required to describe a single character of text.

The letters Mbps mean *megabits per second*. For example, 20 Mbps means that 20 million (20,000,000) bits of data are being transmitted per second over copper or FO cable.

There is no direct correlation between *megahertz* and *megabits*. Different encoding schemes represent information in different ways.

WIRELESS

Structured wiring involves a separate system of conductors and cables that run between controllers, sensors, and all of the other equipment. X10 technology uses the existing wiring in your home. The devices are electronically coupled to the premises wiring in your house. Both systems use wireless remote control devices using infrared technology. Infrared (IR) technology is the heart of most wireless systems, although some operate on RF. RF stands for radio frequency. The main drawback of IR is that you must have a straight unobstructed line of sight between the remote and the equipment being controlled. With RF devices, you do not need a straight, unobstructed line of sight. RF sends out a radio signal that goes around corners and through walls and doors. Both systems can control TVs, radios, sound systems, VCR players, CD players, DVD players, lighting, ceiling paddle fans, garage door openers, small relays that plug into a wall receptacle, remote car locking systems, and so on. Sound familiar? You may not have thought about it, but you are already in the world of home automation.

ZIGBEE

ZigBee is a wireless communication system that uses low-power radio signals. The system is aimed at the wireless personal area networks (WPANs), typical for home automation systems. Over 200 manufacturers around the world have formed an alliance to design devices and equipment that operate on the system specifications found in IEEE Standard 802.15.4. The system does not require hard-wiring.

The ZigBee system uses three different types of devices:

- Coordinator
- Router
- End Device

In the United States, the ZigBee system primarily operates on a 2.4 gigahertz (GHz) band. It also transmits on 915 megahertz (MHz).

Z-WAVE

Z-Wave is a low-power wireless communication system. More than 100 manufacturers have formed an alliance and are designing home automation devices based on the Z-Wave standard. These devices work fine within a range of 100 ft (30 m). The devices are readily available at electronic and home center retail centers. Many of the Z-Wave devices are "plug-in" of the type illustrated in Figure 31-1. Other control devices allow the control of lighting.

In the United States, the Z-Wave system operates on 868.42 megahertz (MHz).

Z-Wave devices may be compatible with X10 devices. Some manufacturers have produced bridges to allow the X10 devices to operate on Z-Wave technology. The manufacturers' instruction manuals provide the necessary information relative to compatibility with other systems.

SUMMARY

For residential wiring, life safety is key. Knowledge of the *NEC*® requirements is of utmost importance. It takes training in that field. It is not a job for the amateur. A novice should not get involved in residential wiring.

Life and property are at stake. You will want to study *Articles 800*, *810*, *820*, *830*, and *840* of the *NEC*.

When it comes to communication systems, a choice that must be made is . . . should it be *wireless* or *wired*? The choice is yours! Most installations will involve both.

After reading this chapter, it should be obvious that other than a very basic installation, communication systems become very complex. Installing communications systems is a highly skilled profession and should be left up to the trained professional.

Home automation systems and home communication systems vary. To learn more, visit the websites of the many manufacturers of communication equipment.

Companies such as AT&T and Comcast generally provide the basics such as television, Internet, and security. When homeowners wish to have whole-home audio, video, complex control, leave the installation up to the professional.

Trade schools have classes on the subject of home automation and communication systems.

There are a number of manufacturers of data communication cables. They have a wealth of technical information on their websites. To name a few: *Honeywell*, *Belden*, *General Cable*, *Superior Essex*, and *Comtran*. For those of you wanting more information, go online and search for data cable and data communication cable. You will be amazed at how much technical information is available out there.

REVIEW

Note: Refer to the *Code* or the plans where necessary.

1. One type of home automation system is referred to as
   a. X10.
   b. X20.
   c. HAS.
2. When using an X10 system, which of the following statements is most correct?
   a. A totally independent wiring system must be installed.
   b. The devices generally are "plug-and-play" (PnP), requiring no additional wiring.
   c. An X10 system is ideally suited for installations in new construction.

3. A home automation system that requires a totally separate wiring system is referred to as
   a. an X10 system.
   b. a structured system.
   c. a DC-powered system.
4. The most recommended "category-rated" cable for wiring from various outlets back to the main "service-center" for the interconnection of computers is
   a. Category 1.
   b. Category 2.
   c. Category 3.
   d. Category 5 or 5e.
5. When installing Category 5–rated cable, the minimum bending radius generally recommended is
   a. 4 times the diameter of the cable.
   b. 6 times the diameter of the cable.
   c. 10 times the diameter of the cable.
6. When installing coaxial cable, the minimum bending radius generally recommended is
   a. 4 times the diameter of the cable.
   b. 6 times the diameter of the cable.
   c. 10 times the diameter of the cable.
7. When selecting cables for a structured wiring system or computer hookups, the best choice is to select a cable that has
   a. untwisted pairs of conductors.
   b. twisted pairs of conductors.
8. Match each of the following terms with its description by entering the number of the correct definition in the blank space provided.

a. RG	______	1. cycles per second
b. RJ	______	2. the smallest unit of measurement in the binary system
c. bps	______	3. registered jack
d. mbps	______	4. fiber optic
e. Hz	______	5. bits per second
f. bit	______	6. shielded twisted pair
g. UTP	______	7. radio grade
h. FO	______	8. megabits per second
i. STP	______	9. unshielded twisted pair

CHAPTER **32**

Standby Power Systems

OBJECTIVES

After studying this chapter, the student should be able to

- understand some of the safety issues concerning optional standby power systems.
- understand the basics of standby power.
- understand the types of standby power systems.
- understand wiring diagrams for portable and standby power systems.
- understand transfer switches, disconnecting means, and sizing recommendations.
- understand the *National Electric Code* requirements for standby power systems.

INTRODUCTION

The text that follows is a general discussion and overview of home-type temporary generator systems. It is not possible in this chapter to cover all the variables and details involved with the many types of standby power systems available in the marketplace today.

SAFETY ALERT

Exhaust gases contain deadly carbon monoxide, the same as an automobile engine. One 5.5-kilowatt generator produces as much carbon monoxide as six idling automobiles. Carbon monoxide can cause severe nausea, fainting, or death. It is particularly dangerous because it is an odorless, colorless, tasteless, nonirritating gas. Typical symptoms of carbon monoxide poisoning are dizziness, headache, light-headedness, vomiting, stomachache, blurred vision, and inability to speak clearly. If you suspect carbon monoxide poisoning, remain active. Do not sit down, lie down, or fall asleep. Breathe fresh air—fast!

Precautions to Follow When Operating a Generator

- Always install equipment such as generators, power inlets, transfer switches, and panelboards that are listed by a Nationally Recognized Testing Laboratory (NRTL).
- Always carefully read, understand, and follow the manufacturer's installation instructions.
- Always turn off the power when hooking up standby systems. Do not work on "live" equipment.
- Do not stand in water or otherwise work on electrical equipment with wet hands.
- Do not touch bare wires.
- Store gasoline only in approved red containers clearly marked "GASOLINE."
- Do not operate indoors.
- Do not operate in the garage.
- Operate a portable generator only outside in open air, not close to windows, doors, or other vents. Exhaust fumes from the generator can infiltrate the house or a neighbor's house.
- Death is just around the corner when the above warnings are not heeded!

WHY STANDBY (TEMPORARY) POWER?

Everyone has experienced a power outage, sometimes more often than one would like. You are left in the dark. Your furnace does not operate. You are concerned about water pipes freezing in the winter. Your refrigerator and freezer are not running. Your sump pump is not operating, and your basement is flooding. Your overhead garage door does not operate. You want the peace of mind that the essential equipment and appliances will continue to function while you are on an extended trip. Or, possibly worst of all, you may miss the most important sports game of the year on TV!

A few questions have to be answered. In your home, which loads are critical and must continue to operate in the event of a utility power outage? You need to know this so you can select a generator panelboard with adequate space for branch circuits. Only you can make this decision. Another question that must be answered is, how large must the generator be to serve the loads that are considered critical in your home? Still another question is, how simple or complicated a standby system do you want?

WHAT TYPES OF STANDBY POWER SYSTEMS ARE AVAILABLE?

The Simplest

Home centers usually carry the simplest and most economical portable generators, as shown in Figure 32-1. These consist of a gasoline-driven

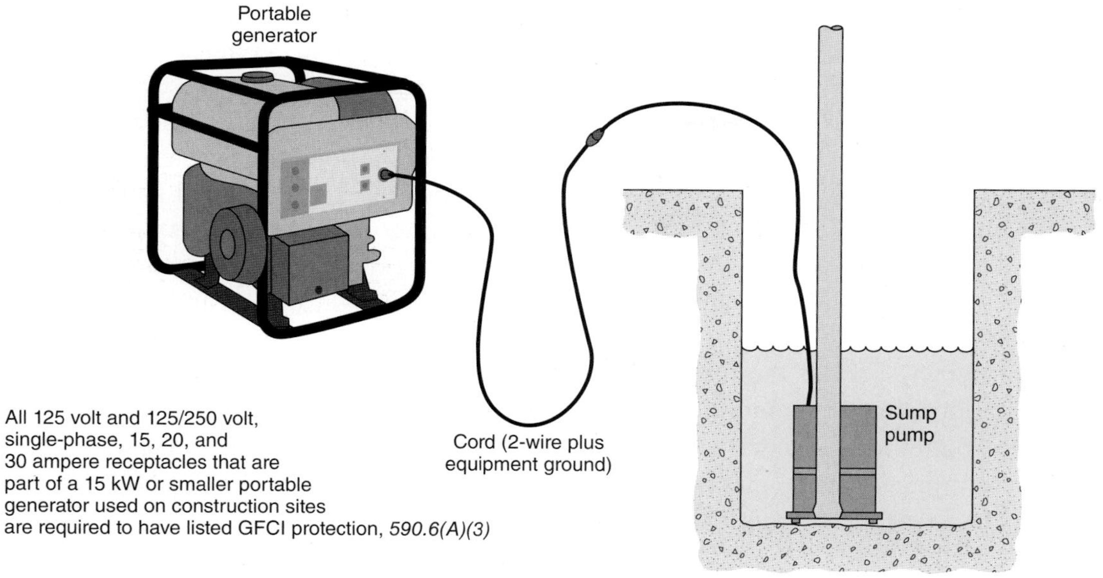

FIGURE 32-1 A portable generator serving standby power to a 120-volt sump pump.

motor/generator set that must be started manually. Some models have the recoil "pull-to-start" feature; others have battery electric start. These are the types construction workers use for temporary power on job sites. Depending on the size of the fuel tank and the load being supplied, these smaller generators might be capable of running for 4 to 8 hours. This type of portable generator is commonly referred to as "backup" power.

These small portable generators consist of a gasoline engine driving a small electrical generator, just like a gasoline engine drives the blades of a lawn mower or snow thrower. Some generator sets have one or more standard 15- or 20-ampere single or duplex grounding-type receptacles, some have GFCI receptacles, and others have 20- and 30-ampere twistlock-type receptacles. Some are rated 125 volts, whereas others are rated 125/250 volts.

These generators are available in sizes up to about 7000 watts. Also available are models with more "bells and whistles" such as oil alerts, adjustable output voltage, larger fuel tanks, and so on.

The procedure for this type is to start up the generator, then plug an extension cord into the receptacle and run it to whatever critical cord-and-plug-connected load needs to operate.

If you are not home when a power outage occurs, you are out of luck!

Receptacles on Portable Generator Sets

Because of the extreme hazards always present for workers on construction sites, particularly when using portable generators for temporary power, *NEC® 590.6(A)(3)* requires that All 125 volt and 125/250 volt, single-phase, 15-, 20-, and 30 ampere receptacle outlets that are a part of a 15 kW or smaller portable generator shall have listed ground-fault circuit interrupter protection for personnel.* The requirements of *Article 590* apply to *Temporary Wiring* and clearly cover construction sites.

SAFETY ALERT

Electricity from generators can also cause electric shock when used at dwellings. Purchase a generator that has GFCI protection. For existing generators without GFCI protection, use a listed portable cord that has GFCI protection.

*Reprinted with permission from NFPA 70-2014.

The Next Step Up

The next step up involves both permanent wiring and cord-and-plug-connected wiring.

This is by far the most popular when it comes to standby power for homes. In the event of a power outage, you must manually start the generator, flip the transfer switching device over from normal power to standby power, and plug in the power "patch" cord (Figure 32-2).

The permanent wiring part of the installation involves connecting specific branch circuits to a separate generator panelboard (usually located right next to the main panelboard), installing and properly connecting a transfer switch (discussed later), and installing a power inlet receptacle (discussed later). The generator panelboard serves those circuits that you have selected as critical to operate in the event of a power outage. This panelboard is sometimes referred to as a generator panelboard, a selected load panelboard, an emergency panelboard (these systems in dwellings almost never meet the definition of an emergency system in *NEC 700.2*), and sometimes a critical load panelboard (Figure 32-3).

The cord-and-plug-connected part of the installation consists of merely a flexible power patch cord that runs between the generator and the power inlet receptacle. This cord and the generator are set up as needed.

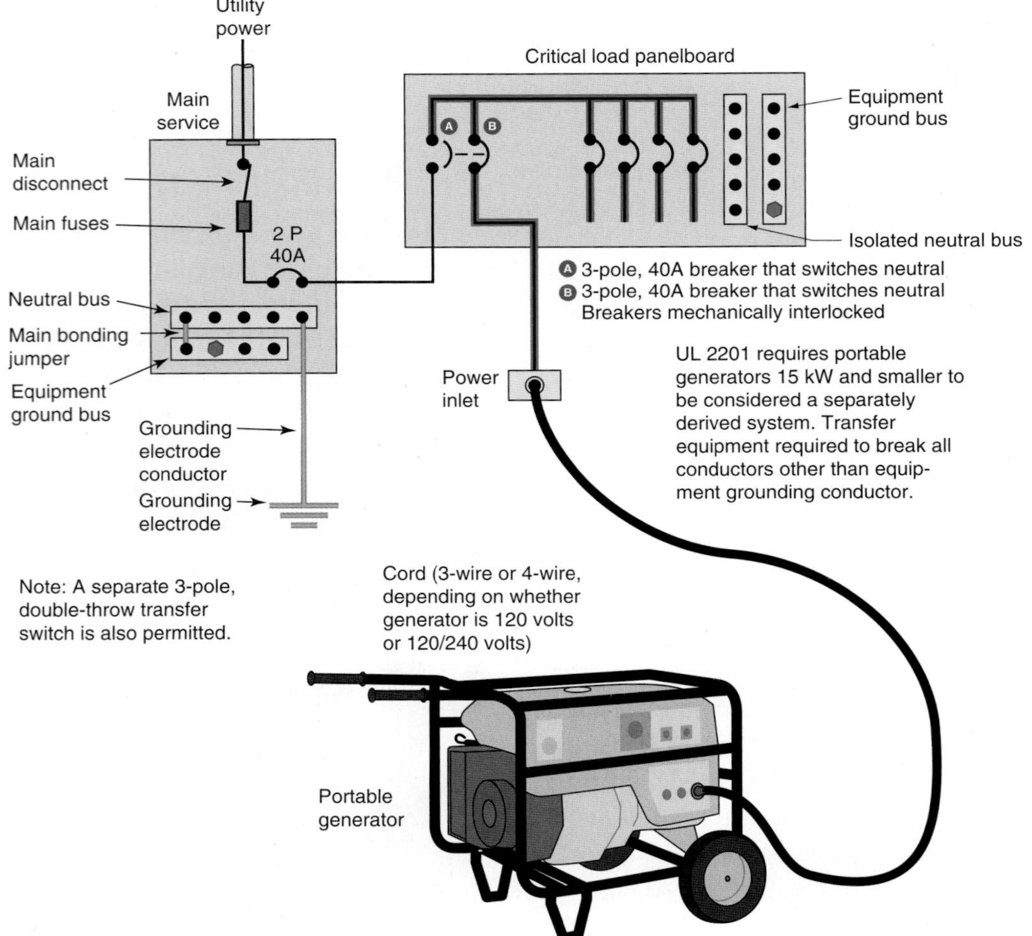

FIGURE 32-2 A portable generator supplies standby power to a critical load panelboard. The circuit breakers in the critical load panelboard must be manually turned to the temporary power position. These breakers are equipped with a mechanical interlock so there will be no feedback of electricity between the generator power and the utility normal power.

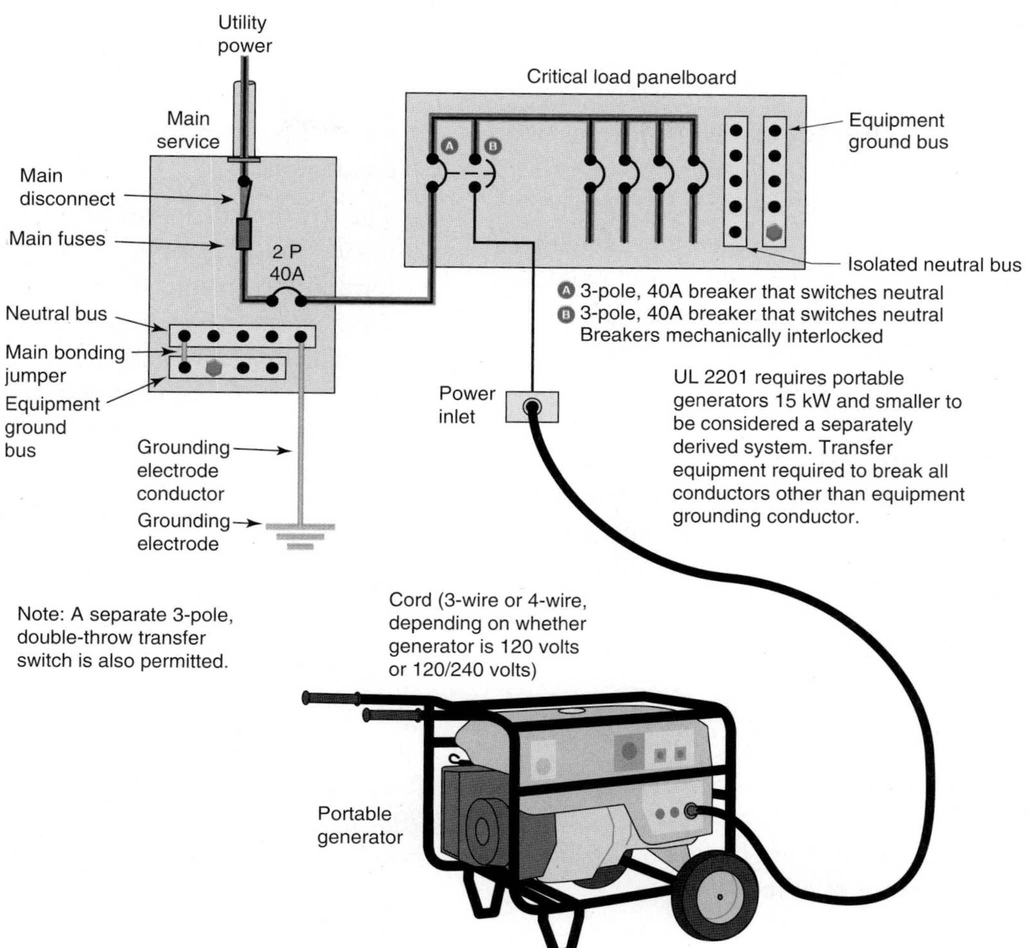

FIGURE 32-3 A critical load panelboard served by normal power from the main service panelboard. A portable generator is shown plugged into the power inlet. To switch to standby power, the circuit breakers in the critical load panelboard must be manually turned to the standby position. These breakers are equipped with a mechanical interlock so there will be no feedback of electricity between the generator power and the utility normal power.

These are gasoline-powered generators that use a flexible 4-wire rubber cord (12 AWG for 5000-watt generators, 10 AWG for 7500-watt generators) that plugs into a female polarized twistlock receptacle on the generator. The other end of the cord plugs into a polarized twistlock receptacle mounted on the outside of the house in a weatherproof power inlet box. This polarized twistlock receptacle is permanently connected to transfer equipment or to a special electrical generator panelboard in which the critical branch circuits are connected. The generator panelboard will have a transfer mechanism plus from 4 to 20 branch-circuit breakers; it all depends on the wattage rating of the generator. Some generator panelboards have the polarized twistlock receptacle mounted as part of and just below the generator panelboard.

During normal operation, this generator panelboard is fed from a 2-pole circuit breaker in the main panelboard. This 2-pole breaker might be rated 30, 40, 50, or 60 amperes, depending on the number of critical branch circuits to be supplied. When utility power is lost, this generator panelboard is fed from the generator. When the electric utility loses power, you must *manually* turn the "transfer switch" from its normal power position to its temporary power position. One manufacturer of electrical equipment furnishes a panelboard that contains two circuit breakers that are mechanically interlocked. These are required to be 3-pole breakers because the neutral from a portable generator must be switched. These breakers serve as the transfer switching means. Only one breaker can be in the ON position at the same time. One of these breakers brings the normal power to the generator panelboard. The second breaker brings the power from the generator to the generator panelboard. When one breaker is turned OFF, the other is turned ON. This panelboard will have four or more branch-circuit breakers.

With this system, the procedure generally is to first plug the 4-wire cord into the polarized twist-lock receptacle on the generator; plug the other end into the power inlet receptacle; and then start the generator, which is always located outdoors. The transfer switch in the generator panelboard is then switched to the GEN position.

When normal power is restored, turn the transfer switch in the generator panelboard back to its normal position; turn off the generator according to the manufacturer's instructions; unplug the power cord from the generator; and finally, unplug the other end of the cord from the power inlet receptacle.

UL Standard Requirements

The UL Safety Standard UL 2201 is titled, *Portable Engine-Generator Assemblies*. This standard covers portable generators that are rated up to 15 kW. The guide card information for the standard in the UL *White Book* in category "Engine Generators for Portable Use (FTCN)" reads,

> "This category covers internal-combustion-engine-driven generators rated 15 kW or less, 250 V or less, which are provided only with receptacle outlets for the ac output circuits. The generators may incorporate alternating- or direct-current generator sections for supplying energy to battery-charging circuits.
>
> When a portable generator is used to supply a building wiring system:
>
> 1. The generator is considered a separately derived system in accordance with ANSI/NFPA 70, *National Electrical Code (NEC)*.
> 2. The generator is intended to be connected through permanently installed Listed transfer equipment that switches all conductors other than the equipment grounding conductor.
> 3. The frame of a listed generator is connected to the equipment grounding conductor and the grounded (neutral) conductor of the generator. When properly connected to a premises or structure, the portable generator will be connected to the premises or structure grounding electrode for its ground reference.
> 4. Portable generators used other than to power building structures are intended to be connected to ground if required by the *NEC*."*

The safety standard adds to or supplements requirements in the *NEC*. Let's look at these rules one by one.

1. "The generator is considered a separately derived system in accordance with ANSI/NFPA 70, *National Electrical Code (NEC)*." The term *separately derived system* as defined in *NEC Article 100* means that the transfer equipment must break all system conductors from the generator, including the neutral. This rule is in place to prevent the equipment grounding conductor from being in parallel with the neutral and carrying neutral current. Normally, this rule would require the generator to be grounded at the source. The *NEC* has special rules for grounding portable generator electrical systems in *NEC 250.34*. See our discussion following list item 4.
2. "The generator is intended to be connected through permanently installed Listed transfer equipment that switches all conductors other than the equipment grounding conductor." The installation of transfer equipment that switches all conductors, including the grounded or neutral conductor, is the deciding factor in determining that the system is in fact a *separately*

derived system. Once again, separately derived systems are generally required to have the neutral grounded (connected to earth) at the source. This is not the case for portable generators. See our discussion following list item 4.

3. "The frame of a Listed generator is connected to the equipment-grounding conductor and the grounded (neutral) conductor of the generator. When properly connected to a premises or structure, the portable generator will be connected to the premises or structure grounding electrode for its ground reference." These connections ensure that a ground-fault return path is established back to the source, which is the generator winding. Be very cautious when selecting a generator. Some manufacturers produce generators with a floating neutral. As a result, it is impossible for an overcurrent device to operate on a ground fault because no circuit for fault current can be established. The connection of the neutral and equipment grounding conductor to the frame of the generator is required in *NEC 250.34(A)(2)* and *(C)*.
4. "Portable generators used other than to power building structures are intended to be connected to ground if required by the *NEC*." The *NEC* contains special requirements for grounding of portable generators in *250.34*. So long as the generator supplies loads only through receptacles mounted on the generator and the equipment grounding conductor is connected to the generator frame, the frame of the portable generator is not required to be connected to a grounding electrode at the generator such as one or more ground rods.

When connected to the service-supplied electrical system by cord-and-plug connection, the equipment grounding conductor serves as a ground or earth reference for the electrical system because the conductor is extended to the windings of the generator.

Important: If you intend to have a GFCI breaker-protected branch circuit transferred to the generator panelboard, be sure to connect this circuit to a GFCI breaker in the generator panelboard.

Because GFCIs and AFCIs require a neutral conductor, make sure the neutral conductor is properly connected between the main panelboard neutral bus and the neutral bus in the critical panelboard.

The neutral bus in the generator panelboard is insulated from the generator panelboard enclosure.

The ground bus in the generator panelboard is bonded to the generator panelboard enclosure.

Standby generators of this type can have run times as long as 16 hours. Again, it depends on the loading. The manufacturer's installation and operating instructions will provide this information.

If you are not home when the power outage occurs, you are out of luck!

The Top of The Line

Standby power served by a permanently installed generator is truly standby power. These types of generator systems are covered under UL Standard 2200, *Stationary Engine Generator Assemblies*.

Standby power is a system that allows you to safely provide electrical power to your home in the event of a power outage. Instead of running extension cords from a portable generator located outside of the house to whatever critical loads (lights, sump pumps, refrigerators, etc.), you select the branch-circuits you feel are critical or essential to satisfactorily operate your home electrical system. These branch circuits are placed in a panelboard that will be transferred to the generator in the event of a power outage from the electric utility. The diagrams in this chapter show this.

The generator for this type of system is permanently installed on a concrete pad at a suitable location outside of the home, convenient to making up all of the electrical connections to serve the selected critical loads. These systems generally run on natural gas (Figure 32-4) or propane (liquid petroleum gas—LPG), and in rare cases on diesel fuel. All the conductors between the generator, the transfer switch, and the generator panelboard installed for the selected critical loads are permanently wired. Note than the power conductors and control conductors are installed in separate raceways to satisfy the requirements in *NEC 725.136*.

The automatic transfer switch monitors the incoming electric utility's voltage. Should a power

Courtesy of Kohler Power Systems.

FIGURE 32-4 A natural gas-fueled generator.

outage occur, the standby generator automatically starts, the transfer switch "transfers," and the selected loads continue to function. When normal power is restored, the transfer switch automatically transfers back to the normal power source, and the generator shuts down.

Because the fuel supply is a permanently installed natural gas line, these systems can run indefinitely in conformance to the manufacturer's specifications.

To be totally automatic, an electronic controller that is part of the transfer switch immediately senses the loss of normal power and "tells" the standby generator to start, and to tell the transfer switch to transfer from normal power to standby power. This controller senses when normal power is restored, allowing the transfer switch to return to normal power and to shut off the standby generator. All of the circuitry is reset, and it is now ready for the next power interruption.

WIRING DIAGRAMS FOR A TYPICAL STANDBY GENERATOR

For simplicity, the diagrams in Figures 32-5 and 32-6 are one-line diagrams. Actual wiring consists of two ungrounded conductors and one grounded neutral conductor. Equipment grounding of all of the components is accomplished through the metal raceways that interconnect the components. See *250.118*.

The neutral bus in a panelboard that serves selected loads must *not* be connected to the metal panelboard enclosure. Connecting the neutral conductor, the grounding electrode conductor, and the equipment grounding conductors together is permitted only in the main service panelboard.

If you were to bond the neutral conductor and the metal enclosures of the equipment (panelboard, transfer switch, and generator) together beyond the main service panelboard, you would create a parallel path.

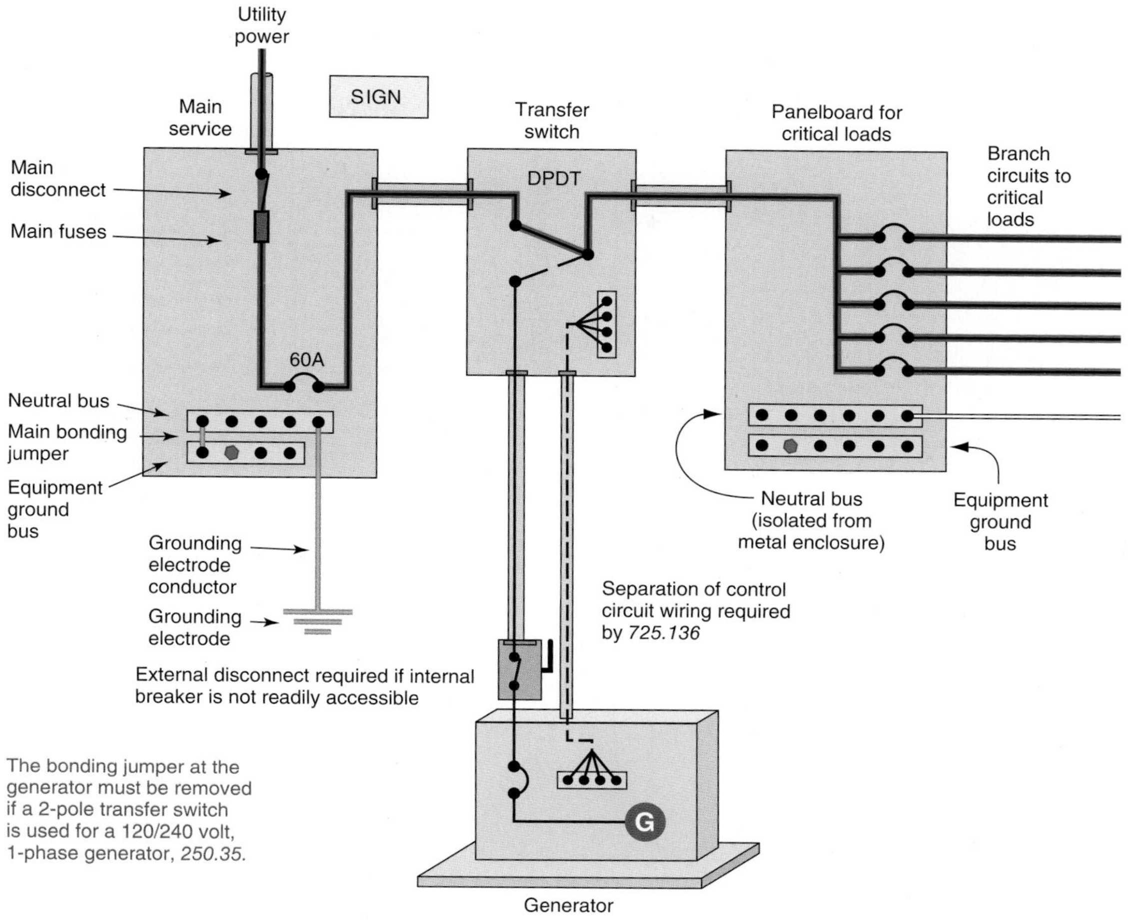

FIGURE 32-5 A natural gas-driven generator connected through a transfer switch. The transfer switch is in the normal power position. Power to the critical load panelboard is through the main service panelboard, then through the transfer switch.

A parallel path means that some of the normal return current and fault current will flow on the grounded neutral conductor, and some will flow on the metal raceways and other enclosures. This is not a good situation! See *NEC 250.6* and *250.24(A)(5)*.

Manufacturers of generators in most cases do not connect the generator neutral conductor lead to the metal frame of the generator. Instead, they connect it to an isolated terminal. Then it is up to you to determine how to connect the generator in compliance with the *NEC* and/or local electrical codes.

Most inspectors will permit the internal neutral bond in a portable generator to remain in place. In fact, when you purchase a portable generator, you should insist the neutral-to-case bond is in place. Without the bond, an overcurrent device cannot function on a ground fault because a return path does not exist. For permanently installed generators, inspectors will generally not permit a bond between the generator neutral and the metal frame of the generator because of the explicit requirements in the *NEC*.

A sign must be placed at the service-entrance main panelboard indicating where the standby generator is located and what type of standby power it is. See *702.8(A)*.

The total time for complete transfer to standby power is approximately 45–60 seconds.

To further give the homeowner assurance that the standby generator will operate when called upon, some systems provide automatic "exercising" of the system periodically, such as once every 7 or 14 days for a run time of 7 to 15 minutes.

FIGURE 32-6 A natural gas-driven generator connected through a transfer switch. The transfer switch is in the standby position, feeding power to the critical load panelboard.

SAFETY ALERT

When an automatic type of standby power system is in place and set in the automatic mode, the engine may crank and start at any time without warning. This would occur when the utility power supply is lost. To prevent possible injury, be alert, aware, and very careful when working on the standby generator equipment or on the transfer switch. Always turn the generator disconnect to the OFF position, then lock out and tag out the switch, warning others not to turn the switch back ON. In the main panelboard, locate the circuit breaker that supplies the transfer equipment, turn it OFF, then lock out and tag out the circuit breaker feeding the transfer switch.

TRANSFER SWITCHES OR EQUIPMENT

A transfer switch or equipment shifts electrical power from the normal utility power source to the standby power source. A transfer switch isolates the utility source from the standby source in such a way that there is no feedback from the generator to the utility's system or vice versa. When normal power is restored, the automatic transfer switch resets itself and is ready for the next power outage.

Transfer switches or equipment for typical residential applications might have ratings of 40 to 200 amperes. Some have wattmeters for balancing generator loads.

For home installations, transfer switches may be 3-pole, double-throw (TPDT) or double-pole, double-throw (DPDT), depending on how the type of system is being installed. As stated above for listed portable generators, a 3-pole, double-throw transfer switch or equipment is required.

For permanently installed generators, either a 2-pole or 3-pole transfer switch or equipment is permitted so long as the generator produced electrical system is coordinated with the transfer equipment.

Two-pole transfer switches or equipment do not break the "neutral" conductor of a 120/240-volt single-phase system. Technically speaking, the *NEC* refers to this type of system as a *nonseparately derived system*. A nonseparately derived system is properly grounded through the grounding electrode system of the normal premises wiring at the service equipment. A system where the neutral is disconnected by the transfer switch is referred to as a *separately derived system* and must be regrounded according to *250.30*. Separately derived systems are common in commercial and industrial installations and are required for connection of portable generators.

If a transfer switch is connected on the line side of the main service disconnect, it must be listed as being suitable for use as service equipment. In Figures 32-5 and 32-6, the transfer switch is not on the line side of the main service disconnect and therefore does not have to be listed for use as service equipment.

A transfer switch must "break-before-make." Without the "break-before-make" feature, a dangerous situation is present and could lead to destruction of the generator, personal injury, or death. If the transfer switch does not separate the utility line from the standby power line, utility workers working on the line could be seriously injured. For low-cost, nonautomatic transfer systems, the transferring means might consist of two 2-pole or 3-pole circuit breakers that are mechanically interlocked so both cannot be ON at the same time.

Capacity and Ratings of Transfer Equipment (*NEC 702.6*)

The rating of the transfer switch must be capable of safely handling the load to be served. Otherwise, the transfer switch could get so hot as to cause a fire. This is nothing new; it is the same hazard as overloading a conductor. Specific rules on the capacity of the generator are dependent on the type of the transfer switch used, as follows:

1. If manual transfer equipment is used, an optional standby system is required to have adequate capacity and rating for the supply of all equipment intended to be operated at one time. The user of the optional standby system is permitted to select the load that is connected to the system.
2. If automatic transfer equipment is used, an optional standby system must comply with parts (a) or (b).
   (a) Full Load. The standby source shall be capable of supplying the full load that is transferred by the automatic transfer equipment.
   (b) Load Management. If a system is employed that will automatically manage the connected load, the standby source must have a capacity sufficient to supply the maximum load that will be connected by the load management system.

A transfer switch must also be capable of safely interrupting the available fault current that the generator or utility is capable of delivering. Listed transfer switches or equipment may be provided with or without integral overcurrent protection. The suitability of listed transfer equipment for interrupting or withstanding short-circuit current is marked on the transfer equipment.

A typical transfer switch might take 10 to 15 seconds to start the generator. This eliminates nuisance start-ups during momentary utility power outages. After start-up, another 10- to 15-second delay is provided to allow the voltage to stabilize. After start-up and warm-up, full transfer takes place. Similar time-delay features are used for the return to normal power. Some systems can do the full transfer in a few seconds.

The UL Product Standard is 1008, *Transfer Switch Equipment*.

See the book *Electrical Grounding and Bonding* published by Cengage Learning for additional information on grounding and bonding of separately derived and nonseparately derived alternate power systems supplied by generators.

DISCONNECTING MEANS

A disconnecting means is required for a generator. *NEC 445.18* states that Generators shall be equipped with a disconnect(s), lockable in the open position in accordance with 110.25 by means of which the generator and all protective devices and control apparatus are able to be disconnected entirely from the circuits supplied by the generator except where both of the following conditions apply:

1. The driving means for the generator can be readily shut down
2. The generator is not arranged to operate in parallel with another generator or other source of voltage.*

NEC 225.34 requires that all disconnecting means shall be grouped.

If an outdoor housed generator set is equipped with a readily accessible disconnecting means that is located within sight of the building or structure supplied, an additional disconnecting means is not required where ungrounded ("hot") conductors serve or pass through the building or structure, *NEC 702.12*. Most electrical inspectors will allow the cord-and-plug connection for a portable generator to serve as the disconnecting means.

Where the normal power main service disconnecting means is located inside of the home, some electrical inspectors will require that a disconnect for the standby power from the generator be installed near the main service equipment.

Having one disconnecting means inside the home and another outside of the home presents quite a challenge for firefighters or others wanting to totally shut off the power to the home. This is why *NEC 702.9(A)* requires that a sign be placed at the service-entrance equipment that indicates the type and location of on-site optional standby power sources.

*Reprinted with permission from NFPA 70-2014.

GROUNDING

Grounding the metal frame of a portable generator is through the equipment grounding conductor in the power cord. Grounding of a hard-wired generator is accomplished by means of a metallic wiring method or other means acceptable by *250.118*.

It is not necessary to connect the frame of a portable generator to a grounding electrode, *250.34(A)*.

CONDUCTOR SIZE FROM STANDBY GENERATOR

The conductors that run from the standby generator to the transfer switch and to the generator panelboard that contains the branch-circuit overcurrent devices shall be sized not less than 115 percent of the generator's nameplate current rating. See *NEC 445.13*. This applies if overcurrent protection of the conductors is not provided at the generator source. If the generator is equipped with an overcurrent device, the standard rules for sizing conductors in *NEC 240.4* apply.

EXAMPLE

A 7500-watt, single-phase, 120/240-volt generator.

Step 1: Calculate the generator's current rating.

$$\text{Amperes} = \frac{\text{Watts}}{\text{Volts}} = \frac{7500}{240} = 31.25 \text{ Amperes}$$

Step 2: 31.25 × 1.15 = 35.94 amperes.

Step 3: Refer to *NEC Table 310.15(B)(16)* to determine the minimum size conductors.

In the 60°C column of *Table 310.15(B)(16)*, we find that an 8 AWG copper conductor has an allowable ampacity of 40 amperes. The discussion on when to use the 60°C and when to use the 75°C columns of *NEC Table 310.15(B)(16)* is covered in Chapter 4.

The neutral conductor is permitted to be sized according to *220.61*, which under certain conditions allows the neutral conductor to be smaller than the phase conductors. This information is found in *NEC 445.13*.

GENERATOR SIZING RECOMMENDATIONS

No matter what type or brand of generator you purchase, you will have to size it properly. There is no better way to do this than to follow the manufacturer's recommendations. Generators for home use are generally rated in watts. The manufacturer of the generator has taken into consideration that watts = volts × amperes. Some manufacturers suggest that after adding up all of the loads to be picked up by the generator, add another 20% capacity for future loads.

As stated above, the type of transfer switch installed determines the minimum capacity required for the generator. If a manual transfer switch or equipment is installed, a generator is required to have capacity for all of the equipment intended to be operated at one time. The user of the optional standby system is permitted to select the load connected to the system. See *NEC 702.6(B)(1)*.

If the generator is supplied with an automatic transfer switch or equipment, it is required to be capable of supplying the full load that is transferred by the automatic transfer equipment. A load management system is permitted to be installed, in which case the generator must be sized not smaller than required to supply the maximum load that the load management system will allow for it to be connected to the equipment. See *NEC 702.6(B)(2)*.

Wattage ratings of appliances vary greatly, as shown in Table 32-1. Resistive loads such as toasters, heaters, and lighting do not have an initial high inrush surge of current. Motors, on the other hand, do have a high inrush surge of starting current, which lasts for only a few seconds until the motor gets up to speed. This inrush must be taken into consideration when selecting a generator. Here are some

TABLE 32-1

Wattage ratings of appliances.

	TYPICAL OPERATING WATTAGE REQUIREMENTS	STARTING WATTAGE REQUIREMENTS
Blender	600	700
Broiler	1600	1600
Central Air Conditioner		
10,000 Btu	1500	4500
20,000 Btu	2500	7500
24,000 Btu	3800	11,000
32,000 Btu	5000	15,000
40,000 Btu	6000	18,000
Coffee Maker	900–1750	900–1750
CD Player	50–100	50–100
Clothes Dryer		
Electric	5000 @ 240 volts	6500
Gas	700	2200
Computer	300–800	300–800
Curlers	50	50

(continues)

TABLE 32-1 (*Continued*)

	TYPICAL OPERATING WATTAGE REQUIREMENTS	STARTING WATTAGE REQUIREMENTS
Dehumidifier	250	350
Dishwasher	1500	2500
Electric		
Drill	250–750	300–900
Fry Pan	1300	1300
Blanket	50–200	50–200
Space Heater	1650	1650
Water Heater	4500–8000 @ 240 volts	4500–8000 @ 240 volts
Electric Range		
6-in. Surface Unit	1500	1500
8-in. Surface Unit	2100	2100
Oven	4500	4500
Lights	Add wattage of bulbs	Add wattage of bulbs
Furnace Fan		
⅛ Horsepower Motor	300	900
⅙ Horsepower Motor	500	1500
¼ Horsepower Motor	600	1800
⅓ Horsepower Motor	700	2100
½ Horsepower Motor	875	2650
Garage Door Opener		
⅓ Horsepower	725	1400
¼ Horsepower	550	1100
Hair Dryers	300–1650	300–1650
Iron	1200	1200
Microwave Oven	700–1500	1000–2300
Radio	15 to 500 (with components)	15 to 500
Refrigerator	700–1000	2200
Security, Home	25–100	25–100
Sump Pump		
⅓ Horsepower	800	2400
½ Horsepower	1050	3150
Television	300	300
Toaster		
4-Slice	1500	1500
2-Slice	950	950
Automatic Washer	1150	3400
Well Pump Motor		
⅓ Horsepower	800	2400
½ Horsepower	1050	3150
1 Horsepower	1920 @ 240 volts	5500
Vacuum Cleaner	1100	1600
VCR	150–250	150–250
Window Fan	200	300

typical *approximate* values for household loads. For heating-type appliances, the typical operating wattage values are used. For motor-operated appliances, the starting wattage values should be used. Verify actual wattage ratings by checking the nameplate on the appliance. Verify that the connected loads do not exceed the generator's marked capacity.

Most generators are capable of handling a momentary "inrush" of an extra 50% of their rating. Here again, check the manufacturer's specifications.

THE *NATIONAL ELECTRICAL CODE* REQUIREMENTS

The *NEC* addresses optional standby systems in *Article 702*. This includes those systems that are permanently installed in their entirety, and those systems that are intended for connection to a premises wiring system from a portable alternate power supply.

- *NEC 702.4*: All of the equipment must have the capacity and rating for the loads that will be supplied by the equipment. The equipment is required to be suitable for the maximum available fault current at its terminals.
- *NEC 702.5*: Transfer equipment must be suitable for the intended use and designed to prevent the interconnection of the generator and the normal utility supply.
- *NEC 702.6*: Some systems, where practical, require audible and visual signal devices to indicate when the normal power system has been transferred to the standby system. The manufacturer of the equipment provides these features. Signal devices are not required for portable standby generators, *702.6, Exception*.
- *NEC 702.7(A)*: A sign is required at the service-entrance equipment to indicate the type and location of the standby generator.
- *NEC 702.7(B):* Where the grounded circuit conductor connected to the optional standby generator is connected to a grounding electrode conductor at a location remote from the standby source, a sign is required at the grounding location that identifies all standby and normal sources connected at that location. For most residential standby systems, this connection is made on the neutral bus in the generator panelboard. This neutral bus in turn has a conductor run to the neutral bus in the main panelboard, where the neutral bus and equipment grounding bus are connected together, and a grounding electrode conductor is connected and run to the grounding electrode of the system. The neutral bus in the generator panelboard is isolated from the panelboard enclosure. The equipment grounding bus in the generator panelboard is bonded to the panelboard enclosure.
- *NEC 702.10*: It is permissible to run standby power conductors and normal power conductors in the same raceway or enclosure.
- *NEC 702.11(B)*: Where a portable optional standby source is used as a nonseparately derived system, the equipment grounding conductor is required to be bonded to the system grounding electrode. The equipment grounding conductor (green or bare) from the standby generator to the generator panelboard is connected to the equipment grounding bus in the generator panelboard. The equipment grounding bus in the generator panelboard is bonded to the generator panelboard enclosure. In turn, the equipment grounding bus in the generator panelboard is bonded back to the main panelboard either by a metallic wiring method, or by a separate equipment grounding conductor. At the main panelboard, the grounding electrode conductor is connected to both the equipment ground bus and the neutral bus.
- *NEC 702.12* applies to outdoor generator sets, both permanently installed and portable, that supply an optional standby system.

  (A) covers permanently installed generators and portable generators greater than 15 kW. ▶Where an outdoor housed generator set is equipped with a readily accessible disconnecting means in accordance with 445.18, and the disconnecting means is located within sight of the building or structure supplied, an additional disconnecting means shall not be required where ungrounded conductors serve or pass through the building or structure. Where the generator supply conductors terminate at a disconnecting means in or on a building or structure, the disconnecting means is required to meet the requirements of 225.36.*

That section permits the disconnecting means to consist of a circuit breaker, molded case switch, general-use switch, snap switch, or other approved means.

(B) covers portable generators rated 15 kW or less. If a portable generator, rated 15 KW or less, is installed using a flanged inlet or other cord and plug type connection, a disconnecting means is not required where ungrounded conductors serve or pass through a building or structure.◀*

*Reprinted with permission from NFPA 70-2014.

Permits

Permanently installed generators require a considerable amount of electrical work and gas line piping. The installation will probably require applying for a permit so that proper inspection by the authorities can be made. Check with your local building official.

Sound Level

Because generators produce a certain level of noise (decibels) when running, you will want to check with your building authorities to make sure the generator you choose is in compliance with local codes relative to any sound ordinance that might be applicable. Sound-level information is provided by manufacturers in their descriptive literature.

REVIEW

Note: Refer to the *Code* or plans where necessary.

1. The basic safety rule when working with electricity is to ________________________

   ____

2. Where would the logical location be for running a portable generator?
3. The best advice to follow is to always use (listed) (cheapest) (smallest) equipment. (Circle the correct answer.)
4. How would you define the term *standby power*? ________________________

   ____

5. Describe in simple terms the three types of standby power systems.

   ____

   ____

   ____

   ____

   ____

6. Is it permitted to ground the neutral conductor of the standby generator to the metal enclosure of the generator, transfer switch, or critical (generator) panelboard? Explain.

   ____

   ____

   ____

   ____

7. Briefly explain the function of a transfer switch.

____

____

____

____

8. When a transfer switch transfers to standby power, the electrical connection inside the switch: (Circle the correct answer.)

   a. maintains connection to the normal power supply as it makes connection to the standby power supply.

   b. breaks the connection to the normal power supply before it makes connection to the standby power supply.

9. A typical transfer switch for residential application is: (Circle the correct answer.)

   a. a 4-pole, double pole switch (TPDT).

   b. a double-pole, double-throw switch (DPDT).

   c. a single-pole, double-throw switch (SPDT).

   d. a double-pole, single-throw switch (DPST).

10. The technical *NEC* definition of a system in which the neutral conductor is not switched is referred to as: (Circle the correct answer.)

    a. a separately derived system.

    b. a nonseparately derived system.

11. The *NEC* requires that a ______________________ means be provided for standby power systems. In the case of portable gen sets, this might be as simple as pulling out the plug on the extension cord that is plugged into the receptacle on the gen set. For permanently installed standby power generators, this might be on the gen set and/or separately provided inside or outside of the home.

12. The conductors running from a permanently installed standby power generator to the transfer switch shall not be less than (100%) (115%) (125%) of the generator's output rating if overcurrent protection of the conductors is not provided at the generator. (Circle the correct answer.)

13. *NEC 590.6(A)(3)* requires that all 125-volt and 125/250-volt, single-phase, 15-, 20-, and 30-ampere receptacle outlets used on a construction site that are a part of a portable generator shall have listed: (Circle the correct answer.)

    a. ground-fault circuit interrupter protection for personnel.

    b. grounding-type receptacles.

    c. twist-lock receptacles.

CHAPTER **33**

Residential Utility-Interactive Photovoltaic Systems

OBJECTIVES

After studying this chapter, you should be able to

- identify the components of a residential utility-interactive solar photovoltaic system.
- recognize the electrical hazards unique to solar photovoltaic systems.
- apply *National Electrical Code* requirements to the installation of a residential utility-interactive solar photovoltaic system.

INTRODUCTION

Electricity from sunlight? Is this possible? This is not only possible, but it has become very practical through the installation of utility-interactive photovoltaic systems. A combination of factors recently has made photovoltaic (PV) systems very popular:

- Increased use of electricity throughout the United States
- Increased costs for electricity production
- Environmental concerns with the use of fossil fuels
- Concern over dependence on foreign sources of energy
- Increased efficiency of photovoltaic systems

Installation of photovoltaic systems is being encouraged by government and electrical utilities through tax incentives and rebates. Some states require utilities to produce defined amounts of electricity through use of renewable resources. Electrical utilities can meet the requirement by construction of large central photovoltaic generating plants or through distributed generation. Generation of electricity by the consumer at the point of use (distributed generation) decreases the need for utility-generated power. Excess electricity will be supplied to the utility grid for use by other customers. Electrical utility companies are required to purchase customer-generated electricity at predetermined rates.

Photovoltaic systems such as these are known as utility interactive. Depending on size, a utility-interactive PV system can supply most or all of a home's electrical load. Photovoltaic systems are being retrofitted to existing homes and even installed on new homes as they are constructed. Battery storage of the generated electricity may also be included, but is not as common.

Electrical Hazards

Electrical work for the installation of photovoltaic systems may seem complex because there are significant differences in the equipment when compared to typical residential wiring. This is the reason *NEC® 690.4(C)* requires that PV equipment be installed only by qualified persons. The term "qualified person" is defined in *NEC Article 100* and means, One who has skills and knowledge related to the construction and operation of the electrical equipment and installations and has received safety training to recognize and avoid the hazards involved.*

First and foremost, a utility-interactive PV system is a supply of electricity for the home, not another load. Photovoltaic modules on the roof are generating electricity when exposed to sunlight. Although the utility disconnect (main circuit breaker) for a house can be turned off, dangerous levels of electricity will still be present on the dwelling as long as the sun is shining. Modules on the roof of a house are connected in series in strings that operate at up to 600 volts.

Electricity generated by the photovoltaic modules is direct current (dc). All conductors and components must be listed for use with dc voltage.

The sun shines for more than three hours at a time, so all conductors/components must be sized for continuous currents (three hours or longer of operation).

Because string conductors of the correct type are permitted to be exposed on the roof of a dwelling, good workmanship is critical. There are open conductors, exposed to the elements, operating at up to 600 volts of continuous current (Vdc) that cannot be turned off! If a short circuit does occur, it is not always obvious. Module short-circuit current is only slightly higher than normal operating current. Contrast this with the short-circuit current from sources such as a utility or even 12 Vdc vehicle batteries. Shorting out these sources results in extremely high fault currents. High-fault currents are easier to detect. High-fault currents will also cause an overcurrent device (fuse or circuit breaker) to open quickly.

The connection of the utility-interactive PV system to the existing service panel is of concern. Service disconnecting means, grounding, and proper labeling must be accomplished to maintain a safe electrical installation.

It is obvious that there are some special considerations for the installation of residential photovoltaic systems. *Article 690* was added to the *NEC* in 1984 to establish minimum electrical standards for the installation of photovoltaic systems. *Chapters 1* through *4* of the *NEC*, along with the requirements of *Article 690*, will apply to residential PV installations. Local jurisdictions may have amendments

*Reprinted with permission from NFPA 70-2014.

to the *NEC* for PV installations. Always check with the local authority having jurisdiction (AHJ) before starting the installation in a new area.

THE BASIC UTILITY-INTERACTIVE PV SYSTEM

Several components are required to convert sunlight into useful amounts of electricity. A basic utility-interactive system will consist of modules, mounting racks, combiner/transition boxes, inverter(s), and several disconnects. A grounding electrode system and connection to the existing service panel will be required. See Figures 33-1 and 33-2 for examples of basic PV system components and arrangements.

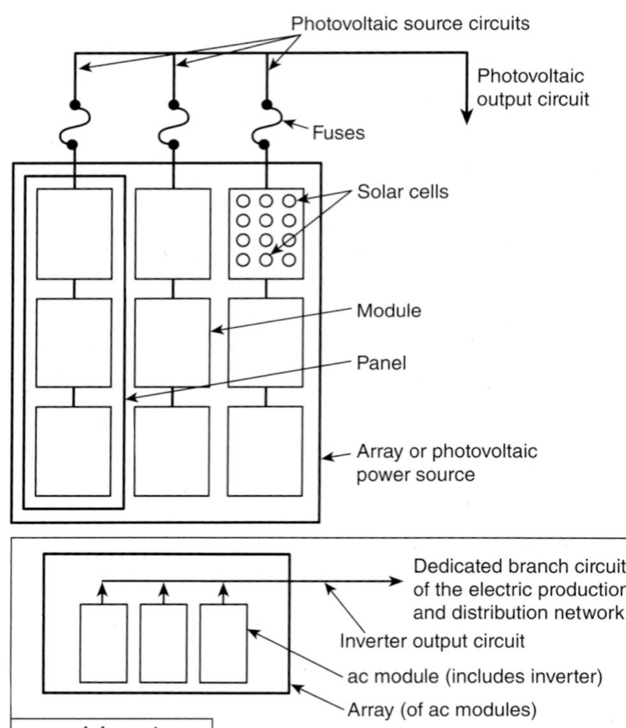

Notes:
1. These diagrams are intended to be a means of identification for photovoltaic system components, circuits, and connections.
2. Disconnecting means required by *Article 690, Part III*, are not shown.
3. System grounding and equipment grounding are not shown. See *Article 690, Part V*.

FIGURE 33-1 Identification of solar photovoltaic system components.

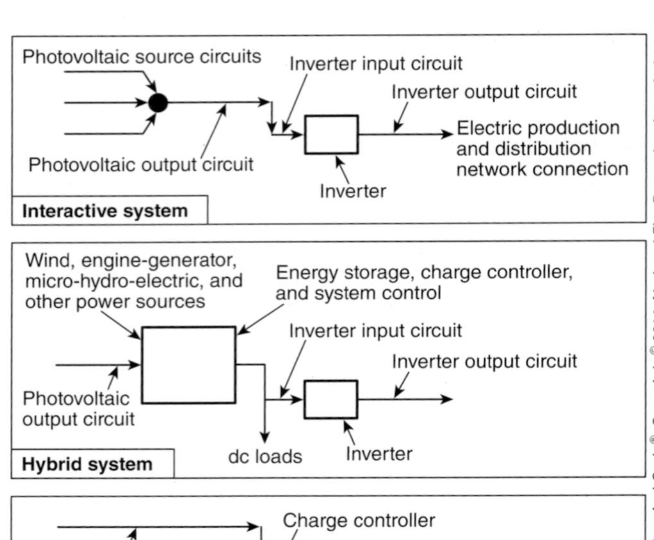

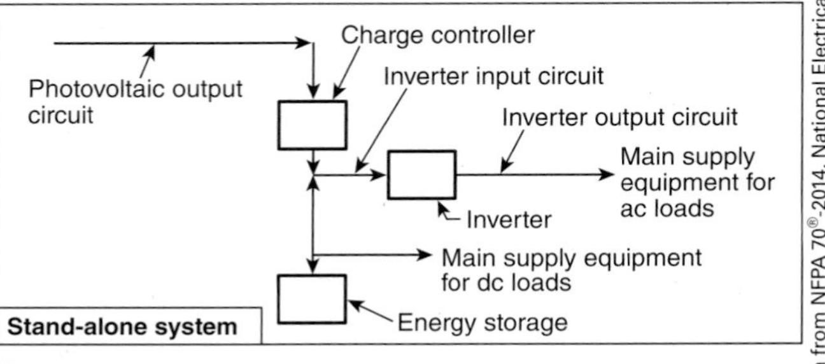

Notes:
1. These diagrams are intended to be a means of identification for photovoltaic system components, circuits, and connections.
2. Disconnecting means and overcurrent protection required by *Article 690* are not shown.
3. System grounding and equipment grounding are not shown. See *Article 690, Part V*.
4. Custom designs occur in each configuration, and some components are optional.

FIGURE 33-2 Identification of solar photovoltaic system components in common system configurations.

Solar Cells, Modules, and Arrays

A photovoltaic module is the basic unit of power production in the system. A module is a manufactured unit made up of many semiconductor PV cells encased in a protective covering and mounted to an aluminum frame. All modules are required by the *NEC* to be listed to nationally recognized standards (*ANSI/UL 1703*). Individual photovoltaic modules are mounted to a support rack that is connected to roof members or, in some cases, to ground mounted structural supports. Individual modules are wired together in a series circuit. Factory-installed leads are provided by the manufacturer for this purpose. Type *USE-2* or *PV* conductors can be spliced to module leads to facilitate circuiting. A separate equipment

FIGURE 33-3 Rooftop array of modules.

grounding conductor is used to ground each module. Most strings will consist of 8 to 15 modules. The quantity of modules in a series circuit will be limited by the combined open circuit voltage of the string. A typical residential array of roof mounted modules is shown in Figure 33-3.

Multiple strings are combined together (in parallel) at the combiner box. This allows a single pair of conductors to deliver current to an inverter. Combiner boxes may be fused or nonfused. Most residential PV systems are made up of three to six strings that can be combined in a single box. String fusing is generally not required for three strings or fewer. Module ratings will determine this, but some designers specify fused combiner boxes even when not required.

Many inverters have the ability to combine several strings at the input terminals. With an inverter like this, the string conductors are routed to the inverter without being combined first. The combiner box would not be required. A junction box, known as a transition box, is used to splice the open string conductors to a wiring method (usually conduit and THWN conductors) for connection of the inverter. The entire assembly of racks, modules and combiner/junction boxes is known as a photovoltaic array. Available space, sunlight, and the ultimate amount of electricity desired will determine the number of modules in the array. Obviously, the larger the system, the more expensive the installation, so the budget must be considered.

Requirements for disconnecting means are provided in *Part III* of *Article 690*. The requirements vary according to the type of arrays that are installed. A disconnect for the ungrounded dc string conductor(s) is generally required before the conductors enter the dwelling. For building-integrated systems, the use of a metallic wiring method such as EMT or FMC is common so that string conductors can be routed through a dwelling unit attic. Metallic raceways provide more protection for the 400–600 Vdc string conductors, which remain energized until the sun goes down. Firefighters responding to a house fire will open the service disconnecting means on arrival. This does not de-energize the PV string conductors if the sun is still shining. Metallic raceways will provide a level of protection for the firefighters who may encounter the dc wiring method when responding to a fire. Use of a metallic raceway eliminates the requirement for an array disconnect on the roof but not at the inverter. The inverter will be energized from two different sources: dc from the array and ac from the utility connection. Disconnecting means must be provided for both sources. Some inverters are manufactured with integral disconnects. A circuit breaker in an adjacent electrical panel can serve as the ac disconnect, but all safety disconnects must be within sight of the inverter and may need to be grouped. The dc grounded conductor is not permitted to be opened by any disconnecting means. The inverter in Figure 33-4 has an integrated dc disconnect.

FIGURE 33-4 A residential utility-interactive inverter.

The Inverter

The inverter is generally required to be installed in a readily accessible location and in a location with sufficient working space per *NEC 110.26(A)*. Utility-interactive inverters are permitted to be mounted in not-readily-accessible locations such as on roofs under several conditions in *690.15(A)*. Larger systems may use two or more inverters with output combined in a separate ac panelboard before supplying the service.

Grounding/bonding of the utility-interactive PV system is required and is usually accomplished at the inverter location. Both the ac and the dc systems contain grounded current carrying conductors. Grounding of the dc conductor is accomplished at the inverter.

A new grounding electrode must be installed (and bonded to the existing premises' grounding electrode) or the existing grounding electrode for the dwelling must be accessed for connection. Correct termination of the grounding electrode conductor, equipment grounding conductors, and grounded ac/dc conductors at the inverter is critical.

Utility-interactive inverters are required to provide ground-fault protection for the array. Proper operation of the ground-fault protection system is dependent on the correct terminations of the grounding and grounded conductors. Inverter design will dictate whether the positive or the negative dc conductor is the grounded current carrying conductor for the array. Inverter size is based on the capacity of the array. Most residential inverters are in the 2 kW to 10 kW range. There are advantages to installing an inverter indoors. Cooler temperatures result in better operating efficiency, but outdoor installation is also common.

Safety Features of Inverters

Just as with modules and combiner boxes, inverters are required to be manufactured to nationally recognized standards. Inverters that are listed by a Nationally Recognized Testing Laboratory (NRTL) to ANSI/UL 1741 have met this standard. One requirement of UL 1741 is known as anti-islanding. Anti-islanding means that an inverter must automatically turn off ac output when utility power is lost. This feature prevents an inverter from supplying electricity into the utility grid when there is an outage. Electrical utility workers risk being electrocuted by inverter-supplied electricity without this.

Connection of Inverters

The two options for connection of the inverter to the electrical service panel are known as a supply-side or a load-side connection. A connection on the utility side of the main disconnect is known as a supply-side point of connection. A load-side connection will supply electricity on the customer side of the main disconnect. Supply-side connection size is not limited by the *NEC*, but a second electrical service is created so the requirements of *NEC Article 230 (Services)* will apply. A load-side connection will have limitations. A dedicated circuit breaker (suitable for backfeed) or fusible disconnect is required. The total supply of current to a panelboard is limited to 120% of the busbar rating. The service panel is supplied by both the utility (through the main circuit breaker) and the inverter (through a back-fed circuit breaker). The sum of the ampere ratings of the main circuit breaker and the inverter circuit breaker cannot exceed the rating of the panelboard bus multiplied by 1.2 (120%). A circuit breaker used for connection of the inverter will have to be located at the opposite end of the bus from the main circuit breaker, to avoid overloading of the bus.

Building-Integrated Photovoltaic Modules

Photovoltaic modules that also serve as an outer protective finish for a building are known as building-integrated photovoltaic (BIPV) modules. This type of module is often installed in the form of roofing tiles intended to blend in with surrounding non-PV roof tiles. See Figure 33-5. BIPV modules are investigated through the listing process for conformance to appropriate fire-resistance and waterproofing standards, which apply to roofing tiles, along with the electrical standards of *UL/ANSI 1703*. A module the size of a roofing tile is obviously much smaller than the standard photovoltaic module. Many more of the smaller modules are required to create strings.

FIGURE 33-5 Building-integrated photovoltaic roof tiles.

Micro-Inverters and ac Photovoltaic Modules

A single small inverter connected to each photovoltaic module is known as a micro-inverter. Instead of connecting multiple modules to a single inverter, each module will have its own attached inverter. The output of each module is connected directly to the micro-inverter with the existing module leads. Multiple micro-inverters are connected in parallel on a single circuit, which then supplies the service panel of the home. Micro-inverters are required to be listed to ANSI/UL 1741 just as with the larger string inverters. Ground-fault protection, anti-islanding, and the other requirements of utility-interactive inverters are applicable. Output current for a single micro-inverter is approximately 0.8 amps. This would permit up to 15 inverters on a single 15-amp circuit (allowing for the continuous current multiplier). The inverter shown in Figure 33-6 is a micro-inverter.

An ac photovoltaic module is essentially a normal dc module with the micro-inverter installed at the factory. The dc conductors are covered and inaccessible on an ac module. The absence of field-installed dc string conductors is a big advantage with both micro-inverters and ac modules. Hazards associated with the dc string conductors are minimized with the use of micro-inverters and eliminated with ac modules. *NEC* requirements for ac circuits connecting micro-inverters/ac modules to service panels are similar to normal branch-circuit rules.

FIGURE 33-6 A utility-interactive micro-inverter.

NATIONAL ELECTRICAL CODE REQUIREMENTS

Article 690, Solar Photovoltaic Systems, of the *NEC* provides the specific requirements for installation of PV systems. In addition, the general requirements of *NEC Chapters 1* through *4* apply, except as modified by *Article 690.* A utility-interactive PV system operates in parallel with the utility (primary source); portions of *Article 705, Interconnected Electric Power Production Sources*, will apply as well. We will look at some of the requirements of *Article 690* in the following section.

Part I. General

Article 690 begins by providing the scope or what is covered by the article and defining many terms used in the article and PV industry. Specific applicable sections of *Article 705* are referenced in *NEC 690.3*. An exception to *NEC 690.3* makes it clear that PV systems installed in hazardous (classified) locations must comply with the applicable portions of *Articles 500* through *516*. Several of the basic rules are covered in *690.4*. The requirement that PV equipment be installed by qualified persons, the arrangement of module connections, and the requirements for listing of equipment, as well as for installation by qualified personnel, are all found in *NEC 690.4*. Ground-fault protection for grounded dc photovoltaic arrays is required, *NEC 690.5*. *Part I* ends with *NEC 690.6*, which provides the rules for the installation of alternating current modules.

Part II. Circuit Requirements

Requirements for PV source and output circuits are found in this section. Voltage and current parameters for design of PV circuits are defined. Maximum dc voltage for a series string of modules is defined as the sum of the rated open-circuit voltage (V_{oc}) of the modules, corrected for the lowest expected ambient temperature. *Table 690.7* must be used if the manufacturer does not provide temperature-correction coefficients in their installation instructions. This table provides multipliers that correspond to the lowest expected ambient temperature. Multiply the sum of the module V_{oc} by the factor from *Table 690.7* to find the maximum voltage. This voltage is used to determine the voltage rating of cables, disconnects, overcurrent devices, and other equipment.

The size of PV system conductors and components is determined by the amount of current that flows through them. *NEC 690.8* provides the method for determining the circuit size and current. The rated module short-circuit current (I_{sc}) is the starting point for both of these calculations. Sunlight may be more intense in the field than at the testing lab, so *690.8(A)* requires the module short-circuit current to be multiplied by 1.25 (125%) to determine maximum circuit current. Module current is continuous (3 hours or more), which is why the maximum circuit current is required to be multiplied by 1.25 (125%) again for sizing of conductors and overcurrent devices, *690.8(B)*. Inverter current output is continuous, so the rated output of the inverter must be multiplied by 1.25 (125%) for sizing of conductors and overcurrent devices. See *690.8(A)(3)* and *690.8(B)(1)*.

Overcurrent Protection. Requirements for photovoltaic system overcurrent protection are found in *690.9*. The general requirement is that all conductors and equipment are to be protected in accordance with *Article 240, Overcurrent Protection*. There is an exception to this general requirement, which is often used. If circuit conductors were sized using the requirements of *690.8(B)* and there are no external sources of parallel currents that exceed the ampacity of the conductors, then overcurrent protection is not required. This exception permits parallel strings of modules to be combined without fuse protection of the individual strings. As long as the sum of the short-circuit currents from parallel strings does not exceed the rating of a faulted module or conductor, no overcurrent protection is required. Consider the following example: an array contains three parallel strings of modules, and short-circuit current from each module/string is 7 amperes. If a short circuit occurs in one of the modules/strings, the maximum fault current would be the sum of the short-circuit currents from the other two strings. In this example, the sum is 14 amperes (7 + 7 = 14). As long as the modules have a rating greater than 14 amperes, the exception to *690.9* permits combination of the strings without fuse protection. A fused combiner box would still be allowed, of course. Many designers specify fused combiner boxes even if not required.

Overcurrent devices are required to be not less than 125 percent of the maximum currents calculated in *690.8(A)*. An exception allows an overcurrent protection assembly that is listed for continuous operation at 100 percent of its rating to be used at that rating. Overcurrent devices, either fuses or circuit breakers, which are used in any dc portion of a PV power system are required to be listed and to have appropriate voltage, current, and interrupt ratings. The overcurrent devices are required to be accessible but are not required to be readily accessible. This allows the overcurrent devices to be mounted at a rooftop location for example.

Rapid Shutdown of PV Systems on Buildings. A new requirement was added to the 2014 *NEC* that provides for rapid shutdown of PV systems that are installed on buildings. It's interesting to note that this requirement is in *Part II* of *Article 690*, which contains circuit requirements, rather than in *Part III*, which contains requirements for the disconnecting means for PV systems. The new requirement is in *690.12* and reads,

▶**Rapid Shutdown of PV Systems on Buildings.** PV system circuits installed on or in buildings shall include a rapid shutdown function that controls specific conductors in accordance with 690.12(1) through (5) as follows.

(1) Requirements for controlled conductors shall apply only to PV system conductors

of more than 1.5 m (5 ft) in length inside a building, or more than 3 m (10 ft) from a PV array.

(2) Controlled conductors shall be limited to not more than 30 volts and 240 volt-amperes within 10 seconds of rapid shutdown initiation.

(3) Voltage and power shall be measured between any two conductors and between any conductor and ground.

(4) The rapid shutdown initiation methods shall be labeled in accordance with 690.56(B).

(5) Equipment that performs the rapid shutdown shall be listed and identified.◀*

As indicated in this requirement, the equipment that performs rapid shutdown is not necessarily a traditional disconnecting means such as a switch or circuit breaker. The 2013 Report on Comments for producing the 2014 *NEC* includes the following "The means for rapid shutdown was a topic of much discussion at the ROP meeting and among the stakeholders during the comment period and it was decided among the stakeholders that the devices and methods of compliance should be left open to the standards process so long as proper markings are provided and that special products developed to meet the requirement be listed and identified for the purpose."

The result of the requirement for "rapid shutdown" will be the elimination of shock hazard to first responders such as firefighters as the voltage of the system must not exceed 30 volts power output no more than 240 VA within 10 seconds of rapid shutdown initiation.

Part III. Disconnecting Means

Means is required to be provided to disconnect all ungrounded dc conductors of a PV source from all other conductors in a building or structure in accordance with *NEC 690.13*. This does not apply to grounded dc conductors. Energized grounded current-carrying conductors are not to be interrupted by the disconnect(s), just as with grounded ac conductors (neutrals). Several disconnects may be required for a typical residential installation. The direct-current circuit generated by the array must have a disconnecting means installed at a readily accessible location outside the building, or if inside, closest to the point of entrance. The disconnect is not permitted to be located in bathrooms. An exception to *690.13(A)(1)* permits the direct-current photovoltaic conductors to enter the building without a disconnecting means if contained in metal raceways or enclosures. See *690.31(G)*.

Photovoltaic equipment, such as an inverter, is required to have a disconnecting means. The disconnects are required to be grouped and identified if the equipment is energized from more than one source. A utility-interactive inverter is connected to two sources of electricity. The array on the roof generates direct current anytime the sun is shining, and the conductors that connect the inverter to the service are energized by the electric utility. Interruption of the array circuit to the inverter will prevent the inverter from supplying alternating current to the service panel, but the conductors are still energized by the utility supply. If the inverter is located within sight of the circuit breaker connection in the service panel, then the circuit breaker can serve as the disconnecting means.

Requirements for the disconnecting means for utility-interactive inverters that are mounted in locations that are not considered readily accessible are found in *NEC 690.15(A)*. These utility-interactive inverters are permitted to be mounted on roofs or other exterior areas that are not readily accessible if they comply with all the following rules:

(1) A dc PV disconnecting means is mounted within sight of or in each inverter.

(2) An ac disconnecting means is mounted within sight of or in each inverter.

(3) The ac output conductors from the inverter and an additional ac disconnecting means for the inverter are required to be installed in a readily accessible location in accordance with *NEC 690.13(A)*.

(4) A permanent plaque or directory, denoting all electric power sources on or in the premises, is installed in compliance with *NEC 705.10* at each service equipment location and at locations of all electric power production sources capable of being interconnected.

*Reprinted with permission from NFPA 70-2014.

Always check with the local authority having jurisdiction (AHJ) and electrical utility company for policies regarding disconnecting means. Some utility companies will require a dedicated utility disconnect.

Part IV. Wiring Methods

Exposed single-conductor cable is permitted to be installed for array interconnection. Only types *USE-2* and listed *PV* wire are permitted, and there are other limitations. If the circuit operates at over 30 volts and is in a readily accessible location, circuit conductors are required to be guarded or installed in a raceway. Direct-current array conductors that enter the structure are required to be installed in metal raceways or enclosures from the point of penetration to the first readily accessible disconnecting means, per *690.31(G)*.

Wiring methods for the inverter output circuit are only limited by the general requirements of *Chapters 1* through *4* of the *NEC*. Inverter wiring to the service is essentially a branch circuit if a load-side connection is used. A supply-side connection will involve the service wiring method limitations of *Article 230*. Junction, pull, and outlet boxes are permitted to be located behind removable modules that are connected with flexible wiring methods. See *690.34*. Ungrounded photovoltaic power systems are permitted if the installation complies with *690.35(A)* through *(G)*.

Part V. Grounding

According to *690.41*, one conductor of a 2-wire PV system over 50 volts shall be grounded. The exception to *690.42* allows the grounded conductor bond to be made by the ground-fault detection device required by *690.5*. The grounded conductor of the dc system can be either the positive or the negative conductor. Inverter design will determine which conductor is grounded, but any PV system over 50 volts is required to be a grounded system unless it is an ungrounded system in compliance with *690.35*.

Exposed non-current-carrying metal parts of module frames and other equipment are required to be grounded. Methods of grounding module frames will vary among different manufacturers. Installation instructions for the module must be referenced. The size of the equipment grounding conductor is determined by *NEC Table 250.122*. When there is no overcurrent device in the circuit, the module/string short-circuit current shall be the assumed overcurrent device size for reference to *Table 250.122* per *690.45*. Increases in equipment grounding conductor size to address voltage drop consideration are not required. The equipment grounding conductor is not permitted to be smaller than 14 AWG.

A grounding electrode system is required for both dc and ac photovoltaic systems. Systems with both ac and dc grounding requirements, such as utility-interactive systems, must meet the requirements of *690.47(C)*. Optional methods of installing a grounding electrode system are permitted. All methods include a bond to the existing premises' grounding electrode system. A conductor that serves as an equipment grounding conductor and the bond between ac and dc systems is permitted by *690.47(C)(3)*.

Additional grounding electrodes are required to be installed in accordance with *250.52* at the location of all ground-and pole-mounted photovoltaic arrays. The grounding electrodes are required to be connected directly to the array frame or structure. A dc grounding electrode conductor is required to be sized according to *250.166*. Additional grounding electrodes are not permitted to be used as a substitute for equipment grounding or bonding conductors.

Part VI. Marking

Photovoltaic modules will have labels marked with specific information as required by *690.51* (dc modules) and *690.52* (ac modules). This information is required so that system designers and installers are able to size the balance of system components. The completed system will have operating parameters (voltage and current) unique to the design. Field labels marked with the specific dc- and ac-operating parameters must be installed per the rules of *690.53* and *690.54*. Buildings with stand-alone or utility-interactive PV systems are required to have a plaque or directory indicating the presence of the system and location(s) of disconnecting means. See *690.56*. Installers of PV systems should be aware that there are marking requirements for specific components throughout *Article 690*.

Part VII. Connection to Other Sources

The rules for connection of the utility-interactive PV system to the electrical service are found in this part of *Article 690*. An inverter or ac module is required to automatically de-energize its output when utility power is lost. An inverter must stay in this de-energized state until utility power is restored. This is known as "anti-islanding" and is a requirement of *690.61* as well as ANSI/UL 1741. A neutral conductor smaller than the ungrounded (phase) conductors is permitted per *690.62*. A smaller neutral for the inverter is only allowed when used for instrumentation or detection purposes. Installation instructions for an inverter will provide the minimum neutral conductor size.

The *National Electrical Code®* permits two methods for connection of a utility-interactive PV system to the electrical service. Requirements for both methods are found in *705.12 Point of Connection*. A connection on the line side of the service disconnecting means is known as a supply-side connection, *705.12(A)*. Size of the photovoltaic system is virtually unlimited, but the rules of *NEC Article 230* will apply to these service conductors. Requirements for load-side connections are found in *705.12(B)*. The photovoltaic system size is limited by the panel bus rating and main circuit breaker size. See *705.12(D)(2)*. A dedicated PV system circuit breaker, suitable for backfeed and positioned at the opposite end of the bus from the main circuit breaker, is a requirement of *705.12(D)(6)*. See Figures 33-7 and 33-8 for a representation of the two connection methods.

Part VIII. Storage Batteries and Part IX. Systems over 600 Volts

Requirements for photovoltaic storage battery systems are found in *Part VIII* of *Article 690*. The provisions of *Article 480, Storage Batteries*, will apply as well. Specific rules for dwelling units are found in *690.71(B)*. The voltage of storage batteries is limited to 50 volts or less unless live parts are not accessible during routine maintenance.

Part IX of *Article 690* requires photovoltaic systems with a maximum dc voltage over 1000 volts dc to comply with *Article 490*. Note that *690.7(C)* limits photovoltaic source and output circuits to 600 volts or less for one- and two-family dwellings.

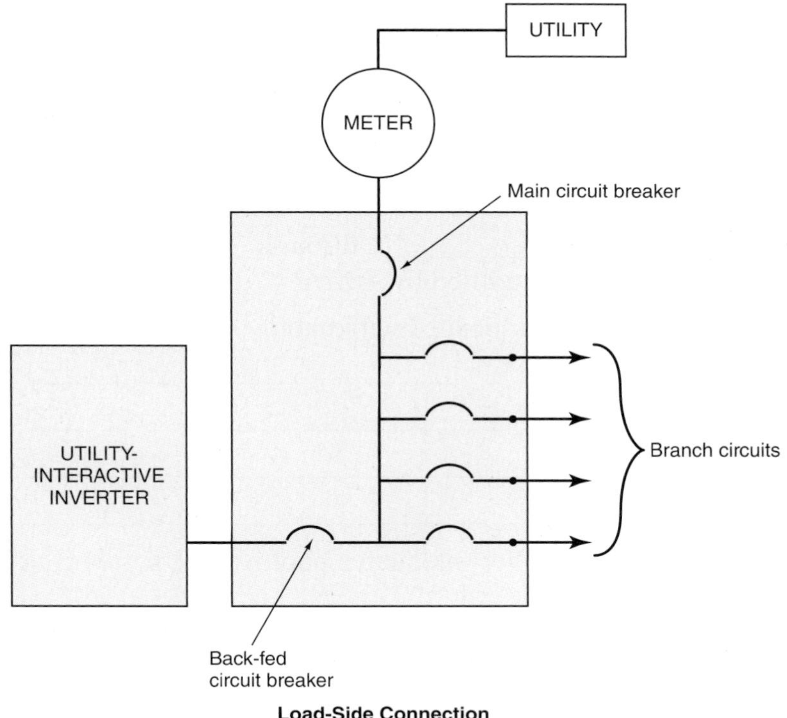

FIGURE 33-7 Load-side connection.

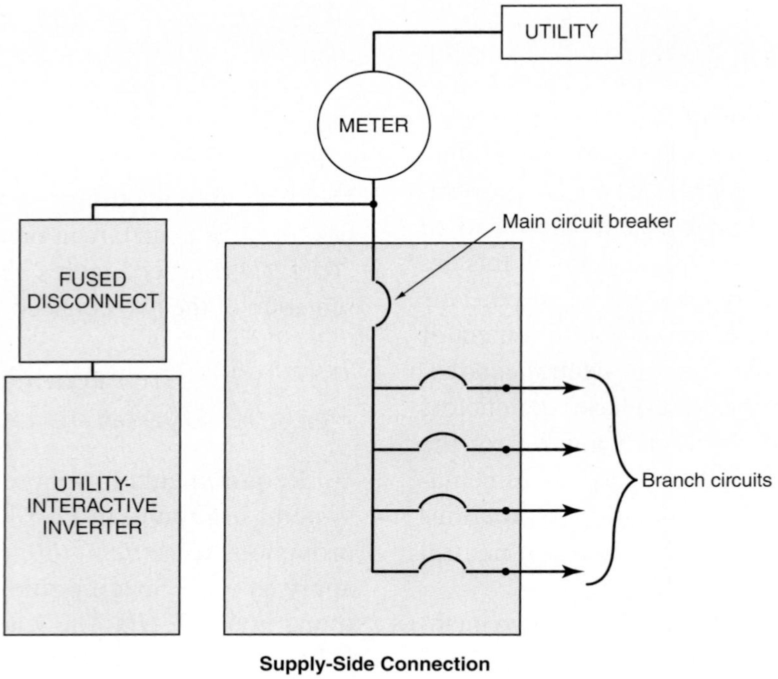

FIGURE 33-8 Supply-side connection.

REVIEW

Note: Refer to the *National Electrical Code* when necessary. Where applicable, responses should be written in complete sentences.

1. *Article* ______________________ of the *National Electrical Code* contains most of the requirements for installation of photovoltaic systems.
2. *NEC Chapters* ______________________ through ______________________ apply to PV installations except as modified by *Article* ______________________.
3. Name four electrical hazards associated with photovoltaic systems:

____

____

____

____

4. Name five components of a utility-interactive photovoltaic system:

____

____

____

____

____

5. Which conductor types are permitted to be installed exposed for array circuits?

____

6. To size conductors and overcurrent devices for source circuits, you must multiply the module string short-circuit current by ____________________.

7. When is a metallic raceway required for photovoltaic source circuits?

____

8. Photovoltaic modules that also serve as an outer protective finish for a building are known as ____________________.

9. Grounding electrode system requirements for a utility-interactive photovoltaic system are found in section ____________________ of the *NEC*.

10. The two methods permitted for connection of a utility-interactive photovoltaic system to a service are known as ____________________ and ____________________.

Specifications for Electrical Work—Single-Family Dwelling

1. **GENERAL:** The "General Clause and Conditions" shall be and are hereby made a part of this division.
2. **SCOPE:** The electrical contractor shall furnish and install a complete electrical system as shown on the drawings and/or in the specifications. Where there is no mention of the responsible party to furnish, install, or wire for a specific item on the electrical drawings, the electrical contractor will be responsible completely for all purchases and labor for a complete operating system for this item.
3. **WORKMANSHIP:** All work shall be executed in a neat and workmanlike manner. All exposed conduits shall be routed parallel or perpendicular to walls and structural members. Junction boxes shall be securely fastened, set true and plumb, and flush with finished surface when wiring method is concealed.
4. **LOCATION OF OUTLETS:** The electrical contractor shall verify location, heights, outlet and switch arrangements, and equipment prior to rough-in. No additions to the contract sum will be permitted for outlets in wrong locations, in conflict with other work, and so on. The owner reserves the right to relocate any device up to 10 ft (3.0 m) prior to rough-in, without any charge by the electrical contractor.
5. **CODES:** The electrical installation is to be in accordance with the latest edition of the *National Electrical Code*, all local electrical codes, and the utility company's requirements.
6. **MATERIALS:** All materials shall be new and shall be listed and bear the appropriate label of Underwriters Laboratories, Inc., or another nationally recognized testing laboratory for the specific purpose. The material shall be of the size and type specified on the drawings and/or in the specifications.
7. **WIRING METHOD:** Wiring, unless otherwise specified, shall be nonmetallic-sheathed cable, armored cable, or electrical metallic tubing (EMT), adequately sized and installed according to the latest edition of the *National Electrical Code* and local ordinances.
8. **PERMITS AND INSPECTION FEES:** The electrical contractor shall pay for all permit fees, plan review fees, license fees, inspection fees, and taxes applicable to the electrical installation and shall be included in the base bid as part of this contract.
9. **TEMPORARY WIRING:** The electrical contractor shall furnish and install all temporary wiring for handheld tools and construction lighting per latest OSHA standards and *Article 590, NEC*, and include all costs in base bid.

10. **WORKSHOP:** Workshop wiring is to be installed in EMT using steel compression gland fittings. Wiring with Type NM cable protected by EMT from the joist space to the outlet, device, or junction box is acceptable as is installation in Type MC cable.

11. **NUMBER OF OUTLETS PER CIRCUIT:** In general, not more than ten (10) lighting and/or receptacle outlets shall be connected to any one lighting branch circuit. Exceptions may be made in the case of low-current-consuming outlets.

12. **CONDUCTOR SIZE:** General lighting branch circuits shall be not smaller than 14 AWG copper protected by 15-ampere overcurrent devices. Small-appliance branch circuits and branch circuits serving bathroom receptacles shall be 12 AWG copper protected by 20-ampere overcurrent devices. Conductors shall be Type THHN. For service-entrance conductors, see Clause 28.

    All other circuits: conductors and overcurrent devices as required by the *Code*.

13. **LOAD BALANCING:** The electrical contractor shall connect all loads, branch circuits, and feeders per the Panelboard Schedule, but shall verify and modify these connections as required to balance connected and calculated loads to within 10% variation.

14. **SPARE CONDUITS:** Furnish and install two empty trade size 1 EMT conduits between workshop and attic for future use.

15. **GUARANTEE OF INSTALLATION:** The electrical contractor shall guarantee all work and materials for a period of one full year after final acceptance by the architect/engineer, electrical inspector, and owner.

16. **APPLIANCE CONNECTIONS:** The electrical contractor shall furnish all wiring materials and make all final electrical connections for all permanently installed appliances such as, but not limited to, the furnace, water heater, water pump, built-in ovens and ranges, food waste disposer, dishwasher, and clothes dryer.

    These appliances are to be furnished by the owner.

17. **CHIMES:** Furnish and install two (2) two-tone door chimes where indicated on the plans, complete with two (2) push buttons and a suitable chime transformer. Allow $150.00 for above items. Chimes and buttons to be selected by owner.

18. **DIMMERS:** Furnish and install dimmer switches where indicated.

19. **EXHAUST FANS:** Furnish, install, and provide connections for all exhaust fans indicated on the plans, including, but not limited to, ducts, louvers, trims, speed controls, and lamps. Included are the recreation room, laundry, rear entry powder room, range hood, and bedroom hall ceiling fans. Allow a sum of $700.00 in the base bid for this. This allowance does not include the two bathroom heat/vent/light units.

20. **LUMINAIRES:** A luminaire allowance of $1500.00 shall be included in the electrical contractor's bid for all surface-mounted luminaires and post lanterns, not including the ceiling paddle fans. Luminaires to be selected by owner. This allowance includes the three (3) medicine cabinets.

    The electrical contractor shall

    a. furnish and install all incandescent and fluorescent recessed luminaires.
    b. install all surface, recessed, track, strip, pendant, and hanging luminaires, and post lanterns.
    c. furnish and install two 2-lamp fluorescent luminaires above the workbench.
    d. furnish and install all lamps except as noted. Fluorescent lamps and incandescent lamps shall be energy-efficient type. Lamps for ceiling paddle fans to be furnished by owner.
    e. furnish and install all porcelain pull-chain and keyless lampholders.

    This luminaire allowance does not include the two bathroom ceiling heat/vent/light units. See Clause 21.

21. **HEAT/VENT/LIGHT CEILING UNITS:** Furnish and install two heat/vent/light units where indicated on the plans, complete with switch assembly, ducts, and louvers required to perform the heating, venting, and lighting operations as recommended by the manufacturer.

22. **PLUG-IN STRIP:** Where noted in the workshop, furnish and install a multioutlet assembly with six outlets.

23. **SWITCHES, RECEPTACLES, AND FACEPLATES:** All flush switches shall be of the quiet ac-rated toggle type. They shall be mounted 46 in. (1.15 m) to center above the finished floor unless otherwise noted.

    Receptacle outlets shall be mounted 12 in. (300 mm) to center above the finished floor unless otherwise noted. All receptacles shall be of the grounding type. Furnish and install, where indicated, ground-fault circuit interrupter (GFCI) receptacles to provide ground-fault circuit protection as required by the *National Electrical Code*. GFCI-type circuit breakers are permitted as an option. All wiring devices are to be provided with ivory handles or faces and shall be trimmed with ivory faceplates except in the kitchen, where chrome-plated steel faceplates shall be used.

    Receptacle outlets, where indicated, shall be of the split-wired design.

24. **TELEVISION OUTLETS:** Furnish and install single-gang, ½-inch, raised plaster covers at each television outlet where noted on the plans. Mount at the same height as receptacle outlets. Furnish and install 75-ohm coaxial cable to each television outlet from a point in the workshop near the main service-entrance switch. Allow 6 ft (1.8 m) of cable in workshop. Furnish and install television plug-in jacks at each location. Faceplates are to match other faceplates in home. All remaining work done by others.

25. **TELEPHONES:** Furnish and install a suitable ½ in. raised single-gang plaster ring at each telephone location, as indicated on the plans.

    Furnish and install four-conductor 18 AWG copper telephone cables from the telephone company's point of demarcation near the service-entrance panelboard to each designated telephone location. Terminate in a proper modular jack, complete with faceplates that match the electrical device faceplates. Allow 6 ft (1.8 m) of cable to hang below ceiling joists. Telephone company to furnish, install, and connect any and all equipment (including grounding connection) up to and including their Standard Network Interface (SNI) device.

    Installation shall be in accordance with any and all applicable *National Electrical Code* and local code regulations.

26. **MAIN SERVICE PANELBOARD:** Furnish and install in the Workshop, where indicated on the plans, one 200-ampere, 120/240-volt, single-phase, 3-wire panelboard with a 200-ampere main circuit breaker. Panelboard shall be series rated for 22,000 amperes available fault current.

    Furnish and install all active and spare breakers as indicated on the Panelboard Schedule.

    Circuit breakers shall be AFCI, GFCI, or dual-function AFCI/GFCI as required by the *NEC*.

27. **SERVICE-ENTRANCE UNDERGROUND LATERAL CONDUCTORS:** To be furnished and installed by the utility. Meter enclosure (pedestal type) to be furnished by the utility and installed by the electrical contractor where indicated on plans. Electrical contractor to furnish and install all panelboards, conduits, fittings, conductors, and other materials required to complete the service-entrance installation from the demarcation point of the utility's equipment to and including the Main Panelboard.

28. **SERVICE-ENTRANCE CONDUCTORS:** Service-entrance conductors supplied by the electrical contractor shall be two 2/0 AWG THHN/THWN phase conductors and one 1 AWG bare neutral conductor. Install trade size 1½ Schedule 40 PVC from Main Panelboard A to the meter pedestal. Install Schedule 80 PVC in areas subject to physical damage.

29. **BONDING AND GROUNDING:** Bond and ground service-entrance equipment in accordance with the latest edition of the *National Electrical Code*, local, and utility code requirements. Install one 4 AWG copper grounding electrode conductor and connect to the reinforcing steel bar(s) in the concrete foundation

in compliance with *NEC 250.52(A)(3)*. The installation of ground rods is optional or as required by the authority having jurisdiction (AHJ).

30. **PANELBOARD B:** Furnish and install in the Recreation Room where indicated on the plans one 30-circuit, 125-ampere, 120/240-volt, single-phase, 3-wire main lug only (MLO) panelboard. Panelboard shall be series rated for 22,000 amperes available fault current.

    Furnish and install all active and spare breakers as indicated on the Panelboard Schedule.

    Circuit breakers shall be AFCI, GFCI, or dual-function AFCI/GFCI as required by the *NEC*.

    Feed panelboard with three 3 AWG THHN or THWN conductors from a 100-ampere, 2-pole breaker in Main Panelboard A. Install conductors in trade size 1 EMT.

31. **CIRCUIT IDENTIFICATION:** All panelboards shall be furnished with typed-card directories with proper designation of the branch-circuit loads, feeder loads, and equipment served. The directories shall be located on the inside of the panelboard cover door in a holder for clear viewing.

32. **SEALING PENETRATIONS:** The electrical contractor shall seal and weatherproof all penetrations through foundations, exterior walls, and roofs.

33. **FINAL SITE CLEAN-UP:** Upon completion of the installation, the electrical contractor shall review and check the entire installation, clean equipment and devices, and remove surplus materials and rubbish from the owner's property, leaving the work in neat and clean order and in complete working condition. The electrical contractor shall be responsible for the removal of any cartons, debris, and rubbish for equipment installed by the electrical contractor, including equipment furnished by the owner or others and removed from the carton by the electrical contractor.

34. **SMOKE DETECTORS:** Furnish and install all smoke detectors and associated wiring per manufacturer's instructions and all codes. See Clause 5. Detectors to be of the ac/dc type. Interconnect to provide simultaneous signaling.

35. **CARBON MONOXIDE DETECTOR:** Furnish and install all carbon monoxide detectors and associated wiring as required by the applicable building code. Detector(s) to be of the ac type with battery backup.

36. **SPECIAL-PURPOSE OUTLETS:** Install, provide, and connect all wiring for all special-purpose outlets. Upon completion of the job, all luminaires and appliances shall be operating properly. See plans and other sections of the specifications for information as to who is to furnish the luminaires and appliances.

37. **AFCI AND GFCI:** Furnish and install ground-fault circuit interrupter (GFCI) and arc-fault circuit interrupter (AFCI) protection as required by the latest edition of the *National Electrical Code*.

38. **CEILING PADDLE FANS:** The electrical contractor shall furnish and install outlet boxes identified as suitable for fan support at all locations where a ceiling suspended paddle fan is likely to be installed. The ceiling-suspended paddle fans, controls, timers, and lamps for same to be furnished by the owner and installed by the electrical contractor.

39. **ELECTRIC VEHICLE CHARGING OUTLET:** Consult with owner to determine whether they will want an outlet installed in the garage for an electric vehicle charger. (This is shown on the plans as "Optional.") Determine the voltage and current requirements for the charger. Update the load calculation for the panelboard where the branch circuit will originate as well as for the service. Make any modification to the electrical system design that is necessary.

Schedule of Special-Purpose Outlets.

Symbol	Description	Volts	Horse-power	Appliance Ampere Rating	Total Appliance Wattage Rating (or VA)	Circuit Ampere Rating	Poles	Wire Size THHN	Circuit Number	Comments
A	Hydromassage tub, Master Bedroom	120	½	10	1200	15	1	14	A9	Connect to separate circuit. Class "A" GFCI circuit breaker.
B	Water pump	240	1	8	1920	20	2	12	A(5–7)	Run circuit to disconnect switch on wall adjacent to pump. Install 10-ampere dual-element fuses as backup protection. Pump controller has integral overload protection.
C	Water heater: top element 4500 W. Bottom element 4500 W.	240	–0–	18.8	4500	30	2	10	A(6–8)	Connected for limited demand. Both elements cannot operate at the same time.
D	Dryer	120/240	120 V ⅙ Motor Only	23.75	5700 Total	30	2	10	B(1–3)	Provide flush-mounted 30 A dryer receptacle.
E	Overhead garage door opener	120	See Unit 16 for explanation of overhead door operator horsepower rating	5.8	696	15	1	14	B14	Unit comes with 3-wire cord. Provide box-cover unit (fuse/switch). Install Fustat Type S fuse, 8 amperes. Unit has integral protection. Fustat fuses are additional backup protection. Connect to garage lighting circuit. Shall be GFCI protected
F	Wall-mounted oven	120/240	–0–	27.5	6600	30	2	10	B(6–8)	
G	Countertop range	120/240	–0–	31	7450	40	2	8	B(2–4)	
H	Food waste disposer	120	⅓	7.2	864	20	1	12	B19	Controlled by SP switch on wall to the right of sink. Provide with both AFCI and GFCI protection.
I	Dishwasher	120	⅓	9.2 Includes 875-W heater	1104	20	1	12	B5	Provide with GFCI protection.
J	Heat/vent/light Master Bedroom bath	120	–0–	12.5	1500	20	1	12	A12	
K	Heat/vent/light Front Bedroom bath	120	–0–	12.5	1500	20	1	12	A11	
L	Attic exhaust fan	120	¼	5.8	696	15	1	14	A10	Run circuit to 4 in. square box. Locate near fan in attic. Provide box-cover unit. Install Fustat Type S fuse, 8 amperes. Unit has integral protection. Fustat fuses provide additional back-up protection.
M	Electric furnace	240	⅓ Motor Only	Motor 3.5 heater 50.7 Total 54.2	13000	70	2	4	A(1–3)	The overcurrent device and branch-circuit conductors shall not be less than 125% of the total load of the heaters and motor. 54.2 × 1.25 = 67.75 [*424.3(B)*].
N	Air conditioner	240	–0–	26.4	6336	45	2	10	A(2–4)	
O	Freezer	120	¼	5.8	696	15	1	14	A13	Install single receptacle outlet.

Appendix

Useful Formulas			
To Find	**Single-Phase**	**Three-Phase**	**Direct Current**
AMPERES when kVA is known	$\frac{\text{kVA} \times 1000}{\text{E}}$	$\frac{\text{kVA} \times 1000}{\text{E} \times 1.732}$	Not applicable
AMPERES when horsepower is known	$\frac{\text{HP} \times 746}{\text{E} \times \text{\% eff.} \times \text{pf}}$	$\frac{\text{HP} \times 746}{\text{E} \times 1.732 \times \text{\% eff.} \times \text{pf}}$	$\frac{\text{HP} \times 746}{\text{E} \times \text{\% eff.}}$
AMPERES when kilowatts is known	$\frac{\text{kW} \times 1000}{\text{E} \times \text{pf}}$	$\frac{\text{kW} \times 1000}{\text{E} \times 1.732 \times \text{pf}}$	$\frac{\text{kW} \times 1000}{\text{E}}$
HORSEPOWER	$\frac{\text{E} \times \text{I} \times \text{\% eff.} \times \text{pf}}{746}$	$\frac{\text{E} \times \text{I}\ 1.732 \times \text{\% eff.} \times \text{pf}}{746}$	$\frac{\text{E} \times \text{I} \times \text{\% eff.}}{746}$
KILOVOLT AMPERES	$\frac{\text{E} \times \text{I}}{1000}$	$\frac{\text{E} \times \text{I} \times 1.732}{1000}$	Not applicable
KILOWATTS	$\frac{\text{E} \times \text{I} \times \text{pf}}{1000}$	$\frac{\text{E} \times \text{I} \times 1.732 \times \text{pf}}{1000}$	$\frac{\text{E} \times \text{I}}{1000}$
VOLT-AMPERES	$\text{E} \times \text{I}$	$\text{E} \times \text{I} \times 1.732 \times \text{pf}$	$\text{E} \times \text{I}$
WATTS	$\text{E} \times \text{I} \times \text{pf}$	$\text{E} \times \text{I} \times 1.732 \times \text{pf}$	$\text{E} \times \text{I}$

$$\text{ENERGY EFFICIENCY} = \frac{\text{Load Horsepower} \times 746}{\text{Load Input kVA} \times 1000}$$

$$\text{POWER FACTOR (pf)} = \frac{\text{Power Consumed}}{\text{Apparent power}} = \frac{\text{W}}{\text{VA}} = \frac{\text{kW}}{\text{kVA}} = \cos\varnothing$$

I = Amperes E = Volts kW = Kilowatts kVA = Kilovolt-amperes

HP = Horsepower % eff. = Percent Efficiency e.g., 90% eff. is 0.90

pf = Power Factor e.g., 95% pf is 0.95

EQUATIONS BASED ON OHM'S LAW:

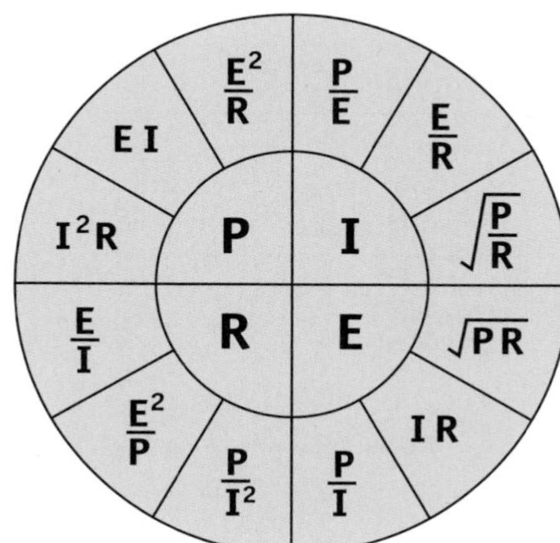

P = POWER, IN WATTS
I = CURRENT, IN AMPERES
R = RESISTANCE, IN OHMS
E = ELECTROMOTIVE FORCE, IN VOLTS

METRIC SYSTEM OF MEASUREMENT

The metric system is known as the International System of Units (SI), taken from the French Le Système International d'Unités. The following conversion table was developed from the latest National Institute of Standards and Technology (NIST) publications. In the table, whenever the slant line (/) is found, say it as "per."

The NIST website is www.nist.gov.

The *NEC* shows both inch-pound and metric units. The term *inch-pound* refers to the way dimensions (i.e., inches, feet, yards), weights (i.e., ounces, pounds), and volume (i.e., cubic inches, cubic feet) are given in the United States. Inch-pound units are also called U.S. Customary units. Whenever possible, use the measurements found in the *NEC* instead of making your own conversions.

Throughout the *NEC*, inch-pound values were converted to metric units in accordance to specific rules established by the National Fire Protection Association.

Soft Metric Conversion

A *soft conversion* is a direct mathematical conversion. A soft conversion is the result of taking a product already designed, manufactured to inch-pound values, then converting those values to metric units. The products dimension does not change in any way. Only the measurement units change. A *soft conversion* is more precise than a *hard conversion*.

Converting an inch-pound measurement to a metric measurement always results in bizarre, fractional values. The same holds true when converting a metric measurement to an inch-pound measurement.

See Example 1 for a soft conversion where there is a real safety issue.

Hard Metric Conversion

A *hard metric conversion* results in a dimension or size that is less precise that a *soft conversion* and does not compromise safety. An inch-pound standard size is replaced with an accepted metric standard size for a particular purpose.

In the *NEC*, to obtain a more convenient numerical expression and eliminate awkward values, an "exact metric conversion" is often rounded (up or down) to some value within acceptable tolerances of the original value, making sure that the final measurement does not violate a "maximum"

EXAMPLE 1

NEC Table 110.26(A)(1) gives the minimum depth of the working space in front of electrical equipment that is likely to require examination, adjustment, servicing, or maintenance while energized. Soft, or more precise conversion of the inch-pound dimension is evident due to the concern of not reducing the working space by conversion:

	Minimum Clear Distance					
Nominal Voltage to Ground	**Condition 1 (Soft Conversion)**	**Hard Conversion**	**Condition 2 (Soft Conversion)**	**Hard Conversion**	**Condition 3 (Soft Conversion)**	**Hard Conversion**
0–150	914 mm (3 ft)	900 mm (35.4 in.)	914 mm (3 ft)	900 mm (35.4 in.)	914 mm (3 ft)	900 mm (35.4 in.)
151–600	914 mm (3 ft)	900 mm (35.4 in.)	1.07 m (3 ft 6 in.)	1 m (39.4 in.)	1.22 m (4 ft)	1.2 m (47.2 in).

or "minimum" *Code* requirement. This is called a *hard conversion*.

A hard metric dimension is also attained by designing a product to metric dimensions. No conversion from inch-pound units is involved. For example, a box is designed to measure 50 × 100 × 100 mm. It's a done deal.

See Example 2 for a hard conversion where safety is not an issue.

EXAMPLE 2

Type NM cable shall be secured at intervals of 4½ ft (1.4 m) and not more than 12 in. (300 mm) from an electrical box. The exact conversion of 4½ ft is 1.3716 m and of 12 in. is 304.8 mm. These values would be cumbersome to use, so the *NEC* rounded off the measurements to obtain convenient numerical expressions.

Multiply This Unit(s)	By This Factor	To Obtain This Unit(s)
acre	4 046.9	square meters (m^2)
acre	43 560	square feet (ft^2)
ampere hour	3 600	coulombs (C)
angstrom	0.1	nanometer (nm)
atmosphere	101.325	kilopascals (kPa)
atmosphere	33.9	feet of water (at 4°C)
atmosphere	29.92	inches of mercury (at 0°C)
atmosphere	0.76	meter of mercury (at 0°C)
atmosphere	0.007 348	ton per square inch
atmosphere	1.058	tons per square foot
atmosphere	1.0333	kilograms per square centimeter
atmosphere	10 333	kilograms per square meter
atmosphere	14.7	pounds per square inch
Bar	100	kilopascals (kPa)
barrel (oil, 42 U.S. gallons)	0.158 987 3	cubic meter (m^3)
barrel (oil, 42 U.S. gallons)	158.987 3	liters (L)
board foot	0.002 359 737	cubic meter (m^3)
bushel	0.035 239 07	cubic meter (m^3)
Btu	778.16	foot-pounds
Btu	252	gram-calories
Btu	0.000 393 1	horsepower-hour
Btu	1 054.8	joules (J)
Btu	1.055 056	kilojoules (kJ)
Btu	0.000 293 1	kilowatt-hour (kWh)
Btu per hour	0.000 393 1	horsepower (hp)
Btu per hour	0.293 071 1	watt (W)
Btu per degree Fahrenheit	1.899 108	kilojoules per kelvin (kJ/K)
Btu per pound	2.326	kilojoules per kilogram (kJ/kg)
Btu per second	1.055 056	kilowatts (kW)
calorie	4.184	joules (J)
calorie, gram	0.003 968 3	Btu
candela per feet squared (cd/ft^2)	10.763 9	candelas per meter squared (cd/m^2)
Note: The former term *candlepower* has been replaced with the term *candela*.		

(*continues*)

(*Continued*)

Multiply This Unit(s)	By This Factor	To Obtain This Unit(s)
candela per meter squared (cd/m^2)	0.092 903	candela per feet squared (cd/ft^2)
candela per meter squared (cd/m^2)	0.291 864	footlambert*
candela per square inch (cd/in.^{2})	1 550.003	candelas per square meter (cd/m^2)
Celsius = (Fahrenheit − 32) × 5/9		
Celsius = (Fahrenheit − 32) × 0.555555		
Celsius = (0.556 × Fahrenheit) − 17.8		
Note: The term *centigrade* was officially discontinued in 1948, and was replaced by the term *Celsius.* The term *centigrade* may still be found in some publications.		
centimeter (cm)	0.032 81	foot (ft)
centimeter (cm)	0.393 7	inch (in.)
centimeter (cm)	0.01	meter (m)
centimeter (cm)	10	millimeters (mm)
centimeter (cm)	393.7	mils
centimeter (cm)	0.010 94	yard
circular mil	0.000 005 067	square centimeter (cm.^{2})
circular mil	0.785 4	square mil
circular mil	0.000 000 785 4	square inch (in.^{2})
cubic centimeter (cm^3)	0.061 02	cubic inch (in.^{3})
cubic foot per second	0.028 316 85	cubic meter per second (m^3/s)
cubic foot per minute	0.000 471 947	cubic meter per second (m^3/s)
cubic foot per minute	0.471 947	liter per second (L/s)
cubic inch (in.^{3})	16.39	cubic centimeters (cm^3)
cubic inch (in.^{3})	0.000 578 7	cubic foot (ft^3)
cubic meter (m^3)	35.31	cubic feet (ft^3)
cubic yard per minute	12.742 58	liters per second (L/s)
cup (c)	0.236 56	liter (L)
decimeter	0.1	meter (m)
decameter	10	meters (m)
Fahrenheit = (9/5 Celsius) + 32		
Fahrenheit = (Celsius × 1.8) + 32		
fathom	1.828 804	meters (m)
fathom	6.0	feet
foot	30.48	centimeters (cm)
foot	12	inches
foot	0.000 304 8	kilometer (km)
foot	0.304 8	meter (m)
foot	0.000 189 4	Mile (statute)
foot	304.8	millimeters (mm)
foot	12 000	mils
foot	0.333 33	yard
foot, cubic (ft^3)	0.028 316 85	cubic meter (m^3)
foot, cubic (ft^3)	28.316 85	liters (L)
foot, board	0.002 359 737	cubic meter (m^3)
cubic feet per second (ft^3/s)	0.028 316 85	cubic meter per second (m^3/s)
cubic feet per minute (ft^3/min)	0.000 471 947	cubic meter per second (m^3/s)
cubic feet per minute (ft^3/min)	0.471 947	liter per second (L/s)

Multiply This Unit(s)	By This Factor	To Obtain This Unit(s)
foot, square (ft^2)	0.092 903	square meter (m^2)
footcandle	10.763 91	lux (lx)
footlambert*	3.426 259	candelas per square meter (cd/m^2)
foot of water	2.988 98	kilopascals (kPa)
foot-pound	0.001 286	Btu
foot-pound-force	1 055.06	joules (J)
foot-pound-force per second	1.355 818	joules (J)
foot-pound-force per second	1.355 818	watts (W)
foot per second	0.304 8	meter per second (m/s)
foot per second squared	0.304 8	meter per second squared (m/s^2)
gallon (U.S. liquid)	3.785 412	liters (L)
gallons per day	3.785 412	liters per day (L/d)
gallons per hour	1.051 50	milliliters per second (mL/s)
gallons per minute	0.063 090 2	liter per second (L/s)
gauss	6.452	lines per square inch
gauss	0.1	millitesla (mT)
gauss	0.000 000 064 52	weber per square inch
grain	64.798 91	milligrams (mg)
gram (g) (a little more than the weight of a paper clip)	0.035 274	ounce (avoirdupois)
gram (g)	0.002 204 6	pound (avoirdupois)
gram per meter (g/m)	3.547 99	pounds per mile (lb/mile)
grams per square meter (g/m^2)	0.003 277 06	ounces per square foot (oz/ft^2)
grams per square meter (g/m^2)	0.029 494	ounces per square yard (oz/yd^2)
gravity (standard acceleration)	9.806 65	meters per second squared (m/s^2)
quart (U.S. liquid)	0.946 352 9	liter (L)
horsepower (550 ft • lbf/s)	0.745 7	kilowatt (kW)
horsepower	745.7	watts (W)
horsepower hours	2.684 520	megajoules (MJ)
inch per second squared (in./s^2)	0.025 4	meter per second squared (m/s^2)
inch	2.54	centimeters (cm)
inch	0.254	decimeter (dm)
inch	0.083 3	feet
inch	0.025 4	meter (m)
inch	25.4	millimeters (mm)
inch	1 000	mils
inch	0.027 78	yard
inch, cubic (in.^{3})	16 387.1	cubic millimeters (mm^3)
inch, cubic (in.^{3})	16.387 06	cubic centimeters (cm^3)
inch, cubic (in.^{3})	645.16	square millimeters (mm^2)
inches of mercury	3.386 38	kilopascals (kPa)
inches of mercury	0.033 42	atmosphere
inches of mercury	1.133	feet of water
inches of water	0.248 84	kilopascal (kPa)
inches of water	0.073 55	inch of mercury

(*continues*)

(*Continued*)

Multiply This Unit(s)	By This Factor	To Obtain This Unit(s)
joule (J)	0.737 562	foot-pound-force (ft•lbf)
kilocandela per meter squared (kcd/m^2)	0.314 159	lambert*
kilogram (kg)	2.204 62	pounds (avoirdupois)
kilogram (kg)	35.274	ounces (avoirdupois)
kilogram per meter (kg/m)	0.671 969	pound per foot (lb/ft)
kilogram per square meter (kg/m^2)	0.204 816	pound per square foot (lb/ft^2)
kilogram-meter-squared (kg•m^2)	23.730 4	pounds-foot squared (lb•ft^2)
kilogram-meter-squared (kg•m^2)	3 417.17	pounds-inch squared (lb•in.^{2})
kilogram per cubic meter (kg/m^3)	0.062 428	pound per cubic foot (lb/ft^3)
kilogram per cubic meter (kg/m^3)	1.685 56	pound per cubic yard (lb/yd^3)
kilogram per second (kg/s)	2.204 62	pounds per second (lb/s)
kilojoule (kJ)	0.947 817	Btu
kilometer (km)	1 000	meters (m)
kilometer (km)	0.621 371	mile (statute)
kilometer (km)	1,000,000	millimeters (mm)
kilometer (km)	1 093.6	yards
kilometer per hour (km/h)	0.621 371	mile per hour (mph)
kilometer squared (km^2)	0.386 101	square mile (mile^2)
kilopound-force per square inch	6.894 757	megapascals (MPa)
kilowatts (kW)	56.921	Btus per minute
kilowatts (kW)	1.341 02	horsepower (hp)
kilowatts (kW)	1 000	watts (W)
kilowatt-hour (kWh)	3 413	Btus
kilowatt-hour (kWh)	3.6	megajoules (MJ)
knots	1.852	kilometers per hour (km/h)
lamberts*	3 183.099	candelas per square meter (cd/m^2)
lamberts*	3.183 01	kilocandelas per square meter (kcd/m^2)
liter (L)	0.035 314 7	cubic foot (ft^3)
liter (L)	0.264 172	gallon (U.S. liquid)
liter (L)	2.113	pints (U.S. liquid)
liter (L)	1.056 69	quarts (U.S. liquid)
liter per second (L/s)	2.118 88	cubic feet per minute (ft^3/min)
liter per second (L/s)	15.850 3	gallons per minute (gal/min)
liter per second (L/s)	951.022	gallons per hour (gal/hr)
lumen per square foot (lm/ft^2)	10.763 9	lux (lx)
lumen per square foot (lm/ft^2)	1.0	footcandles
lumen per square foot (lm/ft^2)	10.763	lumens per square meter (lm/m^2)
lumen per square meter (lm/m^2)	1.0	lux (lx)
lux (lx)	0.092 903	lumen per square foot (foot-candle)
maxwell	10	nanowebers (nWb)
megajoule (MJ)	0.277 778	kilowatt-hours (kWh)
meter (m)	100	centimeters (cm)
meter (m)	0.546 81	fathom
meter (m)	3.2809	feet
meter (m)	39.37	Inches

Multiply This Unit(s)	By This Factor	To Obtain This Unit(s)
meter (m)	0.001	kilometer (km)
meter (m)	0.000 621 4	mile (statute)
meter (m)	1 000	millimeters (mm)
meter (m)	1.093 61	yards
meter, cubic (m^3)	1.307 95	cubic yards (yd^3)
meter, cubic (m^3)	35.314 7	cubic feet (ft^3)
meter, cubic (m^3)	423.776	board feet
meter per second (m/s)	3.280 84	feet per second (ft/s)
meter per second (m/s)	2.236 94	miles per hour (mph)
meter, squared (m^2)	1.195 99	square yards (yd^2)
meter, squared (m^2)	10.763 9	square feet (ft^2)
mho per centimeter (mho/cm)	100	siemens per meter (S/m)
Note: The older term *mho* has been replaced with *siemens*. The term *mho* may still be found in some publications.		
micro inch	0.025 4	micrometer (μm)
mil	0.002 54	centimeters
mil	0.000 083 33	feet
mil	0.001	Inches
mil	0.000 000 025 40	kilometers
mil	0.000 027 78	yards
mil	25.4	micrometers (μm)
mil	0.025 4	millimeter (mm)
miles per hour	1.609 344	kilometers per hour (km/h)
miles per hour	0.447 04	meter per second (m/s)
miles per gallon	0.425 143 7	kilometer per liter (km/L)
miles	1.609 344	kilometers (km)
miles	5 280	feet
miles	1 609	meters (m)
miles	1 760	yards
miles (nautical)	1.852	kilometers (km)
miles squared	2.590 000	kilometers squared (km^2)
millibar	0.1	kilopascal (kPa)
milliliter (mL)	0.061 023 7	cubic inch ($in.^3$)
milliliter (mL)	0.033 814	fluid ounce (U.S.)
millimeter (about the thickness of a dime)	0.1	centimeter (cm)
millimeter (mm)	0.003 280 8	foot
millimeter (mm)	0.039 370 1	Inch
millimeter (mm)	0.001	meter (m)
millimeter (mm)	39.37	mils
millimeter (mm)	0.001 094	yard
millimeter squared (mm^2)	0.001 550	square inch ($in.^2$)
millimeter cubed (mm^3)	0.000 061 023 7	cubic inches ($in.^3$)
millimeter of mercury	0.133 322 4	kilopascal (kPa)
ohm	0.000 001	megohm
ohm	1,000,000	micro ohms
ohm circular mil per foot	1.662 426	nano ohms meter (nΩ•m)

(*continues*)

(*Continued*)

Multiply This Unit(s)	By This Factor	To Obtain This Unit(s)
oersted	79.577 47	amperes per meter (A/m)
ounce (avoirdupois)	28.349 52	grams (g)
ounce (avoirdupois)	0.062 5	pound (avoirdupois)
ounce, fluid	29.573 53	milliliters (mL)
ounce (troy)	31.103 48	grams (g)
ounce per foot squared (oz/ft^2)	305.152	grams per meter squared (g/m^2)
ounces per gallon (U.S. liquid)	7.489 152	grams per liter (g/L)
ounce per yard squared (oz/yd^2)	33.905 7	grams per meter squared (g/m^2)
pica	4.217 5	millimeters (mm)
pint (U.S. liquid)	0.473 176 5	liter (L)
pint (U.S. liquid)	473.177	milliliters (mL)
pound (avoirdupois)	453.592	grams (g)
pound (avoirdupois)	0.453 592	kilograms (kg)
pound (avoirdupois)	16	ounces (avoirdupois)
poundal	0.138 255	newton (N)
pound-foot (lb•ft)	0.138 255	kilogram-meter (kg•m)
pound-foot per second	0.138 255	kilogram-meter per second (kg•m/s)
pound-foot squared (lb•ft^2)	0.042 140 1	kilogram-meter squared (kg•m^2)
pound-force	4.448 222	newtons (N)
pound-force foot	1.355 818	newton-meters (N•m)
pound-force inch	0.112 984 8	newton-meter (N•m)
pound-force per square inch	6.894 757	kilopascals (kPA)
pound-force per square foot	0.047 880 26	kilopascal (kPa)
pound per cubic foot (lb/ft^3)	16.018 46	kilograms per cubic meter (kg/m^3)
pound per foot (lb/ft)	1.488 16	kilograms per meter (kg/m)
pound per foot squared (lb/ft^2)	4.882 43	kilograms per meter squared (kg/m^2)
pound per foot cubed (lb/ft^3)	16.018 5	kilograms per meter cubed (kg/m^3)
pound per gallon (U.S. liquid)	119.826 4	grams per liter (g/L)
pound per second (lb/s)	0.453 592	kilogram per second (kg/s)
pound-inch squared (lb•in.^{2})	292.640	kilograms-millimeter squared (kg•mm^2)
pound per mile	0.281 849	gram per meter (g/m)
pound-square foot (lb•ft^2)	0.042 140 11	kilogram square meter (kg/m^2)
pound per cubic yard (lb/yd^3)	0.593 276	kilogram per cubic meter (kg/m^3)
quart (U.S. liquid)	946.353	milliliters (mL)
square centimeter (cm^2)	197 300	circular mils
square centimeter (cm^2)	0.001 076	square foot (ft^2)
square centimeter (cm^2)	0.155	square inch (in.^{2})
square centimeter (cm^2)	0.000 1	square meter (m^2)
square centimeter (cm^2)	0.000 119 6	square yard (yd^2)
square foot (ft^2)	144	square inches (in.^{2})
square foot (ft^2)	0.092 903 04	square meter (m^2)
square foot (ft^2)	0.111 1	square yard (yd^2)
square inch (in.^{2})	1 273 000	circular mils
square inch (in.^{2})	6.4516	square centimeters (cm^2)
square inch (in.^{2})	0.006 944	square foot (ft^2)

Multiply This Unit(s)	By This Factor	To Obtain This Unit(s)
square inch (in.^{2})	645.16	square millimeters (mm^2)
square inch (in.^{2})	1,000,000	square mils (m^2)
square meter (m^2)	10.764	square feet (ft^2)
square meter (m^2)	1 550	square inches (in.^{2})
square meter (m^2)	0.000 000 386 1	square mile
square meter (m^2)	1.196	square yards (yd^2)
square mil	1.273	circular mils
square mil	0.000 001	square inch (in.^{2})
square mile	2.589 988	square kilometers (km^2)
square millimeter (mm^2)	1 973	circular mils
square yard (yd^2)	0.836 127 4	square meter (m^2)
tablespoon (tbsp)	14.786 75	milliliters (mL)
teaspoon (tsp)	4.928 916 7	milliliters (mL)
therm	105.480 4	megajoules (MJ)
ton (long) (2 240 lb)	1 016.047	kilograms (kg)
ton (long) (2 240 lb)	1.016 047	metric tons (t)
ton, metric	2 204.62	pounds (avoirdupois)
ton, metric	1.102 31	tons, short (2 000 lb)
ton, refrigeration	12 000	Btus per hour
ton, refrigeration	4.716 095 9	horsepower-hours
ton, refrigeration	3.516 85	kilowatts (kW)
ton (short) (2 000 lb)	907.185	kilograms (kg)
ton (short) (2 000 lb)	0.907 185	metric ton (t)
ton per cubic meter (t/m^3)	0.842 778	ton per cubic yard (ton/yd^3)
ton per cubic yard (ton/yd^3)	1.186 55	tons per cubic meter (ton/m^3)
torr	133.322 4	pascals (Pa)
watt (W)	3.412 14	Btus per hour (Btu/hr)
watt (W)	0.001 341	horsepower
watt (W)	0.001	kilowatt (kW)
watt-hour (Wh)	3.413	Btus
watt-hour (Wh)	0.001 341	horsepower-hour
watt-hour (Wh)	0.001	kilowatt-hour (kW/h)
yard	91.44	centimeters (cm)
yard	3	feet
yard	36	Inches
yard	0.000 914 4	kilometer (km)
yard	0.914 4	meter (m)
yard	914.4	millimeters
yard, cubic (yd^3)	0.764 55	cubic meter (m^3)
yard, squared (yd^2)	0.836 127	meter squared (m^2)

*These terms are no longer used, but may still be found in some publications.

ARCHITECTURAL DRAFTING SYMBOLS

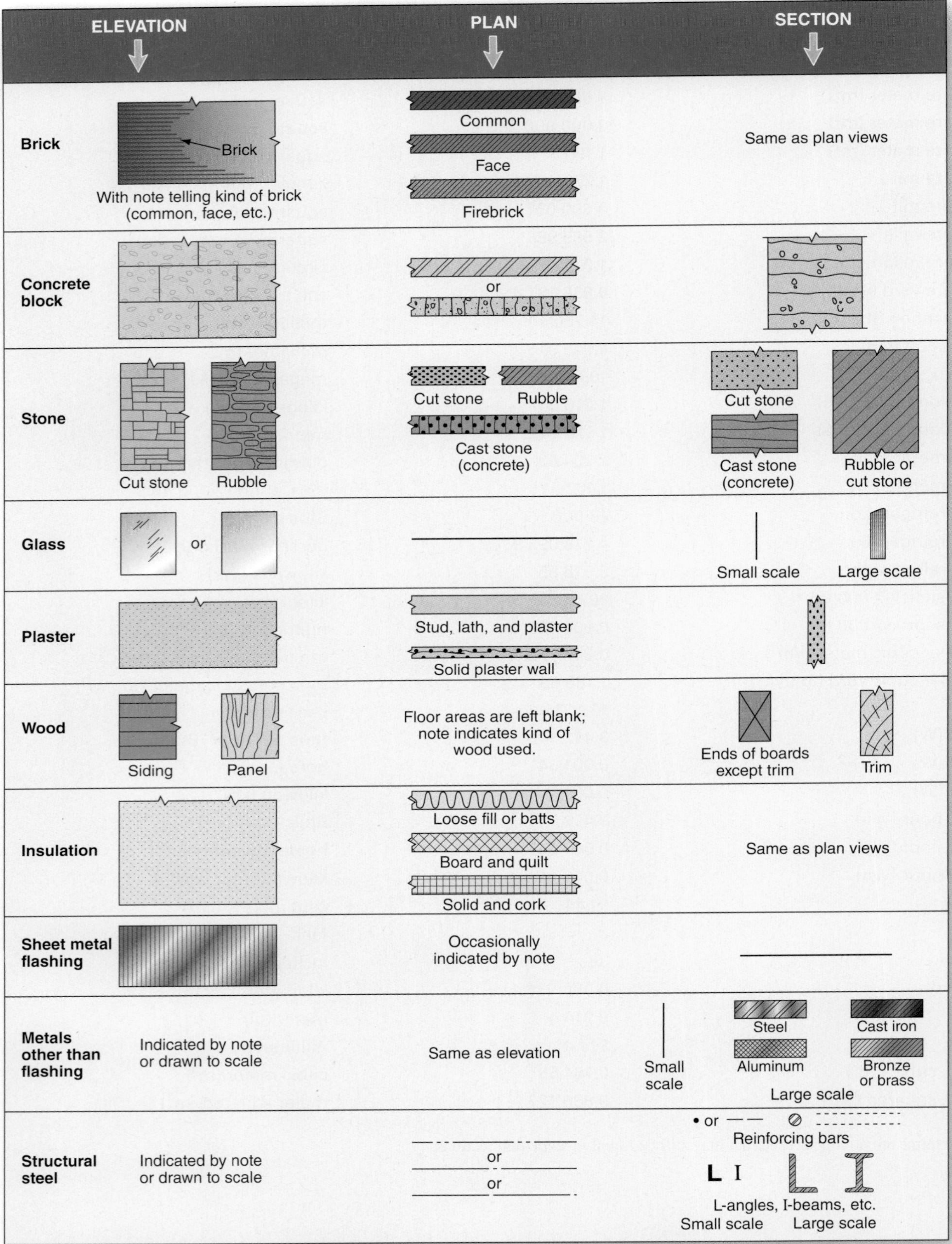

ARCHITECTURAL DRAFTING SYMBOLS (*Continued*)

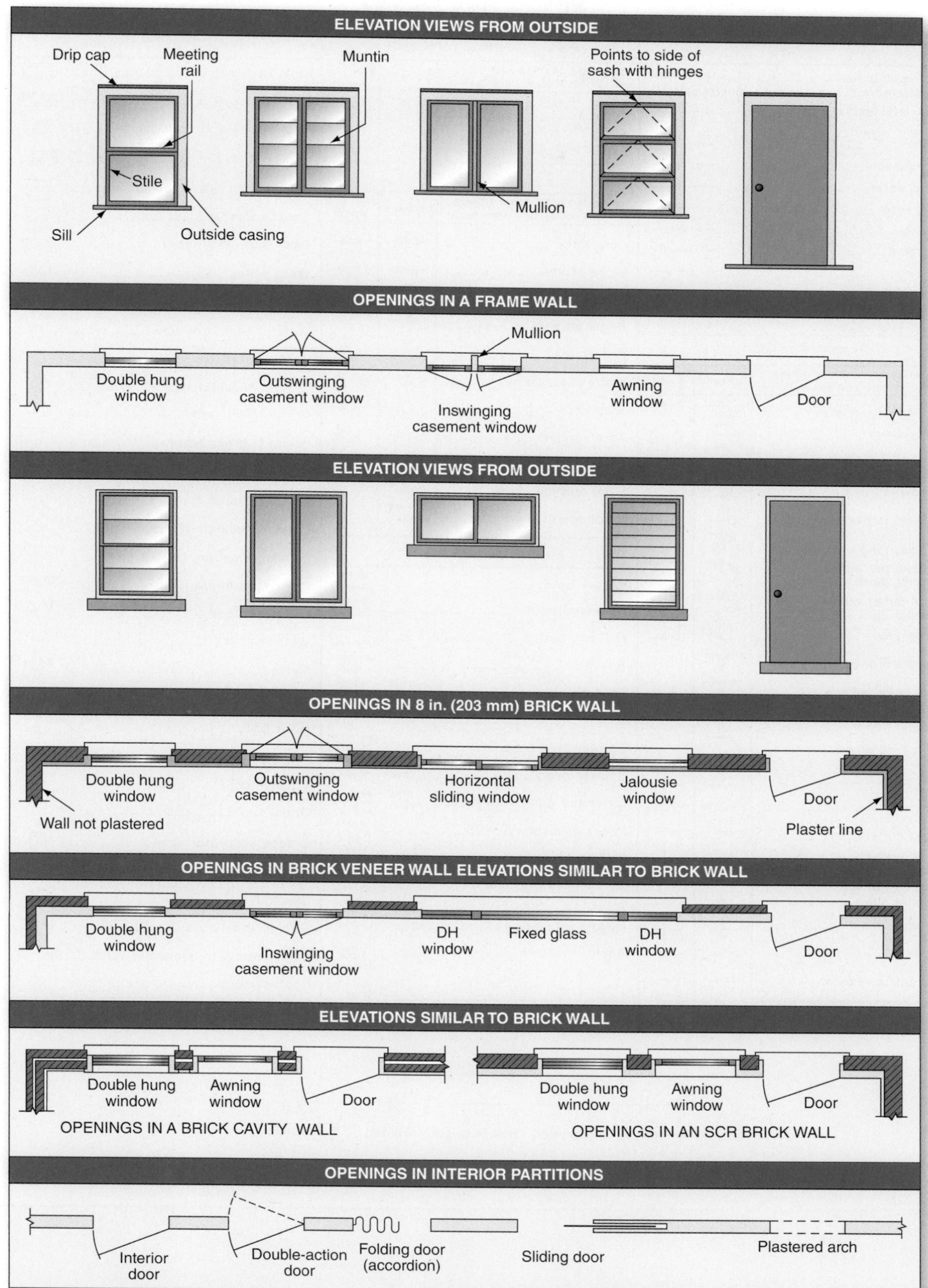

STANDARD SYMBOLS FOR PLUMBING, PIPING, AND VALVES

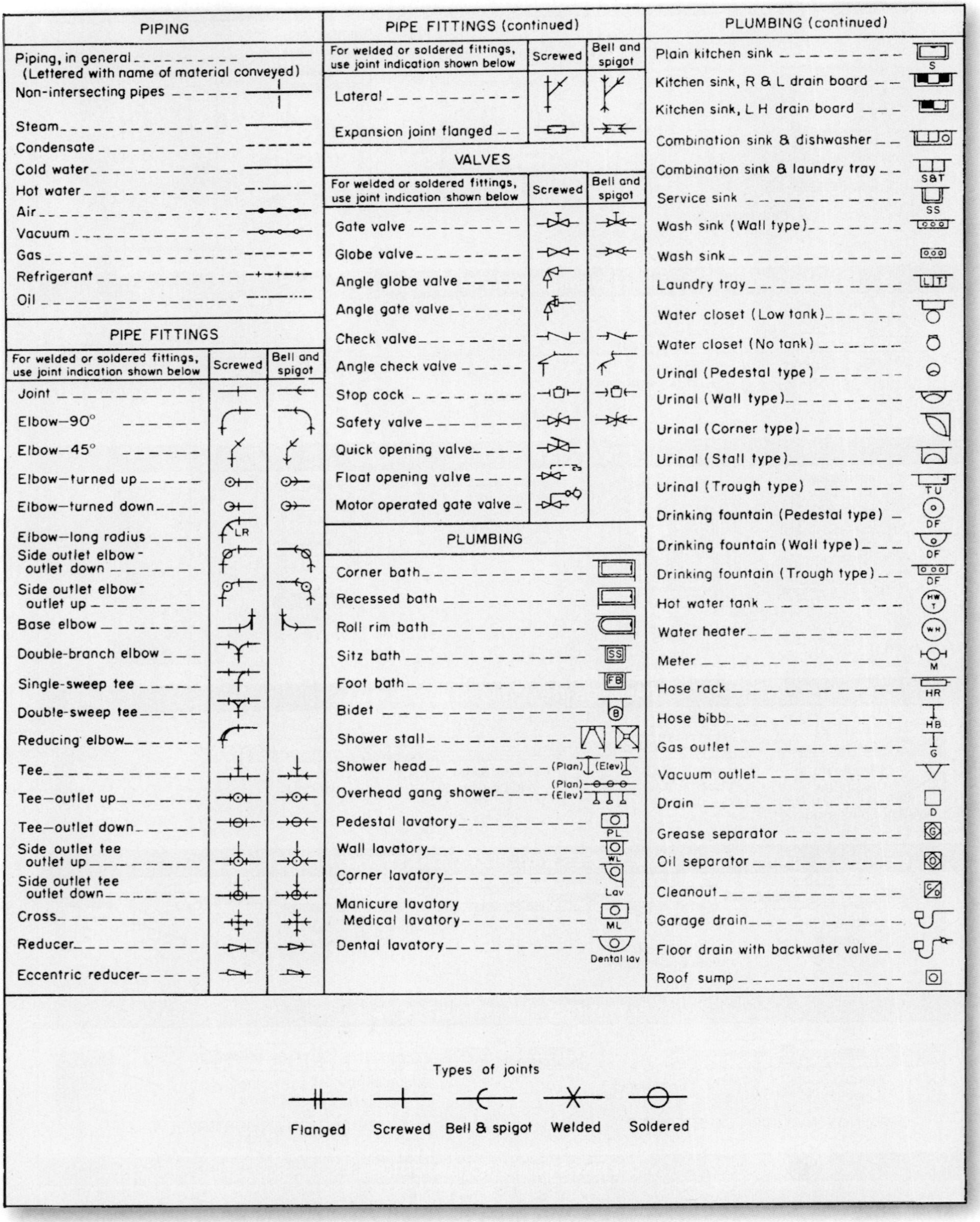

PIPING

	Symbol
Piping, in general (Lettered with name of material conveyed)	
Non-intersecting pipes	
Steam	
Condensate	
Cold water	
Hot water	
Air	
Vacuum	
Gas	
Refrigerant	
Oil	

PIPE FITTINGS

For welded or soldered fittings, use joint indication shown below	Screwed	Bell and spigot
Joint		
Elbow—90°		
Elbow—45°		
Elbow—turned up		
Elbow—turned down		
Elbow—long radius	LR	
Side outlet elbow-outlet down		
Side outlet elbow-outlet up		
Base elbow		
Double-branch elbow		
Single-sweep tee		
Double-sweep tee		
Reducing elbow		
Tee		
Tee—outlet up		
Tee—outlet down		
Side outlet tee outlet up		
Side outlet tee outlet down		
Cross		
Reducer		
Eccentric reducer		

PIPE FITTINGS (continued)

For welded or soldered fittings, use joint indication shown below	Screwed	Bell and spigot
Lateral		
Expansion joint flanged		

VALVES

For welded or soldered fittings, use joint indication shown below	Screwed	Bell and spigot
Gate valve		
Globe valve		
Angle globe valve		
Angle gate valve		
Check valve		
Angle check valve		
Stop cock		
Safety valve		
Quick opening valve		
Float opening valve		
Motor operated gate valve		

PLUMBING

	Symbol
Corner bath	
Recessed bath	
Roll rim bath	
Sitz bath	SS
Foot bath	FB
Bidet	B
Shower stall	
Shower head	(Plan) (Elev)
Overhead gang shower	(Plan) (Elev)
Pedestal lavatory	PL
Wall lavatory	WL
Corner lavatory	Lav
Manicure lavatory Medical lavatory	ML
Dental lavatory	Dental lav

PLUMBING (continued)

	Symbol
Plain kitchen sink	S
Kitchen sink, R & L drain board	
Kitchen sink, L H drain board	
Combination sink & dishwasher	
Combination sink & laundry tray	S&T
Service sink	SS
Wash sink (Wall type)	
Wash sink	
Laundry tray	LT
Water closet (Low tank)	
Water closet (No tank)	
Urinal (Pedestal type)	
Urinal (Wall type)	
Urinal (Corner type)	
Urinal (Stall type)	
Urinal (Trough type)	TU
Drinking fountain (Pedestal type)	DF
Drinking fountain (Wall type)	DF
Drinking fountain (Trough type)	DF
Hot water tank	HW T
Water heater	WH
Meter	M
Hose rack	HR
Hose bibb	HB
Gas outlet	G
Vacuum outlet	
Drain	D
Grease separator	G
Oil separator	O
Cleanout	
Garage drain	
Floor drain with backwater valve	
Roof sump	

Types of joints

Flanged, Screwed, Bell & spigot, Welded, Soldered

SHEET METAL DUCTWORK SYMBOLS

Blank Off. adjustable	Damper automatic M	Damper deflecting	Damper deflecting up
Damper deflecting down	Damper volume PLAN ELEVATION	Duct flow direction	Duct inclined drop D
Duct inclined rise R	Duct section exhaust, return riser to 2nd floor riser to Ist floor	Duct section supply riser to 2nd floor riser to Ist floor	Duct section notation type exhaust K E place kitchen
Duct connection below joist	Fan flexible connection	Vanes	Louver & screen air intake
Ventilator, cowl PLAN ELEVATION	Ventilator, gooseneck PLAN ELEVATION	Ventilator, rainproof PLAN ELEVATION	Ventilator, standard PLAN ELEVATION
SINGLE LINE REPRESENTATION			
Supply S	Return R	Damper & retractor	Anchor PA.
Hanger H	Expansion joint	Louver opening L	Register or grille

SINGLE-FAMILY DWELLING SERVICE-ENTRANCE CALCULATIONS

1. General Lighting Load (*220.12*).

_______________ ft^2 @ 3 VA per ft^2 = _______________ VA

Note: Included in this floor area calculation are all lighting outlets and general-use receptacles. Do not include open porches, garages, or unused or unfinished spaces not adaptable for future use. See *NEC 220.12, Table 220.12,* and *220.14(J).*

2. Minimum Number of 15-ampere Lighting Branch Circuits.

$\frac{\text{Line 1}}{120} = \frac{\rule{3cm}{0.4pt}}{120}$ = _______________ amperes

then, $\frac{\text{amperes}}{15}$ $\frac{\rule{3cm}{0.4pt}}{15}$ = _______________ 15-amperes branch circuits

3. Small-Appliance Load [*210.11(C)(1), 220.52(A),* and *210.52(B)*].

(Minimum of two 20-ampere branch circuits)

_______________ branch circuits @ 1500 VA each = _______________ VA

4. Laundry Branch Circuit [*210.11(C)(2), 220.52(B),* and *210.52(F)*].

(Minimum of one 20-ampere branch circuit)

_______________ branch circuit(s) @ 1500 VA each = _______________ VA

5. Total General Lighting, Small-Appliance, and Laundry Load.

Lines 1 + 3 + 4 = _______________ VA

6. Net Calculated General Lighting, Small-Appliance, and Laundry Loads (less ranges, ovens, and "fastened-in-place" appliances). Apply demand factors from *Table 220.42*.

a. First 3000 VA @ 100% = 3000 VA

b. Line 5 __________ − 3000 = __________ @ 35% = __________ VA

Total a + b = __________ VA

7. Electric Range, Wall-Mounted Ovens, Counter-Mounted Cooking Units (*Table 220.55*). = __________ **VA**

8. Electric Clothes Dryer (*Table 220.51*). = __________ **VA**

9. Electric Furnace (*220.54*).

Air Conditioner, Heat Pump *(Article 440)*.

(Enter largest value, *220.60*) = __________ VA

10. Net Calculated General Lighting, Small-Appliance, Laundry, Ranges, Ovens, Cooktop Units, HVAC.

Lines 6 + 7 + 8 + 9 = __________ VA

11. List "Fastened-in-Place" Appliances *in addition* to Electric Ranges, Electric Clothes Dryers, Electric Space Heating, and Air-Conditioning Equipment.

Appliance		**VA Load**
Water heater:	=	______ VA
Dishwasher:	=	______ VA
Garage door opener:	=	______ VA
Food waste disposer:	=	______ VA
Water pump:	=	______ VA
Gas-fired furnace:	=	______ VA
Sump pump:	=	______ VA
Other: ______	=	______ VA
______	=	______ VA
______	=	______ VA
______	=	______ VA
Total	=	______ VA

12. Apply 75% Demand Factor (*220.53*) if Four or More "Fastened-in-Place" Appliances. If Less Than Four, Figure @ 100%. Do not include electric ranges, electric clothes dryers, electric space heating, or air-conditioning equipment.

Line 11 Total: ______ × 0.75 = ______ VA

13. Total Calculated Load (Lighting, Small-Appliance, Ranges, Dryer, HVAC, "Fastened-in-Place" Appliances).

Line 10 ______ + Line 12 ______ = ______ VA

14. Add 25% of Largest Motor (*220.50* and *430.24*).

______ × 0.25 = ______ VA

Note: The largest motor can be difficult to determine because nothing is in place when service-entrance load calculations are made. It might be an air-conditioning unit or a heat pump. If the dwelling is cooled by an evaporative cooler, the largest motor might be a water pump, a large attic exhaust fan, a large food waste disposer, or a sump pump. The additional 25% of the largest motor is a small portion of the total service-entrance load calculation.

15. Grand Total Line 13 + Line 14. = ______ VA

16. Minimum Ampacity for Ungrounded Service-Entrance Conductors.

$$\text{Amperes} = \frac{\text{Line 15}}{240} = \frac{\_\_\_\_\_\_}{240} = \_\_\_\_\_\_ \text{ amperes}$$

17. Ungrounded Conductor Size (copper). ______ AWG

Note: *310.15(B)(7)* may be used only for 120/240-volt, 3-wire, residential single-phase service-entrance conductors, service lateral conductors, and feeder conductors that carry the entire load of the service to a dwelling unit.

18. Minimum Ampacity for Neutral Service-Entrance Conductor, *220.61* and *310.15(B)(7)*. Do Not Include Straight 240-Volt Loads.

a. Line 6: = __________ VA

b. Line 7: __________ @ 0.70 = __________ VA

c. Line 8: __________ @ 0.70 = __________ VA

d. Line 11: (Include only 120-volt loads.)

__________ __________ VA

__________ __________ VA

__________ __________ VA

__________ __________ VA

__________ __________ VA

Total __________ VA

e. Line d total @ 75% demand factor if four or more per *220.53*, otherwise use 100%.

__________ × 0.75 = __________ VA

f. Add 25% of largest 120-volt motor.

__________ × 0.25 = __________ VA

Total = __________ VA

g. Total a + b + c + e + f. = __________ VA

$$\text{Amperes} = \frac{\text{Line g}}{240} = \frac{\rule{3cm}{0.4pt}}{240} = \rule{2.5cm}{0.4pt}\ \text{amperes}$$

19. Neutral Conductor Size (copper) (*220.61*). __________ AWG

Note: *NEC 310.15(B)(7)* permits the neutral conductor to be smaller than the ungrounded "hot" conductors if the requirements of *215.2, 220.61*, and *230.42* are met. *NEC 220.61* states that a feeder or service neutral load shall be the maximum unbalance of the load determined by *Article 220*. When bare conductors are used with insulated conductors, the conductors' ampacity is based on the lowest temperature rating of the insulated conductors in the raceway, *310.15(B)(4)*. The neutral conductor is not permitted to be smaller than the grounding electrode conductor, *250.24(C)(1)*.

20. Grounding Electrode Conductor Size (copper) (*Table 250.66*). __________ AWG

21. Raceway Size. __________ Trade Size

Obtain dimensional data from *Table 1*, *Table 4*, *Table 5*, and *Table 8*, *Chapter 9*, *NEC*.

SINGLE-FAMILY DWELLING SERVICE-ENTRANCE CALCULATIONS

JOB ______________________________

***Note:* See back of page for line-item instructions.**

1. General Lighting Load ______________ ft^2 @ 3 VA per ft^2 = __________ VA

2. Minimum Number of 15-Ampere Lighting Branch Circuits

$\frac{\text{Line 1}}{120} = \frac{\_\_\_\_}{120} =$ amperes, then $\frac{\text{amperes}}{15} = \frac{\_\_\_\_}{15} =$ 15 ampere lighting branch circuits

3. Small-Appliance Load ______ branch circuits @ 1,500 VA each = __________ VA

4. Laundry Branch Circuit _____ branch circuit(s) @ 1,500 VA each = __________ VA

5. Total General Lighting, Small-Appliance, and Laundry Load Lines 1 + 3 + 4 = __________ VA

6. General Lighting, Small-Appliance, Laundry

a. First 3000 VA @ 100% = __________ VA

b. Line 5 ________ − 3000 = __________ @ 35% = __________ VA

Total a + b = __________ VA

7. Electric Range, Ovens, Counter-Mounted Cooking Units = __________ VA

8. Electric Clothes Dryer = __________ VA

9. Electric Furnace, Air Conditioner, Heat Pump (Enter largest) = __________ VA

10. Add Lines 6 + 7 + 8 + 9 = __________ VA

11. List "Fastened-in-Place" Appliances *in addition* to Electric Ranges, Electric Clothes Dryers, Electric Space Heating, and Air-Conditioning Equipment.

Appliance	**VA Load**
Water heater:	= __________ VA
Dishwasher:	= __________ VA
Garage door opener:	= __________ VA
Food waste disposer:	= __________ VA
Water pump:	= __________ VA
Gas-fired furnace:	= __________ VA
Sump pump:	= __________ VA
Other: __________________	= __________ VA
______________________	= __________ VA
______________________	= __________ VA
Total	= __________ VA

12. Apply 75% Demand Factor (*220.53*) if Four or More "Fastened-in-Place" Appliances. If Less Than Four, Figure @ 100%.

Line 11 Total: ______________ × 0.75 = __________ VA

13. Total Calculated Load (Lighting, Small-Appliance, Ranges, Dryer, HVAC, "Fastened-in-Place" Appliances)

Line 10 __________ + Line 12 __________ = __________ VA

14. Add 25% of Largest Motor (*220.50* and *430.24*).

______________ × 0.25 = __________ VA

15. Grand Total Line 13 + Line 14 = ____________ VA

16. Minimum Ampacity for Ungrounded Service-Entrance Conductors

$$\text{Amperes} = \frac{\text{Line 15}}{240} = \frac{}{120} = \text{\_\_\_\_\_\_\_\_\_\_ amperes}$$

17. Ungrounded Conductor Size (copper) ____________ AWG

18. Minimum Ampacity for Neutral Service-Entrance Conductor

a. Line 6 total: = ____________ VA

b. Line 7: ____________ @ 0.70 = ____________ VA

c. Line 8: ____________ @ 0.70 = ____________ VA

d. Line 11: (Include only 120-volt loads.)

____________ VA ____________ VA

____________ VA ____________ VA

Total VA ____________ VA

e. Line d total @ 75% demand factor if four or more per 220.53; otherwise use 100%.

____________ × 0.75 = ____________ VA

f. Add 25% of largest 120-volt motor. ____________ × 0.25 = ____________ VA

g. Total a + b + c + e + f Total = ____________ VA

19. Neutral Conductor Size (copper) ____________ AWG

20. Grounding Electrode Conductor Size (copper) ____________ AWG

21. Raceway Size ____________ Trade Size (inch)

SINGLE-FAMILY DWELLING SERVICE-ENTRANCE CALCULATIONS

Line-item instructions

1. Included in this floor area calculation are all lighting outlets and general-use receptacles. Do not include open porches, garages, or unused or unfinished spaces not adaptable for future use. See *NEC 220.12*, *Table 220.12*, and *220.14(J)*.
2. No further explanation needed.
3. Minimum of two 20-ampere branch-circuits. See *210.11(C)(1)*, *210.52(B)*, and *220.52(A)*.
4. Minimum of one 20-ampere branch-circuit. See *210.11(C)(2)*, *210.52(F)*, and *220.52(B)*.
5. No further explanation needed.
6. Apply the 100% and 35% demand factors from *Table 220.42*.
7. See *Table 220.55* for load values. Typical household electric ranges come under the 8 kW level.
8. See *Table 220.54* for load values. 5000 VA minimum or nameplate rating, whichever is larger.
9. Enter largest value, *220.60*. Also see *220.51* and *Article 440*.
10. This is the total for General Lighting, Small-Appliance, Laundry, Ranges, Ovens, Cooktop Units, and HVAC.
11. List fastened-in-place appliances. See *220.53*.
12. Apply 75% demand factor. Do not include electric ranges, electric clothes dryers, electric space heating, or air-conditioning equipment. See *220.53*.
13. No further explanation needed.
14. What is the largest motor? The largest motor can be difficult to determine because nothing is in place when service-entrance load calculations are made. It might be an air-conditioning unit or a heat pump. If the dwelling is cooled by an evaporative cooler, the largest motor might be the evaporative cooler motor. It could also be a furnace fan motor, a water pump, a large attic exhaust fan, a sump pump, or a large food waste disposer. Don't waste valuable time quibbling. Enter a large volt-ampere value to be on the safe side.
15. No further explanation needed.
16. No further explanation needed.
17. Shall not be smaller than 100 amperes, *230.42(B)* and *230.79*. See *Table 310.15(B)(16)*. Special conductor sizing might be permitted. *310.15(B)(7)* may be used only for 120/240-volt, 3-wire, residential single-phase service-entrance conductors, service lateral conductors, and feeder conductors that carry the entire load of the service to a dwelling unit. Check whether this is permitted in your locality.
18. See *220.61* and *310.15(B)(7)*. Do not include straight 240-volt loads.
19. See *220.61*. *NEC 310.15(B)(7)* permits the neutral conductor to be smaller than the ungrounded "hot" conductors if the requirement of *215.2*, *220.61*, and *230.42* are met. *NEC 220.61* states that a feeder or service neutral load shall be the maximum unbalance of the load determined by *Article 220*. When bare conductors are used with insulated conductors, the conductors' ampacity is based on the ampacity of the other insulated conductors in the raceway, *310.15(B)(4)*. The neutral conductor must not be smaller than the grounding electrode conductor, *250.24(B)(1)*.
20. See *Table 250.66*. Grounding electrode conductor based on size of service-entrance conductors.
21. Obtain dimensional data from *Table 1*, *Table 4*, *Table 5*, and *Table 8*, *Chapter 9, NEC*.

AFCI/GFCI CHECK-OFF LIST

Instructions:

1) Address/Name or other description of job.

____

2) In AFCI and/or GFCI column, put a check-mark ✓ for the specific room/area if AFCI and/or GFCI protection has been provided. If not required, enter the letters **"NR."**
3) In Comments column, describe how and where AFCI and/or GFCI protection has been provided. For example, AFCI/GFCI breaker in main panel.
4) Certain rooms/areas are not required to have AFCI protection. For example, an attic, in the garage and bathroom. If the attic wiring is part of a branch circuit that does require AFCI protection, then the attic wiring would automatically have AFCI protection.

Room/Area	AFCI Protection for Branch Circuit Wiring	GFCI Protection for Receptacles	Comments
Alcove			
Attic			
Atrium			
Basement-Unfinished Area			
Basement-Finished Area			
Bathroom			
Bedroom			
Closet			
Computer Room			
Crawl Space			
Den			
Dining Room			
Foyer			
Garage			
Great Room			
Hallway			
Family Room			
Kitchen			
Laundry Room			
Library			
Living Room			
Loft			
Office/Study			
Outdoor Lighting Front			
Outdoor Lighting Back			
Pantry			
Pool			
Porch			
Powder Room			
Recreation Room			
Stairwell			
Storage Room			
Sunroom			
Toilet Room			
Utility Room			
Vestibule			
Workshop			

IAEI MEMBERSHIP HAS VALUE

When you join IAEI, you receive exclusive member benefits as well as enjoy the satisfaction of helping to support a broad range of IAEI initiatives that affect everyone who cares about electrical safety.

Along with being part of a community of electrical professionals, your member benefits include:

- Free Code book after three consecutive years of membership ($89.50 value)
- Free Subscription to IAEI print and digital magazine ($112.95 value)
- Free UL White Book ($45.00 value)
- Discounts on IAEI publications
- Discounts on IAEI education training / seminar
- Certification Programs
- Membership Rewards Discount Program

Member Type	1-Year Membership	3-Year Membership
Associate Members	$102.00	$286.00
Inspector Members*	$102.00	$286.00
Student Members**	$78.00	N/A

Contact IAEI customer service department for information on corporate membership categories.

New members, other than students, may choose the multiyear plan when they complete the application form.

MEMBERSHIP APPLICATION | PLEASE PRINT

Name - Last First M.I.

Title

Employer

Address of Applicant

City State or Province ZIP or Postal Code

(Area Code) Telephone Number

Email Date of Birth

How did you hear about IAEI?

Student applicants give school attending** Graduation date

Applicant's Signature

Chapter, where you live or work, if known (Division, where appropriate).

If previous member, give last membership number and last year of membership.

Ray C. Mullin 533800

Endorsed by Endorser's Membership Number

☐ MasterCard ☐ Visa ☐ AMEX ☐ Money Order

☐ Discover ☐ Diners Club ☐ Check

Name on Card

Charge Card Number Expiration Date

Inspector ☐ Associate ☐ Student ☐ Other ☐

Amount Paid $ Specify member type

Inspector Member MUST sign below:

I, ______________________________

meet the qualification for inspector member as described below.

*Inspector members must regularly make electrical inspections for preventing injury to persons or damage to property on behalf of a governmental agency, insurance agency, rating bureau, recognized testing laboratory or electric light and power company.

** Student member must be currently enrolled in an approved college, university, vocational technical school or trade school specializing in electrical training or approved electrical apprenticeship school.

MAIL TO: IAEI
P.O. Box 830848, Richardson
TX 75083-0848
For information call:
(972) 235-1455 (8–5 CST)

ONLINE: Join by scanning the QR Code ▸ or at iaei.org

Section	Chapter No.	Division No.

AC2014

Courtesy of International Association of Electrical Inspectors.

ELECTRICAL TERMS COMMONLY USED IN THE APPLICATION OF THE *NEC*

The following terms are used extensively in the application of the *National Electrical Code*. Many of these definitions are quotes of definitions given by the *NEC*. These have been marked with an *. Recall that definitions that have been revised from the previous edition of the *NEC* are marked with opening and closing red triangles. (▶◀) See Article 100 of the *NEC* for a full listing of *Code* definitions.

When added to the terms related to the electrical industry, they constitute a "cultural literacy" for the electrical construction industry.

A

Accent Lighting. Directional lighting to emphasize a particular object or to draw attention to a part of the field in view.

Iluminación de acento. Iluminación direccional para resaltar un objeto en especial o llamar la atención a una parte del campo de visión.

Accessible (equipment). Admitting close approach: not guarded by locked doors, elevation, or other effective means.*

Accesible (equipo) Que permite que uno se acerque: que no está protegido por puertas cerradas, elevación u otros medios efectivos.

Accessible, Readily. ▶Capable of being reached quickly for operation, renewal, or inspections, without requiring those to whom ready access is requisite to actions such as to use tools, to climb over or remove obstacles or to resort to portable ladders, chairs, and so forth.◀*

Accesible, disponible. Que se puede alcanzar rápidamente para realizar operaciones, renovaciones o inspecciones sin necesitar de aquellos para los cuales el acceso disponible es un requisito para acciones como el uso de herramientas, el salto o la remoción de obstáculos o el uso de escaleras portátiles, sillas y otros objetos similares.

Accessible (wiring methods). Capable of being removed or exposed without damaging the building structure or finish, or not permanently closed in by the structure or finish of the building.*

Accesible (métodos de cableado). Que puede se puede extraer o exponer sin dañar la estructura o el acabado de la edificación o que no está permanentemente encerrado en la estructura o el acabado de la edificación.

Addendum. Modification (change) made to the construction documents (plans and specifications) during the bidding period.

Anexo. Modificación (cambio) hecho a los documentos de construcción (planos y especificaciones) durante el período de oferta.

Adjustment Factor. A multiplier (penalty) that is applied to conductors when there are more than three current-carrying conductors in a raceway or cable. These multipliers are found in *Table 310.15(B)(3)(a)*. An adjustment factor is also referred to as a derating factor.

Factor de ajuste. Un multiplicador (penalización) que se les aplica a los conductores cuando hay más de tres conductores que transmiten corriente en un conducto para cables o en un cable. Estos multiplicadores se encuentran en la *Tabla 310.15(B)(3)(a)*. Al factor de ajuste también se lo denomina factor de reducción.

AL/CU. Terminal marking on switches and receptacles rated 30 amperes and greater, suitable for use with aluminum, copper, and copper-clad aluminum conductors. If not marked, suitable for copper conductors only.

AL/CU. Marca de terminal en interruptores y receptáculos con valor nominal de 30 amperes o más, apto para uso con conductores de aluminio, cobre y aluminio con revestimiento de cobre. Si no está marcado, es apto solo para conductores de cobre.

*Reprinted with permission from NFPA 70-2014.

Alternate Bid. Amount stated in the bid to be added or deducted from the base bid amount proposed for alternate materials and/or methods of construction.

Oferta alternativa. Monto establecido en la oferta a ser agregado o descontado del monto básico de la oferta propuesto para materiales y/o métodos de construcción alternativos.

Ambient Lighting. Lighting throughout an area that produces general illumination.

Iluminación ambiental. Iluminación en toda un área que produce una iluminación general.

Ambient Temperature. The environmental temperature surrounding the object under consideration.

Temperatura ambiente. La temperatura ambiente que rodea al objeto en cuestión.

American National Standards Institute (ANSI). An organization that identifies industrial and public requirements for national consensus standards and coordinates and manages their development, resolves national standards problems, and ensures effective participation in international standardization. ANSI does not itself develop standards. Rather, it facilitates development by establishing consensus among qualified groups. ANSI ensures that the guiding principles—consensus, due process, and openness—are followed.

American National Standards Institute ((Instituto Nacional Estadounidense de Normas, ANSI). Una organización que identifica los requisitos industriales y públicos para establecer estándares nacionales consensuados y coordina y administra su desarrollo, resuelve problemas sobre los estándares nacionales y asegura la participación efectiva en la estandarización internacional. ANSI no desarrolla estándares por sí misma. Sino que facilita el desarrollo al establecer un consenso entre grupos calificados. ANSI se asegura de que se respeten los principios rectores (consenso, debido proceso y apertura).

Ampacity. The maximum current, in amperes, that a conductor can carry continuously under the conditions of use without exceeding its temperature rating.*

Corriente máxima o admisible. La corriente máxima, en amperios, que un conductor puede transportar de forma continua cuando está expuesto a las condiciones de uso sin exceder su escala de temperatura.

Ampere. The measurement of intensity of rate of flow of electrons in an electric circuit. An ampere is the amount of current that will flow through a resistance of 1 ohm under a pressure of 1 volt.

Amperio. La medida de la intensidad de la tasa de flujo de electrones en un circuito eléctrico. Un amperio es la cantidad de corriente que fluirá a través de una resistencia de 1 ohm cuando está sometida a una presión de 1 voltio.

Appliance. Utilization equipment, generally other than industrial, normally built in standardized sizes or types, that is installed or connected as a unit to perform one or more functions such as clothes washing, air conditioning, food mixing, deep frying, and so forth.*

Electrodoméstico. Equipo de uso, generalmente no industrial, armado normalmente en medidas o tipos estandarizados, que se instala o se conecta como unidad para realizar una o más funciones como lavado de ropa, acondicionamiento de aire, mezcla de alimentos, fritura, etc.

Approved. Acceptable to the authority having jurisdiction (AHJ).*

Aprobado. Aceptable para la autoridad que tiene jurisdicción (AHJ).

Arc-Fault Circuit-Interrupter (AFCI). A device intended to provide protection from the effects of arc faults by recognizing characteristics unique to arcing and by functioning to de-energize a circuit when an arcing fault is detected.*

Interruptor de circuito por falla de arco (AFCI). Un dispositivo que tiene el objetivo de proteger de los efectos de las fallas de arco al reconocer las características únicas de los arcos y al activarse para cortar la energía del circuito cuando se detecta una falla de arco.

Architect. One who designs and supervises the construction of buildings or other structures.

Arquitecto. Persona que diseña y supervise la construcción de edificaciones u otras estructuras.

Architectural Drawing. A line drawing showing plan and/or elevation views of the proposed building for the purpose of showing the overall appearance of the building.

Plano arquitectónico. Un dibujo de líneas que muestra vista de plano y/o elevación de la edificación propuesta con el objetivo de mostrar la apariencia global de la edificación.

As-Built Drawings. Contract drawings that reflect changes made during the construction process.

Dibujos As-Built (como está construido). Planos de contrato que reflejan las modificaciones realizadas durante el proceso de construcción.

---

*Reprinted with permission from NFPA 70-2014.

As-Built Plans. When required, a modified set of working drawings that is prepared for a construction project and that includes all variances from the original working drawings that occurred during the project construction.

Planos As-Built (como está construido). Cuando corresponde, un conjunto modificado de planos de trabajo que se prepara para un proyecto de construcción y que incluye todas las variantes de los planos de trabajo originales que se produjeron durante la construcción del proyecto.

Authority Having Jurisdiction (AHJ). The organization, office, or individual responsible for approving equipment, materials, an installation, or a procedure.* An AHJ is usually a governmental body that has legal jurisdiction over electrical installations. The AHJ has the responsibility for making interpretations of the rules, for deciding upon the approval of equipment and materials, and for granting special permission if required. Where a specific interpretation or deviation from the *NEC* is given, get it in writing. See *90.4* of the *NEC*.

Autoridad que tiene jurisdicción (AHJ). La organización, oficina o individuo responsable de aprobar los equipos, los materiales, una instalación o un procedimiento. Una AHJ es por lo general una entidad gubernamental que tiene jurisdicción legal sobre las instalaciones eléctricas. La AHJ tiene la responsabilidad de interpretar las reglas para decidir acerca de la aprobación de equipos y materiales y para otorgar permisos especiales, si son necesarios. Cuando se da una interpretación o desviación específica de *NEC*, obténgala por escrito. Vea *90.4* de *NEC*.

Average Rated Life (lamp). How long it takes to burn out the lamp; for example, a 60-watt lamp is rated 1000 hours. The 1000-hour rating is based on the point in time when 50% of a test batch of lamps burn out and 50% are still burning.

Vida útil nominal promedio (lámpara). Tiempo que le toma quemarse a la lámpara; por ejemplo, una lámpara de 60 vatios tiene una vida útil nominal de 1000 horas. La vida útil nominal de 1000 horas se basa en el punto en el tiempo en el cual el 50% del lote de prueba de lámparas estaban quemadas y el 50% todavía funcionaban.

AWG. The American Wire Gauge. Previously known as the Brown & Sharpe Gauge. The smaller the AWG number, the larger the conductor. There are 40 electrical conductor sizes from 36 AWG through 4/0 AWG. Starting from the smallest size, 36 AWG, each successive size is approximately 1.26 times larger than the previous AWG size. *Table 8* of *Chapter 9* in the *NEC* lists conductor sizes commonly found in the electrical industry. Conductor sizes through 4/0 AWG are shown using the AWG designation. Conductors larger than 4/0 AWG are shown in circular mil area. The 1.26 relationship between sizes is easily confirmed in *Table 8* by checking conductor sizes 4 AWG through 4/0 AWG.

AWG. Calibre de cable estadounidense. Anteriormente se conocía como el calibre de Brown & Sharpe. Cuanto más bajo es el AWG, mayor es el conductor. Hay 40 medidas de conductores eléctricos que van desde 36 AWG a 4/0 AWG. A partir de la menor medida, 36 AWG, cada medida sucesiva es aproximadamente 1,26 veces mayor que la medida AWG anterior. La *Tabla 8* del *Capítulo 9* en *NEC* enumera las medidas de conductores que se encuentran en la industria eléctrica. Las medidas de conductores a través 4/0 AWG se muestran utilizando la designación AWG. Los conductores mayores que 4/0 AWG se muestran en área del círculo de un mil. La relación de 1,26 entre las medidas se confirma fácilmente en la *Tabla 8* al observar las medidas de conductores de 4 AWG a 4/0 AWG.

B

Backlight. Illumination from behind a subject directed substantially parallel to a vertical plane through the optical axis of the camera.

Retroiluminación. Iluminación de fondo que hay detrás de un sujeto que se dirige sustancialmente en paralelo a un plano vertical a través del eje óptico de la cámara.

Ballast. Used for energizing fluorescent lamps. It is constructed of a laminated core and coil windings or electronic components.

Balasto. Se utiliza para proporcionar energía a las lámparas fluorescentes. Se fabrica a partir de un núcleo laminado y bobinados o componentes electrónicos.

Ballast Factor. A measure of the light output (lumens) of a ballast and lamp combination in comparison to an ANSI Standard "reference" ballast operated with the same lamp. It is a measure of how well the ballast performs when compared to the "reference ballast."

Factor de balasto. La medida de luz emitida (lúmenes) de una combinación de balasto y lámpara en comparación con el balasto "de referencia" del estándar ANSI operado con la misma lámpara. Es una medida de la efectividad con la que se desempeña el balasto al compararlo con el "balasto de referencia".

*Reprinted with permission from NFPA 70-2014.

Bare Lamp. A light source with no shielding.

Lámpara descubierta. Una fuente de luz sin cubierta.

Bid. An offer or proposal of a price.

Oferta. Una oferta o una propuesta de precio.

Bid Bond. A written form of security executed by the bidder as principal and by a surety for the purpose of guaranteeing that the bidder will sign the contract, if awarded the contract, for the stated bid amount.

Garantía de oferta. Una forma escrita de garantía firmada por el ofertante como principal y por un garante con el objetivo de garantizar que el ofertante firmará el contrato, si se le otorga, por el monto establecido en la oferta.

Bid Price. The stipulated sum stated in the bidder's bid.

Precio de la oferta. La suma estipulada en la oferta del ofertante.

Black Line Print. Another term for a photocopy or computer-aided drawing (CAD).

Impresión de línea negra. Otro término utilizado para una fotocopia o diseño asistido por computadora (CAD).

Blueprints. A term used to refer to a plan or plans. A photographic print in white on a bright blue ground or blue on a white ground, used especially for copying maps, mechanical drawings, and architects' plans. See text for further explanation.

Planos. Término utilizado para hacer referencia a un plano o planos. Una impresión fotográfica en blanco sobre un fondo azul claro o azul sobre un fondo blanco, utilizada especialmente para copiar mapas, diseños mecánicos y planos arquitectónicos. Consulte las explicaciones adicionales en el texto.

Bonded (bonding). Connected to establish electrical continuity and conductivity.*

Conectado (conexión). Conectado para establecer conductividad y continuidad eléctrica.

Bonding Conductor or Jumper. A reliable conductor to ensure the required electrical conductivity between metal parts required to be electrically connected.*

Conductor de conexión o puente. Conductor seguro para asegurar la conductividad eléctrica requerida entre las partes metálicas que tienen que estar conectadas eléctricamente.

Bonding Jumper, Equipment. The connection between two or more portions of the equipment grounding conductor.*

Puente de conexión, equipo. La conexión entre dos o más partes del conductor de conexión a tierra del equipo.

Bonding Jumper, Main. The connection between the grounded circuit conductor and the equipment grounding conductor at the service.*

Puente de conexión, principal. La conexión entre el conductor del circuito con conexión a tierra y el conductor de conexión a tierra del equipo que está funcionando.

Branch Circuit. The circuit conductors between the final overcurrent device protecting the circuit and the outlet(s).*

Circuito de bifurcación. Los conductores del circuito entre el dispositivo de sobrecorriente que protege al circuito y al(a los) toma(s) de corriente.

Branch Circuit, Appliance. A branch circuit supplying energy to one or more outlets to which appliances are to be connected; such circuits are to have no permanently connected luminaires not a part of an appliance.*

Circuito de bifurcación, electrodoméstico. Un circuito de bifurcación que proporciona energía a uno o más tomas de corriente a los que se van a conectar los electrodomésticos; estos circuitos no deben tener luminarias permanentemente conectadas ni parte de un electrodoméstico.

Branch Circuit, General Purpose. A branch circuit that supplies a number of outlets for lighting and appliances.*

Circuito de bifurcación, propósito general. Un circuito de bifurcación que alimenta a un número de tomas de corriente para iluminación y electrodomésticos.

Branch Circuit, Individual. A branch circuit that supplies only one utilization equipment.*

Circuito de bifurcación, individual. Un circuito de bifurcación que alimenta solo un equipo de uso.

Branch Circuit—Multiwire. A branch circuit consisting of two or more ungrounded conductors having a voltage between them, and a grounded conductor having equal voltage between it and each ungrounded conductor of the circuit, and that is connected to the neutral or grounded conductor of the system.*

Circuito de bifurcación: multialámbrico. Un circuito de bifurcación que tiene dos o más conductores sin conexión a tierra y que tienen un voltaje entre ellos, y un conductor con conexión a tierra que tiene un voltaje igual entre él y cada conductor sin conexión a tierra del circuito, y que está conectado al conductor neutro o con conexión a tierra del sistema.*

*Reprinted with permission from NFPA 70-2014.

Branch-Circuit Overcurrent Protective Device. ▶A device capable of providing protection for service, feeder, and branch circuits and equipment over the full range of overcurrent between its rated current and the interrupting rating. Such devices are provided with interrupting ratings appropriate for the intended use but no less than 5000 amperes.◀*

Dispositivo protector de sobrecorriente del circuito de bifurcación. Un dispositivo capaz de proporcionar protección para los equipos y circuitos de servicio, alimentación y bifurcación en un amplio rango de sobrecorriente entre su corriente nominal y la capacidad de interrupción. A estos dispositivos se les proporcionan capacidades de interrupción para el uso que se pretende, pero que no sea menor que 5000 amperios.

Branch Circuit Selection Current. The value in amperes to be used instead of the rated-load current in determining the ratings of motor branch-circuit conductors, disconnecting means, controllers, and branch-circuit short-circuit and ground-fault protective devices wherever the running overload protective device permits a sustained current greater than the specified percentage of the rated-load current. The value of branch-circuit selection current will always be equal to or greater than the marked rated-load current.* This definition is found in *Article 440*.

Corriente de selección del circuito de bifurcación. El valor en amperios que debe usarse en lugar de la corriente de carga nominal al determinar las capacidades de los conductores de circuito de bifurcación de motor, medios de desconexión, controladores y dispositivos protectores de cortocircuitos en el circuito de bifurcación y fallas en la conexión a tierra siempre que el dispositivo protector de sobrecorriente que está en funcionamiento permita una corriente sostenida mayor que el porcentaje especificado de la corriente de carga nominal. El valor de la corriente de selección del circuito de bifurcación siempre será igual o mayor que la corriente de carga nominal marcada. Esta definición se encuentra en el *Artículo 440*.

Brightness. In common usage, the term "brightness" usually refers to the strength of sensation that results from viewing surfaces or spaces from which light comes to the eye.

Claridad. En su uso común, el término "claridad" generalmente hace referencia a la fuerza de la sensación que se produce al ver superficies o espacios desde donde sale luz y llega a los ojos.

Building Code. The legal requirements set up by the prevailing various governing agencies covering the minimum acceptable requirements for all types of construction.

Código de construcción. Los requisitos legales establecidos por diversas agencias gubernamentales de importancia y que abarca requisitos mínimos aceptables para todos los tipos de construcción.

Building Information Modeling (BIM). A digital representation of physical and functional characteristics of a facility. BIM is a shared knowledge resource for information about a facility forming a reliable basis for decisions during its life-cycle; defined as existing from earliest conception to demolition.

Modelado de información de construcción (BIM). Una representación digital de características físicas y funcionales de una instalación. El BIM es un recurso de conocimiento compartido para información acerca de una instalación que forma una base confiable para las decisiones durante su ciclo de vida útil; se define como existe desde su primera concepción hasta la demolición.

Building Inspector/Official. A qualified government representative authorized to inspect construction for compliance with applicable building codes, regulations, and ordinances.

Inspector/Funcionario de la construcción. Representante gubernamental calificado que está autorizado a inspeccionar construcciones para verificar el cumplimiento de los códigos de construcción, las regulaciones y las ordenanzas aplicables.

Building Permit. A written document issued by the appropriate governmental authority permitting construction to begin on a specific project in accordance with drawings and specifications approved by the governmental authority.

Permiso de construcción. Documento escrito emitido por la autoridad gubernamental correspondiente que autoriza el comienzo de la construcción de un proyecto específico de conformidad con los planos y las especificaciones aprobadas por la autoridad gubernamental.

C

Candela. The international (SI) unit of luminous intensity. Formerly referred to as a "candle," as in "candlepower."

Candela. La unidad internacional (SI) de intensidad luminosa. Antes se conocía como "vela" o "bujía".

Candlepower. A measure of intensity mathematically related to lumens.

Bujía. Medida de intensidad relacionada matemáticamente con los lúmenes.

Carrier. A wave having at least one characteristic that may be varied from a known reference value by modulation.

Portadora. Una onda que tiene al menos una característica que puede variar de un valor de referencia conocido por la modulación.

Carrier Current. The current associated with a carrier wave.

Corriente portadora. La corriente asociada con una onda portadora.

CC. A marking on wire connectors and soldering lugs indicating they are suitable for use with copper-clad aluminum conductors only.

CC. Marca en los conectores de alambre y terminales de soldadura que indica que son aptos para ser usados solo con conductores de aluminio con revestimiento de cobre.

CC/CU. A marking on wire connectors and soldering lugs indicating they are suitable for use with copper or copper-clad aluminum conductors only.

CC/CU. Marca en los conectores de alambre y terminales de soldadura que indica que son aptos para ser usados solo con conductores de cobre o de aluminio con revestimiento de cobre.

Change Order. A written document signed by the owner and the contractor, authorizing a change in the work or an adjustment in the contract sum or the contract time. A contract sum and the contract time may be changed only by a change order.

Orden de modificación. Un documento escrito firmado por el propietario y el contratista en el que se autoriza una modificación en el trabajo o un ajuste en la suma del contrato o en el plazo del contrato. La suma del contrato y el plazo del contrato solo pueden cambiarse con una orden de modificación.

Circuit Breaker. A device designed to open and close a circuit by nonautomatic means and to open the circuit automatically on a predetermined overcurrent without damage to itself when properly applied within its rating.*

Interruptor de circuito. Dispositivo diseñado para abrir y cerrar un circuito por medios no automáticos y para abrir el circuito de forma automática en una sobrecorriente predeterminada sin sufrir daños cuando se lo aplica adecuadamente dentro de su capacidad.

Clothes Closet. A nonhabitable room or space intended primarily for the storage of garments and apparel.*

Armario de ropa. Habitación o espacio no habitable cuyo propósito principal es el almacenamiento de ropa y prendas de vestir.

CO/ALR. Terminal marking on switches and receptacles rated 15 and 20 amperes that are suitable for use with aluminum, copper, and copper-clad aluminum conductors. If not marked, they are suitable for copper or copper-clad conductors only.

CO/ALR. Marca de terminal en interruptores y receptáculos con capacidad de 15 y 20 amperios que son aptos para ser utilizados con conductores de aluminio, cobre y aluminio con revestimiento de cobre. Si no tienen marca, solo son aptos para conductores de cobre o con revestimiento de cobre.

Conductor:

Conductor:

Bare. A conductor having no covering or electrical insulation whatsoever. (See Conductor, Covered.)*

Descubierto. Un conductor que no tiene cubierta ni ningún tipo de aislante eléctrico. (Vea Conductor, cubierto.)

Covered. A conductor encased within material of a composition or thickness that is not recognized by this Code as electrical insulation.* (See Conductor, Bare.)

Cubierto. Conductor revestido por un material de una composición o un grosor que no es reconocido por este Código como aislante eléctrico. (Vea Conductor, descubierto.)

Insulated. A conductor encased within material of a composition and thickness that is recognized by this Code as electrical insulation.*

Aislado. Conductor revestido por un material de una composición o un grosor que es reconocido por este Código como aislante eléctrico.

Connector, Pressure (Solderless). A device that establishes a connection between two or more conductors or between one or more conductors and a terminal by means of mechanical pressure and without the use of solder.*

*Reprinted with permission from NFPA 70-2014.

Conector, presión (sin soldadura). Dispositivo que establece una conexión entre dos o más conductores o entre uno o más conductores y una terminal por medio de presión mecánica y sin el uso de un soldador.

Construction Documents. A term used to represent all drawings, specifications, addenda, and other pertinent construction information associated with the construction of a specific project.

Documentos de construcción. Término utilizado para representar todos los planos, las especificaciones, los anexos y demás información de construcción relevante que estén asociados con la construcción de un proyecto específica.

Construction Drawing. See Drawings.

Plano de construcción. Vea Planos.

Construction Types. (Also see Annex E, *NEC.*)

Tipos de construcción. (Vea también Anexo E, *NEC.*)

Type I. Fire-resistive construction. A building constructed of noncombustible materials such as reinforced concrete, brick, stone, and so on, and having any metal members properly fireproofed, with major structural members designed to withstand collapse and to prevent the spread of fire. Type I construction is subdivided into Types IA and IB.

Tipo I. Construcción con resistencia al fuego. Edificación construida a partir de materiales no combustibles como hormigón reforzado, ladrillos, piedra, etc., y que tiene todos los miembros metálicos resistentes al fuego de la forma adecuada, con miembros estructurales principales que están diseñados para soportar el colapso y para evitar propagación del fuego. La construcción Tipo I se subdivide en Tipos IA e IB.

Type II. Noncombustible construction. A building having all structural members, including walls, floors, and roofs, of noncombustible materials such as reinforced concrete, brick, stone, and so on, and not qualifying as fire-resistive construction. Type II construction is divided into Types IIA, IIB, and IIC.

Tipo II. Construcción no combustible Construcción que tiene todos los miembros estructurales, incluidos las paredes, los pisos y los techos, fabricados con materiales no combustibles como hormigón reforzado, ladrillos, piedras, etc., y que no califica como construcción con resistencia al incendio. La construcción Tipo II está dividida en Tipos IIA, IIB y IIC.

Type III. Construction where the exterior walls are of concrete, masonry, or other noncombustible material. The interior structural members are constructed of any approved materials such as wood or other combustible materials. Type III construction is subdivided into Types IIIA and IIIB.

Tipo III. Construcción donde las paredes exteriores son de hormigón, mampostería u otro material no combustible. Los miembros estructurales internos están construidos con cualquier material aprobado como madera y otros materiales combustibles. La construcción Tipo III se subdivide en Tipos IIIA y IIIB.

Type IV. Construction where the exterior walls are constructed of approved noncombustible materials. The interior structural members are of solid or laminated wood.

Tipo IV. Construcción donde las paredes exteriores están fabricadas con materiales no combustibles aprobados. Los miembros estructurales internos son de madera sólida o laminada.

Type V. Construction where the exterior walls, loadbearing walls, partitions, floors, and roofs are constructed of any approved material. This is wood construction as found in typical residential construction. Type V construction is divided into Types VA and VB.

Tipo V. Construcción donde las paredes exteriores, las paredes que soportan la carga estructural, las particiones, los pisos y los techos están fabricados con cualquier material aprobado. Esta es la construcción de madera como se encuentra en las construcciones residenciales comunes. La construcción Tipo V se divide en Tipos VA y VB.

Continuous Load. A load where the maximum current is expected to continue for 3 hours or more.*

Carga continua. Una carga donde se espera que la corriente máxima continúe durante 3 horas o más.

Contract. An agreement between two or more parties, especially one that is written and enforceable by law.

Contrato. Un acuerdo entre dos o más partes, en especial uno que es escrito y del cual se puede exigir su cumplimiento ante la ley.

Contract Modifications. After the agreement has been signed, any additions, deletions, or modifications of the work to be done are accomplished by change

*Reprinted with permission from NFPA 70-2014.

order, supplemental instruction, and field order. They can be issued at any time during the contract period.

Modificaciones de contrato. Después de que se ha firmado el contrato, toda adición, eliminación o modificación del trabajo que deba hacerse debe realizarse a través de una orden de modificación, instrucción suplementaria y orden de campo. Se pueden emitir en cualquier momento durante el plazo del contrato.

Contractor. A properly licensed individual or company that agrees to furnish labor, materials, equipment, and associated services to perform the work as specified for a specified price.

Contratista. Individuo o compañía con licencia adecuada que se compromete a brindar servicios de mano de obra, materiales, equipos y otros relacionados para llevar a cabo un trabajo especificado a cambio de un precio determinado.

Correction Factor. A multiplier (penalty) that is applied to conductors when the ambient temperature is greater than 86°F (30°C). These multipliers are found in *Table 310.15(B)(2)(a)* of the *NEC*.

Factor de corrección. Un multiplicador (penalización) que se aplica a los conductores cuando la temperatura ambiente es mayor que 86°F (30°C). Estos multiplicadores se encuentran en la *Tabla 310.15(B)(2)(a)* de *NEC*.

CSA. This is the Canadian Standards Association that develops safety and performance standards in Canada for electrical products, similar to but not always identical to those of (Underwriters Laboratories) UL in the United States.

CSA. Esta es la Asociación Canadiense de Estándares (Canadian Standards Association) que desarrolla estándares de seguridad y rendimiento en Canadá para productos eléctricos, similares pero no siempre idénticos a los de (Underwriters Laboratories) UL en los Estados Unidos.

CU. A marking on wire connectors and soldering lugs indicating they are suitable for use with copper conductors only.

CU. Marca que hay en los conectores de alambre o terminales de soldaduras que indica que son aptos para ser usados solo con conductores de cobre.

Current. The flow of electrons through an electrical circuit, measured in amperes.

Corriente. El flujo de electrones a través de un circuito eléctrico que se mide en amperios.

D

Derating Factor. A trade or industry jargon term for Ambient Temperature Correction Factors [*Table 310.15(B)(2)(a)*] for ambient temperatures exceeding 86°F (30°C), Adjustment Factors [*Table 310.15(B)(3)(a)*] that is applied to conductors when there are more than three current-carrying conductors in a raceway or cable.

Factor de reducción. Un término de la jerga de la rama o industria para Factores de corrección de temperatura ambiente [*Tabla 310.15(B)(2)(a)*] para las temperaturas ambiente que exceden 86°F (30°C), los factores de ajuste [*Tabla 310.15(B) (3)(a)*] que se aplican a los conductores cuando hay más de tres conductores que transportan corriente en un conducto para cables o cable.

Details. Plans, elevations, or sections that provide more specific information about a portion of a project component or element than smaller scale drawings.

Detalles. Planos, elevaciones o secciones que brindan información más específica acerca de una parte de un componente o elemento de un proyecto que la que brindan los planos a menor escala.

Device. ►A unit of an electrical system, other than a conductor, that carries or controls electric energy as its principal function.◄*

Dispositivo. Unidad de un sistema eléctrico, que no es un conductor, que transporta o controla energía eléctrica como su función principal.

Diagrams. Nonscaled views showing arrangements of special system components and connections not possible to clearly show in scaled views. A schematic diagram shows circuit components and their electrical connections without regard to actual physical locations. A wiring diagram shows circuit components and the actual electrical connections.

Diagramas. Vistas que no son a escala y que muestran la disposición de componentes y conexiones especiales del sistema que no se pueden mostrar de forma clara en vistas a escala. Un diagrama esquemático muestra los componentes del circuito eléctrico y sus conexiones eléctricas sin importar sus ubicaciones físicas reales. Un diagrama de cableado muestra los componentes del circuito y las conexiones eléctricas reales.

Dimmer. A switch with components that permits variable control of lighting intensity. Some dimmers have electronic components; others have core and coil (transformer) components.

---

*Reprinted with permission from NFPA 70-2014.

Regulador de luz. Interruptor con componentes que permiten el control variable de la intensidad de la luz. Algunos reguladores tienen componentes eléctricos; otros tienen componentes de núcleo y bobina (transformador).

Dimming Ballast. Controls light output of fluorescent lamps.

Balasto de regulación de luz. Controla la cantidad de luz que emiten las lámparas fluorescentes.

Direct Glare. Glare resulting from high luminances or insufficiently shielded light sources in the field of view. Usually associated with bright areas, such as luminaires, ceilings, and windows that are outside the visual task of the region being viewed.

Deslumbramiento directo. Deslumbramiento que se produce por fuentes de alta luminosidad o de luz sin suficiente recubrimiento en el campo de visión. Por lo general, está asociado con áreas luminosas, como las luminarias, los techos y las ventanas que están fuera de la tarea visual de la región que se observa.

Disconnecting Means. A device, or group of devices, or other means by which the conductors of a circuit can be disconnected from their source of supply.*

Medio de desconexión. Dispositivo, o grupo de dispositivos, u otro medio por el cual los conductores de un circuito se pueden desconectar de su fuente de alimentación.

Downlight. A small, direct lighting unit that guides the light downward. It can be recessed, surface-mounted, or suspended.

Aparato de luz intensiva. Unidad pequeña y directa de iluminación que guía la luz hacia abajo. Se puede empotrar, montar en la superficie o suspender.

Drawings. (1) A term used to represent that portion of the contract documents that graphically illustrates the design, location, and dimensions of the components and elements contained in a specific project. (2) A line drawing.

Planos. (1) Término utilizado para representar esa parte de los documentos del contrato que ilustran gráficamente el diseño, la ubicación y las dimensiones de los componentes y elementos que están dentro de un proyecto específico. (2) Gráfico de líneas.

Dry Niche Luminaire. A luminaire intended for installation in the wall of a pool or fountain in a niche that is sealed against the entry of pool water.* This definition is found in *Article 680*.

Luminaria para nicho seco. Luminaria diseñada para ser instalada en la pared de una piscina o fuente en un nicho que está sellado contra la entrada de agua de la piscina. Esta definición se encuentra en el *Artículo 680*.

Dwelling Unit. A single unit, providing complete and independent living facilities for one or more persons, including permanent provisions for living, sleeping, cooking, and sanitation.*

Unidad de vivienda. Una unidad única que brinda instalaciones de vivienda completa e independiente para una o más personas, lo que incluye espacios para vivir, dormir, cocinar y sanitarios.

E

Effective Ground-Fault Current Path. An intentionally constructed, low-impedance electrically conductive path designed and intended to carry current under ground-fault conditions from the point of a ground fault on a wiring system to the electrical supply source and that facilitates the operation of the overcurrent protective device or ground-fault detectors.*

Vía de corriente efectiva para fallas de conexión a tierra. Vía de conducción eléctrica de baja impedancia construida intencionalmente y diseñada con el fin de transportar corriente cuando falla la conexión a tierra desde el punto de la falla de conexión a tierra en un sistema de cableado hasta la fuente de alimentación eléctrica y que facilita el funcionamiento del dispositivo protector de sobrecorriente o de los detectores de fallas de conexión a tierra.

Efficacy. The total amount of light energy (lumens) emitted from a light source divided by the total lamp and ballast power (watts) input, expressed in lumens per watt.

Eficiencia. La cantidad total de energía luminosa (lúmenes) emitida por una fuente de luz dividida por la entrada de energía total (vatios) de la lámpara y el balasto, expresada en lúmenes por vatio.

Elevations. Views of vertical planes, showing components in their vertical relationship, viewed perpendicularly from a selected vertical plane.

Elevaciones. Vistas de planos verticales que muestran componentes en su relación vertical, como se ven de forma perpendicular desde un plano vertical seleccionado.

Equipment. A general term including material, fittings, devices, appliances, luminaires, apparatus, machinery, and the like used as a part of, or in connection with, an electrical installation.*

*Reprinted with permission from NFPA 70-2014.

Equipo. Termino general que incluye materiales, accesorios, dispositivos, electrodomésticos, luminarias, aparatos, maquinarias y similares que se utilizan como parte de una instalación eléctrica o en conexión con ella.

F

Feeder. All circuit conductors between the service equipment, the source of a separately derived system, or other power-supply source and the final branch-circuit overcurrent device.*

Alimentador. Todos los conductores del circuito entre el equipo de servicio, la fuente de un sistema derivado por separado u otra fuente de alimentación eléctrica y el dispositivo de sobrecorriente del circuito de bifurcación.

Fill Light. Supplementary illumination to reduce shadow or contract range.

Iluminación de relleno. Iluminación suplementaria para reducir las sombras o contraer el rango.

Fire Rating. The classification indicating in time (hours) the ability of a structure or component to withstand fire conditions.

Clasificación de incendio. Clasificación que indica en tiempo (horas) la capacidad de una estructura o un componente de tolerar condiciones de incendio.

Flame Detector. A radiant energy-sensing fire detector that detects the radiant energy emitted by a flame.

Detector de incendios. Un detector de incendios que detecta la energía radiante que emite una llama de fuego.

Fluorescent Lamp. A lamp in which electric discharge of ultraviolet energy excites a fluorescing coating (phosphor) and transforms some of that energy to visible light.

Lámpara fluorescente. Lámpara en la cual una descarga eléctrica de energía ultravioleta sale de un recubrimiento fluorescente (fósforo) y transforma parte de esa energía en luz visible.

Footcandle. The unit used to measure how much total light is reaching a surface. One lumen falling on 1 ft^2 of surface produces an illumination of 1 footcandle.

Candela/pie. La unidad utilizada para medir cuánta luz total llega a la superficie. Un lumen que cae en 1 pie^2 de superficie produce una iluminación de 1 candela/pie.

Fully Rated System. All devices installed have an interrupting rating greater than or equal to the specified available fault-current. In a fully rated system, the panelboard short-circuit current rating will be equal to the lowest interrupting rating of any branch-circuit breaker or fuse installed. (See Series- Rated System.)

Sistema completamente clasificado. Todos los dispositivos instalados tienen una capacidad de interrupción que es mayor o igual a la corriente de falla disponible especificada. En un sistema completamente clasificado, la capacidad de corriente de cortocircuito del cuadro terminal será igual a la capacidad de interrupción más baja de cualquier disyuntor o fusible de circuito de bifurcación instalado. (Vea Sistema de clasificación por series.)

Fuse. An overcurrent protective device with a fusible link that operates to open the circuit on an overcurrent condition.

Fusible. Dispositivo protector de sobrecorrientes con una conexión a un fusible que se activa para abrir el circuito cuando se produce una sobrecorriente.

G

General Conditions. A written portion of the contract documents set forth by the owner, stipulating the contractor's minimum acceptable performance requirements including the rights, responsibilities, and relationships of the parties involved in the performance of the contract. General conditions are usually included in the book of specifications but are sometimes found in the architectural drawings.

Condiciones generales. Parte escrita de los documentos de un contrato establecida por el propietario, donde estipula los requisitos mínimos aceptables que debe cumplir el contratista, lo que incluye los derechos, las responsabilidades y las relaciones de las partes involucradas en el cumplimiento del contrato. Por lo general, las condiciones generales se incluyen en el libro de especificaciones, pero a veces se encuentran en los planos arquitectónicos.

General Lighting. Lighting designed to provide a substantially uniform level of illumination throughout an area, exclusive of any provision for special lighting.

Iluminación general. Iluminación diseñada para proporcionar un nivel sustancialmente uniforme de luz en toda un área, exclusivo de cualquier determinación para iluminación especial.

Glare. The sensation produced by luminance within the visual field that is sufficiently greater than the luminance to which the eyes are adapted to cause annoyance, discomfort, or loss of visual performance and visibility.

*Reprinted with permission from NFPA 70-2014.

Deslumbramiento. La sensación producida por una luminaria dentro del campo visual que es lo suficientemente mayor a la luminosidad a la cual están adaptados los ojos como para causar molestia, incomodidad o pérdida de la capacidad visual y visibilidad.

Ground. The earth.*

Tierra. El suelo.

Grounded (Grounding). Connected (connecting) to ground or to a conductive body that extends the ground connection.*

Conectado a tierra (conexión a tierra). Conectado (conexión) a tierra o a un cuerpo conductor que extiende la conexión a tierra.

Grounded, Solidly. Connected to ground without inserting any resistor or impedance device.*

Conectado a tierra, firmemente. Conectado a tierra sin introducir ningún resistor ni dispositivo de impedancia.

Grounded Conductor. A system or circuit conductor that is intentionally grounded.*

Conductor con conexión a tierra. Conductor de sistema o circuito que está conectado a tierra intencionalmente.

Ground Fault. An unintentional, electrically conductive connection between a normally current-carrying conductor of an electrical circuit, and the normally non-current-carrying conductors, metallic enclosures, metallic raceways, metallic equipment, or earth.*

Falla de conexión a tierra. Conexión accidental que conduce electricidad entre un conductor que normalmente transporta corriente de un circuito eléctrico y los conductores que normalmente no transportan corriente, cajas metálicas, conductos eléctricos metálicos, equipos metálicos o tierra.

Ground-Fault Circuit Interrupter (GFCI). A device intended for the protection of personnel that de-energizes a circuit or portion thereof within an established period of time when a current to ground exceeds the values established for a Class A device.*

Interruptor de circuito por falla de conexión a tierra (GFCI). Dispositivo para la protección del personal que interrumpe la alimentación a un circuito o a una porción de un circuito dentro de un período de tiempo establecido cuando una corriente a tierra supera los valores establecidos para un dispositivo de Clase A.

Informational Note. Class A ground-fault circuit interrupters trip when the current to ground is 6 mA or higher and do not trip when the current to ground is less than 4 mA. For further information, see UL 943, Standard for Ground- Fault Circuit Interrupters.*

Nota informativa. Los interruptores de falla de conexión a tierra se activan cuando la corriente a tierra es 6 mA o más y no se activan cuando la corriente es de menos de 4 mA. Para obtener más información, vea UL 943, Estándar para interruptores de circuito por falla de conexión a tierra.

Ground-Fault Current Path. An electrically conductive path from the point of a ground fault on a wiring system through normally non-current-carrying conductors, equipment, or the earth to the electrical supply source.*

Vía de corriente de falla de conexión a tierra. Vía conductora de electricidad que va desde el punto de la falla de conexión a tierra en un sistema de cableado a través de conductores que normalmente no transportan corriente, equipos o la tierra hacia la fuente de alimentación eléctrica.

Ground-Fault Protection of Equipment. A system intended to provide protection of equipment from damaging line-to-ground fault currents by operating to cause a disconnecting means to open all ungrounded conductors of the faulted circuit. This protection is provided at current levels less than those required to protect conductors from damage through the operation of a supply circuit overcurrent device.*

Protección de equipos por falla de conexión a tierra. Sistema que tiene el objetivo de proteger los equipos de los daños producidos por corrientes en las que falla la conexión a tierra mientras operan para generar medios de desconexión para abrir todos los conductores sin conexión a tierra del circuito fallido. Esta protección se proporciona a niveles de corriente menores que los requeridos para proteger los conductores de los daños a través de la activación de un dispositivo de sobrecorriente del circuito de alimentación.

Grounding Conductor, Equipment (EGC). ▶The conductive path(s) that provides a ground-fault current path and connects normally non-current-carrying metal parts of equipment together and to the system grounded conductor or to the grounding electrode conductor, or both.◀*

*Reprinted with permission from NFPA 70-2014.

Conductor de conexión a tierra, equipo (EGC). La(s) vía(s) conductora(s) que proporciona un camino para la corriente con falla de conexión a tierra y conecta las partes metálicas que normalmente no transportan corriente de los equipos y las conecta al conductor con conexión a tierra del sistema o al conductor de electrodo de conexión a tierra o a ambos.

Grounding Electrode. A conducting object through which a direct connection to earth is established.*

Electrodo de conexión a tierra. Objeto conductor a través del cual se establece una conexión directa a tierra.

Grounding Electrode Conductor. A conductor used to connect the system grounded conductor or the equipment to a grounding electrode or to a point on the grounding electrode system.*

Conductor de electrodo de conexión a tierra. Conductor utilizado para conectar el conductor de conexión a tierra del sistema o del equipo a un electrodo de conexión a tierra o a un punto en el sistema de electrodos de conexión a tierra.

H

Habitable Room. A room in a residential occupancy used for living, sleeping, cooking, and eating, but excluding bath, storage and service area, and corridors.

Espacio habitable. Habitación en una ocupación residencial utilizada para vivir, dormir, cocinar y comer, pero que excluye zona de aseo, almacenamiento, servicio y corredores.

HACR. Circuit breakers subjected to additional tests unique to HVAC equipment, have been marked with the letters HACR. The letters stood for Heating, Air Conditioning, and Refrigeration. The HVAC equipment would also be marked with the letters HACR if that is the type of overcurrent protection required by the manufacturer of the HVAC equipment. Today, all standard molded case circuit breakers tested and listed to UL 489 are suitable for use on HVAC equipment. The letters HACR will no doubt disappear from the scene as time goes on.

HACR. Los disyuntores de circuito que están sujetos a evaluaciones adicionales exclusivas para los equipos de HVAC han sido marcados con las letras HACR. La sigla representa calefacción, aire acondicionado y refrigeración. Los equipos de HVAC también tendrán la marca con la sigla HACR si ese es el tipo de protección de sobrecorriente que exige el fabricante de los equipos de HVAC. En la actualidad, todos los interruptores de circuito de caja moldeada estándar que han sido evaluados y enumerados en UL 489 son aptos para su uso en equipos de HVAC. La sigla HACR sin duda desaparecerá de la escena con el paso del tiempo.

Halogen Lamp. An incandescent lamp containing a halogen gas that recycles tungsten back onto the tungsten filament surface. Without the halogen gas, the tungsten would normally be deposited onto the bulb wall.

Lámpara halógena. Lámpara incandescente que contiene un gas halógeno que recicla tungsteno para regresarlo a la superficie del filamento de tungsteno. Sin el gas halógeno, el tungsteno se depositaría normalmente en la pared de la bombilla.

Heat Alarm. A single or multiple station alarm responsive to heat.

Alarma de calor. Alarma de estación simple o múltiple que se activa por calor.

Hermetic Refrigerant Motor-Compressor. A combination consisting of a compressor and motor, both of which are enclosed in the same housing, with no external shaft or shaft seals; the motor operating in the refrigerant. This definition is found in *Article 440*.

Motor compresor refrigerante hermético. Una combinación de un compresor y un motor, los cuales están dentro de la misma carcasa sin eje externo ni juntas de eje; el motor opera en el refrigerante. Esta definición se encuentra en el *Artículo 440*.

High-Intensity Discharge Lamp (HID). A general term for a mercury, metal-halide, or high-pressure sodium lamp.

Lámpara de descarga de alta intensidad (HID). Término general para una lámpara de mercurio, halogenuros metálicos o sodio de alta presión.

High-Pressure Sodium Lamp. A high-intensity discharge light source in which the light is primarily produced by the radiation from sodium vapor.

Lámpara de sodio de alta presión. Fuente de luz de descarga de alta intensidad en la cual la luz es producida principalmente por la radiación emitida por el vapor de sodio.

Horsepower (tools). Horsepower is a measurement of motor torque multiplied by speed. Horsepower also refers to the rate of work (power) an electric motor is capable of delivering. See Torque, Rated Horsepower, and Maximum Developed Horsepower.

Caballos de fuerza (herramientas). El caballo de fuerza es una medida del par de torsión del motor multiplicado por la velocidad. Los caballos de fuerza también se refieren a la tasa de trabajo (fuerza) que puede proporcionar un motor eléctrico. Vea par de torsión, valor nominal en caballos de fuerza y valor máximo desarrollado en caballos de fuerza.

Horsepower, Rated. Rated horsepower is a motor's running torque at its rated running speed. The motor can be run continuously at its rated horsepower without overheating. If the motor is required to give an extra spurt of effort while running, the motor is overloaded and develops extra horsepower to compensate. The most horsepower that can be drawn from a motor to handle this extra effort is its maximum developed horsepower. See Torque, Horsepower, and Maximum Developed Horsepower.

Caballos de fuerza, valor nominal. El valor nominal en caballos de fuerza es el par de torsión de funcionamiento de un motor a su velocidad nominal de funcionamiento. El motor puede funcionar de forma continua a su valor nominal en caballos de fuerza sin sobrecalentarse. Si se necesita que el motor haga un esfuerzo adicional mientras esté en funcionamiento, se sobrecorriente el motor y desarrolla más caballos de fuerza para compensar. El valor más alto en caballos de fuerza que puede producir un motor para manejar este esfuerzo adicional es su valor máximo desarrollado en caballos de fuerza. Vea par de torsión, caballos de fuerza y valor máximo desarrollado en caballos de fuerza.

Horsepower, Maximum Developed. If a motor is required to work harder than its idle speed, it will be overloaded and must develop extra horsepower. The most horsepower that can be expected from a motor to handle this extra effort is referred to as its maximum developed horsepower. See Horsepower, Torque, and Rated Horsepower.

Caballos de fuerza, valor máximo desarrollado. Si se necesita que un motor trabaje con más fuerza que su velocidad de inactividad, se sobrecargará y deberá desarrollar caballos de fuerza adicionales. El valor más alto en caballos de fuerza que se puede esperar de un motor para que maneje este esfuerzo adicional se conoce como su valor máximo desarrollado en caballos de fuerza. Vea caballos de fuerza, par de torsión y valor nominal en caballos de fuerza.

Household Fire Alarm System. A system of devices that produces an alarm signal in the household for the purpose of notifying the occupants of the presence of a fire so that they will evacuate the premises.

Sistema de alarma residencial contra incendios. Sistema de dispositivos que produce una señal de alarma en la residencia con el fin de notificar a los ocupantes acerca de la presencia de un incendio para que evacúen el lugar.

HVAC. A term used by tradesman for Heating, Ventilating and Air-Conditioning.

HVAC. Término utilizado por los comerciantes para referirse a calefacción, ventilación y acondicionamiento de aire.

I

ICC Building Code. One of the families of codes and related publications published by the International Code Council. See www.iccsafe.org for a complete listing of codes produced for the built environment. The International Code Council is an amalgamation of the former International Conference of Building Officials (ICBO), the Building Officials Code Administration (BOCA), and Southern Building Codes Congress International.

Código de construcción del ICC. Una de las familias de códigos y publicaciones relacionadas publicadas por el Consejo Internacional de Códigos (International Code Council, ICC). Visite www.iccsafe.org para consultar un listado completo de los códigos producidos para el entorno construido. El Consejo Internacional de Códigos es una fusión de la ex Conferencia Internacional de Funcionarios de la Construcción (International Conference of Building Officials, ICBO), la Administración del Código de los Funcionarios de la Construcción (Building Officials Code Administration, BOCA) y del Congreso Internacional de Códigos de Construcción del Sur (Southern Building Codes Congress International).

Identified (conductor). The identified conductor is the insulated grounded conductor.

- For sizes 6 AWG or smaller, the insulated grounded conductor is required to be identified by a continuous white or gray outer finish or by three continuous white stripes on other than green insulation along its entire length.
- For sizes larger than 6 AWG, the insulated grounded conductor is required to be identified either by a continuous white or gray outer finish or by three continuous white stripes on other than green insulation along its entire length, or at the time of installation by a distinctive white or gray marking at its terminations that encircles the insulation. See *NEC 200.6*.

Identificado (conductor). El conductor identificado es el conductor aislado con conexión a tierra.

- Para las medidas de 6 AWG o menores, se necesita que se identifique el conductor aislado con conexión a tierra con un acabado externo continuo de color blanco o gris o con tres rayas blancas continuas sobre un aislante que no sea verde a lo largo de toda su longitud.
- Para medidas mayores a 6 AWG, se necesita que se identifique el conductor aislado con conexión a tierra con un acabado externo continuo de color blanco o gris o con tres rayas blancas continuas sobre un aislante que no sea verde a lo largo de toda su longitud, o al momento de la instalación con una marca distintiva en blanco o gris en sus terminaciones que rodee al aislante. Consulte *NEC 200.6*.

Identified (equipment). Recognizable as suitable for the specific purpose, function, use, environment, application, and so forth, where described in a particular *Code* requirement.*

Identificado (equipo). Que se puede reconocer como apto para un fin específico, una función, un uso, un entorno, una aplicación, etc., donde se describa en un requisito de un código en especial.

Identified (terminal). The identification of terminals to which a grounded conductor is to be connected is required to be substantially white in color. The identification of other terminals is required to be of a readily distinguishable different color. See *NEC 200.10*.

Identificado (terminal). La identificación de los terminales a los que se debe conectar un conductor con conexión a tierra debe ser sustancialmente de color blanco. La identificación de otros terminales debe hacerse con un color diferente que sea fácil de diferenciar. Consulte *NEC 200.10*.

IEC. The International Electrotechnical Commission is a worldwide standards organization. These standards differ from those of Underwriters Laboratories. Some electrical equipment might conform to a specific IEC Standard but may or may not conform to the UL Standard for the same item.

IEC. La Comisión Electrotécnica Internacional es una organización de estándares mundiales. Estos estándares difieren de los de Underwriters Laboratories. Algunos equipos eléctricos pueden cumplir un estándar específico de IEC, pero pueden o no cumplir un estándar de UL para el mismo artículo.

Illuminance. The amount of light energy (lumens) distributed over a specific area expressed as footcandles (lumens/ft^2) or lux (lumens/m^2).

Iluminancia. La cantidad de energía de luz (lúmenes) distribuida en un área específica expresada como candela/pie (lúmenes/pie^2) o lux (lúmenes/m^2).

Illumination. The act of illuminating or the state of being illuminated.

Iluminación. El acto de iluminar o el estado de estar iluminado.

Immersion Detection Circuit Interrupter (IDCI). A device integral with grooming appliances that will shut off the appliance when the appliance is dropped in water.

Interruptor de circuito por detección de inmersión (IDCI). Dispositivo integral con aparatos de interrupción que apagarán al aparato si cae al agua.

Incandescent Filament Lamp. A lamp that provides light when a filament is heated to incandescence by an electric current. Incandescent lamps are the oldest form of electric lighting technology.

Lámpara de filamento incandescente. Lámpara que genera luz cuando el filamento se calienta hasta alcanzar la incandescencia gracias a una corriente eléctrica. Las lámparas incandescentes son la forma más antigua de tecnología de iluminación eléctrica.

Indirect Lighting. Lighting by luminaires that distribute 90% to 100% of the emitted light upward.

Luz indirecta. Luz producida por luminarias que distribuye entre el 90% y el 100% de la luz emitida hacia arriba.

Inductive Load. A load that is made up of coiled or wound wire that creates a magnetic field when energized. Transformers, core and coil ballasts, motors, and solenoids are examples of inductive loads.

Carga inductiva. Carga que está compuesta de alambre bobinado o enrollado que crea un campo magnético cuando recibe energía. Los transformadores, los balastos de núcleo y la bobina, los motores y los solenoides son ejemplos de cargas inductivas.

Informational Note. Explanatory material, such as references to other standards, references to related sections of the *NEC*, or information related to a *NEC* rule, is included in the *NEC* in the form of Informational Notes. Such notes are informational only and are not enforceable as requirements of this *Code*.* *See NEC 90.5(C).*

*Reprinted with permission from NFPA 70-2014.

Nota informativa. Material de explicación, como referencias a otros estándares, referencias a secciones relacionadas del *NEC* o información relacionada con una regla *NEC*, que está incluido en *NEC* en forma de notas informativas. Esas notas tienen carácter meramente informativo y no son aplicables como requisitos de este Código. Consulte *NEC 90.5(C)*.

In Sight From (Within Sight From, Within Sight). Where this Code specifies that one equipment shall be "in sight from," "within sight from," or "within sight," and so forth, of another equipment, the specified equipment is to be visible and not more than 50 ft (15 m) distant from the other.*

A la vista desde (dentro de la vista desde, a la vista). Donde este código especifique que un equipo debe estar "a la vista desde", "dentro de la vista desde" o "a la vista", etc., de otro equipo, el equipo indicado deberá ser visible y no estar a más de 50 pies (15 m) de distancia del otro.

Instant Start. A circuit used to start specially designed fluorescent lamps without the aid of a starter. To strike the arc instantly, the circuit utilizes higher open-circuit voltage than is required for the same length preheat lamp.

Encendido instantáneo. Circuito utilizado para encender lámparas fluorescentes especialmente diseñadas sin la ayuda de un arranque. Para activar el arco de forma instantánea, el circuito usa un voltaje más alto de circuito abierto que el que se necesita para la lámpara de precalentamiento de la misma longitud.

International Association of Electrical Inspectors (IAEI). A not-for-profit and educational organization cooperating in the formulation and uniform application of standards for the safe installation and use of electricity, and collecting and disseminating information relative thereto. The IAEI is made up of electrical inspectors, electrical contractors, electrical apprentices, manufacturers, electrical testing laboratories and governmental agencies.

Asociación Internacional de Inspectores Eléctricos (International Association of Electrical Inspectors, IAEI). Organización educativa sin fines de lucro que coopera en la formulación y aplicación uniforme de estándares para la instalación y el uso seguros de la electricidad y en la recolección y divulgación de información relacionada con la misma. La IAEI está formada por inspectores eléctricos, contratistas eléctricos, aprendices eléctricos, fabricantes, laboratorios de pruebas eléctricas y agencias gubernamentales.

Interrupting Rating. The highest current at rated voltage that a device is identified to interrupt under standard test conditions.*

Clasificación de interrupción. La mayor corriente a un voltaje nominal que un dispositivo tiene indicado para interrumpir cuando se encuentra en condiciones de prueba estándar.

Intersystem Bonding Termination. A device that provides a means for connecting intersystem bonding conductors for communications systems to the grounding electrode system.*

Terminación de conexión entre sistemas. Dispositivo que proporciona un medio para conectar conductores de conexión entre sistemas para sistemas de comunicaciones al sistema de electrodos de conexión a tierra.

Ionization Detector. This type of detector triggers an alarm when oxygen and nitrogen particles are ionized in the ionization chamber. The internal circuitry measures a minute amount of electrical current between two plates. When smoke enters the ionization chamber, the current is reduced, triggering the alarm. This type of detector works well for detecting small amounts of smoke such as that resulting from gasoline fires. Consult the manufacturer's literature for more information.

Detector de ionización. Este tipo de detector activa una alarma cuando se ionizan las partículas de oxígeno y nitrógeno en la cámara de ionización. Los circuitos internos miden una diminuta cantidad de corriente eléctrica entre dos placas. Cuando entra humo a la cámara de ionización, la corriente se reduce, lo que activa la alarma. Este tipo de detector funciona bien para detectar pequeñas cantidades de humo como las que se producen en los incendios con gasolina. Consulte la bibliografía del fabricante para obtener más información.

Isolated Ground Receptacle. A grounding-type device in which the equipment ground contact and terminal are electrically isolated from the receptacle mounting means.

Receptáculo aislado con conexión a tierra. Dispositivo de tipo de conexión a tierra en el cual el terminal y el contacto de conexión a tierra del equipo están eléctricamente aislados de los medios de montaje del receptáculo.

K

Kilowatt (kW). One thousand watts equals 1 kilowatt.

Kilovatio (kW). Mil vatios equivalen a 1 kilovatio.

*Reprinted with permission from NFPA 70-2014.

Kilowatt-hour (kWh). One thousand watts of power in 1 hour. One 100-watt lamp burning for 10 hours is 1 kilowatt-hour. Two 500-watt electric heaters operated for 1 hour is 1 kilowatt-hour.

Kilovatio/hora (kWh). Mil vatios de energía en 1 hora. Una lámpara de 100 vatios que está encendida durante 10 horas es 1 kilovatio/hora Dos calentadores eléctricos de 500 vatios que funcionan durante 1 hora es 1 kilovatio/hora.

Kitchen. An area with a sink and permanent provisions for food preparation and cooking.*

Cocina. Zona con un fregadero y suministros permanentes para preparar y cocinar alimentos.

L

Labeled. Equipment or materials to which has been attached a label, symbol, or other identifying mark of an organization that is acceptable to the authority having jurisdiction (AHJ) and concerned with product evaluation that maintains periodic inspection of production of labeled equipment or materials and by whose labeling the manufacturer indicates compliance with appropriate standards or performance in a specified manner.*

Etiquetado. Equipos o materiales a los cuales se les ha colocado una etiqueta, símbolo u otro tipo de marca identificadora de una organización que es aceptable para la autoridad que tiene jurisdicción (AHJ) y que se encarga de la evaluación de productos que mantiene inspecciones periódicas de la producción de equipos o materiales etiquetados y por medio de cuyo etiquetado el fabricante indica que cumple los estándares o el desempeño adecuado de una forma específica.

Labor and Material Payment Bond. A written form of security from a surety (bonding) company to the owner, on behalf of an acceptable prime or main contractor or subcontractor, guaranteeing payment to the owner in the event the contractor fails to pay for all labor, materials, equipment, or services in accordance with the contract. (See Performance Bond and Surety Bond.)

Garantía de pago de materiales y mano de obra. Forma de garantía escrita que emite una compañía afianzadora (garante) para el propietario a nombre de un contratista o subcontratista principal aceptable, donde se garantiza el pago al propietario en caso de que el contratista no pague por la mano de obra, los materiales, los equipos o los servicios de conformidad con el contrato. (Consulte garantía de cumplimiento y fianza.)

Lamp. A generic term for an artificial source of light.

Lámpara. Término genérico para una fuente de luz artificial.

Light. The term generally applied to the visible energy from a source. Light is usually measured in lumens or candlepower. When light strikes a surface, it is either absorbed, reflected, or transmitted.

Luz. Término que se aplica generalmente a la energía visible de una fuente. La luz se mide por lo general en lúmenes o candelas. Cuando la luz choca con la superficie, esta se absorbe, se refleja o se transmite.

Lighting Outlet. An outlet intended for the direct connection of a lampholder or luminaire.*

Toma de luz. Toma de corriente diseñado para una conexión directa de un portalámparas o luminaria.

Listed. Equipment, materials, or services included in a list published by an organization that is acceptable to the authority having jurisdiction (AHJ) and concerned with evaluation of products or services, that maintains periodic inspection of production of listed equipment or materials or periodic evaluation of services, and whose listing states that either the equipment, material, or services meets appropriate designated standards or has been tested and found suitable for a specified purpose.*

Incluidos. Equipos, materiales o servicios que están incluidos en una lista publicada por una organización que es aceptable para la autoridad que tiene jurisdicción (AHJ) y que se encarga de la evaluación de productos o servicios, que mantiene inspecciones periódicas de la producción de los equipos o materiales incluidos o la evaluación periódica de servicios, y cuya inclusión indica que los equipos, materiales o servicios cumplen los estándares adecuados designados o se han probado y se ha determinado que son aptos para un fin específico.

Load. The electric power used by devices connected to an electrical system. Loads can be figured in amperes, volt-amperes, kilovolt-amperes, or kilowatts. Loads can be intermittent, continuous intermittent, periodic, short-time, or varying. See the definition of "Duty" in the *NEC*.

Carga. La energía eléctrica usada por los dispositivos conectados a un sistema eléctrico. Las cargas pueden estar expresadas en amperios, voltios amperios,

*Reprinted with permission from NFPA 70-2014.

kilovoltios amperios o kilovatios. Las cargas pueden ser intermitentes, intermitentes continuas, periódicas, de corto plazo o variantes. Consulte la definición de "Servicio" en *NEC*.

Load Center. A common name for residential panelboards. A load center may not be as deep as a panelboard and generally does not contain relays or other accessories as are available for panelboards. Circuit breakers "plug in" as opposed to the "bolt-in" types used in panelboards. Manufacturers' catalogs will show both load centers and panelboards. The UL standards do not differentiate.

Centro de carga. Nombre común para los paneles residenciales. El centro de carga puede no ser tan profundo como un panel y generalmente no contiene relés ni otros accesorios que están disponibles en los paneles. Los interruptores de circuito están "enchufados" a diferencia de los tipos "atornillados" que se usan en los paneles. Los catálogos de los fabricantes mostrarán los centros de carga y los paneles. Los estándares de UL no los diferencian.

Location, Damp. Locations protected from weather and not subject to saturation with water or other liquids but subject to moderate degrees of moisture. Examples of such locations include partially protected locations under canopies, marquees, roofed open porches, and like locations, and interior locations subject to moderate degrees of moisture, such as some basements, some barns, and some cold-storage warehouses.*

Ubicación, húmeda. Lugares protegidos de las condiciones climáticas y que no están expuestos a la saturación con agua ni otros líquidos, pero que están expuestos a grados moderados de humedad. Algunos ejemplos de este tipo de ubicaciones incluyen ubicaciones que están parcialmente protegidas debajo de doseles, marquesinas, pórticos abiertos con techo y ubicaciones similares, y ubicaciones en el interior que estén expuestas a grados moderados de humedad, como algunos sótanos, algunos establos y algunos depósitos de almacenamiento en frío.

Location, Dry. A location not normally subject to dampness or wetness. A location classified as dry may be temporarily subject to dampness or wetness, as in the case of a building under construction.*

Ubicación, seca. Lugar que normalmente no está expuesto a humedad o superficies mojadas. Una ubicación clasificada como seca puede estar temporalmente expuesta a humedad o superficies mojadas, como en el caso de una edificación que se está construyendo.

Location, Wet. Installations underground or in concrete slabs or masonry in direct contact with the earth; in locations subject to saturation with water or other liquids, such as vehicle washing areas; and in unprotected locations exposed to weather.*

Ubicación, mojada. Instalaciones subterráneas o en losas de hormigón o mampostería en contacto directo con la tierra; en lugares que están expuestos a una saturación con agua u otros líquidos, tales como áreas de lavado de vehículos; y en lugares desprotegidos que están expuestos a las condiciones climáticas.

Locked Rotor Current (LRC). The steady-state current taken from the line with the rotor locked and with rated voltage and frequency applied to the motor.

Corriente con rotor bloqueado (LRC). Corriente de estado estable tomada de la línea con el rotor bloqueado y con un voltaje y una frecuencia nominal aplicados al motor.

Low-Pressure Sodium Lamp. A discharge lamp in which light is produced from sodium gas operating at a partial pressure.

Lámpara de sodio de baja presión. Lámpara de descarga en la que la luz se produce gracias a un gas de sodio que funciona a presión parcial.

Lumens. The SI unit of luminous flux. The units of light energy emitted from the light source.

Lúmenes. Unidad del SI para el flujo luminoso. Las unidades de energía de luz emitidas desde la fuente de luz.

Luminaire. A complete lighting unit consisting of a light source such as a lamp or lamps, together with the parts designed to position the light source and connect it to the power supply. It may also include parts to protect the light source, ballast, or distribute the light. A lampholder itself is not a luminaire.* (Prior to the National Electrical Code adopting the International System (SI) definition of luminaire, the commonly used term in the United States was and in most instances still is "lighting fixture." It will take years for the electrical industry to totally change and feel comfortable with the term luminaire.)

*Reprinted with permission from NFPA 70-2014.

Luminaria. Unidad completa de iluminación que consiste de una fuente de luz como una lámpara o varias, junto con las partes diseñadas para posicionar la fuente de luz y conectarla a la alimentación de energía. También podría incluir partes para proteger la fuente de luz, el balasto o distribuir la luz. Un portalámparas por sí solo no es una luminaria. (Antes de que el Código Eléctrico Nacional adoptara la definición de luminaria del Sistema Internacional (SI), el término utilizado comúnmente en los Estados Unidos era, y en muchos casos aún es, "accesorio de iluminación". Tomará años que la industria eléctrica cambie totalmente y se sienta cómoda con el término luminaria.)

Luminaire Efficiency. The total lumen output of the luminaire divided by the rated lumens of the lamps inside the luminaire.

Eficiencia de la luminaria. La producción total de lúmenes de la luminaria dividida por los lúmenes nominales de las lámparas dentro de la luminaria.

Lux. The SI (International System) unit of illumination. One lumen uniformly distributed over an area 1 square meter in size.

Lux. Unidad de iluminación del SI (Sistema Internacional). Un lumen distribuido de forma uniforme en un área de 1 metro cuadrado.

M

Mandatory Rules. The terms shall or shall not are used when a Code statement is mandatory.

Reglas obligatorias. Los términos "deberá" o "no podrá" se utilizan cuando una afirmación del código es obligatoria.

Maximum Rating of Branch-Circuit Short-Circuit and Ground-Fault Protection. A term used with equipment that has a hermetic motor compressor(s). This is the maximum ampere rating for the equipment's branch-circuit overcurrent protective device. The ampere rating is determined by the manufacturer and is marked on the nameplate of the equipment. The nameplate will also indicate if the overcurrent device can be a circuit breaker, a fuse, or either.

Capacidad máxima de la protección contra fallas de conexión a tierra y cortocircuito en el circuito de bifurcación. Término utilizado con un equipo que tiene compresor(es) de motor hermético. Esta es la capacidad máxima en amperios para el dispositivo protector de sobrecarga del circuito de bifurcación del equipo. La capacidad en amperios está establecida por el fabricante y está marcada en la placa del equipo. La placa también indica si el dispositivo de sobrecarga puede ser un interruptor de circuito, un fusible o cualquiera de los dos.

Mercury Lamp. A high-intensity discharge light source in which radiation from the mercury vapor produces visible light.

Lámpara de mercurio. Fuente de luz de descarga de alta intensidad en la cual la radiación del vapor de mercurio produce luz visible.

Metal-Halide Lamp. A high-intensity discharge light source in which the light is produced by the radiation from mercury together with the halides of metals such as sodium and candium. *Reprinted with permission from NFPA 70-2014.

Lámpara de halogenuros metálicos. Fuente de luz de descarga de alta intensidad en la cual la luz se produce por la radiación del mercurio junto con los halogenuros de metales como sodio y cadmio. *Reproducción autorizada por NFPA 70-2014.

N

National Electrical Code (NEC). The electrical code published by the National Fire Protection Association. This Code provides for practical safeguarding of persons and property from hazards arising from the use of electricity. It does not become law until adopted by federal, state, or local laws and regulations.

Código Eléctrico Nacional (NEC). Código eléctrico publicado por la Asociación Nacional de Protección contra Incendios (National Fire Protection Association). Este código establece la protección práctica de personas y propiedad de peligros que puedan surgir por el uso de la electricidad. No se convierte en ley hasta que sea adoptado por las regulaciones y leyes federales, estatales y locales.

National Electrical Manufacturers Association (NEMA). NEMA is a trade organization made up of many manufacturers of electrical equipment. They develop and promote standards for electrical equipment.

Asociación Nacional de Fabricantes Eléctricos (National Electrical Manufacturers Association, NEMA). NEMA es una organización comercial compuesta de muchos fabricantes de equipos eléctricos. Ellos desarrollan y fomentan estándares para equipos eléctricos.

National Fire Protection Association (NFPA). Located in Quincy, MA. The NFPA is an international Standards Making Organization dedicated to the protection of people from the ravages of fire and electric shock. The NFPA is responsible for developing and writing the National Electrical Code, the Installation of Sprinkler Systems in One- and Two- Family Dwellings and Manufactured Homes, the Life Safety Code, the National Fire Alarm Code, and over 300 other codes, standards, and recommended practices. The NFPA phone number is (800) 344-3555. Its website is www.nfpa.org.

Asociación Nacional de Protección contra Incendios (National Fire Protection Association, NFPA). Ubicada en Quincy, MA. La NFPA es una organización internacional que establece estándares y se dedica a la protección de las personas de los estragos de un incendio o un choque eléctrico. La NFPA es la responsable de desarrollar y redactar el Código Eléctrico Nacional, la instalación de sistemas de aspersores en viviendas de una y dos familias y casas fabricadas, el Código de Seguridad de la Vida, el Código Nacional de Alarmas contra Incendios y más de 300 códigos, estándares y prácticas recomendadas más. El número de teléfono de la NFPA es (800) 344-3555. Su sitio web es www.nfpa.org.

Nationally Recognized Testing Laboratory (NRTL). The term used to define a testing laboratory that has been recognized by OSHA; for example, Underwriters Laboratories (UL), Intertek Testing, and MET Laboratories.

Laboratorio de pruebas reconocido a nivel nacional (Nationally Recognized Testing Laboratory, NRTL). Término utilizado para definir un laboratorio de pruebas que ha sido reconocido por OSHA; por ejemplo, Underwriters Laboratories (UL), Intertek Testing y MET Laboratories.

Neutral Conductor. The conductor connected to the neutral point of a system that is intended to carry current under normal conditions.* In residential wiring, the neutral conductor is the grounded conductor in a circuit. A neutral conductor is always a grounded conductor. A grounded conductor is not always a neutral conductor, such as a "grounded B phase" system discussed in Electrical Wiring—Commercial.

Conductor neutro. Conductor conectado al punto neutro de un sistema diseñado para transportar corriente en condiciones normales. En los cableados residenciales, el conductor neutro es el conductor con conexión a tierra en un circuito. El conductor neutro siempre es un conductor con conexión a tierra. Un conductor con conexión a tierra no siempre es un conductor neutro, como el sistema de "fase B con conexión a tierra" mencionado en Cableado eléctrico—comercial.

Neutral point. The common point on a wye-connection in a polyphase system or midpoint on a single-phase, 3-wire system, or midpoint of a single-phase portion of a 3-phase delta system, or a midpoint of a 3-wire, direct current system.*

Punto neutro. Punto común en una conexión de estrella en un sistema polifásico o el punto medio en un sistema de fase simple de 3 cables, o el punto medio de una parte de fase simple de un sistema delta de 3 fases, o el punto medio de un sistema de corriente directa de 3 cables.

No-Niche Luminaire. A luminaire intended for installation above or below the water without a niche.* The definition is found in *Article 680.*

Luminaria sin nicho. Luminaria diseñada para ser instalada encima o debajo del agua sin un nicho. La definición se encuentra en el *Artículo 680.*

Noncoincidental Loads. Loads that are not likely to be on at the same time. Heating and cooling loads would not operate at the same time. See *220.60* of the *NEC.*

Cargas no coincidentales. Cargas que es probable que no estén encendidas al mismo tiempo. Las cargas de calefacción y refrigeración no funcionarían al mismo tiempo. Consulte *220.60* de *NEC.*

Notations. Words found on plans to describe something.

Anotaciones. Palabras que se encuentran en los planos para describir algo.

O

Occupational Safety and Health Act (OSHA). This is the code of federal regulations developed by the Occupational Safety and Health Administration, U.S. Department of Labor. The electrical regulations are covered in Part 1910, Subpart S. The *NEC* must still be referred to in conjunction with OSHA regulations.

Ley de Seguridad y Salud Ocupacional (Occupational Safety and Health Act, OSHA). Este es el código de regulaciones federales desarrollado por la Administración de Salud y Seguridad Ocupacional del Departamento de Trabajo de los EE. UU. Las regulaciones eléctricas se tratan en la Parte 1910, Subparte S. Todavía se debe hacer referencia a *NEC* en conjunto con las regulaciones de OSHA.

*Reprinted with permission from NFPA 70-2014.

Ohm. A unit of measure for electric resistance. An ohm is the amount of resistance that will allow 1 ampere to flow under a pressure of 1 volt.

Ohmio. Unidad de media para la resistencia eléctrica. Un ohmio es la cantidad de resistencia que permitirá que 1 amperio fluya cuando está sometido a una presión de 1 voltio.

Outlet. A point on the wiring system at which current is taken to supply utilization equipment.*

Toma de corriente. Punto en el sistema de cableado en el cual se toma la corriente para alimentar a un equipo de uso.

Overcurrent. Any current in excess of the rated current of equipment or the ampacity of a conductor. It may result from overload, short circuit, or ground fault.*

Sobrecarga. Toda corriente que exceda la corriente nominal del equipo o la corriente admisible de un conductor. Podría producirse por un exceso de carga, cortocircuito o falla en la conexión a la tierra.

Overcurrent Device. Also referred to as an overcurrent protection device. A form of protection that operates when current exceeds a predetermined value. Common forms of overcurrent devices are circuit breakers, fuses, and thermal overload elements found in motor controllers.

Dispositivo de sobrecorriente. También conocido como dispositivo protector de sobrecorriente. Una forma de protección que opera cuando la corriente excede un valor predeterminado. Las formas comunes de los dispositivos de sobrecorriente son interruptores de circuito, fusibles y elementos de sobrecarga térmica en controladores de motor.

Overload. Operation of equipment in excess of the normal, full-load rating, or of a conductor in excess of rated ampacity that, when it persists for a sufficient length of time, would cause damage or dangerous overheating. A fault, such as a short circuit or a ground fault, is not an overload.*

Sobrecarga. Funcionamiento de un equipo en exceso de la capacidad normal de carga total, o de un conductor en exceso de la corriente admisible nominal, cuando persiste por un período lo suficientemente largo podría causar un daño o un exceso de calentamiento peligroso. Una falla, como un cortocircuito o una falla en la conexión a tierra, no es una sobrecarga.

Owner. An individual or corporation that owns a real property.

Propietario. Individuo o corporación que es dueño de una propiedad real.

P

Panelboard. A single panel or group of panel units designed for assembly in the form of a single panel, including buses and automatic overcurrent devices; equipped with or without switches for the control of light, heat, or power circuits; designed to be placed in a cabinet or cutout box placed in or against a wall or partition and accessible only from the front.*

Panel. Panel único o grupo de paneles diseñados para su ensamblado en forma de un único panel, que incluye buses y dispositivos de sobrecorriente automáticos; equipado con o sin dispositivos para el control de circuitos de luz, calor o energía; diseñado para ser colocado en un gabinete o una caja de corte colocada en una pared o partición o contra ella y accesible solo desde el frente.

Performance Bond. (1) A written form of security from a surety (bonding) company to the owner, on behalf of an acceptable prime or main contractor or subcontractor, guaranteeing payment to the owner in the event the contractor fails to perform all labor, materials, equipment, or services in accordance with the contract. (2) The surety companies generally reserve the right to have the original prime or main or subcontractor remedy any claims before paying on the bond or hiring other contractors.

Garantía de cumplimiento. (1) Forma de garantía escrita que emite una compañía afianzadora (garante) para el propietario a nombre de una prima aceptable, un contratista o subcontratista principal, donde se garantiza el pago al propietario en caso de que el contratista no pague por la totalidad de la mano de obra, los materiales, los equipos o los servicios de conformidad con el contrato. (2) Las compañías afianzadoras generalmente se reservan el derecho a hacer que la prima, el contratista principal o subcontratista original pague todo reclamo antes de pagar la garantía o contratar a otros contratistas.

Permissive Rules. Allowed but not required. Terms such as shall be permitted or shall not be required are used when a Code statement is permissive.

Reglas permisivas. Permitidas, pero no obligatorias. Términos como "estará permitido" o "no se exigirá" se utilizan cuando una afirmación del código es permisiva.

*Reprinted with permission from NFPA 70-2014.

Photoelectric Detector. This type of detector triggers an alarm when light is blocked between its internal light source and sensor. Typically, heavy smoke will block the light. The action of blocking light is similar to that of the safety feature on a garage door operator. Consult the manufacturer's literature for more information.

Detector fotoeléctrico. Este tipo de detector activa una alarma cuando la luz se bloquea entre su fuente de luz interna y el sensor. Por lo general, el humo denso bloqueará la luz. La acción del bloqueo de la luz es similar a la de la función de seguridad de un operador de puerta de garaje. Consulte la bibliografía del fabricante para obtener más información.

Photometry. The pattern and amount of light that is emitted from a luminaire, normally represented as a cross-section through the luminaire distribution pattern.

Fotometría. El patrón y la cantidad de luz que se emite desde una luminaria, normalmente representados como sección transversal a través del patrón de distribución de la luminaria.

Plan. (1) A line drawing (by floor) representing the horizontal geometrical section of the walls of a building. The section (a horizontal plane) is taken at an elevation to include the relative positions of the walls, partitions, windows, doors, chimneys, columns, pilasters, and so on. (2) A plan can be thought of as cutting a horizontal section through a building at an eye level elevation.

Plano. (1) Gráfico de líneas (por piso) que representa la sección geométrica horizontal de las paredes de un edificio. La sección (el plano horizontal) se toma a una elevación para incluir las posiciones relativas de las paredes, particiones, ventanas, puertas, chimeneas, columnas, pilastras, etc. (2) Se puede pensar en un plano como en el corte de una sección horizontal a través de una edificación a la elevación del nivel de los ojos.

Plan Checker. A term sometimes used to describe a building department official who examines the building permit documents.

Inspector de planos. Término que se usa a veces para describir al funcionario del departamento de construcción que examina los documentos del permiso de construcción.

Plans. A term used to represent all drawings, including sections and details and any supplemental drawings, for complete execution of a specific project.

Planos. Término utilizado para representar todos los gráficos, lo que incluye las secciones y los detalles y todo dibujo suplementario, para la ejecución completa de un proyecto específico.

Power Factor. A ratio of actual power (W or kW) being used to apparent power (VA or kVA) being drawn from the power source.

Factor de potencia. Coeficiente de potencia real (W o kW) que se utiliza para potencia aparente (VA o kVA) que se extrae de la fuente de energía.

Power Supply. A source of electrical operating power including the circuits and terminations connecting it to the dependent system components.

Fuente de energía. Fuente de energía eléctrica operativa, lo que incluye los circuitos y las terminaciones que la conectan a los componentes que dependen del sistema.

Preheat. A type of ballast that is easily identified because it has a starter. One type of starter is automatic with two buttons on one end; the other type is a manual On/Off switch that has a momentary "make" position just beyond the "On" position.

Precalentador. Tipo de balasto que es fácil de identificar porque tiene un cebador. Un tipo de cebador es automático con dos botones en un extremo; el otro tipo es un interruptor de encendido/apagado que tiene una posición momentánea de "hacer" justo después de la posición de "encendido".

Preheat Fluorescent Lamp Circuit. A circuit used on fluorescent lamps wherein the electrodes are heated or warmed to a glow stage by an auxiliary switch or starter before the lamps are lighted.

Circuito de precalentamiento de lámpara fluorescente. Circuito usado en las lámparas fluorescentes por medio del cual se calientan los electrodos para colocarlos en una etapa de brillo por efecto de un interruptor auxiliar o cebador antes de que se enciendan las lámparas.

Preliminary Drawings: The drawings that precede the final approved drawings. Usually stamped "PRELIMINARY."

Gráficos preliminares: Los gráficos que preceden a los gráficos finales aprobados. Por lo general, tienen estampado "PRELIMINAR".

Prime Contractor. Any contractor having a contract directly with the owner. Usually the main (general) contractor for a specific project.

Contratista principal. Todo contratista que tiene un contrato directo con el propietario Por lo general, el contratista fundamental (general) para un proyecto específico.

Prints. A term used to refer to a plan or plans.

Impresiones. Término utilizado para hacer referencia a un plano o planos.

Project. A word used to represent the overall scope of work being performed to complete a specific construction job.

Proyecto. Palabra utilizada para hacer referencia al alcance global del trabajo que se realiza para completar un trabajo de construcción específico.

Proposal. A written offer from a bidder to the owner, preferably on a prescribed proposal form, to perform the work and to furnish all labor, materials, equipment and/or service for the prices and terms quoted by the bidder.

Propuesta. Oferta escrita de un ofertante al propietario, preferentemente en un formulario de propuesta prescrito, para realizar el trabajo y brindar toda la mano de obra, los materiales, los equipos y/o el servicio por los precios y términos citados por el ofertante.

Punch List (Inspection List). A list prepared by the owner or his or her authorized representative of items of work requiring immediate corrective or completion action by the contractor.

Lista de puntos pendientes (Lista de inspección). Lista preparada por el propietario o su representante autorizado que incluye los artículos de trabajo que requieren acciones inmediatas de corrección o terminación por parte del contratista.

Q

Qualified Person. One who has skills and knowledge related to the construction and operation of the electrical equipment and installations and has received safety training to recognize and avoid the hazards involved.*

Persona calificada. Persona que tiene las habilidades y el conocimiento relacionados con la construcción y operación de los equipos y las instalaciones eléctricas y que ha recibido capacitación en seguridad para reconocer y evitar los peligros que ello implica.

R

Raceway. ▶An enclosed channel of metallic or nonmetallic materials designed expressly for holding wires, cables, or busbars, with additional functions as permitted in this *Code.*◀*

Conducto de cables. Canal cerrado de materiales metálicos o no metálicos que está diseñado expresamente para el paso de cables, o barras, con funciones adicionales como se permite en virtud de este código.

Rapid Start. The most common type of ballast used today that does not require a starter.

Encendido rápido. El tipo más común de balasto que se utiliza en la actualidad que no requiere un cebador.

Rapid-Start Fluorescent Lamp Circuit. A circuit designed to start lamps by continuously heating or preheating the electrodes. Lamps must be designed for this type of circuit. This is the modern version of the "trigger start" system. In a rapid-start, 2-lamp circuit, one end of each lamp is connected to a separate starting winding.

Circuito de encendido rápido de lámpara fluorescente. Circuito diseñado para el encendido de lámparas al calentar y precalentar de forma continua los electrodos. Las lámparas deben estar diseñadas para este tipo de circuito. Esta es la versión moderna del sistema de "activador de encendido". En un circuito de 2 lámparas de encendido rápido, un extremo de cada lámpara está conectado a un devanado de encendido separado.

Rate of Rise Detector. A device that responds when the temperature rises at a rate exceeding a predetermined value.

Detector térmico. Dispositivo que responde cuando la temperatura sube a una tasa que excede un valor determinado.

Rated-Load Current. The rated-load current for a hermetic refrigerant motor-compressor is the current resulting when the motor-compressor is operated at the rated load, rated voltage, and rated frequency of the equipment it serves.* See *NEC Article 440.*

Corriente de carga nominal. La corriente de carga nominal para un motor compresor refrigerante hermético es la corriente que se genera cuando el motor compresor funciona a la carga nominal, voltaje nominal y frecuencia nominal del equipo al que sirve. Consulte *NEC Artículo 440.*

Receptacle. A receptacle is a contact device installed at the outlet for the connection of an attachment plug. A single receptacle is a single contact device with no other contact device on the same yoke. A multiple receptacle is two or more contact devices on the same yoke.*

Receptáculo. Un receptáculo es un dispositivo de contacto instalado en la toma de corriente para la

*Reprinted with permission from NFPA 70-2014.

conexión de un enchufe. Un receptáculo simple es un dispositivo de contacto simple sin ningún otro dispositivo de contacto en el mismo yugo. Un receptáculo múltiple es un dispositivo de dos o más contactos en el mismo yugo.

Receptacle Outlet. An outlet where one or more receptacles are installed.*

Toma de corriente del receptáculo. Toma de corriente donde hay uno o más receptáculos instalados.

Resistive Load. An electric load that opposes the flow of current. A resistive load does not contain cores or coils of wire. Some examples of resistive loads are the electric heating elements in an electric range, ceiling heat cables, and electric baseboard heaters.

Carga resistiva. Carga eléctrica que se opone al flujo de la corriente. Una carga resistiva no contiene núcleos ni bobinas de cables. Algunos ejemplos de cargas resistivas son los elementos de calefacción eléctrica en una autonomía eléctrica, cables de calefacción del techo y calefactores de rodapié eléctricos.

Root-Mean-Squared (RMS). The square root of the average of the square of the instantaneous values of current or voltage. For example, the RMS value of voltage, line to neutral, in a home is 120 volts. During each electrical cycle, the voltage rises from zero to a peak value (120 × 1.4142 = 169.7 volts), back through zero to a negative peak value, then back to zero. The RMS value is 0.707 of the peak value (169.7 × 0.707 = 120 volts). The root-mean-squared value is what an electrician reads on an ammeter or voltmeter.

Valor eficaz (RMS). La raíz cuadrada del promedio del cuadrado de los valores instantáneos de corriente o voltaje. Por ejemplo, el valor RMS de voltaje, línea a neutro, en un hogar es de 120 voltios. Durante cada ciclo eléctrico, el voltaje se eleva desde cero a un valor pico (120 × 1.4142 = 169.7 voltios), vuelve para pasar el cero y alcanzar un valor pico negativo, luego vuelve a cero. El valor RMS es 0,707 del valor pico (169.7 × 0.707 = 120 voltios). El valor eficaz es el que lee un electricista con un amperímetro o voltímetro.

S

Schedules. Tables or charts that include data about materials, products, and equipment.

Programas. Tablas o cuadros que incluyen información acerca de materiales, productos y equipos.

*Reprinted with permission from NFPA 70-2014.

Scope. A written range of view or action; outlook; hence, room for the exercise of faculties or function, capacity for achievement; all in connection with a designated project.

Alcance. Rango escrito de visión o acción; enfoque; por lo tanto, espacio para el ejercicio de facultades o función, capacidad de logro; todo en conexión con un proyecto designado.

Sections. Views of vertical cuts through and perpendicular to components showing their detailed arrangement.

Secciones. Vistas de cortes verticales y perpendiculares de objetos que muestran su distribución detallada

Series-Rated System. Panelboards are marked with their short-circuit rating in RMS symmetrical amperes. A series-rated panelboard will be determined by the main circuit breaker or fuse, and branch-circuit breaker combination tested in accordance to UL Standard 489. The series rating will be less than or equal to the interrupting rating of the main overcurrent device, and greater than the interrupting rating of the branch-circuit overcurrent devices. (See "Fully Rated System.")

Sistema de capacidad de serie. Los paneles están marcados con sus capacidades de cortocircuito en amperios simétricos de RMS. Un panel con capacidad de series se determinará por el interruptor de circuito principal o fusible, y la combinación de interruptor de circuito de bifurcación probada de conformidad con el estándar 489 de UL. La capacidad de serie será menor o igual a la capacidad de interrupción del dispositivo de sobrecorriente principal y mayor que la capacidad de interrupción de los dispositivos de sobrecorriente del circuito de bifurcación. (Ver "Sistema completamente clasificado".)

Service. The conductors and equipment for delivering energy from the serving utility to the wiring system of the premises served.*

Servicio. Los conductores y los equipos para transportar energía desde el servicio público al sistema de cableado de los lugares que reciben el servicio.

Service Conductors. The conductors from the service point to the service disconnecting means.*

Conductores de servicio. Los conductores desde el punto de servicio hasta el medio de desconexión del servicio.

Service Conductors, Overhead. The overhead conductors between the service point and the first point of connection to the service-entrance conductors at the building or other structure.*

Conductores de servicio, aéreos. Los conductores aéreos entre el punto de servicio y el primer punto de conexión a los conductores de entrada de servicio en la edificación u otra estructura.

Service Conductors, Underground. The underground conductors between the service point and the first point of connection to the service entrance conductors in a terminal box, meter, or other enclosure, inside or outside the building wall.*

Conductores de servicio, subterráneos. Los conductores subterráneos entre el punto de servicio y el primer punto de conexión a los conductores de entrada de servicio en una caja terminal, medidor u otro cerramiento, dentro o fuera de la pared de la edificación.

Informational Note. Where there is no terminal box, meter, or other enclosure with adequate space, the point of connection shall be considered to be the point of entrance of the service conductors into the building.*

Nota informativa. Donde no haya caja terminal, medidor u otro cerramiento con el espacio adecuado, se considerará punto de conexión al punto de entrada de los conductores de servicio dentro del edificio.

Service Drop. The overhead service conductors between the utility electric supply system and the service point.*

Caída de servicio. Los conductores de servicio aéreos entre el sistema de suministro eléctrico público y el punto de servicio.

Service Equipment. The necessary equipment, usually consisting of a circuit breaker or switch and fuses and their accessories, located near the point of entrance of supply conductors to a building or other structure, or an otherwise defined area, and intended to constitute the main control and means of cutoff of the supply.*

Equipo de servicio. El equipo necesario, generalmente consiste de un disyuntor o interruptor de circuito y fusibles y sus accesorios, ubicados cerca del punto de entrada de los conductores de suministro a una edificación u otra estructura, o a otro tipo de área definida, y que está diseñada para constituir el control principal y el medio de corte del suministro.

Service Lateral. The underground conductors between the utility electric supply system and the service point.*

*Reprinted with permission from NFPA 70-2014.

Lateral de servicio. Los conductores subterráneos entre el sistema de suministro eléctrico público y el punto de servicio.

Service Point. The point of connection between the facilities of the serving utility and the premises wiring.*

Punto de servicio. El punto de conexión entre las instalaciones del servicio público y el cableado del lugar.

Informational Note. The service point can be described as the point of demarcation between where the serving utility ends and the premises wiring begins. The serving utility generally specifies the location of the service point based on the conditions of service.*

Nota informativa. Se puede describir al punto de servicio como el punto de demarcación entre donde finaliza el servicio público y comienza el cableado del lugar. El servicio público generalmente especifica la ubicación del punto de servicio en base a las condiciones de servicio.

Shall. Indicates a mandatory requirement.

Debe. Indica un requisito obligatorio.

Short Circuit. A connection between any two or more conductors of an electrical system in such a way as to significantly reduce the impedance of the circuit. The current flow is outside of its intended path, thus the term short circuit. A short circuit is also referred to as a fault.

Cortocircuito. Una conexión entre dos o más conectores de un sistema eléctrico de modo tal que se reduzca significativamente la impedancia del circuito. El flujo de corriente está fuera de la vía pretendida, de ahí proviene el término cortocircuito. El cortocircuito también se conoce como falla.

Short-Circuit Current Rating. The prospective symmetrical fault current at a nominal voltage to which an apparatus or system is able to be connected without sustaining damage exceeding defined acceptance criteria.

Capacidad de corriente de cortocircuito. Corriente prospectiva de falla simétrica a un voltaje nominal al cual un aparato o sistema es capaz de estar conectado sin sufrir daños que excedan los criterios de aceptación definidos.

Should. Indicates a recommendation or that which is advised but not required.

Debería. Indica una recomendación o aquello que se aconseja, pero no se exige.

Site. The place where a structure or group of structures was, or is, to be located (as in construction site).

Sitio. El lugar donde iba a estar o estará ubicada una estructura o un grupo de estructuras (como un sitio de construcción).

Smoke Alarm. A single or multiple station alarm responsive to smoke.

Alarma de humo. Alarma de estación simple o múltiple que responde cuando detecta humo.

Smoke Detector. A device that detects visible or invisible particles of combustion.

Detector de humo. Dispositivo que detecta partículas visibles o invisibles de la combustión.

Sone. A unit of loudness equal to the loudness of a sound of 1 kilohertz at 40 decibels above the threshold of hearing of a given listener.

Sonio. Unidad de sonoridad igual a la sonoridad de un sonido de 1 kilohercio a 40 decibeles por encima del umbral de audición de un oyente dado.

Specifications. Text setting forth details such as description, size, quality, performance, workmanship, and so forth. Specifications that pertain to all of the construction trades involved might be subdivided into "General Conditions" and "Supplemental General Conditions." Further subdividing the specifications might be specific requirements for the various contractors such as electrical, plumbing, heating, masonry, and so forth. Typically, the electrical specifications are found in Division 16.

Especificaciones. Texto que establece detalles como descripción, tamaño, calidad, desempeño, mano de obra, etc. Las especificaciones que pertenecen a todos los oficios de la construcción involucrados pueden ser subdivididas en "Condiciones generales" y "Condiciones generales suplementarias". A su vez, se puede subdividir más las especificaciones en requisitos específicos para los diversos contratistas como eléctricos, de plomería, calefacción, mampostería, etc. Por lo general, las especificaciones eléctricas se encuentran en la División 16.

Split-Wired Receptacle. A receptacle that can be connected to two branch circuits or a multiwire branch circuit. These receptacles may also be used so that one receptacle is live at all times, and the other receptacle is controlled by a switch. The terminals on these receptacles usually have breakaway tabs so the receptacle can be used either as a split-wired receptacle or as a standard receptacle.

Receptáculo de cable partido. Receptáculo que puede ser conectado a dos circuitos de bifurcación o a un circuito de múltiples bifurcaciones. Estos receptáculos también se pueden utilizar para que un receptáculo esté constantemente activado y el otro esté controlado por un interruptor. Los terminales en estos receptáculos tienen por lo general pestañas removibles de modo tal que el receptáculo se pueda usar como receptáculo de cable partido o como receptáculo estándar.

Standard Network Interface (SNI). A device usually installed by the telephone company at the demarcation point where their service leaves off and the customer's service takes over. This is similar to the service point for electrical systems.

Interfaz de red estándar (SNI). Dispositivo generalmente instalado por la compañía telefónica en el punto de demarcación donde el servicio se termina y el servicio del cliente toma el control. Esto es similar al punto de servicio para los sistemas eléctricos.

Starter. A device used in conjunction with a ballast for the purpose of starting an electric discharge lamp.

Cebador. Dispositivo utilizado en conjunto con un balasto con el objetivo de encender una lámpara de descarga eléctrica.

Structure. That which is built or constructed.*

Estructura. Aquella que es construida o armada.

Subcontractor. A qualified subordinate contractor to the prime or main contractor.

Subcontratista. Contratista calificado subordinado al contratista principal o fundamental.

Surety Bond. A legal document issued to ensure the completion of an act by another person.

Contractors usually are required to purchase surety bonds, if they are working on public projects. A surety bond guarantees to one party that another (the contractor) will perform specified acts, usually within a stated period of time. The surety company typically becomes responsible for fulfillment of a contract if the contractor defaults. In the case of a public works project, such as a road, that means that the surety bond protects taxpayers should a contractor go out of business.

Garantía de fianza. Documento legal emitido para asegurar la finalización de un acto por parte de otra persona.

Generalmente se les exige a los contratistas que adquieran garantías de fianza si van a trabajar en proyectos públicos. Una garantía de fianza le garantiza a una parte que la otra (el contratista) realizará los actos especificados, generalmente dentro de un período determinado de tiempo. La compañía afianzadora suele convertirse en la responsable por el cumplimiento del contrato si el contratista incumple. En el caso de un proyecto de obras públicas, como una carretera, esto significa que la garantía de fianza protege a los ciudadanos en caso de que el contratista se declare en bancarrota.

Surface-Mounted Luminaire. A luminaire mounted directly on the ceiling.

Luminaria montada en la superficie. Luminaria montada directamente en el techo.

Surge-Protective Device (SPD). A protective device for limiting transient voltages by diverting or limiting surge current; it also prevents continued flow of follow current while remaining capable of repeating these functions and is designated as follows:*

Dispositivo protector de sobretensión (SPD). Dispositivo protector para limitar los voltajes transitorios al desviar o limitar la corriente de sobretensión; también previene el flujo continuo de corriente de seguimiento mientras que continúa siendo capaz de repetir estas funciones y está designado de la siguiente forma:

Type 1. Permanently connected SPDs intended for installation between the secondary of the service transformer and the line side of the service disconnect overcurrent device.*

Tipo 1. SPD permanentemente conectados diseñados para la instalación entre el secundario del transformador de servicio y el lado de línea del dispositivo de sobrecorriente de desconexión de servicio.

Type 2. Permanently connected SPDs intended for installation on the load side of the service disconnect overcurrent device, including SPDs located at the branch panel.*

Tipo 2. SPD permanentemente conectados diseñados para la instalación del lado de carga del dispositivo de sobrecorriente de desconexión de servicio, lo que incluye los SPD ubicados en el panel de bifurcación.

Type 3. Point of utilization SPDs.*

Tipo 3. SPD del punto de utilización.

Type 4. Component SPDs, including discrete components, as well as assemblies.*

Tipo 4. SPD de componente, lo que incluye componentes separados, al igual que ensamblajes.

Informational Note. For further information on Type 1, Type 2, Type 3, and Type 4 SPDs, see UL 1449, Standard for Surge Protective Devices.*

Nota informativa. Para obtener más información acerca de los SPD de Tipo 1, Tipo 2, Tipo 3 y Tipo 4, consulte UL 1449, estándar para dispositivos protectores de sobretensión.

Suspended (pendant) Luminaire. A luminaire hung from a ceiling by supports.

Luminaria suspendida (colgante). Luminaria que cuelga del techo por medio de soportes.

Switches:

Interruptores:

General-Use Snap Switch. A form of general-use switch constructed so that it can be installed in device boxes or on box covers, or otherwise used in conjunction with wiring systems recognized by this Code.*

Interruptor de encendido de uso general. Forma de interruptor de uso general fabricado de modo tal que pueda instalarse en cajas de dispositivos o en cubiertas de dispositivos o para que pueda usarse junto con sistemas de cableados reconocidos por este código.

General-Use Switch. A switch intended for use in general distribution and branch circuits. It is rated in amperes, and it is capable of interrupting its rated current at its rated voltage.*

Interruptor de uso general. Interruptor diseñado para ser usado en circuitos de distribución general y bifurcación. Está clasificado en amperios y es capaz de interrumpir su corriente nominal a su voltaje nominal.

Motor-Circuit Switch. A switch, rated in horsepower, capable of interrupting the maximum operating overload current of a motor of the same horsepower rating as the switch at the rated voltage.

Interruptor de circuito de motor. Interruptor, clasificado en caballos de fuerza, capaz de interrumpir la corriente máxima de sobrecarga operativa de un motor con la misma capacidad de caballos de fuerza que el interruptor al voltaje nominal.

*Reprinted with permission from NFPA 70-2014.

Symbols. Graphic representations that stand for or represent other things. A symbol is a simple way to show such things as lighting outlets, switches, and receptacles on an electrical plan. The American Institute of Architects has developed a very comprehensive set of symbols that represent just about everything used by all building trades. When an item cannot be shown using a symbol, then a more detailed explanation using a notation or inclusion in the specifications is necessary.

Símbolos. Representaciones gráficas que indican o representan otras cosas. Un símbolo es una forma simple de mostrar cosas como tomas de corriente, interruptores y receptáculos en un plano eléctrico. El Instituto Estadounidense de Arquitectos ha desarrollado un conjunto muy exhaustivo de símbolos que representan casi todo lo que se utilizan en todos los oficios de la construcción. Cuando no se puede mostrar un artículo usando un símbolo, es necesaria una explicación más detallada utilizando una anotación o inclusión en las especificaciones.

T

T & M. An abbreviation for a contracting method called Time and Material. A written agreement between the owner and the contractor wherein payment is based on the contractor's actual cost for labor equipment, materials, and services plus a fixed add-on amount to cover the contractor's overhead and profit.

T y M. Abreviatura de un método de contratación denominado Tiempo y Material. Un contrato escrito entre el propietario y el contratista por medio del cual el pago se basa en el costo real del contratista para el equipo de trabajo, los materiales y servicios más un monto agregado fijo para cubrir los gastos generales y la ganancia del contratista.

Task Lighting. Lighting directed to a specific surface or area that provides illumination for visual tasks.

Iluminación focalizada. Iluminación dirigida a una superficie o área específica que proporciona luz para tareas visuales.

Terminal. A screw or a quick-connect device where a conductor(s) is intended to be connected.

Terminal. Tornillo o dispositivo de conexión rápida donde se debe conectar un conductor.

Thermocouple. A pair of dissimilar conductors so joined at two points that an electromotive force is developed by the thermoelectric effects when the junctions are at different temperatures.

Termocupla. Par de conductores diferentes unidos en dos puntos de modo tal que se desarrolle una fuerza electromotora por los efectos termoeléctricos cuando las uniones tienen temperaturas diferentes.

Thermopile. More than one thermocouple connected together. The connections may be series or parallel, or both.

Termopila. Más de una termocupla conectadas juntas. Las conexiones pueden ser en serie, en paralelo o ambas.

Torque. A measurement of rotation or turning force. Torque is measured in ounce-inches (oz.-in.), ounce-feet (oz.-ft) and pound-feet (lb.-ft). See Horsepower, Rated Horsepower, and Maximum Developed Horsepower.

Par de torsión. Medida de fuerza de rotación o giro. El par de torsión se mide en onza-pulgada (oz-pulg), onza-pie (oz-pie) y libra-pie (lb-pie). Consulte caballos de fuerza, valor nominal en caballos de fuerza y valor máximo desarrollado en caballos de fuerza.

Troffer. A recessed lighting unit, usually long and installed with the opening flush with the ceiling. The term is derived from "trough" and "coffer."

Luminaria empotrada. Unidad de iluminación empotrada, generalmente larga e instalada con la abertura pegada al techo. El término deriva de "en" y "potro".

U

UL. Underwriters Laboratories (UL) is an independent not-for-profit organization that develops standards and tests electrical equipment to these standards.

UL. Underwriters Laboratories (UL) es una organización independiente sin fines de lucro que desarrolla estándares y prueba equipos eléctricos de acuerdo con esos estándares.

UL-Listed. Indicates that an item has been tested and approved to the standards established by UL for that particular item. The UL Listing Mark may appear in various forms, such as the letters UL in a circle. If the product is too small for the marking to be applied to the product, the marking must appear on the smallest unit container in which the product is packaged.

Incluido en la lista de UL. Indica que un artículo ha sido probado y aprobado según los estándares establecidos por UL para ese artículo en particular. La marca de incluido en la lista de UL puede aparecer de varias formas, como las letras UL en un círculo. Si el producto es demasiado pequeño para que se le pueda aplicar la marca, esta debe aparecer en el paquete de la unidad más pequeña en el cual se encuentra el producto.

UL-Recognized. Refers to a product that is incomplete in construction features or limited in performance capabilities. A "Recognized" product is intended to be used as a component part of equipment that has been "listed." A "Recognized" product must not be used by itself. A UL product may contain a number of components that have been "Recognized".

Reconocido por UL. Se refiere a un producto que está incompleto en las características de construcción o limitado en las capacidades de desempeño. Un producto "reconocido" está diseñado para ser usado como parte componente de un equipo que está "incluido en la lista". Un producto "reconocido" no debe usarse por sí solo. Un producto UL puede contener un número de componentes que han sido "reconocidos".

Ungrounded. Not connected to ground or a conductive body that extends the ground connection.*

Sin conexión a tierra. No conectado a la tierra ni a un cuerpo conductor que extienda la conexión a tierra.

Ungrounded Conductor. The conductor of an electrical system that is not intentionally connected to ground. This conductor is often referred to as the "hot" or "live" conductor.

Conductor sin conexión a tierra. Conductor de un sistema eléctrico que no está conectado intencionalmente a la tierra. Este conductor a menudo se conoce como conductor "caliente" o "vivo".

V

Volt. The difference of electric potential between two points of a conductor carrying a constant current of 1 ampere, when the power dissipated between these points is equal to 1 watt. A voltage of 1 volt can push 1 ampere through a resistance of 1 ohm.

Voltio. La diferencia de potencia eléctrica entre dos puntos de un conductor que transporta una corriente constante de 1 amperio, cuando la energía disipada entre estos puntos es igual a un vatio. El voltaje de 1 voltio puede empujar a 1 amperio a través de una resistencia de 1 ohmio.

*Reprinted with permission from NFPA 70-2014.

Voltage (of a circuit). The greatest root-mean-squared (effective) difference of potential between any two conductors of the circuit concerned.*

Voltaje (de un circuito). La mayor diferencia del valor eficaz de potencia entre dos conductores de un circuito determinado.

Voltage (nominal). A nominal value assigned to a circuit or system for the purpose of conveniently designating its voltage class (e.g., 120/240 volts, 480Y/277 volts, 600 volts). The actual voltage at which a circuit operates can vary from the nominal within a range that permits satisfactory operation of equipment.*

Voltaje (nominal). Valor nominal asignado a un circuito o sistema con el fin de designar convenientemente su clase de voltaje (p. ej., 120/240 voltios, 480Y/277 voltios, 600 voltios). El voltaje real al cual funciona un circuito puede variar del nominal dentro de un rango que permite el funcionamiento satisfactorio del equipo.

Voltage Drop. Also referred to as IR drop. Voltage drop is most commonly associated with conductors. A conductor has resistance; when current is flowing through the conductor, a voltage drop will be experienced across the conductor. Voltage drop across a conductor can be calculated using Ohm's law $E = IR$.

Caída de voltaje. También denominada caída de IR. La caída de voltaje se asocia comúnmente con los conductores. Un conductor tiene resistencia; cuando la corriente fluye a través del conductor, se experimentará una caída del voltaje en todo el conductor. La caída del voltaje en el conductor puede calcularse usando la ley del ohmio $E = IR$.

Voltage to Ground. For grounded circuits, the voltage between the given conductor and that point or conductor of the circuit that is grounded; for ungrounded circuits, the greatest voltage between the given conductor and any other conductor of the circuit.*

Voltaje a tierra. Para los circuitos con conexión a tierra, el voltaje entre el conductor determinado y ese punto o conductor del circuito que está conectado a tierra; para circuitos sin conexión a tierra, el mayor voltaje entre el conductor dado y cualquier otro conductor del circuito.

Volt-ampere. A unit of power determined by multiplying the voltage and current in a circuit. A 120-volt circuit carrying 1 ampere is 120 volt-amperes.

Voltio-amperio. Unidad de potencia determinada al multiplicar el voltaje y la corriente en un circuito Un circuito de 120 voltios que transporta 1 amperio es 120 voltio-amperios.

W

Watertight. Constructed so that moisture will not enter the enclosure under specified test conditions.*

Hermético. Construido de modo tal que la humedad no ingrese al recinto cuando está expuesto a condiciones específicas de prueba.

Watt. A measure of true power. A watt is the power required to do work at the rate of 1 joule per second. Wattage is determined by multiplying voltage times amperes times the power factor of the circuit: W = E × I × PF.

Vatio. Medida de potencia verdadera. Un vatio es la potencia necesaria para hacer un trabajo en la tasa de 1 julio por segundo. La potencia en vatios se determina al multiplicar el voltaje por los amperios por el factor de potencia del circuito: W = E × I × PF.

Weatherproof. Constructed or protected so that exposure to the weather will not interfere with successful operation.*

Resistente a la intemperie. Construido o protegido de tal modo que la exposición a los factores climáticos no interfiera con el funcionamiento correcto.

Informational Note. Rainproof, raintight, or watertight equipment can fulfill the requirements for weatherproof where varying weather conditions other than wetness, such as snow, ice, dust, or temperature extremes, are not a factor.*

Nota informativa. Los equipos impermeables o a prueba de lluvia pueden cumplir los requisitos de resistencia a la intemperie cuando no son determinantes otras condiciones meteorológicas no asociadas a la humedad, como la nieve, el hielo, el polvo o las temperaturas extremas.

Wet Niche Luminaire. A luminaire intended for installation in a forming shell mounted in a pool.* This definition is found in *Article 680.*

Luminaria de nicho húmedo. Luminaria diseñada para ser instalada en una carcasa montada en una piscina. Esta definición se encuentra en el *Artículo 680.*

Working Drawing(s). A drawing sufficiently complete with plan and section views, dimensions, details, and notes, so that whatever is shown can be constructed and/or replicated without instructions but subject to clarifications, see drawings.

Gráfico(s) de trabajo. Gráfico suficientemente completo con vistas de plano y sección, dimensiones, detalles y notas, de modo tal que lo que sea que se muestra pueda construirse y/o replicarse sin instrucciones pero sujeto a aclaraciones, consulte gráficos.

Work Order. A written order, signed by the owner or his or her representative, of a contractual status requiring performance by the contractor without negotiation of any sort.

Orden de trabajo. Orden escrita, firmada por el propietario o su representante, de un estado contractual que requiere el cumplimiento del contratista sin negociación de ningún tipo.

Z

Zoning. Restrictions of areas or regions of land within specific geographical areas based on permitted building size, character, and uses as established by governing urban authorities.

Zonificación. Restricciones de áreas o regiones de tierra dentro de zonas geográficas específicas en base a una medida de construcción, el carácter y los usos permitidos según lo establecen las autoridades urbanas gobernantes.

*Reprinted with permission from NFPA 70-2014.

Code Index

Note: **Bold** indicates material is in figures.

A

T

Subject Index

Note: Page numbers in **bold** reference non-text material.

F

H